Lecture Notes in Computer Science 10055

More information about this series at http://www.springer.com/series/7409

Yuan-Fang Li · Wei Hu
Jin Song Dong · Grigoris Antoniou
Zhe Wang · Jun Sun
Yang Liu (Eds.)

Semantic Technology

6th Joint International Conference, JIST 2016
Singapore, Singapore, November 2–4, 2016
Revised Selected Papers

Springer

Editors
Yuan-Fang Li
Information Technology
Monash University
Melbourne, VIC
Australia

Wei Hu
Computer Science and Technology
Nanjing University
Nanjing
China

Jin Song Dong
Computer Science
National University of Singapore
Singapore
Singapore

Grigoris Antoniou
University of Huddersfield
Huddersfield
UK

Zhe Wang
Information and Communication Technology
Griffith University
Brisbane, QLD
Australia

Jun Sun
ISTD
Singapore University of Technology and Design
Singapore
Singapore

Yang Liu
Computer Science and Engineering
Nanyang Technological University
Singapore
Singapore

ISSN 0302-9743 ISSN 1611-3349 (electronic)
Lecture Notes in Computer Science
ISBN 978-3-319-50111-6 ISBN 978-3-319-50112-3 (eBook)
DOI 10.1007/978-3-319-50112-3

Library of Congress Control Number: 2016959173

LNCS Sublibrary: SL3 – Information Systems and Applications, incl. Internet/Web, and HCI

Printed on acid-free paper

This Springer imprint is published by Springer Nature
The registered company is Springer International Publishing AG
The registered company address is: Gewerbestrasse 11, 6330 Cham, Switzerland

Preface

This volume contains the papers presented at JIST 2016: the 6th Joint International Semantic Technology Conference held during November 2–4, 2016, in Singapore. JIST 2016 was co-hosted by National University of Singapore, Nanyang Technological University (Singapore), and Monash University (Australia). JIST is a regional federation of semantic technology-related conferences. It attracts many participants from mainly the Asia Pacific region and often Europe and the USA. The mission of JIST is to bring together researchers in semantic technology research and other areas of semantic-related technologies to present their innovative research results and novel applications.

The main topics of JIST 2016 include ontology and reasoning, linked data, and knowledge graph, among others. JIST 2016 consisted of two keynotes, a main technical track, including (full and short) papers from the research and the in-use tracks, a poster and demo session, a workshop, and two tutorials. There were a total of 34 submissions for the main technical tracks from 17 countries. All papers were reviewed by at least three reviewers and the results were rigorously discussed by the program co-chairs. In all, 16 full papers (47%) and eight short papers were accepted in the technical tracks.

The paper topics are divided into six categories: Ontology and Data Management, Linked Data, Information Retrieval and Knowledge Discovery, RDF and Query, Knowledge Graph, and Applications of Semantic Technologies.

We would like to thank the JIST Steering Committee, Organizing Committee, and Program Committee for their significant contributions. We would also like to especially thank the co-hosts for their support in making JIST 2016 a successful and memorable event. Finally, we would like to express our appreciation to all speakers and participants of JIST 2016. This book is an outcome of their contributions.

November 2016

Yuan-Fang Li
Wei Hu
Jin Song Dong
Grigoris Antoniou
Zhe Wang
Jun Sun
Yang Liu

Organization

Program Committee

Paolo Bouquet	University of Trento, Italy
Nopphadol Chalortham	Silpakorn University, Thailand
C. Chantrapornchai	Kasetsart University, Thailand
Gong Cheng	Nanjing University, China
Paola Di Maio	ISTCS.org/IIT Mandi, India
Stefan Dietze	L3S Research Center, Germany
Dejing Dou	University of Oregon, USA
Jae-Hong Eom	Seoul National University, South Korea
Naoki Fukuta	Shizuoka University, Japan
Volker Haarslev	Concordia University, Canada
Armin Haller	Australian National University, Australia
Masahiro Hamasaki	National Institute of Advanced Industrial Science and Technology (AIST), Japan
Sungkook Han	Wonkwang University, South Korea
Koiti Hasida	AIST, Japan
Wei Hu	Nanjing University, China
Eero Hyvönen	Aalto University, Finland
Ryutaro Ichise	National Institute of Informatics, Japan
Vahid Jalali	Indiana University, USA
Jason Jung	Chung-Ang University, South Korea
Yong-Bin Kang	Monash University, Australia
Takahiro Kawamura	Japan Science and Technology Agency, Japan
Pyung Kim	Jeonju National University of Education, South Korea
Seiji Koide	Ontolonomy, LLC, Japan
Kouji Kozaki	Osaka University, Japan
Seungwoo Lee	KISTII, South Korea
Tony Lee	Saltlux, Inc., South Korea
Yuan-Fang Li	Monash University, Australia
Riichiro Mizoguchi	Japan Advanced Institute of Science and Technology, Japan
Takeshi Morita	Keio University, Japan
Ralf Möller	Universität zu Lübeck, Germany
Shinichi Nagano	Toshiba Corporation, Japan
Ikki Ohmukai	National Institute of Informatics, Japan
Artemis Parvizi	Oxford University Press, UK
Yuzhong Qu	Nanjing University, China
Ulrich Reimer	University of Applied Sciences St. Gallen, Switzerland

Giorgos Stoilos	National Technical University of Athens (NTUA), Greece
Umberto Straccia	ISTI-CNR, Italy
Boontawee Suntisrivaraporn	DTAC, Thailand
Hideaki Takeda	National Institute of Informatics, Japan
Holger Wache	University of Applied Science Northweastern Switzerland, Switzerland
Haofen Wang	East China University of Science and Technology, China
Peng Wang	Southeast University, China
Xin Wang	Tianjin University, China
Zhe Wang	Griffith University, Australia
Krzysztof Wecel	Poznan University of Economics, Poland
Gang Wu	College of Information Science and Engineering, Northeastern University, China
Guohui Xiao	KRDB Research Centre, Free University of Bozen-Bolzano, Italy
Bin Xu	DCST, Tsinghua University, China
Yasunori Yamamoto	Database Center for Life Science, Japan
Xiang Zhang	Southeast University, China
Yuting Zhao	IBM Italy, Italy
Amal Zouaq	Royal Military College of Canada, Canada

Organizing Committee

General Chairs

Jin Song Dong	National University of Singapore, Singapore
Grigoris Antoniou	University of Huddersfield, UK

Program Co-chairs

Yuan-Fang Li	Monash University, Australia
Wei Hu	Nanjing University, China

Publicity Chair

Zhe Wang	Griffith University, Australia

Local Chair

Jun Sun	Singapore University of Technology and Design, Singapore

Workshop Co-chairs

Xin Wang	Tianjing University, China
Hanmin Jung	Korea Institute of Science and Technology Information, Korea

Tutorial Co-chairs

Armin Haller	Australian National University, Australia
Gong Cheng	Nanjing University, China

Poster and Demo Co-chairs

Zhichun Wang	Beijing Normal University, China
Kouji Kozaki	Osaka University, Japan

Finance Chair

Yang Liu	Nanyang Technological University, Singapore

Publicity Chair

Haofen Wang	East China University of Science and Technology, China

Keynotes

Managing Dynamic Ontologies: Belief Revision and Forgetting

Kewen Wang

Griffith University, Brisbane, Australia
k.wang@griffith.edu.au

Ontologies have recently been used in a wide range of practical domains such as e-Science, e-Commerce, medical informatics, bio-informatics, and the Semantic Web. An *ontology* is a formal model of some domain knowledge of the world. It specifies the *formalization* of the domain knowledge as well as the *meaning* (semantics) of the formalization. The Web Ontology Language (OWL), with its latest version, OWL 2, is based on description logics (DLs). Thus, an ontology is often expressed as a knowledge base (KB) in DLs, which consists of both terminological knowledge (or schema information) in the TBox and assertional knowledge (or data information) in the ABox. As with all formal knowledge structures, ontologies are not static, but may evolve over time. Indeed, ontology engineering is described as a life-cycle, which is based on evolving prototypes and specific techniques peculiar to each ontology engineering activity. An important and challenging problem is thus how to effectively and efficiently modify ontologies.

In this talk, we discuss some recent developments and challenges for two paradigms of ontology changes. We focus on model-based approaches.

Knowledge Update: Outdated and incorrect axioms in an ontology have to be eliminated from the ontology and newly formed axioms have to be incorporated into the ontology. In the field of belief change, extensive work has been done on formalising various kinds of changes over logical knowledge bases. In particular, elimination of old knowledge is called contraction and incorporation of new knowledge is called revision. The dominant approach in belief change is the so called AGM framework. Regardless of its wide acceptance, the AGM framework is incompatible with DLs due to its assumption on an underlying logic that includes propositional logic. The incompatibility is the major difficulty in defining DL contraction and revision. Additionally, DL revision is more involved than AGM revision. AGM revision aims to resolve any inconsistency caused while incorporating a new formula. Since a meaningful DL ontology has to be both consistent and coherent (i.e, absence of unsatisfiable concepts), DL revision has to resolve not only inconsistency but also incoherence. Finally, DL contraction and revision should lead to tractable instantiations and at the same time respecting the mathematical properties of AGM contraction and revision.

Forgetting: To support the reuse and combination of ontologies in Semantic Web applications, it is often necessary to obtain smaller ontologies from existing larger ontologies. In particular, applications may require the omission of many terms, e.g., concept names and role names, from an ontology. However, the task of omitting terms from an ontology is challenging because the omission of some terms may affect the relationships between the remaining terms in complex ways. The technique of forgetting provides an effective way for extracting modules from a large ontology.

The Rise of Approximate Ontology Reasoning: Is It Mainstream Yet?

Jeff Z. Pan

University of Aberdeen, Aberdeen, UK

The last five years have seen a growing volume and complexity of ontologies and large-scale linked data available,[1] which present a pressing need for efficient and scalable ontology reasoning services. Major technology vendors are starting to embrace semantic technologies by supporting new standards and integrating with state of the art semantic tools. For example, in their new release 12.1, Oracle Spatial and Graph supports both RDF and OWL2-EL natively,[2] and integrates with an OWL2-DL reasoner (TrOWL) via OWL-DBC.[3]

The second version of the ontology standard OWL (Web Ontology Language) offers a family of ontology languages, including OWL2-DL, the most expressive decidable language in the family, and three tractable sub-languages of OWL2-DL, i.e. OWL2-EL, OWL2-QL and OWL2-RL. Such a two-layered language architecture allows approximate reasoning for OWL2-DL, by approximating OWL2-DL ontologies to those in its tractable sub-languages, so as to exploit efficient and scalable reasoners of the sublanguages. This is motivated by the fact that real-world knowledge and data are hardly perfect or completely digitalised. State of the art approximate reasoners, such as the TrOWL reasoner, can out-perform sound and complete reasoners in time constrained sound-and-complete reasoner competitions, such as the ORE competitions.

In this talk, we will look into how and why approximate reasoners work. Indeed, approximation approaches bring a new dimension – quality, in terms of completeness and soundness of reasoning, into the trade-off between expressiveness and performance, attempting to strike a balance among the three. Once we start to consider such a third dimension, many interesting questions follows: What are the typical approximate reasoning approaches? Should we approximate the input ontology or the input query? Are approximations always finite and unique? Given an ontology and some target queries, are there any best approximations? Why do some approximate reasoning algorithms lose many reasoning results, while others can enjoy high recall? Are approximate reasoning algorithms relevant to optimisations for sound and complete reasoners? Can we extend approximate reasoning algorithm with some post-processing to ensure soundness and completeness? I will discuss many of these questions, in the context of the TrOWL reasoner and related work, and share some thoughts on what approximate reasoning might bring in the near future.

---

1 http://lod-cloud.net/state/.

2 http://download.oracle.com/otndocs/tech/semantic_web/pdf/semtech_datamining_v8.pdf.

3 http://download.oracle.com/otndocs/tech/semantic_web/pdf/trowl_integration_with_orasag.pdf.

Contents

Information Retrieval and Knowledge Discovery

RDF and Query

Knowledge Graph

Applications of Semantic Technologies

Ontology and Data Management

How Can Reasoner Performance of ABox Intensive Ontologies Be Predicted?

Isa Guclu[1], Carlos Bobed[2], Jeff Z. Pan[1](✉), Martin J. Kollingbaum[1], and Yuan-Fang Li[3]

[1] University of Aberdeen, Aberdeen, UK
jeff.z.pan@abdn.ac.uk
[2] University of Zaragoza, Zaragoza, Spain
[3] Monash University, Melbourne, Australia

Abstract. Reasoner performance prediction of ontologies in OWL 2 language has been studied so far from different dimensions. One key aspect of these studies has been the prediction of how much time a particular task for a given ontology will consume. Several approaches have adopted different machine learning techniques to predict time consumption of ontologies already. However, these studies focused on capturing general aspects of the ontologies (i.e., mainly the complexity of their TBoxes), while paying little attention to ABox intensive ontologies. To address this issue, in this paper, we propose to improve the representativeness of ontology metrics by developing new metrics which focus on the ABox features of ontologies. Our experiments show that the proposed metrics contribute to overall prediction accuracy for all ontologies in general without causing side-effects.

Keywords: Semantic web · Ontology reasoning · Prediction · Random forests · Knowledge graph · Practical reasoning

1 Introduction

Semantic technologies have been utilized in various application domains for assisting knowledge management thus far, e.g., data management [13] and software engineering [17]. The worst case complexity 2NEXPTIME-complete [6] of OWL 2 DL, the most expressive profile of OWL 2, constitutes a bottleneck for performance critical environments. Empirical studies show that even the EL profile, with PTIME-complete complexity and less expressiveness, can become too time-consuming [4,11]. To have a scalable environment for implementing semantic technologies, an accurate prediction of ontology time consumption which will guide us about the feasibility of ontology reasoning is needed.

There have been several studies regarding the performance prediction of ontologies. Kang et al. [10] investigated the *hardness category* (categories according to reasoning time) for reasoner-ontology pairs and used machine learning techniques to make a prediction. Using FaCT++ [25], HermiT [5], Pellet [23], and TrOWL [16,18,20,24], they reached high accuracy in terms of hardness category, but not reasoning time.

Y.-F. Li et al. (Eds.): JIST 2016, LNCS 10055, pp. 3–14, 2016.
DOI: 10.1007/978-3-319-50112-3_1

In another study, Kang et al. [12] investigated regression techniques to predict reasoning time. They made experiments using reasoners FaCT++, HermiT, JFact, MORe [21], Pellet and TrOWL with *their syntactic metrics* as features. These metrics are generally effective when there is a balance between TBox axioms and ABox axioms. Our experiments show that accuracy of these metrics decreases as ABox axiom sizes increase. As ABox constitutes the data in an ontology [1,8,27], where TBox constitutes the schema, an approach that can capture the changes in the ABox in a more detailed way is needed to make accurate overall predictions. As observed by Bobed et al. [2], there is an interest in using semantic technologies in mobile devices. In such scenarios, TBox axioms are expected to be more static and the ABox axioms (data) tend to be more frequently changing which necessitates high accuracy in ABox performance prediction. In this paper, we aim to investigate what metrics could help further improve reasoner predictions of ABox intensive ontologies.

Our main contributions can be summarized as follows.

1. We propose an initial set of metrics which estimate the complexity of the TBox concepts and propagates it into the estimated complexity of the ABox.
2. We show that our proposed new metrics for representing the structure of ontologies from the ABox perspective indicate a good research path to improve the accuracy of predicting time consumption of ontology reasoning.

The rest of the paper is as follows. In Sect. 2, we present some related works to place our proposal. In Sect. 3, we define the metrics that we propose in our ongoing work. In Sects. 4 and 5, we explain our experimental settings and the achieved results, respectively. Finally, in Sect. 6, we make some conclusions and draw some future work.

2 Related Work and Background

Ontology metrics, which are features of the ontology expressed numerically or categorically to represent the structure of an ontology, have been effectively utilised in analysing the complexity [28], energy consumption on mobile devices [7], cohesion [26], quality [3] and population task [15] of ontology reasoning.

Kang et al. [10] proposed a set of metrics in 2012 to classify raw reasoning times of ontologies into five large categories: [0 s.–100 ms.], (100 ms.–1 s.], (1 s.–10 s.], (10 s.–100 s.] and (100 s.–∞). Despite the high accuracy of prediction, over an 80%, this approach does not provide actual reasoning time but time categories, which may become obsolete or meaningless according to needs of implementation.

In 2014, Kang et al. [12] extended their work and proposed a new set of metrics to predict actual reasoning time by developing regression models. They extended the previous 27 metrics [10,28] and developed a set of 91 metrics that include 24 ontology-level (ONT) metrics, 15 class-level (CLS) metrics, 22 anonymous class expression (ACE) metrics, and 30 property definition and axiom (PRO) metrics.

While a high number of metrics are usually proposed by researchers, Sazonau et al. [22] proposed instead a local method which involved selecting a *suitable*, small subset of the ontology, and making extrapolation to predict total time consumption of ontology reasoning using the data coming from the processing of such small subset. To do so, they used *Principal Component Analysis* (PCA) [9]. In their experiments, Sazonau et al. [22] observed that 57 of the studied features can be replaced by just one or two features. Using a sample of size of a 10 % of the ontology for reasoning, they argue that they reached good predictions with simple extrapolations. They list advantages of their method as: (1) more accurate performance predictions, (2) not relying on an ontology corpus, (3) not being biased by this corpus, and (4) being able to obtain information about reasoner's behaviour of linear/nonlinear predictability on the corpus. A remarkable contribution of this approach is that it saves researchers from the difficulty/risk of selecting an unbiased corpus [14], which is very difficult while checking the validity of the prediction model and accuracy of the prediction. However, making reasoning with the 10 % of an ontology may not always be applicable especially when the ontology requires high reasoning times.

3 Our Approach

Our claim is that increasing the expressivity of ontology metrics directly helps increasing the accuracy of all the above studies, and enables new studies that target a more feasible implementation environment for semantic technologies.

Part of 91 metrics proposed by Kang et al. [12] are obtained by transforming an ontology into a graph which grasps the relationship between of ABox and TBox axioms. However, their approach calculates the effect of ABox axioms up to a certain extent. It is apparent that connected ABox axioms are more prone to cause more inferences than disconnected ABox axioms. These connections can trigger reasoning time enormously when they come along with a complex TBox. In our work, we have observed that the models trained with this set of 91 metrics begin to lose accuracy in predicting time consumption of ontologies as the ratio between the amount of ABox axioms and TBox axioms increases.

Thus, we propose to include the propagation of the complexity of the TBox into the ABox. To do so, we extend this set of metrics with our 15 *Class Complexity Assertions (CCA)* metrics, which can contribute to performance prediction of ontologies especially when we deal with ontologies which are ABox intensive (i.e., they exhibit a high ABox/TBox ratio). Experiment results and source codes are accessible[1].

3.1 Class Complexity Assertions Metrics

As above mentioned, to capture the interactions between the complexity of the different elements of the TBox and the individuals asserted in the ABox, we have

[1] http://sid.cps.unizar.es/projects/OWL2Predictions/JIST16/.

developed an initial set of features which aim at propagating the complexity of each of the concept expressions in the ontology to the ABox, as well as improving the richness of the TBox metrics.

Thus, let be $N_{CE} = \{CE_i \mid CE_i \in O\}$ with CE any concept expression appearing in any of the logical axioms of the ontology O. For each CE, we estimate its complexity as follows:

$$comp(CE_i) = \frac{height(CE_i) + sigSize(CE_i) + const(CE_i)}{3}$$

with $height(CE_i)$ being the height of the expression as a parsing tree, $sigSize(CE_i)$ being the number of different atomic class names that appear in the expression, and $const(CE_i)$ begin the number of class constructors participating in the class expression.

With this estimation for each CE_i, we calculate the following metrics:

- *TBoxSize:* The count of TBox axioms obtained from OWLAPI.
- *ABoxSize:* The count of ABox axioms obtained from OWLAPI.
- *ABoxTBoxRatio:* The ratio of ABox axioms to TBox axioms.
- *TCCA:* Total amount of estimated complexity of the ontology O (i.e., the class expressions in N_{CE}).

$$TCCA = \sum_{CE_i \in N_{CE}} comp(CE_i)$$

- *AVG_CCA:* Mean estimated complexity of the class expressions in N_{CE}.

$$AVG\_CCA = \frac{TCCA}{|N_{CE}|}$$

- *MAX_CCA:* Maximum estimated complexity of the class expressions in N_{CE}.
- *MIN_CCA:* Minimum estimated complexity of the class expressions in N_{CE}.
- *STD_CCA:* Standard deviation of complexity of the class expressions in N_{CE}.
- *ENT_CCA:* Entropy of the complexity distribution of N_{CE}.

To propagate the complexity of each concept expression to the ABox, we use each of the class assertions as a witness of the complexity of a class expression within the ontology. Then, we aggregate such values to capture what we name the witnessed complexity of the ABox. So, let $Ind_{N_{CE_i}} = \{a \mid a \in Ind(O) \wedge CE_i(a) \in O\}$ the individuals that are explicitly asserted to belong to CE_i. Thus, we define:

- *TWCCA:* Total witnessed complexity of the ABox, which is calculated summing all the products of the estimated complexities of the concept expressions with their *witness individuals*.

$$TWCCA = \sum_{CE_i \in N_{CE}} comp(CE_i) * |Ind_{N_{CE_i}}|$$

- *AVG_WCCA:* Mean witnessed complexity of the ABox of the concept expressions in *O*.

$$AVG\_WCCA = \frac{TWCCA}{|N_{CE}|}$$

- *MAX_WCCA:* Maximum witnessed complexity of a concept expression in *O*.
- *MIN_WCCA:* Minimum witnessed complexity of a concept expression in *O*.
- *STD_WCCA:* Standard deviation of witnessed complexity of the concept expressions in *O*.
- *ENT_WCCA:* Entropy of the witnessed complexity distribution of the concept expressions in *O*.

Note that we apply a Laplace smoothing[2] to include also into the metrics the concept expressions which appear in the ontology but do not have any explicit individual assertion.

4 Experimental Setup

4.1 Evaluation Metrics

R^2, $MAPE$ and $RMSE$ are referred to decide whether our regression model is valid for describing the relation between our metrics and the predictions made by the model. The coefficient of determination (R^2) is a crucial output of regression analysis, indicating to what extent the dependent variable is predictable. For example, a value 0.91 for R^2 means that 91% percent of the variance in Y is predictable from X. Let $y(t)$ be the observed value of y in second t, $\hat{y}(t)$ be the predicted value for y in second t, and $\bar{y}$ be the mean of the observed values, then:

$$R^2 = \frac{\sum_t (\hat{y}(t) - \bar{y})^2}{\sum_t (y(t) - \bar{y})^2} \tag{1}$$

The *Mean Absolute Percentage Error (MAPE)* is a measure of prediction accuracy of a prediction method in statistics that is used to expresses accuracy as a percentage. For calculating the *MAPE* of our prediction model, we will divide the difference of observed and predicted values, divide this by the observed values, and get the average of all observations in the scope. Related to this definition, we define the *Mean Absolute Accuracy Percentage (MAAP)* of our prediction model which is given by (1 - *MAPE*). In this paper, we will refer to *MAAP* to explain the accuracy of a model.

$$MAPE = 100.\frac{\sum_{t=1}^{n} \frac{|\hat{y}(t) - y(t)|}{y(t)}}{n} \tag{2}$$

$$MAAP = 1 - MAPE \tag{3}$$

[2] Adapted from Natural Language Processing, basically, it consists in adding 1 to all the witnessed values of the concept expressions in the ontology.

Finally, the Root Mean Squared Error ($RMSE$) is the square root of the mean/average of the square of all of the error. $RMSE$ represents the sample standard deviation of the differences between observed and predicted values.

$$RMSE = \sqrt{\frac{\sum_{t=1}^{n}(y(t) - \hat{y})^2}{n}} \tag{4}$$

4.2 Data Collection

Reasoner: We have used TrOWL 1.5 for testing EL ontologies as the reasoner to be tested. We deployed *ABox Materialization Task* with TrOWL as our experimental task. In our experiments, we implemented ABox materialization with one thread. We could benefit from parallelization in ABox materialization and it would improve the performance [19] to some extent. As RAM I/O becomes the bottleneck because of the limited bandwidth [19] of the RAM when many worker threads compete for RAM access and this would cause some side-effects in measuring the execution time, we preferred to analyse the performance prediction aspect parallel ABox materialization as future work.

Ontologies: We define an ontology as *ABox-intensive* if the count of ABox axioms in such an ontology is at least 10 times the count of TBox axioms. We made our experiments using ontologies in ORE2014 Reasoner Competition Dataset[3]. From 16,555 ontologies, we have filtered 74 ontologies in EL profile which have the ABox/TBox ratio of at least 10, and created artificial 2779 ABox-intensive ontologies[4] from these ABox-intensive ontologies *randomly* as follows: our method uses the TBox of the original ontology and creates a new ontology using different random subsets of the ABox axioms of the original ontology.

Prediction Model Construction: For predicting the time consumption of ontologies, a random forest based regression model is implemented, using the metrics (predictor variables). Standard 10-fold cross-validation is performed to ensure the generalizability of the model.

5 Results and Evaluation

In our study, we investigated the reasoning performance of a reasoner and ontology characteristics represented by available metrics and our new metrics (CCA). While developing our new metrics, we aimed at capturing the complexity of ontologies without losing accuracy when ABox sizes changed. Our claim is that developing high-quality metrics will increase the accuracy of the prediction

[3] https://zenodo.org/record/10791.

[4] You can find the code of the OntologyChopper at http://sid.cps.unizar.es/projects/OWL2Predictions/JIST16/.

model. Our goal is to make prediction models that can perform on any ontology with high stability using metrics that can represent the ontology with high expressivity.

To ensure the quality of the dataset, we created 2779 artificial ontologies from ORE2014 dataset. To avoid a biased corpus, which would result in misleading generalizations, we generated ontologies with random selection of ABox axioms. We did not put any threshold to cut the experiment, as we wanted to include every result of the dataset without missing any point that could be expressed by the dataset. We believe that wide range of ABox/TBox ratio will help increase the diversity in ontologies.

While working with the quality of the dataset, quality of the feature selection should also be taken into consideration. Inspired from the consistent high accuracy of the Random Forest based regression models in the study of Kang et al. [12], we adopted the same approach. Instead of categorising the time periods, we preferred metrics to give prediction results of time consumption in nanoseconds. We had specified R^2, $MAPE$, and $RMSE$ values as our performance criteria for prediction accuracy.

5.1 Combining 91 Metrics with CCA

In our first set of experiments, we combined 91 metrics with CCA metrics to train the model. We were expecting new CCA metrics would increase the accuracy of prediction, as it contained metrics that would better express the complexity of ABox axioms with TBox axioms. The results obtained in the cross validation procedure for the performance criteria can be seen in Table 1, and in Fig. 1, the $MAPE$ values obtained are visualized.

When we look at the R^2 values, which is indicative to which extent the dependent variable is predictable, we see that both available 91 metrics and combined metrics of 91 metrics and CCA metrics have the values between 97% and 98%. The difference of $RMSE$ values is ≈ 2.5 s. The values of $MAPE$ also show a difference of $\approx 2\%$.

Although this absolute value of a $\approx 2\%$ accuracy increase seems very small, it is a relative improvement of 11% with the first version of our transference metrics, which encourages us to continue to work in this direction improving and extending the definition of such kind of metrics.

Table 1. Contribution of CCA metrics to accuracy of prediction.

	91 Metrics	CCA + 91 Metrics
R^2	0.97607	0.97856
$MAPE$	23.58%	21.10%
$RMSE$	41.4 s	39.1 s

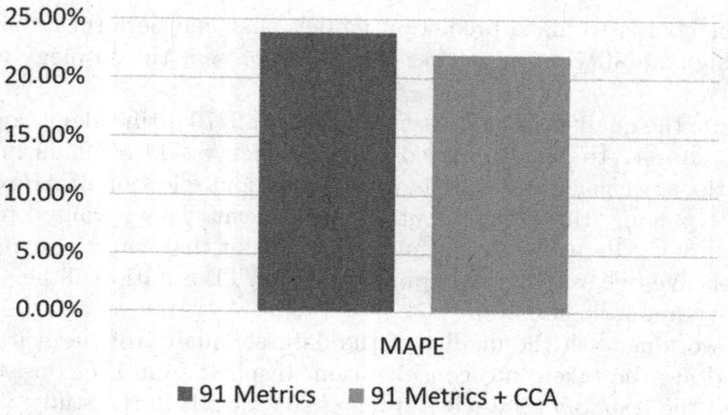

Fig. 1. Change in MAPE when new metrics are added to the prediction model.

5.2 Using CCA Metrics Instead of *some* ABox metrics in 91 Metrics v.1

We searched for the metrics in 91 metrics which are more sensitive to ABox axiom changes. By randomly adding ABox axioms to ontologies, we observed that the change in ABox axioms is highly correlated with some of 91 metrics, i.e., "SOV, CYC, RHLC, IHR, IIR, ITR, IND, aCID, mCID, tCID" [10,28].

In our second set of experiments, we removed the metrics "RHLC, IHR, IIR, ITR, IND, aCID, mCID, tCID" from 91 metrics and replaced with CCA metrics to train the model. The results obtained in the cross validation procedure for the performance criteria can be seen in Table 2, and in Fig. 2, the $MAPE$ values obtained are visualized.

When we look at the R^2 values, we see that both models have the values between 97% and 98%. The difference of $RMSE$ values is $\approx$ 1 s. The values of $MAPE$ show a difference of $\approx$ 4.5%.

The relative improvement in decreasing the average error rate of 91 metrics is about 20% by replacing some of ABox related metrics in 91 metrics with our CCA metrics.

Table 2. Contribution of CCA metrics when replaced with some ABox related metrics in 91 metrics (v.1).

	91 Metrics	CCA + 91 Metrics (v.1)
R^2	0.97607	0.97654
$MAPE$	23.58 %	19.03 %
$RMSE$	41.4 s	41.0 s

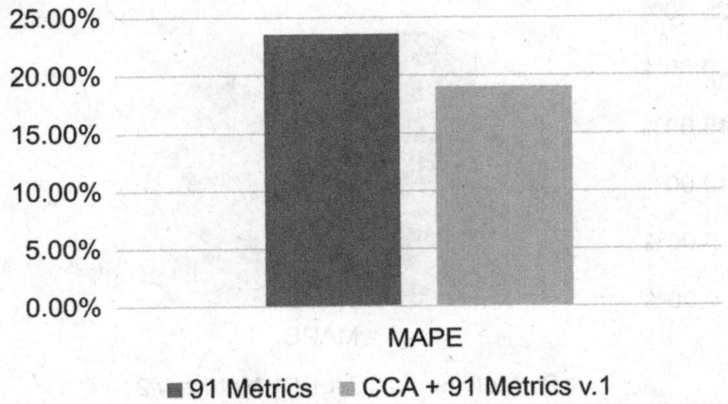

Fig. 2. Change in MAPE when some ABox related metrics in 91 metrics are replaced with CCA metrics (v.1) to the prediction model.

5.3 Using CCA Metrics Instead of *some* ABox metrics in 91 Metrics v.2

In our third case, we removed the metric "CYC" in addition to "RHLC, IHR, IIR, ITR, IND, aCID, mCID, tCID" from 91 metrics and replaced with CCA metrics to train the model. The results obtained in the cross validation procedure for the performance criteria can be seen in Table 3, and in Fig. 3, the $MAPE$ values obtained are visualized.

When we look at the R^2 values, we see that both models have the values between 97% and 98%. The value of $RMSE$ worsened here. The values of $MAPE$ show a difference of ≈ 4%, which is a general improvement but worst than the previous case.

The relative improvement in decreasing the average error rate of 91 metrics is about 18% by replacing some of ABox related metrics in 91 metrics with our CCA metrics.

Table 3. Contribution of CCA metrics when replaced with some ABox related metrics in 91 metrics (v.2).

	91 Metrics	CCA + 91 Metrics (v.2)
R^2	0.97607	0.97449
$MAPE$	23.58%	19.25%
$RMSE$	41.4 s	42.7 s

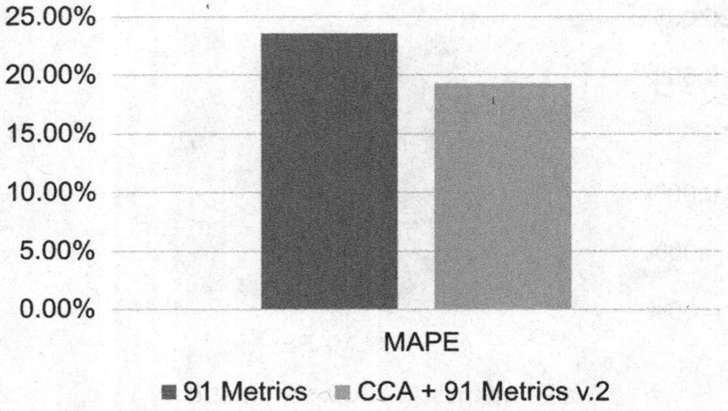

Fig. 3. Change in MAPE when some ABox related metrics in 91 metrics are replaced with CCA metrics (v.2) to the prediction model.

5.4 Evaluation

In our work, we have analysed available metrics and investigated how to bring expressivity of metrics further by developing new metrics to represent ABox axioms (and its interaction with TBox axioms) aspect of ontologies.

According to initial experiments, which compare 91 metrics with combination of 91 metrics and CCA metrics, we observe that adding CCA metrics increases the accuracy of prediction $\approx$ 2.5% and relatively decreasing average error rate $\approx$ 11%.

When we replaced some of the metrics (RHLC, IHR, IIR, ITR, IND, aCID, mCID, tCID) in 91 metrics with CCA metrics, we observe the accuracy of prediction increase $\approx$ 4.5% and relatively, average error rate decrease $\approx$ 20%.

In our third case, we also removed the metric (CYC) in 91 metrics and saw that there was again higher accuracy in prediction but it wasn't as good as the previous model.

Seeing the results above, we conclude that CCA metrics contributes to prediction of ABox-intensive ontologies in our preliminary work. Available metrics (91 metrics proposed by Kang et al. [12]) could grasp the complexity of ontologies to some extent. ABox materialization necessitates new metrics that will represent the interaction of ABox axioms with TBox axioms taking its complexity into account. The weight of ABox axioms in an ontology and their interactions can cause consuming more execution time than expected if their complexity is ignored. We propose our CCA metrics to measure the effect of ABox complexity in performance prediction of ontology reasoning and we want to improve these metrics to measure this aspect of ontologies more effectively. We believe that our study will lead to metrics that are generalizable regardless of the weight of TBox and ABox axioms.

6 Conclusion

Performance prediction of ontology reasoning is a very interesting and challenging topic. In this work, we have started to focus on the performance prediction of ABox-intensive ontologies. We proposed 15 new metrics by extending previous work of Kang et al. [12]. Preliminary results with adding these 15 metrics show slight increase ($\approx 4.5\%$) in the prediction accuracy. And, these results even at the early stages of our research encourage us to continue in this direction. We believe that awareness of the ABox axiom ratio in ontologies and bringing a solution to this change will increase the effectiveness and validity of a performance prediction model.

As future work, firstly, we plan to work on better representation of the interactions between ABox axioms and TBox axioms by developing new metrics. Secondly, we will make experiments with more reasoners on different ontologies that will help understanding the interaction of ABox axioms with TBox axioms in a broader sense. Thirdly, we will use different prediction mechanisms to leverage the contribution of these metrics.

Acknowledgments. This work was partially supported by the EC Marie Curie K-Drive project (286348), the CICYT project (TIN2013-46238-C4-4-R) and the DGA-FSE project.

References

1. Fokoue, A., Meneguzzi, F., Sensoy, M., Pan, J.Z.: Querying linked ontological data through distributed summarization. In: Proceedings of the 26th AAAI Conference on Artificial Intelligence (AAAI2012) (2011)
2. Bobed, C., Yus, R., Bobillo, F., Mena, E.: Semantic reasoning on mobile devices: do androids dream of efficient reasoners? J. Web Semant. **35**(4), 167–183 (2015). ISSN 1570–8268, https://dx.doi.org/10.1016/j.websem.2015.09.002
3. Burton-Jones, A., Storey, V.C., Sugumaran, V., Ahluwalia, P.: A semiotic metrics suite for assessing the quality of ontologies. Data Knowl. Eng. **55**, 84–102 (2005)
4. Dentler, K., Cornet, R., ten Teije, A., de Keizer, N.: Comparison of reasoners for large ontologies in the OWL 2 EL profile. Semant. Web **2**, 71–87 (2011)
5. Glimm, B., Horrocks, I., Motik, B., Stoilos, G., Wang, Z.: Hermit: an OWL 2 reasoner. J. Autom. Reasoning **53**, 245–269 (2014)
6. Grau, B.C., Horrocks, I., Motik, B., Parsia, B., Patel-Schneider, P.F., Sattler, U.: OWL 2: the next step for OWL. J. Web Sem. **6**, 309–322 (2008)
7. Guclu, I., Li, Y.-F., Pan, J.Z., Kollingbaum, M.J.: Predicting energy consumption of ontology reasoning over mobile devices. In: Groth, P., Simperl, E., Gray, A., Sabou, M., Krötzsch, M., Lecue, F., Flöck, F., Gil, Y. (eds.) ISWC 2016. LNCS, vol. 9981, pp. 289–304. Springer, Heidelberg (2016). doi:10.1007/978-3-319-46523-4_18
8. Hogan, A., Pan, J.Z., Polleres, A., Ren, Y.: Scalable OWL 2 reasoning for linked data. In: Polleres, A., d'Amato, C., Arenas, M., Handschuh, S., Kroner, P., Ossowski, S., Patel-Schneider, P. (eds.) Reasoning Web 2011. LNCS, vol. 6848, pp. 250–325. Springer, Heidelberg (2011). doi:10.1007/978-3-642-23032-5_5
9. Jolliffe, I.: Principal Component Analysis. Wiley StatsRef: Statistics Reference Online (2002)

10. Kang, Y.-B., Li, Y.-F., Krishnaswamy, S.: Predicting reasoning performance using ontology metrics. In: Cudré-Mauroux, P., et al. (eds.) ISWC 2012. LNCS, vol. 7649, pp. 198–214. Springer, Heidelberg (2012). doi:10.1007/978-3-642-35176-1_13
11. Kang, Y.-B., Li, Y.-F., Krishnaswamy, S.: A rigorous characterization of classification performance - a tale of four reasoners. In: ORE (2012)
12. Kang, Y.-B., Pan, J.Z., Krishnaswamy, S., Sawangphol, W., Li, Y.-F.: How long will it take? Accurate prediction of ontology reasoning performance. In: AAAI (2014)
13. Li, Y.-F., Kennedy, G., Ngoran, F., Wu, P., Hunter, J.: An ontology-centric architecture for extensible scientific data management systems. Future Gener. Comput. Syst. **29**, 641–653 (2013)
14. Matentzoglu, N., Bail, Ş., Parsia, B.: A corpus of OWL DL ontologies. In: Proceedings DL13 (2013)
15. Maynard, D., Peters, W., Li, Y.: Metrics for evaluation of ontology-based information extraction (2006)
16. Pan, J.Z., Ren, Y., Zhao, Y.: Tractable approximate deduction for OWL. Artificial Intelligence **235**, 95–155
17. Pan, J.Z., Staab, S., Amann, U., Ebert, J., Zhao, Y.: Ontology-Driven Software Development. Springer Publishing Company, Incorporated, Ontology-Driven Software (2012)
18. Pan, J.Z., Thomas, E., Ren, Y., Taylor., S.: Tractable fuzzy and crisp reasoning in ontology applications. In: IEEE Computational Intelligence Magazine (2012)
19. Ren, Y., Pan, J.Z., Lee, K.: Optimising parallel ABox reasoning of EL ontologies. In: Description Logics (2012)
20. Ren, Y., Pan, J.Z., Zhao, Y.: Soundness preserving approximation for TBox reasoning. In: AAAI (2010)
21. Romero, A.A., Grau, B.C., Horrocks, I.: More: modular combination of OWL reasoners for ontology classification. In: SEMWEB (2012)
22. Sazonau, V., Sattler, U., Brown, G.: Predicting performance of OWL reasoners: locally or globally? In: Principles of Knowledge Representation and Reasoning: Proceedings of the Fourteenth International Conference, KR 2014, Vienna, Austria, July 20–24, 2014 (2014)
23. Sirin, E., Parsia, B., Grau, B.C., Kalyanpur, A., Katz, Y.: Pellet: a practical OWL-DL reasoner. J. Web Sem. **5**, 51–53 (2007)
24. Thomas, E., Pan, J.Z., Ren, Y.: TrOWL: tractable OWL 2 reasoning infrastructure. In: Aroyo, L., Antoniou, G., Hyvönen, E., Teije, A., Stuckenschmidt, H., Cabral, L., Tudorache, T. (eds.) ESWC 2010. LNCS, vol. 6089, pp. 431–435. Springer, Heidelberg (2010). doi:10.1007/978-3-642-13489-0_38
25. Tsarkov, D., Horrocks, I.: FaCT++ description logic reasoner: system description. In: Furbach, U., Shankar, N. (eds.) IJCAR 2006. LNCS (LNAI), vol. 4130, pp. 292–297. Springer, Heidelberg (2006). doi:10.1007/11814771_26
26. Yao, H., Orme, A.M., Etzkorn, L.: Cohesion metrics for ontology design and application. J. Comput. Sci. **1**, 107–113 (2005)
27. Yuan Ren, J.Z.P., Lee, K.: Optimising parallel ABox reasoning of el ontologies. In: Proceedings of the 25th International Workshop on Description Logics (DL2012) (2012)
28. Zhang, H., Li, Y.-F., Tan, H.B.K.: Measuring design complexity of semantic web ontologies. J. Syst. Softw. **83**, 803–814 (2010)

Inquiry into RDF and OWL Semantics

Seiji Koide[1,2(✉)] and Hideaki Takeda[1]

[1] National Institute of Informatics, 2-1-2 Hitotsubashi, Chiyoda-ku, Tokyo, Japan
koide@ontolonomy.co.jp

[2] Ontolonomy, LLC., 3-76-3-J901, Mutsukawa, Minami-ku, Yokohama, Japan

Abstract. The purpose of this paper is to present the higher order formalization of RDF and OWL with setting up ontological meta-modeling criteria through the discussion of Russell's *Ramified Type Theory*, which was developed in order to solve *Russell Paradox* appeared at the last stage in the history of set theory. This paper briefly summarize some of set theories, and reviews the RDF and OWL Semantics with higher order classes from the view of Russell's *Principia Mathematica.* Then, a set of criteria is proposed for ontological meta-modeling. Several examples of meta-modeling, including sound ones and unsound ones, are discussed and some of solutions are demonstrated according to the meta-modeling criteria proposed.

Keywords: RDF semantics · OWL semantics · Set theory · Principia mathematica · KIF · Membership loop · Higher order class · Meta-modeling

1 Introduction

The OWL specifications has been split into two parts, Direct Semantics and RDF-based Semantics. The reason of this unhappy partition originates in the different formalizations between OWL DL and RDF. In order to match DL-based semantics to RDF-based semantics, the term of "comprehension conditions (principles)" is introduced into "OWL Semantics and Abstract Syntax, Sect. 5" (Patel-Schneider et al. 2004a) and "OWL 2 Web Ontology Language RDF-Based Semantics (Second Edition)" (Schneider 2014). In addition, "RDF-Compatible Model-Theoretic Semantics" (Patel-Schneider et al. 2004a) states that the *only-if* semantic conditions are necessary to prevent semantic paradoxes with the fourteen comprehension conditions, and OWL2 Appendix[1] mentions "formal inconsistency" instead of "paradox". However, these statements are caused by misunderstanding paradoxes in set theories, and the word "comprehension conditions" is still left in "OWL2 (Second edition)".

Granted that no choice but to introduce some postulates in order to make OWL semantics compatible to RDF semantics, this misunderstanding is partly

[1] http://www.w3.org/TR/2012/REC-owl2-rdf-based-semantics-20121211/#Appendix:_Comprehension_Conditions_.28Informative.29.

Y.-F. Li et al. (Eds.): JIST 2016, LNCS 10055, pp. 15–31, 2016.
DOI: 10.1007/978-3-319-50112-3_2

attributed to RDF semantics itself; indeed, there are not enough explanations in RDF Semantics (Hayes 2004) on the setup of preventing "membership loops". Hayes only claimed that the semantic model distinguishes both properties and classes, regarded as objects, from their extensions, so that this distinction prevents "membership loops". He also stated that the violation of the *axiom of foundation*, which is one of the axioms of standard set theories like Zermelo-Fraenkel (ZF) that forbids infinitely descending chains of membership, does not happen in RDF. However, the mechanism of preventing membership loops is still obscure for readers of RDF Semantics.

In this paper, at Sect. 2, we describe Russell paradox that has roots in Cantor's naive set theory, and summarize the history of set theories for the resolution of the paradox. At Sect. 3, we review RDF Semantics and OWL Semantics with higher order classes. Then, at Sect. 4, we propose a set of criteria for higher order classes and ontological meta-modeling. Those criteria are actually derived from the axioms and principles introduced in *Ramified Type Theory* for the resolution of Russell paradox in Principia Mathematica (PM, Vol.1). At Sect. 5, several examples of meta-modeling, including sound ones and unsound ones, are discussed, and some of solutions are demonstrated according to the meta-modeling criteria presented here.

We call the OWL Full with these criteria *Restricted OWL Full*. Thus, ambiguous word "punning" by W3C is clearly fixed on the meta-modeling with higher order classes and it helps us to deeply understand the semantics and the inference mechanism of RDF/OWL systems that allow ontological meta-modeling.

2 History of Set Theory and Type Theory

The history of set theory is a history of coping with Russell paradox essentially contained by the mathematical concept of set. Russell paradox was resolved in two ways. One is *axiom of separation* by Zermelo and the other is *Ramified Type Theory* by Whitehead and Russell. In the current set theory, a set is discriminated from a class that cannot be a member of set[2]. It is simply stated by Bourbaki (Bourbaki 1966) that there are two sorts of relations, i.e., relations which can make sets and which cannot make sets. This section abstracts set theories by Cantor, Zermelo, type theories by Russell, and the resolution of Russell paradox in Knowledge Interchange Format (KIF 1994).

2.1 RDF Sematics and Sets of Objects

RDF semantics is built on the foundation of set theory as well as every other formal theory in mathematics and logics. In fact, Hayes invokes a set theory named Zermelo-Fraenkel in order to rationalize *membership loops* that do not cause any paradoxes.

[2] A class notion in set theories is different from one in ontology descriptions.

> When classes are introduced in RDFS, they may contain themselves. Such 'membership loops' might seem to violate the *axiom of foundation*, one of the axioms of standard (Zermelo-Fraenkel) set theory, which forbids infinitely descending chains of membership. However, the semantic model given here distinguishes properties and classes considered as objects from their extensions - the sets of object-value pairs which satisfy the property, or things that are 'in' the class - thereby allowing the extension of a property or class to contain the property or class itself without violating the axiom of foundation. (Hayes 2004)

Although the Recommendation claims that the semantic model in RDF semantics does not violate the axiom of foundation on membership loops, yet there are not enough discussions on why and how the membership loops do not violate the axiom of foundation. It only states that $x \neq \mathbb{C}\mathrm{EXT}(x)$ for an object x, where $\mathbb{C}\mathrm{EXT}(x)$ is the class extension of x, and $p \neq \mathbb{I}\mathrm{EXT}(p)$ for a property p, where $\mathbb{I}\mathrm{EXT}(p)$ is the extension of p.

This paper claims the rdfs:Resource and rdfs:Class are proper classes in sets that do not cause paradoxes on sets, because the extension of rdfs:Resource is a totality and the extension of rdfs:Class is also a totality. A totality is not regarded as a set in today's theory. So, the rdfs:Resource and the rdfs:Class should be conceived to be a convention of referring the universal class concept and the universal metaclass concept respectively in the universe of discourse.

2.2 Cantor's Paradox and His Final Legacy

The history of set theories started with Georg Cantor. It is obvious that Cantor assumed members of sets are countable objects and a set is a collection of objects (Cantor 1895). Yet, he actually did not mention about objects, and he clarified the concept of natural numbers based on a (naive) set theory. However, Cantor became to know a paradox (called Cantor's paradox) contained by his own set theory in case of handling the totality of infinite sets including sets of sets[3], and he noticed it in his letter to Dedekind[4] (Cantor 1967). Aczel wrote,

> Cantor's final legacy, beyond the discovery of the transfinite numbers and the continuum hypothesis, was his realization that there could be no set containing everything ... since, given any set, there is a larger set – its set of subsets, the power set. (Aczel 2000)

Today, we know the power set $\wp(a) = \{x \mid x \subseteq a\}$ is too powerful for making sets of infinite cardinality, whereby Cantor discovered the transfinite ordinal numbers, but he also opened a door to lead paradoxes involved by set theories.

---

[3] The universe of natural numbers is factually defined as sets of sets that include the empty set as number zero and powersets of sets as number successors.

[4] In the letter, it is stated that "The system Ω of all numbers is an inconsistent, absolutely infinite multiplicity."

2.3 Comprehension Principle and Russell Paradox

The followings are some of axioms in naive set theory (Boolos 1971).

$$\forall x \forall y \forall z[(z \in x \Leftrightarrow z \in y) \Rightarrow x = y] \tag{1}$$
$$\exists y \forall x[x \in y \Leftrightarrow x \neq x] \tag{2}$$
$$\forall z \forall w \exists y \forall x[x \in y \Leftrightarrow (x = z \vee x = w)] \tag{3}$$
$$\exists y \forall x \exists w[x \in y \Leftrightarrow (x \in w \wedge w \in z)] \tag{4}$$
$$\exists y \forall x[x \in y \Leftrightarrow x = x] \tag{5}$$

The first one is called *axiom of extensionality* (1), and in order respectively, *empty set* (2), *pairing* (3), *union* (4), and *universal* (5). Especially, the empty set $\emptyset$ is defined by the second formula, such that $\emptyset \equiv \{x \mid x \neq x\}$.

The intensional definition of sets such that $\{x \mid \varphi(x)\}$, where $\varphi(x)$ is any formula, is called *comprehension principle* (Kamareddine et al. 2004).

Definition 1 *(Comprehension Principle).*

$$\exists y \forall x[x \in y \Leftrightarrow \varphi(x)] \quad \text{where } y \text{ is not free in } \varphi(x) \tag{6}$$

It was once conceived that the *comprehension principle* was very natural for intensionally defining a set because any (unary) predicate could be applied to a given object in order to determine the membership of an object to a set that holds a property featured by the formula under the law of *excluded middle*; namely, we can determine for any objects whether an object belongs to the set or not. Meanwhile, Russell keenly pointed a paradox in sets. If we take $\varphi(x)$ to $x \notin x$, which intends to denote a set that does not include any membership loops, then it follows that $y = \{x \mid x \notin x\}$ or $\exists y \forall x[x \in y \Leftrightarrow x \notin x]$. Then, in case of instantiating an arbitrary x to y, we obtain a contradiction as follows.

Definition 2 *(Russell Paradox).*

$$\exists y[y \in y \Leftrightarrow y \notin y].$$

Russell revealed that the *comprehension principle* inevitably involves a paradox[5]. J. van Heijenoort stated,

> Bertrand Russell discovered what became known as the Russell paradox in June 1901 [...]. In the letter [to Frege], written more than a year later and hitherto unpublished, he communicates the paradox to Frege. The paradox shook the logicians' world, and the rumbles are still felt today.(From Frege to Gödel in van Heijenoort 1967)

[5] Exactly, Russell pointed the paradox in the expression of functions rather than sets, in the letter to Frege (van Heijenoort 1967).

2.4 Zermelo's Axiom of Separation

To avoid Russell paradox, Ernst Zermelo replaced the *comprehension principle* with the *axiom of separation (aussonderung)* (Kamareddine et al. 2004).

Definition 3 *(Axiom of Separation).*

$$\exists y \forall x[x \in y \Leftrightarrow x \in z \wedge \varphi(x)] \quad \text{where } \ y \text{ does not occur in } \varphi(x) \tag{7}$$

In this case, y must be a proper subset of a set z, of which members satisfy the formula. It may be phrased that the paradox is work-arounded by introducing z distinctively existing. Today, a set theory based on a series of axioms by Zermelo is called Zermelo-Fraenkel (ZF) set theory together with Fraenkel's *axiom of replacement.*

It is worthy to note that in some of modern computer languages a functionality to make a list whose elements are selected from another list so that they satisfy a specified condition is accidentally misnamed *list comprehension*[6]. We claim that this functionality in computer languages should be properly named "list-separation" or "list-selection" in order to avoid misunderstanding between *comprehension principle* in set theories and *list comprehension* in computer languages. As well, the term of "comprehension conditions (principles)", which is strongly associated with paradoxes in sets, in "OWL Semantics and Abstract Syntax, Sect. 5", should be renamed to "list conditions by selection" or something else, in order to avoid unfounded fear of paradoxes associated with *comprehension principle.*

2.5 NBG and Set Theory of KIF 3.0

The Knowledge Interchange Format (KIF) 3.0 theory proposed a special set theory that is based on von Neumann-Bernays-Gödel (NBG) set theory. The Chap. 7 of the reference manual (KIF 1994), which is titled "Sets", starts with the statement below.

> In many applications, it is helpful to talk about sets of objects as objects in their own right, e.g. to specify their cardinality, to talk about subset relationships, and so forth. The formalization of sets of simple objects is a simple matter; but, when we begin to talk about sets of sets, the job becomes difficult due to the threat of paradoxes (like Russell's hypothesized set of all sets that do not contain themselves).
> (Knowledge Interchange Format version 3.0 Reference Manual (KIF 1994))

It is obvious that the KIF Group were worried about involving the paradoxes on sets in the KIF specification. NBG allows us to make sets that include individuals and sets or classes of individuals. The axioms in NBG is roughly separated into two parts; one is for sets, in which a set is expressed using variables with lower letters, and the other is for classes, in which a class is expressed using variables

[6] For example, in Haskell, $[x \uparrow 2 \mid x \leftarrow [1..5]]$ produces $[1, 4, 9, 16, 25]$. (Hutton 2007).

with upper letters. Although both parts of axioms are composed of almost same forms, they are separated owing to the idea that classes are a different sort from sets of individuals in NBG.

KIF distinguishes bounded set (set) from unbounded set (proper class). A bounded set can be a member of a set, but an unbounded set cannot be a member of set. A bounded set is finite. A finite set is bounded, but an infinite set is unbounded. It is consistent to the standard set theories, ZF and NBG. KIF contains one more notion, that is, individuals.

> In KIF, a fundamental distinction is drawn between individuals and sets. A set is a collection of objects. An individual is any object that is not a set. A distinction is also drawn between objects that are bounded and those that are unbounded. This distinction is orthogonal to the distinction between individuals and sets. There are bounded individuals and unbounded individuals. There are bounded sets and unbounded sets. The fundamental relationship among these various types of entities is that of membership. Sets can have members, but individuals cannot. Bounded objects can be members of sets, but unbounded objects cannot. (It is this condition that allows us to avoid the traditional paradoxes of set theory.) (KIF 1994, Sect. 7.1 Basic Concepts)

Although we may have a curiosity what *unbounded individuals* are, there is no explanation in KIF what they are, and the study is beyond the scope of this paper. Russell paradox is described in KIF as follows.

> The paradoxes appear only when we try to define set primitives that are too powerful. We have defined the sentence '(member τ σ)' to be true in exactly those cases when the object denoted by τ is a member of the set denoted by σ, and we might consider defining the term '(setofall τ ϕ)' to mean simply the set of all objects denoted by τ for any assignment of the free variables of τ that satisfies ϕ. Unfortunately, these two definitions quickly lead to paradoxes.
>
> Let $\phi_{\nu/\tau}$ be the result of substituting term τ for all free occurrences of ν in sentence ϕ. Provided that τ is a term not containing any free variables captured in $\phi_{\nu/\tau}$, then the following schema follows from our informal definition. This schema is called the *principle of unrestricted set abstraction.*
>
> $$(\Leftrightarrow \text{(member } \tau \text{ (setofall } \nu\ \phi))\ \phi_{\nu/\tau})$$
>
> (KIF 1994, Sect. 7.4 Paradoxes)

Note that this form is equivalent to the unrestricted *comprehension principle* (6) and it causes Russell paradox.

Instead of this principle, Russell paradox is avoided in KIF by *restricted set abstraction*, in which a set is restricted to bounded sets.

> In the von-Neuman-Gödel-Bernays version of set theory, these paradoxes are avoided by replacing the *principle of unrestricted set abstraction* with the *principle of restricted set abstraction* given above.
>
> $$(\Leftrightarrow \text{(member } \tau \text{ (setofall } \nu\ \phi))\ \text{(and (bounded } \tau)\ \phi_{\nu/\tau}))$$ (*ibid.*)

Note that this form resembles the *axiom of separation* (7) in ZF. Here a bounded object '(bounded τ)' is used instead of the restriction '$x \in z$'.

KIF succeeded to eliminate paradoxes by the concept of bounded set of objects, whereas Russell's *Ramified Type Theory* is much suitable to explain how RDFS can avoid Russell paradox for the correct comprehension of the framework for meta-modeling[7].

2.6 Russell's Ramified Type Theory

Alfred N. Whitehead and Bertrand Russell developed the first type theory, *Ramified Type Theory*, in the epoch-making three-volume books, 'Principia Mathematica' (hereafter *PM* for short). They attempted to solve the Russel paradox together with other paradoxes by capturing them as variations of *vicious circle*.

> An analysis of the paradoxes to be avoided shows that they all result from a certain kind of vicious circle. (*PM, Vol.1*, Introduction, Chap. 2)

They emphasized that statements about "all propositions" are meaningless owing to the totality contained in the statements.

> [...] if we suppose the set to have a total, it will contain members which presuppose this total, then such a set cannot be a total. (*ibid.*)

Therefore, they, first of all, postulate the *vicious-circle principle* in order to avoid paradoxes caused by self-reference.

> The principle which enables us to avoid illegitimate totalities may be stated as follows: "Whatever involves *all* of a collection must not be one of the collection"; or, conversely: "If, provided a certain collection had a total, it would have members only definable in terms of that total, then the said collection has no total." We shall call this the "vicious-circle principle", [...] (*ibid.*)

Whitehead and Russell introduced the idea of propositional function, in which sentences may include variables for not only objects but also functions[8], and value assignments for all variables of objects and functions turn open sentences unambiguous propositions. Thus, functions are also applied to as logical expression. For example, Leibniz equality of $x = y$ is defined as $\forall f[f(x) \Leftrightarrow f(y)]$.

Ramified Types. Types in *PM* have a double hierarchy, that is, (simple) types and orders. The second hierarchy is introduced by regarding also the types of the variables that are bound by a quantifier. Kamareddine, et al. explained the reason using a propositional function $z(\,) \vee \neg z(\,)$, which can involve an arbitrary

[7] The set theory in NBG for individuals and sets can be regarded as a sort of first order logic, and then classes can be regarded as first order. However, RDFS can be regarded as much higher order logic as shown at Sect. 3.

[8] Predicates are functions that return truth value.

proposition for z, then $\forall z \uparrow ()[z() \vee \neg z()]$[9] quantifies over all propositions for z in the universe. We must distinguish a simple proposition $C(a)$ and quantified $\forall z \uparrow ()[z() \vee \neg z()]$. The former does not involve any self-reference but the latter may involve the self reference for z. This problem is solved by dividing types into *orders*. An order is simply a natural number that starts with 0, and in $\forall z \uparrow ()[z() \vee \neg z()]$ we must restrict the form by mentioning the order of propositions. Thus, propositional function of the form $\forall z \uparrow ()^n[z() \vee \neg z()]$ quantifies over only all propositions of order n, and this form has its own order $n+1$.

Definition 4 *(Ramified Types). PM explained ramified types for only unary and binary functions. Kamareddine, et al. extended the definition to n-ary.* (Kamareddine et al. 2004).

1. 0^0 is a ramified type;
2. If $t_1^{a_1}, \ldots, t_n^{a_n}$ are ramified types and $a \in \mathbb{N} > max(a_1, \ldots, a_n)$, then $(t_1^{a_1}, \ldots, t_n^{a_n})^a$ is a ramified type (if $n = 0$ then take $a \geq 0$) ;
3. All ramified types can be constructed using rules 1 and 2.

Note that in $(t_1^{a_1}, \ldots, t_n^{a_n})^a$ we demand that $a > a_i$ for all i. Furthermore, Whitehead and Russell defined *predicative* condition on ramified types.

Definition 5 *(Predicative Types).*

$$(t_1^{a_1}, \ldots, t_n^{a_n})^a \qquad \text{where } a = 0 \text{ if } n = 0, \text{ else } a = 1 + ma\bar{x}(a_1, \ldots, a_n)$$

The followings are some examples of predicative types.

- 0^0;
- $(0^0)^1$;
- $\left((0^0)^1, (0^0)^1\right)^2$;
- $\left((0^0)^1, \left((0^0)^1, (0^0)^1\right)^2\right)^3$.

The above expressions of ramified types are also expressed as follows by Stevens (Stevens 2003) using function symbol F, G, H, and I, and their arguments.

- ${}^0x^0$;
- ${}^1F^{(0/0)}({}^0x^0)$;
- ${}^2H^{(1/(0/0),1/(0/0))}\left({}^1F^{(0/0)}({}^0x^0), {}^1G^{(0/0)}({}^0y^0)\right)$;
- ${}^3I^{(1/(0/0),2/(1/(0/0),1/(0/0)))}\left({}^1F^{(0/0)}({}^0x^0), {}^2H^{(1/(0/0),1/(0/0))}\right)$.

[9] In this paper '↑' is used to indicate the type of variable instead of colon that is usually used in type theory, as a colon is confusing with the notation for namespace in the syntax of Semantic Web.

Here the first 0 in (0/0) stands for the order as argument and the last 0 stands for the type as argument at the individual level. 1 in 1/(0/0) stands for the order as first order as argument. The prefix number of function stands for orders of the form.

Suggested by ramified types, we can put orders to classes on RDF semantics with interpreting a class name as unary predicates in predicate calculus or propositional functions in *PM*. In the next section, we attempt to formalize RDF semantics taking into account of orders in Ramified Type Theory for classes, and claims that RDFS may avoid Russell paradox.

3 Formalization of RDF/OWL Semantics Based on Higher Order Types

3.1 Preliminary Explanations of Notations, Denotations, and Universe of Discourse

Notation. In this formulation, $\mathbb{R}$ stands for the universe of discourse, $\mathcal{P}$ stands for a finite set of logical predicate symbols, $\mathcal{F}$ stands for a finite set of functional symbols, and $\mathcal{V}$ stands for a countable set of vocabularies. Every sentence in this formulation is a triple like $\langle s, p, o \rangle$ composed of words in a vocabulary in $\mathcal{V}$.

Interpretation $\mathcal{I}$ is a mapping from a set of triples and vocabularies $\mathcal{V}$ into the universe of discourse $\mathbb{R}$. Logical symbols, i.e., $\in$ (relation between elements and a set), $\subseteq$ (inclusiveness among sets), $\sqsubseteq$ (sub/super concept of class relation), are used in addition to common logical connectives $\wedge$ (conjunction) and $\vee$ (disjunction). In the domain of RDF and OWL, $\mathcal{F}$ contains only $\mathbb{EXT}(.)$ (extension of property) and $\mathbb{CEXT}(.)$ (extension of class).

There are no variables except for blank nodes in RDF and OWL. So, note that every symbol in RDF and OWL standing for an object is a constant term in logics. However, discussions on sets require variables. Thus, variables for sets are expressed x, y, ... or x_1, x_2, ..., and x_i, y_i. When we indicate variables in logical forms, they may be explicitly expressed with quantifiers $\forall$ or $\exists$.

Tarskian Denotational Semantics. We discriminate sentences and words in sentences from their denotations (Tarski 1946). For example, a word "Tokyo" as logical term *Tokyo* interpreted as a city named Tokyo in Japan, and the denotation is expressed as $\mathcal{I}[\![Tokyo]\!]$ or $Tokyo^{\mathcal{I}}$. In this case, we say "*Tokyo* denotes $Tokyo^{\mathcal{I}}$" or "$Tokyo^{\mathcal{I}}$ is the denotation of *Tokyo*" for $\mathcal{I}[\![Tokyo]\!] = Tokyo^{\mathcal{I}}$. In this representation, $x = x$ is interpreted as $x^{\mathcal{I}} = x^{\mathcal{I}}$ (as tautology), and $a = b$ is interpreted to $a^{\mathcal{I}} = b^{\mathcal{I}}$ (when both denotations are identical). A sentence denotes truth value, of which truth or falsity is decided depending on rules of interpretation and models constructed by ontologists. For example, $New\_York^{\mathcal{I}} = Apple\_City^{\mathcal{I}}$ is true in some case or false in another case.

The interpretation by denotational semantics do not require us to make *unique name assumption*. We represent $a \simeq b$ for owl:sameAs and $a \not\simeq b$ for owl:differentFrom in OWL sentences. Then, the followings hold on different nodes of RDF graph $a^{\mathcal{I}}$ and $b^{\mathcal{I}}$ in OWL[10].

$$\mathcal{I}[\![a \simeq b]\!] \Rightarrow a^{\mathcal{I}} = b^{\mathcal{I}}$$
$$\mathcal{I}[\![a \not\simeq b]\!] \Rightarrow a^{\mathcal{I}} \neq b^{\mathcal{I}}$$

Universe of Discourse by Set Theory. In set theories, a set is extensionally defined by enumerating all members of the set, or intensionally defined by using logical conditions that all members in the set satisfy, which is like *comprehension principle.* However, the expression $x^{\mathcal{I}} = x^{\mathcal{I}}$ is always true in any case, thus the universe of discourse that stands for the totality can be defined as follows,

$$\mathbb{R} \equiv \{x^{\mathcal{I}} \mid x^{\mathcal{I}} = x^{\mathcal{I}}\}.$$

Note that $\mathbb{R}$ as the universe of discourse can contain all denotations (objects in models), and every entity in the universe always belongs to $\mathbb{R}$ because $x^{\mathcal{I}} = x^{\mathcal{I}}$ is always true for any $x^{\mathcal{I}}$. Also note that this form looks like a set but $\mathbb{R}$ is actually not a set but a proper class that contains everything in the universe and cannot be a member of a set.

Property Extension. In this paper, the interpretation of a triple $\langle s, p, o\rangle$ or $\langle s, o\rangle \in \mathrm{EXT}(p)$ is represented as $\langle s^{\mathcal{I}}, o^{\mathcal{I}}\rangle \in \mathbb{E}\mathrm{XT}(p^{\mathcal{I}})$,

$$\mathcal{I}[\![\langle s, o\rangle \in \mathrm{EXT}(p)]\!] \Rightarrow \langle s^{\mathcal{I}}, o^{\mathcal{I}}\rangle \in \mathbb{E}\mathrm{XT}(p^{\mathcal{I}}).$$

$\mathbb{E}\mathrm{XT}(p^{\mathcal{I}})$ is called the (semantic) extension of property $p^{\mathcal{I}}$. $\mathbb{E}\mathrm{XT}(p^{\mathcal{I}})$ is a mapping into the powerset of direct product $\mathbb{R} \times \mathbb{R}$, thereby the arguments $x^{\mathcal{I}}$ and $y^{\mathcal{I}}$ of an ordered pair $\langle x^{\mathcal{I}}, y^{\mathcal{I}}\rangle$ are in $\mathbb{R}$, namely, $(x^{\mathcal{I}} \in \mathbb{R} \wedge y^{\mathcal{I}} \in \mathbb{R})$[11].

Class Extension. We express a triple $\langle s, \mathit{rdf}:\mathit{type}, o\rangle$ as $s \uparrow o$ in this paper, then the class extension can be captured as a set of which members can be interpreted as instances of classes. Namely, for an instance $x^{\mathcal{I}}$ of class $y^{\mathcal{I}}$,

$$\mathcal{I}[\![x \uparrow y]\!] \Rightarrow x^{\mathcal{I}} \Uparrow y^{\mathcal{I}} \equiv \langle x^{\mathcal{I}}, y^{\mathcal{I}}\rangle \in \mathbb{E}\mathrm{XT}(\mathit{rdf}:\mathit{type}^{\mathcal{I}}) \equiv x^{\mathcal{I}} \in \mathbb{C}\mathrm{EXT}(y^{\mathcal{I}}).$$

$\mathbb{C}\mathrm{EXT}(y^{\mathcal{I}})$ is called the (semantic) class extension of class $y^{\mathcal{I}}$.

3.2 Higher Order Classes

We introduce orders into the description of RDFS classes. Namely,

$$^{n}x^{\mathcal{I}} \Uparrow {}^{m}y^{\mathcal{I}} \equiv \langle {}^{n}x^{\mathcal{I}}, {}^{m}y^{\mathcal{I}}\rangle \in \mathbb{E}\mathrm{XT}(\mathit{rdf}:\mathit{type}^{\mathcal{I}}) \equiv {}^{n}x^{\mathcal{I}} \in \mathbb{C}\mathrm{EXT}({}^{m}y^{\mathcal{I}}) \quad \text{where } m > n \geq 0.$$

[10] A question arises in the case of no statements of owl:sameAs and owl:differentFrom for atomic nodes in comparison of two different graphs. We proposed the algorithm named *UNA for atomic objects in the non-UNA condition.* See the motivation and the detail in Koide and Takeda 2011.

[11] See the simple interpretation 3 in RDF Semantics (Hayes 2004).

Here n and m is an order of class x and y, respectively. When exactly $m = n+1$, we call it *predicative* as well as *PM*.

Definition 6 *(Predicative Classes).*

$$^n x^{\mathcal{I}} \Uparrow {}^{n+1}y^{\mathcal{I}} \equiv \langle {}^n x^{\mathcal{I}}, {}^{n+1}y^{\mathcal{I}} \rangle \in \mathbb{E}\mathrm{XT}(\mathit{rdf:type}^{\mathcal{I}}) \text{ where } n \geq 0, \qquad (8)$$

$$^n x^{\mathcal{I}} \Uparrow {}^{n+1}y^{\mathcal{I}} \equiv {}^n x^{\mathcal{I}} \in \mathbb{C}\mathrm{EXT}({}^{n+1}y^{\mathcal{I}}) \text{ where } n \geq 0. \qquad (9)$$

We distinguish $^n x^{\mathcal{I}}$ from $^{n+1}x^{\mathcal{I}}$ as different nodes in an RDF graph, while the x is the same lexical token as IRI or nodeID in $\mathcal{V}$. Therefore, we can obtain the following lemma without violating vicious circle principle,

Lemma 1.

$$^n x^{\mathcal{I}} \Uparrow {}^{n+1}x^{\mathcal{I}} = \langle {}^n x^{\mathcal{I}}, {}^{n+1}x^{\mathcal{I}} \rangle \in \mathbb{E}\mathrm{XT}(\mathit{rdf:type}^{\mathcal{I}}) \text{ where } n \geq 0, \qquad (10)$$

$$^n x^{\mathcal{I}} \Uparrow {}^{n+1}x^{\mathcal{I}} = {}^n x^{\mathcal{I}} \in \mathbb{C}\mathrm{EXT}({}^{n+1}x^{\mathcal{I}}) \text{ where } n \geq 0. \qquad (11)$$

We assume that rdf:type has the same role in the universe from $n = 0$ to ∞. This is actually the same as *axiom of reducibility* (*PM, Vol.1*, Introduction, VI), which is required to enable that there exists a (higher order) function as formally the same as the predicative function that takes individuals as arguments. We extend this principle to every RDF and OWL properties later on.

3.3 Subsumption in Higher Order Classes

The RDFS semantic condition on rdfs:subClassOf contains the following condition that is obtained with extending RDFS-original classes to higher order classes,

$$^n x^{\mathcal{I}} \sqsubseteq {}^m y^{\mathcal{I}} \equiv {}^n x^{\mathcal{I}} \in \mathbb{C} \wedge {}^m y^{\mathcal{I}} \in \mathbb{C} \wedge \mathbb{C}\mathrm{EXT}({}^n x^{\mathcal{I}}) \subseteq \mathbb{C}\mathrm{EXT}({}^m y^{\mathcal{I}}) \text{ where } m, n \geq 1.$$

Here $\sqsubseteq$ represents subclass-superclass relation that is designated by rdfs: subClassOf, and $\mathbb{C}$ may be called the *universal domain of classes*, to which all classes in the universe of discourse belong. The above condition is called *subsumption*.

Then, we introduce a new notion onto the subsumption by higher order classes. If $n = m$ on the above condition, let us call it *interpretable*.

Definition 7 *(Interpretable Class Condition).*

$$^n x^{\mathcal{I}} \sqsubseteq {}^n y^{\mathcal{I}} \equiv {}^n x^{\mathcal{I}} \in \mathbb{C} \wedge {}^n y^{\mathcal{I}} \in \mathbb{C} \wedge \mathbb{C}\mathrm{EXT}({}^n x^{\mathcal{I}}) \subseteq \mathbb{C}\mathrm{EXT}({}^n y^{\mathcal{I}})) \text{ where } n \geq 1. \qquad (12)$$

Namely, both classes related by rdfs:subClassOf must be the same order.

The order n of classes should be greater than zero ($n > 0$), since order 0 is assigned only individuals and an individual cannot have its own extension. Note that this *interpretable class condition* is constructively obtained in accordance with the definition of "being of the same type" *PM* *9.131, in which it is stated u and v "are the same type," if "(1) both are individuals, (2) both are elementary functions taking arguments of the same type". Individuals are interpretable. So, the first order classes are also interpretable. Then, we consider rdfs:subClassOf relation among the first order classes also interpretable. Thus, this procedure may be repeated again and again from order 0 to order n.

3.4 Universal Class in Higher Order Classes

As shown in the entailment rule **rdfs4a** and **rdfs4b** (Hayes 2004), every entity in the universe of discourse is an instance of $\mathit{rdfs\!:\!Resource}^{\mathcal{I}}$,

$$\vdash \forall u^{\mathcal{I}}[u^{\mathcal{I}} \in \mathbb{C}\mathrm{EXT}(\mathit{rdfs\!:\!Resource}^{\mathcal{I}})].$$

RDFS entailment lemma in (Hayes 2004) states that every entity as class is a subclass of the class rdfs:Resource,

$$\vdash \forall c^{\mathcal{I}}[c^{\mathcal{I}} \sqsubseteq \mathit{rdfs\!:\!Resource}^{\mathcal{I}}].$$

We extend these forms for entities and classes to higher order classes in the universe as well as described above.

$$\vdash \forall {}^{n}u^{\mathcal{I}}[{}^{n}u^{\mathcal{I}} \in \mathbb{C}\mathrm{EXT}({}^{n+1}\mathit{rdfs\!:\!Resource}^{\mathcal{I}}) \equiv {}^{n}\mathbb{R}] \qquad \text{where } n \geq 0, \tag{13}$$

$$\vdash \forall {}^{n}c^{\mathcal{I}}[{}^{n}c^{\mathcal{I}} \sqsubseteq {}^{n}\mathit{rdfs\!:\!Resource}^{\mathcal{I}}] \qquad \text{where } n \geq 1. \tag{14}$$

We see that all individuals, which is expressed as ${}^{0}u^{\mathcal{I}}$ belong to ${}^{0}\mathbb{R}$, and all first classes, which is expressed as ${}^{1}u^{\mathcal{I}}$ belong to ${}^{1}\mathbb{R}$, and so forth. Using this extended rule (13), we see that the universe of discourse $\mathbb{R}$ stratifies by orders. Every entity in n-th order universe ${}^{n}\mathbb{R}$ is an instance of $(n+1)$-th order ${}^{n+1}\mathit{rdfs\!:\!Resource}^{\mathcal{I}}$. Therefore, all extensions of ${}^{n}\mathit{rdfs\!:\!Resource}^{\mathcal{I}}$ $(n \geq 1)$ covers all entities in the universe and the union of ${}^{n}\mathbb{R}$ coincides with the universe of discourse $\mathbb{R}$.

Definition 8 *(Universal Class and Stratified Universe).*

$$\bigcup_{i=1 \rightarrow \infty} \mathbb{C}\mathrm{EXT}({}^{i}\mathit{rdfs\!:\!Resource}^{\mathcal{I}}) = \bigcup_{i=0 \rightarrow \infty} {}^{i}\mathbb{R} = \mathbb{R} \tag{15}$$

We abbreviate this form to the following that is described in RDF Semantics.

$$\mathbb{C}\mathrm{EXT}(\mathit{rdfs\!:\!Resource}^{\mathcal{I}}) = \mathbb{R}$$

Thus, $\mathit{rdfs\!:\!Resource}^{\mathcal{I}}$ is appropriate to name *universal class* due to the extension being the universe of discourse $\mathbb{R}$.

3.5 Universal Metaclass in Higher Order Classes

As well as the universe of discourse $\mathbb{R}$, we set up the universal domain of classes in discourse, $\mathbb{C}$, in which all classes in the universe exist. Then, we can define for higher order classes,

$$^{n}c^{\mathcal{I}} \in \mathbb{C}\mathrm{EXT}(^{n+1}\mathit{rdfs\colon Class}^{\mathcal{I}}) \equiv {}^{n}\mathbb{C} \quad \text{where } n \geq 1. \tag{16}$$

Definition 9 *(Universal Metaclass and Stratified Universe of Classes).*

$$\bigcup_{i=2\to\infty} \mathbb{C}\mathrm{EXT}(^{i}\mathit{rdfs\colon Class}^{\mathcal{I}}) = \bigcup_{i=1\to\infty} {}^{i}\mathbb{C} \equiv \mathbb{C} = \mathbb{R}\backslash{}^{0}\mathbb{R} \tag{17}$$

The RDF semantics shows the following condition. It is deemed to be an abbreviation of the universal metaclass (17),

$$\mathbb{C}\mathrm{EXT}(\mathit{rdfs\colon Class}^{\mathcal{I}}) = \mathbb{C}.$$

While $\mathit{rdfs\colon Class}^{\mathcal{I}}$ is appropriate to be called *universal metaclass* as a representative class for the universal domain of classes in discourse, we need to make clear the relation between $\mathbb{R}$ and $\mathbb{C}$ or $\mathit{rdfs\colon Resource}^{\mathcal{I}}$ and $\mathit{rdfs\colon Class}^{\mathcal{I}}$.

From stratified universe (13) and stratified universal domains of classes (16), we obtain the followings,

$$\begin{aligned} {}^{n}\mathit{rdfs\colon Class}^{\mathcal{I}} \in \mathbb{C}\mathrm{EXT}(^{n+1}\mathit{rdfs\colon Resource}^{\mathcal{I}}) \equiv {}^{n}\mathbb{R} \quad &\text{where } n \geq 2,\\ {}^{n}\mathit{rdfs\colon Class}^{\mathcal{I}} \sqsubseteq {}^{n}\mathit{rdfs\colon Resource}^{\mathcal{I}} \quad &\text{where } n \geq 2. \end{aligned}$$

We distinguish ${}^{n}\mathit{rdfs\colon Class}^{\mathcal{I}}$ and ${}^{n+1}\mathit{rdfs\colon Class}^{\mathcal{I}}$ as well as we distinguish ${}^{n}\mathit{rdfs\colon Resource}^{\mathcal{I}}$ and ${}^{n+1}\mathit{rdfs\colon Resource}^{\mathcal{I}}$. Thus, we obtain the followings from (13) and (16),

$$\begin{aligned} {}^{n}\mathit{rdfs\colon Resource}^{\mathcal{I}} \in \mathbb{C}\mathrm{EXT}(^{n+1}\mathit{rdfs\colon Resource}^{\mathcal{I}}) \equiv {}^{n}\mathbb{R} \quad &\text{where } n \geq 1,\\ {}^{n}\mathit{rdfs\colon Resource}^{\mathcal{I}} \in \mathbb{C}\mathrm{EXT}(^{n+1}\mathit{rdfs\colon Class}^{\mathcal{I}}) \equiv {}^{n}\mathbb{C} \quad &\text{where } n \geq 1,\\ {}^{n}\mathit{rdfs\colon Class}^{\mathcal{I}} \in \mathbb{C}\mathrm{EXT}(^{n+1}\mathit{rdfs\colon Class}^{\mathcal{I}}) \equiv {}^{n}\mathbb{C} \quad &\text{where } n \geq 2. \end{aligned}$$

Let us call these complex relations between rdfs:Resource and rdfs:Class *hemi-cross subsumption*, as these equations draw a picture like cross fire but ${}^{n}\mathit{rdfs\colon Resource}^{\mathcal{I}} \not\sqsubseteq {}^{n}\mathit{rdfs\colon Class}^{\mathcal{I}}$.

Thus, if we neglect the orders of classes, the membership loops appear on rdfs:Resource and rdfs:Class, but by seeing the orders, no membership loops exist in the universe.

$$\begin{aligned} \mathit{rdfs\colon Class}^{\mathcal{I}} &\in \mathbb{C}\mathrm{EXT}(\mathit{rdfs\colon Class}^{\mathcal{I}})\\ \mathit{rdfs\colon Resource}^{\mathcal{I}} &\in \mathbb{C}\mathrm{EXT}(\mathit{rdfs\colon Resource}^{\mathcal{I}}) \end{aligned}$$

4 Meta-Modeling Criteria in RDFS and OWL

As shown above, several principles about the order number of classes are addressed to avoid infinite membership loops. Here we set up them as criteria for meta-modeling.

1. (reducible) Every property that is applicable to individuals and the first order classes is applicable to much higher order classes.
2. (predicative) In respect of properties that make the relation of instance and class, i.e., rdf:type, owl:oneOf, rdfs:domain, rdfs:range, etc., the order of class must be plus one to the order of the instances.
3. (interpretable) In respect of properties that make the relation among non-literal objects, the order of arguments must be the same. Note that this principle is not applied to instance objects of Datatype such as strings, numbers, or URLs as datatype.
4. (constructive) Even if we adopt the predicative and the interpretable principles, ambiguous and undecidable entities on orders may still remain. Such a case, the orders must be decidable by ascendingly computing orders from individuals ($n = 0$), and first classes ($n = 1$), or descendingly computable starting at higher orders to lower orders so that the computation terminates at individuals level ($n = 0$).

In RDF-based OWL semantics, the class extension of OWL is defined as follows[12],

$$\mathbb{C}\mathrm{EXT}(c^{\mathcal{I}}) = \{x^{\mathcal{I}} \in \mathbb{R} \mid \langle x^{\mathcal{I}}, c^{\mathcal{I}} \rangle \in \mathbb{E}\mathrm{XT}(\mathit{rdf:type}^{\mathcal{I}})\}.$$

We extend this definition to higher order classes as

$$\mathbb{C}\mathrm{EXT}({}^{n+1}c^{\mathcal{I}}) = \{{}^{n}x^{\mathcal{I}} \in {}^{n}\mathbb{R} \mid \langle {}^{n}x^{\mathcal{I}}, {}^{n+1}c^{\mathcal{I}} \rangle \in \mathbb{E}\mathrm{XT}(\mathit{rdf:type}^{\mathcal{I}})\}. \tag{18}$$

Namely, all individuals ${}^{0}x^{\mathcal{I}}$ in OWL belong to ${}^{0}\mathbb{R}$ of the RDF universe, and all first classes ${}^{1}c^{\mathcal{I}}$ belong to ${}^{1}\mathbb{R}$ of the RDF universe, and so forth.

In the document of RDF-based OWL semantics, a special syntax form is used for sequence of entities, i.e., $\mathbb{S}\mathrm{EQ} \equiv \mathbb{E}\mathrm{XT}(\mathit{rdf:List})$. In this paper, we express the sequence of entities simply $(x, y, \dots)$. So, owl:intersectionOf and owl:unionOf[13] are extended to higher order classes as follows,

$$\begin{aligned} &\langle\, {}^{n}z^{\mathcal{I}}, ({}^{n}c_1^{\mathcal{I}}, \dots, {}^{n}c_m^{\mathcal{I}})\,\rangle \in \mathbb{E}\mathrm{XT}(\mathit{owl:intersectionOf}) \Leftrightarrow \\ &\qquad {}^{n}z^{\mathcal{I}},\ {}^{n}c_1^{\mathcal{I}}, \dots, {}^{n}c_m^{\mathcal{I}} \in {}^{n}\mathbb{C} \wedge \mathbb{C}\mathrm{EXT}({}^{n}z^{\mathcal{I}}) = \bigcap_{i=1\to m} \mathbb{C}\mathrm{EXT}({}^{n}c_i^{\mathcal{I}}), \end{aligned} \tag{19}$$

$$\begin{aligned} &\langle\, {}^{n}z^{\mathcal{I}}, ({}^{n}c_1^{\mathcal{I}}, \dots, {}^{n}c_m^{\mathcal{I}})\,\rangle \in \mathbb{E}\mathrm{XT}(\mathit{owl:unionOf}) \Leftrightarrow \\ &\qquad {}^{n}z^{\mathcal{I}},\ {}^{n}c_1^{\mathcal{I}}, \dots, {}^{n}c_m^{\mathcal{I}} \in {}^{n}\mathbb{C} \wedge \mathbb{C}\mathrm{EXT}({}^{n}z^{\mathcal{I}}) = \bigcup_{i=1\to m} \mathbb{C}\mathrm{EXT}({}^{n}c_i^{\mathcal{I}}). \end{aligned} \tag{20}$$

[12] http://www.w3.org/TR/owl-rdf-based-semantics/#Class_Extensions.

[13] http://www.w3.org/TR/owl2-rdf-based-semantics/#Semantic_Conditions_for_Boolean_Connectives.

As well, the semantic conditions of enumeration is extended as

$$\langle\, {}^{n+1}z^{\mathcal{I}}, ({}^{n}a_1^{\mathcal{I}}, \ldots, {}^{n}a_m^{\mathcal{I}})\,\rangle \in \mathbb{E}\mathrm{XT}(owl\!:\!oneOf) \Leftrightarrow$$
$$ {}^{n+1}z^{\mathcal{I}} \in {}^{n+1}\mathbb{C} \wedge \mathbb{C}\mathrm{EXT}({}^{n+1}z^{\mathcal{I}}) = \{{}^{n}a_1^{\mathcal{I}}, \ldots, {}^{n}a_m^{\mathcal{I}}\}. \quad (21)$$

5 Related Work and Discussion

On Punning. W3C posed six use cases on "punning"[14], but only the first and the second cases in these use cases deserve to discuss in ontological view.

The first case is solved by making $a\!:\!Service$ a meta-class.

$$\boldsymbol{{}^{2}a\!:\!Service\ rdf\!:\!subClassOf\ {}^{2}owl\!:\!Class\ .}$$
$$ {}^{2}a\!:\!Service\ rdf\!:\!type\ {}^{3}owl\!:\!Class\ .$$
$$ {}^{1}a\!:\!Person\ rdf\!:\!type\ {}^{2}owl\!:\!Class\ .$$
$$ {}^{1}s1\ rdf\!:\!type\ {}^{2}a\!:\!Service\ .$$
$$ {}^{1}s1\ a\!:\!input\ {}^{1}a\!:\!Person\ .$$

The first triple shown above is newly added to the original set of the triples, so that the system becomes decidable and $s1$ becomes to be interpreted as the first order class rather than an individual because $n > 1$ for ${}^{n}owl\!:Class$. The last triple must be modified to the form for domain $s1$ and range $a : Person$ *constraints*.

The second use case is a typical quiz for meta-classing. Harry as individual is a eagle, and the eagle as species is in the Red List as endangered species. In the following triples, the first and second triples are newly added to the others, so that they set up $a : Species$ and $a : EndangeredSpecies$ as meta-classes. Then, the case becomes decidable.

$$\boldsymbol{{}^{2}a\!:\!Species\ rdfs\!:\!subClassOf\ {}^{2}owl\!:\!Class.}$$
$$\boldsymbol{{}^{2}a\!:\!EndangeredSpecies\ rdfs\!:\!subClassOf\ {}^{2}a\!:\!Species.}$$
$$ {}^{1}a\!:\!Eagle\ rdf\!:\!type\ {}^{2}owl\!:\!Class.$$
$$ {}^{0}a\!:\!Harry\ rdf\!:\!type\ {}^{1}a\!:\!Eagle.$$
$$ {}^{1}a\!:\!Eagle\ rdf\!:\!type\ {}^{2}a\!:\!Species.$$
$$ {}^{1}a\!:\!Eagle\ rdf\!:\!type\ {}^{2}a\!:\!EndangeredSpecies.$$

Domino-Tilting Puzzle. Motik posed Domino-tilting Puzzle to exemplify undecidable OWL Full (Motik 2007). In this example, GRID is an OWL class and does not interpreted as property. However, this model involves the infinite ascending higher order computation by the resulted stratified form such as ${}^{n}GRID \sqsubseteq \exists rdf\!:type.{}^{n+1}GRID$, starting from an individual GRID $a_{0,0}$ at the coordinate $(0,0)$, and going to $a_{\infty,\infty}$. Then we have no way to terminate the computation.

---

14 http://www.w3.org/2007/OWL/wiki/Punning#Treating_classes_as_instances_of_metaclasses_.28Class_.E2/86/94_Individual.29.

6 Conclusion

We focused on set theories involved in RDF and RDF-based OWL Semantics, and clarified that stratified proper classes such as ${}^{n}rdfs\!:\!Resource$, ${}^{n}rdfs\!:\!Class$ (${}^{n}owl\!:\!Thing$ and ${}^{n}owl\!:\!Class$ as well) do not include membership loops. Then we proposed a set of criteria for meta-modeling that is derived from *Ramified Type Theory* in Principia Mathematica. While it is obvious that unrestricted OWL Full may be undecidable, the proposed meta-modeling criteria is not enough to make meta-modeling computation decidable, even if we fulfill these criteria in meta-modeling as shown in Domino-tilting Puzzle. Let us call such ones *unsound meta-modeling setup*. We need further ways in well-mannered OWL Full meta-modeling so that the systems would be decidable with the computation of higher order classes.

References

Aczel, A.D.: The Mystery of the Aleph, Four Walls Eight Windows (2000)

Boolos, G.S.: The Iterative Conception of Set. J. Philosophy **68–8**, 215–231 (1971)

Bourbaki, N.: Éléments de Mathématique, Chapitres 1 et 2. Hermann (1966)

Cantor, G.: Beiträge zur Begründung der transfiniten Mengenlehre. Mathematische Annalen, Bd.46, S.481-512 (1895). Contributions to the Founding of the Theory of Transfinite Numbers, Dover (1955)

Cantor, G.: Letter to Dedekind (1899) in "From Frege to Gödel A Source Book in Mathematical Logic, 1879–1931". In: van Heijenoort, J. (ed.). Harvard (1967)

Doets, H.C.: Zermelo-Fraenkel Set Theory (2002). http://staff.science.uva.nl/vervoort/AST/ast.pdf

Hutton, G.: Programming in Haskell. Cambridge University Press, New York (2007)

Hayes, P.: RDF Semantics. W3C Recommendation (2004). http://www.w3.org/TR/2004/REC-rdf-mt-20040210/

Kamareddine, F., Laan, T., Nederpelt, R.: A Modern Perspective on Type Theory. Kluwer, New York (2004)

Genesereth, M.R., Fikes, R.E.: Knowledge Interchange Format version 3.0 Reference Manual (1994). http://logic.stanford.edu/kif/Hypertext/kif-manual.html

Koide, S., Takeda, H.: Common Languages for Web Semantics, Evaluation of Novel Approaches to Software Engineering. In: Communications in Computer and Information Science, vol. 230, pp. 148–162. Springer (2011)

Motik, B.: On the properties of metamodeling in OWL. J. Logic Comput. **17**(4), 617–637 (2007). doi:10.1093/logcom/exm027

Patel-Schneider, P.F., Hayes, P., Horrocks, I.: OWL web ontology language semantics and abstract syntax Sect. 5. RDF-Compatible Model-Theoretic Semantics (2004a). http://www.w3.org/TR/owl-semantics/rdfs.html

Patel-Schneider, P.F., Horrocks, I.: OWL web ontology language semantics and abstract syntax Sect. 3. Direct Model-Theoretic Semantics (2004b). http://www.w3.org/TR/owl-semantics/direct.html

Schneider, M.: OWL 2 web ontology language RDF-based semantics, 2nd edn. (2014). http://www.w3.org/TR/owl-rdf-based-semantics/

van Heijenoort, J. (ed.): Russell, B.: Letter to Frege (1902) in "From Frege to Gödel A Source Book in Mathematical Logic, 1879–1931". Harvard (1967)

Whitehead, A.N., Russell, B.: Principia Mathematica, vol. 1. Merchant Books (1910)
Graham, S.: Re-examining Russell's paralysis: ramified type-theory and Wittgenstein's objection to Russell's theory of judgment. J. Bertrand Russell Stud. **23**, 5–26 (2003)
Tarski, A.: Introduction to Logic. Dover (1946/1995). This book is an extended edition of the book title "On Mathematical Logic and Deductive Method," appeared at 1936 in Polish and 1937 in German

Designing of Ontology for Domain Vocabulary on Agriculture Activity Ontology (AAO) and a Lesson Learned

Sungmin Joo[1(✉)], Seiji Koide[2], Hideaki Takeda[1], Daisuke Horyu[3], Akane Takezaki[3], and Tomokazu Yoshida[3]

[1] National Institute of Informatics, 2-1-2 Hitotsubashi, Chiyoda-ku, Tokyo, Japan
{joo,takeda}@nii.ac.jp
[2] Ontolonomy, LLC., 3-76-3-J901, Mutsukawa, Minami-ku, Yokohama, Japan
koide@ontolonomy.co.jp
[3] National Agriculture and Food Research Organization, 1-31-1, Kannondai, Tsukuba, Ibaraki, Japan
{horyu,akane,jones}@affrc.go.jp

Abstract. This paper proposes Agriculture Activity Ontology (AAO) as a basis of the core vocabulary of agricultural activity. Since concepts of agriculture activities are formed by the various context such as purpose, means, crop, and field, we organize the agriculture activity ontology as a hierarchy of concepts discriminated by various properties such as purpose, means, crop and field. The vocabulary of agricultural activity is then defined as the subset of the ontology. Since the ontology is consistent, extendable, and capable of some inferences thanks to Description Logics, so the vocabulary inherits these features. The vocabulary is also linked to existing vocabularies such as AGROVOC. It is expected to use in the data format in the agricultural IT system. The vocabulary is adopted as the part of "the guideline for agriculture activity names for agriculture IT systems" issued by Ministry of Agriculture, Forestry and Fisheries (MAFF), Japan. Also we investigated the usefulness of the ontology as the method for defining the domain vocabulary.

Keywords: Ontology · Agriculture · Agronomic sciences · Knowledge representation · Core vocabulary · Vocabulary management

1 Introduction

The various IT systems have been introduced in farm management to realize better management, i.e., more efficient resource management, finer production control and better product quality. Now data management is indispensable in farm management. Data in farm management is also used in own purpose but the aggregated data is used for statistics, analysis and prediction for area agriculture.

Data in agricultural IT systems is nonetheless not easy to federate and integrate since the languages to describe data are not unified. Terminology in agriculture such as names of activity, equipment, and crop has not been well standardized mainly because agriculture has been *local*. Some of locality comes from

Y.-F. Li et al. (Eds.): JIST 2016, LNCS 10055, pp. 32–46, 2016.
DOI: 10.1007/978-3-319-50112-3_3

diversity of culture and environment and others from the way of business, i.e., farms are small and run independently. But introduction of IT systems changed the situation; farms can be connected beyond the barrier of individual farms, regions, and even culture. But un-unified terminology exists as the problem. Without unified terminology, smooth data exchange cannot be enabled. So standardization of terminology is the key to enhance agriculture with IT systems. We focus on agriculture activity in this paper. Agriculture activity is the most basic element of farm management and also the most difficult to standardize since it is more abstract than other types of terminology like equipment and crop.

In this paper, we investigate the existing vocabulary system for agricultural activities. Then we propose the agriculture activity ontology by paying attention to the linguistic feature of agricultural activities. We also explore reasoning functions with the ontology and web services to utilize the ontology. Finally, we discuss the future directions of the improvement and extension of the ontology.

2 An Existing Resource: AGROVOC

In this section, we survey the features of AGROVOC (a portmanteau of agriculture and vocabulary) [1] as an existing agricultural vocabulary system. AGROVOC is the most well-known vocabulary in agriculture supervised by Food and Agriculture Organization (FAO) of the United Nations. AGROVOC is the thesaurus containing more than 32,000 terms of agriculture, fisheries, food, environment and other related fields. It has international interoperability as it is provided in 21 languages. Each term can have the hierarchical structure with narrower concept and broader concept, and there are 25 top-most concepts such as activities, organisms, location, products and so on. 1,434 narrower concepts are provided in activities which contains the concepts about agriculture activity.

AGROVOC is the well-known vocabulary system which has international interoperability and it contains many terms. However, There are some insufficient features in order to use as the core vocabulary. First of all, the relationship between concepts is not clear. Most of narrower/broader relationship is attached only by considering the pair-wise relationship. Thus hierarchy by these relationships are not so consistent. For instance, *Vegetative propagation* has *Rooting* as the narrower concept. Figure 1 shows the broader concept and the narrower concept of *harvesting. mowing* is located as the narrower concept of *harvesting*, but it is not an appropriate classification considering the general meaning of *mowing*. This kind of problem occurs because AGROVOC is established vocabulary system as the thesaurus. This vague relationship between concepts makes the problem when adding a new term; it is difficult to define the relation with concepts in AGROVOC, i.e., to find the best position to the new term.

In addition, the number of activity names about rice farming, which is important in Asia including Japan, are insufficient. For example, in rice farming, especially *pulling seedlings* and *midseason drainage* which are important activity in a rice paddy, are not contained in AGROVOC.

The ambiguousness of relationship among concepts in the existing vocabulary system of agricultural activities is thus problematic. It is required to clearly

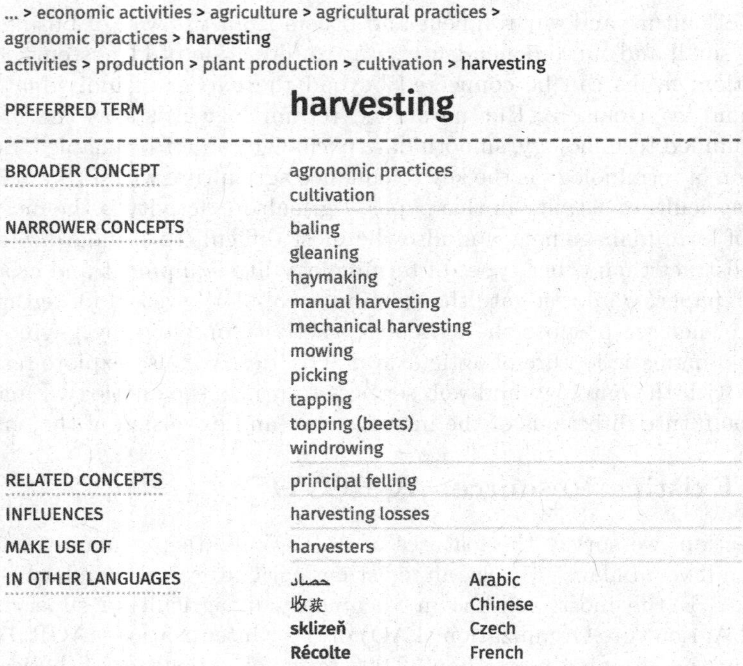

Fig. 1. The broader concept and the narrower concept of *harvesting* in AGROVOC (http://oek1.fao.org/skosmos/agrovoc/en/page/c_3500).

define the relationships among concepts and specify them. In order to solve these problems, this paper suggests the establishment of the ontology for agriculture activity. The ontology can define the clear concepts by separating concepts and representation. Also it can reflect the characteristics of the domain more clearly by structuralizing the relationships.

3 Designing of Agricultural Activity Ontology

This paper describes the Agriculture Activity Ontology (AAO) as the basis of the core vocabulary of agriculture activity, and it provides semantics for agricultural activity names. Also, AAO is formalized by Description Logics in order to define and classify the agricultural activities clearly. Formalization by Description Logics makes it possible to judge the inconsistency and subsumption among concepts, and to enable more logical inferences. The ontology designed by Description Logics can be converted to OWL so that it can processed by computers.

3.1 The Structuralization of the Agricultural Activities

Our strategy to structuralize agricultural activities is the top-down, i.e., starting from the most general activity and expanding it to more specific activities.

The important criteria for the top-down approach is how we can classify more specific concepts consistently. We define more specific concepts by specifying attributes and their values. We furthermore define the general rule for specifying attributes.

We start with the top concept *Agriculture Activity* which denotes all kind of activities related on crop and/or fields. Then we break down the concept into more concrete concepts. When farmers plan or do a certain agricultural activity, the first decision is what for they would take the action, i.e., *purpose* is the first attribute to distinguish agriculture activities. After the purpose is well specified, we use other attributes, i.e., *act* (type of action), *target*, *place*, *means*, *equipment*, and *season* in this order. *Crop* is also introduced so as to define the activity for a specific crop. These eight attributes are used to define the concept and to form the hierarchy of the agricultural activity.

The basic idea of formalization of Agriculture Activity Ontology is that concepts correspond to concepts in Descriptions Logics (DLs) while attributes such as purpose correspond to roles in DLs. By adding the role and the role value, the concepts is defined as the narrower concept of the original concept. If the added role is what is already used in the original concept and the value of the role of the new concept is narrower than that in the original concept, the new concept is also narrower concept of the original concept. It should be noted; not all concepts correspond to terms for farm management since some abstract concepts are introduced just to classify. So we distinguish the abstract concepts not corresponding to terms and concrete concepts corresponding to terms. We call former *category* and the latter *term*. For example, *thinning* and *cutting root* are terms, and their broader concept *activity for uniformity* is the category. The category and the term are succeeding the values of the attributes, and they have the relationship of inclusion.

Now let's look at the ontology in detail. At first, we classify the agriculture activity into two; *crop production activity* which is related to the crop production directly, and *administrative activity* which is related to the farm management. *crop production activity* is classified into the following four activities: *crop growth activity* which is for the purpose of crop growth, *activity for environmental control* which is for the purpose of the environment control, *activity for post production* which is for the purpose of the post production, and *activity for support for crop production* which is for the purpose of the indirect support in the crop production. So as to define narrower concepts, *purpose*, *act*, *target*, *place*, *means*, *equipment*, *season*, *crop* were used as attributes. Classification is conducted by using values of these attributes.

The activity *Seeding* can be defined as follows; First of all, *Seeding* is one of the activities of *Activity for seed propagation*, and *Activity for control of propagation* is the broader concept of *Activity for seed propagation*, *Crop growth activity* is the broader concept of *activity for control of propagation*. *Crop production activity* is the broader concept of *Crop growth activity*. Lastly, the broader concept of *Crop production activity* is *agriculture activity* which is the broadest concept. All these concepts are classified by *purpose* so that *purpose* attribute is used.

Since the values of purpose attributes are hierarchical, activity concepts are hierarchical. *Seed propagation* is the narrower concept of *Control of propagation*, and it then the narrower of *Crop growth*. *Crop growth* is the narrower concept of *Crop production*.

$$Crop\_production\_activity \equiv Agriculture\_activity \sqcap \forall purpose.crop\_production \quad (1)$$

$$Crop\_growth\_activity \equiv Crop\_production\_activity \sqcap \forall purpose.crop\_growth \quad (2)$$

$$Activity\_for\_control\_of\_propagation \equiv Crop\_growth\_activity \sqcap \forall purpose.control\_of\_propagation \quad (3)$$

$$Activity\_for\_seed\_propagation \equiv Activity\_for\_control\_of\_propagation \sqcap \forall purpose.seed\_propagation \quad (4)$$

The formula (1), (2), (3), (4) can be represented as the formula (5).

$$\begin{aligned} Activity\_for\_seed\_propagation \equiv\ & Agriculture\_activity \\ & \sqcap \forall purpose.seed\_propagation \\ & \sqcap \forall purpose.control\_of\_propagation \\ & \sqcap \forall purpose.crop\_growth \\ & \sqcap \forall purpose.crop\_production \end{aligned} \quad (5)$$

Here the purposes of *seed propagation*, *control of propagation*, *crop growth*, *crop production* have the following relation of inclusion by definition.

$$seed\_propagation \sqsubseteq control\_of\_propagation \sqsubseteq crop\_growth \sqsubseteq crop\_production \quad (6)$$

Thus the formula (5) is represented as below.

$$Activity\_for\_seed\_propagation \equiv Agriculture\_activity \sqcap \forall purpose.seed\_propagation \quad (7)$$

On the other hand, *Seeding* is the activity whose *purpose* is *seed propagation*, *place* is *field*, *target* is *seed*, and *act* is *sow*. So it is defined as below.

$$\begin{aligned} Seeding \equiv\ & Activity\_for\_seed\_propagation \\ & \sqcap \forall act.sow \\ & \sqcap \forall target.seed \\ & \sqcap \forall place.field \end{aligned} \quad (8)$$

Therefore, *Seeding* in the agricultural activities is defined like below from the formula (7) and (8).

$$\begin{aligned} Seeding \equiv\ & Agriculture\_activity \\ & \sqcap \forall purpose.seed\_propagation \\ & \sqcap \forall act.sow \\ & \sqcap \forall target.seed \\ & \sqcap \forall place.field \end{aligned} \tag{9}$$

Among *Seeding*, when the *crop* is *rice* and the *place* is *nursery box*, it is classified as *seeding on nursery box*, when the *place* is *paddy field*, it is classified as *direct seeding in flooded paddy field* and when the *place* is *well drained paddy field*, it is classified as *direct seeding in well drained paddy field*. As a result, *Seeding on nursery box*, *Direct seeding in flooded paddy field* and *Direct seeding in well drained paddy field* are in the relationship of siblings, and they are the narrower concept of *seeding*. Here these activities can be represented by the description logic as below.

$$\begin{aligned} Seeding\_on\_nursery\_box \equiv\ & Seeding \\ & \sqcap \forall crop.rice \\ & \sqcap \forall place.nursery\_box \end{aligned} \tag{10}$$

$$\begin{aligned} Direct\_seeding\_in\_flooded\_paddy\_field \equiv\ & Seeding \\ & \sqcap \forall crop.rice \\ & \sqcap \forall place.paddy\_field \end{aligned} \tag{11}$$

$$\begin{aligned} Direct\_seeding\_in\_well\_drained\_paddy\_field \equiv\ & Seeding \\ & \sqcap \forall crop.rice \\ & \sqcap \forall place.well\_drainded\_paddy\_field \end{aligned} \tag{12}$$

The *place* value of *nursery box*, *paddy field* and *well drained paddy field* are defined as a part of *field*.

$$\begin{aligned} nursery\_box &\sqsubseteq field, \\ paddy\_field &\sqsubseteq field, \\ well\_drained\_paddy\_field &\sqsubseteq field \end{aligned} \tag{13}$$

Thus, from the formula (11), (12) and (13), we define the activity *Seeding on nursery box*, *Direct seeding in flooded paddy field* and *Direct seeding in well drained paddy field* as below.

$$
\begin{aligned}
Seeding\_on\_nursery\_box \equiv\ & Agriculture\_activity \\
& \sqcap \forall purpose.seed\_propagation \\
& \sqcap \forall act.sow \\
& \sqcap \forall target.seed \\
& \sqcap \forall crop.rice \\
& \sqcap \forall place.nursery\_box
\end{aligned} \tag{14}
$$

$$
\begin{aligned}
Direct\_seeding\_in\_flooded\_paddy\_field \equiv\ & Agriculture\_activity \\
& \sqcap \forall purpose.seed\_propagation \\
& \sqcap \forall act.sow \\
& \sqcap \forall target.seed \\
& \sqcap \forall crop.rice \\
& \sqcap \forall place.paddy\_field
\end{aligned} \tag{15}
$$

$$
\begin{aligned}
Direct\_seeding\_in\_well\_drained\_paddy\_field \equiv\ & Agriculture\_activity \\
& \sqcap \forall purpose.seed\_propagation \\
& \sqcap \forall act.sow \\
& \sqcap \forall target.seed \\
& \sqcap \forall crop.rice \\
& \sqcap \forall place.well\_drainded\_paddy\_field
\end{aligned} \tag{16}
$$

By combining adding more attributes and subdividing the attribute values, we can flexibly form the hierarchical structure to represent terminology used in agriculture without loosing logical consistency.

3.2 Polysemic Concepts

There are many activities conducted for the multiple purposes in the agricultural activities. The typical case is *Activity for mulching.* One of its purposes is spreading organic matter and other things on the surface of the soil, but there are other purposes; to keep the temperature optimal and controls the temperature, and to refrain weeds. The other example is *Puddling.* It is to plow the bottom of a rice field, but it is also intended to conduct for the purpose of water retention, i.e., preventing from the water leak, and for the purpose of land leveling, i.e., flattening the soil. In our formalization, these concepts are interpreted as polysemic concept and modelled as disjunction of multiple concepts since none of multiple concepts are mandatory rather optional. Here puddling can be expressed with DL as follows;

$$
\begin{aligned}
Puddling \equiv\ & Pulverization \\
& \sqcup Land\_leveling \\
& \sqcup Activity\_for\_water\_retention
\end{aligned} \tag{17}
$$

Now *Puddling* is expanded as follows;

$$\begin{aligned} \mathit{Puddling} \equiv\ & (\mathit{Agriculture\_activity} \\ & \sqcap \forall \mathit{purpose.land\_preparation} \\ & \sqcap \forall \mathit{act.crush} \\ & \sqcap \forall \mathit{place.paddy\_field}) \\ \sqcup\ & (\mathit{Agriculture\_activity} \\ & \sqcap \forall \mathit{purpose.land\_preparation} \\ & \sqcap \forall \mathit{act.level} \\ & \sqcap \forall \mathit{target.field}) \\ \sqcup\ & (\mathit{Agriculture\_activity} \sqcap \forall \mathit{purpose.water\_retention}) \end{aligned} \tag{18}$$

By converting the disjunction form into the formula (18) to the conjunction form, we can infer the formula (19).

$$\begin{aligned} \mathit{Puddling} \sqsupseteq\ & \mathit{Agriculture\_activity} \\ & \sqcap \forall \mathit{purpose.}(\mathit{land\_preparation} \sqcup \mathit{water\_retention}) \\ & \sqcap \forall \mathit{act.}(\mathit{crush} \sqcup \mathit{level}) \\ & \sqcap \forall \mathit{place.paddy\_field} \end{aligned} \tag{19}$$

The polysemic concepts defined with multiple concepts can properly express features for the activities conducted for multiple attributions in the Agriculture Activity Ontology.

3.3 Synonym

There are many synonyms in the vocabulary for agricultural activities. It is easily treated in DL as follows;

$$\mathit{Seeding} \equiv \mathit{Sowing} \tag{20}$$

In addition, expressions in multiple languages are also represented as synonyms. It is important especially for non-English speaking countries[1].

$$\begin{aligned} \mathit{Seeding} &\equiv \mathit{Sowing} \\ &\equiv \text{は種} \end{aligned} \tag{21}$$

So as to correspond to the variety of the vocabulary, the Agriculture Activity Ontology enables to separate the concepts themselves and expressions of the concepts properly.

[1] Indeed, AAO is basically written in Japanese and expressions of concepts and roles in English are optional. But we here provide the English version of AAO for simplicity of explanation.

4 Reasoning by Agriculture Activity Ontology

Generally speaking, the more abstract concepts are, the more difficult it is to define them. The more specific concepts are, the easier it is to take the specific attributes into account. On a specific agriculture activity, it is easy to define the activity with the specific attributes, such as the purpose, the target, the means, etc. Since the purpose of AAO is to keep the agriculture activity terms consistent and well-organized, placing new terms at the appropriate location in the ontology is mandatory. For instance, suppose that we want to add a new term *Making scarecrow.* It is composed of attribute the *purpose* of *pest animal suppression* and attribute the *means* of *physical means*, then the abstract activity *Activity for pest animal suppression by physical means* may become the abstraction of *Making scarecrow*, even if it is not specified explicitly. Furthermore, more abstract *Activity for pest animal suppression* must be the abstraction of *Activity for pest animal suppression by physical means* without attribute the means.

$$\begin{aligned} making\_scarecrow \equiv\ & \forall purpose.pest\_animal\_suppression \\ & \sqcap \forall act.make \\ & \sqcap \forall target.scarecrow \\ & \sqcap \forall means.physical\_means \end{aligned} \tag{22}$$

The question is what is the broader concept of *Making scarecrow* in AAO, and how we can find it. We set up the ontology of attributes with the relationship of inclusion as follows.

$$pest\_animal\_suppression \sqsubseteq biotic\_suppression \sqsubseteq biotic\_control \tag{23}$$

We also set up the hierarchical structure of the agriculture activity as follows.

$$\begin{aligned} & Activity\_for\_pest\_animal\_suppression\_by\_physical\_means \equiv \\ & Activity\_for\_pest\_animal\_suppression \sqcap \forall means.physical\_means \end{aligned} \tag{24}$$

$$\begin{aligned} & Activity\_for\_pest\_animal\_suppression \equiv \\ & Activity\_for\_biotic\_control \sqcap \forall purpose.pest\_animal\_suppression \end{aligned} \tag{25}$$

$$\begin{aligned} Activity\_for\_biotic\_control \equiv\ & activity\_for\_environmental\_control \\ & \sqcap \forall purpose.biotic\_control \end{aligned} \tag{26}$$

Activity for pest animal suppression by physical means is a conjunction of *pest animal suppression* (for purpose) and *physical means* (for means). Thus, there is no contradiction by making *Activity for pest animal suppression by physical means* a broader concept of *Making scarecrow.*

The main task of Description Logics is to compute truth value in subsumption checking [2]. However, it cannot discover subsumers or subsumees for a given subsumee or subsumer in a given ontological hierarchies. Therefore, we introduced

Schank's algorithm for Case-Based Reasoning (CBR) [3] into our OWL [4] reasoning engine named SWCLOS [5,6][2], whereby an appropriate position of a given collection of pairs of attributes and values can be automatically discovered in coherent hierarchies of concepts and their attribute values, starting from a given domain top concept and descending subsuming chains to specific ones. In the systematization of AAO reasoning, the knowledge expressed in DLs is described in OWL. The following shows an example of the formula (12) described in Turtle [7].

```
cavoc.aao:Direct_seeding_in_well_drained_paddy_field a rdfs:Class ;
    rdfs:subClassOf cavoc.aao:Seeding ,
    [ a owl:Restriction ;
      owl:onProperty cavoc:crop ;
      owl:allValuesFrom cavoc:rice ] ,
    [ a owl:Restriction ;
      owl:onProperty cavoc:place ;
      owl:allValuesFrom cavoc:well_drained_paddy_field] .
```

In SWCLOS, we can see the form of any OWL entity in lisp-like expression. The following demonstrates the expression of `cavoc.aao:making_scarecrow` in SWCLOS. Note that it has no subsumer concept defined here. Command **`refine-abstraction-from`** performs Schank's algorithm with parameters, a domain top concept and an entity for the discovery of position.

```
gx(8): (get-form cavoc.aao:Making_scarecrow)
(owl:Class cavoc.aao:Making_scarecrow
  (rdfs:subClassOf
    (owl:Restriction _:g1937
      (owl:onProperty cavoc:purpose)
      (owl:allValuesFrom cavoc:pest_animal_suppression))
    (owl:Restriction _:g1938
      (owl:onProperty cavoc:act)
      (owl:allValuesFrom cavoc:make))
    (owl:Restriction _:g1939
      (owl:onProperty cavoc:target)
      (owl:allValuesFrom cavoc:scarecrow))
    (owl:Restriction _:g1940
      (owl:onProperty cavoc:means)
      (owl:allValuesFrom cavoc:physical_means)))
  (rdfs:label"\"Making scarecrow\"@en"))
gx(9): (refine-abstraction-from
        cavoc.aao:Crop_production_activity cavoc.aao:making_scarecrow)
#<node cavoc.aao:Activity_for_pest_animal_suppression_by_physical_means>
gx(10): (get-form cavoc.aao:making_scarecrow)
(owl:Class cavoc.aao:Making_scarecrow
  (rdfs:subClassOf cavoc.aao:Activity_for_pest_animal_suppression_by_physical_means
    (owl:Restriction _:g1937
      (owl:onProperty cavoc:purpose)
      (owl:allValuesFrom cavoc:pest_animal_suppression))
    (owl:Restriction _:g1938
      (owl:onProperty cavoc:act)
```

[2] SWCLOS is a lisp-based OWL Full processor on top of Common Lisp Object System (CLOS). It is downloadable from https://github.com/SeijiKoide/SWCLOS.

```
    (owl:allValuesFrom cavoc:make))
  (owl:Restriction _:g1939
    (owl:onProperty cavoc:target)
    (owl:allValuesFrom cavoc:scarecrow))
  (owl:Restriction _:g1940
    (owl:onProperty cavoc:means)
    (owl:allValuesFrom cavoc:physical_means)))
(rdfs:label "\"Making scarecrow\"@en"))
```

Here, in `gx(9)` the appropriate position in the hierarchy was decided, and `gx(10)` demonstrated the `cavoc.aao:Making_scarecrow` should be a subclass of `cavoc.aao:Activity_for_pest_animal_suppression_by_physical_means`. Namely, following results were inferred.

$$\begin{aligned} making\_scarecrow \sqsubseteq \{&(\forall purpose.activity\_for\_pest\_animal\_suppression \\ &\sqcap \forall act.make \\ &\sqcap \forall target.scarecrow \\ &\sqcap \forall means.physical\_means) \\ &\sqcap Activity\_for\_pest\_animal\_suppression\_by\_physical\_means\} \end{aligned} \tag{27}$$

5 Web Services Based on Agricultural Activity Ontology

AAO is hosted on CAVOC (Common Agricultural VOCabulary, www.cavoc.org). In this section, we explain the web services of CAVOC based on the agricultural activity ontology.

5.1 Namespace of Agriculture Activity

CAVOC allows browsing and searching concepts of AAO (Fig. 2, www.cavoc.org/aao). The key feature of CAVOC is that it provides URIs for names of agriculture activities. The agriculture activity ontology has unique namespace, and each agriculture activity has URI. Each URI is structured using the http://cavoc.org/aao/ns/1/ namespace, therefore all of the terms and categories are preceded with this URL. Figure 3 is an example of the URI for *Seeding*. In the page, the hierarchical structure is represented in order to indicate the narrower concept, the broader concept, and the relationship between concepts. In addition, it provides the brief natural language explanation of the concept by using values of attributes. The simple interface allows users to browse concepts of AAO through a tree interface, and to search for specific terms.

5.2 Version History

We have developed the agriculture activity ontology with some versions. Table 1 shows the overview of the versions of the agriculture activity ontology. In version 0.94, the first version to open publicly, the concepts were classified by two

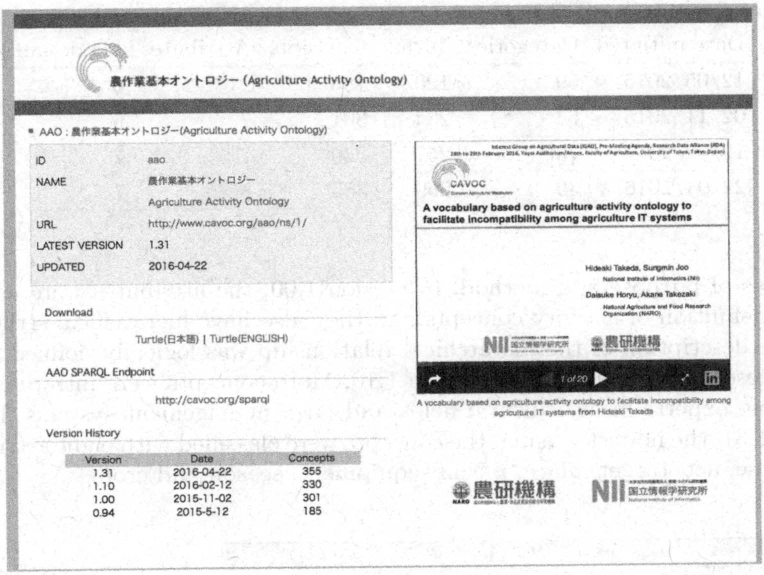

Fig. 2. Main page of AAO (http://cavoc.org/aao).

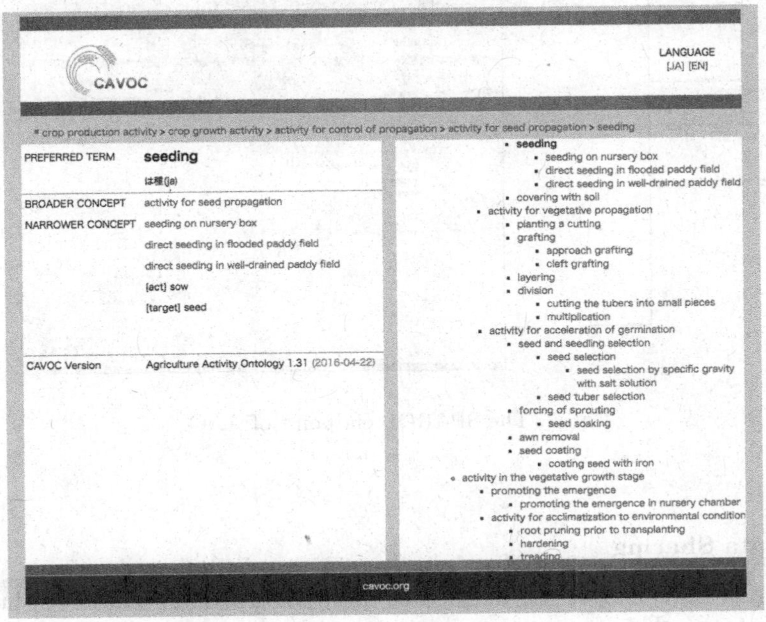

Fig. 3. The namespace of seeding (http://cavoc.org/aao/ns/1/seeding).

Table 1. Listing of the history of AAO version changes

Version	Date initiated	Categories	Terms	Concepts	Attributes	Maximum layer
0.94	12/05/2015	59	126	185	2	6
1.00	02/11/2015	67	234	301	7	6
1.10	12/02/2016	73	257	330	7	7
1.31	22/04/2016	86	269	355	8	7

attributes of purpose and method. In version 1.00, the attributes were used to specify definition of activity concepts and they also have hierarchical structure. Now the description of the hierarchical relationship was logically defined based on the description logic. In the version 1.10, More concepts were introduced by consulting experts in agricultural fields and farm management systems. In the version 1.31, the newest version, the concepts were classified with eight attributes of purpose, act, target, place, means, equipment, season and crop.

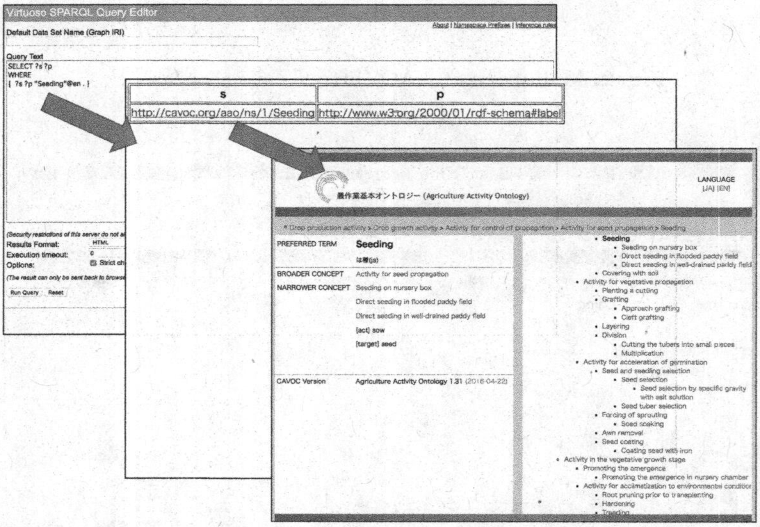

Fig. 4. The SPARQL endpoint of AAO.

5.3 Data Sharing

The data of AAO can be downloaded in the RDF/Turtle formats from http://cavoc.org/aao/. This format is well supported by many semantic tools, and it is possible to convert it into other RDF formats if needed. Also, we provide a SPARQL endpoint for users to explore AAO data using SPARQL queries (Fig. 4).

6 Discussion and Future Work

The agriculture activity ontology ver 1.31, the latest version of the agriculture activity ontology, has 355 concepts which are either categories and terms. It covers most of terms in the national agricultural statistics in Japan. Structuralization with attributes are our own idea so that we need discussion and communication with experts in agriculture more extensively to verify the value of the ontology.

The extension to crop-specific ontologies is one of the important directions of AAO. The scope of the agriculture activity ontology is not just the general terms of agriculture but also covers agriculture activities specialized by crop. The activity specialized in the crop currently contains 10 types of crop such as rice, melon, and other things. By using the attribute of the crop, it is possible to extend to the crop-specific ontologies. We are now developing the crop-based ontologies that can define crop-specific concepts by using crop independent concepts.

There are still issues in Structuralization of the ontology. One of them is *composite* concept. In the agriculture IT systems, there are cases in which the multiple works are managed as a single activity by combining multiple activities. For example, *Raising seedling* is *composite* including*Seeding*, *Fertilization*, *Watering* and other things. However, when the agriculture is planned or implemented, it is managed as *Raising seedling*. We express this activity by combining existing activities. The concept can be expressed with `part-of` relationship, but the simple solution is not always suitable since all of the concepts are not sometimes necessary. We are now considering more appropriate formalization for combination of the relevant activities.

International interoperability is next to do. We have already connections with other activities for agricultural ontologies (for example Crop Ontology Group[3]). Our research has begun from the purpose of establishing the core vocabulary for the field of agriculture of Japan, although it can be independent regardless of language culture so that it can be applied to various languages and cultures. We will improve the ontology in order to develop international core vocabulary.

We designed a domain vocabulary by using ontology based on Description Logics. The design used by the Description Logics can make a concept which has ambiguous meaning classified clearly and deal with the situation when new vocabularies have to be added. The meaning of the concept was defined as suitable attributes for the domain. The structure was constructed to make sense the meaning of the concept by using the value of attributes and it could make the effective processing when the vocabulary lists have to be generated automatically or when the related applications have to be realized. Also it can be used as a tool like dictionary in a namespace for each vocabulary.

3 https://sites.google.com/a/cgxchange.org/cropontologycommunity/.

7 Conclusion

We provide the Agriculture Activity Ontology (AAO) so as to standardize the vocabulary for agricultural activities. By using the ontology, it is possible to define concepts of agriculture activities beyond the linguistic diversity of the vocabulary for agricultural activities. The agriculture activity ontology was adopted as the part of "the guideline for agriculture activity names for agriculture IT systems" issued by Ministry of Agriculture, Forestry and Fisheries (MAFF), Japan in 2016, which is one of the achievements of this study [8]. We are now working to extend our idea to other agriculture domains, i.e., the standardization of vocabulary for agriculture such as the crop, distribution, and agricultural pesticide.

Acknowledgement. This work was supported by Council for Science, Technology and Innovation (CSTI), Cross-ministerial Strategic Innovation Promotion Program (SIP), "Technologies for creating next-generation agriculture, forestry and fisheries" (funding agency: Bio-oriented Technology Research Advancement Institution, NARO).

References

1. AGROVOC Multilingual agricultural thesaurus. Subsequences. J. Mol. Biol. **147**, 195–197 (1981). http://aims.fao.org/vest-registry/vocabularies/agrovoc-multilingual-agricultural-thesaurus
2. Baader, F., et al. (eds.): The Description Logic Handbook. Cambridge University Press, Cambridge (2003)
3. Riesbeck, Christopher K., Roger C. Schank, Inside Case-Based Reasoning, ISBN 0-89856-767-6, LEA (1989)
4. OWL Web Ontology Language Guide, W3C Recommendation 10, February 2004 https://www.w3.org/TR/owl-guide/
5. Koide, S., Takeda, H.: OWL-full reasoning from an object oriented perspective. In: Mizoguchi, R., Shi, Z., Giunchiglia, F. (eds.) ASWC 2006. LNCS, vol. 4185, pp. 263–277. Springer, Heidelberg (2006). doi:10.1007/11836025_27
6. Koide, S.: Theory and implementation of object oriented semantic web language Dr.thesis, Department Informatics School of Multidisciplinary Sciences, The Graduate University for Advanced Studies (SOKENDAI) (2010)
7. RDF 1.1 Turtle, Terse RDF Triple Language, W3C Recommendation 25, February 2014. https://www.w3.org/TR/turtle/
8. Ministry of Agriculture, Fisheries and Forestry (MAFF), Japan: the guideline for agriculture activity names for agriculture IT systems (2016). http://www.kantei.go.jp/jp/singi/it2/senmon_bunka/shiryo/shiryo04.pdf (in japanese)

SQuaRE: A Visual Approach for Ontology-Based Data Access

Michał Blinkiewicz(✉) and Jarosław Bąk(✉)

Institute of Control and Information Engineering, Poznan University of Technology, Piotrowo 3a, 60-965 Poznan, Poland
{michal.blinkiewicz,jaroslaw.bak}@put.poznan.pl

Abstract. We present the SPARQL Query and R2RML mappings Environment (SQuaRE) which provides a visual interface for creating mappings expressed in R2RML. SQuaRE is a web-based tool with easy to use interface that can be applied in the ontology-based data access applications. We describe SQuaRE's main features, its architecture as well as implementation details. We compare SQuaRE with other similar tools and describe our future development plans.

1 Introduction

Ontologies, as a way of expressing knowledge, are becoming more and more popular in various research and practical fields. They allow to define a knowledge base using abstract concepts, properties and relations between them. Ontologies can be expressed in the Web Ontology Language 2 (OWL 2). This is a well-known format of ontologies and most widely used. Ontologies require data to be in a format of RDF[1] triples. Then, using an appropriate reasoner we can obtain new data in the same format. Moreover, we can query such RDF data using SPARQL[2] queries. Nevertheless, ontologies and data need to follow the RDF-based representation. Due to a fact that most of data are stored in different formats, any application of an OWL/OWL2 ontology rises the integration problem between an ontology and stored data. In this case we can transfer our current data format into RDF-based representation and change our software as well as architecture environment or we can create mappings between ontology and our data and then use an appropriate tool that handles such a solution. The first option is very cost-expensive and needs a lot of changes in the current software architecture. The second approach is easier and cheaper. We need to create mappings and then query non-RDF data with SPARQL using ontology, mappings and a tool that enables on-the-fly transfer from non-RDF into RDF data. In this method the most important part is to create appropriate mappings. Currently, a very popular standard for expressing mappings from relational databases to RDF data is W3C's R2RML[3]. The standard allows to use existing relational data in the RDF data model, and then use SPARQL to query such data.

[1] https://www.w3.org/RDF/.
[2] https://www.w3.org/TR/sparql11-overview/.
[3] https://www.w3.org/TR/r2rml/.

Y.-F. Li et al. (Eds.): JIST 2016, LNCS 10055, pp. 47–55, 2016.
DOI: 10.1007/978-3-319-50112-3_4

In this paper we describe SQuaRE, the SPARQL Queries and R2RML mappings Environment, which provides a graphical editor for creating and managing R2RML mappings and for creating and executing SPARQL queries. SQuaRE is a web-based application that simplifies the creation of mappings between a relational database and an ontology. It also enables to test created mappings by defining and executing SPARQL queries. The remainder of this paper is organized as follows. Firstly, we provide preliminary information, then we describe main features of SQuaRE. Next, we present its architecture as well as implementation details. Then, we compare SQuaRE with other similar tools. Finally, we provide conclusions along with future development plans.

2 Preliminaries

SQuaRE is an OBDA-oriented tool which helps inexperienced user to create mappings between a relational database and ontology, and then test those mappings by creating SPARQL queries. Moreover, the tool can be used to write and execute SPARQL queries. In this section we present the main overview of OBDA and R2RML.

Ontology-based Data Access (OBDA) is an approach [8] to separate a user from data sources by means of an ontology which can be perceived as a conceptual view of data. Moreover, by using concepts and relations from the ontology one can define a query in a convenient way. In this case the user operates on a different abstract level than data source. As a result the user defines queries using concepts and relations from the domain of interest and creates complex semantic conditions instead of expressing queries in terms of relational data model. Nevertheless, in order to use OBDA approach with relational data one needs to develop mappings between a relational database and an ontology.

The R2RML recommendation provides a language for expressing mappings from a relational database to RDF datasets. Those mappings allow to view the relational database as a virtual RDF graph. Then, the relational database can be queried using the SPARQL language. Each R2RML mapping is a triples map (an RDF graph) that contains: a logic table (which can be a base table, a view or a valid SQL query); a subject map which defines the subject of all RDF triples that will be generated for a particular logical table row; and a set of predicate-object maps that define the predicates and objects of the generated RDF triples. In order to create R2RML mappings manually, one needs to know about ontologies, RDF, R2RML and SQL at the same time.

SQuaRE tries to overcome the aforementioned issues. The main goal of the tool is to support creation of R2RML mappings and SPARQL queries in a graphical manner. However, at the current state of development SQuaRE supports a graphical editor for R2RML mappings and a text-based interface for creating and executing SPARQL queries.

3 SPARQL Queries and R2RML Mappings Environment

Features. The SQuaRE environment is aimed at providing easy-to-use functions that will support creation and execution of SPARQL queries as well as creation of R2RML mappings. Moreover, SQuaRE allows for management of queries and mappings. A user can save both: mappings and queries for future reference, execution and management. Currently, the tool supports a graphical interface for creating mappings and a text-based interface for creating and executing SPARQL queries. Nevertheless, SQuaRE provides the following useful features:

1. Browsing a relational database – a user can choose a data source and browse its schema. In this view the user sees table names, column names as well as data types stored in each column. An example view of a relational database is shown in Fig. 1.

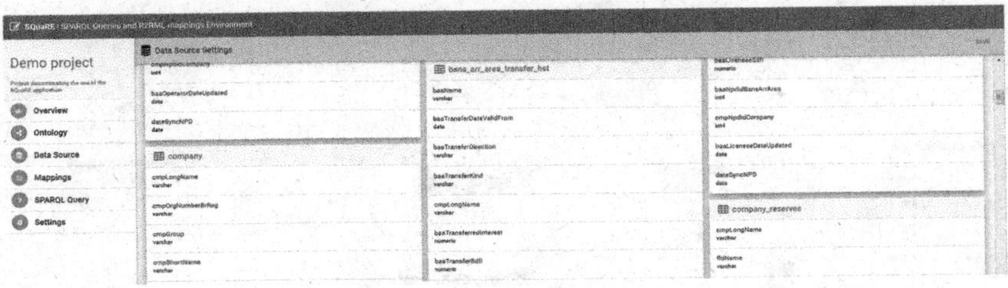

Fig. 1. View of a database schema.

2. Browsing an OWL ontology – a user sees hierarchies of classes, object properties and datatype properties. The user can browse an ontology and search for its elements (Fig. 2).

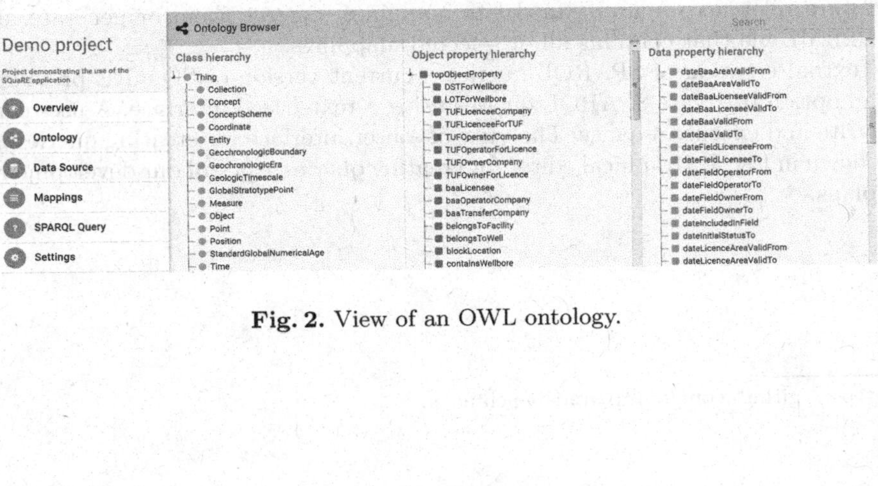

Fig. 2. View of an OWL ontology.

3. Browsing mappings – a list of all created mappings is shown to a user. The user can choose a mapping and then edit it in the mapping creation view (shown in Fig. 3).
4. Graphical creation of R2RML mappings – in a mapping creation view a user can create R2RML mappings. The user needs to choose tables that are going to be mapped. Then, she/he needs to search for an appropriate classes and properties to create mappings using a graphical interface. An example mapping of a table that contains data about companies to the NPD-benchmark ontology[4] is shown in Fig. 3. Classes are represented with an orange background, datatype properties with a green background and object properties with a blue background.

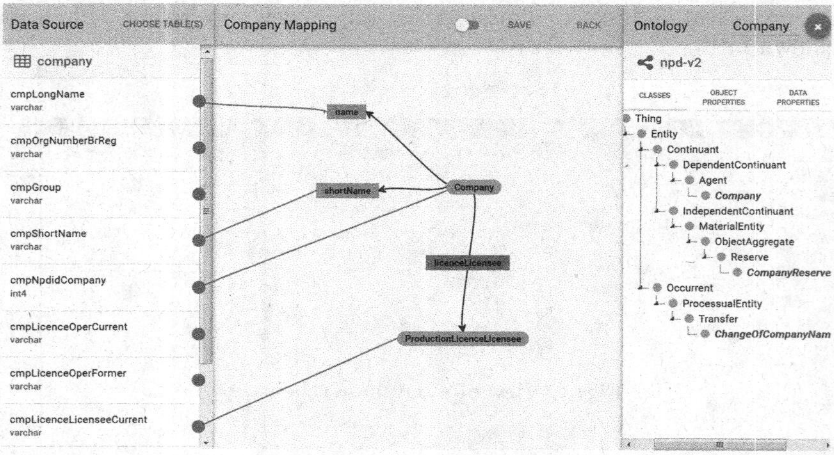

Fig. 3. Creating a mapping.

5. Management of R2RML mappings – each created mapping can be saved for future reference. A user can delete mappings, correct them or generate an R2RML file that contains all or selected mappings.
6. Textual creation of SPARQL queries – current version of SQuaRE provides an option to create SPARQL queries using a text-based interface. A user can write and execute a query. The view of a user interface for creating queries is shown in Fig. 4. Graphical editor for creating queries is one of our development plans.

[4] https://github.com/ontop/npd-benchmark.

7. Management of SPARQL queries – each constructed SPARQL query can be saved and used in future. A user can execute, delete or export a SPARQL query. Moreover, the user can select few queries (or all of them) and generate a separate .txt file that contains their definitions.
8. Execution of SPARQL queries – created SPARQL queries can be executed and results are shown as a table.

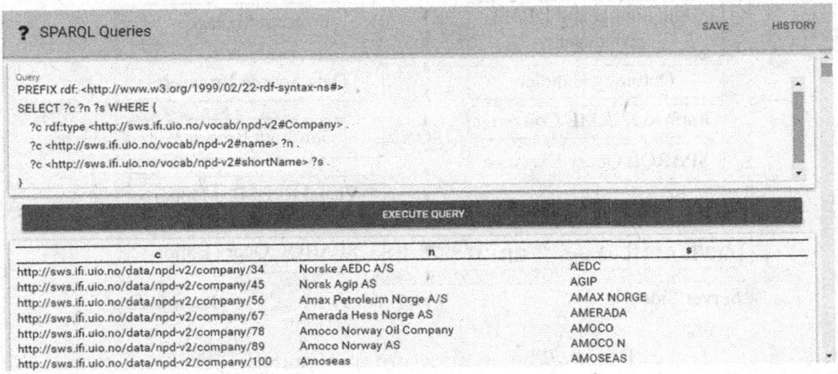

Fig. 4. Defining and executing a SPARQL query.

The aforementioned main list of features provides an intuitive ontology-based access to relational data. Moreover, by exporting functionality (importing features are still in development) a user can use SQuaRE to create mappings and test them by creating SPARQL queries, and then save everything into external files. This allows to import queries and mappings into another tool that supports SPARQL and R2RML.

Architecture and Applied Tools. SQuaRE is developed in Java as a web application. The architecture of SQuaRE is presented in Fig. 5.

The main module, from the user's point of view, is Visual R2RML Mapper which provides tools for visual (graph-based) mappings of relational database metadata, such as table columns, and user provided ontology entities.

Moreover, there are modules responsible for data source configuration and ontology management. The former allows the user to configure a data source by providing DBMS, host location and port, username and password. The latter provides an interface to import an ontology and browse hierarchies of classes and properties.

The server-side modules consist of Data Source and DBMS Manager which manages the user defined data sources and provides JDBC-based access; Ontology Handler for OWL ontology processing; and SPARQL Query Executor which utilizes already defined mappings and allows to execute SPARQL queries in the context of a relational database.

Another key server-side module is Graph to R2RML Converter which is responsible for converting user defined mappings, supplied with client-side visual mapper, in the form of a serialized graph. This is a graph of connections between relational database columns and ontology entities. The graph is then translated into a valid R2RML representation which may be used by the SPARQL Query Executor module.

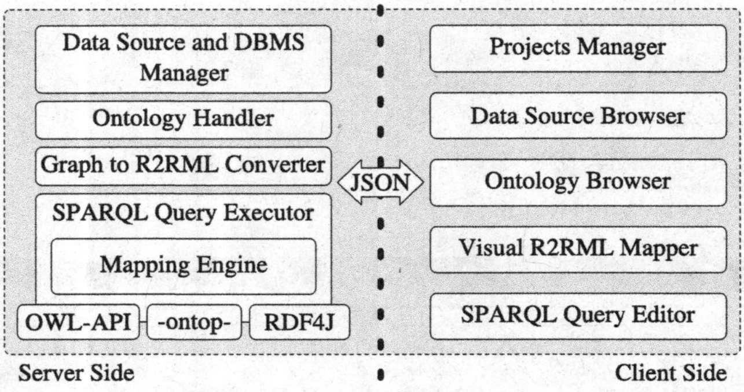

Fig. 5. The architecture of SQuaRE.

SQuaRE applies well-known tools to handle OWL ontologies, relational data and SPARQL queries. The main tools that SQuaRE uses are the following:

- OWL-API [4] – this is the most often used Java library to handle OWL/OWL2 ontologies. It contains a lot of features that are useful when manipulating ontology elements, using reasoner or serialising ontologies.
- The Spring Framework[5] – it is an application framework for the Java platform. Among others, it allows for easy creation of RESTful web services and building backend API.
- -ontop- [2] – it is a platform to query relational databases as virtual RDF graphs using SPARQL. The tool accepts mappings in R2RML and its own OBDA mapping language. SQuaRE uses -ontop- to query relational database using mappings and SPARQL.
- RDF4J[6] – it is a framework for processing RDF data. The tool supports SPARQL in version 1.1 and is used in many third party storage applications. SQuaRE uses it to save all data connected with created mappings and queries.
- Javascript libraries – we use a set of popular Javascript tools like: AngularJS, jQuery, Cytoscape.js, jsPlumb and jsTree.

[5] http://projects.spring.io/spring-framework/.
[6] http://rdf4j.org/.

4 Related Tools

Several tools have been implemented to support a user in defining mappings between data sources and ontologies. We provide the comparison table of the most similar tools to SQuaRE (Table 1).

Table 1. Comparison of main features of SQuaRE and related tools

Features	SQuaRE	OntopPro	Map-On	ODEMapster	Karma	RBA
Visual mappings editor	✓	–	✓	✓	✓	–
Visual SPARQL queries creator	–	–	–	–	–	–
SPARQL queries executor	✓	✓	–	–	–	–
Relational database support	✓	✓	✓	✓	✓	✓
Other data formats support	–	–	–	–	✓	–
Ontology browser	✓	✓	✓	✓	✓	✓
Database schema browser	✓	–	✓	✓	✓	✓
Web-based	✓	–	✓	–	✓	–

SQuaRE, Map-On [9], ODEMapster[7] and Karma [6] are equipped with a visual mapping editor. Map-On provides a graph layout for creating mappings as well as viewing ontologies and databases. ODEMapster supports a tree graphical layout for database schema and ontology. Karma provides a table-like interface for representing data sources and tree layout for visualising an ontology. RBA (R2RML By Assertion) [7] supports a tree layout for displaying databases and ontologies but a user is not able to see a graphical form of mappings. None of compared tools likewise SQuaRE do not have a visual SPARQL query creator. However, only SQuaRE and OntopPro[8] enable SPARQL queries execution against created mappings. All of selected tools are capable to map relational databases but only Karma also supports other data formats (like JSON, CSV etc.). Ontology and database schema browsers are built in all aforementioned tools but OntopPro which is a Protégé plugin does not allow to browse database schema. Furthermore, only SQuaRE, Map-On and Karma are accessible via a Web-based interface.

The aforementioned tools provide different features that overlap in some cases. However, none of them provide the comprehensive functionality for

[7] http://neon-toolkit.org/wiki/ODEMapster.
[8] http://ontop.inf.unibz.it/components/sample-page/.

OBDA-based scenario. SQuaRE provides features for creating and managing of both: R2RML mappings and SPARQL queries. Moreover, it supports users in the execution of queries and presents results in a table-like way. Moreover, we are going to implement support for a graphical creation of SWRL rules [5], which will be another difference to the mentioned tools.

SQuaRE is aimed at providing a simple user interface and easy to use methodology. Nevertheless, it should be perceived as a tool that tries to acquire the best features of other applications and provide them in a graphical way with an easy-to-use interface. The most similar tool at this stage of development is Karma, but without handling SPARQL queries and results in a graphical manner (but Karma provides more mapping methods than SQuaRE and more features regarding data integration). It is worth to notice that SQuaRE is still at the early stage of development whereas most of the tools from the list are being developed in the last few years. Some of them are even discontinued, like ODEMapster or RBA.

5 Summary and Future Work

In this paper we presented the SQuaRE tool which is a web-based environment that provides: (i) creation of R2RML mappings between relational databases and OWL ontologies, and (ii) creation and execution of SPARQL queries. The tool provides a lot of useful features that can be applied in an OBDA-based scenario.

Currently, we are implementing a graph-based method for creating SPARQL queries. In this case we will fully support a graphical environment for handling R2RML and SPARQL. We also plan to include RuQAR [1] to extend reasoning capabilities and provide support for SWRL rules [5]. Moreover, the long term plans are to support other mapping languages, like D2RQ[9] and RML [3]. As a result we will be able to map different data sources like CSV or JSON.

Acknowledgments. This research has been supported by Polish Ministry of Science and Higher Education under grant 04/45/DSPB/0149.

References

1. Bak, J.: Ruqar: reasoning with OWL 2 RL using forward chaining engines. In: ORE (2015)
2. Calvanese, D., Cogrel, B., Komla-Ebri, S., Kontchakov, R., Lanti, D., Rezk, M., Rodriguez-Muro, M., Xiao, G.: Ontop: Answering SPARQL queries over relational databases. Semantic Web, (Preprint), pp. 1–17
3. Dimou, A., Sande, M.V., Colpaert, P., Verborgh, R., Mannens, E., Van de Walle, R.: RML: a generic language for integrated RDF mappings of heterogeneous data. In: Proceedings of the 7th Workshop on Linked Data on the Web, April 2014
4. Horridge, M., Bechhofer, S.: The OWL API: a Java API for OWL ontologies. Semant. Web **2**(1), 11–21 (2011)

[9] http://d2rq.org/.

5. Horrocks, I., Patel-schneider, P.F., Boley, H., Tabet, S., Grosof, B., Dean, M.: SWRL: a semantic web rule language combining OWL and RuleML (2004). Accessed 04 Apr 2013
6. Knoblock, C.A., et al.: Semi-automatically mapping structured sources into the semantic web. In: Simperl, E., Cimiano, P., Polleres, A., Corcho, O., Presutti, V. (eds.) ESWC 2012. LNCS, vol. 7295, pp. 375–390. Springer, Heidelberg (2012). doi:10.1007/978-3-642-30284-8_32
7. Neto, L.E.T., Vidal, V.M.P., Casanova, M.A., Monteiro, J.M.: *R2RML by Assertion*: a semi-automatic tool for generating customised R2RML mappings. In: Cimiano, P., Fernández, M., Lopez, V., Schlobach, S., Völker, J. (eds.) ESWC 2013. LNCS, vol. 7955, pp. 248–252. Springer, Heidelberg (2013). doi:10.1007/978-3-642-41242-4_33
8. Poggi, A., Lembo, D., Calvanese, D., Giacomo, G., Lenzerini, M., Rosati, R.: Linking data to ontologies. In: Spaccapietra, S. (ed.) Journal on Data Semantics X. LNCS, vol. 4900, pp. 133–173. Springer, Heidelberg (2008). doi:10.1007/978-3-540-77688-8_5
9. Siciliaa, Á., Nemirovskib, G., Nolleb, A.: Map-on: A web-based editor for visual ontology mapping. Semantic Web Journal (Preprint), pp. 1–12

Compression Algorithms for Log-Based Recovery in Main-Memory Data Management

Gang Wu(✉), Xianyu Wang, Zeyuan Jiang, Jiawen Cui, and Botao Wang

College of Computer Science and Engineering, Northeastern University, Shenyang, People's Republic of China
wugang@mail.neu.edu.cn, wangxianyu04@gmail.com

Abstract. With the dramatic increases in performance requirement of computer hardware and decreases in its cost in recent years, the relevant research in main-memory database is becoming more and more popular and has a prosperous future. Log-based recovery, which is one of its most important research directions, is a set of problems accompanied by volatile memory. Its problem of stagnation in memory/CPU resulted from the slow I/O speed of non-volatile storage now needs to be addressed urgently. However, there is no specific platform for log-based recovery research. So the study aims to address this issue.

For the specific platform issue, we design and implement a simulation platform called RecoS. RecoS aims at an implementation of recovery sub-system of the main-memory database. It uses cluster substrate to simulate more real data storage and developed interfaces for a variety of recovery strategies. We propose three log compression methods in this paper: (1) the dictionary encoding, (2) the indirectly encoding with no threshold limit and (3) the indirect encoding with a threshold limit. We also adapt ARIES and command logging on the platform, which represents physical and logical logging respectively, focusing on their recovery process and some important details. Regard the recovery platform as the core to investigate the performance of the recovery platform with different load by using different log sets.

Keywords: Main-memory database · Logging · Checkpointing · Failure recovery

1 Background

The development of information technology has a great influence on all walks of life. As a representative one of various techniques, main-memory database has its outstanding advantages in access time and made a significant impact on many applications. It has been one of the important research files in the main- memory database management in recent years. Relying on the background of the main-memory database, this paper mainly studied on[1]:

[1] This paper is partially supported by the National Natural Science Foundation of China No. 61370154 and No. 61332006, and the Fundamental Research Funds for the Central Universities No. N140404009.

Y.-F. Li et al. (Eds.): JIST 2016, LNCS 10055, pp. 56–64, 2016.
DOI: 10.1007/978-3-319-50112-3_5

1. We design and implement a simulation platform called RecoS. RecoS aims at an implementation of recovery sub-system of main-memory database. It uses cluster substrate to simulate realer data storage, and developed interfaces for variety of recovery strategy.
2. We implement the important steps of physical logging and logical logging in RecoS, stressing the detailed difference caused by volatile memory in log, failure recovery and checkpointing between main-database and disk database [1].

2 Overview

RecoS is mainly used to implement the recovery sub-system in MMDB, means three key step, observing log, checkpointing and recovery. The goal of this paper is to implement the platform based on available means independently.

The superstratum of RecoS refers to the recovery part of H-Store, and using Redis as storage in the substratum. [2, 3] RecoS is supposed to simulate and implement some properties and functions such as relational, row store and cluster environment.

2.1 Architecture

RecoS can be a recovery program which is used to compare different strategies with each other. RecoS consists of master nodes which controls the program and Redis instances. Redis includes the cluster which is regarded as storage nodes(the cluster has its own distributed protocols) and the singletons which are regarded as logging record nodes. Using Redis as storage node and logging record node, and control through the master nodes, the Redis instances can be abstracted as a kind of distributed system, it can also abstracted as a kind of distributed system.

Master node: The master node keeps control of the Redis instance and get status through the network connection. It is mainly used to be responsible for all functions except storing data and log, including sending a read/write command, simulated transactions, control timing of logging reading and writing, etc. The master node is aim to implement three functions: mapping table, recovery simulator, transaction simulator. Redis cluster: The main roles Redis cluster is of are data storage and checkpointing. There are some operations for each separate node in the cluster. The connection between master nodes and sub-nodes is the only problem that the super stratum need to pay attention to. Redis logging node: Logging nodes are composed of multiple separate Redis instance. Logging nodes accept the logging of master node and finish the persistence of logging independently. Under the condition that Redis cluster nodes are used to store data, the logging nodes are used to access the log.

2.2 The Implements of Recovery Strategy

The simulate platform stores the logging dispersedly in the cluster. However, there are some differences in the recovery strategy between merging the logging and recovering the logging singly [4].

Physical logging.

(1) Logging format

The platform only record LSN, TxnID, TupleID, and OldValue(NewValue).

(2) Logging record

The logging record are stored in the logging database, then flush the redo logging into the disk of that node. The undo logging will be cleared after committing the transaction.

(3) Logging strategy

The logging can be divided into private logging and group logging. The private logging means a logging chain of a transaction maintenance. The private one will be combined with other logging into a group logging after the commit of transaction [4].

Logical logging: Compared with physical logging, logical logging has differences in format, strategy and recovery. In logging format, the logical logging record LSN, TxnID, SPP, and Params. SPP is the pointer of the store procedure which has been stored. [6] We need to add a kind of string to the logging as SPP. When the string is loaded into the master node to recover the logging, it is supposed to point out the location of stored procedure in the main-memory.

In the logging strategy, we come up with a method which only need to be located in one node. Firstly, we need to number the working nodes in the cluster. If there were two nodes in one stored procedure, we would choose the small one to record the logging.

CheckPointing: The platform uses conforming checkpointing. All the transactions in the checkpointing are in a same status. In order words, all the transactions should be ended with committing, so that we just need to REDO once. This can simplify and accelerate the recovery process [7, 8].

(1) Checkpoints in physical logging
Checkpoints in RecoS use the RDB snapshot of Redis. The system use the SAVE or BGSAVE command on the all nodes in the cluster at a given checkpoint timing. It represents that there are some persisting RDB operation in the background.
(2) Checkpoints in logical logging
Command logging need a transaction conforming checkpointing to cooperate. [9] Transaction conforming checkpointing is similar to delaying the new transactions, waiting for the ongoing transaction has been completed, then making checkpointing.

3 Algorithm

3.1 Recovery

(1) Physical logging recovery

Recovery algorithm for physical logging in RecoS

Algorithm3.1 Recovery algorithm for physical logging

Input: the node number N which to be recovered.
Output: null
for(each node in N):
If (checkPointing) import the node ,and implement the recovery of checkpointing in chapter 3.3.3
Reload the redo_log into main-memory.
Take the redo_log out from master node.
For(all nodes)
Scan all the logging chains excpet redo_log, means scan the txn_redo logging which has not been committed.
Destroy these private txn_redo log.
After mater node have taken all logging of broken nodes, rank the logging according to LSN.
According to the ranked redo logging, do the transaction fragments again.

---

(2) Logical logging recovery

Recovery algorithm for Logical logging in RecoS.

Algorithm3.2 Recovery algorithm for logical logging

Input :the number N of nodes in trouble
Output :null
If(node N is in trouble)
Stop all nodes in the system, if there were a checkpointing, then load the checkpointing and implement the recovery in 3.3.3
Take the redo_log of all nodes, merge it with master node.
Master nodes rank the logging by LSN.
Execute unified recovery.
For(all nodes)
Scan all the logging chains excpet redo_log, and destroy private txn_redo log if possible.

3.2 Logging Compression

Compression is aimed to optimize the space of logging and commit ways. We put forward the coding compression method based on group commit strategy in the commit of logging process [10, 11].

(1) Dictionary encoding for private logging.
Directory encoding for private log

Algorithm3.4 Directory encoding for private log

Input: the time T of commit, the length L of temporary storage region
Output: a set of directory encoding and relative directory
while(in time T || TS is not full)
  TS receive the logging produced by transactions
  Query all dictionary according to txnID.
  If(the txnID is not in the dictionary)
    Record the txnID in the dictionary and then establish a new private logging of this txnID.
    Return dictID, the direction number.
    Replace the txnID to dictID
    Put this logging into the private logging, and add 1 to the temporary storage region.
Submit all private logging chains with the order of the dictionary
Submit the corresponding direction of the group.
Clear the temporary storage region

---

(2) Indirect encoding with threshold limit.
In indirect encoding, suppose that we would like to divide the data into blocks in a same size, if there were less different values in a block, we could compress this block in dictionary encoding, the opposite is not.

(1) Finish a group submission in size L and time T, no matter the temporary storage region is full or not. The logging must be committed when time T out. Set two pointers, p1 and p2, points to the head and tail of the temporary storage region. P1 moves from head to tail, pointing to the position where the logging will be stored. P2 moves from tail to head, pointing to the logging beyond threshold value.
(2) When the logging is to fill the i block, we establish the local dictionary di and a table ti which can only accommodate t value for the range block and the initial value of these table is empty.
(3) When writing logging items into main-memory, we record the TxnID in the table and follow the rules below in block i.
If ti is not full, then write the value into the table, establish an item in the dictionary and write the logging into the block, p1 move to the next position.

If there is such value in ti, write the logging into the block, p1 move to the next position.
If ti is not full and there is not such a value in ti, write the logging into the position that p2 points to, then p2 move to the former position.

(4) If pi point to the next block, then rebuild di + 1and ti + 1, and now ti is useless, return to step2. If pointers meet or time out, end the algorithm, commit the partly logging, the information of pointers in the temporary storage region.

4 Experiments

We use the method of group submission in the experiment. [12] We no longer to consider whether the transaction are rolled back in order to pay attention to the compression of transaction logging encoding. In addition, in a simple submission and submission with the dictionary encoding group, the time occupancy we use for the encoding is not obvious high. Considering the loggings submission, transaction concurrency taking up too much time, the advantages of no-encoding have already been basically eliminated.

4.1 Experimental Environment

The Redis cluster in the experiment contains three nodes. Each node matches to a logging node, logging nodes are directly connected to the master node.

The master node use an 8 GB memory, a disk of 512 GB and Windows8.1 as operating system. Cluster nodes and logging nodes use a memory of 10 GB, a disk of 20 GB, and CentOS6.5 operating system. Nodes in the cluster use Redis3.0 version, other configuration is the same as the default cluster configuration. Logging nodes are separate Redis3.0 instance with the default configuration. Jedis version is 2.7.0.

4.2 Group Commit of Dictionary Encoding

Figure 1 shows the size of L and the relationship of the space taken up by logging and dictionary after dictionary encoding. In this experiment, the number of transaction is fixed to 50, each transaction logging has an average number of 20 and each logging is about 60 B. We will discuss compression effect on these 1000 loggings in this paper.

In Fig. 1, the "uncompressed" part of the logging remains at 69686 bytes, all the stores is loggings. The "compressed" part of the storage is the corresponding loggings and dictionary, the results is about less than 68000 bytes. If the number of logging is overmuch in each group, compression effect not rises linearly, such as compression rate rises steadily from 20 to 500 in abscissa, but fall from 500 to 1000. This illustrates that if the global dictionary contains too many entries, it may affect the final result. At the L = 500, compression rate is about 95.5%.

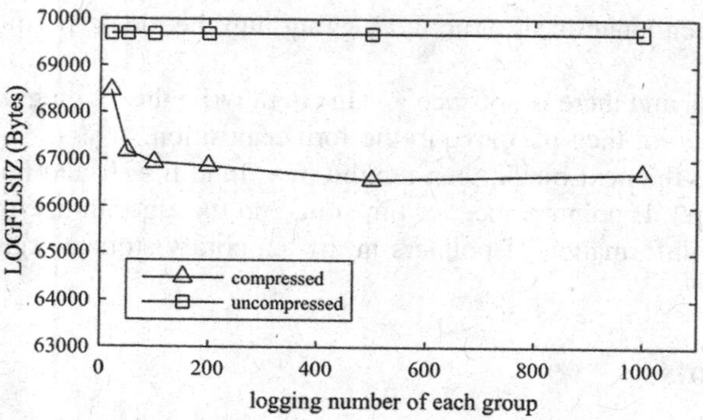

Fig. 1. Group commit with dictionary encoding

4.3 Group Commit of Indirect Encoding

Indirect encoding without threshold limit.
In Fig. 2, the experiment is based on the combination of different transaction number and the average logging each transaction produced. The diagram shows that indirect encoding has advantage on the compression when transactions are more and the average loggings are less.

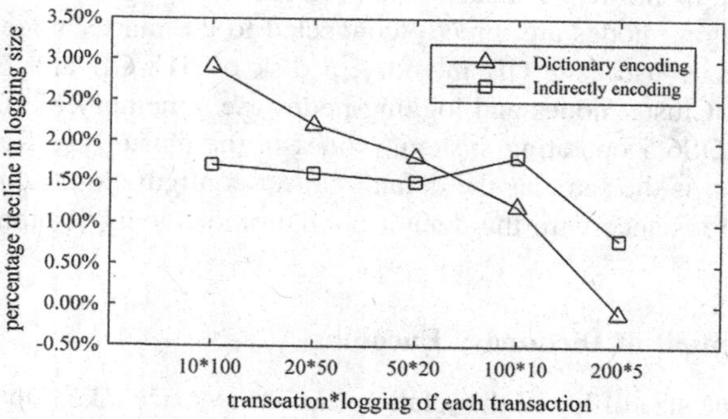

Fig. 2. A comparison between indirect and dictionary encoding on the combination of different transaction number and log per transaction

Indirect encoding with threshold limit.
Encoding with a threshold limit have obvious advantages when the part of to be compressed is lower. If the logging is 60 bytes. The concurrent transactions increase but the compressible part become less. Using the dictionary encoding and indirect encoding are unable to achieve the effect of compression, and indirect encoding with threshold limit doesn't produce dictionary for block, then avoid the overhead of the dictionary, and this makes some effect of compression.

4.4 Comparing of Recovery

We compare the recovering efficiency from no-compression, dictionary encoding and indirect encoding. Regarding the platform as the core, we inspect the performance through different loggings scale in different size of the recovery platform load. As shown in the Fig. 3, there is a list of the recovery time in a scale of 1000, 3000, 6000, 10000, 20000, and 50000.

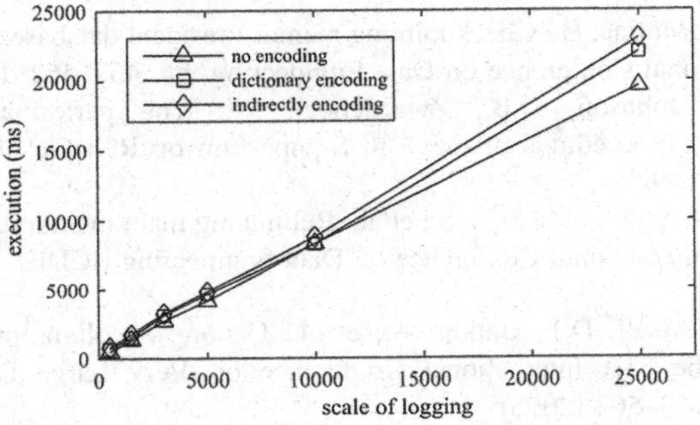

Fig. 3. Comparison with several methods the recovery of submission

5 Conclusions

This article has studied study the traditional and active logging recovery from the availability of the main-memory database. The paper mainly stated the two aspects:

1. In order to study the recovery strategy of main-memory database, especially in the cluster, we design and implement a simulate platform—RecoS. The platform is aimed to separate user's focus from the multifarious database system to the deep study of recovery sub-system. We use Redis to store data and logging, and the client logic program to control the nodes.
2. As to logging recovery, the strategies that are not based on logging will be a main direction of main-memory data recovery development.

References

1. Woo, S., Ho Kim, M., Joon Lee, Y.: Accommodating logical logging under fuzzy checkpointing in main memory databases. In: Proceedings of the International Database Engineering and Applications Symposium. IDEAS 1997. IEEE, pp. 53–62 (1997)
2. Kallman, R., Kimura, H., Natkins, J., et al.: H-store: a high-performance, distributed main memory transaction processing system. Proc. VLDB Endowment **1**(2), 1496–1499 (2008)

3. Stonebraker, M., Madden, S., Abadi, D.J., et al.: The end of an architectural era: (it's time for a complete rewrite). In: Proceedings of the 33rd International Conference on Very Large Data Bases. VLDB Endowment 2007, pp. 1150–1160 (2007)
4. Yao, C., Agrawal, D., Chen, G., et al.: Adaptive logging: optimizing logging and recovery costs in distributed In-memory databases. In: International Conference (2016)
5. Mohan, C., Haderle, D., Lindsay, B., et al.: ARIES: a transaction recovery method supporting fine-granularity locking and partial rollbacks using write-ahead logging. ACM Trans. Database Syst. (TODS) **17**(1), 94–162 (1992)
6. Stonebraker, M., Weisberg, A.: The VoltDB main memory DBMS. IEEE Data Eng. Bull. **36** (2), 21–27 (2013)
7. Salem, K., Garcia-Molina, H.: Checkpointing memory-resident databases. In: Proceedings of the Fifth International Conference on Data Engineering, pp. 452–462. IEEE (1989)
8. Elnozahy, E.N., Johnson, D.B., Zwaenepoel, W.: The performance of consistent checkpointing. In: Proceedings of the 11th Symposium on Reliable Distributed Systems, pp. 39–47. IEEE(1992)
9. Malviya, N., Weisberg, A., Madden, S., et al.: Rethinking main memory OLTP recovery. In: 2014 IEEE 30th International Conference on Data Engineering (ICDE), pp. 604–615. IEEE (2014)
10. Stonebraker, M., Abadi, D.J., Batkin, A., et al.: C-store: a column-oriented DBMS. In: Proceedings of the 31st International Conference on Very Large Data Bases. VLDB Endowment, pp. 553–564 (2005)
11. Abadi, D.J., Boncz, P.A., Harizopoulos, S.: Column-oriented database systems. Proc. VLDB Endowment **2**(2), 1664–1665 (2009)
12. Garcia-Molina, H., Salem, K.: Main memory database systems: an overview. IEEE Trans. Knowl. Data Eng. **4**(6), 509–516 (1992)

Linked Data

An Empirical Study on Property Clustering in Linked Data

Saisai Gong, Haoxuan Li, Wei Hu(✉), and Yuzhong Qu

State Key Laboratory for Novel Software Technology, Nanjing University, Nanjing 210023, China
ssgong.nju@gmail.com, hxli.nju@gmail.com, {whu,yzqu}@nju.edu.cn

Abstract. Properties are used to describe entities, a part of which are likely to be clustered together to constitute an aspect. For example, first name, middle name and last name are usually gathered to describe a person's name. However, existing automated approaches to property clustering remain far from satisfactory for an open domain like Linked Data. In this paper, we firstly investigated the relatedness between properties using five different measures. Then, we employed three clustering algorithms and two combination methods for property clustering. Based on a moderate-sized sample of Linked Data, we empirically studied the property clustering in Linked Data and found that a proper combination of different measures gave rise to the best result. Additionally, we showed how the property clustering can improve user experience in our entity browsing system.

Keywords: Property clustering · Property relatedness · Entity browsing · Empirical study · Linked data

1 Introduction

With the development of Linked Data, billions of RDF triples have been published to describe numerous entities. An entity usually involves multiple aspects and its property-values may focus on different aspects. For instance, graduate from and work at reveal the career information of a person, while parent, spouse and child deliver her family information. Therefore, it is natural to cluster properties into meaningful groups based on the aspects that they intend to describe. Property clustering is useful for many applications such as entity browsing, ontology editing, query completion, etc. It makes the presented information more formatted and understandable and significantly enhances the capability of users to consume the large-scale Linked Data [8].

Take, for example, the case of entity browsing. Many state of the art systems support users to manually cluster properties [13]. But due to the limited energy and knowledge of the users, this type of manual operations is only effective at a small scale. In consideration of an open domain like Linked Data, automated property clustering is needed to solve the scalability issue, but its performance

Y.-F. Li et al. (Eds.): JIST 2016, LNCS 10055, pp. 67–82, 2016.
DOI: 10.1007/978-3-319-50112-3_6

is still far from satisfactory. One reason is the fact that, when browsing entities, the data is multi-sourced and the vocabularies involved are barely predictable, i.e. probably use any vocabularies, thus it is difficult to find out useful patterns among properties in advance and make use of them to guide clustering. Another reason is that the properties used by the entities are largely heterogeneous, which makes identifying similar aspects more difficult and less reliable.

In this paper, we empirically studied the property clustering in Linked Data, which is defined as to automatically assign a subset of property set $\mathbf{P}$ (collected from the descriptions of entities) to a set of disjoint clusters $\{g_1, g_2, \ldots, g_m\}$, such that the properties in a cluster g_k focus on an aspect or a dimension of the content. But, $g_1 \cup g_2 \cup \ldots \cup g_m$ is unnecessarily equal to $\mathbf{P}$.

In order to achieve property clustering, we firstly measured the relatedness of properties from five perspectives: lexical similarity between property names, semantic relatedness between property names, distributional relatedness between properties, range relatedness between properties and overlap of property values. We then employed three widely-used algorithms, namely density-based clustering, hierarchical clustering and spectral clustering. Furthermore, to combine various relatedness measures and clustering results, we developed two combination methods based on linear combination and consensus clustering.

We sampled a moderate-sized dataset of Linked Data for our empirical study and manually built gold standards to assess the relatedness measures, clustering algorithms and combination methods. Moreover, we integrated the property clustering in our entity browsing system called SView[1], in order to observe the use of property clustering in practice. Overall, we tried our best in this study to provide answers to the following questions:

- Q1. What is the most effective measure(s) for measuring the relatedness between properties?
- Q2. What is the most effective algorithm(s) for clustering properties into meaningful groups?
- Q3. Can the combination method(s) improve the property clustering and how largely?
- Q4. Are there any general principles or guidelines for using the property clustering in practice?

In the rest of this paper, we first introduce the measures of property relatedness in Sect. 2. Then, we present the clustering algorithms in Sect. 3 and the combination methods in Sect. 4. Our experimental results are reported in Sect. 5. We show the application of property clustering to entity browsing in Sect. 6 and discuss related work in Sect. 7. Finally, we summarize our findings in Sect. 8 and conclude this paper in Sect. 9.

2 Property Relatedness Measures

In this section, we present five types of relatedness between properties and formalize them as numerical measures. All the measures are assumed non-negative.

[1] http://ws.nju.edu.cn/sview.

2.1 Lexical Similarity Between Property Names

A property is usually associated with several human-readable names, e.g. labels. When the names of two properties share many common characters, it indicates some kind of relatedness between their meanings. For example, both mouth position and mouth elevation describe the mouth information of a river. By detecting this aspect, we took advantage of I-Sub string measure [16] to characterize the lexical similarity. Let p_i, p_j be two properties and l_i, l_j be the names of p_i, p_j respectively, the *lexical similarity* between p_i and p_j, denoted by R_I, is measured by I-Sub as follows:

$$\begin{aligned} R_I(p_i, p_j) &= \textit{I-Sub}(l_i, l_j) \\ &= Comm(l_i, l_j) - Diff(l_i, l_j) + Winkler(l_i, l_j), \end{aligned} \tag{1}$$

where $Comm(l_i, l_j)$ represents the commonality of l_i, l_j, while $Diff(l_i, l_j)$ represents their difference. $Winkler(l_i, l_j)$ is a coefficient to adjust the result. The score of I-Sub was normalized from $[-1, 1]$ to $[0, 1]$; a larger value implies a higher similarity.

2.2 Semantic Relatedness Between Property Names

WordNet provides several semantic relations among concepts/words, e.g. hyperonymy, and it is widely considered as a knowledge base for measuring semantic relatedness between words [3], based on shortest paths, information theory, etc. To measure the WordNet-based relatedness of two properties p_i, p_j, we transformed their property names to the normalized forms by splitting names, removing stop words and stemming. Let l'_i, l'_j be the normalized forms of the names of p_i, p_j, respectively. $|.|$ counts the number of words in a normalized name. The *semantic relatedness* between p_i and p_j, denoted by R_W, is calculated based on WordNet 3.0 as follows:

$$R_W(p_i, p_j) = \min\left(\frac{\sum_{x \in l'_i} \max_{y \in l'_j} Lin(x, y)}{|l'_i|}, \frac{\sum_{y \in l'_j} \max_{x \in l'_i} Lin(x, y)}{|l'_j|}\right), \tag{2}$$

where $Lin(x, y)$ denotes the Lin's WordNet-based word relatedness between x and y [11], which is based on the shortest path and information theory.

2.3 Distributional Relatedness Between Properties

In the area of computational linguistics, distributional relatedness [5] is a measure of word relatedness in distributional semantics, through word co-occurrence in different contexts such as bigrams, sentences or documents. Based on the selected context, the strength of relatedness between co-occurrent words is quantified by using mutual information or other measures. Inspired by this research line, we studied property co-occurrence in use and conceived an entity's RDF description as the context from which co-occurrence is found, i.e. properties are used together to describe the entity. For example, founded by and key person co-occur to describe important people in a company. Specifically, we used the *symmetrical uncertainty coefficient*, denoted by R_U, which computes the strength of relatedness between variables in terms of normalized mutual information. Given two properties p_i, p_j, the *distributional relatedness* between p_i and p_j, denoted by R_U, is calculated using the symmetrical uncertainty coefficient as follows:

$$R_U(p_i, p_j) = 2 \cdot \frac{H(p_i) + H(p_j) - H(p_i, p_j)}{H(p_i) + H(p_j)}, \tag{3}$$

where $H(p_i) = -\sum_{x \in \{p_i, \overline{p_i}\}} P(x) \log P(x)$ obtains the entropy of p_i, $H(p_i, p_j) = -\sum_{x \in \{p_i, \overline{p_i}\}} \sum_{y \in \{p_j, \overline{p_j}\}} P(x, y) \log P(x, y)$ counts the joint entropy of p_i, p_j. The score of R_U is in $[0, 1]$; a higher value indicates a stronger relatedness.

To obtain R_U, we need the probabilities $P(p_i), P(p_i, p_j), P(p_i, \overline{p_j}), P(\overline{p_i}, \overline{p_j})$. Let $\mathbb{G}$ be the RDF dataset from which the probabilities would be estimated and p_i, p_j be two properties. $Res(\mathbb{G})$ denotes the entities appearing in the subject or object position of any RDF triples in $\mathbb{G}$, and $ResDesc(\mathbb{G}, p_i)$ be the entities in $\mathbb{G}$ that is particularly described by p_i (i.e. p_i appears at the predicate position of any RDF triples). The probabilities $P(p_i), P(p_i, p_j), P(p_i, \overline{p_j})$ and $P(\overline{p_i}, \overline{p_j})$ are estimated as follows:

$$P(p_i) = \frac{|ResDesc(\mathbb{G}, p_i)|}{|Res(\mathbb{G})|}, \tag{4}$$

$$P(p_i, p_j) = \frac{|ResDesc(\mathbb{G}, p_i) \cap ResDesc(\mathbb{G}, p_j)|}{|Res(\mathbb{G})|}, \tag{5}$$

$$P(p_i, \overline{p_j}) = \frac{|ResDesc(\mathbb{G}, p_i) \cap (Res(\mathbb{G}) - ResDesc(\mathbb{G}, p_j))|}{|Res(\mathbb{G})|}, \tag{6}$$

$$P(\overline{p_i}, \overline{p_j}) = \frac{|Res(\mathbb{G}) - (ResDesc(\mathbb{G}, p_i) \cup ResDesc(\mathbb{G}, p_j))|}{|Res(\mathbb{G})|}. \tag{7}$$

We leveraged the Billion Triples Challenge (BTC) 2011 dataset[2], a representative subset of the Linked Data to estimate the above probabilities. As different URIs may refer to the same entity, which are called *coreferent URIs*, we firstly found out the coreferent URIs and then merged their RDF descriptions by replacing coreferent URIs with a uniform ID. We used two kinds of ontology semantics owl:sameAs and inverse functional properties, and computed a transitive closure to identify coreferent URIs. More sophisticated coreference resolution algorithms can be found in [9]. This modified dataset would be treated as the RDF dataset $\mathbb{G}$ as described above.

2.4 Range Relatedness Between Properties

Property ranges may be URIs of related types (i.e. classes), which indicate certain kind of relatedness as well. For example, if two properties have the ranges

[2] http://km.aifb.kit.edu/projects/btc-2011/.

delicious food and handicraft respectively, both of them deliver the tourist information of a tourist city. We leveraged the ranges of properties to measure their relatedness, except rdfs:Resource and owl:Thing. If no axioms involving property ranges can be found, we used the conjunction of the types of the property values instead. Let p_i, p_j be two properties and $\mathbf{T}_i, \mathbf{T}_j$ be the sets of ranges for p_i, p_j respectively, the *range relatedness* between p_i, p_j, denoted by R_T, is calculated as follows:

$$R_T(p_i, p_j) = \max_{c_i \in \mathbf{T}_i, c_j \in \mathbf{T}_j} R_W(c_i, c_j), \tag{8}$$

where $R_W(c_i, c_j)$ reuses Eq. (2) to compute the WordNet-based relatedness of the names of c_i, c_j.

Many other relations exist between properties, such as sub-/super-properties or domain relatedness, which indicate the strength of property relatedness as well. For example, both medalist and champion are sub-properties of has participant, and they give the winner information of a sport event participants. A part of super-properties exist in ontology axioms, however, a larger amount of essential super-properties that can be used for property clustering by semantic relatedness are not defined formally in ontologies and thus only latent. For instance, in most ontologies, the three properties length, width and depth do not have a super-property like physical dimension. Besides, the work in [9] observed that property domains are not as useful as ranges. Therefore, we do not consider those relatedness using other relations presently.

2.5 Overlap of Property Values

There may also exist synonymous properties to describe the same entity. For example, both has book and write describe a book written by an author. In this case, common values should be frequently shared by these properties. We used the vector space model (specifically, the TF/IDF model) to represent the values of a property. The text of each property value is collected, e.g. local names of URIs and lexical forms of literals after normalization, and all the terms in the text are used to construct a term frequency vector, where each component corresponds to the number of occurrences of a particular term. Given two properties p_i, p_j, the *overlap of property values* between p_i, p_j, denoted by R_O, is computed by the cosine similarity of the corresponding vectors $\mathbf{v}_i, \mathbf{v}_j$:

$$R_O(p_i, p_j) = \frac{\sum_{k=1}^{n} v_{ik} \cdot v_{jk}}{\sqrt{\sum_{k=1}^{n} v_{ik}^2} \cdot \sqrt{\sum_{k=1}^{n} v_{jk}^2}}, \tag{9}$$

where n is the dimension of the vector space and v_{ik}, v_{jk} are the components of the vectors $\mathbf{v}_i, \mathbf{v}_j$.

3 Property Clustering Algorithms

To work with an arbitrary relatedness measure, we employed the following three well-known clustering algorithms in Weka 3 [15], which use the previously computed relatedness as input and generate a set of property clusters.

DBSCAN, denoted by C_D, finds clusters based on the density of properties in a region. Its key idea is for each property in a cluster, the neighborhood of a given radius (Eps) has to contain at least a minimum number of properties ($MinPts$). In other words, each non-trivial cluster in the result must own at least $MinPts$ properties.

Single linkage clustering, referred to as C_L, is an agglomerative hierarchical algorithm, which repeatedly merges two most related clusters in a bottom-up fashion until meeting some criteria. The single linkage relatedness of two clusters is derived from the two most related properties in the two clusters. The single linkage clustering is terminated when the maximum relatedness of any two clusters is no greater than a threshold θ.

Spectral clustering, denoted by C_S, leverages the spectrum of a relatedness matrix of properties to divide them into clusters. A threshold η for cluster number needs to be pre-defined.

Except the aforementioned parameters and thresholds, we kept the default settings in Weka for the three clustering algorithms.

4 Combination Methods

Combining various relatedness measures helps obtain a property clustering with better accuracy and coverage. There exist two typical methods to conduct combination. One is to combine the measures before clustering, for example, to use a *linear combination* of different relatedness measures for each property pair and carry out a clustering algorithm to produce clusters. Given the five relatedness measures, the combined relatedness, denoted by R_{all}, is defined as follows:

$$\begin{aligned} R_{all}(p_i, p_j) = \omega_1 R_I(p_i, p_j) + \omega_2 R_W(p_i, p_j) + \omega_3 R_U(p_i, p_j) \\ + \omega_4 R_T(p_i, p_j) + \omega_5 R_O(p_i, p_j), \end{aligned} \tag{10}$$

where $\omega_i \in [0, 1]$ denotes the weight coefficient value for a specific property relatedness measure, and $\sum_{i=1}^{5} \omega_i = 1$. In this study, we investigated various values for linearly combining our relatedness measures.

Another method is to first conduct clustering based on individual relatedness measures and then aggregate these individual results using *ensemble clustering*. In this study, we selected *consensus clustering* [1] to realize ensemble clustering. Given a set of individual clusterings corresponding to different relatedness measures, the goal of computing a consensus clustering is to achieve a clustering that minimizes the distance among individual clusterings. The problem of finding an optimal consensus clustering is NP-hard. We implemented CC-Pivot [1], a 3-approximation algorithm, to calculate the consensus clustering.

5 Empirical Study

In this section, we report our study of the relatedness measures, clustering algorithms and combination methods for the property clustering in Linked Data. The source code and sample data for this empirical study are all available at our website[3].

5.1 Dataset

We randomly sampled 20 entities of various types (classes) in Linked Data, each of which is integrated from a DBpedia URI with its coreferent URIs that refer to the same entity using owl:sameAs relations. The finally-selected URIs were required to be accessible via HTTP protocol (to eliminate outdated ones), and have sufficient properties and values, i.e. having more than 50 properties. Overall, the 20 entities involve 12 sources: DBpedia, DBTune, Freebase, GeoNames, LinkedGeoData, LinkedMDB, New York Times, OpenCyc, Project Gutenberg, RDF Book Mashup, The World Factbook and YAGO. We distinguished properties in the forward and backward directions, and considered that different directions of the same property represent different properties. The properties holding a sample entity at the subject position is referred to as the *forward* properties, while at the object position is referred to as the *backward* properties. Table 1 lists the names of the entities with their types and numbers of properties. Note that an entity can have multiple types and we just show an important one.

Table 1. 20 sample entities with their types and numbers of properties

Entity name	Type	#Prop.	Entity name	Type	#Prop.
Hong Kong Airport	Airport	95	Bob Jones University	Institution	130
Michael Nesmith	Artist	145	British Museum	Museum	110
Jeremy Shockey	Athlete	107	Load (album)	MusicalWork	91
Deep Purple	Band	108	Edmund Stoiber	Politician	82
A Clockwork Orange	Book	61	Amazon River	River	142
The Pentagon	Building	110	William H. Holmes	Scientist	65
Baltimore	City	351	Polymelus	Species	51
Adobe Systems	Company	127	Doom II	Software	79
Finland	Country	574	Burlington Township	Township	62
Eyes Wide Shut	Film	99	Barney & Friends	TVShow	79

5.2 Experiment Setup

To observe the prevalence of property clustering in Linked Data and assess the effectiveness of the relatedness measures, clustering algorithms and combination

[3] http://ws.nju.edu.cn/sview/propcluster.zip.

methods, we sought to build for each sample entity a reference clustering that is meaningful, aspect-coherent and compact, so as to compare the algorithmically-generated clustering with the reference ones. Due to the large number of properties (shown in Table 1), it is hard to ask users to manually build the reference clustering. Hence, we did not start from scratch but leveraged existing reasonably good clustering.

Freebase divides properties describing similar aspects into *types* and groups similar types into *domains*[4]. For example, /music/group_member describes the member information of a music group, where group_member is a type and music is a domain. Thus, we invited three PhD candidates in the field of Linked Data to assign each property of a sample entity to the most relevant /domain/type, for example assign a property band member to /music/group_member, and created the reference clustering such that properties are clustered together if they are assigned to the same /domain/type. The average inter-rater agreement score of the 20 entities, measured by Fleiss' κ [6], is 0.895, and the minimum inter-rater agreement score for an entity is 0.814. From the high inter-rater agreement score, we saw that strong agreement exists among the three judges, which guarantees the statistical significance of our empirical study.

By using the reference clusterings as our golden standard, we evaluated the algorithmically-generated property clustering in terms of the following five metrics: *Precision*, *Recall*, *F-Score*, *Rand Index* and *Normalized Mutual Information* (NMI). These metrics are the well-known criteria assessing how well a clustering matches the golden standard. For a clustering π, $S(\pi)$ gives the total number of property pairs in the same clusters:

$$S(\pi) = \{(p_i, p_j) \mid \exists g_k \in \pi,\, p_i, p_j \in g_k,\, i < j\}. \tag{11}$$

Let π_{gs} be a golden standard clustering. The Precision, Recall and F-Score for a computed clustering π w.r.t. π_{gs} are calculated as follows:

$$\text{Precision} = \frac{S(\pi) \cap S(\pi_{gs})}{S(\pi)}, \tag{12}$$

$$\text{Recall} = \frac{S(\pi) \cap S(\pi_{gs})}{S(\pi_{gs})}, \tag{13}$$

$$\text{F-Score} = \frac{2 \cdot \text{Precision} \cdot \text{Recall}}{\text{Precision} + \text{Recall}}. \tag{14}$$

Rand Index penalizes both false positive and false negative decisions in clustering, while NMI can be information-theoretically interpreted. Their values are both rational numbers in [0, 1] range, and a higher value indicates a better clustering. We refer the reader to [17] for the detailed calculation.

5.3 Results of Relatedness Measures and Clustering Algorithms

We clustered the properties of the 20 sample entities by using each relatedness measure and clustering algorithm, and computed the harmonic mean (h-mean)

[4] https://developers.google.com/freebase/guide/basic_concepts.

of Precision, Recall, F-Score, Rand Index and NMI, respectively. To obtain the optimal parameters of a clustering algorithm, we enumerated the values of Eps and $MinPts$ of DBSCAN in $\{0.6, 0.7, 0.8, 0.9\}$ and $\{2, 3, 4, 5\}$ respectively, θ of single linkage in $\{0.6, 0.7, 0.8, 0.9, 1.0\}$, η of spectral clustering from 5 to 20 with 1 interval, and selected the parameters that achieved the highest h-mean of F-Score for the 20 entities. As a result, we set $MinPts = 2, Eps = 0.9, \theta = 0.1$ and $\eta = 5$.

Table 2. Average performance w.r.t. relatedness measures and clustering algorithms

(a) Precision

	C_D	C_L	C_S
R_I	.235	.235	.184
R_W	.215	.215	.198
R_U	.242	.242	.177
R_T	.170	.170	.215
R_O	.247	.247	.188

(b) Recall

	C_D	C_L	C_S
R_I	.273	.273	.449
R_W	.266	.266	.337
R_U	.433	.433	.410
R_T	.381	.381	.329
R_O	.137	.138	.427

(c) F-Score

	C_D	C_L	C_S
R_I	.253	.253	.261
R_W	.238	.238	.250
R_U	.310	.310	.248
R_T	.235	.235	.260
R_O	.176	.177	.261

(d) Rand Index

	C_D	C_L	C_S
R_I	.549	.549	.500
R_W	.672	.672	.584
R_U	.644	.644	.503
R_T	.547	.547	.628
R_O	.709	.708	.516

(e) NMI

	C_D	C_L	C_S
R_I	.387	.387	.229
R_W	.441	.441	.231
R_U	.507	.507	.224
R_T	.364	.364	.255
R_O	.520	.520	.216

Table 2 depicts the result of this experiment, where each row and column represent the relatedness measure and clustering algorithm used, respectively. From the first two columns of each table, we found that the results of DBSCAN (C_D) and single linkage (C_L) are very similar. In fact, when the value of $MinPts$ of DBSCAN is set to 2, DBSCAN is nearly identical to single linkage. From Table 2(a), R_I, R_U and R_O achieved the highest values in Precision by using C_D or C_L. Among them, R_U and R_I also achieved high Recall using either clustering algorithm, as shown in Table 2(b), and thus both of them achieved the highest values in F-Score as shown in Table 2(c). R_T achieved low Precision, but had a relatively good Recall. Although the F-Score of R_T seems good in Table 2(c), the values of R_T in Rand Index and NMI are almost the lowest, as listed in Table 2(d) and (e). To summarize, R_I, R_U and R_O may be the most effective measures for clustering in terms of Precision in general. R_U, R_T and R_I may be the most effective measures in terms of Recall in general. R_U and R_I may be the most effective measures in terms of F-Score. These best measures may vary w.r.t the nature of clustering algorithms used.

From the third column of each table, we saw that the performance of spectral clustering (C_S) is very different from C_D (or C_L). R_T may be the most effective measure for clustering in terms of Precision in general. R_U, R_I and R_O may be the most effective measures in terms of Recall. R_I, R_T and R_O may be the most effective measures in terms of F-Score. Also, we can see that the Rand Index and NMI of C_S are generally much lower than the other two algorithms. These indicate that DBSCAN and single linkage performed better than spectral clustering in our evaluation.

5.4 Results of Combination Methods

The second experiment is to evaluate whether a certain combination method can improve the performance of clustering. In this experiment, we used DBSCAN (C_D) with $MinPts = 2$ and $Eps = 0.9$, and spectral clustering (C_S) with $\eta = 5$ as the clustering algorithms since these parameter settings generally achieved good F-Score according to the results of the previous experiment.

The total number of possible linear combinations of five measures is 26 ($= 2^5 - 1 - 5$), which is large. So we did not investigate all these possible combinations (Recall that we intend to investigate how a proper linear combination improves the performance compared to single measures, not for the best combination on specific datasets). Instead, for C_D we tried to find whether the linear combination of R_I, R_U and R_O (each of which has a relatively high Precision) can improve the performance, and whether R_T or R_W can improve the performance when combined with R_I, R_U and R_O. As a result, seven linear combinations were investigated in Table 3 (C_D), where $\omega_i \in [0,1]$ is a weight coefficient. For C_S, similar to the C_D, we tried to find seven linear combinations as well. Because the values of the measures in Precision are similar, we chose to try the linear combination of R_I, R_T and R_O, each of which has a relatively high F-Score instead of Precision. Finally, seven linear combinations were investigated in Table 3 (C_S), where $\omega_j \in [0,1]$ is also a weight coefficient like ω_i as aforementioned. We enumerated each weight coefficient value ω_i or ω_j from 0 to 1 with 0.05 interval and finally selected the combinations that achieved the highest F-Score in average. Additionally, we investigated the performance of the consensus clustering induced by using the corresponding measures. The average runtime of a clustering algorithm on the 20 entities is in 40 s.

Table 3 shows the harmonic means of Precision, Recall, F-Score, Rand Index and NMI achieved by using single measures for C_D and C_S, linear combinations of various measures and ensemble clustering. For each table, the 6th to 12th rows represent the linear combination with their weight coefficient values, and the 13th to 19th rows represent the performance of consensus clustering induced by using various measures.

For C_D, we observed that the linear combination of different measures may greatly improve the Recall values (about 50% from 0.433 to 0.899 in some cases). However, the linear combination achieved lower Precision as compared with individual measures. Furthermore, it also has a substantial decrease in the values of Rand Index and NMI (about 50% from 0.709 to 0.364 in some cases on Rand Index and about 50% from 0.520 to 0.268 in some cases on NMI). These results may be due to that different measures complemented each other to cover more

Table 3. Comparison on single relatedness measures and two combination methods

Clustering algorithm: C_D	Precision	Recall	F-Score	Rand Index	NMI
R_I	.235	.273	.253	.549	.387
R_W	.215	.266	.238	.672	.441
R_U	.242	.433	.310	.644	.507
R_T	.170	.381	.235	.547	.364
R_O	.247	.137	.176	.709	.520
$.3R_I + .7R_U$	.218	.757	.339	.471	.379
$.5R_I + .5R_O$	.209	.619	.313	.411	.265
$.6R_U + .4R_O$	.214	.716	.330	.477	.375
$.3R_I + .5R_U + .2R_O$	.211	.883	.341	.398	.318
$.3R_I + .5R_U + .1R_T + .1R_O$	.205	.878	.333	.372	.277
$.2R_I + .1R_W + .2R_U + .5R_O$	.216	.790	.339	.438	.344
$.2R_I + .1R_W + .15R_U + .1R_T + .45R_O$	.207	.899	.337	.364	.268
R_I, R_U	.287	.148	.196	.732	.563
R_I, R_O	.331	.051	.089	.744	.566
R_U, R_O	.290	.066	.108	.755	.575
R_I, R_U, R_O	.273	.210	.237	.706	.513
R_I, R_U, R_T, R_O	.292	.102	.151	.744	.560
R_I, R_W, R_U, R_O	.290	.115	165	.726	.548
R_I, R_W, R_U, R_T, R_O	.256	.213	.232	.677	.493
Clustering algorithm: C_S	Precision	Recall	F-Score	Rand Index	NMI
R_I	.184	.449	261	.500	.229
R_W	.198	.337	.250	.584	.231
R_U	.177	.410	.248	.503	.224
R_T	.215	.329	.260	.628	.255
R_O	.188	.427	.261	.516	.216
$.75R_I + .25R_T$	.209	.410	.277	.563	.269
$.8R_T + .2R_O$	.215	.356	.268	.613	.255
$.7R_I + .3R_O$	.194	.449	.271	.520	.242
$.7R_I + .25R_T + .05R_O$	.206	.418	.276	.558	.278
$.2R_I + .5R_U + .2R_T + .1R_O$	.209	.392	.272	.585	.285
$.7R_I + .1R_W + .1R_T + .1R_O$	.191	.464	.271	.497	.249
$.2R_I + .1R_W + .5R_U + .1R_T + .1R_O$	.198	.455	.276	.499	.234
R_I, R_T	.181	.211	.195	.620	.331
R_T, R_O	.237	.167	.196	.705	.376
R_I, R_O	.218	.160	.184	.691	.361
R_I, R_T, R_O	.184	.298	.228	.552	.268
R_I, R_U, R_T, R_O	.200	.151	.172	.681	.361
R_I, R_W, R_T, R_O	.202	.129	.158	.691	.362
R_I, R_W, R_U, R_T, R_O	.207	.213	.210	.662	.320

types of properties describing similar aspects while bringing noises. The ensemble clustering based on various measures has an increase in Precision (0.247 to 0.331 in some cases) without loss of Rand Index and NMI, due to the fact that two properties were assigned to the same cluster by ensemble clustering only if there is sufficient number of individual clustering results. This indicates that, when seeking for a clustering with a high Precision for C_D, the ensemble clustering may be better than the linear combination, while the linear combination tends to find a clustering with a high Recall.

For C_S, we found that the linear combination of different measures cannot improve the performance for C_S. The ensemble clustering based on various measures has an great decrease in Recall (about 60% from 0.449 to 0.184 in some cases). This indicates that the combination methods may not be helpful for improving the performance for C_S. The bad results may be related to the similarity of these measure for C_S.

6 Application to Entity Browsing

For years, it has been a great challenge to provide general users with smart views for browsing interlinked RDF descriptions of entities. As a use scenario of property clustering, we developed an online system called SView for browsing linked entities. It groups and orders property-values of entities by lenses for a neat presentation, and offers various mechanisms for discovering related entities such as exploration based on link patterns and similarity-based entity recommendation. Moreover, users can personalize their browsing experience and they never work alone. They can edit lenses and consolidate entities following their own opinions, and their efforts are alleviated due to crowdsourced contributions from all users. Additionally, SView leverages users' contributions to generate smart views, e.g. global lenses and global viewpoints on entity consolidation. The smart views are, in turn, shared among all users when browsing linked entities.

Figure 1 shows the screenshot for viewing an entity "The Pentagon"[5] in SView. Since this entity contains hundreds of property-values, it is not very readable if there is no appropriate organizing method. To address this issue, SView groups and orders property-values with lenses (e.g. "Building"), which reuse property clustering to describe closely related aspects of an entity and thus to help users capture related information quickly. A weighted set cover problem is formulated and solved to automatically pick up a small number of the lenses that can cover as many relevant properties as possible.

We invited 24 master students in computer science to compare the presentation of 10 DBpedia entities with and without property clustering in SView. The SUS (System Usability Scale) scores in average indicated that nearly 16% improvement can be achieved by using the property clustering (72.85 versus 62.86), and this result is statistically significant ($p < 0.05$).

[5] http://dbpedia.org/resource/The_Pentagon.

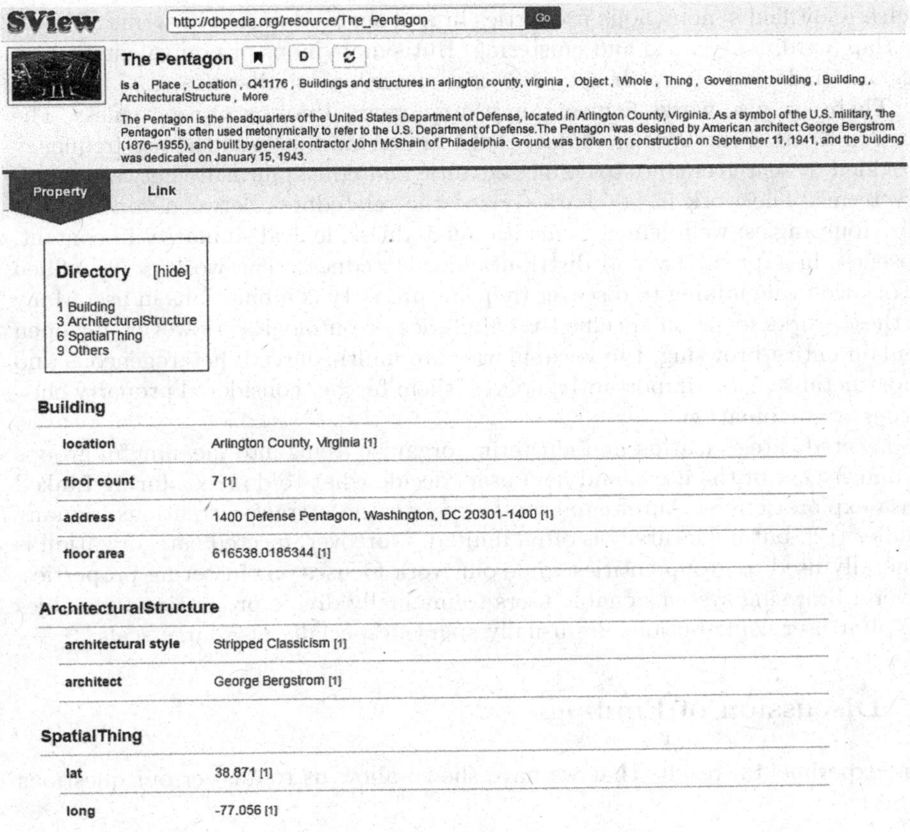

Fig. 1. Screenshot for browsing entities in SView with property clustering

It is worth noting that Wikidata[6] and Freebase[7] also organize related properties in adjacent positions. See Reasonator[8] for an item-type-optimized manner of Wikidata entity browsing.

7 Related Work

Property similarity is a special kind of property relatedness. Existing work that dedicates to property similarity finds synonymous or equivalent properties for applications like ontology mapping [14], entity linkage [10] and query expansion [2,18]. Specifically, the work in [2] leveraged association rule mining to exploit synonymous properties. The work in [18] defined statistical knowledge patterns,

---

[6] http://www.wikidata.org.
[7] http://www.freebase.com.
[8] http://tools.wmflabs.org/reasonator.

which identified synonymous properties in and across datasets in terms of triple overlap, cardinality ratio and clustering. But synonymous or equivalent properties are inadequate to cover the properties describing similar aspects.

There are also works focusing on a more general notion of relatedness. The work in [7] used the Web as its knowledge source and utilized the use frequency provided by search engines to define semantic relatedness measure between ontology terms. The work in [4] characterized the relatedness between vocabularies from four angles: well-defined semantic relatedness, lexical similarity in content, closeness in expressivity and distributional relatedness. The work in [9] refined association rule mining to discover frequent property combinations in use. Many of these works focus on specified vocabularies or ontologies. However, for open domain entity browsing, the vocabularies are multi-sourced, heterogeneous and unpredictable. More importantly, none of them further considered property clustering or combination.

Faceted categorization and clustering organize items into meaningful groups to make sense of the items and help users decide what to do next during Linked Data exploration [8]. Automated facet construction attracts attentions in many studies [12], but its accuracy is often limited. Moreover, faceted categorization is generally used to group entities while our work focuses on clustering properties. Several browsing systems enable users to manually divide properties and values [13], but user contributions are usually sparse, especially at a large scale.

8 Discussion of Findings

The experimental results that we have shown allow us to answer our questions in Sect. 1.

- We empirically evaluated five kinds of relatedness measures between properties: lexical similarity between property names (R_I), semantic relatedness between property names (R_W), distributional relatedness between properties (R_U), range relatedness between properties (R_T) and overlap of property values (R_O). The result of our empirical study is uneven for every measure, which can be explained in two aspects. On one hand, most sample entities have considerable variance due to difference sources and property numbers; on the other hand, there is no measure that can achieve a high value for every clustering algorithm on either of Precision, Recall, Rand Index and NMI. In terms of F-Score, R_I and R_U generally generate the clusterings that are closer to the reference ones in our study than the other measures.
- We empirically evaluated three clustering algorithms: DBSCAN (C_D), single linkage (C_L) and spectral clustering (C_S). From the overall results, C_D is similar to C_L under our parameter settings, and C_S is greatly different from them. The results of C_D, C_L and C_S are uneven for each measure, and that of C_S is relatively stable. However, C_D and C_L usually generate better clustering results.

- We empirically evaluated the linear combination of measures and ensemble clustering using C_D and C_S. For C_D, our empirical study shows that the linear combination of relatedness measures tends to generate a clustering that features a high Recall, while ensemble clustering (consensus clustering) is recommended to use if a high Precision is preferred. For C_S, both of them are of little avail.

However, there are some issues that have not been fully covered during our evaluation:

- There are a diversity of methods to calculate lexical similarity between property names. In our study, we only tried I-Sub based on our previous experience in ontology matching. But it is possible that there is a great difference among the performances of different similarity measures. Additionally, distributional relatedness between properties highly depends on the underlying dataset used for estimation. Leveraging a more appropriate dataset can improve the performance.
- There are not a few well-known clustering algorithms that we have not considered in our evaluation, i.e. non-negative matrix factorization [15], which may achieve better performance.
- The two combination methods can improve C_D significantly, but can work on C_S barely. It is probable that both of the two combination methods have special favorites on clustering algorithms. However, due to the limited sample entities, we have not observed such correlation between the combination methods and the clustering algorithms.
- In our evaluation, properties used for clustering came from multiple sources and they were largely heterogeneous. At present, the clustering algorithms and combination methods have not achieved satisfiable F-Score values. This implies that more sophisticated solutions need to be developed for property clustering in Linked Data.

9 Conclusion

In this paper, we studied the property clustering in Linked Data and evaluated five property relatedness measures, three property clustering algorithms and two combination methods. Our experimental results demonstrated the feasibility of the automated property clustering. We also showed that property clustering can enhance entity browsing in practice.

In future work, we will improve the quality of property clustering by leveraging user feedback and active learning. We will also explore more use scenarios for property clustering in Linked Data.

Acknowledgements. This work is supported by the National Natural Science Foundation of China (No. 61370019). We appreciate our students' participation in the experiments.

References

1. Ailon, N., Charikar, M., Newman, A.: Aggregating inconsistent information: ranking and clustering. J. ACM **55**(5), 23 (2008)
2. Abedjan, Z., Naumann, F.: Synonym analysis for predicate expansion. In: Cimiano, P., Corcho, O., Presutti, V., Hollink, L., Rudolph, S. (eds.) ESWC 2013. LNCS, vol. 7882, pp. 140–154. Springer, Heidelberg (2013). doi:10.1007/978-3-642-38288-8_10
3. Budanitsky, A., Hirst, G.: Evaluating WordNet-based measures of lexical semantic relatedness. Comput. Linguist. **32**(1), 13–47 (2006)
4. Cheng, G., Gong, S., Qu, Y.: An empirical study of vocabulary relatedness and its application to recommender systems. In: Aroyo, L., Welty, C., Alani, H., Taylor, J., Bernstein, A., Kagal, L., Noy, N., Blomqvist, E. (eds.) ISWC 2011, Part I. LNCS, vol. 7031, pp. 98–113. Springer, Heidelberg (2011). doi:10.1007/978-3-642-25073-6_7
5. Evert, S.: Corpora and collocations. In: Lüdeling, L., Kytö, M. (eds.) Corpus Linguistics: An International Handbook, pp. 1212–1248. Mouton de Gruyter, Berlin (2008)
6. Fleiss, J.: Measuring nominal scale agreement among many raters. Psychol. Bull. **76**(5), 378–382 (1971)
7. Gracia, J., Mena, E.: Web-based measure of semantic relatedness. In: Bailey, J., Maier, D., Schewe, K.-D., Thalheim, B., Wang, X.S. (eds.) WISE 2008. LNCS, vol. 5175, pp. 136–150. Springer, Heidelberg (2008). doi:10.1007/978-3-540-85481-4_12
8. Hearst, M.: Clustering versus faceted categories for information exploration. Commun. ACM **49**(4), 59–61 (2006)
9. Hu, W., Jia, C.: A bootstrapping approach to entity linkage on the semantic web. J. Web Semant. **34**, 1–12 (2015)
10. Isele, R., Bizer, C.: Active learning of expressive linkage rules using genetic programming. J. Web Semant. **23**, 2–15 (2013)
11. Lin, D.: An information-theoretic definition of similarity. In: ICML 1998, pp. 296–304. Morgan Kaufmann, San Francisco (1998)
12. Oren, E., Delbru, R., Decker, S.: Extending faceted navigation for RDF data. In: Cruz, I., Decker, S., Allemang, D., Preist, C., Schwabe, D., Mika, P., Uschold, M., Aroyo, L.M. (eds.) ISWC 2006. LNCS, vol. 4273, pp. 559–572. Springer, Heidelberg (2006). doi:10.1007/11926078_40
13. Quan, D., Karger, D.: How to make a semantic web browser. In: WWW 2004, pp. 255–265. ACM, New York (2004)
14. Shvaiko, P., Euzenat, J.: Ontology matching: state of the art and future challenges. IEEE Trans. Knowl. Data Eng. **25**(1), 158–176 (2013)
15. Smith, T., Frank, E.: Introducing machine learning concepts with WEKA. In: Mathé, E., Davis, S. (eds.) Statistical Genomics, pp. 353–378. Springer, Heidelberg (2016)
16. Stoilos, G., Stamou, G., Kollias, S.: A string metric for ontology alignment. In: Gil, Y., Motta, E., Benjamins, V.R., Musen, M.A. (eds.) ISWC 2005. LNCS, vol. 3729, pp. 624–637. Springer, Heidelberg (2005). doi:10.1007/11574620_45
17. Wagner, S., Wagner, D.: Comparing clusterings: an overview. Universität Karlsruhe, Fakultät für Informatik (2007)
18. Zhang, Z., Gentile, A.L., Blomqvist, E., Augenstein, I., Ciravegna, F.: Statistical knowledge patterns: identifying synonymous relations in large linked datasets. In: Alani, H., et al. (eds.) ISWC 2013, Part I. LNCS, vol. 8218, pp. 703–719. Springer, Heidelberg (2013). doi:10.1007/978-3-642-41335-3_44

A MapReduce-Based Approach for Prefix-Based Labeling of Large XML Data

Jinhyun Ahn[1], Dong-Hyuk Im[2], and Hong-Gee Kim[1,3(✉)]

[1] Biomedical Knowledge Engineering Laboratory and Dental Research Institute, Seoul National University, Seoul, South Korea
{jhahncs,hgkim}@snu.ac.kr

[2] Department of Computer and Information Engineering, Hoseo University, Cheonan, South Korea
dhim@hoseo.edu

[3] Institute of Human-Environment Interface Biology, Seoul National University, Seoul, South Korea

Abstract. A massive amount of XML (Extensible Markup Language) data is available on the web, which can be viewed as tree data. One of the fundamental building blocks of information retrieval from tree data is answering structural queries. Various labeling schemes have been suggested for rapid structural query processing. We focus on the prefix-based labeling scheme that labels each node with a concatenation of its parent's label and its child order. This scheme has been adapted in RDF (Resource Description Framework) data management systems that index RDF data in tree by grouping subjects. Recently, a MapReduce-based algorithm for the prefix-based labeling scheme was suggested. We observe that this algorithm fails to keep label size minimized, which makes the prefix-based labeling scheme difficult for massive real-world XML datasets. To address this issue, we propose a MapReduce-based algorithm for prefix-based labeling of XML data that reduces label size by adjusting the order of label assignments based on the structural information of the XML data. Experiments with real-world XML datasets show that the proposed approach is more effective than previous works.

1 Introduction

A large volume of XML (Extensible Markup Language) data from various areas are publicly available on the web, with some examples including DBLP (computer science bibliography), UniprotKB (protein information), SwissProt (protein sequence database), and Treebank (tagged sentences). XML data can be viewed as a tree data model that represents parent-child relationships between elements. XPath [1] is widely used to represent structural queries against XML data. One of the fundamental structural queries requires the determination of whether an ancestor/descendant relationship exists between two given elements. The simplest method to answer such queries is to traverse the tree to determine if a path exists between the two given elements. However, if the elements are far apart, one has to visit many elements. To overcome this disadvantage, labeling

Y.-F. Li et al. (Eds.): JIST 2016, LNCS 10055, pp. 83–98, 2016.
DOI: 10.1007/978-3-319-50112-3_7

schemes have been proposed, in which each node is labeled so that its ancestor/descendant relationship can be determined by considering only those two labels. Interval-based, prefix-based, and prime number-based labeling schemes exist for tree data models. Of these schemes, we focus on the prefix-based labeling scheme that has been extensively utilized in many practical systems [2], including Microsoft SQL Server [3] and SiREN [4]. Note that SiREN is an information retrieval engine for RDF (Resource Description Framework) data that is a graph data model. In SiREN, RDF triples are converted into tree data structures by grouping subjects. The prefix-based labeling scheme is adopted to index the tree. With the popularity of the prefix-based labeling scheme, [5] proposed a MapReduce-based algorithm for prefix-based labeling of XML data to make it more applicable for massive real-world XML datasets (or Linked Open Data[1] if processed like SiREN).

Although previous approaches can label large XML data in parallel, they are not efficient at producing smaller label sizes. This makes it difficult for the prefix-based labeling scheme to handle massive XML data. Previous approaches assign labels to each node in the order presented in the XML file. We note that if we change the label assignment order appropriately, the resultant prefix-based label size is reduced while conforming to the prefix-based labeling scheme. Reducing label size is important because the query processing performance depends on how large the labels are. Obviously, the smaller the labels are, the faster ancestor/descendant relationships can be determined. Therefore, devising a way of generating the smallest label size is very important.

In this paper, we propose a dynamic labeling technique for the prefix-based labeling scheme, designed to produce a smaller label size than previous works. Specifically, the proposed approach extends [6]'s MapReduce-based repetitive prime labeling algorithm, which is similar to prefix-based labeling. The proposed approach adjusts the label assignment order during the labeling process, enabling it to reduce label size. The adjustment of label assignment is based on structural properties of the tree that can be obtained during the labeling process. The proposed approach is implemented on MapReduce, which allows multiple machines to perform labeling in parallel.

The contribution of this paper is summarized as follows:

- We extend the existing MapReduce-based repetitive prime labeling algorithm [6] to perform prefix-based labeling more efficiently than the state-of-the-art prefix-based labeling systems.
- We devise a novel technique that improves the above extension.
- Experiments with real-world XML datasets are conducted to show the effectiveness of the proposed approaches compared with state-of-the-art works.

This paper is organized as follows. Problem definitions and notations are stated in Sect. 2. Section 3 briefly overviews related work. In Sect. 4, we motivate our work by discussing the drawbacks of the state-of-the-art approach. Our approach is demonstrated in Sect. 5. Experimental results are discussed in Sect. 6. The paper is concluded in Sect. 7.

[1] http://linkeddata.org.

2 Preliminarily

We briefly introduce the MapReduce framework, XML data and prefix-based labeling scheme in this section, as these are closely related in our problem setting.

MapReduce is a programming model in a distributed computing environment. A MapReduce-based program consists of a map and reduce phase [7]. First, input data is split into several parts by an InputFormat, each of which is denoted as InputSplit. Each mapper reads each InputSplit and sends a part of InputSplit that is grouped by a map key to reducers. The reducer then processes the received data. All mappers and reducers run on each machine independently in a parallel fashion.

XML data can be viewed as a sequence of *start/end-tags*. The *empty-element-tags* (encoded in `<.../>`) and text contents between *start-tag* and *end-tag* are ignored in this paper to simplify notation. These can be represented in an expanded form consisting explicitly of dummy *start-tags* and *end-tags*. An *element* is a logical component that is a pair of a *start-tag* and a matching *end-tag*. Specifically, XML InputSplit is taken into account in order to model XML data in the context of the MapReduce framework. The formal definition of XML InputSplit is stated in Definition 1.

Definition 1. ***XML InputSplit***: Given an XML file D and a split size b in bytes. An ith InputSplit $I_i(D, b)$ is the sequence of *start/end-tags* split by b bytes, such that

$$I_i(D, b) = (t_j, t_{j+1}, ..., t_{j+n})$$

where t_j is the jth tag in D such that

$$t_j = (name, offset, type)$$

The $name$ denotes the string between two brackets. The position of its first bracket in D is $offset$, which is provided automatically by the MapReduce framework. The $type$ is either $start$ or end. For readability, we sometimes denote $t^{start} = (*, *, start)$ and $t^{end} = (*, *, end)$. If the context is clear, we simply state I_i to indicate $I_i(D, b)$.

Example 1. The start-tag of a root element is represented by (dblp, 0, *start*) if the name of the root element is "dblp". The $offset$ of the first child of the root element is set as 6 since `<dblp>` has six characters.

We implemented an XMLInputFormat to split an XML file into a sequence of XML InputSplits, as TextInputFormat provided by Apache Hadoop cannot be used here because it splits by *characters*. A tag may be split into different XML InputSplit (e.g., `<d` in I_i and `blp/>` in I_{i+1}). Thus, the XMLInputFormat is designed to be split by *brackets* to split into tags correctly.

Definition 2. ***XML Data Labeling Problem***: Given an XML file D, output an injective map L from a set of elements E of D to a set LS of particular labels.

$$L : E \longrightarrow LS$$

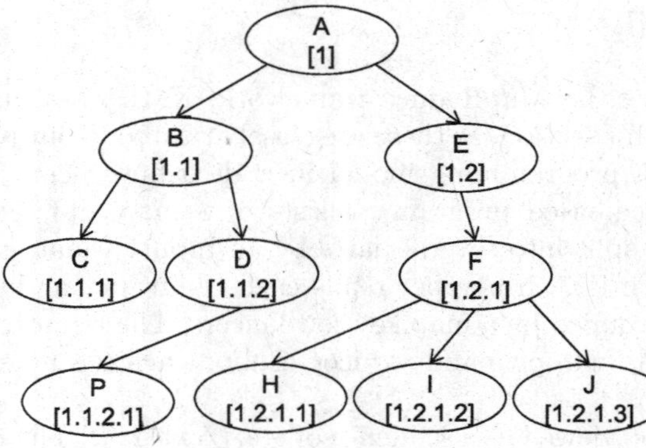

Fig. 1. A tree labeled by the prefix-based labeling scheme

L should be injective because every element $e \in E$ must be labeled and no two elements have the same label. LS is based on the choice of labeling scheme.

In particular, we focus on the conventional prefix-based labeling scheme [8–11], defined in Definition 3.

Definition 3. ***Prefix-based Labeling Scheme***: The prefix-based label $L(e)$ of an element e is defined as follows:

$$L(e) = L(pa(e)).order(e)$$

where $pa(e)$ is the parent of e, period(.) is the delimiter, and $order(e)$ is the child order of e under $pa(e)$. The label size $size(L(e))$ of $L(e)$ is the number of digits in $L(e)$ except for delimiters. To indicate the suffix label $order(e)$, we denote $suffix(L(e))$.

Example 2. See Fig. 1. The root element A has 1 as its prefix-based label. L(E) is 1.2 because we have $L(\text{A}) = 1$ and E is the second child corresponding to $order(\text{E}) = 2$. The label size of L(H) is 4, while the label size of L(E) is 2.

3 Related Work

Answering ancestor/descendant relationships between two given elements in XML data can be achieved by tree traversal. However, this approach is inefficient when the two given elements are far away from each other because it requires visiting all of the intermediate elements in a path connecting them. To avoid visiting numerous nodes, we may maintain additional data for each element. The additional data is called an *index* or *label*. For two given elements a and b, we can determine the ancestor/descendant relationships by applying operations to the labels of a and b. The simplest labeling scheme is to have the label to retain a list of ancestor/descendant elements. However, this approach is inefficient because the space requirement is too large. To overcome the space disadvantage, diverse tree labeling schemes have been proposed [12].

Previous works are classified in two categories: tree labeling schemes and tree labeling algorithms. Tree labeling schemes define *what* the label is, while tree labeling algorithms define *how* to label each element. Note that the focus of this paper is a tree labeling algorithm based on the prefix-based labeling scheme.

Tree Labeling Schemes: The interval-based labeling scheme labels each element with (*start*, *end*, *level*) [13–17]. The *start* and *end* of an element represents the interval that includes all of its child element's intervals recursively. Therefore, the parent/child relationship can be determined from whether or not an interval is included. Additionally, *level* is used for determining whether the child is a direct child. The prime number labeling scheme calculates labels on the basis of the product of prime numbers [18,19]. In [18], after a unique prime number is assigned to every element, the product of the parent element's label and its own prime number (self-label) becomes the element's label. The disadvantage is that a unique prime number must be assigned to every element. If the number of elements increases, the number of prime numbers used will increase accordingly. However, variations have been proposed to overcome this disadvantage. In [19], the prime number recycling method is proposed, in which the self-label becomes the prime number following the prime number of its parent. This has two advantages over [18]. First, the label size is reduced because this scheme reuses prime numbers. Second, the keyword-based search query processing is efficient because self-labels of descendant elements of an element are larger than or equal to the self-label of the element. The prefix-based labeling scheme is a labeling method in which the parent element label is the prefix, and the child's order number is the suffix [8–11]. A delimiter is needed to distinguish between the prefix and the suffix. For example, if the root element's label is 1, its first child element's label is 1.1 and its second child element's label is 1.2, where the period(.) is the delimiter.

Tree Labeling Algorithm: In-memory tree labeling algorithms can be straightforwardly implemented. However, a massive XML file cannot be processed if there is insufficient memory. To solve this, the MapReduce framework has been adapted.

There exists a few studies on XML data labeling algorithms based on MapReduce [5,6]. [5] proposed two MapReduce-based tree labeling algorithms for interval-based labeling and prefix-based labeling schemes. Let's first discuss the interval-based labeling algorithm. During the map phase, *start* and *end* values are assigned to each element in InputSplit. In InputSplit, for a start/end-tag, if there is no corresponding end/start-tag, then *end* (or *start*) value is left empty. These incomplete labels are sent to reducers. At each map step, the label of the last element and the number of elements are recorded on an HDFS (Hadoop Distributed File System) file. When all map steps are completed, all of the information collected in the HDFS file is combined to generate one offset table, which

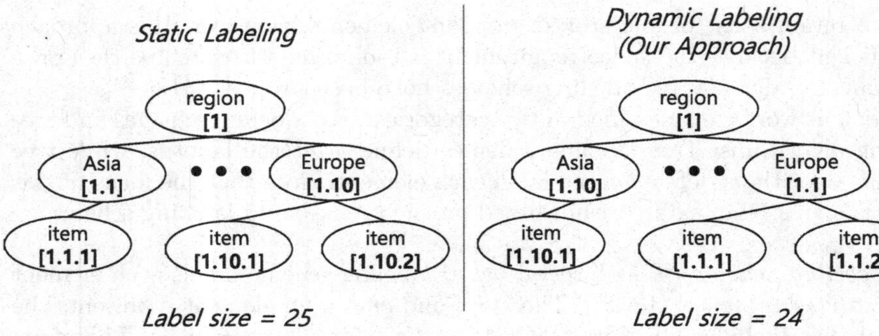

Fig. 2. Label size differs depending on the order of labels for unordered XML data.

can be accessed by all machines. In the reduce step, the offset table is used for completing the incomplete label of each element.

To explain the motivation for our work, the state-of-the-art MapReduce-based prefix-based labeling algorithm and its drawbacks are discussed in details in Sect. 4.

4 Motivation

Regarding the order of child elements, XML data are classified as unordered and ordered. The order of child elements is significant in ordered XML data, whereas this order is not significant in unordered XML data. In most real-world XML data, the order among child elements is unimportant [20,21]. In this regard, this paper focuses on unordered XML data. Specifically, we note that for unordered XML data, the order of label assignments affects the size of the resultant labels, as discussed in [6].

In Fig. 2, two different labeling results are depicted for the same XML data. The same prefix-based labeling scheme is used here. The figure is drawn on the assumption that the Asia element precedes the Europe element in the XML data. We also assume that there are eight elements between Asia and Europe, which are omitted in the figure. Label assignment on the left is by the order of the elements serialized in the XML data. We call it *static* labeling because the order of labels is determined by the order in the XML data. Asia is assigned 1.1 because it is the first child. Europe is assigned 1.10 because it is the 10th child. On the other hand, label assignment on the right is *dynamic* labeling because label order is not determined by the order in the XML data. Here, we have $L(\text{Europe}) = 1.1$ and $L(\text{Asia}) = 1.10$. Asia is assigned 10 because it has fewer descendants than Europe. In *dynamic* labeling, 10 is inherited to one node, whereas it is inherited to two elements in *static* labeling. By changing the label assignment order, we can reduce the label size (e.g., from 25 to 24 in Fig. 2).

The state-of-the-art prefix-based labeling algorithm [5] contains *static* labeling and does not support *dynamic* labeling. InputSplits are created from an input

XML file by a given split size, each of which is then assigned to each mapper. In the map phase, each element with an incomplete prefix-based label is sent to reducers using the name of the element as the map key. In addition, the following information is output from each mapper: the number of not paired end-tags (called `basement`), the sibling order of the last element, and the incomplete label of the last element. The information is stored in an HDFS file to create an offset table that has n rows, where n is the number of InputSplits. All previous rows from 1 to $n-1$ are combined to obtain the nth row in the offset table. In the reduce phase, the label in the offset table is appended to the prefix part, and s suffixes in the incomplete label are removed when s is the `basement` value. Note that a row in the offset table is made of all previous rows. Therefore, the order of the label assignment is the same as the order of InputSplit, which is the same as the order of rows. In other words, the order of label assignment is fixed to the order present in the XML file, which corresponds to *static* labeling.

5 The Proposed Approach

The proposed approach is extended from RepMR [6], which proposes a MapReduce-based XML data labeling algorithm for the repetitive prime labeling scheme [19]. It appears that a simple modification of RepMR can achieve our goal as these two labeling schemes are closely related. We call the simple modification PrxMR. However, we observe that label size can be further reduced using our novel *DCL(dynamic compressed element labeling)* technique. The approach with *DCL* is called PrxMR+. Section 5.1 briefly reviews PrxMR. Section 5.2 discusses the limitations of PrxMR and then explains our alternative PrxMR+.

5.1 PrxMR

PrxMR is a modification of RepMR [6] so that prefix-based labels are generated instead of repetitive prime labels. The overall algorithm is the same; however, the difference between them is the labeling scheme employed. This is possible because prefix-based labels and repetitive prime labels are similar in that labels are based on parent's label and child order. These two schemes deal with child order differently though. An illustrative example of PrxMR is shown in Fig. 3. Upper Tree Labeling is the process of labeling a portion of elements extracted from the root (called an upper tree). The label assignment order is determined based on the number of descendants. The upper tree is stored in an HDFS file for reducers to access. Label population is performed by reducers independently by referring to the shared upper tree.

Upper Tree Labeling. Figure 4 illustrates the Upper Tree Labeling step. For a given split size, we assume that three XML InputSplits are created. One sub-upper tree (denoted as SUP) is extracted for each InputSplit, in which nested elements are not included (e.g., type in SUP_1). Since these nested elements can be labeled locally in the Label Population step. Start/end-tags are depicted in

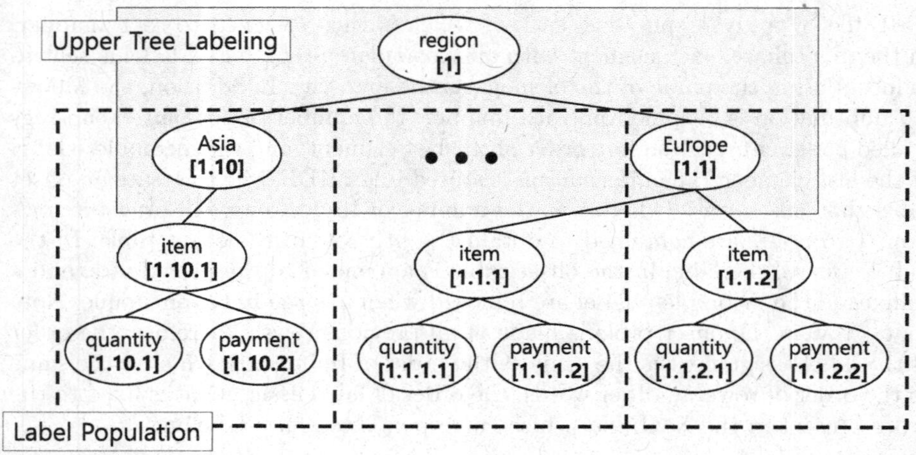

Fig. 3. Conceptual diagram of PrxMR. Prefix-based labels are represented in the bracket below the element name.

dotted rectangles while elements are in solid rectangles. For sub-upper trees (e.g. SUP_2) that have two or more root elements, a *DUMMY* tag becomes the root. These sub-upper trees are merged by paring root elements to create the upper tree (UP) depicted in the right-most part. The number of all descendants including itself is *dsc*, while *sibs* is the number of siblings including itself. Note that the order of book, device, and food elements is changed by *dsc*. The bold rectangle in UP represents a compressed element that corresponds to a sequence of elements, defined in Definition 4.

Definition 4. ***Compressed Element***: A compressed element C in UP is constructed based on a sequence $W \subset I_i$ of tags in an XML InputSplit I_i.

$$W = \{\texttt{<}e_i^{start}\texttt{>}, ..., \texttt{</}e_i^{end}\texttt{>}, \texttt{<}e_j^{start}\texttt{>}, ..., \texttt{</}e_k^{end}\texttt{>}\}$$

such that there exists a start-tag <e> in W if and only if there exists a matching end-tag </e> in W, which means that all tags are paired. The first element (e_i^{start},e_i^{end}) is called *leader*. Consider a sub-sequence $S \subset W$ of *leader*'s sibling elements as follows:

$$S = \{(s^{start}, s^{end}) | level(s^{start}) = level(leader^{start}) \text{ and } s^{start}, s^{end} \in W\}$$

The *level* values are calculated by iterating over W. The $level(leader^{start})$ is initially set to 0. The value increases by 1 when a start-tag is encountered and decreases by 1 for an end-tag. Then C is encoded in a 3-tuple:

$$(leader, sibs, dsc),$$

where $sibs = |S|$ and $dsc = \sum_{s \in S} s.dsc$.

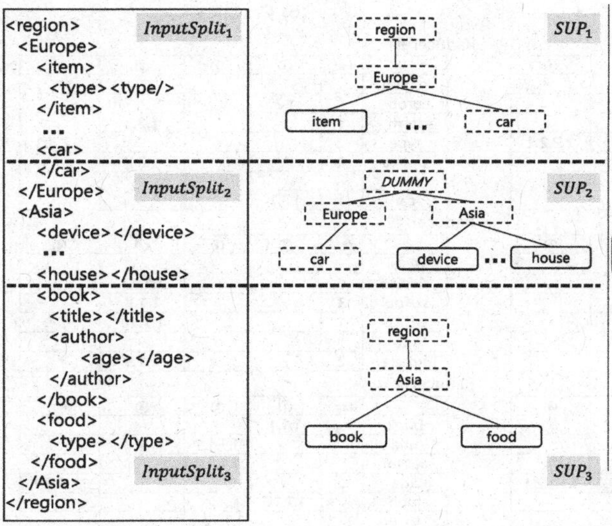

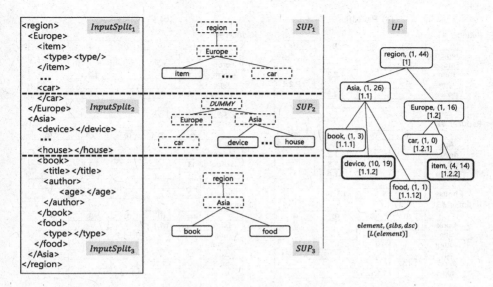

Fig. 4. Example dataflow of Upper Tree Labeling. Sub-upper trees (SUP) are extracted from each InputSplits and then merged into an upper tree (UP). Note that elements that can be labeled locally are not included in UP.

Example 3. (device,10,19) is a compressed element. There exist ten sibling elements in the same level such that devise is the first one and house is the last one. Therefore, food element is assigned 1.1.12 because device.$sibs = 10$.

The order between two (compressed or regular) elements A and B in UP is determined by the average number of descendants of each direct children defined in Eq. 1.

$$order(A) < order(B) \text{ iff } \frac{A.dsc}{A.sibs} \geq \frac{B.dsc}{B.sibs} \tag{1}$$

Example 4. In Fig. 4, we have $order(\textbf{book}) < order(\textbf{device})$ because $\frac{book.dsc}{book.sibs} = \frac{3}{1} = 3$ and $\frac{device.dsc}{device.sibs} = \frac{19}{10} = 1.9$.

Label Population. Figure 5a is an illustrative example of PrxMR that assigns prefix-based labels to each element by referring to UP in Fig. 4. Three reducers are executed because InputSplit id is chosen as the map key, which means that elements in an InputSplit are sent exclusively to a reducer. There are four cases where each element e is labeled by matching to UP. The $offset$ value is utilized to match between elements.: ① If e is found in UP, then take the label in UP. ②: If e is not found in UP, but its parent $pa(e)$ is found in UP, then we automatically obtain $L(e)$ since $pa(e)$ has already been encountered in the same reducer and therefore $L(pa(e))$ is available. For example, in Reducer #3, the title is not in UP, but its parent book is. Therefore, book must have been

(a) PrxMR

Reducer #1

	①	②	③	④
region	1			
Europe	1.2			
Item			**1.2.2**	
type				**1.2.2.1**
police			**1.2.3**	
...				
car	1.2.1			

Reducer #2

	①	②	③	④
Asia	1.1			
device			**1.1.2**	
school			**1.1.3**	
office			**1.1.4**	
...				
house			**1.1.11**	

Reducer #3

	①	②	③	④
book	1.1.1			
title		1.1.1.1		
author		1.1.1.2		
age				1.1.1.2.1
food	1.1.12			
type		1.1.12.1		

(b) PrxMR+

Reducer #1

	dsc.	①	②	③	④
region		1			
Europe		1.2			
Item	**1**			**1.2.3**	
type					**1.2.3.1**
police	**2**			**1.2.2**	
...					
car	1	1.2.1			

Reducer #2

	dsc.	①	②	③	④
Asia		1.1			
device	**0**			**1.1.11**	
school	**3**			**1.1.8**	
office	**1**			**1.1.10**	
...					
house	**2**			**1.1.9**	

Reducer #3

	dsc.	①	②	③	④
book		1.1.1			
title			1.1.1.1		
author			1.1.1.2		
age					1.1.1.2.1
food		1.1.12			
type			1.1.12.1		

Fig. 5. Example dataflow of the Label Population step of PrxMR and PrxMR+. Each element e is assigned $L(e)$ differently in four cases (① ~ ④) by referring to UP in Fig. 4. The number of descendants is represented by *dsc*, which is only exploited by PrxMR+. Elements in compressed nodes are labeled differently, highlighted in boldface.

encountered before title according to the sequence serialized in the XML data. ③: e is the compressed element in UP. For example, device is the compressed element with $L(\text{devise}) = 1.1.2$ from UP. The sub-sequent sibling elements are labeled in the presented order. ④: The other case. For example, in Reducer #3, the age element is automatically labeled because it is the child of the author element that has been labeled by case ②.

5.2 PrxMR+

Although PrxMR successfully reduces the label size, it is still inefficient in the way elements in a compressed element are labeled, corresponding to case ③ in the Label Population step.

See elements in bold in Fig. 5a. The suffix labels increase sequentially in the presented order. For example, in Reducer #2, device, school, office, ... , house elements are assigned suffix labels from 2 to 11. This corresponds to *static* labeling; its inefficiency in terms of the label size was discussed in Sect. 4. We observed that this inefficiency can be overcome by utilizing the information about *dsc* of each sibling elements to adjust suffix labels. In this regard, we devise *DCL* in Definition 5.

Definition 5. *DCL(Dynamic Compressed Element Labeling)*: Given a compressed element C, by Definition 4, we have a sequence W of tags and its

Table 1. Real-world XML Datasets

Name	Elements	Avg. fanout (max)	Avg. depth (max)
treebank	2,437,666	2.2 (51)	7.8 (36)
swissProt	2,977,030	6.6 (342)	3.5 (5)
dblp	3,332,130	9.0 (750)	2.9 (6)
psd7003	21,305,818	3.9 (151)	5.1 (7)
kegg	63,445,091	2.4 (188)	7.3 (10)

sub-sequence S of $C.leader$'s sibling elements. The dsc of $s \in S \subset W$ is obtained as follows:

$$s.dsc = \text{ the number of start-tags in } W \text{ between } s^{start} \text{ and } s^{end}$$

By ordering S in decreasing order of dsc, we obtain a sorted set $DSC(S)$. Then, the label of each element in S is defined as follows:

$$L(s) = L(pa(C.leader)).i$$

such that $i = suffix(L(C.leader)) + (d - 1)$ and s is the dth one in $DSC(S)$.

Example 5. See Fig. 5b. In Reducer #1, we have a compressed element $C = (\text{item}, 4, 14)$ and $S = \{\textbf{item}, ..., \textbf{police}\}$. We obtain $DSC(S) = \{\textbf{police}, ..., \textbf{item}\}$, because item.$dsc$ < police.dsc assuming that they have the minimum and maximum dsc among elements in S. We also have $L(pa(\text{item})) = 1.2$ and $suffix(L(\text{item})) = 2$. Therefore, we have $L(\text{police}) = 1.2.2$ because police is the first element in $DSC(S)$.

6 Performance Study

Experiments with real-world XML datasets were conducted. Our cluster consists of four machines (2.6 GHz CPU, 32 GB RAM). Hadoop 2.7.1 and JDK 1.8 were used for the implementations. The proposed approaches (PrxMR and PrxMR+) were compared with the state-of-the-art approach [5], denoted as STATIC. In the experiments, the split size b was assumed to be 524,288 bytes.

Table 1 lists five datasets obtained from the XML Repository at University of Washington[2]. The number of children of an element is *fanout*. The number of elements from e to root is the *depth* of an element e.

6.1 Labeling Time

Figure 6 shows the labeling time for the five datasets. There is no significant difference between the three approaches for all datasets. Nevertheless, we observe

[2] http://www.cs.washington.edu/research/xmldatasets/www/repository.html.

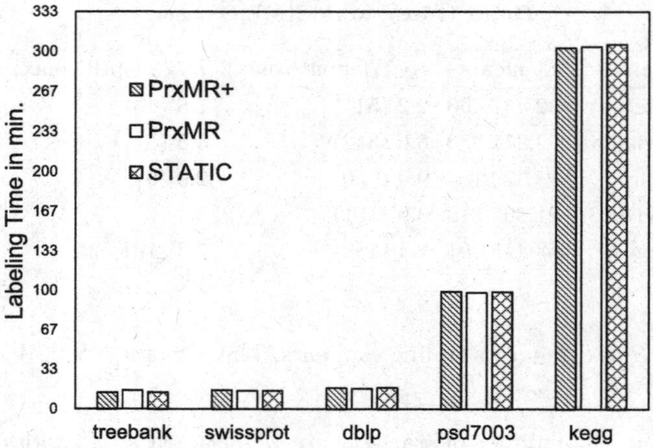

Fig. 6. Labeling time

that our approaches are slightly better than STATIC for larger datasets. This can be understood by the fact that STATIC completes labels in the reduce phase using a slightly complex operation (i.e., calibration operation stated in [5]), whereas our approach employs a very simple operation (i.e., Label Population).

6.2 Label Size

The average label size is calculated in the experiments (Table 2). This measure is more important than the total label size, because prefix matching is performed between individual labels to answer structural queries. For all cases, our approaches generate smaller label size than STATIC. The Upper Tree Labeling technique helps avoid assignments of larger suffix numbers to elements with many descendants, which contributes to reducing label size. Note that label size is the number of digits in decimal representation, which means that even a very small reduction in label size is meaningful. The largest label sizes were seen in treebank even though it is the smallest dataset due to its large depth. According to the prefix-labeling scheme, suffix numbers are appended following descen-

Table 2. Average Label Size

	PrxMR+	PrxMR	STATIC
treebank	**11.656**	11.658	11.678
swissprot	**7.762**	7.765	8.291
dblp	**7.558**	7.559	7.672
psd7003	**9.796**	9.800	9.913
kegg	**10.466**	**10.466**	10.741

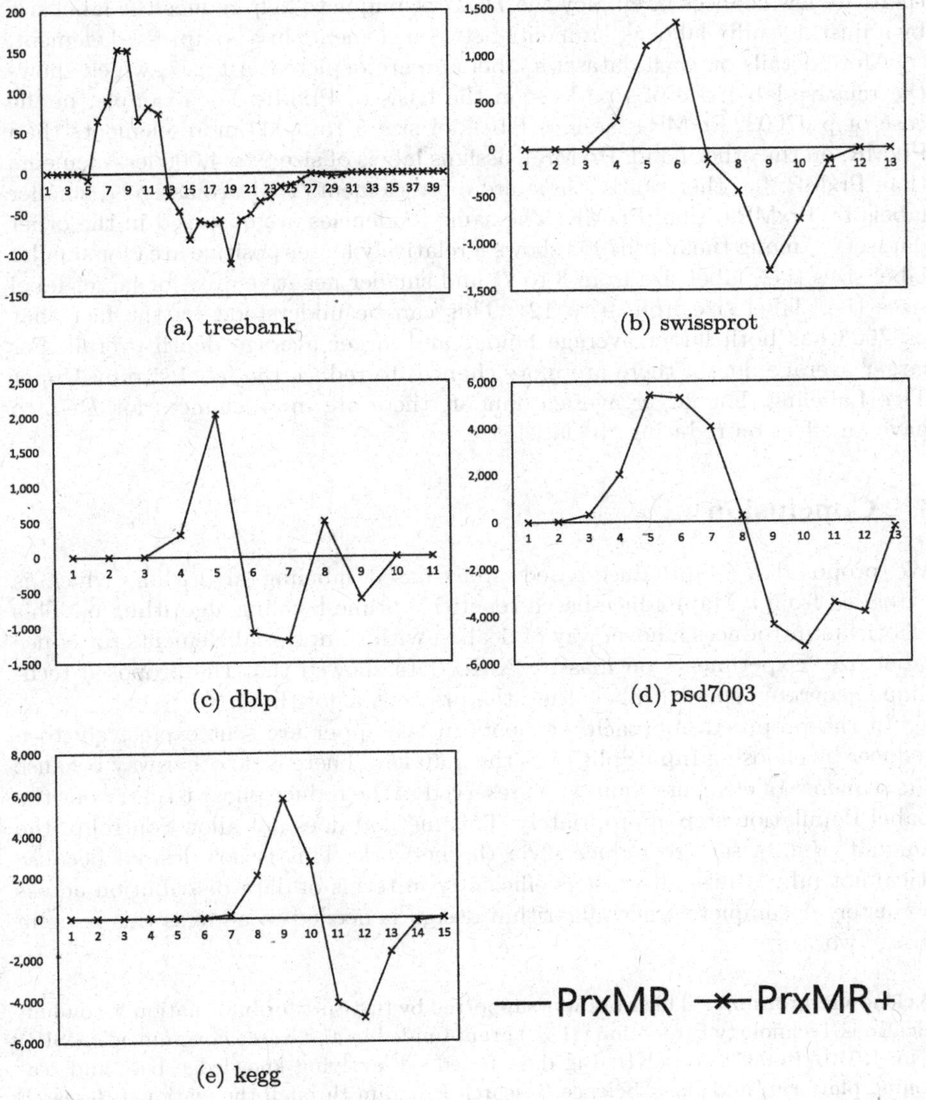

Fig. 7. The relative number of labels for each label size, on the basis of PrxMR. X-axis indicates the label size. For example, individual label sizes of kegg range from 1 to 15. The y-axis indicates the relative number of elements for a label size x by calculating $|\{e|size(L_{\text{PrxMR+}}(e)) = x\}| - |\{e|size(L_{\text{PrxMR}}(e)) = x\}|$, where $L_{\text{PrxMR+}}$ and L_{PrxMR} are the labeling map by PrxMR+ and PrxMR, respectively.

dants. In the case of kegg, PrxMR+ and PrxMR have the same label size. As seen in Table 1, kegg has a relatively small average fanout, which means that there are few chances to employ the *DCL* technique to help reduce the label size by adjusting suffix label assignments between elements in a compressed element.

More details on each dataset's label size are depicted in Fig. 7, which shows the relative label size of PrxMR+ on the basis of PrxMR. For example, in the case of psd7003, PrxMR+ assigns labels of size 5 to 5,000 more elements than PrxMR; on the other hand, PrxMR+ assigns labels of size 9 to 4,000 less elements than PrxMR. In other words, there are more elements that are assigned smaller labels by PrxMR+ than PrxMR. The same tendencies are observed in the other datasets. Among them, psd7003 shows a relatively larger positive area for smaller label sizes (i.e., label size from 3 to 7) and smaller negative area for larger label sizes (i.e., label size from 9 to 12). This can be understood by the fact that psd7003 has both larger average fanout and larger average depth overall. For larger average depth, there are more chances to reduce the label size by Upper Tree Labeling. For larger average fanout, there are more chances for *DCL* to have an effect on reducing the label size.

7 Conclusion

We proposed a MapReduce-based prefix-based labeling algorithm, which is extended from a MapReduce-based repetitive prime labeling algorithm [6]. This algorithm introduces a novel way of dealing with compressed elements to reduce label size. Experiments on massive XML data showed that the proposed technique generated smaller labels than the previous algorithms.

In the proposed approach, elements in a mapper are sent exclusively to a reducer by choosing InputSplit id as the map key. There is no other way because the sequence of elements should be preserved in the reduce phase to carry out the Label Population step appropriately. This method does not allow control of the amount of data sent to reducers via the network. This is not desired because it cannot fully utilize all workers efficiently in terms of data distribution across a cluster. A completely new algorithm design is needed to address this issue in future works.

Acknowledgement. This work was supported by Institute for Information & communications Technology Promotion (IITP) grant funded by the Korea government (MSIP) (No. R0101-16-0054, WiseKB: Big data based self-evolving knowledge base and reasoning platform) and Basic Science Research Program through the National Research Foundation of Korea (NRF) funded by the Ministry of Science, ICT & Future Planning (NRF-2014R1A1A1002236).

References

1. Clark, J., DeRose, S., et al.: XML path language (XPath) (1999)
2. Pal, S., Cseri, I., Seeliger, O., Rys, M., Schaller, G., Yu, W., Tomic, D., Baras, A., Berg, B., Churin, D., et al.: XQuery implementation in a relational database system. In: Proceedings of the 31st International Conference on Very Large Data Bases, VLDB Endowment, pp. 1175–1186 (2005)
3. O'Neil, P., O'Neil, E., Pal, S., Cseri, I., Schaller, G., Westbury, N.: ORDPATHs: insert-friendly XML node labels. In: Proceedings of the 2004 ACM SIGMOD International Conference on Management of Data, pp. 903–908. ACM (2004)
4. Delbru, R., Toupikov, N., Catasta, M., Tummarello, G.: A node indexing scheme for web entity retrieval. In: Aroyo, L., Antoniou, G., Hyvönen, E., Teije, A., Stuckenschmidt, H., Cabral, L., Tudorache, T. (eds.) ESWC 2010. LNCS, vol. 6089, pp. 240–256. Springer, Heidelberg (2010). doi:10.1007/978-3-642-13489-0_17
5. Choi, H., Lee, K.H., Lee, Y.J.: Parallel labeling of massive XML data with mapreduce. J. Supercomputing **67**(2), 408–437 (2014)
6. Ahn, J., Im, D.H., Lee, T., Kim, H.G.: A dynamic and parallel approach for repetitive prime number labeling of XML data with MapReduce. J. Supercomputing (To Appear)
7. Dean, J., Ghemawat, S.: MapReduce: simplified data processing on large clusters. Commun. ACM **51**(1), 107–113 (2008)
8. Xu, L., Ling, T.W., Wu, H., Bao, Z.: DDE: from dewey to a fully dynamic XML labeling scheme. In: SIGMOD. ACM (2009)
9. Tatarinov, I., Viglas, S.D., Beyer, K., Shanmugasundaram, J., Shekita, E., Zhang, C.: Storing and querying ordered XML using a relational database system. In: Proceedings of the 2002 ACM SIGMOD International Conference on Management of Data, pp. 204–215. ACM (2002)
10. Lin, R.-R., Chang, Y.-H., Chao, K.-M.: A compact and efficient labeling scheme for XML documents. In: Meng, W., Feng, L., Bressan, S., Winiwarter, W., Song, W. (eds.) DASFAA 2013. LNCS, vol. 7825, pp. 269–283. Springer, Heidelberg (2013). doi:10.1007/978-3-642-37487-6_22
11. Lu, J., Meng, X., Ling, T.W.: Indexing and querying XML using extended dewey labeling scheme. Data Knowl. Eng. **70**(1), 35–59 (2011)
12. Klaib, A., Joan, L.: Investigation into indexing XML data techniques (2014)
13. Xu, L., Bao, Z., Ling, T.W.: A dynamic labeling scheme using vectors. In: Wagner, R., Revell, N., Pernul, G. (eds.) DEXA 2007. LNCS, vol. 4653, pp. 130–140. Springer, Heidelberg (2007). doi:10.1007/978-3-540-74469-6_14
14. Li, C., Ling, T.W.: QED: a novel quaternary encoding to completely avoid relabeling in XML updates. In: CIKM. ACM (2005)
15. Christophides, V., Karvounarakis, G., Plexousakis, D., Scholl, M., Tourtounis, S.: Optimizing taxonomic semantic web queries using labeling schemes. Web Semant. Sci. Serv. Agents World Wide Web **1**(2), 207–228 (2004)
16. Xu, L., Ling, T.W., Wu, H.: Labeling dynamic XML documents: an order-centric approach. IEEE Trans. Knowl. Data Eng. **24**(1), 100–113 (2012)
17. Subramaniam, S., Haw, S.C., Soon, L.K.: Relab: A subtree based labeling scheme for efficient XML query processing. In: 2014 IEEE 2nd International Symposium on Telecommunication Technologies (ISTT), pp. 121–125. IEEE (2014)
18. Wu, X., Lee, M.L., Hsu, W.: A prime number labeling scheme for dynamic ordered XML trees. In: ICDE (2004)

19. Sun, D.H., Hwang, S.C.: A labeling methods for keyword search over large XML documents. J. KIISE **41**(9), 699–706 (2014)
20. Wang, Y., DeWitt, D.J., Cai, J.Y.: X-Diff: An effective change detection algorithm for XML documents. In: 2003 Proceedings of the 19th International Conference on Data Engineering, pp. 519–530. IEEE (2003)
21. Leonardi, E., Bhowmick, S.S., Madria, S.: XANDY: Detecting changes on large unordered XML documents using relational databases. In: Zhou, L., Ooi, B.C., Meng, X. (eds.) DASFAA 2005. LNCS, vol. 3453, pp. 711–723. Springer, Heidelberg (2005). doi:10.1007/11408079_65

RIKEN MetaDatabase: A Database Platform as a Microcosm of Linked Open Data Cloud in the Life Sciences

Norio Kobayashi[1,2,3(✉)], Kai Lenz[1], and Hiroshi Masuya[2,1]

[1] Advanced Center for Computing and Communication (ACCC), RIKEN, 2-1 Hirosawa, Wako, Saitama 351-0198, Japan
{norio.kobayashi,kai.lenz}@riken.jp

[2] BioResource Center (BRC), RIKEN, 3-1-1, Koyadai, Tsukuba, Ibaraki 305-0074, Japan
hmasuya@brc.riken.jp

[3] RIKEN CLST-JEOL Collaboration Center, RIKEN, 6-7-3 Minatojima-minamimachi, Chuo-ku, Kobe 650-0047, Japan

Abstract. The amount and heterogeneity of life-science datasets published on the Web have considerably increased recently. However, biomedical scientists face numerous serious difficulties in finding, using and publishing useful databases. In order to solve these issues, we developed a Resource Description Framework-based database platform, called RIKEN MetaDatabase, which allows biologists to easily develop, publish and integrate databases. The platform manages metadata of both research data and individual data described with standardised vocabularies and ontologies, and has a simple browser-based graphical user interface for viewing data including tabular and graphical views. The platform was released in April 2015, and 110 databases including mammalian, plant, bioresource and image databases with 21 ontologies have been published through this platform as of July 2016. This paper describes the technical knowledge obtained through the development and operation of RIKEN MetaDatabase as a challenge for accelerating life-science data distribution promotion.

Keywords: Semantic web · Database cloud platform · Database integration · Life sciences

1 Introduction

The life sciences have been developed rapidly and subdivided into specialised study fields. Thus, the study of the life sciences has generated numerous heterogeneous datasets, making it difficult for researchers to find data from this flood of information, use them appropriately in their research and publishing them in a useful way for other researchers. Considering these difficulties, two major issues arise. The first issue is realising rich and useful data integration in a sustainable

Y.-F. Li et al. (Eds.): JIST 2016, LNCS 10055, pp. 99–115, 2016.
DOI: 10.1007/978-3-319-50112-3_8

way: linking data, integrating data systematically using standardised vocabularies, representing semantics and publishing the data location. The second issue involves realising easy, flexible and low-cost operation that allows many data developers and biologists to participate in the process of data integration.

These difficulties also occur within a research institute. RIKEN is the largest Japanese comprehensive science institute, having both large-scale research centres and many small-scale laboratories, which generate large-scale life-science datasets in various fields. The institute is confronted with issues regarding the realisation of collaborative research promotion over different fields within it. Therefore, database infrastructure is required for the publication and promotion of RIKEN's research results. This situation can be presented as a microcosm of the linked open data cloud in the life sciences.

We consider RIKEN's problem described above as a case study of data utilisation in the life sciences. To solve this problem, we developed RIKEN MetaDatabase—which is a database platform based on the Resource Description Framework (RDF), which realises metadata management at low cost, systematic data integration and global publication on the Web. RIKEN MetaDatabase was published in April 2015 with RIKEN's original databases, as well as external databases associated with these databases and ontologies. Here, we discuss the advantages of RDF for solving life-science data distribution and future issues, focusing on RIKEN MetaDatabase implementation, data integration and comparison with other cases.

The rest of this paper is organised as follows. Section 2 presents previous work related to this study. Section 3 discusses the requirement specifications for the database platform. Sections 4, 5 and 6 discuss design issues including functions, workflow of database publication, detailed data view and implementation. Sections 7 and 8 introduce available databases and comprehensively review RIKEN MetaDatabase by introducing concrete database projects. Finally, Sect. 9 concludes this study.

2 Related Work

We here review the existing RDF-based database platforms for life-science data publication and integration related to RIKEN MetaDatabase.

The Harvard Catalyst (https://catalyst.harvard.edu) is an information resource-sharing platform for human health research that enables collaboration among researchers within a group of 31 institutes including Harvard University. RDF-based data integration and federated search among distributed servers are used as the network's mining tool [1]. However, the platform does not aim at hosting the researchers' databases for data integration purposes.

Bio2RDF (http://bio2rdf.org) provides major existing life-science datasets by converting them into the RDF format [2]. The generators of the original data and the informaticians for RDFising are different in this case. Instead, in our approach, the original data generators participate in the data integration by RDFising these data themselves.

BioPortal (http://bioportal.bioontology.org) in the National Center for Biomedical Ontology (NCBO) is a data federation platform based on RDF and OWL, as well as our platform [3]. However, the primary focus of BioPortal is data integration and coordination between ontologies, while our approach focuses on inter-linking data items in the researchers' databases directly.

RDF Portal (http://integbio.jp/rdf/) in the National Bioscience Database Center (NBDC) and our platform apply a common data integration concept, which collects RDF datasets from various study fields. RDF Portal allows researchers from different institutes and universities to combine their RDF datasets. In addition, it provides SPARQL query interfaces for each dataset and across all datasets. Instead, our platform supports both generating and collecting RDF data. Furthermore, RIKEN MetaDatabase provides a data browser that can also be used by non-RDF users.

EBI RDF Portal (https://www.ebi.ac.uk/rdf/) currently hosts six datasets produced by large-scale projects, which are indispensable for data analysis and integration to many bioinformatists. In contrast, our platform is aimed at hosting both large- and middle–small-scale projects and laboratories, all of which want to contribute to the life sciences through the publication of their research-based data.

To generate RDF data, we employ a spreadsheet to describe the raw data and convert them into RDF. A similar tool is OpenRefine (http://openrefine.org/), which can generate RDF data from various source files including a spreadsheet. In OpenRefine, the data that are to be converted into RDF are defined outside the source files. Instead, our spreadsheet includes all the data to be converted into RDF, and the RDF expert can easily recognise the RDF data structure from the spreadsheet.

RightField [4] is a tool for editing life-science data given in spreadsheets by embedding the ontology annotations. A RightField spreadsheet allows a user to select terms from a given ontology dataset which includes subclass relations, individuals, and combinations.

3 Requirement Specifications for the Life-Science Database Platform

3.1 Requirements for Cloud-Based Databases in the Life Sciences

As an in-house database platform for a comprehensive research institute, RIKEN MetaDatabase should support different types and sizes of datasets generated by research projects of all scales. At the same time, RIKEN MetaDatabase should form a uniform knowledge base by including not only RIKEN's internal datasets but also global datasets interlinked on the Web. In addition, it should provide a simple operational workflow, by which biologists can easily participate in global data integration without specialised data integration skills.

Ideally, both biologists and informaticians should cooperate closely for data integration and mining through the database platform. Therefore, we conclude that RIKEN MetaDatabase is required to be a database platform providing well-coordinated datasets, which can be easily integrated into global datasets and used by data scientists. Furthermore, the data publication workflow should be simple so that biologists can operate it easily.

To satisfy these requirements, we have designed the database platform as a cloud-based platform that allows many database developers to deploy their data without management hardware and to ensure significant and flexible computational resources. We also decided to adopt Semantic Web technologies. Easily operable interfaces for data generation and publication were also designed. At present, there is no other database platform that realises cloud-based data and database publication independently of the data types and sizes. In other words, this platform meets various needs of database developers in an organisation or community in a cost-effective way.

3.2 Data Integration

For developing RIKEN MetaDatabase, we aim to realise the following two types of data integrations:

1. Data integration in a specialised research field.
   Data integration to realise a comprehensive dataset across research projects (research organisations or international consortium) to form a unified database. In this case, we assume that research projects bring non-redundant datasets, which may belong to the same data class. Research projects should share the data structure defined using their class, rather than sharing data entities (or instances), to enable unified data handling and management.
2. Data integration among different research fields.
   Data integration among different research fields or for co-operable research to realise mutual data links across datasets. In this case, projects provide related datasets, which belong to different data classes. Here, some data entities may act as links between the different datasets.

Case 1 can be achieved by introducing common or standardised data schemata and ontologies, where each data entity is usually described as an instance of a class or an ontology term. Therefore, data integration is not achieved by providing a direct link between data entities but by sharing a common class and semantic links (RDF properties) defining the data structure. In case 2, data entities from different datasets are directly connected by semantic links, whereas each dataset is described by different specialised data schemata. Data entities themselves are the links that allow the expansive combination of different communities.

Semantic Web employing RDF is a technology that satisfies these two types of integrations simultaneously; case 1 can be realised using the RDF scheme and the web ontology language (OWL), whereas, in case 2, a data linking mechanism can be applied. However, as explained in Sect. 2, most platforms do not satisfy both cases 1 and 2. Therefore, we propose here a novel practical approach to solve this problem.

4 Grand Design of the RDF-based RIKEN MetaDatabase Platform

4.1 RDF Data Structure Suitable for Life-Science Data Integration

Prior to discussing the grand design, we will investigate the data structures in the RIKEN databases. These datasets are represented in tabular form or hosted by a relational database system (data not shown). Therefore, in order to realise a simple and user-friendly database infrastructure system, as described above, we restrict the RDF data handled by RIKEN MetaDatabase to tabular-type database data and tree-type ontology data.

Tabular form used to describe tabular-type database data represents the RIKEN MetaDatabase data that can be easily generated and browsed. Tree form used to describe ontology data represents the concepts and data classes with their conceptual hierarchy to refer to the databases. Using the two kinds of data forms, RIKEN MetaDatabase aims to build a single integrated RDF dataset by managing multiple tabular and tree ontology data individually.

4.2 Tabular Data Model

We introduce a tabular data model for describing RDF data in which all RDF resources are associated with an RDF class. A table is generated for each class of subject instances of RDF triplets. Figure 1 shows the data structure of the RDF data described in tabular form. The presented table is separated into an RDF scheme definition part and a data part.

	1	2	3	4
1	English Attribution	Background strain	name	taxon
2	日本語属性	背景系統	名称	生物種
3	Property URI		rdfs:label	obo:RO_0002162
4	Data type	animal:0000004	rdf:langString	owl:Class
5		animal:0000004_7	"AIZ [Mus musculus molossinus]"@en	NCBITaxon:57486
6		animal:0000004_10	"AKT [Mus musculus musculus]"@en	NCBITaxon:39442
7		animal:0000004_12	"AST [Mus musculus musculus (wagneri)]"@en	NCBITaxon:39442
8		animal:0000004_23	"BFM/2 [Mus musculus domesticus]"@en	NCBITaxon:10092
9		animal:0000004_51	"Car [Mus caroli]"@en	NCBITaxon:10089

Fig. 1. A spreadsheet describing RDF data of class Background strain (http://metadb.riken.jp/db/rikenbrc_mouse/animal_0000004) in tabular form. In this example, the second column includes a list of instances of class Background strain, the third column includes a list of literal values of rdf:langString and the fourth column includes a list of Taxon classes as instances of owl:Class.

Functions of Rows. The RDF scheme definition part is presented in the top four rows in the table. In the first and second rows, English and Japanese column names of the table are displayed on the GUI, respectively. The third and fourth rows describe the properties and classes of the objects of the triplets used to convert the tabular data into RDF, respectively. The fifth and subsequent rows include the data.

Functions of Columns. The first column is a comment column, which is not converted into RDF.

The second column includes a list of instances (resources) of the common class that is the subject of all triplets described in the table, namely a list of subject instances. Using the table coordinates (r, c) to locate the data points, where r is the row and c is the column, $(4, 2)$ contains the data class, $(3, 2)$ is empty, and $(m, 2)$ for $m \geq 5$ are the instances of class $(4, 2)$.

The third and subsequent columns describe the properties and objects for the subjects listed in the second column. $(3, n)$ is a property and $(4, n)$ is a class or a data type of the instances or the literals listed as (m, n), respectively, where $m \geq 5$ and $n \geq 3$. Here, the triplet $(m, 2), (3, n), (m, n)$ is equivalent to the following set of RDF triplets:

$$(m, 2)\ \ (3, n)\ \ (m, n).$$
$$(m, 2)\ \ \texttt{rdf:type}\ \ (4, 2).$$
$$(m, n)\ \ \texttt{rdf:type}\ \ (4, n).$$

where $(4, n)$ is a RDF class, or

$$(m, 2)\ \ (3, n)\ \ (m, n).$$
$$(m, 2)\ \ \texttt{rdf:type}\ \ (4, 2).$$

where $(4, n)$ is a data type and (m, n) is a literal denoted as the form of data type $(4, n)$.

Moreover, each pair of a property and an object class (or data type) in the third and subsequent columns can appear multiple times, in order to describe multiple triplets sharing a common subject, property and object class (or data type).

4.3 Correspondence with the RDF Scheme

In order to manage multiple RDF datasets as databases or ontologies in RIKEN MetaDatabse, we introduce a specialised data category corresponding to the existing RDF scheme elements, as shown in Table 1.

A database is an RDF dataset with tabular data which compose an individual database, and corresponds to an RDF named graph. An ontology is an OWL ontology managed as an RDF named graph. A property and a class are equivalent to an RDF property and an RDF class as an instance of rdf:Property and rdfs:Class, respectively. An instance is limited to an instance i of rdf:Resource explicitly described as triplet i **`rdf:type`** c, where c is an RDF class. The reason we introduce limited instances is to establish data re-usability; when a class is specified, the instances of that class can be accurately obtained without orphan instances, which are not associated with any class.

Table 1. Correspondence between RIKEN MetaDatabase and the RDF scheme

RIKEN MetaDatabase	RDF scheme	Description
database	named graph	an individual dataset with multiple classes
ontology	named graph	an individual ontology written in OWL
property	instance of rdf:Property	equivalent to rdf:Property
class	instance of rdfs:Class	a concept or a rdf:Resource set
instance	instance of class	an instance typed by a class

4.4 RDF Data Generation and Publication

We design a procedure through which users can generate and publish their RDF data. Tree ontology data—usually described in OWL, which can be downloaded from public repositories or generated by an existing ontology editor—can be uploaded directly to the RIKEN MetaDatabase platform and immediately published. On the other hand, for tabular data, we apply a spreadsheet-based workflow, which can be operated by biologists as follows:

Step 1. Generating a spreadsheet. In this step, the user (database developer) describes the spreadsheet as a Microsoft Excel file or Tab-Separated Values (TSV) files, which represents a tabular data model. Using multiple spreadsheets in Microsoft Excel or TSV files, the user can describe a complicated database in which multiple tables are linked in a relational database management system.

Step 2. Generating an RDF dataset. The spreadsheet generated in the previous step is converted into RDF by the user using our application program. The program generates not only the RDF data converted from the raw data but also a structure definition file that describes the order of the columns and the column names.

Step 3. Uploading the RDF dataset. Both the database and the structure definition files are uploaded by a service administrator. The uploaded data are immediately published.

4.5 User Interface for Data Input and Output

RIKEN MetaDatabase employs both a graphical user interface (GUI), working on the user's web browser, and an application programming interface (API), for data input and output.

For data input, a registration interface of RDFised tabular data and tree ontology data is implemented. This function is closed; only the service administrators can operate this function since they should be able to check the uploaded data before publication.

The data publishing function is implemented in both API and GUI. As an API, we use an interface that actuates as a SPARQL endpoint accessible via the HTTP protocol, which is a standardised RDF data access protocol. The GUI works on the user's web browser and displays RDF data in various formats such as tabular and tree formats. In addition, it offers a list of RDF data archives for download and access to the SPARQL endpoint described above with query editor and result display functions.

5 Data Display Functions

To demonstrate the RIKEN management of RDF data, several fixed display forms are prepared as views for each data category. The data are shown using only their multilingual labels rather than their Unified Resource Identifiers (URIs), as a default, but both labels and URIs can be shown to RDF experts. The implemented views are summarised as follows:

List view shows lists of databases and ontologies, and includes a keyword search function to filter those data.
Tree view shows OWL ontologies as trees based on the subclass relationships.
Tabular view shows a list of instances of a specified class.
Card view shows a selected instance.
Download view is used for downloading RDF data archives for each database.
SPARQL search view supports editing queries and result displaying.

By default, the tabular and card views are devised to show RDF data to biologists. We describe these views in detail below.

Tabular View. Tabular view is a special feature of RIKEN MetaDatabase that shows an RDF graph data in tabular form. This view can be generated for each class and shows the name and description of the concerned class, all instances of the class and the triplets whose subject is one of the instances. Moreover, the triplets, whose object is an instance described not only in the database but also in other databases, are shown so that data integration via instances can be realised. A selected RDF class with its instances can be shown in this view. However, using a structure definition file generated from a spreadsheet of tabular data, the column names and column order can be customised.

An example of tabular view is shown in Fig. 2. The first column is a list of instances of the class. The second and subsequent columns form sets, each of which is associated with a property and describes a list of objects of triplets with the same property. Furthermore, the columns are reversely linked to the instances listed in the first column; thus, each column is associated with a property of triplets reversely linked to the instances of the first column.

By default, the name of the first column is the label of the class and those of the second and subsequent columns are the labels of the corresponding properties. However, a structure definition file can be uploaded and the column names can be overwritten by the column names in the structure definition.

The data in each row can be sorted in ascending or descending order as specified by the user. Furthermore, the data in the rows can be filtered by full-text search for humanly readable metadata of the data records using keywords specified by the user for each column.

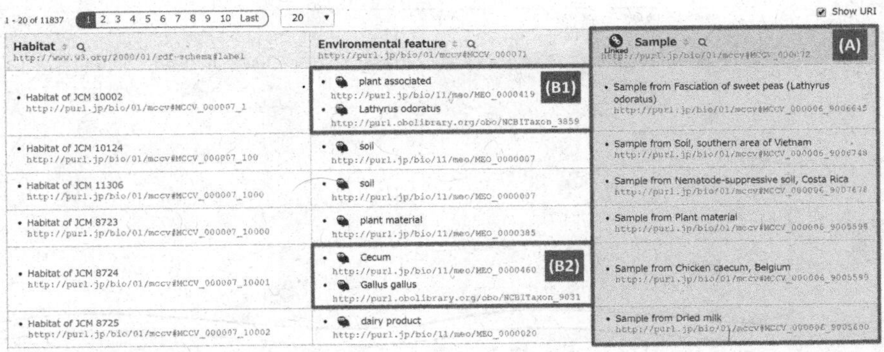

Fig. 2. A snapshot of the tabular view of a class Habitat of the Japan Collection of Microorganisms (JCM) resource database (http://metadb.riken.jp/metadb/db/rikenbrc_jcm_microbe). (A) The third column (class Sample) includes the instances to link to the subject instances of the first column, namely reversely linked instances of class Sample. (B1,B2) Multiple objects with same subject and predicate pairs can be displayed in the form of a list in the corresponding cell.

Card View. Card view, as shown in Fig. 3, is mainly used to show an instance and its triplets, which is linked to other instances or is reversely linked from other instances. In the card view, a user can view a long triplet path by traversing the connected triplets, in a sequence, from the corresponding instance. By default, only triplets including that particular instance are shown. A user can select an instance connected via a triplet to show further triplets having the selected instance, and the new triplets are shown as a new nested card in the original card view.

6 Implementation

Reducing of both development and operational costs is most important for realising persistent database services worldwide while ensuring service stability. In our implementation of RIKEN MetaDatabse, we adopt a simple architecture consisting of two components: (a) a web server providing GUI and (b) an RDF triplet store. The web server provides web pages having data display functions through a data display view, as described above. The RDF data displayed on a view are obtained from the RDF triplet store. In addition, the web server

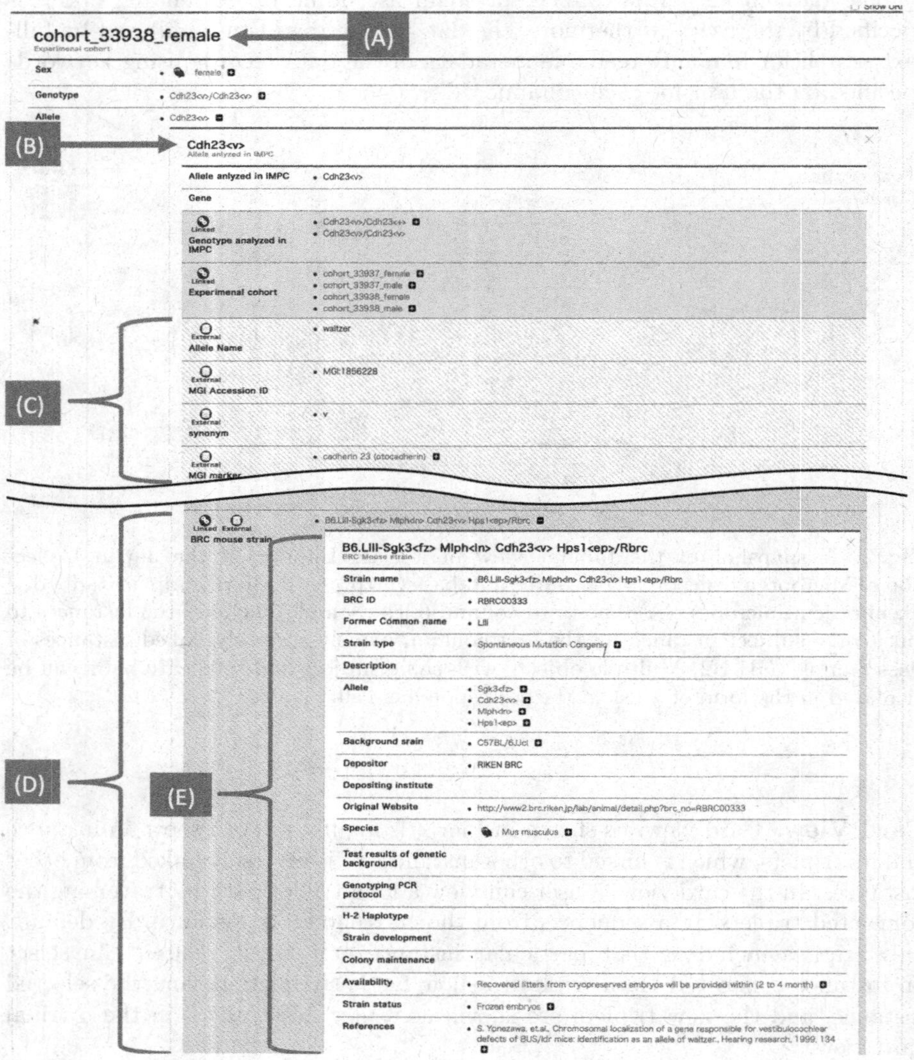

Fig. 3. A snapshot of the card view of an instance of an experimental cohort (http://metadb.riken.jp/db/IMPC_RDF/Cohort) used in the International Mouse Phenotype Consortium (IMPC) database (described in Sect. 8.2 in detail) that links to the instances of other databases. (A) is an instance of the Cohort class of KO mice, (B) represents an allele, *Cdh23-v* (*walther*), carried by the cohort in the IMPC database and described in the URI of the Mouse Genome Informatics (MGI) RDF database, (C) is a list of triplets in the MGI RDF database, (D) is a list of links from the BioResouce Center (BRC) Mouse Strain database and (E) is the detailed information of the mouse strain which has multiple alleles including *Cdh23-v* in the BRC Mouse Strain database.

functions as a SPARQL endpoint for submitting SPARQL queries generated by the web server. In our current platform, we employ the Openlink Virtuoso open-source version 7 as the RDF triplet store, and the web server is implemented as a Java servlet using Apache Tomcat version 8. To ensure stability, portability and continuity of the platform, we deploy these software components on RIKEN's private cloud, called RIKEN Cloud Service (http://cloudinfo.riken.jp), which provides multi-purpose Linux-based virtual machines. The web server (a) is located on a virtual machine connected to the global network. The RDF triplet store (b) is deployed on a specialised virtual machine to realise fast SPARQL operations, which connects a 1 TB flash memory storage via an InfiniBand network where the Virtuoso database directory is located.

7 Available Databases

As of July 2016, 21 public ontologies, including Gene Ontology (GO), Phenotypic Quality Ontology (PATO), NCBI Organismal Classification (NCBITaxon) and Semanticscience Integrated Ontology (SIO) have been selected and published as mirrors. These ontologies refer to 110 databases including 59 of RIKEN's original databases. The remaining 51 databases are external databases that are converted from originally non-RDF databases and linked from RIKEN's databases. In total, RIKEN MetaDatabase carries 148 million triplets, 797 classes, 2.94 million instances and 1,352 properties. The original databases are from various research fields, e.g., FANTOM (mammalian [5]), FOX Hunting (plant [6]), Heavy-atom Database System (protein [7]) and Metadata of BioResouce Center (BRC) resources (bioresources [8–10]).

7.1 Database Directory Service

RIKEN MetaDatabase provides a specialised database and RDF datasets that provide easy access to data. The specialised database is the RIKEN Database Directory, which is a catalogue of RIKEN's databases including the non-RDF databases. The catalogue data are designed to be compatible with the Integbio Database Catalog (http://integbio.jp/dbcatalog/?lang=en), which aims at inter-ministry integration of life-science databases in Japan. In addition, W3C's Health Care and Life Sciences (HCLS) Community Profile data (http://www.w3.org/TR/hcls-dataset/) including statistics data are generated for each database and for entire datasets, and are published as RDF archives and via the SPARQL endpoint. RIKEN MetaDatabase also provides the SPARQL Builder Metadata (http://sparqlbuilder.org/), which are generated and published for more intelligent SPARQL search. The SPARQL Builder Metadata is a profile of the SPARQL endpoint that describes the RDF graph structure. Thus, the SPARQL Builder tool [11] generates a SPARQL query that obtains triplet paths connecting two ontological concepts specified by the user.

8 Discussion

We developed RIKEN MetaDatabase as a cloud-based database platform, which realises Semantic-Web-based data integration with a simplified workflow, implemented through the cooperation of biologists and informaticians. In this section, we comprehensively review our methodology and the development process and operation of RIKEN MetaDatabase from various perspectives.

8.1 Contributions of RIKEN MetaDatabase for Different Types of Users

For Database Publishers. A data publisher is a biologist who has research results, converts the data into RDF and publishes the converted RDF data. The advantages of data generation using a spreadsheet are summarised as follows:

1. new columns can be easily added,
2. text format data such as TSV can be easily imported, and
3. the readability of tabular data is high for humans.

Adding new columns is required for including a triplet corresponding to a new RDF property. This feature is enabled because the RDF is open for adding new data (the open world assumption). Importing text format allows bioinformaticians to input data derived through the existing techniques where life-science data processing is often performed using script languages such as Perl and text data are often used for data exchange, rather than the RDF graph format. Finally, the tabular form is suitable for data typing and data confirmation before publication.

Especially for biologists, this methodology does not solve RDF-specific difficulties such as usage of URIs for data resource identification and selecting suitable vocabularies including ontologies, properties and classes. However, these difficulties are successfully reduced by generating spreadsheet templates in collaboration with informaticians.

Data integration based on RDF is also to the publisher's advantage. Though the integration can be realised by creating semantic links from one publisher's data to another's, as RDF triplets, a more attractive advantage is that data published later may be linked to existing data, which is already integrated by the original publishers. Furthermore, the appropriate data to link new data can be easily discovered through the tabular view without SPARQL.

For database users. Previously, database users had great difficulties in discovering the types of databases available, where these were published and how to use them. RIKEN MetaDatabase collects metadata of databases published by RIKEN and functions as a one-stop-shop of databases. Indeed, these metadata are published as a database catalogue in the RIKEN Database Directory and the HCLS Community Profile using standardised vocabularies, which help users to discover the data.

Furthermore, by employing standards for metadata publication such as RDF and SPARQL, RIKEN MetaDatabase provides standardised API to data access as a SPARQL endpoint. In addition, for users who are not familiar with RDF, it provides intuitive data views such as the tabular data view—which is a popular form for biologists, facilitating data view and operation.

For the RIKEN institute. Since RIKEN has researchers in various fields, including genome, plant, animal, brain, medical, bioresource and informatics, we can handle a wide range of metadata descriptions and bio-medical concepts. Development of a novel ontology is required for new types of research data and concepts. RIKEN easily realises this collaboration among various researchers for an internal collaborative research. Consequently, we are accomplishing the difficult task of the ontology development. We propose that this collaborative metadata integration model should be used in an open environment.

8.2 Contributions of RIKEN MetaDatabase to Inter-labs and Global Data Integration

In this section, we present the main contributions of RIKEN MetaDatabase.

Open data promotion. The development of RIKEN MetaDatabase is a step toward open access to research data. The platform provides easy and interactive access to previously untapped data stored in laboratory records. In addition, RIKEN MetaDatabase facilitates an easy, rapid and cost-effective publication of databases by small laboratories. For example, in the case of ENU-induced Mutations in RIKEN Mutant Mouse Library (http://metadbdev.riken.jp/sandbox/db/BRC-ENU-inducedMutationsInRIKENMutantMouseLibrary), the data developer did not have sufficient expertise and hardware to develop a public database. Using RIKEN MetaDatabase, they easily published their data on the Web. Furthermore, through collaboration with us and the RDF experts in DNA Data Bank of Japan (DDBJ), their data were integrated with data within RIKEN and DDBJ by applying a common data scheme as shown below.

Data coordination referring common data resources. The second contribution is the data coordination between different databases hosted in RIKEN MetaDatabase to share common URI of instances. In order to describe alleles and genes in mice, we applied common gene records (http://metadb.riken.jp/metadb/db/mgi_rdf) imported from the Mouse Genome Informatics (MGI) database (http://www.informatics.jax.org). Here, the MGI approved the publication of the RDF version of the mouse gene records. We have promoted the common use of MGI gene records in RIKEN MetaDatabase. As a result, MGI records are used in multiple databases, such as Metadata of BRC mouse resources and phenotypes (http://metadb.riken.jp/metadb/db/rikenbrc_celle), Metadata of Functional Glycomics with KO mice database (http://metadb.riken.jp/metadb/db/Glycomics_mouse) and the International Mouse Phenotype Consortium (IMPC: http://www.mousephenotype.org) RDF data (http://metadb.riken.jp/metadb/db/IMPC_RDF). In these databases, the data items related to genes are linked to MGI allele or gene records. Through this association, the integrated information, including what public experimental material (mouse strain in this case) are available correspond to the phenotype data published in IMPC, can be obtained, as shown in Fig. 3.

Scheme-level integration across databases. The third contribution is towards scheme-level integration. In the ENU-induced Mutations in RIKEN Mutant Mouse Library, next-generation sequencing (NGS) metadata are described in the common RDF scheme, which is developed by cooperation of DDBJ and RIKEN, based on the broadly used XML scheme for the NGS metadata. Using this scheme, RIKEN plans to develop a unified pipeline to publish NGS metadata on the Web and deposit NGS data as public archives operated by DDBJ. It is expected that this pipeline will promote worldwide sharing of NGS data from RIKEN. Moreover, metadata from the Japan Collection of Microorganisms (JCM) resources (http://metadb.riken.jp/metadb/db/rikenbrc_jcm_microbe) are described based on a common RDF scheme for strains of microorganisms, the Microbial Culture Collection Vocabulary (MCCV: http://bioportal.bioontology.org/ontologies/MCCV), which is used in MicrobeDB.jp (http://microbedb.jp/). The integrated database represents the encyclopaedia of microbes based on metagenome data. By applying the MCCV, basic information on the microbe strains released from JCM can be related to the metagenome data in MicorbeDB.jp. Phenotype data of experimental animals are also integrated by the J-phenome project (http://jphenome.info). J-phenome is a portal of phenotype databases hosted by RIKEN MetaDatabase in which the RDF scheme for the description of animals' phenotypes are unified using common phenotype ontologies such as the Mammalian Phenotype Ontology PATO. The unified scheme contributes to the development of a common application to determine the cross-species relationship between phenotypes using an inter-ontology relationship library produced by machine reasoning [12,13]. In summary, scheme-level integration in RIKEN MetaDatabase contributes to the common use of query, application and workflow pipelines to handle the same (or similar) data across databases.

International collaboration. The integration of multiple datasets in RIKEN MetaDatabase contributes to international collaboration. IMPC is an umbrella of comprehensive phenotyping of mouse mutants [14]. Through the cooperation of this consortium, multiple research centres released measurement data produced from the standardised phenotyping pipeline. As a member of IMPC, RIKEN BRC produced RDF version of IMPC phenotype data including more than 50 million triplets, which are now hosted by RIKEN MetaDatabase. Although the IMPC website provides a rich interface, visualising various phenotype data, the RDF version of the IMPC data presented in RIKEN MetaDatabase can be used by data scientists who want to integrate these phenotype data with other datasets related from different databases. For example, using the SPARQL endpoint of RIKEN MetaDatabase (http://metadb.riken.jp/sparql), a data user can perform a federated query between RIKEN and EBI RDF platform (https://www.ebi.ac.uk/rdf/) to retrieve what phenotype can be expressed when a specific biological pathway is inactivate, utilising the connection of IMPC dataset and Reactome (http://www.reactome.org) dataset.

In summary, using RIKEN MetaDatabase, seamless data integration can be performed from the inner-research institute to the worldwide level.

8.3 Open Issues

RIKEN MetaDatabase is a simple database system and platform built on our private cloud infrastructure. Data generation and publication costs for biologists are reduced, since they do not need to prepare and operate their own server. Since the system does not support data access control, it cannot handle private datasets or datasets under development. However, the system is lightweight and requires only two virtual machines. Therefore, we can build multiple instances of the system on the private cloud for each research project, with each project having its own access control using a firewall. However, the federated SPARQL search is not available yet. Our future work will include the development of such effective federation among these instances, as an ideal database federation model on the global Web.

Since the publication of RIKEN MetaDatabase in April 2015, our efforts for data dissemination are in progress. So far, we have participated in international database projects such as IMPC, for mouse phenotype databases, and W3C's HCLS group, for database profile, so that our published metadata can be easily linked to other published datasets.

To promote the reuse of common URIs, discussion-based cooperation among database developers is currently promoted by forming a working group of representatives from research centres in RIKEN. However, we have not yet implemented the automatic ontology annotation function on the RIKEN MetaDatabase. To address this issue, application of RightField in data construction workflow may prove useful in expansion of Excel spreadsheets, allowing semi-automatic ontology annotation.

9 Conclusions

We discussed the requirement specifications, design, development and operation of a database platform called RIKEN MetaDatabase handled by the comprehensive research institute RIKEN. One of the major difficulties is the practical co-localisation of open data framework RDF and the development of simple data processing methods for biologists. In order to solve these issues, we developed a template spreadsheet for data creation, which is a GUI that realises intuitive data views including tabular view. The database platform is deployed on our private cloud infrastructure and multiple system instances can be generated. Thus far, data integration from different research fields, such as IMPC, has been successfully realised on the platform.

Future work includes the realisation of practical federation among multiple system instances, so that an integrated database can be realised that supports our proposed data views. This will be accomplished by developing an individual database for each research project in a distributed environment and intelligent support for selecting suitable vocabularies for biologists.

References

1. Vasilevsky, N., Johnson, T., Corday, K., Torniai, C., Brush, M., Segerdell, E., Wilson, M., Shaffer, C., Robinson, D., Haendel, M.: Research resources: curating the new eagle-i discovery system. Database (Oxford), 20:2012 (2012)
2. Belleau, F., Nolin, M.A., Tourigny, N., Rigault, P., Morissette, J.: Bio2RDF: towards a mashup to build bioinformatics knowledge systems. J Biomed. Inform. **41**(5), 706–716 (2008)
3. Whetzel, P.L., Noy, N.F., Shah, N.H., Alexander, P.R., Nyulas, C., Tudorache, T., Musen, M.A.: BioPortal: enhanced functionality via new Web services from the National Center for Biomedical Ontology to access and use ontologies in software applications. Nucleic Acids Res. **39**(Web Server issue): W541–W545 (2011)
4. Wolstencroft, K., Owen, S., Horridge, M., Krebs, O., Mueller, W., Snoep, J.L., du Preez, F., Goble, C.: RightField: embedding ontology annotation in spreadsheets. Bioinformatics **27**(14), 2021–2012 (2011)
5. The FANTOM Consortium and the RIKEN PMI and CLST (DGT): A promoter-level mammalian expression atlas. Nature **507**, 462–470 (2014)
6. Ichikawa, T., Nakazawa, M., Kawashima, M., Iizumi, H., Kuroda, H., Kondou, Y., Tsuhara, Y., Suzuki, K., Ishikawa, A., Seki, M., Fujita, M., Motohashi, R., Nagata, N., Takagi, T., Shinozaki, K., Matsui, M.: The FOX hunting system: an alternative gain-of-function gene hunting technique. Plant J. **45**, 974–985 (2006)
7. Sugahara, M., Asada, Y., Shimada, H., Taka, H., Kunishima, N.: HATODAS II: heavy-atom database system with potentiality scoring. J. Appl. Crystallogr. **42**, 540–544 (2009)
8. Yoshiki, A., Ike, F., Mekada, K., Kitaura, Y., Nakata, H., Hiraiwa, N., Mochida, K., Ijuin, M., Kadota, M., Murakami, A., Ogura, A., Abe, K., Moriwaki, K., Obata, Y.: The mouse resources at the RIKEN BioResource center. Exp. Anim. **58**(2), 85–96 (2009)
9. Nakamura, Y.: Bio-resource of human and animal-derived cell materials. Exp. Anim. **59**(1), 1–7 (2010)
10. Yokoyama, K.K., Murata, T., Pan, J., Nakade, K., Kishikawa, S., Ugai, H., Kimura, M., Kujime, Y., Hirose, M., Masuzaki, S., Yamasaki, T., Kurihara, C., Okubo, M., Nakano, Y., Kusa, Y., Yoshikawa, A., Inabe, K., Ueno, K., Obata, Y.: Genetic materials at the gene engineering division. RIKEN BioResource Center. Exp. Anim. **59**(2), 115–124 (2010)
11. Yamaguchi, A., Kozaki, K., Lenz, K., Wu, H., Yamamoto, Y., Kobayashi, N.: Efficiently finding paths between classes to build a SPARQL query for life-science databases. In: Qi, G., Kozaki, K., Pan, J.Z., Yu, S. (eds.) JIST 2015. LNCS, vol. 9544, pp. 321–330. Springer, Heidelberg (2016). doi:10.1007/978-3-319-31676-5_24

12. Hoehndorf, R., Schofield, P.N., Gkoutos, G.V.: PhenomeNET: a whole-phenome approach to disease gene discovery. Nucleic Acids Res. **39**(18), 119 (2011)
13. Robinson, P.N., Khler, S., Oellrich, A.: Sanger Mouse Genetics Project, Wang, K., Mungall, C.J., Lewis, S.E., Washington, N., Bauer, S., Seelow, D., Krawitz, P., Gilissen, C., Haendel, M., Smedley, D.: Improved exome prioritization of disease genes through cross-species phenotype comparison. Genome Res. **24**(2), 340–348 (2014)
14. Dickinson, M.E., Flenniken, A.M., Ji, X., Teboul, L., Wong, M.D., White, J.K., Meehan, T.F., Weninger, W.J., Westerberg, H., Adissu, H., et al.: High-throughput discovery of novel developmental phenotypes. Nature **537**(7621), 508–514 (2016)

A Preliminary Investigation Towards Improving Linked Data Quality Using Distance-Based Outlier Detection

Jeremy Debattista(✉), Christoph Lange, and Sören Auer

University of Bonn and Fraunhofer IAIS, Bonn, Germany
{debattis,langec,auer}@cs.uni-bonn.de

Abstract. With more and more data being published on the Web as Linked Data, Web Data quality is becoming increasingly important. While quite some work has been done with regard to quality assessment of Linked Data, only few works have addressed quality improvement. In this article, we present a preliminary an approach for identifying potentially incorrect RDF statements using distance-based outlier detection. Our method follows a three stage approach, which automates the whole process of finding potentially incorrect statements for a certain property. Our preliminary evaluation shows that a high precision is maintained with different settings.

Keywords: Outlier detection · Data quality · Linked data

1 Introduction

A rationale of the Semantic Web is to provide real-world things, also called *resources*, with descriptions in common data formats that are meaningful to machines. Furthermore, Linked Data emphasises on the reuse and linking of these resources, thus assisting in the growth of the *Web of* (meaningful) *Data*. Schemas, some being lightweight and others being more complex, have been defined for various use cases and application scenarios in order provide structure to the descriptions of semantic resource based on a common understanding. Nevertheless, since linked datasets are usually originating from various structured (e.g. relational databases), semi-structured (e.g. Wikipedia) or unstructured sources (e.g. plain text), a complete and accurate *semantic lifting* process is difficult to attain. Such processes can often contribute to incomplete, misrepresented and noisy data, especially for semi-structured and unstructured sources. Issues caused by these processes can be attributed to the fact that either the knowledge worker is not aware of the various implications of a schema (e.g. incorrectly using inverse functional properties), or because the schema is not well defined (e.g. having an open domain and range for a property). In this article, we are concerned with the latter, aiming to identify potentially incorrect statements in order to improve the quality of a knowledge base.

Y.-F. Li et al. (Eds.): JIST 2016, LNCS 10055, pp. 116–124, 2016.
DOI: 10.1007/978-3-319-50112-3_9

When analysing the schema of the DBpedia dataset we found out that from around 61,000 properties, approximately 59,000 had an undefined domain and range. This means that the type of resources attached to such properties as the subject or the object of an RDF triple can be very generic, i.e. `owl:Thing`. Whilst this is not forbidden, it makes a property ambiguous to use. For example, the property `dbp:author`, whose domain and range are undefined, has instances where the subject is of type `dbo:Book` and the object of type `dbo:Writer`, and other instances where the subject is of type `dbo:Software` and the object of type `dbo:ArtificialSatellite`.

The key research question in this paper is *can distance-based outlier techniques help in identifying quality problems in linked datasets?* In this article we investigate how triples can be clustered together based on their distance. This distance is identified by a semantic similarity measure that takes into consideration the subject type, object type, and the underlying schema. Furthermore, we evaluate complementary aspects of the proposed approach. More specifically, we were interested to see how different settings in our approach affect the precision and recall values.

This article is structured as follows. The state-of-the-art is described in Sect. 2. Our proposed approach is explained in Sect. 3. Experiments of our approach are documented in Sect. 4. Conclusions and an outlook to future work are discussed in Sect. 5.

2 Related Work

Various research efforts have tackled the problem of detecting incorrect RDF statements using different techniques. These include *statistical distribution* [8], *schema enrichment* [9,12] and *crowdsourcing* [1,10]. Outlier detection techniques such as [11] are used to validate the correctness of data literals in RDF statements, which is out of the scope of this research as our approach considers only statements where the subject and object are resources.

Statistical Distribution. Paulheim et al. [8] describe an algorithm based on the statistical distribution of types over properties in order to identify possibly faulty statements. Statistical distribution was used in order to predict the probability of the types used on a particular property, thus with some confidence verify the correctness of a triple statement. Their three step approach first computes the frequency of the predicate and object combination in order to identify those statements that have a low value. Cosine similarity is then used to calculate a confidence score based on the statement's subject type probability and the object type probability. Finally, a threshold value is applied to mark those statements that are potentially incorrect. Our approach uses semantic similarity to identify whether a statement could be a possibly incorrect statement or not, instead of statistical distribution probabilities. Therefore, our similarity approach takes into consideration the semantic topology of types and not their statistical usage.

Schema Enrichment. Schema enrichment is also a popular technique to detect incorrect statements. Töpper et al. [9] enrich a knowledge base schema with additional axioms before detecting incorrect RDF statements in the knowledge base itself. Such an approach requires external knowledge in order to enrich the ontology. Similarly, Zaveri et al. [12] apply a semi-automated schema enrichment technique before detecting incorrect triples.

Crowdsourcing WhoKnows? [10] is a crowdsourcing game where users contribute towards identifying inconsistent, incorrect and doubtful facts in DBpedia. Such crowdsourcing efforts ensure that the quality of a dataset can be improved with more accuracy, as a human assessor can identify such problems even from a subjective point of view. During the evaluation, the users identified 342 triples that were potentially inconsistent from a set of overall 4,051 triples, reporting a precision value of 46%. A similar crowdsourcing effort was undertaken by Acosta et al. in [1]. They used pay-per-hit micro tasks as a means of improving the outcome of crowdsourcing efforts. Their evaluation focuses on checking the correctness of the object values and their data types, and the correctness of interlinking with related external sources, thus making it incomparable to our approach. In contrast to crowdsourcing, our preliminary approach gives a good precision in identifying outliers without the need of any human intervention, in an acceptable time (± 3 min to compute outliers of a 10 K dump). Nonetheless, at some point, human expert intervention would still be required (in our approach) to validate the correctness of the detected outliers, but with any (semi-)automatic learning approaches, human intervention is reduced.

3 Improving Dataset Quality by Detecting Incorrect Statements

The detection and subsequent cleaning of potentially incorrect RDF statements aids in improving the quality of a linked dataset. There were a number of attempts to solve this problem in the best possible manner (cf. Sect. 2). We apply the distance-based outlier technique by Knorr et al. [6] in a Linked Data scenario. Exploiting reservoir sampling and semantic similarity measures, clusters of RDF statements based on the statement' subject and object types are created, thus identifying the potentially incorrect statements. We implemented[1] this approach as a metric for *Luzzu* [2].

3.1 Approach

Following [6], our proposed Linked Data adapted method has three stages: *initial*, *mapping*, and *colouring*. These three stages automate the whole process of finding potentially incorrect statements for a certain property. In the *initial* stage, k (the size of the reservoir) RDF statements are added to a reservoir sampler. Following the initialisations of the constants, the *mapping* stage groups data objects in various cells based on the mapping properties described in [6]. Finally, the *colouring* stage identifies the cells that contain outlier data objects.

[1] The Java code can be found in our GIT repository: https://goo.gl/bGRKxi.

Initial Stage. The initial steps are crucial for achieving a more accurate result, i.e. a better identification of potentially incorrect statements. We start by determining the approximate distance D that is used in the second stage to condition the mapping, and thus the final clustering of RDF statements. The approximate value D is valid for a particular property, i.e. the property whose triples are being assessed. Therefore, two properties (e.g. *dbp:author* and *dbp:saint*, i.e. the patron saint of, e.g., a town) will have different values of D according to the triples, their types, and ultimately the similarity measure chosen. Currently, in our approach we assume that a resource is typed with **only** one class, choosing the most specific type if a resource is multi-typed (e.g. *dbo:Writer* and not *dbo:Person*). Additionally, a threshold fraction p (between 0 and 1) is defined by the user during the initial phase, affecting the number of data objects in a cluster M. Therefore, p can be considered to be a sensitivity function that increases or decreases the amount of data objects in a cluster.

Determining the Approximate Distance. Our approach makes use of reservoir sampling as described in [3]. The rationale is that D is approximated by a sample of the data objects being assessed, to identify the acceptable maximum distance between objects mapped together in a cell, in a quick and automated way. To determine the approximate distance we applied two different implementations (cf. Sect. 4 for their evaluation), one based on a simple sampling of triples and another one based on a modified reservoir sampler, which we call the *type-selective*. From the sample set (for both implementations), a random data object is chosen to be the *host*, and is removed from the sampler. All remaining statements in the sampler are semantically compared with the host individually and their distance values are stored in a list. The median distance is than chosen from the list of distances. We chose the median value over the mean value as a central tendency since the latter can be influenced by outliers.

In the first implementation (simple sampling), the reservoir selects a sample of triples, irrelevantly of their subject and object types. The main limitation is that, irrelevantly of the size of the reservoir, the approximate distance D value can bias towards the more frequent pairs of the subject and object types. Therefore, the sampler might not represent the broad types attached to the particular property being assessed.

In order to attempt to solve the *sampler representation problem*, we propose the *type-selective* reservoir sampler. The proposed reservoir sampler modifies the simple sampler by adding a condition that only one statement with a certain subject type and object type can be added to the reservoir. In other words, when there are two distinct statements with matching subject types and object types, only one of these statements will be added to the reservoir.

Mapping Stage. The mapping stage attends to the clustering of data objects (i.e. RDF statements in our case) in cells. An RDF statement is chosen at random from the whole set of data objects and is placed in a random cell. This is called the *host* cell. Thereafter, every other RDF statement in the dataset is mapped to an appropriate cell by first comparing it to the data object in this host cell.

Semantic Similarity Measure. In order to check if an RDF statement fits in a cell with other similar RDF statements, a semantic similarity measure is used. More specifically, since we are mostly concerned about the distance between two statements, we use a normalised semantic similarity measure. The similarity between two statements S_1 and S_2 is defined as the average of the similarity between the statements' subjects, and the similarity between the statements' objects.

Colouring Stage. After mapping all data objects to the two-dimensional space, the *colouring* process colours cells to identify outlier data objects, based on the process identified in [6]. In [6], the minimum number of objects (M) required in a cell such that data objects are not considered as outliers is calculated as $M = N \cdot (1 - p)$ where N is the total number of data objects, and p is the threshold fraction value determined in the *initial* stage.

4 Experiments and Evaluations

The primary aim of this experiment is to compare if the automatic approach of setting approximate D value gives an advantage over the manual setting. All experiments in this part of the evaluation used the same similarity measure configuration, i.e. Zhou IC [13] with the Mazandu measure [7], as implemented in the Semantic Measures Library & Toolkit [4].

This experiment is split into two sub-experiments. In the first part, we evaluated triple statements in DBpedia with the predicate http://dbpedia.org/property/author using the proposed approach with the p and D parameters manually set to determine the precision and recall values. In the second part of this evaluation we repeat this experiment but the value of D is determined by the two automated approaches described in Sect. 3. For both experiments, p was set to: 0.99, 0.992, 0.994, 0.996, and 0.998.

Sub-experiment #1 – Setting Approximate D Manually. In this manual experiment, the D value for the evaluated property was obtained as an estimate from a manual calculation of the similarity values of the different types. From Fig. 1, we observe that on average our approach achieved around 76% precision. On the other hand, the recall values were low, with an average of 31%. We also observed that increasing the approximate value D does not result in an increasing precision. For example, in Fig. 1 we spot that the precision value for the D value of 0.3335 is greater than that of 0.3555 when p was set to 0.996. When D was set to 0.3555, 39 more outliers were detected, (true positives –7, false positives +42 data objects). This slight change in *true positives* and *false positives* was expected as the data objects cluster with similar data objects whose distance is the smallest. Therefore, the change in D might have moved some objects from one cell to another with the consequence that a previously non-outlier cell is now marked as an outlier, since a number of data objects might have moved to other

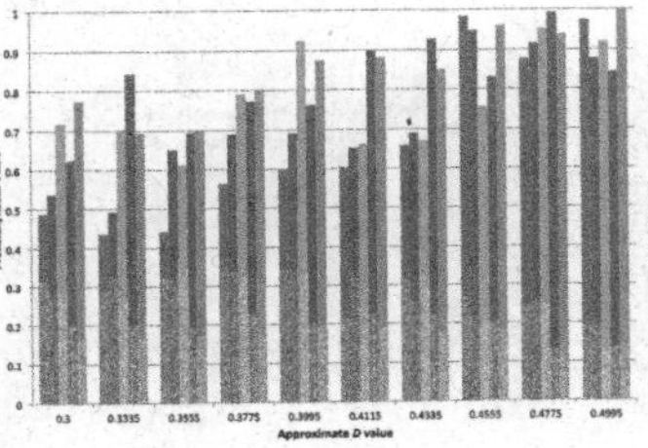

Fig. 1. The precision and recall values for the authors property dump with different values for D and p. The solid bars denote precision values, whilst the striped overlapped bars denote recall.

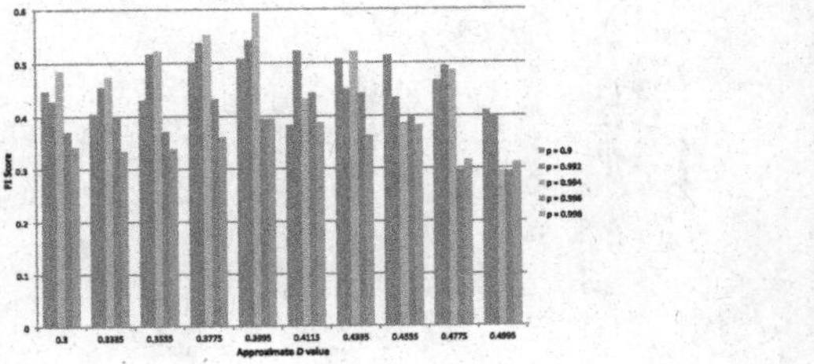

Fig. 2. The F1 score authors property dump with different values for D and p.

cells. Figure 2 represents the F1 score for the authors property dump manual experiment, showing an average of almost 43% for this harmonic mean score.

Sub-experiment #2 – Setting Approximate D Automatically. The same evaluated property was used in this experiment, where first an approximate D value was calculated first using the *simple* reservoir sampler and then using the *type-selective* reservoir sampler. A single host was chosen randomly from these reservoir samplers, together with a starting host location. The choice of a random data object will not affect the precision of the algorithm, as all data objects will be compared and mapped in suitable cells. From Fig. 3 we observe that the *type-selective* sampler outperforms its simpler counterpart for all p values with regard to the precision. One possible reason is due to the low approximate D values identified by the simple reservoir sampler. Low approximate D values mean that less data objects get mapped together in cells, since the approximate distance becomes smaller and data objects will be dispersed throughout the whole 2D space. This means that since less data objects are mapped in the same cell or surrounding cells, it would be more difficult to reach the $M + 1$ quota, and thus more cells will be marked as outliers. Therefore, whilst a low approximate D could lead for a decrease of *false positives* in non-outlier cells, it can also increase of *false negatives* (thus decreasing *true positives*), as objects that should not be marked as outliers could end up in outlier-marked cells. The main factors that affect the approximate D value are (1) the choice of the semantic similarity measure, and (2) the underlying schema (cf. limitations in Sect. 4.1). Furthermore, this approximate D value and the user-defined sensitivity threshold value (p) affect the precision and recall.

Following these experiments, in Fig. 4 we compared the *type-selective* precision and recall results for every p against the manual approach. For this comparison we used the manual scores that got the highest F1 measure for each p value, thus having a balance between the precision and recall. Figure 4 shows that the manual approach performed overall better than the automatic one in terms of

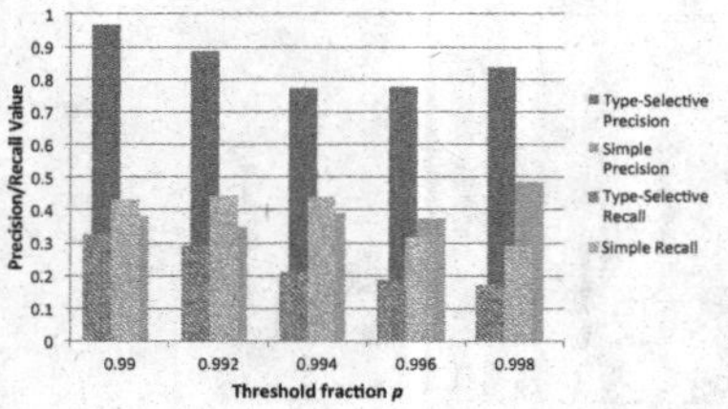

Fig. 3. The precision and recall values for the authors property dump with different values for p and a generated D value.

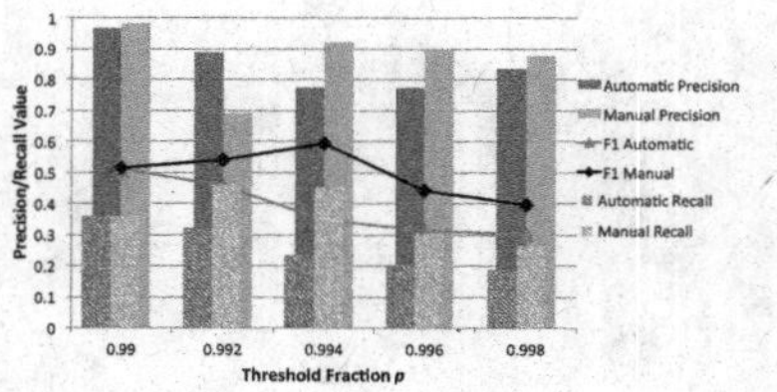

Fig. 4. Precision and recall values for the authors property dump comparing the manual results against the automatic results for multiple values of the fraction p.

the F1 measure. Nevertheless, in most cases, there are no large discrepancies between the two. The automatic approach resulted into a higher approximate D value than the manual approach. The approximation D value for the automatic approach was 0.482147, 0.0826 more than the given manual approximation D value with the highest F1 value (i.e. 0.3995 for threshold fraction p).

4.1 Discussion

The led evaluation is as yet not conclusive, since we only evaluated our approach with one property. This evaluation also showed that our approach produces a low recall value and thus a low F1 measure. A higher recall, without comprising the precision, would have been ideal, as with low recall we are missing a relevant data objects that should have been marked as outliers. One must also note that the choice of a semantic similarity measure will also affect the precision and recall values of such an approach, in a way that its results are the deciding factor where a data objects is mapped.

Nevertheless, our approach has a number of known limitations:

1. the approach is limited to knowledge bases without blank nodes, which can effect the degree of similarity, thus making this approach less robust and generic;
2. the approach does not fully exploit the semantics of typed annotations in linked datasets, since our approach assumes that an instance is a member of only one type, in particular the most specific type assigned to the resource;
3. the evaluated semantic similarity measures are limited to hierarchical '*is-a*' relations that might be more fitting to biomedical ontologies having deep hierarchies;
4. the sampled population might not reflect the actual diverse population of the data objects that have to be clustered in both sampler implementations. Thus, with both implementations we will not achieve the best representative sample, such as that obtained by stratified sampling [5];

5. whilst with the *simple* sampler outliers might occur in the sample population, with the *type-selective* sampler there is a 100% certainty that outlier data objects are present in the sample that determines the approximate D. Knorr et al. [6] had foreseen this problem and whilst suggesting that sampling provides a reasonable starting value for D, it cannot provide a high degree of confidence for D because of the unpredictable occurrence of outliers in the sample.

5 Conclusions

In this article we investigated the possibility of detecting potentially incorrect RDF statements in a dataset using a time and space efficient approach. More specifically, we applied a distance-based clustering technique [6] to identify outliers in a Linked Data scenario. While providing satisfactory results, our approach has a number of limitations that we are currently addressing. However, the preliminary results give us an indication on the research question set in the introduction. In the future, we aim to extend our experiments by using semantic relatedness measures instead of the semantic similarity measures, thus our distance based measure will also consider the semantic relationships between two terms, such as `owl:equivalentClass`.

References

1. Acosta, M., Zaveri, A., Simperl, E., Kontokostas, D., Auer, S., Lehmann, J.: Crowdsourcing linked data quality assessment. In: Alani, H., Kagal, L., Fokoue, A., Groth, P., Biemann, C., Parreira, J.X., Aroyo, L., Noy, N., Welty, C., Janowicz, K. (eds.) ISWC 2013. LNCS, vol. 8219, pp. 260–276. Springer, Heidelberg (2013). doi:10.1007/978-3-642-41338-4_17
2. Debattista, J., Auer, S., Lange, C.: Luzzu - a framework for linked data quality analysis. In: 2016 IEEE International Conference on Semantic Computing, Laguna Hills (2016)
3. Debattista, J., Londoño, S., Lange, C., Auer, S.: Quality assessment of linked datasets using the approximation. In: 12th European Semantic Web Conference Proceedings (2015)
4. Harispe, S., Ranwez, S., Janaqi, S., Montmain, J.: Semantic measures for the comparison of units of language, concepts or entities from text and knowledge base analysis, October 2013. arXiv abs/1310.1285
5. Hausman, J.A., Wise, D.A.: Stratification on endogenous variables and estimation: the gary income maintenance experiment. In: Manski, C.F., McFadden, D.L. (eds.) Structural Analysis of Discrete Data with Econometric Applications. MIT Press, Cambridge (1981)
6. Knorr, E.M., Ng, R.T., Tucakov, V.: Distance-based outliers: algorithms and applications. VLDB J. **8**(3–4), 237–253 (2000)
7. Mazandu, G.K., Mulder, N.J.: A topology-based metric for measuring term similarity in the gene ontology. Adv. Bioinf. **2012**, 1–17 (2012)
8. Paulheim, H., Bizer, C.: Improving the quality of linked data using statistical distributions. Int. J. Semant. Web Inf. Syst. **10**(2), 63–86 (2014)

9. Töpper, G., Knuth, M., Sack, H.: DBpedia ontology enrichment for inconsistency detection. In: Proceedings of the 8th International Conference on Semantic Systems, I-SEMANTICS 2012, pp. 33–40. ACM, New York (2012)
10. Waitelonis, J., Ludwig, N., Knuth, M., Sack, H.: WhoKnows? - evaluating linked data heuristics with a quiz that cleans up DBpedia. Int. J. Interact. Technol. Smart Educ. (ITSE) **8**(3), 236–248 (2011)
11. Wienand, D., Paulheim, H.: Detecting incorrect numerical data in DBpedia. In: Presutti, V., d'Amato, C., Gandon, F., d'Aquin, M., Staab, S., Tordai, A. (eds.) ESWC 2014. LNCS, vol. 8465, pp. 504–518. Springer, Heidelberg (2014). doi:10.1007/978-3-319-07443-6_34
12. Zaveri, A., Kontokostas, D., Sherif, M.A., Bühmann, L., Morsey, M., Auer, S., Lehmann, J.: User-driven quality evaluation of DBpedia. In: Proceedings of the 9th International Conference on Semantic Systems, I-SEMANTICS 2013, pp. 97–104. ACM, New York (2013)
13. Zhou, Z., Wang, Y., Gu, J.: A new model of information content for semantic similarity in wordnet. In: FGCNS 2008 Proceedings of the 2008 Second International Conference on Future Generation Communication and Networking Symposia, vol. 3, pp. 85–89. IEEE Computer Society, December 2008

Information Retrieval and Knowledge Discovery

Linked Data Collection and Analysis Platform for Music Information Retrieval

Yuri Uehara(✉), Takahiro Kawamura, Shusaku Egami, Yuichi Sei, Yasuyuki Tahara, and Akihiko Ohsuga

Graduate School of Information Systems,
University of Electro-Communications, Tokyo, Japan
{uehara.yuri,kawamura,egami.shusaku}@ohsuga.is.uec.ac.jp,
{seiuny,tahara,ohsuga}@uec.ac.jp

Abstract. There has been extensive research on music information retrieval (MIR), such as signal processing, pattern mining, and information retrieval. In such studies, audio features extracted from music are commonly used, but there is no open platform for data collection and analysis of audio features. Therefore, we build the platform for the data collection and analysis for MIR research. On the platform, we represent the music data with Linked Data, which are in a format suitable for computer processing, and also link data fragments to each other. By adopting the Linked Data, the music data will become easier to publish and share, and there is an advantage that complex music analysis will be facilitated. In this paper, we first investigate the frequency of the audio features used in previous studies on MIR for designing the Linked Data schema. Then, we build a platform, that automatically extracts the audio features and music metadata from YouTube URIs designated by users, and adds them to our Linked Data DB. Finally, the sample queries for music analysis and the current record of music registrations in the DB are presented.

Keywords: Linked data · Audio features · Music information retrieval

1 Introduction

Recently, there are a large number of studies on music. Music Information Retrieval (MIR) deals with music on computers and has been studied in various ways [1]. In these studies, audio features extracted from music are frequently used, however, there is no open platform for collecting data including the audio features for music analysis. Therefore, we propose the platform for MIR research in this paper.

On the platform, we used Linked Data format, since it is suitable for complex searches for audio features and songs-related metadata. Also, the music data in the Linked Data can be easily linked to the external databases (DBs) such as DBpedia[1], and then become more valuable when published and shared in

[1] http://wiki.dbpedia.org/.

Y.-F. Li et al. (Eds.): JIST 2016, LNCS 10055, pp. 127–135, 2016.
DOI: 10.1007/978-3-319-50112-3_10

public. Thus, we built a platform, that automatically extracts audio features of music, transforms them to graphs in the Linked Data, and then insert the graphs to our music Linked Data DBs with connections to the external DBs. Note that this platform is designed for music-related researchers and developers, who intend to analyze music information and create their own applications, e.g., recommendation mechanism. Use of a listener is beyond the scope of this paper.

The rest of the paper is organized as follows. Related works in terms of MIR research using audio features and Linked Data are shown in Sect. 2. In Sect. 3, we describe the schema design of music Linked Data. After the system for automatically analyzing audio features is proposed in Sect. 4, some examples of music analysis are described in Sect. 5. Finally, we conclude this paper with future works in Sect. 6.

2 Related Work

There is MusicBrainz, that is an open database about music. The MusicBrainz database[2] has the music data, such as song title, artist name, etc., which are described in Resource Description Framework (RDF). The data are created mainly by participants, and thus there is a mechanism of data registration that requires the verification and approval of the other participants in order to maintain the reliability of the data.

In addition, Music Ontology[3] offers the data model related to music. This provides vocabulary in the RDF model of the MusicBrainz data for describing the relationship of music information.

Also, in our previous research, the music recommendation using Linked Data is proposed in [2]. We constructed Linked Open Data (LOD) by the data retrieved from Last. fm, Yahoo! Local, Twitter, and Lyric Wiki. Then, they proposed a method for recommending songs according to associative relations in the LOD.

However, there is no data of audio features in the MusicBrainz database, the Music Ontology, and our previous LOD set. Therefore, we built a platform that provides audio features combined with music metadata in Linked Data format for open MIR research.

3 Schema Design of Music Information

In this section, designing Linked Data schema, including audio features and music metadata is described.

[2] https://musicbrainz.org/.
[3] http://musicontology.com/.

3.1 Selection of Audio Features

Audio features refer to the characteristics of the music, such as *Tempo* representing the speed of the track, tonality of track and quality of sound, the features used in MIR studies vary. For example, Osmalskyj et al. used *Tempo* and *Loudness* to identify cover songs [3]. Luo et al. used the audio features *Pitch*, *Zero crossing rate*, etc. to detect of common mistakes in novice violin playing [4]. Thus, it is necessary to survey which audio features should be prepared in the music schema.

Thus, we investigated the frequency of the audio features used in previous MIR studies. We collected 114 papers published in the International Society of Music Information retrieval (ISMIR)[4] in 2015, which is the top conference in the field of MIR. Table 1 shows the results.

Table 1. Number of using audio features

Audio Features	Count
Tempo	18
Pitch	10
MFCC	9
Beat	8
Loudness	7
Chord	5
Chroma, Key, Zero crossing rate, Roll off	3
Roughness, Timbre, Low energy, RMS energy, Brightness, Mode, Duration	2
Harmony, Volume, Articulation, Energy ratio, Swing ratio, Spectral irregularity, Inharmonicity, Vibrate, Rhythm, Dynamics	1

We found in Table 1 that *Tempo* representing the speed of music songs has been the most used in many studies, although there are some features which have been used just once. Then, we selected some of the audio features according to the following policies.

1. Features which are similar to each other can be integrated. (Beat, Swing ratio and Rhythm are integrated Tempo.)
2. Features appeared just once in the publications can be ignored. (Harmony, Volume, Articulation, Energy ratio, Spectral irregularity, Inharmonicity, Vibrate, Dynamics are deleted.)
3. Features, which cannot be extracted through a song, can be ignored, since a user's input is assumed to be song by song. Features in a series of songs can be extracted by querying for the resulted DB.(Pitch, Loudness, Chord, Chroma, Timbre, Duration are deleted.)
4. Features should be quantitative in numerical values, and qualitative ones like an emotional feature are not included.(MFCC is deleted.)

[4] http://www.ismir.net/.

3.2 Design of the Schema

Linked Data is an RDF format, in which data fragments are linked by any semantic relations.

We defined original properties for selected audio features, excluding *Key* and *Mode*, since there were no existing properties for them, or the properties are not appropriate for our purpose. In terms of *Key* and *Mode*, there are appropriate properties in the Music Ontology, and thus we used them. Then, we classified properties of audio features into some classes for making them easy to use. Table 2 shows the classes and properties corresponding to the audio features.

Table 2. Class and property of audio features

Class	Property	Audio Features	Count
Tempo	tempo	Tempo	28
Key	key	Key, Mode	5
Timbre	zerocross	Zero crossing rate	3
	rolloff	Roll off	3
	brightness	Brightness	2
Dynamics	rmsenergy	RMS energy	2
	lowenergy	Low energy	2

In Table 2, *Mode* can be included in the *Key* class, thus we used the same property. Based on these definitions, we designed the music schema and built the Linked Data set for music. *Tempo* means speed of the song, *Key* means tonality of song, *Mode* means volume difference of the major chord and a minor chord, *Zero crossing rate* means the rate at which the signal changes from positive to negative or back, *Roll off* means ratio of bass which accounts for 85 percent of the total, *Brightness* means ratio of high-range (more than 1500 Hz), *RMS energy* means the average of the volume (root mean square), *Low energy* means Ratio of sound low in volume. Figure 1 shows part of the Linked Data.

We designed the music schema with the video id (URI) of YouTube. In Fig. 1, the id: dvgZkm1xWPE indicates a song "Viva La Vida" by Coldplay. In the graph, the id node links to the classes of audio features and then links to each audio feature. Also, we added some degrees for categorizing numerical values in the features. The *lowenergy*, the *rmsenergy*, and the *brightness* have a class by 0.1, the *zerocross* has a class by 100, and the *rolloff* has a class by 1000. The *tempo* has tmarks based on tempo values[5], which is a measure of the speed marks: Slow means 39 or less bpm, Largo means 40–49 bpm, Lento means 50–55 bpm, Adagio means 56–62 bpm, Andante means 63–75 bpm, Moderato means 76–95 bpm, Allegretto means 96–119 bpm, Allegro means 120–151 bpm, Vivace

[5] http://www.sii.co.jp/music/try/metronome/01.html.

means 152–175 bpm, Presto means 176–191 bpm, Prestissimo means 192–208 bpm, Fast means over 209 bpm.

In addition, we extended the schema of music metadata, that includes not only song title, artist name, etc. in the Music Ontology, but also lyricist name, cd name for the complex search for music information. Figure 2 shows part of the metadata in our Linked Data. In the graph, the video id node links to the class of metadata, and then links to the detailed value, as well as the graph for the audio features. Also, some nodes such as the artist name are linked to the external DBs like DBpedia.

Figures 1 and 2 are the graphs of "Viva La Vida" by Coldplay, and thus the two graphs can be linked with the video id of YouTube.

4 Music Information Extracting

4.1 System Overview

The system architecture for our music information extraction is shown in Fig. 3, and its workflow is indicated by the number 1 to 11 as follows.

```
1. Download the video data from the YouTube video URI designated by a user in a web browser.
2. Call the MATLAB process that analyzes audio features in the video file.
3. Store the obtained audio features in an RDB, MySQL.
4. Call the RDF create program.
5. Obtain the music information for the video from the YouTube website.
6. Search the music metadata using  Last.fm API.
7. Query the audio features of the video for MySQL.
8. Convert the metadata and audio features to RDF graphs, and store them in an RDF store,
   Virtuoso.
9. Notify the completion to the user.
```

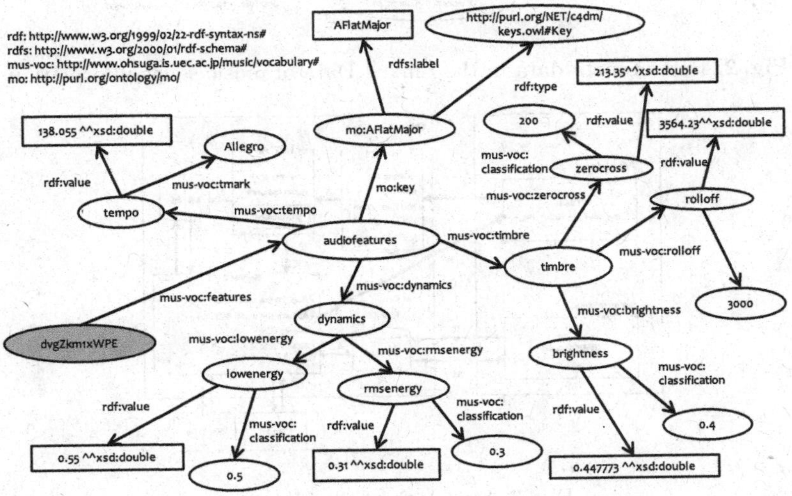

Fig. 1. Part of audio features in the Linked Data of music songs information

```
10. Submit a simple SPARQL query for confirmation.
11. Returns the evidence of the inclusion of new sub-graphs corresponding to the video.
```

Our system obtains videos to analyze audio features from YouTube, and so public users can easily extend the music information on the platform. However, we discard the video files after extracting the audio features, and thus we believe this process does not cause any legal or moral problems.

The workflow is divided into several phases. The first phase is for analyzing the audio features of the YouTube video, and the second phase is for acquiring the metadata of the YouTube video. Then, the third phase converts the metadata and audio features to RDF graphs, and the RDF graphs are stored in Virtuoso

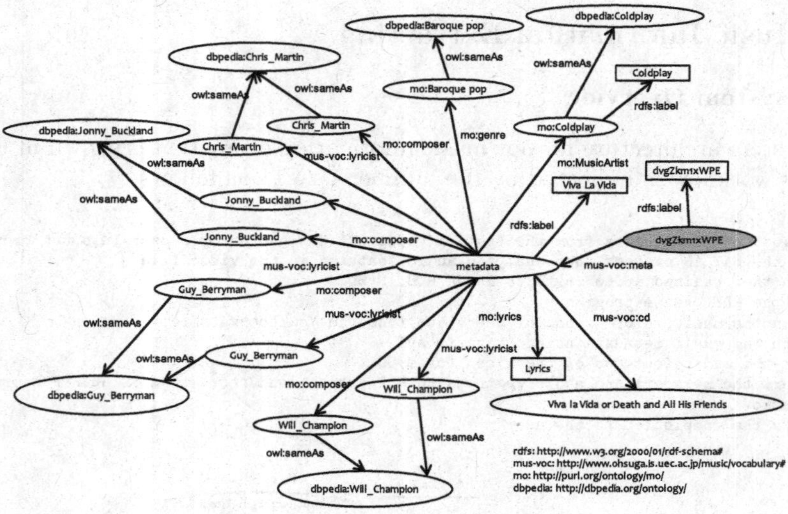

Fig. 2. Part of meta data in the Linked Data of music songs information

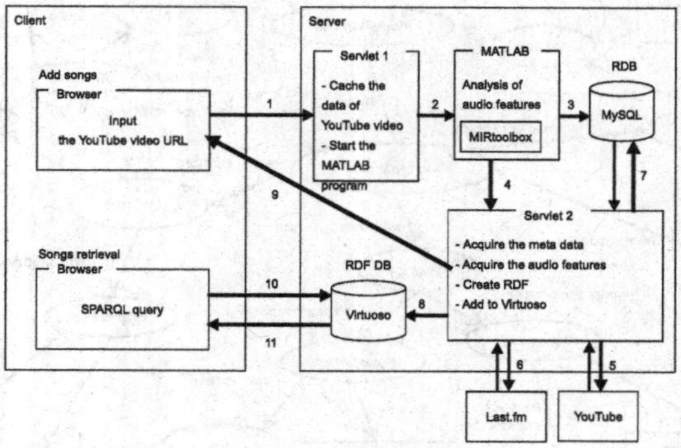

Fig. 3. Structure of the system

database. Then, the final phase is for the confirmation of newly added graphs. We describe the detail of each phase in the following sections.

4.2 Analyzing Audio Features

This phase (1–3 in Fig. 3) obtains and analyzes the audio features described in Sect. 3 from a specified video on the YouTube website. We used MIRtoolbox running on the MATLAB for analysis of the audio features. The MIRtoolbox includes several signal processing algorithms, which are commonly used in MIR.

First, a user inputs a URI of the YouTube video in the input form in a web browser. Then, the extract program downloads and caches the YouTube video data, and then it starts the MATLAB program that analyzes audio features of the video. Finally, the extracted audio features are temporally stored in MySQL database.

4.3 Searching Music Metadata

This phase (4–6 in Fig. 3) acquires music metadata, such as track title and artist name base on schema designed in Sect. 3. First, the video information is obtained from the YouTube. Then, we search the corresponding metadata, such as track title and artist name using Last.fm API for extending the related information.

4.4 Adding Linked Data

This phase (7 and 8 in Fig. 3) adds the audio features and the metadata to the music Linked Data.

First, the RDF create program gets the audio features from MySQL database, then it converts the audio features and the above metadata to RDF graphs based on the schema design. Finally, the RDF graphs are stored in Virtuoso.

4.5 Confirming Results

Figure 4 shows an example that a user added a new music "Doom and Gloom" by The Rolling Stones (the id: 1DWiB7ZuLvI) to the existing graph including "Applause" by Lady Gaga (the id: _bHhpufKRjs) in RDF DB. The two graphs are linked at a classification of the audio feature *Low energy* since they have the similar values in that feature. We visualize the RDF graph using the visualization tool the Visualization of RDF graph by ARC2[6]. Note that part of metadata and audio features are omitted for convenience.

[6] http://www.kanzaki.com/works/2009/pub/graph-draw.

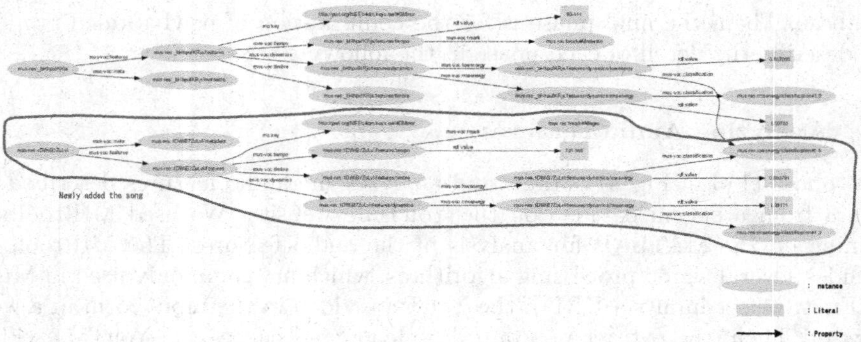

Fig. 4. The graph added new data of music song

5 Example of Music Analysis

The current number of music registered in the platform is 1073 and the number of triples automatically extracted for representing the audio features and the metadata for that music is 20858. The platform is publicly available at http://www.ohsuga.is.uec.ac.jp/music/.

In this section, we show the results of some example queries on the platform, and how the music Linked Data can be used for MIR.In the SPARQL Query, we specified the audio feature *Brightness* of the song "Hello, Goodbye" by The Beatles, and search other songs, in which the value of the *Brightness* is similar to the specified song. As the result, we get 5 songs, in which the *Brightness* has the similar degree in Table 3.

```
[SPARQL Query]
PREFIX mus-voc:<http://www.ohsuga.is.uec.ac.jp/music/vocabulary#>
PREFIX mo:<http://purl.org/ontology/mo/>

SELECT ?artist_x ?title_x ?brightness_x
WHERE { ?metadata rdfs:label ?title .
        ?resource mus-voc:meta ?metadata .
        ?resource mus-voc:features ?features .
        ?features mus-voc:timbre ?timbre .
        ?timbre mus-voc:brightness ?brightnessc .
        ?brightnessc rdf:value ?brightness.
        ?brightnessc_x rdf:value ?brightness_x.
        ?timbre_x mus-voc:brightness ?brightnessc_x .
        ?features_x mus-voc:timbre ?timbre_x.
        ?resource_x mus-voc:features ?features_x .
        ?resource_x mus-voc:meta ?metadata_x .
        ?metadata_x rdfs:label ?title_x .
        ?metadata_x mo:MusicArtist ?MusicArtist_x .
        ?MusicArtist_x rdfs:label ?artist_x .
        FILTER regex(?title,"Hello Goodby") . }
ORDER BY (
  IF( ?brightness < ?brightness_x,
      ?brightness_x - ?brightness,
      ?brightness - ?brightness_x )
) LIMIT 5
```

Table 3. Result of submitting the SPARQL query

artist_x	title_x	brightness_x
The Beatles	Can't Buy Me Love	0.553889
Whitney Houston	Never Give Up	0.559786
Coldplay	Princess Of China Ft. Rihanna	0.560039
Lady Gaga	Judas	0.550279
The Beatles	Penny Lane	0.550221

6 Conclusion and Future Work

In this paper, we proposed a platform for providing audio features and the music metadata to MIR research. Thus, we designed the Linked Data schema for audio features and the metadata. Then, we built the platform for the music data collection and analysis and showed the current status of the dataset and a simple example of its use.

In future, we plan to provide more sophisticated examples and applications of music information analysis, which will encourage the expansion of the music Linked Data to music researchers and developers.

Acknowledgments. This work was supported by JSPS KAKENHI Grant Numbers 16K12411, 16K00419, 16K12533.

References

1. Kitahara, T., Nagano, H.: Advancing Information Sciences through Research on Music: 0. Foreword. IPSJ magazine. Joho Shori **57**(6), 504–505 (2016)
2. Wang, M., Kawamura, T., Sei, Y., Nakagawa, H., Tahara, Y., Ohsuga, A.: Context-aware Music Recommendation with Serendipity Using Semantic Relations. In: Proceedings of 3rd Joint International Semantic Technology Conference, pp. 17–32 (2013)
3. Osmalskyj, J., Foster, P., Dixon, S., Embrechts, J.J.: Combining features for cover song identification. In: Proceedings of the 16th International Society for Music Information Retrieval Conference, pp. 462–468 (2015)
4. Luo, Y.-J., Su, L., Yang, Y.-H., Chi, T.-S.: Real-time music tracking using multiple performances as a reference. In: Proceedings of the 16th International Society for Music Information Retrieval Conference, pp. 357–363 (2015)

Semantic Data Acquisition by Traversing Class–Class Relationships Over Linked Open Data

Atsuko Yamaguchi[1(✉)], Kouji Kozaki[2], Kai Lenz[3], Yasunori Yamamoto[1], Hiroshi Masuya[4,3], and Norio Kobayashi[3,4,5]

[1] Database Center for Life Science (DBCLS), Research Organization of Information and Systems, 178-4-4 Wakashiba, Kashiwa, Chiba 277-0871, Japan
{atsuko,yy}@dbcls.rois.ac.jp

[2] The Institute of Scientific and Industrial Research (ISIR), Osaka University, 8-1 Mihogaoka, Osaka, Ibaraki 567-0047, Japan
kozaki@ei.sanken.osaka-u.ac.jp

[3] Advanced Center for Computing and Communication (ACCC), RIKEN, 2-1 Hirosawa, Wako, Saitama 351-0198, Japan
{kai.lenz,norio.kobayashi}@riken.jp

[4] BioResource Center (BRC), RIKEN, 3-1-1, Koyadai, Tsukuba, Ibaraki 305-0074, Japan
hmasuya@brc.riken.jp

[5] RIKEN CLST-JEOL Collaboration Center, 6-7-3 Minatojima-minamimachi, Chuo-ku, Kobe 650-0047, Japan

Abstract. Linked Open Data (LOD), a powerful mechanism for linking different datasets published on the World Wide Web, is expected to increase the value of data through mashups of various datasets on the Web. One of the important requirements for LOD is to be able to find a path of resources connecting two given classes. Because each class contains many instances, inspecting all of the paths or combinations of the instances results in an explosive increase of computational complexity. To solve this problem, we have proposed an efficient method that obtains and prioritizes a comprehensive set of connections over resources by traversing class–class relationships of interest. To put our method into practice, we have been developing a tool for LOD exploration. In this paper, we introduce the technologies used in the tool, focusing especially on the development of a measure for predicting whether a path of class–class relationships has connected triples or not. Because paths without connected triples can be predicted and removed, using the prediction measure enables us to display more paths from which users can obtain data that interests them.

Keywords: Linked data · Class–class relationships · Data integration · Path finding

Y.-F. Li et al. (Eds.): JIST 2016, LNCS 10055, pp. 136–151, 2016.
DOI: 10.1007/978-3-319-50112-3_11

1 Introduction

An important feature of Linked Data is to provide an efficient mechanism for linking different datasets published on the World Wide Web. The method enables users to mash up different data, and combinations of various datasets are expected to contribute to new innovations [1]. The Linked Open Data (LOD) Cloud (http://lod-cloud.net/) shows the evolution of LOD and many datasets published as Linked Data in a large variety of domains. In government, publishing government data as open data is promoted as an important policy in many countries. In science, open science is strongly promoted because open data publishing is essential for providing evidence for testing new theories. Particularly, in the life sciences, many Resource Description Framework (RDF) databases are published as LOD to provide foundations for data integration towards open science [2–5].

To use these databases published as LOD efficiently, users must be permitted to obtain data in a flexible manner according to their interests. An important case is to find paths of links between instances (resources) whose types are given two classes for integrative data analysis with semantics. For example, when biomedical researchers obtain molecular pathways through their experiments, they want to obtain a set of their IDs in the Reactome database[1], IDs of proteins related them, and their protein names. These paths can be obtained by retrieving chains of properties (links) that connect instances of classes such as Pathway, Protein and Protein Name. In other words, these paths can be obtained by traversing paths of class–class relationships over the LOD.

However, many users have difficulty specifying the appropriate path of class–class relations suitable for their search request, because each RDF dataset has different data schema that must be analyzed. Therefore, the technologies for exploring RDF datasets and displaying the summary of paths of class–class relationships are strongly required. As a method for exploring RDF datasets, we developed a metadata specification called SPARQL Builder Metadata (SBM) and a crawling tool to extract SBM through SPARQL endpoints. In addition, as a means of displaying paths, we developed some graph-based methods for expressing class–class relationships in what we call a class graph. In this paper, of these methods, we focus especially on the proposal of a measure for removing meaningless paths of class–class relationships through which a user cannot obtain any data.

The remainder of this paper is organized as follows: In Sect. 2, we describe existing work related to our approach. Section 3 introduces an application, SPARQL Builder, based on our proposed methods to make the explanations concrete. In addition, we explain two technologies, SBM and class graphs, that support our method, as mentioned above. Section 4 discusses a measure for avoiding meaningless paths and shows the performance of the proposed measure through computational experiments. Discussion and conclusion are provided in Sect. 5.

---

1 https://www.ebi.ac.uk/rdf/services/reactome/.

2 Related Work

Some methods for obtaining paths from LOD have been proposed. To find a path or paths between two given resources, RelFinder [6] uses an algorithm that iteratively finds interim resources (sequentially related triples) by following RDF triples. Our approach differs from that of RelFinder in that ours accepts two classes of interest and finds paths between individuals that belong to each respective class. Because each class often contains thousands of individuals, and multiple paths between any two end individuals are highly probable, there are many cases with which RelFinder's algorithm cannot cope.

Another related application is Visor [7], which enables users to browse RDF datasets in the light of class–class relationships. For Visor, an exploratory search called Multi-Pivot, that extracts concepts and relationships from ontologies of interest to the user, was developed. The extracts are visualized and used for semantic searches among instances (data) associated with ontology terms. However, Visor provide no method for finding an end-to-end path through multiple resources.

As another approach for exploring RDF datasets using classes, a web based tool named Sparklis[2] [8] presents users with lists of classes in its target endpoints and allows them to construct queries through facet-based graphical user interfaces (GUIs). Faceted navigation, such as that in [9], is a very powerful approach to the browsing of datasets and finding specific resources. However, our approach focuses on obtaining comprehensive data having a class–class relationship that interests an user.

Related work regarding class graphs introduced in Subsect. 3.3, includes class association graphs [10]. Although for both class graphs and class association graphs nodes correspond to classes, edges of class association graphs correspond to the number of triples between classes, while edges of class graphs correspond to properties and their statistical values.

3 Data Acquisition Based on Class–Class Relationships

To discuss the technologies that are required to put our approach, which enables a user to understand RDF datasets by traversing class–class relationships, into practice, we first introduce an application called *SPARQL Builder*. The SPARQL Builder system generates a SPARQL query based on a path of class–class relationships over LOD. Then, we briefly explain two important technologies, *SPARQL Builder Metadata (SBM)* and *class graphs*, to realize the system.

3.1 SPARQL Builder

We have been developing a practical LOD search tool called SPARQL Builder for users who are not familiar with the SPARQL language to generate SPARQL

[2] http://www.irisa.fr/LIS/ferre/sparklis/osparklis.html.

queries without knowledge of SPARQL and RDF data schema [11]. The system is based on our proposed method, with which users can obtain their required data flexibly by traversing a path of class–class relationships over LOD.

The most important issue for the application is how to find candidate paths of class–class relationships for a user. In other words, how to compute candidate paths more efficiently and accurately is key for the application. To compute paths efficiently, the system obtains class–class relationships from LOD in advance, stores it as metadata using SBM, and constructs a labeled multigraph named a class graph as mentioned in Subsects. 3.2 and 3.3. In addition, we will discuss in Sect. 4 how to remove unnecessary paths to display paths more accurately.

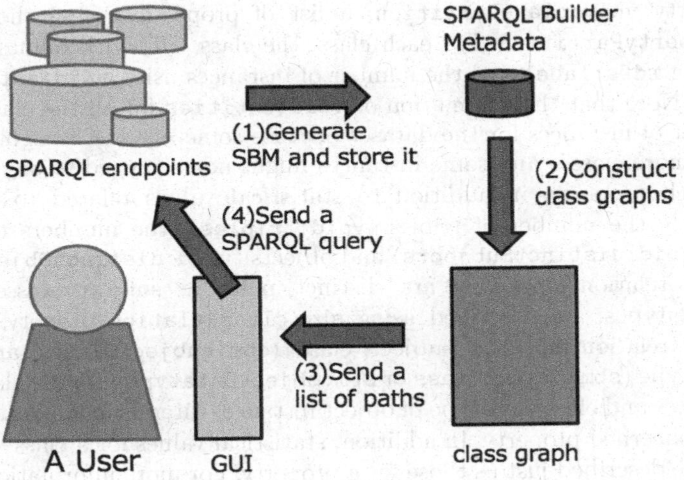

Fig. 1. An overview of the SPARQL builder system.

An overview of the SPARQL Builder system's architecture is shown in Fig. 1. (1) SPARQL Builder manages SBM generated by a crawler to access SPARQL endpoints of LOD in advance. (2) When a user accesses the SPARQL Builder system via a web browser as a GUI and selects two classes from a list of classes shown in the GUI initially, the class graph for the RDF dataset including the selected two classes is constructed using SBM, and possible paths between the selected two classes are computed and sent to the GUI. (3) Then, the SPARQL Builder GUI displays the list of paths in the user's web browser. If the user selects one path from the list, the system generates the SPARQL query corresponding to the path. (4) A user can obtain data by throwing the SPARQL query from the GUI.

Therefore, using the SPARQL Builder GUI, a user can explore RDF datasets of interest by specifying classes and a path. SPARQL Builder supports 38 SPARQL endpoints as of March 2016. Subsect. 3.2 relates to step (1) in Fig. 1. Subsect. 3.3 relats to step (2).

3.2 SBM

SBM is a summary of RDF datasets provided via a SPARQL endpoint. As described above, SBM is designed for describing a summary of class–class relationships in RDF datasets provided by SPARQL endpoints to enable SPARQL Builder to obtains necessary information quickly for computing paths of a sequentially connected class–class relationships.

SBM is defined as an extension of the VoID[3] and SPARQL 1.1 service description[4] with our original vocabulary of name space `sbm:`. For a SPARQL endpoint, SBM consists of summaries of the datasets provided as named graphs in the endpoint. For an individual RDF dataset, the summary for the dataset typed by `void:Dataset` includes a list of classes using the property `void:classPartition`, a list of properties using the property `void:propertyPartition`. For each class, the class URI with human-readable labels using `rdfs:label` and the number of instances using `void:entities` are described. Note that the summation of `void:entities` for all the classes is not the number of instances for the dataset because some instances might be typed to two or more classes and some instances might not be typed to any class.

For each property, in addition to statistical values related to the property such as the number of triples (`void:triples`), the numbers of distinct subjects (`void:distinctSubjects`) and objects (`void:distinctObjects`), and class–class relationships that are distinct pairs of subject classes, object classes/datatypes, are described using `sbm:classRelation` propety. For each class–class relationship, the subject class (`sbm:subjectClass`) and object class/datatype (`sbm:objectClass` or `sbm:objectDatatype`) are the class of subject instances and class/datatype of object instances/literals in triples associated with the concerned property. In addition, statistical values for a class–class relationship are described just as those for a property. For more information, see the web page[5] of the SBM specification.

To obtain SBM from RDF datasets, we implemented a crawler that generates SBM data by throwing lightweight but numerous SPARQL queries to the concerned SPARQL endpoints. The reason we used lightweight queries was to avoid time out response for a query. Because we found through our preliminary study that the size of the result for each query is not larger than the maximum size of result for a SPARQL endpoint, we focused on the response time for each query more than on the intermediate sizes of results by the queries to obtain information required for SBM. We have already crawled 38 endpoints including life-science SPARQL endpoints provided at EBI RDF Platform (https://www.ebi.ac.uk/rdf/platform) [2], Bio2RDF [3] Release 3 (http://download.bio2rdf.org/release/3/release.html) and Database Center for Life Science (http://dbcls.rois.ac.jp/en/services) as of April 2016. The crawled metadata in SBM are available through the web[6].

---

[3] https://www.w3.org/TR/void/.
[4] https://www.w3.org/TR/sparql11-service-description/.
[5] http://www.sparqlbuilder.org/doc/sbm_2015sep/.
[6] http://www.sparqlbuilder.org/sbm/.

Because SPARQL Builder discovers sequentially connected triples associated with a path specified by a user, the comprehensiveness of domain-range information for properties and class declarations for instances are important metadata. As described in [12], many datasets miss domain-range information for properties. In addition, a class is sometimes not typed as a class. Therefore, to make available as many RDF datasets as possible, the crawler extracts not only declared classes explicitly typed by `rdfs:Class` and domain–range relationships using the properties `rdfs:domain` and `rdfs:range`, but also classes expressed implicitly such as a subject or an object for the property `rdfs:subClassOf` and implicit class–class relationships found by triples with subjects and objects typed by some classes. By expanding classes and class–class relationships using such inferences, over 99.9% of instances of the 38 databases crawled are associated with classes.

Although SBM is designed for SPARQL Builder, statistical values gathered by the crawler can be applied to more general cases, such as evaluation of a SPARQL endpoint and RDF datasets in the endpoints. For example, a qualitative categorization with levels from 1 to 3 of SPARQL endpoints based on SBM is introduced in [11]: Level 1 corresponds to datasets that have domain-range declarations for all pf the properties and class declarations for all of the resources. Level 3 corresponds to datasets fpr which none of the resources is typed by any class. The other datasets with neither Level 1 or 3 are Level 2. Datasets with Level 2 can be further evaluated by the ratio of typed resources and the ratio of properties with domain-range declarations.

3.3 Class Graphs

To compute paths between two classes efficiently, we used a specialized graph whose nodes and edges correspond to the classes and the class–class relationships, respectively. We call such a graph a *class graph*.

Formally, a class graph is defined as follows: Given an RDF dataset R, we denote by C the set of all classes in R. A class graph $G_R = (V, E, c, p)$ of R is a directed labeled multigraph defined as follows: V is a $|C|$-sized set of nodes and c is a one-to-one mapping from V to a set of URLs of C. E is a multiset of directed edges between the nodes of V, and p maps E to a set of URLs of predicates in R. To construct E and p from R, we add to E a directed edge e_{pred} from node n_d to n_r, where $c(n_d) = class_d$ and $c(n_r) = class_r$, and define $p(e_{pred}) = pred$ if $pred$ satisfies either of the following two conditions: (1) both the triples "*pred* rdfs:domain $class_d$" and "*pred* rdfs:range $class_r$" exist in R for some classes $class_d$ and $class_r$; (2) there exist three triples "*sub pred ob*", "*sub* rdf:type $class_d$", "*ob* rdf:type $class_r$" in R, where *sub* and *ob* are resources and $class_d$ and $class_r$ are classes.

A class graph can be constructed from SBM efficiently, because SBM include a list of all the classes and a list of all the class–class relationships. V corresponds to a set of objects of the property `void:classPartition`. We can define c by referring to objects of the property `void:class`. Each object of the property `sbm:classRelation` corresponds to edge e in E. From the subject of the property `sbm:classRelation`, we can find the property URI for the class–class relationship. Because inferred classes and class–class relationships are included in SBM as explained in Subsect. 3.2, nodes include inferred classes and edges include inferred class–class relationships as the definition of edges includes condition (2).

Given a class graph G_R, we define a *class path* p from a start class *start* to an end class *end* as a sequence $(n_0, e_1, n_1, e_2, \ldots, n_k)$, where the nodes n_i and edges e_i of G_R satisfy the following conditions: (1) $c(n_0) = start$, $c(n_k) = end$, (2) $c(n_i) \neq end$ for any $i \neq k$, and (3) e_i is a directed edge from n_{i-1} to n_i or from n_i to n_{i-1}. An edge e_i directed from n_{i-1} to n_i or from n_i to n_{i-1} is called *forward* or *reverse* directed, respectively. The length of a class path $(n_0, e_1, \ldots, n_k)$ is defined as k. To compute possible class paths between two classes in a practical time, the maximum length of class paths is given in advance and is currently set as 4.

A class path corresponds to a multi-step class–class relationship to obtain the instances of an end class from the instances of a start class by relating a sequence of predicates $p(e_i)$. By searching the possible class paths from the start class to the end class, we can obtain candidates of connections between data allowing a user to select a class path according to the interests of the user. Class paths between two classes can be found in a practically short time using an algorithm provided in [13].

4 Removal of Empty Paths

4.1 Measure to Remove Empty Paths

Although many paths can be computed efficiently using the technologies introduced in Sect. 3, we found through our preliminary investigation that some class paths have no sequence of instances obtained by traversing triples along the class paths. For example, as in Fig. 2, for a class path $(n_0, e_1, n_1, e_2, n_2)$ with forward edges e_1 and e_2, there might be no two triples of $(r_1\ p(e_1)\ r_2)$ and $(r_2\ p(e_1)\ r_3)$ such that r_1, r_2, and r_3 are instances of classes $c(n_0)$, $c(n_1)$, and $c(n_2)$, respectively. We call such a path an *empty path*. A user cannot obtain any data using an empty path because there are no connected triples from a start class to an end class. For example, if a user selects an empty path from a list of paths displayed in the SPARQL Builder GUI at step (3), the generated SPARQL query has no result at step (4). Therefore, it is important to present a method for removing as many such empty paths as possible from candidate paths.

An exact method for deciding whether a class path is an empty path involves using a SPARQL query with `ASK` corresponding to the path and deciding it is an empty path if the result is "false". Concretely, for a class path $(n_1, e_1, n_2, e_2, \ldots, n_m)$, using the following SPARQL query, we can decide whether a class path is an empty path.

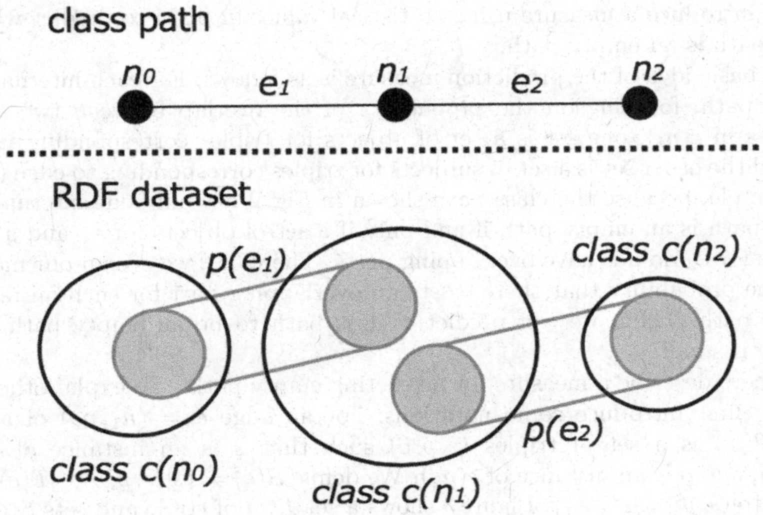

Fig. 2. Example of an empty path. Although two nodes n_0 and n_2 are connected in the class graph, there is no sequence of triples connecting instances in class $c(n_0)$ to those in class $n(n_2)$.

ASK {
  $?r_1$ $p(e_1)$ $?r_2$. (or $?r_2$ $p(e_1)$ $?r_1$. if e_1 is backward)
  $?r_2$ $p(e_2)$ $?r_3$. (or $?r_3$ $p(e_2)$ $?r_2$. if e_2 is backward)
  ...
  $?r_{m-1}$ $p(e_{m-1})$ $?r_m$. (or $?r_m$ $p(e_{m-1})$ $?r_m$. if e_{m-1} is backward)
  $?r_1$ rdf:type $c(n_1)$.
  $?r_2$ rdf:type $c(n_2)$.
  ...
  $?r_m$ rdf:type $c(n_m)$
  }

If a SPARQL endpoint always returned the result for a query quickly, our system could extract only non-empty class paths from a set of class paths to show them to a user. However, using current triple stores, an ASK query sometimes consumes too much time. In addition, because the number of class paths is sometimes very large, the total time for computing a set of non-empty paths from a set of class paths tends to be very long. Furthermore, obtaining results for ASK in advance as we did for SBM is intractable through the Internet from SPARQL endpoints because results for all the paths with length four for all the combinations of classes are required. Therefore, it is not realistic in our approach at the moment to use ASK query to decide whether a class path is empty. In fact, although we had tried to use ASK query for SPARQL Builder to remove empty paths, it had often not worked even for relatively small RDF datasets. Therefore, we now introduce a measure using statistical values in SBM to predict whether a class path is an empty path.

The basic idea of the prediction measure is as follows: For each internal node n_i in a path, focusing on the probability of the overlap between two sets of instances in $c(n_i)$, one set is a set of objects for triples corresponding to edge (e_i), and the other set is a set of subjects for triples corresponding to edge (e_{i+1}). For example, because the class path shown in Fig. 2 has only one internal node n_1, the path is an empty path if and only if a set of objects for e_1 and a set of subjects for e_2 do not have overlapping parts. Therefore, we design our measure to be the probability that there exists an overlapping part for such an internal node in path. Then, we can predict a class path to be an empty path if the measure is small.

We here describe a measure for predicting empty paths. To explain the measure, we first introduce some notations. For an edge $e = (n_1, n_2)$ of a class graph, $T(e)$ is a set of triples (s, p, o) such that s is an instance of $c(n_1)$, $p = p(e)$, and o is an instance of $c(n_2)$. We define $S(e) = \{s|(s, p, o) \in T(e)\}$ and $O(e) = \{o|(s, p, o) \in T(e)\}$. Figure 3 shows a set $T(e)$ of triple and sets $S(e)$ and $O(e)$ of instances in an RDF dataset for an edge e of a class graph.

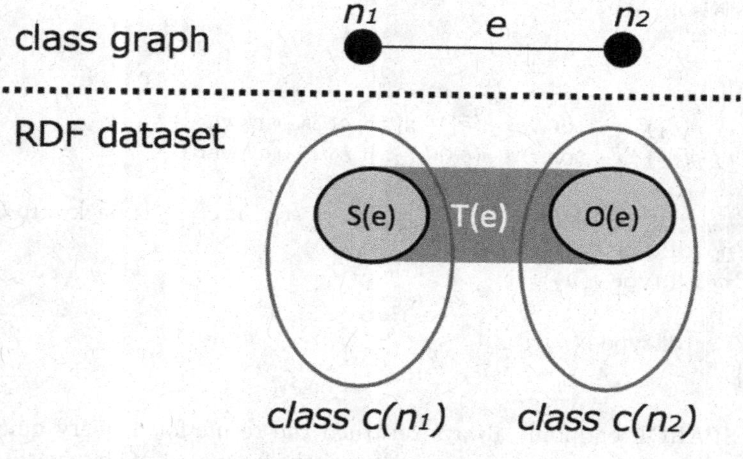

Fig. 3. Sets $T(e)$, $S(e)$, and $O(e)$ in an RDF dataset.

For a node n_i of a class graph, we denote by $|n_i|$ the number of instances in the class $c(n_i)$. Note that $|T(e)|$, $|S(e)|$, and $|O(e)|$ are described using the properties `void:triples`, `void:distinctSubject`, and `void:distinctObject`, respectively, from instances of `sbm:ClassRelation` located in `sbm:PropertyPartition` of SBM. Similarly, we can obtain $|n_i|$ for each n_i from SBM using `void:entities` from instances of `sbm:ClassPartition`.

For a class path $p = (n_0, e_1, n_1, \ldots, n_k)$, $I(p)$ is a set of sequences $(t_1, \ldots, t_k)$ of triples satisfying the following two conditions: (1) For $t_i = (s_i, p_i, o_i)$, if e_i is forward in p, then s_i is an instance of $c(n_{i-1})$, $p(e) = p_i$, and o_i is an instance of $c(n_i)$. If e_i is backward, then s_i is an instance of $c(n_i)$, $p(e) = p_i$, and o_i is an instance of $c(n_{i-1})$. (2) for two triples $t_{i-1} = (s_{i-1}, p_{i-1}, o_{i-1})$ and $t_i = (s_i, p_i, o_i)$, $r_{i-1} = r_i$ where $r_{i-1} = o_{i-1}$ if e_{i-1} is forward, $r_{i-1} = s_{i-1}$ if e_{i-1} is backward, $r_i = s_i$ if e_i is forward, and $r_i = o_i$ if e_i is backward. Note that for an empty path p, $I(p)$ is empty.

We now introduce our proposed measure Pr based on statistical values, such as $|S(e)|$ and $|O(e)|$ for edge e, included by SBM. To simplify the definition, we assume here that all the edges in $p = (n_0, e_1, n_1, \ldots, n_k)$ are forward. We then define $Pr(p) = (1 - (1 - |S(e_2)|/|n_1|)^{|O(e_1)|}) \times (1 - (1 - |S(e_3)|/|n_2|)^{|O(e_2)|}) \times \cdots \times (1 - (1 - |S(e_k)|/|n_{k-1}|)^{|O(e_{k-1})|})$. Note that if there is a backward edge e_i in p, $O(e_i)$ and $S(e_i)$ in the corresponding terms for e_i of the definition above must be changed to $S(e_i)$ and $O(e_i)$, respectively.

$Pr(p)$ is a very rough approximation of the probability of the existence of a triple sequence in $I(p)$, assuming that each instance in $S(e)$ and $O(e)$ is uniformly distributed independently of the other in instances of classes $c(n_1)$ and $c(n_2)$, respectively, for each edge $e = (n_1, n_2)$ appearing in p. Even though in reality, occurrences in instances between $c(n_1)$ and $c(n_2)$ might not be independent, because they are connected by triples, this assumption simplifies the definition of Pr. Under the assumption, if there are two edges e_i and e_{i+1} sharing one node n_i, for each instance in $O(e_i)$, the probability of a connecting instance in $S(e_{i+1})$ is $|S(e_{i+1})|/|n_i|$. Therefore, for all the instances in $O(e_i)$, because the probability of having no instances in $S(e_{i+1})$ is $(1 - |S(e_{i+1})|/|n_i|)^{|O(e_i)|}$, the probability of having at least one instance is $1 - (1 - |S(e_{i+1})|/|n_i|)^{|O(e_i)|}$. Because there are $k - 1$ such nodes in a path $p = (n_0, e_1, n_1, \ldots, n_k)$, Pr can be computed by multiplying the probabilities for the nodes.

Because $|O(e)|$, $|S(e)|$, and $|n_i|$ can be found in SBM as objects for the properties `void:distinctObjects`, `void:distinctSubjects`, void:entities, respectively, a class graph can hold these values with edges and nodes in a class graph when being constructed from SBM. Therefore, $Pr(p)$ can be computed in $O(A(n)B(m)k)$, where $A(n)$ is the look-up time of a node from n nodes, and $B(m)$ is the look-up time of an edge from m edges in a class graph. Because the length k of a path is always limited to four or five for the SPARQL Builder system, and $A(n)$ and $B(m)$ can be almost constant for n and m, respectively, through implementation using an efficient look-up structure such as a hash table, the computation time of $Pr(p)$ can be almost constant. Therefore, even for datasets with very large n and m, $Pr(p)$ can be computed in a short time using SBM.

4.2 Evaluation of the Measure Through Computational Experiment

To investigate the prediction performances of using the measure Pr, which is easily computable from SBM, we checked whether each class path p was an empty path, and we compared the results with the value of $Pr(p)$. We selected

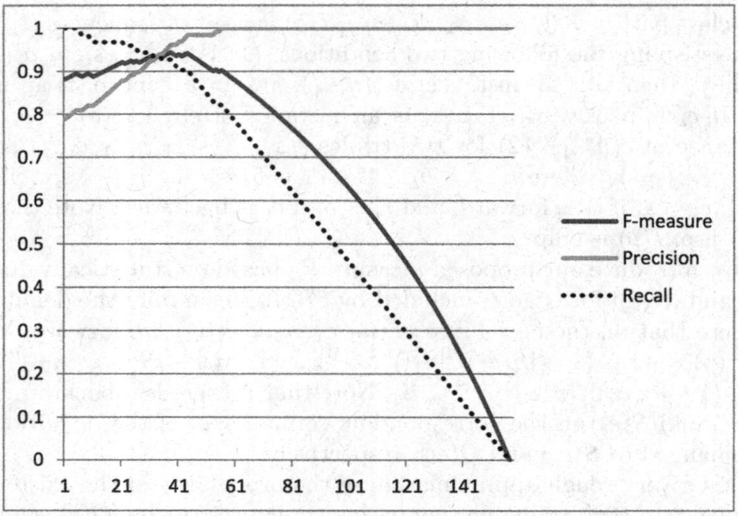

Fig. 4. The distribution of the F-measures with recalls and precisions for an obtained set of class paths by removing n paths of the smaller $Pr(p)$ for the Allie dataset.

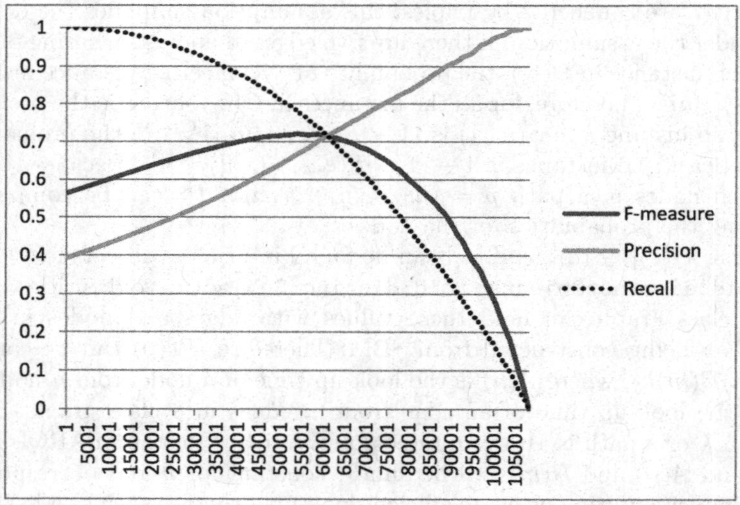

Fig. 5. The distribution of the F-measures with recalls and precisions for an obtained set of class paths by removing n paths of the smaller $Pr(p)$ for the Reactome dataset.

three datasets from Allie [14], Reactome from EBI RDF Platform [2], Affymetrix from Bio2RDF [3] as of April 2016 with various ratios, approximately 0.8, 0.4, and 0.25, respectively, of non-empty paths of all the paths between two classes.

We first processed the datasets to produce metadata written in SBM. We then set a maximum path length of four for the experiment. Then, we computed

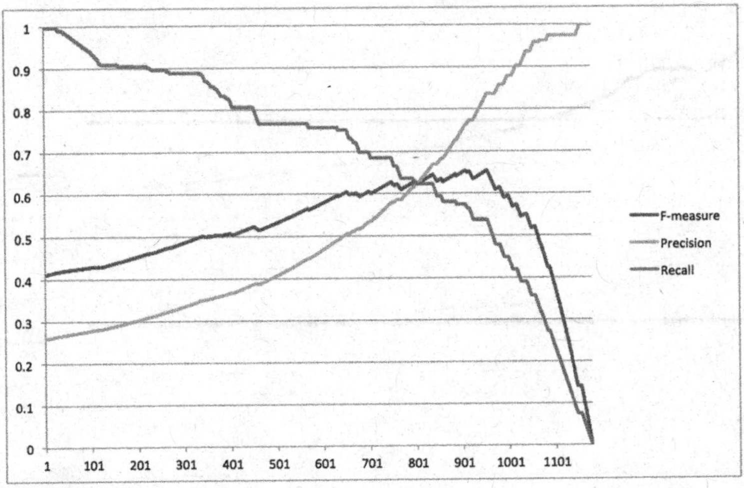

Fig. 6. The distribution of the F-measures with recalls and precisions for an obtained set of class paths by removing n paths of the smaller $Pr(p)$ for the Affymetrix dataset.

all of the class paths between every pair of classes using the algorithm proposed in [13] from the metadata previously computed for each dataset. For each class path p, we then computed $Pr(p)$ using the metadata.

To prepare exact positive and negative sets, i.e., to decide whether each class path $(n_1, e_1, n_2, e_2, \ldots, n_m)$ is an empty path, we downloaded the three datasets and checked all of the class paths with a maximum length of four to determine whether they were empty paths. Concretely, we uploaded these three datasets locally to the open source version of OpenLink Virtuoso 7.20.3215 on a CentOS 5.11 machine with 24 cores of Intel Xeon 2.53 GHz and accessed the triple store with Virtuoso Driver 4.0 for JDBC and Virtuoso Jena Driver 2.6.2 using the `ASK` SPARQL query described as an exact method in Subsect. 4.1.

To evaluate Pr as the measure for predicting an empty path, we aligned all of the paths in ascending order with respect to Pr, and computed the precision, recall, and F-measure of the performance when n paths of the smaller Pr are removed from the original set of paths. Figure 4 shows the distribution of the precisions, recalls, and F-measures for the Allie dataset. The x-axis corresponds to the number of removed paths. The precision for the number n of removed paths is the ratio of non-empty paths to all of the paths after the n paths from the smallest Pr are removed. Similarly, the recall for n is the ratio of non-empty paths to all of the non-empty paths after the n paths are removed. The best F-measure was 0.944 with precision 0.961, and recall 0.927 was found at the threshold 0.389 of Pr.

Similarly, Figs. 5 and 6 show the distribution of the precisions, racalls, and F-measures for the Reactome and Affymetrix datasets. For Reactome, the best F-measure was 0.721 with precision 0.667 and the recall 0.783 found at the

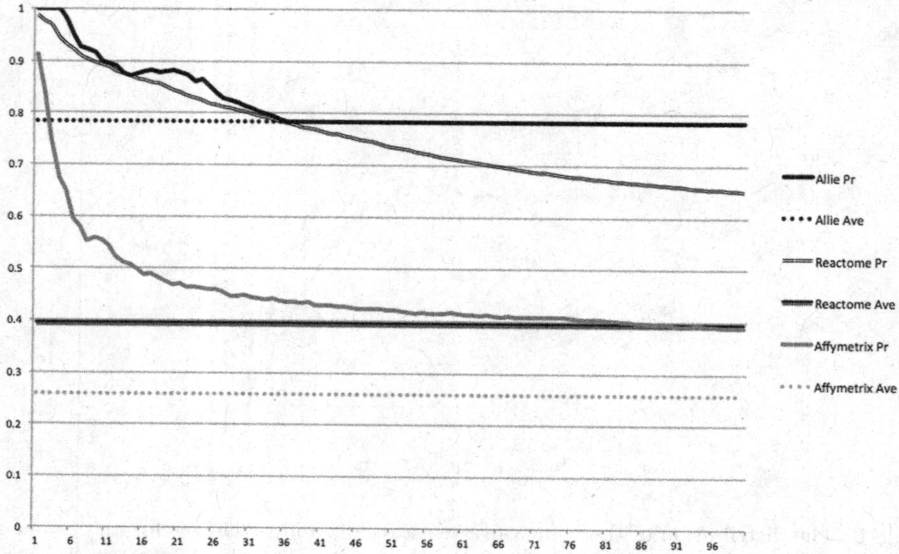

Fig. 7. Performance using Pr. Allie Pr, Reactome Pr and Affymetrix Pr show accuracy rates of the top n ($1 \leq n \leq 100$) paths using Pr, i.e., average ratio of non-empty paths among top n path for every pair of classes, for Allie, Reactome and Affymetrix datasets, respectively. Allie Ave, Reactome Ave and Affymetrix Ave show the average non-empty path ratio for all the paths for Allie, Reactome and Affymetrix, respectively.

threshold 0.000252 of Pr. For Affymetrix, the best F-measure was 0.654 with precision 0.836 and recall 0.538 found at the threshold 0.121 of Pr.

From a practical point of view, because systems such as SPARQL Builder can show only a small number, for example, ten, of class paths at one time as a result of limited screen size, it is important that the top n paths include as many non-empty paths as possible. Therefore, we plotted the average ratio of non-empty paths in the top n paths, sorted using the measure Pr. In other words, we plotted the precision for non-empty paths after all of the paths were removed except for the top n paths. Figure 7 shows a comparison between the average ratios of non-empty paths in the top n paths sorted by Pr and the average ratio of non-empty paths to all paths for Allie, Reactome and Affymetrix datasets. The x-axis corresponds to the number n of paths from the largest Pr. Because the average non-empty path ratios for all of the paths are 0.783, 0.393, and 0.25 for Allie, Reactome and Affymetrix datasets, respectively, we expect it to be easiest for Allie and most difficult for Affymetrix to obtain non-empty paths from all the class paths among the three datasets. For all three dataset, we can see from Fig. 7 that the average ratios of non-empty paths for the top n paths are always more than those for all the class paths. For example, the average ratios of non-empty paths for the top ten paths are 0.9, 0.893, and 0.555 for Allie, Reactome and Affymetrix datasets, respectively.

Especially for the Affymetrix dataset, although the average non-empty path ratio for all of the paths is approximately 0.25, the top five paths have the average accuracy 0.85. The average non-empty path ratio for all of the paths corresponds to the expected accuracy of selecting n paths randomly. The top n paths from one to 100 always obtain better accuracy than the averages for the three dataset. Therefore, we can claim that Pr is useful in removing empty paths from a set of class paths.

5 Discussion and Conclusion

In this study, we discussed a novel LOD exploring methodology and its application to enabling practical LOD data discovery from a SPARQL endpoint. We achieved data acquisition using class paths of interest to a user based on class–class relationships, with paths of traversing class–class relations being computed and used for two classes given by a user. Therefore, our approach requires some technologies including SBM specification, crawling for obtaining SBM, an efficient algorithm for finding class paths, and a measure for predicting empty paths proposed in this paper. These technologies are strongly related to one another in realizing our approach. For example, because an efficient algorithm finds many paths, including empty paths, we need a prediction measure for empty paths. Computing the prediction measure in nearly constant time requires metadata written in SBM designed to provide multiple statistical values in machine-readable form. Obtaining SBM requires an efficient crawler for SBM. Because we showed that the performance of our proposed measure Pr is practically useful, our approach is now ready to be realized as a system such as SPARQL Builder for analysis of LOD.

As for the measure for predicting empty paths based on an approximation of the existence probability using statistical values in SBM, the proposed measure were successfully demonstrated by our evaluation experiment, using three life-science datasets. For the experiment shown in Fig. 7, even though the ratio of the top n paths for Allie and Reactome datasets are very similar, they are very different from that for the Affymetrix dataset. The ratio of the top n paths for the Affymetrix dataset decreases quickly as n increases. This might have resulted from the lowness of the average ratio of non-empty class paths to all class paths for the Affymetrix dataset. However, we believe there to be another reason for the difference of the performance because the performances for Allie and Reactome are similar, even though the average ratios of non-empty class paths to all class paths are more different between Allie and Reactome datasets than between Reactome and Affymetrix. By analyzing datasets in regard to many features, the measure might be able to improve its performance. In addition, although this paper proposed only a measure for predicting empty paths, a measure for estimating the size of data obtained by a path, corresponding to the size of $I(p)$, might also be useful because data are strongly connected through p if $I(p)$ is large.

Our approach has been evaluated using life-science datasets. Although we believe that our approach does not depend on domain-specific techniques, we would like to confirm it by using datasets from another domain. In addition, our future work will include an improvement of a federated search across large-scale SPARQL endpoints by applying our methodology to reduce search space and enhance the effectiveness of data traversing for searches that have not been realized. By introducing a federated search system, SPARQL Builder can support queries using class paths between classes from different datasets.

Acknowledgments. This work was supported by JSPS KAKENHI Grant Number 25280081, 24120002 and the National Bioscience Database Center (NBDC) of the Japan Science and Technology Agency (JST).

References

1. Heath, T., Bizer, C.: Linked Data: Evolving the Web into a Global Data Space. Synthesis Lectures on the Semantic Web: Theory and Technology, 1st edn. 1: 1, 1–136. Morgan & Claypool (2011)
2. Jupp, S., Malone, J., Bolleman, J., Brandizi, M., Davies, M., Garcia, L., Gaulton, A., Gehant, S., Laibe, C., Redaschi, N., Wimalaratne, S.M., Martin, M., Le Novére, N., Parkinson, H., Birney, E., Jenkinson, A.M.: The EBI RDF platform: linked open data for the life sciences. Bioinformatics **30**(9), 1338–1339 (2014)
3. Belleau, F., Nolin, M.A., Tourigny, N., Rigault, P., Morissette, J.: Bio2RDF: towards a mashup to build bioinformatics knowledge systems. J. Biomed. Inf. **41**(5), 706–716 (2008)
4. Redaschi, N., UniProt Consortium: UniProt in RDF: tackling data integration and distributed annotation with the semantic web. Nat. Precedings (2009). doi:10.1038/npre.2009.3193.1
5. Fu, G., Batchelor, C., Dumontier, M., Hastings, J., Willighagen, E., Bolton, E.: PubChemRDF: towards the semantic annotation of PubChem compound and substance databases. J. Cheminformatics **7**(34) (2015). doi:10.1186/s13321-015-0084-4
6. Heim, P., Hellmann, S., Lehmann, J., Lohmann, S., Stegemann, T.: RelFinder: revealing relationships in RDF knowledge bases. In: Chua, T.-S., Kompatsiaris, Y., Mérialdo, B., Haas, W., Thallinger, G., Bailer, W. (eds.) SAMT 2009. LNCS, vol. 5887, pp. 182–187. Springer, Heidelberg (2009). doi:10.1007/978-3-642-10543-2_21
7. Popov, I.O., Schraefel, M.C., Hall, W., Shadbolt, N.: Connecting the dots: a multi-pivot approach to data exploration. In: Aroyo, L., Welty, C., Alani, H., Taylor, J., Bernstein, A., Kagal, L., Noy, N., Blomqvist, E. (eds.) ISWC 2011. LNCS, vol. 7031, pp. 553–568. Springer, Heidelberg (2011). doi:10.1007/978-3-642-25073-6_35
8. Ferré, S.: Sparklis: a SPARQL endpoint explorer for expressive question answering. In: Proceedings of the ISWC 2014 Posters & Demonstrations Track, CEUR Workshop Proceedings 1272, Riva del Garda, Italy (2014)
9. Oren, E., Delbru, R., Decker, S.: Extending faceted navigation for RDF data. In: Cruz, I., Decker, S., Allemang, D., Preist, C., Schwabe, D., Mika, P., Uschold, M., Aroyo, L.M. (eds.) ISWC 2006. LNCS, vol. 4273, pp. 559–572. Springer, Heidelberg (2006). doi:10.1007/11926078_40
10. Qu, Y., Ge, W., Cheng, G., Gao, Z.: Class association structure derived from linked objects. In: Proceedings of the Web Science Conference (WebSci 2009: Society On-Line), Athens, Greece (2009)

11. Yamaguchi, A., Kozaki, K., Lenz, K., Wu, H., Kobayashi, N.: An intelligent SPARQL query builder for exploration of various life-science databases. In: The 3rd International Workshop on Intelligent Exploration of Semantic Data (IESD 2014), CEUR Workshop Proceedings 1279, Riva del Garda, Italy (2014)
12. Villalon, P., Suárez-Figueroa, M.C., Gómez-Pérez, A.: A double classification of common pitfalls in ontologies. In: Workshop on Ontology Quality (OntoQual 2010), Lisbon, Portugal (2010)
13. Yamaguchi, A., Kozaki, K., Lenz, K., Wu, H., Yamamoto, Y., Kobayashi, N.: Efficiently finding paths between classes to build a SPARQL query for life-science databases. In: Qi, G., Kozaki, K., Pan, J.Z., Yu, S. (eds.) JIST 2015. LNCS, vol. 9544, pp. 321–330. Springer, Heidelberg (2016). doi:10.1007/978-3-319-31676-5_24
14. Yamamoto, Y., Yamaguchi, A., Bono, H., Takagi, T.: Allie: a database and a search service of abbreviations and long forms. Database (2011). doi:10.1093/database/bar013

Estimation of Spatio-Temporal Missing Data for Expanding Urban LOD

Shusaku Egami[1(✉)], Takahiro Kawamura[1,2], and Akihiko Ohsuga[1]

[1] Graduate School of Informatics and Engineering,
The University of Electro-Communications, Tokyo, Japan
egami.shusaku@ohsuga.lab.uec.ac.jp, takahiro.kawamura@jst.go.jp,
ohsuga@uec.ac.jp
[2] Japan Science and Technology Agency, Tokyo, Japan

Abstract. The illegal parking of bicycles has been an urban problem in Tokyo and other urban areas. We have sustainably built a Linked Open Data (LOD) relating to the illegal parking of bicycles (IPBLOD) to support the problem solving by raising social awareness. Also, we have estimated and complemented the temporally missing data to enrich the IPBLOD, which consisted of intermittent social-sensor data. However, there are also spatial missing data where a bicycle might be illegally parked, and it is necessary to estimate those data in order to expand the areas. Thus, we propose and evaluate a method for estimating spatially missing data. Specifically, we find stagnation points using computational fluid dynamics (CFD), and we filter the stagnation points based on popularity stakes that are calculated using Linked Data. As a result, a significant difference in between the baseline and our approach was represented using the chi-square test.

Keywords: Linked open data · Urban problem · Illegally parked bicycles

1 Introduction

The illegal parking of bicycles have been an urban problem in Tokyo and other urban areas since the number of bicycles owned in Japan is large. An increased awareness of health problems [1] and energy conservation [2] led to a 2.6-fold increase in bicycle ownership in Japan from 1970 to 2013. In addition to the insufficient availability of bicycle parking spaces and public transportations such as city buses, the inadequate public knowledge on bicycle parking laws, has contributed to this problem. Illegally parked bicycles obstruct vehicles, cause road accidents, encourage theft, and disfigure streets.

In order to address this problem, we believe it would be useful to publish the distribution of illegally parked bicycles as Linked Open Data (LOD). For example, it would serve to visualize illegally parked bicycles, suggest locations for optimal bicycle parking spaces, assist with the removal of illegally parked bicycles, and assist with the urban design. Thus, we built the illegally parked bicycle

Y.-F. Li et al. (Eds.): JIST 2016, LNCS 10055, pp. 152–167, 2016.
DOI: 10.1007/978-3-319-50112-3_12

LOD (IPBLOD) based on social data after designing LOD schema [3]. Furthermore, we estimated and complemented temporal missing data using Bayesian networks, since the temporal missing data of the social sensor data is inevitable. Therefore, IPBLOD became a temporally enriched LOD, and it became possible to suggest the efficient timing of removal of illegally parked bicycles by the city, through the visualization of time-series changes in the distribution of illegally parked bicycles.

However, there are not only temporal missing data, but also spatial missing data where bicycles might be illegally parked. It is necessary to complement the spatial missing data in order to apply IPBLOD to various urban areas. However, it is not satisfied merely by social sensors when collecting observation points of illegally parked bicycles.

In this paper, we propose the method for geographically expanding LOD by estimating spatial missing data. We thought that observation points of illegally parked bicycles have spatial or geographic features common such as road width and building density. Thus, we considered the flow of people as the fluid, and we estimated spatial missing data in such a way as to find stagnation points of the fluid. Specifically, we first simulated airflow in urban area using computational fluid dynamics (CFD) and found stagnation points using stagnation point patterns defined by us. Next, we collected POI information around each of the stagnation points and calculate popularity stakes of the POIs using DBpedia Japanese. Then, we filtered stagnation points if their sum of the popularity stakes of POIs is less than the threshold. We considered the filtered stagnation points as estimated data and added the data to IPBLOD separately from real data. Therefore, our contributions are the geographical expansion of IPBLOD, development of an approach for estimating spatial missing data using CFD and Linked Data, and evaluation of this approach. We aim to collect new accurate data related to estimated observation points from social sensors, by raising social awareness through the visualization of estimated observation points.

The remainder of this paper is organized as follows. In Sect. 2, correcting data, building IPBLOD, estimating temporally missing data, and visualization of IPBLD are presented. In Sect. 3, the approach for estimating spatial missing data using CFD and DBpedia Japanese is described. Also, we evaluate our results. In Sect. 4, related works of data collection and urban LOD are described. Finally, Sect. 5 concludes this paper with future works.

2 Illegally Parked Bicycle LOD

We have sustainably built IPBLOD and applied them to Tokyo and other several urban areas. Managing urban problem data *joining* multiple tables in (distributed) RDBs is troublesome from the aspect of data interoperability and maintenance, since the urban problem is closely related to multiple domains, such as government data, legal data, and social data as we already incorporated POIs and weather data in this application, and also those have different schemata. Thus, Linked Data is a suitable format as the data infrastructure of not only illegally parked bicycles, but also urban problems in general, since Linked Data can have advantages of flexible linkability and schema.

We divided our approach of sustainable LOD construction into the following five steps. Steps (2) to (5) are executed repeatedly as more input data become available.

1. Designing LOD schema
2. Collecting observation data and factor data
3. Building the LOD based on schema
4. Using Bayesian networks to estimate the missing number of illegally parked bicycles at each location
5. Visualizing illegally parked bicycles using LOD.

2.1 Building IPBLOD

First, we designed IPBLOD based on the result of extracting domain requirements from Web articles related to illegally parked bicycles. Figure 1 shows an overview of IPBLOD schema.

Next, we collected tweets containing location information, pictures, hashtags, and the number of illegally parked bicycles. Furthermore, we collected information on POI using Google Places API[1] and Foursquare API[2]. Also, we obtained bicycle parking information from websites of municipalities and in

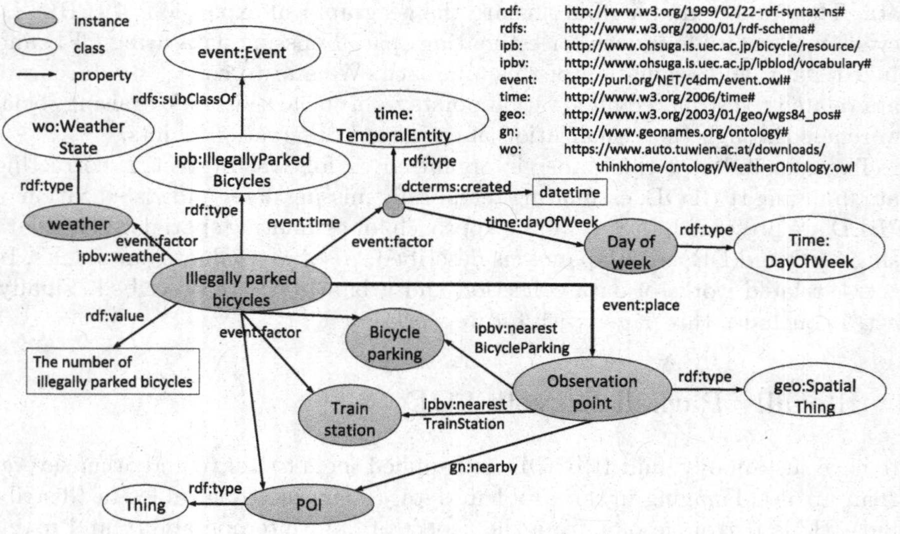

Fig. 1. LOD schema containing instances

[1] https://developers.google.com/places/?hl=en.

[2] https://developer.foursquare.com/.

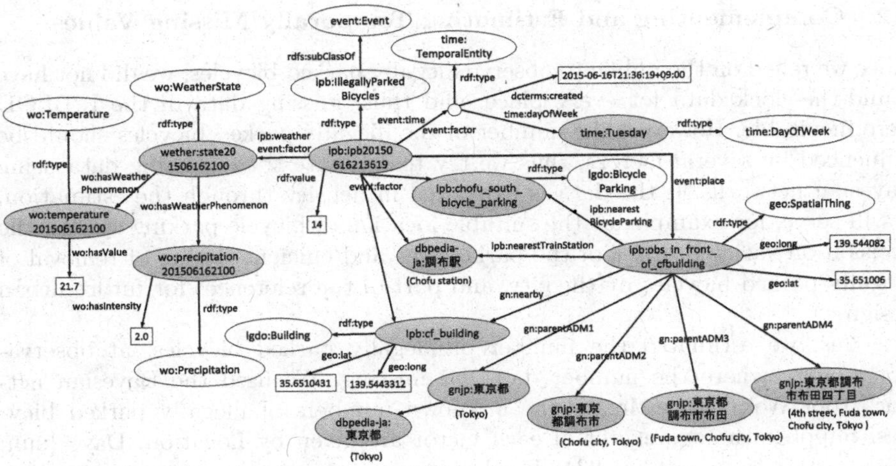

Fig. 2. Part of the integrated LOD

cooperation with the Bureau of General Affair of Tokyo[3]. The Bureau of General Affairs of Tokyo publishes Open Data on bicycle parking areas as CSV. The data contain names, latitudes, longitudes, addresses, capacities, and business hours. More information was collected from municipalities, for example, monthly parking fees and daily parking fees. Also, we retrieved weather information from the website of the Japanese Meteorological Agency (JMA)[4].

The collected data on illegally parked bicycles are converted to LOD based on the designed schema. First, the server program collects tweets containing the particular hash-tags, the location information, and the number of illegally parked bicycles in real time. The number of illegally parked bicycles is extracted from the text of tweets using regular expressions. Next, the server program checks whether there is an existing observation point within a radius of less than 30 m by querying our endpoint[5] using the SPARQL query. If there is no observation point on the IPBLOD, the point is added as a new observation point. In order to add new observation points, the nearest POI information is obtained using Google Places API and Foursquare API. The new observation point is generated based on the name of the nearest POI. It is possible to obtain the types of the POI from Google Places API and Foursquare API. We map the types of POI to classes in LinkedGeoData [14]. Thus, the POI is an instance of classes in LinkedGeoData. However, some POIs do not have a recognized types. Therefore, their types are decided by a keyword search with the name of the POI. Figure 2 shows part of the IPBLOD. The LOD are stored in Virtuoso[6] Open-Source Edition. Also, the RDF data set is published with CC-BY license on our website[7].

---

[3] http://www.soumu.metro.tokyo.jp/30english/index-en.htm.
[4] http://www.jma.go.jp/jma/indexe.html.
[5] http://www.ohsuga.is.uec.ac.jp/sparql.
[6] http://virtuoso.openlinksw.com/.
[7] http://www.ohsuga.is.uec.ac.jp/bicycle/dataset.html.

2.2 Complementing and Estimating Temporally Missing Values

Since we relied on the public to observe illegally parked bicycles, we did not have round-the-clock data for every place, and thus, missing data in the IPBLOD were inevitable. However, the number of the illegally parked bicycles should be influenced by several factors, thus we try to estimate these missing data using Bayesian networks. If the data is expanded in density through the estimation, it will serve, for example, as the suitable location of bicycle parking spaces, the decision on variable prices of the parking fee and efficient timing of removal of illegally parked bicycles by the city, and part of the references for future urban design.

Thus, we estimated the number of illegally parked bicycles, at observation points, where the number data are missing. We used the Bayesian network tool Weka[8] to estimate the unknown numbers of illegally parked bicycles. Suppose the aggregates of each factor are given by Location, Day={sun, mon,...,sat}, Hour={0,1,...,23}, Precipitation={0,1,...}, Temperature={...,-1,0, 1,...}, DailyFee ={0,1,...}, MonthlyFee={0,1,...}, Density={0,1,...}, Commuters={0,1,...}, POIs={0,1}, and Number (of illegally parked bicycles)={1,...,4}, then the observation data are stored as an aggregate O of vectors $o \in Location \times Day \times Hour \times Precipitation \times Temperature \times DailyFee \times MonthlyFee \times Density \times Commuters \times POIs \times Number$. In fact, POIs are divided to 46 elements, for example, restaurant, bar, supermarket, hospital, and school. The number of illegally parked bicycles is classified into four classes by Jenks natural breaks [13], which are often used in Geographic Information Systems (GISs). The range is 0 to 6, 7 to 17, 18 to 3, and 36 to 100. Therefore, the input data is a set O that consists of vectors with 56 elements, and the amount of the input data is 897. We used HillClimber as a search algorithm, and also used Markov blanket classifier. The maximum number of parent nodes was seven. The average estimation accuracy of ten times 10-fold cross validation became 70.9%, after random sampling with a 90 % rate.

Then, we examined the observation data in each observation point from the first observation date to the last observation date. If there are no data at 9 am or 9 pm, we estimated and complemented the number of illegally parked bicycles. Then, we added the estimated number and its probability to IPBLOD as follows.

```
@prefix ipb: <http://www.ohsuga.is.uec.ac.jp/ipblod/
    vocabulary#>
@prefix bicycle: <http://www.ohsuga.is.uec.ac.jp/bicycle/
    resource/>
bicycle:ipb_{observation point}_{datetime}
    ipb:estimatedValue [ rdf:value  "0-7" ;
        ipb:probability  "0.772"^^xsd:double ] .
```

[8] http://www.cs.waikato.ac.nz/ml/weka/.

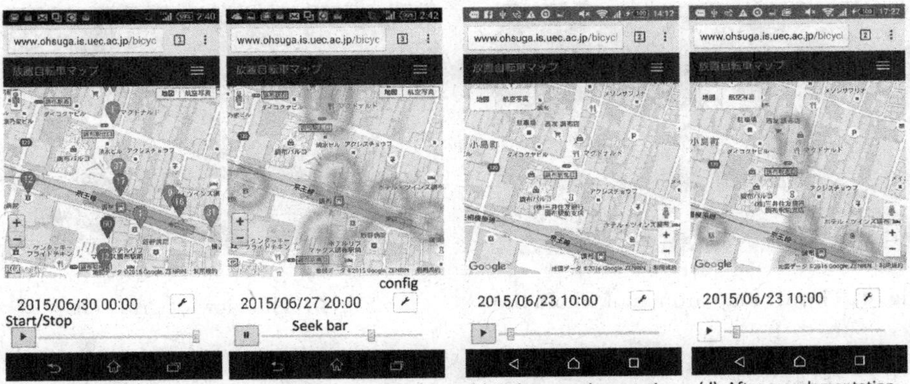

(a). Marker (b). Heatmap (c). Before complementation (d). After complementation

Fig. 3. Screenshots of the visualization application

2.3 Visualization of IPBLOD

Data visualization enables people to intuitively understand data contents. Thus, it can possibly raise the awareness of an issue among local residents. Furthermore, it is expected that we shall collect more urban data. In this section our visualization application of the IPBLOD is described.

As an example of the use of these data, we developed a Web application that visualizes illegally parked bicycles. The application can display time-series changes in the distribution of illegally parked bicycles on a map. Also, the application has a responsive design, so it is possible to use it on various devices such as PCs, smartphones, and tablets. When the start and end times are selected, and the play button is pressed, the time series changes of the distribution of the illegally parked bicycles are displayed. Figure 3(a) and (b) show screenshots of an Android smartphone, on which the Web application is displaying such an animation near Chofu Station in Tokyo using a heatmap and a marker UI.

The IPBLOD contain not only the data collected from Twitter, but also the data estimated by Bayesian networks. Therefore, time-series changes in the distribution of illegally parked bicycles become smoother than before estimating the missing values. Figure 3(c) and (d) show the comparison between the before and after complementation. The time-series changes after complementation are successive, whereas the time-series changes before complementation are intermittent.

3 Estimating Spatial Missing Data

There are spatio-temporal missing data in IPBLOD since the data is collected from social sensors. We estimated and complemented temporal missing data with 70.9% accuracy using Bayesian networks. However, spatial missing data

Fig. 4. The 3D map around of Chofu Sta.

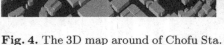

Fig. 5. The view of grid cells

(unobserved points where bicycles might be illegal parked) have not been complemented. In this study, we geographically expand IPBLOD by estimating and complementing the spatial missing data. We consider the flow of people to the fluid, and we find stagnation points of areas around train stations by airflow simulation using 3D maps and CFD. In Japan, since there are generally illegally parked bicycles around train stations, we selected those as our simulation areas. Then, we validate correlation of stagnation points and observation points of illegally parked bicycles. Furthermore, we filter stagnation points using DBpedia Japanese, and we regard these filtered points as new observation points. Therefore, in this section, we describe the hybrid approach using CFD and Linked Data for estimating spatial missing data.

3.1 Finding Stagnation Points Using CFD

There are wind tunnel test and CFD as the methods of airflow simulation. CFD is the method which observes the movements of fluid using computer simulation. A wind tunnel test requires expensive and large equipment, but CFD is easy to experiment with in different environments when using a computer. However, CFD can not produce exact copies of fluid movements, since CFD uses an approximate solution.

We first obtained maps of building from Geospatial Information Authority of Japan[9]. This data consists of 2D polygons. We converted the 2D maps to 3D maps using ArcGIS for Desktop[10]. Since we could not obtain information on the height of the buildings, we set the height of all buildings to 30 m. Figure 4 shows the 3D map around of Chofu Station in Tokyo. Red markers are observation points of illegally parked bicycles. We obtained observation points in a CSV format from the SPARQL endpoint of the IPBLOD, and then we imported the CSV to the 3D map. Next, we simulated the airflow around the station using Airflow Analyst[11], which is a simulation software run on ArcGIS. Figure 5 shows the grid cells which is set as the analysis range. We set the analysis range to

[9] http://www.gsi.go.jp/ENGLISH/index.html.
[10] http://www.esri.com/software/arcgis/arcgis-for-desktop.
[11] http://www.airflowanalyst.com/en/index.php.

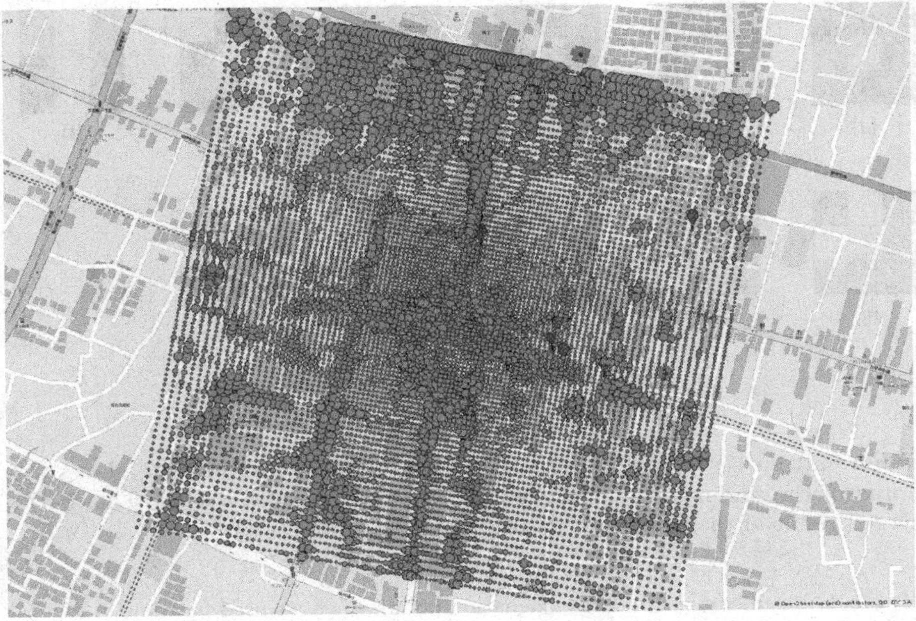

Fig. 6. The result of the airflow simulation

include all observation points around of Chofu Station. In Fig. 5, we selected 700 × 700 square meteres around Chofu Station as the analysis range. Also, the node spacing is 5 m, and the number of nodes is 10,000. The wind direction is set as being parallel to the road of the train station. Since it is considered that people come to the station from four directions in the case of Chofu Station, we simulated the airflow while changing the wind directions such as 11-degree, 109-degree, 190-degree, and 288-degree.

Figure 6 shows the visualization of the average wind velocity based on the results of the simulation, when the wind direction is 11-degree and the wind speed is 5 m/s. The size of the circle means the average wind velocity. We found stagnation points based on this numerical data. A stagnation point is a point where the velocity of the fluid is zero in the flow field. We tried to find stagnation points using patterns in Fig. 7. A black node is a node with average wind velocity $x > 0.1$. The white node is the node which is $x = 0$. The grey node is the node that $0 < x \leq 0.1$. In general, a stagnation point is a white node under these conditions. However, white nodes became buildings in our experiment. Therefore, we defined grey nodes as stagnation points. Table 1 shows the total accuracy of the findings of stagnation points around Chofu Station, Fuchu Station, and Shinjuku Station using the all patterns. The precision is the ratio of stagnation points within a 20-m radius from an observation point. The recall is the ratio of observation points that have stagnation points within a 20-m radius. As the result, the F-measure when we used the pattern (j) became the highest. Hence,

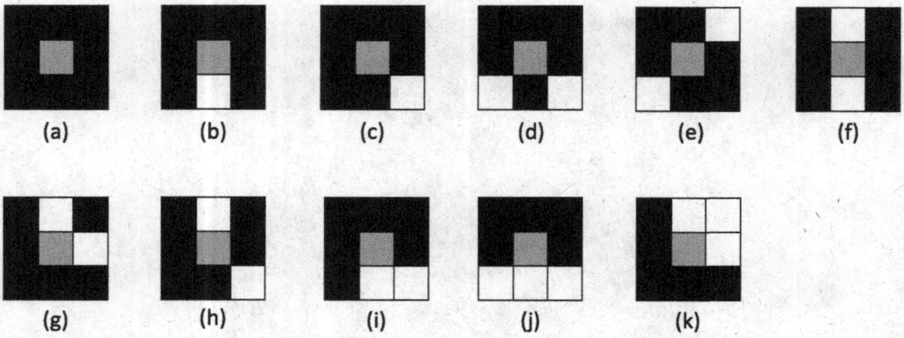

Fig. 7. Patterns of stagnation points

we use pattern (j) to find stagnation points in this study. Figure 8 shows the results of the findings stagnation points around of Chofu Station. This is the merged result of the simulation results of the four directions.

Table 1. Results of the findings stagnation points when we used patterns in Fig. 7

Pattern	Precision	Recall	F-measure
(a)	0.102	0.286	0.150
(b), (c)	0.0833	0.0357	0.0500
(d), (e), (f), (g), (h), (i)	0.000	0.000	0.000
(j)	0.0913	0.429	0.151
(k)	0.0746	0.107	0.0880

3.2 Filtering Stagnation Points Using DBpedia Japanese

We found the stagnation points, but, there were many noise points as can be seen in Fig. 8. We assumed that bicycles tend to be parked illegally at stagnation points having nearby POIs, whose popularity stakes are high. Therefore, we calculated the popularity stakes of the POIs around of the stagnation points and then filtered the stagnation points using the popularity stakes.

We first obtained the POIs information within a 20-m radius from the stagnation points using Google Places API. Then, we calculated the number of links from person resources to POIs on DBpedia Japanese. Also, we mapped the types of POIs to DBpedia Japanese resources. We considered the number of inbound links from person resources as the popularity stakes, and we obtained the number of links from instances of `foaf:Person` to types of POIs. Then, we calculated the sum of the popularity stakes of POIs, and we filtered stagnation points if the sum of the popularity stakes is less the threshold. We varied the threshold from

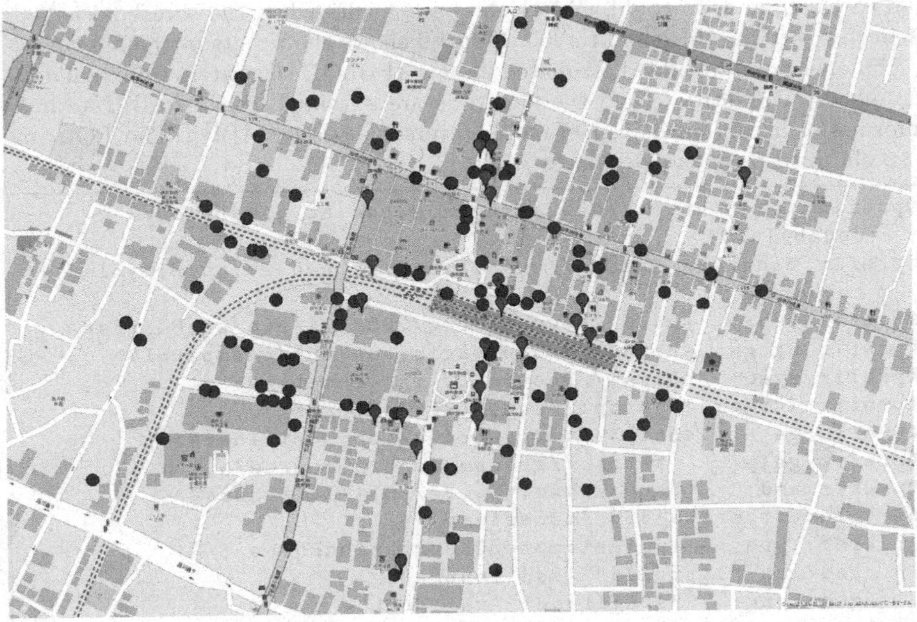

Fig. 8. The stagnation points around of Chofu station

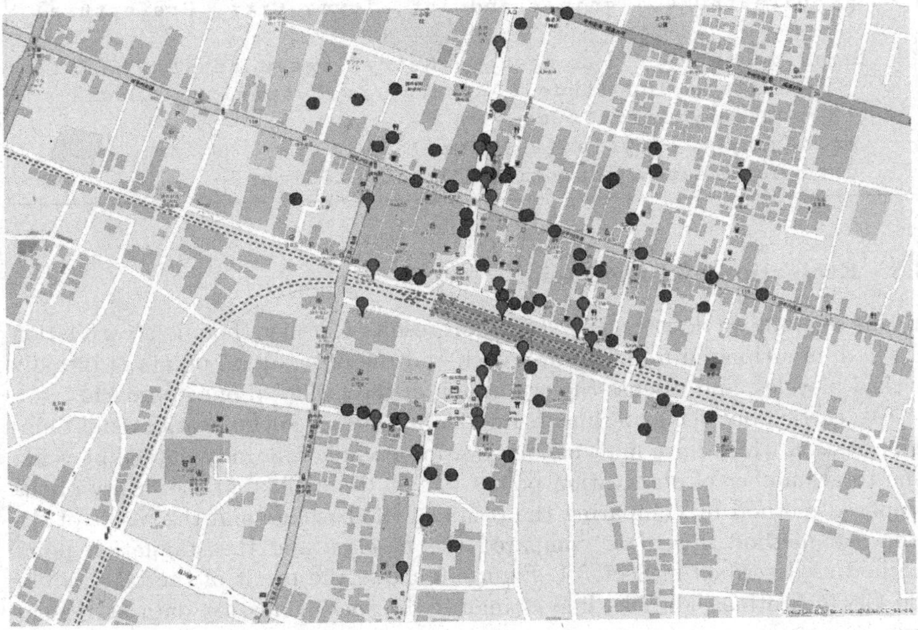

Fig. 9. The filtered stagnation points around of Chofu station

100 to 1,000, and we could achieve the best results when the threshold is 200. Hence, we set the threshold to 200. Figure 9 shows the results of the filtering.

Furthermore, we added estimated data to IPBLOD separately from real data as follows. The latitude and the longitude are obtained from ArcGIS. Address information is obtained from Yahoo! Reverse Geocoder API[12]. The POIs are also obtained from Google Places API.

```
@prefix ipb: <http://www.ohsuga.is.uec.ac.jp/ipblod/
    vocabulary#>
@prefix bicycle: <http://www.ohsuga.is.uec.ac.jp/bicycle/
    resource/>
@prefix geo:   <http://www.w3.org/2003/01/geo/wgs84_pos#> .
@prefix ogcgs: <http://www.opengis.net/ont/geosparql#> .
@prefix ngeo:  <http://geovocab.org/geometry#> .
@prefix dcterms: <http://purl.org/dc/terms/> .
@prefix gn:    <http://www.geonames.org/ontology#> .
@prefix gnjp:    <http://geonames.jp/resource/> .
@prefix xsd:   <http://www.w3.org/2001/XMLSchema#> .
bicycle:estimated_obs_{timestamp}
    rdf:type    ipb:EstimatedObservationPoint ;
    geo:lat "latitude"^^xsd:double ;
    geo:long    "longitude"^^xsd:double ;
    gn:parentADM     gnjp:{Prefecture}
    gn:parentADM2    gnjp:{City, Prefecture} ;
    gn:parentADM3    gnjp:{Town, City, Prefecture} ;
    gn:parentADM4    gnjp:{Land lot, Town, City, Prefecture}
        ;
    ngeo:geometry    [ a ngeo:Geometry; ogcgs:asWKT  "POINT(
        latitude,longitude)"^^<http://www.openlinksw.com/
        schemas/virtrdf#Geometry> . ] ;
    gn:nearby   bicycle:{POI name} ;
    dcterms:created "datetime"^^xsd:dateTime .
```

3.3 Evaluation and Discussion

In this section, we describe the validation results whether there is a correlation of the data estimated from our approach and the observation points of illegally parked bicycles, and discuss the evaluation of the utility of our approach.

We carried out the experiments on Chofu Station, Fuchu Station, and Shinjuku Station which have multiple observation points of illegally parked bicycles. The total number of observation points was 56. We validated the utility of the proposed method by comparing the result of the baseline and the result of the proposed method. First, we compared the baseline and the stagnation point method as described in Sect. 3.1. Figure 10 shows the result of the baseline for the Chofu Station. The baseline estimates the spatial missing data at regular

[12] http://developer.yahoo.co.jp/webapi/map/openlocalplatform/v1/reversegeocoder.html.

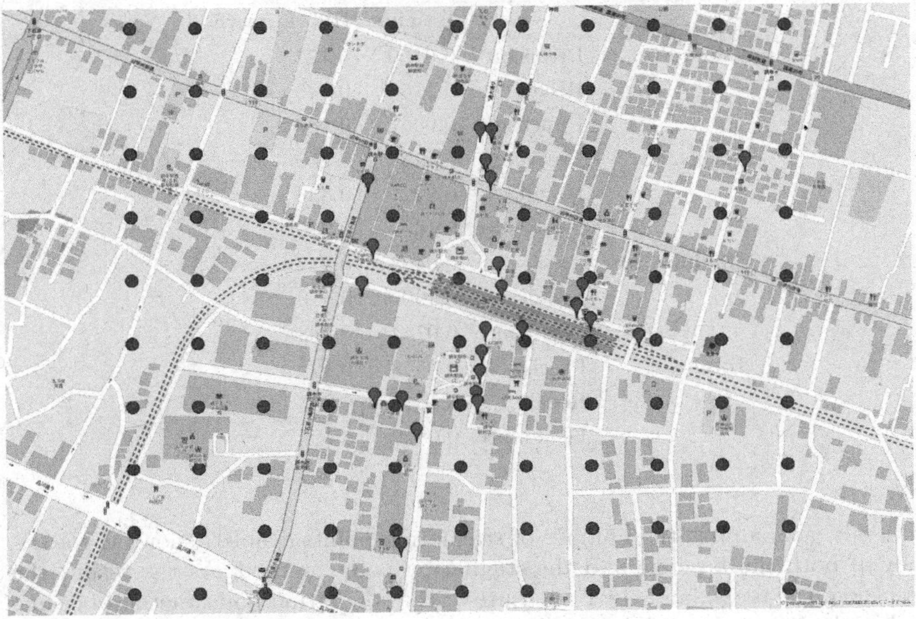

Fig. 10. The stagnation points of baseline

Table 2. Evaluation results of both baseline and stagnation point method

		Baseline	Stagnation point method
Chofu Sta.	Precision	0.0496	0.0726
	Recall	0.231	0.423
	F-measure	0.0816	0.152
Fuchu Sta.	Precision	0.125	0.188
	Recall	0.222	0.333
	F-measure	0.160	0.240
Shinjuku Sta.	Precision	0.0493	0.100
	Recall	0.190	0.476
	F-measure	0.0784	0.165
Total	Precision	0.0550	0.0913
	Recall	0.214	0.429
	F-measure	0.0876	0.151

Table 3. Evaluation results of both baseline and hybrid method

		Baseline	Hybrid method
Chofu Sta.	Precision	0.0469	0.121
	Recall	0.115	0.346
	F-measure	0.0667	0.180
Fuchu Sta.	Precision	0.125	0.250
	Recall	0.222	0.333
	F-measure	0.160	0.286
Shinjuku Sta.	Precision	0.0493	0.117
	Recall	0.190	0.476
	F-measure	0.0784	0.188
Total	Precision	0.0559	0.129
	Recall	0.161	0.393
	F-measure	0.0829	0.194

intervals, as many as the number of stagnation points. Table 2 shows the accuracy of both the baseline and the stagnation point method. As the result, the precision, the recall, and the F-measure of the stagnation point method became higher than the result of the baseline. Also, we validated the utility of the stagnation point method using the chi-square test. The null hypothesis is that there is no difference between the result (recall or precision) of the baseline and the result (recall or precision) of the stagnation point method, and we used a standard level of significance $p < 0.05$. As the result, the p-value of precision was 0.01591, and the p-value of recall was 2.244e-06. Hence, we found that there is a significant difference between the result of baseline and the result of the stagnation point method.

Next, we compared the baseline and the hybrid method (filtering stagnation points). Table 3 shows the accuracy of both the baseline and the hybrid method. As the result, the precision, the recall, and the F-measure of the hybrid method became higher than the result of the baseline. Also, as it is possible to see from Tables 2 and 3, the precision and the F-measure of the hybrid method became higher than the result of the stagnation point method. Therefore, there is the utility of the hybrid method using POIs and DBpedia Japanese. Also, we validated the utility of the hybrid method using the chi-square test. The null hypothesis is that there is no difference between the result of the baseline and the result of the hybrid method, and we used a standard level of significance $p < 0.05$. As the result, the p-value of precision was 7.393e-05, and the p-value of recall was 2.244e-06. Hence, we found that there is a significant difference between the result of the baseline and the result of the hybrid method. Although the accuracy is not high, the data of estimated points are considered to help to collect new data from social sensors.

The accuracy of the estimated data in this study was low for the following reasons. The number of observation points was less. There is a possibility that new observation points are found around the estimated points. Therefore, we should conduct a field survey in the future work; then we should evaluate our approach once again. Also, we could not exclude the stagnation points of the interior of the premises, since we could not obtain the data of building premises. The noise points were caused by this reason.

4 Related Work

In most cases, LOD sets have been built based on existing databases. However, there is little LOD available so far that describes urban problems. Thus, methods for collecting new data to build urban problem LOD are required. Data collection methods for building Open Data include crowdsourcing and gamification. A number of projects have employed these techniques. OpenStreetMap [5] is a project that creates an open map using crowdsourced data. Anyone can edit the map, and the data are published as Open Data. Similarly, FixMyStreet [6] is a platform for reporting regional problems such as road conditions and illegal dumping. Crowdsourcing to collect information in FixMyStreet has meant that regional problems are able to be solved more quickly than ever before. Zook et al. [7] reported a case, where crowdsourcing was used to link published satellite images with OpenStreetMap after the Haitian earthquake. A map of the relief efforts was created, and the data were published as Open Data. Celino et al. [9] proposed an approach for editing and adding Linked Data using a game with a purpose (GWAP) [8] and human computation. However, since the data relating to illegally parked bicycles are time-series data, it is difficult to collect data using these approaches. Therefore, we proposed a method to build IPBLOD while complementing the spatio-temporal missing data.

Also, there have been studies about building Linked Data for cities. Lopez et al. [10] proposed a platform that publishes sensor data as Linked Data. The platform collects streamed data from sensors and publishes Resource Description Framework (RDF) data in real time using IBM InfoSphere Stream and C-SPARQL [11]. The system is used in Dublinked2[13], which is a data portal of Dublin, Ireland, that publishes information about bus routes, delays, and congestion which is updated every 20 s. However, since embedding sensors are costly, this approach is not suitable for our study.

Furthermore, Bischof et al. [4] proposed a method for the collection, complementation, and republishing of data as Linked Data, as with our study. This method collects data from DBpedia [12], Urban Audit[14], United Nations Statistics Division (UNSD)[15], and U.S. Census[16] and then utilizes the similarity among such large Open Data sets on the Web. However, we could not find the

[13] http://www.dublinked.ie/.
[14] http://ec.europa.eu/eurostat/web/cities.
[15] http://unstats.un.org/unsd/default.htm.
[16] http://www.census.gov/.

corresponding data sets and thus could not apply the same approach to our study. Therefore, we estimated temporal missing data using Bayesian network, and we estimated spatial missing data using CFD and DBpedia Japanese.

In other areas, Bogárdi-Mészöly et al. [15] proposed a method for the detection of scenic leaves and blossoms viewing places. The proposed system collects images from Flickr[17], and then the system ranks scenic leaves and blossoms viewing places based on social features and image features. However, since we do have not enough amounts of the observation point's data and their images, this method is not suitable for our study. Furthermore, Hirota et al. [16] proposed a method for estimating missing metadata of images based on the image similarity, the photo-taking condition similarity, and the tag similarity. This method assumes that there is a sufficient amount of data for the estimation as well as Bogárdi-Mészöly et al.

5 Conclusion and Future Work

In this paper, we presented building IPBLOD while complementing temporal missing data, and we described geographically expansion of IPBLOD by estimating the spatial missing data. The mainly technical contribution is the proposal of a hybrid method using CFD and DBpedia Japanese for estimating the spatial missing data in LOD. Also, we evaluated our method using indicators such as precision, recall, and F-measure. Furthermore, we validated the utility of our method using the chi-square test. We expect that it will increase the social awareness of local residents regarding the illegally parked bicycle problem and encourage them to post more data over a wide area, through the visualization of estimated spatial data (new observation points).

In the future, we will estimate spatial missing data in more urban areas, and we will check true-false results to go to estimated points. Also, we will incorporate a new method such as machine learning to solve the problem that was described in Sect. 3.3. Furthermore, we will visualize estimated observation points and will design incentive for social sensors (workers of crowdsourcing), in order to collect more data related to illegally parked bicycles.

Acknowledgments. This work was supported by JSPS KAKENHI Grant Numbers 16K12411, 16K00419, 16K12533.

References

1. Nishi, N.: The 2nd Health Japan 21: goals and challenges. J. Fed. Am. Soc. Exp. Biol. 28(1), 632.19 (2014)
2. Ministry of Internal Affairs, Communications: Current bicycle usage and bicycle-related accident. http://www.soumu.go.jp/main_content/000354710.pdf. Accessed 10 September 2015 (Japanese)

[17] https://www.flickr.com/.

3. Egami, S., Kawamura, T., Ohsuga, A.: Building urban LOD for solving illegally parked bicycles in Tokyo. In: Groth, P., Simperl, E., Gray, A., Sabou, M., Krötzsch, M., Lecue, F., Flöck, F., Gil, Y. (eds.) ISWC 2016. LNCS, vol. 9982, pp. 291–307. Springer, Heidelberg (2016). doi:10.1007/978-3-319-46547-0_28
4. Bischof, S., Martin, C., Polleres, A., Schneider, P.: Collecting, integrating, enriching and republishing open city data as linked data. In: Arenas, M., Corcho, O., Simperl, E., Strohmaier, M., d'Aquin, M., Srinivas, K., Groth, P., Dumontier, M., Heflin, J., Thirunarayan, K., Staab, S. (eds.) ISWC 2015. LNCS, vol. 9367, pp. 57–75. Springer, Heidelberg (2015). doi:10.1007/978-3-319-25010-6_4
5. Haklay, M., Weber, P.: Openstreetmap: User-generated street maps. IEEE Pervasive Comput. **7**(4), 12–18 (2008)
6. King, S.F., Brown, P.: Fix my street or else: using the internet to voice local public service concerns. In: Proceedings of the 1st International Conference on Theory and Practice of Electronic Governance, pp. 72–80 (2007)
7. Zook, M., Graham, M., Shelton, T., Gorman, S.: Volunteered geographic information and crowdsourcing disaster relief: a case study of the Haitian earthquake. World Med. Health Policy **2**(2), 7–33 (2010)
8. von Ahn, L.: Games with a purpose. IEEE Comput. **39**(6), 92–94 (2006)
9. Celino, I., Cerizza, D., Contessa, S., Corubolo, M., Dell' Aglio, D., Valle, E.D., Fumeo, S., Piccinini, F.: Urbanopoly: collection and quality assesment of geo-spatial linked data via a human computation game. In: Proceedings of the 10th Semantic Web Challenge (2012)
10. Lopez, V., Kotoulas, S., Sbodio, M.L., Stephenson, M., Gkoulalas-Divanis, A., Aonghusa, P.M.: QuerioCity: A linked data platform for urban information management. In: Cudré-Mauroux, P., Heflin, J., Sirin, E., Tudorache, T., Euzenat, J., Hauswirth, M., Parreira, J.X., Hendler, J., Schreiber, G., Bernstein, A., Blomqvist, E. (eds.) ISWC 2012. LNCS, vol. 7650, pp. 148–163. Springer, Heidelberg (2012). doi:10.1007/978-3-642-35173-0_10
11. Barbieri, D.F., Ceri, S.: C-SPARQL: SPARQL for continuous querying. In: Proceedings of the 18th International Conference on World Wide Web, pp. 1061–1062 (2012)
12. Auer, S., Bizer, C., Kobilarov, G., Lehmann, J., Cyganiak, R., Ives, Z.: DBpedia: A nucleus for a web of open data. In: Aberer, K., Choi, K.-S., Noy, N., Allemang, D., Lee, K.-I., Nixon, L., Golbeck, J., Mika, P., Maynard, D., Mizoguchi, R., Schreiber, G., Cudré-Mauroux, P. (eds.) ASWC/ISWC -2007. LNCS, vol. 4825, pp. 722–735. Springer, Heidelberg (2007). doi:10.1007/978-3-540-76298-0_52
13. BlackJenks, G.F.: The data model concept in statistical mapping. Int. Yearb. Cartography **7**(1), 186–190 (1967)
14. Stadler, C., Lehmann, J., Höffner, K., Auer, S.: LinkedGeoData: A core for a web of spatialopen data. Semant. Web J. **3**(4), 333–354 (2012)
15. Bogárdi-Mészöly, A., Rövid, A., Yokoyama, S.: Detect scenic leaves and blossoms viewing places from flickr based on social and image features. In: Proceedings of the 5th IIAI International Congress on Advanced Applied Informatics, pp. 1162–1167 (2016)
16. Hirota, M., Fukuta, N., Yokoyama, S., Ishikawa, H.: A robust clustering method for missing metadata in image search results. J. Inf. Process. **20**(3), 537–547 (2012)

RDF and Query

ASPG: Generating OLAP Queries for SPARQL Benchmarking

Xin Wang[1(✉)], Steffen Staab[1,2], and Thanassis Tiropanis[1]

[1] Web and Internet Science Group, University of Southampton, Southampton, UK
xwang@soton.ac.uk

[2] Institute for Web Science and Technology, University of Koblenz-Landau, Mainz, Germany

Abstract. The increasing use of data analytics on Linked Data leads to the requirement for SPARQL engines to efficiently execute Online Analytical Processing (OLAP) queries. While SPARQL 1.1 provides basic constructs, further development on optimising OLAP queries lacks benchmarks that mimic the data distributions found in Link Data. Existing work on OLAP benchmarking for SPARQL has usually adopted queries and data from relational databases, which may not well represent Linked Data. We propose an approach that maps typical OLAP operations to SPARQL and a tool named ASPG to automatically generate OLAP queries from real-world Linked Data. We evaluate ASPG by constructing a benchmark called DBOBfrom the online DBpedia endpoint, and use DBOB to measure the performance of the Virtuoso engine.

Keywords: OLAP · Linked data · Benchmarking · Query generation · SPARQL · DBpedia

1 Introduction

Linked Data principles foster the provisioning and integration of a large amount of heterogeneous distributed datasets [2]. SPARQL 1.1 [11] has introduced aggregations that enable users to do basic analytics. Though limited, SPARQL 1.1 is expressive enough to implement Online Analytical Processing (OLAP) which is an approach to analysing and reporting multidimensional statistics from different perspectives and levels of granularity [3,5].

OLAP contains a rich set of combinations of analytical operations which generate a high workload on SPARQL engines that target the support of analytics queries. In fact, the scalability of SPARQL engines to execute OLAP queries is still rather limited owing further development and optimization. Such optimization and comparison of best developments critically depend on benchmarks that can measure the performance of SPARQL engines on analytic tasks from various perspectives. Several OLAP benchmarks for SPARQL have been proposed. For example Kämpgen and Harth [12] convert queries and data from the Star Schema Benchmark (SSB) to SPARQL and Linked Data using the RDF

Y.-F. Li et al. (Eds.): JIST 2016, LNCS 10055, pp. 171–185, 2016.
DOI: 10.1007/978-3-319-50112-3_13

Data Cube Vocabulary [6]. Since SSB is based on a relational database scenario, its data do not necessarily resemble common Linked Data structures. Another example, the BowlognaBench [8], uses data and queries based on the Bowlogna Ontology [7]. Similar to SSB for SPARQL, BowlognaBench covers a specific scenario which may not represent the heterogeneity and structure of Linked Data.

Görlitz et al. [9] propose a SPARQL query generator called SPLODGE to release benchmarks from pre-defined queries. Following the same direction we present a tool called Analytical SPARQL Generator (ASPG) that generates OLAP queries in SPARQL which can be used to construct benchmarks. ASPG takes an RDF graph as input and selects triples by semi-random walk. Selected triples are parametrised to generate basic graph patterns (BGPs) which are then extended with aggregations that resemble OLAP operations. Queries produced by ASPG are guaranteed to return results from the given RDF graph since they are parametrised from triples in the RDF graph. We construct an analytical benchmark based on DBpedia, referred to as DBpedia OLAP Benchmark (DBOB), using ASPG generated queries. We evaluate Virtuoso[1] using DBOB and present the results.

The remaining sections of this paper are organised as follows: technical details of ASPG are described in Sect. 2; queries and dataset of DBOB are presented in Sect. 3; experiment settings and evaluation result of DBOB are given in Sect. 4, and conclusions are given in Sect. 6. Due to page limit, a complete list of DBOB queries is given in http://xgfd.github.io/ASPG/.

2 Generating OLAP Queries from Linked Data

In this section we discuss the correspondence between typical OLAP operations and SPARQL components, and provide details of generating SPARQL queries from arbitrary RDF graphs that resemble typical OLAP operations.

SPARQL queries consist of basic graph patterns (BGPs) which can be viewed as graphs with variable nodes. When evaluating a BGP against a RDF graph, results are returned if and only if the BGP matches a sub-graph of the given RDF. Consequently, given an arbitrary RDF graph, we can construct BGPs that are guaranteed to return results by parametrising sub-graphs in the RDF. By controlling the structure of sub-graphs we can obtain BGPs that consist of chains or star-shaped triple patterns of arbitrary lengths. We simulate typical OLAP operations by summarising along properties (using GROUP BY) with randomly selected aggregate operations (e.g. SUM, COUNT, AVG etc.). In particular we discuss the challenges to generate queries from RDF graphs that are too large to fit in a single store and describe RDF summarising and sampling techniques to resolve those issues.

2.1 Background of OLAP Operations

OLAP queries operate on a multidimensional data model that is referred to as an OLAP cube. Each data point in the cube is associated with two types of

[1] http://virtuoso.openlinksw.com/.

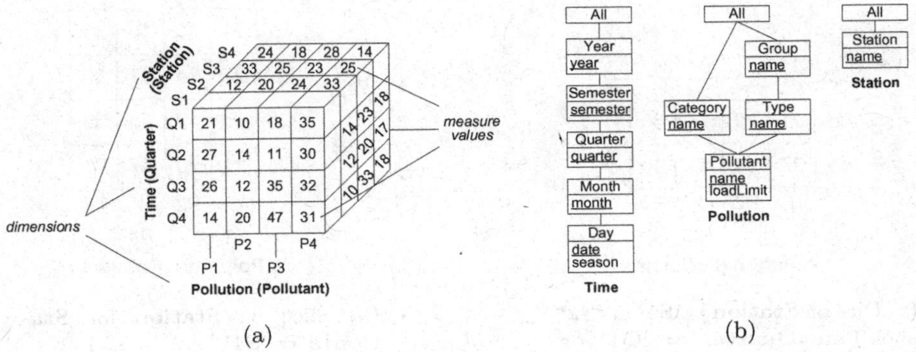

Fig. 1. A three-dimensional cube having dimensions **Time**, **Pollution**, and **Station**, and a measure **concentration**. Dimension hierarchies are shown on the right [4].

attributes, dimensions and measures. Dimensions identify data points and are usually organised hierarchically. Measures represents associated values of a data point and are usually operands of aggregations. An example of OLAP cube is shown in Fig. 1. There is no clear distinction between dimensions and measures. Any set of attributes that uniquely identifies a data point can be viewed as dimensions, and the remaining attributes are measures.

Typical OLAP operations defined on cubes include:

- Dice: Selecting a subset of an OLAP cube (Fig. 2a).
- Slice: Slice is a specific case of dice picking a rectangular subset of a cube by choosing a single value for one of its dimensions (Fig. 2b).
- Roll-up: Aggregating data by climbing up the hierarchy of a dimension (Fig. 2c).
- Drill-down: Aggregating data at a lower level of the dimension hierarchy (Fig. 2d). Drill-down is the reverse operation of roll-up.

In this paper we do not take into account operations that involve multiple OLAP cubes, such as drill-across [4], since multiple RDF graphs can be merged into one graph by taking their union.

Kämpgen et al. [13] describe an approach to map OLAP queries into SPARQL queries with the RDF Data Cube (QB) vocabulary [6]. Since many Linked Data and SPARQL queries do not use QB, we examine the semantics of the above OLAP operations and propose a mapping between OLAP and SPARQL queries that are not limited to specific vocabularies.

2.2 Generating Dice and Slice Queries in SPARQL

Dice and Slice select a subset of an OLAP cube while in SPARQL the same functionality is achieved by BGPs.

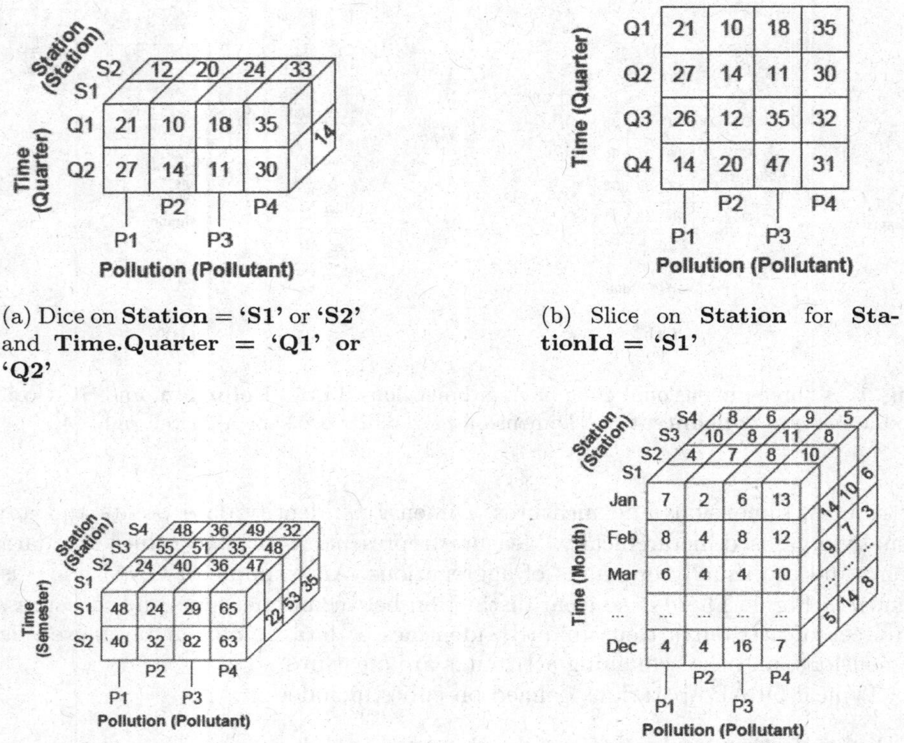

(a) Dice on **Station** = **'S1'** or **'S2'** and **Time.Quarter** = **'Q1'** or **'Q2'**

(b) Slice on **Station** for **StationId** = **'S1'**

(c) Roll-up to the **Semester** level

(d) Drill-down to the **Month** level

Fig. 2. OLAP operations [4].

An OLAP data point and its attributes (dimensions and measures) correspond to a subject and its properties in a RDF graph[2]. Dice selects multiple data points in an OLAP cube, whereas in SPARQL it is analogous to a BGP with optional constraints on object values (using FILTER), as shown below:

Query 1.

```
SELECT ?P ?Q ?S ?concentration
WHERE
{ ?point  :Pollution     ?P ; # FILTER(?P ="P1"|| ?P ="P2")
          :Time          ?Q ; # FILTER(?Q ="Q1"|| ?Q ="Q2")
          :Station       ?S ; # FILTER(?S ="S1"|| ?S ="S2")
          :concentration :?concentration
}
```

[2] Mapping an OLAP data point to a subject is just one intuitive approach. An OLAP data point can be mapped to any RDF term.

Unlike in relational databases (where dimensions are usually keys that are distinguished from measures), we argue that dimensions and measures are indistinguishable in RDF and SPARQL. Thus any BGPs correspond to a valid Dice operation (with Slice as a special case) in OLAP. There may be difficulties to aggregate on certain types of values, since most aggregations in SPARQL are arithmetic. Meanwhile it is always possible to convert an arbitrary type to a numeric. For example a literal can be converted to its length (i.e. STRLEN), and a resource can be converted to its number of occurrences (i.e. COUNT). Queries generated with the above modifications may not be meaningful in a practical sense, they serve the purpose as far as benchmarks are concerned. In the rest of this paper we interchangeably use Dice query and BGP when no confusion is caused.

2.3 Generating Roll-Up and Drill-Down Queries in SPARQL

Roll-up and drill-down group measure values at a specific dimension level and aggregate values in each group using a given aggregate function. Without losing generality, we focus on the mapping of roll-up since drill-down is the reverse operation. In SPARQL Roll-up is achieved by GROUPing BY some variables (i.e. dimensions) in a query and aggregate on other variables (i.e. measures).

Given a Dice query (a BGP basically) that selects entities at the specified dimension levels (i.e. there is a triple pattern matching each of the specified dimension levels), simply GROUPing BY the specified dimension levels and applying an aggregate function on measure values (i.e. any variable object not appeared in GROUP BY) would simulate Roll-up in SPARQL. Taking Query 1 as an example, if we would like to know the concentration of each pollutant at each station averaged over all time points, we would GROUP BY variable *?P* and *?S* and apply AVG on variable *?concentration*, as shown in the query below:

Query 2.

```
SELECT ?P ?S (AVG(?concentration) AS ?avgCon)
WHERE
{ ?point  :Pollution     ?P ;
          :Time          ?Q ;
          :Station       ?S ;
          :concentration :?concentration
} GROUP BY ?P ?S
```

It is worth noticing that GROUPing BY all variables in a BGP does not change the result of the BGP[3]. Thus in SPARQL a Dice query can be trivially extended to a Roll-up query by appending a GROUP BY all variables clause at the end of its BGP.

A more involved case is when we are interested in dimension levels that do not explicitly appear in a Dice query. For example, in Query 2 instead of asking for

[3] It is enough to GROUP BY a subset of all variables that uniquely identifies an entity. Variables excluded from GROUP BY can be selected using the SAMPLE aggregation.

concentration per pollutant, we may ask for the same measure per Category in the Pollution hierarchy (shown in Fig. 1b). Depending on whether the hierarchy (dimension instance as in [4]) is explicitly stated in the RDF graph being queried, we use two different techniques to generate Roll-up queries.

Dimension hierarchy is explicit. Assuming the hierarchy is stated as triples, e.g. in the form

P_i *:rollupTo* C_j

where P_i is an instance of Pollutant, C_j is an instance of Category and :rollupTo states that its object is one level above its subject in the dimension hierarchy, we can add the triple pattern

?P a Pollutant; :rollupTo ?C. ?C a Category.

to Query 2 and GROUP BY *?C* (and *?S*) instead of *?P*.

Dimension hierarchy is absent. In this case values can be manually categorised in SPARQL using an IF expression

rdfTerm IF (boolean cond, rdfTerm expr1, rdfTerm expr2)

where the whole expression evaluates to the value of *expr*1 when *cond* evaluates to *true*, otherwise *expr*2. By nesting IF expressions, we can define a surjective (only) function

$$cat : rdfTerm \rightarrow rdfTerm$$

that maps a value to a category defined by users. For example, assuming both $P1, 2$ belong to $C1$, we can express *cat* in SPARQL as

$$cat(?P) := IF(?P = P1||?P = P2, C1, Other),$$

and convert Query 2 to the following query[4]

Query 3.

```
SELECT ?C ?S (AVG(?concentration) AS ?avgCon)
WHERE
{ ?point :Pollution      ?P ;
         :Time           ?Q ;
         :Station        ?S ;
         :concentration :?concentration
} GROUP BY (cat(?P) AS ?C) ?S
```

[4] SPARQL 1.1 doesn't have the ability to define new functions, and therefore *cat* should be considered as a macro in Query 3.

This technique is more useful to categorise numerics (or elements of totally ordered sets) into different ranges. For example, we can define a *cat* to group numbers into ranges as

$$cat(x) := IF(x <= low, \text{“}Low'', IF(x <= high, \text{“}Medium'', \text{“}High''))$$

where *low* and *high* are numbers.

Given a BGP (i.e. a Dice query), ASPG adopts a naive heuristic to extend it to a Roll-up query: (1) It randomly selects a subset of all variables of the BGP as dimensions, and the remains as measures; (2) All dimensions are used in a GROUP BY clause; (3) If a measure is known to be numerical, it is aggregated using one of the set functions COUNT, MAX, MIN, AVG, SUM, GroupConcat. Otherwise, this measure is firstly converted to a literal with STR and then to an integer with STRLEN, and aggregated using a set function. This procedure is listed below:

```
queryGen(BGP)
    D, M ∈ vars(BGP)
    GroupBy ←“GROUP BY”
    for d ∈ D
        GroupBy ← concat(GroupBy, d)

    SELECT ←“SELECT”
    for m ∈ M
        AGG ∈ {COUNT, MAX, MIN, AVG, SUM, GroupConcat}
        if m is numerical
            SELECT ← concat(SELECT, AGG(m))
        else
            SELECT ← concat(SELECT, AGG(STRLEN(STR(m))))

    query ← concat(SELECT, BGP, GroupBy)
    return query
```

2.4 Generating Basic Graph Patterns

We generate BGPs by replacing nodes in RDF graphs with variables. A RDF graph (or a BGP) can be decomposed into star-shaped or chain-shaped sub graph patterns. Considering a triple (or a triple pattern) as an undirected edge between subject and object, we define the degree of a node as the number of edges connecting to this node. A star-shaped graph pattern has one and only one central node with a degree greater than 1 and all other nodes of degree 1. A chain only has nodes whose degree are no more than 2. We generate a sub-graph from a RDF graph by repeating two steps: (1) select one node in the RDF graph as root, (2) add an edge connected to the root to the sub-graph. A star-shaped graph pattern is generated by selecting the same node as root in every iteration, while a chain is generated as selecting as root the other node in the last added edge in each iteration. We generate a mix of stars and chains by controlling a branching probability of whether to select a different root in each step, as shown in the pseudo code below, where *RDF* is a RDF graph, *T* is a termination predicate function mapping a BGP to a boolean, *p* is branching probability, and *parametrise* maps a non-property IRI to a variable:

```
BGPGen(RDF, T, p)
    BGP ← {}
    root ∈ IRIs(RDF)
    while (!T(BGP))
        E ← getTriples(root)
        e ∈ E
        BGP ← parametrise(e) ∪ BGP
        if (random() < p)
            root ← root
        else
            root ← getObject(e)
    return BGP
```

The termination function is used to control the length of generated BGPs. In this paper we define the termination condition to result in true if either the BGP reaches 10 triple patterns or the longest path in the BGP reaches 5.

The above algorithm guarantees a non-empty result set when evaluating the generated BGP against the source RDF, but there is no guarantee about the size of the result. To avoid BGPs whose result size is too small for aggregation, we evaluate generated BGPs and filter out those whose result size is less than a threshold. This safe guard is not always necessary. Later we present a set of queries generated from DBpedia and none of the BGPs falls below a threshold of 100,000.

Generating BGP with large or remote RDF graphs. When using the above method, one may encounter difficulties when the RDF graph cannot be used as a direct input to *BGPGen*. For example, the graph may be too large to be traversed or it is only available as a SPARQL endpoint. In order to deal with such cases, we adopt techniques that combines ontology and triple sampling to convert large RDF graphs into smaller ones. We describe our techniques using DBpedia as an example, but the techniques can be applied to any graph.

To generate a BGP we need to know the connection between nodes. Such information is often captured in an ontology-like structure of a RDF graph that gather all instance level properties to their classes. For simplicity we still use ontology to refer to such structure. We can issue a SPARQL CONSTRUCT query to recover the ontology (assuming all instances in the RDF belong to some classes, i.e. all *rdf:type* are explicit). However, to construct the whole ontology in one query is likely to end with a time out. Instead, we first retrieve all classes, and then use a script to collect properties between any two classes using the query template below:

Query 4.

```
CONSTRUCT
{ dbo:$1 ?p dbo:$2 }
SELECT DISTINCT ?p
WHERE { [ a dbo:$1] ?p [ a dbo:$2]. }
```

where *$1* and *$2* are replaced by class names (e.g. Person, Event etc.). The ontology is the union of all graphs returned by Query 4. The ontology can be used as the input graph (i.e. the parameter G) in the BGP generation algorithm with some extra care taken. Since all nodes in the ontology are classes, they should all be replaced with variables in generated BGPs. In addition, when following a reflexive property, a new variable should be used as root. For example, *dbo:Person* has a reflexive property *foaf:knows*. When this property is included in a BGP, its subject and object should be two different variables.

Using the ontology instead of the original RDF graph significantly reduces the complexity of BGP generation. However it does not always guarantee that the generated queries have results against the original graphs. For example in DBpedia both *Athlete* and *Artist* are sub-classes of *Person*, an instance of either *Athlete* or *Artist* may also has a *rdf:type* property pointing to *Person*. As a result properties of both *Athlete* and *Artist* are gathered at *Person*. There is a chance that an *Athlete* property and an *Artist* are connected to the same node in a BGP, which may not match any triple in the original graph. This issue can be relieved by gathering properties only to the lowest class of an instance, however doing that in SPARQL is quite cumbersome[5].

When the above method is not applicable (e.g. generating BGPs from DBpedia), we employ triple sampling as an alternative approach to extract subsets of RDF graphs. By repetitively sampling sub-graphs of simple shapes, a more complex and larger sub-graph can be constructed. For example, in ASPG we sample DBpedia using triple chains of length 5, as show in Query 5.

Query 5. Chained triple sampling

```
CONSTRUCT
{
  ?s ?p1 ?n1. ?n1 ?p2 ?n2.
  ?n2 ?p3 ?n3. ?n3 ?p4 ?n4.
  ?n4 ?p5 ?e.
}
WHERE
{
  ?s a dbo:$1. ?e a dbo:$2.
  ?s ?p1 ?n1. ?n1 ?p2 ?n2.
  ?n2 ?p3 ?n3. ?n3 ?p4 ?n4.
  ?n4 ?p5 ?e.
}
```

[5] It requires to calculate the position of an item in a linked list and to identify the maximum item in a set. Refer to https://git.io/vwP0t for more details.

where *$1* and *$2* are replaced by class names. It is left to users to decide how many and what class pairs are used. For example, in the construction of DBOB we use the top 50 classes that have most instances, and it turns out that triple chains sampled by Query 5 intertwine with each other. The result graph is significantly smaller than DBpedia while its structure is rich enough to generate complex queries.

In addition we may also want to identify properties whose ranges are numerics, even it is always possible to convert an arbitrary type to a numeric in SPARQL. Such information enables us to identify variables of numerics to which aggregate functions can be directly applied.

2.5 Complexity Analysis

We examine the time complexity of aggregate functions used in ASPG, namely GROUP BY and set functions COUNT, MAX, MIN, AVG, SUM, GroupConcat (excluding SAMPLE).

GROUP BY can be realised by the application of a higher-order 'map' function on a constant time lower-order function and each set function can be mapped to a higher-order 'fold' function on a constant time arithmetic function. All aggregations used in ASPG have $O(n)$ time complexity, where n is the size of query result regardless of the grouping of the result. We exclude SAMPLE from ASPG since it is a $O(1)$ operation.

We conclude that the time complexity of aggregating on a BGP is linear in the number of aggregate functions and independent of the grouping. In other words, the time complexity of a query (generated by ASPG) can be characterised by its BGP and its number of aggregate functions.

3 DBOB: A Benchmark Constructed with ASPG

In order to evaluate ASPG, we construct an OLAP benchmark named DBOB from DBpedia's online endpoint. DBOB contains 12 queries, of which Q1–3 are real-world queries from online analysis and Q4–12 are generated with ASPG.

Query 4–12 are generated following the steps below:

1. Retrieving the top 50 classes from DBpedia having most instances.
2. Sampling from the DBpedia SPARQL endpoint using chains of length 5 whose endpoints are drawn from instances of the 50 classes.
3. Generating OLAP queries from the RDF graph gained from step 2.
4. Evaluating the query against DBpedia and filtering out those whose result size is less than 100,000.

Due to the page limit the complete list of DBOB queries is available at http://xgfd.github.io/ASPG/.

4 Evaluation

We evaluate ASPG from two perspectives to show that ASPG is able to generate non-trivial queries. Firstly we compare DBOB queries to OLAP4LD-SSB queries [12] with respect to query complexity and types of query patterns. Secondly we use DBOB to evaluate a Virtuoso engine and analyses the result.

4.1 DBOB Quereis Vs. OLAP4LD-SSB Queries

As stated in Sect. 2.5, the time complexity of a query can be decomposed into the complexity of its BGPs and the numbers of aggregate functions. We roughly measure the complexity of a BGP by its number of triple patterns[6] (Table 1).

Table 1. Comparison of DBOB and OLAP4LD-SSB queries.

DBOB	Q1	Q2	Q3	Q4	Q5	Q6	Q7	Q8	Q9	Q10	Q11	Q12
# of triple patterns	10	5	4	4	6	4	4	8	7	2	7	7
# of group by-s	1	1	3	1	1	1	1	1	1	1	1	1
# of set functions	2	3	3	4	5	4	3	8	7	1	7	7
OLAP4LD-SSB												
# of triple patterns	6	6	7	9	8	8	10	10	8	9	11	13
# of group by-s	0	0	0	1	1	1	1	1	1	1	1	1
# of set functions	1	1	1	1	1	1	1	1	1	1	1	1

Comparing to OLAP4LD-SSB, the number of triple patterns of ASPG queries vary a lot, as a result of random sampling. In addition, since ASPG does not focus on the semantic of queries, it can simply add as many aggregate functions as required. The ability of providing triple patterns and aggregate functions on demand makes ASPG a very flexible tool for benchmarking.

4.2 Evaluating Virtuosos with DBOB

We run DBOB on a DBPedia 3.9 endpoint hosted on a machine with the following settings: 4*2.9 GHz CPU, 16 G memory, Ubuntu 14.04.4, Virtuoso opensource 7.1.0.

We use the BSBM query driver[7] to execute all queries with 0 warm up and 20 runs.

---

[6] The complexity of a BGP is also affected by the number of intermediate results in each join. However the later requires detailed statistics to estimate which are not always available.

[7] http://wifo5-03.informatik.uni-mannheim.de/bizer/berlinsparqlbenchmark/spec/BenchmarkRules/index.html#datagenerator.

The evaluation result is shown in Table 2, where QET stands for query execution time in seconds, #Rslt is the query result size before aggregation, #Trpl is the number of triple patterns, and #AF is the number of aggregate functions. We also calculate the correlation between QET and the number of triple patterns, result size and the number of aggregate functions respectively.

Table 2. DBOB evaluation result.

	Q1	Q2	Q3	Q4	Q5	Q6	Q7	Q8	Q9	Q10	Q11	Q12
#Rslt	600	39.4K	10.4K	95.5K	59.1K	65.3K	548.5K	120.8K	258.8K	81.2M	175.5K	5.0M
QET	0.65	1.19	1.31	0.10	0.72	0.31	2.90	0.08	1.43	19.33	2.74	1.73
	Correlation											
	#Rslt		#Trpl		#AF							
QET	0.99		0.07		0.38							

Most queries are finished in no more than 3 s. This may due to that queries with aggregation usually do not need to materialise all intermediate results. In addition we see the correlation between QET and the number of triple patterns is quite low. It is not surprising since QET of BGPs is mainly affected by the number of intermediate results which is not captured by only the number of triple patterns. At the same time the number of aggregate functions shows a higher impact on QET. One possible reason could be the high number of aggregate functions in ASPG queries. Alternatively as the contribution to QET from aggregation is liner to result size, the relatively higher impact from aggregation may just be a side effect of the high correlation between the result size and QET. It may be worth measuring only the execution time of aggregation, however such measure is usually difficult to obtain from outside of query engines.

5 Related Work

We divide related work into two categories: SPARQL query generators and SPARQL benchmarks.

5.1 Related Query Generators

ASPG generates queries from a RDF graph, which is similar to SPLODGE [9]. SPLODGE exploits query characteristics (e.g. join type, query type, variable pattern) and constructs queries from a federated RDF graph. While ASPG focuses on simulating OLAP queries, SPLODGE aims to generate queries for federated benchmarks. Both decompose queries into star-shaped or chained triple patterns. ASPG queries are generated by replacing nodes in a sub-RDF-graph with variables, while SPLODGE queries are generated from linked predicates (i.e. a pair of predicates sharing a common node). SPLODGE queries are not guaranteed to have results, but statistics are used to increase the chance.

FEASIBLE [16] represents a different approach to generate benchmark queries. Instead of generating queries from a RDF graph, it takes existing queries (from query logs) as prototypes and generates similar queries. Comparing to ASPG and SPLODGE, FEASIBLE queries are usually more close to real-world queries.

5.2 Related Benchmarks

To the best of our knowledge only two existing benchmarks are based in an OLAP scenario, namely BowlognaBench [8] and OLAP4LD [12]. We also review a few popular non-OLAP benchmarks.

- Lehigh University Benchmark (LUBM) [10] is designed with focus on inference and reasoning capabilities of RDF engines.
- SP^2Bench [17] has a focus of testing the performance of a variety of SPARQL features.
- The Berlin SPARQL Benchmark (BSBM) [1] mimics a e-commerce scenario and its dataset resembles a relational database.
- DBpedia SPARQL Benchmark (DBPSB) [14] uses (a sub set of) DBpedia as testing data and most used DBpedia queries as testing queries.
- BowlognaBench models an OLAP use case around the Bowlogna Ontology [7] and implements queries such as TopK, Max, Min, Path etc.
- OLAP4LD converts dataset and queries of the Star Schema Benchmark [15] into RDF and SPARQL. It resembles OLAP queries in relational databases.

We compare DBOB with aforementioned benchmarks in Table 3.

Table 3. Comparison of DBOB and existing benchmarks, adapted from [14]. Synthetic stands for artificially generated data; Real stands for real-world data; Mix stands for a mix of the former two types.

	LUMB	SP2Bench	BSBM	DBPSB	OLAP4LD	Bowlogna	DBOB
Dataset type	Synthetic	Synthetic	Synthetic	Real	Synthetic	Synthetic	Real
Query type	Synthetic	Synthetic	Synthetic	Real	Synthetic	Synthetic	Mix
Num. of classes	43	8	8	239	7	76	239
Num. of properties	32	22	51	1200	28	36	1200

6 Conclusions and Future Plan

In this paper we present ASPG that can be used to generate Dice, Slice, Roll-up and Drill-down queries in SPARQL. By exploiting ontologies and triple sampling techniques, ASPG is able to generate queries from large RDF graphs or graphs available as SPARQL endpoints. We further construct a benchmark called DBOB with ASPG and DBpedia to evaluate processing time of OLAP SPARQL queries.

Queries generated by ASPG usually have more complex BGPs compared to real-world queries. Perhaps human users are more likely to issue simple queries and combine their results afterwards, due to the lack of convenient query builders and constraints on query complexity from SPARQL endpoints. We argue that as far as query processing time is concerned, generated queries may give more insight on the performance of SPARQL engines than simple real-world queries. In addition, it is likely that the increasing demand of SPARQL analytics will foster better tools that enable users to generate complex queries. The Roll-up generation heuristic used by ASPG may contribute to the creation of such tools.

Currently ASPG queries only consist of one BGP and randomly selected aggregate functions, while real-world queries may also employ FILTERs and sub-queries (e.g. Q2 and Q3 of DBOB). As a result ASPG queries only represent some basic analytical needs. A future plan is to extend ASPG to generate multiple BGPs and sub queries that covers a broader range of analysis operations.

References

1. Bizer, C., Schultz, A.: The Berlin SPARQL benchmark. Int. J. Semant. Web Inf. Syst. (IJSWIS) - Special Issue on Scalability and Performance of Semantic Web Systems **5**(2), 1–24 (2009)
2. Capadisli, S., Auer, S., Riedl, R.: Linked Statistical Data Analysis. Semantic Web (2013)
3. Chaudhuri, S., Dayal, U.: An overview of data warehousing and OLAP technology. ACM SIGMOD Record **26**(1), 65–74 (1997)
4. Ciferri, C., Ciferri, R., Gómez, L., Schneider, M., Vaisman, A., Zimányi, E.: Cube algebra: a generic user-centric model and query language for OLAP cubes. Int. J. Data Warehous. Min. **9**(2), 39–65 (2013)
5. Codd, E.F., Codd, S.B., Salley, C.T.: Providing OLAP (on-line Analytical Processing) to user-analysts: an IT mandate. Codd Date **32**, 3–5 (1993)
6. Cyganiak, R., Reynolds, D., Tennison, J.: The RDF Data Cube Vocabulary
7. Demartini, G., Enchev, I.: The bowlogna ontology: fostering open curricula and agile knowledge bases for Europe ' s higher education. Landscape **0**, 1–11 (2012)
8. Demartini, G., Enchev, I., Wylot, M., Gapany, J., Cudré-Mauroux, P.: BowlognaBench-Benchmarking RDF analytics. Data-Driven Process Discovery Anal. **116**, 82–102 (2011)
9. Görlitz, O., Thimm, M., Staab, S.: SPLODGE: systematic generation of SPARQL benchmark queries for linked open data. In: Cudré-Mauroux, P., et al. (eds.) ISWC 2012. LNCS, vol. 7649, pp. 116–132. Springer, Heidelberg (2012). doi:10.1007/978-3-642-35176-1_8
10. Guo, Y., Pan, Z., Heflin, J.: LUBM: a benchmark for OWL knowledge base systems. Web Semant. **3**(2–3), 158–182 (2005)
11. Harris, S., Seaborne, A.: SPARQL 1.1 Query Language (2013)
12. Kämpgen, B., Harth, A.: No size fits all – running the star schema benchmark with SPARQL and RDF aggregate views. In: Cimiano, P., Corcho, O., Presutti, V., Hollink, L., Rudolph, S. (eds.) ESWC 2013. LNCS, vol. 7882, pp. 290–304. Springer, Heidelberg (2013). doi:10.1007/978-3-642-38288-8_20

13. Kämpgen, B., ORiain, S., Harth, A.: Interacting with Statistical Linked Data via OLAP Operations. In: Simperl, E., Norton, B., Mladenic, D., Della Valle, E., Fundulaki, I., Passant, A., Troncy, R. (eds.) ESWC 2012. LNCS, vol. 7540, pp. 87–101. Springer, Heidelberg (2015). doi:10.1007/978-3-662-46641-4_7
14. Morsey, M., Lehmann, J., Auer, S., Ngonga Ngomo, A.-C.: DBpedia SPARQL benchmark – performance assessment with real queries on real data. In: Aroyo, L., Welty, C., Alani, H., Taylor, J., Bernstein, A., Kagal, L., Noy, N., Blomqvist, E. (eds.) ISWC 2011. LNCS, vol. 7031, pp. 454–469. Springer, Heidelberg (2011). doi:10.1007/978-3-642-25073-6_29
15. Neil, P.O., Neil, B.O., Chen, X.: Star Schema Benchmark - Revision 3. Technical report, UMass/Boston (2009)
16. Saleem, M., Mehmood, Q., Ngonga Ngomo, A.-C.: FEASIBLE: a feature-based SPARQL benchmark generation framework. In: Arenas, M., et al. (eds.) ISWC 2015. LNCS, vol. 9366, pp. 52–69. Springer, Heidelberg (2015). doi:10.1007/978-3-319-25007-6_4
17. Schmidt, M., Hornung, T., Lausen, G., Pinkel, C.: SP2Bench: a SPARQL performance benchmark. In: Proceedings of the International Conference on Data Engineering, pp. 222–233. IEEE (2009)

Towards Answering Provenance-Enabled SPARQL Queries Over RDF Data Cubes

Kim Ahlstrøm(✉), Katja Hose, and Torben Bach Pedersen

Department of Computer Science, Aalborg University, Aalborg, Denmark
{kah,khose,tbp}@cs.aau.dk

Abstract. The SPARQL 1.1 standard has made it possible to formulate analytical queries in SPARQL. While some approaches have become available for processing analytical queries on RDF data cubes, little attention has been paid to answering provenance-enabled queries over such data. Yet, considering provenance is a prerequisite to being able to validate if a query result is trustworthy. The main challenge for existing triple stores is the way provenance can be encoded in standard triple stores based on context values (named graphs). Hence, in this paper we analyze the suitability of existing triple stores for answering provenance-enabled queries on RDF data cubes, identify their shortcomings, and propose an index to handle the high number of context values that provenance encoding typically entails. Our experimental results using the Star Schema Benchmark show the feasibility and scalability of our index and query evaluation strategies.

1 Introduction

The rapid expansion of the Linked Open Data (LOD) cloud and the introduction of SPARQL 1.1 have created new possibilities for the integration of online data. It is natural to use this vast amount of linked data to answer analytical queries [1]. Several initiatives have already been started to facilitate analytics over the Semantic Web [9,14,18]. When querying data from remote sources, provenance data is essential to ensure that the results are interpreted in a correct manner. Provenance data is not limited to quality control, there are many more uses such as access control, result ranking, query optimization, and provenance filters [19]. Therefore, it is important not to limit the descriptive power of provenance data. Hence, we use the W3C PROV-O vocabulary [20].

The standard way to encode provenance data is using reification. However, due to the verbose nature of reification we use provenance identifiers to link the provenance data as suggested in [13]. The context value of a triple is used to store a provenance identifier, this identifier corresponds to the subject in a provenance triple, thus connecting one or more information triples to a provenance triple.

We observe the problem that standard triple stores, i.e., Jena TDB [2] and RDF4J Native [8], are not designed to support provenance data. Hence, to support it, we either need to find an encoding so that standard triple stores can support it or develop a new type of triple store (see related work in Sect. 8). In this paper we make the following contributions:

Y.-F. Li et al. (Eds.): JIST 2016, LNCS 10055, pp. 186–203, 2016.
DOI: 10.1007/978-3-319-50112-3_14

- Analysis of the suitability of standard triple stores for answering provenance-enabled analytical queries.
- Two query processing strategies to enable provenance-enabled SPARQL queries over RDF data cubes.
- Proposing the Context Index to reduce the number of context values that have to be considered to answer a query.
- Evaluation of our strategies combined with the index using the Star Schema Benchmark.

The rest of this paper is structured as follows: we start with preliminaries in Sect. 2, here we also present our running example. Section 3 presents the baseline strategy for answering provenance-enabled analytical queries. In Sect. 4 we propose our novel Context Index. In Sect. 5 we combine the baseline strategy with the index. Next, in Sect. 6 the materialization strategy is presented and how it can be combined with the Context Index. In Sect. 7 we evaluate the proposed strategies. We conclude with related work in Sect. 8 followed by the conclusion and future work in Sect. 9.

2 Preliminaries

In this section, we present how provenance data is encoded, define provenance-enabled analytical queries, and present our running example of a provenance-enabled analytical query and an RDF data cube.

2.1 Encoding Provenance

Provenance data describes a piece of data, in terms of its origin, how it was created, when it was changed, and who created it. Reification is the standard (W3C) for expressing provenance information about triples. However, using the context value as a provenance identifier is gaining popularity [4, 10, 13, 22, 24]. To further define what provenance is, we need to define an RDF triple.

An *RDF triple* consists of a *subject*, *predicate*, and *object*, where the subject is related to the object through the relationship defined by the predicate. Formally we say a triple t is defined as $t = (s, p, o) \in (U \cup B) \times U \times (U \cup B \cup L)$, where U is a set of IRIs, B is a set of blank nodes, and L is a set of literals. An *RDF graph* contains a set of triples, each graph has a unique identifying IRI. When a triple is contained in an RDF graph, we write the triple as a *quad*, it consists of a subject, predicate, object, and *context value*. The context value contains the unique identifier of the graph.

Given a set of IRIs U, a set of blank nodes B, a set of literals L, and a set of query variables V, a *triple pattern* (TP) is defined as $TP = (s, p, o) \in (U \cup B \cup V) \times (U \cup V) \times (U \cup B \cup L \cup V)$. A *basic graph pattern* (BGP) consists of a set of triple patterns joined via shared query variables.

In this work, we distinguish between information triples, metadata triples, and provenance triples. An *information triple* represents a piece of information.

An example of an information triple is: (ex:kim foaf:name "Kim"), this triple encodes that ex:kim has the name "Kim". A *metadata triple* describes part of the structure of the RDF data cube, e.g., (ex:City ex:rollUpTo ex:Country), this triple describes that the city level rolls up to the country level. A *provenance triple* describes an information triple; in this paper, we use the W3C PROV Ontology [20] (PROV-O) to describe the provenance data.

PROV-O defines provenance based on three core components: *Entities* are defined as conceptual or physical things. In the context of RDF, an entity represents a set of information triples. *Activities* express how entities are created or changed. *Agents* are actors that interact with activities or entities.

When we have a collection of connected provenance triples, we say that the provenance triples constitute a *provenance graph.* Figure 1 illustrates the provenance graph of an information triple using the PROV-O components. The provenance graph has two entities illustrated as ovals. There is one activity, illustrated by a rectangle, it uses one entity and generates another entity, the generated entity is marked as trusted. The agent, illustrated as a pentagon, is associated with the activity.

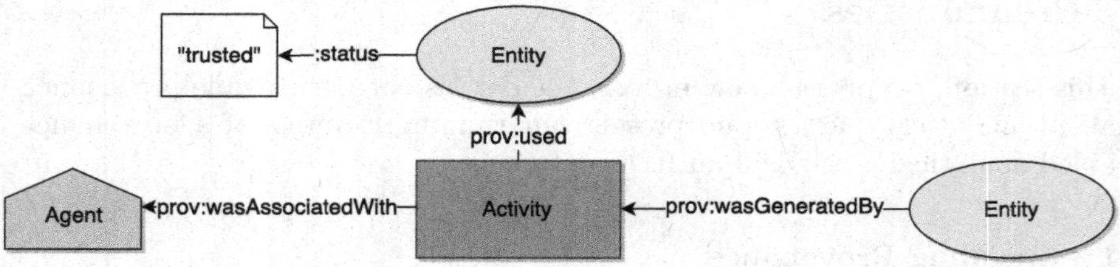

Fig. 1. Provenance graph

Information triples are linked to a provenance graph via an entity. The identifier of the entity is linked with the context value of the information triple. We will show a concrete example of this in the next section.

2.2 Provenance-Enabled RDF Data Cubes

In many ways, an RDF data cube is a traditional cube as defined in the context of relational databases [17]. However, the underlying data is formatted as RDF and instead of a schema an ontology is used; a full definition of the standard ontology for defining RDF data cubes is provided by the W3C [6]. An *RDF data cube* contains *observations*, these are the focus of the desired analysis, e.g., sales of books. An observation has a set of numerical attributes called *measures.* The observations are connected to a set of *dimensions* through a *dimension property.* Dimensions contain hierarchies of *levels* that are connected via *level properties*; each level may contain several attributes.

The RDF data cube in our running example has three dimensions: Date, Shop, and Location, the observations are sales of books and have one measure

:price. The Date dimension has three levels: Day, Month, and Year, the Book dimension has one level: Book, and the Location has three levels: Shop, City, and Country. The levels of these dimensions are described by attributes such as :monthName and :yearNumber. The structure of the RDF data cube is illustrated in Fig. 2. Throughout this paper, we use the QB4OLAP vocabulary[1] to define our cube. We choose this vocabulary because it builds upon the W3C standard vocabulary [6] and in addition defines levels, aggregate functions, cardinalities, and hierarchies. The strategies presented in this paper are not limited to the QB4OLAP vocabulary.

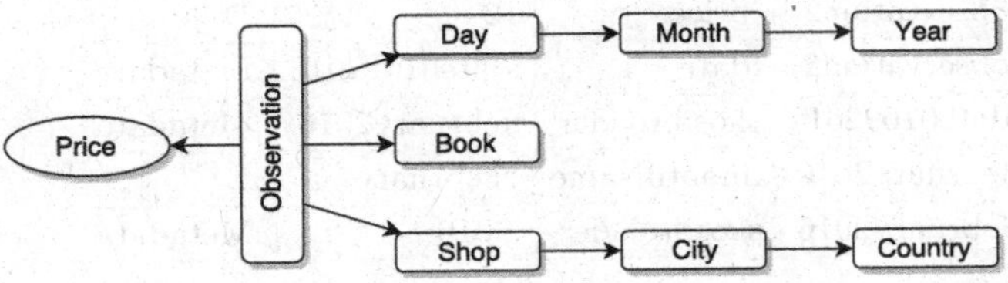

Fig. 2. Structure of the RDF data cube in running example

Table 1 shows a subset of the RDF data cube consisting of information triples and metadata triples; the data describes two book purchases made at two different dates. For brevity, the example does not contain any Book or Shop dimensions, or the QB4OLAP cube definition.

In this paper, we use provenance identifiers to link information triples to a provenance graph [13]. A *provenance identifier* is an IRI that is stored as the context value of the information triple and as the subject of a provenance triple from the corresponding provenance graph. In this example, we use capital letters to represent the IRI of the provenance identifiers, e.g., :A. Not all context values are provenance identifiers, all triples that define part of the cube structure, e.g., (:january2016 skos:broader :2016), are stored in the graph :Metadata.

In Table 2, five provenance graphs are displayed. The provenance graphs describe the provenance identifiers: :A, :B, :C, :D, and :E. These provenance identifiers link the information triples to the provenance graphs. Each provenance graph consists of three provenance triples. The provenance identifier represents the PROV-O entity that has been generated by some activity and the activity "used" some source entity during the generation. The source entities have a status attribute that is either "trusted" or "unknown". All provenance triples are stored in the graph :Provenance.

[1] https://github.com/lorenae/qb4olap/blob/master/rdf/qb4olap.1.2.ttl.

Table 1. The RDF data cube *cube* showing two book sales made on two different dates

Subject	Predicate	Object	Context value
:observation1	:price	7	:A
:observation1	:date	:date31012016	:Metadata
:date31012016	skos:broader	:january2016	:Metadata
:january2016	:monthName	"January"	:B
:january2016	skos:broader	:2016	:Metadata
:2016	:yearNumber	2016	:C
:observation2	:price	12	:D
:observation2	:date	:date01022016	:Metadata
:date01022016	skos:broader	:february2016	:Metadata
:february2016	:monthName	"February"	:E
:february2016	skos:broader	:2016	:Metadata

Table 2. Provenance triples for *cube*

Subject	Predicate	Object	Context value
:A	prov:wasGeneratedBy	:BookExtractor	:Provenance
:BookExtractor	prov:used	:DBpedia	:Provenance
:DBpedia	:status	"trusted"	:Provenance
:B	prov:wasGeneratedBy	:DateExtractor1	:Provenance
:DateExtractor1	prov:used	:DateRepository1	:Provenance
:DateRepository1	:status	"trusted"	:Provenance
:C	prov:wasGeneratedBy	:CalenderExtractor	:Provenance
:CalenderExtractor	prov:used	:CSVFile	:Provenance
:CSVFile	:status	"trusted"	:Provenance
:D	prov:wasGeneratedBy	:WebTableExtractor	:Provenance
:WebTableExtractor	prov:used	:WebTable	:Provenance
:WebTable	:status	"unknown"	:Provenance
:E	prov:wasGeneratedBy	:DateExtractor2	:Provenance
:DateExtractor2	prov:used	:DateRepository2	:Provenance
:DateRepository2	:status	"trusted"	:Provenance

3 Processing Provenance-Enabled Analytical Queries

In this section, we define provenance-enabled analytical queries, propose our baseline strategy called "Native Querying Strategy", and conduct an analysis of shortcomings of standard triple stores in this context.

3.1 Provenance-Enabled Analytical Queries

A *provenance-enabled analytical query* is a SPARQL [12] query that is executed over an RDF data cube, where the data the query is evaluated over is filtered using a provenance query. Similar to [24] we adopt an approach where the analytical part and the provenance part of the query are seperated, both to enhance understandability and for performance reasons. Alternatively, it is possible to rewrite the queries such that they are combined into one, however, this is a complex process [24]. Using SPARQL 1.1 we express the analytical query: "find the average price of sold books per year", see Query 1. Note that in this query we use the skos:broader predicate to roll-up a level along a hierarchy, as defined by the QB4OLAP vocabulary [9].

```
SELECT ?year AVG(?price)
WHERE {
  ?fact :price ?price ;
       :date ?dayLevel .
  ?dayLevel skos:broader ?monthLevel .
  ?monthLevel skos:broader ?yearLevel .
  ?yearLevel :yearNumber ?year .
}
GROUP BY ?year
```

Query 1. Analytical query

A provenance query is characterized by always returning a set of context values, these correspond to the provenance identifiers from the information triples. Query 2 shows a provenance query that finds the context values of all triples that originate from trusted sources.

```
SELECT ?provenanceIdentifiers
FROM :Provenance
WHERE {
  ?provenanceIdentifiers prov:wasGeneratedBy ?activity .
  ?activity prov:used ?entity .
  ?entity :status ?status .
  FILTER (?status ="trusted") .
}
```

Query 2. Provenance query

3.2 Native Querying Strategy

As a baseline strategy, we propose the "Native Querying Strategy", henceforth referred to as "Native". Figure 3 illustrates the steps of this strategy.

Fig. 3. Steps of the native strategy

First, the provenance query (PQ) is executed and the provenance identifiers of all matching provenance graphs are returned. When executing Query 2 over the provenance data presented in Table 2, the following provenance identifiers are returned: `:A`, `:B`, `:C`, and `:E`. Second, the analytical query (AQ) is updated by adding additional FROM clauses, we name this query $AQ^{'}$. For each provenance identifier, we add a FROM clause with the IRI of the provenance identifier. This ensures that only information triples with a provenance graph that matches the provenance query are considered when executing the AQ. Additionally, we need to make sure that we always have a valid cube. Therefore, a FROM clause with the `:Metadata` IRI is always added. In Query 3 we see $AQ^{'}$; four FROM clauses have been added, one for each of the provenance identifiers and one for the cube metadata.

```
SELECT ?year AVG(?price)
FROM :A
FROM :B
FROM :C
FROM :Metadata
WHERE {
  ?fact :price ?price ;
       :date ?dayLevel .
  ?dayLevel skos:broader ?monthLevel .
  ?monthLevel skos:broader ?yearLevel .
  ?yearLevel :yearNumber ?year .
}
GROUP BY ?year
```

Query 3. Updated analytical query ($AQ^{'}$)

Third, we execute $AQ^{'}$ over the RDF data cube. When executed over the example cube in Table 1 the query will evaluate to: "2016, 7". This means that in 2016 the average price for sold books was 7.

3.3 Preliminary Analysis

In this section, we make a preliminary analysis of the Native strategy and how it performs on standard triple stores to determine its strengths and weaknesses. We hypothesize that standard triple stores, i.e. Jena TDB [2] Band RDF4J native [8], are not able to efficiently handle a large number of FROM clauses. This is important, because the Native strategy may have hundreds of FROM clauses, depending on the provenance query and the provenance data.

To test this, we create a small RDF data cube with 68,700 triples, where 65% are information triples, 19% are provenance triples, and 16% are metadata triples. We create three RDF data cubes such that the information triples are distributed over 1000, 500, and 100 provenance identifiers. We execute the same query on the three datasets rewritten such that the number of FROM clauses in the query matches the datasets, i.e., 1000, 500, and 100 FROM clauses. The experiment is conducted on the Jena TDB and RDF4J native triple stores, both stores are created with the GSPO, GPOS, GOSP, SPOG, POSG, and OSPG indices to ensure that the queries are evaluated in an efficient manner. Table 3 shows the results of this experiment. We see that, when the number of FROM

Table 3. Runtime of standard triple stores using the native strategy

	100 FROM clauses	500 FROM clauses	1000 FROM clauses
Jena TDB	1.4 s	7.5 s	13.7 s
RDF4J native	1.0 s	28.0 s	47.0 s

clauses increases, the query evaluation time increases. Based on this observation, we can conclude that the hypothesis holds and it is indeed a problem for standard triple stores to handle a high number of FROM clauses in SPARQL queries.

4 Context Index

To reduce the number of FROM clauses in the AQ, we propose the *Context Index*. Using this index it is possible to reduce the number of provenance identifiers that need to be added as FROM clauses. First, we explain the structure of the index, then how it is used, and last how it is constructed.

4.1 Structure

The Context Index is an unbalanced tree where each node corresponds to a predicate from the RDF data cube except the root node. Each child of the root is a measure or a dimension property; the node is named after the predicate of the corresponding triple. The following nodes are attributes or level properties, again named after the corresponding predicates. Each node that is an attribute is connected to one or more leaf nodes, the leaf nodes are named after provenance identifiers for that specific attribute.

Figure 4 illustrates the index constructed based on the data used in the running example, see Table 1. The root node has two children, the measure `:price` and the dimension property `:date`. This dimension property links to the bottom level of the Date dimension, the Day level. This level does not have any attributes, only a level property. Recall that QB4OLAP uses the predicate `skos:broader` to identify these. The month level has the attribute `:monthName` and the level property `skos:broader`. The year level has the attribute `:yearNumber`. The leaves are the property identifiers from Table 1, such that the provenance identifier of the observations with the predicate `:price` are the leaves of that predicate, i.e., `:A` and `:D`.

4.2 Lookup

The lookup is split into two steps. First, we analyze the analytical query.

The WHERE clause of the analytical query is traversed when path-shaped BGPs are extracted. A *path-shaped BGP* (PSB) is a non-circular chain of triple patterns $\{(s_1, p_1, o_1), (s_2, p_2, o_2), ..., (s_n, p_n, o_n), ..., (s_m, p_m, o_m)\}$, where

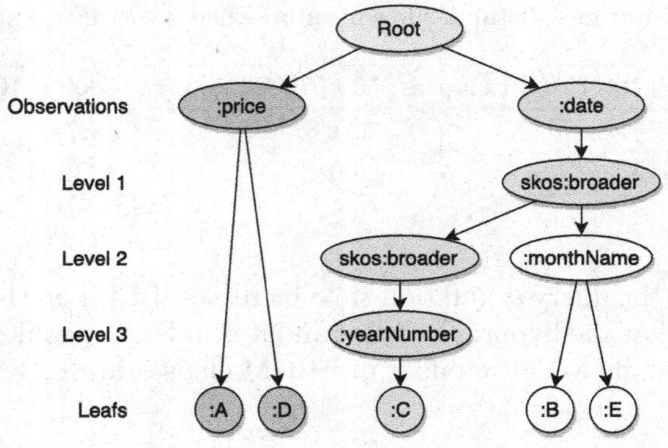

Fig. 4. The context index.

each triple pattern is linked to at most two triple patterns via object-subject joins, such that $o_i = s_{i+1}$.

If we analyze the analytical query, Query 1, introduced above we see that two PSBs are found. They are illustrated in Fig. 5.

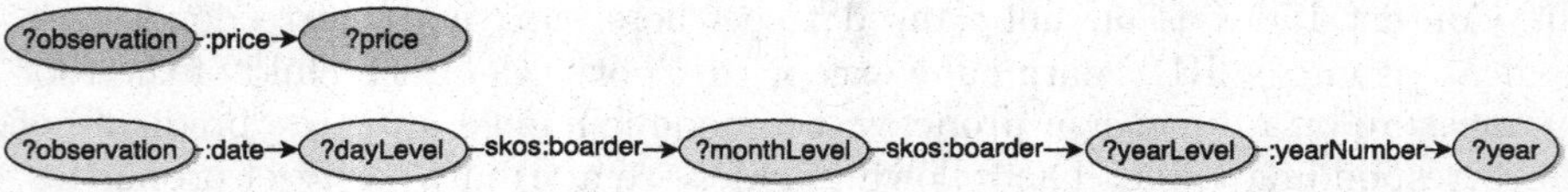

Fig. 5. Two path-shaped basic graph patterns (Color figure online)

Second, we use the PSB to look up the provenance identifiers in the context index. For each PSB, we perform a lookup in the index. By matching the predicates of the triple patterns with the predicates in the index, we identify a set of leaf nodes. This is the set of provenance identifiers that is needed to answer the analytical query. We have color-coded the two PSBs in Fig. 5 in blue and yellow, and correspondingly marked the lookup paths in Fig. 4. The result of the lookup consists of the three provenance identifiers: `:A`, `:D`, and `:C`. This is useful because any provenance identifier that is not found this way, can be discarded. This is because they will not be used when answering the analytical query. Therefore, they can be used to reduce the number of FROM clauses, we will elaborate on this in Sect. 5.

4.3 Construction

Now we explain how the index is built. The index is precomputed and updated when the data is updated. To build the index, a full scan of the RDF data cube is required. First, the RDF data cube definition is traversed; starting from the observations all measures are added as nodes, they are named after the predicate of the specific measure, e.g., `:price`. For each dimension, a new node is created, these nodes are named after the dimension property, e.g., `:date`. Second, in a depth-first manner each dimension is traversed one level at a time. All attributes are added as nodes to their corresponding level and named after the attribute predicate. Similarly, each parent level spawns a new node. This node is named after the level property, in QB4OLAP these are always called `skos:broader`. This continues until all dimensions have been traversed. Third, for each attribute a query is generated and issued that returns the provenance identifiers of that specific attribute. These are added as leafs to the attribute. Figure 4 illustrates a fully constructed context index.

5 Index-Based Native Strategy

In this section, we combine the Native Querying Strategy with the Context Index in order to address the problem of too many FROM clauses, as discussed in Sect. 3.3. In this strategy, we use the fact that the provenance query potentially finds more provenance identifiers than what are actually needed to answer the analytical query. Figure 6 illustrates the steps of the strategy.

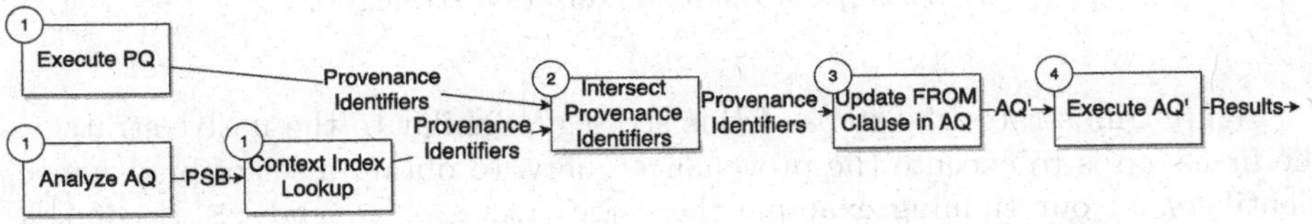

Fig. 6. Steps of the index-based native strategy

First, the provenance query is executed, as described in Sect. 3.2, and a set of provenance identifiers is returned. In our running example, this corresponds to the set: (`:A`, `:B`, `:C`, `:E`). In parallel to this the analytical query is analyzed and the path-shaped BGPs are identified. These are used to look up the set of provenance identifiers in the context index, as described in Sect. 4. This set of provenance identifiers is: (`:A`, `:D`,`:C`). Second, the intersection of these two sets is found by finding common IRIs. We say it is minimum because it contains the minimum set of provenance identifiers that is needed to answer the provenance-enabled analytical query. The intersection of the two aforementioned sets is: (`:A`, `:C`). This means that only the information triples with these provenance identifers are needed to answer the provenance-enabled analytical query. The next two steps are identical to the second and third step in the Native Strategy (Sect. 3.2). Third, the IRIs of the provenance identifiers are added as FROM clauses to the analytical query to produce the updated analytical query. Fourth, the updated analytical query is executed over the RDF data cube.

6 Materialization Strategy

In this section, we propose an additional strategy that relies on materialization. Further, we explain how it is combined with the Context Index.

6.1 Materialization Strategy

In this strategy, we materialize the subset of the RDF data cube based on the provenance identifiers and execute the analytical query over the materialized subset. Obviously, this strategy allows for reuse of the materialized cube when identical queries are issued in the fugure. However, this is not the main benefit of this strategy. As shown in Sect. 3.3, standard triple stores are not able to handle queries with a large number of FROM clauses. By first materializing the sub-cube and then executing the analytical query over that, we split a complex query into two simple queries, which is easier for the triple stores to handle efficiently.

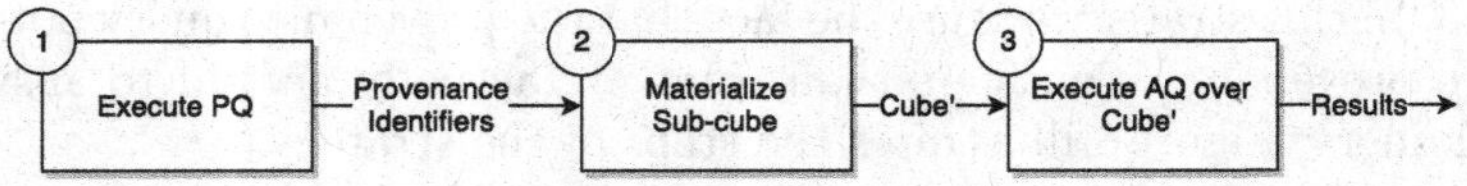

Fig. 7. Steps of the materialization strategy

Figure 7 illustrates the steps of this strategy. Similar to the native strategy, the first step is to execute the provenance query, to obtain a set of provenance identifiers. In our running example these are: `:A`, `:B`, `:C`, and `:E`. Second, a SPARQL CONSTRUCT query is created. This query creates the sub-cube, we call this `Cube'`. In the WHERE clause, we match all triple patterns and in the CONSTRUCT clause we insert the matching triple patterns but change the context value to `:MaterializedSubCube`. For each provenance identifier a FROM clause is created. Again we also add the `:Metadata` graph. This query creates a sub-cube that only contains information triples that satisfy the provenance query. In Query 4 the CONSTRUCT query for `Cube'` is shown. It creates a set of quads where the subject, predicate, and object remain unchanged but the context value is set to `:MaterializedSubCube`.

```
CONSTRUCT {
   GRAPH <:MaterializedSubCube> {?s ?p ?o }
}
FROM <:A>
FROM <:B>
FROM <:C>
FROM <:E>
FROM <:Metadata>
WHERE {
  ?s ?p ?o
}
```

Query 4. Query for materializing the sub-cube (`Cube'`)

The result of the CONSTRUCT query is shown in Table 4.

Table 4. The triples of the sub-cube (`Cube'`)

Subject	Predicate	Object	Context value
:observation1	:price	7	:MaterializedSubCube
:observation1	:date	:date31012016	:MaterializedSubCube
:date31012016	skos:broader	:january2016	:MaterializedSubCube
:january2016	:monthName	"January"	:MaterializedSubCube
:january2016	skos:broader	:2016	:MaterializedSubCube
:2016	:yearNumber	2016	:MaterializedSubCube
:observation2	:date	:date01022016	:MaterializedSubCube
:date01022016	skos:broader	:febuary2016	:MaterializedSubCube
:febuary2016	:monthName	"Febuary"	:MaterializedSubCube
:febuary2016	skos:broader	:2016	:MaterializedSubCube

Third, the analytical query is executed over the materialized sub-cube (`Cube'`). A single FROM clause is added with the identifier `:MaterializedSubCube`. When executed over `Cube'`, the query will evaluate to: "2016, 7".

6.2 Index-Based Materialization Strategy

The materialization strategy can further be improved by combining it with the context index. The steps are illustrated in Fig. 8.

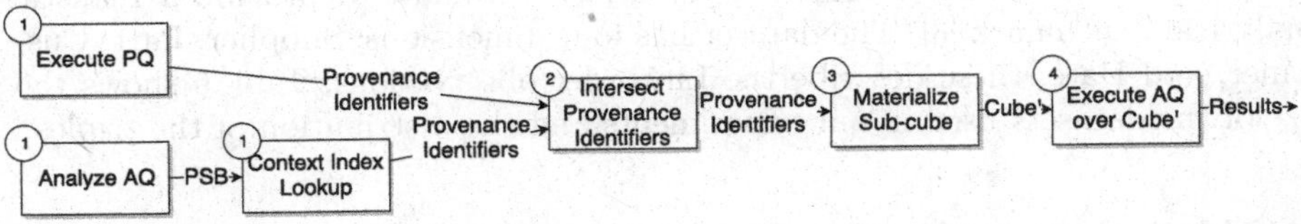

Fig. 8. Steps of the index-based materialization strategy

The first two steps are identical to the first two steps explained in Sect. 5. The last two steps are identical to step two and three from the materialization strategy explained above. Therefore, we will only give an example. First, the provenance query is executed yielding the set (`:A, :B, :C, :E`). The index is used and the set (`:A, :D,:C`) is output. Second, the intersection of the two sets are determined (`:A, :C`). Third, the CONSTRUCT query for the sub-cube is built and executed, resulting in the sub-cube `Cube'`, see Table 5. Fourth, the analytical query is executed over `Cube'`.

Table 5. Materialized cube where index is used

Subject	Predicate	Object	Context value
:observation1	:price	7	:MaterializedSubCube
:observation1	:date	:date31012016	:MaterializedSubCube
:date31012016	skos:broader	:january2016	:MaterializedSubCube
:january2016	skos:broader	:2016	:MaterializedSubCube
:2016	:yearNumber	2016	:MaterializedSubCube
:observation2	:date	:date01022016	:MaterializedSubCube
:date01022016	skos:broader	:febuary2016	:MaterializedSubCube
:febuary2016	skos:broader	:2016	:MaterializedSubCube

7 Experiments

To empirically evaluate our Native and Materialization strategies in combination with the context index, we conduct a series of experiments on the Jena TDB triple store. In order to test the scalability of our strategies we use the Star Schema Benchmark dataset and generate matching provenance data.

7.1 Experimental Environment

Hardware Platform. All experiments were run in a virtual machine with an AMD Opteron (TM) Processor 6274 (dual core), 16 GB of RAM, 500 GB harddisk, running Ubuntu 14.04.4 LTS.

SSB Dataset. For evaluating our strategies we use the Star Schema Benchmark [21]. This dataset refines the TPC-H benchmark to provide a realistic analytical benchmark [3]. The dataset has four dimensions: Supplier, Part, Customer, and Date, these describe the Lineorder observations. Table 6 shows the size of the datasets used in our experiments and the distribution of the triples.

Table 6. Overview of the four different sizes of cubes used in the experiments

Triples	Information triples	Provenance triples	Metadata triples	Unique provenance graphs
1,000,000	744,000	6,000	250,000	30,000
1,800,000	1,300,000	10,000	490,000	60,000
8,000,000	5,588,000	12,000	2,400,000	300,000
13,500,000	8,980,000	20,000	4,500,000	600,000

Provenance Generation. We generate provenance data for the SSB dataset. Each provenance graph consists of between 42 and 72 provenance triples with a varying number of entities, agents, and activities.

We use three different levels of granularity, unique, shared, and split. *Unique* is the finest level of granularity, each information triple is described by its own provenance graph. This means that each information triple has a unique provenance identifier. *Shared* is the coarsest level of granularity; all information triples share the same provenance graph, thus all information triples have the same provenance graph. *Split* has a varying level of granularity. When splitting we select an attribute, e.g., `:monthName`, which encodes the name of the month. Splitting on this attribute means that all dates from a given month, e.g., January, will share the same provenance graph. Depending on the number of distinct values for the attribute, the granularity may vary.

We vary the granularity level on each of the four dimensions and the observations. As default we choose to split the Lineorders by the attribute `:custkey`, this means that all Lineorders made by the same customer share the same provenance graph. All information triples from the same dimension share the same provenance graph, such that all information triples from the supplier dimension have the same provenance graph.

Workload. As workload we consider an analytical query and a set of synthetic provenance queries. The provenance queries we use return a slice of the RDF data cube, such that 10%, 20% ... 90% of the provenance identifiers are selected. This allows us to measure the performance of the strategies in a controlled manner. The slice is designed such that a valid RDF data cube will always be returned, thus ensuring that all provenance-enabled queries are valid over the RDF data cube.

7.2 Results

Analysis. The scalability and performance of our strategies are reported in Fig. 9. The x-axis shows the size of the RDF data cubes in millions of triples and the y-axis shows the execution time in seconds on a logarithmic scale. The provenance query in this experiment has 10% of the provenance identifiers. Note that the native and native+Index strategies are omitted for large cubes due to execution times exceeding one hour. Due to the poor scaling of the Native strategies, it is difficult to make any conclusions. We observe that the materialization strategies are faster than the native strategies, up to two orders of magnitude, if we consider the context index, then even more. We also observe that the Context Index reduces the execution time of both strategies. The index provides a constant improvement of 50% for the Materialization Strategy on all scales. For the Native Strategy this improvement is only 5%.

Figure 10 illustrates how the strategies scale when the number of provenance identifiers increase. On the x-axis is the number of FROM clauses and on the y-axis is the query evaluation time in seconds. In this experiment we use an RDF

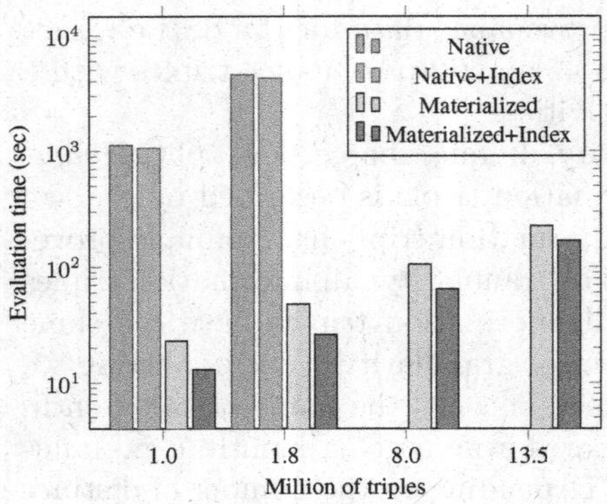

Fig. 9. Query execution over different sizes of RDF data cubes (log scale)

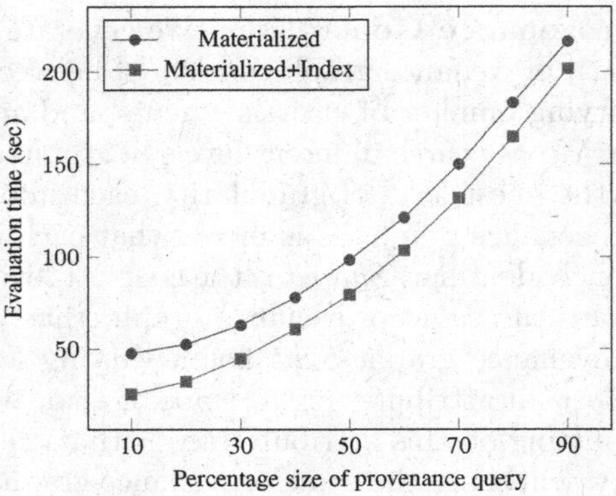

Fig. 10. Query execution with increasing number of FROM clauses

data cube of 1.8 million triples. Due to very large performance difference, the Native strategies have been omitted but they show a similar tendency. On the lower percentage, the Context Index gains a 100% speed up. As the percentage increases the performance of the index relatively decreases. Again this is because the index provides a constant improvement.

Discussion. As expected the Context Index reduces the number of FROM clauses and thereby reduces query time. Because of the synthetic nature of the provenance queries, we see a fixed performance improvement. However, because the index is constructed on load time the index has to be rebuilt when the data is updated. The Materialized strategy benefits the most from the Context Index. The index does not provide a substantial improvement for the Native Strategy, because it suffers from poor support in standard triples stores. The Materialization Strategy is up to 100 times faster than the Native Strategy. The price for this performance improvement lies in the additional storage cost for storing and updating the materialized cube. Reuse of the materialized cube was not part of the evaluation, but we expect that this would further improve this strategy. Combining the Context Index with the Materialized proves to be the best strategy.

8 Related Work

Several custom RDF provenance storage systems have been proposed in the literature. RDFProv [5] is based on a relational store for answering provenance queries over scientific workflows. By using mappings and translation algorithms on-the-fly SPARQL queries can be answered over the relational store. This approach is limited to workflow provenance queries. Our work addresses the problem of combining queries with provenance filters. In our work, we propose strategies for combining analytical queries and provenance queries.

Chebotko et al. [4] optimized Apache HBase to handle a high number of large provenance graphs, this is primarily done by using specialized indexes for select and join operations. Similar to PDFProv querying both the provenance and information triples are not considered.

TriplePROV [23] stores RDF data in molecule templates and custom physical storage models to enable fast retrieval of triples. When queries are evaluated a provenance polynomial is constructed that makes it possible to track the triples used for answering the query. While provenance polynomials are a powerful tool for some tasks, it is not possible to query the provenance polynomials using SPARQL. However, by using the context value to identify provenance triples, we do not suffer from this limitation. These custom RDF storage systems all have in common that their techniques cannot be applied to standard triple stores such as Jena TDB [2] or RDF4J Native [8]. The strategies we suggest are applicable in a storage independent manner.

In this work we build upon the query execution strategies proposed in [24]. However, it is not possible or sensible to directly apply these strategies to answer provenance-enabled analytical queries over RDF data cubes, because important metadata would be discarded. Additionally, unlike our strategies these are not applicable to standard triple stores.

The area of RDF data cubes is in constant growth. While there are several groups working towards how to best combine business intelligence and the Semantic Web [1,7,15,18,19], there is an agreement that this area carries a lot of potential for enabling web analytics. While some works on optimizing the execution of analytical queries [14,16] others work on adding spatial concepts [11]. Our work takes a step further by working towards how quality, security, and traceability can be enabled through provenance. In this paper, we address the problem of efficiently answering provenance-enabled analytical queries over RDF data cubes.

9 Conclusions and Future Work

In this paper, we work towards the problem of answering provenance-enabled analytical queries over RDF data cubes stored in standard triple stores. We observe that the main problem of evaluating such queries is handling SPARQL queries with a high number of FROM clauses. We propose the Native Strategy for answering provenance-enabled analytical queries and present two improvements. The first improvement is the novel Context Index that takes advantage of the cube structure to reduce the number of FROM clauses. The second improvement is the Materialization Strategy, it splits a provenance-enabled analytical query into a construction query and an analytical query, thus avoiding executing complex query over many graphs. Finally, we perform an empirical evaluation of our strategies using the Star Schema Benchmark augmented with provenance data. Our experimental evaluation confirms that the materialization strategy is efficient and scales for large RDF data cubes and the context index provides a consistent improvement for both strategies.

Building on the results of this work, we see a number of possible paths of future work. The current evaluation focuses on scale tests. However, using real-life data would help us further optimize the strategies and the index. Additionally, we would like to apply our strategies in a distributed setting, this which involves for a series of new challenges.

Acknowledgments. This research was partially funded by the Danish Council for Independent Research (DFF) under grant agreement No. DFF-4093-00301.

References

1. Abelló, A., Romero, O., Pedersen, T.B., Berlanga, R., Nebot, V., Aramburu, M.J., Simitsis, A.: Using semantic web technologies for exploratory OLAP: a survey. TKDE **27**(2), 571–588 (2015)
2. Apache software foundation. Jena TDB (3.1.0). https://jena.apache.org/
3. Bog, A., Plattner, H., Zeier, A.: A mixed transaction processing and operational reporting benchmark. ISF **13**(3), 321–335 (2011)
4. Chebotko, A., Abraham, J., Brazier, P., Piazza, A., Kashlev, A., Lu, S.: Storing, indexing and querying large provenance data sets as RDF graphs in apache HBase. In: Services, pp. 1–8 (2013)
5. Chebotko, A., Lu, S., Fei, X., Fotouhi, F.: RDFProv: a relational RDF store for querying and managing scientific workflow provenance. DKE **69**(8), 836–865 (2010)
6. Cyganiak, R., Reynolds, D.: The RDF data cube vocabulary. W3C recommendation, W3C, January 2014. http://www.w3.org/TR/2014/REC-vocab-data-cube-20140116/
7. Deb Nath, R.P., Hose, K., Pedersen, T.B.: Towards a programmable semantic extract-transform-load framework for semantic data warehouses. In: DOLAP, pp. 15–24 (2015)
8. Eclipse RDF4J. RDF4J (2.0.1). http://rdf4j.org/
9. Etcheverry, L., Vaisman, A., Zimányi, E.: Modeling and querying data warehouses on the semantic web using QB4OLAP. In: Bellatreche, L., Mohania, M.K. (eds.) DaWaK 2014. LNCS, vol. 8646, pp. 45–56. Springer, Heidelberg (2014). doi:10.1007/978-3-319-10160-6_5
10. Flouris, G., Fundulaki, I., Pediaditis, P., Theoharis, Y., Christophides, V.: Coloring RDF triples to capture provenance. In: Bernstein, A., Karger, D.R., Heath, T., Feigenbaum, L., Maynard, D., Motta, E., Thirunarayan, K. (eds.) ISWC 2009. LNCS, vol. 5823, pp. 196–212. Springer, Heidelberg (2009). doi:10.1007/978-3-642-04930-9_13
11. Gür, N., Hose, K., Pedersen, T.B., Zimányi, E.: Modeling and querying spatial data warehouses on the semantic web. In: Qi, G., Kozaki, K., Pan, J.Z., Yu, S. (eds.) JIST 2015. LNCS, vol. 9544, pp. 3–22. Springer, Heidelberg (2016). doi:10.1007/978-3-319-31676-5_1
12. Harris, S., Seaborne, A.: SPARQL 1.1 query language. W3C recommendation, W3C, March 2013. http://www.w3.org/TR/2013/REC-sparql11-query-20130321/
13. Hartig, O., Thompson, B.: Foundations of an alternative approach to reification in RDF (2014). CoRR abs/1406.3399

14. Ibragimov, D., Hose, K., Pedersen, T.B., Zimányi, E.: Towards exploratory OLAP over linked open data - a case study. In: BIRTE, pp. 1–18 (2014)
15. Ibragimov, D., Hose, K., Pedersen, T.B., Zimányi, E.: Processing aggregate queries in a federation of SPARQL endpoints. In: Gandon, F., Sabou, M., Sack, H., d'Amato, C., Cudré-Mauroux, P., Zimmermann, A. (eds.) ESWC 2015. LNCS, vol. 9088, pp. 269–285. Springer, Heidelberg (2015). doi:10.1007/978-3-319-18818-8_17
16. Jakobsen, K.A., Andersen, A.B., Hose, K., Pedersen, T.B.: Optimizing RDF data cubes for efficient processing of analytical queries. In: COLD (2015)
17. Jensen, C.S., Pedersen, T.B., Thomsen, C.: Multidimensional Databases and Data Warehousing. Synthesis Lectures on Data Management. Morgan & Claypool Publishers, San Rafael (2010)
18. Jovanovic, P., Romero, O., Simitsis, A., Abelló, A.: ORE: an iterative approach to the design and evolution of multi-dimensional schemas. In: DOLAP, pp. 1–8 (2012)
19. Laborie, S., Ravat, F., Song, J., Teste, O.: Combining business intelligence with semantic web: overview and challenges. In: INFORSID, pp. 99–114 (2015)
20. McGuinness, D., Lebo, T., Sahoo, S.: PROV-o: The PROV ontology. W3C recommendation, W3C, April 2013. http://www.w3.org/TR/2013/REC-prov-o-20130430/
21. O'Neil, P., O'Neil, B., Chen, X.: Star schema benchmark. Technical report, UMass/Boston, June 2019. http://www.cs.umb.edu/~poneil/StarSchemaB.PDF
22. Wang, H., Wu, T., Qi, G., Ruan, T.: On publishing Chinese linked open schema. In: Mika, P., Tudorache, T., Bernstein, A., Welty, C., Knoblock, C., Vrandečić, D., Groth, P., Noy, N., Janowicz, K., Goble, C. (eds.) ISWC 2014. LNCS, vol. 8796, pp. 293–308. Springer, Heidelberg (2014). doi:10.1007/978-3-319-11964-9_19
23. Wylot, M., Cudre-Mauroux, P., Groth, P.: TripleProv: efficient processing of lineage queries in a native RDF store. In: WWW, pp. 455–466 (2014)
24. Wylot, M., Cudre-Mauroux, P., Groth, P.: Executing provenance-enabled queries over web data. In: WWW, pp. 1275–1285 (2015)

Data Analysis of Hierarchical Data for RDF Term Identification

Pieter Heyvaert(✉), Anastasia Dimou, Ruben Verborgh, and Erik Mannens

iMinds – IDLab – Ghent University, Ghent, Belgium
pheyvaer.heyvaert@ugent.be

Abstract. Generating Linked Data based on existing data sources requires the modeling of their information structure. This modeling needs the identification of potential entities, their attributes and the relationships between them and among entities. For databases this identification is not required, because a data schema is always available. However, for other data formats, such as hierarchical data, this is not always the case. Therefore, analysis of the data is required to support RDF term and data type identification. We introduce a tool that performs such an analysis on hierarchical data. It implements the algorithms, Daro and S-Daro, proposed in this paper. Based on our evaluation, we conclude that S-Daro offers a more scalable solution regarding run time, with respect to the dataset size, and provides more complete results.

1 Introduction

Data often originally resides in (semi-)structured formats. Tools [1,2] and mapping languages [3,4] allow to describe how Linked Data, via RDF triples, is generated based on the original data. Information structure modeling [5] (henceforth referred to as 'modeling') is required during the creation of these descriptions. This modeling includes the following tasks: (1) identify the candidate entities, their attributes and the relationships among these entities; (2) generate IRIs for the entities; and (3) define the data type of each attribute, if needed. For RDF these tasks align with RDF term identification. However, they can be fulfilled in different ways, and not every way results in the desired RDF triples. Additionally, current tools come short in fulfilling these tasks (semi-)automatically or do not provide the users with the required information to fulfill them manually. This information includes the data model, keys and data types. Though, this information can be found in the data schema, for hierarchical data the schema is not always available, nor always complete, as opposed to databases. Tools, such as XmlGrid[1] and FreeFormatter[2] for XML data, exist to generate these schemas.

The described research activities were funded by Ghent University, iMinds, the Institute for the Promotion of Innovation by Science and Technology in Flanders (IWT), the Fund for Scientific Research Flanders (FWO Flanders), and the European Union.

[1] http://xmlgrid.net/xml2xsd.html.

[2] http://www.freeformatter.com/xsd-generator.html.

Y.-F. Li et al. (Eds.): JIST 2016, LNCS 10055, pp. 204–212, 2016.
DOI: 10.1007/978-3-319-50112-3_15

However, they do not give all the aforementioned information, such as keys, and the data type information is not fine-grained enough when working with dirty data. Additionally, manually extracting this information is error-prone and time consuming, as the complete data source needs to be analyzed. In this paper, we introduce a tool[3] to obtain the required information of hierarchical data to address the three tasks. The tool implements two algorithms, Daro and S-Daro, to conduct the data analysis in a scalable way, as the dataset can become large. Based on theoretical analysis, the key discovery of Daro is not always complete, while for S-Daro it is. From our evaluation, we conclude that S-Daro has a better run time when the dataset size increases. The remainder of the paper is structured as follows. In Sect. 2, we discuss the related work. In Sect. 3, we explain, using an example, how the data analysis information can be used to fulfill the modeling tasks. In Sect. 4, we explain the two algorithms. In Sect. 5, we elaborate on the evaluation of the two algorithms. Finally, in Sect. 6, we conclude the paper.

2 Related Work

For XML the data model, keys and data types can be described via the XML schema. However, not in all cases is the schema available, nor complete. Tools exist that allow to generate a schema based on an XML file, such as XmlGrid and FreeFormatter. The same is applicable for JSON and the JSON schema [6]. The tool at http://jsonschema.net can be used to generate a JSON schema given a JSON input. These tools provide data model and data type information. However, the latter lacks detail as a single data type is given when certain data fractions might have different data types. Furthermore, these tools lack key discovery.

3 Example: RDF Term Identification Using Data Analysis

In most cases Linked Data is interpreted as a graph structure, as done by RDF, where the nodes (representing the entities and their attributes) are linked using edges (representing the relationships). Using the XML example in Listing 1.1, we execute the three aforementioned tasks (see Sect. 1) of the modeling process to identify the RDF terms, taking into account which information from the data analysis is used to fulfill each task. We aim to give one possible set of declarative statements of how these terms are generated, using the mapping language RML [4], based on the data model, keys and data types. Subsequently, these statements are used to generate RDF triples.

[3] https://github.com/RMLio/data-analysis-cli; available under the MIT license.

```
<person>
  <firstName>John</firstName>
  <lastName>Doe</lastName>
  <car id="0695-77968-33844">
    <brand>Peugeot</brand>
    <purchDate>12-01-2015</purchDate>
  </car>
</person>
<person>
  <firstName>Jane</firstName>
  <lastName>Doe</lastName>
  <car id="0695-77968-33897">
    <brand>Peugeot</brand>
    <purchDate>16-01-2015</purchDate>
  </car>
</person>
</persons>
```

Listing 1.1. XML example with person metadata (**http://ex.com/persons.xml**)

Task 1: Identify Entities, Attributes and Relationships Using Data Model. RDF term identification is required to find the appropriate IRIs, blank nodes, and literals. It is supported by using the *data model*. The tree structure of these data sources allows determining possible entities, their literals and relationships by looking at the XML elements and XML attributes: parent elements (i.e., elements with child elements) are identified as entities (IRIs or blank nodes), and leaf elements (i.e., elements with no child elements) and attributes as the entities corresponding IRIs' or blank nodes' literals. Additionally, if a parent element has a parent element as a child, there exists a relationship between the corresponding entities. In the example, the parent elements are `<person>` and `<car>`. This leads to:

```
1  @prefix rr: <http://www.w3.org/ns/r2rml#> . @prefix rml:
2  <http://semweb.mmlab.be/ns/rml#> . @prefix xsd:
3  <http://www.w3.org/2001/XMLSchema#> .
4
5  <#PersonMapping>
6    rml:logicalSource [
7      rml:source "http://ex.com/persons.xml";
8      rml:referenceFormulation ql:XPath;
9      rml:iterator "/persons/person" ] .
10 <#CarMapping>
11   rml:logicalSource [
12     rml:source "http://ex.com/persons.xml";
13     rml:referenceFormulation ql:XPath;
14     rml:iterator "/persons/person/car" ] .
```

For each parent elements there is a triples map (lines 5 and 10). Each map requires a logical source, which includes the path to the parent element (lines 9 and 14). The leaf elements of `<person>` are `<firstName>` and `<lastName>`. Consequently, they can be identified as literals of the parent element's IRI or blank node, resulting in:

```
1  <#PersonMapping> rr:predicateObjectMap <#PreObjMapFirstName> .
2  <#PreObjMapFirstName> rr:objectMap [ rml:reference "firstName" ] .
3  <#PersonMapping> rr:predicateObjectMap <#PreObjMapLastName> .
4  <#PreObjMapLastName> rr:objectMap [ rml:reference "lastName" ] .
```

A predicate object map, with an object map, is added to the triples map for the `<firstName>` (lines 1 and 2) and `<lastName>` (lines 3 and 4). The same is the case for the parent element `<car>` and its leaf elements `<brand>`, `<purchDate>`, and the attribute `@id`, resulting in:

```
<#CarMapping> rr:predicateObjectMap <#PreObjMapBrand> .
<#PreObjMapBrand> rr:objectMap [ rml:reference "brand" ] .
<#CarMapping> rr:predicateObjectMap <#PreObjMapID> .
<#PreObjMapID> rr:objectMap [ rml:reference "@id" ] .
<#CarMapping> rr:predicateObjectMap <#PreObjMapPurchaseDate> .
<#PreObjMapPurchaseDate> rr:objectMap <#ObjMapPurchaseDate> .
<#ObjMapPurchaseDate> rml:reference "purchDate" .
```

Furthermore, we conclude that there is a relationship between these two entities, because `<car>` is a child element of `<person>`. This is done by adding a new predicate object map to the triples map for `<person>`, together with a parent triples map that refers to the triples map for `<car>`. This results in:

```
<#PersonMapping> rr:predicateObjectMap <#PreObjMapCar> .
<#PreObjMapCar> rr:objectMap [ rr:parentTriplesMap <#CarMapping> ] ] .
```

Task 2: Generate IRIs Using Keys. In most cases the IRIs have a specific structure, and certain elements of this structure are depended on the data. Additionally, each IRI has to represent at most one entity. This can be accomplished by using *keys* as part of the IRIs. Keys are data fractions that have a unique value for each entity in the original data. In the example, a key identified for the persons is `firstName`. A key identified for the cars is `@id`. This results in:

```
<#PersonMapping> rr:subjectMap [ rr:template "http://ex.com/person/{firstName}] .
<#CarMapping> rr:subjectMap [ rr:template"http://ex.com/car/{@id} ] .
```

A subject map is added to the triples map of each element together with a possible template to generate IRIs using the specified keys.

Task 3: Define Data Types. The *data types* of all values are string with exception of the purchase date (`<purchDate>`; lines 7 and 15), which is a date. This results in the following statement, where date data type is added to the object map corresponding with `<purchDate>`.

```
<#ObjMapPurchaseDate> rr:datatype xsd:date .
```

Subsequently, these statements can be used directly or via a tool, e.g., the RMLEditor [1], to provide the predicates to generate the desired triples.

4 Algorithms

Preliminaries. We structure hierarchical data using a *tree*, in which each node has a set of *properties*, regardless of the data format, e.g., XML or JSON. Each property points to one or more children or data values. For the example in Listing 1.1, the properties of `<person>` are given by the paths `firstName`, `lastName` and `car`. N is the set of all nodes in the tree. $\mathcal{P}$ is the set of all multi-level properties of a node. *Multi-level properties* are the properties of a node including all properties of that node's childnode trees. For the example in Listing 1.1, the multi-level properties of `<person>` are given by the paths `firstName`, `lastName`, `car/brand`, `car/id`, `car/purchaseDate`. P is used for a set of properties where $P \subseteq \mathcal{P}$. The value v of a node n for a certain (multi-level) property p is defined as $(n, p, v) \in N \times P \times V$, where V represents all values. Two nodes are distinguishable from each other given a set of properties if for at least one property the values of both nodes are not the same. This is formally given in Eq. 1.

$$\begin{aligned} dist(n, n', P) = \exists p \in P \ \wedge \ &\exists(n, p, v) \in N \times P \times V \ \wedge \\ &\exists(n', p, v') \in N \times P \times V : v \neq v' \end{aligned} \tag{1}$$

Daro. The first algorithm is based on the ROCKER algorithm, which uses a refinement operator for the discovery of keys, proposed by Soru et al. [7]. The operator refines which keys are worth checking, opposed to checking all possible keys. Originally, it was applied for key discovery on RDF datasets. Our version supports hierarchical data sources, and is called 'Data Analysis using the ROCKER Operator' (Daro). It uses a *scoring function* that gives the ratio of the number of nodes that is distinguishable given a set of properties over the total number of nodes ($score(P)$ in Eq. 2). P is a key if $score(P) = 1$, because that means that all nodes are uniquely identifiable using P. Additionally, the function $sortByScore(P)$ returns the properties of P ascendantly ordered using their score, i.e., $\forall p_i, p_j \in P : i \leq j \implies score(p_i) \geq score(p_j)$.

$$score(P) = \frac{|\{n \in N \mid \forall n' \in N : n \neq n' \Rightarrow dist(n, n', P)\}|}{|N|} \tag{2}$$

The refinement operator ($\rho(P)$ in Eq. 3) defines which sets of properties need to be checked next given a set of (previously checked) properties. It requires the properties of $\mathcal{P}$ to be ordered using $sortByScore(\mathcal{P})$.

$$\rho(P) = \begin{cases} \mathcal{P} \text{ if } P = \emptyset, \\ \{P \cup \{p_1\}, \ldots, P \cup \{p_i\}\} : & p'_0 \in sortByScore(P) \wedge \\ & (\exists p_j \in \mathcal{P} : p'_0 = p_j) \wedge (p_i \in \mathcal{P} : i < j) \end{cases} \tag{3}$$

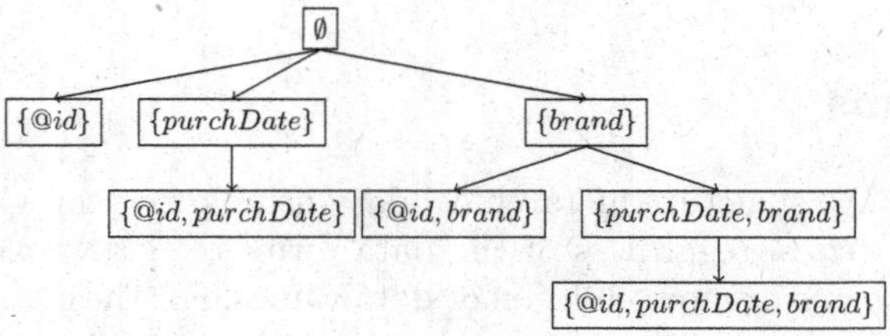

Fig. 1. Complete refinement operator tree for <car> of Listing 1.1

We explain the operator using <car> from Listing 1.1. In Fig. 1 you can see the complete refinement tree for the child elements and attributes of <car>, i.e.,

$\mathcal{P} = \{@id, purchDate, brand\}$. First, we start with an empty set of sets $(= \emptyset)$, because we do not have a set of properties. Applying the operator on the empty set $(= \rho(\emptyset))$ results in the following sets of properties: $\{@id\}$, $\{purchDate\}$ and $\{brand\}$. This is visualized in the second level of the tree. The sets on the third and fourth level of the tree are generated by applying refinement operator on each element on the second and third level, respectively. The theorems and proofs regarding the operator are given in the original work by Soru et al. [7]. Additionally, we created a function to generate the data model and a method that analyzes the values of the properties in order to provide the data types.

Algorithm 1. Daro	**Algorithm 2.** S-Daro
1: $nodes \leftarrow xml.query(nodePath)$	1: $nodes \leftarrow xml.query(nodePath)$
2: $foundKeys \leftarrow [\,]$	2: $trees \leftarrow [\,]$
3: **if** $\neg nodes.isEmpty()$ **then**	3: **if** $\neg nodes.isEmpty()$ **then**
4: $paths \leftarrow getPaths(nodes)$	4: $paths \leftarrow getPaths(nodes)$
5: $buildTreeAndIndex(nodes, paths)$	5: $model \leftarrow getModel(paths)$
6: $model \leftarrow getModel(paths)$	6: $possibleKeys \leftarrow generateKeys(paths)$
7: $paths \leftarrow sortByScore(paths)$	7: **for** $node$ in $nodes$ **do**
8: **if** $score(paths) = 1$ **then**	8: **for** key in $possibleKeys$ **do**
9: $q \leftarrow new\ PriorityQueue()$	9: **if** $\neg key.parent.valid()$ **then**
10: $q.add(\emptyset, 0)$	10: $groups \leftarrow [\,]$
11: **while** $\neg q.isEmpty()$ **do**	11: **for** $path$ in key **do**
12: $P \leftarrow q.pop()$	12: $value \leftarrow node.query(path)$
13: $P' \leftarrow \rho(P)$	13: $analyze(value)$
14: **for** p in P' **do**	14: $tree \leftarrow trees.search(path)$
15: $s \leftarrow score(p)$	15: $group \leftarrow tree.search(value)$
16: **if** $s = 1$ **then**	16: $groups.add(group)$
17: $foundKeys.add(p)$	17: **end for**
18: **else**	18: **if** $groups.hasDupNode()$ **then**
19: $q.add(p, s)$	19: $key.valid(false)$
20: **end if**	20: **end if**
21: **end for**	21: **end if**
22: **end while**	22: **end for**
23: **end if**	23: **end for**
24: **end if**	24: **end if**

The pseudo-code[4] of the algorithm can be found in Algorithm 1. The properties and the nodes are used to build a search tree of the nodes and an index over the values, during which also the data types are determined (line 5). The data model is generated using the properties (line 6). If the score of the set of all properties is 1, then all nodes are unique when taking into account all properties (line 8). Only when this is true, we continue the key discovery.

Keys that are supersets of already found keys will not be returned, because the algorithm only adds set of properties to the queue again when they are not keys. Therefore, the number of found keys might be smaller than the total number of keys. It depends on the data which keys will be found and which keys not, as the refinement operator is based on the scores of the properties. These scores are based on the actual values of the data. However, the algorithm always returns all keys consisting of one property, together with all the keys that contain

---

[4] For brevity, we did not include the code that allows users to determine the data model, keys, and data types separately.

a property that on itself is not a key. The reason is that the empty set (added on line 10) results in checking all possible keys consisting of one property, and properties that are not a key are used to generate new possible keys using the operator (line 19) until a key is found or none can be found.

Key discovery is the most expensive part of the analysis, because the different elements of the data have to be compared. The other elements of the data analysis only require a single pass over the data. However, they are done during the key discovery, because it is needed to iterate over the data in any case.

S-Daro. The second algorithm is called 'Scalable Data Analysis using the ROCKER Operator' (S-Daro). While building upon the ROCKER algorithm, it builds up an index for each property containing all possible values present in that dataset together with the nodes that have this value. Additionally, it does not use the scoring function to lower the run time. Algorithm 2 contains the pseudo code. Using the refinement operator of the previous algorithm, we determine all the possible sets of properties (line 6). They are all possible keys. Additionally, for each set we remember on which other set it was based, if applicable. In the refinement operator tree, this is the set on the lower level to which it connects. We call this set the parent set. A set is only evaluated if the parent set is not valid (i.e., not a key; line 9). If the parent set is valid than the current set stays valid, because the properties of the current set are a superset of the properties of the parent set [8]. If for all properties with those values there is one node (besides the current node) that is present (line 18), than the set is not a key. The current node and that specific node are indistinguishable using these properties.

As opposed to Daro, this algorithm returns all keys, because, besides the keys that were marked valid during checking, also the keys that have a valid parent key are valid keys. Like for Daro, key discovery is the most expensive part of the analysis, because the different elements of the data have to be compared.

5 Evaluation

In this section, we elaborate on the evaluation conducted on Daro and S-Daro. The criterion of the evaluation is the run time, because the algorithms are only useful for practical purposes if they finish within a reasonable amount of time. We have evaluated[5] both algorithms using 4 sets of 240 artificially generated files[6]. These files have between 100 and 30,000 nodes, and have between 6 and 13 properties. Their data is about people and their jobs. In Fig. 2a and b plots of the fitted functions of the run times for both algorithms can be found for 6 and 13 properties, respectively. We see that S-Daro outperforms Daro, when the number of nodes becomes larger. The functions are polynomial of the second

[5] All experiments were conducted on a 64-bit Ubuntu 14.04 machine with 128 GB of RAM and a 24-core 2.40 GHz CPU. Each algorithm was run in a Docker container and was able to use at any moment a maximum of 8 GB of RAM and 1 CPU core.

[6] http://rml.io/data/ISWC16/ph/files.

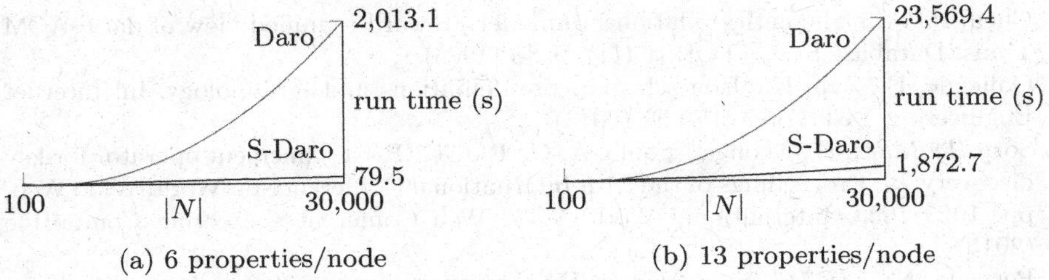

Fig. 2. Daro vs S-Daro

degree for both algorithms. Nevertheless, the function for S-Daro rises slower than for Daro, because the coefficient of the quadratic number of nodes of S-Daro remains small when compared to the coefficient for Daro. However, the coefficient for S-Daro can still be fitted to an exponential function. The reason for this is the exponential growth of possible keys in function of the total number of properties [7]. Therefore, when the number of properties becomes too large even S-Daro might not be able to provide a result within a desired time frame.

6 Conclusion

Our tool implements the two algorithms Daro and S-Daro with support for XML data sources. However, they are applicable to other formats of hierarchical data, such as JSON. Although both algorithms benefit from the refinement operator regarding their run times, the evaluation showed that S-Daro outperforms Daro when the number of nodes becomes larger. Furthermore, the incompleteness of the key discovery of Daro drives the choice towards S-Daro when all keys are required. However, certain use cases might find the results of Daro sufficient.

References

1. Heyvaert, P., Dimou, A., Herregodts, A.-L., Verborgh, R., Schuurman, D., Mannens, E., Walle, R.: RMLEditor: a graph-based mapping editor for linked data mappings. In: Sack, H., Blomqvist, E., d'Aquin, M., Ghidini, C., Ponzetto, S.P., Lange, C. (eds.) ESWC 2016. LNCS, vol. 9678, pp. 709–723. Springer, Heidelberg (2016). doi:10.1007/978-3-319-34129-3_43
2. Pinkel, C., Schwarte, A., Trame, J., Nikolov, A., Bastinos, A.S., Zeuch, T.: DataOps: seamless end-to-end anything-to-RDF data integration. In: Gandon, F., Guéret, C., Villata, S., Breslin, J., Faron-Zucker, C., Zimmermann, A. (eds.) ESWC 2015. LNCS, vol. 9341, pp. 123–127. Springer, Heidelberg (2015). doi:10.1007/978-3-319-25639-9_24
3. Das, S., Sundara, S., Cyganiak, R., R2RML: RDB to RDF mapping language. Working group recommendation, W3C. http://www.w3.org/TR/r2rml/
4. Dimou, A., Sande, M.V., Colpaert, P., Verborgh, R., Mannens, E., Rik Van de Walle, R.M.L.: A generic language for integrated rdf mappings of heterogeneous data. In: Workshop on Linked Data on the Web (2014)

5. Chen, P.P.-S.: The entity-relationship model - toward a unified view of data. ACM Trans. Database Syst. (TODS) **1**(1), 9–36 (1976)
6. Galiegue, F., Zyp, K., Json schema: core definitions and terminology. In: Internet Engineering Task Force (IETF) (2013)
7. Soru, T., Marx, E., Ngonga Ngomo, A.-C.: ROCKER - a refinement operator for key discovery. In: Proceedings of the 24th International Conference on World Wide Web, pp. 1025–1033. International World Wide Web Conferences Steering Committee (2015)
8. Pernelle, N., Saïs, F., Symeonidou, D.: An automatic key discovery approach for data linking. Web Semant. Sci. Serv. Agents WWW **23**, 16–30 (2013)

PIWD: A Plugin-Based Framework for Well-Designed SPARQL

Xiaowang Zhang[1,3,4], Zhenyu Song[1,3], Zhiyong Feng[2,3](✉), and Xin Wang[1,3]

[1] School of Computer Science and Technology, Tianjin University, Tianjin, China
[2] School of Computer Software, Tianjin University, Tianjin 300350, China
zyfeng@tju.edu.cn
[3] Tianjin Key Laboratory of Cognitive Computing and Application, Tianjin, China
[4] Key Laboratory of Computer Network and Information Integration, Southeast University, Ministry of Education, Nanjing 211189, China

Abstract. In the real world datasets (e.g., DBpedia query log), queries built on well-designed patterns containing only AND and OPT operators (for short, WDAO-patterns) account for a large proportion among all SPARQL queries. In this paper, we present a plugin-based framework for all SELECT queries built on WDAO-patterns, named PIWD. The framework is based on a parse tree called *well-designed AND-OPT tree* (for short, WDAO-tree) whose leaves are basic graph patterns (BGP) and inner nodes are the OPT operators. We prove that for any WDAO-pattern, its parse tree can be equivalently transformed into a WDAO-tree. Based on the proposed framework, we can employ any query engine to evaluate BGP for evaluating queries built on WDAO-patterns in a convenient way. Theoretically, we can reduce the query evaluation of WDAO-patterns to subgraph homomorphism as well as BGP since the query evaluation of BGP is equivalent to subgraph homomorphism. Finally, our preliminary experiments on gStore and RDF-3X show that PIWD can answer all queries built on WDAO-patterns effectively and efficiently.

Keywords: SPARQL · BGP · Well-designed patterns · Subgraph homomorphism

1 Introduction

Resource Description Framework (RDF) [23] is the standard data model in the semantic web. RDF describes the relationship of entities or resources using directed labelling graph. RDF has a broad range of applications in the semantic web, social network, bio-informatics, geographical data, etc. [3,28,29]. The standard query language for RDF graphs is SPARQL [19]. Though SPARQL is powerful to express queries over RDF graphs [2], generally, the query evaluation of the full SPARQL is PSPACE-complete [18].

Currently, there are some popular query engines for supporting the full SPARQL such as Jena [7] and Sesame [6]. However, they become not highly

Y.-F. Li et al. (Eds.): JIST 2016, LNCS 10055, pp. 213–228, 2016.
DOI: 10.1007/978-3-319-50112-3_16

efficient when they handle some large RDF datasets [33,34]. Currently, gStore [33,34] and RDF-3X [16] can highly efficiently query large datasets. But gStore and RDF-3X merely provide querying services of BGP. Therefore, it is very necessary to develop a query engine with supporting more expressive queries for large datasets.

Since the OPT operator is the least conventional operator among SPARQL operators [30], it is interesting to investigate those patterns extending BGP with the OPT operator. Let us take a look at the following example.

An RDF example in Table 1 describes the entities of bloggers and blogs. The relationship between a blogger and a blog is revealed in the property of *foaf:maker*. Both blogger and blog have some properties to describe themselves. Triples can be modeled as a directed graph substantially.

Table 1. bloggers.rdf

Subject	Predict	Object
id1	foaf:name	Jon Foobar
id1	rdf:type	foaf:Agent
id1	foaf:weblog	foobar.xx/blog
foobar.xx/blog	dc:title	title
foobar.xx/blog	rdfs:seeAlso	foobar.xx/blog.rdf
foobar.xx/blog.rdf	foaf:maker	id1
foobar.xx/blog.rdf	rdf:type	rss:channel

Example 1. Consider the RDF dataset G storing information in Table 1. Given a BGP $Q = ((?x, \mathit{foaf{:}maker}, ?y) \text{ AND } (?z, \mathit{foaf{:}name}, ?u))$, its evaluation over G is as follows:

$[\![Q]\!]_G =$

?x	?y	?z	?u
foobar.xx/blog.rdf	id1	id1	Jon Foobar

Consider a new pattern Q_1 obtained from Q by adding the OPT operator in the following way:
$Q_1 = (((?x, \mathit{foaf{:}maker}, ?y) \text{ OPT } (?y, \mathit{rdf{:}type}, ?v)) \text{ AND } (?z, \mathit{foaf{:}name}, ?u))$, the evaluation of Q_1 over G is as follows:

$[\![Q_1]\!]_G =$

?x	?y	?v	?z	?u
foobar.xx/blog.rdf	id1	foaf:Agent	id1	Jon Foobar

Consider another pattern $Q_2 = (((?x, \mathit{foaf{:}maker}, ?y) \text{ OPT } (?y, \mathit{rdf{:}type}, ?z)) \text{ AND } (?z, \mathit{foaf{:}name}, ?u))$, the evaluation of Q_2 over G is the empty set, i.e., $[\![Q_2]\!]_G = \emptyset$.

In the above example, Q_1 is a well-designed pattern while Q_2 is not a well-designed pattern [18].

In fact, we investigate that queries built on well-designed patterns are very popular in a real world. For example, in LSQ [20], a Linked Dataset describing SPARQL queries extracted from the logs of four prominent public SPARQL endpoints containing more than one million available queries shown in Table 2, queries built on well-designed patterns are over 70% [9,22].

Table 2. SPARQL logs source in LSQ

Dataset	Date	Triple number
DBpedia	30/04/2010 to 20/07/2010	232,000,000
Linked Geo Data (LGD)	24/11/2010 to 06/07/2011	1,000,000,000
Semantic Web Dog Food (SWDF)	16/05/2014 to 12/11/2014	300,000
British Museum (BM)	08/11/2014 to 01/12/2014	1,400,000

Furthermore, queries with well-designed AND-OPT patterns (for short, WDAO-patterns) are over 99% among all queries with well-designed patterns in LSQ [9,22]. In short, the fragment of WDAO-patterns is a natural extension of BGP in our real world. Therefore, we mainly discuss WDAO-patterns in this paper.

In this paper, we present a plugin-based framework for all SELECT queries built on WDAO-patterns, named PIWD. Within this framework, we can employ any query engine evaluating BGP for evaluating queries built on WDAO-patterns in a convenient way. The main contributions of this paper can be summarized as follows:

- We present a parse tree named *well-designed AND-OPT tree* (for short, WDAO-tree), whose leaves are BGP and all inner nodes are the OPT operator and then prove that for any WDAO-pattern, it can be translated into a WDAO-tree.
- We propose a plugin-based framework named *PIWD* for query evaluation of queries built on WDAO-patterns based on WDAO-tree. Within this framework, a query could be evaluated in the following three steps: (1) translating that query into a WDAO tree T; (2) evaluating all leaves of T via query engines of BGP; and (3) joining all solutions of children to obtain solutions of their parent up to the root.
- We implement the proposed framework PIWD by employing gStore and RDF-3X and evaluate the experiments on LUBM.

The rest of this paper is organized as follows: Sect. 2 briefly introduces the SPARQL, conception of well-designed patterns and OPT normal form. Section 3 defines the well-designed and-opt tree to capture WDAO-patterns. Section 4 presents PIWD and Sect. 5 evaluates experimental results. Section 6 summarizes our related works. Finally, Sect. 7 summarizes this paper.

2 Preliminaries

In this section, we introduce RDF and SPARQL patterns, well-designed patterns, and OPT normal form [18].

2.1 RDF

Let I, B and L be infinite sets of *IRIs*, *blank nodes* and *literals*, respectively. These three sets are pairwise disjoint. We denote the union $I \cup B \cup L$ by U, and elements of $I \cup L$ will be referred to as *constants*.

A triple $(s, p, o) \in (I \cup B) \times I \times (I \cup B \cup L)$ is called an *RDF triple*. A *basic graph pattern* (BGP) is a set of triple patterns.

2.2 Semantics of SPARQL Patterns

The semantics of patterns is defined in terms of sets of so-called *mappings*, which are simply total functions $\mu \colon S \to U$ on some finite set S of variables. We denote the domain S of μ by $\mathrm{dom}(\mu)$.

Now given a graph G and a pattern P, we define the semantics of P on G, denoted by $[\![P]\!]_G$, as a set of mappings, in the following manner.

- If P is a triple pattern (u, v, w), then

$$[\![P]\!]_G := \{\mu \colon \{u, v, w\} \cap V \to U \mid (\mu(u), \mu(v), \mu(w)) \in G\}.$$

  Here, for any mapping μ and any constant $c \in I \cup L$, we agree that $\mu(c)$ equals c itself. In other words, mappings are extended to constants according to the identity mapping.
- If P is of the form P_1 UNION P_2, then $[\![P]\!]_G := [\![P_1]\!]_G \cup [\![P_2]\!]_G$.
- If P is of the form P_1 AND P_2, then $[\![P]\!]_G := [\![P_1]\!]_G \bowtie [\![P_2]\!]_G$, where, for any two sets of mappings Ω_1 and Ω_2, we define

$$\Omega_1 \bowtie \Omega = \{\mu_1 \cup \mu_2 \mid \mu_1 \in \Omega_1 \text{ and } \mu_2 \in \Omega_2 \text{ and } \mu_1 \sim \mu_2\}.$$

  Here, two mappings μ_1 and μ_2 are called *compatible*, denoted by $\mu_1 \sim \mu_2$, if they agree on the intersection of their domains, i.e., if for every variable $?x \in \mathrm{dom}(\mu_1) \cap \mathrm{dom}(\mu_2)$, we have $\mu_1(?x) = \mu_2(?x)$. Note that when μ_1 and μ_2 are compatible, their union $\mu_1 \cup \mu_2$ is a well-defined mapping; this property is used in the formal definition above.
- If P is of the form P_1 OPT P_2, then

$$[\![P]\!]_G := ([\![P_1]\!]_G \bowtie [\![P_2]\!]_G) \cup ([\![P_1]\!]_G \smallsetminus [\![P_2]\!]_G),$$

  where, for any two sets of mappings Ω_1 and Ω_2, we define

$$\Omega_1 \smallsetminus \Omega_2 = \{\mu_1 \in \Omega_1 \mid \neg \exists \mu_2 \in \Omega_2 : \mu_1 \sim \mu_2\}.$$

- If P is of the form SELECT$_S(P_1)$, then $[\![P]\!]_G = \{\mu|_{S \cap \text{dom}(\mu)} \mid \mu \in [\![P_1]\!]_G\}$, where $f|_X$ denotes the standard mathematical notion of restriction of a function f to a subset X of its domain.
- Finally, if P is of the form P_1 FILTER C, then $[\![P]\!]_G := \{\mu \in [\![P_1]\!]_G \mid \mu(C) = \mathit{true}\}$.
  Here, for any mapping μ and constraint C, the evaluation of C on μ, denoted by $\mu(C)$, is defined in terms of a three-valued logic with truth values *true*, *false*, and *error*. Recall that C is a boolean combination of atomic constraints. For a bound constraint bound(?x), we define:

$$\mu(\text{bound}(?x)) = \begin{cases} \mathit{true} & \text{if } ?x \in \text{dom}(\mu); \\ \mathit{false} & \text{otherwise.} \end{cases}$$

For an equality constraint $?x = ?y$, we define:

$$\mu(?x = ?y) = \begin{cases} \mathit{true} & \text{if } ?x, ?y \in \text{dom}(\mu) \text{ and } \mu(?x) = \mu(?y); \\ \mathit{false} & \text{if } ?x, ?y \in \text{dom}(\mu) \text{ and } \mu(?x) \neq \mu(?y); \\ \mathit{error} & \text{otherwise.} \end{cases}$$

Thus, when $?x$ and $?y$ do not both belong to dom(μ), the equality constraint evaluates to *error*. Similarly, for a constant-equality constraint $?x = c$, we define:

$$\mu(?x = c) = \begin{cases} \mathit{true} & \text{if } ?x \in \text{dom}(\mu) \text{ and } \mu(?x) = c; \\ \mathit{false} & \text{if } ?x \in \text{dom}(\mu) \text{ and } \mu(?x) \neq c; \\ \mathit{error} & \text{otherwise.} \end{cases}$$

A boolean combination is then evaluated using the truth tables given in Table 3.

Table 3. Truth tables for the three-valued semantics.

p	q	$p \wedge q$	$p \vee q$
true	*true*	*true*	*true*
true	*false*	*false*	*true*
true	*error*	*error*	*true*
false	*true*	*false*	*true*
false	*false*	*false*	*false*
false	*error*	*false*	*error*
error	*true*	*error*	*true*
error	*false*	*false*	*error*
error	*error*	*error*	*error*

p	$\neg p$
true	*false*
false	*true*
error	*error*

2.3 Well-Designed Pattern

A UNION-*free* pattern P is *well-designed* if the followings hold:

- P is safe;

- for every subpattern Q of form (Q_1 OPT Q_2) of P and for every variable $?x$ occurring in P, the following condition holds:

  If $?x$ occurs both inside Q_2 and outside Q, then it also occurs in Q_1.

Consider the definition of well-designed patterns, some conceptions can be explained as follows:

Remark 1. In the fragment of and-opt patterns, we exclude FILTER and UNION operators and it contains only AND and OPT operators at most. It is obvious that *and-opt* pattern must be UNION-*free* and safe.

We can conclude that WDAO-patterns are decided by variables in subpattern.

- **UNION-*free* Pattern**: P is UNION-free if P is constructed by using only operators AND, OPT, and FILTER. Every graph pattern P is equivalent to a pattern of the form denoted by (P_1 UNION P_2 UNION $\cdots$ UNION P_n). Each P_i $(1 \leq i \leq n)$ is UNION-free.
- **Safe**: If the form of (P FILTER R) holds the condition of $var(R) \subseteq var(P)$, then it is safe.

Note that the OPT operator provides really optional left-outer join due to the weak monotonicity [18]. A SPARQL pattern P is said to be weakly monotone if for every pair of RDF graphs G_1, G_2 such that $G_1 \subseteq G_2$, it holds that $[\![P]\!]_{G_1} \sqsubseteq [\![P]\!]_{G_2}$. In other words, we assume μ_1 represents $[\![P]\!]_{G_1}$, and μ_2 represents $[\![P]\!]_{G_2}$. Then there exists μ' such that $\mu_2 = \mu_1 \cup \mu'$. Weakly monotone is an important property to characterize the satisfiability of SPARQL [31]. For instance, consider the pattern Q_1 in Sect. 1, ($?y$, *rdf:type*, $?v$) are really optional.

2.4 OPT Normal Form

A UNION-free pattern P is in *OPT normal form* [18] if P meets one of the following two conditions:

- P is constructed by using only the AND and FILTER operators;
- $P = (P_1$ OPT $P_2)$ where P_1 and P_2 patterns are in OPT normal form.

For instance, the pattern Q aforementioned in Sect. 1 is in OPT normal form. However, consider the pattern $(((?x, p, ?y)$ OPT $(?x, q, ?z))$ AND $(?x, r, ?z))$ is not in OPT normal form.

3 Well-Designed And-Opt Tree

In this section, we propose the conception of the well-designed and-opt tree (WDAO-tree), any WDAO-pattern can be seen as an WDAO-tree.

3.1 WDAO-tree Structure

Definition 1 (WDAO-tree). *Let P be a well-designed pattern in OPT normal form. A well-designed tree T based on P is a redesigned parse tree, which can be defined as follows:*

- *All inner nodes in T are labeled by the* OPT *operator and leaves are labeled by BGP.*
- *For each subpattern* $(P_1 \text{ OPT } P_2)$ *of P, the well-designed tree* T_1 *of* P_1 *and the well-designed tree* T_2 *of* P_2 *have the same parent node.*

For instance, consider a WDAO-pattern P[1]

$$P = (((p_1 \text{ AND } p_3) \text{ OPT}_2\ p_2) \text{ OPT}_1 \\ ((p_4 \text{ OPT}_4\ p_5) \text{ OPT}_5\ (p_6 \text{ OPT}_6\ p_7))).$$

The WDAO-tree T is shown in Fig. 1. As shown in this example, BGP - $(p_1 \text{ AND } p_3)$ is the exact matching in P, which corresponds to the non-optional pattern. Besides, in WDAO-tree, it is the leftmost leaf in T. We can conclude that the leftmost node in WDAO-tree means the exact matching in well-designed SPARQL query pattern.

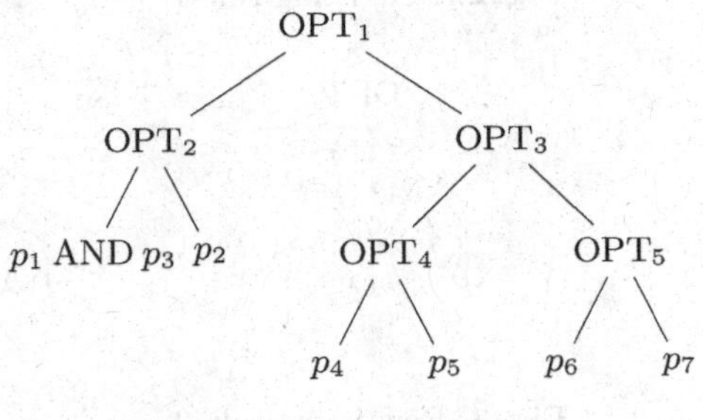

Fig. 1. WDAO-tree

3.2 Rewritting Rules over WDAO-tree

As described in Sect. 1, WDAO-tree does not contain any OPT operator in its leaves. In this sense, patterns as the form of Q_1 in Sect. 1 cannot be transformed into WDAO-tree since it is not OPT normal form.

Proposition 1. *[18, Theorem 4.11] For every UNION-free well-designed pattern P, there exists a pattern Q in OPT normal form such that P and Q are equivalent.*

[1] We give each OPT operator a subscript to differentiate them so that readers understand clearly.

In the proof of Proposition 1, we apply three rewriting rules based on the following equations: let P, Q, R be patterns and C a constraint,

- $(P \,\mathrm{OPT}\, R) \,\mathrm{AND}\, Q \equiv (P \,\mathrm{AND}\, Q) \,\mathrm{OPT}\, R$;
- $P \,\mathrm{AND}\, (Q \,\mathrm{OPT}\, R) \equiv (P \,\mathrm{AND}\, Q) \,\mathrm{OPT}\, R$;
- $(P \,\mathrm{OPT}\, R) \,\mathrm{FILTER}\, C \equiv (P \,\mathrm{FILTER}\, C) \,\mathrm{OPT}\, R$.

Intuitively, this lemma states that AND operator can forward and OPT operator can backward in a well-designed pattern with preserving the semantics. The above three rules can be deployed on a WDAO-tree. For each WDAO-tree T, there exists T' corresponding to T after applying rewriting rules.

Figures 2 and 3 have shown that the process of rewriting rules after generating grammar tree and finally WDAO-tree can be obtained. Clearly, WDAO-tree has less height than the grammar tree.

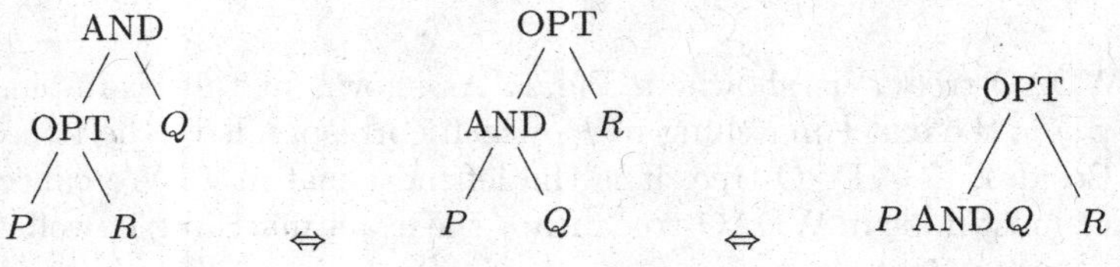

Fig. 2. Rewritting rule-1

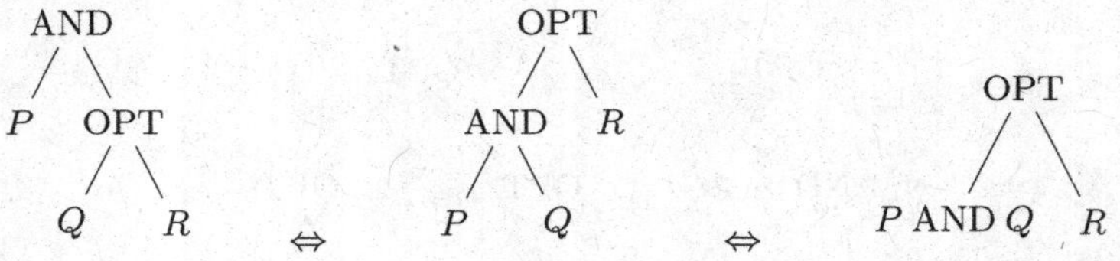

Fig. 3. Rewritting rule-2

3.3 WDAO-tree Construction

Before constructing WDAO-tree, we recognize query patterns and attachments at first. Then we rewrite query patterns by rewritting rules, which leads to a new pattern. Based on this new pattern, we construct WDAO-tree in the principle of Definition 1.

In the process of the WDAO-tree construction, we firstly build the grammar tree of SPARQL patterns, whose inner node is either AND operator or OPT operator. This process is based on recursively putting the left pattern and right pattern of operator in the left node and right node respectively until the pattern does not contain any operator. Then we apply the rewritting rules to the

---

Algorithm 1. rewritting rules

Input: GrammarTree with *Root*;
Output: RewriteTree with *Root*;
 1: **while** not all AND.child IS OPT **do**
 2:   **Procedure** ReWriteRules(*Root*)
 3:   **if** *Root* IS AND **then**
 4:     **if** *Root* IS OPT **then**
 5:       swap(*Root.left*,*Root.right.left*);
 6:       swap(*Root.right*,*Root.right*);
 7:       swap(*Root.left.left*,*Root*);
 8:       swap(*Root.left.right*,*Root.left.right*);
 9:     **end if**
10:     **if** *Root.right* IS OPT **then**
11:       swap(*Root.left*,*Root.left*);
12:       swap(*Root.right*,*Root.left.left*);
13:       swap(*Root.left.left*,*Root*);
14:       swap(*Root.left.right*,*Root.left.right*);
15:     **end if**
16:   **end if**
17:   **Procedure** ReWriteRules(*Root.left*)
18:   **End Procedure**
19:   **Procedure** ReWriteRules(*Root.right*)
20:   **End Procedure**
21: **End Procedure**
22: **end while**
23: **return** *Root*;

---

grammar tree in Algorithm 1 to build rewriting-tree whose only leaf node is single triple pattern. Different rewritting rules are adopted depending on OPT operator are AND operator's left child or right child. Since WDAO-tree's inner nodes only contain AND operators, After getting rewriting-tree, we merge the AND operators only containing leaf child nodes with its child nodes into new nodes in order to get a WDAO-tree.

The WDAO-tree construction can be executed in PTIME. Given a pattern containing n ANDs and m OPTs, the construction of the grammar tree and rewriting tree have $O(n + m)$ time complexity and $O(nm)$ time complexity, respectively. Furthermore, the merge of nodes whose parent is AND has $O(n)$ time complexity.

4 PIWD Demonstration

In this section, we introduce PIWD, which is a plugin-based framework for well-designed SPARQL.

4.1 PIWD Overview

PIWD is written in Java in a 2-tier design shown in Fig. 4. The bottom layer consists of any BGP query framework which is used as a black box for evaluating BGPs. Before answering SPARQL queries, the second layer provides the rewriting process and left-outer join evaluation, which lead to the solutions.

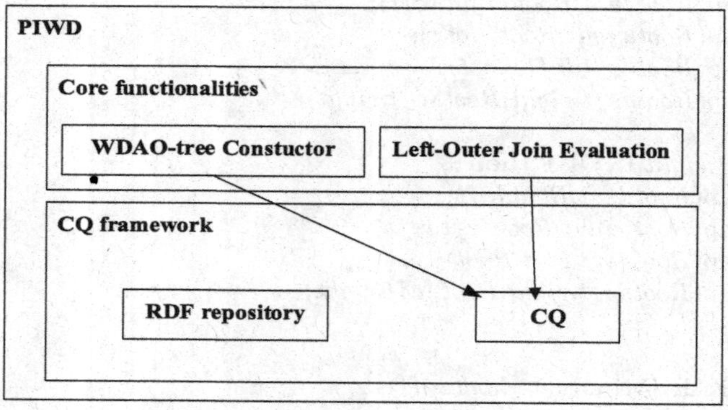

Fig. 4. PIWD architecture

BGP query framework supports both query and RDF data management, such as gStore, RDF-3X and so on, which solve the problem of subgraph isomorphism. PIWD provides the left-outer join between the BGPs. That is, the problem of answering well-designed SPARQL has been transformed into the problem of subgraph isomorphism and left-outer join between triple patterns.

4.2 Answering Queries over PIWD

The query process over PIWD can be described as follows:

Firstly, WDAO-tree is built after rewriting rules on the grammar tree. Secondly, post-order traversal is applied on WDAO-trees. The traversal rule is: If the node is a leaf node without the OPT operator, BGP query framework is deployed on it to answer this query and return solutions which is stored in a stack. If the node is an inner node labeled by the OPT operator, we get the top two elements in the stack and left-outer join them. We repeat this process until all of WDAO-tree's nodes are visited. Finally, only one element in the stack is the final solutions.

In the querying processing, BGP query framework serves as a query engine to support queries from leaves in WDAO-trees. OPT operators take an essential position in the query processing. Users receive optional solutions based on OPT operators which contribute to the semantic abundance degree since optional solutions are considered in this sense. In other words, OPT operators lead to the explosive growth of the solution scale.

The query process is described in Algorithm 2.

Algorithm 2. Query Processing over PIWD

Input: WDAO-tree with *Root*; Prefix *prefix*; *Stack* to store subresults;
Output: Query result *result*;
  1: **Procedure** TraverseTree(*Root*)
  2: **if** *root* is not null **then**
  3:     **Procedure** TraverseTree(*Root* → *Lnode*)
  4:     **End Procedure**
  5:     **Procedure** TraverseTree(*Root* → *Rnode*)
  6:     **End Procedure**
  7:     **if** *node* is not *OPTIONAL* **then**
  8:         *subquery*=AssembleQuery(*prefix*,*node*);
  9:         *subresult*=QueryIngStore(*subquery*);
 10:         Push(*Stack* , *subresult*);
 11:     **else**
 12:         *r*=Pop(*Stack*);
 13:         *l*=Pop(*Stack*);
 14:         *result*=*l* ⋈ *r*;
 15:         Push(*Stack* , *result*);
 16:     **end if**
 17: **end if**
 18: **End Procedure**
 19: *list*=ConvertToList(*Stack*);
 20: **return** *list*;

---

5   Experiments and Evaluations

This section presents our experiments. The purpose of the experiments is to evaluate the performance of different WDAO-patterns.

5.1   Experiments

Implementations and running environment. All experiments were carried out on a machine running Linux, which has one CPU with four cores of 2.40GHz, 32GB memory and 500GB disk storage. All of the algorithms were implemented in Java. gStore [33, 34] and RDF-3X [16] are used as the underlying query engines to handle BGPs. In our experiments, there is no optimization in our OPT operation.

gStore and RDF-3X. Both gStore and RDF-3X are SPARQL query engines for subgraph matching. gStore stores RDF data in disk-based adjacency lists, whose format is *[vID,vLabel,adjList]*, where *vID* is the vertex ID, *vLabel* is the corresponding URI, and *adjList* is the list of its outgoing edges and the corresponding neighbor vertices. gStore converts an RDF graph into a data signature graph by encoding each entity and class vertex. Some different hash functions such as BKDR and AP hash functions are employed to generate signatures, which compose a novel index (called VS*-tree). A filtering rule and efficient search algorithms are developed for subgraph queries over the data signature graph in order to speed up query processing. gStore can answer exact SPARQL queries and queries with wildcards in a uniform manner. RDF-3X engine is a RISC-style architecture for executing SPARQL queries over large repositories of RDF triple. Physical design is workload-independent by creating appropriate indexes over a single giant triples table in RDF-3X. And the query processor is RISC-style by relying mostly on merge joins over sorted index lists. gStore and RDF-3X have good performances in BGPs since their query methods are based on subgraph matching.

Dataset. We used LUBM[2] as the dataset in our experiments to investigate the relationship between query response time and dataset scale. LUBM, which features an ontology for the university domain, is a standard benchmark to evaluate the performance of semantic Web repositories, In our experiments, we used LUBM1, LUBM50, LUBM100, LUBM150 and LUBM200 as our query datasets. The LUBM dataset details in our experiments are shown in Table 4.

Table 4. LUBM Dataset Details

Dataset	Number of triples	RDF NT File Size(bytes)
LUBM1	103,104	14,497,954
LUBM50	6,890,640	979,093,554
LUBM100	13,879,971	1,974,277,612
LUBM150	20,659,276	2,949,441,119
LUBM200	27,643,644	3,954,351,227

SPARQL queries. The queries over LUBM were designed as four different forms, which corresponds to different WDAO-trees. The details of queries are described in Table 5. Clearly, OPT nesting in Q_2 is the most complex among four forms. Furthermore, we build the AND operator in each query.

Table 5. SPARQL queries Details

QueryID	Pattern	OPT amount
Q_1	$(P_1$ AND P_2 AND $P_3)$ OPT P_4	1
Q_2	$((P_1$ AND P_2 AND $P_3)$ OPT $P_4)$ OPT $(P_5$ OPT $P_6)$	3
Q_3	$((P_1$ AND P_2 AND $P_3)$ OPT $P_4)$ OPT P_5	2
Q_4	P_1 OPT $((P_2$ AND P_3 AND $P_4)$ OPT $P_5)$	2

[2] http://swat.cse.lehigh.edu/projects/lubm/.

5.2 Evaluation on PIWD

The variation tendencies of query response time are shown in Tables 6 and 7 and Fig. 5. Query efficiency is decreased with higher response time when OPT nesting becomes more complex. Furthermore, there has been a significant increase in query response time when the dataset scale grows up. For instance, we observe Q_2, which corresponds to the most complex pattern in our four experimental SPARQL patterns. When the dataset is ranging from LUBM100 to LUBM200, its query response time extends more than five times even though the dataset scale extends two times. In this sense, OPT nesting complexity in WDAO-patterns influences query response time especially for large dataset scale.

Table 6. Query Response Time[ms] on gStore

	LUBM1	LUBM50	LUBM100	LUBM150	LUBM200
Q_1	1,101	617,642	1,329,365	2,126,383	2,978,237
Q_2	1,870	1,010,965	2,901,295	6,623,806	10,041,836
Q_3	1,478	637,128	1,359,315	2,191,356	3,068,692
Q_4	1,242	644,155	1,456,232	2,151,811	3,129,246

Table 7. Query Response Time[ms] on RDF-3X

	LUBM1	LUBM50	LUBM100	LUBM150	LUBM200
Q_1	1,231	625,703	1,401,782	2,683,461	3,496,156
Q_2	1,900	1,245,241	2,983,394	7,286,812	10,852,761
Q_3	1,499	640,392	1,427,392	2,703,981	3,672,970
Q_4	1,316	648,825	1,531,547	2,791,152	3,714,042

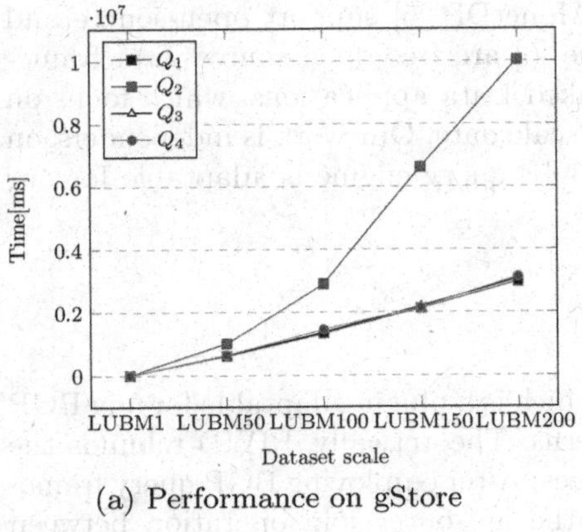

(a) Performance on gStore

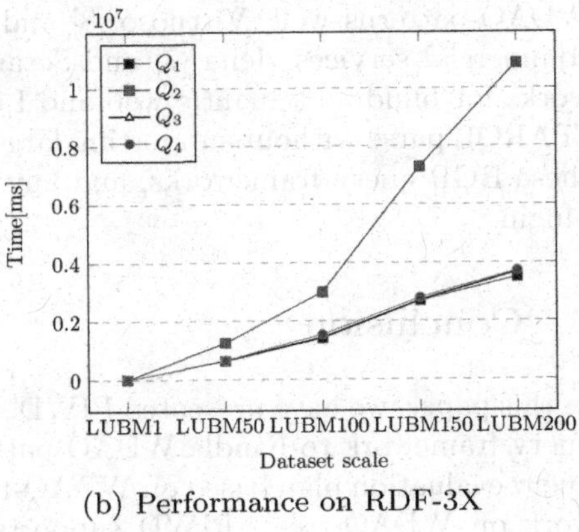

(b) Performance on RDF-3X

Fig. 5. Query response time over LUBM

6 Related Works

In this section, we survey related works in the following three areas: BGP query evaluation algorithms, well-designed SPARQL and BGP query evaluation frameworks.

BGP query algorithms have been developed for many years. Existing algorithms mainly focus on finding all embedding in a single large graph, such as ULLmann [24], VF2 [14], QUICKSI [21], GraphQL [11], SPath [32], STW [25] and TurboIso [10]. Some optimization method has been adapted in these techniques, such as adjusting matching order, pruning out the candidate vertices. However, the evaluation of well-designed SPARQL is not equivalent to the BGP query evaluation problem since there exists inexact matching.

It has been shown that the complexity of the evaluation problem for the well-designed fragment is coNP-complete [18]. The quasi well-designed pattern trees (QWDPTs), which are undirected and ordered, has been proposed [12]. This work aims at the analysis of containment and equivalence of well-designed pattern. Efficient evaluation and semantic optimization of WDPT have been proposed in [4]. Sparm is a tool for SPARQL analysis and manipulation in [13]. Above-mentioned all aim at checking well-designed patterns or complexity analysis without evaluation on well-designed patterns. Our WDAO-tree is different from QWDPTs in structure and it emphasizes reconstructing query plans. The OPT operation optimization has been proposed in [15], which is different from our work since our work aims to handle a plugin in any BGP query engine in order to deal with WDAO-patterns in SPARQL queries.

RDF-3X [16], TripleBit [27], SW-Store [1], Hexastore [26] and gStore [33,34] have high performance in BGPs. RDF-3X create indexes in the form of B+ tree, as well as TripleBit in the form of ID-Chunk. All of them have efficient performance since they concentrate on the design of indexing or storage. However, they can only support exact SPARQL queries, since they replace all literals (in RDF triples) by ids using a mapping dictionary. In other words, they cannot support WDAO-patterns well. Virtuoso [8] and MonetDB [5] support open-source and commercial services. Jena [7] and Sesame [6] are free open source Java frameworks for building semantic web and Linked Data applications, which focus on SPARQL parse without supporting large-scale date. Our work is independent on these BGP query frameworks, and any BGP query engine is adaptable for our plugin.

7 Conclusion

In this paper, we have presented PIWD, which is a plugin adaptable for any BGP query framework to handle WDAO-patterns. Theoretically, PIWD rebuilds the query evaluation plan based on WDAO-trees. After employing BGP query framework on WDAO-trees, PIWD supports the left-outer join operation between triple patterns. Our experiments show that PIWD can deal with complex and multi-level nested WDAO-patterns. In the future, we will further handle other non-well-designed patterns and deal with more operations such as UNION. Besides, we will consider OPT operation optimization to improve efficiency of PIWD and implement our framework on distributed RDF graphs by applying the distributed gStore [17].

Acknowledgments. This work is supported by the programs of the National Key Research and Development Program of China (2016YFB1000603), the National Natural Science Foundation of China (NSFC) (61502336), and the open funding project of Key Laboratory of Computer Network and Information Integration (Southeast University), Ministry of Education (K93-9-2016-05). Xiaowang Zhang is supported by Tianjin Thousand Young Talents Program and the project-sponsored by School of Computer Science and Technology in Tianjin University.

References

1. Abadi, D.J., Marcus, A., Madden, S.R., Hollenbach, K.: SW-store: a vertically partitioned DBMS for semantic web data management. VLDB J. **18**(2), 385–406 (2009)
2. Angles, R., Gutierrez, C.: The expressive power of SPARQL. In: Sheth, A., Staab, S., Dean, M., Paolucci, M., Maynard, D., Finin, T., Thirunarayan, K. (eds.) ISWC 2008. LNCS, vol. 5318, pp. 114–129. Springer, Heidelberg (2008). doi:10.1007/978-3-540-88564-1_8
3. Arenas, M., Pérez, J.: Querying semantic web data with SPARQL. In: Proceedings of SIGMOD 2011, pp. 305–316 (2011)
4. Barcelo, P., Pichler, R., Skritek, S.: Efficient evaluation and approximation of well-designed pattern trees. In: Proceedings of PODS 2015, pp. 131–144 (2015)
5. Boncz, P.A., Zukowski, M., Nes, N.J.: MonetDB/x100: Hyper-pipelining query execution. In: Proceedings of CIDR 2005 (2005)
6. Broekstra, J., Kampman, A., Harmelen, F.: Sesame: a generic architecture for storing and querying RDF and RDF schema. In: Horrocks, I., Hendler, J. (eds.) ISWC 2002. LNCS, vol. 2342, pp. 54–68. Springer, Heidelberg (2002). doi:10.1007/3-540-48005-6_7
7. Carroll, J.J., Dickinson, I., Dollin, C., Reynolds, D., Seaborne, A., Wilkinson, K.: Jena: implementing the semantic web recommendations. In: Proceedings of WWW 2004, pp. 74–83 (2004)
8. Erling, O., Mikhailov, I.: RDF support in the virtuoso DBMS. Studies in Computational Intelligence, pp. 7–24 (2009)
9. Han, X., Feng, Z., Zhang, X., Wang, X., Rao, G.: On the statistical analysis of practical SPARQL queries. In: Proceedings of WebDB 2016, Article 2(2016)
10. Han, W.S., Lee, J., Lee, J.H.: Turbo ISO: Towards ultrafast and robust subgraph isomorphism search in large graph databases. In: Proceedings of SIGMOD 2013, pp. 337–348 (2013)
11. He, H., Singh, A.K.: Query language and access methods for graph databases. In: Aggarwal, C.C., Wang, H. (eds.) Managing and Mining Graph Data. Springer, Heidelberg (2010)
12. Letelier, A., Pérez, J., Pichler, R., Skritek, S.: Static analysis and optimization of semantic web queries. Proceedings of PODS **38**(4), 84–87 (2012)
13. Letelier, A., Pérez, J., Pichler, R., Skritek, S.: SPAM: a SPARQL analysis and manipulation tool. Proc. VLDB **5**(12), 1958–1961 (2012)

14. Luigi, P.C., Pasquale, F., Carlo, S., Mario, V.: A (sub)graph isomorphism algorithm for matching large graphs. IEEE Trans. Pattern Anal. Mach. Intell. **26**(10), 1367–1372 (2004)
15. Medha, A.: Left bit right: for SPARQL join queries with OPTIONAL patterns (left-outer-joins). In: Proceedings of SIGMOD 2015, pp. 1793–1808 (2015)
16. Neumann, T., Weikum, G.: The RDF3X engine for scalable management of RDF data. VLDB J. **19**(1), 91–113 (2010)
17. Peng, P., Zou, L., Özsu, M.T., Chen, L., Zhao, D.: Processing SPARQL queries over distributed RDF graphs. VLDB J. **25**(2), 243–268 (2016)
18. Pérez, J., Arenas, M., Gutierrez, C.: Semantics and complexity of SPARQL. ACM Trans. Database Syst. **34**(3), 30–43 (2009)
19. Prud'hommeaux, E., Seaborne, A.: SPARQL query language for RDF. W3C Recommendation (2008)
20. Saleem, M., Ali, M.I., Hogan, A., Mehmood, Q., Ngomo, A.-C.N.: LSQ: the linked SPARQL queries dataset. In: Arenas, M., et al. (eds.) ISWC 2015. LNCS, vol. 9367, pp. 261–269. Springer, Heidelberg (2015). doi:10.1007/978-3-319-25010-6_15
21. Shang, H., Zhang, Y., Lin, X., Yu, J.X.: Taming verification hardness: an efficient algorithm for testing subgraph isomorphism. Proc. VLDB **1**(1), 364–375 (2008)
22. Song, Z., Feng, Z., Zhang, X., Wang, X., Rao, G.: Efficient approximation of well-designed SPARQL queries. In: Song, S., Tong, Y. (eds.) WAIM 2016. LNCS, vol. 9998, pp. 315–327. Springer, Heidelberg (2016). doi:10.1007/978-3-319-47121-1_27
23. Swick, R.R.: Resource description framework (RDF) model and syntax specification. In: W3C Recommendation (1998)
24. Ullmann, J.R.: An algorithm for subgraph isomorphism. J. ACM **23**(1), 31–42 (1976)
25. Wang, H.: Efficient subgraph matching on billion node graphs. In: Proc. of VLDB 2012, 5: article 9 (2012)
26. Weiss, C., Karras, P., Bernstein, A.: Hexastore: sextuple indexing for semantic web data management. Proc. VLDB **1**, 1008–1019 (2008)
27. Yuan, P., Liu, P., Wu, B., Jin, H., Zhang, W., Liu, L.: Triplebit: a fast and compact system for large scale RDF data. Proc. VLDB **6**(7), 517–528 (2013)
28. Zhang, X., Feng, Z., Wang, X., Rao, G., Wu, W.: Context-free path queries on RDF graphs. In: Groth, P., Simperl, E., Gray, A., Sabou, M., Krötzsch, M., Lecue, F., Flöck, F., Gil, Y. (eds.) ISWC 2016. LNCS, vol. 9981, pp. 632–648. Springer, Heidelberg (2016). doi:10.1007/978-3-319-46523-4_38
29. Zhang, X., Van den Bussche, J.: On the power of SPARQL in expressing navigational queries. Computer J. **58**(11), 2841–2851 (2015)
30. Zhang, X., Van den Bussche, J.: On the primitivity of operators in SPARQL. Inf. Process. Lett. **114**(9), 480–485 (2014)
31. Zhang, X., Van den Bussche, J.: On the satisfiability problem for SPARQL patterns. J. Artif. Intell. Res. **56**, 403–428 (2016)
32. Zhao, P., Han, J.: On graph query optimization in large networks. Proc. VLDB **3**(1–2), 340–351 (2010)
33. Zou, L., Mo, J., Chen, L., Özsu, M.T., Zhao, D.: gStore: answering SPARQL queries via subgraph matching. Proc. VLDB **4**(8), 482–493 (2011)
34. Zou, L., Özsu, M.T., Chen, L., Shen, X., Huang, R., Zhao, D.: gStore: a graph-based SPARQL query engine. VLDB J. **23**(4), 565–590 (2014)

Knowledge Graph

Non-hierarchical Relation Extraction of Chinese Text Based on Scalable Corpus

Xiaoheng Su[1], Hai Wan[1](✉), Ruibin Chen[1], Qi Liu[1], Wenxuan Zhang[1], and Jianfeng Du[2]

[1] School of Data and Computer Science, Sun Yat-sen University, Guangzhou 510006, China
wanhai@mail.sysu.edu.cn,
{suxh8,chenrb6,liuq99,zhangwx26}@mail2.sysu.edu.cn

[2] Guangdong University of Foreign Studies, Guangzhou 510006, China
dududjf@gmail.com

Abstract. As for ontology construction from Chinese text, the non-hierarchical relation extraction is harder than the concept extraction and its extraction effect is still not satisfactory. In this paper, we put forward a scalable corpus model, which uses Tongyici Cilin and word2vec to calculate terms' similarity and add the qualified candidate terms to the corpora. In this way we can expand the scalable corpus while extracting non-hierarchical relations. In turn, the scalable corpus that has been expanded with the new terms will facilitate the non-hierarchical relation extraction further. We carry out the experiment with Chinese texts in the domain of Computer, whose results show that with expansion of the corpus, the extraction effect will be better and better.

Keywords: Relation extraction · Scalable corpus · Chinese text

1 Introduction

Maedche et al. defined the ontology structure as a five tuple: $O := \{C, R, H_C, rel, A_O\}$, where O, C, R and A_O indicated the ontology, the set of concepts, the set of relations and the set of axioms respectively. And H_C denotes a set of hierarchical relations among concepts with inheritance, for example, $H_C(C_1, C_2)$ expresses C_1 is a subconcept of C_2; $rel : R \rightarrow C \times C$ is a function, $rel(R) = (C_1, C_2)$ can also written in $R(C_1, C_2)$, denoting a set of non-hierarchical relations, such as $capital(China, Beijing)$ [1]. Therefore the main tasks of ontology construction from the Chinese text are concept and relation extractions, where the non-hierarchical relation extraction is a harder problem. The difficulty lies in unstructured organization of Chinese text and various relations occurring in Chinese sentences. Aiming at the non-hierarchical relation extraction, two kinds of methods are focused on. One is based on lexical rules matching, which is mainly suitable for English, while not working well in Chinese. The other is based on association rule analysis, which can determine whether

Y.-F. Li et al. (Eds.): JIST 2016, LNCS 10055, pp. 231–238, 2016.
DOI: 10.1007/978-3-319-50112-3_17

there exist relations between two terms, however it is difficult to extract the exact relations. In this case, the paper tries to establish a robust non-hierarchical relation extraction model, making relations extraction in Chinese texts accurate and stable. In the model, firstly initialize the two scalable corpora including the concept corpus and the relation corpus, then use Density Extraction Algorithm proposed in the paper to extract the core concepts from the Chinese text, and seek for high quality of domain terms through core concepts, then add them into scalable corpora, next take advantage of improved association rule analysis to capture term pairs based on similarity analysis including word2vec and Tongyici Cilin, and extract relations after pruning sentences, finally expand corpus again. The rest of this paper is organized as follows: Sect. 2 introduces research status of ontology construction. Section 3 elaborates the main ideas of our method. Section 4 exhibits the experimental design and result analysis. Finally, Sect. 5 concludes our work.

2 Related Work

Nowadays ontology construction has received widespread attention from researchers. Generally, ontology construction algorithms are based on statistical methods and lexical syntactic patterns, and the idea of combining multiple methods has become a mainstream [2]. In the study of the concept extraction, Roberto et al. presented a novel method for filtering uncorrelated terms from candidate terms based on Domain Relevance(DR) and Domain Consensus (DC), their methods made the extracted concepts more representative for consistency of terms in the same domain [3]. On this basis, Haitao He et al. presented a method of the domain concept extraction based on semantic rules and association rules [4]. And for the relation extraction, there are two major approaches, lexical syntactic patterns [5] and association rule analysis [6]. By the limitations of these two methods, Jun Gu et al. proposed a improved association rule to acquire relations, which applied association rule to two terms and a verb at the same time, and could get pretty triples, however it ignored the case that noun expressions act as relations [7]. In order to extract more relations including nouns, Fan Yu raised a set of grammatical rules, which paid attention to the elimination of pseudo verbs and nouns, and experiments proved that this method helped getting subjects, predicates and objects precisely [8]. To improve the quality of extraction, Yufang Zhang et al. integrated similarity analysis based on the context into above methods [9]. Compared with literatures, our method does not require plenty of data to identify the domain terms and maintains the higher F1-value of relation extraction. More important is that we develop a robust extraction model, which can rigorously expand the scalable corpora and find more concepts and non-hierarchical relations from corpora in turn.

3 Extraction Based on Scalable Corpora

The scalable corpus is a set of domain terms that are the most common expressions in this domain and can be divided into the scalable concept corpus and the scalable relation corpus. The scalable concept [1] corpus refers to a set of concepts or instances mainly including nouns and noun phrases, and the scalable relation corpus is composed by a set of relations mainly in verbs or nouns. Our proposed extraction model includes the following two parts. But firstly we need establish a higher purity of initial corpora, this step is introduced in Part 4 detailedly.

3.1 Part One: Expand the Scalable Corpora

Since the scalable corpora are not complete and some domain terms in the document may not be in initial corpora, the primary objective of this part is to expand corpora with key terms of the document preliminarily, which lays the foundation of association rule analysis in the second part. Figure 1 depicts the work flow of the first part.

Firstly, segment the document and extract core concepts with Density Extraction Algorithm, and add them to the concept corpus. Secondly, locate the context of the core concepts and extract nouns and verbs around them. After that, filter out nouns that are not suitable for concepts through TF-IDF, and put the remainders keeping higher scores to the concept corpus. Last, filter out nouns and verbs that are unlikely to be relations with similarity analysis based on existent corpora, and add the remaining verbs or nouns to the relation corpus.

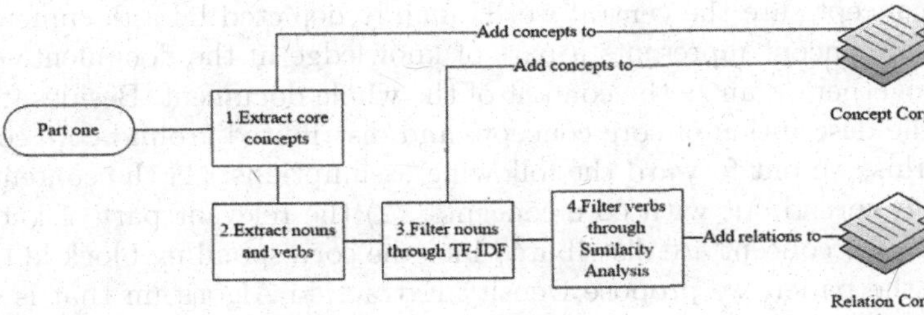

Fig. 1. Flow diagram of the first part

3.2 Part Two: Extract Relations

After the first part, the corpora have filled with key terms of this document, which can be regarded as the clue to extraction of concepts and relations further. The main goal in this part is to extract more non-hierarchical relations and concepts and expand the corpora again. Figure 2 shows the whole process.

Firstly, use the improved Apriori algorithm and the scalable concept corpus to get term pairs which may have the strong relations. Next, prune sentences where term pairs are and extract predicates as candidate relations. Then, filter out the relations and concepts using similarity analysis including word2vec and Tongyici

[1] In the paper, the concept refers to some concept or the instance of some concept.

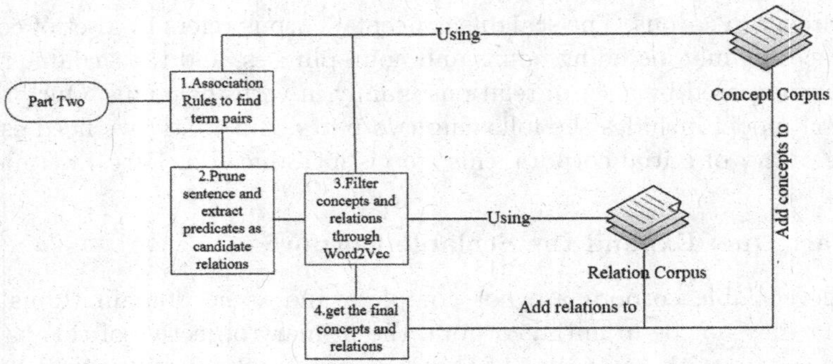

Fig. 2. Flow diagram of the second part

Cilin. If the pair of terms are both in concept corpus, then extract relations with the maximum similarity; if either of the pair of terms is in concept corpus, choose candidate relations whose similarity is greater than the preset threshold as the relation and the other term as the new concept. Finally, obtain the all relations and concepts from the previous step, and add them into corpora respectively.

3.3 Density Extraction Algorithm

The core concepts are the central words mainly depicted by a document. Generally a core concept represents a part of knowledge in the document. All core concepts together organize the content of the whole document. Besides, the document is the description of core concepts and distributed around core concepts. Based on this, we put forward the following assumptions: (1) the content of the document is spread out with core concepts; (2) the relevant part of knowledge about each core concept are distributed into the corresponding block in the document. In the paper, we propose Density Extraction Algorithm that is used to extract the core concepts from a single document based on the line density and uniformity.

Firstly, the document needs to have a segmentation including removing stop words and to be processed in the form of one sentence in one line. Then calculate the line density of each word, with the way that the number of total occurrences is divided by the number of spanning lines. Next, consider uniformity of each word, which refers to uniform distribution of each word in the corresponding block. In order to achieve the word uniformity, we divide the corresponding block of each word into consecutive several bands with equality, at the same time every band is made up with consecutive several rows. To this end each core concept must appear at least specified number of times in all bands. In order to enforce this method perfectly, we need to choose different arguments including the number of bands, the number of consecutive lines in one band and the specified density value according to the length of the specified Chinese document to process.

3.4 Improved Apriori Algorithm

Apriori Algorithm is a kind of realization of association rule analysis, aiming at getting some items that may have correlations between them [11]. The basic idea is that if items often appear in the same sentences, there probably be relations between them. Here we apply it to the term pairs extraction from documents, then find the concrete relations between term pairs.

Firstly, define a terms set $T = \{t_1, t_2, ..., t_m\}$ which contains the basic terms, $S = \{s_1, s_2, ..., s_n\}$ is a sentence set where the s_i expresses the ith sentence in S and is a set that contains some terms in T, $s_i \subseteq T$.

Then define an association rule $A \Rightarrow B$, where $A \subseteq T$, $B \subseteq T$, $A \cap B = \emptyset$. The rule is implemented by support and confidence.

Support of $A \Rightarrow B$ is the ratio of sentences containing both A and B to all sentences referred to as Sup$(A \Rightarrow B)$. It is also the probability $P(A \cup B)$.

Confidence of $A \Rightarrow B$ is the ratio of sentences containing A and B to sentences containing A referred to as Conf$(A \Rightarrow B)$. It is also the probability $P(\frac{A \cup B}{A})$.

In order to get the higher quality of term pairs, we make some improvements in choosing sentences. A sentence must contain at least a term in the concept corpus. From this perspective, it is beneficial to expand the corpora through core concepts beforehand in Part 1.

3.5 Prune Sentences

In a sentence, some nouns are not always used as subject or objects and the verbs are not for predicates. We call these words pseudo nouns and pseudo verbs [8]. For example, noun in preposition + noun. In the paper we take the following methods to prune sentences such that the sentence only includes standard nouns and verbs.

Firstly, merge some consecutive nouns and consecutive verbs because a complete sense of phrase may be split by the segmentation tool. Secondly, as some special words in Chinese which have a collocation with ordinary words will change the part of speech, we need to remove these collocations or change their parts of speech. Next, delete all adjectives and time words and these collocations: preposition + verb, noun + conjunction, conjunction + noun, preposition + noun, noun + particle. Finally, delete all remaining non-nouns and non-verbs.

3.6 Similarity Analysis

In order to expand our concept and relation corpus but no introduction of impurity, we use similarity analysis method including Tongyici Cilin and word2vec to expand our corpora.

The first similarity calculation method is based on synonyms, using the coding and structural features of the extended version of Tongyici Cilin [10]. Tongyici Cilin classification adopts hierarchy system with 5 layers of structure. With increasing of the level, the semantic description is closer and closer, and the similarity and correlation are higher.

Literature [12] presented a kind of semantic method based on context, but selected words must involve cooccurring words in all sentences. Here we find that word2vec is a natural tool for semantic similarity, which converts a word to a vector based on context using deep learning [13]. We compare the terms' similarity by computing the cosine value of their vectors. In addition, word2vec can also compute the similarity between relations.

4 Experiment and Analysis

Purity of the initial corpora is critical to extract the new concepts and relations from the specified domain documents. Initially, We crawls pages in the domain of Computer from Baidu Encyclopedia and extracts concepts and relations according to HTML tags to initialize the corpora. But we find that there is still a few impurities in corpora, so secondary purification is required.

Owing to special nature of Chinese text, we need to carry on the Chinese word segmentation. Here we choose "Jieba"[2] Chinese text segmentation tool. "Jieba" allows user to load a customer's dictionary of segmentation, we combine the relation corpus and the concept corpus as a synthetic term dictionary of "Jieba", to some extent it can solve a part of long tail concepts problem. Then we use the above crawled documents to train a word2vec model. Afterwards pick out 300 concepts from the initial concept corpus manually that can represent the domain of Computer. Then compute the average similarity distance between each term in initial corpora and 300 concepts though word2vec, finally filter the concepts lower than preset threshold that can be viewed as impurities. Similarly for relations. Eventually about 8000 pure concepts and 3000 relations are obtained.

In the paper, the precision rate, recall rate and F1-value are adopted to measure the results of extraction. Specially, we just evaluate the extracted relations from Chinese documents. Here we choose 50 documents from the Chinese version of "Computer Culture" written by June Jamrich Parsons as experiment documents, contents of these documents involve computer memory, operating system, CPU, computer network, database, software engineering and so on.

We mainly discuss two kinds of experiments with different expansion of corpora. One is expansion inside a single document, namely, all documents adopt the same initial corpora. And the other is continuous expansion within all documents. That is, the current document may adopt the corpora produced from the previous document. Here we choose the first 10 documents and compute the assessment values. Concrete results are shown in Fig. 3.

The average F1-value of the first expansion reaches to 76.9 %, even more exciting is that the F1-value of the second expansion reaches to 82.7 %. Besides, for the same document the precision value of the first expansion is generally greater than the second, stable at 85.3 %, and the recall value of the second expansion is commonly greater than the first, as high as 89.6 %. In fact the first expansion just follows the basic procedure depicted by our model, and concepts

[2] https://github.com/fxsjy/jieba.

or relations in the front of the document may not be extracted because initial corpora are smaller and incomplete. Therefore the precision is not too high but stable in the first case. Unlike the first expansion, the second expansion is a continuous process beneficial to find more concepts and relations. For example, a domain concept does not exist in the concept corpus, but after extraction of the last document, the concept is captured and added into the corpus, so in this round, we can extract relative relations and concepts about the concept. We continue to conduct the same experiment in the following 40 documents, the results of the experiments are not quite different.

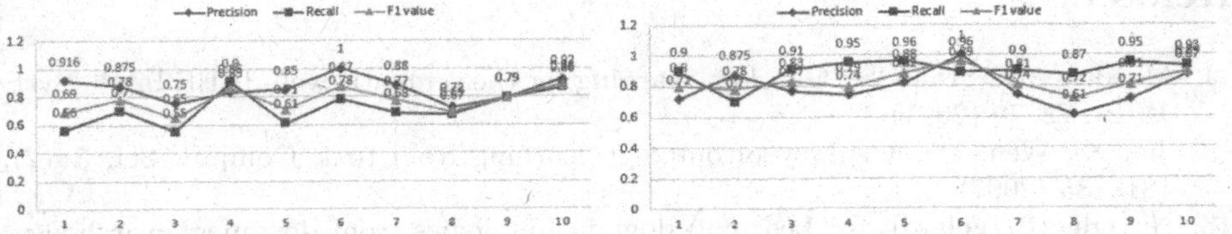

Fig. 3. The first expansion (left) and the second expansion (right)

Here we compare our method with the relevant method [8], which also uses association rule analysis and sentence pruning technique, but adopts different strategies. Detailed comparison is seen in the following table.

Table 1. Comparison with other method

Methods	Precision	Recall	F1
Literature [8] (2013)	77.1 %	71.4 %	72.6 %
Our method	78.6 %	89.6 %	82.7 %

Table 1 shows that all evaluation values of our method are higher than Literature [8] on the same test documents. Although we utilize method of Literature [8], our scalable corpora are expanding while extracting relations, more domain terms not just simple results of text segmentations are extracted.

5 Conclusion

The paper puts forward a new ontology extraction model by setting up two scalable corpora. In this model, expanding corpora and extracting new concepts or relations occur at the same time. Scalable corpora facilitate concept and relation extractions which enrich the corpora in turn. The result of experiment indicates that overall F1-value of this method is 82.7 %, and it is better than related methods. More importantly, our method can perform better and better when extraction continues all the way. Besides, it has a strong portability, that is to say, concepts and relations in other domain can be extracted in the same way without a large number of modification.

Acknowledgments. Hai Wan's research was in part supported by the National Natural Science Foundation of China under grant 61573386, Natural Science Foundation of Guangdong Province under grant 2016A030313292, Guangdong Province Science and Technology Plan projects under grant 2016B030305007, and Sun Yat-sen University Young Teachers Cultivation Project under grant 16lgpy40.

References

1. Maedche, A., Staab, S.: Ontology learning for the semantic web. IEEE Intell. Syst. **16**(2), 72–79 (2001)
2. Jia, X., Wen, D.: A survey of ontology learning from text. Comput. Sci. **34**(2), 181–185 (2007)
3. Navigli, R., Velardi, P.: Learning domain ontologies from document warehouses and dedicated web sites. Comput. Linguist. **30**(2), 151–179 (2004)
4. He, H., Shanhong, Z., et al.: Research on domain ontology the concept extraction based on association rule and semantic rules. J. Jilin Univ. (Info. Sci. Edt.) **32**(06), 657–663 (2014)
5. Hearst, M.A.: Automatic acquisition of hyponyms on large text corpora. In: Proceedings of the 14th International Conference on Computational Linguistics, pp. 539–545, Nantes, France (1992)
6. Buitelaar, P., Daniel, O., et al.: A Protege plug-in for ontology extraction from text based on linguistic analysis. In: Proceedings of the 1st European Semantic Web Symposium (2004)
7. Gu, J., Yan, M., et al.: Research on ontology relation acquisition based on improved association rule. Info. Stud. Theo. Appl. **34**(12), 121–125 (2011)
8. Yu, F., Cheng, H., et al.: Non-hierarchical relations extraction of chinese texts based on grammar rules and improved association rules. Lib. Info. Ser. **57**(22), 126–131 (2013)
9. Zhang, Y., Yang, F., et al.: Study on context based domain ontology the concept extraction and the relation extraction. Appl. Res. Comput. **27**(1), 74–76 (2010)
10. Tian, J., Zhao, W.: The method of word similarity calculation based on synonym word lin. J. Jilin. Univ. **28**(6), 602–608 (2010)
11. Agrawal, R., Ramakrishnan, S.: Fast algorithms for mining association rule in large databases. In: Proceedings of the 20th International Conference on Very Large Data Bases, pp. 487–499. VLDB (1994)
12. Zhang, Y., Yang, F.: Study on context based domain ontology the concept extraction and the relation extraction. Appl. Res. Comput. **27**(1), 74–76 (2010)
13. Mikolov, T., Chen et al.: Efficient estimation of word representations in vector space. arXiv preprint arxiv:1301.3781 (2013)

Entity Linking in Web Tables with Multiple Linked Knowledge Bases

Tianxing Wu(✉), Shengjia Yan, Zhixin Piao, Liang Xu, Ruiming Wang, and Guilin Qi

School of Computer Science and Engineering, Southeast University, Nanjing, China
{wutianxing,sjyan,piaozhx,liang.xu,wangruiming,gqi}@seu.edu.cn

Abstract. The World-Wide Web contains a large scale of valuable relational data, which are embedded in HTML tables (i.e. Web tables). To extract machine-readable knowledge from Web tables, some work tries to annotate the contents of Web tables as RDF triples. One critical step of the annotation is entity linking (EL), which aims to map the string mentions in table cells to their referent entities in a knowledge base (KB). In this paper, we present a new approach for EL in Web tables. Different from previous work, the proposed approach replaces a single KB with multiple linked KBs as the sources of entities to improve the quality of EL. In our approach, we first apply a general graph-based algorithm to EL in Web tables with each single KB. Then, we leverage the existing and newly learned "`sameAs`" relations between the entities from different KBs to help improve the results of EL in the first step. We conduct experiments on the sampled Web tables with Zhishi.me, which consists of three linked encyclopedic KBs. The experimental results show that our approach outperforms the state-of-the-art table's EL methods in different evaluation metrics.

Keywords: Entity linking · Web tables · Linked knowledge bases

1 Introduction

The current World-Wide Web contains a large scale of relational data in the form of HTML tables (i.e. Web tables), which have already been viewed as an important kind of sources for knowledge extraction on the Web. To realize the vision of Semantic Web, various efforts [9–12,19,20] have been made to interpret the implicit semantics of Web tables by annotating their contents as RDF triples. One critical step of such annotation is entity linking (EL), which refers to map the string mentions in table cells to their referent entities in a given knowledge base (KB). For example, in the third column of the Web table in Fig. 1, EL aims to link the string mention "`Michael Jordan`" to the entity "`Michael Jordan (American basketball player)`" in a given KB. Without correct identified entities, the annotation on Web tables is hard to get accurate RDF triples. Thus, in this paper, we focus on studying the problem of EL in Web tables.

Y.-F. Li et al. (Eds.): JIST 2016, LNCS 10055, pp. 239–253, 2016.
DOI: 10.1007/978-3-319-50112-3_18

Team	City	Owner	Arena
Los Angeles Clippers	Los Angeles	Steve Ballmer	Staples Center
Dallas Mavericks	Dallas	Mark Cuban	American Airlines Center
...	...	...	...
Charlotte Hornets	Charlotte	**Michael Jordan**	Time Warner Cable Arena
Chicago Bulls	Chicago	Jerry Reinsdorf	United Center

Fig. 1. An example of web table describing the information of NBA teams

There exist two main problems in previous work for EL in Web tables as follows. **(1)** Many work [9–11,19,21,22] strongly relies on the features based on specific information, such as column headers (e.g. "`Team`", "`City`", etc. in the first row of Fig. 1) in Web tables, entity types in the target KB, and so on. Therefore, it is obvious that these approaches can not work well when the given Web table or KB contains no or few such information. **(2)** Most of the existing approaches [2,9,10,16,19,21,22] only consider linking string mentions in table cells to a single KB, which can not ensure good coverage of EL in Web tables. This problem is also presented in [15] when performing EL in natural language text.

To overcome the above problems, we propose a new general approach for EL in Web tables with multiple linked KBs. The proposed approach contains two steps. We first apply a graph-based algorithm without using any specific information to EL in Web tables with each single KB. Then, we present three heuristic rules leveraging the existing and newly learned "`sameAs`" relations between the entities from different KBs to improve the results of EL in the first step. The second step of our approach can not only reduce the errors generated by EL with each single KB, but also improve the coverage of the EL results. In experiments, we map the string mentions in the cells of sampled Web tables to the entities in Zhishi.me [14], which is the largest Chinese linked open data and composed of three Chinese linked online encyclopedic KBs: Chinese Wikipedia[1], Baidu Baike[2] and Hudong Baike[3]. The evaluation results show that our approach outperforms two state-of-the-art systems (i.e. TabEL [2] and LIEGE [16]) in terms of MRR (i.e. Mean Reciprocal Rank[4]), precision, recall and F1-score.

The rest of this paper is organized as follows. Section 2 outlines some related work. Section 3 introduces the proposed approach in detail. Section 4 presents the experimental results and finally Sect. 5 concludes this work and describes the future work.

[1] https://zh.wikipedia.org.
[2] http://baike.baidu.com.
[3] http://www.baike.com.
[4] https://en.wikipedia.org/wiki/Mean_reciprocal_rank.

2 Related Work

In this section, we review some related work regarding semantic annotation on Web tables, which usually tackles three tasks: entity linking (EL), column type inference and relation extraction between the entities in the same row but different columns. After Cafarella et al. [6] reported that there are more than 150 million Web tables embedded with high-quality relational data, lots of researchers realized that Web tables are important sources that can be used for many applications, such as information extraction and structured data search. Hence, there emerged various work about semantic annotation on Web tables.

Hignette et al. [9] proposed an aggregation approach to annotate the contents of Web tables using vocabularies in the given ontology. It first annotates cells, then columns, finally relations between those columns. Similarly, Syed et al. [19] also presented a pipeline approach, which first infers the types of columns, then links cell values to entities in the given KB, finally selects appropriate relations between columns. Zhang [22] designed a tool called TableMiner for annotating Web tables. TableMiner only focuses on column type inference and EL, and can not extract relations from Web tables. Afterwards, Zhang [21] also proposed some strategies to improve TableMiner. Limaye et al. [10] and Mulwad et al. [11] described two approaches which can respectively jointly model the EL, column type inference and relation extraction tasks for Web tables. The main difference between our approach and these work is that we do not use any specific information for the task of EL, such as column headers and captions of Web tables, entity types in KBs, semantic markups in Web pages, and so on.

There also exists some work in specific scenarios about semantic annotation on Web tables without the step of EL. In the work of Venetis et al. [20], their approach weakens the impacts of EL, and directly infers the types of columns and determines the relationships by the frequency of different patterns in large scale isA and relation databases, which are both built from Web pages but usually unavailable to most of the researchers. Besides, Muñoz et al. [12] proposed an approach to mine RDF triples from Wikipedia tables. In this work, they can directly identify the entities in Wikipedia with internal links and article titles.

The closest work to our approach is done by Shen et al. [16] and Bhagavatula et al. [2]. Shen et al. [16] tried to link the string mentions in list-like Web tables (multiple rows with one column) to the entities in a given KB. Bhagavatula et al. [2] presented TabEL, a table entity linking system, which uses a collective classification technique to collectively disambiguate all mentions in a given Web table. Both of these two work do not use any specific information for EL, and can be applied to any KB. Here, we focus on EL with multiple linked KBs instead of a single KB, in order to improve the quality of EL in Web tables.

3 Approach

In this section, we introduce our proposed approach for entity linking (EL) in Web tables, which consists of two main steps: EL with any single KB and improving EL using "`sameAs`" links between multiple linked KBs.

Team	City	Owner	Arena
Los Angeles Clippers	Los Angeles	Steve Ballmer	Staples Center
Dallas Mavericks	Dallas	Mark Cuban	American Airlines Center
...	...	...	...
Charlotte Hornets	Charlotte	Michael Jordan	Time Warner Cable Arena
Chicago Bulls	Chicago	Jerry Reinsdorf	United Center

related mentions for "Michael Jordan"

Fig. 2. An example of related mentions in the same row or column

3.1 Entity Linking with a Single KB

Candidate Generation. For each string mention in table cells, we first need to identify its candidate referent entities in the given KB. Here, we segment each mention in word level, so each mention can be represented by a set of words. If an entity e in the given KB or one of e's synonyms in BabelNet [13] (a Web-scale multilingual synonym thesaurus) contains at least one word of some mention m, then e is taken as one candidate referent entity of the mention m. For example, the mention "`Charlotte`" has candidate referent entities such as "`Charlotte, North Carolina`", "`Charlotte, Illinois`" and "`Charlotte Hornets`". The results of candidate generation is that each mention may correspond to a set of candidate entities.

Entity Disambiguation. In entity disambiguation, we aim to choose an entity from the candidate set as each mention's referent entity in the given KB. As shown in Fig. 2, we can easily find that mentions in the same row or column tend to be related. In other words, there exists some potential association between any two mentions appearing in the same Web table. Therefore, we choose to jointly disambiguate all the mentions in one table using a graph-based algorithm:

(a) Firstly, for each given table, we build an *Entity Disambiguation Graph* only using mentions and their candidate referent entities as the graph nodes.
(b) Secondly, in each constructed *Entity Disambiguation Graph*, we compute the initial importance of each mention for joint disambiguation and the semantic relatedness between different nodes as the **EL impact factors** to decide whether an entity is the referent entity of a given mention.
(c) Finally, with iterative probability propagation using the **EL impact factors** until convergence, each entity gets its probability to be the referent entity of the given mention and our algorithm makes the EL decisions based on these probabilities.

In the following part of this section, we describe the above three steps in detail.

(a) Building *Entity Disambiguation Graph*. For each given table, we build an *Entity Disambiguation Graph*, which consists of two kinds of nodes and two kinds of edges introduced as follows.

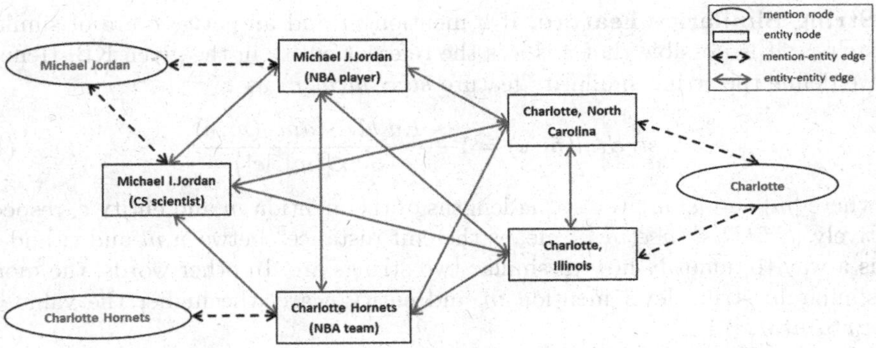

Fig. 3. An example of constructed *Entity Disambiguation Graph*

- **Mention Node**: These nodes refer to the mentions in Web tables.
- **Entity Node**: These nodes represent mentions' candidate referent entities in the given KB.
- **Mention-Entity Edge**: A mention-entity edge is an undirected edge between a mention and one of its candidate referent entities.
- **Entity-Entity Edge**: An entity-entity edge is an undirected edge between entities.

An example of the constructed *Entity Disambiguation Graph* is given in Fig. 3. Due to the limited space, it is only the part of the constructed *Entity Disambiguation Graph* for the Web table in Fig. 2, and lots of nodes and edges are not shown in Fig. 3. Note that each mention such as "`Michael Jordan`" should has a mention-entity edge linking to any of its candidate entities "`Michael J. Jordan (NBA player)`" and "`Michael I. Jordan (CS scientist)`". Entity-entity edges should also be created between all the entity nodes in the graph.

(b) Computing the EL Impact Factors. After constructing the *Entity Disambiguation Graph* for the given Web table, each node or edge is assigned with a probability. For the entity nodes, their probabilities refer to the possibilities of they being the referent entities of mentions, and are initialized as 0 before affected by the **EL impact factors**, which are actually (1) the probabilities of mention nodes, and they can be viewed as the importance of mentions for joint disambiguation; (2) the probabilities assigned to edges, and they are semantic relatedness between nodes. In this paper, we equally treat each mention, so when there exist k mentions in the Web table, the importance of each mention is initialized to $1/k$. Since there are two kinds of edges in each constructed *Entity Disambiguation Graph*, entity-entity edges and mention-entity edges should be respectively associated with the semantic relatedness between entities and that between mentions and entities.

For the semantic relatedness between mentions and entities, we use two features to measure it as follows.

- **String Similarity Feature.** If a mention m and an entity e are of similar strings, it is possible that e is m's the referent entity in the given KB. Hence, we define the string similarity feature $strSim(m, e)$ as

$$strSim(m, e) = 1 - \frac{EditDistance(m, e)}{\max\{|m|, |e|\}} \tag{1}$$

where $|m|$ and $|e|$ are the string lengths of the mention m and entity e, respectively. $EditDistance(m, e)$ means the edit distance[5] between m and e, and it is a way to quantify how dissimilar two strings are. In other words, the more similar in string level mention m and entity e are, the higher the value of $strSim(m, e)$ is.
- **Mention-Entity Context Similarity Feature.** Given a mention m and one of its candidate referent entities e, if they are semantic related, they tend to share similar context. Here, for obtaining the context of the given mention m, we first collect other mentions in the row or column where m locates. Then, we segment each collected mention into a set of words. Finally, we take all the words as the context of m and it is denoted by $menContext(m)$. For the context of the entity e, we first collect all the RDF triples which e exists in, and then segment each object (when e is the subject) or each subject (when e is the object) into a set of words. These words are also treated as e's context $entContext(e)$. To calculate the *mention-entity context similarity feature* $contSim_{me}(m, e)$ between the mention m and the entity e, we apply the Jaccard Similarity[6] as follows:

$$contSim_{me}(m, e) = \frac{|menContext(m) \cap entContext(e)|}{|menContext(m) \cup entContext(e)|} \tag{2}$$

Given a mention m and an entity e, to integrate the *string similarity feature* $strSim(m, e)$ with the *mention-entity context similarity feature* $contSim_{me}(m, e)$, we define the **Mention-Entity Semantic Relatedness** $SR_{me}(m, e)$ as follows:

$$SR_{me}(m, e) = 0.99 \times (\alpha_1 \cdot strSim(m, e) + \beta_1 \cdot contSim_{me}(m, e)) + 0.01 \tag{3}$$

where both α_1 and β_1 are set to 0.5 in this work. $SR_{me}(m, e)$ at least equals 0.01, in order to keep the connectivity of the *Entity Disambiguation Graph* during the subsequent process of probability propagation.

For the semantic relatedness between entities, we also define following two features to measure it.

- **Triple Relation Feature.** If two entities are in the same RDF triple, they are obviously semantic related. Thus, we compute the *triple relation feature* $IsRDF(e_1, e_2)$ between the entity e_1 and the entity e_2 as

$$IsRDF(e_1, e_2) = \begin{cases} 1, & e_1 \textit{ and } e_2 \textit{ are in the same RDF triple} \\ 0, & \textit{otherwise} \end{cases} \tag{4}$$

[5] https://en.wikipedia.org/wiki/Edit_distance.
[6] https://en.wikipedia.org/wiki/Jaccard_index.

– **Entity-Entity Context Similarity Feature.** Similar to the idea introduced in the *mention-entity context similarity feature*, i.e. semantic related entities may be of similar context, and we use the same process for extracting the context of each entity. Given an entity e_1 and an entity e_2, we also use Jaccard Similarity to compute the *entity-entity context similarity feature* $contSim_{ee}(e_1, e_2)$ between their respective context $entContext(e_1)$ and $entContext(e_2)$ as

$$contSim_{ee}(e_1, e_2) = \frac{|entContext(e_1) \cap entContext(e_2)|}{|entContext(e_1) \cup entContext(e_2)|} \tag{5}$$

To acquire the semantic relatedness between an entity e_1 and an entity e_2, we compute the **Entity-Entity Semantic Relatedness** $SR_{ee}(e_1, e_2)$ integrating *triple relation feature* $IsRDF(e_1, e_2)$ with the *entity-entity context similarity feature* $contSim_{ee}(e_1, e_2)$ as follows:

$$SR_{ee}(e_1, e_2) = 0.99 \times (\alpha_2 \cdot IsRDF(e_1, e_2) + \beta_2 \cdot contSim_{ee}(e_1, e_2)) + 0.01 \tag{6}$$

where both α_2 and β_2 are also set to 0.5.

(c) Iterative Probability Propagation. To combine different EL impact factors for the EL decisions, we utilize iterative probability propagation to compute the probabilities associated with entity nodes (i.e. the probabilities for entities to be the referent entities of mentions) until convergence. The detailed process of our proposed iterative probability propagation on each *Entity Disambiguation Graph* is described as follows.

Given an *Entity Disambiguation Graph* $\boldsymbol{G} = (\boldsymbol{V}, \boldsymbol{E})$ containing $\boldsymbol{n}$ nodes (with $\boldsymbol{k}$ mention nodes and $\boldsymbol{l}$ entity nodes), each node is assigned to an integer index from $\boldsymbol{1}$ to $\boldsymbol{n}$. We use these indexes to represent the nodes, and an $n \times n$ adjacency matrix of the *Entity Disambiguation Graph* $\boldsymbol{G}$ is denoted as $\boldsymbol{A}$, where $\boldsymbol{A}_{ij}$ refers to the transition probability from the node i to the node j and $\boldsymbol{A}_{ij} = \boldsymbol{A}_{ji}$. Since the edge between the node i to the node j has been associated with a probability, which is the semantic relatedness (defined in Eqs. 3 and 6) between different nodes, we define $\boldsymbol{A}_{ij}$ as

$$\boldsymbol{A}_{ij} = \begin{cases} \frac{SR_{me}(i,j)}{SR_{me}(i,*)}, & \text{if } i \neq j \text{ and } i \text{ represents a mention node} \\ \gamma \times \frac{SR_{ee}(i,j)}{SR_{ee}(i,*)}, & \text{if } i \neq j \text{ and } i, j \text{ represent two entity nodes} \\ (1-\gamma) \times \frac{SR_{me}(i,j)}{SR_{me}(i,*)}, & \text{if } i \neq j,\ i \text{ is an entity node and } j \text{ is a mention node} \\ 0, & \text{if } i = j \end{cases} \tag{7}$$

where $SR_{me}(i, j)$ is the *mention-entity semantic relatedness* between a mention node and an entity node (defined in Eq. 3), $SR_{me}(i, j) = SR_{me}(j, i)$, $SR_{ee}(i, j)$

is the *entity-entity semantic relatedness* between entity nodes (defined in Eq. 6), $SR_{ee}(i,*)$ means the total *entity-entity semantic relatedness* between i and its adjacent entity nodes and γ is set to 0.5. If i represents a mention node, $SR_{me}(i,*)$ is the total *mention-entity semantic relatedness* between i and all of its adjacent nodes. If i denotes an entity node, $SR_{me}(i,*)$ is the total *mention-entity semantic relatedness* between i and its adjacent mention nodes.

Here, we finally define a notation, i.e. an $n \times 1$ vector $\boldsymbol{r}$ for all the nodes, where $\boldsymbol{r}(i)$ means the probability for the node i to be the referent entity of some mention (if i is an entity node). To compute $\boldsymbol{r}$ with iterative probability propagation, we first set its initial value $\boldsymbol{r}^0$. As introduced before, if the node i is a mention node, $\boldsymbol{r}^0(i)$ is set to the initial importance of i, i.e. $1/k$. If i is an entity node, $\boldsymbol{r}^0(i) = 0$. Then, we update $\boldsymbol{r}$ in the process of iterative probability propagation using other EL impact factors, i.e. *mention-entity semantic relatedness* and *entity-entity semantic relatedness* encoded in the matrix $\boldsymbol{A}$. In this way, the recursive form of $\boldsymbol{r}$ is given as follows:

$$\boldsymbol{r}^{t+1} = ((1-d) \times \frac{\boldsymbol{E}}{n} + d \times \boldsymbol{A}) \times \boldsymbol{r}^t \tag{8}$$

where t is the number of iterations and $\boldsymbol{E}$ is an $n \times n$ square matrix of all 1's. In this formula, to ensure the matrix $\boldsymbol{A}$ is aperiodic to converge, we add a special kind of undirected edges from each node to all the other nodes and give each edge a small transition probability controlled by the damping factor d. In other words, during the process of iterative probability propagation, there exists a probability that the EL impact factors are propagated by neither the defined *mention-entity edges* nor *entity-entity edges*, but the above special kind of edges associated with small transition probabilities. Since the process of iterative probability propagation is similar to the PageRank algorithm [5], we apply the same setting that $d = 0.85$. After the iterative probability propagation, given a mention m and its corresponding set of candidate referent entities $ESet(m) = \{e_1, e_2, ..., e_s\}$, we pick the entity which is of the highest probability in $ESet(m)$ as the referent entity of m.

Different from other methods, our approach for EL in Web tables with a single KB does not rely on any specific information but only general RDF triples. Thus, it can be applied to any KB containing RDF triples in Linking Open Data[7], including DBpedia [1,3], Yago [17,18], Freebase [4], Zhishi.me [14] and etc.

3.2 Improving Entity Linking with Multiple Linked KBs

Entity Linking (EL) in Web tables only with a single KB can not always ensure a good coverage. One solution is to respectively perform the task of EL with different KBs so that we can improve the coverage of the EL results. However, the problem is that there may exist conflicts among the results of EL with different KBs. In this paper, we have done a test on Zhishi.me consisting of three largest Chinese linked online encyclopedic KBs, i.e. Chinese Wikipedia, Baidu Baike and

[7] http://linkeddata.org/.

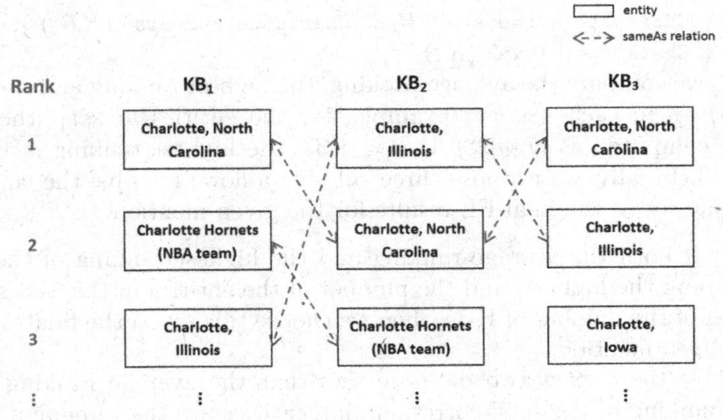

Fig. 4. An example of EL results: the ranking lists of entities in different KBs for the mention "`Charlotte`" in Fig. 2

Hudong Baike. We first apply our proposed approach for EL with a single KB to our extracted Web tables (more than 70 thousand) with Chinese Wikipedia, Baidu Baike and Hudong Baike, respectively. Then, given a mention in a Web table and its identified entities in three KBs, if two identified entities have the "`sameAs`" relation, they can be considered as the same individual, otherwise they are different, i.e. there exists a conflict. According to the statistics, the conflicts exist in totally 38.94 % EL results (one result refers to a mention with its identified entities from different KBs). After that, we observe the EL results of the above test and analyse the reasons for such conflicts among the results of EL with different KBs as follows:

- **Reason 1**: For some KBs, the EL results are really incorrect, that is to say, some potential correct referent entities do not rank the highest.
- **Reason 2**: The "`sameAs`" relations are incomplete between KBs, i.e. there does not exist the "`sameAs`" relations between some equivalent entities from different KBs.

Based on these two reasons, we present our approach in detail to solve the conflicts of EL with different KBs in the following part of this section.

Suppose that there are n different linked KBs. For each given KB, we first apply our proposed approach to EL with a single KB to the given Web table. Then, for each mention, we can get its n ranking lists of referent entities. Afterwards, with the "`sameAs`" relations between entities in different KBs, we group the entities representing the same individual into different sets. For example, in Fig. 4, we can get 4 sets as follows:

(1) Set_1 = {"`Charlotte, North Carolina`" (KB_1), "`Charlotte, North Carolina`" (KB_2), "`Charlotte, North Carolina`" (KB_3)};

(2) Set_2 = {"`Charlotte, Illinois`" (KB_1), "`Charlotte, Illinois`" (KB_2), "`Charlotte, Illinois`" (KB_3)};

(3) $Set_3 = \{$"`Charlotte Hornets`" (KB_1), "`Charlotte Hornets`" (KB_2) $\}$;
(4) $Set_4 = \{$"`Charlotte, Iowa`" $(KB_3)\}$.
After that, we compute the average ranking, the highest ranking and the number of the entities in each set. For example, for the entities in set_1, the average ranking is computed as $(1 + 2 + 1)/3 = 1.33$, the highest ranking is 1 and the number is 3. Finally, we propose three rules as follows to solve the conflicts by choosing one set as the final EL results for the given mention.

- **Rule 1**: If both the average ranking and the highest ranking of the entities in a set rank the highest, and the number of the entities in this set is not less than half of the number of KBs, then we choose this set as the final EL results for the given mention.
- **Rule 2**: If there exist two or more sets that the average ranking and the highest ranking of the sets' corresponding entities are the same and rank the highest, also the number of the entities in each of these sets is not less than half of the number of KBs, then we choose one set at random as the final EL results for the given mention.
- **Rule 3**: If the number of the entities in each set is less than half of the number of KBs, the original EL results of the given mention remain unchanged.

To obtain the global and local optimal EL results at the same time, we consider not only the average ranking and the highest ranking of the entities in each set, but also the number of times that each individual (represented by a set of entities) occurs in different KBs. If the number of the entities in a set is less than half of the number of KBs, it means that the individual represented by these entities is covered by few KBs, so the average ranking is not convincing and it is not reasonable to take this set of entities to solve the conflicts among the results of EL with all the KBs.

According to *Reason 2*, if there exist more "`sameAs`" relations between the entities from different KBs, we may better solve the conflicts with our proposed rules. Here, in order to learn new "`sameAs`" relations, we define three features and train a supervised learning classifier Support Vector Machine (SVM), which is of the best performance in most situations [8]. The proposed features are introduced as follows:

- **Synonym Feature**. This feature tries to detect that whether the strings of two entities may represent synonyms. We input the strings of two entities e_1 and e_2 into BabelNet [13], if these two strings may represent synonyms in BabelNet, the synonym feature $isSyn = 1$, otherwise $isSyn = 0$.
- **String Similarity Feature**. This feature captures the linguistic relatedness between entities. It is denoted by $strSim(e_1, e_2)$, where e_1 and e_2 are entities. We use Eq. 1 to compute this feature using edit distance.
- **Entity-Entity Context Similarity Feature**. For two given entities in different KBs, this feature measures the similarity between the extracted context of entities and is already defined in Eq. 5.

After completing the "`sameAs`" relations with this SVM classifier, we also utilize our proposed rules to decide the final EL results. It is to verify whether the performance is improved compared with that of the rules only using the existing "`sameAs`" relations.

4 Experiments

In this section, we evaluated our approach on the sampled Web tables with three linked KBs (i.e. Chinese Wikipedia, Baidu Baike and Hudong Baike) in Zhishi.me, and compared our approach with two state-of-the-art systems for EL in Web tables and two degenerate versions of our approach.

4.1 Data Set and Evaluation Metrics

Since entity linking in Web tables with multiple linked KBs is a new task, we do not have any existing benchmark. Therefore, we need to generate ground truths by ourselves. We have extracted more than 70 thousand Web tables containing relational data from the Web. We randomly sampled 200 Web tables and invited five graduate students to manually map each string mention in table cells to the entities in each KB of Zhishi.me. The labeled results is based on majority voting and are publicly available[8]. Besides, in order to train the SVM classifier to learn new "`sameAs`" relations between different KBs, we also need to manually generate the labeled data. We first randomly selected 500 existing "`sameAs`" relations as the positive labeled data. Then, we random selected 3,000 entity pairs, each of which consists of the entities from different KBs. Finally, we also asked the five graduate students to label them and 3,000 entity pairs were all labeled as negative.

For each sampled Web table, we performed EL with our approach and the designed comparison methods. We evaluated the results with four metrics, which are Precision, Recall, F1-score and MRR (Mean Reciprocal Rank [7]). F1-score is the harmonic mean of precision and recall. Mean Reciprocal Rank is used for evaluating the quality of the ranking lists. For a mention m, the reciprocal rank in EL is the multiplicative inverse of the rank for m's referent entity. For example, if the correct referent entity of m is in the second place in the ranking lists generated by some EL algorithm, the reciprocal rank is 1/2.

4.2 Comparsion Methods

We compared our approach with the following methods.

- ***TabEL***: TabEL [2] is the current state-of-the-art system for EL in Web tables, and it uses a collective classification technique with several general features to collectively disambiguate all mentions in a given Web table. Besides, any KB can be used for EL in Web tables with TabEL.

[8] https://github.com/jxls080511/MK-EL.

Table 1. The overall EL results evaluated with each single KB

Knowledge base	Approach	Precision	Recall	F1-score	MRR
Chinese Wikipeida	TabEL	0.823	0.809	0.816	0.858
	LIEGE	0.778	0.747	0.762	0.813
	Our-s	0.830	0.797	0.813	0.860
	Our-m-e	0.861	0.821	0.841	0.881
	Our-m-(e+n)	**0.873**	**0.828**	**0.850**	**0.887**
Baidu Baike	TabEL	0.659	0.628	0.643	0.707
	LIEGE	0.629	0.576	0.601	0.670
	Our-s	0.696	0.652	0.673	0.725
	Our-m-e	0.758	0.705	0.731	0.746
	Our-m-(e+n)	**0.774**	**0.727**	**0.750**	**0.776**
Hudong Baike	TabEL	0.681	0.649	0.665	0.780
	LIEGE	0.661	0.632	0.646	0.751
	Our-s	0.708	0.642	0.673	0.768
	Our-m-e	0.729	0.700	0.714	0.787
	Our-m-(e+n)	**0.744**	**0.708**	**0.726**	**0.796**

- ***LIEGE***: LIEGE [16] is a general approach to link the string mentions in list-like Web tables (multiple rows with one column) to the entities in a given KB. It proposes an iterative substitution algorithm with three features to EL in Web lists. This approach can also be applied to EL in Web tables with any KB.
- ***Our-s***: It is a degenerate version of our approach. It only uses our proposed approach for EL with a single KB and does not utilize the rules with "`sameAs`" relations to improve the EL results.
- ***Our-m-e***: It is also a degenerate version of our approach. After performing EL with each single KB, it only uses existing "`sameAs`" relations (without newly learned "`sameAs`" relations) to improve the EL results.

4.3 Result Analysis

In the whole version of our proposed approach (denoted as ***Our-m-(e+n)***), we first apply a graph-based algorithm without using any specific information to EL in Web tables with each single KB, and then we leverage the existing and newly learned "`sameAs`" relations between the entities from different KBs to help improve the results of EL. Table 1 gives the overall results of our approach and the designed comparison methods evaluated with each single KB, and we can see that:

- Our approach for EL with a single KB, i.e. *Our-s*, is comparable to the state-of-the-art system TabEL and outperforms LIEGE, which reflects the effectiveness of our proposed graph-based algorithm.

- *Our-m-e* is always better than *Our-s* in precision, recall, F1-score and MRR. It shows the value of our proposed heuristic rules for improving the EL results of *Our-s*.
- The whole version of our approach, i.e. *Our-m-(e+n)* outperforms all the other comparison methods, which verifies the superiority of our approach for EL in Web tables with multiple linked KBs. Compared with *Our-m-e*, the better performance of *Our-m-(e+n)* demonstrates that the newly learned "`sameAs`" relations are beneficial to solve the conflicts among the EL results of *Our-s* with different KBs.

Besides, we also calculated the precision, recall and F1-score as the evaluation results of our approach (i.e. *Our-m-(e+n)*) on the whole Zhishi.me. The precision is 0.831, recall is **0.903** and F1-score is 0.866. The most important thing is that the recall is significantly improved, which shows EL in Web tables with multiple linked KBs can really ensure a good coverage.

5 Conclusions and Future Work

In this paper, we presented a new approach for EL in Web tables with multiple linked KBs. We first proposed an algorithm based on graph-based iterative probability propagation to perform EL with each single KB. In order to improve the EL results generated by the first step, we then applied three heuristic rules leveraging the existing and newly learned "`sameAs`" relations between the entities from different KBs. The experimental results showed that our approach not only outperforms the designed state-of-the-art comparison methods in different evaluation metrics, but also can use any single KB or linked KBs for EL in the Web tables.

As for the future work, we first will build more benchmarks of other languages for the new task of EL in Web tables with multiple linked KBs and further verify the effectiveness of our approach in other languages, especially English. We then plan to provide APIs or tools as the programming interface of our proposed approach. Finally, we also consider to extend our approach to cross-lingual entity linking in Web tables with multiple linked KBs.

Acknowledgements. This work is supported in part by the National Natural Science Foundation of China (NSFC) under Grant No. 61272378, the 863 Program under Grant No. 2015AA015406 and the Research Innovation Program for College Graduates of Jiangsu Province under Grant No. KYLX16_0295.

References

1. Auer, S., Bizer, C., Kobilarov, G., Lehmann, J., Cyganiak, R., Ives, Z.: DBpedia: a nucleus for a web of open data. In: Aberer, K., et al. (eds.) ASWC/ISWC - 2007. LNCS, vol. 4825, pp. 722–735. Springer, Heidelberg (2007). doi:10.1007/978-3-540-76298-0_52

2. Bhagavatula, C.S., Noraset, T., Downey, D.: TabEL: entity linking in web tables. In: Arenas, M., et al. (eds.) ISWC 2015. LNCS, vol. 9366, pp. 425–441. Springer, Heidelberg (2015). doi:10.1007/978-3-319-25007-6_25
3. Bizer, C., Lehmann, J., Kobilarov, G., Auer, S., Becker, C., Cyganiak, R., Hellmann, S.: Dbpedia-a crystallization point for the web of data. Web Seman. Sci. Serv. Agents WWW **7**(3), 154–165 (2009)
4. Bollacker, K., Evans, C., Paritosh, P., Sturge, T., Taylor, J.: Freebase: a collaboratively created graph database for structuring human knowledge. In: SIGMOD, pp. 1247–1250 (2008)
5. Brin, S., Page, L.: Reprint of: the anatomy of a large-scale hypertextual web search engine. Comput. Netw. **56**(18), 3825–3833 (2012)
6. Cafarella, M.J., Halevy, A., Wang, D.Z., Wu, E., Zhang, Y.: Webtables: exploring the power of tables on the web. PVLDB **1**(1), 538–549 (2008)
7. Craswell, N.: Mean reciprocal rank. In: Liu, L., Özsu, M.T. (eds.) Encyclopedia of Database Systems, p. 1703. Springer, Heidelberg (2009)
8. Fernández-Delgado, M., Cernadas, E., Barro, S., Amorim, D.: Do we need hundreds of classifiers to solve real world classification problems? J. Mach. Learn. Res. **15**(1), 3133–3181 (2014)
9. Hignette, G., Buche, P., Dibie-Barthélemy, J., Haemmerlé, O.: Fuzzy annotation of web data tables driven by a domain ontology. In: Aroyo, L., et al. (eds.) ESWC 2009. LNCS, vol. 5554, pp. 638–653. Springer, Heidelberg (2009). doi:10.1007/978-3-642-02121-3_47
10. Limaye, G., Sarawagi, S., Chakrabarti, S.: Annotating and searching web tables using entities, types and relationships. PVLDB **3**(1–2), 1338–1347 (2010)
11. Mulwad, V., Finin, T., Joshi, A.: Semantic message passing for generating linked data from tables. In: Alani, H., et al. (eds.) ISWC 2013. LNCS, vol. 8218, pp. 363–378. Springer, Heidelberg (2013). doi:10.1007/978-3-642-41335-3_23
12. Muñoz, E., Hogan, A., Mileo, A.: Using linked data to mine RDF from wikipedia's tables. In: WSDM, pp. 533–542 (2014)
13. Navigli, R., Ponzetto, S.P.: Babelnet: building a very large multilingual semantic network. In: ACL, pp. 216–225 (2010)
14. Niu, X., Sun, X., Wang, H., Rong, S., Qi, G., Yu, Y.: Zhishi.me - weaving chinese linking open data. In: Aroyo, L., Welty, C., Alani, H., Taylor, J., Bernstein, A., Kagal, L., Noy, N., Blomqvist, E. (eds.) ISWC 2011. LNCS, vol. 7032, pp. 205–220. Springer, Heidelberg (2011). doi:10.1007/978-3-642-25093-4_14
15. Pereira, B.: Entity linking with multiple knowledge bases: an ontology modularization approach. In: Mika, P., et al. (eds.) ISWC 2014. LNCS, vol. 8797, pp. 513–520. Springer, Heidelberg (2014). doi:10.1007/978-3-319-11915-1_33
16. Shen, W., Wang, J., Luo, P., Wang, M.: Liege: link entities in web lists with knowledge base. In: SIGKDD, pp. 1424–1432 (2012)
17. Suchanek, F.M., Kasneci, G., Weikum, G.: Yago: a core of semantic knowledge. In: WWW, pp. 697–706 (2007)
18. Suchanek, F.M., Kasneci, G., Weikum, G.: Yago: a large ontology from wikipedia and wordnet. Web Seman. Sci. Serv. Agents WWW **6**(3), 203–217 (2008)
19. Syed, Z., Finin, T., Mulwad, V., Joshi, A.: Exploiting a web of semantic data for interpreting tables. In: WebSci, vol. 5 (2010)
20. Venetis, P., Halevy, A., Madhavan, J., Paşca, M., Shen, W., Wu, F., Miao, G., Wu, C.: Recovering semantics of tables on the web. PVLDB **4**(9), 528–538 (2011)

21. Zhang, Z.: Learning with partial data for semantic table interpretation. In: Janowicz, K., Schlobach, S., Lambrix, P., Hyvönen, E. (eds.) EKAW 2014. LNCS (LNAI), vol. 8876, pp. 607–618. Springer, Heidelberg (2014). doi:10.1007/978-3-319-13704-9_45
22. Zhang, Z.: Towards efficient and effective semantic table interpretation. In: Mika, P., et al. (eds.) ISWC 2014. LNCS, vol. 8796, pp. 487–502. Springer, Heidelberg (2014). doi:10.1007/978-3-319-11964-9_31

Towards Multi-target Search of Semantic Association

Xiang Zhang[1,2](✉) and Yulian Lv[3]

[1] School of Computer Science and Engineering, Southeast University, Nanjing, China
x.zhang@seu.edu.cn
[2] Key Laboratory of Data Engineering and Knowledge Services,
Nanjing University, Nanjing, China
[3] College of Software Engineering (Suzhou), Southeast University, Suzhou, China
lvyulian@seu.edu.cn

Abstract. Semantic association represents group relationship among objects in linked data. Searching semantic associations is complicated, which involves the search of multiple objects and the search of their group relationships simultaneously. In this paper, we propose this kind of search as a multi-target search, and we compare it to traditional search tasks, which we classify as single-target search. A novel search model is introduced, and the notion of virtual document is used to extract linguistic information of semantic associations. Multi-target search is finally fulfilled by a PageRank-like ranking scheme and a top-*K* selection policy considering object affinity. Experiments show that our approach is effective in improving retrieval precision on semantic associations.

Keywords: Linked data · Semantic association · Multi-target search

1 Introduction

Linked data provides good practice for connecting and sharing objects by URI and RDF. An important knowledge we can discovered in linked data is the explicit or hidden relationship among objects, which is named as semantic associations. Stated in [1], semantic association is defined as group relationships among multiple objects.

At present, massive semantic associations can be discovered efficiently from large volume of linked data. But few studies have been done on searching these semantic associations and making use of them in a friendly and efficient manner. It was proposed to transform semantic associations into text-based structure and searching based on keywords in [2]. However, this approach still lacks considerations on the group relationships. Traditional keyword-based search only consider the hit of the keywords in single information target in a dataset, which can be a single document or a web page. We name it as single-target search. In our cases, the search model involves the search of multiple objects and the group relationships simultaneously. For example, a search of associations could be a set of keywords: "Tim Berners-Lee", "Ted Nelson" and "Doug Engelbart". Each set of keywords in this search hits an object. User wants to find out the group relationships among these three scientists. Thus, searching semantic associations is a complicated combination of multiple single-target searches.

Y.-F. Li et al. (Eds.): JIST 2016, LNCS 10055, pp. 254–262, 2016.
DOI: 10.1007/978-3-319-50112-3_19

In this paper, we propose a multi-target search model of semantic associations in Sect. 4. The notion of virtual document is used also in Sect. 4 to extract linguistic information of semantic associations. In Sect. 5, multi-target search is fulfilled by a PageRank-style ranking scheme and a top-K selection policy considering object affinity. Experiments are discussed in Sect. 6, showing that our approach is feasible in improving retrieval precision.

2 System Architecture

An overview of the architecture of multi-target search is given in Fig. 1. **Association Miner** discovers meaningful semantic associations in multiple linked data, and then passes the mining results to ***Association Indexer***. The latter uses a ***VDoc Extractor*** to build the linguistic information of each association. Once a user poses a multi-target search on indexed associations, the search will be divided into multiple single-target searches, each performed by a corresponding ***Single-Target Query Processor***. The ***Association Ranker*** will sort the search result based on a ***Static Ranker***, which rank the results using a PageRank-style ranking algorithm to filter out important associations, and also on a ***Top-K Selector***, which utilizes an ***Object Affinity* Assessor** to control the size of result set and makes a reasonable combination of the multiple single-target search into a multi-target one. The final search result will return to user through ***Multi-Target Query Processor***.

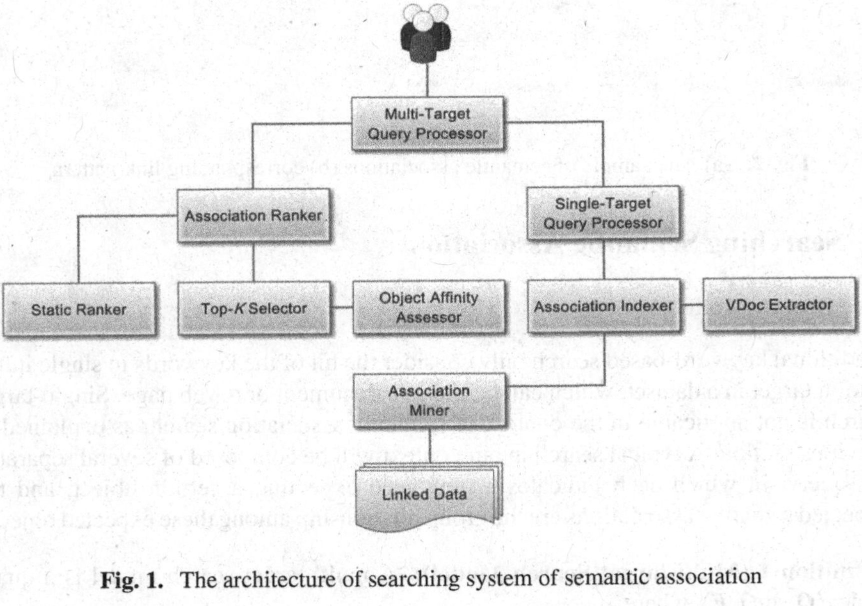

Fig. 1. The architecture of searching system of semantic association

3 Discovering Semantic Associations

Early definitions of semantic association are based on RDF paths between two objects. In [3], Aleman declared that, if two objects are semantically connected by a semantic path then they are semantically associated. In [4], Kochut used defined directionality path to characterize semantic associations. But path-based definition of semantic associations is not able to characterize complex group relationships among multiple objects. Thus, a graph-based definition was proposed in [1], in which a semantic association is a graph instantiation of a frequent subgraph called link pattern.

Figure 2(b) shows a link pattern discovered in real-world linked data. It represents a paper-presenting association among "Author", "Publication" and "Conference". This pattern is discovered by pattern-growth frequent subgraph mining algorithm, assuring the pattern is typical and meaningful. Figure 2(a) presents two semantic associations. One is a paper-presenting event in WWW2004 conference, the other is on ISWC2003 conference. Both are instantiations of the link pattern shown in Fig. 2(b). A discussion was given in [1] on mining link patterns and semantic associations.

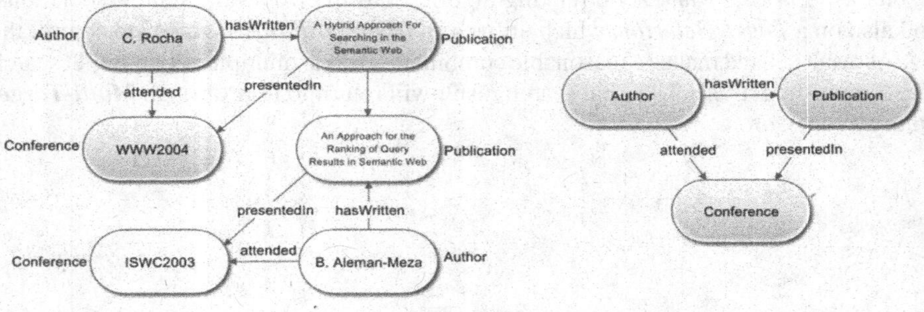

Fig. 2. (a) An example of semantic associations (b) corresponding link pattern

4 Searching Semantic Association

4.1 Search Model

Traditional keyword-based search only consider the hit of the keywords in single information target in a dataset, which can be a single document or a web page. Single-target search is not applicable in the context of semantic association search, as explained in previous section. A typical search in our context will be composed of several separated sub-query, in which each indicates a user need expecting a certain object, and the expected semantic associations are the group relationship among these expected objects.

Definition 1 (Multi-target Search Model): A multi-target search model is a quadruple: $\langle O, A, Q, F \rangle$, where:

(1) O is the set of objects in linked data;
(2) A is a set of semantic associations discovered in linked data;

(3) Q is the query set. A query $q \in Q$ is a set of sub-queries: $q = \{q_1, q_2, \ldots q_k\}$ in which each sub-query is a bag of query keywords, and will be posed to hit a set of objects in O.

(4) $F = \langle F_o, F_Q \rangle$ is the query function comprising two sub-functions. F_o is a function mapping a sub-query to a set of objects. For a sub-query $q_i \in q$, $F_o(q_i) = O_{q_i} \subseteq O$ is the set of hit objects. F_Q is a function mapping a query to a set of semantic associations. For a query $q \in Q$ and $q = \{q_1, q_2, \ldots q_k\}$, $F_o(q) = A_q \subseteq A$ is the research result of q, in which each association contains at least one object from $F_o(q_1)$ to $F_o(q_k)$.

Multi-target search model is a two-step search model. Given a semantic association query containing sub-queries, the first step of the search model aims at finding separated sets of objects using sub-queries. The second step of the model utilizes the sets of objects hit in the first step to find appropriate semantic associations. The sub-queries in multi-target search model enable a search with finer granularity comparing to single-target search model.

4.2 Search Process

The graph structure of the semantic association brings a great barrier to the process of searching. Transforming the graph into text-based structure is a good idea. Therefore, we borrowed the notion of virtual document referred in [5].

Briefly, for a literal node, the description is a collection of words derived from the literal itself; for a URI, it is a collection of words extracted from the local name, *rdfs:label*, *rdfs:comment* and other possible annotations.

Given an object o, its linguistic information is a bag of words defined in Eq. 1:

$$info(o) = \bigcup \{LN(o), RL(o), RC(o), OA(o)\} \quad (1)$$

Here, $LN(o)$ represents the bag of words in the local name of o, $RL(o)$ represents the bag of words in the *rdfs:label* of o, $RC(o)$ represents the bag of the *rdfs:comment* of o and $OA(o)$ represents the bag of the other possible annotations. The operator $\cup$ stands for merging several sets of words together.

When a multi-target search is posed, it will be divided into single-target ones and each single-target search will look for corresponding objects, then a resulted set of associations will be built using hit objects, shown in Algorithm 1.

Algorithm 1 . The process of multi-target search

Input: A query $q = \{q_1, q_2, \dots q_k\}$, in which q_i is a sub-query.

1. for $i = 1$ to k:
   1) find target objects $o_1, o_2, \dots o_x$ that match keywords in q_i;
   2) for each target objects t_j, find semantic associations $a_{j1}, a_{j2}, \dots a_{jy}$ that involve target object o_j; then add these associations into an association set: s_i, which is the result set of the corresponding sub-query.
2. for $i = 1$ to k:
   1) quick sort all associations in each s_i with their internal id;
   2) find s_m, $(1 \le m \le k)$, which contains least semantic associations;
3. for each semantic association a_l in s_m:
   1) for $j = 1$ to k: use binary search to check whether s_j contains a_l;
   2) if every s_j $(j \ne m)$ contains a_l, put a_l into result set R;

Output: Resultset $R = \{a_1, a_2, \dots a_n\}$

5 Ranking Scheme

The time complexity of finding the intersection of semantic association will be $O(h * N \log N)$, when the keywords are frequent words, the amount of hit objects may be very large and response time may be unacceptable for user. So a top-*K* selection policy is used to control the volume of results, and meanwhile to sort objects according to mutual affinity.

A PageRank-like static ranking scheme is firstly performed on resulted objects, to evaluate their importance and authority. For object o_i, its importance is calculated as Eq. 2, in which *N* is total number of objects, $L(o_j)$ is the number of objects that connected with o_j and the value of damping factor *d* is 0.85.

$$PageRank(o_i) = \frac{1-d}{N} + d \sum_{o_j} \frac{PageRank(o_j)}{L(o_j)} \tag{2}$$

Next, for each group of hit objects in a single-target search, top-2 *K* objects are selected according to their PageRank value. Further, top-*K* objects are selected based on the evaluation of their affinity to objects in other groups. For object o_i and o_j, the affinity between them is as Eqs. 3 and 4. Here a_m represents a semantic association.

$$affinity(o_i, o_j) = \sum_{sa_i} \text{connection}(a_m, o_i, o_j) \tag{3}$$

$$\text{connection}(a_m, o_i, o_j) = \begin{cases} 0, & a_m \ \textit{dones't contain } o_i \textit{ or } o_j \\ 1, & a_m \ \textit{contains both } o_i \textit{ and } o_j \end{cases} \tag{4}$$

6 Evaluation

6.1 Dataset

DBpedia is used as dataset in this paper. It is a widely-used linked data on structured information extracted from Wikipedia. As of September 2014, the entire DBpedia dataset contains about 30 billion triples. A part of it is used as our dataset, which includes 54,046,420 triples. 14,608 linked patterns and 47,535,150 semantic associations are discovered from this dataset.

6.2 Evaluation Method

Since we don't find any other similar research about multi-target search of semantic association, we performed experiments comparing our multi-target search model to a pure single-target search model as stated in [2]. Both response time and retrieval precision are evaluated between two search models based on 500,000 randomly selected semantic associations. The size of result is also considered in the experiment because it will affect the time cost for user to understand the results.

A set of keywords are randomly picked to construct 10 variable-length different queries. A human ground truth is built by semantic web experts for the evaluation of precision. For each query, we select top 20 records from two models respectively as a potential set of records. Since the two models may produce the result overlap, the size of set may less than 40. Then we invite experts to pick up the top 10 useful semantic associations these match the query keywords as possible from the set. The statistical information of the amount of hit objects is shown in Table 1, the amount that each group of keywords hit is separated by 'l', the amount of result of each query will be shown in Sect. 6.4

Table 1. Statistical information of hit objects

Query 1	Query 2	Query 3	Query 4	Query 5
179 | 209	11 | 11 | 4	102 | 1 | 13	71 | 120 | 5	1 | 8
Query 6	Query 7	Query 8	Query 9	Query 10
102 | 1 | 6 | 12	7 | 9 | 8	2 | 1	223 | 1	31 | 86

6.3 Evaluation on Response Time and Precision

The horizontal axis in Figs. 3 and 4 present 10 different queries. The vertical axis is the response time and precision of each query in Figs. 3 and 4 respectively. The blue column presents the single-target search model and the red column presents the multi-target search model.

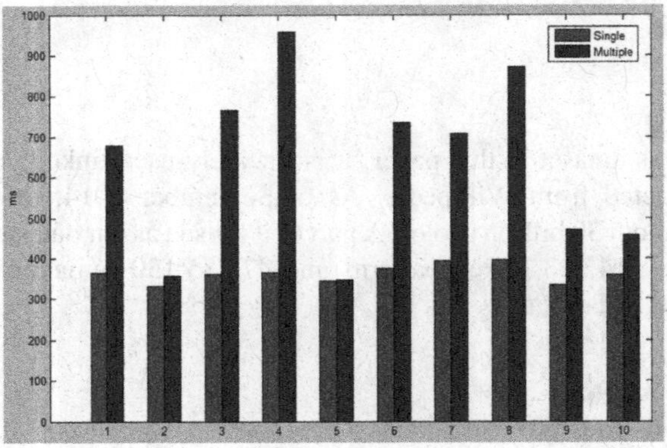

Fig. 3. Evaluation on response time (Color figure online)

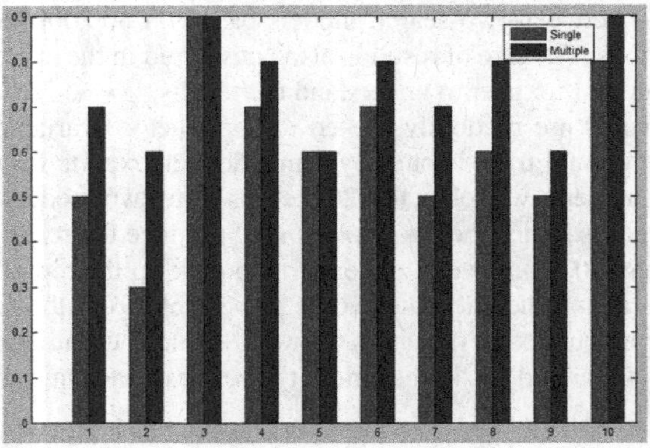

Fig. 4. Evaluation on precision (Color figure online)

Figure 3 shows that, the response time of single-target search model is quite stable, the average response time is about 400 ms, but that of multi-target search vary a lot. When users search with popular keywords, it may hit large quantities of objects, and then get a large size of result. The search process may take a long time. Figure 4 shows that the average precision of single target search model is 61%, the average of multi-target search model is 74%.

6.4 Evaluation on Result Size

Figure 5 clearly presents the significant difference of result sizes with or without top-K selection. The average value of result size with top-K selection is 35, while it is 109 without top-K selection. The result size with top-K selection is only about 12%–72% of that without top-K selection. A top-K selection policy can control the volume of results effectively, and it will also take user less time to distinguish useful information.

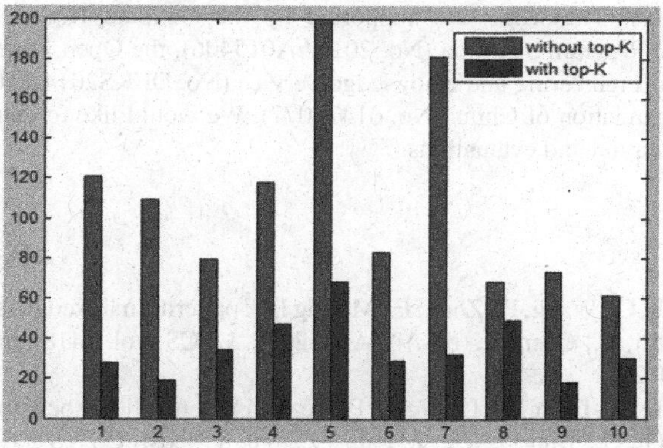

Fig. 5. Evaluation on result size (Color figure online)

7 Related Work

Traditional definition of semantic association is paths connecting two objects. In [3], if two objects are semantically connected by a semantic path then they are semantically associated. In [6], Myungjin proposed a semantic association search methodology that consists of how to find relevant information for a given user's query. In [7], Viswanathan presented a method to find relevant semantic association paths through user-specific intermediate entities.

Virtual document is constructed and used for ontology matching in [8]. The size of virtual document is easy to control, can be applied to search. Therefore we transform the semantic association into text-based structure in [2]. In order to solve the problem of short text, we propose k-step virtual document.

Top-K algorithm is widely used in RDF research. In [9], Tran proposed a novel algorithm for the exploration of top-K matching subgraphs. In [10], Huiying Li present an algorithm for searching top-K answers on RDF data. In our paper, top-K algorithm is used to control the size of result and search space.

8 Conclusions and Future Works

In this paper, a multi-target search model is proposed for searching semantic association. Comparing to single-target search model, our model consider the search of multiple objects and their group relationship simultaneously. The notion of virtual document is used to extract and represent linguistic information of objects and semantic associations. A PageRank-style ranking scheme and a top-K selection policy considering object affinity are used. Experiments show that our approach is feasible in improving retrieval precision.

In our future wok, the type of object will be taken into consideration to achieve higher retrieval accuracy. The search system will be evaluated on its efficiency on large-volume semantic associations.

Acknowledgements. The work was supported by the National High-Tech Research and Development (863) Program of China (No. 2015AA015406), the Open Project of Jiangsu Key Laboratory of Data Engineering and Knowledge Service (No. DEKS2014KT002), and National Natural Science Foundation of China (No. 61472077). We would like to thank Xing Li for his efforts in implementation and evaluations.

References

1. Zhang, X., Zhao, C., Wang, P., Zhou, F.: Mining link patterns in linked data. In: Gao, H., Lim, L., Wang, W., Li, C., Chen, L. (eds.) WAIM 2012. LNCS, vol. 7418, pp. 83–94. Springer, Heidelberg (2012)
2. Wang, C., Zhang, X., Lv, Y., Ji, L., Wang, P.: Searching semantic associations based on virtual document. In: Qi, G., Tang, J., Du, J., Pan, J.Z., Yu, Y. (eds.) CSWS 2013. CCIS, vol. 406, pp. 62–75. Springer, Heidelberg (2013)
3. Aleman-Meza, B., Halaschek-Wiener, C., Arpinar, I.B., Sheth, A.P.: Context-aware semantic association ranking, vol. 1, no. 3, pp. 33–50 (2003)
4. Kochut, K.J., Janik, M.: SPARQLeR: extended SPARQL for semantic association discovery. Semant. Web Res. Appl. **4519**, 145–159 (2007)
5. Le, B.T., Dieng-Kuntz, R., Gandon, F.: On ontology matching problems. In: Proceedings of the International Conference on Enterprize Information Systems, pp. 236–243 (2003)
6. Lee, M., Kim, W.: Semantic association search and rank method based on spreading activation for the semantic web. In: Proceedings of the International Conference on Industrial Engineering and Engineering Management, pp. 523–1527 (2009)
7. Viswanathan, V., Krishnamurthi, I.: Finding relevant semantic association paths through user-specific intermediate entities. Hum. Centric Comput. Inf. Sci. **2**(1), 1–11 (2012)
8. Qu, Y., Hu, W., Cheng, G.: Constructing virtual documents for ontology matching. In: Proceedings of the International Conference on World Wide Web, pp. 23–31 (2006)
9. Tran, T., Wang, H., Rudolph, S., Cimiano, P.: Top-k exploration of query candidates for efficient keyword search on graph-shaped (RDF) data. In Proceedings of the IEEE International Conference on Data Engineering, pp. 405–416 (2009)
10. Li, H., Wang, Y.: Ranked keyword query on semantic web data. In: Proceedings of the International Conference on Fuzzy Systems and Knowledge Discovery, pp. 2285–2289 (2010)

Linking Named Entity in a Question with DBpedia Knowledge Base

Huiying Li(✉) and Jing Shi

School of Computer Science and Engineering, Southeast University, Nanjing 210096, People's Republic of China
{huiyingli,220151530}@seu.edu.cn

Abstract. The emerging Linked Open Data provides an opportunity to answer the natural language question based on knowledge bases (KB). One challenge of the question answering (QA) problem is to link the entity mention in the question with the entity in the existing knowledge base. This study proposes an approach to link entity mention with a DBpedia entity. We propose an entity-centric indexing model to help search candidate entities in KB. After obtaining the candidate entities, we expand the context of the entity mention with WordNet and ConceptNet, we compute the context similarity between the expanded context and the property value of the candidate entity and the popularity of the candidate entity. Finally, we rerank the candidate entities by leveraging these features. Evaluations are performed on DBpedia version 2015, the evaluation tests show that our approach is promising in dealing with linking named entity in DBpedia.

1 Introduction

Linked Open Data (LOD) aims to publish structured data to enable the interlinking of such data and therefore enhance their utility [1]. It shares information that can be read automatically by computers and allows data from different sources to be connected and queried. LOD consists of an unprecedented volume of structured datasets currently amounts to 50 billion facts that are represented as Resource Description Framework (RDF) triples on the web. Recently, many large scale publicly available knowledge bases including DBpedia [2] and YAGO [3] have emerged.

The large amount of Linked Data has become an important resource to support question answering. However, many challenges are encountered in returning a right answer based on the knowledge base for a natural language question (utterance). The following example represents a question and some snippets in DBpedia.

Sample question: "what town was martin luther king assassinated in?".

Sample knowledge base: A snippet of DBpedia dataset is shown in Fig. 1, which lists three different instances named "martin luther king".

Y.-F. Li et al. (Eds.): JIST 2016, LNCS 10055, pp. 263–270, 2016.
DOI: 10.1007/978-3-319-50112-3_20

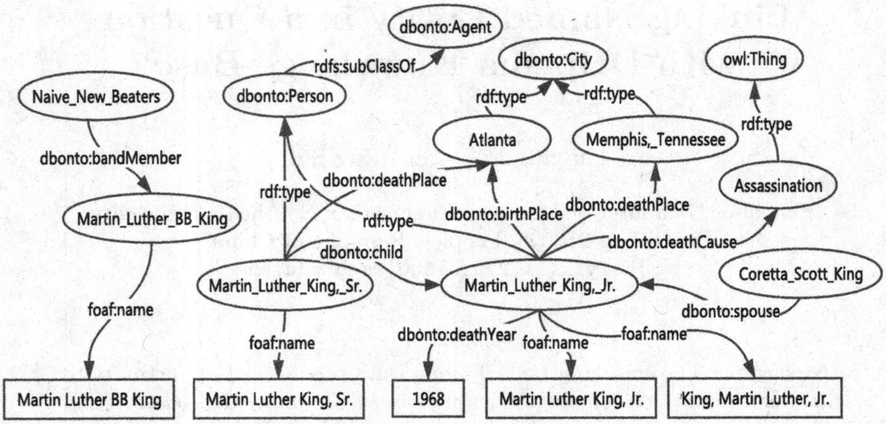

Fig. 1. RDF snippet Example

In examining the sample question and the sample knowledge base, we find that to answer the assassinated town of the civil rights leader Martin Luther King, we should map the queried "martin luther king" to the civil rights leader, then retrieve the assassinated town of this entity from the knowledge base directly.

Hence, the primary task for question answering is to locate the entity mention in the question and link it with an entity in the knowledge base. Locating entity mention (Named Entity Recognition) is out of the scope of our paper. We focus on the entity linking task in this paper. We propose an approach to link the entity mention in a question with an entity in the knowledge base. We index the surface forms for every entity in DBpedia by Lucene. Using this index, we can generate a ranked candidate entity list for each entity mention. Furthermore, the context similarity and entity popularity of the candidate entities are calculated, then we rerank the candidate entities with the combination of these measures.

The remainder of this article is organized as follows: Sect. 2 introduces the related work. Section 3 proposes the method to generate candidate entities. Section 4 presents the named entity linking approach. Section 5 details the experimental results of our approach. Section 6 concludes the study.

2 Related Work

The emergence of large scale knowledge bases like DBpedia and YAGO has spurred great interests in the entity linking task, which maps the textual entity mention to its corresponding entity in the knowledge base. [5] is the first work that considers Wikipedia as an information source for named entity disambiguation. The disambiguation is performed using an SVM kernel that compares the lexical context around the ambiguous named entity to the context of the candidate Wikipedia page. The subsequent work on Wikification [6–8] recognize the global document-level topical coherence of the entities. [6] addresses the entity linking problem through maximizing the agreement between the text of the mention document and the context of the Wikipedia entity, as well as the agreement among the categories associated with the candidate entities. [7] defines the semantic context as a set of unambiguous surface forms in the text, and uses the Normalized Google Distance (NGD) [9] to compute the relatedness. [8] formalizes the Disambiguation to Wikipedia task as an optimization problem with local and global variants, and analyzes their strengths and weaknesses. LINDEN [10] is a framework to link named entities in text with a knowledge base unifying Wikipedia and Word-Net, by leveraging the rich semantic knowledge embedded in the Wikipedia and the taxonomy of the knowledge base. The semantic associativity and semantic similarity are considered based on the constructing of semantic Network. A probabilistic approach is proposed in [11], entity mentions are disambiguated jointly across an entire document by combining a document-level prior of entity co-occurrences with local information captured from mentions and their surrounding context.

3 Candidate Entity Generation

Given an entity mention m, we generate the set of candidate entities E_m in this section. Generally, the candidates in E_m should have the name of the surface form of m. To solve this problem, we need to build an index for all entities in the knowledge base. We also find that the information in the KB is usually entity-centric, and many triples describe the property (attribute and relation) value pairs of an entity. Based on this observation, we consider the entity as the basic index unit.

We group the properties into multiple categories to reduce the fields number and preserve some of the original structure. Totally, we group RDF properties into five fields for a given entity e, and these fields and their values are listed as follows:

- *Name*: The *Name* field collects the name attributes of entity e, and we consider the name attribute as $foaf : name$, $dbonto : alias$, $rdfs : label$ or the attribute ends with "name", "label", or "title".
- *Type*: The *Type* field represents the $rdf : type$ attribute of entity e.
- *Attribute*: The *Attribute* field collects the literal attributes except for the name attributes. The field values are the corresponding attribute values, and the attribute names are also indexed for QA.
- *OutRelation*: The *OutRelation* field collects the object attributes from entity e to other entities e_i. The field values are a list of items, and each item represents an object attribute information for entity e. The item is composed of the object attribute name, the name of entity e_i, and the class and super class type of entity e_i.
- *InRelation*: The *InRelation* field collects the object attributes from other entities e_i to entity e. The field values are a list of items, and each item is composed of the object attribute name, the name of entity e_i, and the class and super class type of entity e_i.

Based on the proposed index method, we index all the entities in KB by Lucene. For each mention m, we look up the index and search for the mention m directly in the *name* field. We take the entity mention as the keyword to perform a search in Lucene. The entities returned by Lucene are considered as candidate entities. Each returned entity is assigned a relevance score by Lucene. We represent this relevance score by $LS(m, e)$, it means the Lucene score for entity e.

For entity mention "martin luther king" in question "what town was martin luther king assassinated in?", three entities in Fig. 1 are returned as candidate entity. The entity with highest Lucene score is *Martin_Luther_BB_King*, then *Martin_Luther_King,_Sr.*, finally *Martin_Luther_King,_Jr.* Based on the context in the question, entity *Martin_Luther_King,_Jr.* should be the linked entity. We find that although our index can help search candidate entities and rank them by Lucene score $LS(m, e)$. It is not enough for named entity linking.

4 Named Entity Linking

In this section, we discuss how to rerank the candidate entities generated in Sect. 3.

4.1 Context Similarity

Our guiding premise is that the property of the target entity should be similar to the context of the entity mention. Given an entity mention m, E_m is the set of candidate entities generated in Sect. 3. Based on the guiding premise, one reranking factor is the cosine similarity between the context of the entity mention and the properties of the candidate entity.

$$CosSim(m, prop) = \frac{m.T \bullet prop.T}{||m.T|| \bullet ||prop.T||} \tag{1}$$

For every property *prop* of candidate entity e, *prop.T* contains the words of the property, the words of the property value, and the words of the value type. Where $m.T$ contains all words occurring around the entity mention in the question. Meanwhile, to expand the context of the entity mention, we import its synonyms and related words with the help of WordNet and ConceptNet. WordNet is used to obtain the synonyms, and ConceptNet is used to obtain the related words and similar words. The *prop.T* and $m.T$ are represented in the standard vector space model, where each component corresponds to a term, and the term weight is the frequency of the term.

$$CS(m, e) = \sum_{prop \in e} CosSim(m, prop) \tag{2}$$

The value of *Context similarity* for each candidate entity e is defined as the sum of the *CosSim* to each property.

4.2 Entity Popularity

Given an entity mention m, E_m is the set of candidate entities. The other reranking factor is the entity popularity. Each $e \in E_m$ containing the same surface form m has different popularity, some entities are rare and some entities are popular for the given surface form. A popular entity means that there are many entities which points to it. Based on the *inRelation* field index in Sect. 3, we can calculate the number of links which point to an entity.

For example, for the entity mention "martin luther king", the entity *Martin_Luther_BB_King* and *Martin_Luther_King,_Sr.* are much rarer than *Martin_Luther_King,_Jr.*, and in most cases when people mention "martin luther king", they mean the civil rights leader rather than other two entities.

Hence, we define the popularity score $PS(e)$ for entity e as:

$$PS(e) = \frac{num(e)}{\sum_{ce \in E_m} num(ce)} \tag{3}$$

where $num(e)$ is the number of links which points to entity e.

4.3 Candiates Reranking

Based on the three score factors introduced above, we can rerank the candidate entities. We consider using a linear reranking function as follows:

$$\hat{e} = \operatorname*{argmax}_{e \in E_m} S(m, e) \tag{4}$$

where $S(m, e)$ is the reranking score for each $e \in E_m$, it can be calculated by:

$$S(m, e) = W \bullet X_m(e) \tag{5}$$

The feature vector $X_m(e)$ is generated for each $e \in E_m$ where $X_m(e) =< LS(m, e), CS(m, e), PS(e) >$. These features affect the reranking score, different features have different degrees of importance. The weight vector W, namely, $W =< w_1, w_2, w_3 >$, which gives different weights for each feature in $X_m(e)$. It can be learned by training on the question answer dataset. Therefore, the candidate entities can be reranked according the their score $S(m, e)$, the $top - 1$ entity $\hat{e}$ is selected as the predicted mapping entity for mention m.

To learn the weight vector W by training on the question answer dataset, we use Support Vector Machine(SVM) based on maximum interval. The SVM is completed by Libsvm toolkit[1].

4.4 Detecting Unlinkable Mention

The disambiguation method discussed above implicitly assumes that DBpedia contains all entities that the entity mention refers to. In practice, there may be

[1] http://www.csie.ntu.edu.tw/~cjlin/libsvm/#nuandone.

contexts where entity mention m refers to an entity e that is not covered in DBpedia, especially when e is not a popular entity. We call this entity mention as unlinkable mention. To deal with the unlinkable mention, we adopt a simple method and learn a threshold τ to validate the top one entity e for entity mention m. If $S(m, e)$ is less than the threshold τ, we consider entity mention m as unlinkable.

5 Experimental Study

We use the WEBQUESTIONS dataset [12], which consists of 5,810 question/answer pairs, as test questions. The questions are split into training and testing sets, which contain 3,778 questions and 2,032 questions, respectively. We learn the weight vector W by training on the training dataset, and test our approach on the testing dataset. The real-world RDF dataset DBpedia 2015 is selected as the KB to answer the questions.

We adopt the evaluation measure precision which is used in most work about entity linking. The precision is calculated as the number of correctly linked entity mentions divided by the total number of all mentions. Since the entities are returned by ranked score, we evaluate the precisions at different k (k means how many entities are returned). The precision of $top - k$ is calculated as the number of entity mentions, which is correctly linked in $top - k$ entities, divided by the total number of all mentions.

Table 1. Feature set effectiveness over the "What" questions

Feature Set	*top*-1		*top*-2		*top*-3	
	Number	Precision	Number	Precision	Number	Precision
LS	452	0.40	522	0.46	560	0.50
LS+CS	512	0.61	563	0.67	589	0.70
LS+PS	508	0.60	536	0.63	538	0.63
LS+CS+PS	545	**0.65**	595	**0.71**	600	**0.72**

To evaluate the effectiveness of our approach on different questions, we consider three typical questions in the testing sets of the WEBQUESTIONS, "What" questions, "Who" questions, and "Where" questions.

Table 1 shows the precisions of the proposed approach on "What" questions. We analyze the effectiveness of different feature sets, it shows the precision and the number of correctly linked mentions obtained by our approach with different feature sets. It can be seen that every feature has a positive impact on the performance of our approach, and with the combination of all features our approach can obtain the best result. The precisions of LS are the results that only Lucene score is considered. The improvement achieved by adding CS (context similarity) feature to LS is greater than that can be achieved by adding PS

(popularity score) feature, which means that the *CS* feature is quite useful to deal with entity linking problem in "What" questions.

It shows the precisions at different *k* ($k = 1$, 2, and 3 respectively). For a tested question, the result is considered to be right when the right target entity is returned in the *top-k* results. The best result is 0.65 when only *top*-1 entity is considered. It is nature the precision increases to 0.72 when *top*-3 entities are considered.

Table 2. Feature set effectiveness over the "Who" questions

Feature Set	*top*-1		*top*-2		*top*-3	
	Number	Precision	Number	Precision	Number	Precision
LS	168	0.63	201	0.75	209	0.78
LS+CS	164	0.62	187	0.70	200	0.75
LS+PS	192	**0.72**	202	**0.76**	209	**0.78**
LS+CS+PS	179	0.67	196	0.74	202	0.76

But the precision of unlinkable mention is only 0.18, it means that the simple threshold setting method is not effective for detecting unlinkable mention. Although the precision of unlinkable mention can be increased by changing the threshold, but it decreases the precision of entity linking.

Table 3. Feature set effectiveness over the "Where" questions

Feature Set	*top*-1		*top*-2		*top*-3	
	Number	Precision	Number	Precision	Number	Precision
LS	138	0.56	169	0.69	180	0.73
LS+CS	143	0.58	167	0.68	173	0.70
LS+PS	172	**0.70**	187	**0.76**	188	**0.76**
LS+CS+PS	154	0.63	176	0.72	177	0.72

Table 2 shows the precisions on "Who" questions. The *PS* feature has a positive impact on the performance of our approach, but the *CS* feature has a slight improve on the performance, sometimes it has a negative impact on the contrary. It leads that our approach only obtains a little improve of precision with the combination of all features.

Table 3 shows the precisions on "Where" questions. The results are similar to those in Table 2, the *PS* feature obtains the best result.

6 Conclusions

In this study, we propose a named entity linking approach on the DBpedia dataset. We set up an entity-centric indexing, candidate entities are returned according to the LS score given an entity mention. Then we consider a set of useful features for entity linking. The CS feature measures the cosine similarity between the context of the mention and the property values of the candidate entity. The PS feature measures the popularity of the candidate entity. Finally, a SVM re-ranker is used to score each candidate entity. The evaluation tests show that our approach is promising in dealing with entity linking problem on Linked Data.

Acknowledgments. The work is supported by the Natural Science Foundation of Jiangsu Province under Grant BK20140643 and the National Natural Science Foundation of China under grant No. 61502095.

References

1. Bizer, C., Heath, T., Berners-Lee, T.: Linked data - the story so far. IJSWIS **5**(3), 1–22 (2009)
2. Zaveri, A., Kontokostas, D., Sherif, M.A., Bühmann, L., Morsey, M., Auer, S., Lehmann, J.: User-driven quality evaluation of DBpedia. In: 9th International Conference on Semantic Systems (I-SEMANTICS 2013), pp. 97–104 (2013)
3. Suchanek, F., Kasneci, G., Weikum, G.: Yago: a large ontology from wikipedia and wordnet. J. Web Semant. **6**(3), 203–217 (2008)
4. Miller, G.A.: WordNet: a lexical database for english. Commun. ACM **38**(11), 39–41 (1995)
5. Bunescu, R., Pasca, M.: Using encyclopedic knowledge for named entity disambiguation. In: Proceedings of the 11th Conference of the European Chapter of the Association for Computational Linguistics, pp. 9–16 (2006)
6. Cucerzan, S.: Large-scale named entity disambiguation based on wikipedia data. In: Proceedings of EMNLP-CoNLL, pp. 708–716 (2007)
7. Milne, D., Witten, I.H.: Learning to link with wikipedia. In: Proceedings of the 17th Conference on Information and Knowledge Management, pp. 509–518 (2008)
8. Ratinov, L., Roth, D., Downey, D., Anderson, M.: Local, global algorithms for disambiguation to wikipedia. In: Proceedings of the 49th Annual Meeting of the Association for Computational Linguistics: Human Language Technologies, pp. 1375-1384 (2011)
9. Cilibrasi, R.L., Vitanyi, P.M.B.: The google similarity distance. IEEE Trans. Knowl. Data Eng. **19**(3), 370–383 (2007)
10. Shen, W., Wang, J., Luo, P., Wang, M.: LINDEN: linking named entities with knowledge base via semantic knowledge. In: Proceedings of WWW, pp. 449–458 (2012)
11. Ganea, O., Ganea, M., Lucchi, A., Eickhoff, C., Hofmann, T.: Probabilistic bag-of-hyperlinks model for entity linking. In: Proceedings of WWW (2016)
12. Berant, J., Chou, A., Frostig, R., Liang, P.: Semantic parsing on Freebase from question-answer pairs. In: Proceedings of the 2013 Conference on Empirical Methods in Natural Language Processing, pp. 1533–1544 (2013)

Applications of Semantic Technologies

Hypercat RDF: Semantic Enrichment for IoT

Ilias Tachmazidis[1(✉)], John Davies[2], Sotiris Batsakis[1], Grigoris Antoniou[1], Alistair Duke[2], and Sandra Stincic Clarke[2]

[1] University of Huddersfield, Huddersfield, UK
i.tachmazidis@hud.ac.uk
[2] British Telecommunications, Ipswich, UK

Abstract. The rapid growth of sensor networks and smart devices has led to the generation of an increasing amount of information. Such information typically originates from various sources and is published in different formats. One of the key prerequisites for the Internet of Things (IoT) is interoperability. The Hypercat specification defines a lightweight JSON-based hypermedia catalogue, and is tailored towards the existing needs of industry. In this work, we propose a semantic enrichment of Hypercat, defining an RDF-based catalogue. We propose an ontology that captures the core of the Hypercat RDF specification and provides a mapping mechanism between existing JSON and proposed RDF properties. Finally, we propose a new type of search, called *Semantic Search*, which allows SPARQL-like queries on top of semantically enriched Hypercat catalogues and discuss how this semantic approach offers advantages over what was previously available.

1 Introduction

In 2014, Innovate UK (the UK's innovation agency) funded the Internet of Things Ecosystem Demonstrator programme. Eight industry-led projects were funded to deliver IoT 'clusters', each centred around a data hub to aggregate and expose data feeds from multiple sensor types.

A major objective of the programme was to address interoperability and this led to Hypercat, a standard for representing and exposing Internet of Things data hub catalogues [6] over web technologies, to improve data discoverability and interoperability. The idea is to enable distributed data repositories (hubs) to be used jointly by applications through making it possible to query their catalogues in a uniform machine-readable format. This enables applications to identify and access the data they need, whatever the data hub in which they are held.

As described in the specification of Hypercat (Beart et al. [3]), this is achieved through employing the same principles on which linked data and the web are built: data accessible through standard web protocols and formats (HTTPS, JSON, REST); the identification of resources through URIs; and the establishment of common, shared semantics for the descriptors of datasets. From this perspective, Hypercat represents a pragmatic starting point to solving the issues

Y.-F. Li et al. (Eds.): JIST 2016, LNCS 10055, pp. 273–286, 2016.
DOI: 10.1007/978-3-319-50112-3_21

of managing multiple data sources, aggregated into multiple data hubs, through linked data and web approaches. It incorporates a lightweight, JSON-based approach based on a technology stack used by a large population of web developers and as such offers a very low barrier to entry.

Each Hypercat catalogue lists and annotates any number of URIs (which typically identify data sources), each having a set of relation-value pairs (metadata) associated with it. In this way, Hypercat allows a server to provide a set of resources to a client, each with a set of semantic annotations. Importantly, there is only a small set of core mandatory metadata relations which a valid Hypercat catalogue must include, thus implementers are free to use any set of annotations to suit their needs. A Hypercat developer community is emerging, with open source tools becoming available[1]. Hypercat provides a standard, machine-processable means for resource discovery, which enables an interoperable ecosystem.

The complexity and diversity of IoT data sets is one of the main reasons why they have emerged as a key use case for linked data and semantic technologies recently[2]. Linked data enable the integration of data into a common, browsable and accessible knowledge graph, while leaving data distributed and managed in different systems, under the control of different contributors. The use of linked data technologies has been effective in many cases where information from different sources needs to be put together in a generic way, to enable a variety of applications, without the need to encode the constraints of the applications in the data model. Semantic web technologies add to this the ability to apply meaningful data models (ontologies) both to improve interoperability between systems, and to enable improved data analysis (see e.g. Lecue et al. [8]). It is therefore natural to consider how the Hypercat specification could be serialised in a semantic language and to investigate the benefits that could accrue from such a materialisation and that is the subject of this paper. One can envisage a more expressive, richer catalogue where data policies/licences, as well as the data flows that relate to them (see d'Aquin et al. [5]) are represented as machine readable information, enabling the implementation of inference rules to support automated reasoning in tasks such as data discovery and policy validation. This can be achieved using Semantic Web technologies [1] such as RDF and OWL that allow for representation of the meaning of data.

In this paper, we propose a semantic enrichment of the Hypercat specification that further increases interoperability by defining an RDF-based catalogue. Catalogue information is published based on a well-defined ontology that: (a) captures the core of the Hypercat RDF specification and (b) provides a mapping mechanism between existing JSON and proposed RDF properties. We describe how existing Hypercat JSON catalogues can be systematically translated into Hypercat RDF catalogues. We then propose a new type of search,

[1] https://hypercatiot.github.io/.

[2] See for example the "Semantic Cities" - http://research.ihost.com/semanticcities14/ - series of workshops.

which allows SPARQL-like queries on top of semantically enriched Hypercat catalogues that capture the semantic hierarchy of classes and properties.

The rest of the paper is organized as follows. Section 2 provides background on Semantic Web technologies. Section 3 presents the current Hypercat 3.00 specification. Section 4 introduces the Hypercat ontology, while Sect. 5 describes the translation of a JSON-based catalogue into an RDF-based catalogue. Section 6 presents the Hypercat RDF specification, while Sect. 7 introduces *Semantic Search*. We conclude in Sect. 8.

2 Background

The Semantic Web [1] evolved out of the Web with the aim to represent Web content in a form that is machine understandable and processable. Today, Web content, in HTML format, retrieved using search engines is typically suitable for human consumption, while content that is generated automatically from databases is usually presented without the original structural information of a given database. Formats such as JSON[3] are used for data exchange in a structured way, enabling machines to parse and generate data. However, even these formats do not address the following problem: the meaning of Web content is not machine-accessible. Semantic Web technologies are used to represent the semantics of Web content in a machine readable form in order to apply intelligent methods (i.e., reasoning) and automate tasks that are currently handled manually by users. Automating tasks is even more important today for the proliferation of connected devices that are part of the *Internet of Things* (IoT) [7].

Machine readable semantics of concepts of an application domain can be defined using Semantic Web standards. Specifically, an *ontology* is an explicit and formal specification of a conceptualization. An ontology consists of definitions of concepts (classes of objects) of the domain and relationships between these concepts (e.g., class hierarchies), and can be defined using the Web Ontology Language *OWL* [2]. OWL is a W3C standard and the current version is OWL 2[4]. Facts about application domain objects and their relations can be asserted using the RDF[5] format. Using RDF, Web resources are connected using a labelled graph representation, and simple ontologies containing descriptions of these resources can be defined using RDF Schema or RDF/S[6].

Using Semantic Web standards such as RDF and OWL, Web resources are represented with machine readable semantics, allowing for automatically interfering implied facts about these resources. Retrieving information represented using RDF format can be achieved using the SPARQL query language[7], which is a W3C standard and the current version is SPARQL 1.1[8]. Furthermore, reasoning and querying can be combined for retrieving not only explicitly asserted

[3] http://www.json.org/.
[4] http://www.w3.org/TR/owl2-overview/.
[5] http://www.w3.org/RDF/.
[6] http://www.w3.org/TR/rdf-schema/.
[7] http://www.w3.org/TR/rdf-sparql-query/.
[8] http://www.w3.org/TR/sparql11-query/.

facts, but also implied facts based on asserted facts and concept definitions and their relations into an ontology. Automatic inference and retrieval is very important when data is voluminous and changes fast (e.g., streaming data) which is a typical case in Internet of Things application scenarios.

3 Hypercat 3.00 Specification

In this section, we provide the basic notions of the Hypercat 3.00 specification[9]. Hypercat is a lightweight JSON-based hypermedia catalogue format for exposing collections of URIs, with each URI having any number of RDF-like triple statements about it.

By definition, a Hypercat catalogue is a file representing an unordered collection of resources on the web, with each item in the catalogue referring to a single resource by its URI. Thus, a Hypercat catalogue may expose a collection of resources, such as data feeds, and provide links to external Hypercat catalogues. Although the definition of a catalogue within a catalogue is not allowed, catalogues may be linked by referring to other catalogue URIs. In addition, a given catalogue may provide metadata about itself and each catalogue item.

The structure of a Hypercat catalogue is defined based on a *Catalogue Object*, which is a JSON object. A given *Catalogue Object* contains the following properties: (a) *items*, which is a list of items (JSON array of zero or more *Item Objects*), and (b) *catalogue-metadata*, which is an array of *Metadata Objects* describing the catalogue object (JSON array of *Metadata Objects*).

An *Item Object* (from the *items* array) is a JSON object, which contains the following properties: (a) *href*, which is an identifier for the resource item (URI as a JSON string), and (b) *item-metadata*, which is an array of *Metadata Objects* describing the resource item (JSON array of *Metadata Objects*).

A *Metadata Object* is a JSON object, which describes a single relationship between the parent object (either the catalogue or catalogue item) and some other entity or concept denoted by a URI, such a relationship is applicable to both the catalogue itself and each catalogue item. A *Metadata Object* contains the following properties: (a) *rel*, which is a relationship between the parent object and a target noun, expressed as a predicate (URI of a relationship as a JSON string), and (b) *val*, the entity (noun) to which the *rel* property applies (JSON string, optionally the URI of a concept or entity).

The structure that is described above constitutes the basic core of any given Hypercat catalogue. However, the Hypercat 3.00 specification defines a far more detailed model compared to the aforementioned description. Thus, in the remainder of this paper, we explore each aspect of the Hypercat 3.00 specification while providing the corresponding semantically enriched solution based on an OWL ontology, which is asserted in RDF format.

---

[9] http://www.hypercat.io/uploads/1/2/4/4/12443814/hypercat_specification_3.00rc1-2016-02-23.pdf.

4 Hypercat Ontology

In this section, we provide the definition of an OWL ontology that captures the aforementioned Hypercat structure, thus providing a translation mechanism from a JSON-based to an RDF-based catalogue. The proposed Hypercat ontology is available with the uri

http://portal.bt-hypercat.com/ontologies/hypercat

and captures the core properties that would enable the development of RDF-based catalogues. Namespaces for the Hypercat ontology can be written prefixing concepts and properties with "hypercat:". Currently, it is part of an IoT Data Hub[10], while as a next step it will be proposed to the Hypercat community for standardization. We believe that providing catalogues in RDF, based on a well-defined ontology, would further increase interoperability and offer intelligent reasoning capabilities.

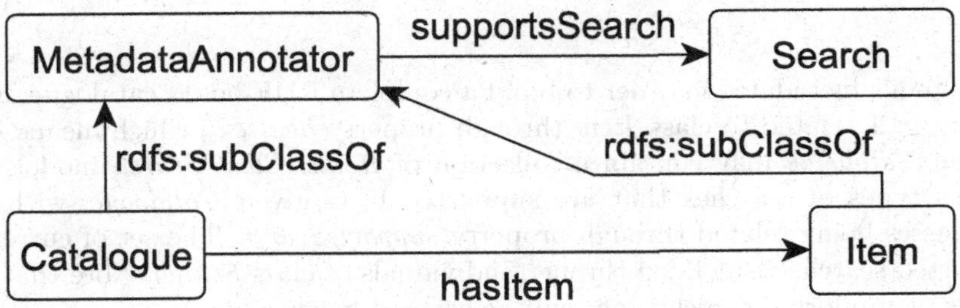

Fig. 1. The hypercat ontology.

The Hypercat ontology consists of a class hierarchy that is depicted in Fig. 1, a range of properties that are included in Tables 1, 2, and 3, and a set of individuals that are described in Table 4. The core hierarchy is rich enough to capture the corresponding constructs of the JSON-based catalogue while providing the flexibility for further extensions.

As described above, a *Metadata Object* is applicable to both the catalogue itself and each catalogue item. Thus, we define class *MetadataAnnotator*, which captures metadata properties that are applicable to both *Catalogue Objects* and *Item Objects.* Note that the Hypercat 3.00 specification defines certain properties as applicable to either *Catalogue Objects* or *Item Objects*, but not both, as such properties cannot be included in the definition of class *MetadataAnnotator*.

Subsequently, class *MetadataAnnotator* has two subclasses, namely class *Catalogue* and class *Item*. In essence, class *Catalogue* models a *Catalogue Object* defining properties that are applicable only to the catalogue's metadata, while class *Item* models an *Item Object* defining properties that are applicable only

[10] http://portal.bt-hypercat.com/.

Table 1. *MetadataAnnotator* properties mapped to existing JSON properties

JSON-based	RDF-based
urn:X-hypercat:rels:hasDescription:en	rdfs:comment
urn:X-hypercat:rels:supportsSearch	hypercat:supportsSearch
urn:X-hypercat:rels:isContentType	hypercat:isContentType
urn:X-hypercat:rels:hasHomepage	hypercat:hasHomepage
urn:X-hypercat:rels:containsContentType	hypercat:containsContentType
urn:X-hypercat:rels:hasLicense	hypercat:hasLicense
urn:X-hypercat:rels:acquireCredential	hypercat:acquireCredential

Table 2. *Catalogue* properties mapped to existing JSON properties

JSON-based	RDF-based
urn:X-hypercat:rels:eventSource	hypercat:eventSource
urn:X-hypercat:rels:hasRobotstxt	hypercat:hasRobotstxt

to an item's metadata. In order to build a complete RDF-based catalogue, class *Catalogue* is related to class *Item* through property *hasItem*, which means that a given *Catalogue* may contain a collection of *Items*. Class *Search* models the various types of searches that are supported by a given *Catalogue*, with the two classes being related through property *supportsSearch*. The set of currently supported searches is defined through individuals of class *Search*. Note that the details of supported search types will be covered below.

5 Hypercat JSON to Hypercat RDF

Prior to exploring each aspect of the Hypercat 3.00 specification, we provide a mapping of existing JSON properties and proposed RDF relations/individuals for each defined class. Tables 1, 2, 3 and 4 can be used in order to create a translator from a JSON-based catalogue to an RDF-based catalogue. Note that this work is in line with recent developments in the Semantic Web community, namely the translation of JSON data to RDF using RML[11].

Table 3. *Item* properties mapped to existing JSON properties

JSON-based	RDF-based
urn:X-hypercat:rels:accessHint	hypercat:accessHint
urn:X-hypercat:rels:lastUpdated	hypercat:lastUpdated

[11] http://rml.io/.

Table 4. *Search* individuals mapped to existing JSON properties

JSON-based	RDF-based
urn:X-hypercat:search:simple	hypercat:SimpleSearch
urn:X-hypercat:search:geobound	hypercat:GeoboundSearch
urn:X-hypercat:search:lexrange	hypercat:LexrangeSearch
urn:X-hypercat:search:multi	hypercat:MultiSearch
urn:X-hypercat:search:prefix	hypercat:PrefixSearch

For example, for the JSON-based catalogue with the uri

http://portal.bt-hypercat.com/cat

the following *catalogue-metadata*, namely metadata about the catalogue itself:

"rel": "urn:X-hypercat:rels:isContentType"
"val": "application/vnd.hypercat.catalogue+json"

will be translated into the following RDF triple, using Table 1:

<http://portal.bt-hypercat.com/cat-rdf>
<http://portal.bt-hypercat.com/ontologies/hypercat#isContentType>
"application/n-triples".

Note that we choose to represent our RDF-based catalogue in N-Triples[12] format. Thus, we define that our RDF-based catalogue is of MIME type "application/n-triples". In addition, the URI of the catalogue needs to be changed from http://portal.bt-hypercat.com/cat to

http://portal.bt-hypercat.com/cat-rdf

since the RDF-based catalogue will be stored in a different location.

This translation pattern applies to all properties of Tables 1 and 2 except for property *supportsSearch* where, for the JSON-based catalogue, both *rel* and *val* contain URIs. Thus, for the JSON-based catalogue http://portal.bt-hypercat.com/cat, the following *catalogue-metadata*:

"rel": "urn:X-hypercat:rels:supportsSearch"
"val": "urn:X-hypercat:search:simple"

will be translated in the following RDF triple, using Tables 1 and 4:

<http://portal.bt-hypercat.com/cat-rdf>
<http://portal.bt-hypercat.com/ontologies/hypercat#supportsSearch>
<http://portal.bt-hypercat.com/ontologies/hypercat#SimpleSearch>.

Note that Table 4 provides the mapping from JSON-based URIs that represent the various types of searches to the RDF-based individuals of class *Search*, which represent semantically the various types of searches.

[12] http://www.w3.org/TR/n-triples/.

Finally, the following *Item Object* (in *items*):

"href":"http://api.bt-hypercat.com/sensors/feeds/UUID"

and "item-metadata" containing

"rel":"urn:X-hypercat:rels:lastUpdated"
"val":"2015-12-01T00:00:00Z"

will be translated in the following RDF triple, using Table 3:

<http://api.bt-hypercat.com/sensors/feeds/UUID>
<http://portal.bt-hypercat.com/ontologies/hypercat#lastUpdated>
"2015-12-01T00:00:00Z".

For properties that are not defined in Tables 1, 2, 3 and 4, Hypercat RDF publishers are encouraged to develop their own OWL ontology by extending the one proposed in this work. Thus, the newly defined ontology would capture the meaning of their catalogue, by defining additional properties, and would enable the full translation of their JSON-based catalogue into an RDF-based catalogue. Alternatively, Hypercat RDF publishers could translate and publish only standardized properties (using Tables 1, 2, 3 and 4). Even though in this case the RDF-based catalogue would contain less information compared to the JSON-based catalogue, it would still be a valid Hypercat RDF catalogue.

In order to fully translate the JSON-based catalogue of the BT Data Hub, an extension of the core ontology has been developed and made available with the uri

http://portal.bt-hypercat.com/ontologies/bt-hypercat

Namespaces for the BT Hypercat ontology can be written prefixing concepts and properties with "bt-hypercat:". In this way, "item metadata" containing

"rel":"urn:X-bt:rels:feedTitle",
"val":"Met Office Datapoint Observations"

that could not be translated using the core ontology, can now be translated in the following RDF triple, using the BT Hypercat ontology:

<http://api.bt-hypercat.com/sensors/feeds/UUID>
<http://portal.bt-hypercat.com/ontologies/bt-hypercat#feed_title>
"Met Office Datapoint Observations".

6 Hypercat RDF Specification

In this section, we examine each aspect of the Hypercat RDF specification by following the structure of (the JSON-based) Hypercat 3.00 specification.

Hypercat File Format Specification: We have already provided a description of the OWL ontology and how an RDF-based catalogue should be developed. In addition, we have presented all standard semantic properties and their correspondence to JSON properties. However, we need to elaborate on several aspects that have not been covered. Thus, each instance of class *MetadataAnnotator* must include the mandatory property *rdfs:comment*, and may include the optional properties *hypercat:isContentType*, *hypercat:hasHomepage*, *hypercat:containsContentType* and *hypercat:supportsSearch*. In addition, each instance of class *Catalogue* must include the mandatory property *hypercat:isContentType*.

As described above, we define each RDF-based catalogue in N-Triples format. Thus, RDF-based catalogues are of MIME type "application/n-triples". In terms of extensibility, we follow the Hypercat 3.00 specification:

- An unknown metadata relationship should be ignored.
- New search method supported by a catalogue server may be added.
- Human readable descriptions may be added in any language.
- Old style catalogues may point to new style and vice versa without version ambiguity.
- A catalogue may contain any number of other properties and classes as developers see fit as long as they are defined in an ontology.

Hypercat Server API Specification: An RDF-based Hypercat server follows the JSON-based Hypercat server specification with several minor adjustments. Every RDF-based Hypercat server must provide a publicly readable "*/cat-rdf*" endpoint serving a Hypercat document asserted in RDF. Requests to an RDF-based Hypercat server, such as insert and delete, can be implemented in a similar fashion as for a JSON-based Hypercat server, while the response will be an RDF-based catalogue instead of a JSON-based catalogue.

A Simple Search Mechanism is implemented in a similar way on top of RDF-based catalogues. For a given RDF-based catalogue http://portal.bt-hypercat.com/cat-rdf, we can advertise that it supports the simple search mechanism by including the following triple:

<http://portal.bt-hypercat.com/cat-rdf>
<http://portal.bt-hypercat.com/ontologies/hypercat#supportsSearch>
<http://portal.bt-hypercat.com/ontologies/hypercat#SimpleSearch>.

All query parameters must be URL encoded and are all optional. If multiple search parameters are supplied, the server must return the intersection of items where search parameters match in a single item, combining parameters with boolean AND.

Simple search supports the following parameters: (a) *s*, which is the N-Triple's subject, (b) *p*, which is the N-Triple's predicate, and (c) *o*, which is the N-Triple's object. Note that each parameter should be inserted in exactly the same form as it would appear in the RDF-based catalogue. For example, we could query a given catalogue based on the following query strings (even though queries must be URL encoded, for readability we present them as plain text):

?s=<http://portal.bt-hypercat.com/ItemID>
?p=<http://portal.bt-hypercat.com/ontologies/hypercat#isContentType>
?p=<http://portal.bt-hypercat.com/ontologies/hypercat#supportsSearch>&
o= <http://portal.bt-hypercat.com/ontologies/hypercat#SimpleSearch>
?o="2015-12-01T00:00:00Z"

Hypercat Subscription: The Hypercat 3.00 specification describes a simple subscription system, providing an API for polling catalogues. A client subscribed to a stream of events from a Hypercat server will receive a stream of events, with each event containing an event name and a body. By first fetching a catalogue and then accumulating catalogue events, a client may keep a synchronised local copy of a given catalogue.

The existing Hypercat subscription mechanism can be used, with minor changes, for an RDF-based catalogue. An RDF-based catalogue which can be used for subscribing to, must be annotated with the property *hypercat:eventsource*. For events concerning a specific catalogue item within a catalogue, the event name is the (unique) N-Triple's subject of all RDF triples for the specified item. Moreover, the event body for an item update event is a set of N-Triples (related to the specified item), while for an item deletion event, the event body is an empty string.

Hypercat Resource Subscription: Hypercat provides the ability to link to resources through URIs. Such resources may contain real-time data required by various applications. Use-cases where client applications require real-time data feeds from devices, hubs or other services are very common in the field of IoT. A possible solution where all data are placed directly into Hypercat catalogues was considered in the past, but was inapplicable due to the simple data model of a JSON-based catalogue. On the other hand, an RDF-based catalogue could serve as a solution for importing data directly into a Hypercat catalogue, given that imported data is semantically enriched and is expressed as N-Triples. However, we believe that such a decision should be part of a wider discussion within the Hypercat community since incorporating data into a catalogue will result in RDF-based catalogues providing a functionality that will not be supported by a JSON-based catalogue.

Hypercat Signing: Hypercat Signing for an RDF-based catalogue remains an open issue as *JSON Web Signature* is not applicable to RDF. Thus, we defer Hypercat signing, based on a well-accepted standard, to future work. However, the same intuition is applicable to an RDF-based catalogue, namely we can create a signature for the entire catalogue or a specific item based on all triples in the catalogue or triples that correspond to a specific item respectively.

Hypercat Security Access Hints: Although systems supporting Hypercat should provide open data with traversable links whenever possible, many systems will potentially provide resources or catalogues only to authenticated clients. Thus, where resources or catalogues are discoverable, but not accessible without authentication, authentication information can be provided to clients.

An item that requires authentication, should point at a machine or human readable description of the authentication method, using property *hypercat:accessHint*. Note that in case multiple *hypercat:accessHint* declarations are present, the client should assume that the resource can be accessed using multiple authentication systems.

Hypercat Security Credential Acquisition: A Hypercat catalogue may support various methods of acquiring access credentials in order to access catalogues and resources. Thus, a catalogue or item, should point at a self-describing web page or resource helping the client acquire credentials, using property *hypercat:acquireCredential*. Note that when multiple *hypercat:acquireCredential* declarations are present, the client should assume that credentials can be acquired in multiple ways.

Hypercat Geographic Bounding Box Search: A geographic search allows for filtering items that fall within a geographic region, which is defined by a bounding box. A given RDF-based catalogue may reuse the following properties:

http://www.w3.org/2003/01/geo/wgs84_pos#lat
http://www.w3.org/2003/01/geo/wgs84_pos#long

which are well-defined by an external ontology. In addition, a given RDF-based catalogue can inform a client that geographic bounding box search is supported, by including the following triple:

<http://portal.bt-hypercat.com/cat-rdf>
<http://portal.bt-hypercat.com/ontologies/hypercat#supportsSearch>
<http://portal.bt-hypercat.com/ontologies/hypercat#GeoboundSearch>.

Geographic search supports the following parameters: (a) *geobound-minlat*, which is the inclusive lower bound of latitude of bounding box, (b) *geobound-maxlat*, which is the inclusive upper bound of latitude of bounding box, (c) *geobound-minlong*, which is the inclusive lower bound of longitude of bounding box, and (d) *geobound-maxlong*, which is the inclusive upper bound of longitude of bounding box. Geographic search queries are submitted by providing the aforementioned parameters.

Hypercat Lexicographic Range Search: Lexicographic search allows searching for items which, when sorted lexicographically, fall between a minimum and maximum. A given RDF-based catalogue can inform a client that lexicographic range search is supported, by including the following triple:

<http://portal.bt-hypercat.com/cat-rdf>
<http://portal.bt-hypercat.com/ontologies/hypercat#supportsSearch>
<http://portal.bt-hypercat.com/ontologies/hypercat#LexrangeSearch>.

The property *hypercat:lastUpdated* could be used for dates and time.

Lexicographic search supports the following parameters: (a) *lexrange-p*, which is the N-Triple's predicate to search on, (b) *lexrange-min*, which is the lower bound of range to return (inclusive), and (c) *lexrange-max*, which is the upper bound of range to return (non-inclusive). Lexicographic search queries are submitted by providing the aforementioned parameters.

Hypercat Robots Exclusion Search: A given Hypercat RDF catalogue can contain information for the client about an associated *robots.txt* file, which is used by websites to communicate with web crawlers and other web robots, using property *hypercat:hasRobotstxt*. Note that if a robots exclusion file is provided, it must be located at the *BASE_URL* of a catalogue and it must be named *robots.txt*, namely N-Triple's object should contain a URI of the form "[BASE_URL]/robots.txt".

Hypercat Multi-Search: Hypercat supports several different search extensions, which are mainly variations of the simple search, and thus, only allow for simple interactions with a catalogue. Combining geographic search and lexicographic range search could be done by submitting two independent queries and then processing the results accordingly. However, such a solution is inefficient.

Multi-search allows a client to combine single or multiple search mechanisms supported by a server so as to retrieve only the items of interest. A given RDF-based catalogue can inform a client that multi-search is supported, by including the following triple:

<http://portal.bt-hypercat.com/cat-rdf>
<http://portal.bt-hypercat.com/ontologies/hypercat#supportsSearch>
<http://portal.bt-hypercat.com/ontologies/hypercat#MultiSearch>.

In order to retain compatibility between JSON-based and RDF-based catalogues, a multi-search query, for a given RDF-based catalogue, could be submitted as a single JSON object. Multi-search supports the following parameters: (a) *query*, which is a JSON string, holding a URL query string as passed to underlying search mechanism, (b) *intersection*, which is a JSON array of objects that could contain *query*, *intersection* or *union*, and (c) *union*, which is a JSON array of objects that could contain *query*, *intersection* or *union*. Searches may be nested to allow complex mixing of union and intersection.

Hypercat Prefix Match Search: Prefix match search allows searching for items where the N-Triple's object specified in the query is a prefix match of the N-Triple's object in a triple describing a catalogue item. As with simple search, any N-Triple's predicate can be used. A given RDF-based catalogue can inform a client that prefix match search is supported, by including the following triple:

<http://portal.bt-hypercat.com/cat-rdf>
<http://portal.bt-hypercat.com/ontologies/hypercat#supportsSearch>
<http://portal.bt-hypercat.com/ontologies/hypercat#PrefixSearch>.

If multiple search parameters are supplied, the server must return the intersection of items where search parameters match a single item, combining parameters with boolean AND. Prefix match search supports the following parameters: (a) *prefix-s*, which is a prefix of the N-Triple's subject, (b) *prefix-p*, which is a prefix of the N-Triple's predicate, and (c) *prefix-o*, which is a prefix of N-Triple's object.

Hypercat Linked Data rel: The newly introduced Hypercat JSON *rel* for specifying that the item at hand is an instance of an RDF class, namely http://www.w3.org/1999/02/22-rdf-syntax-ns#type is part of the RDF concepts vocabulary. Thus, we do not need to include a new property.

Hypercat License rel: Hypercat catalogues or linked resources may be available under a specific license. Thus, in order to allow clients to determine the license under which the data is released, we specify the property *hypercat:hasLicense*, which should point at a machine or human readable version of a license. Where multiple *hypercat:hasLicense* declarations are present, the client should assume that the resource is available under multiple licenses.

7 Semantic Search

Semantic search allows SPARQL-like queries on top of semantically enriched Hypercat catalogues, providing a searching mechanism that captures the underlying semantic hierarchy. Given a query where *rel* (or N-Triple's predicate) is *rdf:type* and *val* (or N-Triple's object) is *bt-hypercat:Feed*, with BT catalogue's ontology defining that *bt-hypercat:SensorFeed* is subclass of *bt-hypercat:Feed*, semantic search will return all catalogue items that are instances of both *bt-hypercat:Feed* and *bt-hypercat:SensorFeed* - this would not be possible without the use of the Hypercat RDF to encode the subclass relationship.

A given JSON-based catalogue can inform a client that semantic search is supported, by including the following *rel val* pair:

"rel": "urn:X-hypercat:rels:supportsSearch"
"val": "urn:X-hypercat:search:semantic"

while a given RDF-based catalogue must include the following triple:

<http://portal.bt-hypercat.com/cat-rdf>
<http://portal.bt-hypercat.com/ontologies/hypercat#supportsSearch>
<http://portal.bt-hypercat.com/ontologies/hypercat#SemanticSearch>.

If multiple search parameters are supplied, the server must bind them to a single triple pattern and run a SPARQL query, including reasoning based on both catalogue's ontology and catalogue itself. Semantic search supports the following parameters for JSON-based (resp. RDF-based) catalogues: (a) *sem-href* (resp. *sem-s*), which is a resource URI (resp. the N-Triple's subject), (b) *sem-val* (resp. *sem-p*), which is a semantic metadata relation (resp. N-Triple's predicate), and (c) *sem-rel* (resp. *sem-o*), which is a semantic metadata value (resp. N-Triple's object).

In terms of implementation, reasoning over the given Hypercat ontology can be performed using standard reasoners such as Pellet[13] and HermiT[14],

[13] http://clarkparsia.com/pellet/.
[14] http://hermit-reasoner.com/.

while querying can be based on the query engine of Apache Jena[15]. Alternatively, an OBDA approach proposed by Botoeva et al. [4] can be considered in order to perform SPARQL queries over JSON data.

8 Conclusion

In this work, we presented a semantic enrichment of the Hypercat specification, which allows the definition of an RDF-based catalogue. We proposed an ontology that captures the core of the Hypercat RDF specification. In addition, we showed how existing JSON-based catalogues can be translated into RDF-based catalogues in an automated fashion. Finally, we proposed a new type of search, called *Semantic Search*, which allows SPARQL-like queries on top of semantically enriched catalogues.

In future work, we plan to propose and standardize the Hypercat RDF specification by working closely with the Hypercat community. In addition, we intend to collaborate with existing partners in order to provide a publicly available converter from JSON-based to RDF-based catalogues, and a publicly available implementation of *Semantic Search*. In this way, richer Hypercat catalogues will provide a higher degree of interoperability.

References

1. Antoniou, G., van Harmelen, F.: A Semantic Web Primer. Cooperative Information Systems, 2nd edn. The MIT Press, Cambridge (2008)
2. Antoniou, G., van Harmelen, F.: Web ontology language: OWL. In: Staab, S., Studer, R. (eds.) Handbook on Ontologies. IHIS, pp. 67–92. Springer, Heidelberg (2004)
3. Beart, O.: Hypercat 3.00 specification (2016)
4. Botoeva, E., Calvanese, D., Cogrel, B., Rezk, M., Guohui Xiao, O., Relational, B.: DBs: a study for MongoDB. In: Proceedings of the 29th International Workshop on Description Logics, Cape Town, South Africa, 22–25 April 2016 (2016)
5. d'Aquin, M., Adamou, A., Daga, E., Liu, S., Thomas, K., Motta, E.: Dealing with diversity in a smart-city datahub. In: ISWC, pp. 68–82 (2014)
6. Davies, J., Fisher, M.: Internet of things - why now? J. Inst. Telecommun. Prof. **7**(3), 36–42 (2015)
7. Gubbi, J., Buyya, R., Marusic, S., Palaniswami, M.: Internet of things (IoT): a vision, architectural elements, and future directions. Future Gener. Comput. Syst. **29**(7), 1645–1660 (2013)
8. Lécué, F., Tucker, R., Bicer, V., Tommasi, P., Tallevi-Diotallevi, S., Sbodio, M.: Predicting severity of road traffic congestion using semantic web technologies. In: Presutti, V., d'Amato, C., Gandon, F., d'Aquin, M., Staab, S., Tordai, A. (eds.) ESWC 2014. LNCS, vol. 8465, pp. 611–627. Springer, Heidelberg (2014). doi:10.1007/978-3-319-07443-6_41

[15] https://jena.apache.org/index.html.

Enabling Spatial OLAP Over Environmental and Farming Data with QB4SOLAP

Nurefşan Gür[1(✉)], Katja Hose[1], Torben Bach Pedersen[1], and Esteban Zimányi[2]

[1] Department of Computer Science, Aalborg University, Aalborg, Denmark
{nurefsan,khose,tbp}@cs.aau.dk

[2] Department of Computer and Decision Engineering, Université Libre de Bruxelles, Bruxelles, Belgium
ezimanyi@ulb.ac.be

Abstract. Governmental organizations and agencies have been making large amounts of spatial data available on the Semantic Web (SW). However, we still lack efficient techniques for analyzing such large amounts of data as we know them from relational database systems, e.g., multidimensional (MD) data warehouses and On-line Analytical Processing (OLAP). A basic prerequisite to enable such advanced analytics is a well-defined schema, which can be defined using the QB4SOLAP vocabulary that provides sufficient context for spatial OLAP (SOLAP). In this paper, we address the challenging problem of MD querying with SOLAP operations on the SW by applying QB4SOLAP to a non-trivial spatial use case based on real-world open governmental data sets across various spatial domains. We describe the process of combining, interpreting, and publishing disparate spatial data sets as a spatial data cube on the SW and show how to query it with SOLAP operators.

1 Introduction

In late 2012, the Danish government joined the Open Data movement by making several raw digital data sets [3] freely available at no charge. These data sets span domains such as environmental data, geospatial data, business data from transport to tourism, fishery, forestry, and agriculture. GovAgriBus Denmark[1] was an initial effort in 2014 to make Danish government Open Data from various domains available as Linked Open Data (LOD) [2] on the Semantic Web in order to pose queries across domains. If the corresponding domains can be related through space and location, spatial attributes of these data sets become particularly interesting as we can derive spatial joins and containment relationships that were not encoded in the original data sets. Danish government organizations and agencies continue publishing data sets for new domains and update the corresponding data sets regularly on a yearly basis, which brings opportunities in querying the expanding spatial data with analytical perspectives on the Semantic Web. Responding to such queries is a complex task, which requires well-defined schemas to facilitate OLAP operations on the Semantic Web. QB4SOLAP [7]

[1] https://datahub.io/dataset/govagribus-denmark.

Y.-F. Li et al. (Eds.): JIST 2016, LNCS 10055, pp. 287–304, 2016.
DOI: 10.1007/978-3-319-50112-3_22

aims to support intelligent multidimensional querying in SPARQL by providing context to SOLAP and its elements on the SW. However, QB4SOLAP has not been applied on complex real-world data yet. This could bring particular challenges with the use of real-world spatial data. In this paper, we address the challenging problem of multidimensional querying with SOLAP operations on the SW by applying QB4SOLAP on real-world open governmental data sets from various domains. These domains span from livestock farming to environment, where many of them have spatial information.

In this paper, we design a spatial data cube schema with data from livestock farming, environment, and geographical domains. Every data set is downloaded from different governmental sources in various formats. The downloaded data is prepared and conciliated with a spatial data cube schema in order to publish it on the SW with QB4SOLAP. We use the common SOLAP operators [8] on the spatial data cube for advanced analytical queries. These analytical queries give perspective on the use case data sets that are linked and published with QB4SOLAP as a unified spatial data cube. Having the use case data sets with spatial attributes also allows us to reveal patterns across the use case domains that were not possible before. We share our experiences with the best practices and methods together with the lessons learned. Finally, we show how to formulate and execute SPARQL queries with individual and nested SOLAP operations.

The remainder of this paper is structured as follows. Section 2 presents the background and motivation, Sect. 3 discusses related work and presents the state-of-the-art spatial data cubes on the SW. Section 4 presents the data sources for the use case while Sect. 5 describes how to annotate and publish the use case data as a spatial data cube on the SW. Section 6 presents SOLAP operators and their SPARQL implementation. Section 7 presents a brief overview of the process and reflects on the problems and improvements. Finally, Sect. 8 concludes the paper with an outlook to future work.

2 Background and Motivation

The Semantic Web supports intelligent querying via SPARQL with active inference and reasoning on the data in addition to capturing its semantics. Linked Open Data on the Semantic Web is an important source to support Business Intelligence (BI). Multidimensional data warehouses and OLAP are advanced analytical tools in analyzing complex BI data. State-of-the-art SW technologies support advanced analytics over *non-spatial* SW data. QB4SOLAP supports intelligent multidimensional querying in SPARQL by providing context to spatial data warehouses and its concepts. Variety of the data is an intriguing concept on both the Semantic Web and in complex BI systems. The variety of the data and heterogeneous representation formats (e.g., CSV, JSON, PDF, XML, and SHP) require underlying conceptualizations and data models with well-defined spatial (and temporal) dependencies, which can be modeled with QB4SOLAP in order to answer complex analytical queries. Complex queries cannot be answered from within one domain alone but span over multiple disciplines and various data sources. As a result, this paper is driven by the motivation of using QB4SOLAP as a proof of concept for spatial data warehouses on the Semantic Web by using open (government) data of various domains from different sources, which creates a non-trivial spatial use case.

3 State of the Art

Data warehouses and OLAP technologies have been successful for analyzing large volumes of data [1], including integrating with external data such as XML [16]. Combining DW/OLAP technologies with RDF data makes RDF data sources more available for interactive analysis. Kämpgen et al. propose an extended model [11] on top of the RDF Data Cube Vocabulary (QB) [4] for interacting with statistical linked data via OLAP operations directly in SPARQL. In OLAP4LD [10], Kämpgen *et al.* suggest enhancing query performance of OLAP operations expressed as SPARQL queries by using RDF aggregate views. The W3C published a list of RDF cube implementations [19]. However, they all have inherent limitations of QB and thus cannot support OLAP dimensions with hierarchies and levels, and built-in aggregate functions.

Etcheverry et al. [6] introduce QB4OLAP as an extended vocabulary based on QB, with a full MD metamodel, supporting OLAP operations directly over RDF data with SPARQL queries. Matei *et al.* [12] use QB and QB4OLAP as a basis to support OLAP queries in Graph Cube [20] with the IGOLAP vocabulary. Jakobsen *et al.* [9] study OLAP query optimization techniques over QB4OLAP data cubes. However, none of these approaches and vocabularies support *spatial* DWs.

QB4SOLAP(v1) [7] is the first attempt to model and query spatial DWs on the SW, and QB4SOLAP(v2) [8] is a foundation for spatial data warehouses and SOLAP operators on the SW, which is currently under submission with completely revised formal semantics of SOLAP operators and SPARQL query generations algorithms. QB4SOLAP is an extension of QB4OLAP with spatial concepts. The QB4OLAP vocabulary is compatible with the QB vocabulary. Therefore QB4SOLAP provides backward compatibility with other statistical or MD data cube vocabularies in addition to providing spatial context for querying with SOLAP. Figure 1 depicts the QB4SOLAP(v2) vocabulary. Capitalized terms with non-italic font represent RDF classes, capitalized terms with italic font represent RDF instances, and non-capitalized terms represent RDF properties. Classes in external vocabularies are depicted in light gray background and font. QB, QB4OLAP, and QB4SOLAP classes are shown with white, light gray, and dark gray backgrounds. Original QB terms are prefixed with `qb:`[2]. QB4OLAP and QB4SOLAP terms are prefixed with `qb4o:`[3] and `qb4so:`[4]. Spatial classes are prefixed with `geo:`[5], where the spatial extension to QB4SOLAP is based on the GeoSPARQL [15] standard from the Open Geospatial Consortium (OGC) for representing and querying geospatial linked data on the SW.

[2] RDF Cube: http://purl.org/linked-data/cube#.
[3] QB4OLAP: http://purl.org/qb4olap/cubes#.
[4] QB4SOLAP: http://w3id.org/qb4solap#.
[5] GeoSPARQL: http://www.opengis.net/ont/geosparql#.

QB4SOLAP is a promising approach for modeling, publishing, and querying spatial data warehouses on the SW. However, it has only been validated with a synthetic use case. Andersen et al. [2] consider publishing/converting open Danish governmental spatial data as Linked Open Data without considering the MD aspects of geospatial data. In this paper, however, we validate QB4SOLAP with a non-trivial use case, which is created as a spatial data cube from open Danish government spatial data. Furthermore, we show how to exploit multidimensional spatial linked data on the SW, which is not solely about adding semantics and linking disparate data sets on the SW, but also about enabling analytical queries by modeling them as spatial data cubes.

4 Source Data

In order create a spatial data cube of livestock holdings in Danish farms, we have gathered data that is published by different agencies in Denmark. We have found these domains to be particularly interesting as they represent a non-trivial use case that covers spatial attributes and measures, which can be modeled in a spatial data cube for multidimensional analysis. In the following we first give a brief overview of the flat data and their sources, and then represent the whole use case data set as a spatial data cube in Sect. 5.

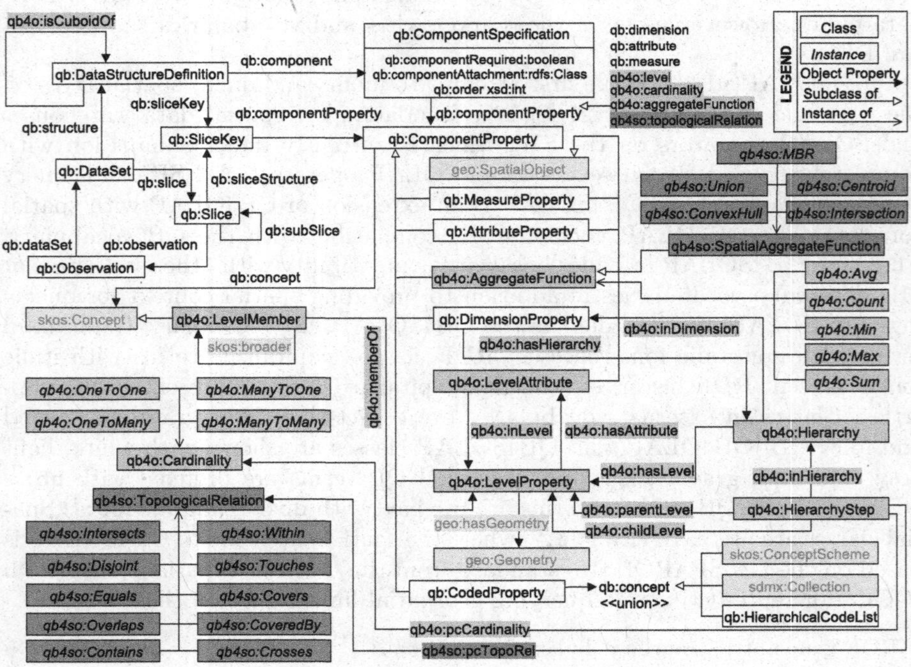

Fig. 1. QB4SOLAP vocabulary

The environmental protection agency under the Ministry of Environment and Food of Denmark regulates the livestock units (DE)[6] per area in order to keep nitrate leaching under control in vulnerable areas. Prohibition rules against the establishment of livestock farms and the siting of animal housing are determined with respect to livestock units and distance to specific natural habitats (e.g., ammonia vulnerable areas, water courses, and water supply facilities etc.) [13].

Livestock Farming (CHR) Data. The Ministry of Environment and Food of Denmark (http://en.mfvm.dk) publishes the central husbandry (livestock) registry (CHR) data, which is the central database used for registration of holdings and animals. We refer to this set of data as *CHR data.* We have downloaded several relevant data sets from http://jordbrugsanalyser.dk in livestock farming domain. The CHR data collection is downloaded in SHP format as 6 data sets, where each data set represents the state of the farms for a year between the years 2010 to 2015. SHP format is used for *shapefiles*, which store geometry information of the spatial features in a data set. In each shapefile, there is information about more than 40,000 farms. In total, the CHR data collection contains around 240,000 records. Farm locations are given as `(X,Y)` point coordinates. Each data set has 24 attributes in which the important ones are: *CHR - Central Husbandry (animal) Registry (holding) number, CVR - Central Company Registry number (owner company of the holding), DE (Livestock unit), Address of the holding (Postnr and Commune), Geographical position of the holding (X and Y coordinates), Different type of normalized herds, Number of animals for each herd, Animal code and label, Animal usage code and purpose.*

Environmental Data. Public environmental data is published on Denmark's environment portal http://www.miljoeportal.dk/, where we can find information about nitrate catchment areas and vulnerable sites. The soil measurements contain data from 2008 to 2015 [14]. We downloaded the data sets in SHP format, which have recently become available on the portal. The files record measurements of the soil quality across Denmark. The environmental data collection contains 3 data sets about nitrogen reduction potentials and phosphor and nitrate classifications of the soil. Temporal validity of the soil measurement data is recorded in the attributes with timestamps. Each data set keeps records of polygon areas. In total, the environmental data collection contains around 30,000 records. Datasets have attribute fields about the area of the polygons, CVR number of the data provider agency or company, responsible person name, etc. The important attributes, which record the soil measurement data are: *Nitrate class type, Nitrogen reduction potentials, Phosphor class type.*

Geographical (Regions) Data. The primary use case data is built around livestock farming (CHR) and environmental data as mentioned above. In sorder to pursue richer analysis upon this use case we enrich the spatiality of the use

6 Livestock units are used to produce statistics describing the number of livestocks in farms.

case data by adding two geographical data sets; parishes and drainage areas of Denmark. These data sets are spatially and topically relevant since we have found pre-aggregated maps created by the Ministry of Environment and Food of Denmark at parish and drainage area levels for livestock farming data. We downloaded parishes and drainage areas of Denmark as SHP files from http://www.geodata-info.dk/. The total number of records of the geographical data collection are 2,300. These data sets have attribute fields such as: *Drainage area name, Parish name, Total area, etc.*

Central Company Registry (CVR) Data. Danish companies, agencies and industries are registered in the Central Company Register (CVR). Every livestock holding is owned by a company and has a CVR number. Environmental data also records the CVR number of the corresponding data provider agencies. Through this CVR number, we can access detailed information of the companies and contact details of the responsible people. This collection allows evaluating interesting queries with the selected domains given above. The CVR data is published at http://cvr.dk and can be accessed via a web service with a Danish social security number log-in. We accessed and downloaded only the data in CSV format that are accredited for publishing. This data includes attributes such as: *Company name, Phone number, and Address etc.*

5 Publishing Spatial Data Cubes with QB4SOLAP

The QB4SOLAP vocabulary allows to define *cube schemas* and *cube instances*. A cube schema defines the structure of a cube as an instance of the class `qb:DataStructureDefinition` in terms of dimension levels, measures, aggregation functions (e.g., SUM, AVG, and COUNT) on measures, spatial aggregation functions[7] on spatial measures, fact-level cardinality relationships, and topological relationships. The properties used to express these relationships are: `qb4o:level`, `qb:measure`, `qb4o:aggregateFunction`[8], `qb4o:cardinality`, and `qb4so:topologicalRelation` respectively (Fig. 1). These schema level metadata are used to define MD data sets in RDF. Cube instances are the members of a cube schema that represent level members, facts and measure values. We describe the cube schema elements in Sect. 5.1 and the cube instances in Sect. 5.2 with their examples.

5.1 GeoFarmHerdState Cube Schema in RDF

As our use case we create a spatial data cube of livestock holdings that we refer to as *GeoFarmHerdState*. The use case data cube is created from the flat data sets of the livestock farming (CHR) data, the environmental data, the

[7] Spatial aggregation functions aggregate two or more spatial objects and return a new spatial object, e.g., union, buffer, and convexHull etc.

[8] SpatialAggregateFunction is a subclass of AggregateFunction. Thus, measures and spatial measures use the same property `qb4o:aggregateFunction`.

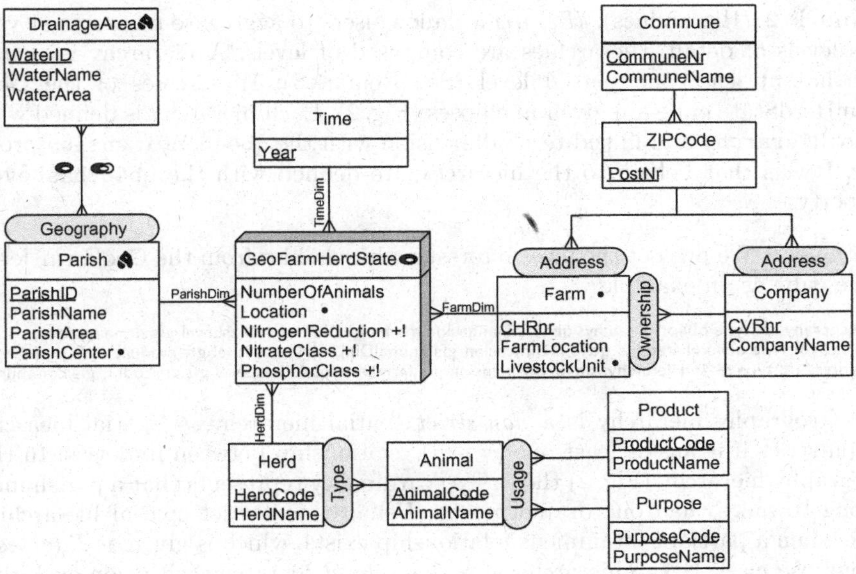

Fig. 2. GeoFarmHerdState – conceptual MD schema of livestock holdings data

geographical (regions) data, and the company registry (CVR) data collections, which are explained in Sect. 4. We create this data cube by thoroughly analyzing the attributes of the flat data sets from the collected relevant domains and conciliating them by foreign keys or by overlaying the SHP files of the spatial data sets and deriving new attributes from the intersected areas. After deriving useful spatial information across use case domains, we generalize the tabular data of the several use case data sets into the GeoFarmHerdState data cube. Figure 2 shows the multidimensional conceptual schema of the GeoFarmHerdState spatial cube. The multidimensional elements of the cube are explained in Remarks 1–7 followed by their examples in RDF. The underlying syntax for RDF examples is given in Turtle. We prefix the schema elements of the GeoFarmHerdState cube with `gfs:`.

Remark 1 (Dimensions). *Dimensions* provide perspectives to analyze the data. The GeoFarmHerdState cube has four dimensions, in which the two of them are *spatial* (FarmDim, ParishDim). All dimensions in the cube are defined with `qb:DimensionProperty`. A dimension is spatial if it has at least one spatial level (See Remark 3). Dimension hierarchies are defined with `qb4o:hasHierarchy` property. Hierarchies and their types are explained later in Remark 2.

Example 1. We give two spatial dimensions as an example.

```
gfs:farmDim rdf:type qb:DimensionProperty; qb4o:hasHierarchy gfs:ownership , gfs:address.
gfs:parishDim rdf:type qb:dimensionProperty; qb4o:hasHierarchy gnw:geography.
```

Remark 2 (Hierarchies). *Hierarchies* allow users to aggregate measures at various levels of detail. Hierarchies are composed of levels. A hierarchy is *spatial* if it has at least one spatial level (See Remark 3). Hierarchies of the GeoFarmHerdState cube are given in ellipses (Fig. 2). Each hierarchy is defined with `qb4o:Hierarchy` and linked to its dimension with the `qb4o:inDimension` property. Levels that belong to the hierarchy are defined with the `qb4o:hasLevel` property.

Example 2. We present the most interesting hierarchies from the GeoFarmHerdState cube as an example.

```
gfs:geogprahy rdf:type qb4o:Hierarchy; qb4o:inDimension gfs:parishDim; qb4o:hasLevel gfs:drainageArea.
gfs:usage rdf:type qb4o:Hierarchy; qb4o:inDimension gfs:animalDim; qb4o:hasLevel gfs:product , gfs:purpose.
gfs:address rdf:type qb4o:Hierarchy; qb4o:inDimension gfs:farmDim; qb4o:hasLevel gfs:zipCode , gfs:commune.
```

The Geography hierarchy is a *non-strict* spatial hierarchy. A spatial hiearchy is non-strict if it has at least one $(n - n)$ relationship between its levels. In the Geography hierarchy (Fig. 2) the $(n - n)$ cardinality represents that a parish may belong to more than one drainage area. Usually, non-strict spatial hierarchies arise when a partial containment relationship exists, which is given as *Intersects* in our use case. Usage hierarchy is a *generalized* hierarchy with non-exclusive paths to splitting levels (Product and Purpose) and has no joining level but the top level *All*. Finally, the Address and Ownership hierarchies are *parallel dependent* hierarchies. Parallel hierarchies arise when a dimension has several hierarchies sharing some levels. Note that the Address hierarchy has different paths from the Company and Farm levels (Fig. 2).

Remark 3 (Levels). *Levels* have a set of attributes (See Remark 4) that describes the characteristics of the level members (See Remark 9). Levels are defined with the `qb4o:LevelProperty` and their attributes are linked with the `qb4o:hasAttribute` property. A level is *spatial* if it has an associated geometry. Therefore, spatial levels have the property `geo:hasGeometry`, which defines the geometry of the spatial level in QB4SOLAP.

Example 3. We present a spatial level (Parish) as an example with its attributes. Attributes and spatial attributes of levels are further described in Remark 4.

```
gfs:parish rdf:type qb4o:LevelProperty; qb4o:hasAttribute gfs:parishID;
    qb4o:hasAttribute gfs:parishName; qb4o:hasAttribute gfs:parishArea;
    qb4o:hasAttribute gfs:parishCenter; geo:hasGeometry gfs:parishPolygon.
```

Note that the Parish level is defined as a spatial level because it has an associated polygon geometry (`gfs:parishPolygon`), which is specified with the `geo:hasGeometry` property. Some other spatial characteristics of the levels can be recorded in the spatial attributes of the level such as the center point of the parish (`gfs:parishCenter`).

Remark 4 (Attributes). Attributes and *spatial* attributes are defined with the `qb4o:LevelAttribute` property and linked to their levels with the qb4o:inLevel property. An attribute is spatial if it is defined over a spatial domain. Attributes are defined as ranging over XSD literals[9] and spatial attributes must be ranging over spatial literals, i.e., well-known text literals (WKT) from OGC schemas[10]. Spatial attributes are a sub-property of the geo:Geometry class. Further, the domain of the spatial attribute should be specified with rdfs:domain, which must be a geometry. Finally, the spatial attribute must be specified as an instance of geo:SpatialObject with the rdfs:subClassOf property. Examples of attributes are given in the following.

Example 4. We present spatial and some non-spatial attributes of the Parish level.

```
gfs:parishID rdf:type qb4o:LevelAttribute; qb4o:inLevel gfs:parish; rdfs:range xsd:positiveInteger.
gfs:parishName rdf:type qb4o:LevelAttribute; qb4o:inLevel gfs:parish; rdfs:range xsd:string.
gfs:parishCenter rdf:type qb4o:LevelAttribute; rdfs:subPropertyOf geo:Geometry; qb4o:inLevel gfs:parish;
    rdfs:domain geo:Point; rdfs:subClassOf geo:SpatialObject; rdfs:range geo:wktLiteral , virtrdf:Geometry.
```

We have mentioned in Remark 3 that spatial levels are defined through their associated geometries, which are not given as a level attribute. For the Parish level we present the following example of the corresponding geometry.

```
gfs:parishPolygon rdf:type geo:Geometry; rdfs:domain geo:MultiSurface;
    rdfs:subClassOf geo:SpatialObject; rdfs:range geo:wktLiteral , virtrdf:Geometry.
```

Remark 5 (Hierarchy Steps). Hierarchy steps define the structure of the hierarchy in relation to its corresponding, levels. A hierarchy step entails a roll-up relation between a lower (child) level and an upper (parent) level with a cardinality. The cardinality $(n-n, 1-n, n-1, n-n)$ relationship describes the number of members in one level that can be related to a member in the other level for both child and parent levels. A hierarchy step is *spatial* if it relates a spatial child level and a spatial parent level, in which case it entails a topological relationship between these spatial levels. Both spatial and non-spatial hierarchy steps are defined as a blank node with the qb4o:HierarchyStep property and linked to their hierarchies with the qb4o:inHierarchy property. The parent and child levels are linked to hierarchy steps with the qb4o:childLevel property and the qb4o:parentLevel property. The cardinality of a hierarchy step is defined by the qb4o:pcCardinality property. And finally, the topological relationship[11] of a hierarchy step is defined by the qb4so:pcTopoRel property.

Example 5. The following illustrates the hierarchy steps of the spatial hierarchy Geography and non-spatial hierarchy Address as it has different paths from child levels Farm and Company.

```
## Geography hierarchy structure ##
_:geography_hs1 rdf:type qb4o:HierarchyStep; qb4o:inHierarchy gfs:geography;
    qb4o:childLevel gfs:parish; qb4o:parentLevel gfs:drainageArea;
    qb4o:pcCardinality qb4o:ManyToMany; qb4so:pcTopoRel qb4so:Intersects, qb4so:Within.
## Address hierarchy structure ##
```

[9] XML Schema Definition: http://www.w3.org/TR/xmlschema11-1/.
[10] OGC Schemas: http://schemas.opengis.net/.
[11] Topological relations are Boolean predicates that specify how two spatial objects are related to each other, e.g., within, intersects, touches, and crosses etc.

```
_:farm_address_hs1 rdf:type qb4o:HierarchyStep; qb4o:inHierarchy gfs:address;
    qb4o:childLevel gfs:farm; qb4o:parentLevel gfs:zipCode;
    qb4o:pcCardinality qb4o:ManyToOne.
_:farm_address_hs2 rdf:type qb4o:HierarchyStep; qb4o:inHierarchy gfs:address;
    qb4o:childLevel gfs:zipCode; qb4o:parentLevel gfs:commune;
    qb4o:pcCardinality qb4o:ManyToOne.
_:company_address_hs1 rdf:type qb4o:HierarchyStep; qb4o:inHierarchy gfs:address;
    qb4o:childLevel gfs:company; qb4o:parentLevel gfs:zipCode;
    qb4o:pcCardinality qb4o:ManyToOne.
_:company_address_hs2 rdf:type qb4o:HierarchyStep; qb4o:inHierarchy gfs:address;
    qb4o:childLevel gfs:zipCode; qb4o:parentLevel gfs:commune;
    qb4o:pcCardinality qb4o:ManyToOne.
```

Remark 6 (Measures). Measures record the values of a phenomena being observed. Measures and *spatial* measures are defined with `qb:MeasureProperty`. A measure is spatial if it is defined over a spatial domain. Similarly to attributes (Remark 4), measures are defined ranging over XSD literals and spatial measures must be ranging over spatial literals.

Example 6. The following shows an example of a spatial measure (Location) and a non-spatial measure (NumberOfAnimals).

```
gfs:location rdf:type qb:MeasureProperty; rdfs:subPropertyOf sdmx-measure:obsValue;
    rdfs:subClassOf geo:SpatialObject; rdfs:domain geo:Point;
    rdfs:range geo:wktLiteral , virtrdf:Geometry.
gfs:numberOfAnimals rdf:type qb:MeasureProperty;
    rdfs:subPropertyOf sdmx-measure:obsValue; rdfs:range xsd:decimal.
```

Remark 7 (Fact). Fact defines the data structure (DSD) of the cube with `qb:DataStructureDefinition`. The dimensions are given as `components` and defined with the `qb4o:level` property as the dimensions are linked to the fact at the lowest granularity level. A fact is *spatial* if it relates two ore more spatial levels. Similarly, measures are given as `components` of the fact and are defined with the `qb:measure` property. Aggregation functions on measures and spatial aggregation functions on spatial measures are also defined in the DSD with `qb4o:aggregateFunction`. Fact-level cardinality relationships and topological relationships are defined with `qb4o:cardinality` and `qb4so:topologicalRelation` in DSD, respectively (Fig. 1).

Example 7. The following shows the data structure definition of the cube GeoFarmHerdState, which is defined with corresponding measures and dimensions.

```
# – GeoFarmHerdState Cube Definition of the Fact FarmHerdState
gfs:GeoFarmHerdState rdf:type qb:DataStructureDefinition;
    # Lowest level for each dimensions in the cube
    qb:component [qb4o:level gfs:herd; qb4o:cardinality qb4o:ManyToOne ];
    qb:component [qb4o:level gfs:time; qb4o:cardinality qb4o:ManyToOne ];
    qb:component [qb4o:level gfs:farm; qb4o:cardinality qb4o:ManyToOne;
        qb4so:topologicalRelation qb4so:Equals ];
    qb:component [qb4o:level gfs:parish; qb4o:cardinality qb4o:ManyToMany;
        qb4so:topologicalRelation qb4so:Within ];
    # Measures in the cube
    qb:component [qb:measure gfs:numberOfAnimals; qb4o:aggregateFunction qb4o:Sum];
    qb:component [qb:measure gfs:location; qb4o:aggregateFunction qb4so:ConvexHull];
    qb:component [qb:measure gfs:nitrogenReduction; qb4o:aggregateFunction qb4o:Avg];
    qb:component [qb:measure gfs:nitrateClass; qb4o:aggregateFunction qb4o:Avg];
    qb:component [qb:measure gfs:phosphorClass; qb4o:aggregateFunction qb4o:Avg].
```

5.2 GeoFarmHerdState Cube Instances in RDF

Cube instances are the members of a cube schema that represent level members and facts (members), which are explained in Remarks 8 and 9 below. We prefix the instances of the GeoFarmHerdState cube with `gfsi:`.

Remark 8 (Fact members). Fact members (i.e., facts of FarmHerdState) are instances of the `qb:Observation` class. Each fact member is related to a set of dimension *base* level members and has a set of measure values. Every fact member has a unique identifier (IRI) which is prefixed with `gfsi:`.

Example 8. The following shows an example of a single fact member, which represents the state of a farm with CHR no. 39679 in the year 2015 that has the herd code 15.

```
gfsi:farm_39679_2015 rdf:type qb:Observation;
## Dimension levels and base level members associated with the fact member
    gfs:herdCode gfsi:herd_15; gfs:year gfsi:year_2015;
    gfs:chrNumber gfsi:farm_39679; gfs:parishID gfsi:parish_8311;
## Measures associated with the fact member
    gfs:numberOfAnimals "100.0"^^xsd:decimal; gfs:nitrateClass "3"^^xsd:integer;
    gfs:nitrogenReduction "0.75"^^xsd:decimal; gfs:phosporClass "3"^^xsd:integer;
    gfs:location "POINT(8.3713 56.7912)"^^geo:wktLiteral.
```

Remark 9 (Level members). Level members are defined with `qb4o:LevelMember`. They are linked to their corresponding levels from the schema with the `qb4o:memberOf` property. For each level member there is a set of attribute values. Due to the roll-up relations between levels of hierarchy steps (Remark 5), the `skos:broader` property relates a child level member to its parent level member.

Example 9. The following shows an example of a child level member in the Parish level and one of its parent level members in the DrainageArea level from the Geography dimension. Figure 3 presents a map snapshot for fact members and level members. Parish level member "Astrup" is highlighted and DrainageArea level member "Mariager Inderfjord" is marked with red borders. Note that Astrup intersects another drainage area "Langerak", therefore it links to two parent level members via `skos:broader`.

```
## Parish level member
gfsi:parish_8648 rdf:type gfs:parish;
    qb4o:memberOf gfs:parish; skos:broader gfsi:water_3710, gfsi:water_159;
    gfs:parishID 8311; gfs:parishName"Astrup"; gfs:parishArea 46,118;
    gfs:parishCenter"POINT(8.2552, 56.8176)"^^geo:wktLiteral;
    gfs:parishPolygon"POLYGON((8.4038 56.7963, 8.3984 56.7721, 8.3689 56.7410, 8.3411 56.7372, 8.3078
    56.7281, 8.2987 56.7601, 8.2563 56.7763, 8.3112 56.8087, 8.3511 56.8137, 8.4038 56.7963))"^^geo:wktLiteral.
## DrainageArea level member
gfsi:water_159 rdf:type gfs:drainageArea;
    qb4o:memberOf gfs:drainageArea; gfs:waterID 159;
    gfs:waterName"Mariager Inderfjord"; gfs:waterArea 267,477;
    gfs:drainageGeo "POLYGON((8.6048 56.9843, 8.5908 56.8969, 8.5707 56.8664,
    8.5975 56.8519, 8.5215 56.8483, 8.3959 56.7625, 8.3938, 56.7340, 8.3613 56.6802,
    8.2584 56.7764, 8.2475 56.7051, 8.2175 56.7232, 8.3121 56.8441, 8.2806 56.8659,
    8.3602 56.9569, 8.4786 56.9713, 8.5474 56.9905, 8.6048 56.9843))"^^geo:wktLiteral.
```

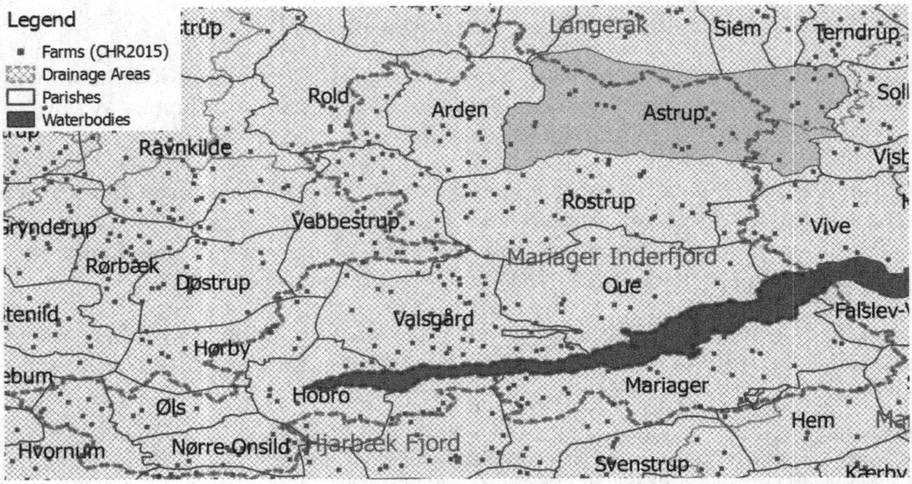

Fig. 3. GeoFarmHerdState – fact members and level members of Example 9 marked

6 SOLAP Operators over GeoFarmHerdState Cube

Spatial OLAP (SOLAP) operates on spatial data cubes. SOLAP increases the analytical capabilities of OLAP by taking into account the spatial information in the cube. SOLAP operators involve spatial conditions or spatial functions. Spatial conditions specify constraints (i.e., spatial Boolean predicates) on the geometries associated to cube members or measures, while spatial functions derive new data from the cube, which can be used, e.g., to derive dynamic spatial hierarchies.

6.1 SOLAP Operators

In what follows we present common SOLAP operators and examples of these operators on GeoFarmHerdState cube.

Remark 10 (S-Slice). The s-slice operator removes a dimension from a cube by choosing single spatial value in a spatial level. It returns a cube with one dimension less.

Example 10. We can perform an s-slice operation in different ways.
1. Slice on farms (state of the farms) of the largest parish.
2. Slice on farms (state of the farms) of the drainage area containing `"POINT(10.43951 55.47006)"`.

The first one applies a spatial function call (for finding the *largest* parish by area) on a spatial level Parish and performs the slice. The second one applies a spatial predicate (for finding where a given point is *within* a particular drainage area) in a spatial level DrainageArea and performs the operation. The corresponding SPARQL queries are:

```
# 1 – s-slice with spatial function #
SELECT ?obs WHERE {
    ?obs rdf:type qb:Observation;
        gfs:parishID ?parish.
    ?parish gfs:parishPolygon ?parishGeo.
# Inner select for finding the largest parish
    { SELECT ?x (MAX(?area) as ?maxArea) WHERE{
    ?obs rdf:type qb:Observation;
        gfs:parishID ?parish.
    ?parish gfs:parishPolygon ?x.
    BIND (bif:st_area(?x) as ?area)}}
FILTER ?parishGeo = ?x) }
```

```
# 2 – s-slice with spatial predicate #
SELECT ?obs WHERE {
    ?obs rdf:type qb:Observation;
        gfs:parishID ?parish.
    ?parish qb4o:memberOf gfs:parish;
        skos:broader ?drainageArea.
    ?drainageArea gfs:drainageGeo ?drainageGeo.
FILTER (bif:st_within("POINT(10.43951 55.47006)",
?drainageGeo)) }
```

Remark 11 (S-Dice). The s-dice operator keeps the cells of the cube that satisfy the spatial predicate over dimension levels, attributes, or measures. It returns a subset of the cube with filtered members of the cube.

Example 11. In the following we show two examples of the s-dice operator.
1. Filter the farms located within 5 Km buffer from the center of a drainage area.
2. Filter the farms located within 2 Km distance from the center of their parish, which are in the nitrate class I areas.

In the first s-dice operation, initially, a spatial function is applied on level members of the DrainageArea level to get the *center* of their polygon geometries. Then, the level members of the Farm level are filtered with a spatial Boolean predicate with respect to the farm locations that are *within* a 5 Km buffer area of the center of the drainage areas. In the second s-dice operation, a spatial function is applied to the spatial measure farm location to get the *distance* of the farms from the center of their parish, which is followed by Boolean predicates; to filter the farms that are less than 2 Km away from the center of their parishes and are on nitrate class I areas.

```
# 1 – s-dice on dimension levels #
SELECT ?obs WHERE {
    ?obs rdf:type qb:Observation;
        gfs:farmID ?farm;
        gfs:parishID ?parish.
    ?farm gfs:farmLocation ?farmGeo.
    ?parish qb4o:memberOf gfs:parish;
        skos:broader ?drainageArea.
    ?drainageArea gfs:waterPolygon ?drainagePoly.
BIND (bif:st_centroid (?drainagePoly) as ?drainageCenter)
FILTER (bif:st_within(?drainageCenter, ?farmGeo, 5)) }
```

```
# 2 – s-dice on measures #
SELECT ?obs WHERE {
    ?obs rdf:type qb:Observation;
        gfs:location ?farmLocation;
        gfs:nitrateClass ?nitClass;
        gfs:parishID ?parish.
    ?parish gfs:parishCenter ?parishCent.
BIND (bif:st_distance (?farmLocation, ?parishCent)
AS ?distance)
FILTER (?distance < 2 && ?nitClass = 1)}
```

Remark 12 (S-Roll-up). The s-roll-up operator aggregates measures of a given cube by using an aggregate function and a spatial function along a spatial dimension's hierarchy. It returns a cube with measures at a coarser granularity for a given dimension.

Example 12. In the following, we present two examples of the s-roll-up operator.
1. Total amount of animals in the farms, which are closest to their parishes' center.
2. Average percentage of nitrogen reduction potentials in the parishes that are within and/or intersect the drainage area "Nibe-Bredning".

In the first s-roll-up operator, measures are aggregated to the Parish level after selecting the farms with respect to their proximity to the center of the parish with a spatial function. In the second s-roll-up operator, measures are aggregated to a specified drainage area ("Nibe-Bredning") at the DrainageArea level. We select all the possible topological cases where a parish intersects or within the drainage area, which means measures from the farms that are outside Nibe-Bredning are also aggregated to the level of this drainage area. In order to prevent this, the query needs to include an s-drill-down operator (Remark 13) to farms from Parish level and apply a spatial Boolean predicate to select the farms *within* the drainage area and then aggregate.

```
# 1 – s-roll-up #
SELECT ?parish (SUM(?animalCount) AS ?totalAnimals)
WHERE { ?obs rdf:type qb:Observation;
    gfs:numberOfAnimals ?animalCount;
    gfs:farmID ?farm;
    gfs:parishID ?parish.
?farm gfs:farmLocation ?farmGeo.
?parish gfs:parishCenter ?parishCent.
# Inner select for finding the
# closest farms to the parish centers #
    {SELECT ?farm1 (MIN(?distance) AS
    ?minDistance) WHERE
    { ?obs rdf:type ab:Observation;
        gfs:farmID ?farm1;
        gfs:parishID ?parish1.
    ?farm1 gfs:farmLocation ?farm1Geo.
    ?parish1 gfs:parishCenter ?parish1Cent.
    BIND (bif:st_distance (?farm1Geo, parish1Cent)
    AS ?distance) } GROUP BY ?farm1 }
    FILTER (?farm = ?farm1 && bif:st_distance
    (?farmGeo, ?parishCent) = ?minDistance )}
GROUP BY ?parish
```

```
# 2 – s-roll-up #
SELECT ?drainageArea (AVG(?nitRed) AS ?avgNitRed)
WHERE { ?obs rdf:type qb:Observation;
    gfs:location ?farmLocation;
    gfs:nitrogenReduction ?nitRed;
    gfs:parishID ?parish.
    ?parish qb4o:memberOf gfs:parish;
        gfs:parishPolygon ?parishGeo;
        skos:broader ?drainageArea.
    ?drainageArea gfs:memberOf gfs:drainageArea;
        gfs:waterPolygon ?drainageGeo;
        gfs:waterName ?drainageName.
FILTER (bif:st_within(?parishGeo, ?drainageGeo)
|| bif:st_intersects(?parishGeo, ?drainageGeo)
&& ?drainageName ="Nibe-Bredning")}
GROUP BY ?drainageArea
```

Remark 13 (S-Drill-down). The s-drill-down operator disaggregates measures of a given cube by using an aggregate function and a spatial function along a spatial dimension's hierarchy. It is the inverse operator of s-roll-up, therefore s-drill-down disaggregates the previously summarized data to a child level in order to obtain measures at a finer granularity.

6.2 Nested SOLAP Operations

A nested set of SOLAP operators can be designed with the pattern ($s\text{–}dice_2(s\text{–}roll\text{–}up_1(\ldots s\text{–}roll\text{–}up_k(s\text{–}slice_1(\ldots s\text{–}slice_n(s\text{–}dice_1(DataCube)))))))$). Initially a sub-cube is selected from the (spatial) data cube with the first s-dice. Afterwards, a number of s-slices can be applied, which is followed by a series of s-roll-ups. Finally, the expression ends with another s-dice for getting the final sub-cube at a coarser granularity by filtering the aggregated measures. In the following, we present a nested SOLAP operation example for the running case GeoFarmHerdState spatial data cube.

Example 13. ($^3s\text{–}roll\text{–}up(^2s\text{–}slice(^1s\text{–}dice(GeoFarmHerdState))))$: This pattern represents a typical nested SOLAP operation that can be paraphrased for the running use case as follows: [1]Filter the farm states located within a 2 Km distance from the center of their parish and [2]slice on the parish which has the most number of topological relations (intersects, within) with a drainage area, [3]average the nitrogen reduction potential of the drainage areas intersecting with the parish.

7 Discussion and Perspectives

In the following, we give and evaluate the steps of our process with respect to the guidelines for publishing governmental linked data [18]. We discuss the particular challenges that we encountered and possible future improvements.

(1) Specification. The first step is to specify the scope of the data by identifying and analyzing the data sources. We identified the data sources for the domains of CHR data, Environmental data, Geographical data, and CVR data as described in Sect. 4. In order to find the correct relations between these domains we had to search documentations (i.e., [13,14]) and acquire knowledge about the domains' interests. As the purpose is to publish open data, the definition of an Open Data license is also required at this level.

(2) Modeling. We used the spatially extended MultiDim model [17] for designing the MD conceptual schema of the use case spatial data cube (Fig. 2) from the collected flat data sets. This process requires good knowledge of spatial data warehouses and its concepts. In order to model the spatial data cube in RDF, QB4SOLAP provides the state-of-the-art semantic spatial data cubes. Therefore, we annotate the designed use case conceptual schema with QB4SOLAP.

Modeling the RDF data with QB4SOLAP provides all the core concepts of spatial data warehouses (i.e., spatial dimensions, spatial levels, and spatial hierarchies) for spatial data on the SW. Therefore, QB4SOLAP conveniently handles the conceptual modeling process of DWs on the SW and clearly describes the certain relations that should be considered during the logical modeling process (e.g., cardinality and topological relationships to create integrity constraints for ER models).

(3) Generation. This step of the overall process involves the most complex tasks. In order to fully generate a spatial data cube in RDF the following subprocesses are performed: transformation and data conciliation.

(3.1) Transformation. The RDF triples were generated with ad-hoc C# code for mapping from relational CSV files to RDF. In total, 12 CSV files were organized based on the relational representation (snowflake schema) of an MD conceptual model such that: We obtained *one* fact table with foreign keys of the related dimension (base) levels and measures, *four* tables with each dimensions' base level, and *seven* tables for the remaining levels along the hierarchies. These 12 tables are related by referential integrity constraints. Every level table also records the level attributes and attribute values. In order to create this relational CSV files, we pursued a number of data conciliation activities as described in the following item.

(3.2) Data Conciliation. Initial data sets are downloaded in different formats i.e., SHP format for CHR, Environmental, and Geographical data; CSV format for CVR data. In order to create the desired relational implementation of the use case spatial data cube, we used the unique identifiers (i.e., CVR and CHR numbers) or utilized spatial joins by joining attributes from one geometry feature to another based on the spatial containment relationship. For instance, we overlaid the point coordinates of the farms from CHR data and polygon coordinates of three environmental data sets in order to intersect and find the soil quality measurements for NitrogenReduction, NitrateClass and PhosphorClass of each farm. Another interesting spatial join is utilized for relating the Parish level members with DrainageArea level members. Since there is an $(n - n)$ cardinality relationship, some parishes intersect with more than one drainage area, thus we used topological relationships (intersects and within) to find the related child and parent level members. For interacting and handling the spatial data, we used QGIS with integration to PostGIS[12]. We used PostgreSQL to create and export relational tables.

The lack of tools for mapping a spatial multidimensional model to the relational model has been an impediment since we have to use topological relationships, where there is an $(n - n)$ cardinality relationship. Therefore, semantic ETL for data warehouses [5] is an important research topic, which requires improvements also for spatial data. Semi-automated tool support of geo-semantic ETL for publishing data warehouses on the SW is a promising improvement for handling the above processes such that the spatial joins can be processed efficiently. Before publishing the final RDF data, a comprehensive data cleansing step is essential for removing redundant columns and cleaning the noise due to unescaped characters, denormalized spatial literals, and encoding problems.

(4) Publication. In order to store and publish the RDF data we chose the Virtuoso Universal Server as a triple store. The details about the SPARQL endpoint can be found on the project page http://extbi.cs.aau.dk/GeoFarmHerdState.

Publication of metadata in Danish and English languages should be completed. Also for enabling efficient discovery of published spatial data cubes, adding an entry of the data in the CKAN repository (`datahub.io`) is required.

(5) Exploitation. The goal of our research is to re-use open government data and publish it as spatial data cubes on the SW for advanced multidimensional analysis. Therefore, we show how to query in SPARQL with SOLAP operators.

We recognize the need for non-expert SW users to write their spatial analytical queries in our high-level SOLAP language instead of the lower-level complex SPARQL language. Thus, a query system with a GUI that can interpret spatial data cube schemas for allowing users to perform high level SOLAP operations is ongoing work. Performing SOLAP queries in SPARQL to work over multiple RDF cubes with *s-drill-across* and supporting spatial aggregation (*s-aggregation*) over spatial measures are other important improvements on exploitation of spatial data cubes.

[12] QGIS: http://www.qgis.org/ PostGIS: http://postgis.net/.

8 Conclusion and Future Work

The need for spatial analytical queries on the Semantic Web increases constantly with regularly published open government data, but there is a lack of effective solutions and efficient models. As a first attempt to publish spatial data cubes from open data, we have shown that the QB4SOLAP vocabulary can be used to link Danish government data that is published in different domains. First, we have studied the use case data sets thoroughly with corresponding regulations and requirements in order to satisfy cross-domain interests (e.g., tracking soil quality in livestock farms and farm animals density on drainage areas etc.). Second, we have conciliated the flat data sets in order to model the MD concepts of a spatial data cube. Third, we described the most popular individual SOLAP operators and a nested SOLAP operation pattern with examples and their SPARQL implementation.

In this paper, the QB4SOLAP vocabulary is validated by a non-trivial spatial use case. As a proof of concept, we showed that linking spatial (governmental open) data on the Semantic Web can be achieved at an advanced level, not solely linking spatial open data on the SW but also modeling this data for advanced analytical queries with SOLAP operations.

Several directions are interesting for future research: developing a geo-semantic ETL tool to support the process of creating spatial data cubes on the SW, a GUI for non-expert users to perform SOLAP operations on SW spatial cubes, extending the use case and implementing advanced SOLAP queries, such that; as s-drill-across and s-aggregation on the SW.

Acknowledgments. This research is partially funded by "The Erasmus Mundus Joint Doctorate in Information Technologies for Business Intelligence Doctoral College (IT4BI-DC)".

References

1. Abelló, A., Romero, O., Pedersen, T.B., Aramburu, M.J., et al.: Using semantic web technologies for exploratory OLAP: a survey. TKDE **27**, 571–588 (2014)
2. Andersen, A.B., Gür, N., Hose, K., Jakobsen, K.A., Pedersen, T.B.: Publishing danish agricultural government data as semantic web data. In: Supnithi, T., Yamaguchi, T., Pan, J.Z., Wuwongse, V., Buranarach, M. (eds.) JIST 2014. LNCS, vol. 8943, pp. 178–186. Springer, Heidelberg (2015). doi:10.1007/978-3-319-15615-6_13
3. Arendt, J.B.: Denmark releases its digital raw material, Ministry of Finance of Denmark, October 2012. http://uk.fm.dk/news/
4. Cyganiak, R., Reynolds, D., Tennison, J.: The RDF Data Cube Vocabulary (2014)
5. Nath, D.R.P., Hose, K., et al.: Towards a Programmable Semantic Extract-Transform-Load Framework for Semantic Data Warehouses. In: DOLAP, pp. 15–24 (2015)
6. Etcheverry, L., Vaisman, A., Zimányi, E.: Modeling and querying data warehouses on the semantic web using QB4OLAP. In: Bellatreche, L., Mohania, M.K. (eds.) DaWaK 2014. LNCS, vol. 8646, pp. 45–56. Springer, Heidelberg (2014). doi:10.1007/978-3-319-10160-6_5

7. Gür, N., Hose, K., Pedersen, T.B., Zimányi, E.: Modeling and querying spatial data warehouses on the semantic web. In: Qi, G., Kozaki, K., Pan, J.Z., Yu, S. (eds.) JIST 2015. LNCS, vol. 9544, pp. 3–22. Springer, Heidelberg (2016). doi:10.1007/978-3-319-31676-5_1
8. Gür, N., Pedersen, T.B., Zimányi, E., Hose, K.: A foundation for spatial data warehouses on the semantic web. Journal paper, under submission (2016)
9. Jakobsen, K.A., Andersen, A.B., Hose, K., Pedersen, T.B.: Optimizing RDF data cubes for efficient processing of analytical queries. In: COLD (2015)
10. Kämpgen, B., Harth, A.: OLAP4LD – a framework for building analysis applications over governmental statistics. In: Presutti, V., Blomqvist, E., Troncy, R., Sack, H., Papadakis, I., Tordai, A. (eds.) ESWC 2014. LNCS, vol. 8798, pp. 389–394. Springer, Heidelberg (2014). doi:10.1007/978-3-319-11955-7_54
11. Kämpgen, B., O'Riain, S., Harth, A.: Interacting with statistical linked data via OLAP operations. In: Simperl, E., Norton, B., Mladenic, D., Della Valle, E., Fundulaki, I., Passant, A., Troncy, R. (eds.) ESWC 2012. LNCS, vol. 7540, pp. 87–101. Springer, Heidelberg (2015). doi:10.1007/978-3-662-46641-4_7
12. Matei, A., Chao, K.-M., Godwin, N.: OLAP for multidimensional semantic web databases. In: Castellanos, M., Dayal, U., Pedersen, T.B., Tatbul, N. (eds.) BIRTE 2013-2014. LNBIP, vol. 206, pp. 81–96. Springer, Heidelberg (2015). doi:10.1007/978-3-662-46839-5_6
13. Danish Ministry of the Environment. Consolidated Act on Livestock Farming Environmental Approvals (2012). http://eng.mst.dk/media
14. Directive, Nitrates: Danish nitrate action programme 2008–2015 regarding the nitrates directive; 91/676/eec. Technical report, Nitrates Directive (2012)
15. Open Geospatial Consortium: GeoSPARQL: A geographic query language for RDF data. W3C Recommendation (2014)
16. Pedersen, D., Riis, K., Pedersen, T.B.: Query optimization for OLAP-XML federations. In: DOLAP, pp. 57–64 (2002)
17. Vaisman, A., Zimányi, E.: Spatial data warehouses. In: Vaisman, A., Zimányi, E. (eds.) Data Warehouse Systems. Design and Implementation. DCSA, pp. 427–473. Springer, Heidelberg (2014)
18. Villazón-Terrazas, B., Vilches-Blázquez, L., Corcho, O., Gómez-Pérez, A.: Methodological guidelines for publishing government linked data. In: Wood, D. (ed.) Linking Government Data, pp. 27–49. Springer, New York (2011)
19. W3C.: Data Cube Implementations (2014). https://www.w3.org/2011/gld/wiki/Data_Cube_Implementations
20. Zhao, P., Li, X., Xin, D., Han, J.: Graph cube: on warehousing and OLAP multidimensional networks. In: SIGMOD, pp. 853–864 (2011)

Classification of News by Topic Using Location Data

Zolzaya Dashdorj[1(✉)], Muhammad Tahir Khan[2], Loris Bozzato[3], and SangKeun Lee[1]

[1] Korea University, 1, 5-ga, Anam-dong, Seongbuk-gu, Seoul, Republic of Korea
{zolzaya,yalphy}@korea.ac.kr
[2] Taiger Singapore, 3 Fusionopolis Place #04-56, Galaxis, Singapore, Singapore
tahir.khan@taiger.com
[3] Fondazione Bruno Kessler, Via Sommarive 18, 38123 Trento, Italy
bozzato@fbk.eu

Abstract. In this work, we will consider news articles to determine geo-localization of their information and classify their topics on the basis of an available open data source: OpenStreetMap (OSM). We propose a knowledge-based conceptual and computational approach that disambiguates place names (i.e., geo-objects and regions) mentioned in news articles in terms of geographic coordinates. The geo-located news articles are analyzed to identify local topics: we found that the mentioned geo-objects are a good proxy to classify news topics.

1 Introduction

Enormous amount of information has been generated over web sources and social platforms by the activities of millions of people worldwide. The quantitative understanding of such information provides great impacts on the analysis of human behaviors and human dynamics on a macro-level [5,6,12]. However, in order to carry out the analysis on a micro level (i.e., provinces and districts), there is a need for geo-localization of information that describes different contexts of events (i.e., emergency vs non emergency) or human activities. Obtaining such useful and explanatory information from web sources and social platforms becomes an emerging interest in the areas of information retrieval, data mining and social network analysis [8].

However, the Natural Language Processing (NLP) successfully recognizes micro level entities (i.e., organization and location) from free text. But, it suffers from the ambiguity of place names mentioned in news in terms of geographic coordinates. To solve such ambiguity problem, previous studies [10,13] rely on geo-referenced texts in social platforms which include geo-coordinates in the meta data. More recent studies have considered to study local topics on social media [9] based on the area-specific term occurrence, estimating area specific scores of terms and occurrences using term frequency, as well as average and standard deviation of the longitude and latitude of raw geotagged information.

Y.-F. Li et al. (Eds.): JIST 2016, LNCS 10055, pp. 305–314, 2016.
DOI: 10.1007/978-3-319-50112-3_23

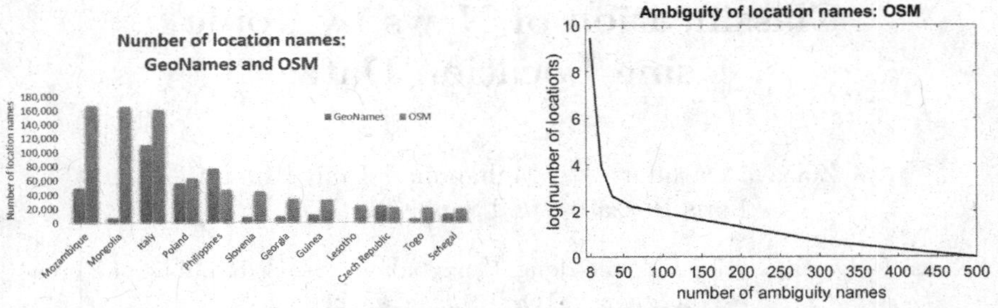

(a) Place Names Over Countries in GeoNames vs OSM

(b) Ambiguous Place Names: OpenStreetMap

Fig. 1. Open spatial data-sources

However, geotagged information in media has not been studied much to determine local topics and no previous work has been done in the identification of local topics from non geo-referenced information on media. The fact that news topics can be extracted from the geo-localization of the article, which has not received an attention: still, researchers have studied topic modeling on news favorably using Latent Dirchlet Allocation (LDA) [2].

In this paper, we propose a news classification model based on location data included in the news by disambiguating the place names mentioned in the news, without an expensive estimation cost on geographical features. In order to deal with ambiguous place names mentioned in news in terms of geographic coordinates, we implement several heuristic disambiguation techniques using well known publicly available geographic gazetteers, namely Geo-names database (GNS) [1] and OpenStreetMap (OSM) [2]. However, GNS has been exploited in few studies [14]. But, it describes a less number of geo-names in some countries, while the OSM is well enriched with more geo-names including also geo-objects (i.e., organizations) that voluntarily collected with the official language of countries and other eight common languages. For example, the number of geo-names collected in GNS covering the country, Mongolia is around 7,017 in total which is almost 22 times smaller than the geo-objects we have collected from OSM. Figure 1(a) shows a comparative number of geo-names across different countries in public data-sources: GNS [3] vs OSM [4].

Thus, we use the OSM which is an open spatial data-source of rich information about geographical objects and features that localized over official administrative divisions of countries. Our research is concentrated on news articles which does not contain any geo-references data and to the best of our knowledge, this research is the first attempt on using OSM for news classification. The potential

[1] http://www.geonames.org.
[2] http://www.openstreetmap.org/.
[3] http://www.geonames.org/statistics.
[4] http://osmstats.neis-one.org/?item=countries.

application of this research is a location based recommender system in mobile computing and social networks.

2 Method Definition

We first build a knowledge base, called *Gazeto*, consisting of place names taxonomy referring to different coordinates, extracted from OSM. We define the term, *place name* in this research as a geographical coordinate associated to an organization or an administrative division. We apply conceptual and geo-computational approaches to web crawled news contents in order to identify the place names associated with the most likelihood that refer to their actual locations. Finally, we evaluate if news topic is predictable based on location features like regions, organizations, categories organization, by classifying the news into local topics as well as global topics. In this context, *local topics* are defined as topics which are identified from the news associated to location features. In the following sections, we will explain in detail each of the steps of our methods.

2.1 Place Names Knowledge Base

Using spatial open data in OSM, we collected the spatial features of 165,179 Points of Interest (POIs) over the entire country of Mongolia including of 378,491 properties (i.e., name, geo-object type, address) by using a spatial tool, OSM2PGSQL[5], and stored in geographical datastore - PostgreSQL[6]. In OSM, geo-objects are described in a pair of key and value e.g., *restaurant* is a sub type of the category *amenity*. The OSM dataset also contains a number of additional features like e.g. email, address, and website. We discarded these features and kept only primary features as described in OSM Wiki website[7] i.e., geo-objects along with their categories, such as sport shop, fast food restaurant and others. Using an *OSMonto* ontology [4], the primary features are refined to 125,398 geo-objects. We added a collection of hierarchical administrative divisions (e.g., countries, provinces, cities) in our *KB*. According to OSM, up to 8 administrative levels can be stored. The relations between geo-objects and administrative divisions are computed by PostGIS [8] functions allowing us to identify the administrative division of a given geo-object or the ascendant administrative division of a given division. We express a place name *LN* for an organization which is expressed by the pair of administrative division and geo-object name that formalized as: ⟨*Country, Province, District, Geo-Object*⟩. In total, we have collected 4,422 place names in our knowledge base. The ambiguity of place names referring to different coordinates is described in Fig. 1(b). Below 500 place names are ambiguous in terms of geo-references.

---

5 http://wiki.openstreetmap.org/wiki/Osm2pgsql.
6 http://www.postgresql.org/.
7 http://wiki.openstreetmap.org/wiki/Map_Features.
8 http://postgis.net/.

2.2 Place Names Disambiguation Methods

To choose a correct location L_j for the ambiguous candidate LN_i, we use heuristics in a similar fashion as [8,14]. However, the study was evaluated on macro-context (i.e., country, city and province). We propose the following heuristic techniques given news articles for place name disambiguation. The comparable baselines are NP and UC approaches.

- **NLP Pos Tagger (NP).** We adopt CRF Classifier—Stanford Named Entity Recognizer (NER) [7] for identifying the entities (i.e., organization and location) from given news articles. We train the sample data using one versus the rest method.
- **Unique Consistency (UC).** We examine if the candidate is non-ambiguous compared to the place names in our KB. Non ambiguous candidate refers to only one geographic coordinate. For entity extraction, we use the NLP package, *LingPipe NER* (Alias-i 2008) [1]: setting all matches phrase without case sensitive in the mapping. This approach will constitute our baseline estimation.
- **Sequence Consistency (SC).** The phrases in a paragraph are ordered and correlated to each other. Most place names are intuitively related to the near place names. We examine if the ascendant of the candidate LN_i is the same as the non ambiguous place name on the left side LN_{i-1} or the right side LN_{i+1} for the disambiguation.
- **Distance Consistency (DC).** The candidate with the minimum distance to non ambiguous place names will be chosen for disambiguation. If there is no non ambiguous place name defined in the news article, the candidate location is disambiguated based on the minimum distance to the central point of all the locations that associated to ambiguous candidates in the news.
- **Category Consistency (CC).** We verify the category of the candidate whether it is defined the same as the most probable category of non ambiguous place names in the news for the disambiguation. Otherwise, we check the similarity between the category of the candidate and the most probable category identified in terms of the parent category.
- **Neighborhood Category Consistency (NCC).** We check here up to 10 nearby geo-objects within the $radius = 5\,\text{m}$ to the candidate if those share the same category as the most probable category of non-ambiguous place names in the news for the disambiguation. Otherwise, the similarity between the category of the nearby geo-objects to the candidate is calculated as the approach CC.

2.3 Classification of Local Topics from News

We classify topics over news based on OSMonto categorical taxonomies. First, we identify the correlation between geo-objects and news topics based on OSMOnto taxonomies, in order to estimate if the news topics are predictable based on such categorical taxonomies using learning algorithms. We then implement SVM multi-class classification using the one-against-all method. The news sets are pre-computed to generate the categorical probability distribution based on geo-object occurrence (P_{poi}) and categorical term occurrence (P_{news}). The correlation is estimated by Bhattacharyya coefficient[9] which is a correlation coefficient between the news category distribution P_{news} and categorical geo-object distribution P_{poi} over the entire news text: $BC(P_{poi}, P_{news}) = \sum_{i=1}^{n} \sqrt{P_{poi}(i) * P_{news}(i)}$, where BC is equal to 0 for a complete mismatch, otherwise 1 for a perfect match. The distributions, P_{poi}, P_{news} are labeled by k class denoted by k-mean. We construct k SVM models where k is the number of classes that describes categories organization and categorical terms based on the most geo-object occurrence and categorical term occurrence in each news, respectively. The vector features are the words in each news. The SVM for class k is constructed using the set of training examples and their desired outputs, (x_i, y_i). The mth SVM is trained with all of the examples in the mth class with positive labels, and all other examples with negative labels. The the decision function is: $x = argmax_{m=1,\ldots,k}((w^m)^T \phi(x) + b^m)$, where x is in the class which has the largest value of the decision function.

2.4 Classification of Global Topics from News

Given news articles where the place names are disambiguated, we obtain global topics comparing with the local topics based on the geo-location features using one of the most popular topic model, latent Dirichlet allocation (LDA) [2]. In this study, we use Fast LDA [11] to evaluate the topic modeling based on perplexity. The perplexity, used by convention in language modeling. The algorithm uses a substantially smaller number of variational parameters, with no dependency on the dimensionality of the dataset. By calculating the perplexity of a test set, we can evaluate the generalization ability of a model, a lower perplexity score indicates better generalization performance. For a dataset D_{test}, the perplexity is: $perplexity(D_{test}) = exp - \frac{\sum_d log(P(w_d))}{\sum_d N_d}$, where w_d are the words in test dataset that are not repeated, N_d is the whole words in test dataset.

3 Experimental Results and Evaluation

3.1 Data Pre-Processing

To the best of our knowledge, there is no ground-truth resource to evaluate our approach. We decided to build our own evaluation set to estimate the performance of our model, sampling 510 english news articles[10] crawled from a daily newspaper site, UB Post[11]. One may notice that the amount of dataset

9 Wiki link: https://en.wikipedia.org/wiki/Bhattacharyya_distance.
10 Dataset: https://www.dropbox.com/sh/sio0goqw2soaavd/AABZHaNdcNC3VAN1XbKgCATPa?dl=0.
11 http://ubpost.mongolnews.mn/.

Table 1. Place names in the knowledge base

# All LN	# Ambiguous LN	# Non-ambiguos LN	# Top categories	# Total categories
9,099	952	3,232	56	721

is not very large: this is due to the cost constraints on geographic computation. The geo-objects which tagged in English language: *name*, *official_name* and *name : en*, are considered. Table 1 shows the total number of place names which populated in our *KB* including 21 provinces, 9 districts and the capital city. 11% of them is ambiguous referring to different coordinates, and the rest is non-ambiguous unique entities.

3.2 Disambiguation of Place Names

We identified a total of 1,847 locations in the set of sampled news. Table 2 shows the performance accuracy for the disambiguation methods. Precision, or Accuracy, is calculated as the number of correctly disambiguated locations divided by the number of locations identified in the news samples. Since we do not have a ground-truth resource to compare our approach, the baseline estimation is considered as an unique consistency (UC) approach. By combining all the methods, we out perform the baseline approach by 53 % more. We analyzed such correlation on the reduced categories which could make a more sense naturally about news topics, such as: *office*, *shop*, *amenity*, *cuisine*, *leisure*, *tourism*, *sport*, *historic*, *nature*, *religion*, *public transportation* and *other*. The correlation result is described in the table, where the (reduced) *BC* is relatively increased for all news. This indicates that the news topics are strongly correlated to the geo-objects mentioned in the text without considering locations.

3.3 Local Topics Identification

To avoid mis-disambiguation of place names in the news, we consider the place names which generated by the reliable ground truth (baseline) methods, UC and NP. This provides unique place names referred to one coordinate only. Given the news text, we estimate the statistical feasibility on news topics prediction based on geo-object types, vice versa. For example, a given news *"KFC is opened near central square in Mongolia as the nations first ever western fast food chain"*, the geo-object KFC is a fast food type of amenity and a business building which are described in OSMonto. Then, the topic of the news is associated to food and business. In fact, we can come up with an assumption that the local topics can be identified by the geo-objects described in the news text.

We pre-process the data by removing stop words in text *D*. We got totally 13,714 number of vocabularies in the text corpus. The BOW model is interpreted as a feature vector of the words given text *D* across each document. The BOW as vector for each news is labeled by the top categories in OSMonto when a geo-object is mentioned in the news text.

Table 2. Disambiguation and identification of place names on news sample

Method	# Count	Accuracy	Average *BC*	Reduced *BC*
NP	1,105 out of 4,231	26%	52%	52%
UC (Baseline)	837	45%	55%	69%
UC + SC	936	51%	53%	68%
UC + DC	1,356	73%	63%	83%
UC + DC + CC	1,382	75%	64%	83%
UC + DC + CC + NCC	1,813	98%	50%	89%

Table 3. SVM classifier

Estimation	Categories Organization - UC	Categories Organization - NP	Categorical Terms
Overall Accuracy	60.84	79.17	70.47
Macro-F1	50.0	100.0	100.0

In OSMonto, there are 385 category combinations in total: using these categories, we would like to classify the news topics. The categories are hierarchically organized into two-levels. For example, a *restaurant* is a sub-type of *amenity*, and a *supermarket* is a sub-type of *shop*. We sampled the dataset containing a geo-object found in articles in order to categorize the news. The result is reported on one against all validation (we chose 70 % for training and 30 % for testing; therefore we performed a random permutation before splitting the dataset for training and testing) in multi-class SVM classifier. The SVM classifier is denoted by a library, LIBSVM [3] which uses a linear kernel with penalty cost C=1. The news are localized based on the most likelihood administrative divisions or the most likelihood organizations. Table 3 shows the accuracy and macro-f1 of multi-class SVM classifier in average, over news categorized by the top categories organization which disambiguated by the baseline techniques UC and NP. Among the experiments, a classifier *Categories Organization - NP* performs well around 79.17 % of accuracy which considers news sets categorized by the top categories based on geo-object occurrence, where the place names have been disambiguated by NP method. This shows the geo-objects in news text as a proxy to predict news topics. But the overall accuracy and Macro-F1 are relatively small on a classifier *Categories Organization - UC* on the same categorized test sets, where we applied the UC method. This might be the impact of the dataset size that generated by UC method which generates a baseline set of news with non-ambiguous place names, around 45 % of geo-localized news. We also tested the classifier *Categorical Terms* on news categorized by the top categories based on the categorical word term occurrence. The result shows a similar accuracy with the classifier on the dataset of categories organization that indicates news topics are strongly correlated with the geo-objects mentioned in the news.

3.4 Global Topics Identification

In this study, we propose to classify news topics over the entire set incorporating with the Fast LDA to obtain lower dimensional feature representation for our subsequent classification task. The perplexity based evaluation is performed on test sets given a certain location features like organizations, categories organization and locations, respectively. In this paper, we took a sample as 10 % of the data for test purpose and used the rest for model training.

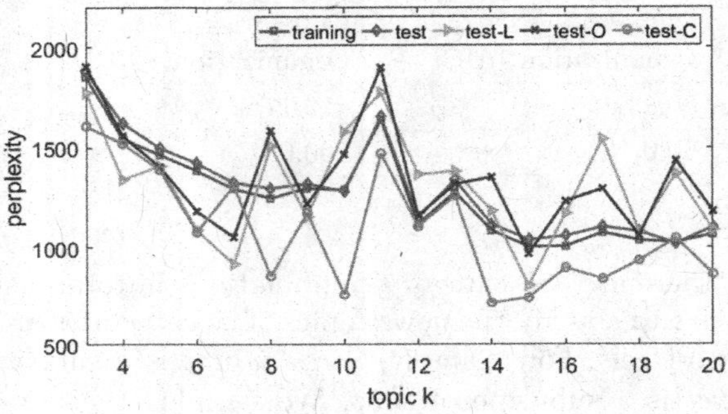

Fig. 2. Perplexity Over Entire and Geo-Located Dataset

Therefore we performed a random permutation before splitting the dataset for training and testing in the classifier. Hyper-parameter is set: use D_k/D to initialize *alpha* where D_k is the number of data points in class k and D is the total number of data points, $\epsilon = 0.01$, where k is the number of topics and V is the vocabulary size, ϵ is the laplacian smoothing parameter. The comparison of average perplexities over the training set, the test set and the classified test sets by location, organization and category organization (i.e., Oyu-tolgoi mining), is presented in Fig. 2. From the comparison, we observed similar perplexities in training and test set, but mildly higher perplexity on the test dataset. But the perplexities on the classified test sets are relatively lower that indicates a better performance. Thus, the global LDA classifier does not work well and it highlights that local topic identification is important to consider. For instance, we obtained a lowest perplexity on *topic-15*, where the model performs better, comparing the perplexities in training set and location feature classified test sets.

4 Conclusion and Discussion

We introduced a knowledge base geo-location recognition model based on OpenStreetMap as it allows us to extract the knowledge about geo-referenced place names in hierarchical administrative divisions. Using the knowledge base, our conceptual and computation methods can identify place names (i.e., organizations and administrative divisions) in free text that can be applied for any language. The place names disambiguation techniques perform reasonably well, around 98%. Therefore, we found that news topics are strongly correlated to the location based features mentioned in news and showed that geo-object is a good proxy to predict news topics, with around 79.17% accuracy (98% accuracy on reduced categories). However, news correlation to locations is highly dependent on the granularity of the city. Our approach could be largely affected by the distribution of geo-objects contributed to OSM, over different countries. In future works, we would like to use other gazetteers (i.e., GeoNames, Open Directory Project) for collecting a sufficient amount of location based features. For topic classification and rating task, we will analyze topics over time changes based on the features like topics importance and topics sentiment.

Acknowledgments. This research was supported by Basic Science Research Program through the National Research Foundation of Korea (NRF) funded by the Ministry of Science, ICT and Future Planning (number 2015R1A2A1A10052665).

References

1. Alias-i. LingPipe 4.1.0 (2008). http://alias-i.com/lingpipe/
2. Blei, D.M., Ng, A.Y., Jordan, M.I.: Latent dirichlet allocation. J. Mach. Learn. Res. **3**, 993–1022 (2003)
3. Chang, C.-C., Lin, C.-J.: LIBSVM: A library for support vector machines. ACM Trans. Intell. Syst. Technol. 2, 27:1–27:27 (2011). http://www.csie.ntu.edu.tw/~cjlin/libsvm
4. Codescu, M., Horsinka, G., Kutz, O., Mossakowski, T., Rau, R.: Osmonto - an ontology of openstreetmap tags. In: State of the map Europe (SOTM-EU) (2011)
5. Dashdorj, Z., Serafini, L., Antonelli, F., Larcher, R.: Semantic enrichment of mobile phone data records. In: MUM, p. 35. ACM (2013)
6. Dashdorj, Z., Sobolevsky, S.: Impact of the spatial context on human communication activity. In: Proceedings of the 2015 ACM UbiComp/ISWC Adjunct 2015, Osaka, Japan, 7–11 September 2015, pp. 1615–1622 (2015)
7. Finkel, J.R., Grenager, T., Manning, C.: Incorporating non-local information into information extraction systems by gibbs sampling. In: Proceedings of the 43nd Annual Meeting of the Association for Computational Linguistics, pp. 363–370 (2005)
8. Inkpen, D., Liu, J., Farzindar, A., Kazemi, F., Ghazi, D.: Detecting and disambiguating locations mentioned in twitter messages. In: Gelbukh, A. (ed.) CICLing 2015. LNCS, vol. 9042, pp. 321–332. Springer, Heidelberg (2015). doi:10.1007/978-3-319-18117-2_24
9. Ishida, K.: Estimation of user location and local topics based on geo-tagged text data on social media. In: 2015 IIAI 4th International Congress on Advanced Applied Informatics (IIAI-AAI), pp. 14–17, July 2015
10. Kinsella, S., Murdock, V., O'Hare, N.: "I'm eating a sandwich in glasgow": Modeling locations with tweets. In: Proceedings of the 3rd International Workshop on Search and Mining User-generated Contents, SMUC 2011. ACM, New York (2011)

11. Shan, H., Banerjee, A.: Mixed-membership naive bayes models. Data Min. Knowl. Disc. **23**(1), 1–62 (2011)
12. Sobolevsky, S., Sitko, I., Grauwin, S., des Combes, R.T., Hawelka, B., Arias, J.M., Ratti, C.: Mining urban performance: Scale-independent classification of cities based on individual economic transactions. CoRR, abs/1405.4301 (2014)
13. Van Laere, O., Quinn, J., Schockaert, S., Dhoedt, B.: Spatially aware term selection for geotagging. IEEE Trans. Knowl. Data Eng. **26**(1), 221–234 (2014)
14. Volz, R., Kleb, J., Mueller, W.: Towards ontology-based disambiguation of geographical identifiers. In: Bouquet, P., Stoermer, H., Tummarello, G., Halpin, H. (eds.) Proceedings of the WWW 2007 Workshop I^3: Entity-Centric Approaches to Information and Knowledge Management on the Web, Canada, CEUR Workshop Proceedings (2007)

Monitoring and Automating Factories Using Semantic Models

Niklas Petersen[1,2](✉), Michael Galkin[1,2,3], Christoph Lange[1,2], Steffen Lohmann[2], and Sören Auer[1,2]

[1] University of Bonn, Bonn, Germany
{petersen,galkin,langec,auer}@cs.uni-bonn.de
[2] Fraunhofer IAIS, Sankt Augustin, Germany
steffen.lohmann@iais.fraunhofer.de
[3] ITMO University, Saint Petersburg, Russia

Abstract. Keeping factories running at any time is a critical task for every manufacturing enterprise. Optimizing the flows of goods and services inside and between factories is a challenge that attracts much attention in research and business. The idea to fully describe a factory in a digital form to improve decision making is called a virtual factory. While promising virtual factory frameworks have been proposed, their semantic models lack depth and suffer from limited expressiveness. We propose an enhanced semantic model of a factory, which enables views spanning from the high level of supply chains to the low level of machines on the shop floor. The model includes a mapping to relational production databases to support federated queries on different legacy systems in use. We evaluate the model in a production line use case, demonstrating that it can be used for typical factory tasks, such as assembly line identification or machine availability checks.

1 Introduction

The Industry 4.0 vision [2] aims at digitizing engineering, production and manufacturing with the goal of (i) a seamless integration of devices, sensors, machines as well as software and IT systems, (ii) increased flexibility thanks to pushing more intelligence from centralized planning systems to the edge, (iii) increased efficiency due to automated data exchange and analysis within the value chain. Currently, much information is isolated within different applications, which prevents efficient access for real-time analytics [8]. The ultimate goal of Industry 4.0 (and related initiatives with different names in different regions, such as Industrie du Futur in France or Industrial Internet in the US) is the creation of a *Smart Factory* [6].

A Smart Factory is defined as a factory that supports people and machines in performing their tasks by providing context-aware information. For instance, the location of information about orders, products, machines, the available work force and the overall factory are rarely available in a unified database and format. The related idea of a *Virtual Factory* [14] proposes a framework that links all

Y.-F. Li et al. (Eds.): JIST 2016, LNCS 10055, pp. 315–330, 2016.
DOI: 10.1007/978-3-319-50112-3_24

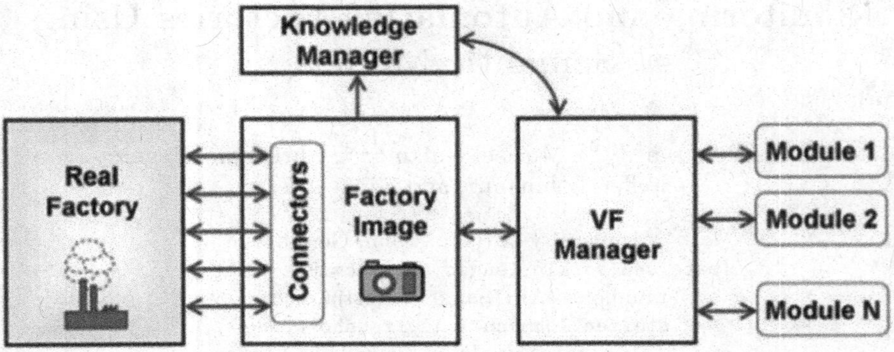

Fig. 1. Virtual factory framework as proposed by [12]

this information together, providing a mirror of the real factory (see Fig. 1) and thus paving the way towards more innovative factory prototyping, assembly line optimization, product design and mass customization [12].

In order to realize such a virtual factory, a number of interoperability challenges need to be solved. These include the identification of relevant data and information, their representation, unified access and interlinking. Of particular importance in this regard is the support of different views on the data (logistics and supply chain, manufacturing, quality control, etc.), the support of different levels of granularity of information representation (operational, strategic, etc.) as well as the access and integration of various data models and structures as used by existing systems and applications (XML, relational, enterprise models, etc.).

An integrated approach that provides a holistic view on an enterprise and its assets (such as factories) has not yet emerged. To fill this gap, we develop the notion of a *Semantic Factory*, employing semantic knowledge representation formalisms and technologies. The rationale is to employ a network of ontologies and vocabularies as a semantic fabric to represent, interlink and integrate the heterogeneous information being distributed in a variety of systems and information sources (e.g. manufacturing execution, quality management and enterprise resource planning systems or sensors, etc.). For creating the ontology, in a first step, we represent the static assets of an enterprise, such as its factories, assembly lines, workforce, etc. We then integrate dynamic information including business processes, shift plans, orders, etc. Finally, we map our model to production databases that contain the respective data for the dynamically changing concepts.

The mappings are performed in a minimally-invasive way, equipping existing systems of record with semantic interfaces (e.g. by using the W3C R2RML standard for mapping relational data to RDF). As a result, information and data in a factory can be integrated in a pay-as-you-go fashion, where mappings as well as the network of Semantic Factory ontologies evolve as required by specific use cases and application scenarios. Examples of use cases are (1) energy management, where the energy consumption is allocated to specific machines, work orders or customers, and (2) tool management, with the goal to minimize the time needed to equip machines with the required tools. In addition to introducing the Semantic Factory model, we demonstrate how decisions within an enterprise can be based on data currently hidden in different legacy systems. This includes the performance of supply chains, the detection of suitable assembly lines and the analysis of assets on a map.

The rest of the paper is structured as follows: We provide motivating examples and derive requirements for an integrated virtual factory representation in Sect. 2. In Sect. 3, we give an overview of related work. The overall architecture and factory ontology as its core model are presented in Sect. 4. In Sect. 5, we describe the implementation. We evaluate the performance of our approach in Sect. 6. The paper is concluded in Sect. 7 with an outlook on future work.

2 Motivating Examples and Requirements

A key motivation of our Semantic Factory model is to establish a holistic and integrated view on an enterprise in order to reduce the overall complexity and improve decision making. This includes the workforce, business processes, machines, shift plans, supply chains, etc. While a lot of this information is already captured by different IT systems, it is rarely accessible in a combined way without investing significant manual effort. Thus, the goal of this work is to make all data that is currently stored in various systems available in a unified model to support users with different roles in decision making.

2.1 Motivating Examples

An example is a factory planner who requires diverse information about order plans, workforce availability and machine maintenance dates. Another example is a machinist who needs to know which tools are to be mounted into which machine, where these tools are located, where the material is stored and what quality control standards are required during the production process. A controller, on the other hand, wants to keep track of the productivity of a factory and get an overview of certain Key Performance Indicators (KPIs). These comprise, in particular, information on the production time and effort required by each machine for each product, such as employee effort and energy consumption data.

To provide all those stakeholders with the information needed to perform their tasks and optimize decision making, we aim at semantically describing as many assets of a factory as possible, taking into account information from different manufacturing systems.

2.2 Requirements

We elicited the following requirements in the context of a research project for a global manufacturing company. The company's objective to gain a better picture of its assets (e.g. machines and factories) led us to develop an ontology that serves as the core element in the overall architecture.

From descriptions of the assets provided by the company and from interviewing domain experts, we gained an overview on typical tasks, processes and problems of each stakeholder. The interviews took place at the company site in multiple meetings, where the company's current IT infrastructure was described in detail. That way, we gathered requirements for the Semantic Factory model step by step:

Semantic Multi-modality. The types of data found in a factory context are diverse. Hence, the representation of various information, including attribute trees, relational, sensor, tabular, graph and entity data, must be supported.

Multi-dimensionality. Information along several dimensions must be represented and captured, such as:

- *Business processes:* Temporal views on diverse business activities are required to judge the success of an enterprise.
- *Spatial hierarchies:* The exploration of assets from a geographical perspective must be possible to increase the findability of said assets and related information.
- *Lifecycle:* Product and business lifecycles must be represented to support strategy management and business innovation.

Multi-granularity. Views on different levels of detail must be provided:

- *Components:* Instant access to sensor and component data must be enabled to support possible intervention measures.
- *Factories:* To decrease the complexity of factories, master and operational data needs to be accessible in a singular view.
- *Organization:* A big picture of all business units is needed, including their hierarchies and responsibilities.

Traceability and Integration. Data and information is currently spread across various systems, such as manufacturing execution systems, quality assurance systems, enterprise resource planning systems, etc. It is important to integrate all relevant information from these systems, while maintaining the systems' record-keeping character. When integrating information from these systems, the provenance of the data must be preserved, and changes to information in the source systems must be reflected in the integrated views, wherever possible in real time.

3 Related Work

There are two categories of related work: (i) existing frameworks that aim to describe factories as completely as possible, and (ii) existing ontologies representing assembly lines.

Terkaj et al. [13] propose a Virtual Factory Data Model represented as an OWL ontology based on the Industry Foundation classes (IFC)[1] standard. The purpose of their ontology is to describe business processes that involve machines requiring specific resources. However, the advantage of modeling each concept twice, once as a class whose instances represent real occurrences (e.g. `IfcProduct`, `IfcProcess`), and then as a class whose instances are supposed to describe generic objects and types (e.g. `IfcTypeProduct` "describes a generic object type that can be related to a geometric or spatial context", `IfcTypeProcess` "describes a generic process type to transform an input into output") is not clearly justified. A significant part of the ontology employs such a logical duplication which is misleading for non-experts. Furthermore, the rationale of proposing property classes such as `VffProcessProperties` to "characterize processes" instead of using object or data properties is not described.

Chen et al. [4] propose a multi-agent framework to monitor and control dynamic production floors. The ontology, serialized in XML, is optimized for the communication between different agents. It describes Radio-Frequency Identification (RFID) tags [16] attached to factory objects and addresses requirements specific to a bike manufacturing use case. Although RFID sensors are an important component of Industry 4.0, the purely XML-based ontology without logical formalisms behind, as they are provided by RDF(S) and OWL, lacks semantics and does not allow for universal and convenient querying.

Büscher et al. [3] introduce the Virtual Production Intelligence platform based on the Condition Based Factory Planning (CBFP) approach. The authors developed an OWL-based CBFP ontology advocating "the decoupling of domain business logic and the technical implementation of a planning system" [3]. The ontology is relatively small, consisting of only five classes. As it is not available online, we consider the CBFP ontology rather abstract and superficial. Detailed evaluations and experiments are not provided, making it hard to assess the practical contribution of the work.

Kim et al. [7] propose an OWL ontology and an information sharing framework to allow collaborative assembly design. The heart of the ontology is the assembly line and its direct environment. The ontology defines assemblies and constraints leveraging capabilities of SWRL and OWL. However, the lack of a published online version prevented us from reusing it. Nevertheless, the conceptual design influenced the one of our ontology, i.e., several concepts in the classes hierarchy and a few properties have been recreated.

---

[1] http://www.buildingsmart-tech.org/specifications/ifc-overview.

Ameri et al. [1] propose the Digital Manufacturing Market (DMM), a semantic web-based framework for agile supply chain deployment. DMM employs the Manufacturing Service Description Language (MSDL) at a semantic level. MSDL is an upper-level ontology expressed in OWL DL. Description Logic is extensively used to characterize supply and demand entities on several levels, such as the supplier, shop, machine and process levels. However, the granularity and ramification (especially for an upper-level ontology) impose restrictions on the usability, i.e., only a domain expert would have enough expertise to create a working model with accompanying queries. Furthermore, the ontology is again not available online, which prevented us from performing a thorough semantic analysis and considering an adaptation of concepts.

Zuehlke [17] introduces the SmartFactory initiative, which comprises best practices from the technical, architectural, planning, security and human dimensions. The initiative is envisioned to define and elaborate on the concept of *factory-of-things* as a vision of future manufacturing. Although semantic services involving ontologies and knowledge bases are claimed to be a part of the concept, the author does neither provide any examples nor references of such ontologies. Therefore, the presented concept is rather an implementation roadmap than a technical contribution.

4 Semantic Factory Architecture and Ontology

Based on the requirements presented in Sect. 2, we designed an architecture for a Semantic Factory application (see Fig. 2). Its core is a factory ontology, which describes real world objects, such as employees, machines and factories, locations of assets and their relations with each other. Operational data, such as information about work orders, machine sensor and process data, is also covered by the factory ontology; in practice, this data is dynamically mapped from the respective databases to the ontology.

The ontology is made available through an RDF triple store. Different applications can execute queries on the data, which is expressed as RDF or available in relational databases. Finally, for geospatial data, an external map provider service is used for drawing, among others, factories on a world map.

The following subsections explain how we developed the factory ontology following the methodology proposed by Uschold et al. [15]. We first defined the purpose and scope of the ontology; then, we captured step-by-step the domain knowledge, conceptualized and formalized the ontology and aligned it with existing ontologies. Finally, we evaluated the ontology by measuring the performance of certain queries for different sized datasets (see Sect. 6).

4.1 Purpose and Scope

The purpose of the ontology is to provide a holistic view of an enterprise. This is realized by implementing the requirements specified in Sect. 2. The intended

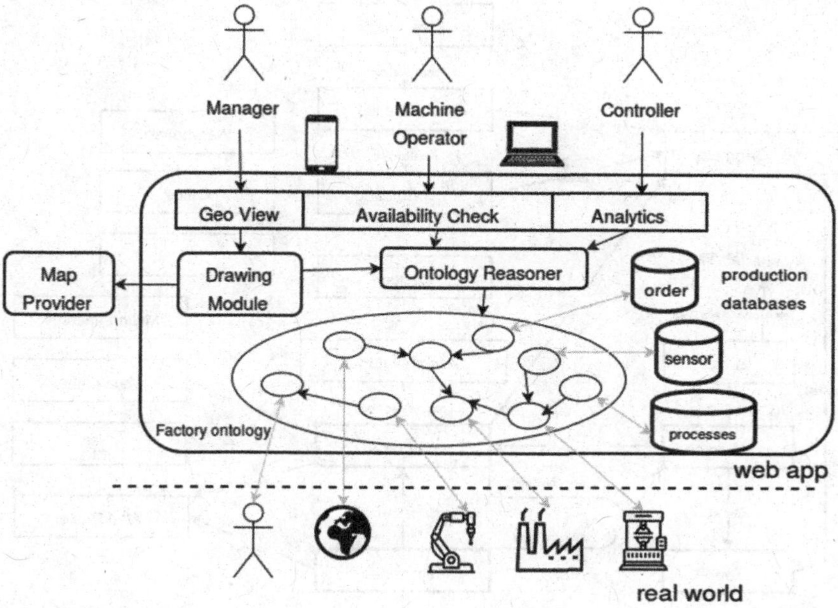

Fig. 2. Semantic Factory architecture

users are different stakeholders of an enterprise, such as managers, machine operators and controllers. Each of them needs different information to effectively and efficiently perform the corresponding tasks and duties. Thus, the ontology enables viewing the factory from different perspectives to support each class of stakeholders in their decision making.

4.2 Capturing Domain Knowledge

We captured the domain knowledge in three ways:

1. The company provided us with descriptive material of the domain, including maps of factories, descriptions of machines and work orders, process information, sensor data and tool knowledge. The types of input material ranged from formatted and unformatted text documents to spreadsheets and SQL dumps.
2. A live demonstration of a particular machine execution was given, including a discussion of further contextual information which was missing in the material. In subsequent meetings, open questions were clarified and concrete use cases of the ontology were discussed.
3. We reviewed relevant existing ontologies with the intention to build upon available conceptualizations and formalizations of domain knowledge.

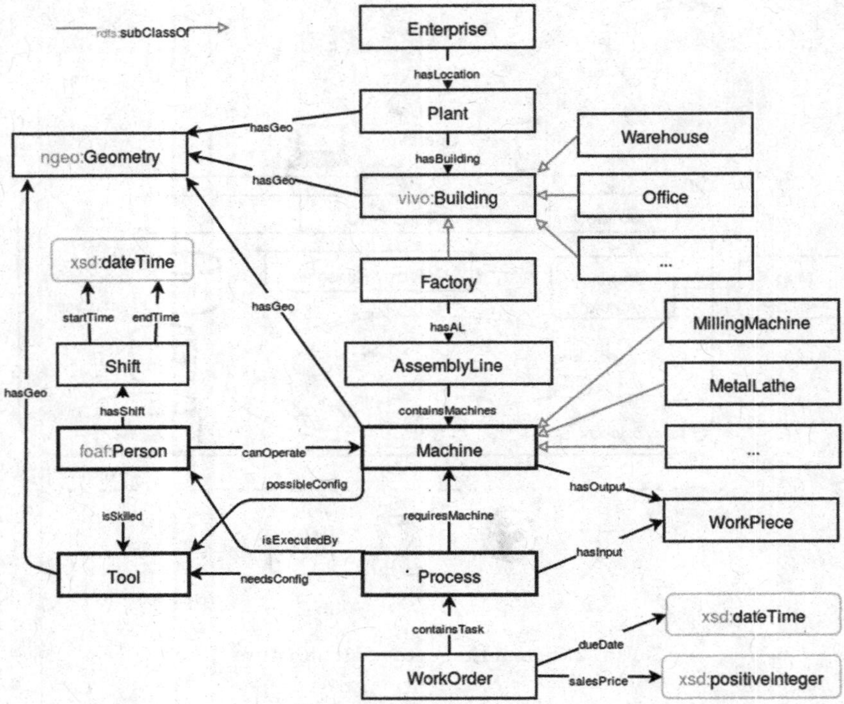

Fig. 3. Core concepts of the factory ontology

4.3 Conceptualizing and Formalizing

The resulting ontology comprises 86 classes, 73 object properties and 142 datatype properties. Since it has been designed to support an industrial project with sensitive business logic descriptions, not the entire ontology could be made publicly available. However, the part of the ontology that can be published is accessible via a permanent URL[2]. In the following, we describe the ontology, starting from the high-level organizational layer to the low-level machine and sensor layers. The core concepts of the ontology are depicted in Fig. 3.

In any concrete scenario, the class `Enterprise` is instantiated to represent the organization to which all further resources belong. Possible instances may be "Volkswagen", "General Electric" or "Samsung". Inter-organizational supply chains could involve multiple enterprises. Different production locations of an enterprise are described using the class `Plant`. A plant can comprise one or many `Buildings`, such as an office building, a factory building or a warehouse building. Typically, we can assume that each building serves a single function, though other configurations can also be represented, as OWL ontologies support

[2] https://w3id.org/i40/smo/.

multi-membership. Therefore, the subclasses of `Building` are not defined to be disjoint in the ontology.

Each `Factory` building may have one or multiple `AssemblyLines`. An assembly line usually consists of a sequence of multiple machines. The classes `MillingMachine` and `MetalLathe` are examples of subclasses of the abstract class `Machine`. Machines can be configured to use certain `Tools` to produce specific work pieces. As an example, a milling machine may be equipped with different lathes or end mills of varying granularity. `WorkPieces` represent everything which is an output of a machine. Once a work piece reaches its final stage of production, it becomes a product and is ready for shipping. Plants, factories and machines may have a representation of their geographical location (`ngeo:Geometry`[3]), which can, for instance, be used by front-end applications to display them on a map. A clear description of the entire capital of an enterprise supports the controllers and managers to keep track of utilized and unutilized assets.

As a next step, we describe the part of the ontology that represents the everyday operation of a factory. Employees are instances of the class `foaf:Person`. Each employee is qualified to operate specific machines (`canOperate`) and skilled to use certain tools (`isSkilled`). The class `WorkOrder` describes orders driven by customers. Each such order contains one or more `Processes`, which need to be executed to fully complete the order. Typical processes may be the configuration of a machine, the execution of a machine, quality control, etc. Each order has a due date and a sales price. The properties `requiresMachine`, `hasInput`, `needsConfig`, `isExecutedBy` are used together with the `Process` class. For example, they define which `Machine` is required for that process together with the needed configuration (tools assembled) and the input `Material`. All these details are required by the machine operator in the event of reconfiguring the machine for a specific order.

4.4 Aligning with Existing Ontologies

The ontology includes concepts and properties from well-known ontologies. The *Semantic Sensor Ontology*[4] provides us with a rich description of sensors, their measurements, devices and related concepts. The workforce and employees are described based on definitions by the *Friend of a friend* (FOAF)[5] ontology. Coordinates of factories and machines are based on the latitude and longitude definitions of the *W3C Geo Vocabulary*[6]. Finally, we reused geometrical concepts, such as the representation of polygons, from the *NeoGeo Geometry Ontology*[7].

[3] Prefixes are defined according to http://prefix.cc.
[4] https://www.w3.org/2005/Incubator/ssn/ssnx/ssn.
[5] http://xmlns.com/foaf/spec/.
[6] https://www.w3.org/2003/01/geo/wgs84_pos.
[7] http://geovocab.org/geometry.html.

5 Implementation and Application

We developed a software system that implements the presented Semantic Factory architecture and ontology and applied it to industry data. In this section, we first describe the front-end and back-end implementation. Then, we illustrate its usage by presenting various SPARQL queries that retrieve information for different production management tasks.

5.1 Front-End

The front-end is realized as a web application to facilitate access from different devices. The decision is motivated by the diversity of IT systems, platforms and devices usually deployed in a factory. The application uses the web framework *AngularJS*[8], which follows the model-view-controller design pattern to separate logic from representation. We created multiple views to address the collected requirements:

The *map view* (Fig. 4a) projects all instances (e.g. buildings, machines) with a geographical representation on a map by making use of the map provider *MapBox API*[9]. This API offers map tiles based on the open geographical database *OpenStreetMap*[10]. The projection itself is realized using the *leaflet.js*[11] JavaScript library. Coordinates in the factory ontology are represented using the *NeoGeo Geometry Ontology*[12] concepts and properties translated into `leaflet` geographical objects to be drawn on the map.

Further information, such as the person currently operating a machine, which order is executed or the status of a machine, is provided in the *machine view* that can be opened from the map view (see Fig. 4b) or independently by the machine operator. Besides static information, the pop-up contains also links to operational views and services. For example, the machine operator can follow the order link to retrieve additional information of that order. Furthermore, based on the tools required for the next machine operation, the "Find available tools" functionality points to the respective geographical location of the tools in the factory. The links "Visualize", "Analyze" and "Predict" point to external pages that provide additional graphical content about the machine, machine usage indicators and prediction dates when machine parts are worn out and need to be replaced.

5.2 Back-End

The back-end consists of a Python web server[13] that supports REST API calls from the front-end. Each request triggers the generation of SPARQL queries

[8] https://angularjs.org.
[9] https://www.mapbox.com/developers/.
[10] http://www.openstreetmap.org.
[11] http://www.leafletjs.com.
[12] http://geovocab.org/geometry.html.
[13] http://flask.pocoo.org.

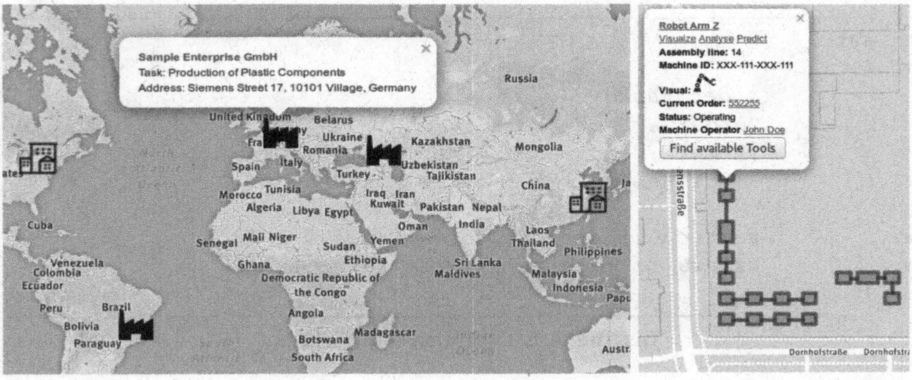

(a) Map View (b) Machine View

Fig. 4. Implementation of the Semantic Factory (geographic views).

executed either on the factory ontology or the production databases. To provide access to the ontology, the Python library *rdflib*[14] is used.

Access to the relational production databases is realized using the D2RQ[15] system. D2RQ provides a generator for creating an RDF mapping file for the database tables and columns, thus preserving the schema of the relational database. Using the mapping file, incoming SPARQL queries are translated ad-hoc into SQL queries and are executed on the respective database. Thus, D2RQ acts as a gateway between the web server and relational databases.

Once the data is obtained, it is returned in JSON data format[16] and processed by the respective front-end controller.

5.3 Factory Queries

In the following, we provide a set of SPARQL queries that demonstrate the usage of the factory ontology.

Order Feasibility Check. Listing 1.1 shows a query that determines if certain machines in the factory are free to use or already scheduled for other production plans. Each factory work order contains a list of tasks to be completed by a different machine. Thus, each needed machine is checked for its availability. Only if all machines are available, the query returns a positive answer such that the work order can be started. The query always checks the current state of the factory.

[14] https://github.com/RDFLib/rdflib.
[15] http://d2rq.org.
[16] http://json.org.

```
ASK
{
        # get tasks
        ?order a :WorkOrder .
        ?order :requiredMachines ?machineList .
        ?machineList rdfs:member ?machine.

        # check if the needed machines are free
        EXISTS { ?machine :isFree false } .
}
```

Listing 1.1. Order feasibility check

Retrieve Geographical Coordinates. Listing 1.2 shows a query to retrieve the machines, their names and coordinates. The outline of a machine is conceived as a polygon, represented as an `rdf:List` of geographical points, each with latitude and longitude, which is linked to the machine using the `ngeo:posList` datatype property. This information is returned to the front-end to be projected on the world map.

```
SELECT ?machine ?label
           (GROUP_CONCAT( ?lat ; separator=";") AS ?lats)
           (GROUP_CONCAT(?long ; separator=";") AS ?longs)
WHERE {
        ?machine rdfs:label ?label .
        ?machine ngeo:posList/rdf:rest*/rdf:first ?point .
        ?point geo:lat ?lat .
        ?point geo:long ?long .
} GROUP BY ?machine ?label
```

Listing 1.2. Retrieve geographical coordinates of machines

Suitable Assembly Lines. Listing 1.3 shows a query to find suitable assembly lines with regard to their sequence. Assembly lines that contain more machines than required but fulfill the correct order are still considered suitable. Thus, certain stations may be skipped within an assembly line.

Suppose, for example, that the machines 2 and 4 are required for an order. Suitable assembly lines include those having the following sequences of machines: `2,4` or `1,2,3,4,5`. Sequences such as `4,2` or `4,3,2` are considered non-suitable.

The query itself works as follows: First, assembly line candidates are filtered (`MINUS`) based on whether they contain the required machines in the work order. Second, of those assembly lines, the position of the needed machines is calculated. This is achieved by preparing the needed order (`?reqSequence`) and then retrieving the position of each machine in that order (?machineSeq). Third, these sequences are concatenated into strings and, finally, it is checked by a regular expression if the sequence is increasing.[17]

```
SELECT ?assemblyLine {
  ?assemblyLine :machineList ?lists .

  # filter all assembly lines with the wrong order
  FILTER REGEX(?seq,"^0*1*2*3*4*5*6*7*8*9*$")

  # concatenate order of the lists into a string
  {SELECT ?lists (GROUP_CONCAT(?machineSeq; separator="")
                                 AS ?seq)

   #Machine Sequence, _sorted_ by required order Sequence
  {SELECT ?lists ?machineSeq {
        {SELECT ?lists ?machineInstance ?machines
        (STRAFTER(STR(?memberProp), "_") AS ?machineSeq)
               {?lists rdfs:member ?machineInstance .
                ?lists ?memberProp ?machineInstance .
                ?machineInstance a ?machines . }}

   # get required Sequence of the Work Order
   {SELECT ?machines (STRAFTER(STR(?prop), "_")
                                AS ?requiredSeq) {
           :sampleOrder :requiredMachines ?orderList .
           ?orderList   ?prop             ?machines .
           FILTER (STRSTARTS(STR(?prop), STR(rdf:_)))}
           ORDER BY ?requiredSeq}

    # Identify Assembly Lines Candidates
  {SELECT ?lists ?assembly
         {?assembly d:machineList ?lists.}}
   MINUS
  {SELECT ?lists {
          ?lists          a                   rdf:Seq .
          ?workOrder   :requiredMachines  ?neededMachineList
              .
          ?neededMachineList  rdfs:member ?machineType .
     FILTER NOT EXISTS{?lists (rdfs:member/a) ?machineType
         .}}}
   } ORDER BY ?lists ?requiredSeq
} GROUP BY ?lists }}
```

Listing 1.3. Find suitable assembly lines

[17] The query is limited to sequences of up to 9 machines but may be extended.

6 Evaluation

We evaluated our factory ontology and software application by testing the performance of the SPARQL queries introduced in Subsect. 5.3. These queries were chosen due to their representativeness in an industrial setting. For that, we prepared multiple datasets, consisting of 10 K, 100 K, 1 M, 2 M and 5 M triples. The datasets contain generated test data based on our factory ontology, such as order information, workforce details, assembly lines, etc.

The queries were executed using the *ARQ* SPARQL processor version 2.13.0[18]. The machine we used for the experiment contains 8 GB of RAM, 256 GB SSD and an Intel i7-3537U CPU with 2.00 GHz.

Figure 5 depicts the results of the performance evaluation. While the growth for the "Retrieve machine coordinates" query of Listing 1.2 is linear, a response time of 25 s in larger datasets is not satisfactory for a front-end application. Thus, large datasets should be split to keep the execution time end-user friendly.

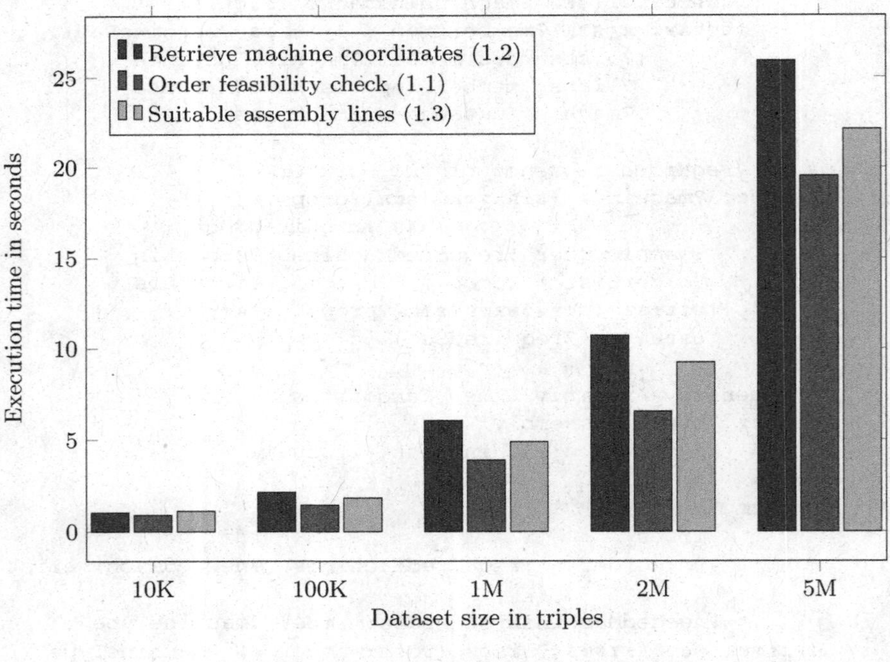

Fig. 5. Query execution performance

Similarly, as in the previous query, the execution time for the "Order Feasibility Check" query of Listing 1.1 grows linearly. While the overall performance

[18] https://jena.apache.org/documentation/query/.

is slightly better, one should nevertheless keep machine status data in an isolated dataset.

Finally, for the large "Suitable Assembly Lines" query of Listing 1.3, it is rather surprising that the execution time is quite similar to the short previous queries. As before, with a linear growth, the performance evolution becomes predictable and it is recommended to split instance data depending on certain services.

Overall, ontology-centered web applications are feasible with a satisfactory performance. As a common rule of thumb, the acceptable response time for complex operations is less than 10 s in order to keep the user's attention on the task [9]. Thus, certain crucial instance data should be kept in different triple stores to stay below that threshold.

However, the `d2rq` system for accessing relational databases reveals performance issues. First experiments using `ontop`[19] to query databases seem to be a promising alternative.

7 Conclusion and Future Work

The use of data-centric approaches in engineering, manufacturing and production is currently a widely discussed topic (cf. Industry 4.0, smart manufacturing or cyber-physical systems initiatives). The complexity of data integration in general is perceived to be one of the major bottlenecks of the field. A key issue in engineering, manufacturing and production is to be able to efficiently and effectively manage factories.

This paper described an ontology and an integration infrastructure to obtain a holistic view of the status of a factory from different perspectives. We see the work presented in this paper as a first step towards establishing an ontology-based integration approach for manufacturing, which is centered around a common information model, but at the same time supports the management of data in a decentralized manner in the existing systems of record. The integration follows a loosely-coupled architecture, where the decentralized data sources are mapped on demand to the factory ontology. The factory ontology is not supposed to be a fixed, monolithic schema, but rather a flexible, evolving and interlinked knowledge fabric. For this purpose, we have developed the collaborative vocabulary development methodology and support environment VoCol [5].

We see a number of directions for future work. In particular, the Semantic Factory approach could be expanded from single factories to an integration approach covering the entire enterprise as well as supply networks (e.g. based on the SCOR model [10]) involving a large number of organizations. Another promising direction of future work is the exploitation of the integrated data for advanced analytics and forecasting [11].

Acknowledgments. This work has been supported by the German Federal Ministry of Education and Research (BMBF) in the context of the projects LUCID (grant no. 01IS14019C), SDI-X (no. 01IS15035C) and Industrial Data Space (no. 01IS15054).

[19] http://ontop.inf.unibz.it.

References

1. Ameri, F., Patil, L.: Digital manufacturing market: a semantic web-based framework for agile supply chain deployment. J. Intell. Manuf. **23**(5), 1817–1832 (2012)
2. Brettel, M., Friederichsen, N., Keller, M., Rosenberg, M.: How virtualization, decentralization and network building change the manufacturing landscape: an industry 4.0 perspective. Int. J. Mech. Ind. Sci. Eng. **8**(1), 37–44 (2014)
3. Büscher, C., Voet, H., Krunke, M., Burggräf, P., Meisen, T., Jeschke, S.: Semantic information modelling for factory planning projects. Procedia CIRP **41**, 478–483 (2016)
4. Chen, R.S., Tu, M.A.: Development of an agent-based system for manufacturing control and coordination with ontology and rfid technology. Expert Syst. Appl. **36**(4), 7581–7593 (2009)
5. Halilaj, L., Petersen, N., Grangel-González, I., Lange, C., Auer, S., Coskun, G., Lohmann, S.: Vocol: an integrated environment to support version-controlled vocabulary development. In: 20th International Conference on Knowledge Engineering and Knowledge Management. Springer Verlag (2016, in print)
6. Hermann, M., Pentek, T., Otto, B.: Design principles for industrie 4.0 scenarios: a literature review. Technische Universität Dortmund, Dortmund (2015)
7. Kim, K.Y., Manley, D.G., Yang, H.: Ontology-based assembly design and information sharing for collaborative product development. Comput. Aided Des. **38**(12), 1233–1250 (2006)
8. Newman, D., Gall, N., Lapkin, A.: Gartner defines enterprise information architecture. Gartner Group (2008)
9. Nielsen, J.: Response times: The 3 important limits. Usability Engineering (1993)
10. Petersen, N., Grangel-González, I., Coskun, G., Auer, S., Frommhold, M., Tramp, S., Lefrançois, M., Zimmermann, A.: SCORVoc: vocabulary-based information integration and exchange in supply networks. In: 10th International Conference on Semantic Computing (ICSC 2016), pp. 132–139. IEEE (2016)
11. Petersen, N., Lange, C., Auer, S., Frommhold, M., Tramp, S.: Towards federated, semantics-based supply chain analytics. In: Abramowicz, W., Alt, R., Franczyk, B. (eds.) BIS 2016. LNBIP, vol. 255, pp. 436–447. Springer, Heidelberg (2016). doi:10.1007/978-3-319-39426-8_34
12. Sacco, M., Pedrazzoli, P., Terkaj, W.: VFF: virtual factory framework. In: Proceedings of 16th International Conference on Concurrent Enterprising (ICE 2010), pp. 21–23. IEEE (2010)
13. Terkaj, W., Urgo, M.: Virtual factory data model to support performance evaluation of production systems. In: Proceedings of the Workshop on Ontology and Semantic Web for Manufacturing (OSEMA 2012). CEUR-WS, vol. 886, pp. 24–27 (2012)
14. Upton, D.: The real virtual factory. Harvard Bus. Rev. **74**(4), 123–133 (1996)
15. Uschold, M., Gruninger, M., et al.: Ontologies: principles, methods and applications. Knowl. Eng. Rev. **11**(2), 93–136 (1996)
16. Want, R.: An introduction to RFID technology. IEEE Pervasive Comput. **5**(1), 25–33 (2006)
17. Zuehlke, D.: SmartFactory–towards a factory-of-things. Annu. Rev. Control **34**(1), 129–138 (2010)

Author Index

Printed in the United States
By Bookmasters

Springer Collected Works in Mathematics

More information about this series at http://www.springer.com/series/11104

I. Schur

Issai Schur

Gesammelte Abhandlungen I

Editors
Alfred Brauer
Hans Rohrbach

Reprint of the 1973 Edition

Springer

Author
Issai Schur (1875 – 1941)
Universität Berlin
Berlin
Germany

Editors
Alfred Brauer (1894 – 1985)
University of North Carolina
Chapel Hill, NC
USA

Hans Rohrbach (1903 – 1993)
Universität Mainz
Mainz
Germany

ISSN 2194-9875
Springer Collected Works in Mathematics
ISBN 978-3-662-48753-2 (Softcover)
978-3-642-61948-9 (Hardcover)

Library of Congress Control Number: 2012954381

Mathematics Subject Classification (2010): 20.XX, 01A75, 40.0X, 15.85

Springer Heidelberg New York Dordrecht London

Printed on acid-free paper

Springer-Verlag GmbH Berlin Heidelberg is part of Springer Science+Business Media
(www.springer.com)

ISSAI SCHUR

GESAMMELTE ABHANDLUNGEN

BAND I

Herausgegeben von

Alfred Brauer und Hans Rohrbach

Springer-Verlag
Berlin · Heidelberg · New York 1973

ISBN-13: 978-3-642-61948-9 e-ISBN-13: 978-3-642-61947-2
DOI: 10.1007/978-3-642-61947-2

Offsetdruck: Julius Beltz, Hemsbach/Bergstr.
Bindearbeiten: Konrad Triltsch, 87 Würzburg

Vorwort

Die Ergebnisse, Methoden und Begriffe, die die mathematische Wissenschaft dem Forscher ISSAI SCHUR verdankt, haben ihre nachhaltige Wirkung bis in die Gegenwart hinein erwiesen und werden sie unverändert beibehalten. Immer wieder wird auf Untersuchungen von SCHUR zurückgegriffen, werden Erkenntnisse von ihm benutzt oder fortgeführt und werden Vermutungen von ihm bestätigt. Daher ist es sehr zu begrüßen, daß sich der Springer-Verlag bereit erklärt hat, die wissenschaftlichen Veröffentlichungen von I. SCHUR als Gesammelte Abhandlungen herauszugeben.

Die Besonderheit des mathematischen Schaffens von SCHUR hat einst MAX PLANCK, als Sekretär der physikalisch-mathematischen Klasse der Preußischen Akademie der Wissenschaften zu Berlin, gut gekennzeichnet. In seiner Erwiderung auf die Antrittsrede von SCHUR bei dessen Aufnahme als ordentliches Mitglied der Akademie am 29. Juni 1922 bezeugte er, daß SCHUR „wie nur wenige Mathematiker die große Abelsche Kunst übe, die Probleme richtig zu formulieren, passend umzuformen, geschickt zu teilen und dann einzeln zu bewältigen".

Zum Gedächtnis an I. SCHUR gab die Schriftleitung der Mathematischen Zeitschrift 1955 einen Gedenkband heraus, aus dessen Vorrede wir folgendes entnehmen (Mathematische Zeitschrift **63,** 1955/56): „Aus Anlaß der 80. Wiederkehr des Tages, an dem Schur in Mohilew am Dnjepr geboren wurde, vereinen sich Freunde und Schüler, um sein Andenken mit diesem Bande der Zeitschrift zu ehren, die er selbst begründet hat. Sie sind in alle Welt zerstreut durch die Katastrophe, in deren Verlauf Schur durch vorzeitige Emeritierung 1935 die Wirkungsstätte verlor, an der seine Vorlesungen drei Jahrzehnte lang Studenten für die Mathematik begeistert hatten ... Möge dieser Band an Schurs Gesamtwerk erinnern und von seiner Fruchtbarkeit zeugen."

Den nachstehend abgedruckten Abhandlungen stellen wir zur Würdigung von I. SCHUR als Forscher, Hochschullehrer und Mensch die Ansprache voran, die der erste der beiden Herausgeber 1960 bei der 150-Jahrfeier der Berliner Universität auf deren Einladung hin Schur gewidmet hat. Diese Würdigung macht insbesondere deutlich, wie sehr die politischen Verhältnisse der dreißiger Jahre in Deutschland die letzten Lebensjahre von SCHUR überschatteten und wie stark seine Schaffenskraft durch Druck und Verfolgung beeinträchtigt wurde. Erst 1939, nach seiner Ankunft in Israel, war er wieder imstande, wissenschaftlich zu arbeiten, doch hat er bis zu seinem Tode (Januar 1941) nichts mehr veröffentlicht.

In seinem Nachlaß sowie in dem seines Sohnes GEORG SCHUR fanden sich mehrere fast fertige Manuskripte. Von diesen sind drei durch M. FEKETE und M. SCHIFFER dem American Journal of Mathematics zur Veröffentlichung eingereicht worden und 1945 bzw. 1947 dort erschienen. Drei weitere Manuskripte bilden den Hauptteil des Anhangs, den die beiden Herausgeber diesen Gesammelten Abhandlungen angefügt haben. Dieser Anhang enthält außerdem die von SCHUR publizierten Aufgaben sowie Ergebnisse von SCHUR in Arbeiten anderer Mathematiker.

Bei der Überarbeitung der im Anhang abgedruckten nachgelassenen Untersuchungen von SCHUR haben uns die Herren RICHARD H. HUDSON (University of South Carolina,

Columbia, S.C.), Rudolf Kochendörffer (Universität Dortmund) und Alfred Stöhr (Freie Universität Berlin) wesentlich unterstützt. Ihnen hierfür auch an dieser Stelle unseren Dank auszusprechen, ist uns ein Bedürfnis. Ebenso danken wir dem Springer-Verlag für die gute Planung, Durchführung und Ausstattung des Gesamtwerks.

Chapel Hill, N.C. und Mainz, Februar 1973 — Alfred Brauer, Hans Rohrbach

Hinweis

Nicht aufgenommen sind die beiden von Frobenius mit Schur verfaßten Abhandlungen. Sie sind abgedruckt in F. G. Frobenius, Gesammelte Abhandlungen. III, 355–386. Berlin-Heidelberg-New York: Springer 1968.

Die Paginierung am oberen Rand jeder Seite entspricht der Originalpaginierung. Die dem Inhaltsverzeichnis dieser Gesammelten Abhandlungen entsprechende fortlaufende Paginierung befindet sich am unteren Rand jeder Seite.

Gedenkrede auf Issai Schur[1]

ALFRED BRAUER

Magnifizenz Professor SCHRÖDER!

Professor REICHARDT!

Meine Damen und Herren!

Den Veranstaltern dieser Tagung möchte ich meinen herzlichsten Dank sagen, daß Sie mich eingeladen haben, hier meines verehrten Lehrers, ISSAI SCHUR, zu gedenken. Ferner möchte ich seiner Magnifizenz, dem Herrn Rektor, für seine mich ehrenden Worte in seiner gestrigen Ansprache bestens danken.

Es ist für mich eine große Freude, an die Stätte zurückzukehren, der ich fast meine ganze wissenschaftliche Ausbildung verdanke und mit der ich fast 20 Jahre als Student, Assistent und Dozent verbunden war.

Es scheint mir eine schöne Idee zu sein, bei einem Jubiläum einer Universität derer zu gedenken, die segensreich an ihr gewirkt haben. Ich persönlich denke heute in Dankbarkeit an alle, die hier als meine Lehrer, als meine Kollegen oder als meine Kommilitonen mir wissenschaftlich und menschlich so viel gegeben haben.

Ich bin stolz darauf, aus dieser Universität hervorgegangen zu sein. Im Frühjahr 1925, noch bevor ich an meiner Dissertation zu arbeiten begonnen hatte, bot Schur mir eine Assistentenstelle bei FELIX KLEIN in Göttingen an. Ich lehnte dieses Angebot ab. Ich zog es vor, als gewöhnlicher Student hier zu bleiben. Allerdings wußte ich genau, was Berlin und insbesondere Schur mir geben konnte. Ich habe diesen Entschluß nie bereut. Alle diejenigen unter den Anwesenden, die damals hier waren, werden mir beipflichten.

ISSAI SCHUR wurde am 10. Januar 1875 als Sohn des Großkaufmanns MOSES SCHUR und seiner Ehefrau GOLDE, geb. LANDAU, zu Mohilew am Dnjepr in Rußland geboren. Seit seinem dreizehnten Lebensjahre lebte er im Hause seiner Schwester und seines Schwagers in Libau, um dort das ausgezeichnete Nicolai-Gymnasium zu besuchen. Das Abitur bestand er als bester Schüler, und er wurde durch Verleihung einer Goldmedaille ausgezeichnet. Ungefähr zu dieser Zeit starb sein Vater, während seine Mutter ein hohes Alter erreichte. Einen ihrer letzten Geburtstage feierte SCHUR durch Widmung einer seiner Arbeiten.

Im Herbst 1894 bezog SCHUR die Universität Berlin. Dort studierte er zunächst Physik, bald aber wandte er sich ganz der Mathematik zu. An der Berliner Universität bestand er am 27. November 1901 die Doktorprüfung summa cum laude. Die Dissertation wurde mit dem Prädikat „egregium“ angenommen. Im Lebenslauf seiner Dissertation nennt SCHUR insbesondere FROBENIUS, FUCHS, HENSEL und SCHWARZ als seine Lehrer. Wie sich später zeigte, war es FROBENIUS, der auf SCHURS Arbeitsweise und mathematisches Interesse den größten Einfluß hatte. Wie damals üblich, hatte er bei der

[1] Rede gehalten am 8. November 1960 auf Einladung der Humboldt-Universität in Berlin anläßlich der SCHUR-Gedenkfeier im Rahmen der 150-Jahrfeier der Universität. (Leicht abgeändert 1971.)

Für Überlassung von Material bin ich der inzwischen verstorbenen Gattin von SCHUR, Frau REGINA SCHUR, seiner Tochter, Frau HILDE ABELIN, Professor M. SCHIFFER und meinem Bruder, RICHARD BRAUER, dankbar, ebenso meiner Frau, HILDE BRAUER, für technische Hilfe.

Doktorprüfung einige von ihm gewählte Thesen zu verteidigen, die in seiner Dissertation abgedruckt sind, ebenso wie die Namen seiner Opponenten. Im Jahre 1903 habilitierte sich SCHUR in Berlin als Privatdozent.

Am 2. September 1906 heiratete Schur Dr. med. REGINA FRUMKIN. Diese Ehe war überaus glücklich. Seine Frau verstand es meisterhaft ihm vieles abzunehmen, damit er sich ganz der Mathematik widmen konnte.

Aus dieser Ehe sind zwei Kinder hervorgegangen, ein Sohn, dem er zu Ehren von FROBENIUS den Vornamen GEORG gab, und eine Tochter HILDE. SCHUR hätte es gern gesehen, wenn sein Sohn, der für Mathematik sehr begabt war, dieses Fach studiert hätte. Dieser aber zog vor, Physik zu studieren, um mit seinem Vater nicht konkurrieren zu müssen. Er bestand noch das Staatsexamen, mußte aber wegen seiner Auswanderung sein Studium aufgeben. In späteren Jahren war er als Versicherungsmathematiker in Israel tätig. Auf seinen Berechnungen beruht die sogenannte National-Versicherung Israels. Sein Interesse für die reine Mathematik kam immer wieder zum Durchbruch. In zwei der in SCHURS Nachlaß gefundenen Arbeiten (als Nr. 81 und 82 der Gesammelten Abhandlungen erstmals veröffentlicht), findet sich ein Beweis seines Sohnes. SCHURS Tochter ist mit einem Arzt, Dr. ABELIN, in Bern verheiratet. Von ihren vier Kindern hat SCHUR die Geburt der ältesten drei noch erlebt. Er hing an diesen Enkelkindern mit großer Liebe.

Im Jahre 1911 wurde SCHUR auf Vorschlag von HAUSDORFF als dessen Nachfolger als planmäßiger außerordentlicher Professor nach Bonn berufen. 1916 kehrte er in der gleichen Stellung nach Berlin zurück. Hier wurde er 1919 ordentlicher Professor. Im Jahre 1922 wählte ihn die Preußische Akademie der Wissenschaft in Berlin zum Mitglied. Ich glaube die Einstellung SCHURS zur Mathematik und zu seinen Arbeiten nicht besser charakterisieren zu können, als wenn ich hier seine Antrittsrede in der Akademie in der öffentlichen Sitzung am Leibniztage 1922 und die Erwiderung von PLANCK als Sekretär der physikalisch-mathematischen Klasse verlese[2]. Ich erinnere mich, daß diese Reden einen großen Eindruck auf die Zuhörer machten.

Auf die Mitgliedschaft in der Akademie ist SCHUR immer besonders stolz gewesen. Er besuchte regelmäßig ihre Sitzungen, und viele seiner Arbeiten sind in ihren Sitzungsberichten erschienen. Später wurde SCHUR noch Korrespondierendes Mitglied der Akademien in Leningrad, Leipzig, Halle und Göttingen.

Die Jahre von 1915–1933 waren in wissenschaftlicher Beziehung äußerst erfolgreich für ihn. Da war es ein entsetzlicher Schlag, als es Ende April 1933 gerüchtweise bekannt wurde, daß SCHUR von seinem Amt beurlaubt werden sollte. Am 1. Mai wurde das Gerücht zur Tatsache. Am Nachmittag dieses Tages suchten ROHRBACH und ich SCHUR auf, um die Hoffnung auszusprechen, daß diese Beurlaubung nur vorübergehend sein würde. Äußerlich war SCHUR völlig ruhig und gefaßt, aber innerlich wurde seine Arbeitskraft durch dieses Ereignis aufs stärkste vermindert. Zwar gelang es den Bemühungen von ERHARD SCHMIDT, die Beurlaubung vom Wintersemester 1933/34 an rückgängig zu machen, da sie auch nach den damaligen Gesetzen ungesetzlich war, weil SCHUR schon vor Ende des ersten Weltkrieges preußischer Beamter gewesen war. Kaum war die Beurlaubung bekannt geworden, als SCHUR ein Angebot von der Universität von Wisconsin in Madison erhielt. Aber er lehnte dieses Angebot ab, da er sich nicht mehr kräftig genug fühlte, in einer anderen Sprache Vorlesungen zu halten.

Nach seiner Wiedereinsetzung durfte SCHUR nur noch ausgewählte Spezialvorlesungen halten. Während der nächsten zwei Jahre wurden ihm immer wieder neue Schwierigkeiten bereitet, bis er sich dem Druck fügte und sich bereit erklärte, sich zum 31. August 1935 emeritieren zu lassen. Hätte er diesen Schritt nicht unternommen, so wäre er bald darauf

[2] Vgl. diese Gesammelten Abhandlungen. II, 413-415.

seines Amtes ohnehin enthoben worden. Denn noch vor Beginn des Wintersemesters 1935/36 wurden die letzten wenigen jüdischen Mitglieder des Lehrkörpers aus ihren Ämtern entfernt.

Für SCHUR ergab sich noch einmal die Möglichkeit einer kurzen Lehrtätigkeit. Im Frühjahr 1936 wurde er von der Eidgenössischen Technischen Hochschule in Zürich eingeladen, eine Reihe von Vorlesungen über die Darstellungstheorie endlicher Gruppen zu halten. Diese Vorlesung wurde von STIEFEL ausgearbeitet und ist im Druck erschienen, aber seit vielen Jahren vergriffen. Sie ist auch heute noch vielleicht die beste Einführung in dieses Gebiet.

Das zwangsweise Ende seiner Lehrtätigkeit im Alter von 61 Jahren bedeutete einen schweren Schlag für SCHUR. Während der kurzen Zeit, in der ROHRBACH dann noch Assistent am Mathematischen Seminar der Berliner Universität war, war es noch möglich, indirekt Bände aus der Bibliothek des Seminars einzusehen. Aber als ROHRBACH diese Stellung verlor und als Assistent nach Göttingen ging, waren wir von der mathematischen Welt mehr und mehr abgeschlossen. Ein Beispiel soll das illustrieren. Als LANDAU im Februar 1938 starb, sollte SCHUR am Grabe eine Gedenkrede halten. Dazu brauchte er einige mathematische Tatsachen, die ihm entfallen waren. Er bat mich zu versuchen, diese aus der Literatur festzustellen. Selbstverständlich war es mir verwehrt, die Bibliothek des Mathematischen Seminars, für deren Aufbau ich jahrelang gearbeitet hatte, zu benutzen. Ich wandte mich mit einem Gesuch an die Preußische Staatsbibliothek. Es wurde mir gestattet, gegen Bezahlung einer Gebühr den Lesesaal dieser Bibliothek für eine Woche zu benutzen. Bücher aber entleihen durfte ich nicht. So konnte ich SCHUR wenigstens einige seiner Fragen beantworten.

In diesen Jahren habe ich SCHUR oft besucht. Die ständigen neuen Bestimmungen, die das Leben aller deutschen Juden mehr und mehr erschweren sollten, führten bei SCHUR zu schweren Depressionen. Er befolgte alle diese Gesetze aufs genaueste. Aber trotzdem geschah es einige Male, daß er, als er mir auf mein Klingeln die Wohnungstür öffnete, erleichtert ausrief: „Ach, Sie sind es und nicht die Gestapo." Häufig war es unmöglich, mit ihm über Mathematik zu sprechen. Gelegentlich diskutierten wir das folgende, auf FROBENIUS zurückgehende Problem, das SCHUR in seiner letzten Berliner Vorlesung etwas behandelt hatte. „Gegeben sind n positive ganze Zahlen $a_1, a_2, \dots, a_n$. Eine Schranke $F(a_1, a_2, \dots, a_n)$ ist zu bestimmen, so daß die Diophantische Gleichung $a_1x_1 + a_2x_2 + \dots + a_nx_n = N$ immer Lösungen in positiven ganzen Zahlen für alle $N > F(a_1, a_2, \dots, a_n)$ hat.

SCHUR stellte sich auf den Standpunkt, daß er nicht mehr das Recht habe, die Resultate dieser gemeinsamen Überlegungen, weder in Deutschland, noch im Ausland, zu veröffentlichen. Auch nachdem wir beide ausgewandert waren, beharrte er auf diesem Standpunkte. Nach langem Hin und Her bat er mich, die Arbeit allein zu publizieren. Er billigte meine Fassung. Fast zwei Jahre nach SCHURS Auswanderung, aber noch wenige Wochen vor seinem Tode, wurde diese Arbeit im November 1940 beim American Journal of Mathematics eingereicht.

Eines Sonntags Morgen im Sommer 1938 erschien Frau SCHUR unerwartet in unserer Wohnung. Sie wollte mich in einer dringenden Angelegenheit um Rat fragen. Sie hatte einen Brief abgefangen, in dem SCHUR in zwei Wochen zu einem Termin bei der Gestapo bestellt wurde. Nun hatte SCHUR mehrmals erklärt, daß er eher Selbstmord begehen würde, als einer Vorladung der Gestapo Folge zu leisten. Frau SCHUR hatte nun den Plan, SCHUR auf Grund eines ärztlichen Attestes sofort in ein Sanatorium zu schicken, da er ja tatsächlich krank war. Ich konnte diesen Plan als einzigen Ausweg nur billigen. SCHUR verließ Berlin und ging für einige Wochen in ein Sanatorium. Frau SCHUR ging mit dem ärztlichen Attest am festgesetzten Termin zur Gestapo. Dort wurde sie nur gefragt, warum sie noch nicht ausgewandert seien. Natürlich wollte die Regierung SCHURS Pension einsparen. Frau SCHUR erklärte, daß sie an ihrer Auswanderung arbeiteten, daß es ihnen aber noch nicht gelungen sei, alle Schwierigkeiten aus dem Weg zu räumen.

Die Hauptschwierigkeit bestand in Folgendem. SCHURS planten, nach Israel auszuwandern und hatten das erforderliche Geld. Aber unglücklicherweise hatte Frau SCHUR eine größere Hypothek auf ein Haus in Litauen geerbt. Auf Grund der litauischen Devisenbestimmungen konnte diese Hypothek nicht zurückgezahlt werden. Es war SCHUR verboten, auf diese Hypothek zu verzichten oder sie an das Deutsche Reich abzutreten. Sie mußte zu seinem sonstigen Vermögen zugerechnet werden, und von der Gesamtsumme war die 25prozentige Reichsfluchtsteuer zu zahlen. Dazu reichte SCHURS Geld nicht aus. Nach einigen Monaten gelang es einen Wohltäter zu finden, der sich bereit erklärte, die notwendige Geldsumme zur Verfügung zu stellen. Natürlich war es für SCHUR sehr schmerzlich, gezwungen zu sein, dieses Geschenk anzunehmen.

Endlich waren alle Schwierigkeiten überwunden, und der Paß wurde erteilt. Eines Tages im Januar 1939 rief Frau SCHUR uns an, um uns mitzuteilen, daß SCHUR am selben Abend in Begleitung einer Krankenschwester nach der Schweiz zu seiner Tochter abreisen würde, da sie selbst erst in einigen Tagen folgen könnte. SCHUR würde meine Frau und mich gern noch einmal sehen. Wenige Stunden später standen wir in seinem Arbeitszimmer, um Abschied von ihm für immer zu nehmen. SCHUR selbst glaubte nicht, daß die Auswanderung glücken würde, obgleich alle amtlichen Bestimmungen aufs genaueste befolgt waren. Aber am nächsten Morgen rief mich Frau SCHUR an, um mir mitzuteilen, daß SCHUR bei seiner Tochter in Bern angekommen sei. Dort blieb er einige Wochen, um dann mit seiner Frau nach Israel auszuwandern.

Es konnte gehofft werden, daß SCHURS Zustand sich in Israel bessern würde. Aber keine wesentliche Besserung trat ein. Als SCHIFFER, der SCHUR von Berlin her kannte, ihn zum ersten Male wiedersah, war er erschüttert. SCHUR war kaum zu bewegen, über Mathematik zu sprechen. Auch weiterhin bestand er darauf, daß er nicht mehr das Recht habe, etwas zu veröffentlichen. Zwar hat er anscheinend im Geheimen etwas gearbeitet, denn in seinem Nachlaß sind einige Arbeiten gefunden worden, die mindestens zum Teil in Tel Aviv entstanden sind. Drei von diesen wurden später von FEKETE und SCHIFFER überarbeitet und unter SCHURS Namen im American Journal of Mathematics veröffentlicht.

Nur einmal gelang es, SCHUR zu bewegen, im Mathematischen Seminar der Universität in Jerusalem zu sprechen. Er begann, einen ausgezeichneten Vortrag zu halten wie in seinen besten Zeiten, so daß die Anwesenden, unter denen TOEPLITZ und SCHIFFER waren, über diese plötzliche Besserung beglückt waren. Er erwähnte in seinem Vortrage Resultate von GRUNSKY und getreu seiner Einstellung, über Menschen Gutes zu sagen, benutzte er diese Gelegenheit zu bemerken, wie sehr er GRUNSKY mathematisch und menschlich schätze. Aber plötzlich bat er um Entschuldigung und setzte sich auf einen Stuhl am Vortragstisch, den Kopf vornüber gelegt, als ob er schwer nachdenken müßte. Nach einigen Minuten stand er auf und beendete seinen Vortrag in der selben klaren und eleganten Weise, als ob nichts gewesen wäre. Später stellte es sich heraus, daß er während des Vortrags einen leichten Herzanfall gehabt hatte. Wenige Monate später, am 10. Januar 1941, seinem 66. Geburtstage, bereitete ein neuer Herzanfall seinem Leben ein Ende.

Lassen Sie mich nun das wissenschaftliche Werk SCHURS betrachten. Es ist im Rahmen eines kurzen Vortrags unmöglich, auf alle seine Arbeiten, wenn auch nur kurz, einzugehen. Es sind weit über 70, ohne die zahlreichen Aufgaben mitzuzählen. Hier sind natürlich die Arbeiten im Journal für die reine und angewandte Mathematik eingeschlossen, als deren Verfasser J. SCHUR genannt ist. Bei den ersten dieser Arbeiten war der Vorname nämlich falsch abgekürzt, und SCHUR hielt es für richtig, dies bei den späteren Arbeiten in dieser Zeitschrift nicht zu ändern. Aber tatsächlich legte er großen Wert darauf, stets als I. SCHUR zitiert zu werden. Trotzdem wird auch heute noch gelegentlich sein Vorname falsch abgekürzt.

Das bloße Verlesen der Titel würde mehr als 20 Minuten beanspruchen, und der bloße Titel besagt häufig wenig. Die meisten der SCHURschen Arbeiten sind bedeutungsvoll und sehr inhaltsreich. In der Besprechung der Arbeit „Bemerkungen zur Theorie der beschränkten Linearformen mit unendlich vielen Veränderlichen", deren bescheidener Titel nicht vermuten läßt, daß sie viele wichtige Resultate enthält, sagt O. TOEPLITZ als Referent im Jahrbuch für die Fortschritte der Mathematik: „Aus der Fülle der Resultate und Methoden, die der Praktiker aus dieser Arbeit zu lernen hat, kann hier nur einiges Wenige hervorgehoben werden." Trotzdem ist die Besprechung eine volle Seite lang.

Die Hauptbedeutung SCHURS liegt in seinen Arbeiten zur Gruppentheorie. Hier setzt er das Werk seines Lehrers FROBENIUS fort, dem neben MOLIEN die Darstellungstheorie der endlichen Gruppen durch Gruppen linearer Substitutionen zu verdanken ist. SCHUR beschränkt sich nicht auf endliche Gruppen.

In seiner Dissertation betrachtet er die Darstellung der vollen linearen Gruppe. Es ist dies eine grundlegende Arbeit, die erst später gebührende Beachtung gefunden hat, z.B. im Buch WEYL, Classical Groups, das SCHUR gewidmet ist. Vorher hatte SCHUR seine Theorie auf die Orthogonale Gruppe ausgedehnt. Diese Arbeit ist übrigens auch in einer anderen Richtung von Bedeutung. Durch SCHUR wurde die Aufmerksamkeit der Mathematiker auf den Hurwitzschen Ansatz der Integration über kompakte Lie-Gruppen gelenkt. Etwas später führte dann HAAR sein Maß für kompakte topologische Gruppen ein, das ja heute für die Analysis von großer Wichtigkeit ist. Es sei auch auf die Bedeutung für die Quantenmechanik hingewiesen.

In einer seiner frühesten gruppentheoretischen Arbeiten gab SCHUR einen elementaren Beweis von Sätzen von BURNSIDE und FROBENIUS. Von besonderem Interesse ist hierbei, daß sich dort zum ersten Male der Begriff findet, den man jetzt Verlagerung nennt. Zwischen 1904 und 1907 dehnte SCHUR die FROBENIUS'sche Idee von Darstellungen von endlichen Gruppen durch lineare Transformationen auf die von Darstellungen durch Kollineationen aus. Wie in den eben erwähnten Arbeiten ist SCHUR hier wieder Vorläufer von modernen Entwicklungen. Hier ist vielleicht die erste Stelle, wo sich Ansätze aus der homologischen Algebra finden. Der spezielle Fall der symmetrischen und alternierenden Gruppe erledigte Fragen, wie sie von KLEIN in anderer Sprache aufgeworfen waren.

In anderen Arbeiten zur Darstellungstheorie ersetzt SCHUR den Körper der komplexen Zahlen durch beliebige Körper der Charakteristik Null. Seine Resultate stehen in engem Zusammenhang zur Theorie der Algebren, und man kann sagen, daß SCHUR vieles aus dieser Theorie in anderer Sprache gekannt hat. Auch hier kann man ihn als einen Vorläufer ansehen. In diesem Zusammenhange sind auch zwei gemeinsame Arbeiten von FROBENIUS und SCHUR zu nennen.

Am bekanntesten ist SCHURS Arbeit „Neue Begründung der Theorie der Gruppencharaktere" geworden. Diese ist auch deswegen wichtig, weil sich in dieser Form die Theorie auf die kompakten Lie-Gruppen ausdehnen läßt. Auf die STIEFEL'sche Ausarbeitung der SCHUR-Vorlesung ist bereits hingewiesen worden.

Neben der Gruppentheorie sind es fast alle Zweige der klassischen Algebra und der Zahlentheorie, die Schur bedeutende neue Resultate oder besonders schöne neue Beweise verdanken, die Theorie der algebraischen Gleichungen, Matrizentheorie und Determinantentheorie, Invariantentheorie, elementare Zahlentheorie, additive Zahlentheorie, analytische Zahlentheorie, Theorie der algebraischen Zahlen, Geometrie der Zahlen, Theorie der Kettenbrüche. Aus der Analysis sind insbesondere die Theorie der Integralgleichungen und die Theorie der Unendlichen Reihen zu nennen.

Für die charakteristischen Wurzeln der Matrizen hat sich SCHUR immer sehr interessiert. In einer seiner ersten Arbeiten gibt er einen einfachen Beweis eines Satzes von FROBENIUS über die charakteristischen Wurzeln vertauschbarer Matrizen. In einer anderen Arbeit, die für die Theorie der Integralgleichungen von Wichtigkeit ist, zeigt SCHUR, daß man zu jeder quadratischen Matrix A mit reellen oder komplexen Elementen eine unitär orthogonale Matrix P so bestimmen kann, daß $\bar{P}' A P$ eine Dreiecksmatrix ist, deren Hauptdiagonale von den charakteristischen Wurzeln gebildet ist. Dieses Resultat wird auch heute noch oft gebraucht.

In einer Reihe von Arbeiten studierte SCHUR die Lage der Wurzeln algebraischer Gleichungen und andere Eigenschaften der Polynome. Insbesondere ist hier die Arbeit „Über das Maximum des absoluten Betrags eines Polynoms in einem gegebenen Intervall" und seine Arbeit „Über algebraische Gleichungen, die nur Wurzeln mit negativem Realteil besitzen" zu nennen. Diese letztere ist in der Zeitschrift für angewandte Mathematik und Mechanik erschienen; in ihr gibt SCHUR einen einfachen Beweis des Kriteriums von HURWITZ. Die bekannte Arbeit seines Doktoranden A. COHN, „Über die Anzahl der Wurzeln einer algebraischen Gleichung in einem Kreise", Mathematische Zeitschrift **14** (1922), 110–148, ist von SCHUR angeregt worden.

Schon früh interessierte sich SCHUR für die Frage der Irreduzibilität und der Galoisschen Gruppe einer algebraischen Gleichung, wie einige seiner Aufgaben zeigen. Später zeigte er unter Benutzung einer Methode von BAUER, daß es leicht ist, Gleichungen ohne Affekt vom Grade n zu finden, falls man eine Primzahl im Intervall $\{\frac{1}{2}n \ldots n\}$ kennt. In berühmt gewordenen Arbeiten bewies SCHUR mit Hilfe eines von ihm wieder gefundenen Satzes von SYLVESTER über die Verteilung der Primzahlen, daß alle Polynome der Form

$$1 + g_1 \frac{x}{1!} + g_2 \frac{x^2}{2!} + \ldots + g_{n-1} \frac{x^{n-1}}{(n-1)!} + \frac{x^n}{n!}$$

mit ganzzahligen g_ν im Körper der rationalen Zahlen irreduzibel sind. Hieraus folgt, daß die Abschnitte der Reihen für e^x und $\cos x$, sowie die Laguerreschen Polynome irreduzibel sind. Die Galoissche Gruppe der Laguerreschen Polynome ist die symmetrische Gruppe, die der Abschnitte der Reihe für e^x, wenn n durch 4 teilbar ist, die alternierende, anderenfalls die symmetrische Gruppe. Für jedes ungerade n erhält Schur Gleichungen, deren Gruppe die alternierende ist.

Es kann hier nicht meine Aufgabe sein, auf alle Arbeiten SCHURS einzugehen. Fast alle von ihnen sind auch heute noch von großer Wichtigkeit und viele waren der Ausgangspunkt von Veröffentlichungen anderer Mathematiker.

Im Jahre 1918 gründete SCHUR zusammen mit LICHTENSTEIN, KNOPP und E. SCHMIDT die Mathematische Zeitschrift, die schnell ein hohes Ansehen gewann. Einige von SCHURS eigenen Arbeiten sind hier erschienen.

Die Verleihung des Doktortitels der Universität Berlin konnte erst dann erfolgen, wenn der Kandidat 200 gedruckte Exemplare der Dissertation eingereicht hatte. Kurz nach dem ersten Weltkrieg erschienen die meisten Dissertationen in Zeitschriften. Aber bald weigerten sich die mathematischen Zeitschriften, Dissertationen zu drucken, und der Kandidat mußte die erheblichen Kosten für einen privaten Druck allein aufbringen. Um hier zu helfen, gründeten SCHUR, E. SCHMIDT, BIEBERBACH und V. MISES die „Schriften des

Mathematischen Seminars und des Instituts für Angewandte Mathematik der Universität Berlin" für die Veröffentlichung von Dissertationen. Die Arbeiten erschienen in einzelnen Heften, die zu Bänden vereinigt wurden. Die Universität kaufte einen Teil der Auflage und verwandte sie zum Tausch mit ausländischen Zeitschriften. Dadurch wurden die Kosten für den Kandidaten erheblich vermindert.

Aber wir feiern heute nicht nur den großen Gelehrten, sondern auch den hervorragenden Hochschullehrer der Berliner Universität.

Als ordentlicher Professor war SCHUR vertraglich verpflichtet, in jedem Semester zwei vierstündige Vorlesungen und ein zweistündiges Seminar zu halten. In den Jahren seiner Hauptwirkungszeit, etwa von 1920–1932, baute er langsam zwei Vorlesungszyklen von je vier Vorlesungen auf, einen in Zahlentheorie und einen in Algebra. Der erste bestand aus Zahlentheorie, Theorie der algebraischen Zahlen, Analytische Zahlentheorie I und II, der zweite aus Determinantentheorie, Algebra, Galoissche Theorie, Invariantentheorie. Gelegentlich wurden die höheren Vorlesungen durch andere ersetzt. Zahlentheorie hat SCHUR im Wintersemester 1920 und dann in jedem zweiten Winter gelesen, zum letzten Male im Wintersemester 1932/33. Algebra las SCHUR in den anderen Wintersemestern. Die elementarere der beiden Vorlesungen fand montags, dienstags, donnerstags und freitags von 10–11, die andere an denselben Tagen von 11–12 statt. Das SCHURsche Seminar war dienstags 5–7 vor dem Mathematischen Kolloquium. Außerdem hielt SCHUR Übungen zur Zahlentheorie, zur Determinantentheorie und zur Algebra donnerstags 6–8 nach der Sitzung der Akademie ab in den Semestern, in denen diese Vorlesung gehalten wurde.

Als Dozent war SCHUR hervorragend. Seine Vorlesungen waren äußerst klar, aber nicht immer leicht und erforderten Mitarbeit. SCHUR verstand es meisterhaft, seine Hörer für den Stoff zu interessieren und ihr Interesse wach zu halten. Es galt damals als selbstverständlich, daß jeder Student, der sich nur irgendwie für Mathematik interessierte, wenigstens eine der SCHUR'schen Vorlesungen hörte, auch wenn sein Hauptinteresse auf anderen Gebieten lag. SCHURS Wirken trug wesentlich dazu bei, daß die Zahl der Studenten der Mathematik in Berlin damals so anwuchs. In seiner elementaren Vorlesung waren oft über 400 Hörer. Im Wintersemester 1930 war die Zahl der Studenten, die SCHURS Zahlentheorie belegen wollten, so groß, daß der zweitgrößte Hörsaal der Universität mit etwas über 500 Sitzen zu klein war. Auf SCHURS Wunsch mußte ich eine Parallelvorlesung für etwa 40 Hörer halten.

SCHUR bereitete jede seiner Vorlesungen aufs sorgfältigste vor. Die Berliner Tradition verbot es einem Dozenten der Mathematik, ein Buch in den Hörsaal mitzubringen. Darüber hinaus erwarteten die Studenten, daß eine Vorlesung nicht ein Abklatsch eines Buches sei. Lange vor Beginn des Semesters arbeitete SCHUR jede Vorlesung schriftlich auf losen Blättern aus. Jede Vorlesung bestand aus einigen Abschnitten, jeder Abschnitt aus einer Reihe von Paragraphen, die alle eine Überschrift hatten. Während der Vorlesung hatte SCHUR die betreffenden Seiten seiner Ausarbeitung in seiner Brusttasche. Er nahm sie aber nur selten heraus, z. B. wenn es sich um eine schwierige Abschätzung in der analytischen Zahlentheorie handelte. Es ist wohl nie vorgekommen, daß SCHUR in einer Vorlesung stecken blieb. Gelegentlich hat es sich ereignet, daß SCHUR ein gewisses Resultat als bewiesen benutzte, bis die Hörer ihn darauf aufmerksam machten, daß es noch nicht bewiesen war, und SCHUR feststellen mußte, daß er einen ganzen Paragraphen übersprungen hatte.

Dank der guten Vorbereitung war SCHUR in der Lage, ziemlich viel Stoff in einer Stunde zu erledigen. Aber auf der anderen Seite setzte er seinen Stolz nicht darin, möglichst viel zu schaffen. Er war mehr daran interessiert, daß seine Hörer ihn verstanden und daß sie durch seine Vorlesung für den Stoff interessiert wurden. Ein gewisses Bild seiner Vorlesungen geben die schon erwähnte Ausarbeitung seiner Züricher Vorlesung durch STIEFEL und die kürzlich erschienene Ausarbeitung seiner Vorlesung über Invariantentheorie durch H. GRUNSKY.

SCHUR sprach ruhig und deutlich in ausgezeichnetem Deutsch ohne jeden Akzent. Niemand konnte auf den Gedanken kommen, daß Deutsch nicht SCHURS Muttersprache war.

In seiner Bescheidenheit hat SCHUR oft erklärt, daß seine Vorlesungen nicht sein Werk seien, sondern zu großen Teilen das seiner Vorgänger, insbesondere das von FROBENIUS. Aber sicher enthielten seine Vorlesungen auch manches, was neu war und vielleicht auch heute noch nicht in der Literatur gefunden werden kann. Von historischem Interesse ist z. B. SCHURS Beweis des Determinantensatzes von MINKOWSKI, den er in seinen Vorlesungen gab. Siehe die Arbeit von H. ROHRBACH „Bemerkungen zu einem Determinantensatz von Minkowski" Jahresbericht D.M.V. **40** (1931), 49–53 (eingegangen Januar 1930). SCHUR benutzt in diesem Beweise die Kreise, die heute die Kreise von GERSHGORIN genannt werden, obgleich dessen Arbeit erst 1931 erschienen ist.

SCHURS Übungen schlossen sich eng an seine Vorlesung an. In der ersten Stunde der Doppelstunde stellte SCHUR etwa 8 Aufgaben zur schriftlichen Bearbeitung. Die eingegangenen Lösungen wurden jeweils in der nächsten Woche in der zweiten Stunde der Doppelstunde von SCHURS Assistenten besprochen. Die Assistenten waren H. RADEMACHER (bis zu seinem Fortgang nach Hamburg), K. LÖWNER (bis zu seinem Fortgang nach Prag) und ich selbst von 1928–1935.

Die ersten ein oder zwei Aufgaben in jeder Aufgabenreihe waren reine Zahlenbeispiele; dann folgten theoretische Aufgaben, von leichteren zu schwereren aufsteigend. Oft handelte es sich um den Beweis spezieller Resultate, für die SCHUR sich immer sehr interessierte. Diese Übungen waren für die Ausbildung vieler der Hörer von großem Einfluß. Ich kann das nicht besser als durch folgende Beispiele zeigen.

Im Wintersemester 1921/22 nahmen H. HOPF, mein Bruder RICHARD und ich selbst an den Übungen zur Algebra teil. Die folgende Aufgabe wurde unter anderen gegeben: Es seien $a_1, a_2, \ldots, a_n$ verschiedene ganze rationale Zahlen. Dann sind die Polynome $P(x) = \{(x-a_1)(x-a_2)\ldots(x-a_n)\}^k+1$ für $k=2$ und $k=4$ im Körper der rationalen Zahlen irreduzibel. Diese Aufgabe hatte SCHUR schon früher im Archiv der Mathematik und Physik gestellt, ohne daß eine Lösung eingegangen war. SCHUR bemerkte, daß der Fall $k=2$ leicht sei, daß er aber noch nie eine Lösung für den Fall $k=4$ erhalten habe, obgleich er diese Aufgabe immer wieder gestellt habe. Das war natürlich ein ungeheurer Ansporn für die Hörer. Ich erinnere mich noch genau, daß mein Bruder und ich während der nächsten Tage in verschiedenen Zimmern mit Hochdruck an der Lösung arbeiteten. Von Zeit zu Zeit verglichen wir unsere Resultate und vereinigten sie. Vor Ende der Frist von 5 Tagen zur Einreichung der Aufgaben hatten wir nicht nur eine Lösung der eigentlichen Aufgabe, sondern auch einige Verallgemeinerungen. Unsere Resultate wurden später in die Aufgabensammlung von PÓLYA-SZEGÖ aufgenommen. Eine andere Lösung der Aufgabe hatte HOPF eingereicht. Durch Vereinigung der beiden Lösungen gelang es uns, die Irreduzibilität auch für $k=8$ zu beweisen (Jahresbericht D. M. V. **35** (1926), 99–113). Die Vermutung, daß diese Polynome für alle $k=2^s$ mit $s>0$ im Körper der rationalen Zahlen irreduzibel sind, wurde erst kürzlich von I. SERES (Acta Math. Acad. Sc. Hungaricae **7** (1956), 151–157) bewiesen. Diese Arbeit ist dem Andenken SCHURS gewidmet.

Auch eine Reihe meiner anderen Arbeiten haben ihren Ursprung direkt oder indirekt in den SCHUR'schen Übungen oder in dem SCHUR'schen Seminar. Für das Seminar wählte

Schur zu Beginn jedes Semesters immer eine Reihe kürzlich erschienener Arbeiten, die zum Vortrag unter die Teilnehmer verteilt wurden. Aufgabe des Assistenten war es, den Studenten bei der Vorbereitung ihres Vortrags zu helfen. R. Remak war für viele Jahre ein ständiger Gast im Schur'schen Seminar, ebenso v. Neumann während der Jahre, die er in Berlin war.

Außerhalb der Zeiten seiner Vorlesungen war Schur nur selten in der Universität. Zwar war er mit Schmidt und Bieberbach einer der drei Direktoren des Mathematischen Seminars. Aber die gesamte Verwaltungsarbeit, insbesondere die Verwaltung der Seminarbibliothek und die Zusammenarbeit mit den Studenten, lag in den Händen der drei Assistenten. Nur einmal in jedem Semester kamen der gesamte Lehrkörper der reinen und angewandten Mathematik und zwei Vertreter der Studenten zusammen, um den Vorlesungsplan für das nächste Semester aufzustellen. Es gab keinerlei Komitees. Die einzige weitere Verpflichtung, die Schur hatte, war, gelegentlich einen Doktoranden oder Staatsexamenskandidaten zu prüfen.

So hatte Schur reichlich Zeit, wissenschaftlich zu arbeiten, und das tat er in größtem Maße. Wer spät abends vom Roseneck kommend den Hohenzollerndamm herunterging, der konnte in Schurs Arbeitszimmer in der ersten Etage Ruhlaer Str. 14 die Schreibtischlampe noch brennen sehen. Wenn Schur nachts nicht schlafen konnte, dann las er im Jahrbuch für die Fortschritte der Mathematik. Als er später von Israel seine Bibliothek notgedrungen zum Verkauf anbieten mußte, und das Institute for Advanced Study in Princeton sich für das Jahrbuch interessierte, sandte Schur noch wenige Wochen vor seinem Tode ein Telegramm, daß das Jahrbuch nicht verkauft werden sollte. Erst nach Schurs Tode erwarb das Institut sein Exemplar.

Schurs hervorstechendste menschliche Eigenschaften waren wohl seine große Bescheidenheit, seine Hilfsbereitschaft und sein menschliches Interesse an seinen Studenten. Er legte großen Wert darauf, daß ihm keine Anerkennung für ein Resultat gegeben wurde, das nicht voll und ganz sein Werk war. Vielleicht würde er manches, was heute ihm zugeschrieben wird, nicht billigen.

Seit E. Jacobsthals Resultaten über die Verteilung der quadratischen Reste und Nichtreste interessierte sich Schur sehr dafür, insbesondere für Sequenzen von solchen. Er vermutete, daß für alle k und alle hinreichend großen Primzahlen Sequenzen von k quadratischen Resten und k quadratischen Nichtresten existierten. Um dieses Resultat zu beweisen, stellte Schur die folgende Vermutung auf. Verteilt man die ganzen rationalen Zahlen $1, 2, \ldots, N$ irgendwie auf zwei Klassen, so enthält für jedes k und alle hinreichend großen N mindestens eine der beiden Klassen eine arithmetische Progression der Länge k. Aber jahrelang war es weder Schur noch einem der vielen Mathematiker, die von dieser Schur'schen Vermutung hörten, gelungen, sie zu beweisen.

An einem Septembertage 1927 besuchten mein Bruder und ich Schur, als unerwartet auch v. Neumann, der gerade von der D. M. V.-Tagung zurückgekommen war, zu Schur kam, um ihm zu erzählen, daß auf der Tagung van der Waerden unter Benutzung eines Vorschlags von Artin einen Beweis der kombinatorischen Vermutung vorgetragen habe und unter dem Titel „Beweis einer Baudetschen Vermutung" veröffentlichen würde. Schur war höchst erfreut, aber nach wenigen Minuten enttäuscht, da er sah, daß durch dieses Resultat seine Vermutung über Sequenzen noch nicht bewiesen war, da es sich nur ergeben würde, daß eine der beiden Klassen, aber nicht welche, für eine gegebene große Primzahl eine Sequenz der Länge k enthalten würde.

Baudet war damals ein unbekannter Göttinger Student, der auch später nie etwas Mathematisches veröffentlicht hat. Auf der anderen Seite war damals Schurs Freund Landau Professor in Göttingen, der natürlich die Vermutung kannte, und Landau pflegte jedem Mathematiker, den er traf, unbewiesene Vermutungen als Aufgabe vorzuschlagen. So ist es höchst wahrscheinlich, daß Baudet direkt oder indirekt von der Vermutung gehört hatte. Es wäre daher verständlich gewesen, wenn Schur vorgeschlagen hätte, daß bei der Veröffentlichung von van der Waerdens Arbeit der Titel geändert würde oder daß in einer Fußnote darauf hingewiesen würde, daß es sich um eine alte Vermutung von Schur handele. Aber dazu war Schur viel zu bescheiden.

Wenige Tage nach dem Besuch bei Schur gelang es mir mit Hilfe das Satzes von van der Waerden, die Schur'sche Vermutung für die quadratischen Reste zu beweisen. Schur wies darauf hin, daß meine Beweismethode auch für Sequenzen von k-ten Potenzresten anwendbar sein müsse. Bald darauf teilte er mir mit, daß er den Satz von van der Waerden so mittels meiner Beweismethode erweitern könne, daß es für hinreichend große N immer mindestens eine Klasse geben müsse, die eine Progression der Länge k und zugleich deren Differenz enthält. Schur wollte, daß ich dieses Resultat in meine Arbeit aufnehme. Ich muß gestehen, daß ich nie auf den Gedanken gekommen war, dieses Resultat auszusprechen oder nur zu vermuten. Schur aber stellte sich auf den Standpunkt, daß sein Beweis nur eine Anwendung meiner Methode wäre und ich ihn daher allein veröffentlichen müsse. Selbstverständlich habe ich diesen Satz immer einen Satz von Schur genannt.

Wenige Wochen später gelang es mir, auch die Schur'sche Vermutung für die quadratischen Nichtreste zu beweisen. Nun erklärte Schur, daß er meine Arbeit in der Berliner Akademie vorlegen würde. Aber einige Tage später teilte Schur mir mit, daß sich eine Schwierigkeit ergeben hätte. Es bestand seit Jahrzehnten die Regel, daß, wenn Mitglieder der Akademie in den Sitzungsberichten Arbeiten veröffentlichten, vor ihren Namen keine Titel angeführt wurden. Dagegen wurden bei Arbeiten von Nichtmitgliedern die Autoren mit ihren Titeln genannt. Ich selbst stand noch vor dem Doktorexamen, hatte daher keinen Titel. Es gelang aber Schur durchzusetzen, daß meine Arbeit zur Veröffentlichung in den Sitzungsberichten ohne einen Titel vor meinem Namen angenommen wurde, obgleich dadurch der Eindruck entstehen konnte, daß ich ein Mitglied der Akademie sein könnte.

Um die Arbeit vorzulegen, hatte Schur eine Inhaltsangabe zu machen, die in den Sitzungsberichten abgedruckt wurde. Auf diese hatte ich keinen Einfluß. Es war typisch für Schur, daß er in ihr die von ihm stammende Verallgemeinerung des Satzes von van der Waerden in keiner Weise erwähnte. Dies alles zeigt Schurs Bescheidenheit und sein Bestreben seine Studenten zu fördern.

Ich hoffe gezeigt zu haben, daß Schur nicht nur ein großer Mathematiker gewesen ist, sondern auch ein Mensch, den alle, die ihn kannten, hoch verehrten. Nicht nur seine Doktoranden, sondern alle, die je bei ihm eine Vorlesung gehört haben, werden seiner stets in Dankbarkeit gedenken.

Inhaltsverzeichnis Band I

1.

Über eine Klasse von Matrizen, die sich einer gegebenen Matrix zuordnen lassen

Dissertation, Berlin 1901

Es seien

$$A = (a_{ik}), \quad B = (b_{ik})$$

zwei Matrizen m-ten Grades, deren Elemente a_{ik} und b_{ik} unabhängige Variabele sind,

$$C = (c_{ik}) = AB$$

die aus ihnen zusammengesetzte Matrix; dann ist

$$c_{ik} = a_{i1} b_{1k} + a_{i2} b_{2k} + \cdots + a_{im} b_{mk}.$$

Im Folgenden sollen Matrizen $T(A)$, deren Grad r beliebig ist, betrachtet werden, welche folgende Eigenschaften besitzen:

1. Die r^2 Elemente der Matrix $T(A)$ sind ganze rationale Funktionen der m^2 Variabelen a_{ik};

2. Geht die Matrix $T(A)$ in $T(B)$ oder $T(C)$ über, wenn man a_{ik} durch b_{ik} oder c_{ik} ersetzt, so soll die Gleichung

$$T(A)\, T(B) = T(C)$$

bestehen.

Eine so beschaffene Matrix $T(A)$ nenne ich eine aus A gebildete invariante Form oder Matrix oder auch kürzer eine invariante Form von A. Der Prozess der Bildung einer solchen Matrix soll eine invariante Operation genannt werden.

Sind speziell sämtliche Elemente von $T(A)$ homogene Funktionen n-ter Dimension der Variabelen a_{ik}, so soll die Operation eine homogene und n die Ordnung derselben heissen.

Es genügt, um einzusehen, dass solche invariante Operationen in der That existieren, $T(A)$ gleich A oder gleich der Determinante $|A|$ der Matrix A zu setzen.

Ist $T(A)$ eine invariante Form von A und P eine beliebige konstante Matrix desselben Grades, wie $T(A)$, deren Determinante nicht Null ist, so besitzt auch die Matrix $P^{-1}\,T(A)\,P$ die Eigenschaften 1. und 2., ist also ebenfalls eine invariante Form von A. Zwei solche invariante Formen bezeichne ich als äquivalent.

Zerfällt $T(A)$ in zwei Matrizen $T_1(A)$ und $T_2(A)$, ist also

$$T(A) = \begin{pmatrix} T_1(A) & \\ & T_2(A) \end{pmatrix},$$

so sind, wie man unmittelbar sieht, auch $T_1(A)$ und $T_2(A)$ invariante Formen von A. Jede invariante Form, die einer zerfallenden äquivalent ist, soll eine zerlegbare oder reduktible Form heissen. Ist dann $T(A)$ in die Formen $T_1(A)$ und $T_2(A)$ zerlegbar, so will ich sagen, $T_1(A)$, resp. $T_2(A)$ sei ein Teiler von $T(A)$ oder in $T(A)$ enthalten. Eine invariante Form, die keiner zerfallenden äquivalent ist, nenne ich eine irreduktible oder primitive Form.

Die Bedeutung der invarianten Operationen besteht im Folgenden: Es sei G eine abstrakte (endliche oder unendliche) Gruppe, $\mathrm{A}, \mathrm{B}, \Gamma, \ldots$ ihre Elemente. Ordnet man dann dem Element A die Matrix A, dem Element B die Matrix B, u. s. w. derart zu, dass dem Element AB die Matrix AB entspricht, so sagt man, dass die Matrizen $A, B, C, \ldots$ die Gruppe G darstellen[1]). (Diese Darstellung soll eine eigentliche oder uneigentliche genannt werden, je nachdem die Determinanten der Matrizen $A, B, C, \ldots$ alle von Null verschieden sind oder nicht). Ist nun $T(A)$ eine invariante Operation, und wendet man dieselbe auf die die Gruppe darstellenden Matrizen $A, B, C, \ldots$ an, so erhält man in den Matrizen

$$T(A),\ T(B),\ T(C),\ \ldots$$

eine neue Darstellung der Gruppe G.

Anders ausgedrückt heisst dies: die Matrizen $T(A)$ bilden, wenn man für A alle Matrizen m-ten Grades einsetzt, eine Gruppe, die der allgemeinen homogenen linearen Gruppe in m Variabelen isomorph ist.

In der vorliegenden Arbeit wird bewiesen, dass zwei invariante Formen $T(A)$ und $T_1(A)$ dann und nur dann äquivalent sind, wenn ihre Spuren, d. h. die Summen der Glieder ihrer Hauptdiagonalen einander gleich sind. Die Anzahl der verschiedenen (nicht äquivalenten) primitiven homogenen Operationen n-ter Ord-

---

1) Vergl. Frobenius: „Ueber die Darstellung der endlichen Gruppen durch lineare Substitutionen", Sitzungsberichte der Akademie der Wissenschaften zu Berlin, 1897, S. 994.

nung ist endlich und zwar gleich der Anzahl k der Zerlegungen der ganzen Zahl n in höchstens m gleiche oder verschiedene positive Summanden. Es werden auch die Grade und die Spuren der primitiven invarianten Formen bestimmt.

Der Beweis dieser Sätze gelingt mir, indem ich zeige, dass aus jeder homogenen invarianten Operation n-ter Ordnung eine Darstellung der symmetrischen Gruppe n-ten Grades durch lineare Substitutionen hervorgeht, und dass umgekehrt jeder solchen Darstellung eine und nur eine invariante Operation entspricht, wenn man äquivalente Operationen als nicht verschieden ansieht.

Die vollständige Bestimmung der invarianten Operationen $T(A)$ war bisher, soweit mir die Litteratur bekannt ist, nur für die beiden folgenden Fälle geleistet worden:

1. Ist $T(A)$ vom Grade 1, also eine Funktion der Variabelen a_{ik}, so muss $T(A)$ einer Potenz der Determinante von A gleich sein [1]).

2. Sind die Elemente der Matrix $T(A)$ homogene lineare Funktionen der a_{ik}, und ist die Determinante $|T(A)|$ nicht identisch Null, so ist, wie Herr Frobenius [2]) bewiesen hat, $T(A)$ einer Matrix der Form

$$\begin{pmatrix} A & & & & \\ & A & & & \\ & & \cdot & & \\ & & & \cdot & \\ & & & & \cdot \\ & & & & & A \end{pmatrix}$$

äquivalent.

Abschnitt I.

§. 1. Im Folgenden soll stets mit E_λ die Einheitsmatrix, mit N_λ die Nullmatrix des Grades λ bezeichnet werden.

Es sei $T(A)$ eine beliebige aus der Matrix m-ten Grades A gebildete invariante Form, r der Grad von $T(A)$. Wir setzen speziell $A = xE_m$, wo x eine willkürliche Veränderliche bezeichnet. Dann ist jedes Element der Matrix $T(xE_m)$ eine ganze rationale Funktion von x. Ist n der höchste vorkommende Grad dieser Funktionen, so lässt sich $T(xE_m)$ in der Form

1) Vergl. Hurwitz, „Zur Invariantentheorie", Math. Ann. Bd. 45, S. 381—404, § 5.

2) „Ueber die Darstellung der endlichen Gruppen durch lineare Substitutionen II". Sitzungsberichte, 1899, S. 482.

$$T(xE_m) = x^n C_0 + x^{n-1} C_1 + x^{n-2} C_2 + \cdots + x C_{n-1} + C_n$$

schreiben, wo $C_0, C_1, C_2, \cdots$ gewisse konstante Matrizen bedeuten. Wenn y eine von x verschiedene Veränderliche ist, so ergeben sich aus der Gleichung

$$T(xE_m)\, T(yE_m) = T(xE_m \cdot yE_m) = T(xyE_m)$$

folgende Beziehungen zwischen den Matrizen $C_0, C_1, C_2, \ldots C_n$:

$$(1) \qquad C_\mu^2 = C_\mu, \quad C_\mu C_\nu = 0 \quad (\mu, \nu = 0, 1, \ldots n, \mu \neq \nu).$$

Gehen wir von der Matrix C_0 aus, für welche die Gleichung $C_0^2 = C_0$ besteht, so lässt sich nach bekannten Sätzen über lineare Substitutionen eine Matrix P von nicht verschwindender Determinante derart bestimmen, dass

$$P^{-1} C_0 P = \begin{pmatrix} 1 & & & & & \\ & 1 & & & & \\ & & \ddots & & & \\ & & & 1 & & \\ & & & & 0 & \\ & & & & & \ddots \\ & & & & & & 0 \end{pmatrix} = \begin{pmatrix} E_{r_0} & \\ & N_{r-r_0} \end{pmatrix}$$

wird, wo r_0 dem Range der Matrix C_0 gleich ist. Setze ich $P^{-1} C_\nu P = C'_\nu$, so folgt aus den Gleichungen

$$C'_0 C'_\nu = C'_\nu C'_0 = 0, \quad (\nu = 1, 2, \ldots n)$$

dass die Matrizen $C'_1, C'_2, \ldots C'_n$ die Form $\begin{pmatrix} N_{r_0} & \\ & D_\nu \end{pmatrix}$ haben, wo $D_1, D_2, \ldots D_n$ Matrizen $r - r_0$-ten Grades sind, für welche die Gleichungen

$$D_\mu^2 = D_\mu, \; D_\mu D_\nu = 0 \quad (\mu, \nu = 1, 2, \ldots n, \mu \neq \nu)$$

bestehen. Also lässt sich eine Matrix Q angeben, so dass

$$Q^{-1} D_1 Q = \begin{pmatrix} E_{r_1} & \\ & N_{r-r_0-r_1} \end{pmatrix}$$

wird, wo r_1 eine gewisse nicht negative ganze Zahl bedeutet. Setze ich jetzt

$$P_1 = \begin{pmatrix} E_{r_1} & \\ & Q \end{pmatrix}$$

und bezeichne die Matrizen $P_1^{-1} C'_\nu P_1$ mit C''_ν, so wird

$$C''_0 = \begin{pmatrix} E_{r_0} & & \\ & N_{r_1} & \\ & & N_{r-r_0-r_1} \end{pmatrix}, \quad C''_1 = \begin{pmatrix} N_{r_0} & & \\ & E_{r_1} & \\ & & N_{r-r_0-r_1} \end{pmatrix},$$

$$C''_\nu = \begin{pmatrix} N_{r_0} & & \\ & N_{r_1} & \\ & & D'_\nu \end{pmatrix}, \quad (\nu = 2, 3, \ldots n)$$

wo die Matrizen D'_ν den Gleichungen

$$D'^2_\mu = D_\mu, \quad D'_\mu D'_\nu = 0 \qquad (\mu, \nu = 2, 3, \ldots n, \ \mu \neq \nu)$$

genügen. Indem wir in derselben Weise weiter schliessen, sehen wir leicht ein, dass man eine Matrix M von nicht verschwindender Determinante bestimmen kann, so dass

$$M^{-1}\, T(xE_m)\, M = \begin{pmatrix} x^n E_{r_0} & & & & & \\ & x^{n-1}E_{r_1} & & & & \\ & & x^{n-2}E_{r_2} & & & \\ & & & \ddots & & \\ & & & & xE_{r_{n-1}} & & \\ & & & & & E_{r_n} & \\ & & & & & & N_s \end{pmatrix}$$

wird, wo $r_0, r_1, \ldots r_n$ und s gewisse nicht negative ganze Zahlen sind, deren Summe gleich r ist.

Setze ich nun $M^{-1}T(A)M = T'(A)$, so kann ich aus den Gleichungen

$$(2) \qquad T'(xA) = T'(xE_m)\,T'(A) = T'(A)\,T'(xE_m)$$

schliessen, dass $T'(A)$ in $n+2$ Matrizen zerfällt, dass also

$$T'(A) = \begin{pmatrix} T_n(A) & & & & & \\ & T_{n-1}(A) & & & & \\ & & \ddots & & & \\ & & & T_1(A) & & \\ & & & & T_0(A) & \\ & & & & & S(A) \end{pmatrix}$$

wird, wobei $T_\lambda(xE_m) = x^\lambda E_{r_{n-\lambda}}$ und $S(xE_m) = N_s = 0$ zu setzen ist. Dies ist leicht direkt einzusehen; es ergiebt sich auch aus folgendem allgemeinen Satze des Herrn A. Voss[1]):

„Ist A eine Matrix, welche in die Teilmatrizen $A_1, A_2, \ldots A_q$ zerfällt, und besitzen je zwei der charakteristischen Gleichungen der Matrizen $A_1, A_2, \ldots A_q$ keine Wurzel gemeinsam, so muss jede mit A vertauschbare Matrix B in derselben Weise, wie A, zerfallen, d. h. es ist

$$B = \begin{pmatrix} B_1 & & & \\ & B_2 & & \\ & & \ddots & \\ & & & B_q \end{pmatrix},$$

wo B_ϱ von demselben Grade ist wie A_ϱ."

In der That ist in unserem Falle $T'(A)$ mit $T'(xE_m)$ ver-

1) Sitzungsberichte der math.-physik. Klasse der Akademie der Wissenschaften zu München, Bd. XIX (1889), S. 283.

tauschbar, und es besitzen die charakteristischen Gleichungen der Matrizen $x^n E_{r_0}$, $x^{n-1} E_{r_1}$, … $x E_{r_{n-1}}$, E_{r_n}, N, keine Wurzel gemeinsam.

Aus den Gleichungen (2) erhalten wir

$$T_\nu(xA) = x^\nu T(A), \quad S(A) = S(E_m) S(A) = 0;$$

die erste dieser Gleichungen lehrt uns, dass die Elemente der Matrix $T_\nu(A)$ ganze homogene Funktionen ν-ter Dimension der m^2 Grössen a_{ik} sind, d. h. dass $T_\nu(A)$ eine homogene invariante Form ν-ter Ordnung ist. Da ferner $T_\nu(E_m) = E_{r_{n-\nu}}$ ist, so sehen wir, dass die Determinante von $T_\nu(A)$ nicht identisch verschwindet.

Wir erhalten daher den Satz:

I. „Jede invariante Form $T(A)$ ist zerlegbar in homogene invariante Formen von nicht identisch verschwindender Determinante und eine Nullmatrix, die natürlich nur dann auftritt, wenn die Determinante von $T(A)$ identisch Null ist".

Ich werde daher im Folgenden nur homogene invariante Formen von nicht verschwindender Determinante betrachten. Für diese ist $T(E_m)$ eine Einheitsmatrix und wegen $AA^{-1} = E_m$

$$T(A^{-1}) = [T(A)]^{-1}. \tag{3}$$

Mithin ist die Determinante von $T(A)$ stets dann von Null verschieden, wenn die Determinante der Matrix A nicht verschwindet.

§ 2. Es möge nun A eine sogenannte Diagonalmatrix sein, d. h. nur in der Hauptdiagonale nicht verschwindende Elemente enthalten, und es sei

$$A = X = \begin{pmatrix} x_1 & & & \\ & x_2 & & \\ & & \ddots & \\ & & & x_m \end{pmatrix},$$

wo $x_1, x_2, \ldots x_m$ unabhängige Veränderliche bedeuten. Dann lässt sich $T(X)$ in der Form

$$T(X) = \Sigma C_{\alpha_1, \alpha_2 \ldots \alpha_m} x_1^{\alpha_1} x_2^{\alpha_2} \ldots x_m^{\alpha_m}$$

schreiben, wo die Summation über alle nicht negativen ganzen Zahlen $\alpha_1, \alpha_2, \ldots \alpha_m$ zu erstrecken ist, für die $\alpha_1 + \alpha_2 + \cdots + \alpha_m$ gleich ist der Ordnung n von $T(A)$, und $C_{\alpha_1, \alpha_2 \ldots \alpha_m}$ gewisse konstante Matrizen bedeuten. Geht X in Y über, wenn ich die Variabelen x_μ durch andere Variabele y_μ ersetze, so erhalte ich aus der Gleichung

$$T(X)\,T(Y) = T(XY) = \Sigma C_{\alpha_1,\,\alpha_2\ldots\alpha_m}\,(x_1y_1)^{\alpha_1}(x_2y_2)^{\alpha_2}\ldots(x_my_m)^{\alpha_m}$$

die Relationen

$$C^2_{\alpha_1,\alpha_2\ldots\alpha_m} = C_{\alpha_1,\alpha_2\ldots\alpha_m},\quad C_{\alpha_1,\alpha_2\ldots\alpha_m}\,C_{\beta_1,\beta_2\ldots\beta_m} = 0\,,$$

wo $\beta_1, \beta_2, \ldots \beta_m$ eine von $\alpha_1, \alpha_2, \ldots \alpha_m$ verschiedene Indicesreihe bedeuten. In genau analoger Weise, wie auf S. 9, schliessen wir, dass es eine Matrix M von nicht verschwindender Determinante giebt von der Beschaffenheit, dass die Matrix $M^{-1}\,T(X)\,M$ in der Form einer Diagonalmatrix erscheint, die in der Hauptdiagonale nur Potenzprodukte $x_1^{\alpha_1}\,x_2^{\alpha_2}\ldots x_m^{\alpha_m}$ n-ter Dimension enthält.

Es seien nun $\omega_1, \omega_2, \ldots \omega_m$ die Wurzeln der charakteristischen Gleichung[1]) der Matrix A. Dann lässt sich bekanntlich, da die Elemente von A Variabele und daher $\omega_1, \omega_2, \ldots \omega_m$ untereinander verschieden sind, eine Matrix P von nicht verschwindender Determinante angeben, welche der Bedingung

$$(4)\qquad P^{-1}AP = \begin{pmatrix}\omega_1 & & & \\ & \omega_2 & & \\ & & \ddots & \\ & & & \omega_m\end{pmatrix} = \Omega$$

genügt. Wendet man auf diese Gleichung die invariante Operation $M^{-1}\,T(A)\,M = T_1(A)$ an, so erhält man vermöge der Formel (3)

$$(5)\qquad [T_1(P)]^{-1}\,T_1(A)\,[T_1(P)] = T_1(\Omega).$$

Da nun $T_1(\Omega)$ eine Diagonalmatrix ist, so folgt hieraus leicht, dass die charakteristischen Wurzeln von $T_1(A)$ oder, was dasselbe ist, von $T(A)$ alle die Form

$$(6)\qquad \omega_1^{\alpha_1}\,\omega_2^{\alpha_2}\ldots\omega_m^{\alpha_m}\qquad (\alpha_1+\alpha_2+\ldots+\alpha_m = n)$$

besitzen.

Ich setze nun

$$|\,xE_m - A\,| = x^m - c_1x^{m-1} + c_2x^{m-2} - \ldots \pm c_m$$

und bezeichne die Matrix

$$\begin{pmatrix} c_1, & -c_2, & c_3, & \ldots\ldots & \pm c_{m-1}, & \mp c_m \\ 1 & 0 & 0 & \ldots\ldots & 0 & 0 \\ 0 & 1 & 0 & \ldots\ldots & 0 & 0 \\ . & . & . & \ldots\ldots & . & . \\ 0 & 0 & 0 & \ldots\ldots & 1 & 0 \end{pmatrix}$$

1) Dafür sage ich auch kürzer: $\omega_1, \omega_2, \ldots \omega_m$ sind die charakteristischen Wurzeln der Matrix.

mit $\overline{A}$; dann ist

$$|xE_m - A| = |xE_m - \overline{A}|$$ [1]).

Daher lässt sich auch eine Matrix Q von nicht verschwindender Determinante bestimmen, welche die Gleichung

$$Q^{-1}AQ = \overline{A}$$

befriedigt. Alsdann ist auch $[T(Q)]^{-1}\, T(A)\, [T(Q)] = T(\overline{A})$; demnach ist die Spur der Matrix $T(A)$ gleich der Spur der Matrix $T(\overline{A})$ und folglich eine ganze rationale Funktion der Koeffizienten $c_1, c_2, \ldots c_m$. Daraus folgt leicht:

II. „Ist A eine beliebige Matrix m-ten Grades, sind $\omega_1, \omega_2, \ldots \omega_m$ die charakteristischen Wurzeln von A und ist $T(A)$ eine invariante Form r-ten Grades und n-ter Ordnung, so sind die r charakteristischen Wurzeln von $T(A)$ Produkte von je n der Wurzeln $\omega_1, \omega_2, \ldots \omega_m$, und die Gesamtheit dieser r Produkte ändert sich nicht, wenn man die ω_μ beliebig untereinander vertauscht".

Es gehört also zu jeder invarianten Operation $T(A)$ eine gewisse homogene symmetrische Funktion der charakteristischen Wurzeln $\omega_1, \omega_2, \ldots \omega_m$ von A, nämlich die Spur der Matrix $T(A)$, als Funktion der ω dargestellt, oder was dasselbe ist, die Spur der Matrix $T(\Omega)$.

Diese symmetrische Funktion will ich die Charakteristik Φ der Operation $T(A)$ nennen.

Offenbar bleibt Φ ungeändert, wenn $T(A)$ durch eine äquivalente Operation ersetzt wird.

Aus dem Satze II ergiebt sich als Korollar:

III. „Die Determinante der Matrix $T(A)$ ist gleich der $\frac{nr}{m}$-ten Potenz der Determinante $|A|$" [2]).

Es ist nämlich

$$|A| = \omega_1\,\omega_2 \ldots \omega_m;$$

ebenso ist $|T(A)|$ gleich dem Produkte der r Grössen (6). Da dieses Produkt eine symmetrische Funktion von $\omega_1, \omega_2, \ldots \omega_m$ ist, so ergiebt sich

1) Vgl. Frobenius, Theorie der linearen Formen mit ganzen Koeffizienten, Crelle's Journal, Bd. 86, S. 146.

2) J. Deruyts, Bulletins de l'Acad. des Sciences de Belgique, Série 3, T. 32 (1896), S. 433.

$$|T(A)| = \Pi \omega_1^{\alpha_1} \omega_2^{\alpha_2} \dots \omega_m^{\alpha_m} = (\omega_1 \omega_2 \dots \omega_m)^e,$$

wo e eine ganze Zahl bedeutet. Setze ich nun

$$\omega_1 = \omega_2 = \dots = \omega_m = \omega,$$

so wird $\omega_1 \omega_2 \dots \omega_m = \omega^m$ und $\omega_1^{\alpha_1} \omega_2^{\alpha_2} \dots \omega_m^{\alpha_m} = \omega^n$, also $\omega^{em} = \omega^{nr}$. Folglich ist $e = \frac{nr}{m}$ und, wie zu beweisen war,

$$|T(A)| = |A|^{\frac{nr}{m}}.$$

Dieser Satz ergiebt sich auch leicht aus dem auf S. 7 zitierten Satz 1.; denn offenbar kann die Determinante $|T(A)|$ selbst als eine invariante Form von A aufgefasst werden.

Eine Matrix P, welche der Gleichung (4) genügt, lässt sich stets dann bestimmen, wenn die Elementarteiler der Matrix A alle linear sind[1]). Dann folgt aber aus der sich aus (4) ergebenden Gleichung (5):

IV. „Ist A eine Matrix, deren Elementarteiler alle linear sind, und ist $T(A)$ eine invariante Operation, so besitzt auch die Matrix $T(A)$ lauter lineare Elementarteiler".

Abschnitt II.

§ 3. Um ein Hülfsmittel zur Bildung invarianter Formen zu erhalten, ist es vorteilhaft, an Stelle der Matrizen die ihnen entsprechenden linearen Substitutionen in Betracht zu ziehen.

Es seien

$$(7) \qquad (y') = (y_1', y_2', \dots y_m'), \; (y'') = (y_1'', y_2'', \dots y_m''), \dots$$

und

$$(8) \qquad (x') = (x_1', x_2', \dots x_m'), \; (x'') = (x_1'', x_2'', \dots x_m''), \dots$$

zwei Reihen von beliebig vielen Variabelensystemen, die untereinander durch die Gleichungen

$$(A) \qquad y_\iota^{(\nu)} = \Sigma a_{\iota\varkappa} x_\varkappa^{(\nu)}, \qquad (\iota, \varkappa = 1, 2, \dots m, \; \nu = 1, 2, \dots)$$

wo die m^2 Grössen $a_{\iota\varkappa}$ als unabhängige Veränderliche zu betrachten sind, verbunden sein mögen. Ferner seien

---

1) Vergl. Muth, Theorie der Elementarteiler, Leipzig, 1899.

(φ) $$\varphi_1(y),\ \varphi_2(y),\ \ldots\ \varphi_r(y)$$

r ganze homogene Funktionen gleicher Dimension der Variabelen (7), welche die Eigenschaft besitzen, sich vermöge der Gleichungen (A) als lineare homogene Funktionen

$(\overline{A})$ $$\varphi_\varrho(y) = \Sigma A_{\varrho\sigma}\varphi_\sigma(x) \qquad (\varrho, \sigma = 1, 2, \ldots r)$$

der aus den Variabelen (8) entsprechend gebildeten Funktionen $\varphi_1(x), \varphi_2(x), \ldots \varphi_r(x)$ darstellen zu lassen; und zwar sollen die r^2 Grössen $A_{\varrho\sigma}$ ganze rationale Funktionen der m^2 Variabelen $a_{\iota\varkappa}$ sein. Ein solches System (φ) nenne ich nach dem Vorgange des Herrn Deruyts [1]) ein transformables System und sage, das System gehöre zu der linearen Substitution $(\overline{A})$.

Es können zu $(\overline{A})$ unendlich viele transformable Systeme gehören: sind $\varphi_\varrho(y)$ und $\psi_\varrho(y)$ zwei solche, so gehört auch das System $c_1\varphi_\varrho(y) + c_2\psi_\varrho(y)$, wo c_1 und c_2 beliebige Konstanten sind, zu der linearen Substitution $(\overline{A})$.

Unter den r Funktionen $\varphi_\varrho(y)$ können gewisse lineare Relationen bestehen. Lässt sich aber ein zu $(\overline{A})$ gehörendes transformables System angeben, dessen Funktionen untereinander linear unabhängig sind [2]), so schliesst man leicht, dass die lineare Substitution $(\overline{A})$ alsdann eine invariante Operation $T(A)$ definiert. Ausserdem muss offenbar in diesem Falle die Determinante $|A_{\varrho\sigma}|$ nicht identisch Null sein.

§ 4. Umgekehrt sei $T(A) = (A_{\varrho\sigma})$ eine beliebige aus $A = (a_{\iota\varkappa})$ gebildete invariante Form. Bezeichnet man die Matrizen

$$\begin{pmatrix} y_1' \, y_1'' \ldots\ldots y_1^{(m)} \\ y_2' \, y_2'' \ldots\ldots y_2^{(m)} \\ \ldots\ldots \\ y_m' \, y_m'' \ldots\ldots y_m^{(m)} \end{pmatrix} \quad \text{und} \quad \begin{pmatrix} x_1' \, x_1'' \ldots x_1^{(m)} \\ x_2' \, x_2'' \ldots x_2^{(m)} \\ \ldots\ldots \\ x_m' \, x_m'' \ldots x_m^{(m)} \end{pmatrix}$$

mit Y, resp. mit X, so lassen sich die Gleichungen

$$y_\iota^{(\mu)} = \Sigma a_{\iota\varkappa} x_\varkappa^{(\mu)} \qquad (\iota, \varkappa, \mu = 1, 2, \ldots m)$$

in die Formel

(9) $$Y = AX$$

zusammenfassen; und daraus folgt: $T(Y) = T(A)\,T(X)$.

1) Bulletins de l'Acad. des Sciences de Belgique, Série 3, T. 32 (1896), S. 82.

2) Ein solches System will ich ein independentes nennen.

Ersetzt man daher in den r Elementen irgend einer Kolonne der Matrix $T(A)$ die Veränderlichen $a_{\iota\varkappa}$ durch $y_\iota^{(\varkappa)}$, so erhält man ein transformables System von r Funktionen, das zu der linearen Substitution $(A_{\varrho\sigma})$ gehört.

Aus (9) ergiebt sich auch, wenn $A'(X', Y')$ die zu $A(X, Y)$ konjugierte Matrix bedeutet, $Y' = X'A'$, und also $T(Y') = T(X')\,T(A')$. Folglich erhält man auch, indem man in den r Elementen einer der Zeilen von $T(A)$ für $a_{\iota\varkappa}$ die Veränderlichen $y_\varkappa^{(\iota)}$ substituiert, ein transformables System. Dasselbe gehört aber nicht mehr zu der linearen Substitution $T(A)$, sondern zu der zu $T(A')$ konjugierten linearen Substitution $T'(A') = \mathrm{T}(A)$.

Man überzeugt sich leicht, dass auch $\mathrm{T}(A)$ eine invariante Form von A ist.

Die so erhaltenen zu $T(A)$, resp. zu $\mathrm{T}(A)$ gehörenden transformablen Systeme können sich alle aus linear abhängigen Funktionen zusammensetzen. Wir wollen zeigen, dass man, wenn die Determinante $|T(A)|$ nicht identisch verschwindet, auch ein independentes System angeben kann, das zu $T(A)$, resp. zu $\mathrm{T}(A)$ gehört.

Zu diesem Zwecke ersetze ich in der ersten Kolonne der Matrix $T(A)$ die Grössen $a_{\iota\varkappa}$ durch die Variabelen $y_\iota^{(\varkappa)}$, in der zweiten Kolonne durch eine andere Reihe von Variabelen $y_\iota^{(m+\varkappa)}$, in der dritten durch $y_\iota^{(2m+\varkappa)}$, u. s. w.

Auf diese Weise erhalte ich r Systeme von je r Funktionen

$$\varphi_{1\varrho}(y),\ \varphi_{2\varrho}(y),\ \ldots\ \varphi_{r\varrho}(y), \qquad (\varrho = 1, 2, \ldots r)$$

von denen jedes ein zu $T(A)$ gehörendes transformables System bildet, wobei jedoch je zwei derselben von verschiedenen Variabelen abhängen. Setze ich nun

$$\varphi_\sigma(y) = \varphi_{\sigma 1}(y) + \varphi_{\sigma 2}(y) + \ldots + \varphi_{\sigma r}(y), \qquad (\sigma = 1, 2, \ldots r)$$

so bilden offenbar auch $\varphi_1(y), \varphi_2(y), \ldots \varphi_r(y)$ ein zu $T(A)$ gehörendes transformables System. Es ist aber leicht zu sehen, dass diese r Funktionen untereinander linear unabhängig sind, wenn $|T(A)| \not\equiv 0$ ist. Denn würde eine Gleichung der Form

$$c_1\varphi_1(y) + c_2\varphi_2(y) + \ldots + c_r\varphi_r(y) = 0$$

bestehen, wo $c_1, c_2, \ldots c_r$ in Bezug auf die y konstante Grössen bedeuten, so würde man leicht schliessen, dass die Gleichungen

$$c_1\varphi_{1\varrho}(y) + c_2\varphi_{2\varrho}(y) + \ldots + c_r\varphi_{r\varrho}(y) = 0$$

für alle Werte der y erfüllt sein müssten.

Man könnte also wieder $y_\iota^{(\varkappa)}, y_\iota^{(m+\varkappa)}, \ldots$ durch $a_{\iota\varkappa}$ ersetzen, ohne dass diese Gleichungen zu gelten aufhören; es wird aber dann $\varphi_{\sigma\varrho}(y) = A_{\sigma\varrho}$, und da $|A_{\sigma\varrho}| \neq 0$ ist, muss $c_1 = c_2 = \ldots = c_r = 0$ sein; daher sind die Funktionen $\varphi_1(y), \varphi_2(y), \ldots$ in der That linear unabhängig.

In genau analoger Weise bildet man auch ein zu $T(A) = T'(A')$ gehörendes independentes System.

§ 5. Besonders zu beachten sind die beiden folgenden transformablen Systeme.

1. Man bilde, wenn ein Variabelensystem $(y_1, y_2, \ldots y_m)$ gegeben ist, die sämtlichen $\binom{m+n-1}{n} = r$ Produkte

$$y_1^{\alpha_1}\, y_2^{\alpha_2} \ldots y_m^{\alpha_m} \tag{10}$$

von je n der Variabelen $y_1, y_2, \ldots y_m$. Diese r Produkte bilden offenbar ein independentes transformables System. Die zugehörige lineare Substitution $(\overline{A})$ nenne ich nach dem Vorgange des Herrn Hurwitz[1]) die n-te Potenztransformation der Substitution (A) und bezeichne sie mit $P_n A$. Um die der linearen Substitution $P_n A$ entsprechende Matrix genauer zu fixieren, müssen wir für die Produkte (10) eine bestimmte Reihenfolge festsetzen; und zwar soll das Produkt $y_1^{\alpha_1}\, y_2^{\alpha_2} \ldots y_m^{\alpha_m}$ vor dem Produkt $y_1^{\beta_1}\, y_2^{\beta_2} \ldots y_m^{\beta_m}$ stehen, wenn die erste nicht verschwindende der Differenzen $\alpha_1 - \beta_1, \alpha_2 - \beta_2, \ldots \alpha_m - \beta_m$ positiv ist. Dies ist die von Gauss in der Theorie der symmetrischen Funktionen eingeführte sogenannte lexikographische Anordnung. Bezeichnen wir die Produkte (10) in dieser Reihenfolge mit $Y_1, Y_2, \ldots Y_r$ und die aus den Variabelen $x_1, x_2, \ldots x_m$ entsprechend gebildeten mit $X_1, X_2, \ldots X_r$, so können wir schreiben

$$(Y_1, Y_2, \ldots Y_r) = P_n A\,(X_1, X_2, \ldots X_r).$$

Werden die Variabelen $Y_1, Y_2, \ldots Y_r$ und $X_1, X_2, \ldots X_r$ in einer anderen Reihenfolge genommen, so kommt dies einer Aehnlichkeitstransformation der Matrix $P_n A$ gleich. Um diese Transformation zu bewerkstelligen, hat man die Zeilen der Matrix in gewisser Weise untereinander zu vertauschen und die nämliche Permutation auch in den Spalten der Matrix auszuführen. Eine solche Aehnlichkeitstransformation einer Matrix soll im Folgenden stets als eine elementare Transformation bezeichnet werden.

---

1) Hurwitz, loc. cit.

Setzt man speziell

$$(A) = (\Omega) = (\omega_\iota \delta_{\iota\varkappa}), \qquad (\delta_{\iota\iota} = 1,\ \delta_{\iota\varkappa} = 0,\ \varkappa \neq \iota)$$

so wird $y_1^{\alpha_1} y_2^{\alpha_2} \ldots y_m^{\alpha_m} = \omega_1^{\alpha_1} \omega_2^{\alpha_2} \ldots \omega_m^{\alpha_m} \cdot x_1^{\alpha_1} x_2^{\alpha_2} \ldots x_m^{\alpha_m}$.

Daraus folgt nach § 2, dass man, wenn $\omega_1, \omega_2, \ldots \omega_m$ die charakteristischen Wurzeln der Matrix A sind, die charakteristischen Wurzeln der n-ten Potenztransformation $P_n A$ erhält, indem man die sämtlichen $\binom{m+n-1}{n}$ Produkte von je n der Grössen $\omega_1, \omega_2, \ldots \omega_m$ bildet[1]).

Die Determinante der Matrix $P_n A$ ist, da hier

$$\frac{nr}{m} = \frac{n}{m}\binom{m+n-1}{n} = \binom{n+m-1}{m}$$

ist, gleich $|A|^{\binom{n+m-1}{m}}$. Dieser Determinantensatz ist zuerst von Herrn G. v. Escherich[2]) allgemein bewiesen worden.

Die Charakteristik der invarianten Form $P_n A$ wird erhalten, indem man die Summe aller $\binom{n+m-1}{n}$ Produkte

$$\omega_1^{\alpha_1} \omega_2^{\alpha_2} \ldots \omega_m^{\alpha_m} \qquad (\alpha_1 + \alpha_2 + \cdots + \alpha_m = n)$$

bildet. Diese symmetrische Funktion wird gewöhnlich die n-te Wronski'sche Funktion Aleph genannt[3]). Ich werde sie im Folgenden stets mit $p_n = p_n(\omega_1, \omega_2, \ldots \omega_m)$ bezeichnen.

2. Ein zweites (independentes) transformables System erhält man, wenn n ($n \leqq m$) Variabelensysteme (7) betrachtet werden, indem man die $\binom{m}{n}$ Determinanten

$$Y_{\iota_1, \iota_2, \ldots \iota_n} = \begin{vmatrix} y'_{\iota_1} & y''_{\iota_1} & \ldots & y^{(n)}_{\iota_1} \\ y'_{\iota_2} & y''_{\iota_2} & \ldots & y^{(n)}_{\iota_2} \\ \cdot & \cdot & \cdot\cdot\cdot & \cdot \\ \cdot & \cdot & \cdot\cdot\cdot & \cdot \\ y'_{\iota_n} & y''_{\iota_n} & \ldots & y^{(n)}_{\iota_n} \end{vmatrix}$$

---

1) Franklin, Amer. Journal XVI, S. 205. Dieser Saz findet sich für den Fall $n = 2$, $m = 2$ schon bei Cayley, Crelle Bd. 50, S. 288.

2) „Ueber einige Determinanten", Monatshefte für Math. und Physik, Bd. 3 (1892), S. 68. — Vergl. auch Hurwitz, loc. cit., sowie W. Anissimoff, Math. Ann. Bd. 51, S. 388. — E. Pascal, „Die Determinanten", Leipzig 1900, bezeichnet diesen als den Scholz'schen Determinantensatz.

3) Vergl. K. Th. Vahlen, Encyclopädie der math. Wissenschaften, Leipzig 1899, Bd. I, S. 465.

bildet, wo die Indices $\iota_1, \iota_2, \ldots \iota_n$ die Zahlen 1, 2, ... m durchlaufen und $\iota_1 < \iota_2 < \ldots < \iota_n$ sein soll. Die zugehörige lineare Substitution nenne ich, wie Herr Hurwitz loc. cit., die n-te Determinantentransformation der linearen Substitution (A) und bezeichne sie mit C_nA [1]). Wir wollen hierbei festsetzen, dass die Determinanten $Y_{\iota_1, \iota_2, \ldots \iota_n}$ in der Weise angeordnet werden, dass $Y_{\iota_1, \iota_2, \ldots \iota_n}$ vor $Y_{\varkappa_1, \varkappa_2, \ldots \varkappa_n}$ stehen soll, wenn die erste nicht verschwindende der Differenzen $\iota_1 - \varkappa_1, \iota_2 - \varkappa_2, \ldots \iota_n - \varkappa_n$ negativ ist. Als Elemente der Matrix C_nA treten die $\binom{m}{n}^2$ Unterdeterminanten n-ten Grades der Matrix $A = (a_{\iota\varkappa})$ auf. Die charakteristischen Wurzeln der Matrix C_nA erhält man, wie man sich leicht überzeugt, indem man die $\binom{m}{n}$ Produkte von je n verschiedenen der charakteristischen Wurzeln $\omega_1, \omega_2, \ldots \omega_m$ der Matrix A bildet [2]). Die Charakteristik der invarianten Operation C_nA ist daher die n-te elementarsymmetrische Funktion c_n der m Grössen $\omega_1, \omega_2, \ldots \omega_m$.

§ 6. Es seien uns jetzt zwei lineare Substitutionen

$$(A) \qquad x_\iota = \Sigma a_{\iota\varkappa} x'_\varkappa \qquad (\iota, \varkappa = 1, 2, \ldots m)$$

$$(B) \qquad y_\lambda = \Sigma b_{\lambda\mu} y'_\mu \qquad (\lambda, \mu = 1, 2, \ldots n)$$

der Grade m und n gegeben. Die mn Produkte $x_\iota y_\lambda$ stellen sich offenbar als lineare homogene Funktionen der mn Produkte $x'_\iota y'_\lambda$ dar. Die auf diese Weise entstehende lineare Substitution nennt Herr Hurwitz loc. cit. die Produkttransformation von A und B und bezeichnet sie mit $A \times B$. Ordnet man die Produkte $x_\iota y_\lambda$ und in entsprechender Weise die Produkte $x'_\iota y'_\lambda$ derart an, dass $x_\iota y_\lambda$ vor $x_{\iota_1} y_{\lambda_1}$ stehen soll, falls die erste nicht verschwindende der Differenzen $\iota - \iota_1$, $\lambda - \lambda_1$ negativ ist, so hat die zu der linearen Substitution $A \times B$ gehörende Matrix die Gestalt

$$(11) \qquad \begin{pmatrix} a_{11} B & a_{12} B & \ldots & a_{1m} B \\ a_{21} B & a_{22} B & \ldots & a_{2m} B \\ \cdot & \cdot & \cdot\,\cdot\,\cdot & \cdot \\ a_{m1} B & a_{m2} B & \ldots & a_{mm} B \end{pmatrix},$$

wo $a_{\iota\varkappa} B$ die Matrix $a_{\iota\varkappa}(b_{\lambda\mu})$ bedeutet. Diese Matrix (11) soll stets unter dem Zeichen $A \times B$ verstanden werden.

---

1) Ist $n > m$, so hat man $C_nA = 0$ zu setzen.

2) G. Rados, Math. Ann. Bd. 48, S. 417; W. H. Metzler, Amer. Journal XVI (1894), S. 131.

Werden die Substitutionen (A) und (B) untereinander vertauscht, so kommt dies einer Vertauschung der Produkte $x_\iota y_\lambda$ und $x'_\iota y'_\lambda$ gleich. Daher geht die Matrix $A \times B$ in die Matrix $B \times A$ durch eine elementare Transformation über.

Bildet man in derselben Weise aus zwei anderen Substitutionen A_1 und B_1 die Matrix $A_1 \times B_1$, so beweist man leicht die Gleichung

$$(12) \qquad (A \times B)(A_1 \times B_1) = A A_1 \times B B_1.$$

Ist speziell $A = E_m$, $B = E_n$, so wird $E_m \times E_n = E_{mn}$; daher ist, wenn $|A|$ und $|B|$ von Null verschieden sind,

$$(13) \qquad (A \times B)^{-1} = (A^{-1} \times B^{-1}).$$

Mit Hülfe dieser Gleichung beweist man leicht:

V. „Sind $\alpha_1, \alpha_2, \ldots \alpha_m$, resp. $\beta_1, \beta_2, \ldots \beta_n$ die charakteristischen Wurzeln der Matrizen A und B, so sind die charakteristischen Wurzeln der Matrix $A \times B$ die mn Produkte $\alpha_\iota \beta_\lambda$"[1]).

In analoger Weise definiert man für beliebig viele Matrizen A, B, C, ... die Produkttransformation $A \times B \times C \times \ldots$ und beweist, wenn A_1, B_1, C_1, ... Matrizen derselben Grade sind, wie A, B, C, ..., die Gleichung

$$(A \times B \times C \times \ldots)(A_1 \times B_1 \times C_1 \times \cdots) = AA_1 \times BB_1 \times CC_1 \times \cdots$$ [2]).

§ 7. Sind nun $T_1(A)$, $T_2(A)$, $T_3(A)$, ... eine Reihe von invarianten Formen, so ist auch die Produkttransformation

$$T(A) = T_1(A) \times T_2(A) \times T_3(A) \times \ldots$$

eine invariante Form von A. In der That ist, wenn B eine zweite Matrix desselben Grades wie A bedeutet, nach (12)

$$\begin{aligned} T(A)\,T(B) &= T_1(A)\,T_1(B) \times T_2(A)\,T_2(B) \times \ldots \\ &= T_1(AB) \times T_2(AB) \times \cdots = T(AB). \end{aligned}$$

Ferner folgt aus dem oben Gesagten, dass wir eine zu $T(A)$ äquivalente Operation erhalten, wenn wir bei der Bildung von $T(A)$ die Reihenfolge der Operationen $T_1(A)$, $T_2(A)$, $T_3(A)$, ...

---

1) Franklin loc. cit. Vergl. auch Frobenius: „Ueber die Composition der Charaktere einer Gruppe", Sitzungsberichte, 1899, S. 330.

2) Die Transformationen $A \times B$ und $C_n A$ sind in neuerer Zeit von Herrn Cyp. Stephanos in seiner Abhandlung: „Sur une extension du calcul des substitutions linéaires", Liouville's Journal, Série 5, T. 6 (1900), S. 73, in eingehender Weise untersucht worden.

beliebig ändern. — Bezeichnen wir die Charakteristik von $T_\nu(A)$ mit Φ_ν, so folgt aus dem Satze V., dass die Charakteristik von $T(A)$ gleich ist dem Produkte $\Phi_1 \Phi_2 \Phi_3 \ldots$

Besonders zu beachten ist der Fall

$$T_1(A) = T_2(A) = T_3(A) = \cdots = A.$$

Dann wird $T(A) = A \times A \times \cdots \times A$; wird A genau n Mal genommen, so will ich diese Operation mit $\Pi_n A$ bezeichnen.

Eine andere Reihe von invarianten Operationen erhalten wir aus den Operationen $C_n A$ und $P_n A$, indem wir dieselben beliebig oft wiederholen; so entstehen die Operationen

$$C_\lambda(C_\mu A),\ C_\lambda(P_\mu A),\ P_\lambda(P_\mu A), \text{ u. s. w.}$$

Ich werde aber im Folgenden nur von der ersten Bildungsweise Gebrauch machen.

Abschnitt III.

§ 8. Jede homogene invariante Matrix n-ter Ordnung $T(A)$ lässt sich in der Form

$$(14) \quad T(A) = \sum \begin{bmatrix} \varkappa_1 \varkappa_2 \ldots \varkappa_n \\ \lambda_1 \lambda_2 \ldots \lambda_n \end{bmatrix} a_{\varkappa_1\lambda_1} a_{\varkappa_2\lambda_2} \ldots a_{\varkappa_n\lambda_n} \quad (\varkappa_\nu, \lambda_\nu = 1, 2, \ldots m)$$

schreiben, wo das Zeichen $\begin{bmatrix} \varkappa_1 \varkappa_2 \ldots \varkappa_n \\ \lambda_1 \lambda_2 \ldots \lambda_n \end{bmatrix}$ eine konstante Matrix bedeutet und

$$(15) \qquad \begin{bmatrix} \varkappa_1 \varkappa_2 \ldots \varkappa_n \\ \lambda_1 \lambda_2 \ldots \lambda_n \end{bmatrix} = \begin{bmatrix} \varkappa'_1 \varkappa'_2 \ldots \varkappa'_n \\ \lambda'_1 \lambda'_2 \ldots \lambda'_n \end{bmatrix}$$

zu setzen ist, wenn die beiden Produkte

$$a_{\varkappa_1\lambda_1} a_{\varkappa_2\lambda_2} \ldots a_{\varkappa_n\lambda_n},\quad a_{\varkappa'_1\lambda'_1} a_{\varkappa'_2\lambda'_2} \ldots a_{\varkappa'_n\lambda'_n}$$

sich nur durch die Reihenfolge der Faktoren von einander unterscheiden. Die Summe ist über alle verschiedenen Produkte von je n der m^2 Variabelen $a_{\varkappa\lambda}$ zu erstrecken. Die Anzahl ihrer Glieder oder die Anzahl der Matrizen $\begin{bmatrix} \varkappa_1 \varkappa_2 \ldots \varkappa_n \\ \lambda_1 \lambda_2 \ldots \lambda_n \end{bmatrix}$ ist demnach gleich $\binom{m^2+n-1}{n}$.

Wie wir in § 2 gesehen haben, lässt sich $T(A)$ so transformieren, dass jeder Diagonalmatrix

$$A = X = (x_\varkappa \, \delta_{\varkappa\lambda}) \qquad (\delta_{\varkappa\varkappa} = 1, \; \delta_{\varkappa\lambda} = 0, \; \varkappa \neq \lambda)$$

wieder eine Diagonalmatrix $T(X)$ entspricht. Durch eine elementare Transformation der Matrix $T(A)$ können wir erreichen, dass die in der Hauptdiagonale von $T(X)$ stehenden Produkte $x_1^{n_1} x_2^{n_2} \ldots x_m^{n_m}$ nach dem Gauss'schen lexikographischen Prinzip angeordnet erscheinen, d. h. dass das Produkt $x_1^{n_1} x_1^{n_2} \ldots x_m^{n_m}$ in einer früheren Zeile zu stehen kommt, als das Produkt $x_1^{n'_1} x_2^{n'_2} \ldots x_m^{n'_m}$, falls die erste nicht verschwindende der Differenzen $n_1 - n'_1, \; n_2 - n'_2, \ldots$ positiv ist. Alsdann werden die gleichen unter diesen Produkten hintereinander stehen.

Es möge nun die Charakteristik der Operation $T(A)$ gleich sein

$$\Phi = \Sigma r_{n_1, n_2, \ldots n_m} \, \omega_1^{n_1} \omega_2^{n_2} \ldots \omega_m^{n_m}. \qquad (n_1 + n_2 + \cdots + n_m = n)$$

Dann wird das Produkt $x_1^{n_1} x_2^{n_2} \ldots x_m^{n_m}$ in der Hauptdiagonale von $T(X)$ genau $r_{n_1, n_2, \ldots n_m}$ Mal auftreten, wobei diese Zahlen auch Null sein können.

Ich bezeichne jede Zeile der Matrix $T(A)$, welche, wenn $A = X$ wird, in der Hauptdiagonale das Glied $x_1^{n_1} x_2^{n_2} \ldots x_m^{n_m}$ enthält, mit $Z_{n_1, n_2, \ldots n_m}$, die entsprechenden Spalten mit $S_{n_1, n_2, \ldots n_m}$. Dann möge die (rechteckige) Teilmatrix von $T(A)$, in der sich die $r_{n_1, n_2, \ldots n_m}$ Zeilen $Z_{n_1, n_2, \ldots n_m}$ mit den $r_{n'_1, n'_2, \ldots n'_m}$ Spalten $S_{n'_1, n'_2, \ldots n'_m}$ schneiden, mit $M^{n_1, n_2, \ldots n_m}_{n'_1, n'_2, \ldots n'_m}$ bezeichnet werden.

Aus den Gleichungen

$$AX = (a_{\varkappa\lambda}) (x_\varkappa \, \delta_{\varkappa\lambda}) = (a_{\varkappa\lambda} \, x_\lambda),$$

$$XA = (x_\varkappa \delta_{\varkappa\lambda}) (a_{\varkappa\lambda}) = (x_\varkappa \, a_{\varkappa\lambda}) \qquad (\varkappa, \lambda = 1, 2, \ldots m)$$

erhalten wir

$$T(A) \; T(X) = T[(a_{\varkappa\lambda} \, x_\lambda)],$$

$$T(X) \; T(A) = T[(x_\varkappa \, a_{\varkappa\lambda})].$$

Ist daher

$$t = \Sigma C^{\varkappa_1, \varkappa_2, \ldots \varkappa_n}_{\lambda_1, \lambda_2, \ldots \lambda_n} \, a_{\varkappa_1 \lambda_1} \, a_{\varkappa_2 \lambda_2} \ldots a_{\varkappa_n \lambda_n} \qquad (\varkappa_\nu, \lambda_\nu = 1, 2, \ldots m)$$

ein Element der Matrix $T(A)$, welches der Teilmatrix $M^{n_1, n_2, \ldots n_m}_{n'_1, n'_2, \ldots n'_m}$ angehört, so können in t nur solche Produkte $a_{\varkappa_1 \lambda_1} a_{\varkappa_2 \lambda_2} \ldots a_{\varkappa_n \lambda_n}$

mit nicht verschwindenden Koeffizienten versehen sein, unter deren Indices $\varkappa$ genau n_1 gleich 1, n_2 gleich 2, ... n_m gleich m, und unter deren Indices λ genau n'_1 gleich 1, n'_2 gleich 2, ... n'_m gleich m sind.

Speziell ist z. B. jedes Element der Teilmatrix $M_{n,\,0,\,0\,\ldots\,0}^{n,\,0,\,0\,\ldots\,0}$ gleich ca_{11}^n, wo c eine Konstante ist, die nur in der Hauptdiagonale von Null verschieden (gleich $+1$) sein kann.

Zugleich folgt hieraus, dass, wenn die Indices $\varkappa$ und λ den genannten Bedingungen genügen, die mit $\begin{bmatrix}\varkappa_1\,\varkappa_2\ldots\,\varkappa_n\\ \lambda_1\,\lambda_2\ldots\,\lambda_n\end{bmatrix}$ bezeichnete Matrix nur in dem der Teilmatrix $M_{n'_1,\,n'_2,\ldots\,n'_m}^{n_1,\,n_2,\ldots\,n_m}$ entsprechenden Rechteck nicht verschwindende Elemente enthalten kann. Diese kleinere, $r_{n_1,n_2,\ldots n_m}$ Zeilen und $r_{n'_1,\,n'_2,\ldots\,n'_m}$ Spalten enthaltende Matrix will ich mit $\begin{bmatrix}\varkappa_1\,\varkappa_2\ldots\varkappa_n\\ \lambda_1\,\lambda_2\ldots\,\lambda_n\end{bmatrix}'$ bezeichnen. Ist $\lambda_\nu = \varkappa_\nu$, so wird dann

$$\begin{bmatrix}\varkappa_1\,\varkappa_2\ldots\varkappa_n\\ \lambda_1\,\lambda_2\ldots\,\lambda_n\end{bmatrix}' = E_{r_{n_1,\,n_2,\ldots\,n_m}},$$

wo E_λ, wie stets, die Einheitsmatrix des Grades λ bedeutet.

Die hier betrachtete Form einer invarianten Matrix $T(A)$ will ich die kanonische nennen. Aus dem Gesagten folgt, dass jede invariante Matrix einer in kanonischer Form erscheinenden äquivalent ist.

§ 9. Es sei nunmehr $B = (b_{\varkappa\lambda})$ eine zweite Matrix m-ten Grades, deren Elemente $b_{\varkappa\lambda}$ unabhängige Variabele sind, und es möge wieder

$$AB = C = (c_{\varkappa\lambda})$$

gesetzt werden. Die Gleichung $T(A)\,T(B) = T(AB)$ ist dann identisch mit der Gleichung

$$\sum_{\varkappa,\lambda,\mu,\nu}\begin{bmatrix}\varkappa_1\,\varkappa_2\ldots\varkappa_n\\ \mu_1\,\mu_2\ldots\,\mu_n\end{bmatrix}\begin{bmatrix}\nu_1\,\nu_2\ldots\nu_n\\ \lambda_1\,\lambda_2\ldots\lambda_n\end{bmatrix} a_{\varkappa_1\mu_1}\,a_{\varkappa_2\mu_2}\ldots a_{\varkappa_n\mu_n}\,b_{\nu_1\lambda_1}\,b_{\nu_2\lambda_2}\ldots b_{\nu_n\lambda_n}$$

$$= \sum_{\varkappa,\lambda}\begin{bmatrix}\varkappa_1\,\varkappa_2\ldots\varkappa_n\\ \lambda_1\,\lambda_2\ldots\lambda_n\end{bmatrix} c_{\varkappa_1\lambda_1}\,c_{\varkappa_2\lambda_2}\ldots c_{\varkappa_n\lambda_n}$$

$$= \sum_{\varkappa,\lambda,\mu}\begin{bmatrix}\varkappa_1\,\varkappa_2\ldots\varkappa_n\\ \lambda_1\,\lambda_2\ldots\lambda_n\end{bmatrix} a_{\varkappa_1\mu_1}\,b_{\mu_1\lambda_1}\,a_{\varkappa_2\mu_2}\,b_{\mu_2\lambda_2}\ldots a_{\varkappa_n\mu_n}\,b_{\mu_n\lambda_n}.$$

Indem wir die Matrizen, welche als Koeffizienten gleicher Produkte (*a*) $$a_{\varkappa_1\mu_1}a_{\varkappa_2\mu_2}\ldots a_{\varkappa_n\mu_n}b_{\nu_1\lambda_1}b_{\nu_2\lambda_2}\ldots b_{\nu_n\lambda_n}$$

auf beiden Seiten dieser Gleichung auftreten, einander gleichsetzen, ersehen wir, dass jedes der Matrizenprodukte

$$\begin{bmatrix}\varkappa_1\varkappa_2\ldots\varkappa_n\\ \mu_1\mu_2\ldots\mu_n\end{bmatrix}\begin{bmatrix}\nu_1\nu_2\ldots\nu_n\\ \lambda_1\lambda_2\ldots\lambda_n\end{bmatrix}$$

einer Summe von Matrizen

(*b*) $$\begin{bmatrix}\alpha_1\alpha_2\cdots\alpha_n\\ \beta_1\beta_2\cdots\beta_n\end{bmatrix}$$

gleich ist; hierbei wird die Matrix (*b*) in der erwähnten Summe so oft auftreten, als der Koeffizient des Gliedes (*a*) in der Entwickelung des Produktes

(*c*) $$c_{\alpha_1\beta_1}c_{\alpha_2\beta_2}\cdots c_{\alpha_n\beta_n}$$

angiebt. Offenbar kann derselbe nur dann von Null verschieden sein, wenn die Gesamtheit der Indices $(\alpha_1, \alpha_2, \ldots \alpha_n)$ und $(\beta_1, \beta_2, \ldots \beta_n)$ mit der Gesamtheit der Indices $(\varkappa_1, \varkappa_2, \ldots \varkappa_n)$, resp. $(\lambda_1, \lambda_2, \ldots \lambda_n)$ übereinstimmt.

Ist dies der Fall, so können wir auf Grund der Gleichung (15) schreiben

$$\begin{bmatrix}\alpha_1\alpha_2\cdots\alpha_n\\ \beta_1\beta_2\cdots\beta_n\end{bmatrix}=\begin{bmatrix}\varkappa_1\ \varkappa_2\ \ldots\ \varkappa_n\\ \lambda_1'\ \lambda_2'\ \ldots\ \lambda_n'\end{bmatrix},$$

wo $\lambda_1', \lambda_2', \ldots \lambda_n'$ eine Permutation der Indices $\lambda_1, \lambda_2, \ldots \lambda_n$ bedeuten. Mithin erhalten wir $\binom{m^2+n-1}{n}^2$ Relationen zwischen den Matrizen $\begin{bmatrix}\varkappa_1\ \varkappa_2\ \ldots\ \varkappa_n\\ \lambda_1\ \lambda_2\ \ldots\ \lambda_n\end{bmatrix}$ von der Gestalt

(I.) $$\begin{bmatrix}\varkappa_1\ \varkappa_2\ \ldots\ \varkappa_n\\ \mu_1\ \mu_2\ \ldots\ \mu_n\end{bmatrix}\begin{bmatrix}\nu_1\ \nu_2\ \ldots\ \nu_n\\ \lambda_1\ \lambda_2\ \ldots\ \lambda_n\end{bmatrix}=\sum C\,{}^{(\varkappa),\,(\mu)}_{(\nu),\,(\lambda')}\begin{bmatrix}\varkappa_1\ \varkappa_2\ \ldots\ \varkappa_n\\ \lambda_2'\ \lambda_2'\ \ldots\ \lambda_n'\end{bmatrix},$$

wo die Summation über alle Permutationen der n Indices $\lambda_1, \lambda_2, \ldots \lambda_n$ zu erstrecken ist, für welche die Produkte

(*d*) $$c_{\varkappa_1\lambda_1'}c_{\varkappa_2\lambda_2'}\ldots c_{\varkappa_n\lambda_n'}$$

verschiedene Werte annehmen, und $C^{(\varkappa),\,(\mu)}_{(\nu),\,(\lambda')}$ den Koeffizienten von $a_{\varkappa_1\mu_1}a_{\varkappa_2\mu_2}\ldots a_{\varkappa_n\mu_n}b_{\nu_1\lambda_1}b_{\nu_2\lambda_2}\ldots b_{\nu_n\lambda_n}$ in der Entwickelung des Produktes (*d*) bedeutet.

Ich will jetzt untersuchen, welche unter den Koeffizienten $C^{(\varkappa),(\mu)}_{(\nu),(\lambda')}$ von Null verschieden sind.

Zufolge der Gleichung (15) kann ich durch Vertauschung der Indicespaare $(\varkappa, \mu)$, resp. (ν, λ) erreichen, dass

$$\mu_1 \leqq \mu_2 \leqq \ldots \leqq \mu_n, \quad \nu_1 \leqq \nu_2 \leqq \ldots \leqq \nu_n$$

wird. Jedes Glied in der Entwickelung von (d) hat die Form $a_{\varkappa_1 \varrho_1} b_{\varrho_1 \lambda'_1} a_{\varkappa_2 \varrho_2} b_{\varrho_2 \lambda'_2} \ldots a_{\varkappa_n \varrho_n} b_{\varrho_n \lambda'_n}$. Soll daher $C^{(\varkappa),(\mu)}_{(\nu),(\lambda')}$ nicht Null sein, so muss die Gesamtheit der Indices μ mit der Gesamtheit der Indices ν übereinstimmen. Ist demnach μ_ϱ nicht für alle $\varrho = 1, 2, \ldots n$ gleich ν_ϱ, so ergiebt sich

$$\text{(I a.)} \qquad \begin{bmatrix} \varkappa_1 \varkappa_2 \ldots \varkappa_n \\ \mu_1 \mu_2 \ldots \mu_n \end{bmatrix} \begin{bmatrix} \nu_1 \nu_2 \ldots \nu_n \\ \lambda_1 \lambda_2 \ldots \lambda_n \end{bmatrix} = 0.$$

Ist dagegen $\mu_\varrho = \nu_\varrho$, so mögen n_1 unter den Indices μ gleich 1 n_2 gleich 2, $\ldots$ n_m gleich m sein.

Dann geht (I.) über in

$$\text{(I b.)} \quad \begin{bmatrix} \varkappa_1 \varkappa_2 \ldots \varkappa_{n_1} \varkappa_{n_1+1} \varkappa_{n_1+2} \ldots \varkappa_{n_1+n_2} \ldots \\ 1\;1 \ldots 1\;\;2\;\;\;2\;\;\ldots 2\;\;\ldots \end{bmatrix} \begin{bmatrix} 1\;1 \ldots 1\;\;2\;\;\;2\;\;\ldots 2\;\;\ldots \\ \lambda_1 \lambda_2 \ldots \lambda_{n_1} \lambda_{n_1+1} \lambda_{n_1+2} \ldots \lambda_{n_1+n_2} \ldots \end{bmatrix}$$

$$= \sum C^{(\varkappa),(\mu)}_{(\mu),(\lambda')} \begin{bmatrix} \varkappa_1 \varkappa_2 \ldots \varkappa_{n_1} \varkappa_{n_1+1} \varkappa_{n_1+2} \ldots \varkappa_{n_1+n_2} \ldots \\ \lambda'_1 \lambda'_2 \ldots \lambda'_{n_1} \lambda'_{n_1+1} \lambda'_{n_1+2} \ldots \lambda'_{n_1+n_2} \ldots \end{bmatrix},$$

wo $C^{(\varkappa),(\mu)}_{(\mu),(\lambda')}$ dem Koeffizienten des Gliedes

$$(a') \quad a_{\varkappa_1 1} b_{1\lambda_1} a_{\varkappa_2 1} b_{1\lambda_2} \ldots a_{\varkappa_{n_1} 1} b_{1\lambda_{n_1}} a_{\varkappa_{n_1+1} 2} b_{2\lambda_{n_1+1}} \ldots a_{\varkappa_{n_1+n_2} 2} b_{2\lambda_{n_1+n_2}} \ldots$$

in der Entwickelung des Produktes

$$(d') \quad c_{\varkappa_1 \lambda'_1} c_{\varkappa_2 \lambda'_2} \ldots c_{\varkappa_{n_1} \lambda'_{n_1}} c_{\varkappa_{n_1+1} \lambda'_{n_1+1}} \ldots c_{\varkappa_{n_1+n_2} \lambda'_{n_1+n_2}} \ldots$$

gleich ist. Ich behaupte, dass in (Ib.) nur diejenigen Matrizen $\begin{bmatrix} \varkappa_1 \varkappa_2 \ldots \varkappa_n \\ \lambda'_1 \lambda'_2 \ldots \lambda'_n \end{bmatrix}$ oder, was dasselbe ist, nur diejenigen Produkte (d') in Betracht zu ziehen sind, für welche die Indices $\lambda'_1, \lambda'_2, \ldots \lambda'_{n_1}$ eine

Permutation der Indices $\lambda_1, \lambda_2, \ldots \lambda_{n_1}$ repräsentieren, ebenso $\lambda'_{n_1+1}, \lambda'_{n_1+2}, \ldots \lambda'_{n_1+n_2}$ eine Permutation der Indices $\lambda_{n_1+1}, \lambda_{n_1+2}, \ldots \lambda_{n_1+n_2}$ u. s. w.

Es sei nämlich

$$(a'') \qquad a_{\varkappa_1 \nu_1} b_{\nu_1 \lambda'_1} a_{\varkappa_2 \nu_2} b_{\nu_2 \lambda'_2} \ldots a_{\varkappa_{n_1} \nu_{n_1}} b_{\nu_{n_1} \lambda'_{n_1}} a_{\varkappa_{n_1+1} \nu_{n_1+1}} b_{\nu_{n_1+1} \lambda'_{n_1+1}} \ldots$$

ein Glied in der Entwickelung von (d'), das dem Produkte (a') gleich ist. Ich nehme an, es sei $n_1 > 0$ und es seien unter den Indices $\varkappa_\alpha$ genau s gleich $\varkappa_1$, etwa die Indices $\varkappa_\alpha, \varkappa_\beta, \ldots \varkappa_\sigma$; dann muss offenbar mindestens einer der Indices $\nu_\alpha, \nu_\beta, \ldots \nu_\sigma$ gleich 1 sein, da sonst der Faktor $a_{\varkappa_1 1}$ in (a'') nicht vorkommen würde. Ist etwa ν_ξ ein solcher Index, so setze ich in (d') den Faktor $c_{\varkappa_\xi \lambda'_\xi}$ an die erste Stelle; dann ist in (a'') $\nu_1 = 1$ zu setzen. Ebenso kann ich allgemein durch passende Umstellung der Faktoren des Produktes (d') erreichen, dass

$$\nu_1 = 1,\ \nu_2 = 1, \ldots \nu_{n_1} = 1,\ \nu_{n_1+1} = 2, \ldots \nu_{n_1+n_2} = 2, \ldots$$

wird. Der Quotient der Produkte (a') und (a'') ist aber dann

$$\frac{b_{1\lambda_1} b_{1\lambda_2} \ldots b_{1\lambda_{n_1}} b_{2\lambda_{n_1+1}} b_{2\lambda_{n_1+2}} \ldots b_{2\lambda_{n_1+n_2}} \ldots}{b_{1\lambda'_1} b_{1\lambda'_2} \ldots b_{1\lambda'_{n_1}} b_{2\lambda'_{n_1+1}} b_{2\lambda'_{n_1+2}} \ldots b_{2\lambda'_{n_1+n_2}} \ldots} = 1,$$

und daraus folgt die Richtigkeit unserer Behauptung.

Umgekehrt enthält offenbar die Entwickelung eines jeden Produktes (d'), dessen Indices λ' der erwähnten Bedingung genügen, ein Glied von der Form (a'); die Koeffizienten $C^{(\varkappa),(\mu)}_{(\mu),(\lambda')}$ in der Gleichung (Ib.) sind daher alle von Null verschieden.

Ich bemerke, dass in der Gleichung (Ib.) die Matrizen $\begin{bmatrix} \varkappa_1 \varkappa_2 \ldots \varkappa_n \\ \lambda_1 \lambda_2 \ldots \lambda_n \end{bmatrix}$, wenn $T(A)$ in der kanonischen Form erscheint, durch die mit $\begin{bmatrix} \varkappa_1 \varkappa_2 \ldots \varkappa_n \\ \lambda_1 \lambda_2 \ldots \lambda_n \end{bmatrix}'$ bezeichneten ersetzt werden können.

§ 10. Die Bestimmung sämtlicher invarianten Operationen n-ter Ordnung erfordert scheinbar die Bestimmung sämtlicher Matrizen $\begin{bmatrix} \varkappa_1 \varkappa_2 \ldots \varkappa_n \\ \lambda_1 \lambda_2 \ldots \lambda_n \end{bmatrix}$, welche den Gleichungen (Ia.) und (Ib.) genügen.

Diese Aufgabe vereinfacht sich aber erheblich, wenn ich an-nehme, dass n nicht grösser ist als der Grad m der Matrix A. Es möge dann in (Ib.) sein

$$n_1 = n_2 = \cdots n_n = 1, \quad n_{n+1} = n_{n+2} = \cdots = n_m = 0.$$

Dann tritt auf der rechten Seite von (Ib.) nur ein Glied auf, und zwar erhalten wir, wenn unter den Indicespaaren

$$(\varkappa_1, \lambda_1),\ (\varkappa_2, \lambda_2),\ \ldots\ (\varkappa_n, \lambda_n)$$

q_1 gleich $(\varkappa', \lambda')$ sind, q_2 gleich $(\varkappa'', \lambda'')$, q_3 gleich $(\varkappa''', \lambda''')$, u. s. w., nach dem Polynomialsatz

$$\text{(II.)} \qquad \begin{bmatrix} \varkappa_1\, \varkappa_2 \ldots \varkappa_n \\ 1\; 2 \ldots n \end{bmatrix} \begin{bmatrix} 1\; 2 \ldots n \\ \lambda_1\, \lambda_2 \ldots \lambda_n \end{bmatrix} = q_1!\, q_2!\, q_3! \ldots \begin{bmatrix} \varkappa_1\, \varkappa_2 \ldots \varkappa_n \\ \lambda_1\, \lambda_2 \ldots \lambda_n \end{bmatrix}.$$

Ist zweitens in (Ib.)

$$\varkappa_1 = 1,\ \varkappa_2 = 2,\ \varkappa_3 = 3, \ldots$$
$$\lambda_1 = 1,\ \lambda_2 = 2,\ \lambda_3 = 3, \ldots,$$

so werden alle Koeffizienten $C^{(\varkappa),\,(\mu)}_{(\mu),\,(\lambda')}$ gleich 1, und wir erhalten

$$\text{(III.)} \qquad \begin{bmatrix} 1 & 2 & \ldots n_1, & n_1+1, \ldots n_1+n_2, \ldots \\ 1 & 1 & \ldots 1, & 2 \quad \ldots \quad 2, \quad \ldots \end{bmatrix} \begin{bmatrix} 1 & 1 & \ldots 1, & 2 \quad \ldots \quad 2, \quad \ldots \\ 1 & 2 & \ldots n_1, & n_1+1, \ldots n_1+n_2, \ldots \end{bmatrix}$$
$$= \sum_{\varrho} \begin{bmatrix} 1 & 2 & \ldots n_1, & n_1+1, \ldots & n_1+n_2, \ldots \\ \varrho_1 & \varrho_2 & \ldots \varrho_{n_1} & \varrho_{n_1+1} \ldots & \varrho_{n_1+n_2} \;\; \ldots \end{bmatrix},$$

wo $\varrho_1, \varrho_2, \ldots \varrho_{n_1}$ alle $n_1!$ Permutationen der n_1 Ziffern $1, 2, \ldots n_1$ durchlaufen, ebenso $\varrho_{n_1+1}, \varrho_{n_1+2}, \ldots \varrho_{n_1+n_2}$ alle $n_2!$ Permutationen der n_2 Ziffern $n_1+1,\ n_2+2, \ldots\ n_1+n_2$, u. s. w.

Besonders zu beachten ist der Fall, dass in (II.) sowohl die Indices $\varkappa_1, \varkappa_2, \ldots \varkappa_n$, als auch die Indices $\lambda_1, \lambda_2, \ldots \lambda_n$, abgesehen von der Reihenfolge, mit den Zahlen $1, 2, \ldots n$ übereinstimmen. Dann wird

$$q_1 = q_2 = q_3 \cdots = 1,$$

und die Gleichungen (II.) besagen in diesem Fall nichts Anderes, als dass die $n!$ Matrizen $\begin{bmatrix} 1\; 2 \ldots n \\ \lambda_1\, \lambda_2 \ldots \lambda_n \end{bmatrix}$ eine Darstellung der symmetrischen Gruppe $\mathfrak{S}_n$ des Grades n bilden, wobei die Matrix $\begin{bmatrix} 1\; 2 \ldots n \\ \lambda_1\, \lambda_2 \ldots \lambda_n \end{bmatrix}$ der Permutation $\begin{pmatrix} 1\; 2 \ldots n \\ \lambda_1\, \lambda_2 \ldots \lambda_n \end{pmatrix}$ der n Ziffern $1, 2, \ldots n$

entspricht. Die so gewonnene Darstellung der Gruppe $\mathfrak{S}_n$ wird im Allgemeinen eine uneigentliche sein (vergl. Einleitung), da die Determinanten der Matrizen $\begin{bmatrix} 1 & 2 \dots & n \\ \lambda_1 & \lambda_2 \dots & \lambda_n \end{bmatrix}$ im Allgemeinen Null sind. Nehme ich aber an, dass $T(A)$ auf die kanonische Form gebracht ist, so werden auch die Matrizen $\begin{bmatrix} 1 & 2 \dots & n \\ \lambda_1 & \lambda_2 \dots & \lambda_n \end{bmatrix}'$ die Gruppe $\mathfrak{S}_n$ darstellen, und diese Darstellung ist eine eigentliche, da $\begin{bmatrix} 1 & 2 \dots & n \\ 1 & 2 \dots & n \end{bmatrix}'$ eine Einheitsmatrix ist.

Diese letztere Darstellung nenne ich die zu der invarianten Operation $T(A)$ gehörende Darstellung der symmetrischen Gruppe n-ten Grades. Der Grad f derselben ist gleich dem Koeffizienten $r_{1, 1, \dots 1, 0, 0 \dots}$ von $\omega_1 \omega_2 \dots \omega_n$ in der Charakteristik Φ von $T(A)$. Es kann nicht $f = 0$ sein, denn daraus würde man, da $\begin{bmatrix} 1\,2 \dots n \\ 1\,2 \dots n \end{bmatrix} = 0$ wäre, vermittelst der Formel (II.) leicht schliessen, dass $\begin{bmatrix} \varkappa_1 \varkappa_2 \dots \varkappa_n \\ \lambda_1 \lambda_2 \dots \lambda_n \end{bmatrix}$ für alle Indices $\varkappa$ und λ Null und also auch $T(A) = 0$ ist.

An Stelle der Zahlen $1, 2, \dots n$ kann ich beliebige andere n von einander verschiedene der Zahlen $1, 2, \dots m$ herausgreifen. Ich erhalte auf diese Weise $\binom{m}{n}$ Darstellungen der Gruppe $\mathfrak{S}_n$. Wir werden aber bald sehen, dass diese $\binom{m}{n}$ Darstellungen der zuerst betrachteten äquivalent sind.

§ 11. Ist

$$n = n_1 + n_2 + n_3 + \dots,$$

und bilde ich alle Permutationen der n Ziffern $1, 2, \dots n$, die nur die n_1 ersten Ziffern untereinander vertauschen, ebenso nur die n_2 folgenden, u. s. w., so bilden diese Permutationen eine Untergruppe der symmetrischen Gruppe $\mathfrak{S}_n$. Diese Untergruppe, deren Ordnung gleich $n_1!\, n_2!\, n_3! \dots$ ist, will ich mit $\mathfrak{S}_{n_1, n_2, n_3, \dots}$ bezeichnen.

Bezeichne ich ferner die Summe alle Matrizen

$$\begin{bmatrix} 1 & 2 & \dots & n \\ \lambda_1 & \lambda_2 & \dots & \lambda_n \end{bmatrix}', \qquad (\lambda_1, \lambda_2, \dots \lambda_n) = (1, 2, \dots n)$$

welche den Permutationen der Gruppe $\mathfrak{S}_{n_1, n_2, \dots}$ entsprechen, mit $H_{n_1, n_2, \dots}$, so lehrt uns die Gleichung (III.), dass

$$\text{(III*.)} \begin{bmatrix} 1\,2 \dots n_1, & n_1{+}1 \dots n_1{+}n_2, & \dots \\ 1\,1 \dots 1, & 2 \quad \dots 2, & \dots \end{bmatrix}' \begin{bmatrix} 1\;1 \dots 1, & 2 \quad \dots 2, & \dots \\ 1\;2 \dots n_1, & n_1{+}1, \dots n_1{+}n_2, & \dots \end{bmatrix}' = H_{n_1, n_2, \dots}$$

ist. — Ist allgemein $\mathfrak{H}$ eine abstrakte Gruppe der Ordnung h, und bilden die Matrizen $S_1, S_2, \dots S_h$ eine Darstellung dieser Gruppe, so ist, wenn

$$S_1 + S_2 + \cdots + S_h = H$$

gesetzt wird, $H^2 = hH$. Daher lässt sich die Darstellung $(S_1, S_2, \dots S_h)$ durch eine ihr äquivalente ersetzen, für welche

$$H = h \begin{pmatrix} 1 & & & & & \\ & 1 & & & & \\ & & \ddots & & & \\ & & & 1 & & \\ & & & & 0 & \\ & & & & & \ddots \\ & & & & & & 0 \end{pmatrix}$$

wird. Mithin ist die Spur der Matrix $\frac{1}{h} H$ stets eine nicht negative ganze Zahl. Diese Zahl will ich als die zu der Darstellung $(S_1, S_2, \dots S_h)$ gehörende Charakteristik der Gruppe $\mathfrak{H}$ bezeichnen.

Zufolge der Formel (III*.) ist also die Spur der Matrix

$$\frac{1}{n_1!\, n_2! \dots} \begin{bmatrix} 1 & 2 & \dots n_1, & n_1+1, \dots \\ 1 & 1 & \dots 1 & 2 \quad \dots \end{bmatrix}' \begin{bmatrix} 1 & 1 & \dots 1 & 2 \quad \dots \\ 1 & 2 & \dots n_1, & n_1+1, \dots \end{bmatrix}'$$

gleich der Charakteristik der Gruppe $\mathfrak{S}_{n_1, n_2 \dots}$ für die durch die invariante Operation $T(A)$ erzeugte Darstellung.

Andererseits ist aber nach Formel (II.)

$$\frac{1}{n_1!\, n_2! \dots} \begin{bmatrix} 1 & 1 & \dots 1 & 2 \quad \dots \\ 1 & 2 & \dots n_1, & n_1+1, \dots \end{bmatrix}' \begin{bmatrix} 1 & 2 & \dots n_1, & n_1+1, \dots \\ 1 & 1 & \dots 1 & 2 \quad \dots \end{bmatrix}'$$

$$= \begin{bmatrix} 1\ 1 \dots 1\ 2 \dots 2 \dots \\ 1\ 1 \dots 1\ 2 \dots 2 \dots \end{bmatrix}' = E_{r_{n_1, n_2, \dots}}.$$

Da nun bekanntlich für zwei beliebige Matrizen P und Q die Spur von PQ gleich ist der Spur von QP, so sehen wir also, dass die erwähnte Charakteristik der Gruppe $\mathfrak{S}_{n_1, n_2, \dots}$ gleich ist der Zahl $r_{n_1, n_2, \dots}$.

§ 12. Die Anzahl der zu betrachtenden Untergruppen $\mathfrak{S}_{n_1, n_2, \dots}$ ist gleich der Anzahl k der Zerlegungen der Zahl n in gleiche oder verschiedene positive Summanden. Diese Zahl k giebt aber auch die Anzahl der Klassen gleichberechtigter Elemente der symmetrischen Gruppe $\mathfrak{S}_n$ an.

Es sei

$$R = \begin{pmatrix} 1 & 2 & \dots & n \\ \lambda_1 & \lambda_2 & \dots & \lambda_n \end{pmatrix}$$

eine Permutation der Gruppe $\mathfrak{S}_n$ und $\chi(R)$ die Spur der Matrix $\begin{bmatrix} 1 & 2 & \dots & n \\ \lambda_1 & \lambda_2 & \dots & \lambda_n \end{bmatrix}'$. Dann bilden die $n!$ Zahlen $\chi(R)$ nach der von Herrn Frobenius[1]) eingeführten Bezeichnung einen (zusammengesetzten) Charakter der Gruppe $\mathfrak{S}_n$; und zwar ist, wenn R und R' gleichberechtigte Permutationen von $\mathfrak{S}_n$ sind, $\chi(R) = \chi(R')$. Sind uns die Zahlen $\chi(R)$ gegeben, so erhalten wir nach dem Vorhergehenden die k Zahlen $r_{n_1, n_2, \dots}$ als gewisse lineare Funktionen der $\chi(R)$. Umgekehrt bestimmen auch die Zahlen $r_{n_1, n_2, \dots}$ den Charakter $\chi(R)$ vollständig.

Es möge nämlich $R = R_{\alpha_1, \alpha_2, \dots}$ aus α_1 Zykeln der Ordnung 1 bestehen, α_2 Zykeln der Ordnung 2, u. s. w. Dann ist

(16) $$\alpha_1 + 2\alpha_2 + 3\alpha_3 \dots = n.$$

Ist $R_{\alpha_1, \alpha_2, \dots}$ in der Untergruppe $\mathfrak{S}_{n_1, n_2, \dots}$ enthalten, so zerfällt es in eine Permutation R_1 unter den n_1 ersten Ziffern 1, 2, ... n_1, eine Permutation R_2 unter den n_2 folgenden, u. s. w. Besteht nun R_ν aus $\alpha_{\nu 1}$ Zykeln der Ordnung 1, $\alpha_{\nu 2}$ Zykeln der Ordnung 2, u. s. w., so ist

(17) $$n_1 = \alpha_{11} + 2\alpha_{12} + 3\alpha_{13} + \cdots, \; n_2 = \alpha_{21} + 2\alpha_{22} + 3\alpha_{23} + \cdots, \cdots$$

(18) $$\alpha_1 = \alpha_{11} + \alpha_{21} + \alpha_{31} + \cdots, \; \alpha_2 = \alpha_{12} + \alpha_{22} + \alpha_{32} + \cdots, \cdots.$$

Die Anzahl der in $\mathfrak{S}_{n_1, n_2, \dots}$ enthaltenen mit $R_{\alpha_1, \alpha_2, \dots}$ gleichberechtigten Permutationen ist dann gleich[2])

(19) $$g^{n_1, n_2 \dots}_{\alpha_1, \alpha_2 \dots} = \sum \frac{n_1!}{1^{\alpha_{11}} \alpha_{11}! \, 2^{\alpha_{12}} \alpha_{12}! \dots} \cdot \frac{n_2!}{1^{\alpha_{21}} \alpha_{21}! \, 2^{\alpha_{22}} \alpha_{22}! \dots} \cdots,$$

1) „Ueber Gruppencharaktere", Sitzungsberichte 1896, S. 985 und „Ueber die Darstellung der endl. Gruppen durch lin. Subst." ibid. 1899, S. 482.

2) Frobenius, „Ueber die Charaktere der symmetrischen Gruppe", Sitzungsber. 1900, S. 516.

wo die Summation über alle Lösungen der Gleichungen (17) und (18) zu erstrecken ist. Da nun $n_1!\, n_2! \ldots r_{n_1, n_2, \ldots}$ gleich der Summe der Spuren aller zu der Gruppe $\mathfrak{S}_{n_1, n_2 \ldots}$ gehörenden Matrizen $\begin{bmatrix} 1\; 2 \ldots n \\ \lambda_1\, \lambda_2 \ldots \lambda_n \end{bmatrix}'$ ist, so erhalten wir

$$(20) \qquad n_1!\, n_2! \ldots r_{n_1, n_2, \ldots} = \sum g^{\,n_1,\, n_2, \ldots}_{\,\alpha_1,\, \alpha_2, \ldots}\; \chi(R_{\alpha_1, \alpha_2, \ldots}),$$

die Summation erstreckt über alle Lösungen der Gleichung (16).

Aus den Formeln (20) lassen sich die Zahlen $\chi(R)$ folgendermassen berechnen.

Es sei, wie auf Seite 17, p_ν die ν-te Wronskische Funktion Aleph, ferner s_ν die ν-te Potenzsumme der m Variabelen $\omega_1, \omega_2, \ldots \omega_m$. Dann besteht die Gleichung

$$(21) \qquad \nu p_\nu = s_1 p_{\nu-1} + s_2 p_{\nu-2} + \ldots + s_\nu p_0 \,^{1)}. \qquad (p_0 = 1)$$

Aus dieser Gleichung erhält man leicht

$$(22) \quad \nu!\, p_\nu = \sum_\mu \frac{\nu!}{\mu_1!\, \mu_2! \ldots} \left(\frac{s_1}{1}\right)^{\mu_1} \left(\frac{s_2}{2}\right)^{\mu_2} \cdots \qquad (\mu_1 + 2\mu_2 + 3\mu_3 + \ldots = \nu).$$

Daher ist, wie man unmittelbar sieht,

$$(23) \qquad n_1!\, n_2! \ldots p_{n_1} p_{n_2} \ldots = \sum_\alpha g^{\,n_1,\, n_2 \ldots}_{\,\alpha_1,\, \alpha_2 \ldots}\; s_1^{\alpha_1} s_2^{\alpha_2} s_3^{\alpha_3} \ldots,$$

wo die Zahlen n_ν, α_ν und $g^{\,n_1,\, n_2 \ldots}_{\,\alpha_1,\, \alpha_2 \ldots}$ dieselbe Bedeutung haben, wie oben. Die Bestimmung der Zahlen $\chi(R)$ aus dem Gleichungssystem (20) ist daher mit der Auflösung der Gleichungen (23) nach den Produkten $s_1^{\alpha_1} s_2^{\alpha_2} \ldots$ identisch. Zur Darstellung dieser Produkte als Funktionen der $p_1, p_2, \ldots$ kann man sich aber der Formel

$$s_\nu = \nu \sum (-1)^{\mu_1 + \mu_2 + \cdots - 1} \frac{(\mu_1 + \mu_2 + \cdots + \mu_\nu - 1)!}{\mu_1!\, \mu_2! \ldots \mu_\nu!} p_1^{\mu_1} p_2^{\mu_2} \ldots p_\nu^{\mu_\nu}$$

$$(\mu_1 + 2\mu_2 + 3\mu_3 + \cdots = \nu)$$

bedienen, die sich aus (21) leicht ergiebt.

---

1) Crocchi, Giorn. di mat. 17 (1879), S. 218.

Man kann aber auch die Zahlen $\chi(R)$ direkt aus der Charakteristik $\Phi = \Sigma r_{n_1, n_2 \ldots} \omega_1^{n_1} \omega_2^{n_2} \ldots$ der invarianten Operation $T(A)$ erhalten. Es ist nämlich

$$\frac{1^{\alpha_1}\alpha_1!\,2^{\alpha_2}\alpha_2!\ldots}{n_1!\,n_2!\ldots}\, g^{n_1, n_2 \ldots}_{\alpha_1, \alpha_2 \ldots} = \sum \frac{\alpha_1!}{\alpha_{11}!\,\alpha_{21}!\,\alpha_{31}!\ldots} \cdot \frac{\alpha_2!}{\alpha_{12}!\,\alpha_{22}!\,\alpha_{32}!\ldots}\cdots$$

gleich dem Koeffizienten von $\omega_1^{n_1}\omega_2^{n_2}\ldots\omega_m^{n_m}$ in der Entwickelung des Produktes

$$(\omega_1 + \omega_2 + \cdots \omega_m)^{\alpha_1}(\omega_1^2 + \omega_2^2 + \cdots + \omega_m^2)^{\alpha_2}\cdots = s_1^{\alpha_1} s_2^{\alpha_2} \ldots \text{ [1]}.$$

Daher ist, wie aus (20) unmittelbar folgt,

$$(24) \quad \Phi = \sum r_{n_1, n_2 \ldots} \omega_1^{n_1} \omega_2^{n_2} \ldots = \sum_\alpha \frac{\chi(R_{\alpha_1, \alpha_2 \ldots})}{\alpha_1!\,\alpha_2!\ldots}\left(\frac{s_1}{1}\right)^{\alpha_1}\left(\frac{s_2}{2}\right)^{\alpha_2}\cdots.$$

Stelle ich also die symmetrische Funktion Φ der m Variabelen $\omega_1, \omega_2, \ldots \omega_m$ als Funktion der Potenzsummen $s_1, s_2, \ldots s_n$ dar, so ist $\chi(R_{\alpha_1, \alpha_2 \ldots})$ nichts Anderes, als der Koeffizient von $s^{\alpha_1} s^{\alpha_2} \ldots$ in dieser Entwickelung, multipliziert mit der ganzen Zahl $1^{\alpha_1}\alpha_1!\,2^{\alpha_2}\alpha_2!\,3^{\alpha_3}\alpha_3!\cdots$.

§ 13. Wir haben gesehen, dass durch die Operation $T(A)$ im Ganzen $\binom{m}{n}$ Darstellungen der symmetrischen Gruppe n-ten Grades erzeugt werden. Da nun die Zahlen $r_{n_1, n_2, \ldots}$ sich bei Vertauschung der Indices nicht ändern und der Charakter $\chi(R)$ durch diese Zahlen vollständig bestimmt ist, so sind die Spuren derjenigen Matrizen in diesen $\binom{m}{n}$ Darstellungen, welche derselben Permutation der Gruppe $\mathfrak{S}_n$ entsprechen, einander gleich. Daraus folgt aber, dass die $\binom{m}{n}$ Darstellungen äquivalent sind.

Denn es besteht folgender Satz:

„Ist $\mathfrak{H}$ eine abstrakte Gruppe, und sind

$$S_1, S_2, \ldots S_h \text{ und } S_1', S_2', \ldots S_h'$$

zwei Darstellungen von $\mathfrak{H}$ durch lineare Substitutionen, wobei S_ν und S_ν' demselben Element von $\mathfrak{H}$ entsprechen, so sind diese

1) Frobenius, loc. cit. S. 517.

beiden Darstellungen dann und nur dann äquivalent, wenn die Spuren von S_ν und S'_ν einander gleich sind" [1]).

Da ferner die Charakteristik von $T(A)$ den Gruppencharakter $\chi(R)$ vollständig bestimmt, so erhalten wir den Satz:

VI. „Sind $T(A)$ und $T_1(A)$ zwei invariante Operationen n-ter Ordnung, so sind die beiden Darstellungen der symmetrischen Gruppe n-ten Grades, welche zu diesen Operationen gehören, dann und nur dann äquivalent, wenn die Charakteristiken von $T(A)$ und $T_1(A)$ einander gleich sind".

Abschnitt IV.

§ 14. Soll die Matrix

$$T(A) = \sum \begin{bmatrix} \varkappa_1, \varkappa_2 \dots \varkappa_n \\ \lambda_1, \lambda_2 \dots \lambda_n \end{bmatrix} a_{\varkappa_1\lambda_1} a_{\varkappa_2\lambda_2} \dots a_{\varkappa_n\lambda_n} \qquad (\varkappa_\nu, \lambda_\nu = 1, 2, \dots m)$$

eine aus $A = (a_{\varkappa\lambda})$ gebildete invariante Matrix sein, so müssen, wie wir gesehen haben, die $\binom{m^2+n-1}{n}$ Matrizen $\begin{bmatrix} \varkappa_1\, \varkappa_2 \dots \varkappa_n \\ \lambda_1\, \lambda_2 \dots \lambda_n \end{bmatrix}$ das System der Gleichungen (I.) befriedigen.

Ich will zeigen, dass man die Anzahl der Gleichungen, deren Bestehen für die Invarianteneigenschaft der Matrix $T(A)$ notwendig und hinreichend ist, bedeutend verkleinern kann. Es mögen nämlich die Matrizen $\begin{bmatrix} \varkappa_1\, \varkappa_2 \dots \varkappa_n \\ \lambda_1\, \lambda_2 \dots \lambda_n \end{bmatrix}$ folgenden beiden Bedingungen genügen:

1. Die Gleichungen (II.) und (III.) sind alle erfüllt.
2. Ist

$$\mu_1 \leqq \mu_2 \leqq \dots \leqq \mu_n, \quad \nu_1 \leqq \nu_2 \leqq \dots \leqq \nu_n,$$

so soll, falls μ_ϱ nicht für alle $\varrho = 1, 2, \dots n$ gleich ν_ϱ ist,

$$(\text{I}^*_{\text{a.}}) \qquad \begin{bmatrix} 1\; 2 \dots n \\ \mu_1\, \mu_2 \dots \mu_n \end{bmatrix} \begin{bmatrix} \nu_1\, \nu_2 \dots \nu_n \\ 1\; 2 \dots n \end{bmatrix} = 0$$

sein. — Ich behaupte, dass alsdann auch alle Gleichungen (Ia.) und (Ib.) bestehen. Es ist nämlich nach (II.)

---

1) Molien, Sitzungsberichte der Naturforschergesellschaft zu Dorpat, 1897, Jahrg. 18, S. 259; Frobenius, Sitzungsberichte der Ak. d. Wiss. zu Berlin, 1899, S. 482.

$$\begin{bmatrix}\varkappa_1\,\varkappa_2\dots\varkappa_n\\ \mu_1\,\mu_2\dots\mu_n\end{bmatrix}\begin{bmatrix}\nu_1\,\nu_2\dots\nu_n\\ \lambda_1\,\lambda_2\dots\lambda_n\end{bmatrix},\quad (\mu_1\leqq\mu_2\leqq\dots\leqq\mu_n,\ \nu_1\leqq\nu_2\leqq\dots\leqq\nu_n)$$

abgesehen von einem gewissen Zahlenfaktor, gleich

$$\begin{bmatrix}\varkappa_1\varkappa_2\dots\varkappa_n\\ 1\;2\dots n\end{bmatrix}\begin{bmatrix}1\;2\;\dots n\\ \mu_1\mu_2\dots\mu_n\end{bmatrix}\begin{bmatrix}\nu_1\nu_2\dots\nu_n\\ 1\;2\;\dots n\end{bmatrix}\begin{bmatrix}1\;2\dots n\\ \lambda_1\lambda_2\dots\lambda_n\end{bmatrix};$$

der mittlere Faktor $\begin{bmatrix}1\;2\;\dots n\\ \mu_1\mu_2\dots\mu_n\end{bmatrix}\begin{bmatrix}\nu_1\nu_2\dots\nu_n\\ 1\;2\;\dots n\end{bmatrix}$ ist aber nach (I$^*_a$.) und (III.) entweder gleich Null oder gleich einer Summe von Matrizen der Form $\begin{bmatrix}1\;2\;\dots n\\ \varrho_1\,\varrho_2\dots\varrho_n\end{bmatrix}$, wo $\varrho_1, \varrho_2, \dots \varrho_n$, abgesehen von der Reihenfolge, den Zahlen $1, 2, \dots n$ gleich sind.

Im ersten Falle erhalten wir die Gleichungen (I a.), im zweiten Falle ergiebt sich durch wiederholte Anwendung der Gleichung (II.) eine Relation der Form

$$\text{(I'.)}\qquad \begin{bmatrix}\varkappa_1\,\varkappa_2\dots\varkappa_n\\ \mu_1\,\mu_2\dots\mu_n\end{bmatrix}\begin{bmatrix}\mu_1\,\mu_2\dots\mu_n\\ \lambda_1\,\lambda_2\dots\lambda_n\end{bmatrix} = \sum \overline{C}^{(\varkappa),\,(\mu)}_{(\mu),\,(\lambda')}\begin{bmatrix}\varkappa_1\,\varkappa_2\dots\varkappa_n\\ \lambda'_1\,\lambda'_2\dots\lambda'_n\end{bmatrix},$$

wo $\overline{C}^{(\varkappa),\,(\mu)}_{(\mu),\,(\lambda')}$ gewisse von der speziellen Wahl der Matrix $T(A)$ unabhängige Konstanten und $\lambda'_1, \lambda'_2, \dots \lambda'_n$ gewisse Permutationen der Indices $\lambda_1, \lambda_2, \dots \lambda_n$ bedeuten.

Wären nun die Konstanten $\overline{C}^{(\varkappa),\,(\mu)}_{(\mu),\,(\lambda')}$ von den in der Gleichung (I.) auftretenden $C^{(\varkappa),\,(\mu)}_{(\mu),\,(\lambda')}$ verschieden, so würden wir lineare Beziehungen zwischen den Matrizen $\begin{bmatrix}\varkappa_1\;\varkappa_2\dots\varkappa_n\\ \lambda_1\;\lambda_2\dots\lambda_n\end{bmatrix}$ erhalten, die für jede invariante Operation $T(A)$ bestehen müssten. Dies ist aber nicht der Fall; denn setzt man

$$T(A) = \Pi_n A = A\times A\times\cdots\times A,$$

so hat jedes Element der Matrix $T(A)$ die Form $a_{\varkappa_1\lambda_1}a_{\varkappa_2\lambda_2}\dots a_{\varkappa_n\lambda_n}$; daher müssen in diesem Fall die $\binom{m^2+n-1}{n}$ Matrizen $\begin{bmatrix}\varkappa_1\,\varkappa_2\dots\varkappa_n\\ \lambda_1\,\lambda_2\dots\lambda_n\end{bmatrix}$ linear unabhängig sein[1]). Also ist $\overline{C}^{(\varkappa),\,(\mu)}_{(\mu),\,(\lambda)} = C^{(\varkappa),\,(\mu)}_{(\mu),\,(\lambda)}$, und

1) Auf Grund dieser Eigenschaft der Operation $\Pi_n A$ können wir die Gleichungen (I.) als die definierenden Gleichungen eines Systems komplexer Haupteinheiten auffassen.

folglich stimmen die Gleichungen (I'.) mit den Gleichungen (I.) überein.

§ 15. Es seien nun

(25) $$S_1, S_2, \ldots S_{n!}$$

$n!$ Matrizen f-ten Grades von nicht verschwindender Determinante, die eine Darstellung der symmetrischen Gruppe n-ten Grades $\mathfrak{S}_n$ bilden. Entspricht hierbei S_ν der Permutation $\begin{pmatrix} 1 & 2 & \ldots & n \\ \varrho_1 & \varrho_2 & \ldots & \varrho_n \end{pmatrix}$ der n Ziffern $1, 2, \ldots n$, so wollen wir S_ν mit $\begin{bmatrix} 1 & 2 & \ldots & n \\ \varrho_1 & \varrho_2 & \ldots & \varrho_n \end{bmatrix}'$ bezeichnen.

Ich denke mir eine Matrix M des Grades fq hingeschrieben, wo $q = \binom{m+n-1}{n}$ zu setzen ist. Sind

$$x_1^{\alpha_1} x_2^{\alpha_2} \ldots x_m^{\alpha_m}, \quad x_1^{\alpha_1'} x_2^{\alpha_2'} \ldots x_m^{\alpha_m'}, \ldots$$

die q verschiedenen Produkte von je n der m Variabelen $x_1, x_2, \ldots x_m$ in der lexikographischen Anordnung, so mögen je f aufeinanderfolgende Zeilen der Matrix M der Reihe nach mit

$$Z_{\alpha_1, \alpha_2, \ldots \alpha_m}, \quad Z_{\alpha_1', \alpha_2', \ldots \alpha_m'}, \ldots,$$

ebenso je f aufeinanderfolgende Spalten mit

$$S_{\alpha_1, \alpha_2, \ldots \alpha_m}, \quad S_{\alpha_1', \alpha_2', \ldots \alpha_m'}, \ldots$$

bezeichnet werden. Ferner sei $Q^{n_1, n_2, \ldots n_m}_{n_1', n_2', \ldots n_m'}$ diejenige Teilmatrix f-ten Grades von M, in der sich die f Zeilen $Z_{n_1, n_2, \ldots n_m}$ mit den f Spalten $S_{n_1', n_2', \ldots n_m'}$ schneiden. — Ich bilde nun $\binom{m^2+n-1}{n}$ Matrizen

$$\begin{bmatrix} \varkappa_1 & \varkappa_2 & \ldots & \varkappa_n \\ \lambda_1 & \lambda_2 & \ldots & \lambda_n \end{bmatrix} \qquad (\varkappa_\nu, \ \lambda_\nu = 1, 2, \ldots m)$$

auf folgende Weise.

Die Matrix M, in der

$$Q^{1, 1, 1, \ldots 1, 0, 0, \ldots 0}_{1, 1, 1, \ldots 1, 0, 0, \ldots 0} = \begin{bmatrix} 1 & 2 & \ldots & n \\ \varrho_1 & \varrho_2 & \ldots & \varrho_n \end{bmatrix}'$$

ist, alle übrigen Elemente Null sind, bezeichne ich mit $\begin{bmatrix} 1 & 2 & \ldots & n \\ \varrho_1 & \varrho_2 & \ldots & \varrho_n \end{bmatrix}$.

Es seien ferner $\mu_1, \mu_2, \ldots \mu_n$ n Indices, unter denen n_1 gleich 1, n_2 gleich 2, ... n_m gleich m sind, und es sei $\mu_1 \leqq \mu_2 \leqq \ldots \leqq \mu_n$. Bezeichne ich mit $H_{n_1, n_2, \ldots n_m}$ die Summe der $n_1!\, n_2! \ldots n_m!$ Matrizen $\begin{bmatrix} 1 & 2 & \ldots & n \\ \varrho_1 & \varrho_2 & \ldots & \varrho_n \end{bmatrix}'$, welche Permutationen der auf Seite 27 mit $\mathfrak{S}_{n_1, n_2, \ldots n_m}$ bezeichneten Untergruppe von $\mathfrak{S}_n$ entsprechen, so bedeute $\begin{bmatrix} 1 & 2 & \ldots & n \\ \mu_1 & \mu_2 & \ldots & \mu_n \end{bmatrix}$ diejenige Matrix M, in der

$$Q^{1, 1, \ldots 1, 0, \ldots 0}_{n_1, n_2, \ldots n_n, \ldots n_m} = \frac{1}{\sqrt{n_1!\, n_2! \ldots n_m!}} H_{n_1, n_2, \ldots n_m} = \begin{bmatrix} 1 & 2 & \ldots & n \\ \mu_1 & \mu_2 & \ldots & \mu_n \end{bmatrix}'$$

ist und alle übrigen Elemente verschwinden. Ebenso setze ich $\begin{bmatrix} \mu_1 & \mu_2 & \ldots & \mu_n \\ 1 & 2 & \ldots & n \end{bmatrix}$ derjenigen Matrix M gleich, in der

$$Q^{n_1, n_2, \ldots n_n, \ldots n_m}_{1, 1, \ldots 1, 0, \ldots 0} = \frac{1}{\sqrt{n_1!\, n_2! \ldots n_m!}} H_{n_1, n_2, \ldots n_m} = \begin{bmatrix} \mu_1 & \mu_2 & \ldots & \mu_n \\ 1 & 2 & \ldots & n \end{bmatrix}'$$

ist und alle übrigen Elemente Null sind. Ist ferner $\mu'_1, \mu'_2, \ldots \mu'_n$ eine Permutation der Indices $\mu_1, \mu_2, \ldots \mu_n$ und $\varrho_1, \varrho_2, \ldots \varrho_n$ die inverse Permutation der n Ziffern $1, 2, \ldots n$, so setze ich

$$\begin{bmatrix} 1 & 2 & \ldots & n \\ \mu'_1 & \mu'_2 & \ldots & \mu'_n \end{bmatrix} = \begin{bmatrix} \varrho_1 & \varrho_2 & \ldots & \varrho_n \\ 1 & 2 & \ldots & n \end{bmatrix} \begin{bmatrix} 1 & 2 & \ldots & n \\ \mu_1 & \mu_2 & \ldots & \mu_n \end{bmatrix}$$

$$\begin{bmatrix} \mu'_1 & \mu'_2 & \ldots & \mu'_n \\ 1 & 2 & \ldots & n \end{bmatrix} = \begin{bmatrix} \mu_1 & \mu_2 & \ldots & \mu_n \\ 1 & 2 & \ldots & n \end{bmatrix} \begin{bmatrix} 1 & 2 & \ldots & n \\ \varrho_1 & \varrho_2 & \ldots & \varrho_n \end{bmatrix}.$$

Die noch zu definierenden Matrizen $\begin{bmatrix} \varkappa_1 & \varkappa_2 & \ldots & \varkappa_n \\ \lambda_1 & \lambda_2 & \ldots & \lambda_n \end{bmatrix}$ bestimme ich dann durch die Gleichungen

$$\begin{bmatrix} \varkappa_1 & \varkappa_2 & \ldots & \varkappa_n \\ 1 & 2 & \ldots & n \end{bmatrix} \begin{bmatrix} 1 & 2 & \ldots & n \\ \lambda_1 & \lambda_2 & \ldots & \lambda_n \end{bmatrix} = q_1!\, q_2! \ldots \begin{bmatrix} \varkappa_1 & \varkappa_2 & \ldots & \varkappa_n \\ \lambda_1 & \lambda_2 & \ldots & \lambda_n \end{bmatrix},$$

wo die Zahlen $q_1, q_2, \ldots$ dieselbe Bedeutung haben, wie in der Gleichung (II.).

Ich behaupte, dass alsdann die Matrix

$$T(A) = \sum_{\varkappa, \lambda} \begin{bmatrix} \varkappa_1 & \varkappa_2 & \ldots & \varkappa_n \\ \lambda_1 & \lambda_2 & \ldots & \lambda_n \end{bmatrix} a_{\varkappa_1 \lambda_1} a_{\varkappa_2 \lambda_2} \ldots a_{\varkappa_n \lambda_n}$$

eine aus $A = (a_{\varkappa\lambda})$ gebildete invariante Matrix ist. Dies ist leicht zu sehen; denn aus der Definition der Matrizen $\begin{bmatrix} \varkappa_1 & \varkappa_2 & \ldots & \varkappa_n \\ \lambda_1 & \lambda_2 & \ldots & \lambda_n \end{bmatrix}$

folgt ohne Weiteres, dass für sie die Gleichungen (II.) und ($\overset{*}{\text{Ia}}$.) bestehen. Es gilt aber auch die Gleichung (III.), denn es ist

$$\begin{bmatrix}1 & 2 & \dots & n\\ \mu_1 & \mu_2 & \dots & \mu_n\end{bmatrix}' \begin{bmatrix}\mu_1 & \mu_2 & \dots & \mu_n\\ 1 & 2 & \dots & n\end{bmatrix}' = \frac{1}{n_1!\, n_2! \dots} H^2_{n_1, n_2, \dots n_m} = H_{n_1, n_2, \dots n_m}$$

und daher

$$\begin{bmatrix}1 & 2 & \dots & n\\ \mu_1 & \mu_2 & \dots & \mu_n\end{bmatrix} \begin{bmatrix}\mu_1 & \mu_2 & \dots & \mu_n\\ 1 & 2 & \dots & n\end{bmatrix} = \sum_{\varrho} \begin{bmatrix}1 & 2 & \dots & n\\ \varrho_1 & \varrho_2 & \dots & \varrho_n\end{bmatrix},$$

wo $\varrho_1, \varrho_2, \dots \varrho_n$ alle Permutationen der Gruppe $\mathfrak{S}_{n_1, n_2, \dots}$ durchlaufen.

Aus dem Bestehen der Gleichungen ($\overset{*}{\text{Ia}}$.), (II.) und (III.) folgt aber nach deu Ausführungen des vorigen Paragraphen, dass $T'(A)$ eine invariante Matrix ist.

Die Determinante der so bestimmten Matrix $T'(A)$ wird im Allgemeinen identisch Null sein. Dann können wir sie aber nach § 1 in eine Nullmatrix und eine invariante Form $T(A)$ von nicht identisch verschwindender Determinante zerfällen. Es ist dann leicht zu sehen, dass die zu der invarianten Form $T(A)$ gehörende Darstellung der symmetrischen Gruppe $\mathfrak{S}_n$ der Darstellung (25), von der wir ausgegangen sind, äquivalent sein muss.

§ 16. Wir beweisen jetzt folgenden Satz:

VII. „Zwei invariante Operationen n-ter Ordnung $T(A)$ und $T_1(A)$, deren Determinanten nicht identisch verschwinden, sind äquivalent, wenn die zu ihnen gehörenden Darstellungen der symmetrischen Gruppe n-ten Grades äquivalent sind".

Wir können annehmen, dass die beiden Matrizen

$$T(A) = \sum \begin{bmatrix}\varkappa_1 & \varkappa_2 & \dots & \varkappa_n\\ \lambda_1 & \lambda_2 & \dots & \lambda_n\end{bmatrix} a_{\varkappa_1 \lambda_1} a_{\varkappa_2 \lambda_2} \dots a_{\varkappa_n \lambda_n}$$

und

$$T_1(A) = \sum \begin{Bmatrix}\varkappa_1 & \varkappa_2 & \dots & \varkappa\\ \lambda_1 & \lambda_2 & \dots & \lambda_n\end{Bmatrix} a_{\varkappa_1 \lambda_1} a_{\varkappa_2 \lambda_2} \dots a_{\varkappa_n \lambda_n}$$

auf die kanonische Form gebracht seien; dann möge das Zeichen $\begin{Bmatrix}\varkappa_1 & \varkappa_2, & \dots & \varkappa_n\\ \lambda_1 & \lambda_2 & \dots & \lambda_n\end{Bmatrix}'$ für $T_1(A)$ die analoge Bedeutung haben, wie das Zeichen $\begin{bmatrix}\varkappa_1 & \varkappa_2 & \dots & \varkappa_n\\ \lambda_1 & \lambda_2 & \dots & \lambda_n\end{bmatrix}'$ für $T(A)$ (vergl. Seite 22).

Infolge der Voraussetzung des Satzes kann ich $T_1(A)$ so transformieren, dass für jede Permutation $\varrho_1, \varrho_2, \dots \varrho_n$ der n Ziffern 1, 2, ... n

(26) $$\begin{Bmatrix}1 & 2 & \dots & n \\ \varrho_1 & \varrho_2 & \dots & \varrho_n\end{Bmatrix}' = \begin{bmatrix}1 & 2 & \dots & n \\ \varrho_1 & \varrho_2 & \dots & \varrho_n\end{bmatrix}'$$

wird, und zwar wird hierbei die kanonische Form der Matrix $T_1(A)$ nicht aufgehoben. Dann wird auch nach Gleichung (III.) für beliebige Indices $\mu_1, \mu_2, \dots \mu_n$, die der Bedingung $\mu_1 \leqq \mu_2 \leqq \dots \leqq \mu_n$ genügen,

(27) $$\begin{Bmatrix}1 & 2 & \dots & n \\ \mu_1 & \mu_2 & \dots & \mu_n\end{Bmatrix}' \begin{Bmatrix}\mu_1 & \mu_2 & \dots & \mu_n \\ 1 & 2 & \dots & n\end{Bmatrix}' = \begin{bmatrix}1 & 2 & \dots & n \\ \mu_1 & \mu_2 & \dots & \mu_n\end{bmatrix}' \begin{bmatrix}\mu_1 & \mu_2 & \dots & \mu_n \\ 1 & 2 & \dots & n\end{bmatrix}' = H_{n_1, n_2, \dots n_m},$$

wo $n_1, n_2, \dots n_m$ dieselbe Bedeutung haben, wie auf Seite 24.

Ausserdem besteht, da nach Satz VI. die Charakteristiken von $T(A)$ und $T_1(A)$ einander gleich sein müssen, die Gleichung

(28) $$\begin{bmatrix}\mu_1 & \mu_2 & \dots & \mu_n \\ \mu_1 & \mu_2 & \dots & \mu_n\end{bmatrix}' = \begin{Bmatrix}\mu_1 & \mu_2 & \dots & \mu_n \\ \mu_1 & \mu_2 & \dots & \mu_n\end{Bmatrix}' = E_g,$$

wo $g = r_{n_1, n_2, \dots n_m}$ den Koeffizienten von $\omega_1^{n_1} \omega_2^{n_2} \dots \omega_m^{n_m}$ in der Charakteristik von $T(A)$ bedeutet. Durch eine simultane Transformation der Matrizen $T(A)$ und $T_1(A)$ kann ich erreichen, dass

(29) $$\frac{1}{n_1!\, n_2! \dots} H_{n_1, n_2, \dots n_m} = \begin{pmatrix}E_g & \\ & N\end{pmatrix}$$

wird, wo N eine Nullmatrix ist; diese Transformation kann so ausgeführt werden, dass die Gleichungen (26) bis (28) bestehen bleiben. Nun ist aber, wie man sich leicht überzeugt,

$$\begin{bmatrix}\mu_1 & \mu_2 & \dots & \mu_n \\ 1 & 2 & \dots & n\end{bmatrix}' H_{n_1, n_2, \dots n_m} = n_1!\, n_2! \dots \begin{bmatrix}\mu_1 & \mu_2 & \dots & \mu_n \\ 1 & 2 & \dots & n\end{bmatrix}',$$

$$H_{n_1, n_2, \dots n_m} \begin{bmatrix}1 & 2 & \dots & n \\ \mu_1 & \mu_2 & \dots & \mu_n\end{bmatrix}' = n_1!\, n_2! \dots \begin{bmatrix}1 & 2 & \dots & n \\ \mu_1 & \mu_2 & \dots & \mu_n\end{bmatrix}'.$$

Aus diesen Gleichungen folgt wegen (29), dass in der Matrix $\begin{bmatrix}1 & 2 & \dots & n \\ \mu_1 & \mu_2 & \dots & \mu_n\end{bmatrix}'$ die $f-g$ letzten Zeilen, in der Matrix $\begin{bmatrix}\mu_1 & \mu_2 & \dots & \mu_n \\ 1 & 2 & \dots & n\end{bmatrix}'$ die $f-g$ letzten Spalten lauter Nullen enthalten. Hierbei bedeutet f den Grad der Matrix $\begin{bmatrix}1 & 2 & \dots & n \\ 1 & 2 & \dots & n\end{bmatrix}'$ oder, was dasselbe ist, den Koeffizienten von $\omega_1\, \omega_2 \dots \omega_n$ in der Charakteristik von $T(A)$. Es sei also

$$\begin{bmatrix}1 & 2 & \dots & n\\ \mu_1 & \mu_2 & \dots & \mu_n\end{bmatrix}' = \begin{pmatrix} c_{11}, c_{12}, \dots c_{1g}\\ \cdot\quad\cdot\quad\cdot\quad\cdot \\ c_{g1}\, c_{g2} \dots c_{gg}\\ 0\;\; 0 \;\dots 0\\ \cdot\quad\cdot\quad\cdot\quad\cdot \\ 0\;\; 0 \;\dots 0\end{pmatrix}, \begin{bmatrix}\mu_1 & \mu_2 & \dots & \mu_n\\ 1 & 2 & \dots & n\end{bmatrix}' = \begin{pmatrix} d_{11}, d_{12}, \dots d_{1g}\, 0 \dots 0\\ d_{21}, d_{22}, \dots d_{2g}\, 0\dots 0\\ \cdot\quad\cdot\quad\cdot\quad\cdot\quad\cdot \\ d_{g1}, d_{g2}, \dots d_{gg}\, 0 \dots 0\end{pmatrix};$$

bezeichne ich die Matrix g-ten Grades (c_{ik}) mit C, die Matrix (d_{ik}) mit D, so ist wegen

$$\begin{bmatrix}\mu_1 & \mu_2 & \dots & \mu_n\\ 1 & 2 & \dots & n\end{bmatrix}\begin{bmatrix}1 & 2 & \dots & n\\ \mu_1 & \mu_2 & \dots & \mu_n\end{bmatrix} = n_1!\, n_2! \dots \begin{bmatrix}\mu_1 & \mu_2 \dots \mu_n\\ \mu_1 & \mu_2 \dots \mu_n\end{bmatrix}$$

$$DC = n_1!\, n_2! \dots E_g.$$

Ebenso ist, wenn C_1 und D_1 für $T_1(A)$ dieselbe Bedeutung haben, wie C und D für $T(A)$,

(30) $$D_1 C_1 = n_1!\, n_2! \dots E_g = DC.$$

Ist jetzt M eine Matrix desselben Grades wie $T(A)$, welche in allen den Teilmatrizen $M^{\alpha_1, \alpha_2, \dots \alpha_m}_{\alpha_1, \alpha_2, \dots \alpha_m}$ von $T(A)$[1]) entsprechenden Rechtecken Einheitsmatrizen, in dem der Teilmatrix $M^{n_1, n_2, \dots n_m}_{n_1, n_2, \dots n_m}$ entsprechenden dagegen die Matrix g-ten Grades $Q = C_1 C^{-1}$, sonst überall Nullen enthält, so ist $|M| = |Q| = |C_1|\,|C^{-1}|$, also wegen (30) von Null verschieden. Ich ersetze $T_1(A)$ durch $M^{-1} T_1(A) M$; dann bleiben die Gleichungen (26) bis (29) ungeändert, es geht aber C_1 über in

$$Q^{-1} C_1 = C C_1^{-1} C_1 = C,$$

D_1 in

$$D_1 Q = D_1 C_1 C^{-1} = n_1!\, n_2! .. \; C^{-1} = D.$$

Es wird also

(31) $$\begin{bmatrix}1 & 2 & \dots & n\\ \mu_1, & \mu_2, & \dots & \mu_n\end{bmatrix}' = \begin{Bmatrix}1 & 2 & \dots & n\\ \mu_1 & \mu_2 & \dots & \mu_n\end{Bmatrix}', \begin{bmatrix}\mu_1 & \mu_2 & \dots & \mu_n\\ 1 & 2 & \dots & n\end{bmatrix}' = \begin{Bmatrix}\mu_1 & \mu_2 & \dots & \mu_n\\ 1 & 2 & \dots & n\end{Bmatrix}.$$

Die analoge Transformation können wir für jede andere Indicesreihe $\mu_1', \mu_2', \dots \mu_n'$ ausführen, ohne dass dadurch die Gleichungen (26) und (31) ihre Gültigkeit verlieren.

---

1) Vgl. Seite 21.

Indem wir dieses Verfahren fortsetzen, können wir erreichen, dass die Gleichungen (31) für jede Wahl der Indices $\mu_1, \mu_2, \ldots \mu_n$ bestehen. In Verbindung mit (26) folgt aber dann nach (II.), dass für jede Wahl der Indices $\varkappa_\nu$ und λ_ν

$$\begin{bmatrix} \varkappa_1 \ \varkappa_2 \ldots \varkappa_n \\ \lambda_1 \ \lambda_2 \ldots \lambda_n \end{bmatrix} = \begin{Bmatrix} \varkappa_1 \ \varkappa_2 \ldots \varkappa_n \\ \lambda_1 \ \lambda_2 \ldots \lambda_n \end{Bmatrix}$$

wird. Ich kann also die beiden Matrizen $T(A)$ und $T_1(A)$ durch ihnen äquivalente Matrizen $T'(A)$, resp. $T_1'(A)$ ersetzen, die einander gleich sind; daraus folgt aber, wie zu beweisen war, dass $T(A)$ und $T_1(A)$ äquivalent sind.

In Verbindung mit Satz VI. ergiebt sich hieraus:

VIII. „Zwei invariante Operationen gleichen Grades sind dann und nur dann äquivalent, wenn ihre Charakteristiken einander gleich sind".

Ich bemerke aber, dass dieser Satz vorläufig nur für den Fall $n \leqq m$ bewiesen ist.

§ 17. Aus den Ausführungen der §§ 15 und 16 ergeben sich leicht folgende Sätze über die Zerlegbarkeit der invarianten Operationen.

IX. „Eine invariante Operation $T(A)$ der Ordnung n ist nur auf eine einzige Art in irreduktible Operationen zerlegbar; sie ist zerlegbar oder nicht, je nachdem die zu ihr gehörende Darstellung der symmetrischen Gruppe n-ten Grades $\mathfrak{S}_n$ reduktibel oder irreduktibel ist".

Es möge nämlich die zu $T(A)$ gehörende Darstellung (S) der Gruppe $\mathfrak{S}_n$ in die beiden anderen (S_1) und (S_2) zerlegbar sein. Dann können wir nach § 15 zwei invariante Operationen von nicht identisch verschwindender Determinante $T_1(A)$ und $T_2(A)$ konstruieren, welche die Darstellungen (S_1), resp. (S_2) erzeugen. Alsdann gehört zu der Operation $\begin{pmatrix} T_1(A) & \\ & T_2(A) \end{pmatrix}$ die mit $\begin{pmatrix} (S_1) & \\ & (S_2) \end{pmatrix}$ zu bezeichnende Darstellung der Gruppe $\mathfrak{S}_n$. Daher müssen nach Satz VII. die Operationen $T(A)$ und $\begin{pmatrix} T_1(A) & \\ & T_2(A) \end{pmatrix}$ äquivalent sein, d. h. $T(A)$ ist in die Operationen $T_1(A)$ und $T_2(A)$ zerlegbar.

Ist aber die Darstellung (S) nicht zerlegbar, so ist unmittelbar ersichtlich, dass auch $T(A)$ nicht zerlegbar sein kann.

Es möge nun $T(A)$ auf zwei verschiedene Arten in irreduktible Operationen zerlegbar sein, etwa in

$$T_1(A),\ T_2(A),\ \ldots\ T_p(A)$$

und

$$T'_1(A),\ T'_2(A),\ \ldots\ T'_q(A).$$

Es gehöre zu $T_\mu(A)$ die Darstellung (S_μ), zu $T'_\nu(A)$ die Darstellung (S'_ν) der Gruppe $\mathfrak{S}_n$. Dann sind auch die Darstellungen (S_μ) und (S'_ν) irreduktibel.

Nun ist aber jede Darstellung einer endlichen Gruppe durch lineare Substitutionen nur auf eine einzige Art in primitive Darstellungen zerlegbar [1]). Daher muss jede Darstellung (S_μ) einer Darstellung (S'_ν) äquivalent sein und umgekehrt. Daraus folgt aber nach Satz VII., dass auch $T_\mu(A)$ und $T'_\nu(A)$ äquivalent sein müssen. Dies ist aber der Inhalt unseres Satzes.

Die Anzahl der verschiedenen primitiven Darstellungen einer endlichen Gruppe $\mathfrak{H}$ durch lineare Substitutionen ist, wie die Herren Molien und Frobenius loc. cit. gezeigt haben, gleich der Anzahl der Klassen gleichberechtigter Elemente, in die die Gruppe $\mathfrak{H}$ zerfällt. Diese Anzahl ist für die symmetrische Gruppe $\mathfrak{S}_n$ gleich der Anzahl k der Zerlegungen der Zahl n in gleiche oder verschiedene positive Summanden. Aus Satz IX. folgt daher, dass die Zahl k auch die Anzahl der verschiedenen irreduktiblen Operationen n-ter Ordnung angiebt.

Abschnitt V.

§ 18. Es entsteht nun die Aufgabe, die Charakteristiken der irreduktiblen invarianten Operationen explicite darzustellen.

Es sei wieder, wie in § 5, $C_\varkappa A$ die $\varkappa$-te aus A gebildete Determinantentransformation, $c_\varkappa$ die $\varkappa$-te elementarsymmetrische Funktion der m charakteristischen Wurzeln ω, $\omega_2, \ldots \omega_m$ der Matrix A. Dann ist, wenn

$$\varkappa_1 + \varkappa_2 + \ldots + \varkappa_\varrho = n \qquad (\varkappa_\nu > 0) \tag{32}$$

ist, $C(A) = C_{\varkappa_1} A \times C_{\varkappa_2} A \times \ldots \times C_{\varkappa_\varrho} A$ eine invariante Operation n-ter Ordnung, deren Charakteristik gleich $c_{\varkappa_1}\, c_{\varkappa_2} \ldots c_{\varkappa_\varrho}$ ist.

Ich nehme an, es sei $\varkappa_1 \geqq \varkappa_2 \geqq \varkappa_3 \geqq \ldots \geqq \varkappa_\varrho$, und es seien unter den Zahlen $\varkappa$ genau α_1 gleich 1, α_2 gleich 2, u. s. w.

---

1) Molien und Frobenius, loc. cit.

Dann ist $1\alpha_1 + 2\alpha_2 + 3\alpha_3 + \cdots = n$; setzen wir

$$\lambda_1 = \alpha_1 + \alpha_2 + \alpha_3 + \cdots, \quad \lambda_2 = \alpha_2 + \alpha_3 + \cdots, \quad \lambda_3 = \alpha_3 + \alpha_4 + \cdots,$$

so wird, wenn σ unter den Zahlen λ positiv sind,

$$(33) \qquad \lambda_1 + \lambda_2 + \cdots + \lambda_\sigma = n. \qquad (\lambda_1 \geqq \lambda_2 \geqq \ldots \geqq \lambda_\sigma)$$

Die Entwickelung von $c_{\varkappa_1} c_{\varkappa_2} c_{\varkappa_3} \ldots$ nach Potenzen von $\omega_1, \omega_2, \ldots \omega_m$ enthält dann, wenn die Produkte $\omega_1^{\mu_1} \omega_2^{\mu_2} \ldots$ nach dem Gauss'schen lexikographischen Prinzip angeordnet werden, als höchstes Glied das Produkt $\omega_1^{\lambda_1} \omega_2^{\lambda_2} \ldots \omega_\sigma^{\lambda_\sigma}$, und dieses Glied tritt nur einmal auf. Denke ich mir die Operation $C(A)$ in ihre irreduktiblen Teile zerlegt, so muss unter diesen eine und nur eine Operation auftreten, deren Charakteristik mit dem höchsten Glied $\omega_1^{\lambda_1} \omega_2^{\lambda_2} \ldots \omega_\sigma^{\lambda_\sigma}$ beginnt. Da nun aber die Anzahl k der verschiedenen irreduktiblen Operationen n-ter Ordnung gleich ist der Anzahl der Lösungen der Gleichung (32), so giebt es für jedes der k Produkte n-ter Dimension

$$\omega_1^{\nu_1} \omega_2^{\nu_2} \ldots \omega_m^{\nu_m} \qquad (\nu_1 \geqq \nu_2 \geqq \nu_3 \geqq \ldots \geqq \nu_m)$$

eine und nur eine irreduktible Operation, deren Charakteristik dieses Produkt als höchstes Glied und zwar mit dem Koeffizienten 1 enthält. Die so bestimmte irreduktible Operation wollen wir mit $T_{\nu_1, \nu_2, \ldots}(A)$, ihre Charakteristik mit $\Phi_{\nu_1, \nu_2, \ldots}$ bezeichnen.

Offenbar muss dann jede invariante Operation, deren Charakteristik mit $\omega_1^{\nu_1} \omega_2^{\nu_2} \ldots$ beginnt, die Operation $T_{\nu_1, \nu_2, \ldots}(A)$ enthalten.

§ 19. Wir wollen die Zerlegung (33) der Zahl n in Summanden die zu der Zerlegung (32) associierte nennen. — Es ist dann leicht zu sehen, dass umgekehrt auch die Zerlegung (32) die zu der Zerlegung (33) associierte ist. Dies sieht man am einfachsten ein, indem man eine Matrix n-ten Grades

$$(34) \qquad \begin{matrix} \pi_{11}, & \pi_{12}, & \ldots & \pi_{1n} \\ \pi_{21}, & \pi_{22}, & \ldots & \pi_{2n} \\ \cdot & \cdot & \ldots & \cdot \\ \pi_{n1}, & \pi_{n2}, & \ldots & \pi_{nn} \end{matrix}$$

betrachtet, in deren ν-ter Zeile die $\varkappa_\nu$ ersten Elemente von Null verschieden, die übrigen gleich Null sind[1]); hierbei hat man, wenn

1) An Stelle der Elemente $\pi_{\iota\varkappa}$ könnte man auch Punkte setzen.

$\varrho < n$ ist, $\varkappa_{\varrho+1} = \varkappa_{\varrho+2} = \cdots = \varkappa_n = 0$ zu setzen. Die Zahlen $\lambda_1, \lambda_2, \lambda_3, \ldots$ geben uns nämlich, wie man leicht sieht, dann an, wie viele Elemente in der ersten, resp. zweiten, dritten, ... Spalte der Matrix von Null verschieden sind; auch hier ist $\lambda_{\sigma+1} = \lambda_{\sigma+2} = \cdots = \lambda_n = 0$ zu setzen. Vertauscht man nun die Zeilen und Spalten der Matrix, so sieht man sofort ein, dass $\varkappa_\nu$ für die Zerlegung (33) die analoge Bedeutung hat, wie λ_ν für die Zerlegung (32).

Es kann eintreten, dass die Zerlegung (32) mit der ihr associierten identisch ist, dass also

$$\lambda_1 = \varkappa_1, \ \lambda_2 = \varkappa_2, \ldots \lambda_\sigma = \lambda_\varrho = \varkappa_\varrho$$

wird. Eine solche Zerlegung der Zahl n will ich eine zweiseitige nennen. Setze ich in diesem Falle

$$a_1 = 2\varkappa_1 - 1, \ a_2 = 2\varkappa_2 - 3, \ldots a_n = 2\varkappa_n - 2n + 1,$$

so wird, wenn a_τ die letzte nicht negative dieser Zahlen ist, $a_1 > a_2 > a_3 > \ldots > a_\tau > 0$ und

$$a_1 + a_2 + \cdots + a_\tau = n. \tag{35}$$

Dies sieht man wieder durch Betrachtung der Matrix (34) leicht ein. Denn a_ν ($\nu = 1, 2, \ldots \tau$) giebt uns die Anzahl der nicht verschwindenden Elemente $\pi_{\iota\varkappa}$ an, welche auf der gebrochenen Linie L_ν liegen, die durch die beiden von $\pi_{\nu\nu}$ ausgehenden horizontal nach rechts und senkrecht nach unten verlaufenden Halbstrahlen gebildet wird. Daher ist die Summe der positiven a_ν gleich der Anzahl der nicht verschwindenden Elemente $\pi_{\iota\varkappa}$, also gleich n.

Es entspricht also jeder zweiseitigen Zerlegung von n in gleiche oder verschiedene positive Summanden eine Zerlegung (35) in verschiedene ungerade Summanden.

Umgekehrt erhält man auch leicht aus jeder Zerlegung (35) eine zweiseitige Zerlegung der Zahl n.

Die Anzahl der zweiseitigen Zerlegungen ist daher gleich der Anzahl der Zerlegungen von n in verschiedene ungerade Summanden.

§ 20. Sind

$$\nu_1 + \nu_2 + \nu_3 + \cdots = n \qquad (\nu_1 \geqq \nu_2 \geqq \nu_3 \geqq \ldots)$$

und

$$\nu'_1 + \nu'_2 + \nu'_3 + \cdots = n \qquad (\nu'_1 \geqq \nu'_2 \geqq \nu'_3 \geqq \ldots)$$

zwei Zerlegungen der Zahl n, so will ich die erste Zerlegung von höherer Ordnung nennen, als die zweite, wenn die erste nicht verschwindende der Differenzen

$$\nu_1 - \nu_1', \quad \nu_2 - \nu_2', \quad \nu_3 - \nu_3', \ldots$$

positiv ist; ich schreibe dafür abgekürzt

$$(\nu_1, \nu_2, \nu_3, \ldots) > (\nu_1', \nu_2', \nu_3', \ldots).$$

Es möge nun

$$\omega_1^{\lambda_1'} \omega_2^{\lambda_2'} \omega_3^{\lambda_3'} \ldots \qquad (\lambda_1' \geqq \lambda_2' \geqq \lambda_3' \geqq \ldots)$$

ein von dem höchsten Glied $\omega_1^{\lambda_1} \omega_2^{\lambda_2} \omega_3^{\lambda_3} \ldots$ verschiedenes Glied in der Entwickelung von $c_{\varkappa_1} c_{\varkappa_2} c_{\varkappa_3} \ldots c_{\varkappa_\varrho}$ sein.

Ist dann

$$(\varkappa_1', \varkappa_2', \varkappa_3', \ldots) \qquad (\varkappa_1' \geqq \varkappa_2' \geqq \varkappa_3' \geqq \ldots)$$

die zu $(\lambda_1', \lambda_2', \lambda_3', \ldots)$ associierte Zerlegung, so behaupte ich, dass

$$(36) \qquad (\varkappa_1', \varkappa_2', \varkappa_3', \ldots) > (\varkappa_1, \varkappa_2, \varkappa_3, \ldots)$$

sein muss. — Es ist nämlich, wenn ich das Produkt $\omega_1 \omega_2 \ldots \omega_\nu$ mit Ω_ν bezeichne,

$$\omega_1^{\lambda_1'} \omega_2^{\lambda_2'} \cdots \omega_n^{\lambda_n'} = \Omega_{\varkappa_1'} \Omega_{\varkappa_2'} \Omega_{\varkappa_3'} \ldots$$

Setze ich

$$\omega_{\varkappa_1+1} = \omega_{\varkappa_1+2} = \cdots = \omega_n = 0,$$

so wird

$$c_{\varkappa_1} c_{\varkappa_2} \ldots c_{\varkappa_\varrho} = \omega_1 \omega_2 \ldots \omega_{\varkappa_1} c'_{\varkappa_2} c'_{\varkappa_3} \ldots c'_{\varkappa_\varrho},$$

wo $c'_\varkappa$ die $\varkappa$-te elementarsymmetrische Funktion der $\varkappa_1$ Grössen $\omega_1, \omega_2, \ldots \omega_{\varkappa_1}$ bedeutet. Daher ist das Produkt $\omega_1^{\lambda_1'} \omega_2^{\lambda_2'} \ldots$ entweder Null oder durch $\Omega_{\varkappa_1}$ teilbar, also $\varkappa_1' \geqq \varkappa_1$. Ist $\varkappa_1' = \varkappa_1$, so muss $\Omega_{\varkappa_2'} \Omega_{\varkappa_3'} \ldots$ in der Entwickelung von $c'_{\varkappa_2} c'_{\varkappa_3} \ldots c'_{\varkappa_\varrho}$ auftreten; wir schliessen dann ebenso, dass $\varkappa_2' \geqq \varkappa_2$ sein muss. Allgemein sehen wir, dass, wenn für irgend ein ν

$$\varkappa_1' = \varkappa_1, \quad \varkappa_2' = \varkappa_2, \ldots \quad \varkappa_{\nu-1}' = \varkappa_{\nu-1}$$

ist, $\varkappa_\nu' \geqq \varkappa_\nu$ sein muss. Das ist aber der Inhalt der Formel (36).

§ 21. Es sei nun $T(A)$ eine invariante Operation n-ter Ordnung und $(S) = (S_1, S_2, \ldots S_n)$ die zu ihr gehörende Darstellung der symmetrischen Gruppe n-ten Grades $\mathfrak{S}_n$. Ich bezeichne die Gesamtheit der Matrizen S, welche den geraden Permutationen von $\mathfrak{S}_n$ entsprechen, mit (R), die Gesamtheit der übrigen mit (Q) und setze abgekürzt

$$(S) = (R, Q). \tag{37}$$

Ersetze ich nun jede Matrix Q durch $-Q$, so stellen auch die Matrizen (R) und $(-Q)$ die Gruppe $\mathfrak{S}_n$ dar. Dann nenne ich jede der Darstellung

$$(S') = (R, -Q) \tag{38}$$

äquivalente Darstellung der Gruppe $\mathfrak{S}_n$ eine zu der Darstellung (37) associierte[1]). Nun gehört aber nach § 17 auch zu der Darstellung (S') eine ganz bestimmte invariante Operation $T'(A)$ von nicht verschwindender Determinante. Ich will dann je zwei Operationen $T_1(A)$ und $T_1'(A)$, die zu $T(A)$, resp. zu $T'(A)$ äquivalent sind, associierte Operationen nennen.

Es gilt nun der Satz:

X. „Ist die Operation $T(A)$ durch die Operation $T_1(A)$ teilbar, so ist jede zu $T(A)$ associierte Operation $T'(A)$ durch die zu $T_1(A)$ associierte teilbar. Daher sind $T(A)$ und $T'(A)$ entweder beide reduktibel oder beide irreduktibel“.

Der Beweis dieses Satzes ergiebt sich unmittelbar aus der Betrachtung der beiden Darstellungen (37) und (38) der Gruppe $\mathfrak{S}_n$. Denn offenbar entspricht jeder Zerlegung von (37) in primitive Darstellungen eine analoge Zerlegung von (38).

§ 22. Die Charakteristiken zweier associierter Operationen $T(A)$ und $T'(A)$ stehen in einem einfachen Zusammenhang. Es seien nämlich

$$(S) = (R, Q), \quad (S') = (R', Q')$$

die zu den invarianten Operationen $T(A)$, resp. $T'(A)$ gehörenden associierten Darstellungen der symmetrischen Gruppe $\mathfrak{S}_n$. Bezeichne ich die Spuren der Matrizen S mit $\chi(S)$, die der Matrizen S' mit $\psi(S')$, so ist offenbar, wenn R_ν und R'_ν, resp. Q_ν und Q'_ν dem selben Element der Gruppe entsprechen,

$$\chi(R_\nu) = \psi(R'_\nu), \quad \chi(Q_\nu) = -\psi(Q'_\nu). \tag{a}$$

1) Vergl. Frobenius, Sitzungsberichte der Ak. der Wiss. zu Berlin, 1900, S. 516.

Umgekehrt sind, wenn diese Gleichungen bestehen, (S) und (S') associierte Darstellungen der Gruppe $\mathfrak{S}_n$.

Bezeichne ich mit Φ und Φ' die Charakteristiken der Operationen $T(A)$, resp. $T'(A)$, so ist nach § 12

$$\Phi = \sum \frac{\chi(S_{\alpha_1, \alpha_2 \ldots})}{1^{\alpha_1} \alpha_1! \, 2^{\alpha_2} \alpha_2! \ldots} s_1^{\alpha_1} s_2^{\alpha_2} \ldots,$$

$$(\alpha_1 + 2\alpha_2 + \ldots = n)$$

$$\Phi' = \sum \frac{\psi(S'_{\alpha_1, \alpha_2 \ldots})}{1^{\alpha_1} \alpha_1! \, 2^{\alpha_2} \alpha_2! \ldots} s_1^{\alpha_1} s_2^{\alpha} \ldots,$$

wo $S_{\alpha_1, \alpha_2 \ldots}$ und $S'_{\alpha_1, \alpha_2 \ldots}$ einer Permutation der n Ziffern $1, 2, \ldots n$ entsprechen, die aus α_1 Zykeln der Ordnung 1 besteht, α_2 Zykeln der Ordnung 2, u. s. w. Nun gehört aber eine solche Permutation zur alternierenden Gruppe oder nicht, je nachdem $(-1)^{\alpha_2 + \alpha_4 + \ldots}$ gleich $+1$ oder gleich -1 ist.

Wegen (a) ist daher

$$\chi(S_{\alpha_1, \alpha_2 \ldots}) = (-1)^{\alpha_2 + \alpha_4 + \ldots} \psi(S'_{\alpha_1, \alpha_2, \ldots}).$$

Wir sehen also, dass die Entwickelung von Φ nach den Potenzsummen s_μ in die Entwickelung von Φ' übergeht, indem man s_μ durch $(-1)^{\mu-1} s_\mu$ ersetzt. Diese Bedingung ist notwendig und hinreichend, damit die zu den Charakteristiken Φ und Φ' gehörenden invarianten Operationen associiert seien.

Daraus schliesst man nach § 7 sofort:

XI. „Sind die $2q$ invarianten Operationen $T_\nu(A)$ und $T'_\nu(A)$ $(\nu = 1, 2, \ldots q)$ einander paarweise associiert, so stellen auch die beiden Produkttransformationen

$$T_1(A) \times T_2(A) \times \ldots \times T_q(A)$$

und

$$T'_1(A) \times T'_2(A) \times \ldots \times T'_q(A)$$

associierte Operationen dar".

Hierbei wird natürlich vorausgesetzt, dass die Summe der Ordnungen der Operationen $T_\nu(A)$ nicht grösser ist, als der Grad m der Matrix A.

Besonders hervorzuheben ist, dass die ν-te Potenztransformation $P_\nu A$ und die ν-te Determinantentransformation $C_\nu A$ zwei associierte Operationen repräsentieren.

Es ist nämlich bekanntlich für die ν-te elementarsymmetrische Funktion

$$c_\nu = \sum \frac{(-1)^{\alpha_2+\alpha_4+\cdots}}{\alpha_1!\,\alpha_2!\,\alpha_3!\ldots} \left(\frac{s_1}{1}\right)^{\alpha_1} \left(\frac{s_2}{2}\right)^{\alpha_2} \left(\frac{s_3}{3}\right)^{\alpha_3} \cdots,$$
$$(\alpha_1 + 2\alpha_2 + 3\alpha \ldots = \nu).$$

Andererseits ist aber, wenn p_ν die ν-te Wronski'sche Funktion Aleph bedeutet (vergl. Seite 30),

$$p_\nu = \sum \frac{1}{\alpha_1!\,\alpha_2!\,\alpha_3!\ldots} \left(\frac{s_1}{1}\right)^{\alpha_1} \left(\frac{s_2}{2}\right)^{\alpha_2} \left(\frac{s_3}{3}\right)^{\alpha_3} \cdots,$$
$$(\alpha_1 + 2\alpha_2 + 3\alpha_3 + \ldots = \nu).$$

Es geht also in der That die Entwickelung der Charakteristik c_ν der Operation $C_\nu A$ in die Charakteristik p_ν der Operation $P_\nu A$ über, wenn man $(-1)^{\mu-1}s_\mu$ für s_μ substituiert.

Allgemein folgt hieraus, dass, wenn die Charakteristik einer invarianten Operation $T(A)$, als Funktion von $c_1, c_2, \ldots c_n$ dargestellt, die Form $F(c_1, c_2, \ldots c_n)$ hat, die Charakteristik der zu $T(A)$ associierten Operation gleich $F(p_1, p_2, \ldots p_n)$ ist.

§ 23. Mit Hülfe dieses Satzes können wir die Charakteristiken der irreduktiblen Operationen $T_{\lambda_1, \lambda_2, \ldots}(A)$ bestimmen.

Wir gehen von der Operation $C_n A$ aus; dieselbe ist irreduktibel, weil es nicht möglich ist, ihre Charakteristik c_n als Summe zweier symmetrischer Funktionen mit positiven ganzen Koeffizienten darzustellen. Folglich ist auch die zu ihr associierte Operation $P_n A$ irreduktibel; und zwar haben wir zu setzen

$$C_n A = T_{1,1\ldots1}(A), \quad P_n A = T_n(A).$$

Jetzt bilde ich die Produkttransformation

$$C_1 A \times C_{n-1} A, \quad (C_1 A = P_1 A = A).$$

Die zu ihr associierte Operation ist $P_1 A \times P_{n-1} A$, deren Charakteristik gleich $p_1 p_{n-1}$ ist. Nun ist aber das höchste Glied von $p_1 p_{n-1}$ gleich ω_1^n; daher enthält $P_1 A \times P_{n-1} A$ die Operation $P\ A$, d. h. sie ist zerlegbar in der Operation $P_n A$ und eine zweite, die ich der Uebersicht wegen mit

$$(39) \qquad P_1 A \times P_{n-1} A - P_n A$$

bezeichnen will. Nach Satz X. muss $C_1 A \times C_{n-1} A$ zerlegbar sein in $C_n A$ und eine zweite Operation

$$(40) \qquad C_1 A \times C_{n-1} A - C_n A,$$

die der Operation (39) associiert ist. Die Charakteristiken von (39) und (40) sind

(39′) $$p_1 p_{n-1} - p_n = \begin{vmatrix} p_{n-1} & p_n \\ p_0 & p_1 \end{vmatrix}, \qquad (p_0 = 1)$$

resp.

(40′) $$c_1 c_{n-1} - c_n = \begin{vmatrix} c_{n-1} & c_n \\ c_0 & c_1 \end{vmatrix} \qquad (c_0 = 1).$$

Ich behaupte nun, dass die Operation (39) und (40) irreduktibel sind. Denn wäre die Operation (40), deren Charakteristik mit dem höchsten Glied $\omega_1^2 \omega_2 \omega_3 \dots \omega_{n-1}$ beginnt, nicht äquivalent $T_{2,1,1,\dots 1}(A)$, so müsste sie $C_n A$, und also die Operation (39) $P_n A$ enthalten. Dies ist aber nicht möglich, weil die symmetrische Funktion (39′) das Glied ω_1^n nicht enthält, sondern mit dem höchsten Glied $\omega_1^{n-1}\omega_2$ beginnt. Es ist daher (39) äquivalent $T_{n-1,1}(A)$.

Ich hebe hervor, dass die Zerlegungen

$$n = n \text{ und } n = 1+1+1+\dots+1,$$

resp.

$$n = (n-1)+1 \text{ und } n = 2+1+1+\dots+1,$$

denen associierte Operationen $T_{\lambda_1, \lambda_2, \dots}(A)$ entsprechen, zu gleicher Zeit selbst associierte Zerlegungen der Zahl n sind.

Allgemein beweise ich:

XII. „Sind

(32) $$\varkappa_1 + \varkappa_2 + \dots + \varkappa_\varrho = n \qquad (\varkappa_1 \geqq \varkappa_2 \geqq \dots \geqq \varkappa_\varrho)$$

und

(33) $$\lambda_1 + \lambda_2 + \dots \lambda_\sigma = n \qquad (\lambda_1 \geqq \lambda_2 \geqq \dots \geqq \lambda_\sigma)$$

zwei associierte Zerlegungen der Zahl n, so sind $T_{\varkappa_1, \varkappa_2, \dots \varkappa_\sigma}(A)$ und $T_{\lambda_1, \lambda_2, \dots \lambda_\sigma}(A)$ associierte Operationen. Die Charakteristik $\Phi_{\lambda_1, \lambda_2, \dots \lambda_\sigma}$ von $T_{\lambda_1, \lambda_2, \dots \lambda_\sigma}(A)$ ist gleich der Determinante

(41) $$\begin{vmatrix} p_{\lambda_1}, & p_{\lambda_1+1}, \dots & p_{\lambda_1+\sigma-1} \\ p_{\lambda_2-1}, & p_{\lambda_2}, \dots & p_{\lambda_2+\sigma-2} \\ \cdot & \cdot & \cdot \\ p_{\lambda_\sigma-\sigma+1}, & p_{\lambda_\sigma-\sigma+2}, \dots & p_{\lambda_\sigma} \end{vmatrix} = \left| p_{\lambda_\alpha - \alpha + \beta} \right|^{\alpha}.$$

Hierin sind alle p_α mit negativem Index gleich Null zu setzen.

Ich nehme an, der erste Teil dieses Satzes sei bereits für alle Zerlegungen $(\lambda_1', \lambda_2', \dots)$ bewiesen, die von niedrigerer Ordnung sind, als die Zerlegung (33). Diese Voraussetzung ist berechtigt, weil sie für die beiden niedrigsten Zerlegungen $n = 1 + 1 + \cdots + 1$ und $n = 2 + 1 + \cdots + 1$ erfüllt ist.

Auf Grund dieser Annahme können wir schliessen, dass, wenn

$$(\mu_1, \mu_2, \ldots) \geqq (\lambda_1, \lambda_2, \ldots) \qquad (\mu_1 \geqq \mu_2 \geqq \ldots)$$

ist, die Operation $P_{\mu_1} A \times P_{\mu_2} A \times \cdots$ keine irreduktible Operation $T_{\lambda_1', \lambda_2', \ldots}(A)$ enthält, für die

$$(\lambda_1', \lambda_2', \ldots) < (\lambda_1, \lambda_2, \ldots)$$

ist. Denn wäre dies nicht der Fall, so müsste, wenn $(\varkappa_1', \varkappa_2', \ldots)$ die zu $(\lambda_1', \lambda_2', \ldots)$ associierte Zerlegung ist, nach unserer Voraussetzung $T_{\varkappa_1', \varkappa_2', \ldots}(A)$ zu $T_{\lambda_1', \lambda_2', \ldots}(A)$ associiert und daher in $C_{\mu_1} A \times C_{\mu_2} A \times \ldots$ enthalten sein. Dann wäre aber $\omega_1^{\varkappa_1'} \omega_2^{\varkappa_2'} \ldots$ ein Glied der Entwickelung von $c_{\mu_1} c_{\mu_2} \ldots$ nach Potenzen von $\omega_1, \omega_2, \ldots \omega_m$. Wir haben aber in § 20 gesehen, dass für jedes solche Glied die zu $(\varkappa_1', \varkappa_2', \ldots)$ associierte Zerlegung von nicht niedrigerer Ordnung sein kann als $(\mu_1, \mu_2, \ldots)$; also müsste $(\lambda_1', \lambda_2', \ldots) \geqq (\mu_1, \mu_2, \ldots) \geqq (\lambda_1, \lambda_2, \ldots)$ sein. Dies ergiebt aber einen Widerspruch.

Ich betrachte nun die Determinante (41). Zunächst überzeugt man sich leicht, dass dieselbe eine ganze homogene Funktion n-ter Dimension der m Grössen $\omega_1, \omega_2, \ldots \omega_m$ ist. Daher enthält dieselbe nur Glieder von der Form

$$p_{\mu_1} p_{\mu_2} p_{\mu_3} \ldots, \qquad (\mu_1 + \mu_2 + \mu_3 + \cdots = n) \tag{42}$$

und zwar ergiebt sich aus der Entwickelung der Determinante (und ihrer Unterdeterminanten) nach den Elementen der ersten Zeile, dass für alle Glieder (42)

$$(\mu_1, \mu_2, \ldots) \geqq (\lambda_1, \lambda_2, \ldots) \tag{43}$$

sein muss. Ich denke mir nun alle positiven Glieder der Determinante zu einem Glied t_1, alle negativen zu einem Glied t_2 vereinigt. Dann wird also

$$t = \left| p_{\lambda_\alpha - \alpha + \beta} \right| = t_1 - t_2.$$

Da jedes der Produkte $p_{\mu_1} p_{\mu_2} \ldots$ die Charakteristik der Operation $P_{\mu_1} A \times P_{\mu_2} A \times \ldots$ ist, so existieren jedenfalls zwei invariante

Operationen $T_1(A)$ und $T_2(A)$, denen die Funktionen t_1, resp. t_2 als Charakteristiken entsprechen. Nun denke ich mir $T_1(A)$ und $T_2(A)$ in ihre irreduktiblen Teile zerlegt. Es möge die Operation $T_{\nu_1,\nu_2,\ldots}(A)$ in $T_1(A)$ genau $h_{\nu_1,\nu_2,\ldots}$ Mal, in $T_2(A)$ genau $k_{\nu_1,\nu_2,\ldots}$ Mal enthalten sein. Dann müssen wegen (43) die Zahlen $h_{\nu_1,\nu_2,\ldots}$ und $k_{\nu_1,\nu_2,\ldots}$ für alle $(\nu_1, \nu_2, \ldots) < (\lambda_1, \lambda_2, \ldots)$ Null sein. Es ist dann

$$t_1 = \sum_{\nu} h_{\nu_1,\nu_2,\ldots}\, \Phi_{\nu_1,\nu_2,\ldots}, \quad t_2 = \sum_{\nu} k_{\nu_1,\nu_2,\ldots}\, \Phi_{\nu_1,\nu_2,\ldots},$$

also

$$t = t_1 - t_2 = \sum (h_{\nu_1,\nu_2,\ldots} - k_{\nu_1,\nu_2,\ldots})\, \Phi_{\nu_1,\nu_2,\ldots}.$$

Nun ist aber, wenn ich

$$l_1' = \lambda_1 + m - 1,\ l_2' = \lambda_2 + m - 2, \ldots l_\sigma' = \lambda_\sigma + m - \sigma,\ l_{\sigma+1}' = m - \sigma - 1, \ldots l_{m-1}' = 1,\ l_m' = 0$$

setze,

$$(44) \qquad |p_{\lambda_\alpha - \alpha + \beta}| = \frac{\left|\omega_\varkappa^{l_\iota'}\right|^{1)}}{\left|\omega_\varkappa^{m-\iota}\right|}. \qquad (\iota, \varkappa = 1, 2, \ldots m)$$

Das höchste Glied in der Entwickelung von t ist also gleich

$$\frac{\omega_1^{l_1'}\,\omega_2^{l_2'} \ldots \omega_m^{l_m'}}{\omega_1^{m-1}\,\omega_2^{m-2} \ldots \omega_m^{0}} = \omega_1^{\lambda_1}\,\omega_2^{\lambda_2} \ldots \omega_\sigma^{\lambda_\sigma}.$$

Folglich erhalten wir

$$t = \sum (h_{\nu_1,\nu_2,\ldots} - k_{\nu_1,\nu_2,\ldots})\, \Phi_{\nu_1,\nu_2,\ldots} = \omega_1^{\lambda_1}\omega_2^{\lambda_2} \ldots \omega_\sigma^{\lambda_\sigma} + \text{niedr. Glieder.}$$

Durch successive Vergleichung der Koeffizienten von ω_1^n, $\omega_1^{n-1}\omega_2$, $\omega_1^{n-2}\omega_2^2, \ldots \omega_1^{\lambda_1}\,\omega_2^{\lambda_2} \ldots \omega_\sigma^{\lambda_\sigma}$ auf beiden Seiten dieser Gleichung (das höchste Glied von $\Phi_{\nu_1,\nu_2,\ldots}$ ist $\omega_1^{\nu_1}\,\omega_2^{\nu_2}\ldots$) erhalte ich also

$$h_{\nu_1,\nu_2,\ldots} - k_{\nu_1,\nu_2,\ldots} = 0 \text{ für alle } (\nu_1, \nu_2, \ldots) \neq (\lambda_1, \lambda_2, \ldots)$$

und $h_{\lambda_1,\lambda_2,\ldots} - k_{\lambda_1,\lambda_2,\ldots} = 1$.

Daher ist die Operation $T_2(A)$ in $T_1(A)$ enthalten; d. h. $T_1(A)$ zerfällt in zwei Operationen $T_2(A)$ und $T(A)$. Es gehört aber $T(A)$ zur Charakteristik

$$t = t_1 - t_2 = \Phi_{\lambda_1,\lambda_2,\ldots}.$$

---

1) Vergl. K. Th. Vahlen, Encyclopädie der math. Wissenschaften, Bd. I, S. 465.

und ist daher äquivalent $T_{\lambda_1, \lambda_2, \ldots}(A)$.

Die zu $T(A)$ associierte Operation hat die Charakteristik $|c_{\lambda_\alpha - \alpha + \beta}|$; das höchste Glied dieser symmetrischen Funktion ist aber, wie man sich leicht überzeugt, gleich $\omega_1^{\varkappa_1} \omega_2^{\varkappa_2} \ldots \omega_\varrho^{\varkappa_\varrho}$ und also sind, wie zu beweisen war, $T_{\lambda_1, \lambda_2, \ldots}(A)$ und $T_{\varkappa_1, \varkappa_2, \ldots}(A)$ associierte Operationen.

Da ferner die Charakteristik von $T_{\varkappa_1, \varkappa_2, \ldots}(A)$ gleich

$$|p_{\varkappa_\gamma - \gamma + \delta}| \qquad (\gamma, \delta = 1, 2, \ldots \varrho)$$

ist, so erhalten wir die Gleichung

$$(45) \qquad |c_{\lambda_\alpha - \alpha + \beta}| = |p_{\varkappa_\gamma - \gamma + \delta}|.$$

Ist speziell (33) eine zweiseitige Zerlegung der Zahl n, so wird $T_{\lambda_1, \lambda_2, \ldots \lambda_\sigma}(A) = T_{\varkappa_1, \varkappa_2, \ldots \varkappa_\varrho}(A)$, also sich selbst associiert. Nach dem Ergebnis des § 19 folgt hieraus:

XIII. „Die Anzahl der irreduktiblen invarianten Operationen n-ter Ordnung, die sich selbst associiert sind, ist gleich der Anzahl der Zerlegungen von n in verschiedene ungerade Summanden“ [1]).

Um den Grad $G_{\lambda_1, \lambda_2, \ldots \lambda_\sigma}$ der Matrix $T_{\lambda_1, \lambda_2, \ldots \lambda_\sigma}(A)$ zu erhalten, genügt es in $\Phi_{\lambda_1, \lambda_2, \ldots \lambda_\sigma}$

$$\omega_1 = \omega_2 = \cdots = \omega_m = 1$$

zu setzen. Dann wird $p_\lambda = \binom{m + \lambda - 1}{\lambda}$, $c_\lambda = \binom{m}{\lambda}$, und wir erhalten:

$$(46) \qquad G_{\lambda_1, \lambda_2, \ldots \lambda_\sigma} = \left| \binom{m + \lambda_\alpha - \alpha + \beta - 1}{\lambda_\alpha - \alpha + \beta} \right| = \left| \binom{m}{\varkappa_\gamma - \gamma + \delta} \right|.$$

Setzt man

$$l_1 = \lambda_1 + \sigma - 1, \; l_2 = \lambda_2 + \sigma - 2, \ldots \; l_\sigma = \lambda_\sigma,$$
$$k_1 = \varkappa_1 + \varrho - 1, \; k_2 = \varkappa_2 + \varrho - 2, \ldots \; k_\varrho = \varkappa_\varrho$$

und bezeichnet mit $\Delta(x_1, x_2, \ldots x_\nu)$ das Differenzenprodukt

$$(x_1 - x_2)(x_1 - x_3) \ldots (x_1 - x_\nu)(x_2 - x_3)(x_2 - x_4) \ldots (x_{\nu-1} - x_\nu),$$

---

1) Vergl. Frobenius, loc. cit. S. 529.

so wird, wie eine leichte Rechnung zeigt,

$$(46') \quad \begin{cases} G_{\lambda_1, \lambda_2, \ldots \lambda_\sigma} = \dfrac{\prod_1^\sigma (m + l_\alpha - \sigma)!}{\prod_1^\sigma l_\alpha! \prod_1^\sigma (m-\alpha)!} \varDelta(l_1, l_2, \ldots l_\sigma) \\[2ex] \qquad = \dfrac{\prod_1^\varrho (m+\alpha-1)!}{\prod_1^\varrho k_\alpha! \prod_1^\varrho (m+\varrho-k_\alpha-1)!} \varDelta(k_1, k_2, \ldots k_\varrho). \end{cases}$$

Einen eleganteren Ausdruck erhält man, indem man λ_ν für $\nu > \sigma$ gleich 0 setzt und $\Phi_{\lambda_1, \lambda_2, \ldots \lambda_\sigma}$, was wegen $p_0 = 1$, $p_{-\alpha} = 0$ gestattet ist, als Determinante m-ten Grades

$$|p_{\lambda_\mu - \mu + \nu}| \qquad (\mu, \nu = 1, 2, \ldots m)$$

schreibt. Setzt man hierin wieder $\omega_\mu = 1$, so ergiebt sich

$$46'') \quad G_{\lambda_1, \lambda_2, \ldots \lambda_\sigma} = \frac{\varDelta(l'_1, l'_2, \ldots l'_m)}{\prod_1^m (m-\mu)!} = \frac{\varDelta(l'_1, l'_2, \ldots l'_m)}{\varDelta(1, 2, \ldots m)},$$

wo $l'_1, l'_2, \ldots l'_m$ dieselbe Bedeutung haben, wie in der Formel (44).

§ 24. Zu jeder der k irreduktiblen Operationen $T_{\lambda_1, \lambda_2, \ldots \lambda_\sigma}(A)$ gehört eine primitive Darstellung der symmetrischen Gruppe $\mathfrak{S}_n$. Die Spuren der darstellenden Matrizen bilden einen einfachen Charakter der Gruppe. Nach § 12 erhält man also die k Charaktere von $\mathfrak{S}_n$, indem man die k Determinanten $|p_{\lambda_\alpha} - \alpha + \beta|$ nach den Potenzsummen s_ν entwickelt; und diese Entwickelungen werden gefunden, indem man für p_λ

$$\Sigma \frac{1}{\alpha_1! \, \alpha_2! \ldots} \left(\frac{s_1}{1}\right)^{\alpha_1} \left(\frac{s_2}{2}\right)^{\alpha_2} \ldots \qquad (\alpha_1 + 2\alpha_2 + \cdots = \lambda)$$

einsetzt. Der Grad $f_{\lambda_1, \lambda_2, \ldots \lambda_\sigma}$ des zu $T_{\lambda_1, \lambda_2, \ldots \lambda_\sigma}(A)$ gehörenden Charakters ist speziell gleich dem Koeffizienten von $\frac{s_1^n}{n!}$ in der Entwickelung von $|p_{\lambda_\alpha - \alpha + \beta}|$; es ist daher

$$(47) \quad f_{\lambda_1, \lambda_2, \ldots \lambda_\sigma} = n! \left| \frac{1}{(\lambda_\alpha - \alpha + \beta)!} \right| = \frac{n!}{l_1! \, l_2! \ldots l_\sigma!} \varDelta(l_1, l_2, \ldots l_\sigma),$$

wo $l_1, l_2, \ldots l_\sigma$ dieselbe Bedeutung haben, wie oben.

Bezeichnet man die k Zerlegungen $n = \lambda_1 + \lambda_2 + \cdots$ in irgend einer Reihenfolge mit (1), (2), ... (k), so mögen die entsprechenden

Operationen $T_{\lambda_1,\lambda_2,\ldots}(A)$ mit $T^{(1)}(A)$, $T^{(2)}(A)$, ... $T^{(k)}(A)$, ihre Charakteristiken mit $\Phi^{(1)}$, $\Phi^{(2)}$, ... $\Phi^{(k)}$, die zugehörigen Charaktere mit $\chi^{(1)}(R)$, $\chi^{(2)}(R)$, ... $\chi^{(k)}(R)$ bezeichnet werden.

Ferner wollen wir die Klassen gleichberechtigter Elemente, in die $\mathfrak{S}_n$ zerfällt, mit (0), (1), ... $(k-1)$, die Anzahl der Elemente der Klasse (ϱ) mit h_ϱ bezeichnen. Bestehen die Permutationen R der Klasse (ϱ) aus α_1 Zykeln der Ordnung 1, α_2 Zykeln der Ordnung 2, u. s. w., so soll $\chi^{(\lambda)}(R) = \chi_\varrho^{(\lambda)}$ und $s_1^{\alpha_1} s_2^{\alpha_2} \ldots = s^{(\varrho)}$ gesetzt werden; es ist dann ferner

$$h_\varrho = \frac{n!}{1^{\alpha_1}\alpha_1!\, 2^{\alpha_2}\alpha_2! \ldots}.$$

Nach Einführung dieser Bezeichnungen lässt sich unser Ergebnis in die Formel

$$(48) \qquad n!\,\Phi^{(\lambda)} = \sum_\varrho h_\varrho\, \chi_\varrho^{(\lambda)} s^{(\varrho)} \qquad (\lambda = 1, 2, \ldots k)$$

zusammenfassen.

Aus diesen k Gleichungen lassen sich umgekehrt die $s^{(\varrho)}$ als lineare Funktionen der $\Phi^{(\lambda)}$ darstellen, und zwar ist

$$(49) \qquad s^{(\varrho)} = \sum_\lambda \chi_\varrho^{(\lambda)} \Phi^{(\lambda)}.$$

Dies folgt leicht aus den Gleichungen[1])

$$(50) \qquad \sum_R \chi^{(\lambda)}(R)\, \chi^{(\lambda)}(R^{-1}) = n!, \quad \sum_R \chi^{(\lambda)}(R)\, \chi^{(\mu)}(R^{-1}) = 0,$$

denen die Gruppencharaktere genügen.

Die Gleichung (49) ist wesentlich identisch mit der von Herrn Frobenius[2]) zur Berechnung der Charaktere der symmetrischen Gruppe benutzten.

§ 25. Besonders einfach ist die Bestimmung der Charaktere $\chi_\varrho^{(\lambda)} = \chi^{(\lambda_1,\lambda_2,\ldots\lambda_\sigma)}(R_{\alpha_1,\alpha_2,\ldots})$ in folgenden Fällen:

1. $\sigma = 2$. Dann sei $\lambda_2 = r$, $\lambda_1 = n - r$, also $r \leqq \frac{n}{2}$.

Aus $\qquad \Phi^{(\lambda)} = \begin{vmatrix} p_{n-r} & p_{n-r+1} \\ p_{r-1} & p_r \end{vmatrix} = p_r p_{n-r} - p_{r-1} p_{n-r+1}$

1) Frobenius: „Ueber Gruppencharaktere", Sitzungsberichte, 1896, S. 985. Für die symmetrische Gruppe ist $\chi^{(\lambda)}(R^{-1}) = \chi^{(\lambda)}(R)$.

2) Sitzungsberichte, 1900, S. 519, Formel (6).

erhält man

$$\chi^{(n-r,r)}(R_{\alpha_1,\alpha_2,\ldots}) = \sum \left[\binom{\alpha_1}{\mu_1} - \binom{\alpha_1}{\mu_1-1}\right]\binom{\alpha_2}{\mu_2}\binom{\alpha_3}{\mu_3}\cdots,$$

wo die Summation über alle positiven Lösungen der Gleichung $\mu_1+2\mu_2+\cdots = r$ zu erstrecken und $\binom{\alpha_1}{\mu_1-1} = 0$ zu setzen ist, wenn $\mu_1 = 0$ ist.

2. $$\lambda_1 = n-r,\ \lambda_2 = \lambda_3 = \cdots = \lambda_\sigma = 1.$$

Dann wird, wegen der aus (45) sich ergebenden Gleichung

$$c_n = |p_{1+\alpha-\beta}|, \qquad (\alpha, \beta = 1, 2, \cdots n)$$

$$\begin{vmatrix} p_{n-r}, & p_{n-r+1}, & \cdots p_{n-1}, & p_n \\ 1 & p_1 & \cdots p_{r-1}, & p_r \\ 0 & 1 & \cdots p_{r-2}, & p_{r-1} \\ \cdot & \cdot & \cdots \quad \cdot & \cdot \\ \cdot & \cdot & \cdots \quad \cdot & \cdot \\ 0 & 1 & \cdots \quad 1 & p_1 \end{vmatrix} = c_r p_{n-r} - c_{r-1} p_{n-r+1} + \cdots \pm p_n;$$

und daraus erhält man

$$\chi^{(n-r,1,1,\ldots)}(R_{\alpha_1,\alpha_2,\ldots}) = \sum_{\mu_1+2\mu_2+\cdots=r} (-1)^{\mu_2+\mu_4+\cdots}\binom{\alpha_1-1}{\mu_1}\binom{\alpha_2}{\mu_2}\cdots;$$

speziell ist $\chi^{(n-r,1,1,\ldots)}(E) = f_{n-r,1,1,\ldots} = \binom{n-1}{r}$.

3. $\lambda_1 = \lambda_2 = \cdots = \lambda_r = 2,\ \lambda_{r+1} = \cdots = \lambda_{n-r} = 1.$

Dann ist die zu $\lambda_1+\lambda_2\cdots = n$ associierte Zerlegung: $n = (n-r)+r$, und daher ist

$$\chi^{(2,2,\ldots 2,1,1\ldots)}(R_{\alpha_1,\alpha_2\ldots}) = (-1)^{\alpha_2+\alpha_4+\cdots}\chi^{(n-r,r)}(R_{\alpha_1,\alpha_2,\ldots}).$$

§ 26. Ist $T(A)$ eine beliebige invariante Operation n-ter Ordnung, so hat man, um die Zahlen $h_{\lambda_1,\lambda_2,\ldots}$ zu bestimmen, die angeben, wie oft die primitive Operation $T_{\lambda_1,\lambda_2,\ldots}(A)$ in $T(A)$ enthalten ist, die Charakteristik

(51) $$\Phi = \sum r_{n_1,n_2,n_3\ldots}\,\omega_1^{n_1}\omega_2^{n_2}\omega_3^{n_3}\cdots \qquad (n_1+n_2+n_3+\cdots = n)$$

von $T(A)$ als (lineare) Funktion der Determinanten $|p_{\lambda_\alpha-\alpha+\beta}|$ darzustellen. Man erhält diese Darstellung in eleganter Form, indem man die Zahlen $r_{n_1, n_2, n_3 \ldots}$ symbolisch durch $r_{n_1} r_{n_2} r_{n_3} \ldots$ ersetzt, wo das Symbol r_ν für $\nu = 0$ gleich 1, für $\nu < 0$ gleich 0 zu setzen ist.

Dann wird nämlich

$$\Phi = \sum \left| r_{\lambda_\alpha - \alpha + \beta} \right| \left| p_{\lambda_\alpha - \alpha + \beta} \right|, \qquad (\alpha, \beta = 1, 2, \ldots \sigma) \tag{52}$$

wo die Summation über alle den Bedingungen

$$\lambda_1 + \lambda_2 + \cdots + \lambda_\sigma = n, \qquad \lambda_1 \geqq \lambda_2 \geqq \lambda_3 \geqq \cdots \geqq \lambda_\sigma > 0$$

genügenden ganzen Zahlen zu erstrecken ist. Diese Formel ergiebt sich, indem man Φ mit der Determinante

$$\left| \omega_\varkappa^{m-\iota} \right| \qquad (\iota, \varkappa = 1, 2, \ldots m)$$

multipliziert und die Gleichung (44) berücksichtigt [1]).

Die Determinanten $\left| r_{\lambda_\alpha - \alpha + \beta} \right|$, deren Werte sich ergeben, indem man in ihrer Entwickelung die symbolischen Produkte wieder durch die Zahlen $r_{n_1, n_2, \ldots}$ ersetzt, sind gleich den gesuchten Zahlen $h_{\lambda_1, \lambda_2, \ldots}$ und müssen demnach nicht negative ganze Zahlen sein. Umgekehrt ist diese Bedingung auch hinreichend, damit eine symmetrische Funktion (51) die Charakteristik einer invarianten Operation sein kann.

Ist speziell $T(A) = \Pi_n A$, so wird $\Phi = c_1^n$, also $r_{n_1, n_2, \ldots} = \dfrac{n!}{n_1! n_2! \ldots}$. Man kann in diesem Falle

$$r_\nu = \frac{(n!)^{\frac{\nu}{n}}}{\nu!}$$

setzen und erhält

$$\left| r_{\lambda_\alpha - \alpha + \beta} \right| = n! \left| \frac{1}{(\lambda_\alpha - \alpha + \beta)!} \right| = f_{\lambda_1, \lambda_2, \ldots}.$$

Es wird also $h_{\lambda_1, \lambda_2, \ldots}$ gleich dem Grade der zur Operation $T_{\lambda_1, \lambda_2, \ldots}(A)$ gehörenden Darstellung der symmetrischen Gruppe $\mathfrak{S}_n$ [2]). Daraus folgt nach den Sätzen des Herrn Frobenius [3]), dass

1) Vergl. Nägelsbach, Crelle's Journal, Bd. 81, S. 281.

2) Im Falle $n = 2$ ist demnach $\Pi_2 A$ in die Operationen $C_2 A$ und $P_2 A$ zerlegbar. Dieser Satz findet sich bereits bei Herrn B. Igel: „Zur Theorie der Determinanten", Monatshefte für Math. und Physik, Bd. 3 (1892), S. 55.

3) Sitzungsberichte, 1896, S. 1343, und 1897, S. 994.

die Darstellung der symmetrischen Gruppe, die durch die Operation $\Pi_n A$ erzeugt wird, der durch die Gruppendeterminante von $\mathfrak{S}_n$ gelieferten äquivalent ist.

§ 27. Aus der Gleichung (52) lässt sich ein neuer Beweis für den Satz XII. ableiten.

Ist $\chi(R)$ der zu der Operation $T(A)$ gehörende (zusammengesetzte) Charakter von $\mathfrak{S}_n$, so ist

$$(52')\ \Phi = \Sigma \left|r_{\lambda_\alpha - \alpha + \beta}\right| \cdot \left|p_{\lambda_\alpha - \alpha + \beta}\right| = \sum_{(\alpha_1 + 2\alpha_2 + \cdots = n)} \frac{\chi(R_{\alpha_1, \alpha_2 \ldots})}{1^{\alpha_1}\alpha_1!\, 2^{\alpha_2}\alpha_2 \ldots} s_1^{\alpha_1} s_2^{\alpha_2} \ldots$$

Diese Gleichung wird eine in den s identische, wenn man in ihr

$$(53)\qquad p_\lambda = \Sigma \frac{1}{\beta_1!\, \beta_2!} \cdots \left(\frac{s_1}{1}\right)^{\beta_1} \left(\frac{s_2}{2}\right)^{\beta_2} \cdots \qquad (\beta_1 + 2\beta_2 + \cdots = \lambda)$$

setzt, und zu ihrem Bestehen ist nur erforderlich, dass die Zahlen $r_{n_1, n_2, \ldots}$ und $\chi(R_{\alpha_1 \alpha_2, \ldots})$ durch die Gleichungen

$$n_1!\, n_2! \ldots r_{n_1, n_2 \ldots} = \Sigma\, g_{\alpha_1, \alpha_2 \ldots}^{n_1, n_2 \ldots}\ \chi(R_{\alpha_1, \alpha_2, \ldots})$$

verknüpft sind, wo die $g_{\alpha_1, \alpha_2 \ldots}^{n_1, n_2 \ldots}$ dieselbe Bedeutung haben, wie auf Seite 29. Diese Gleichungen können aber, wenn ich für $\chi(R_{\alpha_1, \alpha_2, \ldots})$ symbolisch $t_1^{\alpha_1} t_2^{\alpha_2} \ldots$ substituiere, ersetzt werden durch

$$(53')\qquad r_\lambda = \Sigma \frac{1}{\beta_1!\, \beta_2! \ldots} \left(\frac{t_1}{1}\right)^{\beta_1} \left(\frac{t_2}{2}\right)^{\beta_2} \cdots.$$

Nun stimmen aber (53) und (53') abgesehen von den Bezeichnungen überein. Mithin kann ich in (52') p_λ durch r_λ und $s_1^{\alpha_1} s_2^{\alpha_2} \ldots$ durch $\chi(R_{\alpha_1, \alpha_2, \ldots})$ ersetzen.

Ich erhalte dann

$$(54)\qquad \Sigma \left|r_{\lambda_\alpha - \alpha + \beta}\right|^2 = \Sigma \frac{\chi^2(R_{\alpha_1, \alpha_2 \ldots})}{1^{\alpha_1}\alpha_1!\, 2^{\alpha_2}\alpha_2! \ldots}.$$

Soll nun $T(A)$ irreduktibel sein, so muss $\chi(R)$ ein einfacher Charakter sein. Für einen solchen ist aber nach der ersten der Formeln (50) die rechte Seite von (54) gleich 1, und da $\left|r_{\lambda_\alpha - \alpha + \beta}\right|$ ganze Zahlen sind, so müssen sie abgesehen von einer, die gleich ± 1 ist, alle verschwinden. Daher muss die Charakteristik einer irreduktiblen Operation die Gestalt $\pm \left|p_{\lambda_\alpha - \alpha + \beta}\right|$ haben; es ist aber das höchste Glied der Determinante gleich $+\omega_1^{\lambda_1}\omega_2^{\lambda_2}\ldots\omega_{\sigma}^{\lambda_{\sigma}}$, und folglich ist das positive Vorzeichen zu wählen.

Der übrige Teil des Satzes XII. ergiebt sich nun wie in § 23.

§ 28. Die Gleichung (52′) lässt sich, wenn man $\chi(R_{\alpha_1,\alpha_2,\ldots})$ durch $t_1^{\alpha_1}t_2^{\alpha_2}\ldots$ ersetzt, in der Form

$$(52''')\quad \sum |r_{\lambda_\alpha-\alpha+\beta}|\,|p_{\lambda_\alpha-\alpha+\beta}| = \sum \frac{1}{\alpha_1!\,\alpha_2!\ldots}\left(\frac{s_1 t_1}{1}\right)^{\alpha_1}\left(\frac{s_1 t_2}{2}\right)^{\alpha_2}\ldots$$

schreiben. Sind nun $\omega_1', \omega_2', \ldots \omega_{m'}'$ irgendwelche Variabele, für die s_ν', p_ν', c_ν' die analoge Bedeutung haben, wie s_ν, p_ν, c_ν für die Variabelen $\omega_1, \omega_2, \ldots \omega_m$, so wird, wenn man $t_1 = s_1', t_2 = s_2', t_3 = s_3', \ldots$ setzt, $r_\nu = p_\nu'$, dagegen, wenn man $t_1 = s_1', t_2 = -s_2', t_3 = s_3', t_4 = -s_4', \ldots$ setzt, $r_\nu = c_\nu'$.

Demnach erhalten wir aus (52‴) die Formeln

$$(55)\quad \begin{cases} \sum |p'_{\lambda_\alpha-\alpha+\beta}|\,|p_{\lambda_\alpha-\alpha+\beta}| = \sum \dfrac{1}{\alpha_1!\,\alpha_2!\ldots}\left(\dfrac{s_1 s_1'}{1}\right)^{\alpha_1}\left(\dfrac{s_2 s_2'}{2}\right)^{\alpha_2}\ldots \\ \sum |c'_{\lambda_\alpha-\alpha+\beta}|\,|p_{\lambda_\alpha-\alpha+\beta}| = \sum \dfrac{(-1)^{\alpha_2+\alpha_4+\ldots}}{\alpha_1!\,\alpha_2!\ldots}\left(\dfrac{s_1 s_1'}{1}\right)^{\alpha_1}\left(\dfrac{s_2 s_2'}{2}\right)^{\alpha_2}\ldots \end{cases}$$

Die rechten Seiten dieser Gleichungen bedeuten nichts Anderes, als die n-te Wronski'sche, resp. n-te elementarsymmetrische Funktion der mm' Variabelen

$$\omega_\mu\,\omega_{\mu'}', \qquad \mu = 1, 2, \ldots m,\ \mu' = 1, 2, \ldots m'.$$

Abschnitt VI.

§ 29. Die Elemente einer invarianten Matrix n-ter Ordnung $T(A) = (A_{\varrho\sigma})$ sind lineare homogene Funktionen der $\binom{m^2+n-1}{n}$ verschiedenen Produkte $a_{\varkappa_1\lambda_1}\,a_{\varkappa_2\lambda_2}\ldots a_{\varkappa_n\lambda_n}$. Im Allgemeinen werden unter den $A_{\varrho\sigma}$ gewisse lineare Relationen mit konstanten Koeffizienten bestehen, die für alle Werte der $a_{\varkappa\lambda}$ erfüllt sind. Sind aber unter den $A_{\varrho\sigma}$ s, und nicht mehr als s, linear unabhängig, so erkennt man leicht, dass diese Zahl ungeändert bleibt, wenn man an Stelle von $T(A)$ eine äquivalente Matrix $T_1(A) = (A'_{\varrho\sigma})$ betrachtet. — Es besteht nun der Satz:

XIV. „Sind $T^{(1)}(A), T^{(2)}(A), \ldots T^{(k)}(A)$ die verschiedenen irreduktiblen invarianten Formen n-ter Ordnung, so sind ihre sämtlichen $G^{(1)2}+G^{(2)2}+\ldots+G^{(k)2}$ Elemente untereinander linear unabhängig."

Hierbei bedeutet $G^{(\lambda)}$ den Grad der Matrix $T^{(\lambda)}(A)$.

Denn ist $T(A)$ eine beliebige invariante Form n-ter Ordnung, so ist nach dem eben Gesagten die grösste Anzahl s der linear unabhängigen unter ihren Elementen jedenfalls nicht grösser, als die Gesamtanzahl der Elemente der k irreduktiblen invarianten Formen $T^{(\lambda)}(A)$, also $\leqq \Sigma G^{(\lambda)2}$. Setzt man aber $T(A) = \Pi_n A$, so treten in $T(A)$ sämtliche Produkte $a_{\varkappa_1\lambda_1} a_{\varkappa_2\lambda_2} \ldots a_{\varkappa_n\lambda_n}$ als Elemente auf. Würden daher unter den Elementen der $T^{(\lambda)}(A)$ lineare Beziehungen bestehen, so müsste

$$(56) \qquad \sum G^{(\lambda)2} > \binom{m^2+n-1}{n}$$

sein. Setzt man aber in der ersten der Formeln (55) $m' = m$, $\omega_\mu = \omega'_{\mu'} = 1$, so wird $|p'_{\lambda_\alpha - \alpha + \beta}| = |p_{\lambda_\alpha - \alpha + \beta}| = G_{\lambda_1, \lambda_2, \ldots}$ und die rechte Seite wird gleich $\binom{m^2+n-1}{n}$, man erhält also in Widerspruch zu (56)

$$(57) \qquad \sum G^{(\lambda)2} = \binom{m^2+n-1}{n}.$$

Der Satz XIV. ist der spezielle Fall eines allgemeinen Satzes über Systeme komplexer Haupteinheiten, den Herr Molien[1]) in seiner Arbeit: „Ueber Systeme höherer komplexer Zahlen" bewiesen hat, und bildet ein Analogon zu dem Satze des Herrn Frobenius[2]), dass die sämtlichen zu einer endlichen Gruppe h-ter Ordnung gehörenden primitiven Gruppenmatrizen zusammen h unabhängige Variabele enthalten.

Mit Hülfe dieses Satzes sind wir imstande für jede invariante Form $T(A)$ die erwähnte Zahl s zu bestimmen, wenn uns die Charakteristik Φ von $T(A)$ bekannt ist. Zu diesem Zwecke haben wir nur nach § 26 Φ als Funktion der $\Phi^{(\lambda)} = |p_{\lambda_\alpha - \alpha + \beta}|$ darzustellen. Ist

$$\Phi = \sum x_\lambda \Phi^{(\lambda)},$$

und sind unter den Koeffizienten x_λ, die nicht negative ganze

1) Math. Ann. Bd. 41, S. 124.

2) Sitzungsberichte, 1897. S. 994.

Zahlen sind, $x_{\lambda_1}, x_{\lambda_2}, \ldots x_{\lambda_\nu}$ von Null verschieden, so ist die gesuchte Zahl s gleich

$$G^{(\lambda_1)^2} + G^{(\lambda_2)^2} + \cdots + G^{(\lambda_\nu)^2}.$$

Da nun unter den invarianten Operationen die Determinantentransformationen $C_{\varkappa_1} A \times C_{\varkappa_2} A \times \cdots$, $C_\lambda(C_\mu A)$ u. s. w. enthalten sind, so erhalten wir eine Methode, Aufgaben über die lineare Abhängigkeit von Determinanten auf entsprechende Aufgaben aus der Theorie der symmetrischen Funktionen zurückzuführen.

§ 30. Ich will für den Satz XIV. oder, was dasselbe ist, für die Gleichung (57) noch einen zweiten Beweis anführen, der sich auf die Betrachtung einer speziellen invarianten Operation stützt.

Ich habe bereits auf Seite 33 erwähnt, dass die Gleichungen (I.), welche die invarianten Matrizen

(58) $$T(A) = \sum \begin{bmatrix} \varkappa_1 \varkappa_2 \ldots \varkappa_n \\ \lambda_1 \lambda_2 \ldots \lambda_n \end{bmatrix} a_{\varkappa_1 \lambda_1} a_{\varkappa_2 \lambda_2} \ldots a_{\varkappa_n \lambda_n}$$

charakterisieren, als die definierenden Gleichungen eines Systems von $\binom{m^2+n-1}{n}$ komplexen Haupteinheiten aufgefasst werden können.

Es mögen allgemein die r Symbole $e_1, e_2, \ldots e_r$ die Basis eines komplexen Zahlensystems bilden, und zwar sei

(59) $$e_i e_k = \sum \gamma_{iks} e_s \qquad (i, k, s = 1, 2, \ldots r).$$

Dann müssen bekanntlich die r^3 Zahlen γ_{iks} folgenden Bedingungen genügen:

1. es ist

(60) $$\sum_{s=1}^{r} \gamma_{iks} \gamma_{slt} = \sum_{s=1}^{r} \gamma_{kls} \gamma_{ist};$$

2. die Determinanten

$$\left|\sum_{i=1}^{r} \gamma_{iks} x_i\right|, \quad \left|\sum_{k=1}^{r} \gamma_{iks} x_k\right|$$

verschwinden nicht identisch, d. h. nicht für jede Wahl der x_i.

Sind

$$x = x_1 e_1 + x_2 e_2 + \cdots + x_r e_r, \quad u = u_1 e_1 + u_2 e_2 + \cdots + u_r e_r$$

zwei beliebige Zahlen des Systems und setzt man

$$y = xu = y_1 e_1 + y_2 e_2 + \cdots y_r e_r,$$

so stellen sich die Parameter y_i als bilineare Formen der Parameter x_i und u_i dar, und zwar ist

(59′) $$y_s = \sum_{i,k} \gamma_{iks} x_i u_k.$$

Bezeichnet man die Matrix r-ten Grades

(61) $$(\gamma_{iks}) \qquad (i, s = 1, 2, \ldots r)$$

mit Γ_k, so folgt aus den Gleichungen (60), dass die r Matrizen Γ_k das gegebene Zahlensystem „darstellen," d. h. es ist

$$\Gamma_i \Gamma_k = \sum_{n=1}^{r} \gamma_{iks} \Gamma_s$$ [1].

Die Bedingung 2. lässt sich folgendermassen ausdrücken: die r Matrizen (61) sind linear unabhängig und es lassen sich r Zahlen $\varepsilon_1, \varepsilon_2, \ldots \varepsilon_r$ so bestimmen, dass $\sum_1^r \varepsilon_n \Gamma_n$ gleich der Einheitsmatrix r-ten Grades wird [2].

In unserem Falle sind die komplexen Einheiten $e_1, e_2, \ldots e_r$ durch die $\binom{m^2+n-1}{n} = r$ Symbole $\begin{bmatrix} \varkappa_1 \varkappa_2 \ldots \varkappa_n \\ \lambda_1 \lambda_2 \ldots \lambda_n \end{bmatrix}$ zu ersetzen. Die Zahlen γ_{iks} lassen sich folgendermassen charakterisieren. Sind $A = (a_{\varkappa\lambda})$ und $B = (b_{\varkappa\lambda})$ zwei Matrizen m-ten Grades, $C = (c_{\varkappa\lambda})$ die aus ihnen zusammengesetzte Matrix, und ersetzt man in (59′) die Parameter x_i in irgend einer Reihenfolge durch die r Produkte

(a) $$a_{\varkappa_1\lambda_1} a_{\varkappa_2\lambda_2} \ldots a_{\varkappa_n\lambda_n},$$

die Parameter u_i durch die entsprechenden Produkte

(b) $$b_{\varkappa_1\lambda_1} b_{\varkappa_2\lambda_2} \ldots b_{\varkappa_n\lambda_n},$$

so werden die y_i gleich den Produkten

(c) $$c_{\varkappa_1\lambda_1} c_{\varkappa_2\lambda_2} \ldots c_{\varkappa_n\lambda_n}.$$

---

1) Vergl. Study, Encycl. der math. Wissensch. Bd. I, S. 147.

2) Umgekehrt erkennt man leicht, dass die r^3 Zahlen γ_{iks} den Bedingungen 1. und 2. genügen, wenn sich r linear unabhängige Matrizen $A_1, A_2, \ldots A_r$ derart bestimmen lassen, dass die Gleichungen $A_i A_k = \sum_{s=1}^{r} \gamma_{iks} A_s$ bestehen, und dass die Einheitsmatrix sich in der Form $\sum_{i=1}^{r} a_i A_i$ darstellen lässt.

Die Zahlen γ_{iks} sind also durch die Entwickelungen der (c) als lineare Formen der (a) und (b) zu bestimmen.

Ich betrachte jetzt diejenige invariante Matrix $T_n(A)$ des Grades $\binom{m^2+n-1}{n}$, die dadurch entsteht, dass man in (58) für die $\begin{bmatrix}\varkappa_1\,\varkappa_2\ldots\varkappa_n\\ \lambda_1\,\lambda_2\ldots\lambda_n\end{bmatrix}$ die entsprechenden Matrizen $\Gamma_n = (\gamma_{iks})$ einsetzt. Um nun zu entscheiden, in welche irreduktiblen Operationen $T_n(A)$ zerlegbar ist, habe ich die Charakteristik

$$\Phi = \sum r_{n_1, n_2, \ldots}\, \omega_1^{n_1} \omega_2^{n_2} \ldots$$

von $T_n(A)$ zu bestimmen.

Die Zahl $r_{n_1, n_2, \ldots n_m}$ giebt an, wie oft das Produkt $a_{11}^{n_1} a_{22}^{n_2} \ldots a_{mm}^{n_m}$ in der Hauptdiagonale von $T_n(A)$ vorkommt. Der Definition der Zahlen γ_{iks} gemäss habe ich in jeder der Entwickelungen der r Produkte (c) den Koeffizienten von

$$(d) \qquad a_{\varkappa_1 \lambda_1} a_{\varkappa_2 \lambda_2} \ldots a_{\varkappa_n \lambda_n}\, b_{11}^{n_1} b_{22}^{n_2} \ldots b_{mm}^{n_m}$$

zu ermitteln und die Summe dieser Koeffizienten zu bilden. Nun enthält aber die Entwickelung von (c) nur dann ein Glied der Form (d), wenn unter der Indices $\lambda_1, \lambda_2, \ldots \lambda_n$ genau n_1 gleich 1, n_2 gleich $2, \ldots n_m$ gleich m sind; und ist diese Bedingung erfüllt, so ist der gesuchte Koeffizient gleich 1, wie unmittelbar ersichtlich ist. Folglich ist $r_{n_1, n_2, \ldots n_m}$ gleich der Anzahl der verschiedenen Produkte

$$c_{\varkappa_1 1} c_{\varkappa_2 1} \ldots c_{\varkappa_{n_1} 1} c_{\varkappa_{n_1+1} 2} \ldots c_{\varkappa_{n_1+n_2} 2} \ldots, \qquad (\varkappa_\nu = 1, 2, \ldots m)$$

also gleich

$$\binom{m+n_1-1}{n_1}\binom{m+n_2-1}{n_2}\cdots\binom{m+n_m-1}{n_m};$$

demnach ist

$$\Phi = \sum \binom{m+n_1-1}{n_1}\binom{m+n_2-1}{n_2}\cdots\binom{m+n_m-1}{n_m}\omega_1^{n_1}\omega_2^{n_2}\ldots\omega_m^{n_m}.$$

Um nun diese symmetrische Funktion nach der in § 26 angegebenen Methode durch die $\Phi^{(\lambda)}$ darzustellen, hat man die dort eingeführten Symbole r_ν durch die Zahlen $\binom{m+\nu-1}{\nu}$ zu ersetzen; dann wird aber

$$|r_{\lambda_\alpha - \alpha + \beta}| = \left|\binom{m+\lambda_\alpha-\alpha+\beta-1}{\lambda_\alpha-\alpha+\beta}\right| = G_{\lambda_1, \lambda_2, \ldots \lambda_\sigma}.$$

Mithin tritt in der Zerlegung der Operation $T_n(A)$ in ihre irreduktiblen Teile jede irreduktible Operation $T^{(\lambda)}(A)$ genau so oft auf, wie ihr Grad $G^{(\lambda)}$ angiebt[1]). Da nun der Grad von $T_n(A)$ gleich $\binom{m^2+n-1}{n}$ ist, so erhalten wir wieder

$$\sum G^{(\lambda)^2} = \binom{m^2+n-1}{n}.$$

Abschnitt VII.

§ 37. Die in der Einleitung gegebene Definition der invarianten Operationen kann noch verallgemeinert werden. Es seien

$$A = (a_{\varkappa\lambda}),\ B = (b_{\varkappa\lambda}),\ C = (c_{\varkappa\lambda}), \cdots$$

Matrizen der Grade a, resp. $b, c, \ldots,$

$$A_1 = (a'_{\varkappa\lambda}),\ B_1 = (b'_{\varkappa\lambda}),\ C_1' = (c'_{\varkappa\lambda}), \ldots$$

eine zweite Reihe von Matrizen derselben Grade $a, b, c, \ldots$; hierbei sollen die Grössen $a_{\varkappa\lambda}, a'_{\varkappa\lambda}, b_{\varkappa\lambda}, b'_{\varkappa\lambda}, \ldots$ unabhängige Variabele bedeuten. Ich stelle mir die Aufgabe, alle Matrizen $T(A, B, C, \ldots)$ zu bestimmen, deren Elemente ganze rationale Funktionen der $a^2+b^2+c^2+\cdots$ Variabelen $a_{\varkappa\lambda}, b_{\varkappa\lambda}, c_{\varkappa\lambda}, \ldots$ sind, und welche der Bedingung

$$(62)\quad T(AA_1, BB_1, CC_1, \ldots) = T(A, B, C, \ldots)\ T(A_1, B_1, C_1, \ldots)$$

genügen. Ich will dann $T(A, B, C, \ldots)$ eine invariante Matrix oder Form von $A, B, C, \ldots$ nennen. Man definiert für dieselben die Begriffe der Aequivalenz und der Zerlegbarkeit ebenso, wie es in der Einleitung für die aus einer einzigen Matrix A gebildeten invarianten Matrizen geschehen ist. Ebenso soll unter der Cha-

---

1) Nach Formel (55) ergiebt sich

$$\Phi = \sum G^{(\lambda)} \Phi^{(\lambda)} = \sum \frac{1}{\alpha_1!\,\alpha_2!\ldots}\left(\frac{m\,s_1}{1}\right)^{\alpha_1}\left(\frac{m\,s_2}{2}\right)^{\alpha_2}\cdots$$
$$(\alpha_1 + 2\alpha_2 + \ldots = n);$$

die rechte Seite dieser Gleichung ist die Charakteristik der invarianten Operation $P_n(E_m \times A)$; dieselbe ist daher der im Texte betrachteten äquivalent.

rakteristik von $T(A, B, C, \ldots)$ die Summe ihrer charakteristischen Wurzeln verstanden werden.

Es soll gezeigt werden, dass der neu eingeführte Begriff auf den bisher behandelten zurückgeführt werden kann.

Es mögen zunächst nur zwei Matrizen A und B in Betracht gezogen werden. Aus (62) erhalten wir

$$(63)\qquad \left\{\begin{aligned} T(A, B) &= T(A, E_b)\, T(E_a, B) \\ &= T(E_a, B)\, T(A, E_b), \end{aligned}\right.$$

wo E_a und E_b die Einheitsmatrizen der Grade a und b bedeuten.

Die Elemente der Matrix $T(A, E_b)$ sind ganze rationale Funktionen $\varphi(a_{\varkappa\lambda})$ der a^2 Variabelen $a_{\varkappa\lambda}$ allein, und es ist

$$T(A, E_b)\, T(A_1, E_b) = T(AA_1, E_b);$$

folglich ist $T(A, E_b) = \mathrm{T}(A)$ eine aus A gebildete invariante Form; ebenso ist $T(E_a, B) = \mathrm{T}_1(B)$ eine invariante Form von B. Die Gleichungen (63) lehren uns, dass die Matrizen $\mathrm{T}(A)$ und $\mathrm{T}_1(B)$ untereinander vertauschbar sind. Ich denke mir nun eine (konstante) Matrix Q von nicht verschwindender Determinante derart bestimmt, dass $Q^{-1}\mathrm{T}(A)\,Q$ in irreduktible Matrizen

$$T_1(A), T_2(A), \ldots T_p(A),$$

deren Grade $r_1, r_2, \ldots r_p$ sein mögen, zerfällt. Denke ich mir für $Q^{-1}\, T(A, B)\, Q$ wieder $T(A, B)$ geschrieben, so sei

$$T(E_a, B) = \mathrm{T}_1(B) = (M_{\iota\varkappa}), \quad (\iota, \varkappa = 1, 2, \ldots p)$$

wo $M_{\iota\varkappa}$ eine aus r_ι Zeilen und $r_\varkappa$ Spalten bestehende Matrix bedeutet. Aus der Vertauschbarkeit von $\mathrm{T}(A)$ und $\mathrm{T}_1(B)$ folgt

$$T_\iota(A)\, M_{\iota\varkappa} = M_{\iota\varkappa}\, T_\varkappa(A).$$

Sind nun $T_\iota(A)$ und $T_\varkappa(A)$ nicht einander äquivalent, so sind ihre Elemente untereinander linear unabhängig [1]) und daher muss $M_{\iota\varkappa} = 0$ sein.

---

1) Hierbei wird stillschweigend die Voraussetzung gemacht, dass der grösste vorkommende Grad n der Funktionen $\varphi(a_{\varkappa\lambda})$ nicht grösser ist als a. Dies ist insbesondere stets der Fall, wenn die Elemente von $T(A, B)$ lineare Funktionen der Variabelen $b_{\varkappa\lambda}$ und $a_{\varkappa\lambda}$ sind.

Daraus folgt, dass $\mathrm{T}_1(B)$ und also auch $T(A, B) = \mathrm{T}(A)\mathrm{T}_1(B)$ stets dann zerlegbar ist, wenn unter den in $\mathrm{T}(A)$ enthaltenen irreduktiblen Operationen zwei von einander verschieden, d. h. nicht äquivalent sind.

Soll also $T(A, B)$ irreduktibel sein, so müssen die $T_\iota(A)$ alle einander äquivalent sein, und ich kann Q so wählen, dass

$$T_1(A) = T_2(A) = \cdots = T_p(A) = T(A)$$

wird. Alsdann sind alle $M_{\iota\varkappa}$ mit $T(A)$ vertauschbar und müssen, wegen der linearen Unabhängigkeit der Elemente von $T(A)$, alle die Form $\varphi_{\iota\varkappa} E_r$ haben, wo $\varphi_{\iota\varkappa}$ eine Funktion der Variabelen $b_{\varkappa\lambda}$ und r den Grad der Matrix $T(A)$ bedeutet. Bezeichne ich die Matrix p-ten Grades $(\varphi_{\iota\varkappa})$ mit $T'(B)$, so wird

$$T(E_a, B) = \mathrm{T}_1(B) = T'(B) \times E_r,$$

und es muss $T'(B)$ eine invariante Form von B sein.

Andererseits kann ich aber setzen

$$\mathrm{T}(A) = E_p \times T(A),$$

folglich wird

$$T(A, B) = T'(B) \times T(A).$$

Man überzeugt sich nun leicht, dass $T(A, B)$ nicht irreduktibel ist, wenn $T'(B)$ zerlegbar ist.

Wir sehen also, dass sich jede irreduktible Matrix $T(A, B)$ auf die Form $T'(B) \times T(A)$ und also auch auf die Form $T(A) \times T'(B)$ transformieren lässt, wo $T(A)$ und $T'(B)$ primitive invariante Matrizen sind. Umgekehrt ist aber auch jede Produktionsformation $T(A) \times T'(B)$ eine invariante Matrix von A und B und zwar eine irreduktible, wenn es $T(A)$ und $T'(B)$ sind. Dies folgt daraus, dass alsdann die Elemente von $T(A) \times T'(B)$ untereinander linear unabhängig sind.

Ebenso überzeugt man sich leicht, dass, wenn $T(A)$ und $T_1(A)$ oder $T'(B)$ und $T'_1(B)$ verschiedene primitive Operationen sind, auch die Elemente der beiden Matrizen $T(A) \times T'(B)$ und $T_1(A) \times T'_1(B)$ untereinander linear unabhängig sind.

Auf Grund dieses Bemerkung können wir genau dieselbe Schlussweise auch auf den Fall dreier Matrizen A, B, C anwenden. Allgemein ergiebt sich:

XV. „Man gewinnt sämtliche irreduktiblen invarianten Formen $T(A, B, C, \ldots)$, indem man die Produktionsformationen

(64) $$T(A) \times T'(B) \times T''(C) \times \cdots$$

bildet, wo $T(A)$, $T'(B)$, $T''(C), \ldots$ beliebige primitive invariante Operationen bedeuten".

Die Charakteristik der Form (64) ist gleich dem Produkte der Charakteristiken von $T(A)$, $T'(B)$, $T''(C), \ldots$. Bezeichnet man die charakteristischen Wurzeln der Matrizen $A, B, C, \ldots$ mit $\alpha_1, \alpha_2, \ldots \alpha_a$, resp. mit $\beta_1, \beta_2, \ldots \beta_b$; $\gamma_1, \gamma_2, \ldots \gamma_c$ u. s. w., so folgt aus XV., dass die charakteristischen Wurzeln einer jeden invarianten Matrix $T(A, B, C, \ldots)$ Produkte der $\alpha, \beta, \gamma, \ldots$ sind, und dass die Spur von $T(A, B, C, \ldots)$ als Summe von Produkten der Form

$$f_1(\alpha_1, \alpha_2, \ldots \alpha_a) f_2(\beta_1, \beta_2, \ldots \beta_b) f_3(\gamma_1, \gamma_2, \ldots \gamma_c) \ldots$$

dargestellt werden kann, wo $f_1, f_2, f_3, \ldots$ ganze homogene symmetrische Funktionen ihrer Argumente sind.

Man beweist auch leicht den Satz:

XVa. „Jede invariante Form $T(A, B, C, \ldots)$ ist nur auf eine einzige Art in irreduktible Formen zerlegbar. Zwei invariante Formen sind dann und nur dann äquivalent, wenn ihre Charakteristiken übereinstimmen."

Sind speziell die Elemente von $T(A, B, C, \ldots)$ homogene lineare Funktionen der Variabelen $a_{\varkappa\lambda}, b_{\varkappa\lambda}, c_{\varkappa\lambda}, \ldots$, so kommen für die irreduktiblen Formen (64) nur die Matrizen $A, B, C, \ldots$ in Betracht. Daher ist jede so beschaffene Form $T(A, B, C, \ldots)$ — wenn ihre Determinante nicht identisch Null ist — einer Matrix der Gestalt

$$\begin{pmatrix} A_A &&&&&&&& \\ & \ddots &&&&&&& \\ && A_A &&&&&& \\ &&& B_B &&&&& \\ &&&& \ddots &&&& \\ &&&&& B_B &&& \\ &&&&&& C_C && \\ &&&&&&& \ddots & \\ &&&&&&&& C \ddots \end{pmatrix}$$

äquivalent [1]).

§ 32. Die zuletzt gemachte Bemerkung lässt eine wichtige Anwendung zu.

Die bisher abgeleiteten Sätze über die invarianten Operationen $T(A)$ sind unter der Voraussetzung bewiesen worden, dass die Ordnung n von $T(A)$ nicht grösser ist als der Grad m der Matrix

1) Vergl. Frobenius, Sitzungsberichte 1899, S. 482.

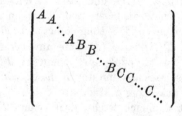

A. Ich will zeigen, dass der Fall $n > m$ sich auf den Fall $n \leqq m$ zurückführen lässt.

Ich bezeichne allgemein die Matrix r-ten Grades $(a_{\iota\varkappa})$, deren Elemente unbestimmt gelassen werden, mit A_r. Dann sei $T(A_n)$ eine beliebige invariante Form n-ter Ordnung; ich denke mir $T(A_n)$ so transformiert, dass $T(X)$ für jede Diagonalmatrix $X = (x_\iota \delta_{\iota\varkappa})$ selbst eine solche wird. Durch Vertauschung der Zeilen und Spalten von $T(A_n)$ kann ich erreichen, dass, wenn m eine gegebene ganze Zahl bedeutet, die kleiner ist als n, die r ersten Diagonalglieder von $T(X)$ nur von $x_1, x_2, \ldots x_m$ abhängen, dagegen die s letzten mindestens eine der Grössen $x_{m+1}, x_{m+2}, \ldots x_n$ als Faktor enthalten. Dann hat $T(A_n)$, wie sich aus den Ausführungen des § 8 ergiebt, die Form

$$T(A_n) = \begin{pmatrix} M_{11} & M_{12} \\ M_{21} & M_{22} \end{pmatrix}, \tag{65}$$

wo die Elemente der Matrix r-ten Grades M_{11} nur von den m^2 Variabelen

$$a_{\iota\varkappa} \qquad (\iota, \varkappa = 1, 2, \ldots m)$$

abhängen, dagegen die Elemente der drei übrigen Matrizen M_{12}, M_{21}, M_{22} alle verschwinden, wenn A_n durch $\begin{pmatrix} A_m & 0 \\ 0 & 0 \end{pmatrix}$ ersetzt wird. Man sieht leicht, dass M_{11} eine aus A_m gebildete invariante Form repräsentiert; ich will dieselbe die durch $T(A_n)$ erzeugte invariante Form $T(A_m)$ nennen.

Man erhält die Charakteristik von $T(A_m)$, indem man in der Charakteristik $\Phi = \sum r_{n_1, n_2, \ldots} \omega_1^{n_1} \omega_2^{n_2} \ldots$ von $T(A_n)$

$$\omega_{m+1} = \omega_{m+2} = \cdots = \omega_n = 0$$

setzt. Speziell hat $T(A_m)$ keine wirkliche Bedeutung, wenn $M_{22} = T(A_n)$ ist: dies tritt dann und nur dann ein, wenn in allen Gliedern $r_{n_1, n_2, \ldots n_n} \; \omega_1^{n_1} \; \omega_2^{n_2} \ldots \omega_n^{n_n}$ von Φ mehr als m Exponenten positiv sind.

Ich beweise nun:

XVI. „Sind $T(A_n)$ und $T_1(A_n)$ äquivalente Operationen, so sind auch die durch sie erzeugten $T(A_m)$ und $T_1(A_m)$ äquivalent."

Es sei nämlich in Analogie mit (65)

$$T(A_n) = \begin{pmatrix} T(A_m), & M_{12} \\ M_{21} & M_{22} \end{pmatrix}, \; T_1(A_n) = \begin{pmatrix} T_1(A_m), & M'_{12} \\ M'_{21} & M'_{22} \end{pmatrix}.$$

Ersetzt man A_n durch $X = (x_\iota \delta_{\iota\varkappa})$, so möge $T(A_m)$ übergehen in X_1, M_{22} in X_2, dann ist $T(X) = \begin{pmatrix} X_1 & \\ & X_2 \end{pmatrix}$, und hierin hängen die Diagonalglieder von X_1 nur von $x_1, x_2, \ldots x_m$ ab, während $X_2 = 0$ wird, wenn $x_{m+1} = x_{m+2} = \ldots x_n = 0$ gesetzt wird. Durch Vertauschung der Zeilen und Spalten von $T_1(A_n)$ kann ich erreichen, dass $T_1(X) = T(X)$ wird. Ist aber dann $T_1(A_n) = S^{-1}\,T(A_n)\,S$, so ist die Matrix S mit $T(X)$ vertauschbar und zerfällt daher, da die charakteristischen Gleichungen der Matrizen X_1 und X_2 offenbar keine Wurzel gemeinsam haben, in zwei Matrizen S_1 und S_2 (vergl. Seite 9), und wir erhalten $T_1(A_m) = S_1^{-1}\,T(A_m)\,S_1$.

§ 33. Es mögen nun

$$T^{(1)}(A_n),\ T^{(2)}(A_n), \ldots\ T^{(k)}(A_n)$$

die sämtlichen verschiedenen primitiven invarianten Formen n-ter Ordnung repräsentieren. Setzt man wieder $A_n = \begin{pmatrix} A_m & 0 \\ 0 & 0 \end{pmatrix}$, so werden diejenigen $T^{(\lambda)}(A)$, deren Charakteristiken mit einem Glied $\omega_1^{\lambda_1}\,\omega_2^{\lambda_2} \ldots \omega_\sigma^{\lambda_\sigma}$ beginnen, in dem $\sigma > m$ ist, alle Null, während die übrigen die invarianten Formen

$$(66) \qquad T^{(1)}(A_m),\ T^{(2)}(A_m), \ldots\ T^{(k')}(A_m)$$

erzeugen mögen. Dann sind offenbar die $T^{(\lambda)}(A_m)$ alle nicht zerlegbar, weil ihre Elemente linear unabhängig sind. Sie enthalten insgesamt $\binom{m^2+n-1}{n} = r$ linear unabhängige Elemente, die lineare homogene Funktionen der $\binom{m^2+n-1}{n}$ Produkte

$$(67) \qquad a_{\varkappa_1\lambda_1}\,a_{\varkappa_2\lambda_2} \ldots a_{\varkappa_n\lambda_n} \qquad (\varkappa_\nu, \lambda_\nu = 1, 2, \ldots m)$$

sind. Denn nach Satz XVI. lässt sich offenbar die durch $\Pi_n A_n$ erzeugte Operation $\Pi_n A_m$, welche die sämtlichen Produkte (67) als Elemente enthält, in die Formen (66), jede mehrmals genommen, zerlegen.

Ersetzt man nun die Produkte (67) durch r unabhängige Variabele $x_1, x_2, \ldots x_r$, so werden, wie man unmittelbar sieht, die r Elemente der k' Matrizen (66) selbst unabhängige Variabele. Es werde dann

$$T^{(\lambda)}(A_m) = (x_{\iota\varkappa}^{(\lambda)}) = X^{(\lambda)};$$

alsdann lassen sich die Variabelen x_ϱ als homogene lineare Funktionen der Variabelen $x_{\iota\varkappa}^{(\lambda)}$ darstellen.

Ist nun B_m eine zweite Matrix m-ten Grades, und setzt man $C_m = A_m B_m$, so möge $X^{(\lambda)}$ in $Y^{(\lambda)}$ oder $Z^{(\lambda)}$ übergehen, wenn man für A_m die Matrix B_m oder C_m substituiert. Die Gleichung $T^{(\lambda)}(A_m)\, T^{(\lambda)}(B_m) = T^{(\lambda)}(C_m)$ besagt dann, dass $X^{(\lambda)} Y^{(\lambda)} = Z^{(\lambda)}$ ist.

Es sei $T(A_m)$ eine beliebige invariante Form n-ter Ordnung, deren Determinante nicht identisch verschwindet. Ersetzt man in ihr die Produkte (67) durch die Variablen x_ϱ, so lassen sich ihre Elemente als homogene lineare Funktionen der r Variabelen $x_{\iota\varkappa}^{(\lambda)}$ darstellen. Es lässt sich aber dann $T(A_m)$ als eine aus den Matrizen

(66′) $$X^{(1)}, X^{(2)}, \ldots X^{(k')}$$

gebildete invariante Form auffassen und ist daher nach dem am Schluss des § 32 gefundenen Ergebnis in die Matrizen (66′), jede eine gewisse Anzahl von Malen genommen, zerlegbar. Soll also $T(A_m)$ speziell selbst irreduktibel sein, so muss sie einer der Formen (66) äquivalent sein [1]). Daher ist die Anzahl der verschiedenen invarianten Operationen n-ter Ordnung $T(A_m)$ gleich k', also gleich der Anzahl der Zerlegungen der ganzen Zahl n in höchstens m gleiche oder verschiedene positive Summanden. Man schliesst jetzt leicht, dass $T(A_m)$ nur auf eine einzige Art in irreduktible Formen zerlegbar ist, und dass diese Zerlegung durch die Charakteristik von $T(A_m)$ eindeutig bestimmt ist (vergl. Satz VIII. und IX.).

§ 34. Ich will jetzt einige Beispiele für die aus mehreren Matrizen $A, B, C, \ldots$ gebildeten invarianten Formen $T(A, B, C, \ldots)$ anführen.

1. Ein solches bietet sich dar, wenn man eine invariante Form $T(A)$ für den Fall betrachtet, dass A in zwei oder mehrere Matrizen niedrigeren Grades zerfällt. Ist nämlich

(68) $$A = \begin{pmatrix} A_1 & & \\ & A_2 & \\ & & \ddots \\ & & & A_q \end{pmatrix},$$

und betrachtet man die Elemente von $A_1, A_2, \ldots A_q$ als unabhängige Variabele, so kann man offenbar $T(A)$ als eine invariante Form der Matrizen $A_1, A_2, \ldots A_q$ ansehen. Daraus ergiebt sich nach Satz XV., dass man jede invariante Form $T(A)$ so trans-

1) Die hier angewandte Beweismethode ist der auf S. 64 zitierten Abhandlung des Herrn Frobenius entlehnt.

formieren kann, dass sie in eine Reihe von Produktransformationen

$$T_1(A_1) \times T_2(A_2) \times \ldots\; T_q(A_q)$$

zerfällt, sobald A die Gestalt (68) besitzt.

2. Von Interesse ist ferner die invariante Form

$$P_n(A \times B \times C) \tag{69}$$

der drei Matrizen A, B und C. Es seien, wie in § 31, $\alpha_1, \alpha_2, \ldots \alpha_a$, resp. $\beta_1, \beta_2, \ldots \beta_b$; $\gamma_1, \gamma_2, \ldots \gamma_c$ die charakteristischen Wurzeln von A, B, C; ich bezeichne mit s_ν, s'_ν, s''_ν die ν-ten Potenzsummen

$$\sum_1^a \alpha_\varrho^\nu,\ \sum_1^b \beta_\sigma^\nu,\ \sum_1^c \gamma_\tau^\nu,$$

mit $\Phi^{(\lambda)}$, resp. mit $\Phi_1^{(\lambda)}$, $\Phi_2^{(\lambda)}$ die Charakteristiken der irreduktiblen invarianten Formen n-ter Ordnung

$$T^{(\lambda)}(A),\ T^{(\lambda)}(B),\ T^{(\lambda)}(C).$$

Die charakteristischen Wurzeln der Matrix $A \times B \times C$ sind die abc Produkte $\alpha_\varrho\, \beta_\sigma\, \gamma_\tau$, die Summe der ν-ten Potenzen dieser Wurzeln ist gleich $s_\nu\, s'_\nu\, s''_\nu$. Daher ist die Charakteristik der Form (69) gleich

$$\Phi = \sum \frac{1}{\varkappa_1!\,\varkappa_2!\ldots}\left(\frac{s_1\, s'_1\, s''_1}{1}\right)^{\varkappa_1}\left(\frac{s_2\, s'_2\, s''_2}{2}\right)^{\varkappa_2}\cdots \qquad (\varkappa_1 + 2\varkappa_2 + \ldots = n)$$

Nach Satz XV. lässt sich (69) in Produktionsformationen

$$T^{(\lambda)}(A) \times T^{(\mu)}(B) \times T^{(\nu)}(C) \tag{70}$$

zerlegen. Ist (70) in (69) $f_{\lambda\mu\nu}$ Mal enthalten, so wird

$$\Phi = \sum \frac{1}{\varkappa_1!\,\varkappa_2!\ldots}\left(\frac{s_1\, s'_1\, s''_1}{1}\right)^{\varkappa_1}\left(\frac{s_2\, s'_2\, s''_2}{2}\right)^{\varkappa_2}\cdots = \sum_{\lambda,\mu,\nu} f_{\lambda\mu\nu}\, \Phi^{(\lambda)}\, \Phi_1^{(\mu)}\, \Phi_2^{(\nu)}. \tag{71}$$

Die Zahlen $f_{\lambda\mu\nu}$ haben eine einfache Bedeutung für die Charaktere der symmetrischen Gruppe n-ten Grades. Es möge nämlich, wie in § 24, der Operation $T^{(\lambda)}(A)$ der Gruppencharakter $\chi^{(\lambda)}(R)$ entsprechen. Ersetzt man in (71) $s_1^{\varkappa_1}\, s_2^{\varkappa_2}\, s_3^{\varkappa_3}\ldots$ durch $\chi^{(\lambda)}(R_{\varkappa_1,\varkappa_2,\varkappa_3\ldots})$, $s_1'^{\varkappa_1}\, s_2'^{\varkappa_2}\, s_3'^{\varkappa_3}\ldots$ durch $\chi^{(\mu)}(R_{\varkappa_1,\varkappa_2,\varkappa_3\ldots})$, wo $R_{\varkappa_1,\varkappa_2,\varkappa_3,\ldots}$ eine Permutation den n Ziffern $1, 2, \ldots n$ bedeutet, die aus $\varkappa_1$ Zykeln der Ordnung 1 besteht, $\varkappa_2$ Zykeln der Ordnung 2, u. s. w., so hat man (vergl. § 27)

$$\Phi^{(\lambda)} = 1,\ \Phi^{(\lambda')} = 0 \qquad (\lambda' \neq \lambda)$$
$$\Phi_1^{(\mu)} = 1,\ \Phi_1^{(\mu')} = 0 \qquad (\mu' \neq \mu)$$

zu setzen; es ist daher

$$\sum \frac{\chi^{(\lambda)}(R_{\varkappa_1, \varkappa_2, \varkappa_3 \ldots})\,\chi^{(\mu)}(R_{\varkappa_1, \varkappa_2, \varkappa_3 \ldots})}{\varkappa_1!\,\varkappa_2!\,\varkappa_3! \ldots} \left(\frac{s_1''}{1}\right)^{\varkappa_1} \left(\frac{s_2''}{2}\right)^{\varkappa_2} \left(\frac{s_3''}{3}\right)^{\varkappa_3} \ldots$$

$$= \sum f_{\lambda\mu\nu}\, \Phi_2^{(\nu)}.$$

Andererseits ist aber

$$\Phi_2^{(\nu)} = \sum \frac{\chi^{(\nu)}(R_{\varkappa_1, \varkappa_2, \varkappa_3 \ldots})}{\varkappa_1!\,\varkappa_2!\,\varkappa_3! \ldots} \left(\frac{s_1''}{1}\right)^{\varkappa_1} \left(\frac{s_2''}{2}\right)^{\varkappa_2} \left(\frac{s_3''}{3}\right)^{\varkappa_3} \ldots,$$

folglich ergiebt sich

$$\chi^{(\lambda)}(R)\,\chi^{(\mu)}(R) = \sum_\nu f_{\lambda\mu\nu}\,\chi^{(\nu)}(R).$$

Diese Formel enthält die Regeln, nach denen, wie sich Herr Frobenius[1]) ausdrückt, die Composition der Gruppencharaktere erfolgt. Die Zahlen $f_{\lambda\mu\nu}$ lassen sich durch die Charaktere darstellen, es ist nämlich[2])

$$f_{\lambda\mu\nu} = \frac{1}{n!} \sum_R \chi^{(\lambda)}(R)\,\chi^{(\mu)}(R)\,\chi^{(\nu)}(R).$$

In derselben Weise findet man, dass die Charakteristik der invarianten Form

$$T^{(\lambda)}(A \times B)$$

gleich ist

$$\sum_{\mu,\nu} f_{\lambda\mu\nu}\, \Phi^{(\mu)}\, \Phi_1^{(\nu)},$$

wo die Zahlen $f_{\lambda\mu\nu}$ dieselbe Bedeutung haben, wie oben.

Abschnitt VIII.

§ 35. In diesem Abschnitt soll eine Methode zur wirklichen Darstellung der primitiven invarianten Formen n-ter Ordnung entwickelt werden.

Zu diesem Zwecke greife ich auf die Ausführungen der §§ 3 und 4 zurück. Es sei $T(A)$ eine beliebige invariante Matrix,

(72) $$\varphi_1(y), \varphi_2(y), \ldots \varphi_r(y)$$

ein zu der linearen Substitution $T(A) = (A_{\varrho\sigma})$ gehörendes trans-

1) „Ueber die Composition der Charaktere einer Gruppe," Sitzungsberichte, 1899, S. 330.

2) ibid. Formel (4).

formables System. Ich nehme an, es seien unter den r Funktionen (72) die s ersten linear unabhängig, die übrigen $r-s=t$ lineare homogene Funktionen dieser s; und zwar sei

$$\varphi_{s+\tau}(y)=\sum_{\sigma=1}^{s} c_{\tau\sigma}\,\varphi_\sigma(y), \qquad (\tau=1,2,\ldots t)$$

wo die st Grössen $c_{\tau\sigma}$ Konstanten bedeuten.

Dann erhalte ich aus $\varphi_\varrho(y)=\sum_1^r A_{\varrho\varrho'}\,\varphi'_{\varrho'}(x)$, wenn

(73) $$A_{\varrho\sigma}+A_{\varrho,s+1}\,c_{1\sigma}+A_{\varrho,s+2}\,c_{2\sigma}+\cdots+A_{\varrho,s+t}\,c_{t\sigma}=B_{\varrho\sigma}$$
$$(\varrho=1,2,\ldots r,\ \sigma=1,2,\ldots s)$$

gesetzt wird,

$$\varphi_\varrho(y)=B_{\varrho1}\,\varphi_1(x)+B_{\varrho2}\,\varphi_2(x)+\cdots+B_{\varrho s}\,\varphi_s(x).$$

Andererseits ist aber

$$\varphi_{s+\tau}(y)=\sum_{\sigma=1}^{s} c_{\tau\sigma}\,\varphi_\sigma(y)=\sum_{\sigma=1}^{s} c_{\tau\sigma}\sum_{\sigma'=1}^{s} B_{\sigma\sigma'}\,\varphi_{\sigma'}(x),$$

und daraus folgt, da $\varphi_1(x),\varphi_2(x),\ldots\varphi_s(x)$ linear unabhängig sein sollen,

(74) $$B_{s+\tau,\sigma}=\sum_{\sigma'=1}^{s} c_{\tau\sigma'}\,B_{\sigma'\sigma}.$$

Ich teile nun die Matrix r-ten Grades $(A_{\varrho\sigma})$ durch zwei aufeinander senkrechte Schnitte, die zwischen der s-ten und $s+1$-ten Zeile, resp. Spalte verlaufen, in vier Teilmatrizen, so dass also

$$(A_{\varrho\sigma})=\begin{pmatrix} M_{11} & M_{12}\\ M_{21} & M_{22}\end{pmatrix}$$

wird. Ferner bezeichne ich das Koeffizientensystem

$$\begin{matrix} c_{11} & c_{12} & \ldots & c_{1s}\\ c_{21} & c_{22} & \ldots & c_{2s}\\ \cdot & \cdot & \cdot & \cdot\\ c_{t1} & c_{t2} & \ldots & c_{ts}\end{matrix}$$

mit C und die Matrix $\begin{pmatrix} E_s & 0\\ C & E_t\end{pmatrix}$ mit L, wo E_s und E_t Einheitsmatrizen bedeuten. Dann ist $|L|=1$ und die zu L inverse Matrix L^{-1} hat die Form $\begin{pmatrix} E_s & 0\\ -C & E_t\end{pmatrix}$.

Nach Einführung dieser Bezeichnungen lassen sich die Gleichungen (73) und (74) in die Matrizenformeln

$$(73')\left\{\begin{array}{l}(B_{\sigma\sigma'}) = \quad M_{11} + M_{12}\,C,\\ (B_{s+\tau,\sigma}) = M_{21} + M_{22}\,C.\end{array}\right. \qquad (\sigma, \sigma' = 1, 2, \ldots s,\ \tau = 1, 2, \ldots t)$$

$$(74') \qquad M_{21} + M_{22}\,C = C\,(M_{11} + M_{12}\,C)$$

zusammenfassen. Daher wird

$$(74'')\left\{\begin{array}{l} L^{-1}\,T(A)\,L = \begin{pmatrix} M_{11} + M_{12}\,C, & M_{12} \\ M_{21} - CM_{11} + CM_{22} - CM_{12}\,C, & M_{22} - CM_{12}\end{pmatrix} \\ \qquad = \begin{pmatrix} M_{11} + M_{12}\,C, & M_{12} \\ 0, & M_{22} - CM_{12}\end{pmatrix}. \end{array}\right.$$

Hierin sind, wie man unmittelbar erkennt, auch die Matrizen $M_{11} + M_{12}\,C$ und $M_{22} - CM_{12}$ invariante Formen der Matrix A.

Wenn daher (vergl. Seite 15) unter den r Elementen einer Kolonne der Matrix $T(A)$ s linear unabhängig, die übrigen $r - s = t$ dagegen lineare Funktionen derselben sind, so lässt sich eine zu $T(A)$ äquivalente Matrix bilden, welche die Gestalt

$$(75) \qquad \begin{pmatrix} T_1(A) & M \\ 0 & T_2(A)\end{pmatrix}$$

hat, wo $T_1(A)$ und $T_2(A)$ wieder invariante Matrizen der Grade s und t sind. Ebenso lässt sich, wenn unter den Elementen einer Zeile von $T(A)$ s linear unabhängig, die übrigen lineare Funktionen derselben sind, nach den Ausführungen des § 4, die zu $T(A')$ konjugierte Matrix $T'(A') = \mathrm{T}(A)$ auf die Form (75), und demnach $T(A)$ auf die Form

$$(75') \qquad \begin{pmatrix} \mathrm{T}_1(A) & 0 \\ M' & \mathrm{T}_2(A)\end{pmatrix}$$

transformieren. Ich bemerke, dass $T(A)$ auch den Matrizen

$$\begin{pmatrix} T_1(A) & \\ & T_2(A)\end{pmatrix}, \text{ resp. } \begin{pmatrix} \mathrm{T}_1(A) & \\ & \mathrm{T}_2(A)\end{pmatrix}$$

äquivalent ist; denn in der That sind die Charakteristiken dieser Matrizen gleich der Charakteristik von $T(A)$.

§ 36. Wird $T(A) = (A_{\varrho\sigma})$ eine Diagonalmatrix, sobald $A = X = (x_\iota\,\delta_{\iota\varkappa})$ gesetzt wird, so wird jedes Element $A_{\varrho\sigma}$, wenn man jedes $a_{\iota\varkappa}$ durch $x_\iota a_{\iota\varkappa} y_\varkappa$ ersetzt, wo x_ι und $y_\varkappa$ beliebige Variabele bedeuten, einen gewissen Faktor

$$x_1^{\alpha_1} x_2^{\alpha_2} \dots x_m^{\alpha_m} y_1^{\beta_1} y_2^{\beta_2} \dots y_m^{\beta_m}$$

heraustreten lassen. Ich will dann sagen, das Element $A_{\varrho\sigma}$ sei vom Gewichte $[\alpha_1, \alpha_2, \dots \alpha_m; \beta_1, \beta_2, \dots \beta_m]$.

Man überzeugt sich leicht, dass, wenn $T_1(A) = (A_{\varrho\sigma})$ eine zu $T(A)$ äquivalente Matrix ist, für welche $T_1(X)$ ebenfalls eine Diagonalmatrix ist, die grösste Anzahl s der linear unabhängigen unter den Elementen von gleichem Gewichte ungeändert bleibt, wenn man $T(A)$ durch $T_1(A)$ ersetzt.

Es seien nun $\lambda_1 + \lambda_2 + \dots + \lambda_\varrho = n$ und $\varkappa_1 + \varkappa_2 + \dots + \varkappa_\varrho = n$ zwei associerte Zerlegungen der Zahl n; dann wissen wir, dass die primitive invariante Form $T_{\lambda_1, \lambda_2, \dots \lambda_\varrho}(A)$ in der Produkttransformation

$$C(A) = C_{\varkappa_1} A \times C_{\varkappa_2} A \times \dots \times C_{\varkappa_\varrho} A = (A_{\varrho\sigma})$$

enthalten sein muss. Bildet man $C(A)$ nach den Vorschriften der §§ 5 und 6, so wird $C(X)$ eine Diagonalmatrix sein und die erste Kolonne von $C(A)$ wird sämtliche Elemente $A_{\varrho\sigma}$ vom Gewichte

$$[\mu_1, \mu_2, \dots \mu_m; \lambda_1, \lambda_2, \dots \lambda_\sigma] \tag{76}$$

enthalten, wo $\mu_1, \mu_2, \dots \mu_m$ beliebige der Bedingung $\mu_1 + \mu_2 + \dots + \mu_m = n$ genügende ganze positive Zahlen bedeuten.

Sind nun unter diesen Elementen s linear unabhängig, die übrigen t lineare Funktionen dieser s, so können wir durch passende Vertauschung der Zeilen und Spalten von $C(A)$ erreichen, dass die erwähnten s Elemente in der ersten Kolonne an erster Stelle stehen.

Geht dadurch $C(A)$ in $C'(A)$ über, so ist auch $C'(X)$ eine Diagonalmatrix, welche, wie $C(X)$, in der ersten Kolonne das Glied $x_1^{\lambda_1} x_2^{\lambda_2} \dots x_\sigma^{\lambda_\sigma}$ enthält. Wir können nun nach dem Ergebnis des vorigen § eine Matrix L von nicht verschwindender Determinante bestimmen, so dass $C_1(A) = L^{-1} C'(A) L$ die Gestalt

$$\begin{pmatrix} T_1(A) & M \\ 0 & T_2(A) \end{pmatrix}$$

annimmt, wo die invariante Matrix $T_1(A)$ vom Grade s ist. Ich behaupte, dass $T_1(A)$ äquivalent $T_{\lambda_1, \lambda_2, \dots \lambda_\sigma}(A)$ ist.

Es ist nämlich (vergl. Formel (74″)) L mit $C'(X)$ vertauschbar, also ist $C_1(X) = C'(X)$, und $T_1(X)$ enthält daher in der ersten Kolonne das Glied $x^{\lambda_1} x^{\lambda_2} \dots x^{\lambda_\sigma}$.

Folglich stehen alle Elemente vom Gewichte (70) der Matrix $C_1(A)$ in der ersten Kolonne von $T_1(A)$ und müssen, da ihre Anzahl gleich s ist, untereinander linear unabhängig sein.

Wäre nun $T_1(A)$ in $T_1'(A)$ und $T_1''(A)$ zerlegbar, so könnte nur eine der Spuren von $T_1'(X)$ und $T_1''(X)$ das Glied $x_1^{\lambda_1} x_2^{\lambda_2} \dots x_\sigma^{\lambda_\sigma}$ enthalten, und man könnte $C_1(A)$ durch eine äquivalente Matrix $C_2(A)$ ersetzen, in der die Anzahl der Elemente vom Gewichte (76) kleiner wäre als s. Dies ist aber nicht möglich, da $C_2(A)$ und $C(A)$ äquivalent wären. Da nun die Charakteristik von $T_1(A)$ mit dem Glied $\omega_1^{\lambda_1} \omega_2^{\lambda_2} \dots \omega_\sigma^{\lambda_\sigma}$ beginnt, so ist, wie zu beweisen war $T_1(A)$ äquivalent $T_{\lambda_1, \lambda_2, \dots \lambda_\sigma}(A)$.

Wir erkennen zugleich, dass die hier betrachtete Zahl s gleich ist dem Grade $G_{\lambda_1, \lambda_2, \dots \lambda_\sigma}$ der Matrix $T_{\lambda_1, \lambda_2, \dots \lambda_\sigma}(A)$.

Man sieht leicht ein, dass diese Zahl auch die grösste Anzahl der linear unabhängigen unter den Elementen der ersten Zeile von $C(A)$ angiebt.

§ 37. Ich will für das eben Gesagte ein Beispiel anführen. Es sei

$$(a) \qquad \begin{matrix} a_{11} & a_{12} & \dots & a_{1m} \\ a_{21} & a_{22} & \dots & a_{2m} \\ \cdot & \cdot & \cdot\cdot\cdot & \cdot \\ a_{p1} & a_{p2} & \dots & a_{pm} \end{matrix}$$

ein System von p Zeilen und m Spalten, und es sei $p < m$.

Die Grössen $a_{\iota\varkappa}$ sollen, wie stets, unabhängige Variabele bedeuten. Aus (a) lassen sich $\binom{m}{p} = m_p$ Determinanten des Grades p bilden und aus diesen $\binom{m_p + n - 1}{n} = r$ verschiedene Produkte von je n.

Ich stelle mir die Aufgabe, zu bestimmen, wie viele unter diesen r Produkten, als Funktionen der $a_{\iota\varkappa}$ betrachtet, linear unabhängig sind.

Zu diesem Zwecke ergänze ich das System (a) durch Hinzufügung weiterer $m - p$ Zeilen zu einem quadratischen System A und bilde die Produkttransformation

$$C_pA \times C_pA \times \cdots \times C_pA = \Pi_n(C_pA).$$

Dann stehen die zu betrachtenden r Produkte alle in der ersten Zeile von $\Pi_n(C_pA)$, und diese Zeile enthält kein von diesen verschiedenes Element. Daher ist die gesuchte Anzahl s gleich dem Grad der irreduktiblen invarianten Form $T_{n,n,\ldots n}(A)$ also ist (vergl. Formel (46'))

$$s = \left|\binom{m}{p-\gamma+\delta}\right| = \left|\binom{m+n-\alpha+\beta-1}{n-\alpha+\beta}\right| \quad \begin{pmatrix}\alpha,\beta = 1,2\cdots p\\ \gamma,\delta = 1,2\cdots n\end{pmatrix}$$

$$= \prod_{\nu=0}^{n-1} \frac{(m+\nu)!\,\nu!}{(m-p+\nu)!\,(p+\nu)!} = \prod_{\lambda=1}^{p} \frac{(m+n-\lambda)!\,(\lambda-1)!}{(n+p-\lambda)!\,(m-\lambda)!}$$

$$= \prod_{\nu=0}^{n-1} \frac{\binom{m+\nu}{p}}{\binom{p+\nu}{p}} = \prod_{\lambda=1}^{p} \frac{\binom{m+n-\lambda}{n}}{\binom{p+n-\lambda}{n}}.$$

Lebenslauf.

Ich, Issai Schur, geboren am 10. Januar 1875 zu Mohilew am Dniepr in Russland, jüdischer Religion, erhielt meine Schulbildung auf dem Nicolai-Gymnasium zu Libau. Michaelis 1894 bezog ich mit dem Zeugniss der Reife die Universität zu Berlin, um mich vornehmlich mathematischen Studien zu widmen, und besuchte die Vorlesungen der Herren Professoren:

Bauschinger, von Bezold, Blasius, Du Bois-Reymond(†), Dilthey, Fischer, Frobenius, Fuchs, Hensel, Hettner, Jastrow, Knoblauch, Kötter, Paulsen, Planck, Rubens, Simmel, Schwarz, Stumpf, Wagner, Warburg.

Ausserdem beteiligte ich mich an den Uebungen des mathematischen Seminars, an den von Herrn Prof. Schwarz geleiteten mathematischen Colloquien, an den mathematisch-physikalischen Uebungen des Herrn Prof. Planck sowie an den unter Leitung des Herrn Prof. Blasius stehenden Uebungen zur Experimentalphysik.

Allen meinen verehrten Lehrern, besonders den Herren Professoren Frobenius, Fuchs, Hensel und Schwarz fühle ich mich zu grossem Danke verpflichtet.

Auch dem Mathematischen Verein der Universität Berlin spreche ich meinen wärmsten Dank aus für die mannigfache Anregung und Belehrung, die ich in seiner Mitte gefunden habe.

Thesen.

I.

Die unvollständige Induktion spielt in der mathematischen Forschung eine wesentliche Rolle.

II.

Es ist nicht möglich, allgemein geltende Grundsätze für die Wertschätzung mathematischer Probleme anzugeben.

III.

In der Zahlentheorie verdient das additive Prinzip nicht geringere Beachtung als das multiplikative.

IV.

Die Annahme, dass alle Naturwissenschaften eine mathematische Behandlungsweise zulassen müssen, ist als eine hypothetische anzusehen.

Opponenten:

Herr stud. math. **L. Lewent,**
Herr Dr. phil. **R. Ziegel,**
Herr Dr. phil. **E. Landau,** Privatdocent a. d. Universität.

2.

Über einen Satz aus der Theorie der vertauschbaren Matrizen

Sitzungsberichte der Preussischen Akademie der Wissenschaften 1902,
Physikalisch-Mathematische Klasse, 120 - 125

Hr. Frobenius hat in seiner Arbeit »Über vertauschbare Matrizen« (Sitzungsberichte 1896) folgenden Satz bewiesen:

»Ist $f(x, y, z, \cdots)$ eine beliebige (ganze rationale) Function der m Variabeln $x, y, z, \cdots$, sind $A, B, C, \cdots$ m Formen, von denen je zwei vertauschbar sind, und sind $a_1, a_2, a_3, \cdots$ (resp. $b_1, b_2, b_3, \cdots$; $c_1, c_2, c_3, \cdots$) die Wurzeln der charakteristischen Gleichung von A (resp. $B, C, \cdots$), so lassen diese Wurzeln sich einander, und zwar unabhängig von der Wahl von f, so zuordnen, dass die Wurzeln der charakteristischen Gleichung der Form $f(A, B, C, \cdots)$ den Grössen $f(a_1, b_1, c_1, \cdots)$, $f(a_2, b_2, c_2, \cdots)$, $f(a_3, b_3, c_3, \cdots)$, $\cdots$ gleich werden.«

Für diesen Satz soll im Folgenden ein einfacher Beweis abgeleitet werden, der von einer Zerlegung der Formen $A, B, C, \cdots$ keinerlei Gebrauch macht.

Es seien

$$\varphi(r) = \prod_{i=1}^{n}(r-a_i),\ \psi(r) = \prod_{i=1}^{n}(r-b_i),\ \chi(r) = \prod_{i=1}^{n}(r-c_i),\ \cdots$$

die charakteristischen Functionen der Formen A, resp. $B, C, \cdots$.

Ich betrachte das Product

$$(1.)\qquad \prod_{\varkappa, \lambda, \mu, \cdots} \{f(x, y, z, \cdots) - f(a_\varkappa, b_\lambda, c_\mu, \cdots)\},$$

wo die Indices $\varkappa, \lambda, \mu, \cdots$ unabhängig von einander die Werthe $1, 2, \cdots n$ durchlaufen.

Der Factor $f(x, y, z, \cdots) - f(a_\varkappa, b_\lambda, c_\mu, \cdots)$ dieses Productes ist gleich

$$\{f(x, b_\lambda, c_\mu, \cdots) - f(a_\varkappa, b_\lambda, c_\mu, \cdots)\} + \{f(x, y, z, \cdots) - f(x, b_\lambda, c_\mu, \cdots)\},$$

also gleich

$$(x-a_\varkappa)\, g_{\varkappa, \lambda, \mu, \cdots} + \{f(x, y, z, \cdots) - f(x, b_\lambda, c_\mu, \cdots)\},$$

wo $g_{\varkappa, \lambda, \mu, \cdots}$ eine ganze rationale Function von x bedeutet.

Daher lässt sich das Theilproduct

$$\prod_{\varkappa=1}^{n} \{f(x, y, z, \cdots) - f(a_\varkappa, b_\lambda, c_\mu, \cdots)\}$$

von (1.) auf die Form

$$h_{\lambda,\mu,\ldots}\,\varphi(x) + h'_{\lambda,\mu,\ldots}\{f(x, y, z, \cdots) - f(x, b_\lambda, c_\mu, \cdots)\}$$

bringen, wo $h_{\lambda,\mu,\ldots} = \prod\limits_{\varkappa=1}^{n} g_{\varkappa,\lambda,\mu,\ldots}$ zu setzen ist, und $h'_{\lambda,\mu,\ldots}$ eine gewisse ganze rationale Function von $x, y, z, \cdots$ bedeutet. Durch weitere Ausführung der Multiplication erhält man für (1.) einen Ausdruck der Form

$$K\varphi(x) + K' \prod_{\lambda,\mu,\cdots} \{f(x, y, z, \cdots) - f(x, b_\lambda, c_\mu, \cdots)\}$$
$$(K' = \prod_{\lambda,\mu,\cdots} h'_{\lambda,\mu,\ldots})$$

und, indem man auf das im zweiten Gliede stehende Product die analogen Schlüsse anwendet, zuletzt die Gleichung

$$(2.)\quad \prod_{\varkappa,\lambda,\mu,\cdots} \{f(x, y, z, \cdots) - f(a_\varkappa, b_\lambda, c_\mu, \cdots)\} = K\varphi(x) + L\psi(y) + M\chi(z) + \cdots;$$

hierin bedeuten $K, L, M, \cdots$ gewisse Functionen der Variabeln $x, y, z, \ldots$.

Ersetzt man, was wegen der Vertauschbarkeit der Matrizen $A, B, C, \ldots$ gestattet ist, in (2.) die Variabeln $x, y, z, \ldots$ durch A, bez. $B, C, \ldots$, so ergibt sich, weil bekanntlich

$$(3.)\quad \varphi(A) = 0, \; \psi(B) = 0, \; \chi(C) = 0, \; \cdots$$

ist, die Gleichung

$$\prod_{\varkappa,\lambda,\mu,\cdots} \{f(A, B, C, \cdots) - f(a_\varkappa, b_\lambda, c_\mu, \cdots)E\} = 0.$$

Wir sehen also, dass die Form $f(A, B, C, \ldots)$ die Gleichung

$$F(r) = \prod_{\varkappa,\lambda,\mu,\cdots} \{r - f(a_\varkappa, b_\lambda, c_\mu, \cdots)\} = 0$$

befriedigt. Ist daher $\Phi(r) = 0$ die Gleichung niedrigsten Grades, der diese Form genügt, so muss $\Phi(r)$ ein Divisor von $F(r)$ sein und folglich die Gestalt

$$\Phi(r) = \prod_{i=1}^{p} \{r - f(a_i, b_i, c_i, \cdots)\}$$

besitzen, wo p den Grad der Function $\Phi(r)$ bedeutet. Hieraus folgt aber nach einem bekannten Satze

$$(4.)\quad |rE - f(A, B, C, \cdots)| = \prod_{i=1}^{p} \{r - f(a_i, b_i, c_i, \cdots)\}^{n_i},$$

wobei $n_1, n_2, \cdots n_p$ gewisse positive ganze Zahlen sind.

Daher hat jede Wurzel der charakteristischen Gleichung der Matrix $f(A, B, C, \cdots)$ die Form $f(a_i, b_i, c_i, \cdots)$.

Dass nun die Zuordnung der Wurzeln $a_i, b_i, c_i, \cdots$ für jede Function f dieselbe ist, sieht man folgendermassen ein.

Mit Hülfe der Gleichungen (1.) lässt sich jede Function von $A, B, C, \cdots$ als Summe

$$\Sigma x_{\varkappa, \lambda, \mu, \ldots} A^\varkappa B^\lambda C^\mu \cdots \qquad (\varkappa, \lambda, \mu, \cdots = 0, 1, 2, \cdots, n-1)$$

darstellen, wo die n^m Grössen $x_{\varkappa, \lambda, \mu \ldots}$ Constanten bedeuten. Man erhält den allgemeinsten Ausdruck für eine Function der Matrizen $A, B, C, \cdots$, indem man diese n^m Grössen als unabhängige Variabele ansieht. Die dieser Form entsprechende Gleichung (4.) wird aber eine in den Variabeln $x_{\varkappa, \lambda, \mu, \ldots}$ identische rationale Gleichung und bleibt demnach für jede specielle Wahl dieser Variabeln bestehen.

Setzt man speciell $f(A, B, C, \cdots) = A$, so geht (4.) über in

$$|rE - A| = \prod_{i=1}^{p} (r - a_i)^{n_i};$$

daher stellen die Wurzeln $a_1, a_2, \cdots a_p$, die i^{te} Wurzel n_i mal gezählt, die sämmtlichen n Wurzeln der charakteristischen Gleichung von A dar. Dasselbe gilt für die Wurzeln $b_1, b_2, \cdots b_p$ in Bezug auf B u. s. w.

Den Grundgedanken des hier durchgeführten Beweises habe ich bereits vor längerer Zeit Hrn. Prof. FROBENIUS mitgetheilt, dessen Rathschlägen ich die Fassung des Beweises in der vorliegenden Form verdanke.

Aus dem Vorhergehenden ergibt sich, dass die Gleichung niedrigsten Grades, der die lineare Function $xA + yB + zC + \cdots$ mit unbestimmten Coefficienten genügt, die Gestalt

$$(1.) \qquad \prod_{i=1}^{r} \{xA + yB + zC + \cdots - (xa_i + yb_i + zc_i + \cdots)E\} = 0$$

besitzt. Es soll nun gezeigt werden, dass auch für jede beliebige Function f die Gleichung

$$\prod_{i=1}^{r} \{f(A, B, C, \cdots) - f(a_i, b_i, c_i, \cdots)E\} = 0$$

besteht.

Setzt man $A - a_i E = A_i$, $B - b_i E = B_i$, $C - c_i E = C_i, \cdots$, so geht (1.) über in

$$(2.) \qquad \prod_{i=1}^{r} (xA_i + yB_i + zC_i + \cdots) = 0.$$

Ich will beweisen, dass die Gleichung (2.) die allgemeinere Gleichung

$$(3.) \qquad \prod_{i=1}^{r} (x_i A_i + y_i B_i + z_i C_i + \cdots) = 0$$

nach sich zieht, wo die mr Grössen $x_i, y_i, z_i, \cdots$ unabhängige Variable bedeuten; oder anders ausgedrückt: aus (2.) folgt, dass jedes aus r Factoren bestehende Product

$$(4.)\qquad A_{\alpha_1} A_{\alpha_2} \cdots B_{\beta_1} B_{\beta_2} \cdots C_{\gamma_1} C_{\gamma_2} \cdots$$

den Werth Null hat, sobald die r Indices $\alpha_1, \alpha_2, \cdots, \beta_1, \beta_2, \cdots, \gamma_1, \gamma_2, \cdots$, abgesehen von der Reihenfolge, den Zahlen $1, 2, \cdots r$ gleich sind.

Es möge (2.), wenn in dem linker Hand stehenden Producte die einander gleichen Factoren zu je einer Potenz vereinigt werden, übergehen in

$$(2'.)\qquad \prod_{i=1}^{s} (xA_i + yB_i + zC_i + \cdots)^{r_i} = 0.$$

Es genügt, zu zeigen, dass sich aus (2'.) die Gleichung

$$(3'.)\qquad \prod_{i=1}^{s} (x_i A_i + y_i B_i + z_i C_i + \cdots)^{r_i} = 0$$

ergibt. Denn in der That lässt sich jedes Product (4.) in der Form

$$(5.)\qquad A_1^{\varkappa_1} B_1^{\lambda_1} C_1^{\mu_1} \cdots A_2^{\varkappa_2} B_2^{\lambda_2} C_2^{\mu_2} \cdots$$

schreiben, wobei die Exponenten $\varkappa_i, \lambda_i, \mu_i, \cdots$ den Bedingungen

$$\varkappa_1 + \lambda_1 + \mu_1 + \cdots = r_1, \quad \varkappa_2 + \lambda_2 + \mu_2 + \cdots = r_2, \cdots$$

zu genügen haben. Das Verschwinden dieser Producte ist aber eine Folge der Gleichung (3'.).

Ferner kann ich ohne Beschränkung der Allgemeinheit voraussetzen, dass die s Wurzeln $a_1, a_2, \cdots a_s$ alle unter einander verschieden sind. Denn es sei für diesen Fall unsere Behauptung bewiesen. Da keine der Grössen

$$x(a_\varkappa - a_\lambda) + y(b_\varkappa - c_\lambda) + z(c_\varkappa - c_\lambda) + \cdots \qquad (\varkappa, \lambda = 1, 2, \cdots s,\ \varkappa \gtrless \lambda)$$

für alle Werthe von $x, y, z, \cdots$ verschwinden soll, lassen sich gewisse Constanten $p, q, \cdots$ derart bestimmen, dass die s Grössen

$$a'_i = a_i + pb_i + qc_i + \cdots \qquad (i = 1, 2, \cdots s)$$

alle unter einander verschieden werden. Ersetzt man A durch die Matrix

$$A' = A + pB + qC + \cdots,$$

so besteht, wenn $A' - a'_i E = A'_i$ gesetzt wird, die Gleichung

$$\prod_{i=1}^{s} (xA'_i + yB_i + zC_i + \cdots)^{r_i} = \prod_{i=1}^{s} \{xA_i + (y+px)\, B_i + (z+qx)\, C_i + \cdots\}^{r_i} = 0.$$

Aus dieser Gleichung folgt nun nach unserer Annahme das Verschwinden von

$$\prod_{i=1}^{s} (x_i A'_i + y_i B_i + z_i C_i + \cdots)^{r_i} = \prod_{i=1}^{s} \{x_i A_i + (y_i + px_i)\, B_i + (z_i + qx_i)\, C_i + \cdots\}^{r_i};$$

daraus ergiebt sich aber auch die Richtigkeit von (3'.), denn die Grössen

$$x_i, y_i + px_i, z_i + qx_i, \cdots$$

stellen zugleich mit $x_i, y_i, z_i, \cdots$ ein System von ms unabhängigen Variabeln dar.

Es seien also die s Wurzeln $a_1, a_2, \cdots a_s$ alle unter einander verschieden. Ist speciell in (5.) $\varkappa_1 + \varkappa_2 + \cdots = r$, mithin $\lambda_i = 0$, $\mu_i = 0, \cdots$, so wird das Product (5.) gleich dem Coefficienten von x^r in der Entwickelung der linken Seite der Gleichung (2'.) nach Potenzen von $x, y, z, \cdots$, also gleich Null.

Ich kann daher annehmen, das Verschwinden der Producte (5.) stehe bereits fest für alle Werthe der Exponenten $\lambda_i, \mu_i, \cdots$, sobald $\Sigma\varkappa_i$ grösser ist als eine gegebene ganze Zahl k. Ich brauche dann nur zu zeigen, dass das Product (5.), das ich mit P bezeichnen will, auch dann verschwindet, wenn $\Sigma\varkappa_i = k$ ist.

Sind in P für ein i nicht alle Exponenten $\lambda_i, \mu_i, \cdots$ gleich Null, so muss

$$A_i P = 0 \tag{6.}$$

sein. Denn ist etwa $\lambda_i > 0$, so ist der zu B_i complementäre Factor von $A_i P$ gleich $A_1^{\varkappa_1} B_1^{\lambda_1} \cdots A_{i-1}^{\varkappa_{i-1}} B_{i-1}^{\lambda_{i-1}} \cdots A_i^{\varkappa_i+1} B_i^{\lambda_i - 1} \cdots$, also gleich Null, weil in diesem Producte die Summe der Exponenten von $A_1, A_2 \cdots$ gleich $k+1$ ist.

Daraus schliesst man sofort, dass $P = 0$ ist, sobald unter den Exponenten $\lambda_1, \lambda_2, \cdots, \mu_1, \mu_2 \cdots$ zwei nicht Null und mit von einander verschiedenen Indices ρ und σ versehen sind. Denn es ist dann wegen (6.)

$$(A_\rho - A_\sigma)P = (a_\sigma - a_\rho)P = 0$$

und mithin, weil $a_\sigma \neq a_\rho$ ist, auch $P = 0$.

Man hat daher nur noch das Verschwinden der Producte

$$P_\rho = A_1^{r_1} A_2^{r_2} \cdots A_{\rho-1}^{r_{\rho-1}} A_\rho^{\varkappa_\rho} B_\rho^{\lambda} C_\rho^{\mu} \cdots A_{\rho+1}^{r_{\rho+1}} \cdots$$

für alle Werthe von $\varkappa_\rho, \lambda, \mu, \cdots$ zu beweisen, die den Bedingungen

$$\varkappa_\rho = k - (r_1 + \cdots r_{\rho-1} + r_{\rho+1} + \cdots) = r_\rho - (\lambda + \mu + \cdots) \geqq 0 \tag{7.}$$

genügen. — Indem man die mit $x^k y^\lambda z^\mu \cdots$ multiplicirte Matrix in der Entwickelung der linken Seite von (2'.) gleich Null setzt, erhält man unter Fortlassung der Producte, deren Verschwinden bereits feststeht, die Gleichung

$$p_1 P_1 + p_2 P_2 + \cdots = 0, \tag{8.}$$

wo die Coefficienten $p_1, p_2, \cdots$ gewisse ganze Zahlen bedeuten, die auch Null sein können. Sind aber für irgend ein ρ die Bedingungen (7.) erfüllt, so ist $p_\rho = \dfrac{r_\rho!}{\varkappa_\rho!\,\lambda!\,\mu!\cdots} > 0$.

Multiplicirt man beide Seiten der Gleichung (8.) mit $A_1 \cdots A_{\varrho-1}$ $A_{\varrho+1} \cdots A_s$, so erhält man wegen (6.)

$$(9.) \qquad A_1 \cdots A_{\varrho-1} A_{\varrho+1} \cdots A_s \cdot P_\varrho = 0.$$

Um sich nun von dem Verschwinden des Productes P_ϱ zu überzeugen, genügt es, dasselbe mit der nicht verschwindenden Constante

$$(a_\varrho - a_1) \cdots (a_\varrho - a_{\varrho-1})(a_\varrho - a_{\varrho+1}) \cdots (a_\varrho - a_s)$$

oder, was dasselbe ist, mit der Matrix

$$R = (A_1 - A_\varrho) \cdots (A_{\varrho-1} - A_\varrho)(A_{\varrho+1} - A_\varrho) \cdots (A_s - A_\varrho)$$

zu multipliciren. Denn es ist

$$R = A_1 \cdots A_{\varrho-1} A_{\varrho+1} \cdots A_s + A_\varrho R',$$

wo R' eine Function von A ist. Daher ist vermöge der Gleichungen (6.) und (9.) $RP_\varrho = 0$ und also auch $P_\varrho = 0$.

Damit ist bewiesen, dass die Producte (5.) alle den Werth Null haben. Diess ist aber der Inhalt der zu beweisenden Gleichung (3).

Diese Gleichung bleibt bestehen, wenn man die Variabeln $x_i, y_i, z_i, \cdots$ durch beliebige, mit $A, B, C, \cdots$ vertauschbare Matrizen ersetzt.

Es sei nun $f(x, y, z, \cdots)$ eine beliebige ganze rationale Function von $x, y, z, \cdots$; dann ist

$$f(x, y, z, \cdots) - f(a_i, b_i, c_i, \cdots) = \{f(x, b_i, c_i, \cdots) - f(a_i, b_i, c_i, \cdots)\}$$
$$+ \{f(x, y, c_i, \cdots) - f(x, b_i, c, \cdots)\} + \{f(x, y, z, \cdots) - f(x, y, c_i, \cdots)\} + \cdots;$$

der erste Summand der rechts stehenden Summe ist aber durch $x - a_i$ theilbar, der zweite Summand durch $y - b_i$, der dritte durch $z - c_i$, u. s. w. Demnach lässt sich die Differenz $f(A, B, C, \cdots) - f(a_i, b_i, c_i, \cdots)E$ in der Form

$$X_i A_i + Y_i B_i + Z_i C_i + \cdots$$

darstellen, wo $X_i, Y_i, Z_i, \cdots$ ganze rationale Functionen von $A, B, C, \cdots$ und also mit $A, B, C, \cdots$ vertauschbar sind. Daher ist, wie zu beweisen war,

$$\prod_{i=1}^{r} \{f(A, B, C, \cdots) - f(a_i, b_i, c_i, \cdots)E\} = 0.$$

Ausgegeben am 20. Februar.

3.

Neuer Beweis eines Satzes über endliche Gruppen

Sitzungsberichte der Preussischen Akademie der Wissenschaften 1902,
Physikalisch-Mathematische Klasse, 1013 - 1019

In seiner Abhandlung »Über auflösbare Gruppen. IV.« (Sitzungsberichte 1901, S. 1216) hat Hr. Frobenius mit Hülfe der Theorie der Gruppencharaktere folgenden Satz bewiesen:

I. »Ist die Gruppe $\mathfrak{G}$ der Ordnung g in der Gruppe $\mathfrak{H}$ der Ordnung $h = gn$ enthalten, sind je zwei Elemente von $\mathfrak{G}$, die in $\mathfrak{H}$ conjugirt sind, auch schon in $\mathfrak{G}$ conjugirt, ist r die Ordnung und m der Index der Commutatorgruppe $\mathfrak{R}$ von $\mathfrak{G}$, und sind m und n theilerfremd, so erzeugen die Elemente von $\mathfrak{H}$, deren Ordnungen in n aufgehen, zusammen mit der Commutatorgruppe von $\mathfrak{H}$ eine charakteristische Untergruppe $\mathfrak{S}$ von $\mathfrak{H}$, deren Ordnung s durch r und n, und deren Index durch m theilbar ist. Sind g und n theilerfremd, so ist $s = rn = \frac{h}{m}$, und die commutative Gruppe $\frac{\mathfrak{H}}{\mathfrak{S}}$ ist der Gruppe $\frac{\mathfrak{G}}{\mathfrak{R}}$ isomorph.«

Für diesen Satz soll im Folgenden ein neuer Beweis abgeleitet werden, der zwar in seinen Grundzügen dem von Hrn. Frobenius gegebenen Beweis nahe verwandt ist, der aber von der Theorie der Gruppencharaktere keinen Gebrauch macht.

Es seien $G_1, G_2, \cdots G_g$ die Elemente von $\mathfrak{G}$, ferner sei

$$\mathfrak{H} = \mathfrak{G}A_1 + \mathfrak{G}A_2 + \cdots + \mathfrak{G}A_n.$$

Dann lässt sich jedes Element P von $\mathfrak{H}$ auf eine und nur eine Weise in der Form $G_\lambda A_\alpha$ darstellen; es möge das Element G_λ mit G_P, das Element A_α mit A_P bezeichnet werden. Ist dann Q ein beliebiges zweites Element von $\mathfrak{H}$, so ergibt sich aus

$$PA_P^{-1} = G_P$$

und

$$A_P Q = G_{A_P Q} A_{A_P Q}$$

die Gleichung

$$PQ = G_P G_{A_P Q} A_{A_P Q}.$$

Hieraus folgt

$$G_P G_{A_P Q} = G_{PQ}.$$

Ersetzt man P durch das Element $A_\alpha P$, so ergibt sich

$$G_{A_\alpha P} G_{A_{A_\alpha P} Q} = G_{A_\alpha PQ};$$

also ist auch

$$\mathfrak{R} G_{A_\alpha P} \cdot \mathfrak{R} G_{A_{A_\alpha P} Q} = \mathfrak{R} G_{A_\alpha PQ}.$$

Bildet man nun auf beiden Seiten dieser Gleichung das Product über $\alpha = 1, 2, \cdots n$ und beachtet, dass die Complexe $\mathfrak{R}G_1, \mathfrak{R}G_2, \cdots \mathfrak{R}G_g$ unter einander vertauschbar sind, so erhält man

$$\text{(1.)} \qquad \mathfrak{R} \prod_\alpha G_{A_\alpha P} \cdot \mathfrak{R} \prod_\alpha G_{A_{A_\alpha P} Q} = \mathfrak{R} \prod_\alpha G_{A_\alpha PQ}.$$

Nun ist aber, weil offenbar das Element $A_{A_\alpha P}$ zugleich mit A_α alle Elemente der Reihe $A_1, A_2, \cdots A_n$ durchläuft,

$$\mathfrak{R} \prod_\alpha G_{A_{A_\alpha P} Q} = \mathfrak{R} \prod_\alpha G_{A_\alpha Q}.$$

Setzt man daher für jedes Element P von $\mathfrak{H}$

$$\mathfrak{R} \prod_\alpha G_{A_\alpha P} = \mathfrak{R}_P,$$

so ergibt sich aus (1.) die Beziehung

$$\text{(2.)} \qquad \mathfrak{R}_P \mathfrak{R}_Q = \mathfrak{R}_{PQ}.$$

Ordnet man also dem Element P von $\mathfrak{H}$ den Complex $\mathfrak{R}_P$ zu, so entspricht für je zwei Elemente P, Q von $\mathfrak{H}$ dem Element PQ der Complex $\mathfrak{R}_P \mathfrak{R}_Q$. Die Complexe $\mathfrak{R}_P$ bilden also gewissermaassen eine Darstellung der Gruppe $\mathfrak{H}$.

Es verdient noch bemerkt zu werden, dass die Complexe $\mathfrak{R}_P$ sich nicht ändern, wenn man anstatt des vollständigen Restsystems $A_1, A_2, \cdots A_n$ von $\mathfrak{H}$ mod. $\mathfrak{G}$ ein anderes vollständiges Restsystem

$$B_1 = S_{A_1} A_1, \qquad B_2 = S_{A_2} A_2, \cdots \qquad B_n = S_{A_n} A_n$$

betrachtet, wo die Elemente $S_{A_1}, S_{A_2}, \cdots S_{A_n}$ alle der Gruppe $\mathfrak{G}$ angehören sollen.

Denn setzt man, wenn $P = G_\mu B_\alpha$ ist, $G_\mu = H_P$ und $B_\alpha = B_P$, so ergibt sich

$$P = G_P A_P = H_P B_P = H_P S_{A_P} A_P,$$

also

$$H_P = G_P S_{A_P}^{-1}.$$

Ferner ist offenbar für jedes Element S von $\mathfrak{G}$

$$G_{SP} = S G_P, \qquad A_{SP} = A_P,$$

also speciell

$$G_{B_\alpha P} = S_{A_\alpha} G_{A_\alpha P}, \qquad A_{B_\alpha P} = A_{A_\alpha P}.$$

Daher erhält man

$$\begin{aligned}\mathfrak{R}\prod_\alpha H_{B_\alpha P} &= \mathfrak{R}\prod_\alpha G_{B_\alpha P} S^{-1}_{A_{B_\alpha P}} \\ &= \mathfrak{R}\prod_\alpha S_{A_\alpha} G_{A_\alpha P} S^{-1}_{A_{A_\alpha P}};\end{aligned}$$

dies ist aber gleich

$$\mathfrak{R}\prod_\alpha G_{A_\alpha P} \cdot \mathfrak{R}\prod_\alpha S_{A_\alpha} S^{-1}_{A_{A_\alpha P}}.$$

Da nun $A_{A_\alpha P}$ zugleich mit A_α alle Elemente $A_1, A_2, \cdots A_n$ durchläuft, so ist

$$\mathfrak{R}\prod_\alpha S_{A_\alpha} S^{-1}_{A_{A_\alpha P}} = \mathfrak{R}\prod_\alpha S_{A_\alpha} S^{-1}_{A_\alpha} = \mathfrak{R},$$

und daher ist in der That, wie zu beweisen war,

$$\mathfrak{R}\prod_\alpha H_{B_\alpha P} = \mathfrak{R}_P.$$

Ich bemerke auch noch Folgendes.

Geht man von einer Gleichung der Form

$$(3.) \qquad \mathfrak{H} = C_1\mathfrak{G} + C_2\mathfrak{G} + \cdots + C_n\mathfrak{G}$$

aus und setzt man, wenn $P = C_\alpha G_\nu$ ist, $C_\alpha = C_P$, $G_\nu = J_P$ und

$$\mathfrak{R}'_P = \mathfrak{R}\prod_\alpha J_{PC_\alpha},$$

so beweist man in ganz analoger Weise, wie es bei dem Beweis der Formel (2.) geschehen ist, dass auch für die Complexe $\mathfrak{R}'_P$ die Beziehung $\mathfrak{R}'_P\mathfrak{R}'_Q = \mathfrak{R}'_{PQ}$ besteht. Man erhält also scheinbar eine neue Darstellung der Gruppe $\mathfrak{H}$ durch Complexe der Gruppe $\frac{\mathfrak{G}}{\mathfrak{R}}$, und es ist von Interesse, zu zeigen, dass die so erhaltenen Complexe $\mathfrak{R}'_P$ von den Complexen $\mathfrak{R}_P$ nicht verschieden sind. Aus (3.) folgt nämlich auch

$$\mathfrak{H} = \mathfrak{G}C_1^{-1} + \mathfrak{G}C_2^{-1} + \cdots + \mathfrak{G}C_n^{-1}.$$

Da es, wie gezeigt wurde, bei der Bildung der Complexe $\mathfrak{R}_P$ auf die Wahl der Elemente $A_1, A_2, \cdots A_n$ innerhalb der Complexe $\mathfrak{G}A_1$, bez. $\mathfrak{G}A_2, \cdots \mathfrak{G}A_n$ nicht ankommt, kann man ohne Beschränkung der Allgemeinheit $C_\alpha^{-1} = A_\alpha$ setzen. Dann folgt aber aus $P = C_P J_P$

$$P^{-1} = J_P^{-1} C_P^{-1} = G_{P^{-1}} A_{P^{-1}},$$

also

$$J_P^{-1} = G_{P^{-1}}.$$

Demnach ist

$$\mathfrak{R}'_P = \mathfrak{R}\prod_\alpha J_{PC_\alpha} = \mathfrak{R}\prod_\alpha G^{-1}_{C_\alpha^{-1}P^{-1}} = \mathfrak{R}\prod_\alpha G^{-1}_{A_\alpha P^{-1}}$$

oder, wenn das zu $\mathfrak{R}_P$ inverse Element der Gruppe $\frac{\mathfrak{G}}{\mathfrak{R}}$ mit $\mathfrak{R}_P^{-1}$ bezeichnet wird,

$$\mathfrak{R}'_P = \mathfrak{R}_{P^{-1}}^{-1}.$$

Es ist aber wegen (2.) $\mathfrak{R}_P\mathfrak{R}_{P^{-1}} = \mathfrak{R}_E = \mathfrak{R}$, also $\mathfrak{R}_{P^{-1}}^{-1} = \mathfrak{R}_P$. Daher ist in der That $\mathfrak{R}'_P = \mathfrak{R}_P$.

Es seien nun

$$(4.) \qquad P_1, P_2, \cdots P_t$$

diejenigen Elemente von $\mathfrak{H}$, für die die Complexe $\mathfrak{R}_{P_1}, \mathfrak{R}_{P_2}, \cdots \mathfrak{R}_{P_t}$ gleich $\mathfrak{R}$ ist; dann bilden die Elemente (4.) eine invariante Untergruppe $\mathfrak{T}$ von $\mathfrak{H}$, und es ist $\frac{\mathfrak{H}}{\mathfrak{T}}$ der commutativen Gruppe isomorph, die von den $\frac{h}{t}$ von einander verschiedenen unter den h Complexen $\mathfrak{R}_P$ gebildet wird. Da ferner diese Complexe alle der Gruppe $\frac{\mathfrak{G}}{\mathfrak{R}}$ der Ordnung m angehören, so ist $\frac{h}{t} \leqq m$.

Das bisher Gesagte gilt für jede beliebige Gruppe $\mathfrak{H}$ in Bezug auf eine ihrer Untergruppen.

Ich mache nun von den Voraussetzungen unseres Satzes Gebrauch.

Es bedeute nunmehr P irgend ein Element von $\mathfrak{G}$, und es mögen die n Elemente

$$A_{A_1P}, A_{A_2P}, \cdots A_{A_nP}$$

mit $A_{\nu_1}, A_{\nu_2}, \cdots A_{\nu_n}$ bezeichnet werden. Dann stimmen, wie bereits mehrfach erwähnt, die n Indices $\nu_1, \nu_2, \cdots \nu_n$, abgesehen von der Reihenfolge, mit den Zahlen $1, 2, \cdots n$ überein. Es möge die Permutation $\begin{pmatrix} 1 & 2 & \cdots & n \\ \nu_1 & \nu_2 & \cdots & \nu_n \end{pmatrix}$ aus den ein- oder mehrgliedrigen Cyklen

$$(\alpha_1, \alpha_2, \cdots \alpha_a), \quad (\beta_1, \beta_2, \cdots \beta_b), \cdots$$

bestehen, so dass $a+b+\cdots = n$ ist. Dann gelten für den ersten Cyklus die Gleichungen

$$A_{\alpha_1}PA_{\alpha_2}^{-1} = G_{A_{\alpha_1}P}, \qquad A_{\alpha_2}PA_{\alpha_3}^{-1} = G_{A_{\alpha_2}P}, \cdots A_{\alpha_a}PA_{\alpha_1}^{-1} = G_{A_{\alpha_a}P}.$$

Hieraus folgt

$$A_{\alpha_1}P^aA_{\alpha_1}^{-1} = G_{A_{\alpha_1}P}G_{A_{\alpha_2}P}\cdots G_{A_{\alpha_a}P}.$$

Man sieht also, dass das rechts stehende Element von $\mathfrak{G}$, das mit G_α bezeichnet werden möge, dem Element P^a von $\mathfrak{G}$ in $\mathfrak{H}$ conjugirt ist. Nach der Voraussetzung unseres Satzes muss sich daher ein Element R von $\mathfrak{G}$ angeben lassen, das der Gleichung $R^{-1}P^aR = G_\alpha$ genügt; folglich ist

$$\mathfrak{R}G_\alpha = \mathfrak{R}P^a.$$

Ebenso zeigt man, dass, wenn

$$G_{A_{\beta_1}P}G_{A_{\beta_2}P}\cdots G_{A_{\beta_b}P} = G_\beta$$

gesetzt wird,

$$\mathfrak{R}G_\beta = \mathfrak{R}P^b$$

ist, u. s. w. Da nun offenbar

$$\mathfrak{R}_P = \mathfrak{R}\prod_\alpha G_{A_\alpha P} = \mathfrak{R}G_\alpha \cdot \mathfrak{R}G_\beta \cdots$$

ist, so erhält man

$$(5.)\qquad \mathfrak{R}_P = \mathfrak{R}P^{a+b+\cdots} = \mathfrak{R}P^n = (\mathfrak{R}P)^n.$$

Nun soll aber n zu der Ordnung m der Gruppe $\frac{\mathfrak{G}}{\mathfrak{R}}$ theilerfremd sein. Daher stimmen wegen (5.) die den g Elementen $G_1, G_2, \cdots G_g$ entsprechenden Complexe $\mathfrak{R}_{G_1}, \mathfrak{R}_{G_2}, \cdots \mathfrak{R}_{G_g}$, abgesehen von der Reihenfolge, mit den Complexen $\mathfrak{R}G_1, \mathfrak{R}G_2, \cdots \mathfrak{R}G_g$ überein. Unter diesen Complexen sind aber genau m von einander verschieden; daher ist die oben erwähnte Zahl $\frac{h}{t}$ nicht kleiner als m und, weil $\frac{h}{t} \leqq m$ ist, gleich m.

Damit ist gezeigt, dass die Untergruppe $\mathfrak{T}$ von $\mathfrak{H}$ vom Index m, also von der Ordnung rn ist. Zugleich ersieht man, dass die Gruppe $\frac{\mathfrak{H}}{\mathfrak{T}}$ der Gruppe $\frac{\mathfrak{G}}{\mathfrak{R}}$ isomorph ist.

Da die Gruppe $\frac{\mathfrak{H}}{\mathfrak{T}}$ eine commutative ist, so enthält die Gruppe $\mathfrak{T}$ die Commutatorgruppe von $\mathfrak{H}$. Sie enthält aber auch jedes Element Q von $\mathfrak{H}$, dessen Ordnung in n aufgeht. Denn aus $Q^n = E$ folgt auch $\mathfrak{R}_Q^n = \mathfrak{R}$; zugleich ist aber auch $\mathfrak{R}_Q^m = \mathfrak{R}$, also, weil n und m theilerfremd sind, $\mathfrak{R}_Q = \mathfrak{R}$; mithin ist das Element Q unter den t Elementen (4.) enthalten.

Daher ist die in unserm Satz erwähnte Gruppe $\mathfrak{S}$ in $\mathfrak{T}$ enthalten, also $\frac{h}{s}$ durch $\frac{h}{t} = m$ theilbar. Man schliesst ferner leicht, dass s durch r und n theilbar ist. Sind insbesondere g und n theilerfremd, so ist $s = rn = t$, also $\mathfrak{S} = \mathfrak{T}$, folglich auch $\frac{\mathfrak{H}}{\mathfrak{S}}$ der Gruppe $\frac{\mathfrak{G}}{\mathfrak{R}}$ isomorph.

Für $r = 1$ erhält man (vergl. a. a. O. § 2) aus dem eben bewiesenen Satze als speciellen Fall den von Hrn. FROBENIUS in seiner Arbeit »Über auflösbare Gruppen. III.« (Sitzungsberichte 1901, S. 849) bewiesenen Satz:

II. »Sind f und g theilerfremde Zahlen, und enthält eine Gruppe $\mathfrak{H}$ der Ordnung fg eine Gruppe $\mathfrak{F}$ der Ordnung f, von deren Elementen nicht zwei in Bezug auf $\mathfrak{H}$ conjugirt sind, so enthält $\mathfrak{H}$ eine und nur eine charakteristische Untergruppe der Ordnung g. Diese wird gebildet von allen Elementen von $\mathfrak{H}$, deren Ordnung in g aufgeht.«

Durch weitere Specialisirung ergibt sich aus diesem Satze:

III. »Enthält die Gruppe $\mathfrak{H}$ der Ordnung $h = gn$ eine aus lauter invarianten Elementen von $\mathfrak{H}$ bestehende Untergruppe $\mathfrak{G}$ der Ordnung g, und sind g und n theilerfremd, so ist $\mathfrak{H}$ das directe Product der Gruppe $\mathfrak{G}$ und einer Gruppe $\mathfrak{N}$ der Ordnung n.«

Dieser Satz lässt sich, wie folgt, direct beweisen.

Es sei wie oben

$$\mathfrak{H} = \mathfrak{G}A_1 + \mathfrak{G}A_2 + \cdots + \mathfrak{G}A_n.$$

Die Complexe

$$P_1 = \mathfrak{G}A_1, \quad P_2 = \mathfrak{G}A_2, \cdots P_n = \mathfrak{G}A_n$$

bilden dann die Gruppe $\frac{\mathfrak{H}}{\mathfrak{G}} = \mathfrak{N}$. Setzt man $A_\nu = A_{P_\nu}$, so besteht für je zwei Elemente P, Q der Gruppe $\mathfrak{N}'$ eine Gleichung der Form

$$A_P A_Q = G_{P,Q} A_{PQ},$$

wo $G_{P,Q}$ ein gewisses Element von $\mathfrak{G}$ bedeutet. Nach dem associativen Gesetz ergibt sich dann für das Element $A_P A_Q A_R$, wo P, Q und R drei beliebige Elemente von $\mathfrak{N}'$ bedeuten, einerseits

$$(A_P A_Q) A_R = G_{P,Q} A_{PQ} A_R = G_{P,Q} G_{PQ,R} A_{PQR},$$

andererseits

$$A_P(A_Q A_R) = A_P G_{Q,R} A_{QR} = G_{Q,R} A_P A_{QR} = G_{Q,R} G_{P,QR} A_{PQR}.$$

Es ist also

$$G_{P,Q} G_{PQ,R} = G_{P,QR} G_{Q,R}.$$

Bildet man nun auf beiden Seiten dieser Gleichung das Product über alle Elemente R von $\mathfrak{N}'$, so erhält man unter Berücksichtigung der Gleichung

$$\prod_R G_{P,QR} = \prod_R G_{P,R},$$

wenn für jedes P das Product $\prod_R G_{P,R}$ mit J_P bezeichnet wird, die Relation

$$G^n_{P,Q} = J_P J_Q J^{-1}_{PQ}. \tag{6.}$$

Da nun n und g theilerfremd sind, lässt sich in der Gruppe $\mathfrak{G}$ für jedes P ein Element K_P bestimmen, das der Bedingung $K^n_P = J_P$ genügt. Setzt man dann

$$B_P = K^{-1}_P A_P,$$

so bilden die n Elemente $B_{P_1}, B_{P_2}, \cdots B_{P_n}$ eine der Gruppe $\mathfrak{N}'$ isomorphe Untergruppe $\mathfrak{N}$ von $\mathfrak{H}$. Denn es ist

$$B_P B_Q B^{-1}_{PQ} = K^{-1}_P K^{-1}_Q K_{PQ} \; A_P A_Q A^{-1}_{PQ} = K^{-1}_P K^{-1}_Q K_{PQ} G_{P,Q}.$$

Bezeichnet man das rechts stehende Element von $\mathfrak{G}$ mit $G'_{P,Q}$, so ist

$$G'^n_{P,Q} = K_P^{-n} K_Q^{-n} K_{PQ}^n G_{P,Q}^n = J_P^{-1} J_Q^{-1} J_{PQ} G_{P,Q}^n .$$

Dies ist aber wegen (6.) gleich E. Da nun andererseits auch $G'^g_{P,Q} = E$ ist, und g und n theilerfremd sind, so ergibt sich $G'_{P,Q} = E$; also ist in der That, wie zu beweisen war, $B_P B_Q = B_{PQ}$.

Es ist nun jedes Element von $\mathfrak{G}$ mit jedem Element von $\mathfrak{N}$ vertauschbar, und diese Gruppen sind auch theilerfremd. Daher ist $\mathfrak{H}$ das directe Product der Gruppen $\mathfrak{G}$ und $\mathfrak{N}$.

Auf einem andern Wege, aber ebenfalls ohne Benutzung der Theorie der Gruppencharaktere, ist der Satz III von Hrn. DE SÉGUIER (Comptes Rendus, T. CXXXV (1902) p. 528; vergl. auch ebenda, T. CXXXIV (1902) p. 692) bewiesen worden. Dagegen ist der von Hrn. DE SÉGUIER a. a. O. veröffentlichte Beweis für den Satz II nicht stichhaltig.

Ausgegeben am 6. November.

4.

Über die Darstellung der endlichen Gruppen durch gebrochene lineare Substitutionen

Journal für die reine und angewandte Mathematik 127, 20 - 50 (1904)

Das Problem der Bestimmung aller endlichen Gruppen linearer Substitutionen bei gegebener Variabelnzahl n ($n > 1$) gehört zu den schwierigsten Problemen der Algebra und hat bis jetzt nur für die binären und ternären Substitutionsgruppen seine vollständige Lösung gefunden.*) Für den allgemeinen Fall ist nur bekannt, daß die Anzahl der in Betracht kommenden Typen von Gruppen eine endliche ist;**) dagegen fehlt noch jede Übersicht über die charakteristischen Eigenschaften dieser Gruppen.

Die Umkehrung dieses Problems bildet in einem gewissen Sinne die Aufgabe: alle Gruppen von höchstens h ganzen oder gebrochenen linearen Substitutionen zu finden, die einer gegebenen endlichen Gruppe $\mathfrak{H}$ der Ordnung h ein- oder mehrstufig isomorph sind, oder auch, wie man sagt, alle Darstellungen der Gruppe $\mathfrak{H}$ durch lineare Substitutionen zu bestimmen.

Diese Aufgabe hat nun im Gegensatz zu der zuerst genannten eine erfolgreiche Behandlung gefunden, indem die Herren *Molien****) und *Frobenius*†) gezeigt haben, daß das Problem der Bestimmung aller Darstellungen der

---

*) Vergl. *A. Wiman*, Encykl. d. math. Wiss., Bd. I, S. 522.

**) *C. Jordan*, dieses Journal, Bd. 84, S. 89.

***) Sitzungsberichte der Naturforscher-Gesellschaft zu Dorpat, 1897, S. 259.

†) „Über Gruppencharaktere", Sitzungsberichte der Berliner Akademie, 1896, S. 985; „Über die Primfaktoren der Gruppendeterminante", ebenda, S. 1343; „Über die Darstellung der endlichen Gruppen durch lineare Substitutionen", ebenda, 1897, S. 994; „Über die Darstellung der endlichen Gruppen durch lineare Substitutionen II.", ebenda, 1899, S. 482. — Diese Abhandlungen werde ich im Folgenden mit „Gruppencharaktere", „Primfaktoren", „D. I." und „D. II." zitieren.

Gruppe $\mathfrak{H}$ durch ganze lineare Substitutionen identisch ist mit dem Problem, die Gruppenmatrix von $\mathfrak{H}$ in nicht weiter zerlegbare Teilmatrizen zu zerfällen. Den ersten und wesentlichsten Schritt zur Lösung dieser Aufgabe bildet, wie sich aus den Untersuchungen des Herrn *Frobenius* ergeben hat, die Zerlegung der Gruppendeterminante von $\mathfrak{H}$ in ihre Primfaktoren oder, was in der Hauptsache dasselbe ist, die Berechnung der Gruppencharaktere von $\mathfrak{H}$.

In der vorliegenden Abhandlung habe ich auseinanderzusetzen versucht, in welcher Weise sich auch das Problem, alle Darstellungen einer gegebenen endlichen Gruppe durch gebrochene lineare Substitutionen zu bestimmen, mit Hilfe der Theorie der Gruppenmatrix und der Gruppencharaktere behandeln läßt.

Ordnet man den h verschiedenen Elementen $A, B, \ldots$ der Gruppe $\mathfrak{H}$ die h linearen Substitutionen von nicht verschwindender Determinante

$$\{A\} \qquad x_\nu = \frac{a_{\nu 1} y_1 + \cdots + a_{\nu, n-1} y_{n-1} + a_{\nu n}}{a_{n1} y_1 + \cdots + a_{n, n-1} y_{n-1} + a_{nn}}, \qquad (\nu = 1, 2, \ldots, n-1)$$

$$\{B\} \qquad x_\nu = \frac{b_{\nu 1} y_1 + \cdots + b_{\nu, n-1} y_{n-1} + b_{\nu n}}{b_{n1} y_1 + \cdots + b_{n, n-1} y_{n-1} + b_{nn}}, \ldots$$

zu, so bilden dieselben eine Darstellung der Gruppe, wenn für je zwei Elemente A, B von $\mathfrak{H}$ die Gleichung $\{A\}\{B\} = \{AB\}$ besteht. Bezeichnet man die Matrizen $(a_{ik}), (b_{ik}), \ldots$ mit $(A), (B), \ldots$, so wird die Matrix $(A)(B)$ sich von der Matrix (AB) nur um einen konstanten Faktor, der auch gleich 1 sein kann, unterscheiden. Umgekehrt entspricht auch jedem System von Matrizen, deren Determinanten sämtlich von Null verschieden sind, und welche die Eigenschaft besitzen, daß für je zwei Elemente A, B der Gruppe eine Gleichung der Form $(A)(B) = r_{A,B}(AB)$ besteht, wo die Größen $r_{A,B}$ Konstanten bedeuten, eine Darstellung der Gruppe durch gebrochene lineare Substitutionen. Ich werde daher im Folgenden jedes so beschaffene System von Matrizen selbst als eine Darstellung der Gruppe durch lineare Substitutionen, und zwar als eine zu dem Zahlensystem $r_{A,B}$ gehörende bezeichnen. Der Grad n der Matrizen $(A), (B), \ldots$ soll der *Grad* der Darstellung genannt werden. Sind insbesondere alle Faktoren $r_{A,B}$ gleich 1, so entspricht dem System der Matrizen $(A), (B), \ldots$ auch eine Darstellung der Gruppe durch die ganzen linearen Substitutionen

$$x_\nu = a_{\nu 1} y_1 + a_{\nu 2} y_2 + \cdots + a_{\nu n} y_n, \qquad (\nu = 1, 2, \ldots, n)$$

$$x_\nu = b_{\nu 1} y_1 + b_{\nu 2} y_2 + \cdots + b_{\nu n} y_n, \ldots .$$

Zwei Darstellungen $(A), (B), \dots$ und $(A'), (B'), \dots$, für die sich h Konstanten $a, b, \dots$ bestimmen lassen, so daß $(A') = a(A)$, $(B') = b(B), \dots$ wird, sind, da ihnen dasselbe System von gebrochenen linearen Substitutionen entspricht, als nicht wesentlich verschieden anzusehen. Zwei solche Darstellungen sollen im Folgenden als einander *assoziiert* bezeichnet werden.

Die Begriffe der *Äquivalenz* und der *Primitivität* sind ebenso wie bei den Darstellungen durch ganze lineare Substitutionen zu definieren (vergl. *Frobenius*, D. II.): Zwei Darstellungen $(A), (B), \dots$ und $(A'), (B'), \dots$ sind einander *äquivalent*, wenn sich eine Matrix P von nicht verschwindender Determinante angeben läßt, so daß

$$(A') = P^{-1}(A)P,\ (B') = P^{-1}(B)P, \dots$$

wird; zwei äquivalente Darstellungen gehören also zu demselben Zahlensystem $r_{A,B}$. Eine Darstellung ist ferner *primitiv*, wenn sich keine ihr äquivalente Darstellung $(A'), (B'), \dots$ der Form

$$(A') = \begin{pmatrix} (A_1), & 0 \\ 0, & (A_2) \end{pmatrix},\quad (B') = \begin{pmatrix} (B_1), & 0 \\ 0, & (B_2) \end{pmatrix}, \dots$$

bestimmen läßt, wo $(A_1), (B_1), \dots$ Matrizen desselben Grades bedeuten.

Für die primitiven Darstellungen einer Gruppe $\mathfrak{H}$ durch gebrochene lineare Substitutionen gilt der analoge Satz wie für die Darstellungen durch ganze lineare Substitutionen (*Frobenius*, Primfaktoren, § 12), daß ihr Grad stets ein Divisor der Ordnung von $\mathfrak{H}$ ist (§ 4).

Die Untersuchung der Darstellungen einer Gruppe durch gebrochene lineare Substitutionen steht in engster Beziehung zu der Theorie der invarianten Elemente der endlichen Gruppen.

Ist $\mathfrak{G}$ eine endliche Gruppe, welche eine aus invarianten Elementen von $\mathfrak{G}$ bestehende Untergruppe $\mathfrak{A}$ derart enthält, daß $\frac{\mathfrak{G}}{\mathfrak{A}}$ der gegebenen Gruppe $\mathfrak{H}$ (einstufig) isomorph ist, so sei

$$\mathfrak{G} = \mathfrak{A}A' + \mathfrak{A}B' + \cdots,$$

und es mögen die Komplexe $\mathfrak{A}A'$, $\mathfrak{A}B', \dots$ den Elementen $A, B, \dots$ der Gruppe $\mathfrak{H}$ entsprechen. — Eine solche Gruppe $\mathfrak{G}$ bezeichne ich im Folgenden als eine *durch die Gruppe $\mathfrak{A}$ ergänzte Gruppe von $\mathfrak{H}$*, wofür ich auch kürzer schreibe, $\mathfrak{G}$ sei eine Gruppe $(\mathfrak{A}, \mathfrak{H})$. — Betrachtet man eine beliebige primitive Darstellung von $\mathfrak{G}$ durch ganze lineare Substitutionen, so besitzt jede einem Element J von $\mathfrak{A}$ entsprechende Matrix die Form $j(E)$, wo j

eine Einheitswurzel und (E) die Einheitsmatrix bedeuten, und die den Elementen $A', B', \ldots$ entsprechenden Matrizen $(A'), (B'), \ldots$ bilden in dem oben definierten Sinne eine (primitive) Darstellung der Gruppe $\mathfrak{H}$ durch gebrochene lineare Substitutionen.

Es lassen sich nun auch ergänzte Gruppen $\mathfrak{G}$ von $\mathfrak{H}$ angeben, die so beschaffen sind, daß jede primitive Darstellung von $\mathfrak{H}$ durch gebrochene lineare Substitutionen (oder eine ihr assoziierte Darstellung) mindestens einer der in der erwähnten Weise durch die Gruppe $\mathfrak{G}$ gelieferten Darstellungen von $\mathfrak{H}$ äquivalent ist. Eine ergänzte Gruppe von $\mathfrak{H}$, welche diese Eigenschaft besitzt, nenne ich eine *hinreichend ergänzte* Gruppe, eine hinreichend ergänzte Gruppe, deren Ordnung möglichst klein ist, eine *Darstellungsgruppe* der Gruppe $\mathfrak{H}$.

Einer gegebenen Gruppe $\mathfrak{H}$ können auch mehrere einander nicht isomorphe Darstellungsgruppen $\mathfrak{G}$ entsprechen. Dagegen sind die (aus invarianten Elementen bestehenden) Untergruppen $\mathfrak{A}$ von $\mathfrak{G}$, für die $\frac{\mathfrak{G}}{\mathfrak{A}} = \mathfrak{H}$ ist, in allen Darstellungsgruppen von $\mathfrak{H}$ einander isomorph. Es entspricht auf diese Weise jeder Gruppe $\mathfrak{H}$ eine wohlbestimmte *Abel*sche Gruppe, die ich als den *Multiplikator* der Gruppe $\mathfrak{H}$ bezeichne. Eine Gruppe, deren Multiplikator die Einheitsgruppe ist, nenne ich eine *abgeschlossene* Gruppe (§§ 1—3).

Man hat demnach, um die sämtlichen primitiven Darstellungen einer gegebenen Gruppe $\mathfrak{H}$ durch gebrochene lineare Substitutionen zu erhalten, in erster Linie eine Darstellungsgruppe $\mathfrak{G}$ von $\mathfrak{H}$ zu bestimmen, was als ein rein gruppentheoretisches Problem aufzufassen ist, und alsdann nach den von Herrn *Frobenius* angegebenen Methoden die primitiven Darstellungen von $\mathfrak{G}$ durch ganze lineare Substitutionen zu untersuchen. Insbesondere läßt sich also für eine abgeschlossene Gruppe jede Darstellung durch gebrochene lineare Substitutionen durch eine Darstellung durch ganze lineare Substitutionen ersetzen.

In § 5 entwickle ich einige Sätze, welche in speziellen Fällen zur Bestimmung des Multiplikators und der Darstellungsgruppen einer gegebenen Gruppe dienen können.

Die hier dargestellte allgemeine Theorie soll in einer später erscheinenden Arbeit weiter verfolgt und auf einige spezielle Klassen von Gruppen angewendet werden.

§ 1.

Es sei $\mathfrak{H}$ eine endliche Gruppe der Ordnung h, deren Elemente $H_0 = E, H_1, \ldots, H_{h-1}$ sind. Es mögen die Matrizen $(H_0), (H_1), \ldots, (H_{h-1})$ eine zu dem Zahlensystem $r_{P,Q}$ gehörende Darstellung von $\mathfrak{H}$ bilden, d. h. den h^2 Gleichungen

$$(1.)\qquad (P)(Q) = r_{P,Q}\,(PQ) \qquad (P, Q = H_0, H_1, \ldots, H_{h-1})$$

genügen. Sind P, Q, R drei beliebige Elemente der Gruppe, so ist die Matrix $(P)(Q)(R)$ nach dem assoziativen Gesetz einerseits gleich

$$r_{P,Q}\,(PQ)(R) = r_{P,Q}\,r_{PQ,R}\,(PQR),$$

andererseits aber auch gleich

$$r_{Q,R}\,(P)(QR) = r_{P,QR}\,r_{Q,R}\,(PQR).$$

Die h^2 Größen $r_{P,Q}$ müssen daher den h^3 Gleichungen

$$(A.)\qquad r_{P,Q}\,r_{PQ,R} = r_{P,QR}\,r_{Q,R} \qquad (P, Q, R = H_0, H_1, \ldots, H_{h-1})$$

genügen.

Umgekehrt läßt sich für jedes System von h^2 nicht verschwindenden Zahlen $r_{P,Q}$, welche den Gleichungen $(A.)$ genügen, eine Darstellung von $\mathfrak{H}$ angeben, die zu dem Zahlensystem $r_{P,Q}$ gehört. Um dies zu beweisen, führe ich h unabhängige Variable $x_{H_0}, x_{H_1}, \ldots, x_{H_{h-1}}$ ein und betrachte die Matrix h-ten Grades

$$X = (r_{PQ^{-1},Q}\, x_{PQ^{-1}}),$$

deren Zeilen und Spalten man erhält, indem man für P und Q der Reihe nach die h Elemente $H_0, H_1, \ldots, H_{h-1}$ setzt. Offenbar läßt sich X in der Form

$$X = \Sigma\,(R)\,x_R$$

schreiben, wo die Summe über alle Elemente von $\mathfrak{H}$ zu erstrecken ist, und das Zeichen (R) eine gewisse Matrix h-ten Grades bedeutet, die man dadurch erhält, daß man in X die Variable x_R gleich 1, die übrigen Variabeln gleich 0 setzt. Ich will zeigen, daß die so erhaltenen h Matrizen $(H_0), (H_1), \ldots, (H_{h-1})$ den Gleichungen (1.) genügen.

Es seien nämlich $y_{H_0}, y_{H_1}, \ldots, y_{H_{h-1}}$ eine andere Reihe von unabhängigen Variabeln, ferner mögen die h Größen $z_{H_0}, z_{H_1}, \ldots, z_{H_{h-1}}$ durch die Gleichungen

$$z_P = \Sigma\, r_{R,S}\, x_R\, y_S, \qquad (P = H_0, H_1, \ldots, H_{h-1})$$

— die Summe über alle Elemente R und S erstreckt, die der Gleichung

$RS = P$ genügen, — bestimmt sein. Es möge X in Y und Z übergehen, wenn man die Variabeln x_P durch y_P und z_P ersetzt. Dann wird

$$XY = (\sum_R r_{PR^{-1},R}\, r_{RQ^{-1},Q}\, x_{PR^{-1}}\, y_{RQ^{-1}}).$$

Nun ist aber nach (*A.*)

$$r_{PR^{-1},RQ^{-1}}\, r_{PQ^{-1},Q} = r_{PR^{-1},R}\, r_{RQ^{-1},Q}.$$

Daher ist

$$\sum_R r_{PR^{-1},R}\, r_{RQ^{-1},Q}\, x_{PR^{-1}}\, y_{RQ^{-1}} = r_{PQ^{-1},Q} \sum_R r_{PR^{-1},RQ^{-1}}\, x_{PR^{-1}}\, y_{RQ^{-1}}.$$

Die rechte Seite dieser Gleichung ist aber gleich $r_{PQ^{-1},Q}\, z_{PQ^{-1}}$. Wir erhalten daher

$$XY = (r_{PQ^{-1},Q}\, z_{PQ^{-1}}) = Z$$

oder

$$\sum_{R,S} (R)(S)\, x_R\, y_S = \sum_T (T)\, z_T = \sum_{R\,S} r_{R,S}\, (RS)\, x_R\, y_S.$$

Es ist daher in der Tat, wie zu beweisen war,

$$(1'.) \qquad (R)(S) = r_{R,S}\,(RS).$$

Da ferner die Matrix (R) sowohl in jeder ihrer Zeilen, als auch in jeder ihrer Spalten ein und nur ein von Null verschiedenes Element enthält, so sind auch die Determinanten $d_{H_0}, d_{H_1}, \ldots, d_{H_{h-1}}$ der Matrizen $(H_0), (H_1), \ldots, (H_{h-1})$ sämtlich von Null verschieden.*) Aus der Gleichung (1'.) ergibt sich noch, indem man auf beiden Seiten die Determinanten bildet,

$$(2.) \qquad r^h_{R,S} = \frac{d_R\, d_S}{d_{RS}}.$$

Wir erhalten den Satz:

I. *Zu einem System von h^2 Zahlen $r_{P,Q}$ gehören dann und nur dann Darstellungen der Gruppe $\mathfrak{H}$, wenn die Zahlen $r_{P,Q}$ den Gleichungen (A.) genügen.*

Das Gleichungssystem (*A.*) besitzt unendlich viele Lösungen. Denn bilden die Zahlen $r_{P,Q}$ eine solche Lösung, und bedeuten $c_{H_0}, c_{H_1}, \ldots, c_{H_{h-1}}$ h beliebige von Null verschiedene Größen, so genügen auch die h^2 Größen

---

*) Ist r die Ordnung des Elementes R von $\mathfrak{H}$, so ist, wie man leicht zeigt,

$$d_R = (-1)^{h-\frac{h}{r}} \prod_S r_{R,S}.$$

$$(3.) \qquad r'_{P,Q} = \frac{c_P c_Q}{c_{PQ}} r_{P,Q}$$

den Gleichungen (*A.*). Zwei Lösungen $r_{P,Q}$ und $r'_{P,Q}$ von (*A.*), für die sich h Größen c_P so bestimmen lassen, daß die Gleichungen (3.) erfüllt sind, sollen als einander *assoziierte* Lösungen oder auch als *assoziierte Zahlensysteme* bezeichnet werden.

Dann können, wie unmittelbar ersichtlich ist, zwei Darstellungen von $\mathfrak{H}$, die zu den Lösungen $r_{P,Q}$ und $\bar{r}_{P,Q}$ von (*A.*) gehören, nur dann einander assoziiert sein (vergl. Einleitung), wenn diese Lösungen einander assoziiert sind. Gehören dagegen zwei Darstellungen von $\mathfrak{H}$ zu zwei einander nicht assoziierten Lösungen von (*A.*), so wollen wir sagen, diese Darstellungen seien von *verschiedenem Typus.*

Denkt man sich nun alle einander assoziierten Lösungen der Gleichungen (*A.*) zu einer Klasse vereinigt, so ist die Anzahl dieser Klassen endlich. Denn sind $d_{H_0}, d_{H_1}, \ldots, d_{H_{h-1}}$ die in der Gleichung (2.) auftretenden Größen, bedeuten $\delta_{H_0}, \delta_{H_1}, \ldots, \delta_{H_{h-1}}$ irgend welche Wurzeln der h Gleichungen $\delta_P^h = d_P$, und ersetzt man die Lösung $r_{P,Q}$ durch die ihr assoziierte Lösung

$$s_{P,Q} = \frac{\delta_P^{-1}\,\delta_Q^{-1}}{\delta_{PQ}^{-1}} r_{P,Q},$$

so ist wegen (2.) $s_{P,Q}^h = 1$. Es enthält also jede Klasse auch solche Lösungen $s_{P,Q}$, für welche die Zahlen $s_{P,Q}$ sämtlich h-te Wurzeln der Einheit sind. Die Anzahl der verschiedenen Klassen ist daher jedenfalls nicht größer als $(h)^{h^2}$.

Ist diese Anzahl gleich m, so mögen die m verschiedenen Klassen von Lösungen mit $K_0, K_1, \ldots, K_{m-1}$ bezeichnet werden. Sind $r_{P,Q}^{(\lambda)}$ und $r_{P,Q}^{(\mu)}$ zwei Lösungen von (*A.*), die den Klassen K_λ und K_μ angehören, so werden auch die Zahlen $r_{P,Q}^{(\lambda)}\, r_{P,Q}^{(\mu)}$ den Gleichungen (*A.*) genügen. Die Klasse K_ν, der diese Lösung angehört, ist offenbar von der speziellen Wahl der Lösungen $r_{P,Q}^{(\lambda)}$ und $r_{P,Q}^{(\mu)}$ innerhalb der Klassen K_λ und K_μ nicht abhängig, also durch die Klassen K_λ und K_μ vollständig bestimmt. Setzt man $K_\nu = K_\lambda K_\mu$, so ist auch $K_\mu K_\lambda = K_\nu$. Ferner folgt, wie man leicht einsieht, aus jeder Gleichung der Form $K_\alpha K_\beta = K_\alpha K_\gamma$ das Übereinstimmen der Klassen K_β und K_γ. Da nun für die festgesetzte Komposition der Klassen auch das assoziative Gesetz erfüllt ist, so definieren die m Klassen $K_0, K_1, \ldots, K_{m-1}$ eine wohlbestimmte *Abel*sche Gruppe $\mathfrak{M}$ der Ordnung m, in der als das

Einheitselement diejenige Klasse K_0 anzusehen ist, der die Lösung $r_{P,Q} = 1$ angehört. Diese Gruppe $\mathfrak{M}$ bezeichne ich als den *Multiplikator* der Gruppe $\mathfrak{H}$.*)

Die Ordnung des Multiplikators von $\mathfrak{H}$ gibt uns also an, wie viele verschiedene Typen von Darstellungen der Gruppe $\mathfrak{H}$ durch gebrochene lineare Substitutionen vorhanden sind.

Die Gleichung (2.) lehrt uns, daß für jede Klasse K_λ die Gleichung $K_\lambda^h = K_0$ besteht. *Die Zahl m kann daher keinen zu h teilerfremden Primfaktor enthalten.*

§ 2.

Es sei nun $\mathfrak{G}$ eine beliebige durch eine *Abel*sche Gruppe $\mathfrak{A}$ ergänzte Gruppe der von uns betrachteten Gruppe $\mathfrak{H}$. Setzt man

$$\mathfrak{G} = \mathfrak{A}G_0 + \mathfrak{A}G_1 + \cdots + \mathfrak{A}G_{h-1}, \qquad (G_0 = E)$$

wobei der Komplex $\mathfrak{A}G_\lambda$ dem Element H_λ der mit $\frac{\mathfrak{G}}{\mathfrak{A}}$ isomorphen Gruppe $\mathfrak{H}$ entsprechen möge, so genügen, wenn die Gruppe $\mathfrak{H}$ durch die h^2 Gleichungen

$$(4.) \qquad H_\lambda H_\mu = H_{\varphi(\lambda,\mu)} \qquad (\lambda, \mu = 0, 1, \ldots, h-1)$$

bestimmt ist, die Elemente $G_0, G_1, \ldots, G_{h-1}$ von $\mathfrak{G}$ h^2 Relationen der Form

$$(5.) \qquad G_\lambda G_\mu = A_{\lambda,\mu} G_{\varphi(\lambda,\mu)}, \qquad (\lambda, \mu = 0, 1, \ldots, h-1;\ A_{0,\lambda} = A_{\lambda,0} = E)$$

wo die $A_{\lambda,\mu}$ Elemente der Untergruppe $\mathfrak{A}$ von $\mathfrak{G}$ bedeuten.

Es seien $A_0 = E, A_1, \ldots, A_{a-1}$ die Elemente,

$$\psi^{(0)}(A), \psi^{(1)}(A), \ldots, \psi^{(a-1)}(A)$$

die a Charaktere der *Abel*schen Gruppe $\mathfrak{A}$. Diese Charaktere lassen sich bekanntlich stets den Elementen $A_0, A_1, \ldots, A_{a-1}$ so zuordnen, daß, wenn $\psi^{(\alpha)}(A) = \psi_{A_\alpha}(A)$ gesetzt wird, für jedes A die Gleichungen

$$(6.) \qquad \psi_{A_\alpha}(A)\,\psi_{A_\beta}(A) = \psi_{A_\alpha A_\beta}(A) \qquad (\alpha, \beta = 0, 1, \ldots, a-1)$$

bestehen.**)

Ist uns nun eine beliebige primitive Darstellung D der Gruppe $\mathfrak{G}$ durch ganze lineare Substitutionen gegeben, bei der dem Element R von $\mathfrak{G}$

---

*) Daß diese Definition des Multiplikators mit der in der Einleitung gegebenen übereinstimmt, wird sich aus dem Folgenden ergeben.

**) Vergl. *Weber*, Lehrbuch der Algebra, Bd. II, Abschnitt II. — Im Folgenden soll stets, wenn die Charaktere einer *Abel*schen Gruppe in der Form $\psi_{A_\alpha}(A)$ geschrieben werden, vorausgesetzt sein, daß die Gleichungen (6.) erfüllt sind.

die Matrix (R) entspricht, so hat für jedes Element A von $\mathfrak{A}$ die Matrix (A) die Form $\psi^{(a)}(A)\cdot(E)$, wo (E) die Einheitsmatrix bedeutet und die Zahlen $\psi^{(a)}(A)$ einen der a Charaktere von $\mathfrak{A}$ repräsentieren. — Man sagt dann (vergl. *Frobenius*, D. II.), die Darstellung D entspreche dem Charakter $\psi^{(a)}(A)$ von $\mathfrak{A}$. — Setzt man

$$\psi^{(a)}(A_{\lambda,\mu}) = r^{(a)}_{H_\lambda, H_\mu}, \tag{7.}$$

so bestehen wegen (5.) die Gleichungen

$$(G_\lambda)(G_\mu) = r^{(a)}_{H_\lambda, H_\mu}(G_{\varphi(\lambda,\mu)}).$$

Die h Matrizen $(G_0), (G_1), \dots, (G_{h-1})$ bilden also eine zu dem Zahlensystem $r^{(a)}_{H_\lambda, H_\mu}$ gehörende Darstellung der Gruppe $\mathfrak{H}$ durch gebrochene lineare Substitutionen. Man erhält aber auch, indem man die sämtlichen dem Charakter $\psi^{(a)}(A)$ von $\mathfrak{A}$ entsprechenden Darstellungen von $\mathfrak{G}$ betrachtet, die sämtlichen zu dem Zahlensystem (7.) gehörenden Darstellungen von $\mathfrak{H}$ durch gebrochene lineare Substitutionen. Denn bilden die Matrizen $(H_0), (H_1), \dots, (H_{h-1})$ eine solche Darstellung, und setzt man

$$(A_\beta G_\lambda) = \psi^{(a)}(A_\beta).(H_\lambda), \quad (\beta = 0, 1, \dots, a-1,\ \lambda = 0, 1, \dots, h-1)$$

so bilden, wie man sich leicht überzeugt, die ah Matrizen $(A_\beta G_\lambda)$ eine dem Charakter $\psi^{(a)}(A)$ entsprechende Darstellung von $\mathfrak{G}$ durch ganze lineare Substitutionen.

Wir erhalten zugleich, da auf Grund der Sätze über die Gruppendeterminante*) jedem der a Charaktere von $\mathfrak{A}$ Darstellungen von $\mathfrak{G}$ entsprechen, in den a Zahlensystemen

$$r^{(a)}_{H_\lambda, H_\mu} \quad (a = 0, 1, \dots, a-1)$$

a Lösungen der Gleichungen (A.).**) Unter diesen a Lösungen werden jedoch im allgemeinen mehrere einander assoziiert sein können, und es fragt sich nun, wie groß die Anzahl m' der verschiedenen Klassen K_λ ist, welche durch die hier betrachteten Lösungen repräsentiert werden; mit

---

*) *Frobenius*, Gruppencharaktere, § 7.

**) Einfacher ergibt sich dies folgendermaßen. Setzt man $A_{\lambda,\mu} = A_{H_\lambda, H_\mu}$, so folgen aus den Gleichungen (5.) auf Grund des assoziativen Gesetzes die Relationen

$$A_{P,Q}\, A_{PQ,R} = A_{P,QR}\, A_{Q,R}; \quad (P, Q, R = H_0, H_1, \dots, H_{h-1})$$

daher ist auch für jeden Charakter $\psi(A)$ von $\mathfrak{A}$

$$\psi(A_{P,Q})\,\psi(A_{PQ,R}) = \psi(A_{P,QR})\,\psi(A_{Q,R}).$$

anderen Worten: wie viele verschiedene Typen von Darstellungen der Gruppe $\mathfrak{H}$ sich aus der Betrachtung der Gruppe $\mathfrak{G}$ ergeben.

Es seien

$$\psi_{B_0}(A),\ \psi_{B_1}(A), \ldots, \psi_{B_{b-1}}(A)$$

diejenigen Charaktere von $\mathfrak{A}$, denen auch lineare Charaktere*) $\chi(R)$ von $\mathfrak{G}$ entsprechen. Dann bilden die Elemente $B_0, B_1, \ldots, B_{b-1}$ eine Untergruppe $\mathfrak{B}$ von $\mathfrak{A}$. Denn entsprechen den Charakteren $\psi_{B_\alpha}(A)$ und $\psi_{B_\beta}(A)$ von $\mathfrak{A}$ zwei lineare Charaktere $\chi^{(\alpha)}(R)$ und $\chi^{(\beta)}(R)$ von $\mathfrak{G}$, so entspricht dem Charakter $\psi_{B_\alpha B_\beta}(A)$ der lineare Charakter $\chi^{(\alpha)}(R)\,\chi^{(\beta)}(R)$. Sollen nun für zwei Elemente B und C von $\mathfrak{A}$ die beiden Zahlensysteme $\psi_B(A_{\lambda,\mu})$ und $\psi_C(A_{\lambda,\mu})$ zwei einander assoziierte Lösungen von (A.) repräsentieren, so müssen sich h Konstanten $\varphi(H_0), \varphi(H_1), \ldots, \varphi(H_{h-1})$ bestimmen lassen, so daß

$$\psi_B(A_{\lambda,\mu}) = \frac{\varphi(H_\lambda)\,\varphi(H_\mu)}{\varphi(H_\lambda H_\mu)}\,\psi_C(A_{\lambda,\mu})$$

oder, was dasselbe ist,

$$\varphi(H_\lambda)\,\varphi(H_\mu) = \psi_{BC^{-1}}(A_{\lambda,\mu})\,\varphi(H_\lambda H_\mu)$$

wird. Setzt man aber dann

$$\chi(A_\alpha G_\lambda) = \psi_{BC^{-1}}(A_\alpha)\,\varphi(H_\lambda),$$

so ergibt sich für je zwei Elemente $R = A_\alpha G_\lambda$ und $S = A_\beta G_\mu$ von $\mathfrak{G}$ die Gleichung $\chi(R)\,\chi(S) = \chi(RS)$, ferner wegen

$$\varphi(H_0)\,\varphi(H_0) = \psi_{BC^{-1}}(A_{0,0})\,\varphi(H_0) = \varphi(H_0),$$

also $\varphi(H_0) = 1$, die Gleichung

$$\chi(A_\alpha) = \psi_{BC^{-1}}(A_\alpha)\,\varphi(H_0) = \psi_{BC^{-1}}(A_\alpha).$$

Daher bilden die ah Zahlen $\chi(A_\alpha G_\lambda)$ einen dem Charakter $\psi_{BC^{-1}}(A)$ entsprechenden linearen Charakter von $\mathfrak{G}$; folglich muß BC^{-1} ein Element von $\mathfrak{B}$ sein. Umgekehrt schließt man leicht, daß diese Bedingung für das Verlangte auch hinreichend ist.

Mithin ist die gesuchte Zahl m' gleich der Anzahl der mod. $\mathfrak{B}$ inkongruenten Elemente von $\mathfrak{A}$, also gleich $\frac{a}{b}$.

---

*) *Frobenius*, Primfaktoren, § 2.

Es sei

$$\mathfrak{A} = \mathfrak{B}A_0 + \mathfrak{B}A_1 + \cdots + \mathfrak{B}A_{m'-1};$$

dann gehören also die m' Lösungen

$$\psi_{A_\alpha}(A_{\lambda,\mu}) = r^{(\alpha)}_{H_\lambda, H_\mu} \qquad (\alpha = 0, 1, \ldots, m'-1)$$

der Gleichungen (A.) m' verschiedenen Klassen $K_0, K_1, \ldots, K_{m'-1}$ an. Ist nun aber $\mathfrak{B}A_\alpha \cdot \mathfrak{B}A_\beta = \mathfrak{B}A_\gamma$, so ist offenbar die Lösung $r^{(\gamma)}_{H_\lambda, H_\mu}$ der Lösung $r^{(\alpha)}_{H_\lambda, H_\mu}\, r^{(\beta)}_{H_\lambda, H_\mu}$ assoziiert, also $K_\gamma = K_\alpha K_\beta$. Daher bilden die Klassen $K_0, K_1, \ldots, K_{m'-1}$ eine der Gruppe $\frac{\mathfrak{A}}{\mathfrak{B}}$ isomorphe Untergruppe $\mathfrak{M}'$ des Multiplikators $\mathfrak{M}$ von $\mathfrak{H}$; demnach ist auch m' ein Divisor der Ordnung m von $\mathfrak{M}$.

Die Zahl m' hat noch eine andere Bedeutung für die Gruppe $\mathfrak{G}$.

Es sei $\mathfrak{K}$ der Kommutator*) der Gruppe $\mathfrak{H}$, $\mathfrak{K}'$ der Kommutator der Gruppe $\mathfrak{G}$ und $\mathfrak{D}$ der größte gemeinsame Teiler der Gruppen $\mathfrak{K}'$ und $\mathfrak{A}$. Es seien r, r' und d die respektiven Ordnungen dieser Gruppen. Dann ist wegen des Isomorphismus der Gruppen $\frac{\mathfrak{G}}{\mathfrak{A}}$ und $\mathfrak{H}$ die Gruppe $\frac{\mathfrak{K}'}{\mathfrak{D}}$ der Gruppe $\mathfrak{K}$ isomorph, also $r' = dr$.

Allgemein besteht für den Kommutator $\mathfrak{K}$ einer Gruppe $\mathfrak{H}$ der Satz: „Ist $\chi(S)$ ein linearer Charakter von $\mathfrak{H}$, so ist für jedes Element R von $\mathfrak{K}$ $\chi(R) = 1$; umgekehrt gehört jedes Element R von $\mathfrak{H}$, welches die Eigenschaft besitzt, daß für jeden linearen Charakter $\chi(S)$ von $\mathfrak{H}$ die Gleichung $\chi(R) = 1$ besteht, dem Kommutator von $\mathfrak{H}$ an.“**)

Diesen Satz wenden wir auf unseren Fall an.

Die Gruppe $\mathfrak{D}$ umfaßt alle Elemente J von $\mathfrak{A}$, die dadurch charakterisiert sind, daß für jeden linearen Charakter $\chi(R)$ von $\mathfrak{G}$ $\chi(J) = 1$ ist, und nur diese Elemente. Nun ist aber für jedes Element A von $\mathfrak{A}$ $\chi(A) = \psi_B(A)$, wo B ein Element von $\mathfrak{B}$ ist. Wir erhalten also alle Elemente von $\mathfrak{D}$, indem wir alle Elemente J von $\mathfrak{A}$ aufsuchen, für die

$$\psi_B(J) = 1 \tag{8.}$$

ist, wo B alle Elemente von $\mathfrak{B}$ durchläuft. Die Gesamtheit der Elemente J, welche den b Bedingungen (8.) genügen, bildet aber bekanntlich eine

---

*) Der Begriff des Kommutators einer Gruppe ist von *Dedekind* (Math. Ann. Bd. 48, S. 548) eingeführt worden. Vergl. auch *Frobenius*, Primfaktoren, § 2.

**) *Frobenius*, a. a. O.

der Gruppe $\frac{\mathfrak{A}}{\mathfrak{B}}$ isomorphe Untergruppe von $\mathfrak{A}$, die Herr *Weber* a. a. O. als die zu $\mathfrak{B}$ *reziproke* Untergruppe von $\mathfrak{A}$ bezeichnet. Es ist daher auch $\mathfrak{D}$ der Gruppe $\mathfrak{M}'$ isomorph, und es ist

$$m' = \frac{a}{b} = d = \frac{r'}{r}.$$

Wir erhalten den für das Folgende wichtigen Satz:

II. *In jeder durch eine Gruppe $\mathfrak{A}$ ergänzten Gruppe $\mathfrak{G}$ von $\mathfrak{H}$ ist der größte gemeinsame Teiler des Kommutators von $\mathfrak{G}$ und der Untergruppe $\mathfrak{A}$ von $\mathfrak{G}$ einer Untergruppe des Multiplikators von $\mathfrak{H}$ isomorph.*

Soll insbesondere $\mathfrak{G}$ eine hinreichend ergänzte Gruppe von $\mathfrak{H}$ sein, so müssen unter den aus der Betrachtung der Darstellungen von $\mathfrak{G}$ durch ganze lineare Substitutionen sich ergebenden Darstellungen von $\mathfrak{H}$ durch gebrochene lineare Substitutionen die sämtlichen m möglichen Typen vertreten sein. Es muß also für jede solche Gruppe $m' = d = m$, folglich auch insbesondere a durch m teilbar sein. Aus dem auf Seite 28 Gesagten folgt auch, daß umgekehrt $\mathfrak{G}$ stets dann eine hinreichend ergänzte Gruppe von $\mathfrak{H}$ ist, wenn $d = m$ ist.

§ 3.

Es soll nun gezeigt werden, daß sich in der Tat für jede Gruppe $\mathfrak{H}$ hinreichend ergänzte Gruppen, und zwar auch solche von der Ordnung mh, angeben lassen.

Wir führen h erzeugende Elemente $Q_0, Q_1, \ldots, Q_{h-1}$ ein und setzen fest, daß die h^2 Elemente

$$(9.)\qquad Q_\lambda\, Q_\mu\, Q^{-1}_{\varphi(\lambda,\mu)} = J_{H_\lambda, H_\mu}, \qquad (\lambda, \mu = 0, 1, \ldots, h-1)$$

wo der Index $\varphi(\lambda, \mu)$ durch die Gleichung (4.) zu definieren ist, mit den h erzeugenden Elementen vertauschbar sein sollen; dann sind die Elemente J_{H_λ, H_μ} auch unter einander vertauschbar. Die h^3 Gleichungen

$$(10.)\qquad Q_\nu \cdot J_{H_\lambda, H_\mu} = J_{H_\lambda, H_\mu} \cdot Q_\nu \qquad (\lambda, \mu, \nu = 0, 1, \ldots, h-1)$$

definieren alsdann eine gewisse unendliche Gruppe $\mathfrak{K}'$; die Elemente J_{H_λ, H_μ} erzeugen eine gewisse unendliche *Abel*sche Gruppe $\mathfrak{M}'$, die in $\mathfrak{K}'$ als invariante Untergruppe enthalten ist. Infolge des assoziativen Gesetzes ergeben sich aus (9.) und (10.) für die J_{H_λ, H_μ} die h^3 Relationen

$$(B.)\qquad J_{P,Q}\, J_{PQ,R} = J_{P,QR}\, J_{Q,R}. \qquad (P, Q, R = H_0, H_1, \ldots, H_{h-1})$$

Jedes Element von $\mathfrak{K}'$ läßt sich auf die Form JQ_λ bringen, wo J der Gruppe $\mathfrak{N}'$ angehört. Es kann sich nicht für $\lambda \neq \mu$ eine Gleichung der Form $JQ_\lambda = J'Q_\mu$ ergeben, da alsdann auch $H_\lambda = H_\mu$ folgen würde, weil doch auch die Elemente H_λ, für die Q_λ eingesetzt, den Bedingungen (9.) und (10.) genügen. Es resultiert aber auch für die Elemente J_{H_λ, H_μ} aus (9.) und (10.) keine Beziehung der Form

$$\prod_{\lambda,\mu} J_{H_\lambda, H_\mu}^{l_{\lambda,\mu}} = E,$$

die sich nicht schon aus den Gleichungen (B.) ergeben würde. Denn wäre dies der Fall, so könnte nicht jedes System von Zahlen r_{H_λ, H_μ}, die den Relationen

(A.) $$r_{P,Q}\, r_{PQ,R} = r_{P,QR}\, r_{Q,R} \qquad (P, Q, R = H_0, H_1, \ldots, H_{h-1})$$

genügen, die Gleichung

(11.) $$\prod_{\lambda,\mu} r_{H_\lambda, H_\mu}^{l_{\lambda,\mu}} = 1$$

erfüllen.*) Wir haben aber in § 1 gesehen, daß sich für jede beliebige Lösung $r_{P,Q}$ der Gleichungen (A.) h Matrizen $(H_0), (H_1), \ldots, (H_{h-1})$ angeben lassen, für welche die Bedingungen $(H_\lambda)(H_\mu) = r_{H_\lambda, H_\mu}(H_{\varphi(\lambda,\mu)})$ bestehen. Die Matrizen (H_λ) genügen jedoch, für die Q_λ eingesetzt, den Bedingungen (9.) und (10.). Daraus würde aber folgen, daß die Zahlen r_{H_λ, H_μ} die Gleichung (11.) befriedigen, was auf einen Widerspruch führt.

Die h^3 Gleichungen (B.) können daher als ein vollständiges System von definierenden Relationen**) für die durch die Elemente J_{H_λ, H_μ} erzeugte *Abel*sche Gruppe $\mathfrak{N}'$ angesehen werden.

Bezeichnet man die $h^2 = p$ Elemente J_{H_λ, H_μ} in irgend einer Reihenfolge mit $X_1, X_2, \ldots, X_p$, so lassen sich die Gleichungen (B.) in der Form

(12.) $$X_1^{\alpha_{\lambda 1}} X_2^{\alpha_{\lambda 2}} \ldots X_p^{\alpha_{\lambda p}} = E \qquad (\lambda = 1, 2, \ldots, n,\ n = h^3)$$

schreiben, wo die Exponenten $\alpha_{\lambda\mu}$ gleich $1, -1$ oder 0 sind.

Es gilt nun bekanntlich folgender Satz:***)

„Genügen p unter einander vertauschbare Elemente $X_1, X_2, \ldots, X_p$ n Relationen der Form (12.), ist $p-s$ der Rang, sind $e_1, e_2, \ldots, e_\varrho$ diejenigen

---

*) Vergl. den am Schluß dieser Seite angeführten Satz.

**) Vergl. *Dyck*, Math. Ann., Bd. 20, S. 1.

***) Vergl. *Frobenius* und *Stickelberger*, dieses Journal, Bd. 86, S. 217.

Elementarteiler des Exponentensystems $(\alpha_{\lambda\mu})$, die größer sind als 1, so läßt sich die durch die Elemente $X_1, X_2, \ldots, X_p$ erzeugte *Abel*sche Gruppe $\mathfrak{N}'$ darstellen als das direkte Produkt einer endlichen Gruppe $\mathfrak{N}$, deren Invarianten die Zahlen $e_1, e_2, \ldots, e_\varrho$ sind, und einer unendlichen Gruppe $\mathfrak{N}''$ des Ranges s, in der kein Element außer dem Hauptelement eine endliche Periode besitzt.*) Mit anderen Worten: es lassen sich in der Gruppe $\mathfrak{N}'$ $\varrho + s$ unabhängige Basiselemente

$$Y_\alpha = X_1^{s_{\alpha 1}} \ldots X_p^{s_{\alpha p}}, \qquad (\alpha = 1, 2, \ldots, \varrho)$$

$$Z_\beta = X_1^{t_{\beta 1}} \ldots X_p^{t_{\beta p}} \qquad (\beta = 1, 2, \ldots, s)$$

angeben, wo Y_α zu dem Exponenten e_α gehört, während für Z_β keinerlei Bedingungen bestehen. – Es möge etwa

$$X_\nu = Y_1^{a_{\nu 1}} \ldots Y_\varrho^{a_{\nu\varrho}} . Z_1^{b_{\nu 1}} \ldots Z_s^{b_{\nu s}}$$

sein. Sind dann $x_1, x_2, \ldots, x_p$ p Zahlen, die den Gleichungen

$$(13.) \qquad x_1^{\alpha_{\lambda 1}} x_2^{\alpha_{\lambda 2}} \ldots x_p^{\alpha_{\lambda p}} = 1 \qquad (\lambda = 1, 2, \ldots, n)$$

genügen, und setzt man

$$y_\alpha = x_1^{s_{\alpha 1}} \ldots x_p^{s_{\alpha p}}, \quad z_\beta = x_1^{t_{\beta 1}} \ldots x_p^{t_{\beta p}}, \qquad \binom{\alpha = 1, 2, \ldots, \varrho}{\beta = 1, 2, \ldots, s}$$

so genügen die ϱ Größen y_α den Gleichungen $y_\alpha^{e_\alpha} = 1$ und es ist

$$(14.) \qquad x_\nu = y_1^{a_{\nu 1}} \ldots y_\varrho^{a_{\nu\varrho}} . z_1^{b_{\nu 1}} \ldots z_s^{b_{\nu s}} .$$

Umgekehrt erhält man die sämtlichen Lösungen der Gleichungen (13.), indem man in (14.) für $y_1, \ldots, y_\varrho$ beliebige Wurzeln der Gleichungen

$$y_1^{e_1} = 1, \ldots, y_\varrho^{e_\varrho} = 1$$

und für $z_1, \ldots, z_s$ beliebige (von Null verschiedene) Größen einsetzt."

Unterwirft man nun die Elemente $X_1, \ldots, X_p$ der Gruppe $\mathfrak{K}'$, also auch die diese Gruppe erzeugenden Elemente $Q_0, Q_1, \ldots, Q_{h-1}$, den s Bedingungen

$$(15.) \qquad Z_\beta = X_1^{t_{\beta 1}} \ldots X_p^{t_{\beta p}} = E, \qquad (\beta = 1, 2, \ldots, s)$$

so definieren die Relationen (9.), (10.) und (15.) eine endliche Gruppe $\mathfrak{K}$,

---

*) Ist kein Elementarteiler von $(\alpha_{\lambda\mu})$ größer als 1, so hat man $\mathfrak{N} = E$ zu setzen. Die Gruppe $\mathfrak{N}$ ist eindeutig bestimmt; es ist dies die umfassendste in $\mathfrak{N}'$ enthaltene endliche Gruppe. Dagegen läßt sich, falls nicht $\mathfrak{N} = E$ oder $s = 0$ ist, die Gruppe $\mathfrak{N}''$ auf verschiedene Arten wählen.

in der die Elemente J_{H_λ, H_μ} die *Abel*sche Gruppe $\mathfrak{N}$ der Ordnung $\overline{m} = e_1 e_2 \dots e_\varrho$ erzeugen. Die Gruppe $\mathfrak{R}$ ist dann als eine ergänzte Gruppe $(\mathfrak{N}, \mathfrak{H})$ von $\mathfrak{H}$ anzusehen.

Ich will zeigen, daß $s = h$, daß $\mathfrak{N}$ dem Multiplikator von $\mathfrak{H}$ isomorph und $\mathfrak{R}$ eine hinreichend ergänzte Gruppe von $\mathfrak{H}$ ist.

Setzt man, wenn X_ν das Element $J_{P,Q}$ bedeutet, $x_\nu = r_{P,Q}$, so ist das Gleichungssystem (13.), abgesehen von der Schreibweise, mit dem Gleichungssystem $(A.)$ identisch. Es möge ferner

$$y_\alpha = x_1^{s_{\alpha 1}} \dots x_p^{s_{\alpha p}} = f_\alpha(r_{R,S}), \quad z_\beta = x_1^{t_{\beta 1}} \dots x_p^{t_{\beta p}} = g_\beta(r_{R,S})^{*)}$$

und

$$y_1^{a_{\nu 1}} \dots y_\varrho^{a_{\nu\varrho}} = F_{P,Q}(y_\alpha), \quad z_1^{b_{\nu 1}} \dots z_s^{b_{\nu s}} = G_{P,Q}(z_\beta)$$

gesetzt werden, sodaß

$$(16.) \qquad r_{P,Q} = F_{P,Q}(y_\alpha)\, G_{P,Q}(z_\beta)$$

wird.

Wäre nun $s < h$, so müßten sich, wenn $c_{H_0}, c_{H_1}, \dots, c_{H_{h-1}}$ h beliebige Variable sind, die h^2 Funktionen $r_{P,Q} = \frac{c_P c_Q}{c_{PQ}}$ dieser Variabeln, die doch eine Lösung von $(A.)$ bilden, und also auch die h Ausdrücke

$$\prod_Q \left(\frac{c_P c_Q}{c_{PQ}}\right) = c_P^h$$

durch $s < h$ Variable $z_1, z_2, \dots, z_s$ ausdrücken lassen, was nicht möglich ist. Es kann aber auch nicht $s > h$ sein. Es mögen nämlich für eine beliebige Lösung $r_{P,Q}$ von $(A.)$ die h Größen $d_{H_0}, d_{H_1}, \dots, d_{H_{h-1}}$ dieselbe Bedeutung haben wie in § 1. Nimmt man dann zu den Gleichungen $(A.)$ noch die h Bedingungen

$$d_{H_0} = 1, d_{H_1} = 1, \dots, d_{H_{h-1}} = 1$$

hinzu, so werden die Zahlen $r_{P,Q}$, wie aus der Formel (2.) folgt, h-te Wurzeln der Einheit. Dies wäre aber nicht möglich, wenn die allgemeinste Lösung des Gleichungssystems $(A.)$ $s > h$ unbestimmte Größen $z_1, z_2, \dots, z_s$ enthielte. Es ist daher in der Tat $s = h$.

Ferner wird jeder der ϱ Ausdrücke $f_\alpha(r_{R,S})$, wenn für $r_{R,S}$ die Ausdrücke $\frac{c_R c_S}{c_{RS}}$ eingesetzt werden, identisch, d. h. für jede Wahl der Größen c_P, gleich 1. Denn andernfalls würde sich in der Gleichung

---

*) Entsprechend hat man zu setzen $Y_\alpha = f_\alpha(J_{R,S})$, $Z_\beta = g_\beta(J_{R,S})$.

$$\left\{f_\alpha\left(\frac{c_R c_S}{c_{RS}}\right)\right\}^{e_\alpha} = 1,$$

die, wie auch die Größen c_P gewählt sein mögen, erfüllt ist, eine Bedingung für diese Größen ergeben. Es ist daher stets

$$(17.)\quad \frac{c_P c_Q}{c_{PQ}} = F_{P,Q}\left\{f_\alpha\left(\frac{c_R c_S}{c_{RS}}\right)\right\} G_{P,Q}\left\{g_\beta\left(\frac{c_R c_S}{c_{RS}}\right)\right\} = G_{P,Q}\left\{g_\beta\left(\frac{c_R c_S}{c_{RS}}\right)\right\}.$$

Es lassen sich aber auch für jede beliebige Wahl der Größen $z_1, z_2, \dots, z_h$ die Größen $a_{H_0}, a_{H_1}, \dots, a_{H_{h-1}}$ so bestimmen, daß $z_\beta = g_\beta\left(\frac{a_R a_S}{a_{RS}}\right)$ wird. Um dies darzutun, genügt es zu zeigen, daß die h Funktionen $\bar{g}_\beta = g_\beta\left(\frac{c_R c_S}{c_{RS}}\right)$ der h Variabeln c_P von einander unabhängig sind. Dies ist aber in der Tat der Fall; denn würde eine Gleichung der Form $\varphi(\bar{g}_1, \bar{g}_2, \dots, \bar{g}_h) = 0$ bestehen, so müßten, da, wie wir in § 1 gesehen haben, jede Lösung von $(A.)$ sich auf die Form $s_{P,Q}\frac{c_P c_Q}{c_{PQ}}$ bringen läßt, wo die Zahlen $s_{P,Q}$ h-te Einheitswurzeln sind, die Größen

$$z_\beta^h = g_\beta(r_{R,S}^h) = g_\beta\left(\frac{c_R^h c_S^h}{c_{RS}^h}\right)$$

dieser Gleichung für jede Lösung $r_{P,Q}$ genügen. Dies ist aber nicht möglich, weil die Größen z_β unbestimmt bleiben sollen.

Daher sind zwei Lösungen

$$r_{P,Q} = F_{P,Q}(y_\alpha)\, G_{P,Q}(z_\beta)$$

und

$$r'_{P,Q} = F_{P,Q}(y'_\alpha)\, G_{P,Q}(z'_\beta)$$

dann und nur dann einander assoziiert, wenn $y'_\alpha = y_\alpha$ ist. Denn lassen sich die h Größen a_P so bestimmen, daß $r'_{P,Q} = \frac{a_P a_Q}{a_{PQ}} r_{P,Q}$ wird, so ergiebt sich

$$y'_\alpha y_\alpha^{-1} = f_\alpha(r'_{R,S})\, f_\alpha(r_{R,S}^{-1}) = f_\alpha(r'_{R,S}\, r_{R,S}^{-1}) = f_\alpha\left(\frac{a_R a_S}{a_{RS}}\right) = 1,$$

also $y'_\alpha = y_\alpha$. Sind aber diese Bedingungen erfüllt, so bestimme man die Größen a_P so, daß

$$z'_\alpha z_\beta^{-1} = g_\beta\left(\frac{a_R a_S}{a_{RS}}\right)$$

wird. Dann erhält man wegen (17.)

$$r'_{P,Q} = F_{P,Q}(y'_\alpha)\, G_{P,Q}(z_\beta)\, G_{P,Q}(z'_\beta z_\beta^{-1}) = \frac{a_P a_Q}{a_{PQ}} r_{P,Q}.$$

Es ist daher jede Lösung von $(A.)$ einer der $\overline{m} = e_1 e_2 \ldots e_\varrho$ Lösungen

$$(18.) \qquad r_{P,Q} = F_{P,Q}(y_\alpha)$$

assoziiert, die man dadurch erhält, daß man für $y_1, \ldots, y_\varrho$ beliebige Wurzeln der Gleichungen $y_1^{e_1} = 1, \ldots, y_\varrho^{e_\varrho} = 1$ setzt. Unter diesen $\overline{m}$ Lösungen sind dagegen nicht zwei einander assoziiert.

Folglich ist $\overline{m} = m$ und die Gruppe $\mathfrak{K}$ von der Ordnung mh.

Für jeden der m Charaktere $\psi(J)$ der Gruppe $\mathfrak{N}$ stimmt das System der Zahlen $\psi(J_{P,Q})$ mit einem der m Zahlensysteme (18.) überein. Da nun dieselben m verschiedenen Klassen von Lösungen der Gleichungen $(A.)$ angehören, so ist für die ergänzte Gruppe $\mathfrak{K} = (\mathfrak{N}, \mathfrak{H})$ von $\mathfrak{H}$ die im vorigen Paragraphen betrachtete Zahl m' gleich m. Daher ist $\mathfrak{K}$, wie zu beweisen war, eine hinreichend ergänzte Gruppe von $\mathfrak{H}$.

Zugleich sehen wir, daß $\mathfrak{N}$ dem Multiplikator $\mathfrak{M}$ von $\mathfrak{H}$ isomorph ist, und erhalten eine Methode zur Bestimmung der Invarianten von $\mathfrak{M}$.

Da nun die Ordnung einer hinreichend ergänzten Gruppe von $\mathfrak{H}$ nach dem am Schluß des vorigen Paragraphen Gesagten nicht kleiner sein kann als mh, so erkennen wir auch, daß die hier konstruierte Gruppe $\mathfrak{K}$ eine Darstellungsgruppe von $\mathfrak{H}$ repräsentiert.

Die Definition der Gruppe $\mathfrak{K}$ vermöge der Gleichungen (9.), (10.) und (15.) hängt wesentlich von der Wahl der Elemente $Z_1, \ldots, Z_h$ innerhalb der Gruppe $\mathfrak{N}'$ ab. Da man nun diese Elemente auf mehrere verschiedene Arten auswählen kann, so können im allgemeinen mehrere einander nicht isomorphe Darstellungsgruppen $\mathfrak{G} = (\mathfrak{A}, \mathfrak{H})$ existieren. Sie sind aber alle von der Ordnung mh, und in jeder von ihnen ist die Untergruppe $\mathfrak{A}$ dem Multiplikator von $\mathfrak{H}$ isomorph.

Es kann aber noch mehr bewiesen werden.

Ich bezeichne als einen linearen Charakter der unendlichen Gruppe $\mathfrak{K}'$ jedes System von nicht verschwindenden Größen $\chi(A)$, die sich den Elementen A von $\mathfrak{K}'$ in der Weise zuordnen lassen, daß für je zwei Elemente A und B von $\mathfrak{K}'$ die Gleichung

$$(19.) \qquad \chi(A)\chi(B) = \chi(AB)$$

besteht. Ist dann $\mathfrak{T}$ der Kommutator von $\mathfrak{K}'$, d. h. diejenige Untergruppe von $\mathfrak{K}'$, die durch die Elemente der Form $ABA^{-1}B^{-1}$ erzeugt wird, so lassen sich die Elemente von $\mathfrak{T}$ in genau analoger Weise, wie es für endliche

Gruppen der Fall ist,*) folgendermaßen charakterisieren: ein Element T von $\mathfrak{K}'$ gehört dann und nur dann dem Kommutator $\mathfrak{T}$ von $\mathfrak{K}'$ an, wenn für jeden linearen Charakter $\chi(A)$ von $\mathfrak{K}'$ die Gleichung $\chi(T) = 1$ besteht.

Auf Grund dieses Kriteriums läßt sich der größte gemeinsame Teiler der Gruppen $\mathfrak{T}$ und $\mathfrak{N}'$ leicht bestimmen.

Offenbar erhält man alle linearen Charaktere $\chi(A)$ von $\mathfrak{K}'$, indem man die h Zahlen $\chi(Q_0), \chi(Q_1), \ldots, \chi(Q_{h-1})$ beliebig wählt und die übrigen Größen $\chi(A)$ vermöge der Gleichungen (19.) bestimmt. Ist nun für einen beliebigen Charakter $\chi(Q_\lambda) = c_{H_\lambda}$, so wird $\chi(J_{P,Q}) = \frac{c_P c_Q}{c_{PQ}}$ und

$$\chi(Y_\alpha) = \chi\{f_\alpha(J_{R,S})\} = f_\alpha\{\chi(J_{R,S})\} = f_\alpha\left(\frac{c_R c_S}{c_{RS}}\right).$$

Dies ist aber, wie wir oben gesehen haben, gleich 1. Daher sind alle Elemente $Y_1, Y_2, \ldots, Y_\varrho$ und also auch die durch diese Elemente erzeugte Gruppe $\mathfrak{N}$ in $\mathfrak{T}$ enthalten. Dagegen gehört kein Element von $\mathfrak{N}''$ (außer dem Hauptelement) der Gruppe $\mathfrak{T}$ an. Denn wäre

$$J = Z_1^{k_1} Z_2^{k_2} \ldots Z_h^{k_h}$$

ein Element von $\mathfrak{T}$, so müßte für jeden linearen Charakter $\chi(A)$ von $\mathfrak{K}'$

$$\chi(J) = [\chi(Z_1)]^{k_1} [\chi(Z_2)]^{k_2} \ldots [\chi(Z_h)]^{k_h} = 1$$

sein. Es ist aber, wenn wieder $\chi(J_{P,Q}) = \frac{c_P c_Q}{c_{PQ}}$ ist,

$$\chi(Z_\beta) = \chi\{g_\beta(J_{R,S})\} = g_\beta\{\chi(J_{R,S})\} = g_\beta\left(\frac{c_R c_S}{c_{RS}}\right).$$

Wir haben aber gesehen, daß die Zahlen c_P so gewählt werden können, daß die h Größen $g_\beta\left(\frac{c_R c_S}{c_{RS}}\right)$ beliebig vorgeschriebene Werte $z_1, z_2, \ldots, z_h$ annehmen. Es kann daher, wenn nicht alle Exponenten $k_1, k_2, \ldots, k_h$ gleich 0 sind, $\chi(J)$ nicht stets gleich 1 sein.

Mithin ist der größte gemeinsame Teiler von $\mathfrak{T}$ und $\mathfrak{N}'$ nichts anderes als die dem Multiplikator $\mathfrak{M}$ von $\mathfrak{H}$ isomorphe Gruppe $\mathfrak{N}$ der Ordnung m. Der Kommutator $\mathfrak{T}$ von $\mathfrak{K}'$ ist daher eine endliche Gruppe, die offenbar als eine durch die Gruppe $\mathfrak{N} = \mathfrak{M}$ ergänzte Gruppe des Kommutators $\mathfrak{K}$ von $\mathfrak{H}$

*) Vergl. den auf S. 30 angeführten Satz. — Es läßt sich leicht einsehen, daß dieser Satz für jede durch endlich viele Elemente erzeugbare Gruppe bestehen bleibt.

aufzufassen ist. Ihre Ordnung ist, falls r die Ordnung von $\mathfrak{R}$ bedeutet, gleich mr.

Es sei nun $\mathfrak{L}$ eine beliebige Darstellungsgruppe von $\mathfrak{H}$, d. h. eine durch die Gruppe $\mathfrak{M}$ ergänzte Gruppe von $\mathfrak{H}$, deren Kommutator $\mathfrak{T}'$ die ganze Gruppe $\mathfrak{M}$ enthält. Es sei

$$\mathfrak{L} = \mathfrak{M}L_0 + \mathfrak{M}L_1 + \cdots + \mathfrak{M}L_{h-1},$$

wo der Komplex $\mathfrak{M}L_\lambda$ dem Elemente H_λ der mit $\frac{\mathfrak{L}}{\mathfrak{M}}$ isomorphen Gruppe $\mathfrak{H}$ entsprechen möge. Dann ist

$$L_\lambda L_\mu L^{-1}_{\varphi(\lambda,\mu)} = M_{H_\lambda, H_\mu}$$

ein Element von $\mathfrak{M}$, also ein invariantes Element von $\mathfrak{L}$. Aus der Annahme, daß der Kommutator $\mathfrak{T}'$ von $\mathfrak{L}$ alle Elemente von $\mathfrak{M}$ enthalten soll, ergibt sich unmittelbar, daß die Elemente $L_0, L_1, \ldots, L_{h-1}$ die ganze Gruppe $\mathfrak{L}$ erzeugen.*) Die Elemente $L_0, L_1, \ldots, L_{h-1}$ genügen aber, für die Q_λ eingesetzt, den die unendliche Gruppe $\mathfrak{K}'$ definierenden Bedingungen (9.) und (10.). Daraus folgt,**) daß die Gruppe $\mathfrak{K}'$ der Gruppe $\mathfrak{L}$ mehrstufig isomorph ist, und zwar derart, daß jedem Element von $\mathfrak{K}'$ nur ein Element von $\mathfrak{L}$ entspricht. Mithin ist auch der Kommutator $\mathfrak{T}$ von $\mathfrak{K}'$ dem Kommutator $\mathfrak{T}'$ von $\mathfrak{L}$ in derselben Weise mehrstufig isomorph. Da aber die Ordnung von $\mathfrak{T}'$ gleich der Ordnung mr von $\mathfrak{T}$ ist, so ist der zwischen $\mathfrak{T}$ und $\mathfrak{T}'$ bestehende Isomorphismus ein einstufiger. Der Kommutator jeder beliebigen Darstellungsgruppe ist demnach der wohlbestimmten Gruppe $\mathfrak{T}$ isomorph.

Wir erhalten den Satz:

III. *Die Kommutatorgruppen je zweier Darstellungsgruppen einer gegebenen Gruppe sind einander isomorph.*

Ist insbesondere $\mathfrak{R} = \mathfrak{H}$, also $r = h$, so stimmt auch jede Darstellungsgruppe von $\mathfrak{H}$ mit ihrer Kommutatorgruppe überein. Aus III. ergibt sich demnach:

IV. *Umfaßt der Kommutator einer Gruppe $\mathfrak{H}$ alle Elemente von $\mathfrak{H}$, so besitzt $\mathfrak{H}$ nur eine Darstellungsgruppe.****)

---

*) Dies folgt daraus, daß jedes Kommutatorelement von $\mathfrak{L}$ die Form $L_\lambda L_\mu L_\lambda^{-1} L_\mu^{-1}$ hat.

**) Vergl. *Dyck* a. a. O. und *Burnside,* Theory of Groups of finite Order, § 182.

***) Die genauere Bestimmung der Anzahl der einander nicht isomorphen Darstellungsgruppen einer Gruppe $\mathfrak{H}$, deren Kommutator nicht alle Elemente von $\mathfrak{H}$ umfaßt, werde ich in einer späteren Arbeit durchführen.

§ 4.

In den bisherigen Betrachtungen ist das Hauptgewicht auf die Untersuchung der sämtlichen Zahlensysteme $r_{P,Q}$ gelegt worden, zu denen Darstellungen der Gruppe $\mathfrak{H}$ durch gebrochene lineare Substitutionen gehören können. Die andere Frage: „wie lassen sich die verschiedenen zu demselben Zahlensystem $r_{P,Q}$ gehörenden primitiven Darstellungen der Gruppe $\mathfrak{H}$ bestimmen?" läßt sich ohne Einführung neuer Hilfsmittel auf Grund der Sätze des Herrn *Frobenius* beantworten.

Es sei wieder wie in § 2

$$\mathfrak{G} = \mathfrak{A} G_0 + \mathfrak{A} G_1 + \cdots + \mathfrak{A} G_{h-1} \qquad (G_0 = E)$$

eine beliebige durch die *Abel*sche Gruppe $\mathfrak{A}$ ergänzte Gruppe von $\mathfrak{H}$; es mögen A_α, $A_{\lambda,\mu}$, $\psi^{(\alpha)}(A)$ und $r^{(\alpha)}_{H_\lambda, H_\mu} = \psi^{(\alpha)}(A_{\lambda,\mu})$ dieselbe Bedeutung haben wie dort. Dann ist, wie wir gesehen haben, die Untersuchung der sämtlichen zu dem Zahlensystem $r^{(\alpha)}_{H_\lambda, H_\mu}$ gehörenden primitiven Darstellungen der Gruppe $\mathfrak{H}$ durch gebrochene lineare Substitutionen identisch mit der Bestimmung der verschiedenen dem Charakter $\psi^{(\alpha)}(A)$ von $\mathfrak{A}$ entsprechenden primitiven Darstellungen der Gruppe $\mathfrak{G}$ durch ganze lineare Substitutionen oder, was dasselbe ist, mit der Bestimmung der diesen Darstellungen entsprechenden zur Gruppe $\mathfrak{G}$ gehörenden Matrizen.*)

Für diese gelten nun folgende Sätze:**)

„Bedeuten die ah Größen $x_{A_\beta G_\lambda}$ $(\beta = 0, 1, \ldots, a-1;\ \lambda = 0, 1, \ldots, h-1)$ unabhängige Variable, so werden die sämtlichen dem Charakter $\psi^{(\alpha)}(A)$ entsprechenden, zur Gruppe $\mathfrak{G}$ gehörigen primitiven Matrizen dadurch erhalten, daß man die zur Gruppe gehörige Matrix h-ten Grades

$$Y^{(\alpha)} = \Big(\sum_\beta \psi^{(\alpha)}(A_\beta)\, x_{A_\beta G_\lambda G_\mu^{-1}}\Big) \qquad (\lambda, \mu = 0, 1, \ldots, h-1)$$

in primitive Teilmatrizen zerlegt. Sind $\Phi_1, \Phi_2, \ldots, \Phi_{l_\alpha}$ die l_α verschiedenen Primfaktoren der Determinante $Q^{(\alpha)}$ von $Y^{(\alpha)}$, ist f_λ der Grad des Primfaktors Φ_λ, so ist

$$Q^{(\alpha)} = \Phi_1^{f_1} \Phi_2^{f_2} \ldots \Phi_{l_\alpha}^{f_{l_\alpha}}, \tag{20.}$$

also

$$f_1^2 + f_2^2 + \cdots + f_{l_\alpha}^2 = h. \tag{21.}$$

---

*) *Frobenius*, D. I.

**) *Frobenius*, Über Relationen zwischen den Charakteren einer Gruppe und denen ihrer Untergruppen, §§ 2 und 3. Sitzungsberichte 1898, S. 501.

Sind ferner $\chi^{(1)}(S), \chi^{(2)}(S), \ldots, \chi^{(l_a)}(S)$ die den l_a Primfaktoren Φ_λ entsprechenden Charaktere von $\mathfrak{G}$, so ist, wenn $S = A_\beta G_\lambda$ ist,

$$(22.) \qquad f_1\chi^{(1)}(S)+f_2\chi^{(2)}(S)+\cdots+f_{l_a}\chi^{(l_a)}(S) = h\,\psi^{(a)}(A_\beta)\,\varepsilon_\lambda,$$

wo ε_λ gleich 1 oder gleich 0 ist, je nachdem $\lambda = 0$ oder > 0 ist."

Setzt man nun, wenn R irgend ein Element von $\mathfrak{G}$ bedeutet,

$$(23.) \qquad y_R^{(a)} = \sum_\beta \psi^{(a)}(A_\beta)\, x_{A_\beta R},$$

so läßt sich $Y^{(a)}$ in der Form

$$Y^{(a)} = (y^{(a)}_{G_\lambda G_\mu^{-1}})$$

schreiben. Ferner wird, wie leicht ersichtlich ist, für jedes Element A von $\mathfrak{A}$

$$(24.) \qquad y^{(a)}_{AR} = \psi^{(a)}(A^{-1})\, y^{(a)}_R.$$

Bezeichnet man das Element H_μ^{-1} von $\mathfrak{H}$ mit $H_{\mu'}$, so erhält man

$$G_\mu G_{\mu'} = A_{\mu,\mu'} G_0 = A_{\mu,\mu'},$$

also $G_\mu^{-1} = A^{-1}_{\mu,\mu'} G_{\mu'}$. Daher ist, wenn $H_\lambda H_\mu^{-1} = H_\varrho$ gesetzt wird,

$$G_\lambda G_\mu^{-1} = A^{-1}_{\mu,\mu'} A_{\lambda,\mu'} G_\varrho.$$

Bezeichnet man nun den Ausdruck $y^{(a)}_{G_\lambda}$ mit z_{H_λ}, so wird wegen (24.)

$$y^{(a)}_{G_\lambda G_\mu^{-1}} = \psi^{(a)}(A_{\mu,\mu'})\,\psi^{(a)}(A^{-1}_{\lambda,\mu'})\, z_{H_\lambda H_\mu^{-1}} = \frac{r^{(a)}_{H_\mu, H_\mu^{-1}}}{r^{(a)}_{H_\lambda, H_\mu^{-1}}}\, z_{H_\lambda H_\mu^{-1}}.$$

Da nun das Zahlensystem $r^{(a)}_{H_\lambda, H_\mu}$ dem Gleichungssystem (A.) genügt und, wegen $G_0 = E$, $r^{(a)}_{H_\lambda, E} = 1$ ist, so ist

$$r^{(a)}_{H_\lambda H_\mu^{-1}, H_\mu}\, r^{(a)}_{H_\lambda, H_\mu^{-1}} = r^{(a)}_{H_\lambda H_\mu^{-1}, E}\, r^{(a)}_{H_\mu, H_\mu^{-1}} = r^{(a)}_{H_\mu, H_\mu^{-1}}.$$

Daher läßt sich die Matrix $Y^{(a)}$ in der Form

$$(25.) \qquad Y^{(a)} = (r^{(a)}_{PQ^{-1}, Q}\, z_{PQ^{-1}}) \qquad (P, Q = H_0, H_1, \ldots, H_{h-1})$$

schreiben. Wir sehen also, daß $Y^{(a)}$ sich auch als die in § 1 dem Zahlensystem $r^{(a)}_{P,Q}$ zugeordnete Matrix h-ten Grades ansehen läßt.

Nimmt man an, $\mathfrak{G} = \mathfrak{K}$ sei eine Darstellungsgruppe von $\mathfrak{H}$, so ist jedes System von Zahlen $r_{P,Q}$, die den Gleichungen (A.) genügen, einem der $a = m$ Zahlensysteme $r^{(a)}_{P,Q}$ assoziiert. Aus dem zitierten Satze läßt sich daher schließen:

V. *Ist $r_{P,Q}$ eine beliebige Lösung der Gleichungen (A.), sind*

$$x_{H_0}, x_{H_1}, \ldots, x_{H_{h-1}}$$

unabhängige Variable, und setzt man

$$(r_{PQ^{-1},Q}\, x_{PQ^{-1}}) = \Sigma (R)\, x_R, \qquad (R = H_0, H_1, \ldots, H_{h-1})$$

so ist jede zu dem Zahlensystem $r_{P,Q}$ gehörende primitive Darstellung der Gruppe $\mathfrak{H}$ einer der primitiven Darstellungen äquivalent, in welche die durch die Matrizen (R) gebildete Darstellung zerlegt werden kann.

Ich will nun zeigen, wie sich die Anzahl l_a der Primfaktoren von $Q^{(a)}$ oder, was dasselbe ist, die Anzahl der verschiedenen dem Charakter $\psi^{(a)}(A)$ von $\mathfrak{A}$ entsprechenden primitiven Darstellungen von $\mathfrak{G}$ genauer bestimmen läßt. Die Methode, die hier zur Bestimmung von l_a angewandt wird, ist der von Herrn *Frobenius**) zur Bestimmung der Anzahl der verschiedenen Primfaktoren der Gruppendeterminante benutzten Methode genau nachgebildet.

Für jeden der l_a dem Charakter $\psi^{(a)}(A)$ entsprechenden Charaktere $\chi^{(\varrho)}(S)$ von $\mathfrak{G}$ ist

$$\chi^{(\varrho)}(A_\beta G_\lambda) = \psi^{(a)}(A_\beta)\, \chi^{(\varrho)}(G_\lambda).$$

Da nun

$$\begin{aligned} ah = \sum_S \chi^{(\varrho)}(S^{-1})\, \chi^{(\varrho)}(S) &= \sum_{\beta,\lambda} \chi^{(\varrho)}(A_\beta^{-1} G_\lambda^{-1})\, \chi^{(\varrho)}(A_\beta G_\lambda) \\ &= \sum_{\beta,\lambda} \psi^{(a)}(A_\beta^{-1} A_\beta)\, \chi^{(\varrho)}(G_\lambda^{-1})\, \chi^{(\varrho)}(G_\lambda) = a \sum_\lambda \chi^{(\varrho)}(G_\lambda^{-1})\, \chi^{(\varrho)}(G_\lambda) \end{aligned}$$

ist,**) so erhält man

$$\sum_\lambda \chi^{(\varrho)}(G_\lambda^{-1})\, \chi^{(\varrho)}(G_\lambda) = h. \tag{26.}$$

Ferner ergibt sich aus der für jeden Charakter f-ten Grades bestehenden Formel***)

$$\sum_S \chi(G_\lambda^{-1} S^{-1} G_\lambda S) = \frac{ah}{f}\, \chi(G_\lambda^{-1})\, \chi(G_\lambda)$$

die Gleichung

$$\sum_{\mu=0}^{h-1} \chi^{(\varrho)}(G_\lambda^{-1} G_\mu^{-1} G_\lambda G_\mu) = \frac{h}{f_\varrho}\, \chi^{(\varrho)}(G_\lambda^{-1})\, \chi^{(\varrho)}(G_\lambda).$$

*) Primfaktoren, § 7.

**) *Frobenius*, Gruppencharaktere, § 5, Formel (10.).

***) ebenda, Formel (5.).

Es ist daher wegen (26.)

$$\sum_{\lambda,\mu} \frac{f_\varrho}{h} \chi^{(\varrho)}(G_\lambda^{-1} G_\mu^{-1} G_\lambda G_\mu) = h$$

und

$$\sum_{\varrho=1}^{l_\alpha} \sum_{\lambda,\mu} \frac{f_\varrho}{h} \chi^{(\varrho)}(G_\lambda^{-1} G_\mu^{-1} G_\lambda G_\mu) = h l_\alpha$$

oder auch

$$h l_\alpha = \sum_{\lambda,\mu} \sum_{\varrho=1}^{l_\alpha} \frac{f_\varrho}{h} \chi^{(\varrho)}(G_\lambda^{-1} G_\mu^{-1} G_\lambda G_\mu).$$

Es sei nun

$$G_\lambda^{-1} G_\mu^{-1} G_\lambda G_\mu = F_{\lambda,\mu} G_\nu,$$

wo $F_{\lambda,\mu}$ ein Element von $\mathfrak{A}$ bedeutet. Dann ist nach (22.)

$$\sum_{\varrho=1}^{l_\alpha} \frac{f_\varrho}{h} \chi^{(\varrho)}(G_\lambda^{-1} G_\mu^{-1} G_\lambda G_\mu) = \psi^{(\alpha)}(F_{\lambda,\mu})\, \varepsilon_\nu.$$

Wir erhalten daher

$$(27.) \qquad h l_\alpha = \sum_{\lambda=0}^{h-1} \sum_{\mu} \psi^{(\alpha)}(F_{\lambda,\mu}),$$

wo μ diejenigen der Zahlen $0, 1, \ldots, h-1$ durchläuft, für die $G_\lambda^{-1} G_\mu^{-1} G_\lambda G_\mu$ ein Element von $\mathfrak{A}$ ist, oder was dasselbe ist, für die H_μ mit H_λ vertauschbar ist. Aus (27.) läßt sich l_α folgendermaßen bestimmen.

Es mögen die h Elemente $H_0, H_1, \ldots, H_{h-1}$ von $\mathfrak{H}$ in k Klassen konjugierter Elemente zerfallen, die mit $(0), (1), \ldots, (k-1)$ bezeichnet werden mögen; es möge die Klasse (ϱ) genau h_ϱ Elemente enthalten, das Zeichen (0) die Hauptklasse (das Element E) bedeuten. Zwei Elemente G_λ und G_μ können offenbar nur dann in $\mathfrak{G}$ konjugiert sein, wenn es H_λ und H_μ in $\mathfrak{H}$ sind. Es läßt sich aber auch, was an einer späteren Stelle die Betrachtung vereinfacht und hier gleich festgesetzt werden soll, das vollständige Restsystem $G_0, G_1, \ldots, G_{h-1}$ von $\mathfrak{G}$ mod. $\mathfrak{A}$ so wählen, daß G_λ und G_μ stets dann in $\mathfrak{G}$ konjugiert sind, wenn H_λ und H_μ einer Klasse (ϱ) angehören. Sind nämlich $H_{\alpha_1}, H_{\alpha_2}, \ldots$ die mit H_1 konjugierten Elemente von $\mathfrak{H}$, so enthält jeder der Komplexe $\mathfrak{A} G_{\alpha_1}, \mathfrak{A} G_{\alpha_2}, \ldots$ mindestens ein mit G_1 konjugiertes Element; man bezeichne dann diese Elemente mit $G_{\alpha_1}, G_{\alpha_2}, \ldots$. Ebenso verfahre man, wenn etwa H_2 ein mit H_1 nicht konjugiertes Element ist, in bezug auf G_2, u. s. w. Gehören dann etwa $H_0, H_1, \ldots, H_{k-1}$ den k verschiedenen Klassen

$$(0), (1), \ldots, (k-1)$$

an, so sind in $\mathfrak{G}$ unter den Elementen $G_0, G_1, \ldots, G_{h-1}$ genau $h_0 = 1$ konjugiert G_0, genau h_1 konjugiert G_1, u. s. w.

Sind nun $H_{\mu_1}, H_{\mu_2}, \ldots$ die mit dem Elemente H_λ der Klasse (ϱ) vertauschbaren Elemente von $\mathfrak{H}$, so ist

$$G_\lambda G_{\mu_\sigma} = F_{\lambda, \mu_\sigma} G_{\mu_\sigma} G_\lambda. \qquad \left(\sigma = 1, 2, \ldots, \frac{h}{h_\varrho}\right)$$

Es seien unter den $\frac{h}{h_\varrho}$ Elementen F_{λ, μ_σ} genau a_λ unter einander verschieden. Dann bilden diese a_λ Elemente offenbar eine Untergruppe $\mathfrak{A}_\lambda$ von $\mathfrak{A}$; und man sieht auch leicht ein, daß a_λ ein Divisor von $\frac{h}{h_\varrho}$ sein muß. Ferner zeigt man leicht, daß die Gruppe $\mathfrak{A}_\lambda$ sich nicht ändert, wenn G_λ durch ein konjugiertes Element G_{λ_1} ersetzt wird. Man erhält auf diese Weise k Untergruppen

$$\mathfrak{A}_0 (= E), \mathfrak{A}_1, \ldots, \mathfrak{A}_{k-1}$$

von $\mathfrak{A}$, die beziehungsweise den k Klassen $(0), (1), \ldots, (k-1)$ von $\mathfrak{H}$ zugeordnet sind.*) Ferner bilden diejenigen Elemente B von $\mathfrak{A}$, für die $\psi^{(\alpha)}(B) = 1$ ist, eine Untergruppe $\mathfrak{B}_\alpha$ von $\mathfrak{A}$. Es sei $\mathfrak{D}_{\alpha\varrho}$ der größte gemeinsame Teiler der Gruppen $\mathfrak{A}_\varrho$ und $\mathfrak{B}_\alpha$, $d_{\alpha\varrho}$ die Ordnung von $\mathfrak{D}_{\alpha\varrho}$. Dann erhalten wir in (27.), wenn H_λ der Klasse (ϱ) angehört,

$$\sum_\mu \psi^{(\alpha)}(F_{\lambda,\mu}) = \frac{h}{a_\varrho h_\varrho} \sum_J \psi^{(\alpha)}(J) = \frac{h\, d_{\alpha\varrho}}{h_\varrho a_\varrho} \sum_{J'} \psi^{(\alpha)}(J'),$$

wo J alle Elemente von $\mathfrak{A}_\varrho$, J' ein vollständiges Restsystem von $\mathfrak{A}_\varrho$ mod. $\mathfrak{D}_{\alpha\varrho}$ durchläuft. Nun bilden aber die $\frac{a_\varrho}{d_{\alpha\varrho}}$ Zahlen $\psi^{(\alpha)}(J')$ einen Charakter der Gruppe $\frac{\mathfrak{A}_\varrho}{\mathfrak{D}_{\alpha\varrho}}$, und da $\psi^{(\alpha)}(J')$ nur dann gleich 1 sein soll, wenn J' der Gruppe $\mathfrak{D}_{\alpha\varrho}$ angehört, so erhalten wir nach einem bekannten Satze

$$\sum \psi^{(\alpha)}(J') = 0,$$

außer wenn $d_{\alpha\varrho} = a_\varrho$, d. h. $\mathfrak{A}_\varrho$ in $\mathfrak{B}_\alpha$ enthalten ist. Es kann daher gesetzt werden

---

*) Ist etwa $\mathfrak{A} = \mathfrak{A}_\varrho + \mathfrak{A}_\varrho B_2 + \cdots + \mathfrak{A}_\varrho B_{\frac{a}{a_\varrho}}$, so entsprechen der Klasse (ϱ) von $\mathfrak{H}$ in $\mathfrak{G}$ $\frac{a}{a_\varrho}$ verschiedene Klassen von je $a_\varrho h_\varrho$ konjugierten Elementen, die durch die Elemente $G_\varrho, B_2 G_\varrho, \ldots, B_{\frac{a}{a_\varrho}} G_\varrho$ repräsentiert werden.

$$(28.) \qquad \sum_{\mu} \psi^{(\alpha)}(F_{\lambda,\mu}) = \frac{h}{h_\varrho}\left[\frac{d_{\alpha\varrho}}{a_\varrho}\right],$$

wo $[x]$, wie gewöhnlich, die größte ganze Zahl bedeutet, die $\leqq x$ ist. Da die rechte Seite der Gleichung (28.) nur von der Klasse (ϱ) abhängt, der das Element H_λ angehört, so erhalten wir aus (27.)

$$l_\alpha = \sum_{\varrho=0}^{k-1}\left[\frac{d_{\alpha\varrho}}{a_\varrho}\right],$$

d. h. l_α ist gleich der Anzahl der Klassen (ϱ) von $\mathfrak{H}$, für die $\mathfrak{A}_\varrho$ in $\mathfrak{B}_\alpha$ enthalten ist.

Diesem Ergebnis läßt sich auch folgende Fassung geben:

VI. *Ist $r_{P,Q}$ irgend eine Lösung der Gleichungen (A.), so besitzen gewisse l ($l \geqq 1$)*) unter den k Klassen konjugierter Elemente von $\mathfrak{H}$ die Eigenschaft, daß, sobald das Element P einer dieser l Klassen angehört und Q ein beliebiges mit P vertauschbares Element ist, $r_{P,Q} = r_{Q,P}$ ist. Diese Zahl l gibt zugleich die Anzahl der verschiedenen (einander nicht äquivalenten)**) zu dem Zahlensystem $r_{P,Q}$ gehörenden primitiven Darstellungen von $\mathfrak{H}$ an.*

Die Grade der Charaktere einer Gruppe sind, wie Herr *Frobenius* (Primfaktoren, § 12) bewiesen hat, Divisoren der Ordnung der Gruppe. In genau analoger Weise läßt sich beweisen:

VII. *Enthält die Gruppe $\mathfrak{G}$ der Ordnung ah eine aus invarianten Elementen von $\mathfrak{G}$ bestehende Untergruppe der Ordnung a, so geht der Grad eines jeden Charakters nicht nur in ah, sondern auch in h auf.*

Hieraus folgt unmittelbar:

VIIa. *Der Grad einer jeden primitiven Darstellung einer Gruppe $\mathfrak{H}$ durch (ganze oder gebrochene) lineare Substitutionen ist ein Divisor der Ordnung von $\mathfrak{H}$.*

Ich beweise den Satz VII folgendermaßen:

Setzt man für die ah Variabeln x_S, wo S alle Elemente von $\mathfrak{G}$ durchläuft, fest, daß für je zwei Elemente S und T $x_{ST} = x_{TS}$ sein soll, so wird für je zwei konjugierte Elemente S und S' der Gruppe $x_S = x_{S'}$. Dagegen sollen, wenn k' die Anzahl der Klassen konjugierter Elemente von $\mathfrak{G}$ und $S, S', S'', \ldots$ k' Elemente von $\mathfrak{G}$ sind, von denen nicht zwei einander konju-

*) Die Hauptklasse (0) besitzt stets diese Eigenschaft.

**) Dagegen können, wenn der Kommutator von $\mathfrak{H}$ nicht gleich $\mathfrak{H}$ ist, unter diesen l primitiven Darstellungen auch mehrere einander assoziiert sein.

giert sind, die Größen $x_S, x_{S'}, x_{S''}, \ldots$ als unabhängige Variable angesehen werden. Es wird dann für zwei beliebige Elemente S und T auch

$$y_{ST}^{(\alpha)} = \sum_\beta \psi^{(\alpha)}(A_\beta)\, x_{A_\beta ST} = \sum_\beta \psi^{(\alpha)}(A_\beta)\, x_{A_\beta TS} = y_{TS}^{(\alpha)}.$$

Daher wird, wenn H_λ und H_μ zwei konjugierte Elemente von $\mathfrak{H}$, also nach unserer Festsetzung G_λ und G_μ konjugierte Elemente von $\mathfrak{G}$ sind, $y_{G_\lambda}^{(\alpha)} = y_{G_\mu}^{(\alpha)}$ oder $z_{H_\lambda} = z_{H_\mu}$. Man bezeichne, wenn $H_\lambda, H_{\lambda_1}, \ldots$ die h_ϱ Elemente der Klasse (ϱ) von $\mathfrak{H}$ sind, die einander gleichen Größen $z_{H_\lambda}, z_{H_{\lambda_1}}, \ldots$ alle mit z_ϱ. Ist nun (ϱ) eine Klasse, für die $\mathfrak{A}_\varrho$ nicht in $\mathfrak{B}_\alpha$ enthalten ist, so lassen sich Elemente A von $\mathfrak{A}_\varrho$ angeben, für die $\psi^{(\alpha)}(A) \neq 1$ ist; es ist dann, weil G_λ und AG_λ konjugierte Elemente von $\mathfrak{G}$ sind, nach (24.)

$$z_\varrho = y_{G_\lambda}^{(\alpha)} = y_{AG_\lambda}^{(\alpha)} = \psi^{(\alpha)}(A^{-1})\, y_{G_\lambda}^{(\alpha)} = 0.$$

Ist dagegen (ϱ) eine der l_α Klassen $(\varrho_0) = (0), (\varrho_1), \ldots$, für die $\mathfrak{A}_\varrho$ in $\mathfrak{B}_\alpha$ enthalten ist, so ist

$$z_\varrho = y_{G_\lambda}^{(\alpha)} = a_\varrho \sum_B \psi^{(\alpha)}(B)\, x_{BG_\lambda}, \qquad (\varrho = \varrho_0, \varrho_1, \ldots)$$

wo B ein vollständiges Restsystem von $\mathfrak{A}$ mod. $\mathfrak{A}_\varrho$ durchläuft, von Null verschieden, weil unter den $\frac{a}{a_\varrho}$ Elementen BG_λ nicht zwei konjugiert, die Variabeln x_{BG_λ} also alle von einander verschieden sind. Zugleich sehen wir, daß die l_α Größen $z_{\varrho_0}, z_{\varrho_1}, \ldots$, die nicht Null sind, als unabhängige Variable angesehen werden können; denn in der Tat enthalten diese Größen, als Funktionen der x_S betrachtet, keine dieser Variabeln gemeinsam.

Nun sind aber die Primfaktoren von $Q^{(\alpha)}$ auch Primfaktoren der Gruppendeterminante von $\mathfrak{G}$. Daher wird (*Frobenius*, Primfaktoren, § 6), wenn

$$\xi_\lambda = \frac{1}{f_\lambda} \sum_S \chi^{(\lambda)}(S)\, x_S$$

gesetzt wird, unter den über die Variabeln x_S gemachten Voraussetzungen

$$\Phi_\lambda = \xi_\lambda^{f_\lambda}.$$

Es ist aber, wenn $\chi^{(\lambda)}(G_\mu)$ mit $\chi_\varrho^{(\lambda)}$ bezeichnet wird, sobald H_μ der Klasse (ϱ) von $\mathfrak{H}$ angehört,

$$\xi_\lambda = \frac{1}{f_\lambda} \sum_{\beta, \mu} \psi^{(\alpha)}(A_\beta)\, \chi^{(\lambda)}(G_\mu)\, x_{A_\beta G_\mu}$$
$$= \frac{1}{f_\lambda} \sum_\mu \chi^{(\lambda)}(G_\mu)\, y_{G_\mu}^{(\alpha)} = \frac{1}{f_\lambda} \sum_{\varrho=0}^{k-1} h_\varrho \chi_\varrho^{(\lambda)} z_\varrho.$$

Aus der Gleichung

$$Q^{(\alpha)} = \prod_{\lambda=0}^{l_\alpha} \xi_\lambda^{f_\lambda^2}$$

folgt dann, da (vergl. Formel (25.)) $Q^{(\alpha)}$ eine ganze rationale Funktion der l_α Variabeln $z_{\varrho_0}, z_{\varrho_1}, \ldots$ ist, deren Koeffizienten ganze algebraische Zahlen sind, da ferner $\frac{h_0\chi_0^{(\lambda)}}{f_\lambda} = 1$ ist, nach einem bekannten Satze, daß die l_α Größen

$$\frac{h_\varrho \chi_\varrho^{(\lambda)}}{f_\lambda} \qquad (\varrho = \varrho_0, \varrho_1, \ldots)$$

ganze algebraische Zahlen sind. Dies ist dann, weil für die übrigen $k - l_\alpha$ Klassen offenbar $\chi_\varrho^{(\lambda)} = 0$ ist, auch für alle $\varrho = 0, 1, \ldots, k-1$ der Fall. Daraus ergibt sich aber, wenn $\chi_{\varrho'}^{(\lambda)}$ den zu $\chi_\varrho^{(\lambda)}$ konjugiert komplexen Wert bedeutet, aus (26.), daß auch

$$\frac{h}{f_\lambda} = \sum_\varrho \frac{h_\varrho \chi_\varrho^{(\lambda)}}{f_\lambda} \chi_{\varrho'}^{(\lambda)}$$

eine ganze algebraische, und also auch ganze rationale Zahl ist. Das ist aber der zu beweisende Satz.*)

§ 5.

Die Bestimmung des Multiplikators und einer Darstellungsgruppe einer gegebenen Gruppe $\mathfrak{H}$ nach den in §§ 1 und 3 angegebenen Methoden führt auf sehr umständliche Rechnungen. Im Folgenden sollen einige Sätze abgeleitet werden, die in vielen Fällen diese Rechnungen zu umgehen ermöglichen.

---

*) Der Satz VII läßt sich, wie Herr *Frobenius* die Güte hatte mir mitzuteilen, einfacher folgendermaßen beweisen.

Es seien (0), (1), ... die k' Klassen konjugierter Elemente, in die die $ah = g$ Elemente der Gruppe $\mathfrak{G}$ zerfallen, es enthalte die α-te Klasse g_α Elemente, und es bedeute (α') die zu (α) inverse Klasse. Sei $e_{\alpha\beta} = 0$, wenn $\beta \neq \alpha'$ ist, und $e_{\alpha\alpha'} = 1$; ferner sei (*Frobenius*, Gruppencharaktere, § 4) $\frac{g p_{\alpha\beta}}{g_\alpha}$ die Anzahl der Lösungen der Gleichung $SR = RSAB$, wo A ein bestimmtes Element der α-ten Klasse ist, wo B die g_β Elemente der β-ten Klasse durchläuft, und R und S unabhängig von einander die g Elemente von $\mathfrak{G}$ durchlaufen. Es ist dann (ebenda, § 3, (12)), wenn f der Grad eines Charakters von $\mathfrak{G}$ ist, $f^2 p_{\alpha\beta'} - g g_\alpha e_{\alpha\beta'} | = 0$, also auch $\left| \frac{g}{g_\alpha} p_{\alpha\beta'} - \frac{g^2}{f^2} e_{\alpha\beta'} \right| = 0$. Folglich sind die k' Wurzeln $\left(\frac{g}{f}\right)^2$ dieser Gleichung ganze Zahlen.

Jede Darstellungsgruppe $\mathfrak{K}$ von $\mathfrak{H}$ ist durch folgende Eigenschaften charakterisiert:

1. $\mathfrak{K}$ enthält eine aus invarianten Elementen von $\mathfrak{K}$ bestehende Untergruppe $\mathfrak{M}$, und es ist $\frac{\mathfrak{K}}{\mathfrak{M}}$ der Gruppe $\mathfrak{H}$ isomorph;

2. der Kommutator von $\mathfrak{K}$ enthält alle Elemente von $\mathfrak{M}$;

3. es gibt keine Gruppe $\mathfrak{G}$, welche die Eigenschaften 1. und 2. besitzt, und deren Ordnung größer ist als die Ordnung der Gruppe $\mathfrak{K}$.

Die folgenden Sätze dienen dazu, eine obere Grenze für die Ordnung m des Multiplikators von $\mathfrak{H}$ abzuleiten. Dieses ist oft hinreichend für die Bestimmung einer Darstellungsgruppe von $\mathfrak{H}$; denn wissen wir etwa, daß m nicht größer ist als eine gewisse ganze Zahl $\overline{m}$, und kennen wir eine Gruppe $\mathfrak{K}$ der Ordnung $\overline{m}h$, welche die Eigenschaften 1. und 2. besitzt, so schließen wir auf Grund des Satzes II, daß $\overline{m} = m$, und daß $\mathfrak{K}$ eine Darstellungsgruppe von $\mathfrak{H}$ ist.

Ich stelle zunächst folgende Betrachtung an.

Die Bedingung 2. besagt, daß für jeden linearen Charakter $\chi(R)$ von $\mathfrak{K}$ $\chi(J) = 1$ sein muß, sobald J ein Element von $\mathfrak{M}$ ist.

Es sei $\mathfrak{K} = \mathfrak{M}Q_0 + \mathfrak{M}Q_1 + \cdots + \mathfrak{M}Q_{h-1}$, wo $\mathfrak{M}Q_\lambda$ dem Element H_λ der mit $\frac{\mathfrak{K}}{\mathfrak{M}}$ isomorphen Gruppe $\mathfrak{H}$ entsprechen möge, es seien $M_0, M_1, \dots, M_{m-1}$ die Elemente,

$$\psi^{(\mu)}(J) = \psi_{M_\mu}(J) \qquad (\mu = 0, 1, \dots, m-1)$$

die m Charaktere der *Abel*schen Gruppe $\mathfrak{M}$.

Betrachtet man irgend eine primitive Darstellung f-ten Grades von $\mathfrak{K}$ durch ganze lineare Substitutionen, die dem Charakter $\psi(J) = \psi_M(J)$ von $\mathfrak{M}$ entspricht, so ist, wenn (J) die dem Element J von $\mathfrak{M}$ entsprechende

---

Bedeutet $q_{\alpha\beta}$ die Anzahl der Lösungen der Gleichung $QP = PQAB$, wenn man P und Q unabhängig von einander die h Elemente $G_0, G_1, \dots, G_{h-1}$ durchlaufen läßt, so ist $\frac{g p_{\alpha\beta}}{g_\alpha} = a^2 q_{\alpha\beta}$. Denn man erhält aus jeder Lösung P, Q a^2 verschiedene Lösungen von $SR = RSAB$, indem man für R die a Elemente $PA_0, PA_1, \dots, PA_{a-1}$, für S unabhängig hiervon die a Elemente $QA_0, QA_1, \dots, QA_{a-1}$ setzt. Mithin ist

$$\left| q_{\alpha\beta'} - \frac{g^2}{a^2 f^2} e_{\alpha\beta'} \right| = 0,$$

also sind die k' Wurzeln $\left(\frac{g}{af}\right)^2 = \left(\frac{h}{f}\right)^2$ dieser Gleichung ganze Zahlen.

Matrix ist, $(J) = \psi_M(J).(E)$. Nun bilden aber die Determinanten der die Gruppe darstellenden Matrizen einen linearen Charakter von $\mathfrak{K}$. Folglich muß für jedes J

$$\{\psi(J)\}^f = \psi_{M^f}(J) = 1,$$

und also auch $M^f = E$ sein. *Daher ist f durch die Ordnung des Elementes M teilbar.* Entsprechen nun etwa dem Charakter $\psi(J)$ von $\mathfrak{M}$ im Ganzen l Charaktere der Grade $f_1, f_2, \ldots, f_l$, so ist nach Formel (21.)

$$f_1^2 + f_2^2 + \cdots + f_l^2 = h.$$

Mithin muß h durch das Quadrat der Ordnung des Elementes M teilbar sein. Da dies für jedes Element M von $\mathfrak{M}$ gilt, so folgt hieraus:

VIII. *Das Quadrat der größten Invariante (also auch jeder Invariante) des Multiplikators von $\mathfrak{H}$ ist ein Divisor der Ordnung h von $\mathfrak{H}$.*

Ist also eine Primzahl p nur in der ersten Potenz in h enthalten, so ist m durch p nicht teilbar. *Speziell sind alle Gruppen von quadratfreier Ordnung abgeschlossene Gruppen.*

Dieser Satz ist in einem viel allgemeineren Satze enthalten.

Bilden die Elemente $H_0, H_1, \ldots, H_{s-1}$ eine Untergruppe $\mathfrak{S}$ von $\mathfrak{H}$, so bilden die in den s Komplexen $\mathfrak{M}Q_0, \mathfrak{M}Q_1, \ldots, \mathfrak{M}Q_{s-1}$ enthaltenen Elemente von $\mathfrak{K}$ eine Untergruppe $\mathfrak{S}'$ der Ordnung ms von $\mathfrak{K}$, die als eine durch die Gruppe $\mathfrak{M}$ ergänzte Gruppe von $\mathfrak{S}$ anzusehen ist. Es sei $\mathfrak{R}'$ der Kommutator von $\mathfrak{S}'$, $\mathfrak{D}$ der größte gemeinsame Teiler von $\mathfrak{M}$ und $\mathfrak{R}'$. Es möge ferner

$$\mathfrak{K} = \mathfrak{S}'A_0 + \mathfrak{S}'A_1 + \cdots + \mathfrak{S}'A_{n-1}$$

sein, sodaß sich jedes Element P von $\mathfrak{K}$ in eindeutiger Weise auf die Form $S'_\alpha A_\nu$ bringen läßt, wo S'_α ein Element von $\mathfrak{S}'$ bedeutet Setzt man dann $S'_\alpha = S'_P$ und bezeichnet den Komplex

$$\mathfrak{R}' \prod_{\nu=0}^{n-1} S'_{A_\nu P}$$

mit $\mathfrak{R}'_P$, so besteht, wie ich in meiner Arbeit „Neuer Beweis eines Satzes über endliche Gruppen"*) gezeigt habe, für je zwei Elemente P und Q von $\mathfrak{K}$ die Beziehung $\mathfrak{R}'_P \mathfrak{R}'_Q = \mathfrak{R}'_{PQ}$; ferner ist für jedes Element P von $\mathfrak{K}$, das

*) Sitzungsberichte der Berliner Akademie, 1902, S. 1013.

dem Kommutator von $\mathfrak{K}$ angehört, $\mathfrak{R}'_P = \mathfrak{R}'$. Ist nun $P = J$ ein Element von $\mathfrak{S}'$, das in $\mathfrak{M}$ enthalten ist, so ist für jedes ν $A_\nu J = JA_\nu$, also $S'_{A_\nu, J} = J$; daraus folgt aber, daß $\mathfrak{R}'_J = \mathfrak{R}' J^n$ ist. Es ist aber zugleich J ein Element des Kommutators von $\mathfrak{K}$, also $\mathfrak{R}'_J = \mathfrak{R}'$. Wir erhalten also für jedes Element J die Gleichung

$$J^n \mathfrak{R}' = \mathfrak{R}',$$

welche besagt, daß die n-te Potenz jedes Elementes von $\mathfrak{M}$ in $\mathfrak{R}'$, und also auch in $\mathfrak{D}$ enthalten ist.

Ist insbesondere die Ordnung j des Elementes J zu n teilerfremd, so ist, weil $J^j = E$ und J^n Elemente von $\mathfrak{D}$ sind, J selbst in $\mathfrak{D}$ enthalten. Berücksichtigen wir noch, daß $\mathfrak{D}$ nach Satz II einer Untergruppe des Multiplikators von $\mathfrak{S}$ isomorph ist, so erhalten wir den Satz:

IX. *Enthält $\mathfrak{H}$ eine Untergruppe $\mathfrak{S}$ von der Ordnung s und dem Index n, und bedeutet $\mathfrak{M}^{(n)}$ diejenige Untergruppe des Multiplikators $\mathfrak{M}$ von $\mathfrak{H}$, die alle Elemente von $\mathfrak{M}$ umfaßt, deren Ordnungen zu n teilerfremd sind, so ist $\mathfrak{M}^{(n)}$ einer Untergruppe des Multiplikators von $\mathfrak{S}$ isomorph.*

Ist speziell s' ein Divisor von h, der zu n teilerfremd ist, so bilden die Elemente von $\mathfrak{M}$, deren Ordnungen in s' aufgehen, eine in $\mathfrak{M}^{(n)}$ enthaltene Untergruppe von $\mathfrak{M}$. Aus IX ergibt sich daher insbesondere:

X. *Ist p^α die höchste Potenz der Primzahl p, die in h aufgeht, und ist $\mathfrak{S}$ eine Untergruppe von $\mathfrak{H}$, deren Ordnung durch p^α teilbar ist, so ist die Ordnung m des Multiplikators von $\mathfrak{H}$ durch keine höhere Potenz von p teilbar, als dies für die Ordnung m' des Multiplikators von $\mathfrak{S}$ der Fall ist.*

Man wird daher, um zu untersuchen, durch welche Potenz von p die Zahl m teilbar ist, zunächst die Multiplikatoren derjenigen Untergruppen von $\mathfrak{H}$, deren Ordnungen durch p^α teilbar sind, zu bestimmen suchen. Ist speziell für eine dieser Gruppen m' zu p teilerfremd, so ist es auch für m der Fall.

Ich will noch bemerken, daß, wenn s und n teilerfremd sind, die oben betrachtete Gruppe $\mathfrak{D}$ mit der Gruppe $\mathfrak{M}^{(n)}$ übereinstimmen muß. Denn da $\mathfrak{D}$ einer Untergruppe des Multiplikators von $\mathfrak{S}$ isomorph ist, so geht nach Satz VIII die Ordnung eines jeden Elementes von $\mathfrak{D}$ in s auf. Daher ist $\mathfrak{D}$ wegen der von uns gemachten Annahme in $\mathfrak{M}^{(n)}$ enthalten. Zugleich ist aber auch $\mathfrak{M}^{(n)}$ eine Untergruppe von $\mathfrak{D}$; folglich ist in der Tat $\mathfrak{D} = \mathfrak{M}^{(n)}$.

Aus X ergibt sich fast unmittelbar:

XI. *Besitzt eine Gruppe* $\mathfrak{H}$ *die Eigenschaft, daß jede ihrer Untergruppen von Primzahlpotenzordnung eine zyklische Gruppe ist, so ist* $\mathfrak{H}$ *eine abgeschlossene Gruppe.*

Um dies zu beweisen, genügt es auf Grund des Satzes X zu zeigen, daß jede zyklische Gruppe $\mathfrak{C}$ **eine abgeschlossene Gruppe ist. Dies ist aber leicht einzusehen. Denn ist** $\mathfrak{G} = (\mathfrak{A}, \mathfrak{C})$ **eine beliebige durch eine** *Abel*sche **Gruppe** $\mathfrak{A}$ **ergänzte Gruppe von** $\mathfrak{C}$, **so ist, weil** $\mathfrak{A}$ **aus invarianten Elementen von** $\mathfrak{G}$ **besteht und** $\frac{\mathfrak{G}}{\mathfrak{A}}$ **eine zyklische Gruppe ist,** $\mathfrak{G}$ **eine** *Abel*sche **Gruppe; ihr Kommutator also gleich** E. **Daraus folgt aber, daß der Multiplikator von** $\mathfrak{C}$ **in der Tat gleich** E **ist.**

5.

Zur Theorie der vertauschbaren Matrizen

Journal für die reine und angewandte Mathematik 130, 66 - 76 (1905)

Bekanntlich lassen sich unter den Matrizen

$$(a_{\varkappa\lambda}) \qquad (\varkappa, \lambda = 1, 2, \ldots n)$$

des Grades n auf unendlich viele Arten n^2 linear unabhängige angeben. Jedes solche System $A_0, A_1, \ldots A_{n^2-1}$ besitzt dann die Eigenschaft, daß sich jede andere Matrix n-ten Grades A in der Form

$$A = c_0 A_0 + c_1 A_1 + \cdots + c_{n^2-1} A_{n^2-1}$$

darstellen läßt. Da nun für $n > 1$ nicht je zwei Matrizen vertauschbar sind, so kann es für $n > 1$ kein System von n^2 linear unabhängigen Matrizen n-ten Grades geben, von denen je zwei vertauschbar sind. Es entsteht auf diese Weise die Aufgabe, die größtmögliche Anzahl linear unabhängiger, unter einander vertauschbarer Matrizen n-ten Grades zu bestimmen.

Es möge die gesuchte Zahl mit $v_n + 1$ bezeichnet werden. Sind dann

$$A_0, A_1, \ldots A_{v_n}$$

$v_n + 1$ vertauschbare Matrizen, die linear unabhängig sind, so muß offenbar jedes der Produkte $A_\beta A_\gamma$ in der Form

$$A_\beta A_\gamma = \sum_{\alpha=0}^{v_n} c_{\alpha\beta\gamma} A_\alpha$$

darstellbar sein, wo die $c_{\alpha\beta\gamma}$ gewisse Konstanten sind. Die Gesamtheit der Matrizen

$$x_0 A_0 + x_1 A_1 + \cdots + x_{v_n} A_{v_n}$$

mit beliebigen Koeffizienten x_ν bildet daher nach der von Herrn *Frobenius**)

---

*) Theorie der hyperkomplexen Größen, Sitzungsberichte der Berliner Akademie, 1903, S. 504.

angewandten Bezeichnung eine kommutative Gruppe der Ordnung $v_n + 1$. Die oben gestellte Aufgabe läßt sich mithin auch folgendermaßen formulieren: „Welches ist die größtmögliche Ordnung einer kommutativen Gruppe von Matrizen n-ten Grades?"

Im folgenden wird bewiesen:

I. *Die Ordnung einer kommutativen Gruppe von Matrizen n-ten Grades ist höchstens gleich* $\left[\frac{n^2}{4}\right]+1$.

Hier bedeutet $\left[\frac{n^2}{4}\right]$, wie üblich, die größte ganze Zahl, die $\frac{n^2}{4}$ nicht übertrifft, d. h. für gerades n die Zahl $\frac{n^2}{4}$, für ungerades n die Zahl $\frac{n^2-1}{4}$.

Sind in einer Gruppe $\mathfrak{H}$ von Matrizen n-ten Grades zwei Teilgruppen $\mathfrak{H}_1$ und $\mathfrak{H}_2$ enthalten, die außer der Null kein Element gemeinsam haben, und läßt sich jedes Element von $\mathfrak{H}$ als die Summe eines Elementes von $\mathfrak{H}_1$ und eines Elementes von $\mathfrak{H}_2$ darstellen, so möge $\mathfrak{H}$ als die *Summe**) der Gruppen $\mathfrak{H}_1$ und $\mathfrak{H}_2$ bezeichnet werden. Die aus den Matrizen xE, wo E die Einheitsmatrix n-ten Grades bedeutet, bestehende Gruppe werde mit $\mathfrak{E}_n$ bezeichnet. Ferner soll eine Gruppe eine *Wurzelgruppe***) genannt werden, wenn jedes Element der Gruppe, zu einer gewissen Potenz erhoben, gleich Null wird.

Es gilt dann der weitere Satz:

II. *Jede kommutative Gruppe von Matrizen n-ten Grades, deren Ordnung gleich* $\left[\frac{n^2}{4}\right]+1$ *ist, ist für* $n>3$ *gleich der Summe der Gruppe* $\mathfrak{E}_n$ *und einer Wurzelgruppe der Ordnung* $\left[\frac{n^2}{4}\right]$, *in der das Produkt von je zwei Elementen gleich Null ist.*

Auf Grund dieses Satzes genügt es daher (für $n>3$), die kommutativen Wurzelgruppen der Ordnung $\left[\frac{n^2}{4}\right]$ zu bestimmen.

Kommutative Wurzelgruppen dieser Ordnung lassen sich für jedes n leicht angeben.

Ist $n=2m$, so bildet die Gesamtheit der Matrizen $(a_{\varkappa\lambda})$, in denen für

$$\varkappa \leqq m \qquad \text{oder} \qquad \lambda > m$$

*) Vergl. *Frobenius,* Theorie der hyperkomplexen Größen II, ebenda S. 634.
**) *Frobenius,* a. a. O. S. 635.

die Größe $a_{\varkappa\lambda}$ gleich Null ist, während die übrigen m^2 Größen $a_{\varkappa\lambda}$ beliebig sind, eine solche Gruppe, die ich mit $\mathfrak{A}_n$ bezeichnen will. Ist $n=2m+1$, so kann man (für $m>0$) zwei kommutative Gruppen der Ordnung

$$m^2+m=\left[\frac{n^2}{4}\right]$$

angeben. Die eine Gruppe umfaßt alle Matrizen $(a_{\varkappa\lambda})$, in denen $a_{\varkappa\lambda}=0$ ist, falls

$$\varkappa\leqq m \qquad \text{oder} \qquad \lambda>m$$

ist, die andere Gruppe umfaßt alle Matrizen $(a_{\varkappa\lambda})$, in denen $a_{\varkappa\lambda}=0$ ist, sobald

$$\varkappa\leqq m+1 \qquad \text{oder} \qquad \lambda>m+1$$

ist. Diese beiden Gruppen bezeichne ich mit $\mathfrak{A}_n$ und $\mathfrak{A}'_n$.

Ich nenne ferner allgemein zwei Gruppen $\mathfrak{G}$ und $\mathfrak{H}$ von Matrizen n-ten Grades, deren Ordnungen einander gleich sind, *äquivalent*, wenn sich eine Matrix P von nicht verschwindender Determinante bestimmen läßt, so daß für jedes Element A von $\mathfrak{G}$ die Matrix PAP^{-1} in $\mathfrak{H}$*) enthalten ist. Dann läßt sich der Satz II noch folgendermaßen präzisieren:

III. *Ist $n\neq 3$, so ist jede kommutative Wurzelgruppe der Ordnung $\left[\frac{n^2}{4}\right]$ von Matrizen n-ten Grades für gerades n der Gruppe $\mathfrak{A}_n$, für ungerades n entweder der Gruppe $\mathfrak{A}_n$ oder der Gruppe $\mathfrak{A}'_n$**) äquivalent. — Für $n=3$ ist eine kommutative Wurzelgruppe der Ordnung $\left[\frac{3^2}{4}\right]=2$ entweder der Gruppe $\mathfrak{A}_3$ oder der Gruppe $\mathfrak{A}'_3$ oder auch der durch die Matrizen der Form*

$$\begin{pmatrix} 0 & 0 & 0 \\ a & 0 & 0 \\ b & a & 0 \end{pmatrix}$$

gebildeten Gruppe äquivalent.

Ich gehe nun an den Beweis der hier ausgesprochenen Sätze.

Da für jedes n die kommutative Gruppe $\mathfrak{E}_n+\mathfrak{A}_n$ die Ordnung $\left[\frac{n^2}{4}\right]+1$ besitzt, so kann die oben definierte Zahl v_n nicht kleiner sein als $\left[\frac{n^2}{4}\right]$.

Es sei nun $\mathfrak{M}'$ eine kommutative Gruppe von Matrizen n-ten Grades, deren Ordnung $m+1$ nicht kleiner als $\left[\frac{n^2}{4}\right]+1$ ist, und es möge, was ohne

---

*) Ich werde im folgenden die Gruppe $\mathfrak{H}$ dann kurz mit $P\mathfrak{G}P^{-1}$ bezeichnen.

**) Die Gruppen $\mathfrak{A}_n$ und $\mathfrak{A}'_n$ sind, wie man leicht zeigt, nicht äquivalent.

Beschränkung der Allgemeinheit angenommen werden darf, $\mathfrak{M}'$ die Einheitsmatrix E enthalten.

Ich nehme zunächst an, daß die charakteristischen Wurzeln jeder Matrix A' von $\mathfrak{M}'$ einander gleich sind. Dann läßt sich A' auf die Form

$$A' = A + \alpha E$$

bringen, wo $A^n = 0$ ist. Die Gesamtheit der Matrizen A bildet dann eine Wurzelgruppe $\mathfrak{M}$ der Ordnung m*), und $\mathfrak{M}'$ läßt sich als Summe der Gruppen $\mathfrak{E}_n$ und $\mathfrak{M}$ auffassen.

Es genügt, die Gruppe $\mathfrak{M}$ für sich zu behandeln.

Nun ist offenbar $v_1 = 0 = \left[\frac{1^2}{4}\right]$. Es sei bereits bewiesen, daß es für $n' < n$ keine kommutative Wurzelgruppe von Matrizen des Grades n' gibt, deren Ordnung größer ist als $\left[\frac{n'^2}{4}\right]$.

Nach einem von Herrn *Cartan***) herrührenden Satze läßt sich in jeder Wurzelgruppe ein von Null verschiedenes Element angeben, das mit jedem Element der Gruppe multipliziert, Null ergibt. Es sei M' eine Matrix der Wurzelgruppe $\mathfrak{M}$, welche diese Eigenschaft besitzt. Ich denke mir M' so gewählt, daß der Rang r dieser Matrix möglichst groß wird. Da $M'^2 = 0$ ist, muß bekanntlich $2r \leqq n$ sein; es sei $n = 2r + s$. Man kann dann stets eine Matrix P von nicht verschwindender Determinante so wählen, daß $PM'P^{-1} = M$ die Form

$$M = \begin{pmatrix} N_{rr} & N_{rs} & N_{rr} \\ N_{sr} & N_{ss} & N_{sr} \\ E_r & N_{rs} & N_{rr} \end{pmatrix}$$

annimmt, wo E_r die Einheitsmatrix r-ten Grades, $N_{\varkappa\lambda}$ allgemein die $\varkappa$ Zeilen und λ Spalten enthaltende Nullmatrix bedeutet.***)

Jedes Element X der mit $\mathfrak{M}$ äquivalenten Gruppe $\mathfrak{M}_1 = P\mathfrak{M}P^{-1}$ hat dann, weil $XM = MX = 0$ ist, die Form

$$X = \begin{pmatrix} N_{rs} & N_{rs} & N_{rr} \\ A & B & N_{sr} \\ C & D & N_{rr} \end{pmatrix},$$

*) Dies gilt auch dann, wenn die Gruppe $\mathfrak{M}'$ nicht kommutativ ist.

**) Sur les groupes bilinéaires et les systèmes des nombres complexes, Ann. de Toulouse, tome XII, 1898. — Vergl. auch *Frobenius*, a. a. O. S. 639.

***) Vergl. z. B. *Muth*, Theorie der Elementarteiler, S. 152 ff.

wo A, B, C, D Matrizen sind, die bezw. s, s, r, r Zeilen und r, s, r, s Spalten enthalten. Ich setze abgekürzt

$$X = \begin{pmatrix} 0 & 0 & 0 \\ A & B & 0 \\ C & D & 0 \end{pmatrix}.$$

Nun soll aber

$$m \geqq \left[\frac{(2r+s)^2}{4}\right] = r^2 + rs + \left[\frac{s^2}{4}\right]$$

sein. Die Gesamtheit der Matrizen B bildet eine kommutative Wurzelgruppe von Matrizen des Grades s, deren Ordnung b nach unserer Annahme nicht größer sein kann als $\left[\frac{s^2}{4}\right]$. Es sei ferner c die größte Anzahl linear unabhängiger unter den Matrizen C; dann ist $c \leqq r^2$. Man kann nun, wie leicht einzusehen ist, eine Basis der Gruppe $\mathfrak{M}_1$ auf folgende Weise wählen: Zuerst hat man

$$b + c' \leqq b + c \leqq r^2 + \left[\frac{s^2}{4}\right]$$

Elemente der Form

$$(1.)\qquad X_\alpha = \begin{pmatrix} 0 & 0 & 0 \\ A_\alpha & B_\alpha & 0 \\ C_\alpha & D_\alpha & 0 \end{pmatrix}, \qquad (\alpha = 1, 2, \ldots b + c')$$

dann

$$g = m - b - c' \geqq rs$$

Elemente der Form

$$(2.)\qquad Y_\beta = \begin{pmatrix} 0 & 0 & 0 \\ A'_\beta & 0 & 0 \\ 0 & D'_\beta & 0 \end{pmatrix}. \qquad (\beta = 1, 2, \ldots g)$$

Es seien unter den Matrizen A'_β genau g_1 linear unabhängig, und zwar seien es etwa die g_1 ersten. Dann lassen sich an Stelle der Basiselemente (2.) die folgenden wählen:

$$(3.)\qquad Y_\beta = \begin{pmatrix} 0 & 0 & 0 \\ A'_\beta & 0 & 0 \\ 0 & D'_\beta & 0 \end{pmatrix}, \qquad (\beta = 1, 2, \ldots g_1)$$

$$(4.)\qquad Z_\gamma = \begin{pmatrix} 0 & 0 & 0 \\ 0 & 0 & 0 \\ 0 & D''_\gamma & 0 \end{pmatrix}, \qquad (\gamma=1,2,\ldots g-g_1$$

wo die D''_γ gewisse Matrizen mit r Zeilen und s Spalten sind.

Man denke sich nun die $b+c'+g_1$ Matrizen A_α und A'_β neben einander geschrieben. Es entsteht auf diese Weise eine Matrix K, die s Zeilen und $(b+c'+g_1)r$ Spalten enthält. Es sei k $(0 \leqq k \leqq s)$ der Rang der Matrix K.

Wir dürfen annehmen, daß bereits mit Hilfe der k letzten Zeilen von K eine nicht verschwindende Determinante k-ten Grades gebildet werden kann. Dies können wir stets durch eine passende Vertauschung der mittleren s Zeilen und Spalten der Matrizen von $\mathfrak{M}$ erreichen, wodurch die Gruppe $\mathfrak{M}$ in eine ihr äquivalente übergeht, während M ungeändert bleibt.

Es werde $s-k=l$ gesetzt und es seien

$$a_{\varrho 1}, \ldots a_{\varrho l}, \qquad b_{\varrho 1}, \ldots b_{\varrho k}$$

die Elemente der ϱ-ten Spalte von K. Dann lassen sich bekanntlich stets kl Konstanten

$$(5.)\qquad \begin{cases} t_{11}, \ldots t_{1k} \\ \cdot\;\cdot\;\cdot\;\cdot\;\cdot \\ t_{l1}, \ldots t_{lk} \end{cases}$$

bestimmen, so daß für jedes ϱ

$$a_{\varrho\alpha} = \sum_{\beta=1}^{k} t_{\alpha\beta}\, b_{\varrho\beta} \qquad (\alpha=1,2,\ldots l)$$

wird.

Wir bezeichnen nun die Matrix (5.) mit T, die Matrix s-ten Grades

$$\begin{pmatrix} E_l & -T \\ 0 & E_k \end{pmatrix}$$

mit S und die Matrix n-ten Grades

$$\begin{pmatrix} E_r & 0 & 0 \\ 0 & S & 0 \\ 0 & 0 & E_r \end{pmatrix}$$

mit Q. Ersetzt man dann die Matrizen

$$X = \begin{pmatrix} 0 & 0 & 0 \\ A & B & 0 \\ C & D & 0 \end{pmatrix}$$

von $\mathfrak{M}$ durch

$$QXQ^{-1}=\begin{pmatrix}0 & 0 & 0\\ SA & SBS^{-1} & 0\\ C & DS^{-1} & 0\end{pmatrix},$$

so enthalten die Matrizen SA, wie man ohne Mühe einsieht, in den ersten l Zeilen lauter Nullen.

Wir können annehmen, daß bereits die Matrizen A diese Eigenschaft besitzen. Dann sind also alle $a_{\varrho\alpha}$ gleich Null.

Nun ergibt sich aber aus

$$Z_\gamma X=\begin{pmatrix}0 & 0 & 0\\ 0 & 0 & 0\\ D''_\gamma A & D''_\gamma B & 0\end{pmatrix}=XZ_\gamma=\begin{pmatrix}0 & 0 & 0\\ 0 & 0 & 0\\ 0 & 0 & 0\end{pmatrix},$$

daß für jedes A

$$D''_\gamma A=0$$

ist. Daher ist auch

$$(6.)\qquad D''_\gamma K=0.$$

Weil aber der Rang der in den letzten k Zeilen von K stehenden Matrix gleich k ist, ergibt sich aus (6.), daß in D''_γ die letzten k Spalten lauter Nullen enthalten müssen. Es kommen also speziell in A'_β höchstens rk, in D''_γ höchstens rl von Null verschiedene Elemente vor. Folglich ist

$$g_1\leqq rk,\qquad g-g_1\leqq rl,$$

also $g\leqq rs$. Da nun aber $g\geqq rs$ sein soll, ergibt sich

$$g_1=rk,\qquad g-g_1=rl,\qquad b=\left[\frac{s^2}{4}\right],\qquad c'=c=r^2.$$

Damit ist zugleich bewiesen, daß m unter der über $\mathfrak{M}'$ gemachten Voraussetzung nicht größer sein kann als $\left[\frac{n^2}{4}\right]$.

Wir können aber noch mehr schließen.

Da die rk Matrizen A'_β linear unabhängig sind, und jedes A in den ersten l Zeilen lauter Nullen enthält, muß sich jede Matrix A als lineare homogene Verbindung der A'_β darstellen lassen. Wir können daher die $r^2+\left[\frac{s^2}{4}\right]$ ersten Elemente unserer Basis so abändern, daß die A_α sämtlich gleich Null werden, d. h. wir können ohne Beschränkung der Allgemeinheit annehmen, daß

$$X_\alpha = \begin{pmatrix} 0 & 0 & 0 \\ 0 & B_\alpha & 0 \\ C_\alpha & D_\alpha & 0 \end{pmatrix} \tag{7.}$$

ist. Es wird dann

$$X_\alpha Y_\beta = \begin{pmatrix} 0 & 0 & 0 \\ B_\alpha A'_\beta & 0 & 0 \\ D_\alpha A'_\beta & 0 & 0 \end{pmatrix} = Y_\beta X_\alpha = \begin{pmatrix} 0 & 0 & 0 \\ 0 & 0 & 0 \\ 0 & D'_\beta B_\alpha & 0 \end{pmatrix},$$

also ist $D_\alpha A'_\beta = 0$ und folglich auch $D_\alpha K = 0$. Daraus ergibt sich aber, daß die k letzten Spalten von D_α lauter Nullen enthalten müssen. Da nun die rl Matrizen D''_γ linear unabhängig sind, muß sich jedes D_α als lineare homogene Verbindung der D''_γ darstellen lassen.

Wir können daher unsere Basiselemente so transformieren, daß die D_α sämtlich Null werden. Es sei also bereits

$$X_\alpha = \begin{pmatrix} 0 & 0 & 0 \\ 0 & B_\alpha & 0 \\ C_\alpha & 0 & 0 \end{pmatrix}.$$

Nun erzeugen aber die B_α für sich eine Wurzelgruppe von Matrizen des Grades s, deren Ordnung gleich $\left[\frac{s^2}{4}\right]$ ist. Es muß sich also, falls $\left[\frac{s^2}{4}\right]$ nicht gleich Null ist, nach dem oben erwähnten Satze des Herrn *Cartan* eine lineare Verbindung

$$M_1 = \begin{pmatrix} 0 & 0 & 0 \\ 0 & B & 0 \\ C & 0 & 0 \end{pmatrix}$$

der X_α angeben lassen, so daß B von Null verschieden ist und, mit jedem B_α multipliziert, Null ergibt. Es wäre aber dann, wie leicht zu sehen ist,

$$M_1 X_\alpha = 0, \qquad M_1 Y_\beta = 0, \qquad M_1 Z_\gamma = 0,$$

also wäre M_1 ein Element von $\mathfrak{M}_1$, das, mit jedem Element der Gruppe multipliziert, Null ergibt. Es besitzt dann für jedes x auch

$$xM + M_1 = \begin{pmatrix} 0 & 0 & 0 \\ 0 & B & 0 \\ xE_r + C & 0 & 0 \end{pmatrix}$$

dieselbe Eigenschaft. Man kann aber x so wählen, daß die Determinante von $xE_r + C$ nicht Null wird. Ist daher $\left[\frac{s^2}{4}\right]$ nicht Null, also B von der Nullmatrix verschieden, so wäre der Rang von $xM + M_1$ größer als r, was unserer über die Zahl r gemachten Voraussetzung widerspricht.

Es muß also $s = 0$ oder $s = 1$ sein.

Ist $s = 0$, so ist $\mathfrak{M}_1$ mit der Gruppe $\mathfrak{A}_n = \mathfrak{A}_{2r}$ identisch.

Es sei demnach $s = 1$, also $n = 2m + 1$. Dann ist wegen $k + l = s = 1$ entweder $k = 1$, $l = 0$ oder $k = 0$, $l = 1$.

Es sei zunächst $k = 1$, $l = 0$. Dann haben die $rk = r$ linear unabhängigen Matrizen A'_β die Form

$$A'_\beta = (a_{\beta 1},\ a_{\beta 2}, \ldots a_{\beta r});$$

ebenso sei

$$D'_\beta = \begin{pmatrix} d_{\beta 1} \\ d_{\beta 2} \\ \vdots \\ d_{\beta r} \end{pmatrix}.$$

Aus $Y_\varkappa Y_\lambda = Y_\lambda Y_\varkappa$ ergibt sich

$$D'_\varkappa A'_\lambda = D'_\lambda A'_\varkappa$$

oder

$$(8.)\qquad d_{\varkappa\varrho} a_{\lambda\sigma} = d_{\lambda\varrho} a_{\varkappa\sigma}. \qquad (\varkappa, \lambda, \varrho, \sigma = 1, 2, \ldots r)$$

Ist nun $r > 1$, so können die Determinanten

$$\begin{vmatrix} a_{\varkappa\sigma} & a_{\lambda\sigma} \\ a_{\varkappa\tau} & a_{\lambda\tau} \end{vmatrix}$$

nicht sämtlich verschwinden, da sonst die r Matrizen A'_β nicht linear unabhängig wären. Es sei etwa

$$\begin{vmatrix} a_{11} & a_{21} \\ a_{12} & a_{22} \end{vmatrix} \neq 0.$$

Dann ergibt sich aus (8.) für $\varrho = 1, 2, \ldots r$

$$d_{1\varrho} a_{21} = d_{2\varrho} a_{11}, \qquad d_{1\varrho} a_{22} = d_{2\varrho} a_{12},$$

also $d_{1\varrho} = d_{2\varrho} = 0$, d. h. $D'_1 = D'_2 = 0$. Man schließt dann auch für einen von 1 und 2 verschiedenen Index μ wegen

$$D'_\mu A'_1 = D'_1 A'_\mu = 0,$$

daß $D'_\mu = 0$ ist. Die Gruppe $\mathfrak{M}_1$ ist daher identisch mit der in der Einleitung definierten Gruppe $\mathfrak{A}_n$.

Ist aber $k=0$, $l=1$, so stimmt $\mathfrak{M}_1$ direkt mit der Gruppe $\mathfrak{A}'_n$ überein.

Der Fall $r=1$ oder $n=3$ bildet eine Ausnahme. In diesem Falle hat, falls $k=1$, $l=0$ und D'_1 nicht gleich Null ist, die von uns betrachtete Basis die Form

$$X_1 = \begin{pmatrix} 0 & 0 & 0 \\ 0 & 0 & 0 \\ x & 0 & 0 \end{pmatrix}, \qquad Y_1 = \begin{pmatrix} 0 & 0 & 0 \\ y & 0 & 0 \\ 0 & z & 0 \end{pmatrix},$$

wo x, y, z von Null verschiedene Zahlen sind. Setzt man dann

$$R = \begin{pmatrix} yz & 0 & 0 \\ 0 & z & 0 \\ 0 & 0 & 1 \end{pmatrix},$$

so wird

$$R X_1 R^{-1} = \begin{pmatrix} 0 & 0 & 0 \\ 0 & 0 & 0 \\ xy^{-1}z^{-1} & 0 & 0 \end{pmatrix}, \qquad R Y_1 R^{-1} = \begin{pmatrix} 0 & 0 & 0 \\ 1 & 0 & 0 \\ 0 & 1 & 0 \end{pmatrix}.$$

Die Gruppe $\mathfrak{M}$ ist daher äquivalent der durch die Matrizen

$$\begin{pmatrix} 0 & 0 & 0 \\ a & 0 & 0 \\ b & a & 0 \end{pmatrix}$$

gebildeten Gruppe.

Damit ist der Satz III vollständig bewiesen.

Es bleibt uns noch übrig, den Fall zu behandeln, daß die Gruppe $\mathfrak{M}'$ der Ordnung $m+1 \geqq \left[\frac{n^2}{4}\right]+1$, von der wir ausgegangen sind, auch solche Matrizen enthält, deren charakteristische Gleichungen zwei verschiedene Wurzeln besitzen. Es sei A eine solche Matrix. Dann läßt sich bekanntlich eine Matrix P so wählen, daß $A_1 = PAP^{-1}$ die Form

$$A_1 = \begin{pmatrix} G & 0 \\ 0 & H \end{pmatrix}$$

annimmt, wo G und H zwei Matrizen gewisser Grade g und h bedeuten, deren charakteristische Gleichungen keine Wurzel gemeinsam haben. Da aber jede Matrix X der Gruppe $P\mathfrak{M}'P^{-1}$ mit A_1 vertauschbar ist, muß nach einem bekannten Satze X die Form

$$X=\begin{pmatrix} Y & 0 \\ 0 & Z \end{pmatrix}$$

haben, wo Y und Z die Grade g und h besitzen.

Es sei bereits allgemein bewiesen, daß für $n'<n$ die Zahl $v_{n'}$ gleich $\left[\frac{n'^2}{4}\right]$ ist. Die Matrizen Y bilden dann eine kommutative Gruppe, deren Ordnung k nicht größer ist als $\left[\frac{g^2}{4}\right]+1$; die Matrizen Z bilden ebenso eine Gruppe, deren Ordnung l höchstens gleich $\left[\frac{h^2}{4}\right]+1$ ist. Es ist aber offenbar

$$m+1\leqq k+l.$$

Man erhält also

$$\left[\frac{(g+h)^2}{4}\right]+1\leqq m+1\leqq k+l\leqq\left[\frac{g^2}{4}\right]+\left[\frac{h^2}{4}\right]+2.$$

Das ist aber nur dann möglich, wenn $gh\leqq 2$ ist, also nur für $n=2$ und $n=3$. Aber auch für diese beiden Fälle ergibt sich, daß m nicht größer sein kann als $\left[\frac{2^2}{4}\right]=1$, bezw. $\left[\frac{3^2}{4}\right]=2$.

Für $n>3$ kommt nur der zuerst behandelte Fall in Betracht. Damit sind auch die Sätze I und II vollständig bewiesen.

Für $n=2$ kommen noch die Gruppen hinzu, die der Gruppe

$$\begin{pmatrix} a & 0 \\ 0 & b \end{pmatrix} \quad (a \text{ und } b \text{ beliebig})$$

äquivalent sind, für $n=3$ die den Gruppen

$$\begin{pmatrix} a & 0 & 0 \\ 0 & b & 0 \\ 0 & 0 & c \end{pmatrix} \quad \text{oder} \quad \begin{pmatrix} a & 0 & 0 \\ b & a & 0 \\ 0 & 0 & c \end{pmatrix} \qquad (a,\ b \text{ und } c \text{ beliebig})$$

äquivalenten Gruppen.

6.
Über eine Klasse von endlichen Gruppen linearer Substitutionen

Sitzungsberichte der Preussischen Akademie der Wissenschaften 1905,
Physikalisch-Mathematische Klasse, 77 - 91

In seiner Abhandlung »Mémoire sur les équations différentielles linéaires à intégrale algébrique« (Journal für Mathematik, Bd. 84, S. 89) hat Hr. JORDAN einen fundamentalen Satz aufgestellt, der sich folgendermaßen formulieren läßt:

Jede endliche Gruppe $\mathfrak{G}$ homogener linearer Substitutionen in n Variabeln enthält eine invariante ABELsche Untergruppe $\mathfrak{F}$ von der Eigenschaft, daß der Quotient λ der Ordnungen von $\mathfrak{G}$ und $\mathfrak{F}$ kleiner ist als eine gewisse allein von n abhängende Zahl.

Während nun im allgemeinen allein für die Zahl λ eine obere Grenze[1] existiert, dagegen die Ordnung der Gruppe $\mathfrak{G}$ noch beliebig großer Werte fähig ist, hört dies auf, der Fall zu sein, sobald nur solche Gruppen $\mathfrak{G}$ in Betracht gezogen werden, bei denen die Spur[2] jeder linearen Substitution einem vorgeschriebenen algebraischen Zahlkörper K angehört. Es läßt sich sogar eine allein durch den Zahlkörper K und die Anzahl n der Variabeln bestimmte Zahl angeben, die als das kleinste gemeinsame Vielfache der Ordnungen aller in Betracht kommenden Gruppen $\mathfrak{G}$ erscheint.

§ 1.

Es soll zunächst angenommen werden, daß der vorgeschriebene Zahlkörper K mit dem Bereich Ω der rationalen Zahlen übereinstimme.

---

[1] Der JORDANsche Beweis liefert keine Methode, eine explizite obere Grenze für die Zahl λ zu bestimmen. Eine solche obere Grenze hat erst in neuerer Zeit Hr. BLICHFELDT (Transactions of the Am. Math. Society, Bd. 4 (1903), S. 387 und Bd. 5 (1904), S. 310) für eine allgemeine Klasse von Gruppen angegeben, die er als primitive Gruppen bezeichnet.

[2] Unter der Spur der linearen Substitution $x_\varkappa = \sum_\lambda a_{\varkappa\lambda} x'_\lambda$ versteht man bekanntlich die Zahl $\sum_\varkappa a_{\varkappa\varkappa}$.

Es gilt dann der Satz:

I. *Ist $\mathfrak{G}$ eine endliche Gruppe homogener linearer Substitutionen in n Variabeln und ist die Spur jeder Substitution von $\mathfrak{G}$ eine (ganze) rationale Zahl, so ist die Ordnung g der Gruppe $\mathfrak{G}$ ein Divisor der Zahl*

$$M_n = \prod_p p^{\left[\frac{n}{p-1}\right]+\left[\frac{n}{p(p-1)}\right]+\left[\frac{n}{p^2(p-1)}\right]+\cdots} \qquad (p = 2, 3, 5, \cdots).$$

Hierbei bedeutet $[a]$ die größte ganze Zahl $\leqq a$, ferner soll p die Reihe der Primzahlen soweit durchlaufen, bis das Produkt von selbst abbricht, d. h. bis zur größten Primzahl, welche $\leqq n+1$ ist.

Der Beweis ergibt sich sehr einfach mit Hilfe der von Hrn. Frobenius begründeten Theorie der Gruppencharaktere.

Es sei nämlich p eine Primzahl, p^m die höchste Potenz von p, die in g aufgeht, und $\mathfrak{P}$ eine Untergruppe der Ordnung p^m von $\mathfrak{G}$. Ist dann P eine Substitution der Gruppe $\mathfrak{P}$, so genügen die charakteristischen Wurzeln:

$$(1) \qquad \omega_1, \omega_2, \cdots, \omega_n$$

von P der Gleichung $x^{p^m} = 1$. Bedeutet nun ρ eine primitive p^mte Einheitswurzel, und kommt die Einheitswurzel ρ^μ unter den Größen (1) genau x_μ Mal vor, so ist die Spur $\chi(P)$ von P gleich

$$\chi(P) = \sum_{\mu=0}^{p^m-1} x_\mu \rho^\mu.$$

Nun soll aber $\chi(P)$ rational sein. Hierfür ist bekanntlich notwendig und hinreichend, daß, falls $p^{m-1} = q$ gesetzt wird,

$$x_q = x_{2q} = \cdots = x_{(p-1)q}$$
$$x_\lambda = x_{\lambda+q} = x_{\lambda+2q} = \cdots = x_{\lambda+(p-1)q} \qquad (\lambda = 1, 2, \cdots q-1)$$

sei. Da nun

$$\rho^q + \rho^{2q} + \cdots + \rho^{(p-1)q} = -1$$
$$\rho^\lambda + \rho^{\lambda+q} + \rho^{\lambda+2q} + \cdots + \rho^{\lambda+(p-1)q} = 0 \qquad (\lambda = 1, 2, \cdots q-1)$$

ist, so erhält man

$$n = x_0 + (p-1)x_q + px_1 + \cdots + px_{q-1}$$

und

$$\chi(P) = x_0 - x_q.$$

Hieraus folgt aber, wenn $y = x_q + x_1 + \cdots + x_{q-1}$ gesetzt wird,

$$\chi(P) = n - py.$$

Ferner ist

$$n - (p-1)y = x_0 + x_1 + \cdots + x_{q-1},$$

also $y \leqq \frac{n}{p-1}$. Bedeutet daher ν die ganze Zahl $\left[\frac{n}{p-1}\right]$, so kommen für $\chi(P)$ nur die Werte

$$n\,,\ n-p\,,\ n-2p\,,\ \cdots,\ n-\nu p$$

in Betracht.

Es mögen nun unter den Spuren der p^m Substitutionen von $\mathfrak{P}$ genau l_α den Wert $n-\alpha p$ besitzen; hierbei ist $l_0 = 1$ zu setzen, da in einer endlichen Gruppe nur die Spur der identischen Substitution E gleich n ist. Es ist dann

$$l_0 + l_1 + \cdots + l_\nu = p^m.$$

Nun bilden aber die p^m Zahlen $[\chi(P)]^\lambda$ für jedes positive ganzzahlige λ einen (zusammengesetzten) Charakter der Gruppe $\mathfrak{G}$.[1] Daher ist die Summe der p^m Zahlen $[\chi(P)]^\lambda$, d. h. die Zahl

$$l_0 n^\lambda + l_1(n-p)^\lambda + \cdots + l_\nu(n-\nu p)^\lambda$$

eine durch p^m teilbare ganze Zahl.[2] Wir erhalten mithin die Kongruenzen

$$\begin{array}{l} l_0 \quad + l_1 + \cdots \qquad + l_\nu \equiv 0 \\ l_0 n + l_1(n-p) + \cdots + l_\nu(n-\nu p) \equiv 0 \\ \cdots\cdots\cdots\cdots\cdots\cdots \qquad (\text{mod. } p^m) \\ l_0 n^\nu + l_1(n-p)^\nu + \cdots + l_\nu(n-\nu p)^\nu \equiv 0. \end{array}$$

Hieraus folgt aber in bekannter Weise, daß

$$l_\alpha \prod_\beta \{n-\alpha p-(n-\beta p)\} \qquad (\beta = 0, 1, \cdots \nu, \ \beta \neq \alpha)$$

durch p^m teilbar ist. Für $\alpha = 0$ ergibt sich, daß p^m ein Divisor der Zahl $p^\nu \nu!$ sein muß. Da aber

$$p^{\left[\frac{\nu}{p}\right]+\left[\frac{\nu}{p^2}\right]+\cdots}$$

die höchste Potenz von p ist, die in $\nu!$ aufgeht, so muß p^m ein Divisor der Zahl

$$p^{\nu+\left[\frac{\nu}{p}\right]+\left[\frac{\nu}{p^2}\right]+\cdots} = p^{\left[\frac{n}{p-1}\right]+\left[\frac{n}{p(p-1)}\right]+\left[\frac{n}{p^2(p-1)}\right]+\cdots}$$

sein.

Hieraus folgt aber unmittelbar, wie zu beweisen war, daß die Ordnung g der Gruppe $\mathfrak{G}$ ein Divisor der Zahl M_n ist.

In dem Satz I ist folgender von Hrn. Minkowski in seiner Arbeit »Zur Theorie der positiven quadratischen Formen« (Journal für Mathematik, Bd. 101, S. 196) bewiesener Satz als spezieller Fall enthalten:

»Die Anzahl der ganzzahligen Transformationen einer positiven quadratischen Form mit n Variabeln (und von nicht verschwindender Determinante) in sich selbst ist ein Divisor der Zahl M_n.«[3]

Umgekehrt folgt, wie noch hervorgehoben werden soll, aus dem Minkowskischen Resultat der Satz I für den speziellen Fall, daß die Koeffizienten aller Substitutionen

---

[1] Frobenius, Sitzungsberichte 1899, S. 330.
[2] Frobenius, Sitzungsberichte 1896, S. 717.
[3] Hr. Minkowski bezeichnet die Zahl M_n mit $\overline{n}|$.

$$x_\varkappa = \sum a_{\varkappa\lambda}^{(\alpha)} x'_\lambda$$

der Gruppe $\mathfrak{G}$ ganze Zahlen sind. Denn alsdann läßt jede dieser Substitutionen die positive quadratische Form von nicht verschwindender Determinante

$$f = \sum_\alpha \sum_\varkappa (a_{\varkappa 1}^{(\alpha)} x_1 + a_{\varkappa 2}^{(\alpha)} x_2 + \cdots + a_{\varkappa n}^{(\alpha)} x_n)^2$$

ungeändert.[1] Daher ist $\mathfrak{G}$ eine Untergruppe der Gruppe $\mathfrak{G}'$ aller ganzzahligen Transformationen der Form f in sich selbst, und da die Ordnung von $\mathfrak{G}'$ in M_n aufgeht, so ist dies auch für die Ordnung von $\mathfrak{G}$ der Fall.

Wie Hr. Minkowski a. a. O. gezeigt hat, lassen sich für jedes n positive quadratische Formen mit n Variabeln angeben, für welche die Anzahl der ganzzahligen Transformationen in sich selbst genau durch dieselbe Potenz der Primzahl p teilbar ist, wie die Zahl M_n.

Hieraus folgt unmittelbar, daß die Zahl M_n das kleinste gemeinsame Vielfache der Ordnungen aller endlichen Gruppen linearer Substitutionen in n Variabeln mit rationalen Spuren repräsentiert.

Vergleicht man dieses Ergebnis mit dem Resultate des Hrn. Minkowski, so könnte die Vermutung entstehen, daß sich jede endliche Gruppe linearer Substitutionen mit rationalen Spuren durch eine Transformation der Variabeln, wodurch ja die Spur jeder Substitution ungeändert bleibt, in eine (ihr ähnliche) Gruppe linearer Substitutionen mit rationalen Koeffizienten überführen läßt. Dies ist jedoch keineswegs der Fall, wie man an dem Beispiel der durch die Substitutionen

$$(A) \qquad x_1 = ix'_1, \quad x_2 = -ix'_2$$
$$(B) \qquad x_1 = x'_2, \quad x_2 = -x'_1$$

erzeugten Gruppe der Ordnung 8, der sogenannten Quaternionengruppe, erkennt.

Diese Gruppe besitzt zwar rationale Spuren, läßt sich aber durch eine Transformation der Variabeln nicht einmal in eine Gruppe reeller Substitutionen überführen. In der Tat seien

$$(A_1) \qquad x_1 = \alpha x'_1 + \beta x'_2, \quad x_2 = \gamma x'_1 + \delta x'_2$$
$$(B_1) \qquad x_1 = \lambda x'_1 + \mu x'_2, \quad x_2 = \nu x'_1 + \rho x'_2$$

zwei reelle Substitutionen, die durch eine passend gewählte Transformation der Variabeln in A und B übergehen mögen. Es müßten dann die Spuren und die Determinanten von A_1, B_1, A_1B_1 mit derjenigen von A, B, AB übereinstimmen. Daraus folgt

[1] Vgl. A. Loewy, Comptes Rendus, 1896, S. 168 und E. H. Moore, Math. Ann. Bd. 50, S. 213.

$$\alpha+\delta=0, \quad \alpha\delta-\beta\gamma=1$$
$$\lambda+\rho=0, \quad \lambda\rho-\mu\nu=1$$
$$\alpha\lambda+\beta\nu+\gamma\mu+\delta\rho=0.$$

Eine leichte Rechnung ergibt

$$(\alpha\mu-\beta\lambda)^2+\beta^2+\mu^2=0.$$

Daher müßte $\beta=0$ sein, was wegen

$$\alpha\delta-\beta\gamma=-\alpha^2-\beta\gamma=1$$

für ein reelles α nicht möglich ist.

§ 2.

Es sei nun $K=\Omega(\varkappa)$ ein beliebig gegebener, durch die algebraische Zahl $\varkappa$ bestimmter Zahlkörper des Grades k.

Ist dann $\mathfrak{G}$ eine endliche Gruppe linearer Substitutionen in n Variabeln, deren Spuren sämtlich dem Körper K angehören, so will ich im folgenden kurz sagen, $\mathfrak{G}$ sei eine Gruppe $\mathfrak{G}_n^{(\varkappa)}$.

Es ist zunächst leicht zu sehen, daß die Ordnung g einer solchen Gruppe $\mathfrak{G}$ eine gewisse, allein durch K und n bestimmte Zahl nicht übersteigen kann. Ist nämlich $\zeta(R)$ die Spur der Substitution R von $\mathfrak{G}$ und sind

$$\zeta(R), \quad \zeta^{(1)}(R), \cdots, \quad \zeta^{(k-1)}(R)$$

diejenigen Zahlen, die aus $\zeta(R)$ dadurch hervorgehen, daß man darin die Größe $\varkappa$ durch die k konjugierten algebraischen Größen ersetzt, so bilden für jedes λ die g Zahlen $\zeta^{(\lambda)}(R)$ einen Charakter der Gruppe $\mathfrak{G}$, daher auch die Zahlen

$$\xi(R)=\zeta(R)+\zeta^{(1)}(R)+\cdots+\zeta^{(k-1)}(R).$$

Es läßt sich folglich eine der Gruppe $\mathfrak{G}$ isomorphe Gruppe linearer Substitutionen in kn Variabeln angeben, worin die Spur der der Substitution R von $\mathfrak{G}$ entsprechenden Substitution den Wert $\xi(R)$ hat. Da diese Zahlen aber rational sind, so muß die Ordnung g der Gruppe nach Satz I ein Divisor der Zahl M_{kn} sein.

Die sich so ergebende obere Grenze für die Ordnungen der Gruppen $\mathfrak{G}_n^{(\varkappa)}$ ist aber im allgemeinen erheblich größer als das kleinste gemeinsame Vielfache dieser Ordnungen, zu dessen Bestimmung erst die folgende Betrachtung führt.

Man bezeichne, wenn ρ eine primitive λ^{te} Einheitswurzel ist, den Körper $\Omega(\rho)$ mit $\Omega^{(\lambda)}$. Ferner sei, wenn p eine gegebene Primzahl ist, $K^{(\mu)}$ der größte gemeinsame Divisor der beiden Körper K und $\Omega^{(p^\mu)}$, $k^{(\mu)}$ der Grad von $K^{(\mu)}$. Es ist dann offenbar

$$k^{(1)} \leqq k^{(2)} \leqq k^{(3)} \leqq \cdots.$$

Da nun die Zahlen $k^{(1)}$, $k^{(2)}, \dots$ sämtlich in dem Grad k von K aufgehen, so muß unter ihnen eine größte vorhanden sein; es sei dies die Zahl $k^{(m_p)}$. Dann ist also

$$k^{(m_p+a)} = k^{(m_p)} \qquad (a = 0, 1, 2, \dots)$$

und, falls $m_p > 1$ ist,

$$k^{(m_p-1)} < k^{(m_p)}.$$

Da ferner der Körper $K^{(m_p)}$ offenbar jeden der Körper $K^{(\mu)}$ enthält, so ist $K^{(\mu)}$ nichts anderes als der größte gemeinsame Divisor der beiden Körper $K^{(m_p)}$ und $\Omega^{(p^\mu)}$.

Es sei nun zunächst $p > 2$. Dann ist jeder der Körper $\Omega^{(p^\mu)}$ ein zyklischer Körper des Grades $p^{\mu-1}(p-1)$, und es entspricht jedem Teiler d dieser Zahl ein und nur ein Divisor des Grades d von $\Omega^{(p^\mu)}$, der ebenfalls ein zyklischer Körper ist. Es ist nun leicht zu sehen, daß die Zahlen $k^{(1)}$, $k^{(2)}, \dots k^{(m_p)}$ die Form haben müssen

$$(2) \qquad k^{(1)} = \frac{p-1}{t_p}, \quad k^{(2)} = \frac{p(p-1)}{t_p}, \dots, \quad k^{(m_p)} = \frac{p^{m_p-1}(p-1)}{t_p},$$

wo t_p einen Divisor von $p-1$ bedeutet.[1] Denn es ist dies jedenfalls richtig für $m_p = 1$. Ist ferner $m_p > 1$, so sei der Divisor $k^{(m_p)}$ von $p^{m_p-1}(p-1) = \frac{p^\nu(p-1)}{t_p}$, wo t_p in $p-1$ aufgeht. Wäre nun $\nu < m_p - 1$, so würde der in $\Omega^{(p^{m_p})}$ enthaltene Körper $\Omega^{(p^{m_p-1})}$ einen Divisor des Grades $k^{(m_p)}$ besitzen, der mit dem Körper $K^{(m_p)}$ übereinstimmen müßte, und es könnte nicht $k^{(m_p-1)} < k^{(m_p)}$ sein. Daher ist $k^{(m_p)} = \frac{p^{m_p-1}(p-1)}{t_p}$. Ist nun $\mu < m_p$, so sind $K^{(m_p)}$ und $\Omega^{(p^\mu)}$ zwei Divisoren des zyklischen Körpers $\Omega^{(p^{m_p})}$, der Grad $k^{(\mu)}$ ihres größten gemeinsamen Divisors muß daher gleich sein dem größten gemeinsamen Teiler ihrer Grade $\frac{p^{m_p-1}(p-1)}{t_p}$ und $p^{\mu-1}(p-1)$. Daher ist in der Tat $k^{(\mu)} = \frac{p^{\mu-1}(p-1)}{t_p}$.

Es sei nun $p = 2$. Der Körper $\Omega^{(2^\mu)}$ besitzt dann für $\mu > 2$ nur drei Divisoren, die nicht in $\Omega^{(2^{\mu-1})}$ enthalten sind, nämlich, wenn σ eine primitive $2^{\mu\text{te}}$ Einheitswurzel ist, die Körper

$$(3) \qquad \Omega^{(2^\mu)} = \Omega(\sigma), \quad \Omega_1^{(2^\mu)} = \Omega(\sigma + \sigma^{-1}), \quad \Omega_2^{(2^\mu)} = \Omega(\sigma - \sigma^{-1})$$

der Grade $2^{\mu-1}$, $2^{\mu-2}$ und $2^{\mu-2}$.

---

[1] Ist $K^{(\mu)}$ für jedes μ gleich Ω, so hat man also $m_p = 1$, $t_p = p-1$ zu setzen.

Ist nun zunächst $m_2 = 1$ oder $m_2 = 2$, so ist jedenfalls $K^{(m_2)}$ gleich $\Omega^{(2)} = \Omega$ oder gleich $\Omega^{(2^2)}$. Ist dagegen $m_2 > 2$, so muß $K^{(m_2)}$, da dieser Körper in $\Omega^{(2^{m_2})}$, aber nicht in $\Omega^{(2^{m_2-1})}$ enthalten ist, mit einem der drei Körper (3) übereinstimmen.

Man setze nun $t_2 = 1$, falls $K^{(m_2)} = \Omega^{(2^{m_2})}$ ist, was für $m_2 = 1$ und $m_2 = 2$ jedenfalls eintritt, dagegen $t_2 = 2$, falls $K^{(m_2)}$ mit einem der beiden Körper $\Omega_1^{(2^{m_2})}$ oder $\Omega_2^{(2^{m_2})}$ zusammenfällt. Es wird dann

$$(4) \qquad k^{(1)} = 1\,,\ k^{(2)} = \frac{2}{t_2}\,,\ k^{(3)} = \frac{2^2}{t_2}\,, \cdots,\ k^{(m_2)} = \frac{2^{m_2-1}}{t_2}\,.$$

Denn ist erstens $t_2 = 1$, so ist für $\mu \leqq m_2$ der Körper $K^{(\mu)}$ der größte gemeinsame Divisor der beiden Körper $\Omega^{(2^\mu)}$ und $K^{(m_2)} = \Omega^{(2^{m_2})}$, also gleich $\Omega^{(2^\mu)}$, und folglich ist $k^{(\mu)} = 2^{\mu-1} = \dfrac{2^{\mu-1}}{t_2}$. Ist ferner $t_2 = 2$, so ist $k^{(m_2)} = 2^{m_2-2} = \dfrac{2^{m_2-1}}{t_2}$; ferner ist für $\mu < m_2$ der Körper $K^{(\mu)}$, als der größte gemeinsame Divisor von $\Omega^{(2^\mu)}$ und $\Omega_1^{(2^{m_2})}$, bzw. $\Omega_2^{(2^{m_2})}$, falls $\mu = 1$ oder $= 2$ ist, gleich Ω, falls $\mu > 2$ ist, gleich dem Körper $\Omega_1^{(2^\mu)}$. Daher ist in der Tat

$$k^{(1)} = 1\,,\ k^{(\mu)} = 2^{\mu-2} = \frac{2^{\mu-1}}{t_2}\,. \qquad (\mu > 1)$$

Es gilt nun der Satz:

II. *Es sei $K = \Omega(x)$ ein algebraischer Zahlkörper, für den die Zahlen m_p und t_p in der geschilderten Weise bestimmt seien. Dann ist das kleinste gemeinsame Vielfache der Ordnungen aller endlichen Gruppen homogener linearer Substitutionen in n Variabeln, deren Spuren sämtlich dem Körper K angehören, gleich der Zahl*

$$M_n^{(x)} = 2^{n - \left[\frac{n}{t_2}\right]} \prod_p p^{m_p \left[\frac{n}{t_p}\right] + \left[\frac{n}{pt_p}\right] + \left[\frac{n}{p^2 t_p}\right] + \cdots}. \qquad (p = 2, 3, 5, \ldots)$$

Hierbei ist das Produkt über alle Primzahlen p zu erstrecken, für welche $t_p \leqq n$ ist, was offenbar nur für endlich viele Primzahlen der Fall ist.

§ 3.

Um den Beweis des Satzes II vorzubereiten, schicke ich folgende Betrachtung voraus.

Eine Gruppe $\mathfrak{G}$ wird bekanntlich zerlegbar oder reduzibel genannt, wenn sie durch eine passend gewählte Transformation der Variabeln in eine ihr ähnliche Gruppe $\mathfrak{G}'$ übergeführt werden kann, in der das Koeffizientensystem jeder Substitution die Form hat

$$\begin{pmatrix} a_{r1} \dots a_{1r} & 0 \dots 0 \\ \dots & \dots \\ a_{r1} \dots a_{1r} & 0 \dots 0 \\ 0 \dots 0 & b_{1s} \dots b_{1s} \\ \dots & \dots \\ 0 \dots 0 & b_{s1} \dots b_{ss} \end{pmatrix},$$

wobei die Zahlen r und s für alle Substitutionen von $\mathfrak{G}'$ dieselben Werte haben. Der Kürze halber will ich hier eine Gruppe $\mathfrak{G} = \mathfrak{G}_n^{(\kappa)}$ im Körper K zerlegbar nennen, wenn in jeder Substitution von $\mathfrak{G}'$ die Teilspuren $\Sigma a_{\kappa\kappa}$ und $\Sigma b_{\lambda\lambda}$ einzeln dem Körper K angehören. Sind dann unter den Substitutionen $(a_{\kappa\lambda})$ im ganzen g', unter den Substitutionen $(b_{\kappa\lambda})$ im ganzen g'' voneinander verschieden, so bilden die ersteren eine Gruppe $\mathfrak{G}_r^{(\kappa)}$ der Ordnung g', die letzteren eine Gruppe $\mathfrak{G}_s^{(\kappa)}$ der Ordnung g''; ferner sind diese Gruppen der Gruppe $\mathfrak{G}$ ein- oder mehrstufig isomorph, und es ist die Ordnung der Gruppe $\mathfrak{G}$ höchstens gleich $g'g''$.

Es sei nun $\mathfrak{G}$ eine Gruppe $\mathfrak{G}_n^{(\kappa)}$ von der Ordnung g, die im Körper K nicht zerlegbar ist. Die g Spuren $\zeta(R)$ der Substitutionen R von $\mathfrak{G}$ bilden dann einen im allgemeinen zusammengesetzten Charakter der Gruppe $\mathfrak{G}$. Es seien nun

$$\chi^{(0)}(R),\ \chi^{(1)}(R),\ \chi^{(2)}(R), \cdots$$

die einfachen Charaktere von $\mathfrak{G}$. Ist dann ρ eine primitive g^{te} Einheitswurzel, so gehören die Zahlen $\chi^{(\lambda)}(R)$ dem Körper $\Omega(\rho)$ an; ferner genügt ρ im Körper K einer irreducibeln Gleichung $F(x) = 0$, deren Grad genau gleich $\frac{\varphi(g)}{d}$ ist, falls d der Grad des größten gemeinsamen Divisors von K und $\Omega(\rho)$ ist. Ersetzt man nun für ein gegebenes λ in den g Zahlen $\chi^{(\lambda)}(R)$ die Größe ρ durch die Wurzeln der Gleichung $F(x) = 0$, so entstehen höchstens $\frac{\varphi(g)}{d}$ verschiedene Systeme von je g Zahlen

$$(5) \qquad \chi^{(\lambda)}(R),\ \chi_1^{(\lambda)}(R),\ \chi_2^{(\lambda)}(R), \cdots.$$

Die g Zahlen $\chi_\alpha^{(\lambda)}(R)$ bilden dann für jedes α wieder einen einfachen Charakter der Gruppe $\mathfrak{G}$. Die Charaktere (5) mögen als die zu $\chi^{(\lambda)}(R)$ relativ konjugierten Charaktere bezeichnet werden.

Enthält nun der Charakter $\zeta(R)$ den einfachen Charakter $\chi^{(\lambda)}(R)$ genau z_λ mal, so ist

$$\zeta(R) = \sum_\lambda z_\lambda \chi^{(\lambda)}(R)$$

und (vgl. Frobenius, Sitzungsberichte 1896, S. 717)

$$(6) \qquad g z_\lambda = \sum_R \zeta(R^{-1}) \chi^{(\lambda)}(R),$$

wo R alle Substitutionen von $\mathfrak{G}$ durchläuft. Da nun die Zahlen $\zeta(R^{-1})$ dem Körper K angehören sollen, so folgt aus (6), daß jeder der zu $\chi^{(\lambda)}(R)$ relativ konjugierten Charaktere in $\zeta(R)$ genau z_λ mal enthalten ist. Es möge nun für den Charakter $\chi^{(\lambda)}(R) = \chi(R)$ die Zahl $z_\lambda > 0$ sein. Sind dann

$$\chi(R), \chi_1(R), \cdots, \chi_{r-1}(R)$$

die sämtlichen verschiedenen zu $\chi(R)$ relativ konjugierten Charaktere, und setzt man

$$\zeta(R) = \chi(R) + \chi_1(R) + \cdots + \chi_{r-1}(R) + \zeta_1(R),$$

so gehören sowohl die Zahlen

$$\xi(R) = \chi(R) + \chi_1(R) + \cdots + \chi_{r-1}(R)$$

als auch die Zahlen $\zeta_1(R)$ dem Körper K an. Wären nun die Zahlen $\zeta_1(R)$ nicht sämtlich gleich Null, so würden sie einen Charakter der Gruppe $\mathfrak{G}$ bilden. Da auch die Größen $\xi(R)$ einen solchen repräsentieren, so würde $\mathfrak{G}$ einer anderen Gruppe ähnlich sein, die in zwei Gruppen linearer Substitutionen zerfällt, deren Spuren die Zahlen $\xi(R)$ und $\zeta_1(R)$ sind, d. h. $\mathfrak{G}$ würde entgegen der gemachten Annahme im Körper K zerlegbar sein.

Wir sehen also, daß der einer im Körper K nicht zerlegbaren Gruppe entsprechende Charakter $\zeta(R)$ die Form haben muß

$$\zeta(R) = \chi(R) + \chi_1(R) + \cdots + \chi_{r-1}(R),$$

wo $\chi(R), \chi_1(R), \cdots \chi_{r-1}(R)$ die sämtlichen zu einem einfachen Charakter relativ konjugierten Charaktere repräsentieren. Ist dann $r = 1$, so ist $\mathfrak{G}$ eine irreduzible, d. h. im Bereich aller Zahlen nicht zerlegbare Gruppe. Ist dagegen $r > 1$, so ist $\mathfrak{G}$ einer Gruppe $\mathfrak{G}'$ ähnlich, deren Substitutionen Koeffizientensysteme der Form

$$\begin{pmatrix} A & 0 & \cdots & 0 \\ 0 & A_1 & \cdots & 0 \\ . & . & \cdots & . \\ 0 & 0 & \cdots & A_{r-1} \end{pmatrix}$$

besitzen, wo $A, A_1, \cdots A_{r-1}$ Matrizen des Grades $\frac{n}{r}$ sind. Die g Matrizen A_α sind dann für jedes α voneinander verschieden und bilden eine der Gruppe $\mathfrak{G}$ einstufig isomorphe irreduzible Gruppe.

§ 4.

Wir beweisen nun folgenden Hilfssatz:

III. *Ist $\mathfrak{A}$ eine* Abelsche *Gruppe linearer Substitutionen in n Variabeln, deren Ordnung a eine Potenz der Primzahl p ist, und ist die Spur* *jeder Substitution von $\mathfrak{A}$ eine Größe des Körpers K, so ist a für $p>2$ höchstens gleich*

$$N_{n,p}=p^{m_p\left[\frac{n}{t_p}\right]},$$

für $p=2$ höchstens gleich

$$N_{n,p}=2^{(m_2-t_2)\left[\frac{n}{t_2}\right]+n}.$$

Dieser Satz ist für $n=1$ leicht zu bestätigen. Denn es sind dann die Spuren aller Substitutionen der Gruppe $a$te Einheitswurzeln, die als dem Körper K angehörende Größen im Körper $K^{(m_p)}$ enthalten sein müssen. Ist nun zunächst $p=2$, $t_2=2$, so enthält $K^{(m_2)}$ nur die Einheitswurzeln $+1$ und -1 und daher ist a höchstens gleich $2=2^{(m_2-2)\left[\frac{1}{2}\right]+1}$. Ist ferner $p\geqq 2$ und $t_p=1$, so enthält $K^{(m_p)}$ alle Einheitswurzeln des Grades p^{m_p}, aber keine primitive p^{m_p+1}te Einheitswurzel, folglich ist $a\leqq p^{m_p}$. Diese Zahl wird aber unter der gemachten Voraussetzung in der Tat gleich $N_{1,p}$. Ist endlich $p>2$ und $t_p>1$, so enthält $K^{(m_p)}$ nur die $a$te Einheitswurzel $+1$ und es wird $a=1=p^{m_p\left[\frac{1}{t_p}\right]}$.

Ich nehme nun an, der Satz III sei bereits für Abelsche Gruppen mit weniger als n Variabeln bewiesen.

Ist dann $\mathfrak{A}$ eine im Körper K zerlegbare Gruppe, so lassen sich zwei der Gruppe $\mathfrak{A}$ ein- oder mehrstufig isomorphe Gruppen $\mathfrak{A}'$ und $\mathfrak{A}''$ linearer Substitutionen in $r<n$ und $s=n-r<n$ Variabeln angeben, deren Spuren dem Körper K angehören, so daß a höchstens gleich wird dem Produkte der Ordnungen a' und a'' dieser Gruppen. Es ist aber nach Voraussetzung

$$a'\leqq N_{r,p}, \quad a''\leqq N_{s,p},$$

also

$$a\leqq N_{r,p}N_{s,p};$$

die Zahl $N_{r,p}N_{s,p}$ ist aber, wie man sofort sieht, höchstens gleich $N_{r+s,p}=N_{n,p}$.

Es sei daher $\mathfrak{A}$ eine im Körper K nicht zerlegbare Gruppe. Dann ist nach dem Ergebnis des § 3, da eine kommutative Gruppe linearer Substitutionen bekanntlich nur dann irreduzibel ist, wenn die Anzahl der Variabeln gleich 1 ist, $\mathfrak{A}$ einer Gruppe ähnlich, in der die Koeffizientenmatrix der Substitution R die Form hat

$$\begin{pmatrix}\psi(R) & 0 & \cdots 0\\ 0 & \psi_1(R) & \cdots 0\\ . & . & \cdots .\\ 0 & 0 & \psi_{n-1}(R)\end{pmatrix},$$

wo die a Zahlen $\psi(R)$ einen (einfachen) Charakter der Abelschen Gruppe $\mathfrak{A}$ bilden, und $\psi_1(R), \cdots \psi_{n-1}(R)$ die übrigen diesem Charakter relativ konjugierten Charaktere bedeuten. Da nun aber die a Einheitswurzeln $\psi(R)$ eine der Gruppe $\mathfrak{A}$ einstufig isomorphe Gruppe bilden, so muß $\mathfrak{A}$ eine zyklische Gruppe sein, und ist $a = p^{\mu}$, so kommen unter den $\psi(R)$ auch primitive $p^{\mu\text{te}}$ Einheitswurzeln vor. Daher wird

$$n = \frac{p^{\mu-1}(p-1)}{k^{(\mu)}},$$

falls $k^{(\mu)}$ wie früher den Grad des größten gemeinsamen Divisors von K und $\Omega^{(p^{\mu})}$ bedeutet.

Ist nun $p = 2$ und $\mu = 1$, so wird $n = \dfrac{1}{k^{(1)}} = 1$, ein Fall, den wir schon erledigt haben.

Ist ferner $\mu \leqq m_p$ und für $p = 2$ noch $\mu > 1$, so wird (vgl. die Formeln (2) und (4)) $n = t_p$ und $N_{n,p}$ in allen Fällen gleich p^{m_p}, also in der Tat $a = p^{\mu} \leqq N_{n,p}$.

Ist endlich $\mu > m_p$, so wird

$$n = \frac{p^{\mu-1}(p-1)}{k^{(m_p)}} = p^{\mu-m_p} . t_p$$

und $N_{n,p}$ für $p \geqq 2$ gleich $p^{m_p \cdot p^{\mu-m_p}}$. Da aber offenbar stets $\mu \leqq m_p p^{\mu-m_p}$ ist, so wird auch hier $a = p^{\mu} \leqq N_{n,p}$.

§ 5.

Wir kommen nun zum Beweise des Satzes II.

Der Beweis stützt sich auf einen von Hrn. Blichfeldt[1] bewiesenen Satz, der folgendermaßen lautet:

»Ist $\mathfrak{G}$ eine endliche Gruppe linearer Substitutionen in n Variabeln, deren Ordnung g eine Primzahlpotenz ist, so läßt sich $\mathfrak{G}$ durch eine Transformation der Variabeln in eine Gruppe $\mathfrak{G}'$ überführen, deren Substitutionen die Form haben

$$x_{\kappa} = \alpha_{\kappa} x'_{\lambda_{\kappa}}, \qquad (\kappa = 1, 2, \cdots, n)$$

wo $\alpha_1, \alpha_2, \cdots \alpha_{\kappa}$ gewisse Konstanten sind und $\lambda_1, \lambda_2, \cdots \lambda_n$ abgesehen von der Reihenfolge mit den Zahlen $1, 2, \cdots, n$ übereinstimmen.«

Betrachtet man in $\mathfrak{G}'$ alle Substitutionen der Form

$$x_{\kappa} = \alpha_{\kappa} x'_{\kappa},$$

so bilden diese eine invariante Abelsche Untergruppe $\mathfrak{G}'$, deren Ordnung gleich f sei. Ferner bilden die verschiedenen den Substitutionen von $\mathfrak{G}'$ entsprechenden Permutationen

$$x_{\kappa} = x'_{\lambda_{\kappa}}$$

[1] Transactions of the American Mathematical Society, Bd. 5 (1904), S. 313.

eine der Gruppe $\mathfrak{G}$ ein- oder mehrstufig isomorphe Gruppe, deren Ordnung d ein Divisor von $n!$ ist, und es ist $g = fd$. Es gilt daher der Satz:

Ist $\mathfrak{G}$ eine endliche Gruppe linearer Substitutionen in n Variabeln, deren Ordnung eine Primzahlpotenz ist, so ist die Ordnung von $\mathfrak{G}$ eine Zahl der Form fd, wo f die Ordnung einer invarianten Abelschen Untergruppe von $\mathfrak{G}$ angibt und d ein Divisor von $n!$ ist.[1]

Um nun zunächst zu zeigen, daß die Ordnung einer Gruppe $\mathfrak{G}_n^{(\varkappa)}$ ein Divisor der Zahl $M_n^{(\varkappa)}$ ist, genügt es offenbar, nachzuweisen, daß die Ordnung g einer solchen Gruppe $\mathfrak{G}$, falls diese Zahl eine Potenz der Primzahl p ist, für $p > 2$ höchstens gleich

$$M_{n,p} = p^{m_p \left[\frac{n}{t_p}\right] + \left[\frac{n}{pt_p}\right] + \left[\frac{n}{p^2 t_p}\right] + \cdots}$$

für $p = 2$ höchstens gleich

$$M_{n,2} = 2^{n - \left[\frac{n}{t_2}\right] + m_2 \left[\frac{n}{t_2}\right] + \left[\frac{n}{2t_2}\right] + \cdots}$$

sein kann.

Es sei nun zunächst $p = 2$. Ist dann $g = fd$, wo f die Ordnung einer Abelschen Untergruppe von $\mathfrak{G}$ und d ein Divisor von $n!$ ist, so ist

$$d \leqq 2^{\left[\frac{n}{2}\right] + \left[\frac{n}{4}\right] + \cdots}$$

und wegen III

$$f \leqq 2^{(m_2 - t_2)\left[\frac{n}{t_2}\right] + n},$$

also ist für $t_2 = 1$

$$g = fd \leqq 2^{m_2 n + \left[\frac{n}{2}\right] + \left[\frac{n}{4}\right] + \cdots}$$

und für $t_2 = 2$

$$g = fd \leqq 2^{n - 2\left[\frac{n}{2}\right] + \left[\frac{n}{2}\right] + m_2\left[\frac{n}{2}\right] + \left[\frac{n}{4}\right] + \cdots}.$$

Die sich so für g ergebenden oberen Grenzen für g sind aber in jedem der beiden Fälle gleich $M_{n,2}$.

Es sei also $p > 2$. Ist zunächst $n = 1$, so wird $\mathfrak{G}$ eine Abelsche Gruppe; die Ordnung einer solchen ist aber höchstens gleich $N_{1,p} = M_{1,p}$.

Ich nehme nun an, es sei schon für $r < n$ gezeigt, daß die Ordnung einer Gruppe $\mathfrak{G}_r^{(\varkappa)}$, falls diese Zahl eine Potenz von p ist, höchstens gleich $M_{r,p}$ ist.

Ist nun die zu untersuchende Gruppe $\mathfrak{G}$ in n Variabeln im Körper K in zwei Gruppen mit $r < n$ und $s = n - r < n$ Variabeln zerlegbar, so schließt man in analoger Weise wie in § 4, daß $g \leqq M_{r,p}\, M_{s,p}$ ist. Dieses Produkt ist aber höchstens gleich $M_{r+s,p} = M_{n,p}$.

---

[1] Vgl. Blichfeldt, a. a. O. S. 320.

Es sei also $\mathfrak{G}$ eine im Körper K nicht zerlegbare Gruppe. Ist dann $\zeta(R)$ die Spur der Substitution R von $\mathfrak{G}$, so sei

$$\zeta(R) = \chi(R) + \chi_1(R) + \cdots + \chi_{r-1}(R),$$

wo $\chi(R), \chi_1(R), \cdots \chi_{r-1}(R)$ gewisse r relativ konjugierte einfache Charaktere von $\mathfrak{G}$ sind. Die Gruppe $\mathfrak{G}$ ist dann einstufig isomorph einer irreduziblen Gruppe $\mathfrak{H}$ linearer Substitutionen in $\chi(E) = \frac{n}{r}$ Variabeln, deren Spuren die Werte $\chi(R)$ haben. Es läßt sich nun in $\mathfrak{H}$, und also auch in $\mathfrak{G}$, eine Abelsche Untergruppe der Ordnung f angeben, so daß $g = fd$ wird, wo d ein Divisor von $\left(\frac{n}{r}\right)!$ ist. Da nun

$$d \leqq p^{\left[\frac{n}{rp}\right] + \left[\frac{n}{rp^2}\right] + \cdots},$$

ferner wegen III die Zahl f höchstens gleich $p^{m_p \left[\frac{n}{t_p}\right]}$ ist, so erhält man

$$g \leqq p^{m_p \left[\frac{n}{t_p}\right] + \left[\frac{n}{pr}\right] + \left[\frac{n}{p^2 r}\right] + \cdots}.$$

Es ist aber leicht zu sehen, daß die Zahl r mindestens gleich t_p sein muß. Denn ist P ein invariantes Element der Ordnung p von G, so ist, weil $\chi(R)$ ein einfacher Charakter der Gruppe $\mathfrak{G}$ ist,

$$\chi(P) = \rho \cdot \chi(E) = \rho \cdot \frac{n}{r},$$

wo ρ eine primitive p^{te} Einheitswurzel ist. Ferner sind unter den Zahlen

$$\chi(P), \chi_1(P), \cdots, \chi_{r-1}(P)$$

die sämtlichen der Zahl $\rho \cdot \frac{n}{r}$ relativ konjugierten Größen enthalten. Da nun aber der größte gemeinsame Divisor der beiden Körper K und $\Omega^{(p)} = \Omega(\rho)$ gleich $\frac{p-1}{t_p}$ ist, so genügt $\rho \cdot \frac{n}{r}$ im Körper K einer irreduziblen Gleichung des Grades t_p. Daher ist in der Tat $r \geqq t_p$ und folglich ist, wie zu beweisen ist,

$$g \leqq p^{m_p \left[\frac{n}{t_p}\right] + \left[\frac{n}{pr}\right] + \left[\frac{n}{p^2 r}\right] + \cdots} \leqq p^{m_p \left[\frac{n}{t_p}\right] + \left[\frac{n}{pt_p}\right] + \left[\frac{n}{p^2 t_p}\right] + \cdots}.$$

§ 6.

Es bleibt uns noch übrig, zu zeigen, daß die Zahl $M_n^{(\varkappa)}$, von der wir nachgewiesen haben, daß sie durch die Ordnung jeder Gruppe $\mathfrak{G}_n^{(\varkappa)}$ teilbar ist, auch wirklich das kleinste gemeinsame Vielfache der Ordnungen aller dieser Gruppen repräsentiert.

Um diesen Nachweis zu führen, hat man für jede Primzahl p eine Gruppe $\mathfrak{G}_n^{(\varkappa)}$ anzugeben, deren Ordnung durch dieselbe Potenz $M_{n,p}$ von p teilbar ist wie die Zahl $M_n^{(\varkappa)}$.

Es sei zunächst $p > 2$ oder $p = 2$ und $t_2 = 1$. Man setze dann

$$\left[\frac{n}{t_p}\right] = \nu\,,\ n = t_p\nu + r\,.$$

Ist nun ρ eine primitive p^{m_p}te Einheitswurzel, so betrachte man die Gesamtheit der Substitutionen der Form

$$(7.)\qquad x_1 = \rho^{\alpha_1} x'_{\lambda_1}\,,\ x_2 = \rho^{\alpha_2} x'_{\lambda_2}\,,\ \cdots\ x_\nu = \rho^{\alpha_\nu} x'_{\lambda_\nu}\,,$$

wo $\alpha_1, \alpha_2, \cdots \alpha_\nu$ unabhängig voneinander die Zahlen $0, 1, \ldots p^{m_p}-1$, ferner $\lambda_1, \lambda_2, \cdots \lambda_\nu$ alle $\nu!$ Permutationen der Ziffern $1, 2, \ldots \nu$ durchlaufen. Die Substitutionen (7.) bilden dann eine Gruppe $\mathfrak{H}$ der Ordnung $p^{m_p\nu}\nu!$

Sind ferner $\rho, \rho', \cdots \rho^{(t_p-1)}$ die zu ρ (in bezug auf K) relativ konjugierten Zahlen, und ersetzt man in der Koeffizientenmatrix A einer beliebigen Substitution von $\mathfrak{H}$ die Einheitswurzel ρ durch $\rho', \cdots \rho^{(t_p-1)}$, so mögen die Matrizen $A_1, \cdots A_{t_p-1}$ entstehen. Ist dann E_r die Matrix r^{ten} Grades

$$\begin{pmatrix} 1 & 0 & \cdots & 0 \\ 0 & 1 & \cdots & 0 \\ . & . & \cdots & . \\ 0 & 0 & \cdots & 1 \end{pmatrix},$$

so bilden die linearen Substitutionen in n Variabeln mit den Koeffizientensystemen

$$\begin{pmatrix} E_r & 0 & 0 & \cdots & 0 \\ 0 & A & 0 & \cdots & 0 \\ 0 & 0 & A_1 & \cdots & 0 \\ . & . & . & \cdots & . \\ 0 & 0 & 0 & \cdots & A_{t_p-1} \end{pmatrix}$$

eine Gruppe der Ordnung $p^{m_p\nu}\nu!$, in der die Spur jeder Substitution dem Körper K angehört. Die Zahl $p^{m_p\nu}\nu!$ ist aber genau durch die Potenz $M_{n,p}$ von p teilbar.

Etwas weniger einfach ist die Behandlung des Falles $p = 2$, $t_2 = 2$. Es sei dann

$$\left[\frac{n}{2}\right] = \nu\,,\ n = 2\nu + r \qquad (r = 0 \text{ oder } 1).$$

Ist ferner σ eine primitive 2^{m_2}te Einheitswurzel, so setze man $\tau = \sigma^{-1}$ oder $-\sigma^{-1}$, je nachdem der größte gemeinsame Divisor von K und $\Omega(\sigma)$ gleich $\Omega(\sigma + \sigma^{-1})$ oder gleich $\Omega(\sigma - \sigma^{-1})$ ist. Betrachtet man dann für $r = 0$ alle Substitutionen der Form

$$x_{11} = \sigma^{\alpha_1} x'_{\lambda_1\beta_1}\,,\ x_{12} = \tau^{\alpha_1} x'_{\lambda_1\gamma_1}\,,\ \cdots,\ x_{\nu 1} = \sigma^{\alpha_\nu} x'_{\lambda_\nu\beta_\nu}\,,\ x_{\nu 2} = \tau^{\alpha_\nu} x'_{\lambda_\nu\gamma_\nu}\,,$$

für $r = 1$ alle Substitutionen der Form

$$x = \pm x',\ x_{11} = \sigma^{\alpha_1} x'_{\lambda_1 \beta_1},\ x_{12} = \tau^{\alpha_1} x'_{\lambda_1 \gamma_1},\ \cdots,\ x_{\nu 1} = \sigma^{\alpha_\nu} x'_{\lambda_\nu \beta_\nu},\ x_{\nu 2} = \tau^{\alpha_\nu} x'_{\lambda_\nu \gamma_\nu},$$

wo $\alpha_1, \alpha_2, \cdots \alpha_\nu$ unabhängig voneinander die Werte $0, 1, \cdots, 2^{m_2} - 1$, die Indices $\lambda_1, \lambda_2, \cdots \lambda_\nu$ alle Permutationen der Zahlen $1, 2, \cdots \nu$ durchlaufen, endlich $\beta_\varkappa, \gamma_\varkappa$ abgesehen von der Reihenfolge die Zahlen 1 und 2 bedeuten, so bilden dieselben eine endliche Gruppe linearer Substitutionen in n Variabeln mit Spuren aus dem Körper K. Die Ordnung dieser Gruppe ist für $r = 0$ gleich $2^{\nu m_2 + \nu} \nu!$, für $r = 1$ gleich $2 . 2^{\nu m_2 + \nu} \nu!$ Man sieht aber leicht ein, daß diese Zahlen genau durch dieselbe Potenz von 2 teilbar sind wie die Zahl $M_n^{(\varkappa)}$.

Damit ist der Satz II vollständig bewiesen.

Ausgegeben am 19. Januar.

7.

Neue Begründung der Theorie der Gruppencharaktere

Sitzungsberichte der Preussischen Akademie der Wissenschaften 1905, Physikalisch-Mathematische Klasse, 406 - 432

Die vorliegende Arbeit enthält eine durchaus elementare Einführung in die von Hrn. Frobenius begründete Theorie der Gruppencharaktere[1], die auch als die Lehre von der Darstellung der endlichen Gruppen durch lineare homogene Substitutionen bezeichnet werden kann.

Eine elementare Begründung dieser Theorie ist zwar in neuerer Zeit bereits von Hrn. Burnside[2] gegeben worden. Hr. Burnside macht jedoch noch von einem dem Gegenstand im Grunde fernliegenden Hilfsmittel, nämlich dem Begriff der Hermiteschen Formen, Gebrauch. Ich halte es daher nicht für überflüssig, eine neue Darstellung der Frobeniusschen Theorie mitzuteilen, die mit noch einfacheren Hilfsmitteln operiert.

Zum Verständnis des Folgenden ist aus der Theorie der linearen Substitutionen im wesentlichen nur die Kenntnis der Anfangsgründe des Kalküls der Matrizen erforderlich. Abgesehen von den rein formalen Regeln dieses Kalküls werden nur noch zwei übrigens sehr leicht zu beweisende Sätze als bekannt vorausgesetzt, die der besseren Übersicht wegen hier angeführt werden mögen:

a) Ist P eine Matrix mit m Zeilen und n Spalten, und sind A und B zwei Matrizen der Grade m und n, deren Determinanten nicht verschwinden, so besitzen die beiden Matrizen P und APB denselben Rang.

b) Ist P eine Matrix mit m Zeilen und n Spalten, deren Rang gleich r ist, so lassen sich zwei Matrizen A und B der Grade m und n von nicht verschwindenden Determinanten bestimmen, so daß in der Matrix $APB = (q_{\alpha\beta})$ die r Koeffizienten $q_{11}, q_{22}, \cdots, q_{rr}$ gleich 1, die übrigen Koeffizienten gleich 0 sind.

---

[1] Sitzungsberichte 1896, S. 985 und S. 1343, ferner 1897, S. 994 und 1899, S. 482.

[2] Acta Mathematica, Bd. 28 (1904), S. 369, und Proceedings of the London Mathematical Society, Ser. 2, Vol. 1 (1904), S. 117.

Der zuletzt angeführte Satz ist identisch mit dem bekannten Theorem, welches besagt, daß eine bilineare Form $f = \sum_{\alpha=1}^{m} \sum_{\beta=1}^{n} p_{\alpha\beta} x_\alpha y_\beta$ vom Range r sich auf die Gestalt $f = u_1 v_1 + u_2 v_2 + \cdots + u_r v_r$ bringen läßt, wo $u_1, u_2, \cdots, u_r$ und $v_1, v_2, \cdots, v_r$ linear unabhängige lineare homogene Funktionen der m Variabeln x_α und der n Variabeln y_β bedeuten.

Die eigentliche Theorie der Gruppencharaktere wird in den §§ 1—5 entwickelt.

Der § 6 enthält eine Anwendung auf die Theorie der charakteristischen Einheiten der endlichen Gruppen[1]; die in diesem Paragraphen durchgeführte Untersuchung verdanke ich einer Anregung des Hrn. Frobenius.

§ 1.

Es sei $\mathfrak{H}$ eine endliche Gruppe der Ordnung h, deren Elemente $H_0, H_1, \cdots, H_{h-1}$ sind.

Sind dann

$$(H_0), (H_1), \cdots, (H_{h-1}) \tag{1.}$$

h Matrizen (lineare Substitutionen) n^{ten} Grades, welche den h^2 Relationen

$$(R)(S) = (RS) \qquad (R, S = H_0, H_1, \cdots, H_{h-1})$$

genügen, so bezeichnen wir das System der Matrizen (1.) als eine *Darstellung* der Gruppe $\mathfrak{H}$. Hierbei brauchen die Matrizen (1.) nicht voneinander verschieden zu sein.

Sind ferner

$$x_{H_0}, x_{H_1}, \cdots, x_{H_{h-1}}$$

h unabhängige Variable, so soll die Matrix

$$X = \sum_R (R) x_R \qquad (R = H_0, H_1, \cdots, H_{h-1})$$

die der Darstellung entsprechende *Gruppenmatrix* genannt werden. Die Zahl n bezeichnen wir als den *Grad* der Darstellung oder der Gruppenmatrix.

Eine Gruppenmatrix $X = (x_{\alpha\beta})$ kann durch folgende Eigenschaften charakterisiert werden:

1. Die Koeffizienten $x_{\alpha\beta}$ sind lineare homogene Funktionen der Variabeln $x_{H_0}, x_{H_1}, \cdots, x_{H_{h-1}}$.

2. Ist $y_{H_0}, y_{H_1}, \cdots, y_{H_{h-1}}$ ein zweites System von h unabhängigen Variabeln, setzt man ferner

$$z_R = \sum_S x_S y_{S^{-1}R} \qquad (R, S = H_0, H_1, \cdots, H_{h-1}),$$

[1] Frobenius, Sitzungsberichte 1903, S. 328.

und geht X in Y oder Z über, falls in X die Variabeln x_R durch y_R oder z_R ersetzt werden, so soll die Gleichung $Z = XY$ bestehen.

Bedeutet E das Hauptelement der Gruppe, so ist offenbar

$$(E)X = X.$$

Daher ist die Determinante $|X|$ der Matrix X dann und nur dann identisch gleich 0, wenn die Determinante der Matrix (E) verschwindet. Ist ferner die Determinante $|X|$ nicht identisch gleich 0, so ist $(E) = E_n$ die Einheitsmatrix n^{ten} Grades. Da ferner für jedes Element R der Gruppe $R^h = E$, also auch $(R)^h = (E)$ ist, so sind, falls $|X| \not\equiv 0$ ist, die Determinanten aller h Substitutionen (R) von Null verschieden.

Ist X eine Gruppenmatrix und A eine konstante Matrix[1], deren Determinante nicht verschwindet, so ist auch $X' = A^{-1}XA$ eine Gruppenmatrix. Jede auf diese Weise aus X hervorgehende Gruppenmatrix X' bezeichnen wir als eine der Gruppenmatrix X *äquivalente* Gruppenmatrix. Ebenso nennen wir die der Matrix X' entsprechende Darstellung der Gruppe $\mathfrak{H}$ der aus X hervorgehenden Darstellung äquivalent.

Eine Gruppenmatrix X soll *reduzibel* genannt werden, wenn sich eine ihr äquivalente Gruppenmatrix angeben läßt, welche die Form

$$\begin{pmatrix} X_1 & 0 \\ U & X_2 \end{pmatrix}$$

besitzt. Hierbei bedeuten X_1 und X_2 zwei quadratische Matrizen gewisser Grade $r > 0$ und $s > 0$, während U eine Matrix mit s Zeilen und r Spalten ist. — Jede der Matrizen X_1 und X_2 besitzt dann ebenfalls die Eigenschaften einer Gruppenmatrix.

Eine nicht reduzible Gruppenmatrix nennen wir *irreduzibel.*[2]

In analoger Weise nennen wir eine Darstellung der Gruppe *reduzibel* oder *irreduzibel*, je nachdem die zugehörige Gruppenmatrix reduzibel oder irreduzibel ist.

Für die Äquivalenz der Gruppenmatrizen gelten folgende unmittelbar zu beweisende Regeln:

1. Ist X' der Gruppenmatrix X äquivalent, so ist auch X der Gruppenmatrix X' äquivalent.

2. Zwei Gruppenmatrizen, die einer dritten äquivalent sind, sind auch untereinander äquivalent.

3. Eine Gruppenmatrix X geht in eine ihr äquivalente Gruppenmatrix über, wenn die Zeilen und Spalten von X in derselben Weise permutiert werden. Daher ist z. B. die Gruppenmatrix $\begin{pmatrix} X_2 & U \\ 0 & X_1 \end{pmatrix}$ reduzibel.

---

[1] D. h. eine Matrix mit konstanten, von den x_R unabhängigen Koeffizienten.

[2] Strenggenommen, müßte auch die Funktion 0 der Variabeln x_R als eine irreduzible Gruppenmatrix angesehen werden. Dieser triviale Fall soll aber, wenn im folgenden von einer irreduziblen Gruppenmatrix gesprochen wird, als ausgeschlossen gelten.

4. Sind zwei Gruppenmatrizen äquivalent, so sind ihre Determinanten, ihre Rangzahlen und ihre Spuren einander gleich. — Hierbei ist unter der *Spur* einer Matrix $(a_{\alpha\beta})$ die Größe $\sum_\alpha a_{\alpha\alpha}$ zu verstehen.

5. Sind $X = (x_{\alpha\beta})$ und $X' = (x'_{\alpha\beta})$ zwei äquivalente Gruppenmatrizen, so ist die Anzahl der linear unabhängigen unter den Funktionen $x_{\alpha\beta}$ der h Variabeln x_R gleich der analogen Anzahl für die Funktionen $x'_{\alpha\beta}$.

Die zuletzt genannte Regel geht daraus hervor, daß die $x_{\alpha\beta}$ als lineare homogene Funktionen der $x'_{\alpha\beta}$ und umgekehrt die $x'_{\alpha\beta}$ als lineare homogene Funktionen der $x_{\alpha\beta}$ darstellbar sind.

§ 2.

Es sollen zunächst einige Eigenschaften der irreduziblen Gruppenmatrizen abgeleitet werden.

Die Grundlage der Untersuchung bildet folgender Satz, der auch in der Burnside'schen Darstellung der Theorie eine wichtige Rolle spielt:

I. *Es seien X und X' zwei irreduzible Gruppenmatrizen der Grade f und f'. Ist dann P eine konstante Matrix mit f Zeilen und f' Spalten, für die die Gleichung*

$$XP = PX'$$

besteht, so ist entweder $P = 0$ oder es sind X und X' äquivalent, und P ist eine quadratische Matrix des Grades $f = f'$ von nicht verschwindender Determinante.

Es sei nämlich P von Null verschieden und $r > 0$ der Rang von P. Es werde $f-r = s$, $f'-r = t$ gesetzt. Man bestimme dann (vgl. Einleitung) zwei Matrizen A und B der Grade f und f', deren Determinanten nicht verschwinden, so daß die Matrix $APB = Q$ die Gestalt

$$\begin{pmatrix} E_r & N_{rt} \\ N_{sr} & N_{st} \end{pmatrix}$$

annimmt; hierbei soll allgemein $E_\varkappa$ die Einheitsmatrix $\varkappa^{\text{ten}}$ Grades bedeuten, während unter $N_{\varkappa\lambda}$ die $\varkappa$ Zeilen und λ Spalten enthaltende Nullmatrix verstanden werden soll. Setzt man nun

$$AXA^{-1} = X_1 \quad , \qquad B^{-1}X'B = X'_1,$$

so wird

$$X_1 Q = QX'_1. \tag{2.}$$

Schreibt man nun X_1 und X'_1 in der Form

$$X_1 = \begin{pmatrix} X_{rr} & X_{rt} \\ X_{sr} & X_{st} \end{pmatrix} \qquad X'_1 = \begin{pmatrix} X'_{rr} & X'_{rt} \\ X'_{sr} & X'_{st} \end{pmatrix},$$

wo $X_{\varkappa\lambda}$ und $X'_{\varkappa\lambda}$ Matrizen mit $\varkappa$ Zeilen und λ Spalten bedeuten, so folgt aus (2.)

$$\begin{pmatrix} X_{rr} & N_{rt} \\ X_{sr} & N_{st} \end{pmatrix} = \begin{pmatrix} X'_{rr} & X'_{rt} \\ N_{sr} & N_{st} \end{pmatrix},$$

d. h. es ist

$$X_{sr} = 0 \quad , \qquad X'_{rt} = 0 .$$

Wäre nun $r < f$ oder $r < f'$, so würde sich ergeben, daß X oder X' reduzibel ist. Daher muß, falls P nicht 0 sein soll, $r = f = f'$ sein. Dann wird aber P eine quadratische Matrix des Grades f von nicht verschwindender Determinante, und die Gruppenmatrizen X und X' sind wegen $P^{-1}XP = X'$ äquivalent.

Aus I folgt:

II. *Ist X eine irreduzible Gruppenmatrix des Grades f, so muß jede mit X vertauschbare konstante Matrix P die Form aE_f besitzen.*

In der Tat sei a eine Wurzel der Gleichung

$$|P - xE_f| = 0 .$$

Dann wird $P - aE_f$ eine Matrix von verschwindender Determinante, für die die Gleichung

$$X(P - aE_f) = (P - aE_f)X$$

besteht. Daher muß nach Satz I die Matrix $P - aE_f$ gleich 0, also $P = aE_f$ sein.

Setzt man wie früher $X = \sum_R (R)x_R$, so ist wegen

$$(E)X = X(E) = X$$

die Matrix (E) mit X vertauschbar. Folglich ist (E) von der Form aE_f, wobei die Zahl a, da $(E)^2 = (E)$ ist, nur die Werte 0 oder 1 besitzen kann. Wäre nun $a = 0$, so würde sich $X = 0$ ergeben, ein Fall, den wir ausgeschlossen haben. Daher ist bei einer irreduziblen Gruppenmatrix stets $(E) = E_f$ und ihre Determinante nicht identisch 0.

Ist ferner $\mathfrak{H}$ insbesondere eine Abelsche Gruppe, so ist jede der Matrizen (R) mit X vertauschbar und also von der Form ρE_f. Hierbei muß die Größe ρ, da $(R)^h = E_f$ ist, eine h^{te} Einheitswurzel sein. Wäre nun $f > 1$, so würde sich ergeben, daß X reduzibel ist. Hieraus folgt:

III. *Eine zu einer Abelschen Gruppe gehörende Gruppenmatrix ist dann und nur dann irreduzibel, wenn ihr Grad gleich 1 ist.*

Wir beweisen nun den wichtigen Satz:

IV. *Es sei*

$$X = (x_{\alpha\beta}) \qquad (\alpha, \beta = 1, 2, \cdots, f),$$

eine irreduzible Gruppenmatrix des Grades f. Ist dann

$$x_{\alpha\beta} = \sum_R a^R_{\alpha\beta} x_R \qquad (R = H_0, H_1, \cdots, H_{h-1}),$$

so bestehen die Gleichungen

$$\text{(I.)} \qquad \sum_R a^{R^{-1}}_{\alpha\beta} a^R_{\gamma\delta} = \frac{h}{f} e_{\alpha\delta} e_{\gamma\beta} \qquad (\alpha, \beta, \gamma, \delta = 1, 2, \cdots, f)$$

wo $e_{\varrho\sigma}$ *gleich* 1 *oder gleich* 0 *zu setzen ist, je nachdem* ρ *gleich* σ *oder von* σ *verschieden ist. Ist ferner*

$$X' = (x'_{\varkappa\lambda}) \qquad (\varkappa, \lambda = 1, 2, \cdots, f')$$

eine der Gruppenmatrix X nicht äquivalente irreduzible Gruppenmatrix des Grades f' *und ist*

$$x'_{\varkappa\lambda} = \sum_R b^R_{\varkappa\lambda} x_R \qquad (R = H_0, H_1, \cdots, H_{h-1}),$$

so gelten die $(ff')^2$ *Relationen*

$$\text{(II.)} \qquad \sum_R a^{R^{-1}}_{\alpha\beta} b^R_{\varkappa\lambda} = 0 \,[1] \qquad (\alpha, \beta = 1, 2, \cdots, f; \ \varkappa, \lambda = 1, 2, \cdots, f').$$

Es mögen nämlich die Matrizen $(a^R_{\alpha\beta})$ und $(b^R_{\varkappa\lambda})$ mit A_R und B_R bezeichnet werden. Dann ist für je zwei Elemente R und S der Gruppe

$$A_R A_S = A_{RS} \quad , \qquad B_R B_S = B_{RS} .$$

Es sei nun $U = (u_{\alpha\beta})$ eine Matrix f^{ten} Grades, deren Koeffizienten $u_{\alpha\beta}$ beliebige Konstanten sind. Setzt man dann

$$(3.) \qquad V = \sum_R A_{R^{-1}} U A_R ,$$

so ist für jedes Element S der Gruppe

$$A_{S^{-1}} V A_S = \sum_R A_{S^{-1}R^{-1}} U A_{RS} ;$$

da nun das Element RS zugleich mit R alle Elemente der Gruppe durchläuft, ferner $S^{-1}R^{-1} = (RS)^{-1}$ ist, so ist die rechts stehende Summe gleich V. Folglich erhält man

$$A_{S^{-1}} V A_S = V$$

und, weil $A_{S^{-1}} A_S = A_E = E_f$ ist, $VA_S = A_S V$. Daher ist V mit allen A_S und also auch mit X vertauschbar und muß mithin die Form vE_f besitzen. Die Größe v ist jedenfalls eine lineare homogene Funktion der $u_{\alpha\beta}$. Setzt man

$$v = \sum_{\beta,\gamma} c_{\beta\gamma} u_{\beta\gamma} ,$$

---

[1] Die Relationen (I.) und (II.) sind in den bisherigen Darstellungen der Theorie nicht ausdrücklich angegeben. Sie sind aber implizite in den für die Gruppencharaktere geltenden Formeln enthalten.

so ergibt sich aus (3.)

$$\sum_R \sum_{\beta,\gamma} a_{\alpha\beta}^{R^{-1}} u_{\beta\gamma} a_{\gamma\delta}^{R} = e_{\alpha\delta} \sum_{\beta,\gamma} c_{\beta\gamma} u_{\beta\gamma}.$$

Daher ist

$$(4.) \qquad \sum_R a_{\alpha\beta}^{R^{-1}} a_{\gamma\delta}^{R} = e_{\alpha\delta} c_{\beta\gamma}.$$

Setzt man speziell $\delta = \alpha$ und bildet die Summe über $\alpha = 1, 2, \cdots, f$, so erhält man

$$f c_{\beta\gamma} = \sum_R \sum_\alpha a_{\gamma\alpha}^{R} a_{\alpha\beta}^{R^{-1}}.$$

Die linke Seite wird aber, weil die Gleichung $A_R A_{R^{-1}} = A_E = E_f$ mit den Relationen

$$\sum_\alpha a_{\gamma\alpha}^{R} a_{\alpha\beta}^{R^{-1}} = a_{\gamma\beta}^{E} = e_{\gamma\beta}$$

identisch ist, gleich $h e_{\gamma\beta}$; daher ist $c_{\beta\gamma} = \frac{h}{f} e_{\gamma\beta}$. Setzt man dies in (4.) ein, so erhält man die Relationen (I.).

Bildet man ebenso, falls jetzt $U = (u_{\alpha\varkappa})$ eine Matrix mit f Zeilen und f' Spalten bedeutet, deren Koeffizienten beliebige Größen sind, die Matrix

$$V = \sum_R A_{R^{-1}} U B_R,$$

so ergibt sich auf demselben Wege die Gleichung $XV = VX'$. Da nun X und X' nicht äquivalent sein sollen, so muß $V = 0$ sein. Folglich ist

$$\sum_R \sum_{\beta,\varkappa} a_{\alpha\beta}^{R^{-1}} u_{\beta\varkappa} b_{\varkappa\lambda}^{R} = 0.$$

Hieraus folgen, da die $u_{\beta\varkappa}$ beliebiger Werte fähig sind, die Relationen (II.).

Aus den eben bewiesenen Relationen ergibt sich leicht einer der Hauptsätze der Theorie:

V. *Sind*

$$X = (x_{\alpha\beta}) \quad , \quad X' = (x'_{\varkappa\lambda}), \cdots$$

beliebig viele zu der Gruppe $\mathfrak{H}$ *gehörende irreduzible Gruppenmatrizen der Grade* $f, f', \cdots$, *von denen nicht zwei äquivalent sind, so sind die* $f^2 + f'^2 + \cdots$ *linearen homogenen Funktionen der Variabeln* x_R *untereinander linear unabhängig.*

In der Tat möge eine Gleichung der Form

$$\sum_{\alpha,\beta} c_{\alpha\beta} x_{\alpha\beta} + \sum_{\varkappa,\lambda} c'_{\varkappa\lambda} x'_{\varkappa\lambda} + \cdots = 0$$

bestehen. Dann ergibt sich, wenn $a_{\alpha\beta}^{R}$ und $b_{\varkappa\lambda}^{R}$ dieselbe Bedeutung haben wie früher, für jedes R die Gleichung

$$\sum_{\alpha,\beta} c_{\alpha\beta} a_{\alpha\beta}^{R} + \sum_{\varkappa,\lambda} c'_{\varkappa\lambda} b_{\varkappa\lambda}^{R} + \cdots = 0.$$

Multipliziert man nun diese Gleichung mit $a^{R^{-1}}_{\gamma\delta}$ und bildet die Summe über $R = H_0, H_1, \cdots, H_{h-1}$, so erhält man wegen (I.) und (II.)

$$\frac{h}{f}\sum_{\alpha,\beta} c_{\alpha\beta}\, e_{\alpha\delta}\, e_{\gamma\beta} = 0\,.$$

Die linke Seite dieser Gleichung ist aber gleich $\frac{h}{f} c_{\delta\gamma}$. Folglich sind alle Koeffizienten $c_{\alpha\beta}$ gleich 0. Ebenso zeigt man, daß die f'^2 Koeffizienten $c'_{\varkappa\lambda}$ verschwinden müssen, usw.

Es gilt ferner der Satz:

VI. *Die Determinante einer irreduziblen Gruppenmatrix ist eine irreduzible Funktion der h Variabeln x_R. Ferner sind zwei irreduzible Gruppenmatrizen dann und nur dann äquivalent, wenn ihre Determinanten einander gleich sind.*

Denn ist $X = (x_{\alpha\beta})$ eine irreduzible Gruppenmatrix des Grades f, so bilden die Koeffizienten $x_{\alpha\beta}$ nach Satz V ein System von f^2 linear unabhängigen Funktionen der Variabeln x_R. Bedeuten nun die f^2 Größen $u_{\alpha\beta}$ beliebige Variable, so kann man die x_R als lineare homogene Funktionen der $u_{\alpha\beta}$ derart bestimmen, daß $x_{\alpha\beta} = u_{\alpha\beta}$ wird. Wäre nun die Determinante $|x_{\alpha\beta}|$ eine zerlegbare Funktion der Variabeln x_R, so würde sich ergeben, daß die Determinante $|u_{\alpha\beta}|$ als ein Produkt von zwei ganzen rationalen Funktionen der $u_{\alpha\beta}$ darstellbar ist, was bekanntlich nicht der Fall ist. — Sind ferner $X = (x_{\alpha\beta})$ und $X' = (x'_{\varkappa\lambda})$ zwei nicht äquivalente irreduzible Gruppenmatrizen, so kann man die Variabeln x_R so wählen, daß die Koeffizienten $x_{\alpha\beta}$ und $x'_{\varkappa\lambda}$ beliebig vorgeschriebene Werte annehmen. Daher können die Determinanten von X und X' nicht einander gleich sein.

Die Relationen (I.) und (II.) lassen sich noch verallgemeinern.

Es seien S und T zwei feste Elemente der Gruppe. Ersetzt man in (I.) die Indizes α und γ durch ρ und σ, multipliziert dann die Gleichung mit $a^S_{\alpha\rho}\, a^T_{\gamma\sigma}$ und bildet die Summe über ρ und σ, so ergibt sich

$$\text{(III.)}\qquad \sum_R a^{SR^{-1}}_{\alpha\beta}\, a^{TR}_{\gamma\delta} = \frac{h}{f}\, a^S_{\alpha\delta}\, a^T_{\gamma\beta}\,.$$

Diese Formel kann noch anders geschrieben werden. Ersetzt man nämlich links R durch $T^{-1}R^{-1}$, so geht die Gleichung über in

$$\text{(III}'.\text{)}\qquad \sum_R a^{SRT}_{\alpha\beta}\, a^{R^{-1}}_{\gamma\delta} = \frac{h}{f}\, a^S_{\alpha\delta}\, a^T_{\gamma\beta}\,.$$

Speziell ergibt sich aus (III.) für $T = E$ die Relation

$$\text{(IV.)}\qquad \sum_R a^{SR^{-1}}_{\alpha\beta}\, a^{R}_{\gamma\delta} = \frac{h}{f}\, a^S_{\alpha\delta}\, e_{\gamma\beta}\,.$$

In analoger Weise erhält man aus der Gleichung (II.) die Formel

$$\text{(V.)} \qquad \sum_R a_{\alpha\beta}^{SR^{-1}} b_{\varkappa\lambda}^{TR} = 0.$$

Auf eine andere Verallgemeinerung der Relationen (I.) hat mich Hr. Frobenius aufmerksam gemacht:

Es sei

$$X_1 = \sum_R (c_{\alpha\beta}^R)\, x_R$$

eine der Gruppenmatrix X äquivalente Gruppenmatrix, und es möge die Matrix $P = (p_{\alpha\beta})$ der Gleichung $X_1 = P^{-1}XP$ genügen. Setzt man $P^{-1} = (q_{\alpha\beta})$, so ist also

$$c_{\gamma\delta}^R = \sum_{\varrho,\sigma} q_{\gamma\varrho}\, a_{\varrho\sigma}^R\, p_{\sigma\delta}.$$

Es ergibt sich dann

$$\sum_R a_{\alpha\beta}^{R^{-1}} c_{\gamma\delta}^R = \sum_{\varrho,\sigma} q_{\gamma\varrho}\, p_{\sigma\delta} \sum_R a_{\alpha\beta}^{R^{-1}} a_{\varrho\sigma}^R = \frac{h}{f} \sum_{\varrho,\sigma} q_{\gamma\varrho}\, p_{\sigma\delta}\, e_{\alpha\sigma}\, e_{\varrho\beta}.$$

Daher ist

$$\sum_R a_{\alpha\beta}^{R^{-1}} c_{\gamma\delta}^R = \frac{h}{f}\, p_{\alpha\delta}\, q_{\gamma\beta}.$$

Diese Gleichung lehrt uns (was auch aus Satz II leicht hervorgeht), daß die Matrix P durch die Bedingung $X_1 = P^{-1}XP$ bis auf einen konstanten Faktor eindeutig bestimmt ist, und liefert eine explizite Methode zur Berechnung der Koeffizienten $p_{\alpha\beta}$.

Mit Hilfe dieser Formel (oder auch auf Grund des Satzes V) beweist man leicht:

1. Stimmen die hf Koeffizienten $c_{\alpha\alpha}^R$ mit den hf Koeffizienten $a_{\alpha\alpha}^R$ überein, so ist P eine Diagonalmatrix $(p_\alpha e_{\alpha\beta})$ und es ist

$$c_{\alpha\beta}^R = \frac{p_\beta}{p_\alpha}\, a_{\alpha\beta}^R.$$

2. Bestehen für ein festes γ die hf Gleichungen

$$c_{\gamma\delta}^R = a_{\gamma\delta}^R \qquad (\delta = 1, 2, \cdots, f, \quad R = H_0, H_1, \cdots, H_{h-1}),$$

so sind alle Koeffizienten $c_{\alpha\beta}^R$ den Koeffizienten $a_{\alpha\beta}^R$ gleich.

§ 3.

Wir wenden uns nun zur Betrachtung der reduziblen Gruppenmatrizen. An erster Stelle beweisen wir den in dieser Form zuerst von Hrn. Maschke (Math. Ann. Bd. 52, S. 363) aufgestellten Satz:

VII. *Ist die Gruppenmatrix X der Gruppenmatrix*

$$X' = \begin{pmatrix} X_1 & 0 \\ U & X_2 \end{pmatrix}$$

äquivalent, so ist sie auch der Gruppenmatrix

$$X'' = \begin{pmatrix} X_1 & 0 \\ 0 & X_2 \end{pmatrix}$$

äquivalent.

Es braucht offenbar nur bewiesen zu werden, daß sich eine konstante Matrix P von nicht verschwindender Determinante bestimmen läßt, so daß

$$X'' = P^{-1} X' P$$

wird. Es seien nun r und s die Grade der Gruppenmatrizen X_1 und X_2, und es werde, falls $X' = \sum_R (R)\, x_R$ ist,

$$(R) = \begin{pmatrix} A_R & 0 \\ C_R & D_R \end{pmatrix}$$

gesetzt, wo A_R und D_R quadratische Matrizen der Grade r und s, dagegen C_R eine rechteckige Matrix mit s Zeilen und r Spalten bedeutet. Es liegt nahe, die Substitution P in der Form

$$P = \begin{pmatrix} E_r & 0 \\ F & E_s \end{pmatrix}$$

anzusetzen und zu fragen, ob sich die s Zeilen und r Spalten enthaltende Matrix F so bestimmen läßt, daß für jedes R

$$P^{-1}(R)P = \begin{pmatrix} A_R & 0 \\ 0 & D_R \end{pmatrix}$$

wird. Als notwendige und hinreichende Bedingung hierfür erhält man leicht

(5.) $$C_R + D_R F = F A_R .$$

Daß sich nun in der Tat eine Matrix F angeben läßt, die dieser Bedingung genügt, läßt sich folgendermaßen einsehen.

Da für je zwei Elemente R und S der Gruppe die Gleichung $(R)(S) = (RS)$ gilt, so bestehen die Relationen

(6.) $$A_R A_S = A_{RS}\,, \qquad D_R D_S = D_{RS}\,,$$

(7.) $$C_R A_S + D_R C_S = C_{RS}\,.$$

Man multipliziere nun die Gleichung (7.) rechts mit $A_{S^{-1}}$ und bilde die Summe über $S = H_0, H_1, \cdots, H_{h-1}$. Setzt man dann noch

$$\frac{1}{h} \sum_S C_S A_{S^{-1}} = F'\,,$$

so ergibt sich wegen $A_S A_{S^{-1}} = A_E$

$$h\, C_R A_E + D_R \cdot h F' = \sum_S C_{RS} A_{S^{-1}}\,.$$

Ersetzt man nun in der rechts stehenden Summe S durch $R^{-1}S$, also S^{-1} durch $S^{-1}R$, so ergibt sich wegen (6.)

$$\begin{aligned}\sum_S C_{RS} A_{S^{-1}} &= \sum_S C_S A_{S^{-1}R} \\ &= \sum_S C_S A_{S^{-1}} A_R \\ &= h F' \cdot A_R.\end{aligned}$$

Daher ist

$$C_R A_E + D_R F' = F' A_R. \tag{8.}$$

Ist nun die Determinante von X_1 nicht 0, so ist $A_E = E_r$; daher genügt $F = F'$ der Gleichung (5.). In jedem Falle wird diese Gleichung durch die Matrix

$$F = F' - D_E C_E$$

befriedigt. Denn setzt man in (7.) $S = E$, so erhält man

$$C_R A_E + D_R C_E = C_R, \tag{9.}$$

und hieraus folgt durch Multiplikation mit A_S, daß $D_R C_E A_S = 0$ ist. Daher ist

$$F A_R = F' A_R - D_E C_E A_R = F' A_R.$$

Andererseits ist

$$C_R + D_R F = C_R - D_R C_E + D_R F'.$$

Die rechte Seite dieser Gleichung ist aber wegen (8.) und (9.) gleich $C_R A_E + D_R F' = F' A_R$. Folglich ist in der Tat, wie zu beweisen ist, $C_R + D_R F = F A_R$.[1]

[1] Hr. Maschke beweist den Satz nur für den allein wichtigen Fall, daß die Determinanten der Substitutionen (R) von 0 verschieden sind, und stützt sich hierbei auf den zuerst von den HH. A. Loewy (Comptes Rendus 1896, S. 168) und E. H. Moore (Math. Ann. Bd. 50, S. 213) bewiesenen Satz, daß sich für jede endliche Gruppe linearer Substitutionen von nicht verschwindenden Determinanten eine positive Hermitesche Form angeben läßt, deren Determinante nicht Null ist, und die bei allen Substitutionen der Gruppe ungeändert bleibt.

Auf den hier angegebenen elementaren Beweis bin ich durch die folgende Mitteilung des Hrn. Frobenius geführt worden, die eine Vereinfachung und Präzisierung des Maschkeschen Beweises enthält:

Es sei $f = \sum_{\alpha,\beta} h_{\alpha\beta} \bar{x}_\alpha x_\beta$ eine positive Hermitesche Form von nicht verschwindender Determinante, die durch alle Substitutionen $Q = (R) = \begin{pmatrix} A_R & 0 \\ C_R & D_R \end{pmatrix}$ der Gruppe in sich transformiert wird. Ist dann $Q' = \begin{pmatrix} A'_R & C'_R \\ 0 & D'_R \end{pmatrix}$ die zu Q konjugierte und konjugiert komplexe Substitution, so besteht, falls H die Matrix $(h_{\alpha\beta})$ bedeutet, die Gleichung $Q'HQ = H$. Schreibt man nun entsprechend H in der Form $\begin{pmatrix} J & K \\ L & M \end{pmatrix}$, so erhält man die Gleichungen

$$D'_R L A_R + D'_R M C_R = L \quad , \qquad D'_R M D_R = M.$$

Nun ist — und dies ist der springende Punkt des Beweises — in einer positiven Hermiteschen Form von nicht verschwindender Determinante jede Hauptunterdeter-

Ehe ich in der Untersuchung weitergehe, will ich als eine einfache Folgerung des Satzes VII einen bekannten Satz über Matrizen ableiten.

Es sei J eine Matrix des Grades n, die der Gleichung $J^2 = J$ genügt. Ist nun r der Rang von J, so lassen sich zwei Substitutionen A und B von nicht verschwindenden Determinanten angeben, so daß $AJB = K$ die Form

$$K = \begin{pmatrix} E_r & 0 \\ 0 & 0 \end{pmatrix}$$

annimmt. Setzt man dann $B^{-1}JB = J'$, so wird $KJ' = K$. Hieraus folgt leicht, daß J' die Form

$$\begin{pmatrix} E_r & 0 \\ C & D \end{pmatrix}$$

besitzt. Hierbei muß, da r auch der Rang von J' ist, $D = 0$ sein. Nun läßt sich aber J als eine Darstellung der allein aus dem Hauptelement E bestehenden endlichen Gruppe auffassen. Daher muß J nach Satz VII auch der Matrix

$$(10.) \qquad J'' = \begin{pmatrix} E_r & 0 \\ 0 & 0 \end{pmatrix}$$

ähnlich sein.

Da ferner auch die Spuren von J und J'' übereinstimmen, die Spur von J'' aber gleich r ist, so ergibt sich zugleich, daß für jede Matrix, die der Gleichung $J^2 = J$ genügt, Spur und Rangzahl einander gleich sind.

Es sei nun $X = \sum_R (R)x_R$ eine zu der Gruppe $\mathfrak{H}$ gehörende Gruppenmatrix von verschwindender Determinante. Es werde $(E) = J$ gesetzt. Ist nun r der Rang von J, so läßt sich, wie wir gesehen haben, eine Matrix P bestimmen, so daß $J'' = P^{-1}JP$ die Form (10.) annimmt. Setzt man nun $P^{-1}XP = X''$, so ergibt sich aus der Gleichung $J''X''J'' = X''$, daß X'' die Form

$$\begin{pmatrix} X_1 & 0 \\ 0 & 0 \end{pmatrix}$$

besitzt, wo X_1 eine Gruppenmatrix des Grades r ist, deren Determinante, weil X_1 für $x_E = 1$, $x_R = 0\ (R \neq E)$ gleich E_r wird, von Null verschieden ist. Hieraus folgt:

---

minante von Null verschieden, und mithin ist die Determinante von M nicht Null. Durch Elimination von D'_R aus den obigen Gleichungen ergibt sich

$$C_R + M^{-1}LA_R = D_R M^{-1}L.$$

Daher genügt die Matrix $F = -M^{-1}L$ der Gleichung (5.) des Textes.

VIII. *Ist X eine Gruppenmatrix vom Range r, so ist X einer Gruppenmatrix der Form*

$$\begin{pmatrix} X_1 & 0 \\ 0 & 0 \end{pmatrix}$$

äquivalent, wo X_1 eine Gruppenmatrix des Grades r von nicht verschwindender Determinante ist.

Man zeigt auch leicht, daß, wenn X noch einer zweiten Gruppenmatrix $\begin{pmatrix} X_1' & 0 \\ 0 & 0 \end{pmatrix}$ äquivalent ist, wo X_1' ebenfalls von nicht verschwindender Determinante ist, X_1 und X_1' äquivalent sein müssen.

Allgemeiner ergibt sich aus dem Satze VII leicht:

IX. *Jede Gruppenmatrix X des Grades n und des Ranges r ist einer Gruppenmatrix äquivalent, welche die Form*

$$(11.)\qquad \begin{pmatrix} X_1 & 0 & \cdots & 0 & 0 \\ 0 & X_2 & \cdots & 0 & 0 \\ . & . & \cdots & . & . \\ 0 & 0 & \cdots & X_m & 0 \\ 0 & 0 & \cdots & 0 & N_{n-r} \end{pmatrix}$$

hat, wobei $X_1, X_2, \cdots, X_m$ irreduzible Gruppenmatrizen bedeuten, und N_{n-r} die Nullmatrix des Grades $n-r$ ist.

Ist insbesondere $\mathfrak{H}$ eine Abelsche Gruppe, so ist jede zu $\mathfrak{H}$ gehörende irreduzible Gruppenmatrix vom Grade 1. Aus IX folgt daher speziell, daß jede Darstellung einer Abelschen Gruppe der Ordnung h durch lineare Substitutionen von nicht verschwindenden Determinanten einer Darstellung äquivalent ist, deren Substitutionen in der Hauptdiagonale h^{te} Einheitswurzeln, sonst überall Nullen enthalten. Für den weiteren speziellen Fall der zyklischen Gruppe ergibt sich hieraus der bekannte Satz, daß jede *periodische* Substitution A, d. h. jede Substitution des Grades n, für die eine Gleichung der Form $A^h = E_n$ besteht, einer anderen ähnlich ist, unter deren Koeffizienten die in der Hauptdiagonale stehenden h^{te} Einheitswurzeln, die übrigen aber gleich Null sind. Hieraus folgt auch, daß die Spur jeder periodischen Substitution als eine Summe von Einheitswurzeln darstellbar ist.

Ist wieder $\mathfrak{H}$ eine beliebige endliche Gruppe und $X = \sum_S (S) x_S$ eine zu $\mathfrak{H}$ gehörende Gruppenmatrix des Ranges r, so ergibt sich aus dem Satz VIII. leicht, daß, wenn s die Ordnung des Elementes S bedeutet, die Spur der Substitution (S) einer Summe von r Einheitswurzeln des Grades s gleich ist.

Ferner gilt der Satz:

X. *Ist eine Gruppenmatrix X, deren Determinante nicht verschwindet, zwei Gruppenmatrizen*

$$\begin{pmatrix} X_1 & 0 & \cdots & 0 \\ 0 & X_2 & \cdots & 0 \\ . & . & \cdots & . \\ 0 & 0 & \cdots & X_m \end{pmatrix}, \quad \begin{pmatrix} X'_1 & 0 & \cdots & 0 \\ 0 & X'_2 & \cdots & 0 \\ . & . & \cdots & . \\ 0 & 0 & \cdots & X'_{m'} \end{pmatrix}$$

äquivalent, wo die X_λ und X'_μ irreduzibel sind, so muß $m = m'$ sein und die Gruppenmatrizen $X_1, X_2, \cdots, X_m$ müssen, abgesehen von der Reihenfolge, der Gruppenmatrizen $X'_1, X'_2, \cdots, X'_m$ äquivalent sein.[1]

Es seien nämlich Φ, Φ_λ und Φ'_μ die Determinanten von X, X_λ und X'_μ. Dann sind nach Satz VI die Funktionen Φ_λ und Φ'_μ irreduzibel, ferner ist

$$\Phi = \Phi_1 \Phi_2 \cdots \Phi_m = \Phi'_1 \Phi'_2 \cdots \Phi'_{m'}.$$

Da nun bekanntlich eine Funktion Φ nur auf eine Weise in Primfaktoren zerlegbar ist, so muß $m = m'$ sein, ferner müssen die Funktionen $\Phi_1, \Phi_2, \cdots, \Phi_m$, abgesehen von konstanten Faktoren, in einer gewissen Reihenfolge mit den Funktionen $\Phi'_1, \Phi'_2, \cdots, \Phi'_m$ übereinstimmen. Beachtet man jedoch, daß die Determinante einer Gruppenmatrix, sofern sie nicht verschwindet, für das Wertsystem $x_E = 1$, $x_R = 0\,(R \neq E)$ gleich 1 wird, so ergibt sich, daß die Funktionen Φ_λ, abgesehen von der Reihenfolge, den Funktionen Φ'_μ direkt gleich sein müssen. Nach Satz VI folgt hieraus aber, daß die Gruppenmatrizen X_λ, abgesehen von der Reihenfolge, den Gruppenmatrizen X'_μ äquivalent sind.

Betrachtet man nun zwei äquivalente irreduzible Gruppenmatrizen als nicht wesentlich voneinander verschieden, so folgt aus IX und X, daß jeder Gruppenmatrix X ein wohlbestimmtes System von irreduziblen Gruppenmatrizen $X_1, X_2, \cdots, X_m$ entspricht, so daß, falls n den Grad und r den Rang von X bedeutet, X der Gruppenmatrix (11.) äquivalent ist. Die Gruppenmatrizen $X_1, X_2, \cdots X_m$ mögen nun als die *irreduziblen Bestandteile* von X bezeichnet werden.

Die irreduziblen Bestandteile einer Gruppenmatrix lassen sich ferner stets so wählen, daß je zwei äquivalente unter ihnen einander gleich werden. Kommt dann X_λ genau r_λ mal vor, so wollen wir r_λ den *Index* von X_λ nennen.[2] Es gehört daher zu jeder Gruppenmatrix X ein gewisses System von l irreduziblen Gruppenmatrizen, von denen nicht zwei äquivalent sind, und ein gewisses System von l Indizes $r_1, r_2, \cdots, r_l$, die wir kurz die Indizes der Gruppenmatrix X nennen.

---

[1] Dieser Satz ist als spezieller Fall in einem allgemeinen von Hrn. A. Loewy (Transactions of the Amer. Mathematical Society, Bd. 4, S. 44) bewiesenen Satze enthalten.

[2] Bedeuten Φ und Φ_λ die Determinanten von X und X_λ und ist Φ nicht gleich 0, so gibt r_λ den Exponenten der höchsten Potenz der irreduziblen Funktion Φ_λ an, die in Φ aufgeht.

Es gilt dann der Satz:

XI. *Es sei* $X = (x_{\alpha\beta})$ *eine Gruppenmatrix von nicht verschwindender Determinante, es seien* $r_1, r_2, \cdots, r_l$ *die Indizes von* X *und* $f_1, f_2, \cdots, f_l$ *die Grade der zugehörigen irreduziblen Gruppenmatrizen. Dann ist* $f_1^2 + f_2^2 + \cdots + f_l^2$ *gleich der Anzahl* g *der linear unabhängigen unter den Funktionen* $x_{\alpha\beta}$ *der* h *Variabeln* x_R. *Ferner ist* $r_1^2 + r_2^2 + \cdots + r_l^2$ *gleich der Anzahl* v *der linear unabhängigen unter den mit* X *vertauschbaren konstanten Matrizen* P. *Endlich ist* l *gleich der Anzahl* v' *der linear unabhängigen unter den konstanten Matrizen, die mit* X *und zugleich mit allen eben charakterisierten Matrizen* P *vertauschbar sind.*

Es seien nämlich $X_1, X_2, \cdots, X_m$ die irreduziblen Bestandteile von X, so daß X der Gruppenmatrix

$$X' = \begin{pmatrix} X_1 & 0 & \cdots & 0 \\ 0 & X_2 & \cdots & 0 \\ \cdot & \cdot & \cdots & \cdot \\ 0 & 0 & \cdots & X_m \end{pmatrix}$$

äquivalent ist. Dann sind unter den Koeffizienten von X' nach Satz V genau $f_1^2 + f_2^2 + \cdots + f_l^2$ linear unabhängig. Daher ist in der Tat $g = f_1^2 + f_2^2 + \cdots + f_l^2$.

Um noch die beiden letzten Behauptungen unseres Satzes zu beweisen, nehmen wir, was ohne Beschränkung der Allgemeinheit geschehen darf, an, die Matrizen $X_1, X_2, \cdots, X_m$ seien so gewählt, daß die r_1 ersten, die r_2 folgenden usw. einander gleich werden. Es mögen dann die Grade von $X_1, X_2, \cdots, X_m$ fortlaufend mit $s_1, s_2, \cdots, s_m$ bezeichnet werden.

Die Zahlen v und v' bleiben offenbar ungeändert, wenn man X durch X' ersetzt. — Es sei nun P eine mit X' vertauschbare konstante Matrix. Wir können dann P in der Form

$$P = \begin{pmatrix} P_{11} & P_{12} & \cdots & P_{1m} \\ P_{21} & P_{22} & \cdots & P_{2m} \\ \cdot & \cdot & \cdots & \cdot \\ P_{m1} & P_{m2} & \cdots & P_{mm} \end{pmatrix}$$

schreiben, wo $P_{\alpha\beta}$ eine Matrix mit s_α Zeilen und s_β Spalten bedeutet. Aus $X'P = PX'$ folgt dann

$$X_\alpha P_{\alpha\beta} = P_{\alpha\beta} X_\beta .$$

Sind nun X_α und X_β nicht einander gleich, also auch nicht äquivalent, so ist $P_{\alpha\beta} = 0$; ist ferner $X_\alpha = X_\beta$, so muß $P_{\alpha\beta}$ die Form aF_α besitzen, wo $F_\alpha = E_{s_\alpha}$ die Einheitsmatrix des Grades $s_\alpha = s_\beta$ bedeutet. Daher hat P die Form

$$P = \begin{pmatrix} p_{11}F_1 \cdots p_{1r_1}F_1 & 0 & \cdots & 0 & \cdots \\ \cdot \quad \cdots \quad \cdot & \cdot & \cdots & \cdot & \cdots \\ p_{r_1 1}F_1 \cdots p_{r_1 r_1}F_1 & 0 & \cdots & 0 & \cdots \\ 0 \quad \cdots \quad 0 & q_{11}F_{r_1+1} & \cdots & q_{1r_2}F_{r_1+1} & \cdots \\ \cdot \quad \cdots \quad \cdot & \cdot & \cdots & \cdot & \cdots \\ 0 \quad \cdots \quad 0 & q_{r_2 1}F_{r_1+1} & \cdots & q_{r_2 r_2}F_{r_1+1} & \cdots \\ \cdot \quad \cdots \quad \cdot & \cdot & \cdots & \cdot & \cdots \end{pmatrix}.$$

Umgekehrt ist jede Matrix P von dieser Form mit X vertauschbar. Daher ist v gleich der Anzahl der in P willkürlich bleibenden Koeffizienten $p_{\alpha\beta}, q_{\gamma\delta}, \cdots$, d. h. es ist in der Tat $v = r_1^2 + r_2^2 + \cdots + r_l^2$.

Soll ferner die konstante Matrix Q mit X und zugleich auch mit allen P vertauschbar sein, so muß zunächst Q dieselbe Form haben wie P. Es möge etwa Q aus P dadurch hervorgehen, daß für $p_{\alpha\beta}$, $q_{\gamma\delta}, \cdots$ die Größen $p'_{\alpha\beta}, q'_{\gamma\delta}, \cdots$ gesetzt werden. Soll nun Q noch mit allen P vertauschbar sein, so muß, wie man leicht sieht, die Matrix $(p'_{\alpha\beta})$ mit allen Matrizen $(p_{\alpha\beta})$, ebenso $(q'_{\gamma\delta})$ mit allen Matrizen $(q_{\gamma\delta})$ vertauschbar sein, usw. Hieraus folgt aber, daß die Matrizen $(p'_{\alpha\beta})$, $(q'_{\gamma\delta}), \cdots$ sich von den Einheitsmatrizen der Grade $r_1, r_2, \cdots$ nur um konstante Faktoren unterscheiden. Diese Bedingung ist auch offenbar hinreichend dafür, daß Q mit X und zugleich mit allen P vertauschbar sei. Die Anzahl v der linear unabhängigen unter den Q ist aber gleich der Anzahl der in Q willkürlich bleibenden Koeffizienten. Da diese Anzahl gleich l ist, so ist, wie zu beweisen ist, $v' = l$.

§ 4.

Aus dem Satz V geht bereits leicht hervor, daß die Anzahl der zu der Gruppe $\mathfrak{H}$ gehörenden, einander nicht äquivalenten irreduziblen Gruppenmatrizen endlich und zwar höchstens gleich h ist.

Um nun die sämtlichen verschiedenen irreduziblen Gruppenmatrizen zu charakterisieren und ihre genaue Anzahl zu bestimmen, betrachten wir die spezielle Gruppenmatrix h^{ten} Grades von nicht verschwindender Determinante

$$X = (x_{PQ^{-1}}),$$

deren Zeilen und Spalten man erhält, indem man für P und Q der Reihe nach die Elemente $H_0, H_1, \cdots, H_{h-1}$ der Gruppe setzt. Diese Gruppenmatrix entspricht der bekannten Darstellung von $\mathfrak{H}$ als Gruppe regulärer Permutationen und möge daher als die *reguläre* Gruppenmatrix bezeichnet werden.

Es seien nun unter den irreduziblen Bestandteilen von X im ganzen k einander nicht äquivalent, etwa die Gruppenmatrizen $X_0, X_1, \cdots, X_{k-1}$. Es sei ferner $f_\varkappa$ der Grad von $X_\varkappa$ und $e_\varkappa$ der zugehörige Index, so daß also, wenn die Determinanten von $X, X_0, \cdots, X_{k-1}$ mit Θ, $\Phi_0, \cdots, \Phi_{k-1}$ bezeichnet werden,

$$\Theta = \Phi_0^{e_0} \Phi_1^{e_1} \cdots \Phi_{k-1}^{e_{k-1}}$$

wird. Es gilt dann der Satz:

XII. *Die Zahlen $f_\varkappa$ und $e_\varkappa$ sind einander gleich. Ferner ist die Zahl k gleich der Anzahl der Klassen konjugierter Elemente der Gruppe.*

Der Beweis ergibt sich auf Grund des Satzes XI sehr einfach. Zunächst ist offenbar

$$(12.) \qquad e_0 f_0 + e_1 f_1 + \cdots + e_{k-1} f_{k-1} = h.$$

Da ferner unter den Koeffizienten der Matrix X genau h linear unabhängig sind, so ist

$$(13.) \qquad f_0^2 + f_1^2 + \cdots + f_{k-1}^2 = h.$$

Wir betrachten nun die mit X vertauschbaren Matrizen

$$Y = (y_{P,Q}) \qquad (P, Q = H_0, H_1, \cdots, H_{h-1}),$$

deren Koeffizienten von der x_R unabhängig sind. Man zeigt leicht, daß Y dann und nur dann mit X vertauschbar ist, wenn

$$y_{P,Q} = y_{RP,RQ} \qquad (P, Q, R = H_0, H_1, \cdots, H_{h-1})$$

ist. Setzt man nun

$$y_{R,E} = y_R,$$

so wird $y_{P,Q} = y_{Q^{-1}P}$, also

$$Y = (y_{Q^{-1}P}).$$

Die Anzahl der in Y willkürlich bleibenden Koeffizienten, die uns die Anzahl der linear unabhängigen unter den Y angibt, ist nun gleich h. Folglich ist nach Satz XI

$$(14.) \qquad e_0^2 + e_1^2 + \cdots + e_{k-1}^2 = h.$$

Aus den Gleichungen (12.), (13.) und (14.) folgt aber

$$(e_0 - f_0)^2 + (e_1 - f_1)^2 + \cdots + (e_{k-1} - f_{k-1})^2 = 0,$$

folglich ist in der Tat $e_\varkappa = f_\varkappa$.

Um nun die Anzahl k genauer zu bestimmen, haben wir noch die konstanten Matrizen

$$Z = (z_{P,Q})$$

zu betrachten, die mit X und zugleich mit allen Y vertauschbar sind. Es ergibt sich, daß

$$z_{P,Q} = z_{RP,RQ} = z_{PR,QR} \qquad (P, Q, R = H_0, H_1, \cdots, H_{h-1})$$

sein muß. Setzt man $z_{R,E} = z_R$, so wird

$$z_{P,Q} = z_{Q^{-1}P} = z_{PQ^{-1}}.$$

Ersetzt man P durch PQ, so erhält man

$$(15.) \qquad z_P = z_{Q^{-1}PQ}.$$

Umgekehrt ist jede Matrix $Z = (z_{Q^{-1}P})$, deren Koeffizienten $z_{H_0}, z_{H_1}, \cdots, z_{H_{k-1}}$ den Bedingungen (15.) genügen, mit X und auch mit Y vertauschbar. Hierbei wird also z_P dann und nur dann gleich z_R, wenn R auf die Form $Q^{-1}PQ$ gebracht werden kann, d. h. wenn P und R konjugierte Elemente sind. Daher ist die Anzahl der in Z willkürlichen bleibenden Koeffizienten, die nach Satz XI mit k übereinstimmen muß, gleich der Anzahl der Klassen konjugierter Elemente, in die die Elemente der Gruppe zerfallen.

Aus der Gleichung (13.) ergibt sich leicht der wichtige Satz:

XIII. *Die Anzahl der zu der Gruppe $\mathfrak{H}$ gehörenden, einander nicht äquivalenten irreduziblen Gruppenmatrizen (Darstellungen) ist genau gleich der Anzahl k der Klassen konjugierter Elemente der Gruppe.*

Zunächst folgt aus dem Vorhergehenden, daß mindestens k nicht äquivalente irreduzible Gruppenmatrizen existieren, nämlich die k irreduziblen Bestandteile $X_0, X_1, \cdots, X_{k-1}$ der regulären Gruppenmatrix. Es muß aber jede andere irreduzible Gruppenmatrix X' einer dieser k Gruppenmatrizen äquivalent sein. Denn wäre dies nicht der Fall, so müßten, falls f' den Grad von X' bedeutet, die

$$f_0^2 + f_1^2 + \cdots + f_{k-1}^2 + f'^2 = h + f'^2$$

Koeffizienten der $k+1$ irreduziblen Gruppenmatrizen $X_0, X_1, \cdots, X_{k-1}$, X' untereinander linear unabhängig sein. Dies ist jedoch nicht möglich, da nicht mehr als h linear unabhängige lineare homogene Funktionen der h Variabeln x_R existieren können.

§ 5.

Bilden die Substitutionen $(H_0), (H_1), \cdots, (H_{k-1})$ eine Darstellung der Gruppe $\mathfrak{H}$, und ist $\chi(R)$ die Spur der Substitution (R), so nennt man das System der h Zahlen $\chi(R)$ den der Darstellung oder der zugehörigen Gruppenmatrix X entsprechenden *Charakter*. Insbesondere bezeichnet man jeden einer irreduziblen Darstellung entsprechenden Charakter als einen *einfachen Charakter* der Gruppe.

Da nun zwei äquivalenten Darstellungen derselbe Charakter entspricht, so ist die Anzahl der einfachen Charaktere der Gruppe gleich der Anzahl k der nicht äquivalenten irreduziblen Gruppenmatrizen X_0, $X_1, \cdots, X_{k-1}$. Die zugehörigen Charaktere sollen mit

$$\chi^{(0)}(R), \chi^{(1)}(R), \cdots, \chi^{(k-1)}(R)$$

bezeichnet werden.

Die Zahl $\chi(E) = r$ gibt den Rang der Matrix (E) und also auch den Rang der Gruppenmatrix X an. Diese Zahl soll der *Grad* des Charakters $\chi(R)$ genannt werden. Insbesondere ist der Grad

$$\chi^{(\varrho)}(E) = f_\varrho$$

des einfachen Charakters $\chi^{(\varrho)}(R)$ gleich dem Grade der zugehörigen Gruppenmatrix X_ϱ.

Allgemeiner ist $\chi(R)$ eine Summe von r Einheitswurzeln. Ist

$$\chi(R) = \rho_1 + \rho_2 + \cdots + \rho_{r-1},$$

so ist

$$\chi(R^{-1}) = \rho_1^{-1} + \rho_2^{-1} + \cdots + \rho_{r-1}^{-1};$$

daher sind $\chi(R)$ und $\chi(R^{-1})$ konjugiert komplexe Größen. Ersetzt man ferner in den Matrizen (R) jeden Koeffizienten durch den konjugiert komplexen Wert, so bilden auch die so entstehenden Matrizen eine Darstellung der Gruppe. Ist daher $\chi(R)$ ein Charakter, so bilden auch die Zahlen $\chi'(R) = \chi(R^{-1})$ einen solchen. Die Charaktere $\chi(R)$ und $\chi'(R)$ werden *inverse* Charaktere genannt. Man schließt auch leicht, daß, wenn $\chi^{(\varrho)}(R)$ ein einfacher Charakter ist, der inverse Charakter $\chi^{(\varrho)}(R^{-1}) = \chi^{(\varrho')}(R)$ ebenfalls ein einfacher Charakter ist.

Da ferner für je zwei Matrizen (R) und (S) die beiden Produkte $(R)(S)$ und $(S)(R)$ dieselben Spuren besitzen, so ist für je zwei Elemente R und S der Gruppe

$$\text{(VI.)} \qquad \chi(RS) = \chi(SR).$$

Ersetzt man hierin R durch $S^{-1}R$, so ergibt sich

$$\text{(VI'.)} \qquad \chi(R) = \chi(S^{-1}RS).$$

Gehören daher zwei Elemente P und R derselben Klasse konjugierter Elemente an, so ist stets $\chi(P) = \chi(R)$.

Ist unter den irreduziblen Bestandteilen der Matrix X die irreduzible Gruppenmatrix X_ϱ genau $r_\varrho \geqq 0$ mal enthalten, so ist offenbar

$$(16.) \qquad \chi(R) = r_0\chi^{(0)}(R) + r_1\chi^{(1)}(R) + \cdots + r_{k-1}\chi^{(k-1)}(R).$$

Bedeutet X speziell die reguläre Gruppenmatrix, so wird, da die Spur von $X = (x_{PQ^{-1}})$ gleich hx_E ist,

$$\chi(R) = h\varepsilon_R,$$

wo ε_R gleich 1 oder gleich 0 zu setzen ist, je nachdem R gleich E oder von E verschieden ist. Da ferner für $X = (x_{PQ^{-1}})$ die Zahl $r_\varkappa$ gleich $f_\varkappa$ ist, so folgt aus (16.)

(VII.) $$\sum_{\varrho=0}^{k-1} f_\varrho \chi^{(\varrho)}(R) = h\varepsilon_R .$$

Es seien ferner ρ und σ zwei verschiedene Zahlen der Reihe $0, 1, \cdots, k-1$. Setzt man

$$X_\varrho = \sum_R (a^R_{\alpha\beta}) x_R , \qquad X_\sigma = \sum_R (b^R_{\varkappa\lambda}) x_R , \qquad \begin{pmatrix} \alpha, \beta = 1, 2, \cdots, f_\varrho \\ \varkappa, \lambda = 1, 2, \cdots, f_\sigma \end{pmatrix}$$

so wird

$$\chi^{(\varrho)}(R) = \sum_\alpha a^R_{\alpha\alpha} , \qquad \chi^{(\sigma)}(R) = \sum_\varkappa b^R_{\varkappa\varkappa} .$$

Aus der Gleichung (vgl. Formel (IV.))

$$\sum_R a^{SR^{-1}}_{\alpha\alpha} a^R_{\beta\beta} = \frac{h}{f_\varrho} a^S_{\alpha\beta} e_{\beta\alpha}$$

ergibt sich dann, indem man die Summe über α und β bildet,

(VIII.) $$\sum_R \chi^{(\varrho)}(SR^{-1}) \chi^{(\varrho)}(R) = \frac{h}{f_\varrho} \chi^{(\varrho)}(S) .$$

Speziell erhält man für $S = E$ die Formel

(IX.) $$\sum_R \chi^{(\varrho)}(R^{-1}) \chi^{(\varrho)}(R) = h .$$

Aus der Gleichung (vgl. Formel (III.))

$$\sum_R a^{SR^{-1}}_{\alpha\beta} a^{TR}_{\beta\alpha} = \frac{h}{f_\varrho} a^S_{\alpha\alpha} a^T_{\beta\beta}$$

ergibt sich ferner, indem man über α und β summiert und beachtet, daß $\sum_\beta a^P_{\alpha\beta} a^Q_{\beta\alpha} = a^{PQ}_{\alpha\alpha}$ ist,

(X.) $$\chi^{(\varrho)}(S) \chi^{(\varrho)}(T) = \frac{f_\varrho}{h} \sum_R \chi^{(\varrho)}(SR^{-1}TR) .$$

In analoger Weise erhält man aus der Formel (V.) die Gleichung

(XI.) $$\sum_R \chi^{(\varrho)}(SR^{-1}) \chi^{(\sigma)}(R) = 0 \qquad (\rho \neq \sigma);$$

hieraus ergibt sich speziell für $S = E$

(XII.) $$\sum_R \chi^{(\varrho)}(R^{-1}) \chi^{(\sigma)}(R) = 0 .$$

Bildet man in der Gleichung (X.) die Summe über $\rho = 0, 1, \cdots, k-1$, so folgt auf Grund der Formel (VII.)

$$\sum_\varrho \chi^{(\varrho)}(S) \chi^{(\varrho)}(T) = \sum_R \varepsilon_{SR^{-1}TR} .$$

Sind nun S und T^{-1} zwei nicht konjugierte Elemente der Gruppe, so wird die Gleichung $SR^{-1}TR = E$ durch kein Element R befriedigt, daher ist in diesem Fall

$$\text{(XIII.)} \qquad \sum_{\varrho} \chi^{(\varrho)}(S)\chi^{(\varrho)}(T) = 0\,.$$

Gehören dagegen S und T^{-1} einer Klasse konjugierter Elemente an, die aus h_S Elementen besteht, so besitzt die Gleichung $SR^{-1}TR = E$ genau $\frac{h}{h_S}$ Lösungen R. Ferner wird dann $\chi^{(\varrho)}(T) = \chi^{(\varrho)}(S^{-1})$; daher besteht die Gleichung

$$\text{(XIV.)} \qquad \sum_{\varrho} \chi^{(\varrho)}(S)\chi^{(\varrho)}(S^{-1}) = \frac{h}{h_S}\,.$$

Die Relationen (VI.) bis (XIV.) bilden die Grundlage der Theorie der Gruppencharaktere. Die wichtigsten unter ihnen sind die Formeln (VI.), (VII.), (IX.), (X.) und (XII.). Die übrigen lassen sich aus diesen durch eine einfache Rechnung ableiten.

Es sei nun X eine beliebige Gruppenmatrix, für die die Zahlen $\chi(R)$ und r_ϱ dieselbe Bedeutung haben wie früher. Multipliziert man dann die Gleichung (16.) mit $\chi^{(\varrho)}(R^{-1})$ und bildet die Summe über $R = H_0, H_1, \cdots, H_{h-1}$, so ergibt sich auf Grund der Relationen (IX.) und (XII.)

$$hr_\varrho = \sum_R \chi(R)\chi^{(\varrho)}(R^{-1})\,.$$

Daher sind die Zahlen r_ϱ und folglich auch die irreduziblen Bestandteile der Gruppenmatrix X allein durch den Charakter $\chi(R)$ bestimmt.

Hieraus ergibt sich der fundamentale Satz:

XIV. *Zwei Darstellungen der Gruppe $\mathfrak{H}$ durch lineare Substitutionen von nicht verschwindenden Determinanten sind dann und nur dann äquivalent, wenn ihnen derselbe Charakter entspricht. Allgemeiner sind zwei Darstellungen dann und nur dann äquivalent, wenn sie denselben Grad und denselben Charakter besitzen.*

Ferner besteht der Satz:

XV. *Der Grad jeder irreduziblen Darstellung der Gruppe $\mathfrak{H}$ ist ein Divisor der Ordnung der Gruppe.*

Dies folgt aus der Gleichung (VIII.), die man auch in der Form

$$\sum_R \chi^{(\varrho)}(R)\left\{\frac{h}{f_\varrho}\varepsilon_{SR^{-1}} - \chi^{(\varrho)}(SR^{-1})\right\} = 0$$

schreiben kann. Da nämlich die h Zahlen $\chi^{(\varrho)}(R)$ nicht sämtlich gleich 0 sind[1], so muß die Determinante h^{ten} Grades

[1] Die Zahl $\chi^{(\varrho)}(E) = f_\varrho$ ist jedenfalls nicht Null.

$$\left| \frac{h}{f_\varrho} \varepsilon_{PQ^{-1}} - \chi^{(\varrho)}(PQ^{-1}) \right| \qquad (P, Q = H_0, H_1, \cdots, H_{h-1})$$

verschwinden. Beachtet man noch, daß die Größen $\chi^{(\varrho)}(R)$ ganze algebraische Zahlen sind, so ergibt sich, daß die Zahl $\frac{h}{f_\varrho}$ einer Gleichung der Form

$$x^h + c_1 x^{h-1} + \cdots + c_h = 0$$

mit ganzen algebraischen Koeffizienten genügt. Folglich ist die Zahl $\frac{h}{f_\varrho}$ eine ganze algebraische und also, da sie rational ist, eine ganze rationale Zahl.

§ 6.

Als eine *charakteristische Einheit* der Gruppe $\mathfrak{H}$ bezeichnet man ein System von h Größen

$$a_{H_0}, a_{H_1}, \cdots, a_{H_{h-1}},$$

für die die h Relationen

$$\sum_R a_{SR^{-1}} a_R = a_S \qquad (R, S = H_0, H_1, \cdots, H_{h-1})$$

bestehen. Ebenso nennt man dann auch die Matrix h^{ten} Grades $A = (a_{PQ^{-1}})$ eine Einheit. Damit die Zahlen a_R eine Einheit bilden, ist notwendig und hinreichend, daß die Gleichung $A^2 = A$ erfüllt sei.

Ist ferner $X = (x_{PQ^{-1}})$, so nennt man die Matrix $\bar{X} = (x_{Q^{-1}P})$ die zu X *antistrophe* Matrix. Setzt man $Y = (y_{PQ^{-1}})$ und $\bar{Y} = (y_{Q^{-1}P})$, so sind X und $\bar{Y}$ vertauschbar, ferner ist, falls $XY = Z$ wird, die zu Z antistrophe Matrix $\bar{Z}$ gleich $\bar{Y}\bar{X}$.

Ist nun $A = (a_{PQ^{-1}})$ eine Einheit, $\bar{A}$ die antistrophe Matrix, so ist auch $\bar{A}^2 = \bar{A}$, ferner besitzt die Matrix $X' = \bar{A}X$ die Eigenschaften einer Gruppenmatrix. Ihre Spur ist gleich

$$\sum_{R,S} a_{S^{-1}R^{-1}S}\, x_R.$$

Daher bilden die Zahlen

$$\varphi(R) = \sum_S a_{S^{-1}RS}$$

einen Charakter des Grades ha_E, den man als den *durch die Einheit* a_R *bestimmten Charakter* bezeichnet. Ist $\varphi(R)$ ein einfacher Charakter, so heißt die Einheit eine *primitive*.

Kennt man eine den einfachen Charakter $\chi(R)$ bestimmende Einheit $A = (a_{PQ^{-1}})$, so kann man auch eine dem Charakter $\chi(R)$ entsprechende irreduzible Gruppenmatrix X_1 konstruieren. Hierzu hat man, wenn b_P die zu a_P konjugiert komplexe Größe bedeutet und $\overline{B} = (b_{Q^{-1}P})$ ist, eine Matrix h^{ten} Grades L zu bestimmen, so daß

$$L^{-1}\overline{B}L = \begin{pmatrix} E_f & 0 \\ 0 & 0 \end{pmatrix} \qquad (f = ha_E = hb_E)$$

wird. Dann hat $L^{-1}\overline{B}XL = (L^{-1}\overline{B}L)(L^{-1}XL) = (L^{-1}XL)(L^{-1}\overline{B}L)$ die Form

$$\begin{pmatrix} X_1 & 0 \\ 0 & 0 \end{pmatrix},$$

wo die Gruppenmatrix X_1 die verlangte Eigenschaft besitzt.

Es soll hier umgekehrt gezeigt werden, wie man die sämtlichen den einfachen Charakter $\chi(R)$ bestimmenden primitiven Einheiten zu berechnen hat, falls es auf irgendeinem Wege gelungen ist, eine dem Charakter $\chi(R)$ entsprechende Darstellung der Gruppe zu bestimmen.

Es seien

$$X_1 = (x_{\alpha\beta}) \quad , \qquad X_1' = (x'_{\kappa\lambda})$$

zwei nicht äquivalente irreduzible Gruppenmatrizen der Grade f und f', ferner $\chi(R)$ und $\chi'(R)$ die ihnen entsprechenden Charaktere. Ist dann

$$x_{\alpha\beta} = \sum_R a^R_{\alpha\beta} x_R \quad , \qquad x'_{\kappa\lambda} = \sum_R b^R_{\kappa\lambda} x^R,$$

so mögen die Matrizen h^{ten} Grades

$$\left(\frac{f}{h} a^{PQ^{-1}}_{\alpha\beta}\right) \quad , \qquad \left(\frac{f'}{h} b^{PQ^{-1}}_{\kappa\lambda}\right)$$

mit $A_{\alpha\beta}$ und $B_{\kappa\lambda}$, ferner die Matrizen

$$\sum_\alpha A_{\alpha\alpha} = \left\{\frac{f}{h}\chi(PQ^{-1})\right\} \quad , \qquad \sum_\kappa B_{\kappa\kappa} = \left\{\frac{f'}{h}\chi'(PQ^{-1})\right\}$$

mit J und J' bezeichnet werden.

Die früher erhaltenen Relationen (III.), (IV.) und (V.) lassen sich dann in den Formeln

$$A_{\alpha\beta}A_{\gamma\delta} = e_{\gamma\beta}A_{\alpha\delta},$$

$$(17.) \qquad A_{\alpha\beta}\overline{A}_{\gamma\delta} = \left(\frac{h}{f} a^P_{\alpha\delta} a^{Q^{-1}}_{\gamma\beta}\right),$$

$$A_{\alpha\beta}B_{\kappa\lambda} = A_{\alpha\beta}\overline{B}_{\kappa\lambda} = 0$$

zusammenfassen; hierbei bedeuten $\overline{A}_{\alpha\beta}$ und $\overline{B}_{\kappa\lambda}$ die zu $A_{\alpha\beta}$ und $B_{\kappa\lambda}$ antistrophen Matrizen. Ebenso sind die Relationen (VIII.) und (XI.) identisch mit den Gleichungen

$$J^2 = J \quad , \qquad JJ' = 0.$$

Die f^2 Matrizen $A_{\alpha\beta}$ sind untereinander linear unabhängig; denn aus

$$\sum_{\alpha,\beta} c_{\alpha\beta} A_{\alpha\beta} = 0$$

folgt wegen (17.), indem man links mit $A_{\gamma\gamma}$, rechts mit $A_{\delta\delta}$ multipliziert, $c_{\gamma\delta}A_{\gamma\delta} = 0$, also $c_{\gamma\delta} = 0$.

Es werde nun

$$\varphi_{\alpha\beta}(R) = \frac{f}{h}\sum_S a_{\alpha\beta}^{S^{-1}RS}$$

gesetzt. Dann ist

$$\varphi_{\alpha\beta}(R) = e_{\alpha\beta}\chi(R).$$

In der Tat ist

$$\begin{aligned}\sum_S a_{\alpha\beta}^{S^{-1}RS} &= \sum_S \sum_{\sigma,\varrho} a_{\alpha\varrho}^{S^{-1}} a_{\varrho\sigma}^{R} a_{\sigma\beta}^{S} \\ &= \sum_{\sigma,\varrho} a_{\varrho\sigma}^{R} \sum_S a_{\alpha\varrho}^{S^{-1}} a_{\sigma\beta}^{S} \\ &= \frac{h}{f}\sum_{\varrho,\sigma} a_{\varrho\sigma}^{R} e_{\alpha\beta} e_{\sigma\varrho} \\ &= \frac{h}{f} e_{\alpha\beta} \sum_{\varrho} a_{\varrho\varrho}^{R}.\end{aligned}$$

Die rechte Seite ist aber, wie zu beweisen ist, gleich $\frac{h}{f} e_{\alpha\beta}\chi(R)$.

Die Gleichung $A_{\alpha\alpha}^2 = A_{\alpha\alpha}$ lehrt uns insbesondere, daß die h Größen $\frac{f}{h} a_{\alpha\alpha}^R$ eine Einheit bilden. Da der durch sie bestimmte Charakter gleich $\varphi_{\alpha\alpha}(R) = \chi(R)$ ist, so ergibt sich:

XVI. *Bilden die h Substitutionen f^{ten} Grades $(a_{\alpha\beta}^R)$ eine irreduzible Darstellung der Gruppe, die dem Charakter $\chi(R)$ entspricht, so repräsentieren für jedes α die h Größen $\frac{f}{h} a_{\alpha\alpha}^R$ eine den Charakter $\chi(R)$ bestimmende charakteristische Einheit.*

Ferner gilt der Satz:

XVII. *Ist $A = (a_{PQ^{-1}})$ eine den Charakter $\chi(R)$ bestimmende (primitive) Einheit, so lassen sich (auf eine und nur eine Weise) f^2 Größen $l_{\alpha\beta}$ berechnen, so daß*

$$a_R = \frac{f}{h}\sum_{\alpha,\beta} l_{\alpha\beta} a_{\alpha\beta}^R$$

wird. Die Matrix $\Lambda = (l_{\alpha\beta})$ genügt dann der Gleichung $\Lambda^2 = \Lambda$ und ist vom Range 1.

Ist nämlich wie früher $X = (x_{PQ^{-1}})$ die reguläre Gruppenmatrix, so besitzen $\bar{A}X$ und $\bar{A}_{11}X$ dieselbe Spur $\sum_R \chi(R^{-1})x_R$. Daher muß sich eine Matrix h^{ten} Grades K bestimmen lassen, so daß

$$\bar{A}X = K^{-1}\bar{A}_{11}XK$$

wird. Speziell ergibt sich für $X = E_h$ die Gleichung $\bar{A} = K^{-1}\bar{A}_{11}K$. Setzt man nun $X = J$, so erhält man, da, wie leicht zu beweisen, $\bar{A}_{11}J = \bar{A}_{11}$ ist,

$$J\bar{A} = \bar{A}J = K^{-1}\bar{A}_{11}K = \bar{A}.$$

Daher ist

$$(18.)\qquad \sum_S \chi(RS^{-1})a_S = \frac{h}{f}a_R.$$

Man setze nun

$$\sum_S a_S a_{\beta\alpha}^{S^{-1}} = l_{\alpha\beta}.$$

Dann wird

$$\sum_{\alpha,\beta} l_{\alpha\beta} a_{\alpha\beta}^R = \sum_S a_S \sum_{\alpha,\beta} a_{\beta\alpha}^{S^{-1}} a_{\alpha\beta}^R.$$

Nun ist aber

$$\sum_{\alpha,\beta} a_{\alpha\beta}^R a_{\beta\alpha}^{S^{-1}} = \sum_\alpha a_{\alpha\alpha}^{RS^{-1}} = \chi(RS^{-1}).$$

Daher ist wegen (18.)

$$\sum_{\alpha,\beta} l_{\alpha\beta} a_{\alpha\beta}^R = \frac{h}{f}a_R,$$

also

$$a_R = \frac{f}{h}\sum_{\alpha,\beta} l_{\alpha\beta} a_{\alpha\beta}^R.$$

Diese Gleichung kann auch in der Form

$$A = \sum_{\alpha,\beta} l_{\alpha\beta} A_{\alpha\beta}$$

geschrieben werden. Es ergibt sich nun

$$A^2 = \sum_{\alpha,\beta,\gamma,\delta} l_{\alpha\beta} l_{\gamma\delta} A_{\alpha\beta} A_{\gamma\delta} = \sum_{\alpha,\beta,\delta} l_{\alpha\beta} l_{\beta\delta} A_{\alpha\delta}.$$

Da andererseits

$$A^2 = A = \sum_{\alpha,\delta} l_{\alpha\delta} A_{\alpha\delta}$$

ist, die Matrizen $A_{\alpha\beta}$ ferner linear unabhängig sind, so ergibt sich

$$\sum_{\alpha,\delta} l_{\alpha\beta} l_{\beta\delta} = l_{\alpha\delta},$$

also in der Tat $\Lambda^2 = \Lambda$. Außerdem wird

$$\sum_S a_{S^{-1}RS} = \chi(R) = \sum_{\alpha,\beta} l_{\alpha\beta}\varphi_{\alpha\beta}(R) = \chi(R)\sum_\alpha l_{\alpha\alpha}.$$

Daher ist die Spur $\sum_\alpha l_{\alpha\alpha}$ oder, was hier dasselbe ist, der Rang der Matrix Λ gleich 1.

Aus der eben durchgeführten Rechnung ergibt sich zugleich, daß eine Matrix $A = \sum_{\alpha,\beta} l_{\alpha\beta} A_{\alpha\beta}$ dann und nur dann eine Einheit bildet, wenn $\Lambda = (l_{\alpha\beta})$ der Gleichung $\Lambda^2 = \Lambda$ genügt, und daß der alsdann durch A bestimmte Charakter gleich $\chi(R)\sum_\alpha l_{\alpha\alpha}$ ist. Soll A eine primitive Einheit sein, so kommt noch hinzu, daß $\sum_\alpha l_{\alpha\alpha} = 1$, also Λ vom Range 1 sein muß.

Es ist noch folgendes zu bemerken.

Da die Determinanten $\begin{vmatrix} l_{\alpha\beta} & l_{\alpha\delta} \\ l_{\gamma\beta} & l_{\gamma\delta} \end{vmatrix}$ sämtlich verschwinden, so lassen sich $2n$ Größen k'_α und k_β bestimmen, so daß $l_{\alpha\beta} = k'_\alpha k_\beta$ wird. Die Bedingung $\sum_\alpha l_{\alpha\alpha} = 1$ besagt dann, daß $\sum_\alpha k'_\alpha k_\alpha = 1$ ist. Sind umgekehrt k'_α und k_β irgendwelche $2n$ Größen, zwischen denen die Relation $\sum k'_\alpha k_\alpha = 1$ besteht, und setzt man $l_{\alpha\beta} = k'_\alpha k_\beta$, so genügt die Matrix $\Lambda = (l_{\alpha\beta})$ der Gleichung $\Lambda^2 = \Lambda$ und ist vom Range 1.

Man erhält daher die sämtlichen den Charakter $\chi(R)$ bestimmenden Einheiten $A = (a_{PQ^{-1}})$, indem man auf alle möglichen Arten $2n$ Größen k'_α und k_β bestimmt, die der Gleichung $\sum_\alpha k'_\alpha k_\alpha = 1$ genügen, und

$$A = \sum_{\alpha,\beta} k'_\alpha k_\beta A_{\alpha\beta}$$

setzt.

Schreibt man ferner für k_α das Zeichen $k_{\alpha 1}$, so lassen sich, wie man leicht schließt, $n(n-1)$ Größen

$$k_{\alpha 2},\ k_{\alpha 3},\ \cdots,\ k_{\alpha n}$$

bestimmen, so daß

$$\sum_\beta k'_\beta k_{\beta 2} = 0,\ \cdots,\ \sum_\beta k'_\beta k_{\beta n} = 0$$

wird, und daß außerdem die Determinante $|k_{\alpha\beta}|$ von Null verschieden ist. Setzt man dann $K = (k_{\alpha\beta})$ und $K^{-1} = (k'_{\alpha\beta})$, so wird $k'_{1\beta} = k'_\beta$. Bezeichnet man nun die Matrix $(a^R_{\alpha\beta})$ mit A_R und setzt

$$K^{-1} A_R K = (c^R_{\alpha\beta})$$

so wird

$$c^R_{11} = \sum_{\alpha,\beta} k'_{1\alpha} k_{\beta 1} a^R_{\alpha\beta} = \frac{h}{f} a_R.$$

Hieraus folgt, daß, wenn die Zahlen a_R eine den Charakter $\chi(R)$ bestimmende primitive Einheit bilden, man stets eine dem Charakter $\chi(R)$ entsprechende Darstellung angeben kann, deren Substitutionen in der ersten Zeile der Hauptdiagonale die Größen $\frac{h}{f} a_R$ enthalten.

Es seien allgemeiner

$$A_1 = (a^{(1)}_{PQ^{-1}}),\ A_2 = (a^{(2)}_{PQ^{-1}}),\ \cdots,\ A_f = (a^{(f)}_{PQ^{-1}})$$

irgendwelche f den Charakter $\chi(R)$ bestimmende Einheiten, für die noch die Gleichungen

$$A_\lambda A_\mu = 0 \qquad (\lambda \neq \mu)$$

bestehen. Es sei nun

$$A_\lambda = \sum_{\alpha,\beta} k'_{\lambda\alpha} k_{\beta\lambda} A_{\alpha\beta},$$

so daß $\sum_{\alpha} k'_{\lambda\alpha} k_{\alpha\lambda} = 1$ ist. Dann ergibt sich aus $A_\lambda A_\mu = 0$ leicht

$$\sum_{\alpha} k'_{\lambda\alpha} k_{\alpha\mu} = 0 \qquad (\lambda \neq \mu).$$

Setzt man daher $K = (k_{\alpha\beta})$, so wird $|k_{\alpha\beta}| \neq 0$ und $K^{-1} = (k'_{\alpha\beta})$; ferner erhält man, falls wie früher $K^{-1} A_R K = (c^R_{\alpha\beta})$ ist,

$$c^R_{\alpha\alpha} = \frac{h}{f} a^{(\alpha)}_R .$$

Sind daher $A_1, A_2, \cdots, A_f$ irgendwelche f Einheiten der betrachteten Art, so läßt sich eine dem Charakter $\chi(R)$ entsprechende Darstellung der Gruppe angeben, deren Substitutionen in den Hauptdiagonalen die Größen

$$\frac{h}{f} a^{(1)}_R, \quad \frac{h}{f} a^{(2)}_R, \quad \cdots, \quad \frac{h}{f} a^{(f)}_R$$

enthalten.

Ausgegeben am 13. April.

8.

Über vertauschbare lineare Differentialausdrücke

Sitzungsberichte der Berliner Mathematischen Gesellschaft 4, 2 - 8 (1905)

Bei der Zusammensetzung der linearen homogenen Differentialausdrücke gilt bekanntlich das kommutative Prinzip im allgemeinen nicht, d. h. sind

$$P(y) = p_0 \frac{d^n y}{dx^n} + p_1 \frac{d^{n-1} y}{dx^{n-1}} + \cdots + p_n y,$$

$$Q(y) = q_0 \frac{d^m y}{dx^m} + q_1 \frac{d^{m-1} y}{dx^{m-1}} + \cdots + q_m y$$

zwei solche Ausdrücke, und bezeichnet man mit $PQ(y)$ denjenigen Ausdruck, der entsteht, indem man in $P(y)$ für y den Ausdruck $Q(y)$ einsetzt, so ist nicht stets $PQ(y) = QP(y)$; vielmehr bedingt das Bestehen dieser Gleichung eine Reihe von Beziehungen zwischen den Funktionen p_λ und q_λ von x.

Es entsteht auf diese Weise die Aufgabe, zu entscheiden, mit welchen Differentialausdrücken $Q(y)$ ein gegebener Ausdruck $P(y)$ vertauschbar ist.[1])

Im folgenden sollen einige Beiträge zu dieser Frage geliefert werden. Insbesondere wird der Satz bewiesen:

I. *Ist $P(y)$ nicht von der Form cy, wo c eine Konstante bedeutet, so sind je zwei mit $P(y)$ vertauschbare lineare homogene Differentialausdrücke auch unter einander vertauschbar.*

1. Ist $P(y) = py$, wo p nicht konstant ist, so zeigt man leicht, daß die mit $P(y)$ vertauschbaren Ausdrücke die Form qy besitzen. Da nun diese Ausdrücke in der Tat sämtlich unter einander vertauschbar sind, so hat man den Satz I nur für den Fall zu beweisen, daß die Ordnung n von $P(y)$ größer als 0 ist.

Der Beweis ergibt sich verhältnismäßig einfach, wenn man den Begriff des linearen homogenen Differentialausdrucks etwas allgemeiner faßt.

Es seien

$$p_0, p_1, p_2, \ldots \qquad (p_0 \neq 0)$$

unendlich viele Funktionen von x und n eine beliebige ganze Zahl, die positiv, negativ oder Null sein kann. Wir fassen dann die Gesamtheit der Funktionen p_λ in Verbindung mit der Zahl n in der symbolischen Formel

$$P = p_0 D^n + p_1 D^{n-1} + p_2 D^{n-2} + \cdots = \sum_0^\infty p_\lambda D^{n-\lambda}$$

---

1) Vergl. Wallenberg, Archiv der Mathematik und Physik, Bd. 4 (1903), S. 252.

zusammen und nennen P einen *symbolischen Differentialausdruck* der *Ordnung* n; $p_0, p_1, p_2, \ldots$ sollen die Koeffizienten von P heißen. Die Zeichen D^n, $D^{n-1}, \ldots$, ebenso wie das Zeichen $+$ sollen hierbei lediglich die Bedeutung von Symbolen haben, die zur Veranschaulichung der Aufeinanderfolge der Funktionen p_λ dienen. Ist ein Koeffizient p_λ gleich 0, so soll das Glied $p_\lambda D^{n-\lambda}$ nicht besonders geschrieben werden, so daß also z. B. $p_0 D^n$ denjenigen Ausdruck bedeuten soll, dessen Ordnung gleich n ist, und dessen Koeffizienten p_0, 0, 0, ... sind. Insbesondere soll $1 \cdot D^n$ auch kurz mit D^n bezeichnet werden. Ist ferner $n \geqq 0$ und

$$P = p_0 D^n + p_1 D^{n-1} + \cdots + p_n D^0,$$

so nenne ich P einen endlichen Differentialausdruck und bezeichne mit $P(y)$ den linearen homogenen Differentialausdruck

$$p_0 \frac{d^n y}{dx^n} + p_1 \frac{d^{n-1} y}{dx^{n-1}} + \cdots + p_n y.$$

2. Für die hier eingeführten symbolischen Differentialausdrücke sollen nun einige Operationsregeln definiert werden.

a) Sind P und Q zwei Ausdrücke, so soll $P = Q$ dann und nur dann gesetzt werden, wenn die Ordnungen und die einzelnen Koeffizienten von P und Q übereinstimmen.

b) Ist $P = \sum\limits_0^\infty p_\lambda D^{n-\lambda}$ und c eine Konstante, so verstehe ich unter cP den Ausdruck $\sum\limits_0^\infty (cp_\lambda) D^{n-\lambda}$. — Der Ausdruck $(-1)P$ soll mit $-P$ bezeichnet werden.

c) Sind

$$P = \sum_0^\infty p_\lambda D^{n-\lambda}, \qquad Q = \sum_0^\infty q_\lambda D^{m-\lambda} \tag{1}$$

zwei Ausdrücke der Ordnungen n und m, und ist r die größere der Zahlen n und m, so bezeichne ich mit $P + Q$ den Ausdruck

$$(p_{n-r} + q_{m-r}) D^r + (p_{n-r+1} + q_{m-r+1}) D^{r-1} + \cdots,$$

wobei p_μ und q_μ für $\mu < 0$ gleich Null zu setzen sind. — Ist $Q = -P$, so setze ich $P + Q = 0$.

d) Mit $P - Q$ bezeichne ich den Ausdruck $P + (-Q)$.

e) Unter dem „Produkt" PQ der beiden Ausdrücke (1) verstehe ich den Ausdruck

$$R = \sum_0^\infty r_\lambda D^{m+n-\lambda},$$

wo

$$r_\lambda = \sum_{\alpha=0}^{\lambda} \sum_{\beta=0}^{\lambda-\alpha} \binom{n-\beta}{\lambda-\alpha-\beta} p_\beta q_\alpha^{(\lambda-\alpha-\beta)} \quad \left(q_\alpha^{(\lambda-\alpha-\beta)} = \frac{d^{\lambda-\alpha-\beta} q_\alpha}{dx^{\lambda-\alpha-\beta}}\right) \tag{2}$$

zu setzen ist.

3. Man zeigt nun ohne Mühe, daß für die eben definierte „Addition“ und „Multiplikation“ die Regeln gelten:

$$P+Q=Q+P,\quad P+(Q+R)=(P+Q)+R,$$
$$P(Q+R)=PQ+PR,\quad (Q+R)P=QP+RP,$$
$$P(QR)=(PQ)R.$$

Man kann daher in bekannter Weise auch für mehrere Ausdrücke $P, P_1 \ldots, P_{m-1}$ die Zeichen

$$P+P_1+\cdots+P_{m-1},\quad PP_1\ldots P_{m-1}$$

und auch das Zeichen P^m einführen.

In Bezug auf die Multiplikation bemerke ich noch folgendes: sind P und Q endliche Ausdrücke, so ist auch $R=PQ$ ein endlicher Ausdruck, und der diesem Ausdruck entsprechende lineare homogene Differentialausdruck $R(y)$ ist nichts anderes, als der aus $P(y)$ und $Q(y)$ zusammengesetzte Differentialausdruck. — Ferner hebe ich noch hervor, daß stets

$$PD^0=D^0P=P$$

ist.

4. Aus der Gleichung (2) ergibt sich für die Koeffizienten r_λ von PQ

$$r_0=p_0q_0,\quad r_1=np_0q_0'+p_1q_0+p_0q_1,$$
$$r_\lambda=p_\lambda q_0+p_0q_\lambda+f_\lambda(p_0,p_1,\ldots p_{\lambda-1},q_0,q_1,\ldots q_{\lambda-1}),\qquad (\lambda>0)$$

wo f_λ eine gewisse ganze rationale Differentialfunktion ist.

Hieraus folgt unmittelbar:

a) Sind P und R gegeben, so kann man einen Ausdruck Q auf eine und nur eine Weise so bestimmen, daß $PQ=R$ wird. Denn setzt man Q als einen Ausdruck an, dessen Ordnung gleich ist der Differenz der Ordnungen von R und P, so bestimmen sich die Koeffizienten von Q sukzessive aus linearen Gleichungen.

b) Sind ebenso Q und R gegeben, so ist P eindeutig bestimmt durch die Bedingung, daß $PQ=R$ sein soll.

c) Aus $PQ=PQ_1$ folgt, daß $Q=Q_1$, ebenso aus $PQ=P_1Q$, daß $P=P_1$ ist.

5. Aus dem eben Gesagten ergibt sich, daß man zu jedem symbolischen Differentialausdruck P der Ordnung n einen wohlbestimmten Ausdruck P_1 der Ordnung $-n$ angeben kann, so daß $PP_1=D^0$ wird. Diesen Ausdruck P_1 will ich mit P^{-1} bezeichnen.[1])

Aus $PP^{-1}=D^0$ ergibt sich

$$P^{-1}PP^{-1}=P^{-1}D^0=D^0P^{-1},$$

also

$$P^{-1}P=D^0.$$

Ferner ist für zwei Ausdrücke P und Q

$$PQQ^{-1}P^{-1}=PD^0P^{-1}=PP^{-1}=D^0,$$

---

1) Der Ausdruck P^{-1} ist für den Fall, daß $P=p_0D^n+p_1D^{n-1}+\cdots+p_nD^0$ ist, bereits von Herrn S. Pincherle (Math. Ann., Bd. 49, S. 378) betrachtet worden.

also

$$(PQ)^{-1} = Q^{-1}P^{-1}.$$

Daraus folgt für jedes positive ganzzahlige m

$$(P^m)^{-1} = (P^{-1})^m.$$

Diesen Ausdruck bezeichne ich mit P^{-m}. Es ist dann offenbar für beliebige positive oder negative ganzzahlige k und l

$$P^k P^l = P^{k+l},$$

wobei unter P^0 der Ausdruck D^0 zu verstehen ist.

Daher sind die positiven und negativen „Potenzen" von P untereinander vertauschbar.

6. Es sei m eine positive ganze Zahl und $P = \sum_0^\infty p_\lambda D^{n-\lambda}$. Dann ist P^m von der Ordnung nm. Setzt man

$$P^m = \sum_0^\infty f_{m\lambda} D^{mn-\lambda},$$

so wird, wie man durch den Schluß von m auf $m+1$ leicht zeigt,

$$f_{m0} = p_0^m, \quad f_{m\lambda} = mp_0^{m-1}p_\lambda + g_{m\lambda}(p_0, p_1, \cdots, p_{\lambda-1}), \qquad (\lambda > 0) \tag{3}$$

wo $g_{m\lambda}(p_0, p_1, \ldots, p_{\lambda-1})$ eine ganze rationale Differentialfunktion von $p_0, p_1, \ldots, p_{\lambda-1}$ bedeutet.

Es sei nun $R = \sum_0^\infty r_\lambda D^{r-\lambda}$ ein Ausdruck von der Ordnung r. Wir fragen: läßt sich, falls m eine gegebene positive ganze Zahl ist, P so bestimmen, daß $P^m = R$ wird?

Zunächst muß r durch m teilbar sein. Ist aber $r = mn$ und setzt man $P = \sum_0^\infty p_\lambda D^{n-\lambda}$, so lehren die Gleichungen (3), daß wegen $p_0^m = r_0$ der Koeffizient p_0 bis auf eine $m$te Einheitswurzel bestimmt ist, und daß, falls p_0 fixiert wird, die übrigen Koeffizienten $p_1, p_2, \ldots$ sich sukzessive aus linearen Gleichungen in eindeutiger Weise berechnen lassen.

Hieraus folgt, daß die Gleichung $P^m = R$, falls r durch m teilbar ist, durch genau m verschiedene symbolische Differentialausdrücke der Ordnung $\frac{r}{m}$ befriedigt wird. Bezeichnet man einen dieser Ausdrücke mit $R^{\frac{1}{n}}$, und ist ϱ eine primitive $m$te Einheitswurzel, so repräsentieren offenbar die m Ausdrücke

$$R^{\frac{1}{m}}, \ \varrho R^{\frac{1}{m}}, \ldots, \ \varrho^{m-1} R^{\frac{1}{m}}$$

die sämtlichen Lösungen der Gleichung $P^m = R$.

7. Es sei nun $P = \sum_0^\infty p_\lambda D^{n-\lambda}$ ein Ausdruck der Ordnung $n > 0$. Bedeutet dann $p_0^{\frac{1}{n}}$ einen eindeutig fixierten Wert von $\sqrt[n]{p_0}$, so existiert nach dem eben Gezeigten ein und nur ein Ausdruck Q der Ordnung 1, der der Gleichung $Q^n = P$ genügt und die Gestalt hat

$$Q = p_0^{\frac{1}{n}} D + q_1 D^0 + q_2 D^{-1} + \cdots.$$

Diesen Ausdruck bezeichne ich mit $P^{\frac{1}{n}}$; ferner setze ich für ein beliebiges ganzzahliges m

$$\left(P^{\frac{1}{n}}\right)^m = P^{\frac{m}{n}}.$$

Dieser Ausdruck ist dann von der Ordnung m. Aus der Berechnungsweise der Koeffizienten von $P^{\frac{1}{n}}$ folgt auch, daß der Koeffizient von $D^{m-\lambda}$ in $P^{\frac{m}{n}}$ eine ganze rationale Differentialfunktion von $p_0^{\frac{1}{n}}$, p_0^{-1}, $p_1, \ldots p_\lambda$ ist.

Die Ausdrücke $P^{\frac{m}{n}}$ sind nun als die Potenzen von $P^{\frac{1}{n}}$ unter einander und wegen $P^{\frac{n}{n}} = P$ auch mit P vertauschbar.

Man kann noch allgemeinere Ausdrücke angeben, die mit P vertauschbar sind. Es seien nämlich $c_0 \neq 0$, $c_1, c_2, \cdots$ unendlich viele Konstanten. Ich verstehe dann unter

$$Q = c_0 P^{\frac{m}{n}} + c_1 P^{\frac{m-1}{n}} + \cdots + c_\lambda P^{\frac{m-\lambda}{n}} + \cdots \tag{4}$$

denjenigen symbolischen Differentialausdruck der Ordnung m, in dem der Koeffizient von $D^{m-\lambda}$ übereinstimmt mit dem Koeffizienten von $D^{m-\lambda}$ in der endlichen Summe $c_0 P^{\frac{m}{n}} + c_1 P^{\frac{m-1}{n}} + \cdots + c_\lambda P^{\frac{m-\lambda}{n}}$.

Man zeigt nun leicht, daß, wenn

$$R = d_0 P^{\frac{r}{n}} + d_1 P^{\frac{r-1}{n}} + \cdots + d_\lambda P^{\frac{r-\lambda}{n}} + \cdots$$

ein zweiter Ausdruck dieser Form ist,

$$QR = e_0 P^{\frac{m+r}{n}} + e_1 P^{\frac{m+r-1}{n}} + \cdots + e_\lambda P^{\frac{m+r-\lambda}{n}} + \cdots$$

wird, wo $e_\lambda = e_0 d_\lambda + e_1 d_{\lambda-1} + \cdots + c_\lambda d_0$ ist. Daher sind die Ausdrücke von der Form (4) unter einander und, weil P selbst unter ihnen enthalten ist, auch mit P vertauschbar.

8. Ich will nun zeigen, daß jeder mit P vertauschbare Ausdruck Q sich auf die Form (4) bringen läßt.

Es sei nämlich $Q = \sum\limits_0^\infty q_\lambda D^{m-\lambda}$ mit P vertauschbar. Dann folgt aus $PQ = QP$ durch Vergleichen der Koeffizienten von D^{m+n-1} auf beiden Seiten

$$np_0 q_0' + p_1 q_0 + p_0 q_1 = mq_0 p_0' + q_1 p_0 + q_0 p_1,$$

also $np_0 q_0' = mq_0 p_0'$. Folglich ist

$$q_0 = c_0 p^{\frac{m}{n}},$$

wo c_0 eine Konstante ist, was auch für $m = 0$ richtig bleibt. Hieraus ergibt sich insbesondere, daß, wenn Q und R zwei mit P vertauschbare Ausdrücke derselben Ordnung m sind, sich stets eine Konstante c so bestimmen läßt, daß die Ordnung des ebenfalls mit P vertauschbaren Ausdrucks $Q - cR$ höchstens gleich $m - 1$ wird.

Nun kennen wir aber für jedes m einen mit P vertauschbaren Ausdruck der Ordnung m, nämlich den Ausdruck $P^{\frac{m}{n}}$. Daher lassen sich sukzessive Konstanten $c_0, c_1, \ldots, c_\lambda$ bestimmen, so daß die mit P vertauschbaren Ausdrücke

$$Q_1 = Q - c_0 P^{\frac{m}{n}}, \quad Q_2 = Q_1 - c_1 P^{\frac{m-1}{n}}, \quad \cdots, \quad Q_{\lambda+1} = Q_\lambda - c_\lambda P^{\frac{m-\lambda}{n}}$$

höchstens von den Ordnungen $m - 1, m - 2, \cdots, m - \lambda - 1$ sind. Da dies nun für jedes noch so große λ gilt, so kann man offenbar direkt

$$Q = c_0 P^{\frac{m}{n}} + c_1 P^{\frac{m-1}{n}} + \cdots + c_\lambda P^{\frac{m-\lambda}{n}} + \cdots$$

setzen.

Hieraus folgt aber, daß die mit P vertauschbaren Ausdrücke auch unter einander vertauschbar sind, worin der Satz I als spezieller Fall enthalten ist.

9. Aus dem auf S. 4 über die Koeffizienten von $P^{\frac{m}{n}}$ Gesagten ergibt sich unmittelbar:

II. *Sind*

$$P(y) = p_0 \frac{d^n y}{dx^n} + p_1 \frac{d^{n-1} y}{dx^{n-1}} + \cdots + p_n y, \qquad (n > 0)$$

$$Q(y) = q_0 \frac{d^m y}{dx^m} + q_1 \frac{d^{m-1} y}{dx^{m-1}} + \cdots + q_m y$$

zwei vertauschbare lineare homogene Differentialausdrücke, so ist q_λ eine ganze rationale Differentialfunktion von $p_0^{\frac{1}{n}}, p_0^{-1}, p_1, \ldots, p_\lambda$.

Ist also insbesondere $p_0 = 1$ und sind $p_1, p_2, \cdots, p_n$ eindeutige analytische oder speziell rationale Funktionen von x, so sind auch die Koeffizienten jedes mit $P(y)$ vertauschbaren Differentialausdrucks eindeutige analytische, bezw. rationale Funktionen von x.

10. Verlangt man, daß die Differentialausdrücke $P(y)$ und $Q(y)$ der Ordnungen n und m vertauschbar sein sollen, so ergeben sich aus der Gleichung $PQ(y) = QP(y)$ genau $n + m$ Differentialgleichungen zwischen den $n + m + 2$ Funktionen $p_0, p_1, \ldots, p_n, q_0, q_1, \ldots, q_m$. Sieht man zunächst $p_0, p_1, \ldots, p_n$ als gegeben an, so erscheinen die ersten $m + 1$ dieser Beziehungen als Differentialgleichungen für $q_0, q_1, \ldots, q_m$. Aus dem Satz II folgt, daß diese Differentialgleichungen in exakter Form integrierbar sind. Setzt man dann die allgemeinen Integralausdrücke für $q_0, q_1, \ldots, q_m$ in die $n - 1$ übrig bleibenden Relationen ein, so ergeben sich $n - 1$ algebraische Differentialgleichungen für die Funktionen $p_0, p_1, \ldots, p_n$, in die noch gewisse Integrationskonstanten eingehen.

Den Charakter dieser $n - 1$ Relationen erkennt man unter Benutzung der früher betrachteten symbolischen Differentialausdrücke auf folgende Weise.

Es sei $P = p_0 D^n + p_1 D^{n-1} + \cdots + p_n D^0$ ein endlicher Ausdruck der Ordnung $n > 0$. Dann hat der Ausdruck $P^{\frac{m}{n}}$ $(m \geqq 0)$ die Form

$$P^{\frac{m}{n}} = R_m + S_m,$$

wo R_m ein wohlbestimmter endlicher Ausdruck der Ordnung m, dagegen S_m höchstens von der Ordnung -1 ist; für $m \equiv 0 \pmod{n}$ hat man S_m gleich 0 zu setzen. Aus $PP^{\frac{m}{n}} - P^{\frac{m}{n}}P = 0$ folgt

$$PR_m - R_m P = S_m P - PS_m.$$

Bezeichnet man daher den wohlbestimmten endlichen Ausdruck $PR_m - R_m P$ mit T_m, so ist $T_m = S_m P - PS_m$, falls m durch n teilbar ist, gleich 0, sonst aber ein Ausdruck, dessen Ordnung höchstens gleich $n - 2$ ist.

Aus dem in Nr. 8 Bewiesenen folgt nun, daß jeder mit P vertauschbare endliche Ausdruck Q der Ordnung m die Form

$$Q = c_0 R_m + c_1 R_{m-1} + \cdots + c_m R_0 \qquad (c_0 \neq 0) \tag{5}$$

besitzen muß, wo $c_0, c_1, \ldots, c_m$ Konstanten sind.

Die Bedingung dafür, daß P mit einem endlichen Ausdruck der Ordnung m vertauschbar sein soll, lautet daher: Es müssen sich $m + 1$ Konstanten $c_0 \neq 0, c_1, \ldots, c_m$ so bestimmen lassen, daß der Ausdruck (5) mit P vertauschbar wird, oder was dasselbe ist, daß die $n - 1$ Koeffizienten von

$$c_0 T_m + c_1 T_{m-1} + \cdots + c_m T_0$$

gleich Null werden. Wegen $T_{\nu n} = 0$ bleiben hierbei die $\left[\frac{m}{n}\right] + 1$ Konstanten c_λ, für welche $\lambda \equiv m \pmod{n}$ ist, ganz willkürlich. Ist insbesondere $m \equiv 0 \pmod{n}$, so genügt es $c_0 = 1$, $c_1 = c_2 = \cdots = c_m = 0$ zu setzen.

Auf die genauere Berechnung der durch P eindeutig bestimmten Ausdrücke R_m und T_m gehe ich hier nicht weiter ein.

9.

Arithmetische Untersuchungen über endliche Gruppen linearer Substitutionen

Sitzungsberichte der Preussischen Akademie der Wissenschaften 1906,
Physikalisch-Mathematische Klasse 164 - 184

Eine endliche Gruppe $\mathfrak{H}$ der Ordnung h, deren Elemente in k Klassen konjugierter Elemente zerfallen, besitzt, wie Hr. Frobenius gezeigt hat, genau k einfache Gruppencharaktere $\chi^{(0)}(R), \chi^{(1)}(R), \cdots, \chi^{(k-1)}(R)$. Jedem Charakter $\chi^{(\varkappa)}(R)$ entspricht eine irreduzible Darstellung der Gruppe $\mathfrak{H}$ durch lineare Substitutionen (Matrizen), in der die dem Element R von $\mathfrak{H}$ entsprechende Substitution A_R die Spur $\chi^{(\varkappa)}(R)$ besitzt. Die verschiedenen unter den h Substitutionen A_R bilden eine der Gruppe $\mathfrak{H}$ ein- oder mehrstufig isomorphe Gruppe $\mathfrak{H}^{(\varkappa)}$ von linearen Substitutionen. Jede andere irreduzible Gruppe von höchstens h linearen Substitutionen, die der Gruppe $\mathfrak{H}$ ein- oder mehrstufig isomorph ist, muß einer der h Gruppen $\mathfrak{H}^{(\varkappa)}$ äquivalent, d. h. durch eine Transformation der Variabeln in $\mathfrak{H}^{(\varkappa)}$ überführbar sein.

Kommt unter den der Gruppe $\mathfrak{H}^{(\varkappa)}$ äquivalenten Gruppen eine Gruppe $\mathfrak{G}$ von linearen Substitutionen vor, deren Koeffizienten sämtlich einem Zahlkörper K angehören, so wollen wir diese Eigenschaft der Gruppe $\mathfrak{H}^{(\varkappa)}$ dadurch kennzeichnen, daß wir sagen: $\mathfrak{H}^{(\varkappa)}$ *ist einer im Körper K rationalen Gruppe äquivalent* oder auch $\mathfrak{H}^{(\varkappa)}$ *ist im Körper K rational darstellbar.*

Nach einem Ergebnis des Hrn. Frobenius läßt sich der Körper K auch stets als ein algebraischer Körper $\Omega(\mu)$ wählen; hierbei bedeutet Ω den Bereich der rationalen Zahlen.

Es entsteht nun die Aufgabe zu untersuchen, durch welche Eigenschaften die algebraischen Körper, in denen die Gruppe $\mathfrak{H}^{(\varkappa)}$ rational darstellbar ist, charakterisiert sind, und insbesondere den kleinsten in Betracht kommenden Grad eines solchen Körpers zu bestimmen.

Diese Aufgabe läßt sich noch etwas verallgemeinern. Es sei P ein gegebener Rationalitätsbereich. Es kann dann gefragt werden, in welchen algebraischen Körpern $\mathrm{P}(\mu)$ über P die Gruppe $\mathfrak{H}^{(\varkappa)}$ rational darstellbar ist.

Ein solcher Körper $\mathrm{P}(\mu)$ muß zunächst alle Zahlen $\chi^{(\varkappa)}(R)$ enthalten. Wenn also die k Zahlen $\chi^{(\varkappa)}(R)$ einen durch die Zahl χ bestimmten Körper $\Omega(\chi)$ erzeugen, und χ im Körper P einer irreduziblen Gleichung des Grades l genügt, so muß der Grad n des Körpers $\mathrm{P}(\mu)$ (in bezug auf P) durch l teilbar sein. Ist nun der kleinste in Betracht kommende Grad n gleich ml, so will ich die Zahl m den *Index* der Gruppe $\mathfrak{H}^{(\varkappa)}$ oder auch des Charakters $\chi^{(\varkappa)}(R)$ in bezug auf den Körper P nennen.

Die Zahl m ist als vollständig bestimmt anzusehen, wenn neben dem Körper P und der Kompositionstabelle für die Elemente der Gruppe $\mathfrak{H}$ noch die k Zahlen $\chi^{(\varkappa)}(R)$ gegeben sind. Die genaue Berechnung von m allein unter Benutzung dieser Daten scheint jedoch mit erheblichen Schwierigkeiten verbunden zu sein.

Die vorliegende Arbeit liefert einige Beiträge zur Lösung dieser Aufgabe.

Insbesondere wird gezeigt: enthält eine Untergruppe $\mathfrak{S}$ von $\mathfrak{H}^{(\varkappa)}$ einen irreduziblen Bestandteil, der im Körper P rational darstellbar ist, genau r mal, so muß r durch m teilbar sein; speziell ist m ein Divisor der Zahl $f_\varkappa = \chi^{(\varkappa)}(E)$.

Dieser allgemeine Satz enthält als speziellen Fall ein von Hrn. Burnside[1] vor kurzem auf anderem Wege gewonnenes wichtiges Resultat, das sich folgendermaßen aussprechen läßt: die Gruppe $\mathfrak{H}^{(\varkappa)}$ läßt sich stets im Körper der h^{ten} Einheitswurzeln rational darstellen, wenn keine ganze Zahl $m' > 1$ existiert, so daß die charakteristische Determinante jeder Substitution von $\mathfrak{H}^{(\varkappa)}$ die m'^{te} Potenz einer rationalen Funktion wird.

Als weitere Folgerung aus unserem Satz erwähne ich noch hier, daß jede auflösbare Gruppe $\mathfrak{H}^{(\varkappa)}$ im Körper der h^{ten} Einheitswurzeln rational darstellbar ist.

Der Index m des Charakters $\chi^{(\varkappa)}(R)$ in bezug auf den Körper P läßt noch eine andere Deutung zu.

Man bezeichne eine im Körper P rationale Gruppe von linearen Substitutionen als *in* P *irreduzibel*, wenn sich keine ihr äquivalente, in P rationale Gruppe angeben läßt, deren Substitutionen Koeffizientenmatrizen der Form $\begin{pmatrix} A & 0 \\ 0 & B \end{pmatrix}$ besitzen, wo die den verschiedenen Substitutionen der Gruppe entsprechenden Matrizen A denselben Grad haben sollen. Haben dann die Zahlen m und l für den Charakter

---

[1] Proceedings of the London Mathematical Society, Series 2, Vol. 3 (1905), S. 239.

$\chi^{(\kappa)}(R)$ dieselbe Bedeutung wie früher, so entspricht dem Charakter $\chi^{(\kappa)}(R)$ eine im Körper P irreduzible Gruppe $\mathfrak{G}^{(\kappa)}$ von linearen Substitutionen, die im Bereiche aller Zahlen in ml irreduzible Bestandteile zerfällt, von denen je m einander äquivalent sind, während die l einander nicht äquivalenten irreduziblen Bestandteile den l mit $\chi^{(\kappa)}(R)$ in bezug auf P konjugierten Charakteren von $\mathfrak{H}$ entsprechen. Die Anzahl der verschiedenen (nicht äquivalenten) im Körper P irreduziblen Darstellungen der Gruppe $\mathfrak{H}$ durch lineare Substitutionen ist gleich der Anzahl der in bezug auf den Körper P nicht konjugierten Charaktere von $\mathfrak{H}$.

§ 1.

Es seien $H_0, H_1, \cdots, H_{h-1}$ die Elemente der gegebenen Gruppe $\mathfrak{H}$, und es sei $\mathfrak{G}$ eine der Gruppe $\mathfrak{H}$ isomorphe Gruppe von höchstens h linearen Substitutionen. Entspricht dem Element R von $\mathfrak{H}$ in der Gruppe $\mathfrak{G}$ die Substitution mit der Koeffizientenmatrix A_R und bedeuten $x_{H_0}, x_{H_1}, \cdots, x_{H_{h-1}}$ unabhängige Variable, so wird die Matrix

$$X = \sum_R A_R x_R \qquad (R = H_0, H_1, \cdots, H_{h-1})$$

eine zu $\mathfrak{H}$ gehörige Gruppenmatrix genannt. Zwei Gruppenmatrizen X und X' desselben Grades heißen äquivalent, wenn sich eine konstante Matrix[1] P von nicht verschwindender Determinante bestimmen läßt, so daß $X' = P^{-1}XP$ wird. Sind die Koeffizienten aller Matrizen A_R Größen eines gegebenen Zahlkörpers P, so soll die Matrix X *im Körper* P *rational* heißen. Ferner bezeichnen wir eine (in P rationale) Gruppenmatrix X als in P reduzibel oder irreduzibel, je nachdem die ihr entsprechende Gruppe $\mathfrak{G}$ in P reduzibel oder irreduzibel ist (vgl. Einleitung). Ist $\zeta(R)$ die Spur der Matrix A_R, so nennen wir das System der h Zahlen $\zeta(R)$ *den Charakter der Gruppenmatrix* X.

Es gilt nun der Satz:

I. *Es seien X und X' zwei im Körper* P *irreduzible Gruppenmatrizen der Grade f und f'. Ist dann P eine konstante Matrix mit f Zeilen und f' Spalten, deren Koeffizienten dem Körper* P *angehören, und besteht die Gleichung*

$$XP = PX',$$

so ist entweder $P = 0$ oder sind X und X' äquivalent, und P ist eine quadratische Matrix des Grades $f = f'$ von nicht verschwindender Determinante.

---

[1] D. h. eine Matrix mit konstanten, von den x_R unabhängigen Koeffizienten.

Der Beweis dieses Satzes ist in genau derselben Weise zu führen wie für den Satz I meiner Arbeit »Neue Begründung der Theorie der Gruppencharaktere«.[1]

Es sei nämlich r der Rang der Matrix P; man bestimme, was stets möglich ist, zwei Matrizen A und B der Grade f und f' mit Koeffizienten aus dem Rationalitätsbereiche P, deren Determinanten von Null verschieden sind, so daß die Matrix APB die Gestalt

$$\begin{pmatrix} E_r & 0 \\ 0 & 0 \end{pmatrix}$$

annimmt; hierin bedeutet E_r die Einheitsmatrix des Grades r. Dann ergibt sich aus der Gleichung $XP = PX'$, daß

$$AXA^{-1} = \begin{pmatrix} X_{rr} & X_{r,f-r} \\ 0 & X_{f-r,f-r} \end{pmatrix}, \qquad B^{-1}X'B = \begin{pmatrix} X_{rr} & 0 \\ X'_{f'-r,r} & X'_{f'-r,f'-r} \end{pmatrix}$$

wird, wo die $X_{\alpha\beta}$ und $X'_{\alpha\beta}$ gewisse Matrizen mit α Zeilen und β Spalten bedeuten, die offenbar im Körper P rational sind. Die Gruppenmatrizen X und X' sind dann auch (vgl. B., Satz VII) den in P rationalen Gruppenmatrizen

$$\begin{pmatrix} X_{rr} & 0 \\ 0 & X_{f-r,f-r} \end{pmatrix}, \text{ bzw. } \begin{pmatrix} X_{rr} & 0 \\ 0 & X'_{f'-r,f'-r} \end{pmatrix}$$

äquivalent. Wäre nun $0 < r < f$ oder $0 < r < f'$, so würde sich ergeben, daß X oder X' in P reduzibel ist. Ist daher P nicht gleich 0, also $r > 0$, so muß $r = f = f'$ sein; dann ist aber P eine quadratische Matrix von nicht verschwindender Determinante, und die Gruppenmatrizen X und X' sind wegen $P^{-1}XP = X'$ äquivalent.

Aus I folgt:

II. *Ist X eine im Körper* P *irreduzible Gruppenmatrix und P eine mit X vertauschbare konstante Matrix, deren Koeffizienten dem Körper* P *angehören, so muß die charakteristische Determinante $|xE-P|$ der Matrix P Potenz einer in* P *irreduziblen Funktion $\phi(x)$ sein und es besteht die Gleichung $\phi(P) = 0$.*

Es sei nämlich

$$|xE-P| = \varphi(x)\varphi_1(x)\varphi_2(x)\cdots,$$

wo $\phi(x), \phi_1(x), \phi_2(x), \cdots$ im Körper P irreduzible Funktionen sind. Dann ist $\phi(P)$ eine in P rationale Matrix von verschwindender Determinante, die mit X vertauschbar ist. Folglich muß nach Satz I die Gleichung $\phi(P) = 0$ bestehen. Ebenso ist $\phi_1(P) = 0$, $\phi_2(P) = 0, \ldots$. Ist daher ω eine Wurzel der Gleichung $|xE-P| = 0$, so wird $\phi(\omega) = 0$, $\phi_1(\omega) = 0$, $\phi_2(\omega) = 0, \ldots$. Die in P irreduziblen Funktionen $\phi(x)$,

[1] Sitzungsberichte 1905, S. 406. — Im folgenden wird diese Arbeit kurz mit B. zitiert.

$\phi_1(x), \phi_2(x), \cdots$ sind demnach nicht relativ prim zueinander und müssen folglich einander gleich sein.

Um nun zu entscheiden, ob eine vorgelegte im Körper P rationale Gruppenmatrix $X = \sum_R A_R x_R$ des Grades n im Körper P reduzibel ist oder nicht, kann man folgendermaßen schließen. Man bestimme die allgemeinste Matrix $V = (v_{\alpha\beta})$, die mit allen h Matrizen A_R vertauschbar ist.[1] Unter den n^2 Koeffizienten $v_{\alpha\beta}$ bleiben gewisse q willkürlich; man kann auch q Parameter $v_1, v_2, \cdots, v_q$ wählen, so daß jede der Größen $v_{\alpha\beta}$ als lineare homogene Funktion von $v_1, v_2, \cdots, v_q$ mit in P rationalen Koeffizienten darstellbar ist. Zieht man nun für die n^2 Größen $v_{\alpha\beta}$ oder, was dasselbe ist, für die q Größen $v_1, v_2, \cdots v_q$ nur Zahlen des Körpers P in Betracht, so läßt sich die Gleichung $|v_{\alpha\beta}| = 0$ als eine diophantische Gleichung im Körper P mit q Unbekannten $v_1, v_2, \cdots v_q$ ansehen. Ist nun X in P irreduzibel, so läßt sich diese diophantische Gleichung nur durch das System $v_1 = 0, v_2 = 0, \cdots v_q = 0$ befriedigen. Ist dagegen X in P reduzibel, so genügen der Gleichung $|v_{\alpha\beta}| = 0$ auch Größen $v_1, v_2, \cdots v_q$ des Körpers P, die nicht sämtlich 0 sind. Denn ist X der in P rationalen Gruppenmatrix

$$X' = \begin{pmatrix} X_1 & 0 \\ 0 & X_2 \end{pmatrix}$$

äquivalent, wo X_1 und X_2 die Grade r und $n-r$ besitzen, so kann man auch eine in P rationale Matrix Q wählen, so daß $X = Q^{-1}X'Q$ wird. Nun ist X' mit der Matrix

$$P' = \begin{pmatrix} E_r & 0 \\ 0 & 0 \end{pmatrix},$$

also X mit der Matrix $Q^{-1}P'Q = P$ vertauschbar. Die Matrix P ist aber in P rational und von der Determinante 0, ohne daß alle Koeffizienten von P verschwinden. — Wir sehen also, daß man, um zu entscheiden, ob eine gegebene in P rationale Gruppenmatrix in P reduzibel ist oder nicht, nur eine einzige diophantische Gleichung im Körper P zu untersuchen hat.

Es sei nun wieder $X = \sum_R A_R x_R$ eine im Körper P irreduzible Gruppenmatrix des Grades n. Bezeichnet man die einfachen Charaktere der Gruppe $\mathfrak{H}$ mit $\chi^{(0)}(R), \chi^{(1)}(R), \cdots, \chi^{(k-1)}(R)$ und mit $\xi(R)$ die Spur der Matrix A_R, so wird

$$\xi(R) = \sum_\varkappa r_\varkappa \chi^{(\varkappa)}(R),$$

wo die $r_\varkappa$ gewisse nicht negative ganze Zahlen sind.

[1] Es genügt $V = \sum_R A_{R^{-1}} U A_R$ zu bilden, wo U eine Matrix mit unbestimmten Koeffizienten ist.

Es seien nun $y_{H_0}, y_{H_1}, \cdots, y_{H_{k-1}}$ irgendwelche Größen des Körpers P, die nur der Bedingung unterworfen sind, daß stets $y_R = y_S$ sein soll, wenn R und S konjugierte Elemente der Gruppe $\mathfrak{H}$ sind. Die Matrix

$$Y = \sum_R A_R y_R$$

ist dann eine mit X vertauschbare Matrix, deren Koeffizienten sämtlich dem Körper P angehören. Setzt man ferner

$$\eta_\varkappa = \sum_\varkappa \chi^{(\varkappa)}(R) y_R,$$

so wird (vgl. Frobenius, Sitzungsberichte 1896, S. 1361)

$$|xE - Y| = \Pi\left(x - \frac{\eta_\varkappa}{f_\varkappa}\right)^{r_\varkappa f_\varkappa},$$

wo $f_\varkappa = \chi^{(\varkappa)}(E)$ den Grad des Charakters $\chi^{(\varkappa)}(R)$ bedeutet. Da nun nach Satz II die Funktion $|xE - Y|$ Potenz einer in P irreduziblen Funktion sein muß, so schließt man leicht, daß diejenigen unter den Zahlen $r_\varkappa$, die von Null verschieden sind, einander gleich sein müssen; ist etwa $r_\lambda = r_{\lambda_1} = \cdots = r_{\lambda_{l-1}} = m > 0$, so müssen außerdem die Größen $\eta_\lambda, \eta_{\lambda_1}, \cdots, \eta_{\lambda_{l-1}}$ die Wurzeln einer im Körper P irreduziblen Gleichung des Grades l sein. Hierfür können wir auch sagen: es müssen die l Charaktere $\chi^{(\lambda)}(R), \chi^{(\lambda_1)}(R), \cdots, \chi^{(\lambda_{l-1})}(R)$ die sämtlichen zu einem (einfachen) Charakter in Bezug auf P konjugierten Charaktere sein. — Wir setzen zur Abkürzung

$$\chi^{(\lambda)}(R) = \chi(R), \chi^{(\lambda_1)}(R) = \chi_1(R), \cdots, \chi^{(\lambda_{l-1})}(R) = \chi_{l-1}(R),$$

so daß

$$\xi(R) = m\{\chi(R) + \chi_1(R) + \cdots + \chi_{l-1}(R)\}$$

wird.

Es sei nun

$$X' = \sum B_R x_R$$

eine zweite in P irreduzible Gruppenmatrix des Grades n', die der Gruppenmatrix X nicht äquivalent ist. Ist $\xi'(R)$ die Spur der Matrix B_R, so ist

$$\xi'(R) = m'\{\chi'(R) + \chi'_1(R) + \cdots + \chi'_{l'-1}(R)\},$$

wo $\chi'(R), \cdots, \chi'_{l'-1}(R)$ die sämtlichen zu einem gewissen einfachen Charakter $\chi'(R)$ von $\mathfrak{H}$ in bezug auf P konjugierten Charaktere sind. Ist dann U eine Matrix mit n Zeilen und n' Spalten, deren Koeffizienten unbestimmt bleibende rationale Zahlen sind, so genügt die in P rationale Matrix

$$P = \sum_R A_{R^{-1}} U B_R$$

den h Gleichungen $A_S P = P B_S$, also auch der Gleichung $XP = PX'$. Da nun X und X' nicht äquivalent sein sollen, so muß nach Satz I die Matrix P gleich 0 sein. Setzt man nun

$$A_R = (a^R_{\alpha\beta}),\quad B_R = (b^R_{\varkappa\lambda}),\qquad \left\{\begin{matrix}\alpha,\beta = 1, 2, \dots n\\ \varkappa,\lambda = 1, 2, \dots n'\end{matrix}\right\}$$

so ergeben sich, weil die Koeffizienten von U beliebige rationale Zahlen bedeuten können, aus $P = 0$ die Relationen

$$\Sigma\, a^{R^{-1}}_{\alpha\beta}\, b^R_{\varkappa\lambda} = 0 .$$

Speziell wird

$$\sum_\alpha \sum_\varkappa \sum_R a^{R^{-1}}_{\alpha\alpha}\, b^R_{\varkappa\varkappa} = 0 .$$

Da nun

$$\xi(R) = \sum_\alpha a^R_{\alpha\alpha}\,,\qquad \xi'(R) = \sum_\varkappa b^R_{\varkappa\varkappa}$$

ist, so wird also

$$\sum_R \xi(R^{-1})\,\xi'(R) = 0 .$$

Diese Gleichung besagt aber (vgl. B., S. 426), daß die $l + l'$ Charaktere $\chi(R), \cdots \chi_{l-1}(R), \chi'(R), \cdots \chi'_{l'-1}(R)$ sämtlich untereinander verschieden sind.

Betrachtet man nun zwei im Körper P rationale Gruppenmatrizen, die einander äquivalent sind, als nicht voneinander verschieden, so erkennt man, daß zu einem einfachen Charakter $\chi(R)$ von $\mathfrak{H}$ nur eine in P irreduzible Gruppenmatrix gehören kann, die, im Bereiche aller Zahlen in irreduzible Bestandteile zerfällt, eine irreduzible Gruppenmatrix Z mit dem Charakter $\chi(R)$ enthält. Daß nun in der Tat zu jedem einfachen Charakter $\chi(R)$ eine in P irreduzible Gruppenmatrix gehört, erkennt man folgendermaßen. Man betrachte speziell die *reguläre* Gruppenmatrix

$$\overline{X} = (x_{PQ^{-1}}) \qquad (P, Q = H_0, H_1, \cdots, H_{h-1})$$

des Grades h. Der Charakter $\xi(R)$ dieser Gruppenmatrix ist gleich

$$\xi(R) = \Sigma f_\varkappa \chi^{(\varkappa)}(R),$$

enthält also jeden einfachen Charakter $\chi^{(\varkappa)}$, und zwar genau $f_\varkappa$ mal. Nun ist $\overline{X}$ gewiß im Körper P rational. Denkt man sich $\overline{X}$ im Körper P in irreduzible Bestandteile $X_0, X_1, X_2, \cdots$ zerlegt, so muß für mindestens eine der Gruppenmatrizen $X_0, X_1, X_2, \cdots$ der ihr entsprechende Charakter $\xi(R)$ auch den einfachen Charakter $\chi(R)$ enthalten.

Fassen wir die gewonnenen Resultate zusammen, so erhalten wir den Satz:

III. *Die Anzahl der verschiedenen (nicht äquivalenten) im Körper* P *irreduziblen Gruppenmatrizen* $X_0, X_1, X_2, \cdots$, *die zur Gruppe* $\mathfrak{H}$ *gehören, ist gleich der Anzahl der in bezug auf* P *nicht konjugierten einfachen Charaktere von* $\mathfrak{H}$. *Denkt man sich die Gruppenmatrizen* $X_0, X_1, X_2, \cdots$ *im Bereiche aller Zahlen in irreduzible Bestandteile zerlegt, so enthalten je zwei keinen irreduziblen Bestandteil gemeinsam. Der Charakter einer jeden der Gruppenmatrizen* $X_0, X_1, X_2, \cdots$ *hat die Form*

$$m\{\chi(R)+\chi_1(R)+\cdots+\chi_{l-1}(R)\},$$

wo $\chi(R), \chi_1(R), \cdots \chi_{l-1}(R)$ *die sämtlichen zu einem einfachen Charakter* $\chi(R)$ *in bezug auf* P *konjugierten Charaktere von* $\mathfrak{H}$ *sind.*

Die durch den Charakter $\chi(R)$ und den Körper P eindeutig bestimmte Zahl m soll nun der *Index* von $\chi(R)$ in bezug auf P genannt werden. Offenbar ist m zugleich auch der Index der zu $\chi(R)$ konjugierten Charaktere $\chi_1(R), \cdots, \chi_{l-1}(R)$.

Es gilt nun der Satz:

IV. *Ist* $\overline{X}$ *eine beliebige in* P *rationale Gruppenmatrix, deren Charakter den einfachen Charakter* $\chi(R)$ *genau* r *mal enthält, so ist* r *durch den Index* m *des Charakters* $\chi(R)$ *in bezug auf* P *teilbar.*

Denn denkt man sich die Gruppenmatrix $\overline{X}$ im Körper P in irreduzible Bestandteile zerlegt, so möge die dem Charakter $\chi(R)$ entsprechende in P irreduzible Gruppenmatrix X genau t mal vorkommen. Da der Charakter $\chi(R)$ in dem Charakter von X genau m mal enthalten ist, so muß $r = tm$ sein.

Wählt man für $\overline{X}$ wieder die reguläre Gruppenmatrix $(x_{PQ^{-1}})$, so ergibt sich:

V. *Der Index* m *des Charakters* $\chi(R)$ *ist ein Divisor der Zahl* $f = \chi(E)$.

§ 2.

Es sei wie früher $X = \Sigma A_R x_R$ eine in P irreduzible Gruppenmatrix, deren Charakter gleich

$$m\{\chi(R)+\chi_1(R)+\cdots+\chi_{l-1}(R)\}$$

ist. Der Grad der Matrix X ist dann gleich mlf, wo $f = \chi(E)$ ist. Die allgemeinste mit X vertauschbare konstante Matrix V läßt sich dann (vgl. die Anmerkung auf S. 168) in der Form

$$V = \sum_R A_{R^{-1}} U A_R$$

darstellen, wo $U = (u_{\alpha\beta})$ eine Matrix des Grades mlf mit unbestimmten Koeffizienten ist. Die charakteristische Determinante $|xE - V|$ von V ist dann (vgl. B. S. 420) die f^{te} Potenz einer ganzen rationalen Funktion des Grades ml, deren Wurzeln untereinander verschieden sind. Man kann daher die Größen $u_{\alpha\beta}$ auch als rationale Zahlen so wählen, daß die ml Wurzeln der Gleichung $|xE-V| = 0$ voneinander verschieden bleiben. Dann wird aber, weil V eine im Körper P rationale mit X vertauschbare Matrix wird, nach Satz II

$$|xE-V| = \{\varphi(x)\}^f,$$

wo $\phi(x)$ eine im Körper P irreduzible Funktion des Grades ml ist. Da zugleich $\phi(V) = 0$ ist, so sind die Elementarteiler der Determinante $|xE-V|$ sämtlich linear. Ist nun

$$\varphi(x) = x^{ml} - a_1 x^{ml-1} - \cdots - a_{ml} = (x-\mu_0)(x-\mu_1)\cdots(x-\mu_{ml-1})$$

und setzt man

$$M = \begin{pmatrix} a_1 & a_2 & a_3 & \cdots & a_{ml-1} & a_{ml} \\ 1 & 0 & 0 & \cdots & 0 & 0 \\ 0 & 1 & 0 & \cdots & 0 & 0 \\ . & . & . & \cdots & . & . \\ 0 & 0 & 0 & \cdots & 1 & 0 \end{pmatrix}, \qquad \mathbf{M} = \begin{pmatrix} \mu_0 & 0 & \cdots & 0 \\ 0 & \mu_1 & \cdots & 0 \\ . & . & \cdots & \\ 0 & 0 & \cdots & \mu_{ml-1} \end{pmatrix},$$

so wird (vgl. Frobenius, Crelles Journal Bd. 86, S. 146)

$$|xE-M| = |xE-\mathbf{M}| = \varphi(x).$$

Daher sind die Matrizen M und $\mathbf{M}$ einander ähnlich. Ferner ist die Matrix

$$T = \begin{pmatrix} M & 0 & \cdots & 0 \\ 0 & M & \cdots & 0 \\ . & . & \cdots & . \\ 0 & 0 & \cdots & M \end{pmatrix}$$

des Grades mlf der Matrix V ähnlich. Man kann daher auch eine in P rationale Matrix Q von nicht verschwindender Determinante wählen, so daß

$$Q^{-1}VQ = T$$

wird. Setzt man nun

$$Q^{-1}XQ = X' = \sum_R B_R x_R,$$

so ist jede der Matrizen B_R mit T vertauschbar. Es werde nun B_R in der Form

$$B_R = (B_{\alpha\beta}) \qquad (\alpha, \beta = 1, 2, \cdots f)$$

geschrieben, wo $B_{\alpha\beta}$ eine Matrix des Grades ml bedeutet. Aus

$$B_R Q = Q B_R$$

folgt dann, daß $B_{\alpha\beta}$ mit M vertauschbar ist. Da nun die charakteristischen Wurzeln von M untereinander verschieden sind, so muß $B_{\alpha\beta} = g_{\alpha\beta}(M)$ sein, wo $g_{\alpha\beta}(t)$ eine ganze rationale Funktion des Grades $ml-1$ von t ist, deren ml Koeffizienten aus linearen Gleichungen mit in P rationalen Koeffizienten zu bestimmen sind, und folglich selbst dem Körper P angehören. Da ferner die Matrizen M und M ähnlich sind, so schließt man auch sofort, daß sich eine Matrix L des Grades mlf von nicht verschwindender Determinante angeben läßt, so daß

$$L^{-1} B_R L = C_R = (\mathrm{B}_{\alpha\beta})$$

wird, wo $\mathrm{B}_{\alpha\beta} = g_{\alpha\beta}(\mathrm{M})$ wird.

Bezeichnet man nun die Matrix f^{ten} Grades

$$\{g_{\alpha\beta}(\mu_\varkappa)\} \qquad (\alpha, \beta = 1, 2, \cdots f)$$

mit $D_R^{(\varkappa)}$ und setzt

$$Z^{(\varkappa)} = \sum_R D_R^{(\varkappa)} x_R,$$

so zerfällt die Matrix $L^{-1} X' L$, wie leicht ersichtlich ist, in die ml Gruppenmatrizen $Z^{(0)}, Z^{(1)}, \ldots Z^{(ml-1)}$ des Grades f. Da nun X im Bereiche aller Zahlen in genau ml irreduzible Bestandteile zerfällt, so müssen die Gruppenmatrizen $Z^{(0)}, Z^{(1)}, \ldots Z^{(ml-1)}$ irreduzibel sein; ferner müssen unter ihnen je m einander äquivalent sein, während die l einander nicht äquivalenten den Charakteren $\chi(R), \chi_1(R), \cdots \chi_{l-1}(R)$ entsprechen.

Es ergibt sich insbesondere, daß die zu dem Charakter $\chi(R)$ gehörende irreduzible Gruppenmatrix Z des Grades f in einem algebraischen Körper $\mathrm{P}(\mu)$ des Grades ml über P rational darstellbar ist.

Es sei umgekehrt eine zu dem Charakter $\chi(R)$ gehörende, im Bereiche aller Zahlen irreduzible Gruppenmatrix

$$Z = \sum D_R x_R$$

des Grades f bekannt, in der die Koeffizienten der Matrizen D_R einem algebraischen Körper $\mathrm{P}(\nu)$ des Grades q über P angehören. Ferner sei

$$D_R = \{g_{\alpha\beta}(\nu)\} \qquad (\alpha, \beta = 1, 2, \cdots f),$$

wo $g_{\alpha\beta}(t)$ eine gewisse ganze rationale Funktion von t mit Koeffizienten aus dem Körper P ist. Es sei

$$\psi(x) = x^q - b_1 x^{q-1} - \cdots - b_q = 0$$

die in P irreduzible Gleichung, der ν genügt. Setzt man dann

$$N = \begin{pmatrix} b_1 & b_2 & b_3 & \cdots & b_{q-1} & b_q \\ 1 & 0 & 0 & \cdots & 0 & 0 \\ 0 & 1 & 0 & \cdots & 0 & 0 \\ . & . & . & \cdots & . & . \\ 0 & 0 & 0 & \cdots & 1 & 0 \end{pmatrix},$$

so wird $\psi(N) = 0$. Bezeichnet man nun die Matrix

$$\{g_{\alpha\beta}(N)\}$$

des Grades nf mit F_R, so wird offenbar, da für je zwei Elemente R und S von $\mathfrak{H}$ die Gleichung $D_R D_S = D_{RS}$ besteht, auch $F_R F_S = F_{RS}$. Daher ist

$$\bar{X} = \sum_R F_R x_R$$

eine zur Gruppe $\mathfrak{H}$ gehörige Gruppenmatrix, die im Körper P rational ist. Sind $\nu_0, \nu_1 \cdots \nu_{q-1}$ die Wurzeln der Gleichung $\psi(x) = 0$ und setzt man

$$\sum_\alpha g_{\alpha\alpha}(\nu_\lambda) = \chi_\lambda(R),$$

so wird die Spur der Matrix F_R gleich

$$\chi_0(R) + \chi_1(R) + \cdots + \chi_{q-1}(R).$$

Da nun die Charaktere $\chi_0(R), \cdots \chi_{q-1}(R)$ in bezug auf den Körper P dem Charakter $\chi(R)$ konjugiert sind, so sind die in P irreduziblen Bestandteile, in die die Gruppenmatrix $\bar{X}$ im Körper P zerfällt, sämtlich der früher betrachteten Matrix X äquivalent. Daher ist der Grad qf der Matrix $\bar{X}$ durch den Grad mlf der Matrix X teilbar; folglich ist auch die Zahl q durch die Zahl ml teilbar.[1]

Wir erhalten den Satz:

VI. *Ist $\chi(R)$ ein einfacher Charakter der Gruppe $\mathfrak{H}$, dessen Index in bezug auf den Körper P gleich m ist, und ist der durch die h Zahlen $\chi(R)$ und die Zahlen von P erzeugte Körper P(χ), als algebraischer Körper über P betrachtet, vom Grade l, so ist der kleinste Grad eines Körpers P(μ) über P, in dem sich die zu $\chi(R)$ gehörende irreduzible Gruppenmatrix Z des Grades $f = \chi(E)$ rational darstellen läßt, gleich ml. Läßt sich Z in einem algebraischen Körper P(ν) des Grades q über P rational darstellen, so muß q durch ml teilbar sein.*

Es sei noch erwähnt, daß der Körper P(μ) des Grades ml keineswegs eindeutig bestimmt zu sein braucht. Es kann vielmehr Z auch in zwei Körpern P(μ) und P(μ') rational darstellbar sein, die nur den Körper P(χ) als gemeinsamen Teiler enthalten. So ist z. B. die durch die Substitutionen

$$A = \begin{pmatrix} i & 0 \\ 0 & -i \end{pmatrix}, \quad B = \begin{pmatrix} 0 & 1 \\ -1 & 0 \end{pmatrix}$$

erzeugte Gruppe der Ordnung 8, deren Index in bezug auf den Körper Ω der rationalen Zahlen gleich 2 ist, nicht nur im Körper $\Omega(i)$, sondern auch in jedem Körper $\Omega(\sqrt{-n})$ rational darstellbar, sobald n eine ganze Zahl ist, die einer Summe von drei Quadraten gleich ist.

[1] Eine ähnliche Betrachtung findet sich in der Abhandlung des Hrn. Dickson: *On the reducibility of linear groups,* Transactions of the Amer. Math. Soc. Bd. 4 (1903), S. 434. Das in dieser Arbeit aufgestellte Resultat kann aber, wie leicht zu sehen, nicht allgemein richtig sein.

Aus dem früher bewiesenen Satz IV folgt unmittelbar:

VII. *Ist* P′ *ein Zahlkörper, der den Körper* P *enthält, so ist der Index* m *des Charakters* $\chi(R)$ *in bezug auf den Körper* P *durch den Index* m' *von* $\chi(R)$ *in bezug auf den Körper* P′ *teilbar.*

Da insbesondere jeder Körper P den Körper Ω enthält, so ist der Index m von $\chi(R)$ in bezug auf einen beliebigen Körper P ein Divisor des Index m_0 von $\chi(R)$ in bezug auf den Körper Ω.

Es sei nun speziell P′ ein algebraischer Körper P(λ) über P, der im Körper P(χ) enthalten ist. Genügt λ einer in P irreduziblen Gleichung des Grades a, so genügt χ im Körper P′ einer irreduziblen Gleichung des Grades $\frac{l}{a} = l'$. Nun läßt sich aber die zu $\chi(R)$ gehörende Gruppenmatrix Z des Grades f in einem algebraischen Körper P′(μ') des Grades $m'l'$ über P′ rational darstellen. Als algebraischer Körper über P betrachtet, besitzt aber P′(μ') den Grad $am'l' = m'l$. Nach Satz VI ist daher $m'l$ durch ml, also m' durch m teilbar. Da andererseits m durch m' teilbar ist, so muß $m' = m$ sein.

VIII. *Ist* P′ *ein algebraischer Körper über* P, *der in dem Körper* P(χ) *enthalten ist, so ist der Index des Charakters* $\chi(R)$ *in bezug auf* P′ *gleich dem Index von* $\chi(R)$ *in bezug auf* P.

Insbesondere ist also der Index m_0 des Charakters $\chi(R)$ in bezug auf den Körper Ω zugleich auch der Index von $\chi(R)$ in bezug auf den Körper $\Omega(\chi)$.

Bemerkenswert ist noch der Fall, daß P der Körper der reellen Zahlen ist. Dann ist die zu $\chi(R)$ gehörende irreduzible Gruppenmatrix Z im Körper P(i), also in einem Körper des Grades 2 über P, rational darstellbar. Daher ist in diesem Falle die Zahl ml ein Divisor von 2; demnach sind nur drei Fälle möglich:

$$l = 1,\ m = 1 \quad ; \quad l = 1,\ m = 2 \quad ; \quad l = 2,\ m = 1.$$

Die im Gebiete der reellen Zahlen irreduziblen Gruppen linearer Substitutionen lassen sich also in drei Arten teilen: 1. solche, die im Gebiete aller Zahlen irreduzibel sind; 2. solche, die im Gebiete aller Zahlen in zwei äquivalente irreduzible Gruppen zerfallen; 3. solche, die im Gebiete aller Zahlen in zwei nicht äquivalente (konjugiert komplexe) irreduzible Gruppen zerfallen (vgl. A. Loewy, *Über die Reduzibilität der reellen Gruppen linearer homogener Substitutionen*, Transactions of the American Math. Soc. Bd. 4, S. 171).

§ 3.

Es sei nun $\mathfrak{S}$ eine Untergruppe der Ordnung s von $\mathfrak{H}$. Ist dann

$$X = \sum_R A_R x_R$$

eine zur Gruppe $\mathfrak{H}$ gehörende Gruppenmatrix, so wird

$$Y = \sum_S A_S x_S,$$

wo S alle Elemente der Untergruppe $\mathfrak{S}$ durchläuft, eine zur Gruppe $\mathfrak{S}$ gehörige Gruppenmatrix. Sind daher

$$\psi^{(0)}(S), \psi^{(1)}(S), \cdots \psi^{(q-1)}(S)$$

die einfachen Charaktere von $\mathfrak{S}$ und ist $\zeta(R)$ der Charakter der Gruppenmatrix X, so wird

$$\zeta(S) = \sum_{\lambda=0}^{q-1} r_\lambda \psi^{(\lambda)}(S),$$

wo die durch die Gleichungen

$$r_\lambda = \frac{1}{s} \sum_S \zeta(S^{-1}) \psi^{(\lambda)}(S)$$

bestimmten Zahlen r_λ nicht negative ganze Zahlen sind. Wir wollen dann der Kürze wegen sagen, der Charakter $\zeta(R)$ enthalte den Charakter $\psi^{(\lambda)}(S)$ der Untergruppe r_λ Mal.

Ist ferner

$$U = \sum_S B_S x_S$$

eine zur Gruppe $\mathfrak{S}$ gehörige Gruppenmatrix des Grades b mit dem Charakter $\eta(S)$, und setzt man für je zwei Elemente P und Q von $\mathfrak{H}$

$$U_{P,Q} = \sum_S B_S x_{PSQ^{-1}},$$

so wird, falls

$$\mathfrak{H} = \mathfrak{S}A_0 + \mathfrak{S}A_1 + \cdots + \mathfrak{S}A_{n-1} \qquad \left(n = \frac{h}{s}\right)$$

ist, die Matrix

$$X = (U_{P,Q}) \qquad (P, Q = A_0, A_1, \cdots A_{n-1})$$

des Grades bn eine zur Gruppe $\mathfrak{H}$ gehörende Gruppenmatrix. Denkt man sich die Gruppenmatrix X im Bereich aller Zahlen in irreduzible Bestandteile zerlegt, so kommt der dem einfachen Charakter $\chi(R)$ von $\mathfrak{H}$ entsprechende irreduzible Bestandteil genau $a = \sum_{\lambda=0}^{q-1} r_\lambda t_\lambda$ vor, falls

$$\chi(S) = \sum_{\lambda=0}^{q-1} r_\lambda \psi^{(\lambda)}(S) \quad , \qquad \eta(S) = \sum_{\lambda=0}^{q-1} t_\lambda \psi^{(\lambda)}(S)$$

ist (Frobenius, Sitzungsberichte 1898, S. 501).

Wir können nun folgenden allgemeinen Satz beweisen:

IX. *Es sei* P *ein gegebener Zahlkörper, es sei m der Index des Charakters $\chi(R)$ von $\mathfrak{H}$ und m' der Index des Charakters $\psi(S)$ der Untergruppe $\mathfrak{S}$ in bezug auf den Körper* P. *Man bezeichne mit*

$$\psi_0(S) = \psi(S) \ , \quad \psi_1(S), \cdots, \quad \psi_{d-1}(S)$$

die in bezug auf P *zu* $\psi(S)$ *konjugierten Charaktere von* $\mathfrak{S}$. *Enthält dann* $\chi(R)$ *den Charakter* $\psi_\delta(S)$ *von* $\mathfrak{S}$ *genau* r_δ *Mal, so muß m ein Divisor der Zahl* $m'(r_0+r_1+\cdots+r_{d-1})$ *sein.*

Man wähle nämlich für die oben betrachtete Gruppenmatrix U von $\mathfrak{S}$ die dem Charakter $\psi(S)$ entsprechende im Körper P irreduzible Gruppenmatrix. Dann wird der Charakter $\eta(S)$ von U gleich

$$\eta(S) = m'\{\psi_0(S)+\psi_1(S)+\cdots+\psi_{d-1}(S)\}.$$

Die Gruppenmatrix X ist dann ebenfalls in P rational und die ihr entsprechende Zahl a wird gleich $m'r_0+m'r_1+\cdots+m'r_{d-1}$. Nach Satz IV ist diese Zahl daher durch m teilbar.

Sind insbesondere für den Charakter $\psi(S)$ die Zahlen m' und d gleich 1, so muß die Zahl $r=r_0$ durch die Zahl m teilbar sein.

IXa. *Besitzt der Charakter* $\psi(S)$ *der Untergruppe* $\mathfrak{S}$ *von* $\mathfrak{H}$ *die Eigenschaft, daß die ihm entsprechende im Bereiche aller Zahlen irreduzible Gruppenmatrix von* $\mathfrak{S}$ *im Körper* P *rational darstellbar ist, und enthält der Charakter* $\chi(R)$ *von* $\mathfrak{H}$ *den Charakter* $\psi(S)$ *von* $\mathfrak{S}$ *genau r Mal, so ist r durch den Index m von* $\chi(R)$ *in bezug auf den Körper* P *teilbar.*

Der hier betrachtete Fall tritt insbesondere ein, wenn $\psi(S)$ den Hauptcharakter von $\mathfrak{S}$ bedeutet, d. h. wenn $\psi(S)=1$ ist. Denkt man sich daher für alle Untergruppen $\mathfrak{S}$ von $\mathfrak{H}$ die Zahlen $r=\frac{1}{s}\sum_S \chi(S)$ gebildet, so müssen diese Zahlen durch den Index m von $\chi(R)$ in bezug auf jeden Körper P teilbar sein. Insbesondere ergibt sich für den durch die Zahlen $\chi(R)$ bestimmten Zahlkörper $\mathrm{P}=\Omega(\chi)$:

X. *Ist* $\chi(R)$ *ein einfacher Charakter der Gruppe* $\mathfrak{H}$, *der den Hauptcharakter einer Untergruppe* $\mathfrak{S}$ *genau r Mal enthält, und sind die den verschiedenen Untergruppen* $\mathfrak{S}$ *von* $\mathfrak{H}$ *entsprechenden Zahlen r ohne gemeinsamen Divisor, so ist die zu* $\chi(R)$ *gehörende im Körper aller Zahlen irreduzible Gruppenmatrix Z im Körper* $\Omega(\chi)$ *rational darstellbar.*

Dieser Fall tritt insbesondere ein, wenn eine der Zahlen r gleich 1 ist. Das sich so ergebende spezielle Resultat ist auf anderem Wege von Hrn. Frobenius (Sitzungsberichte 1903, S. 328) durch Betrachtung der charakteristischen Einheiten der Gruppen gefunden worden. In seiner in der Einleitung zitierten Arbeit hat Hr. Burnside dieses Resultat, offenbar ohne die Frobeniussche Untersuchung zu kennen, von neuem abgeleitet.

Um eine Anwendung des Satzes X zu geben, will ich untersuchen, für welche Gruppen $\mathfrak{H}$ es eintreten kann, daß der Index m_0 eines einfachen Charakters $\chi(R)$ in bezug auf den Körper Ω, oder was dasselbe ist, in bezug auf den Körper $\Omega(\chi)$ den größten zulässigen Wert

$f = \chi(E)$ annimmt. Hierbei nehme ich an, was keine Beschränkung der Allgemeinheit bedeutet, daß die dem Charakter $\chi(R)$ entsprechende der Gruppe $\mathfrak{H}$ isomorphe Gruppe $\mathfrak{G}$ von linearen Substitutionen von der Ordnung h sein soll, so daß der zwischen $\mathfrak{H}$ und $\mathfrak{G}$ bestehende Isomorphismus ein einstufiger wird. Notwendig und hinreichend hierfür ist, daß die Zahl $\chi(R)$ für jedes von E verschiedene Element R nicht gleich f sein soll. Ist dann $\mathfrak{S}$ eine Untergruppe der Ordnung $s > 1$ von $\mathfrak{H}$, und enthält $\chi(R)$ den Hauptcharakter von $\mathfrak{S}$ genau r mal, so muß $r < f$ sein. Denn wäre $r = f$, so würde für jedes Element S von $\mathfrak{S}$ die Zahl $\chi(S)$ gleich f sein. Soll daher $m_0 = f$ sein, so muß, da r durch m_0 teilbar ist, $r = 0$ sein. Durch Betrachtung der zyklischen Untergruppen $\mathfrak{S}$ von $\mathfrak{H}$ ergibt sich insbesondere, daß die charakteristischen Wurzeln jeder Substitution von $\mathfrak{G}$, die identische Substitution E ausgenommen, von 1 verschieden sind. Nun hat Hr. Burnside[1] folgenden Satz bewiesen:

»Ist $\mathfrak{G}$ eine Gruppe linearer Substitutionen, in der die charakteristischen Wurzeln jeder von E verschiedenen Substitution von 1 verschieden sind, so muß jede Untergruppe $\mathfrak{S}$ von $\mathfrak{G}$, deren Ordnung eine Primzahlpotenz p^α ist, entweder zyklisch oder vom »Quaternionentypus«, d. h. der durch die Gleichungen

$$A^{2^{\alpha-1}} = E\,, \qquad B^2 = A^{2^{\alpha-2}}\,, \qquad B^{-1}AB = A^{-1}$$

definierten Gruppe isomorph sein. Im letzteren Falle zerfällt die Untergruppe $\mathfrak{S}$ der Ordnung 2^α in irreduzible Bestandteile vom Grade 2.«

Soll also $m_0 = f$ sein, so muß $\mathfrak{H}$ eine Gruppe sein, deren Untergruppen von Primzahlpotenzordnung entweder zyklisch oder vom Quaternionentypus sind.

Ist insbesondere f eine ungerade Zahl, so müssen alle Untergruppen von Primzahlpotenzordnung zyklisch sein. In diesem Falle ist $\mathfrak{H}$ bekanntlich eine auflösbare Gruppe.

Ist nun speziell f eine Primzahl p, so ist m_0 als Divisor von p entweder gleich 1 oder gleich p. Es ergibt sich der Satz:

XI. *Ist $\mathfrak{G}$ eine irreduzible Gruppe linearer Substitutionen in p Variabeln, wo p eine ungerade Primzahl ist, und ist $\mathfrak{G}$ nicht eine auflösbare Gruppe, in der jede Untergruppe von Primzahlpotenzordnung zyklisch ist, so läßt sich $\mathfrak{G}$ in dem durch die Spuren der Substitutionen von $\mathfrak{G}$ bestimmten Rationalitätsbereich rational darstellen.*

Die auflösbaren Gruppen $\mathfrak{G}$ der hier erwähnten Art wären noch besonders zu betrachten. Ich habe bis jetzt nicht entscheiden können, welche unter diesen Gruppen eine wirkliche Ausnahme bilden. Die Beantwortung dieser Frage scheint schwierig zu sein. Man betrachte

[1] The Messenger of Mathematics, 1905, S. 51.

z. B.; wenn ρ eine primitive Einheitswurzel des Grades 21 bedeutet, die durch die Substitutionen

$$A = \begin{pmatrix} \rho & 0 & 0 \\ 0 & \rho^4 & 0 \\ 0 & 0 & \rho^{16} \end{pmatrix}, \qquad B = \begin{pmatrix} 0 & 0 & 1 \\ 1 & 0 & 0 \\ 0 & \rho^7 & 0 \end{pmatrix}$$

erzeugte Gruppe $\mathfrak{G}$ der Ordnung 63. Es läßt sich leicht zeigen, daß $\mathfrak{G}$ dann und nur dann in dem durch die Zahl $\rho + \rho^4 + \rho^{16}$ bestimmten Zahlkörper rational darstellbar ist, wenn sich eine ganze rationale Funktion $f(\rho)$ mit rationalen Koeffizienten bestimmen läßt, die der Gleichung

$$\rho^7 = f(\rho)\ f(\rho^4)\ f(\rho^{16})$$

genügt.

§ 4.

Von besonderem Interesse ist noch der spezielle Fall, daß der unserer Betrachtung zugrunde liegende Körper P mit dem durch eine primitive h^{te} Einheitswurzel ρ bestimmten Körper $\Omega(\rho)$ übereinstimmt.

Es sei $\chi(R)$ ein einfacher Charakter des Grades f von $\mathfrak{H}$, dessen Index in bezug auf den Körper $\Omega(\rho)$ gleich m ist. Dann ist die zu $\chi(R)$ gehörende irreduzible Gruppenmatrix

$$Z = \sum_R A_R x_R$$

des Grades f in einem algebraischen Körper $\mathrm{P}(\mu)$ des Grades m über $\mathrm{P} = \Omega(\rho)$ rational darstellbar.

Ist nun $\mathfrak{S}$ eine Untergruppe der Ordnung s von $\mathfrak{H}$, welche die Eigenschaft besitzt, daß jede zu $\mathfrak{S}$ isomorphe Gruppe linearer Substitutionen im Körper der s^{ten} Einheitswurzeln und also auch im Körper $\Omega(\rho)$ rational darstellbar ist, und enthält der Charakter $\chi(R)$ den einfachen Charakter $\psi(S)$ von $\mathfrak{S}$ genau $r$mal, so muß nach Satz IXa die Zahl r durch m teilbar sein. Der hier betrachtete Fall tritt jedenfalls ein, wenn $\mathfrak{S}$ die durch ein Element P der Ordnung s erzeugte zyklische Gruppe ist. Ist σ eine $r$fache Wurzel der Gleichung $|xE - A_P| = 0$, so bilden die Zahlen

$$1, \sigma, \sigma^2, \cdots, \sigma^{s-1}$$

einen Charakter $\psi(S)$ von $\mathfrak{S}$, der in dem Charakter $\chi(R)$ von $\mathfrak{H}$ genau $r$mal enthalten ist. Daher muß die Zahl r durch m teilbar sein. Diesem Resultat läßt sich folgende Fassung geben:

XII. *Es sei $\mathfrak{G}$ eine (im Bereiche aller Zahlen) irreduzible Gruppe linearer Substitutionen der Ordnung h und es sei $\mathrm{P} = \Omega(\rho)$ der Körper der h^{ten} Einheitswurzeln. Läßt sich $\mathfrak{G}$ in einem algebraischen Körper $\mathrm{P}(\mu)$ des Grades m über P, aber in keinem Körper niedrigeren Grades über P rational darstellen, so muß die charakteristische Determinante jeder Substitution von $\mathfrak{G}$ die m^{te} Potenz einer ganzen rationalen Funktion sein.*

Läßt sich daher keine Zahl $m' > 1$ angeben, so daß für alle Substitutionen R von $\mathfrak{G}$ die Funktionen $|xE-R|$ sämtlich m'^{te} Potenzen werden, so muß $m = 1$ sein; die Gruppe $\mathfrak{G}$ läßt sich dann also im Körper $\Omega(\rho)$ rational darstellen. Dies ist das in der Einleitung erwähnte Resultat des Hrn. Burnside.

Es verdient noch hervorgehoben zu werden, daß aus dem Vorhandensein einer solchen Zahl $m' > 1$ nicht etwa, wie vermutet werden könnte[1], die Reduzibilität der Gruppe $\mathfrak{G}$ folgt. Man betrachte z. B. die durch die Substitutionen

$$A = \begin{pmatrix} 1 & 0 & 0 & 0 \\ 0 & -1 & 0 & 0 \\ 0 & 0 & 1 & 0 \\ 0 & 0 & 0 & -1 \end{pmatrix}, \quad B = \begin{pmatrix} 1 & 0 & 0 & 0 \\ 0 & 1 & 0 & 0 \\ 0 & 0 & -1 & 0 \\ 0 & 0 & 0 & -1 \end{pmatrix},$$

$$C = \begin{pmatrix} 0 & 1 & 0 & 0 \\ 1 & 0 & 0 & 0 \\ 0 & 0 & 0 & 1 \\ 0 & 0 & 1 & 0 \end{pmatrix}, \quad D = \begin{pmatrix} 0 & 0 & 0 & 1 \\ 0 & 0 & 1 & 0 \\ 0 & 1 & 0 & 0 \\ 1 & 0 & 0 & 0 \end{pmatrix}$$

erzeugte Gruppe der Ordnung 32. Unter den charakteristischen Determinanten der Substitutionen dieser Gruppen kommen nur die Funktionen

$$(x-1)^4, \quad (x+1)^4, \quad (x^2-1)^2, \quad (x^2+1)^2$$

vor, so daß hier $m' = 2$ gesetzt werden kann. Die Gruppe ist aber dennoch irreduzibel.

Von Interesse ist noch folgender Satz:

XIII. *Der Index m eines Charakters $\chi(R)$ des Grades f in bezug auf den Körper $\Omega(\rho)$ ist ein gemeinsamer Divisor der Zahlen f und $\frac{h}{f}$.*[2]

In der Tat ist offenbar $\chi(R) = m\phi(R)$, wo $\phi(R)$ eine Summe von $\frac{f}{m}$ Einheitswurzeln ist. Nun besteht (vgl. B., S. 425) für jedes Element S von $\mathfrak{H}$ die Relation

$$\sum_R \chi(SR^{-1})\chi(R) = \frac{h}{f}\chi(S).$$

Daher ist

$$\sum_R \varphi(SR^{-1})\varphi(R) = \frac{h}{fm}\varphi(S).$$

Setzt man nun ε_R gleich 1, falls R gleich E ist, und $\varepsilon_R = 0$, falls R von E verschieden ist, so kann man diese Gleichung auch in der Gestalt

$$\sum_R \varphi(R)\left\{\varphi(SR^{-1}) - \frac{h}{fm}\varepsilon_{SR^{-1}}\right\} = 0$$

[1] Vgl. die in der Einleitung zitierte Arbeit des Hrn. Burnside (S. 252).

[2] Ist also m durch die α^{te} Potenz einer Primzahl p teilbar, so muß h durch $p^{2\alpha}$ teilbar sein. Man kann noch zeigen, daß h durch $p^{2\alpha+1}$ teilbar sein muß.

schreiben. Da die Zahlen $\phi(R)$ nicht sämtlich 0 sind, so muß die Determinante h^{ten} Grades

$$\left|\varphi(RS^{-1}) - \frac{h}{fm}\varepsilon_{RS^{-1}}\right|$$

gleich 0 sein. Da nun die Zahlen $\phi(R)$ ganze algebraische Zahlen sind, so ergibt sich, daß die Zahl $\frac{h}{fm}$ einer Gleichung der Form

$$x^h + c_1 x^{h-1} + \cdots + c_h = 0$$

mit ganzen algebraischen Koeffizienten genügt. Folglich ist die Zahl $\frac{h}{fm}$ eine ganze algebraische und also eine ganze rationale Zahl.

Zu erwähnen ist noch, daß die Zahl m kleiner als f sein muß, sobald $f > 1$ ist. Denn wäre $m = f$, so müßte jede der Substitutionen A_P nur eine charakteristische Wurzel σ besitzen, daher wäre $A_P = \sigma E$. Ist nun $f > 1$, so wäre die Gruppenmatrix Z reduzibel.

Es gilt ferner der Satz:

XIV. *Jede auflösbare Gruppe $\mathfrak{G}$ von linearen Substitutionen der Ordnung h ist im Körper der h^{ten} Einheitswurzeln rational darstellbar.*

Dieser Satz ist eine unmittelbare Folge eines anderen Satzes:

XV. *Ist $\mathfrak{G}$ eine irreduzible Gruppe von linearen Substitutionen der Ordnung h und ist $\mathfrak{S}$ eine invariante Untergruppe der Ordnung $\frac{h}{n}$ von $\mathfrak{G}$, wo n eine Primzahl ist, so ist die Gruppe $\mathfrak{S}$ entweder im Bereiche aller Zahlen irreduzibel, oder sie zerfällt in n verschiedene (nicht äquivalente) irreduzible Bestandteile desselben Grades.*

Es mögen nämlich zunächst die irreduziblen Bestandteile von $\mathfrak{S}$, deren Anzahl gleich q sei, einander äquivalent sein. Wir können dann $\mathfrak{G}$ durch eine äquivalente Gruppe ersetzen, in der jede Substitution S von $\mathfrak{S}$ die Gestalt

$$S = \begin{pmatrix} S_1 & 0 & \cdots & 0 \\ 0 & S_1 & \cdots & 0 \\ . & . & \cdots & . \\ 0 & 0 & \cdots & S_1 \end{pmatrix}$$

annimmt, wo die Substitutionen S_1 eine der Gruppe $\mathfrak{S}$ isomorphe irreduzible Gruppe $\mathfrak{S}_1$ bilden. Ist f der Grad der Matrix S, so ist hierbei die Matrix $\mathfrak{S}_1$ vom Grade $\frac{f}{q}$. Es sei nun

$$\mathfrak{G} = \mathfrak{S} + \mathfrak{S}P + \mathfrak{S}P^2 + \cdots + \mathfrak{S}P^{n-1}.$$

Dann ist für irgend eine Substitution S von $\mathfrak{S}$ die Substitution

$$P^{-1}SP = S' = \begin{pmatrix} S_1' & 0 & \cdots & 0 \\ 0 & S_1' & \cdots & 0 \\ . & . & \cdots & . \\ 0 & 0 & \cdots & S_1' \end{pmatrix}$$

wieder in $\mathfrak{S}$ enthalten, und wir erhalten einen Automorphismus der Gruppe $\mathfrak{S}$, wenn wir dem Element S das Element S' zuordnen. Man schreibe nun P in der Form

$$P = \begin{pmatrix} P_{11} & \cdots & P_{1q} \\ \cdot & \cdots & \cdot \\ P_{q1} & \cdots & P_{qq} \end{pmatrix},$$

wo die $P_{\alpha\beta}$ gewisse Matrizen des Grades $\frac{f}{q}$ sind. Dann wird

$$S_1 P_{\alpha\beta} = P_{\alpha\beta} S'_1 .$$

Da nun die Matrizen S'_1 eine der Gruppe $\mathfrak{S}_1$ isomorphe irreduzible Gruppe bilden, so müssen sich die Matrizen $P_{\alpha\beta}$ untereinander nur um konstante Faktoren unterscheiden (vgl. B., S. 414), d. h. es ist $P_{\alpha\beta} = c_{\alpha\beta} Q_1$, wo Q_1 eine gewisse Matrix des Grades $\frac{f}{q}$ und $c_{\alpha\beta}$ eine Konstante ist. Es ist nun unmittelbar zu sehen, daß, wenn E_1 die Einheitsmatrix des Grades $\frac{f}{q}$ bedeutet, die Matrix

$$C = \begin{pmatrix} c_{11} E_1 & \cdots & c_{1q} E_1 \\ \cdot & \cdots & \cdot \\ c_{q1} E_1 & \cdots & c_{q1} E_1 \end{pmatrix}$$

mit allen Substitutionen S und auch mit P vertauschbar wird. Daher ist C mit allen Substitutionen der Gruppe $\mathfrak{G}$ vertauschbar und muß folglich, da $\mathfrak{G}$ irreduzibel sein soll, die Form cE besitzen; demnach muß

$$c_{11} = c_{22} = \cdots = c_{qq} = c \quad , \quad c_{\alpha\beta} = 0 \qquad (\alpha \neq \beta)$$

sein. Wäre nun $q > 1$, so würde $\mathfrak{G}$ zerfallen.

Es möge nun die Gruppe $\mathfrak{S}$ mindestens zwei einander nicht äquivalente irreduzible Bestandteile besitzen. Dann läßt sich $\mathfrak{G}$ durch eine äquivalente Gruppe ersetzen, in der jede Substitution S die Form

$$S = \begin{pmatrix} S_1 & 0 & \cdots & 0 \\ 0 & S_2 & \cdots & 0 \\ \cdot & \cdot & \cdots & \\ 0 & 0 & \cdots & S_t \end{pmatrix}$$

annimmt, wo die Matrizen S_α eine der Gruppe $\mathfrak{S}$ isomorphe Gruppe $\mathfrak{S}_\alpha$ bilden, die in einander äquivalente irreduzible Bestandteile zerfällt, während von den Gruppen $\mathfrak{S}_1, \mathfrak{S}_2, \cdots, \mathfrak{S}_t$ je zwei keinen irreduziblen Bestandteil gemeinsam haben.

Ist wieder

$$P^{-1} S P = S' = \begin{pmatrix} S'_1 & 0 & \cdots & 0 \\ 0 & S'_2 & \cdots & 0 \\ \cdot & \cdot & \cdots & \cdot \\ 0 & 0 & \cdots & S'_t \end{pmatrix},$$

so bilden auch die Substitutionen S'_α eine der Gruppe $\mathfrak{S}$ isomorphe Gruppe $\mathfrak{S}'_\alpha$, wobei dem Element S von $\mathfrak{S}$ das Element S'_α von $\mathfrak{S}'_\alpha$ entspricht. Ferner zerfällt offenbar auch $\mathfrak{S}'_\alpha$ in einander äquivalente irreduzible Bestandteile, und von den Gruppen $\mathfrak{S}'_1, \mathfrak{S}'_2, \cdots \mathfrak{S}'_t$ enthalten wieder je zwei keinen irreduziblen Bestandteil gemeinsam. Ist f_α der Grad der Matrix S_α, so werde P in der Form

$$P = \begin{pmatrix} P_{11} & \cdots & P_{1t} \\ \cdot & \cdots & \cdot \\ P_{t1} & \cdots & P_{tt} \end{pmatrix}$$

geschrieben, wo $P_{\alpha\beta}$ eine Matrix mit f_α Zeilen und f_β Spalten ist. Es wird dann wegen $SP = PS'$

$$S_\alpha P_{\alpha\beta} = P_{\alpha\beta} S'_\beta.$$

Enthalten nun die Gruppen $\mathfrak{S}_\alpha$ und $\mathfrak{S}'_\beta$ keinen irreduziblen Bestandteil gemeinsam, so muß, wie man leicht schließt, $P_{\alpha\beta} = 0$ sein. Ferner kann für jedes α nur ein $\beta = \alpha'$ und für ein β nur ein $\alpha = \beta''$ vorhanden sein, so daß $P_{\alpha\beta} \neq 0$ wird. Aus dem Nichtverschwinden der Determinante von P folgt dann, daß $P_{\alpha\alpha'}$ eine quadratische Matrix von nicht verschwindender Determinante sein muß, so daß $\mathfrak{S}_\alpha$ und $\mathfrak{S}'_{\alpha'}$ äquivalente Gruppen werden. Der Substitution P entspricht nun eine Permutation Q zwischen t Ziffern $1, 2, \cdots t$, die die Ziffer α in die Ziffer α' überführt. Man sieht auch sofort ein, daß Q ein Zyklus der Ordnung t sein muß, da andernfalls die Gruppe $\mathfrak{G}$ zerfallen würde. Es ist daher P^t die erste Potenz von P, die in $\mathfrak{S}$ enthalten ist; folglich muß $t = n$ sein. Man kann auch ohne Beschränkung der Allgemeinheit annehmen, daß

$$\mathfrak{S}_2 = \mathfrak{S}'_1,\ \mathfrak{S}_3 = \mathfrak{S}'_2,\ \cdots\ \mathfrak{S}_n = \mathfrak{S}'_{n-1},\ \mathfrak{S}_1 = \mathfrak{S}'_n$$

ist. — Daß nun die Gruppe $\mathfrak{S}_1$ und folglich auch die Gruppen $\mathfrak{S}_2$, $\mathfrak{S}_3, \cdots \mathfrak{S}_n$ irreduzibel sein müssen, kann man folgendermaßen schließen.

Man bezeichne die Spur der Substitution R von $\mathfrak{G}$ mit $\chi(R)$, ferner die Spur der Substitution S_α mit $\psi_\alpha(S)$. Dann wird für jedes Element S von $\mathfrak{S}$

$$\chi(S) = \psi_1(S) + \psi_2(S) + \cdots + \psi_n(S).$$

Ferner ist

$$\sum_S \psi_\alpha(S^{-1})\psi_\beta(S) = 0, \qquad (\alpha \neq \beta)$$

wo die Summation über alle Elemente der Gruppe $\mathfrak{S}$ zu erstrecken ist. Daher ist

$$\sum_S \chi(S^{-1})\chi(S) = \sum_S \{\psi_1(S^{-1})\psi_1(S) + \cdots + \psi_n(S^{-1})\psi_n(S)\}.$$

Die links stehende Summe ist aber offenbar gleich

$$n \sum_S \psi_1(S^{-1})\psi_1(S)$$

Zerfällt nun $\mathfrak{S}_1$ in r irreduzible Bestandteile, so wird

$$\sum_S \psi_1(S^{-1})\psi_1(S) = r^2\frac{h}{n},$$

also

$$\sum_S \chi(S^{-1})\chi(S) = r^2 h.$$

Da nun wegen der Irreduzibilität von $\mathfrak{G}$

$$\sum_R \chi(R^{-1})\chi(R) = h$$

ist, wo R alle Elemente von $\mathfrak{G}$ durchläuft, so muß $r^2 h \leqq h$, also $r = 1$ sein.

Aus dem eben bewiesenen Satz XV ergibt sich nun unser Satz XIV folgendermaßen.

Es genügt offenbar anzunehmen, daß $\mathfrak{G}$ irreduzibel ist. Es sei nun $\mathfrak{G}$ eine auflösbare Gruppe möglichst kleiner Ordnung h, die sich nicht im Körper $\Omega(\rho)$ der h^{ten} Einheitswurzeln rational darstellen läßt. Dann ist der Index m der Gruppe $\mathfrak{G}$ in bezug auf den Körper $\Omega(\rho)$ größer als 1. Wählt man dann in $\mathfrak{G}$ eine invariante Untergruppe $\mathfrak{S}$ der Ordnung $\frac{h}{n}$, wo n eine Primzahl ist, so ist $\mathfrak{S}$ eine auflösbare Gruppe, deren Ordnung kleiner als h ist. Daher würde sich jede zu $\mathfrak{S}$ isomorphe Gruppe linearer Substitutionen im Körper $\Omega(\rho)$ rational darstellen lassen. Folglich müßte die Gruppe $\mathfrak{S}$ jeden ihrer irreduziblen Bestandteile mindestens m mal enthalten. Nach Satz XV muß aber $\mathfrak{S}$ jeden irreduziblen Bestandteil nur einmal enthalten. Die Annahme $m > 1$ führt daher auf einen Widerspruch.

Aus dem Vorhergehenden ergibt sich der für die Anwendungen wichtige Satz:

XVI. *Es sei $\mathfrak{G}$ eine irreduzible Gruppe linearer Substitutionen der Ordnung h, und es sei $\mathfrak{S}$ eine Untergruppe von $\mathfrak{G}$, die auflösbar ist, und die einen ihrer irreduziblen Bestandteile genau r mal enthält. Sind dann die den verschiedenen Untergruppen $\mathfrak{S}$ und den verschiedenen irreduziblen Bestandteilen von $\mathfrak{S}$ entsprechenden Zahlen r ohne gemeinsamen Teiler, so ist $\mathfrak{G}$ im Körper der h^{ten} Einheitswurzeln rational darstellbar.*

Es sei noch erwähnt, daß bis jetzt überhaupt keine Gruppe $\mathfrak{G}$ linearer Substitutionen der Ordnung h bekannt ist, die sich nicht im Körper der h^{ten} Einheitswurzeln rational darstellen läßt.

Ausgegeben am 8. Februar.

10.

Untersuchungen über die Darstellung der endlichen Gruppen durch gebrochene lineare Substitutionen

Journal für die reine und angewandte Mathematik 132, 85 - 137 (1907)

Will man die sämtlichen Darstellungen einer gegebenen endlichen Gruppe $\mathfrak{H}$ durch gebrochene lineare Substitutionen bestimmen, so hat man, wie ich in meiner Arbeit „Über die Darstellung der endlichen Gruppen durch gebrochene lineare Substitutionen"*) gezeigt habe, in erster Linie eine *Darstellungsgruppe* $\mathfrak{K}$ von $\mathfrak{H}$ zu berechnen. Eine solche Gruppe $\mathfrak{K}$ ist durch folgende Eigenschaften charakterisiert:

1. $\mathfrak{K}$ enthält eine aus invarianten Elementen von $\mathfrak{K}$ bestehende Untergruppe $\mathfrak{M}$, und es ist $\frac{\mathfrak{K}}{\mathfrak{M}}$ der Gruppe $\mathfrak{H}$ isomorph;

2. der Kommutator von $\mathfrak{K}$ enthält alle Elemente von $\mathfrak{M}$;

3. es gibt keine Gruppe, welche die Eigenschaften 1 und 2 besitzt, und deren Ordnung größer ist als die Ordnung der Gruppe $\mathfrak{K}$.

Einer Gruppe $\mathfrak{H}$ können auch mehrere Darstellungsgruppen entsprechen, dagegen ist die *Abel*sche Gruppe $\mathfrak{M}$, die ich (D., S. 23) den *Multiplikator* von $\mathfrak{H}$ nenne, eindeutig bestimmt.

Für die Diskussion der Darstellungen von $\mathfrak{H}$ durch gebrochene lineare Substitutionen genügt es, irgend eine Darstellungsgruppe von $\mathfrak{H}$ zu kennen. Dagegen ist es gruppentheoretisch von Interesse, eine Übersicht über die verschiedenen (nicht isomorphen) Darstellungsgruppen von $\mathfrak{H}$ zu gewinnen und die Anzahl dieser Gruppen genauer zu bestimmen. Diese Aufgabe wird in § 1 der vorliegenden Arbeit behandelt. Es wird insbesondere für die gesuchte Anzahl eine obere Schranke abgeleitet, die bei einer allgemeinen

*) Dieses Journal Bd. 127, S. 20. — Im folgenden mit D. zitiert.

Klasse von Gruppen $\mathfrak{H}$, nämlich bei den vollkommenen Gruppen,*) stets erreicht wird.

Kennt man für die gegebene Gruppe $\mathfrak{H}$ eine Gruppe $\mathfrak{R}$, welche die Eigenschaften 1 und 2 besitzt, so ist es im allgemeinen mit erheblichen Schwierigkeiten verbunden, zu entscheiden, ob für sie auch die Eigenschaft 3 besteht. In § 2 leite ich ein Kriterium ab, das in vielen speziellen Fällen die Entscheidung dieser Frage erleichtert. Insbesondere ergibt sich der für die Anwendungen nützliche Satz, daß jede *abgeschlossene* Gruppe $\mathfrak{R}$, welche die Eigenschaften 1 und 2 aufweist, stets eine Darstellungsgruppe von $\mathfrak{H}$ ist. — Hierbei verstehe ich (D., S. 23) unter einer abgeschlossenen Gruppe eine Gruppe, deren Multiplikator die Ordnung 1 besitzt, die also als ihre eigene Darstellungsgruppe erscheint.

Die von mir in D., § 3, angegebene Methode zur Berechnung des Multiplikators einer Gruppe $\mathfrak{H}$ geht von der Betrachtung des allgemeinen Kompositionsschemas für die Elemente der Gruppe aus. In § 3 dieser Arbeit zeige ich, daß es zur Lösung dieser Aufgabe auch genügt, irgend ein vollständiges System von definierenden Relationen zwischen erzeugenden Elementen der Gruppe $\mathfrak{H}$ zu betrachten. Auf Grund dieser Methode bestimme ich in § 4 die Multiplikatorgruppen einer Reihe spezieller Gruppen, darunter auch der *Abel*schen Gruppen.

In § 5 stelle ich die Darstellungsgruppen der bekannten endlichen Gruppen auf, die man durch Betrachtung der binären linearen Substitutionen, deren Koeffizienten *Galois*sche Imaginäre sind, erhält. Die Bestimmung der Grade der sämtlichen irreduziblen Darstellungen dieser Gruppen durch ganze oder gebrochene lineare Substitutionen (mit beliebigen Koeffizienten) ergibt sich alsdann durch Berechnung der Charaktere ihrer Darstellungsgruppen (§ 6).

§ 1.

Es sei

$$\mathfrak{H} = H_0 + H_1 + \cdots + H_{h-1} \qquad (H_0 = E)$$

eine endliche Gruppe der Ordnung h, die durch die h^2 Relationen

$$H_\lambda H_\mu = H_{\varphi(\lambda,\mu)} \qquad (\lambda, \mu = 0, 1, \ldots h-1)$$

*) Unter einer vollkommenen Gruppe versteht man eine Gruppe, die nur kogrediente Isomorphismen in sich zuläßt und keine invarianten Elemente, außer dem Hauptelement E, enthält (vgl. *Hölder*, Math. Ann., Bd. 46, S. 321).

bestimmt ist. Eine Gruppe $\mathfrak{G}$ bezeichne ich dann als eine durch die *Abel*sche Gruppe

$$\mathfrak{A} = A_0 + A_1 + \cdots + A_{a-1}$$

ergänzte Gruppe von $\mathfrak{H}$ oder auch als eine Gruppe $(\mathfrak{A}, \mathfrak{H})$, wenn $\mathfrak{G}$ die Elemente von $\mathfrak{A}$ als invariante Elemente enthält und $\frac{\mathfrak{G}}{\mathfrak{A}}$ der Gruppe $\mathfrak{H}$ isomorph ist (D., S. 22). Setzt man

$$\mathfrak{G} = \mathfrak{A}G_0 + \mathfrak{A}G_1 + \cdots + \mathfrak{A}G_{h-1},$$

wobei der Komplex $\mathfrak{A}G_\lambda$ dem Element H_λ der mit $\frac{\mathfrak{G}}{\mathfrak{A}}$ isomorphen Gruppe $\mathfrak{H}$ entsprechen möge, so besteht für die Elemente $G_0, G_1, \ldots G_{h-1}$ ein System von h^2 Relationen

$$(1.) \qquad G_\lambda G_\mu = A_{\lambda,\mu} G_{\varphi(\lambda,\mu)}, \qquad (\lambda, \mu = 0, 1, \ldots h-1)$$

wo die $A_{\lambda,\mu}$ gewisse Elemente der Untergruppe $\mathfrak{A}$ von $\mathfrak{G}$ bedeuten. Die Gruppe $\mathfrak{G}$ ist als vollständig bestimmt anzusehen, wenn die h^2 Elemente $A_{\lambda,\mu}$ gegeben sind; ich will daher sagen, die Gruppe $\mathfrak{G}$ *entspreche dem Elementensystem* $A_{\lambda,\mu}$. Setzt man noch $A_{\lambda,\mu} = A_{H_\lambda, H_\mu}$, so genügen diese Elemente, wie sich aus dem assoziativen Gesetz ergibt, den h^3 Relationen

$$(2.) \qquad A_{P,Q} A_{PQ,R} = A_{P,QR} A_{Q,R}. \qquad (P, Q, R = H_0, H_1, \ldots H_{h-1})$$

Sind aber umgekehrt für irgend ein System von h^2 Elementen $A_{\lambda,\mu} = A_{H_\lambda, H_\mu}$ der Gruppe $\mathfrak{A}$ diese Bedingungen erfüllt, so bilden die ah Elemente

$$(3.) \qquad A_\alpha G_\lambda, \qquad (\alpha = 0, 1, \ldots a-1, \quad \lambda = 0, 1, \ldots h-1)$$

wenn vorausgesetzt wird, daß $G_0, G_1, \ldots G_{h-1}$ mit allen Elementen von $\mathfrak{A}$ vertauschbar sein und den Relationen (1.) genügen sollen, eine Gruppe $\mathfrak{G}$ der Ordnung ah. Denn alsdann sind für die Komposition der ah Elemente (3.) alle Bedingungen der Gruppenbildung erfüllt. Diese Gruppe $\mathfrak{G}$ ist dann eine gewisse durch die Gruppe $\mathfrak{A}$ ergänzte Gruppe von $\mathfrak{H}$.

Sind $C_0, C_1, \ldots C_{h-1}$ irgend welche Elemente von $\mathfrak{A}$ und setzt man

$$\bar{A}_{\lambda,\mu} = C_\lambda C_\mu C_{\varphi(\lambda,\mu)}^{-1} A_{\lambda,\mu},$$

so genügen auch die Elemente $\bar{A}_{\lambda,\mu} = \bar{A}_{H_\lambda, H_\mu}$ den Gleichungen (2.). Die beiden Elementensysteme $A_{\lambda,\mu}$ und $\bar{A}_{\lambda,\mu}$ sollen als einander *assoziiert* bezeichnet werden. Offenbar sind zwei Gruppen $(\mathfrak{A}, \mathfrak{H})$, die assoziierten Elementen-systemen $A_{\lambda,\mu}$ und $\bar{A}_{\lambda,\mu}$ entsprechen, als nicht von einander verschieden anzusehen. Denn ersetzt man das vollständige Restsystem $G_0, G_1, \ldots G_{h-1}$ von $\mathfrak{G}$ mod. $\mathfrak{A}$ durch das vollständige Restsystem $C_0 G_0, C_1 G_1, \ldots C_{h-1} G_{h-1}$, so treten an Stelle der Elemente $A_{\lambda,\mu}$ die Elemente $\bar{A}_{\lambda,\mu}$.

Sind ferner $A_{\lambda,\mu}$ und $A'_{\lambda,\mu}$ zwei Elementensysteme von $\mathfrak{A}$, denen Gruppen $(\mathfrak{A}, \mathfrak{H})$ entsprechen, so entspricht auch dem System der Elemente $A_{\lambda,\mu} A'_{\lambda,\mu}$ eine solche Gruppe. Denn auch diese Elemente genügen den Gleichungen (2.).

Es sei nun

$$\mathfrak{B} = B_0 + B_1 + \cdots + B_{a-1}$$

ebenso wie $\mathfrak{A}$ eine *Abel*sche Gruppe der Ordnung a. Sollen dann zwei durch die Gruppen $\mathfrak{A}$ und $\mathfrak{B}$ ergänzte Gruppen von $\mathfrak{H}$

$$\mathfrak{G} = \mathfrak{A} G_0 + \mathfrak{A} G_1 + \cdots + \mathfrak{A} G_{h-1}$$

und

$$\mathfrak{G}' = \mathfrak{B} G'_0 + \mathfrak{B} G'_1 + \cdots + \mathfrak{B} G'_{h-1},$$

die durch die Gleichungen

$$G_\lambda G_\mu = A_{\lambda,\mu} G_{\varphi(\lambda,\mu)}, \quad G'_\lambda G'_\mu = B_{\lambda,\mu} G'_{\varphi(\lambda,\mu)}$$

bestimmt sind, einander isomorph sein, so sind drei verschiedene Arten von Isomorphismen zu unterscheiden, die ich als Isomorphismen erster, zweiter und dritter Art bezeichnen will:

1. Es entspricht jedem Element des Komplexes $\mathfrak{A} G_\lambda$ ein Element des Komplexes $\mathfrak{B} G'_\lambda$. — Dann müssen die Gruppen $\mathfrak{A}$ und $\mathfrak{B}$ isomorph sein. Man kann auch das vollständige Restsystem $G'_0, G'_1, \ldots G'_{h-1}$ von $\mathfrak{G}'$ mod. $\mathfrak{B}$ so wählen, daß dem Element G_λ von $\mathfrak{G}$ das Element G'_λ von $\mathfrak{G}'$ entspricht; dann entspricht dem Element $A_{\lambda,\mu}$ das Element $B_{\lambda,\mu}$. Läßt sich umgekehrt ein Isomorphismus $\binom{A_\alpha}{B_\alpha}$ zwischen $\mathfrak{A}$ und $\mathfrak{B}$ angeben, der das Element $A_{\lambda,\mu}$ von $\mathfrak{A}$ in das Element $B_{\lambda,\mu}$ von $\mathfrak{B}$ überführt, so erhält man, indem man dem Element $A_\alpha G_\lambda$ von $\mathfrak{G}$ das Element $B_\alpha G'_\lambda$ von $\mathfrak{G}'$ zuordnet, einen Isomorphismus zwischen den Gruppen $\mathfrak{G}$ und $\mathfrak{G}'$.

2. Es entspricht jedem Element A_α von $\mathfrak{A}$ ein Element $\bar{B}_\alpha$ von $\mathfrak{B}$, ferner dem Element G_λ ein Element des Komplexes $\mathfrak{B} G'_{\chi(\lambda)}$, wo aber der Index $\chi(\lambda)$ nicht für jedes λ gleich λ ist. — Dann erhält man offenbar,

indem man dem Element H_λ von $\mathfrak{H}$ das Element $H_{\chi(\lambda)}$ entsprechen läßt, einen Automorphismus $\mathsf{H} = \binom{H_\lambda}{H_{\chi(\lambda)}}$ von $\mathfrak{H}$.*)

3. Es entsprechen Elementen von $\mathfrak{A}$ auch solche Elemente von $\mathfrak{G}'$, die nicht in $\mathfrak{B}$ enthalten sind. — Dann brauchen $\mathfrak{A}$ und $\mathfrak{B}$ nicht isomorph zu sein. Ein solcher Isomorphismus zwischen $\mathfrak{G}$ und $\mathfrak{G}'$ kann aber nur dann bestehen, wenn $\mathfrak{G}$ und $\mathfrak{G}'$ auch invariante Elemente enthalten, die nicht in $\mathfrak{A}$ und $\mathfrak{B}$ vorkommen. Dieser Fall kann also insbesondere nicht eintreten, wenn die Gruppe $\mathfrak{H}$ keine invarianten Elemente (außer E) enthält.

Über die Isomorphismen zweiter Art ist noch folgendes zu bemerken: Zunächst kann man durch passende Wahl des vollständigen Restsystems $G'_0, G'_1, \dots G'_{h-1}$ erreichen, daß dem Elemente G_λ von $\mathfrak{G}$ das Element $G'_{\chi(\lambda)}$ von $\mathfrak{G}'$ entspricht; dem Element $A_{\lambda,\mu}$ von $\mathfrak{A}$ korrespondiert dann das Element $B_{\chi(\lambda),\chi(\mu)} = \overline{B}_{\lambda,\mu}$ von $\mathfrak{B}$. Es sei nun speziell H ein innerer Automorphismus von $\mathfrak{H}$; dann läßt sich also ein Element H_ϱ von $\mathfrak{H}$ angeben, so daß $H_{\chi(\lambda)} = H_\varrho^{-1} H_\lambda H_\varrho$ wird. Entsprechend wird

$$G_\varrho'^{-1} G'_\lambda G'_\varrho = C_\lambda G'_{\chi(\lambda)},$$

wo C_λ ein gewisses Element von $\mathfrak{B}$ bedeutet. Dann erhalten wir aber auch einen Isomorphismus *erster* Art zwischen den Gruppen $\mathfrak{G}$ und $\mathfrak{G}'$, indem wir dem Element $A_\alpha G_\lambda$ von $\mathfrak{G}$ das Element $\overline{B}_\alpha C_\lambda^{-1} G'_\lambda$ von $\mathfrak{G}'$ zuordnen. Denn es entspricht alsdann dem Element

$$(A_\alpha G_\lambda)(A_\beta G_\mu) = A_\alpha A_\beta A_{\lambda,\mu} G_{\varphi(\lambda,\mu)}$$

von $\mathfrak{G}$ in $\mathfrak{G}'$ das Element

$$\overline{B}_\alpha \overline{B}_\beta \overline{B}_{\lambda,\mu} C^{-1}_{\varphi(\lambda,\mu)} G'_{\varphi(\lambda,\mu)} = \overline{B}_\alpha \overline{B}_\beta \overline{B}_{\lambda,\mu} G'_\varrho G'_{\chi\{\varphi(\lambda,\mu)\}} G_\varrho'^{-1};$$

dies ist aber, weil $\chi\{\varphi(\lambda,\mu)\} = \varphi\{\chi(\lambda),\chi(\mu)\}$ ist, gleich

$$\overline{B}_\alpha \overline{B}_\beta G'_\varrho G'_{\chi(\lambda)} G'_{\chi(\mu)} G_\varrho'^{-1} = (\overline{B}_\alpha C_\lambda^{-1} G'_\lambda)(\overline{B}_\beta C_\mu^{-1} G'_\mu).$$

Man hat daher neben den Isomorphismen erster Art nur solche Isomorphismen zweiter Art zwischen $\mathfrak{G}$ und $\mathfrak{G}'$ zu berücksichtigen, die auf äußere Automorphismen H von $\mathfrak{H}$ führen.

---

*) Nach dem Vorgange des Herrn *Frobenius* (Sitzungsberichte der Berl. Akad. 1901, S. 1324) sollen die Isomorphismen einer Gruppe in sich als *Automorphismen* der Gruppe, die kogredienten Isomorphismen in sich als *innere*, die kontragredienten als *äußere* Automorphismen bezeichnet werden.

Ist demnach $\mathfrak{H}$ speziell eine vollkommene Gruppe, so genügt es, um nachzuweisen, daß zwei ergänzte Gruppen $\mathfrak{G}$ und $\mathfrak{G}'$ von $\mathfrak{H}$ nicht isomorph sind, sich auf die Betrachtung der Isomorphismen erster Art zu beschränken.

Nach diesen vorbereitenden Bemerkungen wende ich mich zu der Behandlung der Aufgabe: *wie erhält man aus einer gegebenen Darstellungsgruppe $\mathfrak{K}$ von $\mathfrak{H}$ die etwa noch existierenden anderen Darstellungsgruppen?*

Es sei neben der Darstellungsgruppe $\mathfrak{K} = (\mathfrak{N}, \mathfrak{H})$ noch eine zweite Darstellungsgruppe $\mathfrak{K}' = (\mathfrak{N}', \mathfrak{H})$ bekannt. Dann sind die *Abel*schen Gruppen $\mathfrak{N}$ und $\mathfrak{N}'$ dem Multiplikator $\mathfrak{M}$ von $\mathfrak{H}$ isomorph; wir wollen sie daher beide mit $\mathfrak{M}$ bezeichnen. Es sei

$$\mathfrak{K} = \mathfrak{M} Q_0 + \mathfrak{M} Q_1 + \cdots + \mathfrak{M} Q_{h-1},$$
$$\mathfrak{K}' = \mathfrak{M} Q'_0 + \mathfrak{M} Q'_1 + \cdots + \mathfrak{M} Q'_{h-1},$$

ferner sei

$$Q_\lambda Q_\mu = J_{\lambda,\mu} Q_{\varphi(\lambda,\mu)}, \qquad Q'_\lambda Q'_\mu = J'_{\lambda,\mu} Q'_{\varphi(\lambda,\mu)},$$

wo $J_{\lambda,\mu}$ und $J'_{\lambda,\mu}$ Elemente von $\mathfrak{M}$ bedeuten.

Sind dann $M_0 = E, M_1, \ldots M_{m-1}$ die Elemente, ferner

$$\psi_{M_\varrho}(J) \qquad (\varrho = 0, 1, \ldots m-1;\ \psi_{M_\varrho}(J)\,\psi_{M_\sigma}(J) = \psi_{M_\varrho M_\sigma}(J))$$

die m Charaktere der *Abel*schen Gruppe $\mathfrak{M}$, so repräsentieren die m Zahlensysteme

$$\psi_{M_\varrho}(J_{\lambda,\mu}) = r^{(\varrho)}_{H_\lambda, H_\mu}$$

m Lösungen der Gleichungen

$$\text{(4.)} \qquad r_{P,Q}\, r_{PQ,R} = r_{P,QR}\, r_{Q,R}, \qquad (P, Q, R = H_0, H_1, \ldots H_{h-1})$$

von denen nicht zwei assoziiert sind (vgl. D., §§ 1 und 2). Daher muß jede andere Lösung der Gleichungen (4.) einer dieser m Lösungen assoziiert sein. Dasselbe ist für die m Zahlensysteme $\psi_{M_\varrho}(J'_{\lambda,\mu})$ der Fall. Folglich muß jedem M_ϱ ein anderes wohlbestimmtes Element $\overline{M}_\varrho$ von $\mathfrak{M}$ entsprechen, so daß die Zahlensysteme $\psi_{\overline{M}_\varrho}(J'_{\lambda,\mu})$ und $\psi_{M_\varrho}(J_{\lambda,\mu})$ einander assoziiert werden. Dann ist aber auch für je zwei Indizes ϱ und σ das System der h^2 Zahlen

$$\psi_{\overline{M}_\varrho}(J'_{\lambda,\mu})\,\psi_{\overline{M}_\sigma}(J'_{\lambda,\mu}) = \psi_{\overline{M}_\varrho \overline{M}_\sigma}(J'_{\lambda,\mu})$$

dem Zahlensystem

$$\psi_{M_\varrho}(J_{\lambda,\mu})\,\psi_{M_\sigma}(J_{\lambda,\mu}) = \psi_{M_\varrho M_\sigma}(J_{\lambda,\mu})$$

assoziiert. Daher erhalten wir einen Automorphismus A der Gruppe $\mathfrak{M}$, indem wir dem Element M_ϱ das Element $\overline{M}_\varrho$ zuordnen.

Es läßt sich dann bekanntlich ein anderer (eindeutig bestimmter) Automorphismus $\mathrm{B} = \left(\frac{J}{\overline{J}}\right)$ von $\mathfrak{M}$ angeben, so daß

$$\psi_{\overline{M}_\varrho}(J) = \psi_{M_\varrho}(\overline{J})$$

wird. Es möge B das Element $J'_{\lambda,\mu}$ in $J''_{\lambda,\mu}$ überführen. Dann ist zunächst die Gruppe $\mathfrak{K}'$ der dem Elementensystem $J''_{\lambda,\mu}$ entsprechenden Gruppe

$$\mathfrak{K}'' = \mathfrak{M}Q''_0 + \mathfrak{M}Q''_1 + \cdots + \mathfrak{M}Q''_{h-1} \qquad (Q''_\lambda Q''_\mu = J''_{\lambda,\mu} Q''_{\varphi(\lambda,\mu)})$$

isomorph. Ferner ist für diese Gruppe das Zahlensystem

$$\psi_{M_\varrho}(J''_{\lambda,\mu}) = \psi_{\overline{M}_\varrho}(J'_{\lambda,\mu})$$

dem Zahlensystem $\psi_{M_\varrho}(J_{\lambda,\mu})$ assoziiert, d. h. es lassen sich h Größen c_0, $c_1, \ldots c_{h-1}$ bestimmen, so daß

$$\psi_{M_\varrho}(J''_{\lambda,\mu}) = \frac{c_\lambda c_\mu}{c_{\varphi(\lambda,\mu)}}\,\psi_{M_\varrho}(J_{\lambda,\mu})$$

oder, was dasselbe ist,

$$\psi_{M_\varrho}(J''_{\lambda,\mu} J^{-1}_{\lambda,\mu}) = \frac{c_\lambda c_\mu}{c_{\varphi(\lambda,\mu)}}$$

wird. Betrachtet man jetzt die dem Elementensystem

$$J''_{\lambda,\mu} J^{-1}_{\lambda,\mu} = C_{\lambda,\mu}$$

entsprechende durch die Gruppe $\mathfrak{M}$ ergänzte Gruppe von $\mathfrak{H}$

$$\mathfrak{G} = \mathfrak{M}G_0 + \mathfrak{M}G_1 + \cdots + \mathfrak{M}G_{h-1}, \qquad (G_\lambda G_\mu = C_{\lambda,\mu} G_{\varphi(\lambda,\mu)})$$

so ist für diese Gruppe jedes der m Zahlensysteme $\psi_{M_\varrho}(C_{\lambda,\mu})$ dem Zahlensystem $\psi_{M_0}(C_{\lambda,\mu}) = 1$ assoziiert. Hieraus folgt aber nach den Ergebnissen von D., § 2, daß der Kommutator der Gruppe $\mathfrak{G}$ kein Element von $\mathfrak{M}$ außer dem Hauptelement E enthält.

Umgekehrt schließt man leicht, daß, wenn $\mathfrak{G}$ irgend eine Gruppe $(\mathfrak{M}, \mathfrak{H})$ ist, die durch das Elementensystem $C_{\lambda,\mu}$ bestimmt und deren Kommutator zu $\mathfrak{M}$ teilerfremd ist, die dem Elementensystem $J_{\lambda,\mu}\,C_{\lambda,\mu}$ entsprechende

Gruppe $(\mathfrak{M}, \mathfrak{H})$ eine Darstellungsgruppe $\mathfrak{K}''$ von $\mathfrak{H}$ ist. Denn es sind dann unter den m Zahlensystemen

$$\psi_{M_\varrho}(J_{\lambda,\mu} C_{\lambda,\mu}) \qquad (\varrho = 0, 1, \ldots m-1)$$

nicht zwei einander assoziiert. — Hierbei führen offenbar zwei einander assoziierte Elementensysteme $C_{\lambda,\mu}$ und $C'_{\lambda,\mu}$ von $\mathfrak{M}$ auf dieselbe Gruppe $\mathfrak{K}''$.

Wir gelangen also zu folgendem Resultat:

Es sei $\mathfrak{K}$ eine durch die Elemente $J_{\lambda,\mu}$ bestimmte Darstellungsgruppe von $\mathfrak{H}$; es seien ferner höchstens n Systeme

$$C^{(0)}_{\lambda,\mu}, C^{(1)}_{\lambda,\mu}, \ldots C^{(n-1)}_{\lambda,\mu}$$

von je h^2 Elementen der Gruppe $\mathfrak{M}$ vorhanden, von denen nicht zwei einander assoziiert sind, und die so beschaffen sind, daß die Kommutatorgruppen der ihnen entsprechenden n Gruppen $(\mathfrak{M}, \mathfrak{H})$

$$\mathfrak{G}^{(\nu)} = \mathfrak{M} G^{(\nu)}_0 + \mathfrak{M} G^{(\nu)}_1 + \cdots + \mathfrak{M} G^{(\nu)}_{h-1} \quad (G^{(\nu)}_\lambda G^{(\nu)}_\mu = C^{(\nu)}_{\lambda,\mu} G^{(\nu)}_{\varphi(\lambda,\mu)})$$

zu $\mathfrak{M}$ teilerfremd sind. Dann ist jede Darstellungsgruppe von $\mathfrak{H}$ einer der n durch die Gleichungen

$$Q^{(\nu)}_\lambda Q^{(\nu)}_\mu = J_{\lambda,\mu} C^{(\nu)}_{\lambda,\mu} Q^{(\nu)}_{\varphi(\lambda,\mu)} \qquad (\nu = 0, 1, \ldots n-1)$$

bestimmten Darstellungsgruppen

$$\mathfrak{K}^{(\nu)} = \mathfrak{M} Q^{(\nu)}_0 + \mathfrak{M} Q^{(\nu)}_1 + \cdots + \mathfrak{M} Q^{(\nu)}_{h-1}$$

isomorph.

Ehe ich zur Berechnung der hier definierten Zahl n schreite, will ich zeigen, daß zwischen je zweien der Gruppen $\mathfrak{K}^{(\nu)}$ kein Isomorphismus erster Art bestehen kann. Denn es möge sich etwa zwischen $\mathfrak{K}^{(\varrho)}$ und $\mathfrak{K}^{(\sigma)}$ ein solcher Isomorphismus angeben lassen. Dann muß auch ein Automorphismus $\mathsf{A} = \binom{J}{J'}$ von $\mathfrak{M}$ existieren, der das Element $J_{\lambda,\mu} C^{(\varrho)}_{\lambda,\mu}$ in ein Element der Form

$$J_{\lambda,\mu} C^{(\sigma)}_{\lambda,\mu} C_\lambda C_\mu C^{-1}_{\varphi(\lambda,\mu)} = J_{\lambda,\mu} F_{\lambda,\mu}$$

überführt, wo die C_λ gewisse Elemente von $\mathfrak{M}$ bedeuten. Es wird dann, wenn A das Element $J_{\lambda,\mu}$ in $J'_{\lambda,\mu}$, das Element $C^{(\varrho)}_{\lambda,\mu}$ in $C'_{\lambda,\mu}$ überführt,

$$J_{\lambda,\mu} F_{\lambda,\mu} = J'_{\lambda,\mu} C'_{\lambda,\mu}.$$

Folglich ist auch für jeden Charakter $\psi_M(J)$ von $\mathfrak{M}$

$$(5.)\qquad \psi_M(J_{\lambda,\mu})\,\psi_M(F_{\lambda,\mu}) = \psi_M(J'_{\lambda,\mu})\,\psi_M(C'_{\lambda,\mu}).$$

Nun ist aber, wenn $\psi_M(J') = \chi(J)$ gesetzt wird, auch $\chi(J)$ ein Charakter von $\mathfrak{M}$; es sei etwa $\chi(J) = \psi_{\bar{M}}(J)$. Dann folgt aber aus (5.), weil sowohl die Ausdrücke $\psi_M(F_{\lambda,\mu})$ als auch die Ausdrücke

$$\psi_M(C'_{\lambda,\mu}) = \psi_{\bar{M}}(C^{(\varrho)}_{\lambda,\mu})$$

sich auf die Form $\frac{c_\lambda c_\mu}{c_{\varphi(\lambda,\mu)}}$ bringen lassen, daß die Zahlensysteme $\psi_M(J_{\lambda,\mu})$ und

$$\psi_M(J'_{\lambda,\mu}) = \psi_{\bar{M}}(J_{\lambda,\mu})$$

einander assoziiert sind. Daher muß $\bar{M} = M$ und also auch $J'_{\lambda,\mu} = J_{\lambda,\mu}$ sein. Da nun aber die Kommutatorgruppe von $\mathfrak{K}$ alle Elemente von $\mathfrak{M}$ enthält und offenbar jedes Element von $\mathfrak{M}$, das ein Produkt von Kommutatorelementen der Gruppe $\mathfrak{K}$ ist, sich auch als Produkt der Elemente $J_{\lambda,\mu}$ darstellen läßt, so erzeugen die Elemente $J_{\lambda,\mu}$ die ganze Gruppe $\mathfrak{M}$. Daher muß, weil $J'_{\lambda,\mu} = J_{\lambda,\mu}$ ist, A der identische Automorphismus sein. Demnach ist $C'_{\lambda,\mu} = C^{(\varrho)}_{\lambda,\mu} = F_{\lambda,\mu}$, d. h. die Elementensysteme $C^{(\varrho)}_{\lambda,\mu}$ und $C^{(\sigma)}_{\lambda,\mu}$ sind einander assoziiert. Dies erfordert aber, daß $\varrho = \sigma$ wird.

Ist daher $\mathfrak{H}$ speziell eine vollkommene Gruppe, so sind unter den n Gruppen $\mathfrak{K}^{(\nu)}$ nicht zwei isomorph. — Dagegen können, wenn $\mathfrak{H}$ keine vollkommene Gruppe ist, zwischen den Gruppen $\mathfrak{K}^{(\nu)}$ noch Isomorphismen zweiter oder auch dritter Art bestehen, die von Fall zu Fall besonders zu untersuchen sind.

Es soll nun gezeigt werden, wie man die n Elementensysteme $C^{(\nu)}_{\lambda,\mu}$ zu bestimmen hat.

Es sei

$$\mathfrak{G} = \mathfrak{M}G_0 + \mathfrak{M}G_1 + \cdots + \mathfrak{M}G_{h-1} \qquad (G_\lambda G_\mu = c_{\lambda,\mu}\, G_{\varphi(\lambda,\mu)})$$

eine der n Gruppen $\mathfrak{G}^{(\nu)}$. Der Voraussetzung nach repräsentieren die Elemente $R'_0, R'_1, \ldots R'_{r-1}$ des Kommutators $\mathfrak{R}'$ von $\mathfrak{G}$ genau r mod. $\mathfrak{M}$ inkongruente Elemente von $\mathfrak{G}$. Da wir das Elementensystem $C_{\lambda,\mu}$ durch ein ihm assoziiertes System, oder was dasselbe ist, jedes Element G_λ durch ein anderes Element des Komplexes $\mathfrak{M}G_\lambda$ ersetzen dürfen, können wir annehmen, daß die Elemente $R'_0, R'_1, \ldots R'_{r-1}$ unter den h Elementen G_λ vorkommen.

Es sei nun $\mathfrak{R}$ der Kommutator von $\mathfrak{H}$ und

$$\mathfrak{H} = \mathfrak{R}T_0 + \mathfrak{R}T_1 + \cdots + \mathfrak{R}T_{s-1}. \qquad (rs = h,\ T_0 = E)$$

Bezeichnet man die Komplexe $\mathfrak{R}T_0, \mathfrak{R}T_1, \ldots \mathfrak{R}T_{s-1}$, die die *Abel*sche Gruppe $\mathfrak{S} = \frac{\mathfrak{H}}{\mathfrak{R}}$ bilden, mit $S_0, S_1, \ldots S_{s-1}$, ferner, wenn $T_\sigma = H_\lambda$ ist, das Element G_λ von $\mathfrak{G}$ mit T'_σ und den Komplex $\mathfrak{R}'T'_\sigma$ mit S'_σ, so wird, falls noch

$$S_\varrho S_\sigma = S_{\chi(\varrho,\sigma)} \qquad (\varrho, \sigma = 0, 1, \ldots s-1)$$

ist,

$$(6.) \qquad S'_\varrho S'_\sigma = S'_\sigma S'_\varrho = D_{\varrho,\sigma} S'_{\chi(\varrho,\sigma)};$$

hierbei bedeutet $D_{\varrho,\sigma} = D_{\sigma,\varrho} = D_{S_\varrho, S_\sigma}$ ein gewisses Element von $\mathfrak{M}$.

Man zeigt nun ohne Mühe, daß durch die s^2 Elemente $D_{\varrho,\sigma}$ alle h^2 Elemente $C_{\lambda,\mu}$ mitbestimmt sind, ferner, daß die h^3 aus dem assoziativen Gesetz hervorgehenden (den Gleichungen (2.) analogen) Bedingungsgleichungen für die Elemente $C_{\lambda,\mu}$ sich auf die s^3 Relationen

$$(7.) \qquad D_{S,T} D_{ST,U} = D_{S,TU} D_{T,U} \qquad (S, T, U = S_0, S_1, \ldots S_{s-1})$$

reduzieren. Daraus folgt, daß die Bestimmung der Elementensysteme $C_{\lambda,\mu}$ identisch ist mit der Bestimmung der verschiedenen (einander nicht assoziierten) Systeme $D_{S_\varrho, S_\sigma} = D_{S_\sigma, S_\varrho}$, die den Bedingungen (7.) genügen, oder, was dasselbe ist, die so beschaffen sind, daß die Gleichungen (6.) eine durch die Gruppe $\mathfrak{M}$ ergänzte *Abel*sche Gruppe von $\mathfrak{S}$ definieren.

Es seien nun $\varepsilon_1, \varepsilon_2, \ldots \varepsilon_k$ die Invarianten der *Abel*schen Gruppe $\mathfrak{S}$, ferner mögen die Elemente $S_1, S_2, \ldots S_k$ eine Basis von $\mathfrak{S}$ bilden, wobei das Basiselement $S_\varkappa$ zu dem Exponenten $\varepsilon_\varkappa$ gehören möge. Ist dann

$$S_\varrho = S_1^{a_{\varrho 1}} S_2^{a_{\varrho 2}} \ldots S_k^{a_{\varrho k}}, \qquad (\varrho = k+1, k+2, \ldots s-1)$$

so lassen sich die Elemente $T'_{k+1}, T'_{k+2}, \ldots T'_{s-1}$ innerhalb der Komplexe $\mathfrak{M}T'_{k+1}, \mathfrak{M}T'_{k+2}, \ldots \mathfrak{M}T'_{s-1}$ so wählen, daß auch

$$S'_\varrho = S_1'^{a_{\varrho 1}} S_2'^{a_{\varrho 2}} \ldots S_k'^{a_{\varrho k}} \qquad (\varrho = k+1, k+2, \ldots s-1)$$

wird. Dann sind die s^2 Elemente $D_{\varrho,\sigma}$ vollständig bestimmt, wenn man die k Elemente $D_1, D_2, \ldots D_k$ kennt, für die

$$S_1'^{\varepsilon_1} = D_1\mathfrak{R}',\ S_2'^{\varepsilon_2} = D_2\mathfrak{R}', \ldots S_k'^{\varepsilon_k} = D_k\mathfrak{R}'$$

ist.

Die Elemente $D_1, D_2, \dots D_k$ unterliegen aber keiner einschränkenden Bedingung; denn man überzeugt sich leicht, daß bei jeder Wahl derselben das ihnen entsprechende Elementensystem D_{s_ϱ, s_σ} den Bedingungen (7.) genügt. Es entspricht daher auch jeder beliebigen Wahl von k Elementen $D_1, D_2, \dots D_k$ innerhalb der Gruppe $\mathfrak{M}$ ein System von Elementen $C_{\lambda,\mu}$, welches die verlangte Eigenschaft besitzt.

Über die Wahl der Elemente $T'_1, T'_2, \dots T'_k$ innerhalb der Komplexe $\mathfrak{M}T'_1, \mathfrak{M}T'_2, \dots \mathfrak{M}T'_k$ ist jedoch noch nicht verfügt worden. Sind daher $M_1, M_2, \dots M_k$ beliebige k Elemente von $\mathfrak{M}$, so entspricht den beiden Systemen $D_1, D_2, \dots D_k$ und $D_1 M_1^{\varepsilon_1}, D_2 M_2^{\varepsilon_2}, \dots D_k M_k^{\varepsilon_k}$ dieselbe Gruppe $\mathfrak{G}$.

Wir sehen also, daß die gesuchte Anzahl n gleich ist der Anzahl der verschiedenen Möglichkeiten, innerhalb der Gruppe $\mathfrak{M}$ Elemente $D_1, D_2, \dots D_k$ auszuwählen, wobei zwei Systeme $D_1, D_2, \dots D_k$ und $D'_1, D'_2, \dots D'_k$ als nicht verschieden anzusehen sind, wenn sich k Elemente $M_1, M_2, \dots M_k$ bestimmen lassen, so daß

$$D'_1 = D_1 M_1^{\varepsilon_1},\ D'_2 = D_2 M_2^{\varepsilon_2}, \dots D' = D_k M_k^{\varepsilon_k}$$

wird.

Auf Grund dieses Ergebnisses läßt sich n leicht explizit angeben, wenn man die Invarianten $e_1, e_2, \dots e_l$ der *Abel*schen Gruppe $\mathfrak{M}$ als bekannt annimmt. Es mögen nämlich alsdann $F_1, F_2, \dots F_l$ eine Basis von $\mathfrak{M}$ bilden, und es sei e_λ die Ordnung von F_λ. Setzt man dann

$$D_\alpha = F_1^{d_{\alpha 1}} F_2^{d_{\alpha 2}} \dots F_l^{d_{\alpha l}}, \qquad (\alpha = 1, 2, \dots k)$$

so ersieht man, daß n gleich ist der Anzahl der verschiedenen Möglichkeiten, kl Zahlen $d_{\alpha\beta}$ derart zu wählen, daß $d_{\alpha\beta}$ einen der Werte $0, 1, \dots e_\beta - 1$ annimmt, wobei zwei Systeme $d_{\alpha\beta}$ und $d'_{\alpha\beta}$ als nicht verschieden zu gelten haben, falls $d'_{\alpha\beta} - d_{\alpha\beta}$ sich auf die Form $x\varepsilon_\alpha + y e_\beta$ bringen läßt. Hieraus ergibt sich aber für n der Wert

$$n = \prod_{\alpha,\beta} (\varepsilon_\alpha, e_\beta), \qquad (\alpha = 1, 2, \dots k;\ \beta = 1, 2, \dots l)$$

wo $(\varepsilon_\alpha, e_\beta)$ den größten gemeinsamen Teiler der Zahlen ε_α und e_β bedeutet. Wir erhalten also den Satz:

I. *Ist* $\mathfrak{R}$ *der Kommutator der Gruppe* $\mathfrak{H}$, *sind* $\varepsilon_1, \varepsilon_2, \dots \varepsilon_k$ *die Invarianten der Abelschen Gruppe* $\frac{\mathfrak{H}}{\mathfrak{R}}$ *und* $e_1, e_2, \dots e_l$ *die Invarianten des Multiplikators* $\mathfrak{M}$ *von* $\mathfrak{H}$, *so ist die Anzahl der verschiedenen Darstellungsgruppen* *von $\mathfrak{H}$ höchstens gleich $\prod\limits_{\alpha,\beta}(\varepsilon_\alpha, e_\beta)$ und stets genau gleich dieser Zahl, wenn $\mathfrak{H}$ eine vollkommene Gruppe ist.*

Speziell ergibt sich hieraus:

II. *Sind die Ordnungen der Gruppe $\frac{\mathfrak{H}}{\mathfrak{N}}$ und $\mathfrak{M}$ teilerfremde Zahlen, so besitzt $\mathfrak{H}$ nur eine Darstellungsgruppe.*

Dieser Fall tritt insbesondere ein, wenn der Kommutator von $\mathfrak{H}$ mit der Gruppe $\mathfrak{H}$ übereinstimmt, ein Resultat, das ich in D., § 3, auf anderem Wege abgeleitet habe.

§ 2.

In enger Beziehung zu der eben durchgeführten Untersuchung steht die allgemeinere Aufgabe: wenn eine *Abel*sche Gruppe $\mathfrak{A}$ und eine beliebige Gruppe $\mathfrak{H}$ gegeben sind, alle durch die Gruppe $\mathfrak{A}$ ergänzten Gruppen von $\mathfrak{H}$ zu bestimmen. — An dieser Stelle will ich mich damit begnügen, diese Aufgabe unter Hinzunahme der Forderung zu behandeln, daß die Kommutatorgruppen der zu bestimmenden Gruppen alle Elemente der Untergruppe $\mathfrak{A}$ enthalten sollen.*)

Es sei

$$\mathfrak{L} = \mathfrak{A}L_0 + \mathfrak{A}L_1 + \cdots + \mathfrak{A}L_{h-1}$$

---

*) Die Bedeutung dieser Gruppen beruht auf folgendem. — Man habe irgend eine endliche Gruppe $\mathfrak{H}$ von h gebrochenen linearen Substitutionen

$$x_\varkappa = \frac{a_{\varkappa 1}^{(\alpha)} y_1 + \cdots + a_{\varkappa, n-1}^{(\alpha)} y_{n-1} + a_{\varkappa, n}^{(\alpha)}}{a_{n1}^{(\alpha)} y_1 + \cdots + a_{n, n-1}^{(\alpha)} y_{n-1} + a_{n, n}^{(\alpha)}}. \qquad (\varkappa = 1, 2, \ldots n-1)$$

Man betrachte die h Matrizen n-ten Grades $A_\alpha = (a_{\varkappa\lambda}^{(\alpha)})$. Dann erzeugen die Matrizen $A_\alpha A_\beta A_\alpha^{-1} A_\beta^{-1}$ eine endliche Gruppe $\mathfrak{R}'$; die in $\mathfrak{R}'$ enthaltenen Matrizen der Form

$$\begin{pmatrix} \varrho & 0 & \ldots & 0 \\ 0 & \varrho & \ldots & 0 \\ \cdot & \cdot & \cdot & \cdot \\ 0 & 0 & \ldots & \varrho \end{pmatrix}$$

bilden eine *Abel*sche Gruppe $\mathfrak{A}$, deren Ordnung a sei. Man kann dann h Konstanten r_α bestimmen, so daß die Matrizen $r_\alpha A_\alpha$ eine endliche Gruppe der Ordnung ah erzeugen (vgl. D., §§ 2 und 3). Jede der so entstehenden Gruppen ist dann eine durch die Gruppe $\mathfrak{A}$ ergänzte Gruppe von $\mathfrak{H}$, die die im Texte verlangte Eigenschaft besitzt.

eine solche Gruppe; man bezeichne die Ordnung der *Abel*schen Gruppe $\mathfrak{A}$ mit a, ferner ihre a Charaktere mit

$$\chi^{(0)}(A),\ \chi^{(1)}(A), \ldots \chi^{(a-1)}(A).$$

Ist dann $\mathfrak{K}$ eine Darstellungsgruppe von $\mathfrak{H}$, für die die Zeichen M_ϱ, $J_{\lambda,\mu}$ und $\psi_{M_\varrho}(J)$ dieselbe Bedeutung haben wie früher, und ist noch

$$L_\lambda L_\mu = A_{\lambda,\mu} L_{\varphi(\lambda,\mu)}, \qquad (\lambda,\mu=0,1,\ldots h-1)$$

wo die $A_{\lambda,\mu}$ gewisse Elemente der Gruppe $\mathfrak{A}$ bedeuten, so sind die a Zahlensysteme

$$(8.) \qquad \chi^{(\alpha)}(A_{\lambda,\mu}) \qquad (\alpha=0,1,\ldots a-1)$$

wegen der über die Gruppe $\mathfrak{L}$ gemachten Voraussetzung genau a verschiedenen unter den m Zahlensystemen

$$\psi_{M_\varrho}(J_{\lambda,\mu}) \qquad (\varrho=0,1,\ldots m-1)$$

assoziiert (D., § 2). Es seien dies die Zahlensysteme

$$(9.) \qquad \psi_{M_0}(J_{\lambda,\mu}),\ \psi_{M_1}(J_{\lambda,\mu}), \ldots \psi_{M_{a-1}}(J_{\lambda,\mu}).$$

Dann bilden die a Elemente $M_0, M_1, \ldots M_{a-1}$ eine der Gruppe $\mathfrak{A}$ isomorphe Untergruppe $\mathfrak{M}'$ von $\mathfrak{M}$. Betrachtet man dann die zu $\mathfrak{M}'$ reziproke Untergruppe von $\mathfrak{M}$, d. h. diejenige Gruppe $\mathfrak{N}$, die aus allen den a Bedingungen

$$\psi_{M_\alpha}(N) = 1 \qquad (\alpha=0,1,\ldots a-1)$$

genügenden Elementen N von $\mathfrak{M}$ besteht, so ist die Gruppe $\mathfrak{M}'$, und also auch $\mathfrak{A}$ der Gruppe $\frac{\mathfrak{M}}{\mathfrak{N}}$ isomorph. Wir können daher ohne Beschränkung der Allgemeinheit annehmen, daß $\mathfrak{A}$ die Gruppe $\frac{\mathfrak{M}}{\mathfrak{N}}$ selbst bedeute.

Ist dann etwa

$$\mathfrak{M} = \mathfrak{N} K_0 + \mathfrak{N} K_1 + \cdots + \mathfrak{N} K_{a-1},$$

und bedeutet das Element A_β den Komplex $\mathfrak{N} K_\beta$, so repräsentieren die Wertsysteme

$$\chi^{(\alpha)}(A_\beta) = \chi^{(\alpha)}(\mathfrak{N} K_\beta) = \psi_{M_\alpha}(K_\beta) \qquad (\alpha,\beta=0,1,\ldots a-1)$$

die a Charaktere von $\mathfrak{A} = \frac{\mathfrak{M}}{\mathfrak{N}}$. Setzt man noch

$$J_{\lambda,\mu} = N_{\lambda,\mu} K_{\lambda,\mu}, \quad A_{\lambda,\mu} = \mathfrak{N} K'_{\lambda,\mu},$$

wo $N_{\lambda,\mu}$ ein Element von $\mathfrak{N}$, dagegen $K_{\lambda,\mu}$ und $K'_{\lambda,\mu}$ gewisse unter den Elementen $K_0, K_1, \ldots K_{a-1}$ bedeuten sollen, so stimmen die Zahlensysteme (8.) und (9.) mit den Zahlensystemen

$$\chi^{(\alpha)}(\mathfrak{N} K'_{\lambda,\mu}),$$

bzw.

$$\chi^{(\alpha)}(\mathfrak{N} K_{\lambda,\mu})$$

überein. Man schließt nun in ganz analoger Weise, wie es auf S. 91 bei der Betrachtung zweier Darstellungsgruppen von $\mathfrak{H}$ geschehen ist, daß sich ein Automorphismus der Gruppe $\frac{\mathfrak{M}}{\mathfrak{N}}$ angeben läßt, der das System der Elemente $\mathfrak{N} K'_{\lambda,\mu}$ von $\frac{\mathfrak{M}}{\mathfrak{N}}$ in ein Elementensystem $\mathfrak{N} K''_{\lambda,\mu}$ überführt, so daß das Zahlensystem $\chi^{(\alpha)}(\mathfrak{N} K''_{\lambda,\mu})$ für jedes α dem Zahlensystem $\chi^{(\alpha)}(\mathfrak{N} K_{\lambda,\mu})$ assoziiert wird. Es entspricht dann dem System der Elemente $\mathfrak{N} K''_{\lambda,\mu}$ eine durch die Gruppe $\mathfrak{A} = \frac{\mathfrak{M}}{\mathfrak{N}}$ ergänzte Gruppe $\mathfrak{L}'$ von $\mathfrak{H}$, welche der Gruppe $\mathfrak{L}$ isomorph ist, und zwar derart, daß zwischen $\mathfrak{L}$ und $\mathfrak{L}'$ ein Isomorphismus erster Art besteht. Ferner entspricht dem System der Elemente

$$(\mathfrak{N} K''_{\lambda,\mu})(\mathfrak{N} K_{\lambda,\mu})^{-1} = B_{\lambda,\mu}$$

eine durch die Gruppe $\mathfrak{A}$ ergänzte Gruppe

$$\mathfrak{G} = \mathfrak{A} G_0 + \mathfrak{A} G_1 + \cdots + \mathfrak{A} G_{h-1} \qquad (G_\lambda G_\mu = B_{\lambda,\mu} G_{\varphi(\lambda,\mu)})$$

von $\mathfrak{H}$, welche die Eigenschaft besitzt, daß ihr Kommutator $\mathfrak{R}'$ kein Element von $\mathfrak{A}$ (außer dem Element E) enthält.

Es mögen nun die Elemente $T_1, T_2, \ldots T_k$, bzw. die Zahlen $\varepsilon_1, \varepsilon_2, \ldots \varepsilon_k$ dieselbe Bedeutung haben wie auf S. 94; ferner setze man, falls $T_\lambda = H_\mu$ ist, $G_\mu = T'_\lambda$. Man kann dann durch passende Wahl der Elemente $G_0, G_1, \ldots G_{h-1}$ innerhalb der Komplexe $\mathfrak{A} G_0, \mathfrak{A} G_1, \ldots \mathfrak{A} G_{h-1}$ erreichen, daß die sämtlichen Elemente $B_{\lambda,\mu}$ als vollständig bestimmt erscheinen, sobald man die k Elemente $B_1, B_2, \ldots B_k$ von $\mathfrak{A}$ kennt, die den Bedingungen

$$(\mathfrak{R}' T'_1)^{\varepsilon_1} = B_1 \mathfrak{R}', \; (\mathfrak{R}' T'_2)^{\varepsilon_2} = B_2 \mathfrak{R}', \ldots (\mathfrak{R}' T'_k)^{\varepsilon_k} = B_k \mathfrak{R}'$$

genügen.

Es sei nun etwa

$$B_{\lambda,\mu} = B_1^{\beta_1} B_2^{\beta_2} \ldots B_k^{\beta_k};$$

ferner sei $B_\varkappa = \mathfrak{N} V_\varkappa$ und $B_{\lambda,\mu} = \mathfrak{N} V_{\lambda,\mu}$, wo $V_\varkappa$ und $V_{\lambda,\mu}$ gewisse unter den Elementen $K_0, K_1, \ldots K_{a-1}$ bedeuten. Setzt man dann für die auf S. 94 betrachteten Elemente $D_1, D_2, \ldots D_k$ von $\mathfrak{M}$ die Elemente $V_1, V_2, \ldots V_k$, so bestimmen sich die dort auftretenden Elemente $C_{\lambda,\mu}$ mit Hilfe der $V_\varkappa$ in genau derselben Weise wie die $B_{\lambda,\mu}$ mit Hilfe der $B_\varkappa$, wobei nur zu beachten ist, daß der Exponent $\beta_\varkappa$ modulo der Ordnung des Elementes $B_\varkappa$ von $\frac{\mathfrak{M}}{\mathfrak{N}}$ reduziert werden kann, diese Ordnung aber nicht mit der Ordnung des Elementes $V_\varkappa$ von $\mathfrak{M}$ übereinzustimmen braucht. Jedenfalls ergibt sich aber

$$C_{\lambda,\mu} = \overline{N}_{\lambda,\mu} V_1^{\beta_1} V_2^{\beta_2} \ldots V_k^{\beta_k},$$

folglich auch

$$C_{\lambda,\mu} = N'_{\lambda,\mu} V_{\lambda,\mu},$$

wo $\overline{N}_{\lambda,\mu}$ und $N'_{\lambda,\mu}$ Elemente von $\mathfrak{N}$ bedeuten. Man bilde nun das System der Elemente

$$J_{\lambda,\mu} C_{\lambda,\mu} = N_{\lambda,\mu} N'_{\lambda,\mu} K_{\lambda,\mu} V_{\lambda,\mu} = N''_{\lambda,\mu} K''_{\lambda,\mu},$$

wo auch $N''_{\lambda,\mu}$ ein Element von $\mathfrak{N}$ bedeutet. Diesem Elementensystem entspricht dann eine Darstellungsgruppe $\mathfrak{K}'$ von $\mathfrak{H}$, ferner ist $\frac{\mathfrak{K}'}{\mathfrak{N}}$ eine durch die Gruppe $\frac{\mathfrak{M}}{\mathfrak{N}}$ ergänzte Gruppe von $\mathfrak{H}$, die ebenso wie die Gruppe $\mathfrak{L}'$ durch die Elemente $\mathfrak{N} K''_{\lambda,\mu}$ von $\frac{\mathfrak{M}}{\mathfrak{N}}$ bestimmt ist. Folglich ist $\mathfrak{L}'$ und also auch $\mathfrak{L}$ der Gruppe $\frac{\mathfrak{K}'}{\mathfrak{N}}$ isomorph.

Umgekehrt schließt man leicht, daß, wenn $\mathfrak{K}$ eine beliebige Darstellungsgruppe von $\mathfrak{H}$ ist, die Gruppe $\frac{\mathfrak{K}}{\mathfrak{N}}$ eine durch die Gruppe $\frac{\mathfrak{M}}{\mathfrak{N}}$ ergänzte Gruppe von $\mathfrak{H}$ ist, deren Kommutator alle Elemente von $\frac{\mathfrak{M}}{\mathfrak{N}}$ umfaßt.

Wir können daher den Satz aussprechen:

III. *Sind $\mathfrak{K}, \mathfrak{K}', \mathfrak{K}'', \ldots$ die verschiedenen Darstellungsgruppen von $\mathfrak{H}$, so erhält man die sämtlichen durch Abelsche Gruppen $\mathfrak{A}$ ergänzten Gruppen von $\mathfrak{H}$, deren Kommutatorgruppen alle Elemente von $\mathfrak{A}$ enthalten, indem man die verschiedenen Untergruppen $\mathfrak{N}, \mathfrak{N}', \ldots$ des Multiplikators von $\mathfrak{H}$ aufsucht und die Gruppen*

$$\frac{\mathfrak{K}}{\mathfrak{N}}, \quad \frac{\mathfrak{K}'}{\mathfrak{N}}, \quad \frac{\mathfrak{K}''}{\mathfrak{N}}, \ldots$$
$$\frac{\mathfrak{K}}{\mathfrak{N}'}, \quad \frac{\mathfrak{K}'}{\mathfrak{N}'}, \quad \frac{\mathfrak{K}''}{\mathfrak{N}'}, \ldots$$
$$\cdot \quad \cdot \quad \cdot \quad \cdot \quad \cdot \quad \cdot \quad \cdot$$

bildet.

Aus diesem Satze läßt sich eine für die Anwendungen wichtige Folgerung ziehen.

IV. *Es sei $\mathfrak{L}$ eine durch die Abelsche Gruppe $\mathfrak{A}$ ergänzte Gruppe von $\mathfrak{H}$, deren Kommutator alle Elemente von $\mathfrak{A}$ enthält; es sei ferner a die Ordnung der Gruppe $\mathfrak{A}$ und m' die Ordnung des Multiplikators von $\mathfrak{L}$. Dann ist am' durch die Ordnung m des Multiplikators $\mathfrak{M}$ von $\mathfrak{H}$ teilbar. Läßt sich insbesondere eine Darstellungsgruppe $\mathfrak{Q}$ von $\mathfrak{L}$ angeben, die zugleich eine ergänzte Gruppe von $\mathfrak{H}$ ist, so ist $\mathfrak{Q}$ auch eine Darstellungsgruppe von $\mathfrak{H}$ und es ist $am'=m$. Ist speziell $\mathfrak{L}$ eine abgeschlossene Gruppe, so ist $\mathfrak{L}$ selbst eine Darstellungsgruppe von $\mathfrak{H}$.*

Ist nämlich r die Ordnung des Kommutators von $\mathfrak{H}$, so besitzt der Kommutator von $\mathfrak{L}$ die Ordnung ar, der Kommutator jeder Darstellungsgruppe $\mathfrak{Q}$ von $\mathfrak{L}$ die Ordnung $m'ar$. Nun läßt sich aber nach Satz III eine Darstellungsgruppe $\mathfrak{K}$ von $\mathfrak{H}$ und eine Untergruppe $\mathfrak{N}$ von $\mathfrak{M}$ angeben, so daß $\mathfrak{L}$ der Gruppe $\frac{\mathfrak{K}}{\mathfrak{N}}$ isomorph ist. Es ist aber dann $\mathfrak{K}$ eine ergänzte Gruppe von $\mathfrak{L}$, und da der Kommutator von $\mathfrak{K}$ die Ordnung $\frac{m}{a}\cdot ar$ hat, so ist (nach D., Satz II) die Zahl m' durch $\frac{m}{a}$, d. h. $m'a$ durch m teilbar. Ist insbesondere $\mathfrak{Q}$ auch eine ergänzte Gruppe von $\mathfrak{H}$, so ist umgekehrt auch m durch $m'a$ teilbar, also ist $m'a=m$ und folglich ist $\mathfrak{Q}$ eine Darstellungsgruppe von $\mathfrak{H}$. Dies tritt speziell ein, wenn $m'=1$ ist; denn alsdann ist $\mathfrak{Q}=\mathfrak{L}$, also jedenfalls eine ergänzte Gruppe von $\mathfrak{H}$.

§ 3.

In D., § 3, habe ich eine allgemeine Methode zur Bestimmung des Multiplikators und einer Darstellungsgruppe einer vorgeschriebenen Gruppe $\mathfrak{H}$ angegeben. Diese Methode läßt sich nun, wie jetzt gezeigt werden soll, derart modifizieren, daß sich die erforderlichen Rechnungen in den meisten speziellen Fällen bedeutend einfacher gestalten.

Ich will zunächst die a. a. O. gewonnenen Resultate kurz rekapitulieren.

Es sei wie immer

$$H_\lambda H_\mu = H_{\varphi(\lambda,\mu)} \qquad (\lambda, \mu = 0, 1, \ldots h-1)$$

das Kompositionsschema der Gruppe $\mathfrak{H}$. Man führe h erzeugende Elemente

$Q_0, Q_1, \ldots Q_{h-1}$ ein und setze fest, daß die h^2 Elemente

$$Q_\lambda\, Q_\mu\, Q^{-1}_{\varphi(\lambda,\mu)} = J_{H_\lambda, H_\mu} \qquad (\lambda, \mu = 0, 1, \ldots h-1)$$

mit den h erzeugenden Elementen vertauschbar sein sollen. Die h^3 Gleichungen

$$Q_\nu\, J_{H_\lambda, H_\mu} = J_{H_\lambda, H_\mu}\, Q_\nu \qquad (\lambda, \mu, \nu = 0, 1, \ldots h-1)$$

definieren alsdann eine unendliche Gruppe $\mathfrak{K}'$; die Elemente J_{H_λ, H_μ} erzeugen eine unendliche *Abel*sche Gruppe $\mathfrak{N}'$, die durch das System der h^3 Relationen

$$J_{P,Q}\, J_{PQ,R} = J_{P,QR}\, J_{Q,R} \qquad (P, Q, R = H_0, H_1, \ldots H_{h-1})$$

vollständig definiert erscheint. Die Gruppe $\mathfrak{N}'$ läßt sich dann als das direkte Produkt einer unendlichen *Abel*schen Gruppe $\mathfrak{N}''$ vom Range h, in der kein Element außer dem Hauptelement E eine endliche Periode besitzt, und einer endlichen Gruppe $\mathfrak{N}$ darstellen. Die Gruppe $\mathfrak{N}$ — die umfassendste in $\mathfrak{N}'$ enthaltene endliche Gruppe — ist dem Multiplikator $\mathfrak{M}$ von $\mathfrak{H}$ isomorph und repräsentiert zugleich den größten gemeinsamen Teiler der Untergruppe $\mathfrak{N}'$ und des Kommutators $\mathfrak{T}$ der Gruppe $\mathfrak{K}'$.

Ich nehme nun an, die Gruppe $\mathfrak{H}$ möge sich durch n Elemente

$$S_1 = H_1,\; S_2 = H_2, \ldots S_n = H_n$$

erzeugen lassen, für die ein System

$$f_\varkappa(S_\nu) = E \qquad (\varkappa = 1, 2, \ldots q)$$

von q die Gruppe $\mathfrak{H}$ vollständig definierenden Relationen bekannt ist;*) hierbei bedeutet $f_\varkappa(S_\nu)$ ein gewisses Produkt der S_ν mit positiven oder negativen Exponenten.

Bezeichnet man nun die den Elementen $H_1, H_2, \ldots H_n$ entsprechenden Elemente $Q_1, Q_2, \ldots Q_n$ von $\mathfrak{K}'$ mit $T_1, T_2, \ldots T_n$, so erzeugen diese n Elemente eine Untergruppe $\mathfrak{G}$ von $\mathfrak{K}'$, die durch die qn Relationen

$$(10.) \qquad T_\lambda \cdot f_\varkappa(T_\nu) = f_\varkappa(T_\nu) \cdot T_\lambda \qquad (\lambda = 1, 2, \ldots n;\; \varkappa = 1, 2, \ldots q)$$

definiert werden kann.

Die q Elemente

$$J_1 = f_1(T_\nu),\; J_2 = f_2(T_\nu), \ldots J_q = f_q(T_\nu)$$

---

*) Vgl. *Dyck,* Math. Ann., Bd. 20, S. 1.

gehören sämtlich der Untergruppe $\mathfrak{N}'$ von $\mathfrak{K}'$ an; sie mögen die Untergruppe $\mathfrak{B}'$ von $\mathfrak{G}$ erzeugen. Ist etwa H_λ gleich dem Produkt $g_\lambda(S_\nu)$ der n Elemente S_ν*) und bezeichnet man das Element $g_\lambda(T_\nu)$ von $\mathfrak{G}$ mit G_λ, so lassen sich offenbar die h^2 Elemente

$$G_\lambda G_\mu G_{\varphi(\lambda,\mu)}^{-1} = F_{H_\lambda, H_\mu} \qquad (\lambda, \mu = 0, 1, \ldots h-1)$$

als Produkte der $J_\varkappa$ darstellen. Umgekehrt lassen sich auch die Elemente $J_\varkappa$, wie unmittelbar ersichtlich ist, durch die F_{H_λ, H_μ} ausdrücken.

Aus den für die F_{H_λ, H_μ} bestehenden Gleichungen

$$(11.) \qquad F_{P,Q} F_{PQ,R} = F_{P,QR} F_{Q,R}$$

werden sich dann gewisse s unabhängige Gleichungen von der Form

$$(12.) \qquad J_1^{\beta_{\sigma 1}} J_2^{\beta_{\sigma 2}} \ldots J_q^{\beta_{\sigma q}} = E \qquad (\sigma = 1, 2, \ldots s)$$

ergeben, die den Gleichungen (11.) vollständig äquivalent sind und sich daher als ein System von definierenden Relationen für die durch die $J_\varkappa$ erzeugte *Abel*sche Gruppe $\mathfrak{B}'$ ansehen lassen.

Nun hat jedes der Elemente Q_λ von $\mathfrak{K}'$ die Form $C_\lambda G_\lambda$, wo C_λ ein gewisses Element von $\mathfrak{N}'$ bedeutet, und speziell die n Elemente $C_1, C_2, \ldots C_n$ gleich E zu setzen sind. Es wird dann

$$J_{H_\lambda, H_\mu} = C_\lambda C_\mu C_{\varphi(\lambda,\mu)}^{-1} F_{H_\lambda, H_\mu}.$$

Daher läßt sich die Gruppe $\mathfrak{N}'$ durch die $h-n$ Elemente $C_0, C_{n+1}, \ldots C_{h-1}$ und die q Elemente $J_1, J_2, \ldots J_q$ erzeugen.

Es möge jetzt (vgl. den D., S. 32, zitierten Satz) die *Abel*sche Gruppe $\mathfrak{B}'$ das direkte Produkt einer endlichen Gruppe $\mathfrak{B}$ und einer unendlichen Gruppe $\mathfrak{B}''$ vom Range k sein, in der kein Element außer dem Hauptelement eine endliche Periode besitzt.

Es kann dann nicht $k < n$ sein, weil sonst die Untergruppe $\mathfrak{N}''$ von $\mathfrak{N}'$ höchstens vom Range $h-n+k < h$ wäre, während wir doch wissen, daß dieser Rang genau gleich h ist. Daß aber k auch nicht größer sein kann als n, also gleich n sein muß, läßt sich etwa folgendermaßen einsehen.

Man setze, wenn $H_\alpha H_\beta H_\gamma \ldots = H_\sigma$ ist,

$$G_\alpha G_\beta G_\gamma \ldots = F_{H_\alpha, H_\beta, H_\gamma, \ldots} G_\sigma.$$

---

*) Hierbei soll für $\lambda = 1, 2, \ldots n$ das Produkt $g_\lambda(S_\nu)$ gleich S_λ sein.

Dann ist das Element $F_{H_\alpha, H_\beta, H_\gamma, \ldots}$ ein gewisses Produkt der F_{H_λ, H_μ}, und zwar ergibt sich aus dem assoziativen Gesetz für irgend welche p Elemente $A, B, \ldots N, P$ von $\mathfrak{H}$ die Beziehung

$$(13.)\qquad F_{A, B, \ldots N, P} = F_{A, B, \ldots N} F_{A, B, \ldots N, P},$$

die zur Berechnung der $F_{H_\alpha, H_\beta, H_\gamma, \ldots}$ dienen kann. Bezeichnet man nun das Produkt

$$\prod_R F_{P, R} \qquad (R = H_0, H_1, \ldots H_{h-1})$$

mit F_P, so ist, wie ich zeigen will,

$$(14.)\qquad F^h_{A, B, \ldots N, P} = F_A F_B \ldots F_N F_P F^{-1}_{AB \ldots NP}.$$

Für zwei Elemente A, B folgt dies nämlich leicht direkt aus (11.), indem man $P = A$, $Q = B$ setzt und auf beiden Seiten das Produkt über alle Elemente R von $\mathfrak{H}$ bildet. Es möge daher die Formel (14.) für weniger als p Elemente bereits als richtig gelten. Dann folgt aber aus (13.)

$$\begin{aligned} F^h_{A, B, \ldots N, P} &= F^h_{A, B, \ldots N} F^h_{AB \ldots N, P} \\ &= F_A F_B \ldots F_N F^{-1}_{AB \ldots N} F_{AB \ldots N} F_P F^{-1}_{AB \ldots NP}. \end{aligned}$$

Dies ist aber die zu beweisende Formel (14.).

Nun soll aber $G_\lambda = g_\lambda(T_\nu)$ sein. Ist daher $g_\lambda(T_\nu) = T_\alpha T_\beta T_\gamma \ldots$, so ist also

$$F_{S_\alpha, S_\beta, S_\gamma, \ldots} = E;$$

folglich ist wegen (14.)

$$(15.)\qquad F_{S_\alpha S_\beta S_\gamma \ldots} = F_{H_\lambda} = F_{S_\alpha} F_{S_\beta} F_{S_\gamma \ldots}.$$

Daher läßt sich jedes der Elemente F_{H_λ} durch die n Elemente $F_{S_1}, F_{S_2}, \ldots F_{S_n}$ ausdrücken. Aus (14.) folgt aber dann, daß sich auch die h-ten Potenzen aller $F_{A, B}$ und also auch aller Elemente von $\mathfrak{B}'$ als Produkte von Potenzen der n Elemente F_{S_ν} darstellen lassen, was offenbar für $k > n$ nicht möglich wäre.

Daher ist der Rang k der Gruppe $\mathfrak{B}''$ genau gleich n.*)

Es ist ferner leicht einzusehen, daß die Gruppe $\mathfrak{B}$, die als die umfassendste in $\mathfrak{B}'$ enthaltene endliche Gruppe definiert werden kann, gleich $\mathfrak{N}$ ist.

---

*) Zugleich ergibt sich, daß die n Elemente F_{S_ν} von einander abhängig sind und keine endliche Periode besitzen.

Da nämlich jedes Kommutatorelement von $\mathfrak{K}'$ die Form $Q_\lambda Q_\mu Q_\lambda^{-1} Q_\mu^{-1}$ hat und dieses Element auch gleich $G_\lambda G_\mu G_\lambda^{-1} G_\mu^{-1}$ ist, so ist der Kommutator $\mathfrak{T}$ von $\mathfrak{K}'$ zugleich auch der Kommutator von $\mathfrak{G}$. Nun enthält aber $\mathfrak{T}$ die ganze endliche Gruppe $\mathfrak{N}$. Folglich ist $\mathfrak{N}$ eine Untergruppe von $\mathfrak{B}'$ und also auch von $\mathfrak{B}$; da andererseits $\mathfrak{B}$ als eine endliche Untergruppe von $\mathfrak{N}'$ auch in $\mathfrak{N}$ enthalten sein muß, so ist in der Tat $\mathfrak{B}=\mathfrak{N}$.

Wir schließen hieraus, da $\mathfrak{N}$ dem Multiplikator $\mathfrak{M}$ von $\mathfrak{H}$ isomorph ist, daß das System $(\beta_{\sigma\varkappa})$ der Exponenten von (12.) genau vom Range $q-n$ ist, ferner daß diejenigen Elementarteiler dieses Systems, die größer sind als 1, mit den Invarianten $e_1, e_2, \ldots e_l$ der *Abel*schen Gruppe $\mathfrak{M}$ übereinstimmen. Insbesondere ist $m=e_1 e_2 \ldots e_l$ der größte gemeinsame Teiler aller Unterdeterminanten $(q-n)$-ten Grades des Systems $(\beta_{\sigma\varkappa})$.

Aus diesem Resultat ergibt sich noch eine für die Anwendungen nützliche Bemerkung.

Da die Relationen (12.) die Untergruppe $\mathfrak{B}'$ von $\mathfrak{G}$ vollständig definieren, so muß sich jede aus den Gleichungen (10.) resultierende Beziehung zwischen den Elementen $J_\varkappa = f_\varkappa(T_\nu)$ auch aus den Gleichungen (12.), d. h. durch Potenzieren und Multiplizieren dieser Gleichungen, ableiten lassen; oder, genauer gesprochen: ergibt sich eine Gleichung

$$J_1^{\gamma_1} J_2^{\gamma_2} \ldots J_q^{\gamma_q} = E,$$

so muß

$$\gamma_\varkappa = \sum_{\sigma=1}^{s} a_\sigma \beta_{\sigma\varkappa} \qquad (\varkappa=1,2,\ldots q)$$

sein, wo die a_σ gewisse ganze Zahlen bedeuten. Hat man daher aus den Gleichungen (10.) auf irgend einem Wege s' Relationen der Form

$$J_1^{\gamma_{\sigma 1}} J_2^{\gamma_{\sigma 2}} \ldots J_q^{\gamma_{\sigma q}} = E \qquad (\sigma=1,2,\ldots s')$$

gewonnen, so müssen die Elementarteiler des Systems $(\gamma_{\sigma\varkappa})$ nach einem bekannten Satze über Systeme mit ganzzahligen Koeffizienten*) durch die entsprechenden Elementarteiler $e_1, e_2, \ldots e_l$ des Systems $(\beta_{\sigma\varkappa})$ teilbar sein. Ist insbesondere der Rang von $(\gamma_{\sigma\varkappa})$ gleich $q-n$, so ist der größte gemeinsame Teiler $\overline{m}$ aller Unterdeterminanten des Grades $q-n$ von $(\gamma_{\sigma\varkappa})$ durch m teilbar. Man erhält auf diese Weise eine obere Grenze für die gesuchte Zahl m.

---

*) Vgl. *Frobenius*, dieses Journal Bd. 88, S. 96, und Sitzungsber. der Berl. Akad., 1894, S. 31; ferner *Hensel*, dieses Journal Bd. 114, S. 109.

Man wird in vielen Fällen mit Erfolg noch anders verfahren können. Man unterwerfe, wenn $s_1, s_2, \dots s_n$ die Ordnungen der Elemente $S_1, S_2, \dots S_n$ von $\mathfrak{H}$ sind, die n die Gruppe $\mathfrak{G}$ erzeugenden Elemente $T_1, T_2, \dots T_n$ noch den n Bedingungen

$$(16.)\qquad T_1^{s_1} = E,\quad T_2^{s_2} = E, \dots T_n^{s_n} = E.$$

Die durch die Gleichungen (10.) und (16.) definierte Gruppe $\mathfrak{G}'$ ist dann, wie ich zeigen will, eine endliche Gruppe.

Es sei nämlich $\mathfrak{S}$ eine beliebige Untergruppe von $\mathfrak{H}$ und es sei

$$\mathfrak{H} = \mathfrak{S}A_0 + \mathfrak{S}A_1 + \cdots + \mathfrak{S}A_{t-1}.$$

Dann ergibt sich, wenn P ein Element von $\mathfrak{S}$ ist, aus der Gleichung

$$F_{P,Q} F_{PQ,A_i} = F_{P,QA_i} F_{Q,A_i},$$

indem man auf beiden Seiten das Produkt über alle Elemente Q von $\mathfrak{S}$ bildet,

$$\prod_Q F_{P,Q} \cdot \prod_Q F_{PQ,A_i} = \prod_Q F_{P,QA_i} \cdot \prod_Q F_{Q,A_i}.$$

Nun ist aber offenbar

$$\prod_Q F_{PQ,A_i} = \prod_Q F_{Q,A_i};$$

folglich ist

$$\prod_Q F_{P,Q} = \prod_Q F_{P,QA_i},$$

und daraus ergibt sich

$$F_P = \prod_R F_{P,R} = \{\prod_Q F_{P,Q}\}^t. \qquad (R = H_0, H_1, \dots H_{h-1})$$

Wendet man dies auf die aus den Potenzen von S_ν bestehende zyklische Gruppe $\mathfrak{S}$ an, so erhält man

$$F_{S_\nu} = F'^{\frac{h}{s_\nu}}_{S_\nu},$$

wo

$$F'_{S_\nu} = \prod_{\sigma=0}^{s_\nu-1} F_{S_\nu, S_\nu^\sigma}$$

zu setzen ist. Fügt man nun zu den Gleichungen (10.) noch die Gleichungen (16.) oder, was offenbar dasselbe ist, die Gleichungen

$$F'_{S_\nu} = E$$

hinzu, so wird auch $F_{S_\nu} = E$; daraus folgt aber wegen (14.) und (15.), daß die h-ten Potenzen aller $F_{A,B}$ und also auch aller q Elemente $J_\varkappa$ gleich E werden. Dies besagt aber, daß die Gruppe $\mathfrak{G}'$ in der Tat eine endliche Gruppe ist, und zwar eine Gruppe, deren Ordnung keinen zu h teilerfremden Primfaktor enthält.

Ich will nun zeigen, daß der Kommutator von $\mathfrak{G}'$ genau von der Ordnung mr ist, falls r wie früher die Ordnung des Kommutators von $\mathfrak{H}$ bedeutet.

Es mögen nämlich die Elemente $F'_{s_1}, F'_{s_2}, \dots F'_{s_n}$ von $\mathfrak{G}$ die Untergruppe $\mathfrak{F}$ von $\mathfrak{G}$ erzeugen. Dann ist offenbar $\mathfrak{G}'$ der Gruppe $\frac{\mathfrak{G}}{\mathfrak{F}}$ isomorph, die durch die verschiedenen unter den Komplexen $\mathfrak{F}R$ gebildet wird, falls R alle Elemente von $\mathfrak{G}$ durchläuft. Nun besitzt aber kein Element von $\mathfrak{F}$ (außer E) eine endliche Periode; denn wäre dies der Fall, so würde sich wegen $F_{s_\nu} = F'^{\frac{h}{s_\nu}}_{s_\nu}$ eine Bedingungsgleichung für die n Elemente F_{s_ν} ergeben, was, wie wir auf S. 103 gesehen haben, nicht möglich ist. Daher ist $\mathfrak{F}$ zu der Gruppe $\mathfrak{B} = \mathfrak{R}$ und also auch zu dem Kommutator $\mathfrak{T}$ von $\mathfrak{G}$ teilerfremd. Sind nun

$$R_0, R_1, \dots R_{mr-1}$$

die Elemente der Gruppe $\mathfrak{T}$, so sind die Komplexe

$$\mathfrak{F}R_0, \mathfrak{F}R_1, \dots \mathfrak{F}R_{mr-1}$$

von einander verschieden. Diese Komplexe bilden aber offenbar den Kommutator von $\frac{\mathfrak{G}}{\mathfrak{F}}$. Daher ist auch die Ordnung des Kommutators von $\mathfrak{G}'$ gleich mr.*)

Es gelingt nun vielfach, die Ordnung und die Konstitution der Gruppe $\mathfrak{G}'$ direkt aus den sie definierenden Relationen zu ermitteln. Man erhält dann den Multiplikator von $\mathfrak{H}$ als den größten gemeinsamen Teiler des Kommutators von $\mathfrak{G}'$ und der durch die Elemente $J_\varkappa = f_\varkappa(T_\nu)$ erzeugten Gruppe $\mathfrak{A}$. Hierbei ist noch zu bemerken, daß, wenn $A_1, A_2, \dots A_n$ irgend welche Elemente von $\mathfrak{A}$ sind, es auch genügt, an Stelle der Gruppe $\mathfrak{G}'$ die

*) Leichter läßt sich das einsehen, indem man zeigt, daß $\mathfrak{G}'$ eine *hinreichend* ergänzte Gruppe von $\mathfrak{H}$ ist (vgl. D., S. 23 und S. 31), d. h. daß man durch Betrachtung der Darstellungen von $\mathfrak{G}'$ durch ganze lineare Substitutionen die sämtlichen Darstellungen von $\mathfrak{H}$ durch gebrochene lineare Substitutionen erhält.

durch die Elemente $A_1 T_1, A_2 T_2, \ldots A_n T_n$ erzeugte Untergruppe von $\mathfrak{G}'$ zu betrachten. Denn offenbar stimmt der Kommutator dieser Untergruppe mit dem Kommutator von $\mathfrak{G}'$ überein.

§ 4.

Zur Illustration der im vorigen Paragraphen entwickelten Methoden mögen folgende Beispiele dienen:

1. $\mathfrak{H}$ ist eine zyklische Gruppe der Ordnung h. — Eine solche ist durch eine einzige Gleichung $S^h = E$ definiert. Die Gruppe $\mathfrak{G}'$ ist durch die Gleichung $T^h = E$ zu definieren und stimmt also mit $\mathfrak{H}$ überein. Es ist hier daher $m = 1$ (vgl. auch D., S. 50).

2. Ich betrachte ferner die durch die Gleichungen

$$(17.)\qquad S_1^{2^t} = E, \quad S_2^2 = S_1^{2^{t-1}}, \quad S_2^{-1} S_1 S_2 = S_1^{-1} \qquad (t \geqq 2)$$

definierte Gruppe $\mathfrak{Q}_{t+1}$ der Ordnung 2^{t+1}. — Die Gruppe $\mathfrak{G}$ ist dann zu definieren durch

$$T_1^{2^t} = J_1, \quad T_2^2 = J_2 T_1^{2^{t-1}}, \quad T_2^{-1} T_1 T_2 = J_3 T_1^{-1},$$

wo J_1, J_2, J_3 mit T_1 und T_2 vertauschbar sein sollen. Aus der zweiten dieser Gleichungen folgt, daß $T_1^{2^{t-1}}$ mit T_2 vertauschbar ist. Erhebt man daher beide Seiten der letzten Gleichung zur 2^{t-1}-ten Potenz, so ergibt sich

$$T_1^{2^{t-1}} = J_3^{2^{t-1}} T_1^{-2^{t-1}}$$

oder

$$(18.)\qquad J_1^{-1} J_3^{2^{t-1}} = E.$$

Die oben betrachtete Zahl $q - n$ ist hier aber gleich $3 - 2 = 1$, ferner ist der größte gemeinsame Teiler der Exponenten in der Gleichung (18.) gleich 1. Daher ist auch in diesem Falle $m = 1$.

Die hier betrachtete Gruppe der Ordnung 2^{t+1} ist dadurch charakterisiert, daß sie nur ein Element der Ordnung 2 enthält und keine zyklische Gruppe ist.*) Man bezeichnet sie passend als die Gruppe der Ordnung 2^{t+1} vom Quaternionentypus. Diese Gruppe ist also eine abgeschlossene Gruppe.

---

*) Vgl. *Burnside*, Theory of Groups of finite Order, § 63.

3. Eine abgeschlossene Gruppe ist auch die durch die Gleichungen

$$(19.)\qquad S_1^{2^t}=E,\quad S_2^2=E,\quad S_2^{-1}S_1S_2=S_1^{-1+2^{t-1}}\qquad (t\geqq 3)$$

definierte Gruppe $\mathfrak{Q}'_{t+1}$ der Ordnung 2^{t+1}.*) Denn die Gruppe $\mathfrak{G}$ ist hier zu definieren durch

$$T_1^{2^t}=J_1,\quad T_2^2=J_2,\quad T_2^{-1}T_1T_2=J_3T_1^{-1+2^{t-1}},$$

wo die $J_\varkappa$ wieder mit T_1 und T_2 vertauschbar sein sollen. Aus der letzten Gleichung folgt

$$T_2^{-2}T_1T_2^2=T_1=J_3\cdot J_3^{-1+2^{t-1}}T_1^{(-1+2^{t-1})^2}=J_3^{2^{t-1}}J_1^{-1+2^{t-2}}T_1,$$

also

$$J_1^{-1+2^{t-2}}J_3^{2^{t-1}}=E.$$

Da auch hier $q-n=1$ ist, ferner wegen $t\geqq 3$ der größte gemeinsame Teiler der Exponenten in der eben gewonnenen Gleichung gleich 1 ist, so ist in der Tat $m=1$.

Aus den unter 1., 2. und 3. erhaltenen Ergebnissen folgt in Verbindung mit dem Satz X meiner früheren Arbeit:

V. *Ist p^a die höchste Potenz der Primzahl p, die in der Ordnung h der Gruppe $\mathfrak{H}$ aufgeht, und sind die Untergruppen der Ordnung p^a von $\mathfrak{H}$ entweder zyklische Gruppen oder auch, wenn $p=2$ ist, vom Typus* 2. *oder* 3., *so ist die Ordnung des Multiplikators von $\mathfrak{H}$ nicht durch p teilbar.***)

4. Es sei nun $\mathfrak{D}_t$ die Diedergruppe der Ordnung $2^t\,(t\geqq 3)$, d. h. die durch die Gleichungen

$$P_1^{2^{t-1}}=E,\; P_2^2=E,\; P_2^{-1}P_1P_2=P_1^{-1}$$

definierte Gruppe. — Der Multiplikator und die Darstellungsgruppen dieser Gruppe können durch folgende Überlegungen bestimmt werden. Man betrachte nämlich die durch die Gleichungen (17.) definierte Gruppe $\mathfrak{Q}_{t+1}$ der

---

*) Vgl. *Burnside*, a. a. O., § 65.

**) Ich will noch auf eine andere Klasse von Gruppen von Primzahlpotenzordnung hinweisen, die abgeschlossene Gruppen sind. Es sind dies die durch die Gleichungen

$$P^{p^\mu}=E,\quad Q^{p^\nu}=E,\quad Q^{-1}PQ=P^{1+p^{\mu-\nu}}$$

definierten Gruppen der Ordnung $p^{\mu+\nu}$ (vgl. *Burnside*, a. a. O., § 67). Hierbei kann ν eine beliebige positive ganze Zahl bedeuten, während μ für $p>2$ größer als ν, für $p=2$ größer als $\nu+1$ sein muß.

Ordnung 2^{t+1}. Das Element $S_1^{2^{t-1}} = M$ dieser Gruppe ist ein invariantes Element, ferner ist M eine Potenz des Kommutatorelementes $S_2^{-1} S_1 S_2 S_1^{-1} = S_1^{-2}$, also in dem Kommutator von $\mathfrak{Q}_{t+1}$ enthalten. Bezeichnet man nun die Gruppe $E + M$ mit $\mathfrak{M}$, so ist $\frac{\mathfrak{Q}_{t+1}}{\mathfrak{M}}$ der Gruppe $\mathfrak{D}_t$ isomorph. Da aber $\mathfrak{Q}_{t+1}$ eine abgeschlossene Gruppe ist, so ergibt sich nach Satz IV unmittelbar, daß $\mathfrak{Q}_{t+1}$ eine Darstellungsgruppe von $\mathfrak{D}_t$ ist. Daher ist auch $\mathfrak{M}$ der Multiplikator von $\mathfrak{D}_t$, also ist für die Diedergruppe $m = 2$. Ebenso wie $\mathfrak{Q}_{t+1}$ sind ferner auch die Gruppen $\mathfrak{Q}'_{t+1}$ und $\mathfrak{D}_{t+1}$ Darstellungsgruppen von $\mathfrak{D}_t$. Andere Darstellungsgruppen besitzt aber die Diedergruppe $\mathfrak{D}_t$ nicht. Der Kommutator $\mathfrak{R}$ von $\mathfrak{D}_t$ ist nämlich die durch das Element P_1^2 erzeugte Gruppe der Ordnung 2^{t-2}, ferner ist $\frac{\mathfrak{D}_t}{\mathfrak{R}}$ eine *Abel*sche Gruppe, deren Invarianten $\varepsilon_1 = 2$, $\varepsilon_2 = 2$ sind. Daher ist hier (wegen $m = 2$) die in § 1 betrachtete Zahl n gleich 4. Stellt man aber nach dem dort angegebenen Verfahren, etwa von der Gruppe $\mathfrak{Q}_{t+1}$ ausgehend, die Darstellungsgruppen von $\mathfrak{D}_t$ auf, so ergibt sich außer den drei Gruppen $\mathfrak{Q}_{t+1}$, $\mathfrak{Q}'_{t+1}$ und $\mathfrak{D}_{t+1}$ noch als vierte die durch die Gleichungen

$$R_1^{2^t} = E, \quad R_2^2 = R_1^{2^{t-1}}, \quad R_2^{-1} R_1 R_2 = R_1^{-1+2^{t-1}}$$

zu definierende Gruppe. Setzt man aber $S_1 = R_1$, $S_2 = R_1 R_2$, so genügen diese Elemente den Gleichungen (19.); daher ist die neu hinzukommende Gruppe der Gruppe $\mathfrak{Q}'_{t+1}$ isomorph.

5. Um eine weitere Anwendung unserer allgemeinen Resultate zu geben, will ich diejenigen Gruppen behandeln, die sich als direkte Produkte zweier oder mehrerer Gruppen darstellen lassen. — Ich beweise zunächst folgenden Satz:

VI. *Es seien $\mathfrak{B}$ und $\mathfrak{W}$ zwei endliche Gruppen, deren Kommutatorgruppen mit $\mathfrak{R}$ und $\mathfrak{S}$, und deren Multiplikatorgruppen mit $\mathfrak{C}$ und $\mathfrak{D}$ bezeichnet werden mögen. Ferner seien die Abelschen Gruppen $\frac{\mathfrak{B}}{\mathfrak{R}}$ und $\frac{\mathfrak{W}}{\mathfrak{S}}$ die direkten Produkte zyklischer Gruppen der Ordnungen $\eta_1, \eta_2, \ldots \eta_k$ und $\zeta_1, \zeta_2, \ldots \zeta_l$. Dann ist der Multiplikator der Gruppe $\mathfrak{B} \times \mathfrak{W}$ das direkte Produkt der Gruppen $\mathfrak{C}$, $\mathfrak{D}$ und der kl zyklischen Gruppen der Ordnungen*

$$(\eta_1, \zeta_1), (\eta_1, \zeta_2), \ldots (\eta_2, \zeta_1), (\eta_2, \zeta_2), \ldots (\eta_k, \zeta_l).$$

Hier bedeutet das Zeichen $\mathfrak{B} \times \mathfrak{W}$ das direkte Produkt der Gruppen $\mathfrak{B}$ und $\mathfrak{W}$ und das Zeichen $(\eta_\varkappa, \zeta_\lambda)$ wie früher den größten gemeinsamen Teiler der Zahlen $\eta_\varkappa$ und ζ_λ.

Es sei nämlich

$$\mathfrak{B} = V_0 + V_1 + \cdots + V_{r-1},$$
$$\mathfrak{W} = W_0 + W_1 + \cdots + W_{s-1},$$

ferner sei

$$(20.) \qquad V_\varkappa V_\lambda = V_{\varphi(\varkappa,\lambda)}, \quad W_\varrho W_\sigma = W_{\psi(\varrho,\sigma)}, \qquad \begin{pmatrix} \varkappa, \lambda = 0, 1, \ldots r-1 \\ \varrho, \sigma = 0, 1, \ldots s-1 \end{pmatrix}$$

wo $\varphi(\varkappa, \lambda)$ und $\psi(\varrho, \sigma)$ gewisse Indizes der Reihe $0, 1, \ldots r-1$, bzw. $0, 1, \ldots s-1$ bedeuten.

Die Gruppe $\mathfrak{H} = \mathfrak{B} \times \mathfrak{W}$ kann dann durch die Gleichungen (20.) und die rs Relationen

$$W_\varrho V_\varkappa = V_\varkappa W_\varrho \qquad (\varkappa = 0, 1, \ldots r-1, \ \varrho = 0, 1, \ldots, s-1)$$

definiert werden. Die Gruppe $\mathfrak{G}$ des vorigen Paragraphen ist dann zu definieren durch die Gleichungen

$$(21.) \qquad X_\varkappa X_\lambda = K_{\varkappa,\lambda} X_{\varphi(\varkappa,\lambda)},$$
$$(22.) \qquad Y_\varrho Y_\sigma = L_{\varrho,\sigma} Y_{\psi(\varrho,\sigma)}, \qquad \begin{pmatrix} \varkappa, \lambda = 0, 1, \ldots r-1 \\ \varrho, \sigma = 0, 1, \ldots s-1 \end{pmatrix}$$
$$(23.) \qquad Y_\varrho X_\varkappa = J_{\varkappa,\varrho} X_\varkappa Y_\varrho,$$

wo die Elemente $K_{\varkappa,\lambda}$, $L_{\varrho,\sigma}$ und $J_{\varkappa,\varrho}$ mit den erzeugenden Elementen $X_\varkappa$ und Y_ϱ vertauschbar sein sollen. Schreibt man noch ausführlicher

$$K_{\varkappa,\lambda} = K_{V_\varkappa, V_\lambda}, \quad L_{\varrho,\sigma} = L_{W_\varrho, W_\sigma}, \quad J_{\varkappa,\varrho} = J_{V_\varkappa, W_\varrho},$$

so ergeben sich zunächst aus (21.) und (22.) auf Grund des assoziativen Gesetzes die Relationen

$$(24.) \qquad K_{A,B} K_{AB,C} = K_{A,BC} K_{B,C}, \qquad (A, B, C = V_0, V_1, \ldots V_{r-1})$$
$$(25.) \qquad L_{P,Q} L_{PQ,R} = L_{P,QR} L_{Q,R}. \qquad (P, Q, R = W_0, W_1, \ldots W_{s-1})$$

Ferner erhält man aus (23.), wie eine leichte Rechnung zeigt, die Gleichungen

$$(26.) \qquad J_{A,P} J_{B,P} = J_{AB,P},$$
$$(27.) \qquad J_{A,P} J_{A,Q} = J_{A,PQ}.$$

Bezeichnet man nun das Element $X_\varkappa Y_\varrho$ mit $G_{\varkappa,\varrho}$, so wird

$$G_{\varkappa,\varrho} G_{\lambda,\sigma} = F_{\varkappa,\lambda,\varrho,\sigma} G_{\varphi(\varkappa,\lambda), \psi(\varrho,\sigma)},$$

wo

$$F_{\varkappa,\lambda,\varrho,\sigma} = K_{\varkappa,\lambda} L_{\varrho,\sigma} J_{\lambda,\varrho}$$

zu setzen ist.

Man überzeugt sich nun leicht, daß für diese Elemente die sämtlichen $(rs)^3$ aus dem assoziativen Gesetz hervorgehenden (den Gleichungen (11.) entsprechenden) Bedingungen erfüllt sind, sofern die Elemente

$$(28.) \qquad K_{A,B}, \quad L_{P,Q}, \quad J_{A,P}$$

den Gleichungen (24.) bis (27.) genügen. Daher ist die durch die Elemente (28.) erzeugte *Abel*sche Gruppe $\mathfrak{B}'$ durch die Gleichungen (24.) bis (27.) vollständig definiert. Erzeugen nun die Elemente $K_{A,B}$ die Gruppe $\mathfrak{C}'$, die Elemente $L_{P,Q}$ die Gruppe $\mathfrak{D}'$, endlich die Elemente $J_{A,P}$ die Gruppe $\mathfrak{J}$, so ist $\mathfrak{B}'$ offenbar das direkte Produkt der Gruppen $\mathfrak{C}'$, $\mathfrak{D}'$ und $\mathfrak{J}$. Ferner ist die umfassendste in $\mathfrak{B}'$ enthaltene endliche Gruppe $\mathfrak{B}$ das direkte Produkt der größten in $\mathfrak{C}'$, $\mathfrak{D}'$ und $\mathfrak{J}$ enthaltenen endlichen Gruppen. Die beiden ersteren sind nun, wie aus dem früheren folgt, den Multiplikatorgruppen $\mathfrak{C}$ und $\mathfrak{D}$ von $\mathfrak{V}$ und $\mathfrak{W}$ isomorph und mögen selbst mit $\mathfrak{C}$ und $\mathfrak{D}$ bezeichnet werden.

Die Gruppe $\mathfrak{J}$ ist aber, wie ich zeigen will, selbst eine endliche Gruppe, die als das direkte Produkt der kl zyklischen Gruppen der Ordnungen $(\eta_\varkappa, \zeta_\lambda)$ anzusehen ist.

In der Tat besagen die Gleichungen (26.), daß die r unter einander vertauschbaren Elemente

$$J_{A,P} \qquad (A = V_0, V_1, \dots V_{r-1})$$

für jedes P eine der Gruppe $\mathfrak{V}$ (ein- oder mehrstufig) isomorphe Gruppe bilden. Daraus folgt sofort, daß $J_{A,P} = E$ ist, sobald A dem Kommutator $\mathfrak{R}$ von $\mathfrak{V}$ angehört, und daß $J_{A,P} = J_{B,P}$ ist, falls A und B mod. $\mathfrak{R}$ kongruent sind. Ebenso ergibt sich aus (27.), daß für jedes A die Gleichung $J_{A,P} = E$ gilt, wenn P dem Kommutator $\mathfrak{S}$ von $\mathfrak{W}$ angehört, und daß $J_{A,P} = J_{A,Q}$ ist, falls PQ^{-1} in $\mathfrak{S}$ vorkommt. Es sei nun

$$\mathfrak{V} = \mathfrak{R} C_0 + \mathfrak{R} C_1 + \cdots + \mathfrak{R} C_{a-1},$$
$$\mathfrak{W} = \mathfrak{S} D_0 + \mathfrak{S} D_1 + \cdots + \mathfrak{S} D_{b-1}.$$

Man setze dann noch $\mathfrak{R} C_\alpha = R_\alpha$, $\mathfrak{S} D_\beta = S_\beta$ und

$$J_{C_\alpha, D_\beta} = J_{R_\alpha S_\beta}.$$

Dann wird offenbar $J_{A,P} = J_{R_\alpha, S_\beta}$, falls A dem Komplex R_α und P dem Komplex S_β angehört. Ferner reduzieren sich die Gleichungen (26.) und (27.)

auf die Gleichungen

(29.) $$J_{R,S} J_{R',S} = J_{RR',S}, \qquad (R, R' = R_0, R_1, \dots R_{a-1})$$

(30.) $$J_{R,S} J_{R,S'} = J_{R,SS'}. \qquad (S, S' = S_0, S_1, \dots S_{b-1})$$

Es mögen nun die Komplexe $R_1, R_2, \dots R_k$ eine Basis der *Abel*schen Gruppe $\frac{\mathfrak{B}}{\mathfrak{R}}$, die Komplexe $S_1, S_2, \dots S_l$ eine Basis der Gruppe $\frac{\mathfrak{W}}{\mathfrak{S}}$ bilden, und zwar sei hierbei $\eta_\varkappa$ die Ordnung von $R_\varkappa$ und ζ_λ die Ordnung von S_λ. Dann folgt aus (29.) und (30.), daß die ab Elemente $J_{R,S}$ sich als Produkte der kl Elemente $J_{\varkappa,\lambda} = J_{R_\varkappa, S_\lambda}$ darstellen lassen, ferner erhält man

$$J_{\varkappa,\lambda}^{\eta_\varkappa} = E, \qquad J_{\varkappa,\lambda}^{\zeta_\lambda} = E,$$

also

(31.) $$J_{\varkappa,\lambda}^{(\eta_\varkappa, \zeta_\lambda)} = E.$$

Es ist aber leicht zu sehen, daß, wenn diese Gleichungen bestehen, auch die allgemeineren Beziehungen (29.) und (30.) sämtlich erfüllt sind. Daher kann die *Abel*sche Gruppe $\mathfrak{J}$ durch die Gleichungen (31.) definiert werden und ist demnach in der Tat das direkte Produkt der kl zyklischen Gruppen der Ordnungen $(\eta_\varkappa, \zeta_\lambda)$.

Da nun das direkte Produkt $\mathfrak{B}$ der Gruppen $\mathfrak{C}, \mathfrak{D}$ und $\mathfrak{J}$ dem Multiplikator von $\mathfrak{H} = \mathfrak{B} \times \mathfrak{W}$ isomorph ist, so ist unser Satz bewiesen.

Kennt man zwei Darstellungsgruppen von $\mathfrak{B}$ und $\mathfrak{W}$, so ist es auch leicht, eine Darstellungsgruppe von $\mathfrak{B} \times \mathfrak{W}$ anzugeben. Sind nämlich

$$\mathfrak{B}' = \mathfrak{C} V_0' + \mathfrak{C} V_1' + \cdots + \mathfrak{C} V_{r-1}'$$

und

$$\mathfrak{W}' = \mathfrak{D} W_0' + \mathfrak{D} W_1' + \cdots + \mathfrak{D} W_{s-1}'$$

Darstellungsgruppen von $\mathfrak{B}$ und $\mathfrak{W}$, so erzeugen die $r+s$ Elemente $V_0', \dots V_{r-1}'$, $W_0', \dots W_{s-1}'$, falls festgesetzt wird, daß die rs Elemente

$$W_\varrho' V_\varkappa' W_\varrho'^{-1} V_\varkappa'^{-1} = J_{\varkappa,\lambda}$$

mit den $r+s$ erzeugenden Elementen vertauschbar sein sollen, eine Darstellungsgruppe von $\mathfrak{B} \times \mathfrak{W}$.

Beachtet man noch, daß der Kommutator $\mathfrak{T}$ von $\mathfrak{H} = \mathfrak{B} \times \mathfrak{W}$ das direkte Produkt der Gruppen $\mathfrak{R}$ und $\mathfrak{S}$, ferner daß die Gruppe $\frac{\mathfrak{H}}{\mathfrak{T}}$ das direkte Produkt der Gruppen $\frac{\mathfrak{B}}{\mathfrak{R}}$ und $\frac{\mathfrak{W}}{\mathfrak{S}}$ ist, so folgert man leicht aus unserem Satz VI durch den Schluß von n auf $n+1$ den allgemeineren Satz:

VII. *Es sei* $\mathfrak{H}$ *das direkte Produkt der n Gruppen* $\mathfrak{H}_1, \mathfrak{H}_2, \ldots \mathfrak{H}_n$. *Es möge* $\mathfrak{R}_\nu$ *der Kommutator,* $\mathfrak{M}_\nu$ *der Multiplikator der Gruppe* $\mathfrak{H}_\nu$ *sein; ferner sei die Abelsche Gruppe* $\frac{\mathfrak{H}_\nu}{\mathfrak{R}_\nu}$ *das direkte Produkt von* k_ν *zyklischen Gruppen der Ordnungen* $\varepsilon_{\nu 1}, \varepsilon_{\nu 2}, \ldots \varepsilon_{\nu k_\nu}$. *Dann ist der Multiplikator von* $\mathfrak{H}$ *das direkte Produkt der Gruppen* $\mathfrak{M}_1, \mathfrak{M}_2, \ldots \mathfrak{M}_n$ *und der* $\sum_{\mu<\nu} k_\mu k_\nu$ *zyklischen Gruppen der Ordnungen*

$$(\varepsilon_{11}, \varepsilon_{21}), (\varepsilon_{11}, \varepsilon_{22}), \ldots (\varepsilon_{1k_1}, \varepsilon_{nk_n}), (\varepsilon_{21}, \varepsilon_{31}), \ldots (\varepsilon_{n-1, k_{n-1}}, \varepsilon_{n, k_n}).$$

Für *Abel*sche Gruppen ergibt sich hieraus unter Berücksichtigung des Umstandes, daß jede zyklische Gruppe eine abgeschlossene Gruppe ist:

VIII. *Ist* $\mathfrak{H}$ *eine Abelsche Gruppe und sind* $\varepsilon_1, \varepsilon_2, \ldots \varepsilon_n$ *die Ordnungen der Elemente irgend einer Basis von* $\mathfrak{H}$, *so ist der Multiplikator von* $\mathfrak{H}$ *das direkte Produkt von* $\binom{n}{2}$ *zyklischen Gruppen der Ordnungen*

$$(\varepsilon_1, \varepsilon_2), (\varepsilon_1, \varepsilon_3), \ldots (\varepsilon_1, \varepsilon_n), (\varepsilon_2, \varepsilon_3), \ldots (\varepsilon_{n-1}, \varepsilon_n).$$

Sind also

$$p^{\alpha_1}, p^{\alpha_2}, \ldots p^{\alpha_k}, \qquad (\alpha_1 \geqq \alpha_2 \geqq \cdots \geqq \alpha_k)$$

$$q^{\beta_1}, q^{\beta_2}, \ldots q^{\beta_l}, \qquad (\beta_1 \geqq \beta_2 \geqq \cdots \geqq \beta_l)$$

$$\cdot \quad \cdot \quad \cdots \cdot$$

die Primzahlinvarianten einer *Abel*schen Gruppe, so ist die Ordnung ihres Multiplikators gleich

$$p^{\alpha_2 + 2\alpha_3 + \cdots + (k-1)\alpha_k} \cdot q^{\beta_2 + 2\beta_3 + \cdots + (l-1)\beta_l} \cdots,$$

die Ordnung jeder ihrer Darstellungsgruppen also gleich

$$p^{\alpha_1 + 2\alpha_2 + \cdots + k\alpha_k} \cdot q^{\beta_1 + 2\beta_2 + \cdots + l\beta_l} \cdots.$$

§ 5.

Mit Hilfe der im vorigen entwickelten Sätze gelingt es nun verhältnismäßig leicht, die Darstellungsgruppen der Gruppen binärer linearer Substitutionen, deren Koeffizienten Größen eines beliebigen *Galois*schen Feldes*) sind, genau zu berechnen.

Es sei p^n die Ordnung des gegebenen *Galois*schen Feldes, das ich nach dem Vorgange des Herrn *Dickson* mit $GF[p^n]$ bezeichnen will.

*) Vgl. *Dickson*, Linear Groups with an Exposition of the *Galois* Field Theory, Leipzig, 1901.

Es werden hier folgende drei Gruppen untersucht werden:

1. Die Gruppe $\mathfrak{F}_{p^n}$, die durch die gebrochenen linearen Substitutionen $\xi = \frac{\alpha\eta + \beta}{\gamma\eta + \delta}$ gebildet wird, deren Determinante $\alpha\delta - \beta\gamma$ gleich 1 ist. — Die Gruppe $\mathfrak{F}_{p^n}$ ist für $p = 2$ von der Ordnung $2^n(2^{2n} - 1)$, für eine ungerade Primzahl p von der Ordnung $\frac{p^n(p^{2n} - 1)}{2}$ und ist für $p^n > 3$ eine einfache Gruppe.

2. Die Gruppe $\mathfrak{L}_{p^n}$, die durch die ganzen linearen Substitutionen

$$\xi_1 = \alpha\eta_1 + \beta\eta_2, \quad \xi_2 = \gamma\eta_1 + \delta\eta_2$$

gebildet wird, deren Determinante gleich 1 ist. — Die Ordnung der Gruppe $\mathfrak{L}_{p^n}$ ist gleich $p^n(p^{2n} - 1)$. Ist $p = 2$, so ist $\mathfrak{L}_{p^n}$ der Gruppe $\mathfrak{F}_{p^n}$ isomorph. Es genügt daher, die Gruppe $\mathfrak{F}_{p^n}$ nur für den Fall $p > 2$ zu betrachten.

3. Die Gruppe $\mathfrak{H}_{p^n}$ der gebrochenen linearen Substitutionen $\xi = \frac{\alpha\eta + \beta}{\gamma\eta + \delta}$ von nicht verschwindender Determinante. — Die Ordnung dieser Gruppe ist gleich $p^n(p^{2n} - 1)$. Für $p = 2$ ist ferner $\mathfrak{H}_{p^n}$ der Gruppe $\mathfrak{L}_{p^n}$ isomorph; es soll daher auch $\mathfrak{H}_{p^n}$ nur für den Fall $p > 2$ betrachtet werden.

Ehe ich an die Bestimmung der Darstellungsgruppen dieser drei Gruppen gehe, schicke ich noch eine Hilfsbetrachtung voraus.

Es sei allgemein $\mathfrak{H}$ eine Gruppe der Ordnung h, ferner sei p eine Primzahl und $h = p^n r$, wo $n > 0$ und r zu p teilerfremd sein soll. Ich nehme an, die in $\mathfrak{H}$ enthaltenen Gruppen der Ordnung p^n seien *Abel*sche Gruppen. Man bezeichne, wenn $\mathfrak{P}$ eine dieser Gruppen ist, die Elemente einer Basis von $\mathfrak{P}$ mit $P_0, P_1, \dots P_{k-1}$.

Es sei nun A ein mit der Gruppe $\mathfrak{P}$ vertauschbares Element von $\mathfrak{H}$, so daß für jedes Element P von $\mathfrak{P}$ auch das Element $A^{-1}PA = P'$ in $\mathfrak{P}$ enthalten ist. Der Automorphismus $\binom{P}{P'}$ von $\mathfrak{P}$ ist dann vollständig bestimmt, wenn die k Elemente

$$P'_\varrho = A^{-1} P_\varrho A = P_0^{a_{\varrho 0}} P_1^{a_{\varrho 1}} \dots P_{k-1}^{a_{\varrho\, k-1}} \qquad (\varrho = 0, 1, \dots k-1)$$

bekannt sind. Ich will nun zeigen: *soll die Ordnung* m *des Multiplikators* $\mathfrak{M}$ *von* $\mathfrak{H}$ *durch* p *teilbar sein können, so muß die Bedingungskongruenz*

$$(32.) \qquad |a_{\varrho\sigma} - x\delta_{\varrho\sigma}| \equiv 0 \pmod{p} \qquad (\delta_{\varrho\varrho} = 1,\ \delta_{\varrho\sigma} = 0 \text{ für } \varrho \neq \sigma)$$

*ein Paar reziproker Wurzeln besitzen.**)

*) Die Wurzeln der Kongruenz (32.) können etwa als Größen des *Galois*schen Feldes $GF[p^{k!}]$ aufgefaßt werden.

Es sei nämlich $\mathfrak{K}$ eine Darstellungsgruppe von $\mathfrak{H}$. Es möge dem Element R von $\mathfrak{H}$ der Komplex $\mathfrak{M}K_R$ der mit $\mathfrak{H}$ isomorphen Gruppe $\frac{\mathfrak{K}}{\mathfrak{M}}$ entsprechen. Durchläuft dann P die p^n Elemente von $\mathfrak{P}$, so bilden die in den Komplexen $\mathfrak{M}K_P$ enthaltenen $p^n m$ Elemente eine Untergruppe $\mathfrak{P}'$ von $\mathfrak{K}$, die als eine durch die Gruppe $\mathfrak{M}$ ergänzte Gruppe von $\mathfrak{P}$ anzusehen ist. Der Kommutator $\mathfrak{N}'$ von $\mathfrak{P}'$ wird dann durch die p^{2n} Elemente

$$K_P K_Q K_P^{-1} K_Q^{-1} = J_{P,Q}$$

erzeugt, wo P und Q die sämtlichen Elemente von $\mathfrak{P}$ durchlaufen. Hierbei gehört, da ja $PQP^{-1}Q^{-1} = E$ ist, das Element $J_{P,Q}$ der Untergruppe $\mathfrak{M}$ von $\mathfrak{P}'$ an. Aus den Gleichungen

$$K_P K_Q = J_{P,Q} K_Q K_P$$

erhält man nun für die $J_{P,Q}$ ohne Mühe folgende Beziehungen:

(33.) $$J_{P,Q} J_{Q,P} = E,$$

(34.) $$J_{P,Q} J_{P,R} = J_{P,QR},$$

(35.) $$J_{P',Q'} = J_{P,Q}.$$

Hierbei bedeutet auch R ein beliebiges Element von $\mathfrak{P}$. — Aus den Gleichungen (33.) und (34.) ergibt sich nun unmittelbar, daß jedes der Elemente $J_{P,Q}$ als Produkt der k^2 Elemente $J_{\varrho\sigma} = J_{P_\varrho, P_\sigma}$ darstellbar ist, und zwar ist, wenn $P = P_0^{x_0} P_1^{x_1} \dots P_{k-1}^{x_{k-1}}$ und $Q = P_0^{y_0} P_1^{y_1} \dots P_{k-1}^{y_{k-1}}$ gesetzt wird,

$$J_{P,Q} = \prod_{\varrho,\sigma} J_{\varrho\sigma}^{x_\varrho y_\sigma}. \qquad (\varrho, \sigma = 0, 1, \dots k-1)$$

Beachtet man noch, daß $J_{\sigma\varrho} = J_{\varrho\sigma}^{-1}$ ist, so kann man diese Gleichung auch in der Form

$$J_{P,Q} = \prod_{\varrho<\sigma} J_{\varrho\sigma}^{x_\varrho y_\sigma - x_\sigma y_\varrho}$$

schreiben. Aus den Relationen (35.) ergeben sich insbesondere die $\binom{k}{2}$ Formeln

(35'.) $$J_{\varkappa\lambda} = \prod_{\varrho<\sigma} J_{\varrho\sigma}^{a_{\varkappa\varrho} a_{\lambda\sigma} - a_{\varkappa\sigma} a_{\lambda\varrho}}. \qquad (\varkappa < \lambda)$$

Bezeichnet man nun die Matrix $\binom{k}{2}$-ten Grades der Determinanten zweiten Grades $\begin{vmatrix} a_{\varkappa\varrho} & a_{\varkappa\sigma} \\ a_{\lambda\varrho} & a_{\lambda\sigma} \end{vmatrix}$ mit $(c_{\alpha\beta})$, ferner die Determinante $|c_{\alpha\beta} - \delta_{\alpha\beta}|$ mit d, so ergibt sich aus (35'.), wie man leicht sieht, für jedes der Elemente $J_{\varkappa\lambda}$ die Gleichung

$$J_{\varkappa\lambda}^d = E;$$

folglich ist auch allgemeiner $J_{P,Q}^d = E$. Nun lehren aber die Gleichungen (34.), daß auch $J_{P,Q}^{p^n} = E$ ist. Ist daher d zu p teilerfremd, so ergibt sich $J_{P,Q} = E$. Nun ist aber, wie ich D., S. 49, gezeigt habe, die r-te Potenz jedes Elementes J von $\mathfrak{M}$ in $\mathfrak{N}'$ enthalten. Soll daher die Ordnung m von $\mathfrak{M}$ durch p teilbar sein, so muß $d \equiv 0$ (mod. p) sein, da andernfalls $J^r = E$ wäre, was für $m \equiv 0$ (mod. p) nicht möglich ist. — Die Bedingung $d \equiv 0$ (mod. p) besagt nun, daß eine der $\binom{k}{2}$ Wurzeln der Kongruenz

$$|c_{\alpha\beta} - x\delta_{\alpha\beta}| \equiv 0 \quad (\text{mod. } p)$$

gleich 1 ist. Diese Wurzeln sind aber bekanntlich,*) wenn $w_0, w_1, \ldots w_{k-1}$ die k Wurzeln der Kongruenz (32.) bedeuten, die Größen $w_0 w_1, w_0 w_2, \ldots w_{k-2} w_{k-1}$. Demnach kann in der Tat, wie zu beweisen ist, m nur dann durch p teilbar sein, wenn mindestens eines der Produkte $w_\varrho w_\sigma (\varrho \neq \sigma)$ gleich 1 ist.

Ich kehre nun zur Betrachtung der Gruppen $\mathfrak{F}_{p^n}$, $\mathfrak{L}_{p^n}$ und $\mathfrak{H}_{p^n}$ zurück.

Im folgenden soll mit $\left(\frac{\alpha\beta}{\gamma\delta}\right)$ die gebrochene lineare Substitution $\xi = \frac{\alpha\eta + \beta}{\gamma\eta + \delta}$, mit $\binom{\alpha\beta}{\gamma\delta}$ die entsprechende ganze lineare Substitution bezeichnet werden, ferner soll v eine primitive Wurzel des Feldes $GF[p^n]$ bedeuten.**)

Ich behandle zuerst die Gruppen $\mathfrak{F}_{p^n} (p > 2)$ und $\mathfrak{L}_{p^n} (p \geqq 2)$.***)

Die Gruppe $\mathfrak{L}_{p^n}$ enthält, wenn $p > 2$ ist, das von $E = \begin{pmatrix} 1 & 0 \\ 0 & 1 \end{pmatrix}$ verschiedene invariante Element $F = \begin{pmatrix} -1 & 0 \\ 0 & -1 \end{pmatrix}$, und es ist, wenn die Gruppe $E + F$ mit $\mathfrak{A}$ bezeichnet wird, $\frac{\mathfrak{L}_{p^n}}{\mathfrak{A}}$ der Gruppe $\mathfrak{F}_{p^n}$ isomorph. Ferner ist F ein Kommutatorelement von $\mathfrak{L}_{p^n}$; denn bedeuten α und β zwei der Gleichung $\alpha^2 + \beta^2 = -1$ genügende Zahlen des Feldes $GF[p^n]$, so wird

$$(36.) \qquad \begin{pmatrix} \alpha & \beta \\ \beta & -\alpha \end{pmatrix} \begin{pmatrix} 0 & 1 \\ -1 & 0 \end{pmatrix} \begin{pmatrix} \alpha & \beta \\ \beta & -\alpha \end{pmatrix}^{-1} \begin{pmatrix} 0 & 1 \\ -1 & 0 \end{pmatrix}^{-1} = F.$$

---

*) Vgl. *G. Rados*, Math. Ann., Bd. 48, S. 417.

**) Betrachtet man die Gruppen $\mathfrak{F}_{p^n}$ und $\mathfrak{H}_{p^n}$ als Gruppen von Permutationen der $p^n + 1$ Symbole $\infty, 0, 1, v, \ldots v^{p^n - 2}$, so hat man der Substitution $\left(\frac{\alpha\,\beta}{\gamma\,\delta}\right)$ die Permutation $\eta = \frac{\alpha\xi + \gamma}{\beta\xi + \delta}$ zuzuordnen.

***) Für das im folgenden über die Konstitution dieser Gruppen angeführte verweise ich auf *Dickson*, Linear Groups, Cap. X und XII.

Bezeichnet man daher die Ordnung des Multiplikators von $\mathfrak{F}_{p^n}$ mit m, die des Multiplikators von $\mathfrak{L}_{p^n}$ mit m', so ist auf Grund unserer allgemeinen Resultate m durch 2 und $2m'$ durch m teilbar.

Ich werde nun zeigen, daß $\mathfrak{L}_{p^n}$ für $p \geqq 2$ mit Ausnahme der Fälle $p^n=4$ und $p^n=9$ eine abgeschlossene Gruppe ist. Es folgt dann unmittelbar, daß m für $p^n \neq 9$ gleich 2, und daß $\mathfrak{L}_{p^n}$ eine Darstellungsgruppe von $\mathfrak{F}_{p^n}$ ist. Dagegen ist, wie ich gleich bemerken will, m für $p^n=9$ gleich 6 und m' gleich 3.

Die Gruppe $\mathfrak{L}_{p^n}$ enthält nämlich Elemente der Ordnung p^n-1 und auch solche der Ordnung p^n+1. Ist daher q^a die höchste in der Ordnung $l=p^n(p^{2n}-1)$ von $\mathfrak{L}_{p^n}$ aufgehende Potenz einer zu $2p$ teilerfremden Primzahl q, so ist jede Untergruppe der Ordnung q^a von $\mathfrak{L}_{p^n}$ eine zyklische Gruppe. Mithin ist nach Satz V die gesuchte Zahl m' nicht durch q teilbar. Ist ferner $p>2$, so ist $F=\begin{pmatrix}-1 & 0\\ 0 & -1\end{pmatrix}$ das einzige Element der Ordnung 2, das in $\mathfrak{L}_{p^n}$ vorkommt. Denn soll die Substitution $R=\begin{pmatrix}\varkappa & \lambda\\ \mu & \nu\end{pmatrix}$ von der Ordnung 2 sein, so muß

$$\begin{pmatrix}\varkappa & \lambda\\ \mu & \nu\end{pmatrix}=\begin{pmatrix}\varkappa & \lambda\\ \mu & \nu\end{pmatrix}^{-1}=\begin{pmatrix}\nu, & -\lambda\\ -\mu & \varkappa\end{pmatrix},$$

also $\lambda=\mu=0$, $\varkappa=\nu$, ferner $\varkappa^2=1$, also $\varkappa=\pm 1$ sein; da nun R nicht gleich E sein kann, so muß $R=F$ sein. Beachtet man noch, daß die in der Gleichung (36.) auftretenden Substitutionen $\begin{pmatrix}\alpha & \beta\\ \beta & -\alpha\end{pmatrix}$ und $\begin{pmatrix}0 & 1\\ -1 & 0\end{pmatrix}$ eine nicht zyklische Gruppe der Ordnung 8 erzeugen, so ergibt sich, daß, wenn $p>2$ und 2^b die höchste in l aufgehende Potenz von 2 ist, die Untergruppen der Ordnung 2^b von $\mathfrak{L}_{p^n}$ vom Quaternionentypus sind. Demnach ist nach Satz V die Zahl m' für $p>2$ auch nicht durch 2 teilbar. Folglich ist in jedem Falle $m'=p^k$.

Die Untergruppen der Ordnung p^n von $\mathfrak{L}_{p^n}$ sind nun *Abel*sche Gruppen, deren Invarianten sämtlich gleich p sind. Daher ist*) $k \leqq \binom{n}{2}$.

Ist demnach $n=1$, so wird $k=0$ und also in der Tat $m'=1$.

Es sei daher $n>1$. Ich betrachte dann die Untergruppe $\mathfrak{P}$ der Ordnung p^n von $\mathfrak{L}_{p^n}$, die aus allen Substitutionen der Form $\begin{pmatrix}1 & 0\\ \gamma & 1\end{pmatrix}$ besteht. Eine Basis dieser Gruppe bilden die Substitutionen

$$P_0=\begin{pmatrix}1 & 0\\ 1 & 1\end{pmatrix},\ P_1=\begin{pmatrix}1 & 0\\ v^2 & 1\end{pmatrix}, \ldots P_{n-1}=\begin{pmatrix}1 & 0\\ v^{2(n-1)} & 1\end{pmatrix}.$$

*) Vgl. D., Satz X, und den Satz VIII dieser Arbeit.

Da nämlich die Potenzen

$$(37.)\qquad v^2,\ v^{2p},\ v^{2p^2}, \dots v^{2p^{n-1}}$$

von v^2 unter einander verschieden sind, so genügt v^2 einer mod. p irreduziblen Kongruenz des Grades n, die wir in der Form

$$(38.)\qquad x^n - a_{n-1}x^{n-1} - a_{n-2}x^{n-2} - \cdots - a_1 x - a_0 \equiv 0 \pmod{p}$$

schreiben können. Daher läßt sich jede Größe γ des Feldes $GF[p^n]$ auf eine und nur eine Weise in der Form

$$\gamma = c_0 + c_1 v^2 + c_2 v^4 + \cdots + c_{n-1} v^{2(n-1)}$$

darstellen, wo $c_0, c_1, \dots c_{n-1}$ Zahlen der Reihe $0, 1, \dots p-1$ bedeuten. Da aber alsdann

$$\begin{pmatrix} 1 & 0 \\ \gamma & 1 \end{pmatrix} = P_0^{c_0} P_1^{c_1} \dots P_{n-1}^{c_{n-1}}$$

wird, so bilden $P_0, P_1, \dots P_{n-1}$ in der Tat eine Basis der Gruppe $\mathfrak{P}$.

Nun ist die Substitution

$$A = \begin{pmatrix} v & 0 \\ 0 & v^{-1} \end{pmatrix}$$

mit $\mathfrak{P}$ vertauschbar, und zwar erhält man insbesondere

$$A^{-1}P_0 A = P_1,\quad A^{-1}P_1 A = P_2,\ \dots\ A^{-1}P_{n-2}A = P_{n-1},\quad A^{-1}P_{n-1}A = P_0^{a_0} P_1^{a_1} \dots P_{n-1}^{a_{n-1}}.$$

Soll daher m' durch p teilbar sein, so muß nach dem auf S. 114 angegebenen Hilfssatz die Kongruenz

$$\begin{vmatrix} -x, & 1, & 0, & \dots & 0, & 0 \\ 0, & -x, & 1, & \dots & 0, & 0 \\ \cdot & \cdot & \cdot & \cdot & \cdot & \cdot \\ 0, & 0, & 0, & \dots & -x, & 1 \\ a_0, & a_1, & a_2, & \dots & a_{n-2}, & a_{n-1}-x \end{vmatrix} \equiv 0 \pmod{p}$$

ein Paar reziproker Wurzeln besitzen. Die links stehende Determinante ist aber gleich

$$(-1)^n (x^n - a_{n-1}x^{n-1} - a_{n-2}x^{n-2} - \cdots - a_1 x - a_0),$$

also abgesehen vom Vorzeichen gleich der linken Seite der Kongruenz (38.). Die Wurzeln dieser Kongruenz sind die n Größen (37.). Soll daher $m' \equiv 0$

(mod. p) sein, so muß eine Gleichung der Form

$$v^{2(p^\varrho + p^\sigma)} = 1$$

bestehen, wo $0 \leqq \sigma < \varrho \leqq n-1$ ist. Da nun v zum Exponenten $p^n - 1$ gehört, so erfordert dies, daß $2(p^{\varrho-\sigma}+1)$ durch p^n-1 teilbar sei. Zwei solche Zahlen ϱ und σ lassen sich aber nur dann bestimmen, wenn p^n gleich 4 oder gleich 9 ist. — Daher ist in der Tat für $p^n \neq 4$ und $\neq 9$ die Zahl $m' = 1$.

Es ist noch folgendes zu bemerken. Für $p^n > 3$ ist $\mathfrak{F}_{p^n}$ eine einfache Gruppe und besitzt als solche nur eine Darstellungsgruppe. Aber auch der Fall $p^n = 3$ bildet keine Ausnahme. Denn die Gruppe $\mathfrak{F}_3$*) besitzt die Ordnung 12, ihr Kommutator die Ordnung 4; nach Satz II hat daher, da $\frac{12}{4} = 3$ zu der Ordnung $m = 2$ des Multiplikators von $\mathfrak{F}_3$ teilerfremd ist, auch $\mathfrak{F}_3$ nur eine Darstellungsgruppe.

Wir erhalten den Satz:

IX. *Die Darstellungsgruppe der Gruppe* $\mathfrak{F}_{p^n} (p > 2)$ *ist für* $p^n \neq 9$ *die Gruppe* $\mathfrak{L}_{p^n}$. *Die Gruppe* $\mathfrak{L}_{p^n} (p \geqq 2)$ *ist, wenn* p^n *von* 4 *und von* 9 *verschieden ist, eine abgeschlossene Gruppe.*

Der Ausnahmefall $p^n = 4$ erledigt sich leicht. Denn die Gruppe $\mathfrak{L}_4$ ist der Gruppe $\mathfrak{F}_5$ isomorph;**) daher ist die Darstellungsgruppe von $\mathfrak{L}_4$ die Gruppe $\mathfrak{L}_5$.

Von größerem Interesse ist der Fall $p^n = 9$. Da stets $m' \leqq p^{\binom{n}{2}}$, also $m \leqq 2p^{\binom{n}{2}}$ ist, so kommen hier für m nur die Werte 2 und 6 in betracht. Wäre nun $m = 2$, so wäre $\mathfrak{L}_9$ die Darstellungsgruppe von $\mathfrak{F}_9$. Der Grad einer jeden irreduziblen Darstellung von $\mathfrak{F}_9$ durch ganze oder gebrochene lineare Substitutionen müßte daher unter den Graden $f = \chi(E)$ der Charaktere von $\mathfrak{L}_9$ enthalten sein. Nun ist aber die Gruppe $\mathfrak{F}_9$ der Ordnung $\frac{9(9^2-1)}{2} = 360$ der alternierenden Gruppe $\mathfrak{A}_6$ des Grades 6 isomorph.***) Diese Gruppe ist, wie zuerst Herr *Wiman*†) nachgewiesen hat, einer von Herrn *Valentiner*††)

---

*) Die Gruppe $\mathfrak{F}_3$ ist der alternierenden Gruppe des Grades 4 isomorph.

**) Diese beiden Gruppen der Ordnung 60 sind der alternierenden Gruppe des Grades 5 isomorph. Vgl. *Dickson*, Linear Groups, S. 309.

***) Vgl. *Dickson*, a. a. O.

†) Math. Ann., Bd. 47, S. 531.

††) Schriften der Dänischen Akademie, Serie 6, Bd. V.

aufgestellten (irreduziblen) Gruppe ternärer linearer Substitutionen isomorph. Daher müßte eine der eben genannten Zahlen f gleich 3 sein. Es ist aber $f=\chi(E)$, falls $\chi(F)=-\chi(E)$ ist, durch 2 teilbar (vgl. D., S. 48). Ist dagegen $\chi(F)=\chi(E)$, so ist f auch der Grad eines Charakters der Gruppe $\mathfrak{F}_9$, oder was dasselbe ist, der Gruppe $\mathfrak{A}_6$. Unter den Graden der von Herrn *Frobenius**) bestimmten Charaktere der alternierenden Gruppe $\mathfrak{A}_6$ kommt aber die Zahl 3 nicht vor. Daher kann $\mathfrak{L}_9$ nicht die Darstellungsgruppe von $\mathfrak{F}_9$ sein. Folglich ist $m=6$, und da m' durch $\frac{m}{2}=3$ teilbar und nicht größer als 3 ist, $m'=3$.**) Hieraus folgt, daß die Darstellungsgruppe $\mathfrak{L}_9'$ von $\mathfrak{F}_9$ von der Ordnung 6.360 ist und zugleich die Darstellungsgruppe von $\mathfrak{L}_9$ repräsentiert.

Auf die genauere Bestimmung der Gruppe $\mathfrak{L}_9'$ gehe ich hier nicht ein, da ich in einer später erscheinenden Arbeit die Darstellungen der symmetrischen und alternierenden Gruppen durch gebrochene lineare Substitutionen ausführlich behandeln werde.

Aus dem Satze IX ergibt sich auch, wie ich noch hinzufügen will, unmittelbar, daß die Gruppe $\mathfrak{L}_{p^n}$ und das direkte Produkt der Gruppe $\mathfrak{F}_{p^n}$ und einer Gruppe der Ordnung 2 die einzigen Gruppen $\mathfrak{G}$ der Ordnung $p^n(p^{2n}-1)$ sind, die eine invariante Untergruppe $\mathfrak{A}$ der Ordnung 2 derart enthalten, daß $\frac{\mathfrak{G}}{\mathfrak{A}}$ der Gruppe $\mathfrak{F}_{p^n}$ isomorph wird.***) Denn enthält der Kommutator $\mathfrak{K}$ von $\mathfrak{G}$ die Gruppe $\mathfrak{A}$, so wird $\mathfrak{G}$ für $p^n \neq 9$ der Darstellungsgruppe von $\mathfrak{F}_{p^n}$, für $p^n=9$, falls $\mathfrak{M}'$ den Multiplikator von $\mathfrak{L}_9$ bedeutet, nach Satz III der Gruppe $\frac{\mathfrak{L}_9'}{\mathfrak{M}'}$ isomorph; in jedem Falle wird also $\mathfrak{G}$ der Gruppe $\mathfrak{L}_{p^n}$ isomorph. Ist aber $\mathfrak{K}$ zu $\mathfrak{A}$ teilerfremd, so wird für $p^n>3$ die Gruppe $\mathfrak{K}$ isomorph $\mathfrak{F}_{p^n}$, also ist $\mathfrak{G}$ das direkte Produkt von $\mathfrak{A}$ und $\mathfrak{F}_{p^n}$. Daß auch der Fall $p^n=3$ keine Ausnahme bildet, ergibt sich wieder aus dem Umstand, daß der Index des Kommutators von $\mathfrak{F}_3$ zu der Ordnung des Multiplikators von $\mathfrak{F}_3$ teilerfremd ist.

Ich betrachte nun die Gruppe $\mathfrak{H}_{p^n}$ $(p>2)$.

---

*) Sitzungsberichte der Berl. Akad., 1899, S. 337, und 1901, S. 303.

**) Aus dem Gesagten ergibt sich zugleich, daß die Valentinergruppe sich nicht als Gruppe von weniger als 3.360 ganzen linearen Substitutionen schreiben läßt. Dies ist zuerst von Herrn *Wiman* a. a. O. nachgewiesen worden.

***) Die Vermutung, daß dieser Satz für die Gruppe $\mathfrak{F}_p$ besteht, hat bereits Herr *Frobenius,* Sitzungsberichte der Berl. Akad., 1902, S. 365, ausgesprochen.

Diese Gruppe enthält die Gruppe $\mathfrak{F}_{p^n}$ als Untergruppe vom Index 2, und zwar ist $\mathfrak{F}_{p^n}$ der Kommutator der Gruppe.

Ist nun $p^n \neq 9$, so ist der Multiplikator von $\mathfrak{F}_{p^n}$ von der Ordnung 2. Daher ist nach D., Satz IX, die Ordnung m'' des Multiplikators von $\mathfrak{H}_{p^n}$ durch keine ungerade Primzahl teilbar; es ist also $m'' = 2^\lambda$. Nun sind aber, wenn 2^b die höchste in der Ordnung $p^n(p^{2n}-1)$ von $\mathfrak{H}_{p^n}$ aufgehende Potenz von 2 ist, die Untergruppen der Ordnung 2^b von $\mathfrak{H}_{p^n}$ Diedergruppen. Da der Multiplikator einer solchen Gruppe, wie auf S. 109 gezeigt worden ist, die Ordnung 2 besitzt, so ist (vgl. D., Satz X) $2^\lambda \leqq 2$, also m'' gleich 1 oder gleich 2.

Ich betrachte nun die Gruppe $\mathfrak{G}_{p^n}$, die aus allen ganzen linearen Substitutionen $\binom{\alpha\ \beta}{\gamma\ \delta}$ von nicht verschwindender Determinante (mit Koeffizienten aus dem Felde $GF[p^n]$) besteht. Diese Gruppe enthält $p^n - 1$ invariante Elemente $\binom{v^\varkappa\ 0}{0\ v^\varkappa}$, die eine zyklische Gruppe $\mathfrak{C}$ bilden, und es ist $\frac{\mathfrak{G}_{p^n}}{\mathfrak{C}}$ der Gruppe $\mathfrak{H}_{p^n}$ isomorph. Ferner ist der Kommutator von $\mathfrak{G}_{p^n}$ die Gruppe $\mathfrak{L}_{p^n}$. Die Ordnung dieser Gruppe ist aber gleich dem zweifachen Wert der Ordnung des Kommutators von $\mathfrak{H}_{p^n}$. Daher ist nach D., Satz II, die Zahl m'' gerade; folglich ist m'' für $p^n \neq 9$ gleich 2.

Auch der Fall $p^n = 9$*) spielt hier keine Ausnahmerolle. Denn da der Multiplikator von $\mathfrak{F}_9$ die Ordnung 6 besitzt und m'' gerade ist, so sieht man leicht ein, daß für m'' nur die Werte 2 und 6 in Betracht kommen. Es kann aber nicht $m'' \equiv 0 \pmod{3}$ sein. Denn man betrachte die Untergruppe $\mathfrak{P}$ der Ordnung 9 von $\mathfrak{H}_9$, die aus den Substitutionen $\binom{1\ 0}{\gamma\ 1}$ besteht. Als Basis dieser Gruppe kann man die Substitutionen

$$P_0 = \binom{1\ 0}{1\ 1}, \quad P_1 = \binom{1\ 0}{v\ 1}$$

wählen. Setzt man ferner $A = \binom{v\ 0}{0\ 1}$, so wird

$$A^{-1} P_0 A = \binom{1\ 0}{v\ 1} = P_1, \quad A^{-1} P_1 A = \binom{1\ \ 0}{v^2\ 1} = P_0^{b_0} P_1^{b_1},$$

wo die Exponenten b_0 und b_1 durch die (irreduzible) Kongruenz

$$x^2 \equiv b_1 x + b_0 \pmod{3},$$

*) Die Gruppe $\mathfrak{H}_9$, deren Ordnung gleich $720 = 6!$ ist, und die eine der alternierenden Gruppen 6. Grades isomorphe invariante Untergruppe enthält, ist bekanntlich *nicht* der symmetrischen Gruppe des Grades 6 isomorph.

der v genügt, bestimmt sind. Wäre nun $m'' \equiv 0 \pmod{3}$, so müßte auf Grund des früher abgeleiteten Hilfssatzes die Kongruenz

$$\begin{vmatrix} -x, & 1 \\ b_0, & b_1 - x \end{vmatrix} = x^2 - b_1 x - b_0 \equiv 0 \pmod{3}$$

reziproke Wurzeln besitzen. Dies ist aber nicht der Fall, da die Wurzeln dieser Kongruenz die Größen v und v^3 sind, und v zum Exponenten $3^2 - 1 = 8$ gehört. Daher ist auch für $p^n = 9$ die Zahl m'' gleich 2.

Nach den Ergebnissen des § 1 müssen sich, da der Index des Kommutators von $\mathfrak{H}_{p^n}$ gleich 2 und auch m'' gleich 2 ist, für die Gruppe $\mathfrak{H}_{p^n}$ zwei Darstellungsgruppen der Ordnung $2p^n(p^{2n}-1)$ angeben lassen, zwischen denen kein Isomorphismus erster Art besteht.

Diese beiden Gruppen lassen sich folgendermaßen charakterisieren.

Es möge w eine primitive Wurzel des *Galois*schen Feldes $GF[p^{2n}]$ bedeuten; ferner sei $p^n - 1 = 2^r q$, wo q ungerade ist, und

$$u = w^{\frac{(p^n+1)q}{2}}.$$

Dann ist u^2 eine Größe des Feldes $GF[p^n]$, das durch die Null und die $p^n - 1$ Potenzen von $w^{p^n+1} = v$ gebildet wird, ferner ist $u^{2^r} = -1$. Betrachtet man nun die Substitutionen

$$U = \begin{pmatrix} u & 0 \\ 0 & u^{-1} \end{pmatrix}, \quad U' = \begin{pmatrix} u^{2^{r-1}+1} & 0 \\ 0 & u^{2^{r-1}-1} \end{pmatrix},$$

so bilden die in den Komplexen $\mathfrak{L}_{p^n}$ und $U\mathfrak{L}_{p^n}$, bzw. $\mathfrak{L}_{p^n}$ und $U'\mathfrak{L}_{p^n}$ enthaltenen Substitutionen je eine Gruppe der Ordnung $2p^n(p^{2n}-1)$. Die so entstehenden Gruppen

$$\mathfrak{K}_{p^n} = \mathfrak{L}_{p^n} + U\mathfrak{L}_{p^n},$$
$$\mathfrak{K}'_{p^n} = \mathfrak{L}_{p^n} + U'\mathfrak{L}_{p^n}$$

sind die beiden gesuchten Darstellungsgruppen von $\mathfrak{H}_{p^n}$.*)

Daß diese Gruppen einander nicht isomorph sind, sieht man am einfachsten folgendermaßen ein.**)

*) Für $p^n \equiv -1 \pmod{4}$ ist $\mathfrak{K}'_{p^n}$ die Gruppe der Substitutionen $\begin{pmatrix} \alpha & \beta \\ \gamma & \delta \end{pmatrix}$ mit Koeffizienten aus dem Felde $GF[p^n]$, deren Determinanten $\alpha\delta - \beta\gamma$ die Werte ± 1 haben.

**) Für $n = 1$ folgt dies bereits aus dem Umstand, daß $\mathfrak{H}_p$ eine vollkommene Gruppe ist (vgl. *Hölder*, a. a. O., § 11).

In der Gruppe $\mathfrak{K}'_{p^n}$ kommen mehrere Elemente der Ordnung 2 vor, z. B. die Substitution $F=\begin{pmatrix}-1 & 0\\ 0 & -1\end{pmatrix}$ und

$$U'\begin{pmatrix}0 & 1\\ -1 & 0\end{pmatrix}=\begin{pmatrix}0 & u^{2^{r-1}+1}\\ -u^{2^{r-1}-1} & 0\end{pmatrix}.$$

Dagegen ist in $\mathfrak{K}_{p^n}$ die Substitution F das einzige Element der Ordnung 2. Daß dies für die Untergruppe $\mathfrak{L}_{p^n}$ dieser Gruppe der Fall ist, ist bereits erwähnt worden. Es kann aber auch nicht eine Substitution der Form

$$\begin{pmatrix}u & 0\\ 0 & u^{-1}\end{pmatrix}\begin{pmatrix}\alpha & \beta\\ \gamma & \delta\end{pmatrix} \qquad (\alpha\delta-\beta\gamma=1)$$

die Ordnung 2 besitzen. Denn es müßte dann sein

$$\begin{pmatrix}u & 0\\ 0 & u^{-1}\end{pmatrix}\begin{pmatrix}\alpha & \beta\\ \gamma & \delta\end{pmatrix}=\begin{pmatrix}\delta & -\beta\\ -\gamma & \alpha\end{pmatrix}\begin{pmatrix}u^{-1} & 0\\ 0 & u\end{pmatrix};$$

hieraus würde folgen

$$\beta=0,\ \gamma=0,\quad \delta=\alpha u^2,$$

also wegen $\alpha\delta-\beta\gamma=1$ noch $\alpha^2u^2=1$. Eine solche Gleichung kann aber nicht bestehen, weil $u^2=v^q$ ein quadratischer Nichtrest des Feldes $GF[p^n]$ ist.

Aus dem eben Gesagten folgt zugleich, daß, wenn 2^c die höchste in $2p^n(p^{2n}-1)$ aufgehende Potenz von 2 ist, die Untergruppen der Ordnung 2^c der Gruppe $\mathfrak{K}_{p^n}$ vom Quaternionentypus sind. Dagegen sind, wie sich zeigen läßt, die Untergruppen der Ordnung 2^c von $\mathfrak{K}'_{p^n}$ von dem in § 4 unter 3. erwähnten Typus.

Hieraus schließt man leicht, daß die Gruppen $\mathfrak{K}_{p^n}$ und $\mathfrak{K}'_{p^n}$ abgeschlossene Gruppen sind.

Zuletzt bemerke ich noch, daß auch die Gruppe $\mathfrak{G}_{p^n}$ der Ordnung $p^n(p^n-1)(p^{2n}-1)$ stets (für $p\geqq 2$) eine abgeschlossene Gruppe ist.

§ 6.

Handelt es sich nun darum, die sämtlichen Darstellungen der Gruppen $\mathfrak{F}_{p^n}$, $\mathfrak{L}_{p^n}$ und $\mathfrak{H}_{p^n}$ durch ganze oder gebrochene lineare Substitutionen zu bestimmen, so hat man auf Grund der im vorigen Paragraphen gewonnenen Resultate nur die Darstellungen der Gruppe $\mathfrak{L}_{p^n}$ und einer der beiden Darstellungsgruppen von $\mathfrak{H}_{p^n}$, etwa der Gruppe $\mathfrak{K}_{p^n}$, durch ganze lineare

Substitutionen zu betrachten. — Eine wesentliche Ausnahme bildet für die Gruppen $\mathfrak{F}_{p^n}$ und $\mathfrak{L}_{p^n}$ nur der Fall $p^n = 9$, da in diesem Falle $\mathfrak{L}_9$ nicht die Darstellungsgruppe von $\mathfrak{F}_9$ und auch nicht eine abgeschlossene Gruppe ist.

Um nun eine genaue Übersicht über die verschiedenen Gruppen ganzer linearer Substitutionen, die den Gruppen $\mathfrak{L}_{p^n}$ und $\mathfrak{K}_{p^n}$ isomorph sind, zu geben, will ich im folgenden die Charaktere dieser beiden Gruppen berechnen.

Es mögen zunächst einige von Herrn *Frobenius**) herrührende Definitionen und Sätze vorausgeschickt werden:

„Es sei $\mathfrak{H}$ eine endliche Gruppe der Ordnung h, deren Elemente in k Klassen konjugierter Elemente zerfallen. Es möge die Klasse, der das Element R angehört, mit (R) bezeichnet werden, und es sei h_R die Anzahl der in (R) vorkommenden Elemente. Die Gruppe $\mathfrak{H}$ besitzt dann k einfache Charaktere

$$\chi^{(0)}(R),\ \chi^{(1)}(R), \dots \chi^{(k-1)}(R)$$

und es bestehen die Relationen

$$\chi^{(\varkappa)}(Q^{-1}RQ) = \chi^{(\varkappa)}(R),$$

$$\text{(39.)} \qquad \sum_R \chi^{(\varkappa)}(R^{-1})\,\chi^{(\varkappa)}(R) = h, \qquad \sum_R \chi^{(\varkappa)}(R^{-1})\,\chi^{(\lambda)}(R) = 0, \qquad (\varkappa \neq \lambda)$$

$$\text{(40.)} \qquad \sum_{\varkappa=0}^{k-1} \chi^{(\varkappa)}(R)\,\chi^{(\varkappa)}(S^{-1}) = \frac{h}{h_R}\,\varepsilon_{R,S};$$

hierbei ist in den Formeln (39.) die Summation über alle Elemente R der Gruppe zu erstrecken, ferner hat man in (40.) die Größe $\varepsilon_{R,S}$ gleich 1 oder gleich 0 zu setzen, je nachdem R und S einander konjugiert sind oder nicht.

Ein System von h Zahlen $\chi(R)$ wird als ein zusammengesetzter Charakter von $\mathfrak{H}$ bezeichnet, wenn sich k ganze rationale Zahlen $r_\varkappa$ bestimmen lassen, so daß für jedes R

$$\chi(R) = \sum_{\varkappa=0}^{k-1} r_\varkappa \chi^{(\varkappa)}(R)$$

wird; ist insbesondere jede der Zahlen $r_\varkappa$ nicht negativ, so nennt man $\chi(R)$ einen eigentlichen Charakter der Gruppe. Für den Charakter $\chi(R)$ gelten die Formeln

$$\sum_R \chi(R)\,\chi^{(\varkappa)}(R^{-1}) = h r_\varkappa, \qquad \sum_R \chi(R^{-1})\,\chi(R) = h\,(r_0^2 + r_1^2 \cdots + r_{k-1}^2).$$

*) Sitzungsberichte der Berl. Akad., 1896, S. 985, 1898, S. 501 und 1899, S. 330.

Ein zusammengesetzter Charakter $\chi(R)$ ist dann und nur dann ein einfacher Charakter, wenn die ganze rationale Zahl $\chi(E)$ positiv und $\sum_R \chi(R^{-1})\chi(R) = h$ ist.

Sind ferner $\chi(R)$ und $\chi_1(R)$ zwei einfache oder zusammengesetzte Charaktere der Gruppe, so bilden auch die h Zahlen $\chi(R)\chi_1(R)$ einen solchen.

Ist endlich $\mathfrak{Q}$ eine Untergruppe der Ordnung q von $\mathfrak{H}$ und ist $\varphi(Q)$ ein eigentlicher Charakter der Gruppe $\mathfrak{Q}$, so erhält man einen eigentlichen Charakter $\chi(R)$ der Gruppe $\mathfrak{H}$, indem man

$$\chi(R) = \frac{h}{q h_R} \sum_Q \varphi(Q) \tag{41.}$$

setzt; hierbei ist die Summe über alle h_R Elemente Q der Klasse (R) zu erstrecken und $\varphi(Q) = 0$ zu setzen, falls Q nicht in $\mathfrak{Q}$ enthalten ist."

Ich will hier noch eine Formel ableiten, die sich auf Gruppen mit invarianten Elementen bezieht und eine Verallgemeinerung der Formel (40.) repräsentiert.

Es sei nämlich $\mathfrak{G}$ eine Gruppe der Ordnung g, die eine aus a invarianten Elementen von $\mathfrak{G}$ bestehende Untergruppe $\mathfrak{A}$ enthält. Es möge nun die Gruppe $\mathfrak{G}$ genau k Klassen konjugierter Elemente

$$(R_0),\ (R_1), \ldots (R_{k-1})$$

enthalten von der Beschaffenheit, daß kein Element einer dieser Klassen einem Element einer anderen Klasse derselben Reihe mod. $\mathfrak{A}$ kongruent wird; es sei g_ϱ die Anzahl der in (R_ϱ) enthaltenen Elemente. Ist nun R ein Element der Klasse (R_ϱ), so mögen a_ϱ verschiedene Elemente von (R_ϱ), etwa

$$A_0 R,\ A_1 R, \ldots A_{a_\varrho - 1} R$$

dem Komplex $\mathfrak{A}R$ angehören. Dann bilden (vgl. D., S. 43) die Elemente $A_0, A_1, \ldots A_{a_\varrho-1}$ eine Untergruppe $\mathfrak{A}_\varrho$ von $\mathfrak{A}$, die von der Wahl des Elementes R innerhalb der Klasse (R_ϱ) unabhängig ist. Ist nun $\psi^{(\alpha)}(A)$ einer der a Charaktere der *Abel*schen Gruppe $\mathfrak{A}$ und bilden diejenigen Elemente B von $\mathfrak{A}$, für die $\psi^{(\alpha)}(B) = 1$ ist, die Untergruppe $\mathfrak{B}_\alpha$ von $\mathfrak{A}$, so mögen $\mathfrak{B}_\alpha$ und $\mathfrak{A}_\varrho$ genau $d_{\alpha\varrho}$ Elemente gemeinsam haben. Dann ist, wie ich D., S. 44, gezeigt habe, die Anzahl l_α derjenigen (einfachen) Charaktere

$$\chi^{(1)}(R),\ \chi^{(2)}(R), \ldots \chi^{(l_\alpha)}(R) \tag{42.}$$

von $\mathfrak{G}$, für die

$$\chi^{(\lambda)}(AR) = \psi^{(a)}(A)\,\chi^{(\lambda)}(R)$$

ist, durch die Gleichung

$$(43.)\qquad l_a = \sum_{\varrho=0}^{k-1} \left[\frac{d_{a\varrho}}{a_\varrho}\right]$$

bestimmt, wo $\left[\frac{d_{a\varrho}}{a_\varrho}\right]$ gleich 0 oder gleich 1 zu setzen ist, je nachdem $d_{a\varrho} < a_\varrho$ oder $= a_\varrho$ wird.

Für die l_a Charaktere (42.) bestehen nun, wie ich zeigen will, noch folgende Relationen: ist R ein Element von $\mathfrak{G}$, das einem der g_ϱ Elemente der Klasse (R_ϱ) mod. $\mathfrak{A}$ kongruent ist, so ist

$$(44.)\qquad \sum_{\lambda=1}^{l_a} \chi^{(\lambda)}(R^{-1})\,\chi^{(\lambda)}(R) = \frac{g\,a_\varrho}{g_\varrho\,a}\left[\frac{d_{a\varrho}}{a_\varrho}\right],$$

sind ferner R und S zwei Elemente von $\mathfrak{G}$, für die kein Element des Komplexes $\mathfrak{A}R$ einem Element des Komplexes $\mathfrak{A}S$ konjugiert ist, so ist

$$(45.)\qquad \sum_{\lambda=1}^{l_a} \chi^{(\lambda)}(R^{-1})\chi^{(\lambda)}(S) = 0.$$

Bei dem Beweise der Formel (44.) kann offenbar angenommen werden, daß R selbst der Klasse (R_ϱ) angehört; denn die linke Seite dieser Gleichung bleibt ungeändert, wenn R durch irgend ein Element des Komplexes $\mathfrak{A}R$ ersetzt wird. — Nun ist aber, falls $\chi^{(\lambda)}(E) = f_\lambda$ gesetzt wird,

$$(46.)\qquad \chi^{(\lambda)}(P)\chi^{(\lambda)}(Q) = \frac{f_\lambda}{g}\sum_T \chi^{(\lambda)}(PT^{-1}QT),$$

wo T alle Elemente von $\mathfrak{G}$ durchläuft, ferner ist (vgl. D., Formel (23.))

$$(47.)\qquad \sum_{\lambda=1}^{l_a} f_\lambda\chi^{(\lambda)}(P) = \frac{g}{a}\,\psi^{(a)}(P) \quad \text{oder} \quad = 0,$$

je nachdem P in $\mathfrak{A}$ enthalten ist oder nicht. Aus (46.) ergibt sich nun

$$\sum_{\lambda=1}^{l_a} \chi^{(\lambda)}(R^{-1})\chi^{(\lambda)}(R) = \frac{1}{g}\sum_T \sum_{\lambda=1}^{l_a} f_\lambda\chi^{(\lambda)}(R^{-1}T^{-1}RT).$$

Nach (47.) ist hier die Teilsumme

$$\sum_{\lambda=1}^{l_a} f_\lambda\chi^{(\lambda)}(R^{-1}T^{-1}RT),$$

falls $T^{-1}RT = JR$ gesetzt wird, gleich $\frac{g}{a}\psi^{(\alpha)}(J)$, wenn J in $\mathfrak{A}_\varrho$ enthalten ist, sonst aber gleich 0. Ferner wird die Gleichung $T^{-1}RT = JR$ für jedes Element J von $\mathfrak{A}_\varrho$ durch genau $\frac{g}{g_\varrho}$ verschiedene Elemente T befriedigt. Daher ist

$$(44'.)\qquad \sum_{\lambda=1}^{l_a} \chi^{(\lambda)}(R^{-1})\chi^{(\lambda)}(R) = \frac{1}{g}\cdot\frac{g}{g_\varrho}\cdot\frac{g}{a}\sum_J \psi^{(\alpha)}(J),$$

wo J alle Elemente von $\mathfrak{A}_\varrho$ durchläuft. Die Summe $\sum\limits_J \psi^{(\alpha)}(J)$ ist aber, wie ich D., S. 43, genauer ausgeführt habe, gleich $a_\varrho\left[\frac{d_{\alpha\varrho}}{a_\varrho}\right]$. Setzt man dies in die Gleichung (44'.) ein, so ergibt sich die Formel (44.). — Ebenso erhält man aus (46.)

$$\sum_{\lambda=1}^{l_a} \chi^{(\lambda)}(R^{-1})\chi^{(\lambda)}(S) = \frac{1}{g}\sum_T \sum_{\lambda=1}^{l_a} f_\lambda \chi^{(\lambda)}(R^{-1}T^{-1}ST).$$

Die rechte Seite dieser Gleichung ist nach (47.) gleich 0, weil auf Grund der über R und S gemachten Voraussetzung das Element $R^{-1}T^{-1}ST$ für kein Element T von $\mathfrak{G}$ in $\mathfrak{A}$ enthalten sein kann; damit ist auch die Formel (45.) bewiesen.

Ich wende mich nun zur Berechnung der Charaktere der Gruppen $\mathfrak{L}_{p^n}$ und $\mathfrak{K}_{p^n}$. Hierbei setze ich überall zur Abkürzung $s = p^n$.

1. *Die Gruppe $\mathfrak{L}_s$ für den Fall $s \equiv 1$ (mod. 2).*

Setzt man

$$E = \begin{pmatrix}1 & 0\\ 0 & 1\end{pmatrix},\quad F = \begin{pmatrix}-1 & 0\\ 0 & -1\end{pmatrix},\quad P = \begin{pmatrix}1 & 0\\ 1 & 1\end{pmatrix},\quad Q = \begin{pmatrix}1 & 0\\ v & 1\end{pmatrix},\quad A = \begin{pmatrix}v & 0\\ 0 & v^{-1}\end{pmatrix}$$

und versteht unter B ein Element der Ordnung $s+1$ der Gruppe, so zerfallen die $s(s^2-1)$ Elemente von $\mathfrak{L}_s$ in die $s+4$ Klassen

$$(E),\ (F),\ (P),\ (Q),\ (FP),\ (FQ),$$

$$(A),\ (A^2), \ldots (A^{\frac{s-3}{2}}),\qquad (B),\ (B^2), \ldots (B^{\frac{s-1}{2}}),$$

und zwar enthalten die ersten beiden dieser Klassen je ein Element, die 4 folgenden je $\frac{s^2-1}{2}$ Elemente, während jede der $\frac{s-3}{2}$ Klassen (A^α) aus $s(s+1)$, jede der $\frac{s-1}{2}$ Klassen (B^b) aus $s(s-1)$ Elementen besteht.*) Ist ferner

$$\varepsilon = (-1)^{\frac{s-1}{2}},$$

so stimmt für $\varepsilon = +1$ jede Klasse (R) mit der inversen Klasse (R^{-1}) überein; dagegen ist dies, falls $\varepsilon = -1$ ist, für alle Klassen mit Ausnahme der 4 Klassen (P), (Q), (FP), (FQ) der Fall, und zwar sind dann (P) und (Q) bzw. (FP) und (FQ) inverse Klassen. Endlich gehören im allgemeinen R und FR zwei verschiedenen Klassen an; eine Ausnahme tritt, falls $\varepsilon = +1$ ist, nur für die Elemente der Klasse $(A^{\frac{s-1}{4}})$, falls $\varepsilon = -1$ ist, nur für die Elemente der Klasse $(B^{\frac{s+1}{4}})$ ein.

Hieraus folgt auf Grund der Formel (43.), daß die Gruppe $\mathfrak{L}_s$ genau $\frac{s+4-1}{2} + 1 = \frac{s+5}{2}$ (einfache) Charaktere $\chi(R)$ besitzt, für die

$$\chi(F) = \chi(E) \tag{48.}$$

ist, und $\frac{s+4-1}{2} = \frac{s+3}{2}$ Charaktere, für die

$$\chi(F) = -\chi(E) \tag{49.}$$

ist. Ich will den Charakter $\chi(R)$ als einen Charakter erster oder zweiter Art bezeichnen, je nachdem die Gleichung (48.) oder (49.) besteht.

Die $s+4$ Charaktere von $\mathfrak{L}_s$ lassen sich in folgender Tabelle zusammenfassen:

	1	1	$\frac{s-3}{2}$	$\frac{s-1}{2}$	2	2
$\chi(E)$	1	s	$s+1$	$s-1$	$\frac{1}{2}(s+1)$	$\frac{1}{2}(s-1)$
$\chi(F)$	1	s	$(-1)^a(s+1)$	$(-1)^\beta(s-1)$	$\frac{\varepsilon}{2}(s+1)$	$-\frac{\varepsilon}{2}(s-1)$
$\chi(P)$	1	0	1	-1	$\frac{1}{2}(1 \pm \sqrt{\varepsilon s})$	$\frac{1}{2}(-1 \pm \sqrt{\varepsilon s})$
$\chi(Q)$	1	0	1	-1	$\frac{1}{2}(1 \mp \sqrt{\varepsilon s})$	$\frac{1}{2}(-1 \mp \sqrt{\varepsilon s})$
$\chi(A^a)$	1	1	$\rho^{\alpha a} + \rho^{-\alpha a}$	0	$(-1)^a$	0
$\chi(B^b)$	1	-1	0	$-(\sigma^{\beta b} + \sigma^{-\beta b})$	0	$-(-1)^b$.

*) Vgl. *Dickson*, Linear Groups, Cap. XII.

Hierin bedeuten ϱ und σ primitive Einheitswurzeln der Grade $s-1$, bzw. $s+1$; für a und α sind die Zahlen $1, 2, \ldots \frac{s-3}{2}$, für b und β die Zahlen $1, 2, \ldots \frac{s-1}{2}$ zu setzen. Die Werte der Charaktere für die hier fehlenden Klassen (FP) und (FQ) ergeben sich vermöge der Formeln

$$\chi(FP) = \frac{\chi(F)}{\chi(E)}\chi(P), \qquad \chi(FQ) = \frac{\chi(F)}{\chi(E)}\chi(Q).$$

Die vorstehende Tabelle habe ich auf folgendem Wege berechnet.

Der in der ersten Kolonne stehende Charakter ist der Hauptcharakter $\chi^{(0)}(R) = 1$. Die $1 + \frac{s-3}{2}$ Charaktere der Grade s und $s+1$ ergeben sich durch Betrachtung der Untergruppe $\mathfrak{Q}$ der Ordnung $s(s-1)$ von $\mathfrak{L}_s$, die aus allen Substitutionen der Form $\begin{pmatrix} \varkappa & 0 \\ \lambda & \varkappa^{-1} \end{pmatrix}$ besteht. Unter den Substitutionen dieser Untergruppe gehören je $\frac{s-1}{2}$ den Klassen (P), (Q), (FP), (FQ) an, nämlich bzw. die Substitutionen

$$\begin{pmatrix} 1 & 0 \\ \mu & 1 \end{pmatrix}, \begin{pmatrix} 1 & 0 \\ \nu\mu & 1 \end{pmatrix}, \begin{pmatrix} -1 & 0 \\ -\mu & -1 \end{pmatrix}, \begin{pmatrix} -1 & 0 \\ -\nu\mu & -1 \end{pmatrix},$$

wo für μ die $\frac{s-1}{2}$ quadratischen Reste $1, \nu^2, \ldots \nu^{s-3}$ des Feldes $GF[p^n]$ zu setzen sind. Ferner gehören, wenn a eine der Zahlen $1, 2, \ldots \frac{s-3}{2}$ bedeutet, die $2s$ Substitutionen

$$\begin{pmatrix} \nu^a & 0 \\ \lambda & \nu^{-a} \end{pmatrix}, \begin{pmatrix} \nu^{-a} & 0 \\ \lambda & \nu^a \end{pmatrix} \qquad (\lambda = 0, 1, \nu, \nu^2, \ldots \nu^{s-2})$$

der Klasse (A^a) an. Außerdem enthält $\mathfrak{Q}$ noch die Substitutionen E und F. Dagegen besitzen die $\frac{s-1}{2}$ Klassen (B^b) in $\mathfrak{Q}$ keinen Repräsentanten.

Man erhält nun, wie unmittelbar ersichtlich ist, einen linearen Charakter $\psi^{(\alpha)}(R)$ von $\mathfrak{Q}$, indem man, falls $R = \begin{pmatrix} \nu^d & 0 \\ \lambda & \nu^{-d} \end{pmatrix}$ ist,

$$\psi^{(\alpha)}(R) = \varrho^{\alpha d} \qquad (\alpha = 0, 1, \ldots s-2)$$

setzt. Mit Hilfe der Formel (41.) ergibt sich dann für jeden der Charaktere $\psi^{(\alpha)}(R)$ von $\mathfrak{Q}$ folgender eigentliche zusammengesetzte Charakter $\zeta_\alpha(R)$ von $\mathfrak{L}_s$:

$$\zeta_\alpha(E) = s+1, \qquad \zeta_\alpha(F) = (-1)^\alpha (s+1),$$
$$\zeta_\alpha(P) = \zeta_\alpha(Q) = 1, \qquad \zeta_\alpha(FP) = \zeta_\alpha(FQ) = (-1)^\alpha,$$
$$\zeta_\alpha(A^a) = \varrho^{\alpha a} + \varrho^{-\alpha a}, \qquad \zeta_\alpha(B^b) = 0.$$

Hat nun α einen der Werte $1, 2, \ldots \frac{s-3}{2}$, so ergibt die Rechnung $\sum_R \zeta_\alpha(R^{-1})\,\zeta_\alpha(R) = s(s^2-1)$; daher ist $\zeta_\alpha(R)$ ein einfacher Charakter der Gruppe. Dies gibt uns die $\frac{s-3}{2}$ Charaktere des Grades $s+1$ unserer Tabelle; sie mögen noch mit $\chi^{(2)}(R), \chi^{(3)}(R), \ldots \chi^{\left(\frac{s-1}{2}\right)}(R)$ bezeichnet werden. Ist dagegen $\alpha = 0$, so ist, wie man ebenso zeigt,

$$\chi^{(1)}(R) = \zeta_0(R) - \chi^{(0)}(R) = \zeta_0(R) - 1$$

ein einfacher Charakter. Dies ist der in der zweiten Kolonne der Tabelle angegebene Charakter.

Für $\alpha = \frac{s-1}{2}$ erhält man noch den zusammengesetzten Charakter $\zeta_{\frac{s-1}{2}}(R) = \zeta(R)$:

$$\begin{gathered}\zeta(E) = s+1, \qquad \zeta(F) = \varepsilon(s+1),\\ \zeta(P) = \zeta(Q) = 1, \qquad \zeta(FP) = \zeta(FQ) = \varepsilon,\\ \zeta(A^a) = 2(-1)^a, \qquad \zeta(B^b) = 0.\end{gathered}$$

Da, wie man leicht findet, $\sum_R \zeta(R^{-1})\,\zeta(R) = 2s(s^2-1)$ ist, so ist $\zeta(R)$ eine Summe von zwei verschiedenen einfachen Charakteren $\xi_1(R)$ und $\xi_2(R)$, und zwar ist, weil $\zeta(F) = \varepsilon\zeta(E)$ ist, auch

$$\xi_1(F) = \varepsilon\,\xi_1(E), \qquad \xi_2(F) = \varepsilon\,\xi_2(E). \tag{50.}$$

Ich betrachte nun die durch die Potenzen von B gebildete zyklische Untergruppe der Ordnung $s+1$. Dem Charakter

$$\psi^{(\beta)}(B^e) = \sigma^{\beta e} \qquad (e=0, 1, \ldots s)$$

dieser Untergruppe entspricht, wie man leicht zeigt, folgender eigentliche Charakter $\varphi_\beta(R)$ von $\mathfrak{L}_s$:

$$\begin{gathered}\varphi_\beta(E) = s(s-1), \qquad \varphi_\beta(F) = (-1)^\beta s(s-1),\\ \varphi_\beta(P) = \varphi_\beta(Q) = \varphi_\beta(FP) = \varphi_\beta(FQ) = 0,\\ \varphi_\beta(A^a) = 0, \qquad \varphi_\beta(B^b) = \sigma^{\beta b} + \sigma^{-\beta b}.\end{gathered}$$

Man bilde nun für $\beta = 1, 2, \ldots \frac{s+1}{2}$ den zusammengesetzten Charakter

$$\vartheta_\beta(R) = \chi^{(1)}(R)\,\zeta_\beta(R) - \zeta_\beta(R) - \varphi_\beta(R),$$

der aber nicht mehr ein eigentlicher Charakter zu sein braucht. Es ergibt sich:

$$\vartheta_\beta(E)=s-1, \qquad \vartheta_\beta(F)=(-1)^\beta(s-1),$$
$$\vartheta_\beta(P)=\vartheta_\beta(Q)=-1, \qquad \vartheta_\beta(FP)=\vartheta_\beta(FQ)=-(-1)^\beta,$$
$$\vartheta_\beta(A^a)=0, \qquad \vartheta_\beta(B^b)=-(\sigma^{\beta b}+\sigma^{-\beta b}).$$

Man überzeugt sich nun leicht, daß für $\beta=1,2,\ldots\frac{s-1}{2}$

$$\sum_R \vartheta_\beta(R^{-1})\,\vartheta_\beta(R)=s(s^2-1)$$

wird. Da außerdem $\vartheta_\beta(E)>0$ ist, so sind daher $\vartheta_1(R), \vartheta_2(R),\ldots \vartheta_{\frac{s-1}{2}}(R)$ einfache Charaktere, die, wie unmittelbar ersichtlich ist, unter einander verschieden sind. Es sind das die $\frac{s-1}{2}$ Charaktere des Grades $s-1$ unserer Tabelle; sie mögen auch mit $\chi^{(\frac{s+1}{2})}(R),\ldots \chi^{(s-1)}(R)$ bezeichnet werden.

Für $\beta=\frac{s+1}{2}$ erhält man noch den zusammengesetzten Charakter $\vartheta_{\frac{s+1}{2}}(R)=\vartheta(R)$:

$$\vartheta(E)=s-1, \qquad \vartheta(F)=-\varepsilon(s-1),$$
$$\vartheta(P)=\vartheta(Q)=-1, \qquad \vartheta(FP)=\vartheta(FQ)=\varepsilon,$$
$$\vartheta(A^a)=0, \qquad \vartheta(B^b)=-2(-1)^b.$$

Die Rechnung ergibt ferner

$$\sum_R \vartheta(R^{-1})\,\vartheta(R)=2s(s^2-1), \qquad \sum_R \vartheta(R^{-1})\chi^{(\gamma)}(R)=0. \qquad {\scriptstyle(\gamma=0,1,\ldots s-1)}$$

Daher ist $\vartheta(R)$ die Summe oder die Differenz von zwei verschiedenen einfachen Charakteren $\eta_1(R)$ und $\eta_2(R)$, die von den bereits erhaltenen s Charakteren $\chi^{(\gamma)}(R)$ verschieden sind. Da außerdem $\vartheta(F)=-\varepsilon\,\vartheta(E)$ ist, so muß, wie leicht zu ersehen ist, auch

$$\eta_1(F)=-\varepsilon\eta_1(E), \qquad \eta_2(F)=-\varepsilon\eta_2(E)$$

sein; folglich sind $\eta_1(R)$ und $\eta_2(R)$ wegen (50.) auch von den Charakteren $\xi_1(R)$ und $\xi_2(R)$ verschieden. Endlich findet man noch

$$\sum_R \zeta(R^{-1})\chi^{(\gamma)}(R)=0; \qquad {\scriptstyle(\gamma=0,1,\ldots s-1)}$$

daher sind auch die Charaktere $\xi_1(R)$ und $\xi_2(R)$ von den s Charakteren $\chi^{(\nu)}(R)$ verschieden.

Es sind nun im ganzen $s+4$ einfache Charaktere zu berechnen; wir haben bereits s erhalten, folglich fehlen noch 4 und das müssen die Charaktere $\xi_1(R)$, $\xi_2(R)$, $\eta_1(R)$, $\eta_2(R)$ sein.

Die Berechnung dieser 4 Charaktere ergibt sich durch Anwendung der Formeln (44.) und (45.) auf unseren Fall.

Zunächst zeige ich, daß $\vartheta(R)$ notwendig die Summe der beiden Charaktere $\eta_1(R)$ und $\eta_2(R)$ sein muß. Ist nämlich $\varepsilon=+1$, so sind $\eta_1(R)$ und $\eta_2(R)$ Charaktere zweiter Art. Unter den s Charakteren $\chi^{(\nu)}(R)$ sind ferner solche zweiter Art

$$\chi^{(2)}(R), \chi^{(4)}(R), \ldots \chi^{\left(\frac{s-1}{2}\right)}(R), \quad \chi^{\left(\frac{s+1}{2}\right)}(R), \chi^{\left(\frac{s+5}{2}\right)}(R), \ldots \chi^{(s-2)}(R).$$

Es muß aber nach (44.) die Summe der Quadrate der Grade aller Charaktere zweiter Art gleich $\frac{s(s^2-1)}{2}$ sein, daher ist

$$\frac{s-1}{4}(s+1)^2+\frac{s-1}{4}(s-1)^2+\eta_1^2(E)+\eta_2^2(E)=\frac{s(s^2-1)}{2},$$

also

$$\eta_1^2(E)+\eta_2^2(E)=\frac{(s-1)^2}{2}.$$

Für $\varepsilon=-1$ erhält man ebenso durch Betrachtung der Charaktere erster Art

$$1+s^2+\frac{s-3}{4}(s+1)^2+\frac{s-3}{4}(s-1)^2+\eta_1^2(E)+\eta_2^2(E)=\frac{s(s^2-1)}{2},$$

also wieder

$$\eta_1^2(E)+\eta_2^2(E)=\frac{(s-1)^2}{2}.$$

Wäre nun $\vartheta(R)=\eta_1(R)-\eta_2(R)$, also $\eta_1(E)-\eta_2(E)=s-1$, so würde sich $\eta_1(E)\,\eta_2(E)=-\frac{(s-1)^2}{2}$ ergeben, was nicht möglich ist, da $\eta_1(E)$ und $\eta_2(E)$ positive ganze Zahlen sind. Daher ist in der Tat

$$\vartheta(R)=\eta_1(R)+\eta_2(R).$$

In ähnlicher Weise lassen sich aus der Formel (44.) oder (45.) auch die genauen Werte der Ausdrücke $\xi_1^2(R)+\xi_2^2(R)$ und $\eta_1^2(R)+\eta_2^2(R)$ für $R=E, P, Q, A^a, B^b$ berechnen. Da man außerdem die Werte von $\xi_1(R)+\xi_2(R)$

und $\eta_1(R)+\eta_2(R)$ kennt, so erhält man für jedes der Zahlenpaare $\xi_1(R)$ und $\xi_2(R)$ bzw. $\eta_1(R)$ und $\eta_2(R)$ eine quadratische Gleichung. Es ergibt sich hierbei, daß für $R=E, A^a, B^b$ die Gleichungen $\xi_1(R)=\xi_2(R)$ und $\eta_1(R)=\eta_2(R)$ bestehen. Eine einfache Überlegung lehrt dann auch, in welcher Weise noch die Wurzeln der den Elementen P und Q entsprechenden quadratischen Gleichungen auf die Charaktere $\xi_1(R)$ und $\xi_2(R)$ bzw. $\eta_1(R)$ und $\eta_2(R)$ zu verteilen sind. Die auf diesem Wege berechneten Werte der 4 Charaktere sind den beiden letzten Kolonnen unserer Tabelle zu entnehmen.

Ich füge noch folgendes hinzu.

Die $\frac{s+5}{2}$ Charaktere der Tabelle, für die $\chi(F)=\chi(E)$ ist, repräsentieren zugleich, wenn je zwei Elemente R und FR als nicht von einander verschieden angesehen werden, die Charaktere der Gruppe $\mathfrak{F}_s$.*)

Da ferner die Gruppe $\mathfrak{L}_s$ für $s\neq 9$ die Darstellungsgruppe von $\mathfrak{F}_s$ ist, so lehrt uns die Tabelle, daß es im ganzen $\frac{s+3}{2}$ verschiedene irreduzible Gruppen ganzer linearer Substitutionen gibt, die der Gruppe $\mathfrak{F}_s$ isomorph sind, und zwar hat man eine Gruppe des Grades s,**) $\frac{s-4-\varepsilon}{4}$ Gruppen des Grades $s+1$, $\frac{s-2+\varepsilon}{4}$ Gruppen des Grades $s-1$ und 2 Gruppen des Grades $\frac{s+\varepsilon}{2}$. Ebenso gibt es (für $s\neq 9$) im ganzen $\frac{s+3}{2}$ verschiedene irreduzible Gruppen gebrochener linearer Substitutionen, die der Gruppe $\mathfrak{F}_s$ isomorph sind, und die sich nicht als Gruppen von $\frac{s(s^2-1)}{2}$ ganzen linearen Substitutionen schreiben lassen. Unter diesen Gruppen haben $\frac{s-2+\varepsilon}{4}$ Gruppen den Grad $s+1$, $\frac{s-\varepsilon}{4}$ den Grad $s-1$ und 2 den Grad $\frac{s-\varepsilon}{2}$.

Die hier erwähnten der Gruppe $\mathfrak{F}_s$ isomorphen Gruppen der Grade $\frac{s+1}{2}$ und $\frac{s-1}{2}$ sind für den Fall $s=p$ zuerst von Herrn *Klein* aus der Transformationstheorie der elliptischen Funktionen erhalten worden (Math. Ann., Bd. 15, S. 275).

---

*) Für den Fall $s=p$ sind diese Charaktere bereits von Herrn *Frobenius*, Sitzungsberichte der Berl. Akad., 1896, S. 1021, angegeben worden.

**) Diese Gruppe ergibt sich unmittelbar aus der auf S. 116 erwähnten Darstellung von $\mathfrak{F}_s$ als Permutationsgruppe von $s+1$ Symbolen.

2. Die Gruppe $\mathfrak{L}_s$ für den Fall $s = 2^n$.

Die $s(s^2-1)$ Elemente dieser Gruppe zerfallen in $s+1$ Klassen konjugierter Elemente. Setzt man nämlich, wie bei $s \equiv 1 \pmod 2$,

$$E = \begin{pmatrix} 1 & 0 \\ 0 & 1 \end{pmatrix}, \quad P = \begin{pmatrix} 1 & 0 \\ 1 & 1 \end{pmatrix}, \quad A = \begin{pmatrix} v & 0 \\ 0 & v^{-1} \end{pmatrix}$$

und bezeichnet mit B irgend eine Substitution der Ordnung $s+1$ der Gruppe, so sind dies die Klassen

$$(E), (P), (A), (A^2), \ldots (A^{\frac{s-2}{2}}), (B), (B^2), \ldots (B^{\frac{s}{2}}),$$

und zwar enthält die erste dieser Klassen ein Element, die zweite s^2-1 Elemente, ferner besteht jede der $\frac{s-2}{2}$ Klassen (A^a) aus $s(s+1)$, jede der Klassen (B^b) aus $s(s-1)$ Elementen. Jede dieser Klassen stimmt mit der inversen Klasse überein.*)

Die $s+1$ Charaktere der Gruppe $\mathfrak{L}_{2^n}$ ergeben sich in ganz analoger Weise, wie die s mit $\chi^{(0)}(R), \chi^{(1)}(R), \ldots \chi^{(s-1)}(R)$ bezeichneten Charaktere der Gruppe $\mathfrak{L}_s$ für den Fall $s \equiv 1 \pmod 2$. Man erhält folgende Tabelle:

	1	1	$\frac{s-2}{2}$	$\frac{s}{2}$
$\chi(E)$	1	s	$s+1$	$s-1$
$\chi(P)$	1	0	1	-1
$\chi(A^a)$	1	1	$\varrho^{\alpha a}+\varrho^{-\alpha a}$	0
$\chi(B^b)$	1	-1	0	$-(\sigma^{\beta b}+\sigma^{-\beta b})$.

Hier bedeuten wieder ϱ und σ primitive Einheitswurzeln der Grade $s-1$ und $s+1$. Für a und α sind die Werte $1, 2, \ldots \frac{s-2}{2}$, für b und β die Werte $1, 2, \ldots \frac{s}{2}$ zu setzen.

Da die Gruppe $\mathfrak{L}_{2^n}$ (für $2^n \neq 4$) eine abgeschlossene Gruppe ist, so ergibt sich aus unserer Tabelle, daß es im ganzen 2^n verschiedene irreduzible Gruppen gebrochener linearer Substitutionen gibt, die der Gruppe $\mathfrak{L}_{2^n}$ isomorph sind, und zwar hat man eine Gruppe des Grades 2^n, ferner $2^{n-1}-1$ Gruppen des Grades 2^n+1 und 2^{n-1} Gruppen des Grades 2^n-1. Alle

*) Vgl. *Dickson*, Linear Groups, a. a. O.

diese Gruppen lassen sich auch als Gruppen von $2^n(2^{2n}-1)$ ganzen linearen Substitutionen schreiben. Der niedrigste in Betracht kommende Grad ist gleich 2^n-1.*)

Der Ausnahmefall $s=4$ erledigt sich dadurch, daß man an Stelle der Gruppe $\mathfrak{L}_4$ die ihr isomorphe Gruppe $\mathfrak{F}_5$ betrachtet.

3. *Die Darstellungsgruppe $\mathfrak{K}_s$ der Gruppe $\mathfrak{H}_s$ (s ungerade).*

Die Gruppe $\mathfrak{K}_s$, deren genaue Definition auf S. 122 gegeben worden ist, enthält, wie man ohne Mühe erkennt, Substitutionen der Ordnung $2(s-1)$ und auch solche der Ordnung $2(s+1)$. Bezeichnet man wie früher mit E, F, P die Substitutionen

$$\begin{pmatrix}1 & 0\\ 0 & 1\end{pmatrix}, \begin{pmatrix}-1 & 0\\ 0 & -1\end{pmatrix}, \begin{pmatrix}1 & 0\\ 1 & 1\end{pmatrix}$$

der Untergruppe $\mathfrak{L}_s$ von $\mathfrak{K}_s$ und mit A und B zwei Elemente der Ordnungen $2(s-1)$ und $2(s+1)$, so wird $A^{s-1}=B^{s+1}=F$, ferner zerfallen die $2s(s^2-1)$ Elemente von $\mathfrak{K}_s$ in die $2s+2$ Klassen konjugierter Elemente

$$(E), (F), (P), (FP), (A), (A^2), \dots (A^{s-2}), (B), (B^2), \dots (B^s);$$

hierbei enthalten die ersten beiden dieser Klassen je ein Element, die beiden folgenden je s^2-1 Elemente, während jede der Klassen (A^a) aus $s(s+1)$, jede der Klassen (B^b) aus $s(s-1)$ Elementen besteht. Ferner sind R und R^{-1} für jedes Element R der Gruppe einander konjugiert. Die Elemente R und FR gehören im allgemeinen verschiedenen Klassen an; eine Ausnahme bilden nur die Elemente der beiden Klassen $(A^{\frac{s-1}{2}})$ und $(B^{\frac{s+1}{2}})$. Daher hat, wie aus Formel (43.) folgt, unsere Gruppe $s+2$ (einfache) Charaktere $\chi(R)$, für die $\chi(F)=\chi(E)$ ist, und s Charaktere, für die $\chi(F)=-\chi(E)$ ist. — Außerdem besitzt die Gruppe, da ihr Kommutator $\mathfrak{L}_s$ vom Index 2 ist, außer dem Hauptcharakter $\chi^{(0)}(R)=1$ noch einen linearen Charakter $\zeta(R)$, und zwar erhält man:

---

*) Für $s=8$ ist $\mathfrak{L}_s$ die bekannte zuerst von Herrn *Cole* (American Journal, Bd. XV, S. 303) angegebene einfache Gruppe der Ordnung 504. Daß diese Gruppe sich nicht als Gruppe gebrochener linearer Substitutionen von weniger als 6 Variabeln darstellen läßt, hat bereits Herr *Wiman* (Göttinger Nachrichten, 1897, S. 62) nachgewiesen. Die Frage, ob hierzu 6 Variable genügen oder nicht, läßt Herr *Wiman* aber a. a. O. noch unentschieden; unser Resultat ergibt, daß diese Frage im bejahenden Sinne zu beantworten ist.

$$\zeta(E)=\zeta(F)=\zeta(P)=\zeta(FP)=1,$$

$$\zeta(A^a)=(-1)^a,\quad \zeta(B^b)=(-1)^b. \qquad (a=1,2,\ldots s-2,\ b=1,2,\ldots s)$$

Die übrigen $2s$ Charaktere der Gruppe $\mathfrak{K}_s$ lassen sich in ähnlicher Weise berechnen, wie die Charaktere der Gruppe $\mathfrak{L}_s$; sie sind in folgender Tabelle zusammengestellt:

	1	1	1	1	$s-2$	s
$\chi(E)$	1	1	s	s	$s+1$	$s-1$
$\chi(F)$	1	1	s	s	$(-1)^\alpha(s+1)$	$(-1)^\beta(s-1)$
$\chi(P)$	1	1	0	0	1	-1
$\chi(A^a)$	1	$(-1)^a$	1	$(-1)^a$	$\varrho^{a\alpha}+\varrho^{-a\alpha}$	0
$\chi(B^b)$	1	$(-1)^b$	-1	$-(-1)^b$	0	$-(\sigma^{\beta b}+\sigma^{-\beta b})$.

Hier bedeutet ϱ eine primitive Einheitswurzel des Grades $2(s-1)$ und σ eine primitive Einheitswurzel des Grades $2(s+1)$. Für die Indizes a und α sind die Werte $1, 2, \ldots s-2$, für b und β die Werte $1, 2, \ldots s$ zu setzen. Die Werte der Charaktere für die noch fehlende Klasse (FP) ergeben sich aus der Relation

$$\chi(FP)=\frac{\chi(F)}{\chi(E)}\chi(P).$$

Aus dieser Tabelle können auch unmittelbar die Charaktere der Gruppe $\mathfrak{H}_s$ abgelesen werden. Man hat hierbei nur diejenigen $s+2$ Charaktere $\chi(R)$ zu berücksichtigen, für die $\chi(F)=\chi(E)$ ist, und je zwei Elemente R und FR als nicht von einander verschieden anzusehen.

Zu jedem der $2s+2$ Charaktere von $\mathfrak{K}_s$ gehört eine dieser Gruppe isomorphe irreduzible Gruppe ganzer linearer Substitutionen. Betrachtet man die diesen Gruppen entsprechenden Gruppen gebrochener linearer Substitutionen, so sind die letzteren der Gruppe $\mathfrak{H}_s$ isomorph. Hierbei ist jedoch zu beachten, daß man für zwei Charaktere $\chi(R)$ und $\chi'(R)$, zwischen denen die Beziehungen

$$\chi'(R)=\zeta(R)\chi(R)$$

bestehen, wie leicht ersichtlich ist, auf dieselbe Gruppe gebrochener linearer Substitutionen geführt wird. Eine genauere Diskussion unserer Tabelle ergibt nun folgendes Resultat: Der Gruppe $\mathfrak{H}_s$ sind im ganzen $s+1$ verschiedene irreduzible Gruppen gebrochener linearer Substitutionen isomorph. Hierunter hat eine den Grad s, ferner haben $\frac{s-1}{2}$ den Grad $s+1$ und $\frac{s+1}{2}$ den Grad $s-1$. Bedeutet noch ε wie früher die Zahl $(-1)^{\frac{s-1}{2}}$, so lassen sich unter diesen Gruppen die Gruppe des Grades s, ferner $\frac{s-2+\varepsilon}{4}$ Gruppen des Grades $s+1$ und $\frac{s-\varepsilon}{4}$ Gruppen des Grades $s-1$ auch als Gruppen von $s(s^2-1)$ ganzen linearen Substitutionen schreiben. Dagegen ist dies für $\frac{s-\varepsilon}{4}$ Gruppen des Grades $s+1$ und für $\frac{s+2+\varepsilon}{4}$ Gruppen des Grades $s-1$ nicht der Fall. Der kleinste in Betracht kommende Grad einer der Gruppe $\mathfrak{H}_s$ isomorphen Gruppe linearer Substitutionen ist gleich $s-1$.

11.

Über die Darstellung der symmetrischen Gruppe durch lineare homogene Substitutionen

Sitzungsberichte der Preussischen Akademie der Wissenschaften 1908,
Physikalisch-Mathematische Klasse, 664 - 678

Eine genaue Übersicht über die irreduziblen Gruppen linearer homogener Substitutionen $\mathfrak{G}_n$, die der symmetrischen Gruppe n^{ten} Grades $\mathfrak{S}_n$ isomorph sind, hat zuerst Hr. Frobenius durch Bestimmung der Charaktere von $\mathfrak{S}_n$ gewonnen[1]. Die Charaktere von $\mathfrak{S}_n$ habe ich später in meiner Dissertation[2] noch auf einem anderen Wege erhalten; zugleich habe ich gezeigt, daß man die wirkliche Konstruktion der Gruppen $\mathfrak{G}_n$ auf die Berechnung gewisser Determinantenrelationen zurückführen kann (D., § 36). Eine weitere Methode zur Berechnung der Charaktere von $\mathfrak{S}_n$ und der Gruppen $\mathfrak{G}_n$ hat Hr. Frobenius in seiner Arbeit *Über die charakteristischen Einheiten der symmetrischen Gruppe*[3] angegeben. In dieser Arbeit hat Hr. Frobenius auch zuerst den Satz ausgesprochen, daß jede der Gruppen $\mathfrak{G}_n$ bei passender Wahl der Variabeln als eine Gruppe mit rationalen Koeffizienten geschrieben werden kann[4].

In der vorliegenden Arbeit soll nun genauer gezeigt werden, daß sich jede der irreduziblen Gruppen $\mathfrak{G}_n$ bei geeigneter Wahl der Variabeln auch als eine Gruppe mit *ganzzahligen* rationalen Koeffizienten darstellen läßt. Da nun jede Gruppe linearer Substitutionen, die der Gruppe $\mathfrak{S}_n$ isomorph ist, als eine endliche Gruppe vollständig reduzibel ist, so ergibt sich zugleich der Satz:

Jede Gruppe linearer homogener Substitutionen, die der symmetrischen Gruppe n^{ten} Grades isomorph ist, läßt sich durch eine lineare Transformation der Variabeln in eine Gruppe mit ganzzahligen rationalen Koeffizienten überführen.

---

[1] Sitzungsberichte 1900, S. 516.

[2] *Über eine Klasse von Matrizen, die sich einer gegebenen Matrix zuordnen lassen*, Berlin 1901. — Im folgenden mit D. zitiert.

[3] Sitzungsberichte 1903, S. 328.

[4] Dieses Resultat ergibt sich auch ohne weiteres aus den Betrachtungen des § 36 meiner Dissertation.

§ 1.

Es mögen zunächst einige Bemerkungen über die Charaktere der symmetrischen Gruppe $\mathfrak{S}_n$ vorausgeschickt werden.

Man habe eine Gruppe $\mathfrak{G}$ linearer homogener Substitutionen von nicht verschwindenden Determinanten, die der Gruppe $\mathfrak{S}_n$ isomorph ist. Entspricht dann die Substitution A von $\mathfrak{G}$ einer Permutation R von $\mathfrak{S}_n$, die in α_1 Zykeln der Ordnung 1, ferner α_2 Zykeln der Ordnung 2 zerfällt usw., so sei $\chi(R) = \chi_{\alpha_1, \alpha_2, \ldots, \alpha_n}$ die Spur der Substitution A. Die Gesamtheit der Zahlen $\chi(R)$ wird als der *Charakter* der Gruppe $\mathfrak{G}$ bezeichnet. Bedeuten ferner $s_1, s_2, \cdots, s_n$ unabhängige Variable, so nenne ich die Funktion

$$\Phi = \sum \frac{\chi_{\alpha_1, \alpha_2, \cdots, \alpha_n}}{\alpha_1!\,\alpha_2! \cdots \alpha_n!} \left(\frac{s_1}{1}\right)^{\alpha_1} \left(\frac{s_2}{2}\right)^{\alpha_2} \cdots \left(\frac{s_n}{n}\right)^{\alpha_n}$$

die *Charakteristik* der Gruppe $\mathfrak{G}$; hierbei ist die Summation über alle nicht negativen ganzzahligen Lösungen der Gleichung

$$\alpha_1 + 2\alpha_2 + \cdots + n\alpha_n = n$$

zu erstrecken. Durch die Charakteristik Φ ist die Gruppe $\mathfrak{G}$, wenn äquivalente (ähnliche) Gruppen als nicht voneinander verschieden gelten, eindeutig bestimmt. Der Koeffizient $f = \chi_{n, 0, \ldots, 0}$ von $\frac{s_1^n}{n!}$ gibt die Anzahl der Variabeln oder den *Grad* der Gruppe $\mathfrak{G}$ an.

Die Anzahl der nicht äquivalenten irreduziblen Gruppen $\mathfrak{G}$, die der Gruppe $\mathfrak{S}_n$ isomorph sind, ist gleich der Anzahl k der Zerlegungen

$$(1.) \qquad n = \lambda_1 + \lambda_2 + \cdots + \lambda_\varrho \qquad (\lambda_1 \leqq \lambda_2 \leqq \cdots \leqq \lambda_\varrho)$$

der Zahl n in positive ganzzahlige Summanden. Die der Zerlegung (1) entsprechende irreduzible Gruppe werde mit $\mathfrak{G}_{\lambda_1, \lambda_2, \ldots, \lambda_\varrho}$ bezeichnet. Die Charakteristik $\Phi_{\lambda_1, \lambda_2, \ldots, \lambda_\varrho}$ dieser Gruppe läßt sich folgendermaßen bestimmen. Man bezeichne mit p_ν die Funktion

$$p_\nu = \sum \frac{1}{\alpha_1!\,\alpha_2! \cdots \alpha_\nu!} \left(\frac{s_1}{1}\right)^{\alpha_1} \left(\frac{s_2}{2}\right)^{\alpha_2} \cdots \left(\frac{s_\nu}{\nu}\right)^{\alpha_\nu} \qquad (\alpha_1 + 2\alpha_2 + \cdots + \nu\alpha_\nu = \nu)$$

und setze noch $p_0 = 1, p_{-1} = p_{-2} = \cdots = 0$. Dann wird (vgl. D., § 23)

$$(2.) \qquad \Phi_{\lambda_1, \lambda_2, \ldots, \lambda_\varrho} = \begin{vmatrix} p_{\lambda_1}, & p_{\lambda_1 - 1}, & \cdots\, p_{\lambda_1 - \varrho + 1} \\ p_{\lambda_2 + 1}, & p_{\lambda_2}, & \cdots\, p_{\lambda_2 - \varrho + 2} \\ \cdot & \cdot & \cdots\quad \cdot \\ p_{\lambda_\varrho + \varrho - 1}, & p_{\lambda_\varrho + \varrho - 2}, & \cdots\, p_{\lambda_\varrho} \end{vmatrix}.$$

Der Grad $f_{\lambda_1, \lambda_2, \ldots, \lambda_\varrho}$ der Gruppe $\mathfrak{G}_{\lambda_1, \lambda_2, \ldots, \lambda_\varrho}$ ist gleich

$$f_{\lambda_1, \lambda_2, \ldots, \lambda_\varrho} = \frac{n!}{\lambda_1!\,(\lambda_2 + 1)! \cdots (\lambda_\varrho + \rho - 1)!} \prod_{\alpha < \beta} (\lambda_\beta - \lambda_\alpha + \beta - \alpha).$$

Differentiiert man auf beiden Seiten der Gleichung (2) nach s_1 und beachtet, daß $\frac{\partial p_\nu}{\partial s_1} = p_{\nu-1}$ ist, so erhält man rechts

$$\begin{vmatrix} p_{\lambda_1-1}, & p_{\lambda_1-2}, & \cdots p_{\lambda_1-\varrho} \\ p_{\lambda_2+1}, & p_{\lambda_2}, & \cdots p_{\lambda_2-\varrho+2} \\ \cdot & \cdot & \cdots \quad \cdot \\ p_{\lambda_\varrho+\varrho-1}, & p_{\lambda_\varrho+\varrho-2}, & \cdots p_{\lambda_\varrho} \end{vmatrix} + \begin{vmatrix} p_{\lambda_1}, & p_{\lambda_1-1}, & \cdots, p_{\lambda_1-\varrho+1} \\ p_{\lambda_2}, & p_{\lambda_2-1}, & \cdots, p_{\lambda_2-\varrho+1} \\ \cdot & \cdot & \cdots \quad \cdot \\ p_{\lambda_\varrho+\varrho-1}, & p_{\lambda_\varrho+\varrho-2}, & \cdots, p_{\lambda_\varrho} \end{vmatrix} + \cdots.$$

Daher ist

$$\frac{\partial \Phi_{\lambda_1,\lambda_2,\ldots,\lambda_\varrho}}{\partial s_1} = \Phi_{\lambda_1-1,\lambda_2,\ldots,\lambda_\varrho} + \Phi_{\lambda_1,\lambda_2-1,\ldots,\lambda_\varrho} + \cdots + \Phi_{\lambda_1,\lambda_2,\ldots,\lambda_\varrho-1}.$$

Vergleicht man auf beiden Seiten dieser Gleichung die Koeffizienten von s_1^{n-1}, so erhält man die für das folgende wichtige Rekursionsformel

$$(3.) \qquad f_{\lambda_1,\lambda_2,\ldots,\lambda_\varrho} = f_{\lambda_1-1,\lambda_2,\ldots,\lambda_\varrho} + f_{\lambda_1,\lambda_2-1,\ldots,\lambda_\varrho} + \cdots + f_{\lambda_1,\lambda_2,\ldots,\lambda_\varrho-1}.$$

Hierbei hat man rechts, falls $\lambda_\nu = \lambda_{\nu-1}$ ist, den νten Summanden gleich 0 zu setzen, außerdem ist für $\lambda_1 = 1$ unter $f_{0,\lambda_2,\ldots,\lambda_\varrho}$ die Zahl $f_{\lambda_2,\ldots,\lambda_\varrho}$ zu verstehen.

Die der Zahl n entsprechenden k Charakteristiken $\Phi_{\lambda_1,\lambda_2,\ldots,\lambda_\varrho}$ denke ich mir nach dem lexikographischen Prinzip angeordnet, d. h. es soll $\Phi_{\lambda_1,\lambda_2,\ldots,\lambda_\varrho}$ vor $\Phi_{\mu_1,\mu_2,\ldots,\mu_\sigma}$ stehen, wenn die erste nicht verschwindende unter den Differenzen

$$\lambda_\varrho - \mu_\sigma, \lambda_{\varrho-1} - \mu_{\sigma-1}, \cdots$$

positiv ist. Die Anordnung ist also folgende:

$$\Phi_n, \Phi_{1,n-1}, \Phi_{2,n-2}, \Phi_{1,1,n-2}, \Phi_{3,n-3}, \Phi_{1,2,n-3}, \ldots.$$

Das αte Glied dieser Reihe soll mit $\Phi^{(\alpha)}$, die zu $\Phi^{(\alpha)}$ gehörende irreduzible Gruppe mit $\mathfrak{G}^{(\alpha)}$ bezeichnet werden. Ist $\alpha < \beta$, so soll $\mathfrak{G}^{(\alpha)}$ auch von *höherer Ordnung* als $\mathfrak{G}^{(\beta)}$ heißen. Nach demselben Prinzip ordnen wir auch die k Produkte $p_{\lambda_1} p_{\lambda_2} \cdots p_{\lambda_\varrho}$ an, sie sollen entsprechend mit $p^{(1)}, p^{(2)}, \cdots$ bezeichnet werden. Dann wird (vgl. D., § 23)

$$p^{(\alpha)} = \Phi^{(\alpha)} + \sum_{\beta=1}^{\alpha-1} c_\beta \Phi^{(\beta)},$$

wo die Koeffizienten c_β nicht negative ganze Zahlen sind. Hieraus folgt, daß, wenn $\mathfrak{G}$ eine der Gruppe $\mathfrak{S}_n$ isomorphe Gruppe linearer Substitutionen mit der Charakteristik $p_{\lambda_1} p_{\lambda_2} \cdots p_{\lambda_\varrho}$ ist, unter den in $\mathfrak{G}$ enthaltenen irreduziblen Gruppen die Gruppe $\mathfrak{G}_{\lambda_1,\lambda_2,\ldots,\lambda_\varrho}$ genau einmal vorkommt, während die übrigen von höherer Ordnung sind.

§ 2.

Eine der Gruppe $\mathfrak{S}_n$ isomorphe Gruppe $\mathfrak{G}$ mit der Charakteristik $p_{\lambda_1} p_{\lambda_2} \cdots p_{\lambda_\varrho}$ kann am einfachsten folgendermaßen hergestellt werden.

Es sei allgemein $\varphi(x_1, x_2, \cdots, x_n)$ irgendeine (ganze rationale) Funktion der Variabeln $x_1, x_2, \cdots, x_n$. Unterwirft man in der Funktion φ die n Variabeln allen $h = n!$ Permutationen von $\mathfrak{S}_n$, so mögen m verschiedene Funktionen

$$(4.)\qquad \varphi_1, \varphi_2, \cdots, \varphi_m$$

entstehen. Jeder Permutation R von $\mathfrak{S}_n$ entspricht dann eine Permutation R' der m Funktionen (4), und diese Permutationen der φ_μ bilden eine mit $\mathfrak{S}_n$ isomorphe Gruppe $\mathfrak{G}$, die auch als Gruppe linearer homogener Substitutionen aufgefaßt werden kann. Die Spur $\chi(R)$ von R' gibt dann an, wie viele unter den Funktionen (4) ungeändert bleiben, wenn die Variabeln die Permutation R erleiden. Die Zahl $\chi(R)$ ist bekanntlich eindeutig bestimmt durch die Permutation R und die Untergruppe $\mathfrak{A}$ von $\mathfrak{S}_n$, bei deren Permutationen die Funktion φ ungeändert bleibt. Man setze

$$\mathfrak{S}_n = \mathfrak{A}S_1 + \mathfrak{A}S_2 + \cdots + \mathfrak{A}S_m$$

und nehme an, daß φ durch die Permutationen des Komplexes $\mathfrak{A}S_\mu$ in φ_μ übergeführt wird. Dann läßt φ_μ nur die Permutationen der Gruppe $S_\mu^{-1}\mathfrak{A}S_\mu$ zu. Daher gibt $\chi(R)$ an, wie viele unter den Gruppen

$$S_1^{-1}\mathfrak{A}S_1,\ S_2^{-1}\mathfrak{A}S_2,\ \cdots,\ S_m^{-1}\mathfrak{A}S_m$$

das Element R enthalten. Bildet man für alle h Permutationen S von $\mathfrak{S}_n$ die Gruppen $S^{-1}\mathfrak{A}S$, so enthalten unter diesen Gruppen, wenn a die Ordnung von $\mathfrak{A}$ ist, genau $a\chi(R)$ die Permutation R. Dies kann man auch anders ausdrücken: unter den h Elementen SRS^{-1} kommen $a\chi(R)$ in $\mathfrak{A}$ vor. Sind nun

$$(5.)\qquad R_1, R_2, \cdots, R_{h_\varrho}$$

die zu R konjugierten Elemente von $\mathfrak{S}_n$, so werden unter den Elementen SRS^{-1} genau $\dfrac{h}{h_\varrho}$ gleich R_ν. Ist demnach a_ϱ die Anzahl der Elemente (5), die in $\mathfrak{A}$ enthalten sind, so wird $a\chi(R) = \dfrac{h}{h_\varrho}\, a_\varrho$, also

$$\chi(R) = \frac{h a_\varrho}{a h_\varrho}\,.$$ [1]

Man denke sich nun die Funktion φ so gewählt, daß $\mathfrak{A}$ die Untergruppe $\mathfrak{S}_{\lambda_1, \lambda_2, \ldots, \lambda_\varrho}$ der Ordnung $\lambda_1!\,\lambda_2! \cdots \lambda_\varrho!$ von $\mathfrak{S}_n$ wird, deren Per-

[1] Vgl. Frobenius, *Über Relationen zwischen den Charakteren einer Gruppe und denen ihrer Untergruppen*, Sitzungsberichte 1898, S. 501.

mutationen die λ_1 ersten Ziffern untereinander vertauschen, ferner die λ_2 folgenden usf. Dann wird (vgl. D., § 12) die durch den Charakter $\chi(R)$ bestimmte Charakteristik Φ von $\mathfrak{G}$ gleich $p_{\lambda_1} p_{\lambda_2} \cdots p_{\lambda_\varrho}$. In diesem Fall soll die Gruppe $\mathfrak{G}$ auch mit $\mathfrak{P}_{\lambda_1, \lambda_2, \ldots, \lambda_\varrho}$ bezeichnet werden.

§ 3.

Es erweist sich als nützlich, im folgenden von den Bezeichnungen der Theorie der Moduln Gebrauch zu machen.

Unter einem Modul verstehe ich hier ein System M von homogenen Formen derselben Ordnung in n Variabeln $x_1, x_2, \cdots x_n$, das durch folgende Eigenschaft charakterisiert ist: sind φ und ψ zwei Formen von M und bedeuten a und b zwei beliebige Konstanten, so enthält M auch die Form $a\varphi + b\psi$. In M lassen sich dann gewisse m linear unabhängige Formen $\varphi_1, \varphi_2, \cdots, \varphi_m$ angeben, so daß die Gesamtheit der Formen von M übereinstimmt mit der Gesamtheit der linearen Verbindungen $a_1\varphi_1 + a_2\varphi_2 + \cdots + a_m\varphi_m$ mit konstanten Koeffizienten a_1, $a_2, \cdots, a_m$. Die Zahl m nennt man die *Ordnung*, die Funktionen φ_1, $\varphi_2, \cdots, \varphi_m$ eine *Basis* des Moduls M. Multipliziert man alle Elemente von M mit irgendeiner Form η, so entsteht ein neuer Modul, der mit ηM bezeichnet werden soll. Der Modul M soll ein *symmetrischer* Modul heißen, wenn für jede Form φ von M auch alle aus φ durch Vertauschung der Variabeln hervorgehenden Formen in M enthalten sind. Hat man mehrere Formen $\varphi, \psi, \cdots$ derselben Ordnung und geht φ bei den $n!$ Vertauschungen der Variabeln in $\varphi', \varphi'', \cdots$ über, ebenso ψ in $\psi', \psi'', \cdots$ usw., so bildet die Gesamtheit der linearen Verbindungen

$$a'\varphi' + a''\varphi'' + \cdots + b'\psi' + b''\psi'' + \cdots$$

einen symmetrischen Modul M. Wir wollen dann sagen: M sei der *durch die Formen* $\varphi, \psi, \cdots$ *erzeugte symmetrische Modul.*

Jedem symmetrischen Modul M in n Variabeln entspricht eine Gruppe $\mathfrak{M}$ linearer homogener Substitutionen, die der Gruppe $\mathfrak{S}_n$ isomorph ist. Bilden nämlich $\varphi_1, \varphi_2, \cdots, \varphi_m$ eine Basis von M und geht φ_μ durch die Permutation R der Variabeln in $\bar{\varphi}_\mu$ über, so ist nach Voraussetzung $\bar{\varphi}_\mu$ in M enthalten und folglich

$$\bar{\varphi}_\mu = a_{\mu 1}\varphi_1 + a_{\mu 2}\varphi_2 + \cdots + a_{\mu m}\varphi_m,$$

wo die $a_{\mu\nu}$ gewisse Konstanten sind. Die so entstehenden $n!$ Substitutionen $A = (a_{\mu\nu})$ bilden dann eine der Gruppe $\mathfrak{S}_n$ isomorphe Gruppe $\mathfrak{M}$, die wir *die Gruppe des Moduls* oder auch genauer die *zur Basis* $\varphi_1, \varphi_2, \cdots, \varphi_m$ *gehörende Gruppe des Moduls* nennen. Wählt man an Stelle von $\varphi_1, \varphi_2, \cdots, \varphi_m$ eine andere Basis von M, so geht $\mathfrak{M}$ in eine äquivalente Gruppe $\mathfrak{M}'$ über.

Man habe nun r Formen derselben Ordnung $\psi_1, \psi_2, \cdots, \psi_r$ in den n Variabeln $x_1, x_2, \cdots, x_n$, die nicht notwendig linear unabhängig sein sollen. Es sei bekannt, daß, wenn ψ_ϱ bei irgendeiner Permutation R der Variabeln in $\bar{\psi}_\varrho$ übergeht, sich Konstanten $b_{\varrho\sigma}$ bestimmen lassen, so daß

$$\bar{\psi}_\varrho = b_{\varrho 1}\psi_1 + b_{\varrho 2}\psi_2 + \cdots + b_{\varrho r}\psi_r$$

wird; man bezeichne die Substitution $(b_{\varrho\sigma})$ mit B. Die Gesamtheit der Ausdrücke $c_1\psi_1 + c_2\psi_2 + \cdots + c_r\psi_r$ mit konstanten Koeffizienten c_ϱ bildet dann einen symmetrischen Modul M, dessen Ordnung m angibt, wie viele unter den Funktionen $\psi_1, \psi_2, \cdots, \psi_r$ linear unabhängig sind. Man wähle nun eine Matrix $Q = (q_{\varrho\sigma})$ von nicht verschwindender Determinante, so daß unter den r Funktionen

$$\varphi_\varrho = q_{\varrho 1}\psi_1 + q_{\varrho 2}\psi_2 + \cdots + q_{\varrho r}\psi_r$$

die ersten m untereinander linear unabhängig, die letzten $r-m$ dagegen gleich Null werden. Dann erhält die Matrix QBQ^{-1} die Form

$$QBQ^{-1} = \begin{pmatrix} A & C \\ 0 & D \end{pmatrix}, \tag{6.}$$

wo A eine Matrix des Grades m bedeutet. Die so entstehenden $n!$ Matrixen A bestimmen dann, wie man leicht sieht, die zur Basis $\varphi_1, \varphi_2, \cdots, \varphi_m$ gehörende Gruppe des Moduls M. Bilden insbesondere die $n!$ Substitutionen B eine mit $\mathfrak{S}_n$ isomorphe Gruppe $\mathfrak{N}$, so folgt aus der Gleichung (6), daß *die Gruppe des Moduls* M *keinen irreduziblen Bestandteil enthält, der nicht auch in der Gruppe* $\mathfrak{N}$ *enthalten ist.*

Es sei wieder M ein beliebiger symmetrischer Modul der Ordnung m in n Variabeln, und es sei A ein Teilmodul der Ordnung a von M, der ebenfalls symmetrisch ist. Sieht man dann zwei Funktionen von M, deren Differenz in A enthalten ist, als nicht voneinander verschieden an, so erscheint M, mod. A betrachtet, gewissermaßen als ein symmetrischer Modul der Ordnung $m - a = r$. Der so entstehende Relativmodul P soll der *zu* A *komplementäre Modul* heißen. In M lassen sich ferner r Formen $\eta_1, \eta_2, \cdots, \eta_r$ bestimmen, so daß jede Form φ von M mod. A einer linearen Verbindung $c_1\eta_1 + c_2\eta_2 + \cdots + c_r\eta_r$ kongruent wird. Die Funktionen $\eta_1, \eta_2, \cdots, \eta_r$ bilden dann, wie wir sagen wollen, eine *Basis des komplementären Moduls* P. Geht η_ϱ durch die Permutation R der Variabeln in $\bar{\eta}_\varrho$ über, so wird

$$\bar{\eta}_\varrho \equiv c_{\varrho 1}\eta_1 + c_{\varrho 2}\eta_2 + \cdots + c_{\varrho r}\eta_r \pmod{\mathrm{A}};$$

hierbei bedeuten die $c_{\varrho\sigma}$ gewisse Konstanten. Die den verschiedenen Permutationen von $\mathfrak{S}_n$ entsprechenden Substitutionen $(c_{\varrho\sigma})$ bilden eine mit $\mathfrak{S}_n$ isomorphe Gruppe $\mathfrak{R}$, die *Gruppe des komplementären Moduls* P. Diese Gruppe $\mathfrak{R}$ läßt sich bekanntlich auch folgendermaßen charakterisieren: man bestimme eine Basis $\xi_1, \xi_2, \cdots, \xi_a$ des Moduls A und bezeichne mit $\mathfrak{A}$ die zu dieser Basis gehörende Gruppe von A; wählt man als Basis von M die m Formen $\xi_1, \cdots, \xi_a, \eta_1, \cdots, \eta_r$, so erhält die dieser Basis entsprechende Gruppe $\mathfrak{M}$ von M die Gestalt

$$\mathfrak{M} = \begin{pmatrix} \mathfrak{A} & 0 \\ \mathfrak{C} & \mathfrak{R} \end{pmatrix}.$$

Hieraus folgt zugleich, daß jeder irreduzible Bestandteil von $\mathfrak{M}$ entweder in $\mathfrak{A}$ oder in $\mathfrak{R}$ enthalten ist.

§ 4.

Man betrachte nun den speziellen symmetrischen Modul Γ in den n Variabeln $x_1, x_2, \cdots, x_n$, der durch die Funktion

$$\varphi = x_1 x_2 \cdots x_{p-1}(x_p + x_{p+1} + \cdots + x_n)$$

erzeugt wird. Hierbei soll p irgendeine ganze Zahl bedeuten, die nicht größer ist als n; für $p = 1$ setze man

$$\varphi = x_1 + x_2 + \cdots + x_n.$$

Unter $C^{(p)}_{\nu_1, \nu_2, \ldots, \nu_r}$ verstehe man für $r \geqq p$ die Summe aller Produkte von je p verschiedenen aus der Reihe der Variabeln $x_{\nu_1}, x_{\nu_2}, \cdots, x_{\nu_r}$, dagegen sei $C^{(p)}_{\nu_1, \nu_2, \ldots, \nu_r} = 0$ für $r < p$. Ist dann

$$\alpha_1, \alpha_2, \cdots, \alpha_p, \beta_1, \beta_2, \cdots, \beta_{n-p}$$

irgendeine Anordnung der Indizes $1, 2, \cdots, n$, so gilt die Formel

$$x_{\alpha_1} x_{\alpha_2} \cdots x_{\alpha_p} \equiv (-1)^p C^{(p)}_{\beta_1, \beta_2, \ldots, \beta_{n-p}} \pmod{\Gamma}.$$

Es genügt offenbar die Formel

$$(7.) \qquad x_1 x_2 \cdots x_p \equiv (-1)^p C^{(p)}_{p+1, p+2, \ldots, n} \pmod{\Gamma}$$

zu beweisen.

Diese Formel ist für $p = 1$ unmittelbar evident; sie sei für die Zahl $p-1$ bereits bewiesen. Ist dann Γ′ der durch die Funktion

$$x_2 \cdots x_{p-1}(x_p + x_{p+1} + \cdots + x_n)$$

erzeugte symmetrische Modul in den $n-1$ Variabeln $x_2, \cdots, x_n$, so wird also

$$(8.) \qquad x_2 \cdots x_p \equiv (-1)^{p-1} C^{(p-1)}_{p+1, \ldots, n} \pmod{\Gamma'}.$$

Ist aber φ in Γ′ enthalten, so kommt $x_1\varphi$ in Γ vor. Daher folgt aus (8)

$$x_1 x_2 \cdots x_p \equiv (-1)^{p-1} x_1 C^{(p-1)}_{p+1, \ldots, n} \pmod{\Gamma}.$$

Ebenso ist für jedes α aus der Reihe $1, 2, \cdots, p$

$$x_1 x_2 \cdots x_p \equiv (-1)^{p-1} x_\alpha C^{(p-1)}_{p+1, \ldots, n} \pmod{\Gamma}$$

und folglich

$$(9.) \qquad p x_1 x_2 \cdots x_p \equiv (-1)^{p-1}(x_1 + x_2 + \cdots + x_p) C^{(p-1)}_{p+1, \ldots, n} \pmod{\Gamma}.$$

Ist nun $n-p<p-1$, so folgt aus dieser Formel, daß $x_1 x_2 \cdots x_p$ in Γ enthalten ist, in Übereinstimmung mit (7). Ist aber $n-p \geqq p-1$, so wähle man $p-1$ Indizes $\gamma_1, \cdots, \gamma_{p-1}$ aus der Reihe $p+1, \cdots, n$ und bezeichne die übrigbleibenden $q=n-2p+1$ Indizes mit $\delta_1, \cdots, \delta_q$. Dann ist jedenfalls

$$x_{\gamma_1} \cdots x_{\gamma_{p-1}} (x_{\delta_1} + \cdots + x_{\delta_q} + x_1 + \cdots + x_p) \equiv 0 \text{ (mod. } \Gamma).$$

Summiert man über alle $\binom{n-p}{p-1}$ Kombinationen $\gamma_1, \cdots, \gamma_{p-1}$, so ergibt sich, wie man leicht sieht,

$$(x_1 + x_2 + \cdots + x_p) C^{(p-1)}_{p+1, \ldots, n} + p\, C^{(p)}_{p+1, \ldots, n} \equiv 0 \text{ (mod. } \Gamma).$$

In Verbindung mit (9) ergibt sich hieraus

$$p x_1 x_2 \cdots x_p \equiv (-1)^p p\, C^{(p)}_{p+1, \ldots, n} \text{ (mod. } \Gamma).$$

Dividiert man durch p, so erhält man die zu beweisende Formel (7).

§ 5.

Um nun die zu der Zerlegung

$$n = \lambda_1 + \lambda_2 + \cdots + \lambda_\varrho \qquad (\lambda_1 \leqq \lambda_2 \leqq \cdots \leqq \lambda_\varrho)$$

gehörende Gruppe $\mathfrak{G} = \mathfrak{G}_{\lambda_1, \lambda_2, \ldots, \lambda_\varrho}$ als Gruppe mit ganzzahligen Koeffizienten darzustellen, wende ich ein Verfahren an, das einer bekannten Methode, die zur Herstellung der speziellen Gruppe $\mathfrak{G}_{1, n-1}$ des Grades $n-1$ dient, nachgebildet ist.

Wir betrachten das Potenzprodukt

$$X = x_1^{\alpha_1} x_2^{\alpha_2} \cdots x_n^{\alpha_n},$$

in dem die λ_1 ersten Exponenten gleich $\rho-1$, die λ_2 folgenden gleich $\rho-2$, usw., die λ_ϱ letzten also gleich 0 sind[1]. Die aus X durch Vertauschung der Variabeln hervorgehenden

$$N = \frac{n!}{\lambda_1! \lambda_2! \cdots \lambda_\varrho!}$$

Produkte mögen mit

$$(10.) \qquad X^{(0)}, X^{(1)}, \ldots, X^{(N-1)}$$

bezeichnet werden. Da X nur bei den Permutationen der Untergruppe $\mathfrak{S}_{\lambda_1, \lambda_2, \ldots, \lambda_\varrho}$ von $\mathfrak{S}_n$ ungeändert bleibt, so erleiden die Funktionen (10) nach dem Ergebnis des § 2 bei den Vertauschungen der Variabeln die Substitutionen der Gruppe $\mathfrak{P}_{\lambda_1, \lambda_2, \ldots, \lambda_\varrho}$, deren Charakteristik gleich

[1] Der triviale Fall $\varrho = 1$, der auf die Darstellung der Gruppe $\mathfrak{S}_n$ durch $n!$ Einsen führt, soll im folgenden ausgeschlossen werden.

$p_{\lambda_1} p_{\lambda_2} \cdots p_{\lambda_\varrho}$ ist. Diese Gruppe erscheint zugleich als die zur Basis (10) gehörende Gruppe des durch die Funktion X erzeugten symmetrischen Moduls

$$\mathrm{M} = \mathrm{M}^{(\lambda_1, \lambda_2, \cdots, \lambda_\varrho)}.$$

Daher enthält die Gruppe dieses Moduls die zu untersuchende Gruppe $\mathfrak{G}_{\lambda_1, \lambda_2, \ldots, \lambda_\varrho}$ als irreduziblen Bestandteil.

Der Modul M enthält weiter die $m = \varrho - 1$ Ausdrücke

$$Y_1 = \frac{X}{x_{\lambda_1}}(x_{\lambda_1} + x_{\lambda_1+1} + \cdots + x_{\lambda_1+\lambda_2})$$

$$Y_2 = \frac{X}{x_{\lambda_1+\lambda_2}}(x_{\lambda_1+\lambda_2} + x_{\lambda_1+\lambda_2+1} + \cdots + x_{\lambda_1+\lambda_2+\lambda_3})$$

$$\cdots\cdots\cdots\cdots\cdots\cdots\cdots\cdots\cdots\cdots$$

$$Y_m = \frac{X}{x_{\lambda_1+\lambda_2+\cdots+\lambda_m}}(x_{\lambda_1+\lambda_2+\cdots+\lambda_m} + x_{\lambda_1+\lambda_2+\cdots+\lambda_m+1} + \cdots + x_{\lambda_1+\lambda_2+\cdots+\lambda_\varrho})$$

und folglich auch den durch diese Ausdrücke erzeugten symmetrischen Modul

$$\mathrm{A} = \mathrm{A}^{(\lambda_1, \lambda_2, \cdots, \lambda_\varrho)}.$$

Die Funktion Y_ν bleibt, wie man leicht sieht, nur bei den Permutationen der Untergruppe

$$\mathfrak{S}_{\lambda_1, \cdots, \lambda_{\nu-1}, \lambda_\nu-1, \lambda_{\nu+1}+1, \lambda_{\nu+2}, \cdots, \lambda_\varrho}$$

ungeändert. Die aus Y_ν durch Vertauschung der Variabeln hervorgehenden N_ν verschiedenen Ausdrücke erleiden daher bei den $n!$ Permutationen der Variabeln die Substitutionen der Gruppe

$$(11.) \qquad \mathfrak{P}_{\lambda_1, \cdots, \lambda_{\nu-1}, \lambda_\nu-1, \lambda_{\nu+1}+1, \lambda_{\nu+2}, \cdots, \lambda_\varrho}.$$

Die Charakteristik

$$p_{\lambda_1} \cdots p_{\lambda_{\nu-1}} p_{\lambda_\nu-1} p_{\lambda_{\nu+1}+1} p_{\lambda_{\nu+2}} \cdots p_{\lambda_\varrho}$$

dieser Gruppe ist aber von höherer Ordnung als die Charakteristik $p_{\lambda_1} p_{\lambda_2} \cdots p_{\lambda_\varrho}$. Nach dem am Schluß des § 1 Gesagten ist daher jeder irreduzible Bestandteil $\mathfrak{G}_{\mu_1, \mu_2, \ldots \mu_\tau}$ der Gruppe (11) von höherer Ordnung als die Gruppe $\mathfrak{G}_{\lambda_1, \lambda_2, \ldots, \lambda_\varrho}$. Denkt man sich nun die $N_1 + N_2 + \cdots + N_m$ Ausdrücke gebildet, die aus $Y_1, Y_2, \cdots, Y_m$ durch Vertauschung der Variabeln hervorgehen, so erfahren diese Ausdrücke bei den $n!$ Permutationen der Variabeln die Substitutionen einer Gruppe $\mathfrak{H}$, die in die m Gruppen (11) zerfällt. Folglich ist die irreduzible Gruppe $\mathfrak{G}$ in $\mathfrak{H}$ nicht enthalten. Aus dem auf S. 669 Gesagten ergibt sich also, daß auch die Gruppe $\mathfrak{A}$ des symmetrischen Teilmoduls A von M die Gruppe $\mathfrak{G}$ nicht enthält. Hieraus schließen wir, daß $\mathfrak{G}$ unter den irreduziblen Bestandteilen der Gruppe $\mathfrak{R}$ des zu A komplementären Moduls P vorkommen muß. Die Ordnung r des Moduls P ist daher nicht kleiner als der Grad

$$f = f_{\lambda_1, \lambda_2, \cdots, \lambda_\varrho}$$

von $\mathfrak{G}$.

Können wir daher f Funktionen $X_1, X_2, \cdots, X_f$ in M angeben von der Art, daß jedes der Produkte (10) mod. A einer linearen Verbindung

$$c_1 X_1 + c_2 X_2 + \cdots + c_f X_f$$

kongruent wird, so muß $r = f$ sein; die Funktionen $X_1, X_2, \cdots, X_f$ bilden dann eine Basis von P, und die zu dieser Basis gehörende Gruppe $\mathfrak{R}$ kann gleich $\mathfrak{G}$ gesetzt werden. Gelingt es insbesondere für die Funktionen $X_1, X_2, \cdots, X_f$ gewisse unter den Potenzprodukten (10) zu wählen und nachzuweisen, daß die Koeffizienten $c_1, c_2, \cdots, c_f$ sämtlich ganzzahlig werden, so werden auch die Koeffizienten der zugehörigen Gruppe $\mathfrak{R} = \mathfrak{G}$ ganze Zahlen.

Man gelangt nun zu einem solchen System von Potenzprodukten $X_1, X_2, \cdots, X_f$ auf folgendem Wege.

Man setze

$$F_{n-1}^{(2)} = 1\,, \qquad F_{n-1}^{(1,1)} = x_{n-1}\,,$$

$$F_{n-2}^{(3)} = 1\,, \qquad F_{n-2}^{(1,2)} = x_{n-2} F_{n-1}^{(2)} + F_{n-1}^{(1,1)}\,, \qquad F_{n-2}^{(1,1,1)} = x_{n-2}^2 F_{n-1}^{(1,1)}$$

usw., zuletzt sei

$$(12.) \quad F^{(\lambda_1, \lambda_2, \cdots, \lambda_\varrho)} = x_1^m F_2^{(\lambda_1 - 1, \lambda_2, \cdots, \lambda_\varrho)} + x_1^{m-1} F_2^{(\lambda_1, \lambda_2 - 1, \cdots, \lambda_\varrho)} + \cdots + x_1 F_2^{(\lambda_1, \lambda_2, \cdots, \lambda_{\varrho-1} - 1, \lambda_\varrho)} + F_2^{(\lambda_1, \lambda_2, \cdots, \lambda_\varrho - 1)}.$$

Hierbei ist rechts für $\lambda_1 = 1$

$$F_2^{(\lambda_1 - 1, \lambda_2, \cdots, \lambda_\varrho)} = F_2^{(\lambda_2, \cdots, \lambda_\varrho)}$$

und, wenn für ein $\varkappa$ die Indizes $\lambda_{\varkappa-1}$ und $\lambda_\varkappa$ einander gleich sind,

$$F_2^{(\lambda_1, \cdots, \lambda_{\varkappa-1}, \lambda_\varkappa - 1, \cdots, \lambda_\varrho)} = 0$$

zu setzen. Auf Grund der Rekursionsformel (12) wird also für jedes System von ϱ Indizes

$$0 < \lambda_1 \leqq \lambda_2 \leqq \cdots \leqq \lambda_\varrho$$

eine wohlbestimmte ganze rationale Funktion $F^{(\lambda_1, \lambda_2, \cdots, \lambda_\varrho)}$ von $\lambda_1 + \lambda_2 + \cdots + \lambda_\varrho$ Variabeln bestimmt, die in bezug auf die erste Variable vom Grade $m = \varrho - 1$ ist, die letzte Variable aber nicht explizite enthält. So ist z. B.

$$F^{(1, n-1)} = x_1 + x_2 + \cdots + x_{n-1}\,,$$

$$F^{(2, n-2)} = x_1(x_2 + \cdots + x_{n-1}) + x_2(x_3 + \cdots + x_{n-1}) + \cdots + x_{n-3}(x_{n-2} + x_{n-1})\,.$$

Aus der Bildungsweise der Funktion $F = F^{(\lambda_1, \lambda_2, \cdots, \lambda_\varrho)}$ geht nun unmittelbar hervor, daß F eine Summe von gewissen f' Potenzprodukten aus der Reihe (10) ist. *Diese Zahl f' ist aber gleich dem Grad $f = f_{\lambda_1, \lambda_2, \cdots, \lambda_\varrho}$ der Gruppe* $\mathfrak{G}$. Denn dies ist richtig für $n = 2$; nimmt man aber unsere

Behauptung für die Funktionen $F^{(\mu_1,\mu_2,\cdots)}$, bei denen die Summe der Indizes $\mu_1, \mu_2, \cdots$ gleich $n-1$ ist, als bewiesen an, so ergibt sich aus (12)

$$f' = f_{\lambda_1-1,\lambda_2,\ldots,\lambda_\varrho} + f_{\lambda_1,\lambda_2-1,\ldots,\lambda_\varrho} + \cdots + f_{\lambda_1,\lambda_2,\ldots,\lambda_\varrho-1}.$$

Die rechts stehende Summe ist aber nach Formel (3) gleich $f_{\lambda_1,\lambda_2,\ldots,\lambda_\varrho}$.

Es soll nun bewiesen werden, daß die f Glieder der Summe F der oben gestellten Forderung genügen:

I. *Jedes der Potenzprodukte $X^{(\alpha)}$ ist mod.* A *kongruent einer ganzzahligen linearen Verbindung der f Potenzprodukte $X_1, X_2, \cdots, X_f$.*

Der Beweis dieses Satzes erfordert ein genaueres Studium des Moduls A und der Funktion F.

§ 6.

Man bilde mit Hilfe der $n-1$ Variabeln

$$(13.) \qquad x_1, x_2, \cdots, x_{\nu-1}, x_{\nu+1}, \cdots, x_n,$$

die der Zerlegung

$$n-1 = (\lambda_1 - 1) + \lambda_2 + \cdots + \lambda_\varrho$$

entsprechenden symmetrischen Moduln

$$\mathrm{M}_\nu = \mathrm{M}^{(\lambda_1-1,\lambda_2,\ldots,\lambda_\varrho)}, \qquad \mathrm{A}_\nu = \mathrm{A}^{(\lambda_1-1,\lambda_2,\ldots,\lambda_\varrho)}$$

in derselben Weise wie früher die Moduln M und A. Dann ist unmittelbar ersichtlich, daß M *den Modul* $x_\nu^m \mathrm{M}_\nu$ *und ebenso* A *den Modul* $x_\nu^m \mathrm{A}_\nu$ *als Teilmodul enthält.* Ferner besitzt A noch folgende Eigenschaft:

II. *Setzt man* $\mu = \lambda_2 - \lambda_1 + 1$ *und versteht unter* $\alpha_1, \alpha_2, \cdots, \alpha_\mu$ *irgendwelche* μ *Indizes aus der Reihe* $1, 2, \cdots, n$, *so ist jedes Potenzprodukt* $X^{(\alpha)}$, *das den Faktor* $x_{\alpha_1}^{m-1} x_{\alpha_2}^{m-1} \cdots x_{\alpha_\mu}^{m-1}$ *enthält, mod.* A *kongruent einer ganzzahligen linearen Verbindung gewisser Potenzprodukte, von denen jedes einen der Faktoren* $x_{\alpha_1}^m, x_{\alpha_2}^m, \cdots, x_{\alpha_\mu}^m$ *enthält.*

Man betrachte nämlich den Ausdruck Y_1, der auch in der Form

$$Y = \frac{X}{x_1 x_2 \cdots x_{\lambda_1}} x_1 x_2 \cdots x_{\lambda_1-1} (x_{\lambda_1} + x_{\lambda_1+1} + \cdots + x_{\lambda_1+\lambda_2})$$

geschrieben werden kann. Setzt man

$$Z = \frac{X}{x_1 x_2 \cdots x_{\lambda_1}} = x_1^{m-1} x_2^{m-1} \cdots x_{\lambda_1+\lambda_2}^{m-1} x_{\lambda_1+\lambda_2+1}^{m-2} \cdots$$

und bezeichnet mit Γ den symmetrischen Modul in den Variabeln $x_1, x_2 \cdots, x_{\lambda_1+\lambda_2}$, der durch die Funktion

$$x_1 x_2 \cdots x_{\lambda_1-1} (x_{\lambda_1} + x_{\lambda_1+1} + \cdots + x_{\lambda_1+\lambda_2})$$

erzeugt wird, so enthält A offenbar den ganzen Modul $Z\Gamma$. Nun ist aber nach § 4

$$x_{\lambda_2-\lambda_1+2} x_{\lambda_2-\lambda_1+3} \cdots x_{\lambda_2+1} \equiv (-1)^{\lambda_1} C^{(\lambda_1)}_{1,2,\ldots,\lambda_2-\lambda_1+1,\lambda_2+2,\ldots,\lambda_2+\lambda_1} \pmod{\Gamma};$$

folglich ist

(14.) $Z \cdot x_{\lambda_2-\lambda_1+2}\, x_{\lambda_2-\lambda_1+3} \cdots x_{\lambda_2+1} \equiv (-1)^{\lambda_1} Z C^{(\lambda_1)}_{1,2,\ldots,\lambda_2-\lambda_1+1,\lambda_2+2,\ldots\lambda_2+\lambda_1}$ (mod. A).

Das links stehende Produkt enthält nun den Faktor $x_1^{m-1} x_2^{m-1} \cdots x_\mu^{m-1}$, ferner ist jedes der $\binom{\lambda_2}{\lambda_1}$ Glieder der rechts stehenden Summe durch mindestens eine der Potenzen $x_1^m, x_2^m, \cdots, x_\mu^m$ teilbar. Da nun die Formel (14) richtig bleibt, wenn die Variabeln $x_1, x_2, \cdots, x_n$ irgendwie permutiert werden, so ist unsere Behauptung als bewiesen anzusehen.

Wir betrachten jetzt die Funktion F. Man bezeichne mit A_ν den Koeffizienten von x_ν^m in F, ferner sei F_ν die Funktion $F^{(\lambda_1-1,\lambda_2,\cdots,\lambda_\varrho)}$ der $n-1$ Variabeln (13) und $D_\nu = F_\nu - A_\nu$. Dann gilt folgende Regel:

III. *Für* $\nu = 1, 2, \cdots, \mu$ *ist* $D_\nu = 0$, *also* $A_\nu = F_\nu$. *Ist aber* $\nu > \mu$, *so wird* D_ν *eine Summe von Potenzprodukten, von denen jedes einen Faktor* $x_{\alpha_1}^{m-1} x_{\alpha_2}^{m-1} \cdots x_{\alpha_\mu}^{m-1}$ *enthält; hierbei bedeuten* $\alpha_1, \alpha_2, \cdots \alpha_\mu$ *Indizes aus der Reihe* $1, 2, \cdots, \nu-1$.

Der Beweis ist mit Hilfe der Rekursionsformel (12) zu führen. Man bezeichne zur Abkürzung die Funktion

$$F_2^{(\lambda_1,\cdots,\lambda_\varkappa-1,\lambda_\varkappa+1,\cdots\lambda_\varrho)}$$

der $n-1$ Variabeln $x_2, \cdots x_n$ mit $F^{(\varkappa)}$, so daß also

$$F = x_1^m F^{(1)} + x_1^{m-1} F^{(2)} + \cdots + x_1 F^{(m)} + F^{(m+1)}$$

wird. Dann ist zunächst, wie zu beweisen ist,

$$A_1 = F^{(1)} = F_1.$$

Ist nun $\nu > 1$ und bedeutet $A_\nu^{(\varkappa)}$ den Koeffizienten von x_ν^m in $F^{(\varkappa)}$, so wird

$$A_\nu = x_1^m A_\nu^{(1)} + x_1^{m-1} A_\nu^{(2)} + \cdots + x_1 A_\nu^{(m)} + A_\nu^{(m+1)}.$$

Hierbei ist zu beachten, daß $A_\nu^{(1)}$ für $\lambda_1 = 1$ gleich 0 zu setzen ist, da dann $F^{(1)} = F_2^{(\lambda_2,\lambda_3,\cdots,\lambda_\varrho)}$ keine der Variabeln in der m^{ten} Potenz enthält. Die mit Hilfe der $n-2$ Variabeln

$$x_2, \cdots, x_{\nu-1}, x_{\nu+1}, \cdots x_n$$

gebildeten Funktionen

$$F^{(\lambda_1-2,\lambda_2,\cdots\lambda_n)}, \quad F^{(\lambda_1-1,\lambda_2\cdots,\lambda_\varkappa-1,\lambda_\varkappa+1,\cdots,\lambda_n)}$$

bezeichne man mit $F_\nu^{(1)}$ und $F_\nu^{(\varkappa)}$ und setze noch $D_\nu^{(\varkappa)} = F_\nu^{(\varkappa)} - A_\nu^{(\varkappa)}$. Für $\lambda_1 = 1$ hat man hierbei $F_\nu^{(1)}$, und also auch $D_\nu^{(1)}$, gleich 0 zu setzen. Ebenso soll, entsprechend einer früher gemachten Festsetzung, $F_\nu^{(\varkappa)} = 0$ sein, wenn für $\varkappa > 2$ die Indizes $\lambda_{\varkappa-1}$ und $\lambda_\varkappa$ einander gleich werden; in diesem Fall wird zugleich $F^{(\varkappa)}$ und also auch $D_\nu^{(\varkappa)}$ gleich 0. Für $\lambda_1 = \lambda_2$ ist aber $F^{(2)} = A_\nu^{(2)} = 0$, dagegen $F_\nu^{(2)} = D_\nu^{(2)}$ von Null verschieden. Nun wird auf Grund unserer Rekursionsformel

$$F_\nu = x_1^m F_\nu^{(1)} + x_1^{m-1} F_\nu^{(2)} + \cdots + x_1 F_\nu^{(m)} + F_\nu^{(m+1)},$$

mithin auch

(15.) $$D_\nu = x_1^m D_\nu^{(1)} + x_1^{m-1} D_\nu^{(2)} + \cdots + x_1 D_\nu^{(m)} + D_\nu^{(m+1)}.$$

Wir nehmen nun den Satz II, der für $n = 2$ leicht zu verifizieren ist, für die Funktionen $F^{(\varkappa)}$ der $n-1$ Variabeln $x_2, \cdots, x_n$ als bereits bewiesen an. Tritt an Stelle der Zahl $\mu = \lambda_2 - \lambda_1 + 1$ bei $F^{(\varkappa)}$ die Zahl $\mu_\varkappa$, so wird für $\lambda_1 > 1$

$$\mu_1 = \mu + 1,\ \mu_2 = \mu - 1,\ \mu_3 = \mu_4 = \cdots = \mu_\varrho = \mu;$$

ist aber $\lambda_1 = 1$, so benutzen wir nur die Gleichungen

$$\mu_2 = \mu - 1,\ \mu_3 = \mu_4 = \cdots = \mu_\varrho = \mu.$$

Ist nun $\nu \leqq \mu$, so wird $\nu - 1 \leqq \mu_\varkappa$ und, da x_ν bei den Funktionen $F^{(\varkappa)}$ als die $(\nu-1)^{\text{te}}$ Variable erscheint, so ist nach Voraussetzung

$$D_\nu^{(1)} = D_\nu^{(2)} = \cdots = D_\nu^{(m+1)} = 0,$$

also in der Tat $D_\nu = 0$, wie zu beweisen war. Ferner ist noch für $\nu = \mu + 1$

$$D_{\mu+1}^{(1)} = D_{\mu+1}^{(3)} = \cdots = D_{\mu+1}^{(m+1)} = 0,$$

also

(16.) $$D_{\mu+1} = x_1^{m-1} D_{\mu+1}^{(2)}.$$

Ist nun $\lambda_1 = \lambda_2$, also $\mu = 1$, so haben wir nur zu zeigen, daß $D_{\mu+1}$ eine Summe von Potenzprodukten ist, von denen jedes den Faktor x_1^{m-1} enthält, und dies wird durch die Formel (16) in Evidenz gesetzt. Wird aber $\lambda_2 > \lambda_1$, so dürfen wir schließen, daß $D_{\mu+1}^{(2)}$ eine Summe von Potenzprodukten ist, von denen jedes den Faktor $x_2^{m-1} x_3^{m-1} \cdots x_\mu^{m-1}$ enthält; daher enthält jedes Glied von $D_{\mu+1}$ den Faktor $x_1^{m-1} x_2^{m-1} \cdots x_\mu^{m-1}$. Es sei nun $\nu > \mu + 1$; dann wird jeder der Ausdrücke

$$D_\nu^{(1)},\ D_\nu^{(3)},\ \cdots D_\nu^{(m+1)}$$

entweder 0 oder eine Summe von Potenzprodukten, von denen jedes einen Faktor

$$x_{\beta_1}^{m-1} x_{\beta_2}^{m-1} \cdots x_{\beta_\mu}^{m-1}$$

enthält. Ferner wird $x_1^{m-1} D_\nu^{(2)}$ in jedem Fall eine Summe, in der jedes Glied einen Faktor

$$x_1^{m-1} x_{\beta_1}^{m-1} \cdots x_{\beta_{\mu-1}}^{m-1}$$

enthält. Hierbei bedeuten $\beta_1, \beta_2, \cdots, \beta_\mu$ Indizes aus der Reihe 2, 3, $\cdots, \nu-1$. Die Formel (15) lehrt uns dann, daß der Ausdruck D_ν in der Tat die behauptete Eigenschaft besitzt.

Der Beweis des Satzes I gestaltet sich nun folgendermaßen.

Man verstehe unter T die Gesamtheit aller Funktionen des Moduls M, die mod. A ganzzahligen linearen Verbindungen der Glieder $X_1, X_2, \cdots, X_f$

von F kongruent sind. Wir haben also zu zeigen, daß T alle Potenzprodukte $X^{(\alpha)}$ enthält. Nun nehmen wir unseren Satz, der für $n = 2$ unmittelbar zu bestätigen ist, für weniger als n Variable als bereits bewiesen an. Sind dann

$$U_\nu^{(0)}, U_\nu^{(1)}, U_\nu^{(2)}, \cdots$$

die Potenzprodukte der Variabeln (13), die die Basis des Moduls M_ν bilden, so ist jedes dieser Produkte mod. A_ν einer ganzzahligen linearen Verbindung der Glieder der Summe F_ν kongruent. Zeigen wir daher, daß in T alle Glieder der Summe $x_\nu^m F_\nu$ vorkommen, so können wir aus der Tatsache, daß A den Modul $x_\nu^m \mathrm{A}_\nu$ enthält, unmittelbar schließen, daß T auch alle Potenzprodukte

$$x_\nu^m U_\nu^{(0)}, x_\nu^m U_\nu^{(1)}, x_\nu^m U_\nu^{(2)}, \cdots,$$

d. h. alle durch x_ν^m teilbaren Potenzprodukte $X^{(\alpha)}$ enthält.

Ist nun $\nu \leqq \mu$, so enthält T gewiß alle Glieder der Summe $x_\nu^m F_\nu$, weil diese Glieder wegen III auch in der Summe F vorkommen. Also kommen in T auch alle Produkte $X^{(\alpha)}$ vor, die einen der Faktoren $x_1^m, x_2^m, \cdots, x_\mu^m$ enthalten. Es sei für $\nu > \mu$ schon gezeigt, daß T alle Produkte $X^{(\alpha)}$ enthält, die durch eine der Potenzen $x_1^m, x_2^m, \cdots, x_{\nu-1}^m$ teilbar sind. Auf Grund des Satzes II können wir dann schließen, daß T auch alle Produkte $X^{(\alpha)}$ enthält, die einen Faktor der Form $x_{\alpha_1}^{m-1} x_{\alpha_2}^{m-1} \cdots x_{\alpha_\mu}^{m-1}$ enthalten, wobei $\alpha_1, \alpha_2, \cdots, \alpha_\mu$ Indizes aus der Reihe $1, 2, \cdots, \nu-1$ bedeuten. Folglich kommen wegen III in T auch alle Glieder der Summe $x_\nu^m D_\nu$ und mithin auch alle Glieder der Summe

$$x_\nu^m F_\nu = x_\nu^m A_\nu + x_\nu^m D_\nu$$

vor. — Damit ist aber der Satz I bewiesen.

Wir können den Satz aussprechen:

IV. *Um die der symmetrischen Gruppe n^{ten} Grades isomorphe irreduzible Substitutionsgruppe $\mathfrak{G}_{\lambda_1, \lambda_2, \cdots, \lambda_\varrho}$ zu konstruieren, bilde man den symmetrischen Modul* A *und den Ausdruck $F = F^{(\lambda_1, \lambda_2, \cdots, \lambda_\varrho)}$, der eine Summe von f Potenzprodukten $X_1 . X_2, \cdots, X_f$ der Variabeln $x_1, x_2, \cdots, x_n$ ist. Führt die Permutation R der Variabeln das Potenzprodukt X_α in $\bar{X}_\alpha$ über, so lassen sich ganze Zahlen $c_{\alpha\beta}$ in eindeutiger Weise berechnen, so daß*

$$\bar{X}_\alpha \equiv c_{\alpha 1} X_1 + c_{\alpha 2} X_2 + \cdots + c_{\alpha f} X_f \pmod{\mathrm{A}}$$

wird. Die so entstandenen $n!$ Substitutionen $(c_{\alpha\beta})$ bilden dann die Gruppe $\mathfrak{G}_{\lambda_1, \lambda_2, \cdots, \lambda_\varrho}$.

Die Koeffizienten $c_{\alpha\beta}$ sind als bekannt anzusehen, sobald es gelingt, die $N - f$ von $X_1, X_2, \cdots X_f$ verschiedenen Potenzprodukte $X^{(\alpha)}$ mod. A durch $X_1, X_2, \cdots, X_f$ darzustellen. Man hat also im ganzen nur $(N-f)f$ Koeffizienten zu berechnen.

§ 7.

Die Rechnung gestaltet sich besonders einfach für den Fall

$$\lambda_1 = 1, \lambda_2 = 1, \cdots, \lambda_{\rho-1} = 1, \lambda_\rho = n - \rho + 1.$$

Die zu betrachtenden Produkte $X^{(\alpha)}$ sind in diesem Fall die Produkte

$$(17.)\qquad x_{\alpha_1}^m x_{\alpha_2}^{m-1} \cdots x_{\alpha_m}. \qquad (\alpha_\nu = 1, 2, \cdots n, m = \rho - 1)$$

Die $f = \binom{n-1}{m}$ Potenzprodukte $X_1, X_2, \cdots, X_f$ sind die Produkte

$$(18.)\qquad x_{\gamma_1}^m x_{\gamma_2}^{m-1} \cdots x_{\gamma_m}. \qquad (\gamma_1 < \gamma_2 < \cdots < \gamma_m \leqq n-1)$$

Um alle Produkte (17) mod. A durch die Produkte (18) darzustellen, hat man nur folgende Regeln zu benutzen: Sind zunächst $\beta_1, \beta_2, \cdots, \beta_m$ die Indizes $\alpha_1, \alpha_2, \cdots, \alpha_m$, nach zunehmender Größe geordnet, so ist

$$(19.)\qquad x_{\alpha_1}^m x_{\alpha_2}^{m-1} \cdots x_{\alpha_m} \equiv \pm x_{\beta_1}^m x_{\beta_2}^{m-1} \cdots x_{\beta_m} \pmod{A};$$

hier ist das Pluszeichen oder das Minuszeichen zu nehmen, je nachdem die Permutation

$$\begin{pmatrix} \beta_1 \beta_2 \cdots \beta_m \\ \alpha_1 \alpha_2 \cdots \alpha_m \end{pmatrix}$$

gerade oder ungerade ist. Wählt man ferner irgendeine Anordnung

$$\mu_1, \mu_2, \cdots, \mu_{m-1}, \nu_1, \nu_2, \cdots, \nu_{n-m}$$

der Zahlen $1, 2, \cdots, n-1$, so wird

$$(20.)\qquad x_{\mu_1}^m x_{\mu_2}^{m-1} \cdots x_{\mu_{m-1}}^2 x_n \equiv -x_{\mu_1}^m x_{\mu_2}^{m-1} \cdots x_{\mu_{m-1}}^2 (x_{\nu_1} + x_{\nu_2} + \cdots + x_{\nu_{n-m}}) \pmod{A}.$$

Durch Kombination der Formeln (19) und (20) kann man leicht jedes der Produkte (17) mod. A als lineare Verbindung der Produkte (18) ausdrücken. Es ergibt sich zugleich, daß für die Koeffizienten $c_{\alpha\beta}$ der zugehörigen Gruppe $\mathfrak{G}_{\lambda_1, \lambda_2, \cdots, \lambda_\rho}$ in dem hier betrachteten Fall nur die Werte $0, 1$ und -1 in Betracht kommen. Ob dies auch allgemein der Fall ist, habe ich bis jetzt nicht entscheiden können.

12.

Neuer Beweis eines Satzes von W. Burnside

Jahresbericht der Deutschen Mathematiker-Vereinigung 17, 171 - 176 (1908)

Zu den wichtigsten neueren Resultaten über endliche Gruppen gehört der von W. Burnside[1]) im Jahre 1900 veröffentlichte Satz:

Jede transitive Permutationsgruppe in p Symbolen ist, wenn p eine Primzahl ist, entweder auflösbar oder zweifach transitiv.

Burnside hat diesen Satz, der als eine wesentliche Ergänzung der Galoisschen Sätze über Gleichungen von Primzahlgrad anzusehen ist, zunächst durch Betrachtungen aus der Theorie der Gruppencharaktere gewonnen. Später hat Burnside einen zweiten Beweis[2]) angegeben, bei dem er zwar von der Theorie der Gruppencharaktere keinen Gebrauch macht, aber noch gewisse Zerlegungssätze aus der Lehre von den periodischen linearen Substitutionen benutzt.

Im folgenden soll ein neuer Beweis des Satzes mitgeteilt werden, der mit wesentlich elementareren Hilfsmitteln operiert und sich auch für Vorlesungszwecke zu eignen scheint.

1. Der Ausgangspunkt meines Beweises ist ein ganz ähnlicher wie bei Herrn Burnside.

Ist $\mathfrak{G}$ eine Gruppe von Permutationen in n Symbolen $0, 1, \ldots, n-1$, so nenne ich die Form

$$F = \sum_{\varkappa, \lambda}^{n-1} {}_0 \, a_{\varkappa\lambda} x_\varkappa y_\lambda$$

eine bilineare Invariante der Gruppe, wenn F jede Permutation A der Gruppe zuläßt, d. h. wenn F ungeändert bleibt, falls die Variabeln $x_0, x_1, \ldots, x_{n-1}$ und $y_0, y_1, \ldots, y_{n-1}$ *gleichzeitig* der Permutation A unterworfen werden. Führt nun A die Ziffer α in α' über, so läßt F die Permutation A stets und nur dann zu, wenn für je zwei Indizes $\varkappa$ und λ

$$a_{\varkappa\lambda} = a_{\varkappa'\lambda'}$$

ist. Hieraus ergibt sich unmittelbar eine notwendige und hinreichende Bedingung für die zweifache Transitivität der Gruppe $\mathfrak{G}$: es muß näm-

---

1) *On some properties of groups of odd order*, Proceedings of the London Math. Soc., Vol. XXXIII, p. 174.

2) *On simply transitive groups of prime degree*, Quarterly Journal, Vol. 37 (1906), p. 215.

lich, wenn $\mathfrak{G}$ zweifach transitiv sein soll, in jeder bilinearen Invariante F von $\mathfrak{G}$

$$a_{11} = a_{22} = \cdots = a_{nn},$$

$$a_{12} = a_{13} = \cdots = a_{1n} = a_{21} = \cdots = a_{n-1,\,n}$$

sein. Dagegen besitzt eine nur einfach transitive Gruppe auch bilineare Invarianten F, deren Koeffizienten nicht diesen Bedingungen genügen. Setzt man nun

$$E = \sum_{\varkappa=0}^{n-1} x_\varkappa y_\varkappa,$$

$$J = \sum_{\varkappa,\,\lambda}^{n-1}{}_0\, x_\varkappa y_\lambda,$$

so erscheint eine zweifach transitive Gruppe $\mathfrak{G}$ dadurch charakterisiert, daß jede bilineare Invariante von $\mathfrak{G}$ die Gestalt $aE + bJ$ besitzt, wo a und b Konstanten sind. Die bilinearen Formen $aE + bJ$ gehören aber zu jeder Gruppe als Invarianten.

Es ist noch folgendes zu bemerken. Sind

$$F = \sum_{\varkappa,\,\lambda} a_{\varkappa\lambda} x_\varkappa y_\lambda, \qquad G = \sum_{\varkappa,\,\lambda} b_{\varkappa\lambda} x_\varkappa y_\lambda$$

zwei bilineare Invarianten von $\mathfrak{G}$, und setzt man

$$c_{\varkappa\lambda} = \sum_{\mu=0}^{n-1} a_{\varkappa\mu} b_{\mu\lambda},$$

so ist, wie man leicht erkennt, auch die Bilinearform

$$H = \sum_{\varkappa,\,\lambda} c_{\varkappa\lambda} x_\varkappa y_\lambda$$

eine Invariante von $\mathfrak{G}$. Diese Form H bezeichnet man als das symbolische Produkt FG. Insbesondere ist auch, wenn das aus λ Faktoren bestehende „Produkt" $FF \ldots F$ mit F^λ bezeichnet wird, zugleich mit F jede Form

(1) $$a_0 F^0 + a_1 F + a_2 F^2 + \cdots + a_m F^m, \qquad (F^0 = E)$$

wo $a_0, a_1, \ldots, a_m$ Konstanten sind, eine Invariante von $\mathfrak{G}$. Bedeutet nun $\varphi(x)$ die ganze rationale Funktion $a_0 + a_1 x + \cdots + a_m x^m$, so soll die Form (1) kurz mit $\varphi(F)$ bezeichnet werden.

2. Es sei nun $n = p$ eine Primzahl und $\mathfrak{G}$ eine transitive Gruppe. Eine solche Gruppe enthält nach dem Cauchy-Sylowschen Satze einen Zyklus P der Ordnung p, und es kann ohne Beschränkung der Allgemeinheit

$$P = (0,\ 1,\ \ldots,\ p-1)$$

angenommen werden. Soll nun $F = \sum_{\varkappa,\lambda} a_{\varkappa\lambda} x_\varkappa y_\lambda$ eine bilineare Invariante von $\mathfrak{G}$ sein, so muß F insbesondere auch die Permutation P zulassen. Daher muß F, wie leicht ersichtlich ist, die Gestalt

$$F = a_0 \sum_{\varkappa=0}^{p-1} x_\varkappa y_\varkappa + a_1 \sum_{\varkappa=0}^{p-1} x_\varkappa y_{\varkappa+1} + a_2 \sum_{\varkappa=0}^{p-1} x_\varkappa y_{\varkappa+2} + \cdots + a_{p-1} \sum_{\varkappa=0}^{p-1} x_\varkappa y_{\varkappa+p-1}$$

besitzen. Hierbei ist $y_\alpha = y_\beta$ zu setzen, falls $\alpha \equiv \beta \pmod{p}$ ist.

Man bezeichne nun die Form

$$\sum_{\varkappa=0}^{p-1} x_\varkappa y_{\varkappa+1}$$

mit R; dann wird, wie man leicht erkennt,

$$\sum_{\varkappa=0}^{p-1} x_\varkappa y_{\varkappa+\lambda} = R^\lambda$$

und

$$R^p = E.$$

Die bilineare Form F kann daher auch in der Gestalt

$$F = a_0 E + a_1 R + a_2 R^2 + \cdots + a_{p-1} R^{p-1}$$

geschrieben werden. Jede bilineare Invariante von $\mathfrak{G}$ ist mithin eine ganze rationale Funktion $\varphi(R)$ von R.

Die Bedingung dafür, daß F auch bei den übrigen Permutationen der Gruppe $\mathfrak{G}$ ungeändert bleiben soll, besteht nun lediglich darin, daß gewisse unter den Koeffizienten a_λ einander gleich werden. Man kann sich daher auch auf die Betrachtung von bilinearen Invarianten F beschränken, bei denen die Koeffizienten a_λ rationale Zahlen sind.

3. Es sei nun $\mathfrak{G}$ nicht zweifach transitiv. Dann gehört zu $\mathfrak{G}$ mindestens eine Invariante

$$F = \varphi(R) = a_0 E + a_1 R + \cdots + a_{p-1} R^{p-1}$$

mit rationalen Koeffizienten, die nicht die Form $aE + bJ$ besitzt. Berücksichtigt man, daß

$$J = E + R + R^2 + \cdots + R^{p-1}$$

ist, so erkennt man, daß die $p-1$ Koeffizienten $a_1, a_2, \ldots, a_{p-1}$ nicht alle einander gleich sein dürfen. Ist also ϱ eine primitive p-te Einheitswurzel, so wird die algebraische Größe $\varphi(\varrho)$ eine nicht rationale Zahl. Diese Größe möge einer irreduziblen Gleichung des Grades

$$e = \frac{p-1}{f} \qquad (e > 1)$$

genügen. Dann läßt sich nach einem bekannten Satz der Kreisteilungslehre eine ganze rationale Funktion $\psi(x)$ bestimmen, so daß $\psi[\varphi(\varrho)]$ eine f-gliedrige Gaußsche Periode wird.[1]) Es sei etwa

$$\psi[\varphi(\varrho)] = \varrho + \varrho^{\gamma} + \varrho^{\gamma^2} + \cdots + \varrho^{\gamma^{f-1}},$$

wo γ mod. p zum Exponenten f gehört. Diese Gleichung besagt nichts anderes, als daß

$$x + x^{\gamma} + x^{\gamma^2} + \cdots + x^{\gamma^{f-1}} = \psi[\varphi(x)] + \chi(x)\{x^0 + x + x^2 + \cdots + x^{p-1}\}$$

wird, wo $\chi(x)$ eine gewisse ganze rationale Funktion bedeutet. Diese Gleichung bleibt auch richtig, wenn man hierin für x die bilineare Form R einführt. Dann wird, da offenbar $R^{\lambda}J = J$ ist,

$$\chi(R)\{R^0 + R + \cdots + R^{p-1}\} = \chi(R) \cdot J = cJ,$$

wo c die Summe der Koeffizienten von $\chi(R)$ bedeutet. Da nun

$$\psi[\varphi(R)] = \psi(F)$$

und cJ Invarianten von $\mathfrak{G}$ sind, so ergibt sich, daß auch

$$H = R + R^{\gamma} + R^{\gamma^2} + \cdots + R^{\gamma^{f-1}} = \psi(F) + cJ$$

eine Invariante von $\mathfrak{G}$ ist.

4. Wir untersuchen nun, bei welchen Permutationen

$$A = \begin{pmatrix} 0 & 1 & \ldots & p-1 \\ \alpha_0 & \alpha_1 & \ldots & \alpha_{p-1} \end{pmatrix}$$

die Bilinearform H ungeändert bleibt. Da

$$H = \sum_{\varkappa=0}^{p-1} x_{\varkappa}(y_{\varkappa+1} + y_{\varkappa+\gamma} + y_{\varkappa+\gamma^2} + \cdots + y_{\varkappa+\gamma^{f-1}})$$
$$= \sum_{\varkappa=0}^{p-1} x_{\alpha_\varkappa}(y_{\alpha_\varkappa+1} + y_{\alpha_\varkappa+\gamma} + y_{\alpha_\varkappa+\gamma^2} + \cdots + y_{\alpha_\varkappa+\gamma^{f-1}})$$

ist und H durch die Permutation A in

$$\sum_{\varkappa=0}^{p-1} x_{\alpha_\varkappa}(y_{\alpha_{\varkappa+1}} + y_{\alpha_{\varkappa+\gamma}} + y_{\alpha_{\varkappa+\gamma^2}} + \cdots + y_{\alpha_{\varkappa+\gamma^{f-1}}})$$

übergeführt wird, so müssen für jeden Wert von $\varkappa$ die Zahlen

$$\alpha_{\varkappa+1},\ \alpha_{\varkappa+\gamma},\ \alpha_{\varkappa+\gamma^2},\ \ldots,\ \alpha_{\varkappa+\gamma^{f-1}}$$

mod. p betrachtet, abgesehen von der Reihenfolge, mit den Zahlen

$$\alpha_\varkappa + 1,\ \alpha_\varkappa + \gamma,\ \alpha_\varkappa + \gamma^2,\ \ldots,\ \alpha_\varkappa + \gamma^{f-1}$$

übereinstimmen.

---

1) Vgl. etwa H. Weber, *Lehrbuch der Algebra*, Bd. 1 (zweite Auflage), § 175.

Es sei nun

$$g(x) = c_0 + c_1 x + \cdots + c_k x^k$$

eine ganze rationale Funktion des Grades $k < p$ mit ganzzahligen Koeffizienten, die den p Kongruenzen

$$g(\varkappa) \equiv \alpha_\varkappa \pmod{p} \qquad (\varkappa = 0, 1, \ldots, p-1)$$

genügt; z. B. ist

$$g(x) = -\sum_{\varkappa=0}^{p-1} \alpha_\varkappa \frac{x(x-1)(x-2)\ldots(x-p+1)}{x-\varkappa}$$

eine solche Funktion. Dann wird wegen der erwähnten Eigenschaft der Zahlen $\alpha_\varkappa$ für jeden Exponenten r

$$(2) \qquad \sum_{\mu=0}^{f-1} \{g(\varkappa) + \gamma^\mu\}^r \equiv \sum_{\mu=0}^{f-1} \{g(\varkappa + \gamma^\mu)\}^r \pmod{p}.$$

Man beachte nun, daß wegen $\gamma^{\mu f} \equiv 1 \pmod{p}$

$$x^f - 1 \equiv (x-1)(x-\gamma)(x-\gamma^2)\ldots(x-\gamma^{f-1}) \pmod{p}$$

ist. Hieraus folgt, daß

$$s_\lambda = 1 + \gamma^\lambda + \gamma^{2\lambda} + \cdots + \gamma^{(f-1)\lambda}$$

kongruent f oder kongruent 0 mod. p ist, je nachdem λ durch f teilbar ist oder nicht. Ist daher $0 < r < f$, so wird

$$(3) \qquad \sum_{\mu=0}^{f-1} \{g(\varkappa) + \gamma^\mu\}^r = \sum_{\lambda=0}^{r} \binom{r}{\lambda} s_\lambda \{g(\varkappa)\}^{r-\lambda} \equiv f\{g(\varkappa)\}^r \pmod{p}.$$

Man setze ferner

$$\{g(x)\}^r = h(x).$$

Dann wird nach der Taylorschen Formel

$$h(x+y) = h(x) + y\frac{h'(x)}{1!} + y^2 \frac{h''(x)}{2!} + \cdots,$$

wobei zu beachten ist, daß $\frac{h^{(\lambda)}(x)}{\lambda!}$ eine Funktion mit ganzzahligen Koeffizienten ist. Daher erhalten wir

$$\sum_{\mu=0}^{f-1} h(\varkappa + \gamma^\mu) = f h(\varkappa) + s_1 \frac{h'(\varkappa)}{1!} + s_2 \frac{h''(\varkappa)}{2!} + \cdots$$

$$\equiv f\left\{h(\varkappa) + \frac{h^{(f)}(\varkappa)}{f!} + \frac{h^{(2f)}(\varkappa)}{(2f)!} + \cdots\right\} \pmod{p}.$$

Aus (2) und (3) ergibt sich daher für jedes $\varkappa$ die Kongruenz

$$\frac{h^{(f)}(\varkappa)}{f!} + \frac{h^{(2f)}(\varkappa)}{(2f)!} + \cdots \equiv 0 \pmod{p}.$$

Es muß mithin die Funktion

$$\frac{h^{(f)}(x)}{f!} + \frac{h^{(2f)}(x)}{(2f)!} + \cdots$$

mod. p durch $x^p - x$ teilbar sein. Da nun der Grad dieser Funktion, falls sie nicht identisch verschwindet und $kr < p$ ist, gleich $kr - f$ wird, so ergibt sich, daß die Zahl $kr - f$ für $r = 1, 2, \ldots, f-1$ entweder negativ oder größer als $p - 1 - f$ ist; insbesondere ist $k - f$ wegen $k < p$ negativ, also $k < f$.[1]) Hieraus folgt aber leicht, daß $k = 1$ sein muß. Denn es sei s die kleinste Zahl, die der Bedingung

$$ks \geqq f$$

genügt. Wäre $k > 1$, so müßte $s < f$, also nach unserem Ergebnis

$$ks > p - 1$$

sein. Daher wäre

$$k(s-1) > p - 1 - k > p - 1 - f.$$

Wegen $e = \frac{p-1}{f} > 1$ ist aber $p - 1 \geqq 2f$, also $p - 1 - f \geqq f$. Folglich würde sich

$$k(s-1) > f$$

ergeben. Dies widerspricht aber der über s gemachten Voraussetzung.

Folglich ist $g(x)$ eine lineare Funktion, d. h. jede Substitution A der Gruppe $\mathfrak{G}$ hat die Gestalt

$$A = \begin{pmatrix} \varkappa \\ a + b\varkappa \end{pmatrix}.$$

Eine solche Gruppe ist aber nach dem bekannten Galoisschen Satz stets auflösbar.

1) Für $f = 1$ ist $\alpha_{\varkappa+1} \equiv \alpha_\varkappa + 1$, also $\alpha_\varkappa \equiv \alpha_0 + \varkappa$, d. h. $k = 1$.

13.

Über die charakteristischen Wurzeln einer linearen Substitution mit einer Anwendung auf die Theorie der Integralgleichungen

Mathematische Annalen 66, 488 - 510 (1909)

Ist eine algebraische Gleichung in der Form

$$\begin{vmatrix} a_{11} - x, & a_{12} & , \cdots, & a_{1n} \\ a_{21} & , a_{22} - x, & \cdots, & a_{2n} \\ \cdot & \cdot & \cdot & \cdot \\ a_{n1} & , a_{n2} & , \cdots, & a_{nn} - x \end{vmatrix} = 0$$

gegeben und sind $\omega_1, \omega_2, \cdots, \omega_n$ ihre Wurzeln, so läßt sich, wie Herr A. Hirsch*) gezeigt hat, in sehr einfacher Weise eine obere Grenze für die absoluten Beträge der ω_ν angeben; bedeutet nämlich a die größte unter den n^2 Zahlen $|a_{\varkappa\lambda}|$, so ist

$$|\omega_\nu| \leqq na.$$

In der vorliegenden Arbeit möchte ich auf einige andere, wesentlich schärfere Ungleichungen aufmerksam machen, die für die absoluten Beträge der ω_ν bestehen. Die einfachste und zugleich wichtigste unter ihnen ist die Ungleichung

$$\sum_{\nu=1}^{n} |\omega_\nu|^2 \leqq \sum_{\varkappa,\lambda} |a_{\varkappa\lambda}|^2.$$

Indem ich noch angebe, welche Bedeutung hier das Auftreten des Gleichheitszeichens besitzt, erhalte ich einen Satz (Satz II), der eine Reihe von bekannten Resultaten über Hermitesche Formen und orthogonalen Substitutionen als spezielle Fälle umschließt.

---

*) „Sur les racines d'une équation fondamentale", Acta Mathematica, Bd. 25, S. 367.

Der zweite Abschnitt enthält eine Anwendung meiner algebraischen Resultate auf die Theorie der linearen homogenen Integralgleichung

$$\lambda \int_a^b K(s, t)\, \varphi(t)\, dt = \varphi(s).$$

Einem von Herrn E. Schmidt für reelle symmetrische Kerne $K(s, t)$ angegebenen Verfahren folgend, zeige ich, wie man auch für einen allgemeinen Kern den Begriff der Ordnung eines Eigenwerts λ auf elementarem Wege ohne Benutzung der Fredholmschen ganzen transzendenten Funktion begründen kann. Zugleich ergibt sich unter allgemeinen Voraussetzungen über den Kern $K(s, t)$ ein Beweis für die absolute Konvergenz der Reihe $\sum \frac{1}{\lambda^2}$, in der jeder Eigenwert λ so oft zu schreiben ist, wie seine Ordnung angibt.

Abschnitt I.

Über die charakteristischen Wurzeln einer linearen Substitution.

§ 1.

Im folgenden soll stets die zu einer Größe a konjugiert komplexe Zahl mit $\bar{a}$ bezeichnet werden. Ebenso soll, wenn $P = (p_{\varkappa\lambda})$ eine Matrix (lineare homogene Substitution) ist, unter $\bar{P}$ die mit Hilfe der Größen $\bar{p}_{\varkappa\lambda}$ gebildete Matrix verstanden werden. Mit P' bezeichne ich, wie üblich, die zu P konjugierte (transponierte) Matrix. Genügt P der Gleichung

$$\bar{P}'P = E, \tag{1}$$

wo $E = (e_{\varkappa\lambda})$ die Einheitsmatrix ist, d. h. wird

$$\sum_\nu \bar{p}_{\nu\varkappa} p_{\nu\lambda} = e_{\varkappa\lambda}, \qquad (e_{\varkappa\varkappa} = 1,\ e_{12} = e_{13} = \cdots = 0)$$

so nenne ich nach dem Vorgange von Herrn L. Autonne*) P eine *unitäre* oder auch eine *unitär orthogonale* Matrix. Die Bedeutung der Gleichung (1) ist bekanntlich die, daß die Hermitesche Einheitsform

$$x_1\bar{x}_1 + x_2\bar{x}_2 + \cdots$$

ungeändert bleibt, wenn $x_\varkappa$ durch $\sum_\lambda p_{\varkappa\lambda} x_\lambda$ (und also $\bar{x}_\varkappa$ durch $\sum_\lambda \bar{p}_{\varkappa\lambda}\bar{x}_\lambda$)

*) „Sur l'Hermitien", Rendiconti del Circolo Matematico di Palermo, T. 16 (1902), S. 104.

ersetzt wird. Sind die Koeffizienten von P reell, so wird P eine gewöhnliche (reelle) orthogonale Matrix.

Es sei nun

$$A = (a_{\varkappa\lambda}) \qquad (\varkappa, \lambda = 1, 2, \cdots, n)$$

eine Matrix mit beliebigen reellen oder komplexen Koeffizienten. Die charakteristischen Wurzeln von A, d. h. die Wurzeln der Gleichung

$$(2) \qquad |A - xE| = \begin{vmatrix} a_{11} - x, & a_{12} & , \cdots, & a_{1n} \\ a_{21} & , a_{22} - x, & \cdots, & a_{2n} \\ \cdot & \cdot & \cdots & \cdot \\ a_{n1} & , a_{n2} & , \cdots, & a_{nn} - x \end{vmatrix} = 0,$$

seien $\omega_1, \omega_2, \cdots, \omega_n$. Dann gilt folgender Satz:

I. *Es läßt sich stets eine unitär orthogonale Matrix* P *bestimmen, sodaß die Matrix*

$$\bar{P}' A P = P^{-1} A P = \mathsf{A}$$

die Form

$$\mathsf{A} = \begin{pmatrix} \omega_1 & 0 & 0 \cdots & 0 \\ c_{21} & \omega_2 & 0 \cdots & 0 \\ c_{31} & c_{32} & \omega_3 \cdots & 0 \\ \cdot & \cdot & \cdot \cdots & \cdot \\ c_{n1} & c_{n2} & c_{n3} \cdots & \omega_n \end{pmatrix}$$

erhält.

Der Beweis ist leicht zu erbringen.*) Da nämlich ω_1 eine Wurzel der Gleichung (2) ist, so können wir n Zahlen $q_1, q_2, \cdots, q_n$ bestimmen, die den n Gleichungen

$$\sum_{\lambda=1}^{n} a_{\lambda\varkappa} q_\lambda = \omega_1 q_\varkappa \qquad (\varkappa = 1, 2, \cdots, n)$$

genügen und nicht sämtlich Null sind. Die Summe $\sum_{\varkappa=1}^{n} \bar{q}_\varkappa q_\varkappa = q$ ist dann eine positive Zahl. Setzt man

$$q_{\varkappa 1} = \frac{q_\varkappa}{\sqrt{q}},$$

so wird auch

$$\sum_{\lambda=1}^{n} a_{\lambda\varkappa} q_{\lambda 1} = \omega_1 q_{\varkappa 1}$$

und

$$\sum_{\lambda=1}^{n} \bar{q}_{\lambda 1} q_{\lambda 1} = 1.$$

*) Vergl. L. Stickelberger, „Über reelle orthogonale Substitutionen", Programm der polyt. Schule, Zürich, 1877, und L. Autonne, a. a. O., S. 119.

Man bestimme nun in bekannter Weise $n(n-1)$ Zahlen

$$\begin{matrix} q_{12}, & q_{13}, & \cdots, & q_{1n} \\ q_{22}, & q_{23}, & \cdots, & q_{2n} \\ \cdot & \cdot & \cdots & \cdot \\ q_{n2}, & q_{n3}, & \cdots, & q_{nn} \end{matrix}$$

so, daß

$$\sum_{\nu=1}^{n} \bar{q}_{\nu\varkappa} q_{\nu\lambda} = e_{\varkappa\lambda} \qquad (\varkappa, \lambda = 1, 2, \cdots, n)$$

wird. Dann ist die Matrix

$$Q = (q_{\varkappa\lambda})$$

eine unitäre Matrix, außerdem wird

$$Q' A Q'^{-1} = \bar{Q}^{-1} A \bar{Q} = \begin{pmatrix} \omega_1 & 0 & 0 & \cdots & 0 \\ b_{21} & b_{22} & b_{23} & \cdots & b_{2n} \\ b_{31} & b_{32} & b_{33} & \cdots & b_{3n} \\ \cdot & \cdot & \cdot & \cdots & \cdot \\ b_{n1} & b_{n2} & b_{n3} & \cdots & b_{nn} \end{pmatrix},$$

wo die $b_{\varkappa\lambda}$ gewisse Zahlen bedeuten. Die charakteristischen Wurzeln der Matrix

$$B = \begin{pmatrix} b_{22} & b_{23} & \cdots & b_{2n} \\ b_{32} & b_{33} & \cdots & b_{3n} \\ \cdot & \cdot & \cdots & \cdot \\ b_{n2} & b_{n3} & \cdots & b_{nn} \end{pmatrix}$$

sind dann die Größen $\omega_2, \omega_3, \cdots, \omega_n$. Nimmt man nun den zu beweisenden Satz, der für $n = 1$ evident ist, für Matrizen mit $n - 1$ Zeilen und Spalten als richtig an, so läßt sich eine unitäre Matrix

$$R = \begin{pmatrix} r_{22} & r_{23} & \cdots & r_{2n} \\ r_{32} & r_{33} & \cdots & r_{3n} \\ \cdot & \cdot & \cdots & \cdot \\ r_{n2} & r_{n3} & \cdots & r_{nn} \end{pmatrix}$$

bestimmen, sodaß

$$R^{-1} B R = \begin{pmatrix} \omega_2 & 0 & \cdots & 0 \\ c_{32} & \omega_3 & \cdots & 0 \\ \cdot & \cdot & \cdots & \cdot \\ c_{n2} & c_{n3} & \cdots & \omega_n \end{pmatrix}$$

wird. Setzt man dann

$$S = \begin{pmatrix} 1 & 0 & \cdots & 0 \\ 0 & r_{22} & \cdots & r_{2n} \\ \cdot & \cdot & \cdots & \cdot \\ 0 & r_{n2} & \cdots & r_{nn} \end{pmatrix}$$

so wird $P = \overline{Q}S$ eine unitär orthogonale Matrix, die den Bedingungen unseres Satzes genügt.*)

Der Satz I läßt sich auch folgendermaßen aussprechen:

I*. *Für jede lineare homogene Substitution A in n Variabeln $x_1, x_2, \cdots, x_n$ lassen sich n Linearformen $y_1, y_2, \cdots, y_n$ angeben, die folgenden Bedingungen genügen:*

1. *es ist*

$$y_1\overline{y}_1 + y_2\overline{y}_2 + \cdots + y_n\overline{y}_n = x_1\overline{x}_1 + x_2\overline{x}_2 + \cdots + x_n\overline{x}_n;$$

2. *die Linearform $y_\varkappa$ geht durch Anwendung der Substitution A in eine Form über, die sich linear und homogen durch $y_1, y_2, \cdots, y_\varkappa$ ausdrücken läßt.*

§ 2.

Aus

$$\overline{P}'AP = \mathsf{A} \tag{3}$$

folgt

$$\overline{P}'\overline{A}'P = \overline{\mathsf{A}}'. \tag{3'}$$

Daher ist wegen $\overline{P}'P = E$

$$\overline{P}'A\overline{A}'P = \mathsf{A}\overline{\mathsf{A}}'.$$

Es ist aber $\overline{P}' = P^{-1}$, folglich besitzen die Matrizen $A\overline{A}'$ und $\mathsf{A}\overline{\mathsf{A}}'$ als ähnliche Matrizen dieselbe Spur, und da die Spur einer Matrix der Form $P\overline{P}'$, wie man leicht sieht, nichts anderes ist als die Summe der Quadrate der absoluten Beträge aller Koeffizienten von P, so erhalten wir

$$\sum_{\varkappa,\lambda} |a_{\varkappa\lambda}|^2 = \sum_{\nu} |\omega_\nu|^2 + \sum_{\varkappa > \lambda} |c_{\varkappa\lambda}|^2.$$

Insbesondere ergibt sich

$$\sum_{\nu} |\omega_\nu|^2 \leqq \sum_{\varkappa,\lambda} |a_{\varkappa\lambda}|^2.$$

Das Gleichheitszeichen steht hier stets und nur dann, wenn die Zahlen $c_{\varkappa\lambda}$ sämtlich gleich Null sind. Dies gibt uns den Satz:

II. *Ist $A = (a_{\varkappa\lambda})$ eine beliebige lineare homogene Substitution in n Variabeln mit den charakteristischen Wurzeln $\omega_1, \omega_2, \cdots, \omega_n$, so ist*

$$\sum_{\nu} |\omega_\nu|^2 \leqq \sum_{\varkappa,\lambda} |a_{\varkappa\lambda}|^2. \tag{4}$$

Das Gleichheitszeichen gilt hier stets und nur dann, wenn sich A durch eine unitär orthogonale Transformation der Variabeln auf die Diagonalform

*) Sind die $a_{\varkappa\lambda}$ und die ω_ν reell, so kann P auch als reelle (orthogonale) Matrix gewählt werden.

$$\Delta = \begin{pmatrix} \omega_1 & 0 & \cdots & 0 \\ 0 & \omega_2 & \cdots & 0 \\ . & . & \cdots & . \\ 0 & 0 & \cdots & \omega_n \end{pmatrix}$$

bringen läßt.

Es ergibt sich zugleich das merkwürdige Resultat, daß, sobald

$$\sum_\nu |\omega_\nu|^2 = \sum_{\varkappa, \lambda} |a_{\varkappa\lambda}|^2$$

ist, die Determinante $|A - xE|$ lauter lineare Elementarteiler besitzen muß.

Dieses Resultat läßt sich noch etwas verallgemeinern.

Sind nämlich die Elementarteiler von $|A - xE|$ sämtlich linear, so läßt sich eine Matrix Q von nicht verschwindender Determinante bestimmen, sodaß

$$Q^{-1} A Q = \Delta$$

wird. Dann wird

$$\bar{Q}' \bar{A}' \bar{Q}'^{-1} = \bar{\Delta}',$$

also

$$\Delta \bar{\Delta}' = Q^{-1} A Q \bar{Q}' \bar{A}' \bar{Q}'^{-1}.$$

Die Spur $\sum_\nu |\omega_\nu|^2$ von $\Delta\bar{\Delta}'$ ist daher gleich der Spur der rechts stehenden Matrix oder, was dasselbe ist, gleich der Spur der Matrix

$$A Q \bar{Q}' \bar{A}' \bar{Q}'^{-1} Q^{-1} = A H \bar{A}' H^{-1},$$

wobei $H = Q\bar{Q}'$ zu setzen ist. Hierbei ist H die Matrix einer positiv definiten Hermiteschen Form von nicht verschwindender Determinante oder, wie man auch kurz sagt, eine positive Hermitesche Matrix.

Läßt sich umgekehrt eine derartige Matrix H so bestimmen, daß die Spur von $A H \bar{A}' H^{-1}$ gleich $\sum_\nu |\omega_\nu|^2$ wird, so wähle man, was stets möglich ist, eine Matrix Q, für die $H = Q\bar{Q}'$ wird. Dann ist $\sum_\nu |\omega_\nu|^2$ zugleich auch die Spur von

$$Q^{-1} A Q \cdot \bar{Q}' \bar{A}' \bar{Q}'^{-1} = B\bar{B}',$$

wo $B = Q^{-1} A Q$ zu setzen ist. Die charakteristischen Wurzeln der zu A ähnlichen Matrix $B = (b_{\varkappa\lambda})$ sind aber wieder die Größen $\omega_1, \omega_2, \cdots, \omega_n$ und, da für B

$$\sum_\nu |\omega_\nu|^2 = \sum_{\varkappa, \lambda} |b_{\varkappa\lambda}|^2$$

wird, so ist B und folglich auch A der Diagonalmatrix Δ ähnlich.

Es ergibt sich auf diese Weise der Satz:

III. *Die Elementarteiler der charakteristischen Determinante einer Matrix* $A = (a_{\varkappa\lambda})$ *mit den charakteristischen Wurzeln* $\omega_1, \omega_2, \cdots, \omega_n$ *sind dann und nur dann sämtlich linear, wenn sich eine positive Hermitesche Matrix* $H=(h_{\varkappa\lambda})$ *so bestimmen läßt, daß*

$$\sum_\nu |\omega_\nu|^2 = \sum_{\varkappa,\lambda,\mu,\nu} h_{\lambda\nu} h'_{\mu\varkappa} a_{\varkappa\lambda} \bar{a}_{\mu\nu}$$

wird; hierbei sollen die Zahlen $h'_{\mu\varkappa}$ *die Koeffizienten von* H^{-1} *bedeuten. Allgemein ist bei einer beliebigen Matrix* A *für jede positive Hermitesche Matrix* $H=(h_{\varkappa\lambda})$

$$\sum_\nu |\omega_\nu|^2 \leqq \sum_{\varkappa,\lambda,\mu,\nu} h_{\lambda\nu} h'_{\mu\varkappa} a_{\varkappa\lambda} \bar{a}_{\mu\nu}.$$

§ 3.

Setzt man

$$a_{\varkappa\lambda} = r_{\varkappa\lambda} + i s_{\varkappa\lambda}, \quad \omega_\nu = \varrho_\nu + i\sigma_\nu,$$

so geht die Ungleichung (4) über in

$$\sum_\nu (\varrho_\nu^2 + \sigma_\nu^2) \leqq \sum_{\varkappa,\lambda} (r_{\varkappa\lambda}^2 + s_{\varkappa\lambda}^2). \tag{5}$$

Andererseits ist aber, da $\omega_1^2, \omega_2^2, \cdots, \omega_n^2$ die charakteristischen Wurzeln von A^2 sind, $\sum\limits_\nu \omega_\nu^2$ die Spur von A^2, also wird

$$\sum_\nu \omega_\nu^2 = \sum_{\varkappa,\lambda} a_{\varkappa\lambda} a_{\lambda\varkappa},$$

oder

$$\sum_\nu (\varrho_\nu^2 - \sigma_\nu^2 + 2i\varrho_\nu\sigma_\nu) = \sum_{\varkappa,\lambda} (r_{\varkappa\lambda} r_{\lambda\varkappa} - s_{\varkappa\lambda} s_{\lambda\varkappa} + 2i r_{\varkappa\lambda} s_{\lambda\varkappa}).$$

Daher ist

$$\sum_\nu (\varrho_\nu^2 - \sigma_\nu^2) = \sum_{\varkappa,\lambda} (r_{\varkappa\lambda} r_{\lambda\varkappa} - s_{\varkappa\lambda} s_{\lambda\varkappa}). \tag{6}$$

Aus (5) und (6) ergibt sich

$$2\sum_\nu \varrho_\nu^2 \leqq \sum_{\varkappa,\lambda} (r_{\varkappa\lambda}^2 + r_{\varkappa\lambda} r_{\lambda\varkappa}) + \sum_{\varkappa,\lambda} (s_{\varkappa\lambda}^2 - s_{\varkappa\lambda} s_{\lambda\varkappa}),$$

$$2\sum_\nu \sigma_\nu^2 \leqq \sum_{\varkappa,\lambda} (r_{\varkappa\lambda}^2 - r_{\varkappa\lambda} r_{\lambda\varkappa}) + \sum_{\varkappa,\lambda} (s_{\varkappa\lambda}^2 + s_{\varkappa\lambda} s_{\lambda\varkappa}).$$

Diese Ungleichungen lassen sich auch, wie eine einfache Betrachtung lehrt, auf folgende Form bringen:

$$\sum_\nu \varrho_\nu^2 \leqq \sum_{\varkappa,\lambda} \left(\frac{r_{\varkappa\lambda} + r_{\lambda\varkappa}}{2}\right)^2 + \sum_{\varkappa,\lambda} \left(\frac{s_{\varkappa\lambda} - s_{\lambda\varkappa}}{2}\right)^2,$$

$$\sum_\nu \sigma_\nu^2 \leqq \sum_{\varkappa,\lambda} \left(\frac{r_{\varkappa\lambda} - r_{\lambda\varkappa}}{2}\right)^2 + \sum_{\varkappa,\lambda} \left(\frac{s_{\varkappa\lambda} + s_{\varkappa\lambda}}{2}\right)^2.$$

oder, was dasselbe ist,

$$\sum_{\nu} \varrho_\nu^2 \leqq \sum_{\varkappa,\lambda} \left|\frac{a_{\varkappa\lambda} + \bar{a}_{\lambda\varkappa}}{2}\right|^2, \tag{7}$$

$$\sum_{\nu} \sigma_\nu^2 \leqq \sum_{\varkappa,\lambda} \left|\frac{a_{\varkappa\lambda} - \bar{a}_{\lambda\varkappa}}{2}\right|^2. \tag{8}$$

Man kann diese Formeln auch anders beweisen. Aus (3) und (3′) folgt

$$\frac{1}{2}\,\bar{P}'(A \pm \bar{A}')P = \frac{1}{2}(\mathsf{A} \pm \bar{\mathsf{A}}').$$

Hieraus schließt man wie früher, daß die Summe der Quadrate der absoluten Beträge aller Koeffizienten der Matrix $\frac{1}{2}(A \pm \bar{A}')$ gleich sein muß der analog gebildeten Summe für die Matrix $\frac{1}{2}(\mathsf{A} \pm \bar{\mathsf{A}}')$. Dies gibt uns aber

$$\sum_{\varkappa,\lambda} \left|\frac{a_{\varkappa\lambda} \pm \bar{a}_{\lambda\varkappa}}{2}\right|^2 = \sum_{\nu} \left|\frac{\omega_\nu \pm \bar{\omega}_\nu}{2}\right|^2 + \frac{1}{2} \sum_{\varkappa > \lambda} |c_{\varkappa\lambda}|^2,$$

woraus dann wieder die Ungleichungen (7) und (8) folgen.

Es sei noch hervorgehoben, daß, wenn in einer der Ungleichungen (4), (7) und (8) das Gleichheitszeichen steht, dies auch in den beiden anderen der Fall sein muß.

§ 4.

Bezeichnet man mit a die größte der Zahlen $|a_{\varkappa\lambda}|$, mit b die größte der Zahlen $\left|\frac{a_{\varkappa\lambda} + \bar{a}_{\lambda\varkappa}}{2}\right|$ und mit c die größte der Zahlen $\left|\frac{a_{\varkappa\lambda} - \bar{a}_{\lambda\varkappa}}{2}\right|$, so folgt aus den Ungleichungen (4), (7) und (8)

$$\sum_{\nu} |\omega_\nu|^2 \leqq n^2 a^2, \qquad \sum_{\nu} \varrho_\nu^2 \leqq n^2 b^2, \qquad \sum_{\nu} \sigma_\nu^2 \leqq n^2 c^2.$$

Speziell wird also

$$|\omega_\nu| \leqq na, \qquad |\varrho_\nu| \leqq nb, \qquad |\sigma_\nu| \leqq nc.$$

Diese Ungleichungen sind bereits von Herrn A. Hirsch a. a. O. abgeleitet worden. Sind ferner die Zahlen $a_{\varkappa\lambda}$ sämtlich reell, so wird $a_{\varkappa\varkappa} - \bar{a}_{\varkappa\varkappa} = 0$; daher ergibt sich inbesondere aus (8)

$$\sum_{\nu} \sigma_\nu^2 \leqq n(n-1)\,c^2.$$

Da außerdem jedes von Null verschiedene σ_ν^2 mindestens zweimal vorkommen muß, so wird

$$\sigma_\nu^2 \leqq \frac{n(n-1)}{2} c^2,$$

also

$$|\sigma_\nu| \leqq c\sqrt{\frac{n(n-1)}{2}}.$$

Diese Ungleichung hat zuerst Herr I. Bendixson*) auf einem anderen Wege erhalten.

Ich möchte noch eine Bemerkung an die Hirschsche Ungleichung

$$|\omega_\nu| \leqq na$$

anknüpfen. Soll hier für eine Wurzel $\omega_\nu = \omega$ das Gleichheitszeichen gelten, so ergibt sich aus dem hier Bewiesenen, daß folgende Bedingungen erfüllt sein müssen:

1. Unter den n Wurzeln $\omega_1, \omega_2, \cdots, \omega_n$ sind $n-1$ gleich 0.

2. Die Koeffizienten $a_{\varkappa\lambda}$ sind alle von gleichem absoluten Betrage a und der absolute Betrag der Summe $\sum_\varkappa a_{\varkappa\varkappa}$ ist gleich na.

3. Die Elementarteiler von $|A - xE|$ sind sämtlich linear, woraus in Verbindung mit 1. folgt, daß A vom Range 1 sein muß.

Hieraus schließt man ohne Mühe, daß die Zahlen $a_{\varkappa\lambda}$ von der Form

$$a_{\varkappa\lambda} = ae^{i(\varphi + \varphi_\varkappa - \varphi_\lambda)}$$

sein müssen, wo $\varphi, \varphi_1, \varphi_2, \cdots, \varphi_n$ reelle Größen bedeuten.

§ 5.

Ehe ich weiter gehe, möchte ich noch darauf aufmerksam machen, daß in dem Satz II der bekannte Hadamardsche Satz über den Maximalwert einer Determinante**) enthalten ist.

Die Determinante D der Matrix $A = (a_{\varkappa\lambda})$ ist nämlich nichts anderes als das Produkt $\omega_1 \omega_2 \cdots \omega_n$. Nach dem Satz über das arithmetische und das geometrische Mittel wird daher

$$(9) \qquad |D|^2 = |\omega_1|^2 |\omega_2|^2 \cdots |\omega_n|^2 \leqq \left(\frac{|\omega_1|^2 + |\omega_2|^2 + \cdots + |\omega_n|^2}{n}\right)^n \leqq \left(\frac{n^2 a^2}{n}\right)^n = n^n a^{2n}.$$

Folglich ist

$$(10) \qquad |D| \leqq n^{\frac{n}{2}} a^n.$$

---

*) „Sur les racines d'une équation fondamentale", Acta Mathematica Bd. 25, S. 359. — Herr A. Hirsch hat noch gezeigt, daß die Bendixsonsche Ungleichung auch dann gilt, wenn nur vorausgesetzt wird, daß die $a_{\varkappa\lambda} + a_{\lambda\varkappa}$ reell sind.

**) „Résolution d'une question relative aux déterminants". Bulletin des sciences mathématiques, 1893, S. 240.

Dies ist die Hadamardsche Ungleichung. Soll $|D| = n^{\frac{n}{2}} a^n$ sein, so muß nach Hadamard weiter

$$|a_{\varkappa\lambda}| = a,$$

$$p_{\varkappa\lambda} = \sum_{\nu} a_{\varkappa\nu} \bar{a}_{\lambda\nu} = n a\, e_{\varkappa\lambda}$$

sein. Dies ergibt sich hier folgendermaßen. Gilt in (10) das Gleichheitszeichen, so müssen auch die Ungleichungen (9) und (4) in Gleichungen übergehen. Daher muß

$$(11) \qquad |\omega_1| = |\omega_2| = \cdots = |\omega_n| = a\sqrt{n}, \qquad |a_{\varkappa\lambda}| = a$$

werden. Außerdem muß, da

$$\sum_{\nu} |\omega_\nu|^2 = \sum_{\varkappa,\lambda} |a_{\varkappa\lambda}|^2$$

wird, die Matrix A der Diagonalmatrix Δ, und zugleich $A\bar{A}'$ der Matrix $\Delta\bar{\Delta}'$ ähnlich sein. Es wird aber wegen (11)

$$\Delta\bar{\Delta}' = n a E.$$

Folglich ist auch, da diese Matrix nur sich selbst ähnlich ist,

$$A\bar{A}' = (p_{\varkappa\lambda}) = n a E$$

d. h. $p_{\varkappa\lambda} = n a\, e_{\varkappa\lambda}$.

§ 6.

Bezeichnet man die mit Hilfe der $\binom{n}{\nu}^2$ Unterdeterminanten ν^{ten} Grades von A gebildete Matrix mit $C_\nu A$, so werden bekanntlich*) die charakteristischen Wurzeln von $C_\nu A$ die $\binom{n}{\nu}$ Größen

$$\omega_{\alpha_1}\omega_{\alpha_2}\cdots\omega_{\alpha_\nu} \qquad (\alpha_1 < \alpha_2 < \cdots < \alpha_\nu).$$

Daher ist wegen (4)

$$(12) \qquad \sum |\omega_{\alpha_1}\omega_{\alpha_2}\cdots\omega_{\alpha_\nu}|^2 \leqq \sum |D_{\alpha\beta}^{(\nu)}|^2,$$

wo $D_{\alpha\beta}^{(\nu)}$ alle Unterdeterminanten ν^{ten} Grades von A durchläuft. Die rechts stehende Zahl ist nichts anderes als die Spur von $(C_\nu A)(C_\nu \bar{A})'$. Es ist aber allgemein

$$(C_\nu P)' = C_\nu P', \qquad (C_\nu P)(C_\nu Q) = C_\nu(PQ).$$

Daher ist $\sum |D_{\alpha\beta}^{(\nu)}|^2$ die Spur von $C_\nu(A\bar{A}')$. Bezeichnet man wie früher das allgemeine Element $\sum_{\mu} a_{\varkappa\mu}\bar{a}_{\lambda\mu}$ von $A\bar{A}'$ mit $p_{\varkappa\lambda}$, so wird also

*) G. Rados, „Zur Theorie der adjungirten Substitutionen", Math. Annalen, Bd. 48, S. 417.

$$(13)\quad \begin{aligned} &\sum |\omega_{\alpha_1}\omega_{\alpha_2}\cdots\omega_{\alpha_\nu}|^2 \\ \leqq &\sum \begin{vmatrix} p_{\alpha_1\alpha_1}, & p_{\alpha_1\alpha_2}, & \cdots, & p_{\alpha_1\alpha_\nu} \\ p_{\alpha_2\alpha_1}, & p_{\alpha_2\alpha_2}, & \cdots, & p_{\alpha_2\alpha_\nu} \\ \cdot & \cdot & \cdot & \cdot \\ p_{\alpha_\nu\alpha_1}, & p_{\alpha_\nu\alpha_2}, & \cdots, & p_{\alpha_\nu\alpha_\nu} \end{vmatrix} \qquad (\alpha_1<\alpha_2<\cdots<\alpha_\nu). \end{aligned}$$

Nun ist aber $A\bar{A}'$ die Matrix der positiven Hermiteschen Form

$$\sum_\varkappa (a_{1\varkappa}x_1 + a_{2\varkappa}x_2 + \cdots + a_{n\varkappa}x_n)(\bar{a}_{1\varkappa}\bar{x}_1 + \bar{a}_{2\varkappa}\bar{x}_2 + \cdots + \bar{a}_{n\varkappa}\bar{x}_n),$$

folglich erscheint auch jede Hauptunterdeterminante von $A\bar{A}'$ als die Determinante einer positiven Hermiteschen Form. Nach einem von Hadamard a. a. O. bewiesenen Satze ist aber eine solche Determinante höchstens gleich dem Produkt der Koeffizienten ihrer Hauptdiagonale.*) Aus (13) ergibt sich daher

$$(13')\quad \sum |\omega_{\alpha_1}\omega_{\alpha_2}\cdots\omega_{\alpha_\nu}|^2 \leqq \sum p_{\alpha_1\alpha_1}\,p_{\alpha_2\alpha_2}\cdots p_{\alpha_\nu\alpha_\nu} \qquad (\alpha_1<\alpha_2<\cdots<\alpha_\nu).$$

Diese Formeln lassen sich als eine Verallgemeinerung des eben erwähnten Hadamardschen Determinantensatzes auffassen.

§ 7.

Aus der Formel (4) lassen sich noch andere interessante Ungleichungen ableiten.

Zunächst kann man, indem man beachtet, daß $\omega_1^r, \omega_2^r, \cdots, \omega_n^r$ die charakteristischen Wurzeln von A^r sind, eine obere Grenze für $\sum_\nu |\omega_\nu|^{2r}$ ableiten.

Von größerem Interesse scheint mir folgende Bemerkung zu sein.

Man habe neben der Matrix noch eine andere Matrix $B = (b_{\mu\nu})$ mit m Zeilen und Spalten. Die charakteristischen Wurzeln von B seien $\eta_1, \eta_2, \cdots, \eta_m$. Dann ist

$$\sum_{\varkappa=1}^{n} a_{\varkappa\varkappa} = \sum_{\varkappa=1}^{n} \omega_\varkappa, \qquad \sum_{\mu=1}^{m} b_{\mu\mu} = \sum_{\mu=1}^{m} \eta_\mu.$$

Ferner ist:

$$\begin{aligned} \sum_{\varkappa,\mu} |\omega_\varkappa + \eta_\mu|^2 &= \sum_{\varkappa,\mu} (\omega_\varkappa\bar{\omega}_\varkappa + \eta_\mu\bar{\eta}_\mu + \omega_\varkappa\bar{\eta}_\mu + \bar{\omega}_\varkappa\eta_\mu) \\ &= m\sum_\varkappa |\omega_\varkappa|^2 + n\sum_\mu |\eta_\mu|^2 + \sum_{\varkappa,\mu} (a_{\varkappa\varkappa}\bar{b}_{\mu\mu} + \bar{a}_{\varkappa\varkappa}b_{\mu\mu}). \end{aligned}$$

---

*) Vergl. auch E. Fischer, „Über den Hadamardschen Determinantensatz", Archiv der Mathematik und Physik, dritte Reihe, Bd. 13, S. 32.

Andererseits ist

$$\sum_{\varkappa,\mu}|a_{\varkappa\varkappa}+b_{\mu\mu}|^2+m\sum_{\varkappa\neq\lambda}|a_{\varkappa\lambda}|^2+n\sum_{\mu\neq\nu}|b_{\mu\nu}|^2 \qquad \begin{pmatrix}\varkappa,\ \lambda=1,2,\cdots,n\\ \mu,\ \nu=1,2,\cdots,m\end{pmatrix}$$

gleich

$$m\sum_{\varkappa,\lambda}|a_{\varkappa\lambda}|^2+n\sum_{\mu,\nu}|b_{\mu\nu}|^2+\sum_{\varkappa,\mu}(a_{\varkappa\varkappa}\bar{b}_{\mu\mu}+\bar{a}_{\varkappa\varkappa}b_{\mu\mu}).$$

Daher gilt stets die Ungleichung

$$\sum_{\varkappa,\mu}|\omega_\varkappa+\eta_\mu|^2\leqq\sum_{\varkappa,\mu}|a_{\varkappa\varkappa}+b_{\mu\mu}|^2+m\sum_{\varkappa\neq\lambda}|a_{\varkappa\lambda}|^2+n\sum_{\mu\neq\nu}|b_{\mu\nu}|^2.$$

Setzt man insbesondere $B=-A$, so kann $\eta_\mu=-\omega_\mu$ angenommen werden, und man erhält

$$\sum_{\varkappa<\mu}|\omega_\varkappa-\omega_\mu|^2\leqq\sum_{\varkappa<\mu}|a_{\varkappa\varkappa}-a_{\mu\mu}|^2+n\sum_{\varkappa\neq\lambda}|a_{\varkappa\lambda}|^2.$$

Hieraus läßt sich sofort eine obere Grenze für den absoluten Betrag der Diskriminante d der Gleichung $|A-xE|=0$ ableiten. Aus

$$d=\prod_{\varkappa<\mu}(\omega_\varkappa-\omega_\mu)^2$$

folgt nämlich

$$|d|^{\frac{2}{n(n-1)}}\leqq\frac{2}{n(n-1)}\sum_{\varkappa<\mu}|\omega_\varkappa-\omega_\mu|^2,$$

daher ist

$$|d|^{\frac{2}{n(n-1)}}\leqq\frac{2}{n(n-1)}\sum_{\varkappa<\mu}|a_{\varkappa\varkappa}-a_{\mu\mu}|^2+\frac{2}{n-1}\sum_{\varkappa\neq\lambda}|a_{\varkappa\lambda}|^2.$$

§ 8.

Setzt man speziell

$$A=\begin{pmatrix}-a_1, & -a_2x_2, & \cdots, & -a_{n-1}x_{n-1}, & -a_nx_n\\ x_2^{-1}, & 0, & \cdots, & 0, & 0\\ 0, & x_3^{-1}x_2, & \cdots, & 0, & 0\\ \cdot & \cdot & \cdot & \cdot & \cdot\\ 0, & 0, & \cdots, & x_n^{-1}x_{n-1}, & 0\end{pmatrix},$$

wo die a_ν beliebige, die x_ν von Null verschiedene Zahlen bedeuten, so wird

(14) $$x^n+a_1x^{n-1}+a_2x^{n-2}+\cdots+a_n=0$$

die charakteristische Gleichung von A. Ist daher

$$b_\nu=|a_\nu|^2,\quad p_\nu=|x_\nu|^2,$$

so genügen die Wurzeln $\omega_1, \omega_2, \cdots, \omega_n$ der Gleichung (14) der Ungleichung

$$\sum_\nu |\omega_\nu|^2 \leqq b_1 + b_2 p_2 + \cdots + b_n p_n + \frac{1}{p_2} + \frac{p_2}{p_3} + \cdots + \frac{p_{n-1}}{p_n}.\text{*)}$$

Diese Formel gilt für alle positiven Zahlen $p_2, p_3, \cdots, p_n$. Für

$$p_2 = p, \quad p_3 = p^2, \quad \cdots, \quad p_n = p^{n-1}$$

erhält man die einfachere Formel

$$\sum_\nu |\omega_\nu|^2 \leqq b_1 + b_2 p + \cdots + b_n p^{n-1} + \frac{n-1}{p}.$$

Allgemeiner folgt aus (12) die Ungleichung

$$\sum_{\alpha_1 < \alpha_2 < \cdots} |\omega_{\alpha_1} \omega_{\alpha_2} \cdots \omega_{\alpha_\nu}|^2 \leqq \binom{n-2}{\nu-1} \sum_{\mu=1}^{n-1} b_\mu p^{\mu-\nu} + \binom{n-1}{\nu-1} b_n p^{n-\nu} + \binom{n-1}{\nu} p^{-\nu},$$

die wieder für jedes positive p richtig ist. Nimmt man insbesondere a_n als von Null verschieden an, so wird für $p = b_n^{-\frac{1}{n}}$

$$\sum_{\alpha_1 < \alpha_2 < \cdots} |\omega_{\alpha_1} \omega_{\alpha_2} \cdots \omega_{\alpha_\nu}|^2 \leqq \binom{n-2}{\nu-1} \sum_{\mu=1}^{n-1} b_\mu b_n^{\frac{\nu-\mu}{n}} + \binom{n}{\nu} b_n^{\frac{\nu}{n}}.$$

Der direkte Beweis dieser für jede algebraische Gleichung geltenden Formeln scheint schwierig zu sein.

§ 9.

Aus dem Satz II lassen sich einige bekannte Sätze über lineare Substitutionen als spezielle Folgerungen ableiten.

Es sei zunächst $A = (a_{\varkappa\lambda})$ eine Hermitesche Matrix, also $a_{\varkappa\lambda} = \bar{a}_{\lambda\varkappa}$. Dann wird

$$\sum_\nu |\omega_\nu|^2 \leqq \sum_{\varkappa,\lambda} a_{\varkappa\lambda} \bar{a}_{\lambda\varkappa} = \sum_{\varkappa,\lambda} a_{\varkappa\lambda} a_{\lambda\varkappa} = \sum_\nu \omega_\nu^2.$$

Da aber andererseits $\sum_\nu \omega_\nu^2 \leqq \sum_\nu |\omega_\nu|^2$ ist, so muß $\sum_\nu \omega_\nu^2 = \sum_\nu |\omega_\nu|^2$ sein, folglich müssen die ω_ν reell sein. Zugleich ergibt sich $\sum_\nu |\omega_\nu|^2 = \sum_{\varkappa,\lambda} |a_{\varkappa\lambda}|^2$, und dies liefert uns die bekannte Tatsache, daß jede Hermitesche Form

---

*) Es läßt sich zeigen, daß hier für $n > 2$ das Gleichheitszeichen nur dann stehen kann, wenn $a_1 = a_2 = \cdots = a_{n-1} = 0$ ist.

durch eine unitär orthogonale Transformation der Variabeln auf die Gestalt $\omega_1 y_1 \bar{y}_1 + \omega_2 y_2 \bar{y}_2 + \cdots + \omega_n y_n \bar{y}_n$ gebracht werden kann.*)

Ist ferner A eine unitär orthogonale Matrix, so wird $A\bar{A}' = E$. Die Spur $\sum_{\varkappa,\lambda} |a_{\varkappa\lambda}|^2$ von $A\bar{A}'$ wird dann gleich der Spur von E, also gleich n. Außerdem ist der absolute Betrag der Determinante D von A (wegen $A\bar{A}' = E$) gleich 1. Da aber $D = \omega_1 \omega_2 \cdots \omega_n$ ist, so ergibt sich

$$1 = (|\omega_1|^2 |\omega_2|^2 \cdots |\omega_n|^2)^{\frac{1}{n}} \leqq \frac{|\omega_1|^2 + |\omega_2|^2 + \cdots + |\omega_n|^2}{n} \leqq \frac{\sum_{\varkappa,\lambda} |a_{\varkappa\lambda}|^2}{n} = 1.$$

Daher muß

$$1 = (|\omega_1|^2 |\omega_2|^2 \cdots |\omega_n|^2)^{\frac{1}{n}} = \frac{|\omega_1|^2 + |\omega_2|^2 + \cdots + |\omega_n|^2}{n}$$

werden; dies erfordert, daß

$$|\omega_1|^2 = |\omega_2|^2 = \cdots = |\omega_n|^2 = 1$$

wird. Daher liegen die charakteristischen Wurzeln einer unitär orthogonalen Substitution A auf dem Einheitskreis. Zugleich ergibt sich aus $\sum_{\nu} |\omega_\nu|^2 = \sum_{\varkappa,\lambda} |a_{\varkappa\lambda}|^2$, daß A durch eine unitär orthogonale Transformation der Variabeln auf die Diagonalform gebracht werden kann. — Diese beiden Sätze sind zuerst von Herrn Frobenius**) aufgestellt worden.

Abschnitt II.

Eine Anwendung auf die Theorie der Integralgleichungen.

§ 10.

Es sei $K(s, t)$ eine für $a \leqq s \leqq b$, $a \leqq t \leqq b$ definierte, reelle oder komplexe Funktion der reellen Variabelen s und t, die im ganzen Defi-

*) Diese Ableitung der Grundeigenschaften der Hermiteschen Formen ist gewiß weniger einfach als der sonst übliche Beweis, der sich gewissermaßen direkt auf den hier mit I bezeichneten Satz stützt. Aber es schien mir von Interesse, darauf hinzuweisen, daß sich diese Eigenschaften unmittelbar aus der die Hermiteschen Formen eindeutig kennzeichnenden Gleichung

$$\sum_{\varkappa,\lambda} a_{\varkappa\lambda} \bar{a}_{\varkappa\lambda} = \sum_{\varkappa,\lambda} a_{\varkappa\lambda} a_{\lambda\varkappa}$$

ablesen lassen.

**) „Über die principale Transformation der Thetafunktionen mehrerer Variabeln", Journal für die reine und angewandte Mathematik, Bd. 95, S. 264. — Vergl. auch A. Loewy, „Über bilineare Formen mit konjugiert imaginären Variabeln", Nova Acta Acad. Leop.-Carol., Bd. 71, S. 379.

nitionsbereich stetig sein soll.*) Eine für $a \leqq s \leqq b$ stetige Funktion $\varphi(s)$ heißt dann nach Herrn Hilbert eine *zum Eigenwert λ gehörende Eigenfunktion des Kerns* $K(s,t)$, wenn

$$\varphi(s) = \lambda \int_a^b K(s,t)\,\varphi(t)\,dt \tag{15}$$

ist. Herr Fredholm**) hat eine gewisse, durch $K(s,t)$ eindeutig definierte ganze transzendente Funktion

$$D(x) = 1 + d_1 x + d_2 x^2 + \cdots$$

angegeben, die nur für die Eigenwerte λ des Kerns $K(s,t)$ verschwindet. Ist λ eine n-fache Nullstelle von $D(x)$, so heißt n die *Ordnung* des Eigenwerts λ. Diese Zahl n hat noch folgende Bedeutung: *es lassen sich n, aber nicht mehr als n, linear unabhängige (stetige) Funktionen* $\varphi_1(s), \varphi_2(s), \cdots, \varphi_n(s)$ *bestimmen, für die*

$$\int_a^b K(s,t)\,\varphi_\alpha(t)\,dt = c_{\alpha 1}\,\varphi_1(s) + c_{\alpha 2}\,\varphi_2(s) + \cdots + c_{\alpha,\alpha-1}\,\varphi_{\alpha-1}(s) + \frac{1}{\lambda}\,\varphi_\alpha(s),$$

wird, wo die $c_{\alpha\beta}$ *Konstanten sind.****)

Es ist nun von Interesse, auf elementarem Wege ohne Einführung der Fredholmschen Funktion $D(x)$ nachzuweisen, daß sich zu jedem Eigenwert λ eine ganze Zahl n angeben läßt, die durch die zuletzt erwähnte Eigenschaft charakterisiert ist. Man erhält auf diese Weise eine neue einfache Definition des Begriffs „Ordnung eines Eigenwerts". Für reelle symmetrische Kerne hat diese Aufgabe bereits Herr E. Schmidt†) gelöst. Die folgende Betrachtung schließt sich auch aufs engste dem Schmidtschen Gedankengang an.

---

*) Um einen einfachen Fall vor Augen zu haben, beschränke ich mich im folgenden auf die Betrachtung stetiger Funktionen. Es würde aber genügen anzunehmen, daß die hier vorkommenden Funktionen gewissen leicht zu formulierenden Bedingungen der Integrabilität genügen.

**) „Sur une classe d'équations fonctionnelles", Acta Mathematica, Bd. 27, S. 365.

***) J. Plemelj, „Zur Theorie der Fredholmschen Funktionalgleichung", Monatshefte für Math. und Phys., Jahrgang XV (1904), S. 93. Vergl. auch E. Goursat, „Recherches sur les équations intégrales linéaires", Annales de la faculté des sciences de l'université de Toulouse, 1908, S. 6. — Daß die Ordnung n von λ die *größte* Zahl ist, welche die im Text genannte Eigenschaft besitzt, wird erst in der Goursatschen Arbeit (§ 44) ausdrücklich hervorgehoben und bewiesen. Es läßt sich dies aber auch ohne Mühe aus den Ausführungen auf S. 119 der Plemeljschen Arbeit folgern.

†) „Zur Theorie der linearen und nichtlinearen Integralgleichungen", I. Teil, Math. Annalen, Bd. 63, S. 443.

§ 11.

Wir bezeichnen im folgenden überall, wenn $\varphi(s)$ eine beliebige Funktion der reellen Variabeln s ist, die konjugiert komplexe Funktion mit $\overline{\varphi}(s)$. Sind $\varphi_1(s), \varphi_2(s), \cdots, \varphi_n(s)$ im Intervall $a \leqq s \leqq b$ stetige Funktionen und ist

$$\int_a^b \varphi_\alpha(s)\,\overline{\varphi}_\beta(s)\,ds = e_{\alpha\beta}, \qquad (\alpha, \beta = 1, 2, \cdots, n) \tag{16}$$

wo die $e_{\alpha\beta}$ in derselben Weise wie in § 1 zu erklären sind, so sage ich, *die Funktionen* $\varphi_1(s), \varphi_2(s), \cdots, \varphi_n(s)$ *bilden ein unitär orthogonales System.*

Es gilt dann folgende Regel: *Sind n linear unabhängige, für $a \leqq s \leqq b$ stetige Funktionen $\psi_1(s), \psi_2(s), \cdots, \psi_n(s)$ gegeben, so lassen sich stets n Funktionen*

$$\varphi_\alpha(s) = \sum_{\beta=1}^{n} p_{\alpha\beta}\,\psi_\beta(s)$$

mit konstanten Koeffizienten $p_{\alpha\beta}$ bestimmen, die ein unitär orthogonales System bilden.

Der Beweis dieses Satzes, der für reelle Funktionen zuerst von Herrn E. Schmidt (a. a. O., § 3) ausgesprochen worden ist, läßt sich prinzipiell am einfachsten folgendermaßen führen. Setzt man

$$\int_a^b \psi_\alpha(s)\,\overline{\psi}_\beta(s)\,ds = h_{\alpha\beta},$$

so wird

$$\sum_{\alpha,\beta} h_{\alpha\beta}\,x_\alpha\,\overline{x}_\beta = \int_a^b [x_1\,\psi_1(s) + \cdots + x_n\,\psi_n(s)]\,[\overline{x}_1\,\overline{\psi}_1(s) + \cdots + \overline{x}_n\,\overline{\psi}_n(s)]\,ds$$

eine positive Hermitesche Form von nicht verschwindender Determinante.*) Die Konstanten $p_{\alpha\beta}$ sind nun so zu bestimmen, daß

$$\int_a^b \varphi_\alpha(s)\,\overline{\varphi}_\beta(s)\,ds = \sum_{\gamma,\delta} p_{\alpha\gamma}\,\overline{p}_{\beta\delta}\,h_{\gamma\delta} = e_{\alpha\beta}$$

wird. Hierzu hat man nur n Linearformen

$$y_\gamma = p_{1\gamma}x_1 + p_{2\gamma}x_2 + \cdots + p_{n\gamma}x_n$$

zu berechnen, für die

$$\sum_{\gamma,\delta} h_{\gamma\delta}\,y_\gamma\,\overline{y}_\delta = x_1\overline{x}_1 + x_2\overline{x}_2 + \cdots + x_n\overline{x}_n$$

wird, was bekanntlich stets möglich ist.

---

*) Vergl. E. Fischer, a. a. O., S. 38.

Bilden ferner $\varphi_1(s), \varphi_2(s), \cdots, \varphi_n(s)$ ein unitär orthogonales System, und ist $f(s)$ eine beliebige für $a \leqq s \leqq b$ stetige Funktion, so setze man

$$\int_a^b f(s)\overline{\varphi}_\alpha(s)\,ds = A_\alpha,$$

also

$$\int_a^b \bar{f}(s)\varphi_\alpha(s)\,ds = \bar{A}_\alpha,$$

Dann wird

$$\int_a^b \Big[f(t) - \sum_{\alpha=1}^n A_\alpha \varphi_\alpha(t)\Big]\Big[\bar{f}(t) - \sum_{\alpha=1}^n \bar{A}_\alpha \overline{\varphi}_\alpha(t)\Big]dt$$

$$= \int_a^b f(t)\bar{f}(t)\,dt - 2\sum_{\alpha=1}^n A_\alpha \bar{A}_\alpha + \sum_{\alpha,\beta} A_\alpha \bar{A}_\beta \int_a^b \varphi_\alpha(t)\overline{\varphi}_\beta(t)\,dt$$

$$= \int_a^b f(t)\bar{f}(t)\,dt - \sum_{\alpha=1}^n A_\alpha \bar{A}_\alpha.$$

Daher ist stets

$$\sum_{\alpha=1}^n A_\alpha \bar{A}_\alpha \leqq \int_a^b f(t)\bar{f}(t)\,dt.$$

Diese Ungleichung soll nach dem Vorgange von E. Schmidt (a. a. O., § 1) als die Besselsche Ungleichung bezeichnet werden.

§ 12.

Zur Abkürzung möge für jede Funktion $f(s)$

$$\int_a^b K(s,t)f(t)\,dt = K(f)$$

gesetzt werden. Wird für n linear unabhängige stetige Funktionen $\varphi_1, \varphi_2, \cdots, \varphi_n$

$$K(\varphi_\alpha) = \sum_{\beta=1}^n a_{\alpha\beta}\varphi_\beta \qquad (\alpha = 1, 2, \cdots, n),$$

so sagen wir: $\varphi_1, \varphi_2, \cdots, \varphi_n$ *erfahren bei Anwendung der Operation* K *die Substitution* $A = (a_{\alpha\beta})$ oder auch *die Funktionen* $\varphi_1, \varphi_2, \cdots, \varphi_n$ *bilden ein invariantes System des Kerns* $K(s, t)$, *das zur Substitution* A *gehört.* Jede in einem invarianten System vorkommende Funktion nenne ich eine *Hauptfunktion**) des Kerns. Die charakteristische Determinante $|A - xE|$

*) „Fonction principale" bei E. Goursat, a. a. O.

der Substitution A soll kurz die *charakteristische Funktion des invarianten Systems* $\varphi_1, \varphi_2, \cdots, \varphi_n$ heißen. Treten an Stelle von $\varphi_1, \varphi_2, \cdots, \varphi_n$ irgend welche n lineare homogene Verbindungen $\psi_1, \psi_2, \cdots, \psi_n$ der φ_α (mit konstanten Koeffizienten), die untereinander linear unabhängig sind, so bilden auch die Funktionen $\psi_1, \psi_2, \cdots, \psi_n$ ein invariantes System des Kerns und die diesem System entsprechende Substitution B ist eine zu A ähnliche Substitution. Man kann hierbei die ψ_α so wählen, daß B irgend eine vorgeschriebene zu A ähnliche Substitution wird. Hieraus folgt zugleich auf Grund der Lehre von den linearen Substitutionen, daß sich für eine charakteristische Wurzel ω von A, die genau m Elementarteiler der Determinante $|A-xE|$ zum Verschwinden bringt, m und nicht mehr als m linear unabhängige Funktionen ψ_μ der Form

$$\psi_\mu = c_{\mu 1}\varphi_1 + c_{\mu 2}\varphi_2 + \cdots + c_{\mu n}\varphi_n \qquad (\mu = 1, 2, \cdots, m)$$

angeben lassen, sodaß

$$K(\psi_\mu) = \omega\psi_\mu$$

wird.*) Ist ω speziell von Null verschieden, so wird $\lambda = \frac{1}{\omega}$ ein Eigenwert des Kerns $K(s, t)$.

Hat man zwei invariante Systeme $\varphi_1, \varphi_2, \cdots, \varphi_n$ und $\psi_1, \psi_2, \cdots, \psi_p$ des Kerns $K(s, t)$ und besitzen die zugehörigen charakteristischen Funktionen keinen Teiler gemeinsam, so müssen, wie in bekannter Weise geschlossen wird, die $n+p$ Funktionen $\varphi_1, \varphi_2, \cdots, \varphi_n; \psi_1, \psi_2, \cdots, \psi_p$ untereinander linear unabhängig sein.

Von Wichtigkeit ist auch noch folgende Bemerkung. Es seien m Funktionen $\varphi_1, \varphi_2, \cdots, \varphi_m$ gegeben, unter denen nur r linear unabhängig sind, etwa die Funktionen $\varphi_1, \varphi_2, \cdots, \varphi_r$. Ist dann bekannt, daß für die m Funktionen $\varphi_1, \varphi_2, \cdots, \varphi_m$ Gleichungen der Form

$$K(\varphi_\mu) = \sum_{\nu=1}^{m} c_{\mu\nu}\varphi_\nu$$

bestehen, wo die $c_{\mu\nu}$ Konstanten sind, so erfahren offenbar die r linear unabhängigen Funktionen $\varphi_1, \varphi_2, \cdots, \varphi_r$ bei Anwendung der Operation K eine gewisse lineare Substitution A, sie bilden also ein invariantes System des Kerns $K(s, t)$. Es gilt dann die leicht zu beweisende Regel: *Die charakteristische Funktion* $|A-xE|$ *des Systems* $\varphi_1, \varphi_2, \cdots, \varphi_r$ *ist ein Divisor der charakteristischen Determinante der Substitution* $C = (c_{\mu\nu})$.

---

*) Vergl. etwa S. Pincherle und U. Amaldi, „Le operazioni distributive" (Bologna, 1901), Kap. IV.

§ 13.

Ich beweise nun folgenden Satz:

IV. *Bilden die n Funktionen* $\psi_1, \psi_2, \cdots, \psi_n$ *ein invariantes System des Kerns* $K(s,t)$ *und ist*

$$(\omega_1 - x)(\omega_2 - x)\cdots(\omega_n - x) \tag{17}$$

die charakteristische Funktion dieses invarianten Systems, so ist stets

$$\sum_{\nu=1}^{n} |\omega_\nu|^2 \leqq \int_a^b\int_a^b |K(s,t)|^2\, ds\, dt. \tag{18}$$

Man bestimme nämlich n lineare homogene Verbindungen $\varphi_1, \varphi_2, \cdots, \varphi_n$ der ψ_α, sodaß die φ_α ein unitär orthogonales System bilden. Dann sind $\varphi_1, \varphi_2, \cdots, \varphi_n$ von selbst linear unabhängige Funktionen, und wir erhalten in diesen Funktionen ein invariantes System des Kerns $K(s,t)$, dessen charakteristische Funktion wieder die Funktion (17) ist. Es sei

$$K(\varphi_\alpha) = \int_a^b K(s,t)\,\varphi_\alpha(t)\,dt = \sum_{\beta=1}^{n} a_{\alpha\beta}\varphi_\beta(s). \tag{19}$$

Setzt man für ein gegebenes s in der Besselschen Ungleichung

$$f(t) = \overline{K}(s,t),$$

wo $\overline{K}(s,t)$ die zu $K(s,t)$ konjugiert komplexe Funktion bedeutet, so wird

$$A_\alpha = \int_a^b \overline{K}(s,t)\,\overline{\varphi}_\alpha(t)\,dt = \sum_{\beta=1}^{n} \overline{a}_{\alpha\beta}\,\overline{\varphi}_\beta(s).$$

Daher ist

$$\sum_{\alpha,\beta,\gamma} a_{\alpha\beta}\,\overline{a}_{\alpha\gamma}\,\varphi_\beta(s)\,\overline{\varphi}_\gamma(s) \leqq \int_a^b K(s,t)\,\overline{K}(s,t)\,dt.$$

Integriert man nun nach s von a bis b, so erhält man wegen (16)

$$\sum_{\alpha,\beta} a_{\alpha\beta}\,\overline{a}_{\alpha\beta} = \sum_{\alpha,\beta} |a_{\alpha\beta}|^2 \leqq \int_a^b\int_a^b K(s,t)\,\overline{K}(s,t)\,ds\,dt.\text{*)} \tag{20}$$

Da aber $\omega_1, \omega_2, \cdots, \omega_n$ die charakteristischen Wurzeln der Substitution $A = (a_{\alpha\beta})$ sind, so ist

$$\sum_{\nu=1}^{n} |\omega_\nu|^2 \leqq \sum_{\alpha,\beta} |a_{\alpha\beta}|^2. \tag{21}$$

Aus (20) und (21) ergibt sich die zu beweisende Ungleichung (18)

*) Vergl. E. Schmidt, a. a. O., § 5.

Ich füge noch folgendes hinzu. Ist $Q = (q_{\alpha\beta})$ irgend eine unitär orthogonale Substitution, und setzt man

$$\chi_\alpha(s) = \sum_{\beta=1}^{n} q_{\alpha\beta}\,\varphi_\beta(s),$$

so bilden auch die Funktionen $\chi_1, \chi_2, \cdots, \chi_n$ ein unitär orthogonales System und es wird

$$K(\chi_\alpha) = \sum_{\beta=1}^{n} c_{\alpha\beta}\,\chi_\beta,$$

wo die Matrix $C = (c_{\alpha\beta})$ aus der Gleichung

$$C = QAQ^{-1} = QA\overline{Q}'$$

zu berechnen ist. Aus dem Satz I ergibt sich daher, daß man jedes invariante System mit der charakteristischen Funktion (17) durch ein *unitär orthogonales System* $\chi_1, \chi_2, \cdots, \chi_n$ ersetzen kann, für das

$$K(\chi_\alpha) = c_{\alpha 1}\chi_1 + c_{\alpha 2}\chi_2 + \cdots + c_{\alpha,\alpha-1}\chi_{\alpha-1} + \omega_\alpha\chi_\alpha$$

wird.

Ist speziell

$$\sum_{\nu=1}^{n} |\omega_\nu|^2 = \sum_{\alpha,\beta} |a_{\alpha\beta}|^2, \tag{22}$$

so werden die Konstanten $c_{\alpha\beta}$ von selbst gleich Null, so daß sich

$$K(\chi_\alpha) = \omega_\alpha\chi_\alpha$$

ergibt. Dies tritt z. B. stets ein, wenn $K(s, t)$ ein reeller, symmetrischer oder allgemeiner ein *Hermitescher Kern* ist, d. h. wenn $K(s, t) = \overline{K}(s, t)$ wird. Denn aus der Gleichung (19) folgt, indem man mit $\overline{\varphi}_\gamma(s)$ multipliziert und nach s zwischen a und b integriert,

$$a_{\alpha\gamma} = \int_a^b\!\!\int_a^b K(s, t)\,\varphi_\alpha(t)\,\overline{\varphi}_\gamma(s)\,dt\,ds.$$

Ist nun $K(s, t) = \overline{K}(t, s)$, so ergibt sich hieraus $a_{\alpha\gamma} = \overline{a}_{\gamma\alpha}$, d. h. $A = (a_{\alpha\beta})$ wird in diesem Fall eine Hermitesche Matrix; für eine solche ist aber die Bedingung (22) gewiß erfüllt. Zugleich ergibt sich, daß die Eigenwerte eines Hermiteschen Kerns sämtlich reell sind.

§ 14.

Es sei nun

$$\lambda = \frac{1}{\omega}$$

ein Eigenwert des Kerns $K(s, t)$. Für jede zu λ gehörende Eigenfunktion φ wird dann

$$K(\varphi) = \omega\varphi.$$

Die Funktion φ erscheint also als eine invariante Funktion des Kerns, und die zugehörige charakteristische Funktion ist $\omega - x$. Ich betrachte nun allgemeiner diejenigen invarianten Systeme $\varphi_1, \varphi_2, \cdots, \varphi_n$ des Kerns, deren charakteristische Funktion gleich $(\omega - x)^n$ wird. Dann wird nach Satz IV

$$n|\omega|^2 \leqq k,$$

wo k das in (18) auftretende Integral bedeuten soll. Also ist $n \leqq k|\lambda|^2$, d. h. n ist unterhalb einer endlichen Grenze gelegen. *Die größte Zahl n, für die sich ein invariantes System $\varphi_1, \varphi_2, \cdots, \varphi_n$ des Kerns mit der charakteristischen Funktion $\left(\frac{1}{\lambda} - x\right)^n$ angeben läßt, soll nun die Ordnung des Eigenwerts λ heißen.* Auf Grund des in § 10 erwähnten Plemelj-Goursatschen Satzes ist diese Zahl n zugleich auch die Ordnung des Eigenwerts im Fredholmschen Sinne.

Es seien ferner

$$\lambda_1 = \frac{1}{\omega_1},\ \lambda_2 = \frac{1}{\omega_2},\ \ldots,\ \lambda_p = \frac{1}{\omega_p}$$

p verschiedene Eigenwerte des Kerns $K(s, t)$, die Ordnung von λ_α sei n_α. Bilden dann für ein gegebenes ϱ die n_ϱ Funktionen

$$\varphi_{\varrho 1},\ \varphi_{\varrho 2},\ \cdots,\ \varphi_{\varrho n_\varrho} \tag{23}$$

ein invariantes System des Kerns mit der charakteristischen Funktion $(\omega_\varrho - x)^{n_\varrho}$, so erhalten wir in den

$$n = n_1 + n_2 + \cdots + n_p$$

Funktionen (23) n linear unabhängige Funktionen, die als ein invariantes System mit der charakteristischen Funktion

$$(\omega_1 - x)^{n_1}(\omega_2 - x)^{n_2} \cdots (\omega_p - x)^{n_p}$$

erscheinen. Folglich ist nach (18)

$$n_1|\omega_1|^2 + n_2|\omega_2|^2 + \cdots + n_p|\omega_p|^2 \leqq k$$

oder, was dasselbe ist,

$$\frac{n_1}{|\lambda_1|^2} + \frac{n_2}{|\lambda_2|^2} + \cdots + \frac{n_p}{|\lambda_p|^2} \leqq k.$$

Dies lehrt uns aber (vergl. E. Schmidt, a. a. O., § 5), daß die Eigenwerte des Kerns $K(s, t)$ keine Häufungsstelle im Endlichen besitzen. Zugleich ergibt sich, daß *die Reihe*

$$\sum \frac{1}{|\lambda|^2},$$

erstreckt über alle Eigenwerte λ des Kerns, wobei jeder Eigenwert so oft zu schreiben ist, wie seine Ordnung angibt, konvergent und zwar kleiner als die Zahl k ist. Hieraus folgt auch, daß die Fredholmsche ganze transzendente Funktion $D(x)$, deren Geschlecht bekanntlich höchstens gleich 2 ist*), eine Produktzerlegung der Form

$$D(x) = e^{ax+bx^2} \prod \left(1 - \frac{x}{\lambda}\right) e^{\frac{x}{\lambda}}$$

aufweist. Dieses Resultat scheint für die hier betrachteten allgemeinen Kerne neu zu sein.

§ 15.

Unsere Betrachtung bedarf noch einer Ergänzung. Ist nämlich n wieder die Ordnung eines Eigenwerts $\lambda = \frac{1}{\omega}$ und bilden $\varphi_1, \varphi_2, \cdots, \varphi_n$ ein invariantes System mit der charakteristischen Funktion $(\omega - x)^n$, so fragt es sich: inwiefern sind die Funktionen $\varphi_1, \varphi_2, \cdots, \varphi_n$ und die zu gehörige lineare Substitution $A = (a_{\alpha\beta})$, für die

$$K(\varphi_\alpha) = \sum_{\beta=1}^{n} a_{\alpha\beta} \varphi_\beta$$

wird, durch den Eigenwert λ eindeutig bestimmt?

Diese Frage ist leicht zu entscheiden.

Hat man nämlich neben $\varphi_1, \varphi_2, \cdots, \varphi_n$ noch ein zweites invariantes System $\psi_1, \psi_2, \cdots, \psi_n$ mit der charakteristischen Funktion $(\omega - x)^n$, so sei

$$K(\psi_\alpha) = \sum_{\beta=1}^{n} b_{\alpha\beta} \psi_\beta.$$

Die Substitution $(b_{\alpha\beta})$ möge mit B bezeichnet werden. Sind dann unter den $2n$ Funktionen

(24) $$\varphi_1, \varphi_2, \cdots, \varphi_n; \quad \psi_1, \psi_2, \cdots, \psi_n$$

r linear unabhängig, etwa $\chi_1, \chi_2, \cdots, \chi_r$, so muß zunächst, weil doch die φ_α (und ebenso die ψ_α) linear unabhängig sein sollen, $r \geqq n$ sein. Ferner bilden die Funktionen $\chi_1, \chi_2, \cdots, \chi_r$ wieder ein invariantes System des Kerns. Nach der am Schluß des § 12 gemachten Bemerkung muß aber die charakteristische Funktion dieses Systems ein Divisor der charakteristischen Determinante $(\omega - x)^{2n}$ der Substitution

$$\begin{pmatrix} A & 0 \\ 0 & B \end{pmatrix}$$

sein, und also gleich $(\omega - x)^r$ werden. Hieraus folgt aber, daß $r = n$ sein muß, da sonst n nicht als die Ordnung des Eigenwerts λ zu kenn-

*) Vergl. Plemelj, a. a. O., S. 101 und T. Lalesco, „Sur l'ordre de la fonction entière $D(\lambda)$ de Fredholm", Comptes Rendus, 25. November 1907, S. 906.

zeichnen wäre. Demnach ist jede der Funktionen $\psi_1, \psi_2, \cdots, \psi_n$ eine lineare homogene Verbindung der Funktionen $\varphi_1, \varphi_2, \cdots, \varphi_n$ mit konstanten Koeffizienten. Zugleich ergibt sich, daß die Substitution B der Substitution A ähnlich sein muß. Sieht man ähnliche Substitutionen als nicht voneinander verschieden an, so erscheint A als eine durch λ eindeutig charakterisierte Substitution. Die Anzahl der zu λ gehörenden linear unabhängigen Eigenfunktionen des Kerns $K(s,t)$ ist dann nichts anderes als die Anzahl der Elementarteiler der Determinante $|A - xE|$.

Die Funktionen $\varphi_1, \varphi_2, \cdots, \varphi_n$ kann man passend als ein *vollständiges zum Eigenwert λ gehörendes invariantes System* bezeichnen.

Allgemeiner schließt man in ähnlicher Weise:

Man habe irgend ein invariantes System $\psi_1, \psi_2, \cdots, \psi_p$ des Kerns $K(s,t)$, die zugehörige charakteristische Funktion sei

$$(\omega_1 - x)^{p_1}(\omega_2 - x)^{p_2} \cdots (\omega_m - x)^{p_m}, \qquad (p_1 + p_2 + \cdots + p_m = p)$$

wo $\omega_1, \omega_2, \cdots, \omega_m$ voneinander verschieden sein sollen. Jede nicht verschwindende unter den Zahlen ω_μ ist dann der reziproke Wert eines Eigenwerts λ_μ des Kerns, und ist n_μ die Ordnung dieses Eigenwerts, so ist $p_\mu \leqq n_\mu$. Sind insbesondere alle m Zahlen $\omega_1, \omega_2, \cdots, \omega_m$ von Null verschieden und bilden die Funktionen

$$\varphi_{\mu 1}, \varphi_{\mu 2}, \cdots, \varphi_{\mu n_\mu} \qquad (\mu = 1, 2, \cdots, m) \tag{25}$$

ein vollständiges zu λ_μ gehörendes invariantes System, so läßt sich jede der p Funktionen $\psi_1, \psi_2, \cdots, \psi_p$ durch die $n_1 + n_2 + \cdots + n_m$ Funktionen (25) *linear und homogen mit konstanten Koeffizienten darstellen.*

14.

Beiträge zur Theorie der Gruppen linearer homogener Substitutionen

American Mathematical Society Transactions 10, 159 - 175 (1909)

In seiner Arbeit *Über die vollständig reduciblen Gruppen, die zu einer Gruppe linearer homogener Substitutionen gehören* † hat Herr A. LOEWY eine wichtige neue Art der Reduktion einer Gruppe $\mathfrak{G}$ linearer homogener Substitutionen angegeben, die er als die Zerlegung der Gruppe unter Hervorhebung der zu ihr gehörenden auf einander folgenden grössten vollständig reduziblen Gruppen bezeichnet. Diese Zerlegung kann man, wenn die Koeffizienten aller Substitutionen von $\mathfrak{G}$ einem gegebenen Zahlkörper Ω angehören, auch innerhalb des Körpers Ω durchführen. Auf diese Weise erhält Herr LOEWY ein System von gewissen zu $\mathfrak{G}$ gehörenden Gruppen $\mathfrak{A}_1, \mathfrak{A}_2, \cdots, \mathfrak{A}_\mu$ mit Koeffizienten aus dem Körper Ω, die inbezug auf Ω vollständig reduzibel sind, und er zeigt auch, dass die Gruppen $\mathfrak{A}_1, \mathfrak{A}_2, \cdots, \mathfrak{A}_\mu$, wenn man äquivalente (ähnliche) Gruppen als nicht von einander verschieden ansieht, in der gegebenen Reihenfolge als eindeutig bestimmt zu betrachten sind.

In der vorliegenden Arbeit soll nun eine, wie ich glaube, nicht unwesentliche Ergänzung der LOEWY'schen Resultate mitgeteilt werden. Es wird nämlich gezeigt, dass die Gruppen $\mathfrak{A}_1, \mathfrak{A}_2, \cdots, \mathfrak{A}_\mu$, wenn wieder äquivalente Gruppen als nicht von einander verschieden gelten, von der Wahl des Körpers Ω gänzlich unabhängig sind; oder genauer: legt man der Betrachtung an Stelle des Körpers Ω einen anderen Körper Ω' zugrunde, der ebenfalls alle Substitutionskoeffizienten von $\mathfrak{G}$ umfasst, und sind $\mathfrak{A}'_1, \mathfrak{A}'_2, \cdots, \mathfrak{A}'_{\mu'}$ die zu $\mathfrak{G}$ gehörenden auf einander folgenden grössten vollständig reduziblen Gruppen inbezug auf Ω', so ist $\mu' = \mu$, ferner sind $\mathfrak{A}_\alpha$ und $\mathfrak{A}'_\alpha$ äquivalente Gruppen.‡

Zur Begründung dieses Resultats bedarf es einer eingehenderen Untersuchung der in einem Körper Ω irreduziblen Gruppen linearer homogener Substitutionen. In §§ 2–4 stelle ich für diese Gruppen eine Reihe von Sätzen auf, die

* Presented to the Society (Chicago) April 17, 1908.

† Transactions of the American Mathematical Society, vol. 6 (1905), p. 504. Vergl. auch L. STICKELBERGER, *Zur Theorie der vollständig reduciblen Gruppen, die zu einer Gruppe linearer homogener Substitutionen gehören*, ibid., vol. 7 (1906), p. 509.

‡ Für den Fall, dass $\mathfrak{G}$ eine zyklische Gruppe ist, die aus den Potenzen einer Substitution besteht, hat diesen Satz bereits Herr J. WIRTH in seiner Dissertation *Über die Elementarteiler einer linearen homogenen Substitution* (Freiburg i. Br., 1906) bewiesen.

ganz analog sind den Sätzen, die ich für den speziellen Fall der endlichen Gruppen bereits in meiner Arbeit *Arithmetische Untersuchungen über endliche Gruppen linearer Substitutionen* * bewiesen habe. Dass diese Satze auch für unendliche Gruppen ihre Geltung behalten, beruht in erster Linie auf dem Umstand, dass jede Gruppe linearer Substitutionen, die inbezug auf einen gegebenen Körper Ω irreduzibel ist, im Bereich aller Zahlen vollständig reduzibel ist.† Auf diese bemerkenswerte Tatsache hat bereits Herr TABER am Schluss seiner Arbeit *Sur les groupes réductibles de transformations linéaires et homogènes* ‡ aufmerksam gemacht. Nimmt man diesen Satz als bekannt an, so lassen sich auch meine anderen Sätze, insbesondere der hier hauptsächlich in Betracht kommende Satz X leichter beweisen. Es schien mir aber von Interesse, die Untersuchung in der Weise durchzuführen, dass sich zugleich ein neuer Beweis für den TABER'schen Satz ergiebt. Während Herr TABER beim Beweis seines Satzes von der Theorie der hyperkomplexen Grössen Gebrauch macht, stützt sich meine Beweisführung auf einen einfachen Hilfssatz (Satz II), der sich leicht direkt begründen lässt.

§ 1.

Unter einer Gruppe linearer Substitutionen (Matrizen) des *Grades g* verstehe ich im folgenden ein beliebiges endliches oder auch unendliches System $\mathfrak{G}$ von linearen homogenen Substitutionen in g Variabeln, wenn das Produkt von je zwei Substitutionen von $\mathfrak{G}$ wieder in $\mathfrak{G}$ enthalten ist. Sind die Koeffizienten aller Substitutionen von $\mathfrak{G}$ Zahlen eines gegebenen Zahlkörpers Ω, so sage ich kurz, $\mathfrak{G}$ sei eine *in Ω rationale Gruppe.* Ebenso verstehe ich unter einer *in Ω rationalen Matrix* eine Matrix, deren Koeffizienten dem Körper Ω angehören. Geht nun eine in Ω rationale Gruppe $\mathfrak{G}$ durch die lineare Transformation T der Variabeln in die ihr äquivalente Gruppe $\mathfrak{G}' \doteq T\mathfrak{G}T^{-1}$ über, und weiss man, dass $\mathfrak{G}'$ wieder in Ω rational ist, so kann man offenbar T auch so wählen, dass die Koeffizienten von T ebenfalls in Ω enthalten sind.

Es seien nun $\mathfrak{G}$ und $\mathfrak{H}$ zwei Gruppen linearer Substitutionen der Grade g und h, die entweder einander isomorph oder allgemeiner einer dritten Gruppe homomorph § sind. Lässt sich dann eine Matrix P mit g Zeilen und h Kolonnen oder auch eine Matrix P mit h Zeilen und g Kolonnen bestimmen, sodass für je zwei entsprechende Substitutionen (Matrizen) G und H der beiden Gruppen $\mathfrak{G}$ und $\mathfrak{H}$ die Gleichung $GP = PH$, bezw. die Gleichung $PG = HP$

* Sitzungsberichte der Berliner Akademie, Februar 1906, p. 164. Im folgenden wird diese Arbeit kurz mit A. zitiert.

† Für den Fall, das Ω den Bereich aller reellen Zahlen bedeutet, hat diesen Satz Herr A. LOEWY in seiner Arbeit *Über die Reducibilität der reellen Gruppen linearer homogener Substitutionen*, Transactions of the American Mathematical Society, vol. 4 (1903), p. 171, bewiesen.

‡ Comptes Rendus de l'Académie des Sciences, April 1906, p. 948.

§ Vergl. G. FROBENIUS und I. SCHUR, *Über die Äquivalenz der Gruppen linearer Substitutionen*, Sitzungsberichte der Berliner Akademie, Februar 1906, p. 209.

besteht, so will ich $\mathfrak{G}$ und $\mathfrak{H}$ als *verkettete* Gruppen bezeichnen; ich sage auch kürzer, die Matrix P genüge der Gleichung

$$(1) \qquad \mathfrak{G}P = P\mathfrak{H} \quad \text{oder} \quad P\mathfrak{G} = \mathfrak{H}P.$$

Natürlich soll hierbei nicht $P = 0$ sein. — Sind insbesondere die beiden Gruppen $\mathfrak{G}$ und $\mathfrak{H}$ in einem Körper Ω rational und weiss man, dass eine von Null verschiedene Matrix P existiert, die einer der Gleichungen (1) genügt, so kann man offenbar auch eine in Ω rationale Matrix wählen, die dieselbe Bedingung erfüllt. Sind ferner $\mathfrak{G}'$ und $\mathfrak{H}'$ zwei zu $\mathfrak{G}$ und $\mathfrak{H}$ äquivalente Gruppen, von denen bekannt ist, dass sie untereinander verkettet sind, so schliesst man leicht, dass auch $\mathfrak{G}$ und $\mathfrak{H}$ verkettet sind. Dieser Fall tritt insbesondere ein, wenn die Gruppen $\mathfrak{G}'$ und $\mathfrak{H}'$ die spezielle Form

$$\mathfrak{G}' = \begin{pmatrix} \mathfrak{A} & 0 \\ \mathfrak{C} & \mathfrak{D} \end{pmatrix}, \qquad \mathfrak{H}' = \begin{pmatrix} \mathfrak{A} & 0 \\ 0 & \mathfrak{E} \end{pmatrix}$$

aufweisen, wo $\mathfrak{A}$ eine gewisse Gruppe des Grades a bedeuten soll. Denn versteht man alsdann unter P' die Matrix

$$P' = (p_{\kappa\lambda}) \qquad (\kappa = 1, 2, \cdots, h;\ \lambda = 1, 2, \cdots, g),$$

wo $p_{11}, p_{22}, \cdots, p_{aa}$ gleich 1, dagegen alle übrigen $p_{\kappa\lambda}$ gleich 0 sind, so wird, wie man leicht findet,

$$P'\mathfrak{G}' = \mathfrak{H}'P' = \begin{pmatrix} \mathfrak{A} & 0 \\ 0 & 0 \end{pmatrix}.$$

Daher sind in diesem Fall $\mathfrak{G}'$ und $\mathfrak{H}'$, und folglich auch $\mathfrak{G}$ und $\mathfrak{H}$ verkettete Gruppen.

Es gilt ferner der für das folgende wichtige Satz:

I. *Es seien $\mathfrak{G}$ und $\mathfrak{H}$ zwei in einem Körper Ω rationale Gruppen linearer Substitutionen der Grade g und h, die entweder isomorph oder einer dritten Gruppe homomorph sind, und es sei $\mathfrak{G}$ insbesondere im Körper Ω irreduzibel. Ist dann P eine in Ω rationale Matrix, die der Gleichung $\mathfrak{G}P = P\mathfrak{H}$ oder der Gleichung $P\mathfrak{G} = \mathfrak{H}P$ genügt, so ist entweder $P = 0$ oder es ist P vom Range g. Im letzteren Falle enthält $\mathfrak{H}$ die Gruppe $\mathfrak{G}$ als irreduziblen Bestandteil inbezug auf Ω.*

Der Beweis dieses Satzes ist genau ebenso zu führen, wie der Beweis des Satzes I meiner mit A. zitierten Arbeit.

Aus dem Satze I ergiebt sich sofort:

I*. *Ist $\mathfrak{G}$ eine im Körper Ω irreduzible Gruppe des Grades g und $\mathfrak{H}$ eine in Ω rationale Gruppe des Grades h, die mit $\mathfrak{G}$ verkettet ist, so muss $h \geqq g$ sein. Ist speziell $h = g$, so muss $\mathfrak{H}$ mit $\mathfrak{G}$ äquivalent sein.*

Um die Untersuchung später nicht unterbrechen zu müssen, schicke ich noch folgende Bemerkung voraus.

Es sei

$$\phi(x) = x^r - c_1 x^{r-1} - c_2 x^{r-2} - \cdots - c_r = 0$$

eine Gleichung mit Koeffizienten aus dem Körper Ω, die in Ω irreduzibel ist; die Wurzeln der Gleichung seien $\rho = \rho_0, \rho_1, \rho_2, \cdots, \rho_{r-1}$. Setzt man

$$R = \begin{bmatrix} c_1 & c_2 & \cdots & c_{r-1} & c_r \\ 1 & 0 & \cdots & 0 & 0 \\ 0 & 1 & \cdots & 0 & 0 \\ \cdot & \cdot & \cdot & \cdot & \cdot \\ 0 & 0 & \cdots & 1 & 0 \end{bmatrix}, \quad P = \begin{bmatrix} \rho_0 & 0 & \cdots & 0 \\ 0 & \rho_1 & \cdots & 0 \\ \cdot & \cdot & \cdot & \cdot \\ 0 & 0 & \cdots & \rho_{r-1} \end{bmatrix}$$

und

$$Q = \begin{bmatrix} \rho_0^{r-1} & \rho_1^{r-1} & \cdots & \rho_{r-1}^{r-1} \\ \rho_0^{r-2} & \rho_1^{r-2} & \cdots & \rho_{r-1}^{r-2} \\ \cdot & \cdot & \cdot & \cdot \\ 1 & 1 & \cdots & 1 \end{bmatrix},$$

so wird, wie man leicht erkennt,

$$RQ = \begin{bmatrix} \rho_0^{r} & \rho_1^{r} & \cdots & \rho_{r-1}^{r} \\ \rho_0^{r-1} & \rho_1^{r-1} & \cdots & \rho_{r-1}^{r-1} \\ \cdot & \cdot & \cdot & \cdot \\ \rho_0 & \rho_1 & \cdots & \rho_{r-1} \end{bmatrix} = QP$$

Also ist, da die Determinante der Matrix Q nicht 0 ist,

(2) $$Q^{-1}RQ = P.$$

Hieraus folgt sofort, dass R der Gleichung

(3) $$R^r - c_1 R^{r-1} - c_2 R^{r-2} - \cdots - c_r E = 0$$

genügt, wo E die Einheitsmatrix bedeutet. — Man habe nun eine Gruppe $\mathfrak{F}$ linearer Substitutionen des Grades f, die im Körper $\Omega(\rho)$ rational ist. Es sei

$$F = \{\phi_{\kappa\lambda}(\rho)\}$$

die Koeffizientenmatrix irgend einer Substitution von $\mathfrak{F}$. Ersetzt man in jeder der in Ω rationalen Funktionen $\phi_{\kappa\lambda}(\rho)$ die Zahl ρ durch die zu ρ konjugierten Zahlen $\rho_1, \rho_2, \cdots, \rho_{r-1}$, so mögen die (zu $\mathfrak{F}$ isomorphen) Gruppen $\mathfrak{F}_1, \mathfrak{F}_2, \cdots, \mathfrak{F}_{r-1}$ entstehen. Substituiert man weiter in $\phi_{\kappa\lambda}(\rho)$ für ρ die Matrix R, so bilden auch die Matrizen

$$\bar{F} = \{\phi_{\kappa\lambda}(R)\},$$

deren Grad gleich rf ist, eine mit $\mathfrak{F}$ isomorphe Gruppe $\bar{\mathfrak{F}}$. Dies folgt daraus, dass jede für ρ bestehende Relation mit Koeffizienten aus dem Körper Ω wegen der Gleichung (3) richtig bleibt, wenn die Zahl ρ durch die Matrix R ersetzt wird.* Ist nun

$$T = \begin{vmatrix} Q & 0 & \cdots & 0 \\ 0 & Q & \cdots & 0 \\ . & . & . & . \\ 0 & 0 & \cdots & Q \end{vmatrix}$$

die Matrix des Grades rf, die durch f-malige Aneinanderreihung der Matrix Q entsteht, so wird $T^{-1}\overline{\mathfrak{F}}\,T$, wie aus (2) leicht folgt, eine Gruppe, die durch eine einfache Vertauschung der Zeilen und Spalten in die Gruppe

$$\mathfrak{F}' = \begin{vmatrix} \mathfrak{F} & 0 & \cdots & 0 \\ 0 & \mathfrak{F}_1 & \cdots & 0 \\ . & . & . & . \\ 0 & 0 & \cdots & \mathfrak{F}_{r-1} \end{vmatrix}$$

übergeht. *Daher ist die Gruppe $\mathfrak{F}'$ mit der im Körper Ω rationalen Gruppe $\overline{\mathfrak{F}}$ äquivalent.*†

§ 2.

Ist $\mathfrak{G}$ eine im Bereiche Z aller Zahlen irreduzible Gruppe linearer Substitutionen, so hat bekanntlich jede Matrix P, die mit allen Substitutionen (Matrizen) von $\mathfrak{G}$ vertauschbar ist,‡ die Form cE, wo c eine Konstante ist. Weiss man aber umgekehrt, dass jede mit $\mathfrak{G}$ vertauschbare Matrix die Form cE hat, so folgt im allgemeinen noch keineswegs, dass $\mathfrak{G}$ im Bereiche Z irreduzibel ist. Es gilt jedoch der Satz:

II. *Ist $\mathfrak{G}$ eine in einem gegebenen Zahlkörper Ω irreduzible Gruppe und weiss man, dass jede mit $\mathfrak{G}$ vertauschbare Matrix die Form cE hat, so ist $\mathfrak{G}$ auch im Bereiche Z aller Zahlen irreduzibel.*

Es sei nämlich $\mathfrak{G}$ in Z reduzibel. Dann kann man bekanntlich, wenn g den Grad von $\mathfrak{G}$ bedeutet, jedenfalls g^2 Konstanten $k_{\alpha\beta}$, die nicht sämmtlich Null sind, bestimmen, sodass für jede Substitution $A = (a_{\alpha\beta})$ von $\mathfrak{G}$ die Gleichung

$$\sum_{\alpha,\beta} k_{\beta\alpha} a_{\alpha\beta} = 0 \tag{4}$$

besteht. Ich will nun zeigen, dass unter den über $\mathfrak{G}$ gemachten Voraussetzungen aus dem Bestehen der Relationen (4) auch die Reduzibilität von $\mathfrak{G}$ inbezug auf Ω geschlossen werden kann, was dann auf einen Widerspruch führt.

Der folgende Beweis ist eine fast wörtliche Nachbildung des Beweises, den Herr FROBENIUS und der Verfasser in der bereits zitierten Arbeit für den

* Vergl. A., p. 172.

† Vergl. L. E. DICKSON, *On the reducibility of linear groups*, Transactions of the American Mathematical Society, vol. 4 (1903), p. 434.

‡ Im folgenden sage ich kurz: die Matrix P ist mit der Gruppe $\mathfrak{G}$ vertauschbar.

bekannten BURNSIDE'schen Fundamentalsatz über irreduzible Gruppen angegeben haben. Nur der grösseren Deutlichkeit wegen soll der Beweis hier ausführlich mitgeteilt werden.

Zunächst ist zu beachten, dass die Konstanten $k_{\alpha\beta}$, da sie sich aus linearen Gleichungen mit Koeffizienten aus dem Körper Ω bestimmen, auch als Zahlen des Körpers Ω angenommen werden können. Man denke sich nun alle in Ω rationalen Matrizen $K = (k_{\alpha\beta})$ ins Auge gefasst, deren Koeffizienten $k_{\alpha\beta}$ den Gleichungen (4) genügen. Unter ihnen seien höchstens s linear unabhängig, etwa $K_1, K_2, \cdots, K_s$. Dann lässt sich jede andere solche Matrix K als lineare homogene Verbindung von $K_1, K_2, \cdots, K_s$ mit in Ω rationalen Koeffizienten darstellen. Nun schliesst man aber sofort, dass auch KA dieselbe Eigenschaft besitzt wie K; daher muss sich KA auf die Form $\sum_\sigma r_\sigma K_\sigma$ bringen lassen, wo die r_σ gewisse Zahlen des Körpers Ω bedeuten. Speziell sei

$$K_\rho A = \sum_{\sigma=1}^{s} r_{\rho\sigma} K_\sigma \qquad (\rho = 1, 2, \cdots, s).$$

Die Matrizen $R = (r_{\rho\sigma})$ bilden dann eine in Ω rationale Gruppe $\mathfrak{R}$, die mit $\mathfrak{G}$ homomorph ist. Diese Gruppe $\mathfrak{R}$ kann auch inbezug auf Ω reduzibel sein. In diesem Fall lassen sich $t < s$ linear unabhängige lineare Verbindungen $L_1, L_2, \cdots, L_t$ der K_σ mit in Ω rationalen Koeffizienten angeben, sodass

$$L_\rho A = \sum_{\sigma=1}^{t} s_{\rho\sigma} L_\sigma \qquad (\rho = 1, 2, \cdots, t) \tag{5}$$

wird und die (in Ω rationalen) Matrizen $S = (s_{\rho\sigma})$ eine inbezug auf Ω **irreduzible** Gruppe $\mathfrak{S}$ bilden.—Ist aber $\mathfrak{R}$ in Ω irreduzibel, so denken wir uns $t = s$ und $\mathfrak{S} = \mathfrak{R}$ gesetzt.—Bezeichnet man nun, wenn $L_\tau = (l^{(\rho)}_{\alpha\beta})$ ist, mit P_α die Matrix

$$P_\alpha = \begin{pmatrix} l^{(1)}_{\alpha 1} & l^{(1)}_{\alpha 2} & \cdots & l^{(1)}_{\alpha g} \\ \cdot & \cdot & \cdot & \cdot \\ l^{(t)}_{\alpha 1} & l^{(t)}_{\alpha 2} & \cdots & l^{(t)}_{\alpha g} \end{pmatrix},$$

so kann man die Gleichungen (5) auch in der Form

$$P_\alpha A = S P_\alpha \tag{6}$$

schreiben. Die P_α sind aber gewiss nicht alle gleich Null, da sonst auch die L_ρ alle Null wären. Folglich sind $\mathfrak{G}$ und $\mathfrak{S}$ verkettete Gruppen, und da sie beide in Ω irreduzibel sind, so müssen sie nach Satz I einander äquivalent sein. Durch passende Wahl von $L_1, L_2, \cdots, L_t$ kann daher auch erreicht werden, dass $S = A$ wird. Die Gleichungen (6) gehen dann über in die Gleichungen $P_\alpha A = A P_\alpha$, d. h. P_α wird mit $\mathfrak{G}$ vertauschbar. Es muss daher auf Grund unserer Annahme über $\mathfrak{G}$ die Matrix P_α die Form $k_\alpha E$ haben; es wird also

$$l^{(\gamma)}_{\alpha\beta} = k_\alpha e_{\gamma\beta} \qquad (\alpha, \beta, \gamma = 1, 2, \cdots, g),$$

wo $e_{\gamma\beta}$ gleich 1 oder gleich 0 zu setzen ist, je nachdem $\gamma = \beta$ oder $\gamma \neq \beta$ ist. Es soll doch aber

$$\sum_{\alpha,\beta} l^{(\gamma)}_{\beta\alpha} a_{\alpha\beta} = 0$$

sein; folglich wird

$$\sum_{\beta} a_{\gamma\beta} k_\beta = 0.$$

Wären nun die Konstanten k_β, wie das unsere Annahme, dass die L_γ linear unabhängig sein sollen, erfordert, nicht sämtlich 0, so würden diese Gleichungen besagen, dass die Gruppe $\mathfrak{G}$ mit der ihr homomorphen Gruppe, die aus lauter Nullen besteht, verkettet ist. Dies ist aber nach Satz I nicht möglich.

§ 3.

Allgemein gilt der Satz:

III. *Ist $\mathfrak{G}$ eine im Körper Ω irreduzible Gruppe des Grades g und P eine mit $\mathfrak{G}$ vertauschbare Matrix, deren Koeffizienten dem Körper Ω angehören, so muss die charakteristische Determinante $|xE-P|$ der Matrix P Potenz einer in Ω irreduziblen Funktion $\phi(x)$ sein und es besteht die Gleichung $\phi(P)=0$.*

Dieser Satz ist dem Satz II meiner mit A. zitierten Arbeit völlig analog und ist ebenso wie dort zu beweisen.

Nun ist die Bedingung dafür, dass die Matrix $P=(p_{\alpha\beta})$ mit der Gruppe $\mathfrak{G}$ vertauschbar sein soll, identisch mit einer Reihe von linearen homogenen Gleichungen für die $p_{\alpha\beta}$ mit Koeffizienten aus dem Körper Ω. Denkt man sich daher die allgemeinste mit $\mathfrak{G}$ vertauschbare Matrix P bestimmt, so erscheinen die $p_{\alpha\beta}$ als lineare homogene Verbindungen von gewissen s Parametern $v_1, v_2, \cdots, v_s$ mit in Ω rationalen Koeffizienten. Sind nun unter den charakteristischen Wurzeln dieser allgemeinsten Matrix P genau r unter einander verschieden, so lassen nach einem bekannten Satz der Algebra für die Parameter v_α auch spezielle rationale Werte einsetzen, so dass die so entstehende spezielle Matrix P ebenfalls genau r verschiedene charakteristische Wurzeln besitzt. Da aber P alsdann eine in Ω rationale Matrix wird, so muss nach Satz III

$$|xE-P| = [\phi(x)]^f$$

sein, wo $\phi(x)$ eine in Ω irreduzible Funktion des Grades r bedeutet und $f=g/r$ zu setzen ist. Ist nun $r=1$, so folgt aus der Gleichung $\phi(P)=0$, dass P die Form cE hat und man schliesst auch sofort, dass die allgemeinste mit $\mathfrak{G}$ vertauschbare Matrix ebenfalls diese Form haben muss. In diesem Fall ist daher nach Satz II die Gruppe $\mathfrak{G}$ auch im Bereiche Z aller Zahlen irreduzibel. Ist aber $r>1$ und

$$\phi(x) = (x-\rho)(x-\rho_1)\cdots(x-\rho_{r-1}),$$

so folgt aus der Gleichung $\phi(P)=0$, weil die ρ_a unter einander verschieden sind, in bekannter Weise, dass die Elementarteiler von $|xE-P|$ sämtlich linear sind.* Hieraus schliesst man leicht, dass sich eine Matrix R von nicht verschwindender Determinante mit Koeffizienten, die dem Körper $\Omega(\rho)$ angehören, bestimmen lässt, so dass

$$P = RPR^{-1} = \begin{pmatrix} M & 0 \\ 0 & N \end{pmatrix}$$

wird; hierin soll $M = \rho E_f$ sein, wo E_f die Einheitsmatrix des Grades f bedeutet, während unter N eine gewisse in $\Omega(\rho)$ rationale Matrix des Grades $rf-f$ zu verstehen ist, unter deren charakteristischen Wurzeln alsdann nur die Grössen $\rho_1, \rho_2, \cdots, \rho_{r-1}$ (jede f Mal) vorkommen. Man führe nun an Stelle von $\mathfrak{G}$ die ihr äquivalente, in $\Omega(\rho)$ rationale Gruppe

$$\mathfrak{G}' = R\mathfrak{G}R^{-1}$$

ein. Da nun $\mathfrak{G}'$ mit der Matrix P' vertauschbar ist und die Matrizen M und N keine charakteristische Wurzel gemeinsam haben, so ergiebt sich nach einem bekannten Satz, dass $\mathfrak{G}'$ die Form

$$\mathfrak{G}' = \begin{pmatrix} \mathfrak{F} & 0 \\ 0 & \mathfrak{H} \end{pmatrix}$$

haben muss, wo $\mathfrak{F}$ insbesondere eine Gruppe des Grades f ist. Es mögen nun $\mathfrak{F}_1, \mathfrak{F}_2, \cdots, \mathfrak{F}_{r-1}, \overline{\mathfrak{F}}$ und $\mathfrak{F}'$ dieselbe Bedeutung haben wie in § 1. Dann sind (vergl. § 1) $\mathfrak{G}'$ and $\mathfrak{F}'$ unter einander verkettete Gruppen. Daher sind auch die ihnen äquivalenten in Ω rationalen Gruppen $\mathfrak{G}$ und $\overline{\mathfrak{F}}$ verkettet; da aber $\mathfrak{G}$ in Ω irreduzibel und $\overline{\mathfrak{F}}$ von gleichem Grade $g = rf$ wie $\mathfrak{G}$ ist, so müssen $\mathfrak{G}$ und $\overline{\mathfrak{F}}$ nach Satz I* äquivalente Gruppen sein. Hieraus folgt aber, dass $\mathfrak{G}$ auch der Gruppe

$$\mathfrak{F}' = \begin{bmatrix} \mathfrak{F} & 0 & \cdots & 0 \\ 0 & \mathfrak{F}_1 & \cdots & 0 \\ \cdot & \cdot & \cdot & \cdot \\ 0 & 0 & \cdots & \mathfrak{F}_{r-1} \end{bmatrix}$$

äquivalent ist.

Die Gruppe $\mathfrak{F}$ ist jedenfalls im Körper $\Omega(\rho)$ irreduzibel.† Denn wäre dies nicht der Fall, so würde man sofort schliessen können, dass die aus $\mathfrak{F}$ hervorgehende in Ω rationale Gruppe $\overline{\mathfrak{F}}$ inbezug auf Ω reduzibel ist; dies ist aber nicht möglich, da $\overline{\mathfrak{F}}$ mit $\mathfrak{G}$ äquivalent ist. Es ist aber $\mathfrak{F}$ *auch im Körper Z aller Zahlen irreduzibel.* Denn wäre die im Körper $\Omega(\rho)$ irreduzible Gruppe $\mathfrak{F}$ in Z reduzibel, so müsste sich nach dem Vorhergehenden eine mit $\mathfrak{F}$ vertauschbare Matrix L angeben lassen, die mindestens zwei verschiedene charak-

* Vergl. A., p. 172.

† Vergl. L. E. Dickson, loc. cit.

teristische Wurzeln w und w' besitzt. Versteht man dann unter $w_1, w_2, \cdots, w_{r-1}$ irgend welche $r-1$ Grössen, die unter einander und auch von w und w' verschieden sind, und setzt man

$$L_1 = w_1 E_f, \ L_2 = w_2 E_f, \ \cdots, \ L_{r-1} = w_{r-1} E_f,$$

so wird die Matrix

$$P_1 = \begin{pmatrix} L & 0 & \cdots & 0 \\ 0 & L_1 & \cdots & 0 \\ \cdot & \cdot & \cdot & \cdot \\ 0 & 0 & \cdots & L_{r-1} \end{pmatrix}$$

mit der Gruppe $\mathfrak{F}'$ vertauschbar. Eine zu P_1 ähnliche Matrix P wäre dann mit der Gruppe $\mathfrak{G}$ vertauschbar. Diese Matrix würde jedoch ebenso wie P_1 mindestens $r+1$ verschiedene charakteristische Wurzeln $w, w', w_1, \cdots, w_{r-1}$ besitzen. Dies widerspricht aber der Annahme, die über die Zahl r gemacht worden ist. — Ebenso zeigt man, dass auch die Gruppen $\mathfrak{F}_1, \mathfrak{F}_2, \cdots, \mathfrak{F}_{r-1}$ im Bereiche Z irreduzibel sind.

Wir erhalten den Satz:

IV. *Jede in einem Körper Ω irreduzible Gruppe $\mathfrak{G}$ linearer Substitutionen ist im Bereich aller Zahlen vollständig reduzibel. Hat die allgemeinste mit $\mathfrak{G}$ vertauschbare Matrix genau r verschiedene charakteristische Wurzeln, so zerfällt $\mathfrak{G}$ im Bereich aller Zahlen in r irreduzible Gruppen desselben Grades. Diese r Gruppen lassen sich auch in r konjugierten algebraischen Körpern über Ω rational darstellen.*

Ferner gilt der Satz:

V. *Zwei in einem Körper Ω irreduzible Gruppen $\mathfrak{G}$ und $\mathfrak{G}_1$, die entweder isomorph oder einer dritten Gruppe homomorph sind, sind dann und nur dann äquivalent, wenn sie im Bereiche Z aller Zahlen einen irreduziblen Bestandteil gemeinsam haben.*

Denn enthalten $\mathfrak{G}$ und $\mathfrak{G}_1$ in Z einen irreduziblen Bestandteil gemeinsam, so sind sie als vollständig reduzible Gruppen unter einander verkettet und folglich nach Satz I* auch äquivalent.

§ 4.

Die zu der Gruppe $\mathfrak{G}$ gehörenden im Bereiche Z irreduziblen Gruppen $\mathfrak{F}, \mathfrak{F}_1, \cdots, \mathfrak{F}_{r-1}$ brauchen keineswegs in dem Sinne von einander verschieden zu sein, dass nicht zwei von ihnen einander äquivalent sind. Es gilt aber der Satz:

VI. *Es möge die im Körper Ω irreduzible Gruppe $\mathfrak{G}$ im Bereiche Z aller Zahlen in die irreduziblen Gruppen $\mathfrak{F}, \mathfrak{F}_1, \cdots, \mathfrak{F}_{r-1}$ zerfallen. Sind dann unter diesen Gruppen genau l vorhanden, von denen nicht zwei einander äquivalent sind, so ist $r/l = m$ eine ganze Zahl und es sind unter den Gruppen $\mathfrak{F}, \mathfrak{F}_1, \cdots, \mathfrak{F}_{r-1}$ je m einander äquivalent.*

Dies ergiebt sich fast unmittelbar aus dem Satz, der besagt, dass zwei in Z irreduzible einander isomorphe Gruppen dann und nur dann einander äquivalent sind, wenn je zwei einander entsprechende Substitutionen der beiden Gruppen dieselbe Spur besitzen.* — Es mögen nämlich die Spuren der Substitutionen der Gruppe $\mathfrak{F}$ zusammen mit den Zahlen von Ω einen Körper Ω' erzeugen. Da nun $\mathfrak{F}$ im Körper $\Omega(\rho)$ rational darstellbar ist, so ist Ω' jedenfalls ein Teilkörper von $\Omega(\rho)$ und daher bekanntlich selbst ein algebraischer Körper über Ω. Es möge etwa Ω' aus Ω durch Adjunktion der Grösse $\chi = \chi(\rho)$ von $\Omega(\rho)$ hervorgehen, so dass also $\Omega' = \Omega(\chi)$ wird. Dann erzeugen die Spuren der Substitutionen der Gruppe $\mathfrak{F}_\lambda$, die aus $\mathfrak{F}$ durch die Permutation $\rho | \rho_\lambda$ hervorgeht, zusammen mit den Zahlen von Ω den durch die Zahl $\chi_\lambda = \chi(\rho_\lambda)$ bestimmten Körper $\Omega(\chi_\lambda)$. Auf Grund des erwähnten Satzes über die Äquivalenz zweier irreduzibler Gruppen schliesst man sofort, dass $\mathfrak{F}_\kappa$ und $\mathfrak{F}_\lambda$ dann und nur dann äquivalente Gruppen sind, wenn $\chi_\kappa = \chi_\lambda$ wird. Genügt aber χ im Körper Ω einer irreduziblen Gleichung des Grades l, so ist bekanntlich l ein Divisor des Grades r des algebraischen Körpers $\Omega(\rho)$ über Ω und es sind unter den Zahlen $\chi, \chi_1, \cdots, \chi_{r-1}$ genau l von einander verschieden und je $m = r/l$ einander gleich. Hieraus folgt aber unser Satz.

Man beweist ferner ganz ähnlich wie bei der analogen Betrachtung A., p. 173:

VII. *Lässt sich die Gruppe $\mathfrak{F}$ in einem algebraischen Körper $\Omega(\sigma)$ des Grades s über Ω rational darstellen, so muss s durch r teilbar sein.*

Der früher bestimmte Körper $\Omega(\rho)$ ist also ein algebraischer Körper *kleinsten Grades* über Ω, in dem die Gruppe $\mathfrak{F}$ rational darstellbar ist.

Ebenso wie für endliche Gruppen (vergl. A., p. 171) gilt auch allgemein der Satz:

VIII. *Die Zahl $m = r/l$ ist ein Divisor des Grades f der Gruppe $\mathfrak{F}$.*

Der Beweis lässt sich mit Hilfe einer allgemeinen Satzes führen, der auch an und für sich von Wichtigkeit ist:

IX. *Man habe eine Gruppe $\mathfrak{H}$ linearer Substitutionen des Grades h, die im Bereiche Z aller Zahlen vollständig reduzibel ist. Unter den irreduziblen Bestandteilen von $\mathfrak{H}$ mögen genau l vorhanden sein, von denen nicht zwei einander äquivalent sind, etwa die Gruppen $\mathfrak{F}_0, \mathfrak{F}_1, \cdots, \mathfrak{F}_{l-1}$; hierbei soll keine dieser Gruppen aus lauter Nullen bestehen. Ist dann f_λ der Grad der Gruppe $\mathfrak{F}_\lambda$, so sind unter den Substitutionen von $\mathfrak{H}$ genau*

$$n = f_0^2 + f_1^2 + \cdots + f_{l-1}^2$$

linear unabhängig. Genügen etwa die Substitutionen $H_1, H_2, \cdots, H_n$ von $\mathfrak{H}$ dieser Bedingung, so sei $H = z_1 H_1 + z_2 H_2 + \cdots + z_n H_n$ die allgemeine Substitution von $\mathfrak{H}$ und speziell sei

* G. Frobenius und I. Schur, loc. cit., Satz II.

(7) $$H_\alpha H_\beta = \sum_{\gamma=1}^{n} a_{\gamma\alpha\beta} H_\gamma .$$

Bezeichnet man dann mit S_α die Matrix

$$S_\alpha = (a_{\kappa\alpha\lambda}) \qquad (\kappa, \lambda = 1, 2, \cdots, n),$$

so bilden die Matrizen $S = z_1 S_1 + z_2 S_2 + \cdots + z_n S_n$ eine der Gruppe $\mathfrak{H}$ isomorphe Gruppe $\mathfrak{S}$, die ebenfalls vollständig reduzibel ist. Unter den irreduziblen Bestandteilen von $\mathfrak{S}$ kommen wieder nur die l Gruppen $\mathfrak{F}_0, \mathfrak{F}_1, \cdots, \mathfrak{F}_{l-1}$ vor, und zwar enthält $\mathfrak{S}$ die Gruppe $\mathfrak{F}_\lambda$ genau f_λ Mal.

Auf die Tatsache, dass die Gruppe $\mathfrak{S}$ zugleich mit der Gruppe $\mathfrak{H}$ vollständig reduzibel ist, hat bereits Herr TABER in seiner in der Einleitung zitierten Arbeit aufmerksam gemacht. Was hier also hinzugefügt wird, ist nur die Bestimmung der irreduziblen Bestandteile von $\mathfrak{S}$. Ich bemerke noch, dass unser Satz für den besonderen Fall, dass die Gruppe $\mathfrak{H}$ irreduzibel ist, bereits von Herrn BURNSIDE.* bewiesen worden ist.

Der Beweis des Satzes IX ergiebt sich folgendermassen.

Es möge der Substitution H_α von $\mathfrak{H}$ in $\mathfrak{F}_\lambda$ die Substitution $F_\alpha^{(\lambda)}$ entsprechen. Sind dann $x_1, x_2, \cdots, x_n$ unabhängige Variable und setzt man

$$X^{(\lambda)} = x_1 F_1^{(\lambda)} + x_2 F_2^{(\lambda)} + \cdots + x_n F_n^{(\lambda)},$$

so sind, weil unter den irreduziblen Gruppen $\mathfrak{F}_0, \mathfrak{F}_1, \cdots, \mathfrak{F}_{l-1}$ nicht zwei einander äquivalent sind, die $f_0^2 + f_1^2 + \cdots + f_{l-1}^2$ Koeffizienten der l Matrizen $X^{(0)}, X^{(1)}, \cdots, X^{(l-1)}$, als Funktionen der x_ν betrachtet, linear unabhängig.† Andererseits sind unter den h^2 Koeffizienten der Matrix

$$x_1 H_1 + x_2 H_2 + \cdots + x_n H_n$$

genau n linear unabhängig. Hieraus schliesst man wegen der über $\mathfrak{H}$ gemachten Annahmen sofort, dass $n = f_0^2 + f_1^2 + \cdots + f_{l-1}^2$ sein muss. Man erkennt auch leicht, dass das durch die Gleichungen (7) definierte System hyperkomplexer Grössen nach der von Herrn FROBENIUS‡ eingeführten Bezeichnungsweise ein DEDEKIND'scher System ist, d. h. es muss die Determinante D der n^2 Grössen

$$p_{\alpha\beta} = \sum_{\kappa, \lambda} a_{\kappa\lambda\kappa} a_{\lambda\alpha\beta} \qquad (\alpha, \beta = 1, 2, \cdots, n)$$

* *On the arithmetical nature of the coefficients in a group of linear substitutions of finite order* (second paper), Proceedings of the London Mathematical Society, ser. 2, vol. 4 (1906), p. 1.

† Vergl. G. FROBENIUS und I. SCHUR, loc. cit., Satz I.

‡ *Theorie der hyperkomplexen Grössen*, Sitzungsberichte der Berliner Akademie, 1903, p. 504 und p. 634.

von Null verschieden sein.* Daher ist die durch die Matrizen

$$x_1 S_1 + x_2 S_2 + \cdots + x_n S_n$$

gebildete Gruppe $\mathfrak{S}$ eine vollständig reduzible Gruppe, die jeden ihrer irreduziblen Bestandteile genau so oft enthält wie sein Grad angiebt.† Dasselbe gilt natürlich auch für die mit $\mathfrak{H}$ isomorphe Gruppe $\mathfrak{S}$. Nun ist aber auch

$$F_\alpha^{(\lambda)} F_\beta^{(\lambda)} = \sum_\gamma a_{\gamma\alpha\beta} F_\gamma^{(\lambda)}. \tag{8}$$

Setzt man

$$F_\alpha^{(\lambda)} = (k_{\mu\nu}^{(\alpha)}) \qquad (\mu, \nu = 1, 2, \cdots, f_\lambda)$$

und bezeichnet mit P_ν die Matrix

$$P_\nu = \begin{vmatrix} k_{1\nu}^{(1)} & k_{1\nu}^{(2)} & \ldots & k_{1\nu}^{(n)} \\ . & . & . & . \\ k_{f_\lambda \nu}^{(1)} & k_{f_\lambda \nu}^{(2)} & \ldots & k_{f_\lambda \nu}^{(n)} \end{vmatrix},$$

so lassen sich die Gleichungen (8) auch in der Form

$$F_\alpha^{(\lambda)} P_\nu = P_\nu S_\alpha$$

schreiben. Da nun die P_ν nicht sämtlich 0 sein können, so besagen diese Gleichungen (vergl. § 1), dass die Gruppe $\mathfrak{S}$ mit der Gruppe $\mathfrak{F}_\lambda$ verkettet ist. Nach Satz I muss daher $\mathfrak{F}_\lambda$ ein irreduzibler Bestandteil von $\mathfrak{S}$ sein und ist nach dem Gesagten genau f_λ Mal in $\mathfrak{S}$ enthalten.‡ Da dies für jedes λ gilt, so erhält man $f_0 + f_1 + \cdots + f_{l-1}$ irreduzible Bestandteile von $\mathfrak{S}$, die aneinander gereiht eine Gruppe des Grades $f_0^2 + f_1^2 + \cdots + f_{l-1}^2 = n$ ergeben. Da aber n der Grad von $\mathfrak{S}$ ist, so kann $\mathfrak{S}$ keinen weiteren irreduziblen Bestandteil enthalten. §

Ist nun speziell $\mathfrak{H} = \mathfrak{G}$, wie früher, eine im Körper Ω irreduzible Gruppe, so wird $f_0 = f_1 = \cdots = f_{l-1} = f$, also $n = lf^2$. Ferner gehören die Zahlen $a_{\gamma\alpha\beta}$ in diesem Fall ebenfalls den Körper Ω an. Denkt man sich nun die in Ω ratio-

* Vergl. TABER, loc. cit. — Herr TABER hat noch die wichtige Bemerkung hinzugefügt, dass für jede beliebige Gruppe $\mathfrak{H}$ die Determinante D dann und nur dann verschwindet, wenn die mit Hilfe der Spuren $q_{\alpha\beta}$ der Matrizen $H_\alpha H_\beta$ gebildete Determinante $\Delta = |q_{\alpha\beta}|$ gleich Null ist. Es lässt sich auch, wie bei dieser Gelegenheit hervorgehoben sei, zeigen, dass *die Determinanten D und Δ stets denselben Rang besitzen.*

† Dass die zu einem System hyperkomplexer Grössen mit nicht verschwindender Determinante D gehörende Gruppe $\mathfrak{S}$ diese Eigenschaft besitzt, hat zuerst Herr MOLIEN in seiner Arbeit *Über Systeme höherer complexer Zahlen*, Mathematische Annalen, Bd. 41 (1893), p. 83, gezeigt. Vergl. auch CARTAN, *Sur les groupes bilinéaires et les systèmes de nombres complexes*, Annales de Toulouse, t. XII (1898), p. 1 und FROBENIUS, loc. cit.

‡ Vergl. auch FROBENIUS, loc. cit., p. 536.

§ Erzeugen die Spuren der Substitutionen der (vollständig reduziblen) Gruppe $\mathfrak{H}$ den Rationalitätsbereich Ω', so gehören auch die Zahlen $a_{\gamma\alpha\beta}$ dem Körper Ω' an (vergl. BURNSIDE, loc. cit., p. 4). Auf Grund dieser Bemerkung kann man weiter schliessen, dass die Gruppe $\mathfrak{H}$ in einem algebraischen Körper über Ω' rational darstellbar ist.

nale Gruppe $\mathfrak{S}$ inbezug auf Ω in irreduzible Bestandteile $\mathfrak{G}_1, \mathfrak{G}_2, \cdots, \mathfrak{G}_p$ zerlegt, so muss jede dieser Gruppen der Gruppe $\mathfrak{G}$ äquivalent sein. Denn wäre dies für die Gruppe $\mathfrak{G}_\lambda$ nicht der Fall, so wären nach Satz V die irreduziblen Bestandteile von $\mathfrak{G}_\lambda$ inbezug auf den Körper Z aller Zahlen von denjenigen der Gruppe $\mathfrak{G}$ wesentlich verschieden, folglich müsste auch die Gruppe $\mathfrak{S}$, im Körper Z zerlegt, einen irreduziblen Bestandteil enhalten, der keiner der Gruppen $\mathfrak{F}, \mathfrak{F}_1, \cdots, \mathfrak{F}_{r-1}$ äquivalent ist. Dies widerspricht aber dem soeben Bewiesenen. Daher ist

$$n = lf^2 = pg = pmlf,$$

und hieraus folgt, dass die Zahl m in der Tat ein Divisor der Zahl f ist.

§ 5.

Es sei nun $\mathfrak{K}$ eine beliebige im Körper Ω rationale Gruppe linearer Substitutionen des Grades k. Diese Gruppe denken wir uns zunächst im Körper Ω in irreduzible Bestandteile zerlegt, es sei etwa

$$\begin{bmatrix} \mathfrak{G}_{11} & 0 & \cdots & 0 \\ \mathfrak{G}_{21} & \mathfrak{G}_{22} & \cdots & 0 \\ \cdot & \cdot & \cdot & \cdot \\ \mathfrak{G}_{p1} & \mathfrak{G}_{p2} & \cdots & \mathfrak{G}_{pp} \end{bmatrix}$$

ein in Ω rationale, mit $\mathfrak{K}$ äquivalente Gruppe, wo $\mathfrak{G}_{11}, \mathfrak{G}_{22}, \cdots, \mathfrak{G}_{pp}$ in Ω irreduzibel sind. Es möge ferner die Gruppe $\mathfrak{G}_{\alpha\alpha}$, im Bereiche Z aller Zahlen zerlegt, die irreduziblen Bestandteile $\mathfrak{F}_{\alpha 1}, \mathfrak{F}_{\alpha 2}, \cdots, \mathfrak{F}_{\alpha r_\alpha}$ aufweisen. Dann repräsentieren die sich so ergebenden $r_1 + r_2 + \cdots + r_p$ Gruppen $\mathfrak{F}_{\alpha\beta}$ in ihrer Gesamtheit die irreduziblen Bestandteile der Gruppe $\mathfrak{K}$. Auf Grund der früher gewonnenen Resultate kann man nun den Satz aussprechen :

X. *Man habe eine Gruppe $\mathfrak{K}$ linearer homogener Substitutionen, die in einem Zahlkörper Ω rational ist. Ist dann $\mathfrak{F}$ einer der irreduziblen Bestandteile, in die $\mathfrak{K}$ im Bereiche aller Zahlen zerfällt, so erzeugen die Spuren der Substitutionen von $\mathfrak{F}$ zusammen mit den Zahlen von Ω einen algebraischen Körper über Ω, ferner lässt sich $\mathfrak{F}$ in einem algebraischen Körper $\Omega(\rho)$ über Ω rational darstellen. Wählt man den Grad r des Körpers $\Omega(\rho)$ möglichst klein, so seien $\rho_1, \rho_2, \cdots, \rho_{r-1}$ die zu ρ konjugierten Zahlen und es möge $\mathfrak{F}$ in $\mathfrak{F}_\lambda$ übergehen, wenn in allen Substitutionskoeffizienten von $\mathfrak{F}$ die Zahl ρ durch ρ_λ ersetzt wird. Dann ist die Gruppe*

$$\mathfrak{F} = \begin{bmatrix} \mathfrak{F} & 0 & \cdots & 0 \\ 0 & \mathfrak{F}_1 & \cdots & 0 \\ \cdot & \cdot & \cdot & \cdot \\ 0 & 0 & \cdots & \mathfrak{F}_{r-1} \end{bmatrix}$$

einer in Ω rationalen und irreduziblen Gruppe $\mathfrak{G}$ äquivalent, die in $\mathfrak{K}$ als irreduzibler Bestandteil inbezug auf Ω enthalten ist.

Ich stelle nun folgende Betrachtung an. Es sei

$$(K) \qquad \bar{x}_\kappa = \sum_{\lambda=1}^{k} c_{\kappa\lambda} x_\lambda$$

eine beliebige Substitution von $\mathfrak{K}$. Ist $\phi(x_1, x_2, \cdots, x_n)$ irgend eine homogene Form der Variabeln $x_1, x_2, \cdots, x_n$, so möge

$$\phi(\bar{x}_1, \bar{x}_2, \cdots, \bar{x}_n) = \bar{\phi}(x_1, x_2, \cdots, x_n)$$

gesetzt werden. Man denke sich nun eine der Gruppe $\mathfrak{K}$ homomorphe Gruppe $\mathfrak{A}$ des Grades a gegeben, in der der Substitution K von $\mathfrak{K}$ die Substitution $A = (a_{\alpha\beta})$ entsprechen möge. Es kann dann eintreten, dass sich gewisse a, nicht notwendig linear unabhängige, Formen $\phi_1, \phi_2, \cdots, \phi_a$ gleicher Ordnung angeben lassen, so dass

$$(9) \qquad \bar{\phi}_\alpha = \sum_{\beta=1}^{a} a_{\alpha\beta} \phi_\beta$$

wird. Ist dies für jede Substitution K von $\mathfrak{K}$ der Fall, so will ich sagen: *$\phi_1, \phi_2, \cdots, \phi_a$ bilden ein zur Gruppe $\mathfrak{A}$ gehörendes Invariantensystem von $\mathfrak{K}$.* Sind insbesondere die Formen $\phi_1, \phi_2, \cdots, \phi_a$ unter einander linear unabhängig, und weiss man, dass für jede Substitution K von $\mathfrak{K}$ Gleichungen der Form (9) bestehen, so sind die Koeffizienten $a_{\alpha\beta}$ eindeutig bestimmt. Die Substitutionen $A = (a_{\alpha\beta})$ bilden dann von selbst eine der Gruppe $\mathfrak{K}$ homomorphe Gruppe. In diesem Falle sage ich auch: *$\phi_1, \phi_2, \cdots, \phi_a$ bilden ein Invariantensystem von $\mathfrak{K}$ mit der Transformationsgruppe $\mathfrak{A}$.* Für die Gleichungen (9) schreibe ich in jedem Falle auch kürzer

$$(\bar{\phi}_\alpha) = \mathfrak{A}(\phi_\alpha).$$

Es gilt nun der für viele Anwendungen nützliche Satz:

XI. *Es sei $\mathfrak{A}$ eine im Körper Ω irreduzible Gruppe, die der in Ω rationalen Gruppe $\mathfrak{K}$ homomorph ist. Hat man dann ein zur Gruppe $\mathfrak{A}$ gehörendes Invariantensystem $\phi_1, \phi_2, \cdots, \phi_a$ von $\mathfrak{K}$ mit Koeffizienten aus dem Körper Ω, so sind die Formen $\phi_1, \phi_2, \cdots, \phi_a$ entweder alle Null oder sie sind unter einander linear unabhängig.*

Dies ist leicht zu beweisen. Denn es seien die Formen ϕ_α nicht alle Null; man bezeichne die Anzahl der linear unabhängigen unter ihnen mit r. Dann lassen sich also r linear unabhängige Formen $\psi_1, \psi_2, \cdots, \psi_r$ mit Koeffizienten aus dem Körper Ω angeben, so dass Gleichungen der Form

$$(10) \qquad \psi_\rho = \sum_{\alpha=1}^{a} p_{\rho\alpha} \phi_\alpha, \qquad \phi_\alpha = \sum_{\rho=1}^{r} q_{\alpha\rho} \psi_\rho$$

bestehen, wo die $p_{\rho\alpha}$ und $q_{\alpha\rho}$ Zahlen von Ω sind. Aus (10) folgt, wenn

$$r_{\rho\sigma} = \sum_{\alpha,\,\beta=1}^{a} p_{\rho\alpha} a_{\alpha\beta} q_{\beta\sigma}$$

gesetzt wird,

$$\bar{\psi}_\rho = \sum_{\sigma=1}^{r} r_{\rho\sigma} \psi_\sigma .$$

Bezeichnet man daher die durch die Substitutionen $R = (r_{\rho\sigma})$ gebildete Gruppe mit $\mathfrak{R}$, so erscheinen die Formen $\psi_1, \psi_2, \cdots, \psi_r$ als ein Invariantensystem von $\mathfrak{K}$ mit der Transformationsgruppe $\mathfrak{R}$. Andererseits wird aber

$$\bar{\phi}_\alpha = \sum_\rho q_{\alpha\rho} \bar{\psi}_\rho = \sum_{\rho,\,\sigma} q_{\alpha\rho} r_{\rho\sigma} \psi_\sigma = \sum_\beta a_{\alpha\beta} \phi_\beta = \sum_{\beta,\,\sigma} a_{\alpha\beta} q_{\beta\sigma} \psi_\sigma ,$$

also ist

$$\sum_{\rho=1}^{r} q_{\alpha\rho} r_{\rho\sigma} = \sum_{\beta=1}^{a} a_{\alpha\beta} q_{\beta\sigma} .$$

Setzt man die Matrix $(q_{\alpha\rho})$ gleich Q, so wird auf Grund der in § 1 festgesetzten Bezeichnungsweise

$$Q\mathfrak{R} = \mathfrak{A} Q .$$

Da nun $\mathfrak{A}$ in Ω irreduzibel ist, so muss nach Satz I der Rang von Q gleich a sein. Dies erfordert aber, dass $r = a$ wird.*

Es sei nun $\mathfrak{F}$ wie früher ein irreduziblen Bestandteil des Grades f von $\mathfrak{K}$ inbezug auf den Bereich aller Zahlen. Ich nehme insbesondere an, dass sich f (nicht identisch verschwindende) lineare homogene Formen $u_1, u_2, \cdots, u_f$ der x_κ angeben lassen, die ein zu $\mathfrak{F}$ gehörendes Invariantensystem von $\mathfrak{K}$ bilden; es sei auch noch verlangt, dass $u_1, \cdots, u_f$ von gewissen gegebenen a linearen homogenen Formen $y_1, y_2, \cdots, y_a$ mit in Ω rationalen Koeffizienten linear unabhängig sein sollen. Hat nun $\mathfrak{F}$ diese Eigenschaft, so gilt dies auch für jede mit $\mathfrak{F}$ äquivalente Gruppe. Denkt man sich insbesondere $\mathfrak{F}$ durch eine äquivalente Gruppe ersetzt, die in dem oben genannten Körper $\Omega(\rho)$ rational ist, so lassen sich die Linearformen $u_1, u_2, \cdots, u_f$ auch so wählen, dass ihre Koeffizienten ebenfalls Zahlen von $\Omega(\rho)$ werden, ohne dass die Linearformen aufhören den früher gestellten Forderungen zu genügen. Geht dann $u_\alpha = u_{\alpha 0}$ in $u_{\alpha\nu}$ über, wenn ρ durch ρ_ν ersetzt wird, so bilden die f Linearformen $u_{1\nu}$, $u_{2\nu}, \cdots, u_{f\nu}$ ein zur Gruppe $\mathfrak{F}_\nu$, ferner die $rf = g$ Linearformen $u_{\alpha\nu}$ zusammengenommen ein zur Gruppe $\mathfrak{F}'$ gehörendes Invariantensystem von $\mathfrak{K}$.

Setzt man nun weiter

$$(11) \qquad z_{\alpha\mu} = \sum_{\nu=0}^{r-1} \rho_\nu^\mu u_{\alpha\nu} \quad (\alpha = 1, 2, \cdots, f;\ \mu = 0, 1, \cdots, r-1),$$

* Nimmt man $\mathfrak{A}$ nicht als irreduzibel an, so ergiebt sich, dass $\mathfrak{A}$ einer in Ω rationalen Gruppe der Form

$$\begin{pmatrix} \mathfrak{R} & \mathfrak{Q} \\ 0 & \mathfrak{P} \end{pmatrix}$$

äquivalent ist.

so werden die Koeffizienten dieser $rf = g$ Linearformen der x_κ, als symmetrische Funktionen der ρ_ν, Zahlen des Körpers Ω.* Ist ferner T die Koeffizientenmatrix der Formen $z_{a\mu}$, als Funktionen der $u_{a\nu}$ aufgefasst, und setzt man

$$T\mathfrak{F}' T^{-1} = \mathfrak{G},$$

so erkennt man sofort, dass die $z_{a\mu}$ ein zur Gruppe $\mathfrak{G}$ gehörendes Invariantensystem von $\mathfrak{K}$ bilden. Die Gruppe $\mathfrak{G}$ ist aber (vergl. § 1) eine in Ω rationale und irreduzible Gruppe. Folglich müssen nach Satz XI die Linearformen $z_{a\mu}$, da sie offenbar nicht alle Null sein können, *unter einander linear unabhängig sein.*

§ 6.

Ich wende mich nun zur Betrachtung des in der Einleitung erwähnten Satzes von Herrn A. Loewy.

Der Gedankengang des Herrn Loewy lässt sich folgendermassen formulieren. Man habe eine im Körper Ω rationale und in diesem Körper vollständig reduzible Gruppe $\mathfrak{A}$ des Grades a, die zu unserer Gruppe $\mathfrak{K}$ homomorph ist. Es mögen sich ferner a unabhängige Linearformen $y_1, y_2, \cdots, y_a$ der x_κ mit Koeffizienten aus dem Körper Ω bestimmen lassen, die ein Invariantensystem von $\mathfrak{K}$ mit der Transformationsgruppe $\mathfrak{A}$ bilden. Wir verlangen weiter, dass die Gruppe $\mathfrak{A}$ noch folgender Bedingung genügen soll: es soll sich keine in Ω irreduzible mit $\mathfrak{K}$ homomorphe Gruppe $\mathfrak{G}$ des grades g angeben lassen, zu der ein lineares Invariantensystem $z_1, z_2, \cdots, z_g$ von $\mathfrak{K}$ mit Koeffizienten aus Ω in der Weise gehört, dass die $a + g$ Linearformen $y_1, y_2, \cdots, y_a, z_1, \cdots, z_g$ unter einander linear unabhängig werden. Dann beweist Herr Loewy:

"Ist $\overline{\mathfrak{A}}$ eine zweite Gruppe, welche genau denselben Forderungen genügt, wie die Gruppe $\mathfrak{A}$, so sind $\mathfrak{A}$ und $\overline{\mathfrak{A}}$ äquivalente Gruppen. Bestimmt man ferner, was jedenfalls möglich ist, zwei zu $\mathfrak{K}$ äquivalente in Ω rationale Gruppen $\mathfrak{K}'$ und $\overline{\mathfrak{K}}$ der Form

$$\mathfrak{K}' = \begin{pmatrix} \mathfrak{A} & 0 \\ \mathfrak{C} & \mathfrak{D} \end{pmatrix}, \quad \overline{\mathfrak{K}} = \begin{pmatrix} \overline{\mathfrak{A}} & 0 \\ \overline{\mathfrak{C}} & \overline{\mathfrak{D}} \end{pmatrix},$$

so sind auch $\mathfrak{D}$ und $\overline{\mathfrak{D}}$ äquivalente Gruppen."†

Man kann daher $\mathfrak{A}$ als die grösste zu $\mathfrak{K}$ gehörende vollständig reduzible Gruppe inbezug auf Ω bezeichnen.

Tritt nun an Stelle des Körpers Ω ein anderer Körper Ω', der ebenso wie Ω alle Koeffizienten der Substitutionen von $\mathfrak{K}$ umfasst, so existiert auch inbezug auf Ω' eine zu $\mathfrak{K}$ gehörende grösste vollständig reduzible Gruppe $\mathfrak{A}'$.

---

* Vergl. L. E. Dickson, loc. cit.

† Herr Loewy macht in seiner Arbeit noch die Voraussetzung, dass die Determinanten der Substitutionen von $\mathfrak{K}$ nicht verschwinden, und dass die reziproke Substitution jeder Substitution von $\mathfrak{K}$ wieder in $\mathfrak{K}$ enthalten ist. Das von Herrn Stickelberger (vergl. das in der Einleitung angeführte Zitat) mitgeteilte Beweisverfahren gestattet aber, den Loewy'schen Satz auch für eine allgemeine Gruppe $\mathfrak{K}$ zu beweisen.

Ich will nun zeigen, dass $\mathfrak{A}$ und $\mathfrak{A}'$ äquivalente Gruppen sind.

Es genügt offenbar, dies allein für den Fall zu beweisen, dass Ω' den Bereich Z aller Zahlen bedeutet. Der Beweis gestaltet sich nun folgendermassen.

Da jede im Körper Ω irreduzible Gruppe im Körper Z vollständig reduzibel ist, so ist die in Ω vollständig reduzible Gruppe $\mathfrak{A}$ in Z ebenfalls vollständig reduzibel. Wir haben also nur zu zeigen, dass $\mathfrak{A}$ auch, im Körper Z betrachtet, die Eigenschaften einer grössten vollständig reduziblen Gruppe besitzt.

Wäre dies nun nicht der Fall, so müsste sich eine zu $\mathfrak{K}$ homomorphe in Z irreduzible Gruppe $\mathfrak{F}$ des Grades f angeben lassen, zu der ein lineares Invariantensystem $u_1, u_2, \cdots, u_f$ gehört, so dass die $a+f$ Linearformen $y_1, \cdots, y_a$, $u_1, \cdots, u_f$ unter einander linear unabhängig sind. Die Gruppe $\mathfrak{F}$ wird dann ein irreduzibler Bestandteil von $\mathfrak{K}$ inbezug auf Z. Man bestimme nun, wie im vorigen Paragraphe, die aus $\mathfrak{F}$ hervorgehende inbezug auf Ω irreduzible Gruppe $\mathfrak{G}$. Dann gehört auch zu dieser, wie wir gesehen haben, ein lineares Invariantensystem $z_{a\mu}$ mit in Ω rationalen Koeffizienten. Da aber $\mathfrak{A}$ inbezug auf Ω eine grösste vollständig reduzible Gruppe sein soll, so dürften die Linearformen $z_{a\mu}$ und $y_1, y_2, \cdots, y_a$ nicht unter einander linear unabhängig sein. Hieraus folgt aber (wegen der Irreduzibilität von $\mathfrak{G}$ inbezug auf Ω) nach einem von Herrn Stickelberger, loc. cit., bewiesenen Satze, dass j e d e der Funktionen $z_{a\mu}$ durch $y_1, y_2, \cdots, y_a$ linear darstellbar sein muss. Es ist aber, wie aus den Gleichungen (11) hervorgeht, jede der Funktionen $u_1, u_2, \cdots, u_f$ eine lineare homogene Form der $z_{a\mu}$ und wäre folglich auch durch $y_1, y_2, \cdots, y_a$ linear und homogen darstellbar. Dies führt aber auf einen Widerspruch.

Ist wie früher

$$\mathfrak{K}' = \begin{pmatrix} \mathfrak{A} & 0 \\ \mathfrak{C} & \mathfrak{D} \end{pmatrix}$$

eine zu $\mathfrak{K}$ äquivalente in Ω rationale Gruppe, so gehört auch zur Gruppe $\mathfrak{D}$ inbezug auf Ω eine grösste vollständig reduzible Gruppe u. s. f. So erhält Herr Loewy die zu $\mathfrak{K}$ inbezug auf Ω gehörenden auf einander folgenden grössten vollständig reduziblen Gruppen $\mathfrak{A}, \mathfrak{A}_1, \cdots, \mathfrak{A}_\mu$. Aus unserer Betrachtung ergiebt sich aber, dass diese Gruppen in dem in der Einleitung genauer erläuterten Sinne von der Wahl des Körpers Ω unabhängig sind.

Man kann dieses Resultat auch folgendermassen aussprechen:

XII. *Ist $\mathfrak{K}$ eine in einem gegebenen Zahlkörper Ω rationale Gruppe linearer homogener Substitutionen und zerlegt man $\mathfrak{K}$ im Bereich aller Zahlen unter Hervorhebung der zu $\mathfrak{K}$ gehörenden auf einander folgenden grössten vollständig reduziblen Gruppen, so ist jede dieser Gruppen im Körper Ω rational darstellbar.*

Berlin, *im März* 1908.

15.

Zur Theorie der linearen homogenen Integralgleichungen

Mathematische Annalen 67, 306 - 339 (1909)

In seiner Inaugural-Dissertation hat Herr E. Schmidt*) den Hilbertschen Fundamentalsatz, der besagt, daß jede Integralgleichung

$$\varphi(s) = \lambda \int_a^b K(s, t)\, \varphi(t)\, dt$$

mit reellem stetigem symmetrischem Kern $K(s, t)$ mindestens einen Eigenwert λ besitzt,**) auf besonders einfachem Wege bewiesen. Herr Schmidt gelangt zu diesem Satz, indem er einen (den kleinsten) Eigenwert γ_1 des iterierten Kerns $K^2(s, t)$ mit Hilfe eines Grenzverfahrens, das der Gräffeschen Methode in der Algebra eng verwandt ist, gewissermaßen explizite berechnet. Da nun aber dieses Verfahren nur *einen* Eigenwert von $K^2(s, t)$ liefert, so entsteht die Frage: wie berechnet man die übrigen Eigenwerte

$$\gamma_2 \leqq \gamma_3 \leqq \cdots$$

des Kerns $K^2(s, t)$?

Auf funktionentheoretischem Wege hat Herr A. Kneser***) diese Aufgabe gelöst, indem er gezeigt hat, daß der Eigenwert γ_n, wenn $\gamma_1, \gamma_2, \cdots, \gamma_{n-1}$ bereits bekannt sind, als der Grenzwert eines gewissen Ausdrucks dargestellt werden kann.

---

*) „Entwicklung willkürlicher Funktionen nach Systemen vorgeschriebener" (Göttingen, 1905); vergl. auch Math. Annalen, Bd. 63, S. 433.

**) Hilbert, „Grundzüge einer allgemeinen Theorie der linearen Integralgleichungen", Erste Mitteilung, Göttinger Nachrichten 1904, S. 49.

***) „Ein Beitrag zur Theorie der Integralgleichungen", Rendiconti del Circolo Matematico di Palermo, T. XXII (1906), S. 233. — Herr Kneser geht übrigens im § 3 seiner Arbeit von der irrtümlichen Annahme aus, daß die (stets reellen) Eigenwerte eines reellen symmetrischen Kerns immer positiv sind. Die Formeln, die er in diesem Paragraphen für den Kern $K(s, t)$ entwickelt, gelten im allgemeinen erst für den iterierten Kern $K^2(s, t)$.

Zu einer anderen, wie ich glaube, einfacheren Lösung gelangt man aber, wenn man die Analogie mit der Algebra weiter verfolgt. Hat man nämlich eine algebraische Gleichung

$$x^m + c_1 x^{m-1} + \cdots = 0,$$

von der wir der Einfachheit wegen annehmen, daß ihre Wurzeln

$$x_1 \geqq x_2 \geqq \cdots \geqq x_m$$

positiv sind, so liefert die Gräffesche Näherungsmethode zunächst nur die Wurzel x_1. Will man nach demselben Verfahren die übrigen Wurzeln erhalten, so betrachtet man für jedes $n \leqq m$ die Gleichung

$$x^{\binom{m}{n}} + c_1^{(n)} x^{\binom{m}{n}-1} + \cdots = 0,$$

der die $\binom{m}{n}$ Produkte

$$x_{\alpha_1} x_{\alpha_2} \cdots x_{\alpha_n} \qquad (\alpha_1 < \alpha_2 < \cdots < \alpha_n)$$

genügen. Auf diese Gleichung angewandt, ergibt dasselbe Näherungsverfahren das Produkt $x_1 x_2 \cdots x_n$; durch Bildung des Quotienten $\frac{x_1 x_2 \cdots x_n}{x_1 x_2 \cdots x_{n-1}}$ erhält man dann x_n.

Ähnlich kann man auch in der Theorie der Integralgleichungen verfahren, wenn es gelingt, zu jeder Integralgleichung

$$\varphi(s) = \lambda \int_a^b K(s, t)\, \varphi(t)\, dt$$

mit den Eigenwerten

$$\lambda_1, \lambda_2, \ldots$$

eine andere Integralgleichung anzugeben, deren Eigenwerte bei vorgeschriebenem n die Produkte

$$\lambda_{\alpha_1} \lambda_{\alpha_2} \cdots \lambda_{\alpha_n} \qquad (\alpha_1 < \alpha_2 < \cdots < \alpha_n)$$

sind, und deren Kern, wenn $K(s, t)$ reell und symmetrisch ist, ebenfalls dieser Bedingung genügt. Diese Eigenschaft besitzt, wie hier gezeigt wird, die Integralgleichung

$$\Phi(s_1, s_2, \cdots, s_n) = \mu \int_a^b \int_a^b \cdots \int_a^b K\begin{pmatrix} s_1, s_2, \cdots, s_n \\ t_1, t_2, \cdots, t_n \end{pmatrix} \Phi(t_1, t_2, \cdots, t_n)\, dt_1\, dt_2 \cdots dt_n,$$

in der, ähnlich wie bei Herrn Fredholm*),

$$K\begin{pmatrix} s_1, s_2, \cdots, s_n \\ t_1, t_2, \ldots, t_n \end{pmatrix} = \frac{1}{n!} \begin{vmatrix} K(s_1, t_1), & K(s_1, t_2), & \cdots, & K(s_1, t_n) \\ K(s_2, t_1), & K(s_2, t_2), & \cdots, & K(s_2, t_n) \\ \cdot & \cdot & \cdot & \cdot \\ K(s_n, t_1), & K(s_n, t_2), & \cdots, & K(s_n, t_n) \end{vmatrix}$$

*) „Sur une classe d'équations fonctionnelles", Acta Mathematica, Bd. 27, S. 365.

zu setzen ist. Diese durch $K(s,t)$ bestimmte Funktion von $2n$ Variabeln $s_1, \cdots, s_n, t_1, \cdots, t_n$ nenne ich den *zu* $K(s,t)$ *assoziierten Kern des Grades n.**) Ist $K(s,t)$ speziell ein reeller symmetrischer Kern, so erhält man durch Anwendung des Schmidtschen Verfahrens auf die zu $K(s,t)$ assoziierten Kerne ohne Hinzunahme irgend welcher neuer Hilfsmittel jeden Eigenwert γ_n von $K^2(s,t)$ als den Grenzwert eines Ausdrucks, in dem nur die von Herrn Schmidt zur Berechnung von γ_1 benutzten Größen $U_{2\nu}$ auftreten. Zugleich ergibt sich ein explizites Verfahren zur Bestimmung der zu γ_n gehörenden Eigenfunktionen des Kerns $K^2(s,t)$.

Die im folgenden (Abschnitt II) entwickelten Sätze über die zu einem gegebenen Kern assoziierten Kerne lassen sich mit Hilfe der Fredholmschen ganzen transzendenten Funktion ohne Mühe beweisen. Doch schien es mir von Interesse, neben diesen Beweisen noch andere anzugeben, die nur von dem Begriff der Ordnung eines Eigenwerts Gebrauch machen, einem Begriff, der sich auch auf elementarem Wege begründen läßt.**) Diese Beweismethode erfordert zwar etwas längere Rechnungen, sie hat aber den Vorzug, daß sie in gewissen Fällen, in denen die Fredholmsche Theorie versagt, noch ihre Gültigkeit behält. Doch habe ich mich der Einfachheit wegen damit begnügt, nur von stetigen Kernen zu sprechen. Die Beschränkung auf reelle symmetrische Kerne würde in den beiden ersten Abschnitten für die Beweisführung keine wesentliche Vereinfachung bedeuten.

Abschnitt I.

Vorbereitende Betrachtungen.

§ 1.

Es sei $K(s,t)$ eine reelle oder komplexe Funktion der reellen Variabeln s und t, die in dem Gebiet $a \leqq s \leqq b$, $a \leqq t \leqq b$ definiert und stetig ist. Eine von Null verschiedene Zahl λ heißt dann bekanntlich ein *Eigenwert des Kerns* $K(s,t)$, wenn sich eine nicht identisch verschwindende, für $a \leqq s \leqq b$ stetige Funktion $\varphi(s)$ bestimmen läßt, die der Gleichung

$$\varphi(s) = \lambda \int_a^b K(s,t)\, \varphi(t)\, dt$$

---

*) Diesem Begriff dürfte in der Theorie der Integralgleichungen eine ähnliche Bedeutung zukommen, wie dem Begriff der assoziierten Differentialgleichungen in der Theorie der linearen Differentialgleichungen (vergl. L. Schlesinger, Handbuch der Theorie der linearen Differentialgleichungen, Bd. II, Teil I, S. 127).

**) Vergl. meine Arbeit „Über die charakteristischen Wurzeln einer linearen Substitution mit einer Anwendung auf die Theorie der Integralgleichungen", Math. Annalen, Bd. 66, S. 488. — Im folgenden wird diese Arbeit kurz mit I zitiert.

genügt; $\varphi(s)$ ist dann eine zu λ gehörende *Eigenfunktion* des Kerns $K(s,t)$. Jedem Eigenwert λ entspricht eine positive ganze Zahl n, die *Ordnung* von λ, die folgendermaßen charakterisiert werden kann: es lassen sich n, aber nicht mehr als n, linear unabhängige stetige Funktionen $\varphi_1(s), \cdots, \varphi_n(s)$ angeben, für die

$$\int_a^b K(s,t)\,\varphi_\alpha(t)\,dt = \sum_{\beta=1}^{n} a_{\alpha\beta}\,\varphi_\beta(s) \qquad (\alpha = 1, 2, \cdots, n)$$

wird; hierbei sollen die $a_{\alpha\beta}$ Konstanten bedeuten, die der Bedingung

$$\begin{vmatrix} a_{11}-x, & a_{12}, & \cdots, & a_{1n} \\ a_{21}, & a_{22}-x, & \cdots, & a_{2n} \\ \cdot & \cdot & \cdot & \cdot \\ a_{n1}, & a_{n2}, & \cdots, & a_{nn}-x \end{vmatrix} = \left(\frac{1}{\lambda} - x\right)^n$$

genügen*). Ich sage dann: Die Funktionen $\varphi_1(s), \cdots, \varphi_n(s)$ bilden ein zu λ *gehörendes, vollständiges invariantes System* des Kerns $K(s,t)$. Es gilt nun folgende Regel**):

Hat man neben den Funktionen $\varphi_\alpha(s)$ noch n andere Funktionen $\psi_1(s), \cdots, \psi_n(s)$, für die ebenfalls

$$\int_a^b K(s,t)\,\psi_\alpha(t)\,dt = \sum_{\beta=1}^{n} a_{\alpha\beta}\,\psi_\beta(s) \qquad (\alpha = 1, 2, \cdots, n)$$

wird, so ist jede der Funktionen $\psi_\alpha(s)$ linear und homogen mit konstanten Koeffizienten durch die Funktionen $\varphi_\alpha(s)$ darstellbar.

Man kann die Funktionen $\varphi_\alpha(s)$ auch so wählen, daß $a_{\alpha\beta} = 0$ wird, sobald $\alpha < \beta$ ist. Dann wird also

$$\int_a^b K(s,t)\,\varphi_\alpha(t)\,dt = a_{\alpha 1}\,\varphi_1(s) + \cdots + a_{\alpha,\alpha-1}\,\varphi_{\alpha-1}(s) + \frac{1}{\lambda}\,\varphi_\alpha(s).$$

Denkt man sich nun alle verschiedenen Eigenwerte des Kerns $K(s,t)$ ins Auge gefaßt und bestimmt zu jedem ein zugehöriges vollständiges invariantes System von der zuletzt erwähnten Art, so erhält man ein endliches oder unendliches System von Funktionen

$$\varphi_1(s),\ \varphi_2(s),\ \cdots,$$

die folgende Eigenschaften besitzen:

1. Im Intervall $a \leq s \leq b$ ist jede der Funktionen stetig, und für jedes m sind $\varphi_1(s), \varphi_2(s), \cdots, \varphi_m(s)$ untereinander linear unabhängig.

---

*) Vergl. I, § 14.

**) Vergl. I, § 15.

2. Es ist für jedes α

$$\int_a^b K(s,t)\,\varphi_\alpha(t)\,dt = c_{\alpha 1}\,\varphi_1(s) + \cdots + c_{\alpha,\alpha-1}\,\varphi_{\alpha-1}(s) + c_{\alpha\alpha}\,\varphi_\alpha(s),$$

wo die $c_{\alpha\beta}$ Konstanten bedeuten und insbesondere alle $c_{\alpha\alpha}$ von Null verschieden sind.

3. Bilden die Funktionen $\psi_1(s), \cdots, \psi_n(s)$ ein zu irgend einem Eigenwert gehörendes vollständiges invariantes System des Kerns $K(s,t)$, so ist jede Funktion $\psi_\alpha(s)$ durch gewisse endlich viele unter den Funktionen $\varphi_\alpha(s)$ linear und homogen mit konstanten Koeffizienten darstellbar.

Jedes System von Funktionen $\varphi_1(s), \varphi_2(s), \cdots$, die diesen drei Bedingungen genügen, bezeichne ich als ein *vollständiges System von Hauptfunktionen* des Kerns $K(s,t)$*). Ist ein solches System bekannt, so erhält man in den Zahlen

$$\lambda_1 = \frac{1}{c_{11}}, \quad \lambda_2 = \frac{1}{c_{22}}, \cdots \tag{1}$$

die sämtlichen Eigenwerte des Kerns, und zwar jeden so oft, wie seine Ordnung angibt. Die Reihe

$$\frac{1}{|\lambda_1|^2} + \frac{1}{|\lambda_2|^2} + \cdots$$

ist dann konvergent und ihr Wert ist höchstens gleich

$$\int_a^b\int_a^b |K(s,t)|^2\,ds\,dt.^{**)}$$

Ferner folgt aus der Bedingung 3 genauer: Sind $\chi_1(s), \cdots, \chi_r(s)$ beliebige stetige Funktionen, für welche Gleichungen der Form

$$\int_a^b K(s,t)\,\chi_\varrho(t)\,dt = \sum_{\sigma=1}^{r} d_{\varrho\sigma}\,\chi_\sigma(s) \qquad (\varrho = 1, 2, \cdots, r)$$

bestehen, wo die $d_{\varrho\sigma}$ Konstanten bedeuten, deren Determinante $|d_{\varrho\sigma}|$ nicht verschwindet, so sind die $\chi_\varrho(s)$ durch gewisse endlich viele unter den Funktionen $\varphi_1(s), \varphi_2(s), \cdots$ linear und homogen mit konstanten Koeffizienten darstellbar. Sind insbesondere $\chi_1(s), \cdots, \chi_r(s)$ nicht alle identisch Null, so kommen unter den Wurzeln der Gleichung

$$|d_{\varrho\sigma} - x e_{\varrho\sigma}| = 0 \qquad (e_{\varrho\varrho} = 1,\ e_{12} = e_{13} = \cdots = 0) \tag{2}$$

die reziproken Werte gewisser Eigenwerte des Kerns vor, also gewisse der Zahlen $c_{\alpha\alpha}$. Wählt man dann N so groß, daß unter den Zahlen

*) Ein solches System existiert natürlich nur dann, wenn der Kern $K(s,t)$ mindestens einen Eigenwert besitzt.

**) Vergl. I, § 14.

$$c_{N+1,N+1},\ c_{N+2,N+2},\ \cdots$$

keine Wurzel von (2) mehr vorkommt, so ist jede der Funktionen $\chi_\varrho(s)$ eine lineare homogene Verbindung der N Funktionen

$$\varphi_1(s),\ \varphi_2(s),\ \cdots,\ \varphi_N(s).*)$$

§ 2.

Um im folgenden auch von der Fredholmschen Theorie Gebrauch machen zu können, empfiehlt es sich, ein Hauptresultat dieser Theorie etwas anders zu formulieren, als dies gewöhnlich geschieht.

Hat der Ausdruck $K\begin{pmatrix} s_1, s_2, \cdots, s_n \\ t_1, t_2, \cdots, t_n \end{pmatrix}$ dieselbe Bedeutung wie in der Einleitung, und setzt man

$$A_n = \int_a^b \int_a^b \cdots \int_a^b K\begin{pmatrix} s_1, s_2, \cdots, s_n \\ s_1, s_2, \cdots, s_n \end{pmatrix} ds_1\, ds_2 \cdots ds_n,$$

so wird bekanntlich

$$D(x) = 1 - A_1 x + A_2 x^2 - \cdots$$

eine ganze transzendente Funktion, die nur für die Zahlen (1) verschwindet, und die Produktzerlegung dieser Funktion hat die Form**)

$$(3) \qquad D(x) = e^{-Ax - Bx^2} \prod_{\alpha=1}^{\infty} \left(1 - \frac{x}{\lambda_\alpha}\right) e^{\frac{x}{\lambda_\alpha}}.$$

Ist ferner $K^m(s,t)$ der m^{te} iterierte Kern, d. h.

$$K^m(s,t) = \int_a^b \int_a^b \cdots \int_a^b K(s, r_1)\, K(r_1, r_2) \cdots K(r_{m-1}, t)\, dr_1\, dr_2 \cdots dr_{m-1},$$

und setzt man

$$U_m = \int_a^b K^m(s,s)\, ds,$$

so wird***)

$$(4) \quad A_n = \sum \frac{(-1)^{\nu_2+\nu_4+\cdots}}{\nu_1!\,\nu_2!\cdots\nu_n!} \left(\frac{U_1}{1}\right)^{\nu_1} \left(\frac{U_2}{2}\right)^{\nu_2} \cdots \left(\frac{U_n}{n}\right)^{\nu_n} \quad (\nu_1 + 2\nu_2 + \cdots + n\nu_n = n),$$

*) Alles dies folgt aus den Ausführungen des § 15 meiner mit I zitierten Arbeit.

**) Vergl. E. Goursat, „Recherches sur les équations intégrales linéaires", Annales de la Faculté des Sciences de l'Université de Toulouse, S. 96, und I, § 14.

***) Vergl. E. Goursat, a. a. O., S. 95, wo diese Formel besonders einfach bewiesen wird.

und hieraus folgt für genügend kleine Werte von $|x|$ die bekannte Relation

$$D(x) = e^{-\frac{U_1}{1}x - \frac{U_2}{2}x^2 - \frac{U_3}{3}x^3 - \cdots} \tag{5}$$

Da nun andererseits, wenn für $m > 1$

$$l_m = \sum_\alpha \frac{1}{\lambda_\alpha^m}$$

gesetzt wird,

$$\prod_\alpha \left(1 - \frac{x}{\lambda_\alpha}\right) e^{\frac{x}{\lambda_\alpha}} = e^{-\frac{l_2}{2}x^2 - \frac{l_3}{3}x^3 - \cdots}$$

ist, so erhält man aus (3) und (5) durch Koeffizientenvergleichung

$$A = U_1, \quad B = \frac{1}{2}(U_2 - l_2) \tag{6}$$

und für $m \geqq 3$

$$U_m = l_m.\text{*)} \tag{7}$$

Umgekehrt folgt aus den Gleichungen (7), wenn A und B aus den Gleichungen (6) bestimmt werden, die Formel (3).

Kennt man daher ein System von Zahlen

$$\lambda_1, \lambda_2, \cdots,$$

für welche die Reihe $\sum_\alpha \frac{1}{\lambda_\alpha^2}$ *absolut konvergent ist, so genügt es zu zeigen, daß für* $m \geqq 3$

$$\sum_\alpha \frac{1}{\lambda_\alpha^m} = \int_a^b K^m(s, s)\, ds$$

ist, um behaupten zu können, daß die Zahlen $\lambda_1, \lambda_2, \cdots$ *die Gesamtheit der Eigenwerte des Kerns* $K(s, t)$ *repräsentieren.*

Alles bis jetzt Gesagte gilt auch dann, wenn s und t Punkte in einem n-dimensionalen Raume bedeuten und an Stelle des einfachen Integrals $\int_a^b$ ein n-faches Integral tritt.

§ 3.

Es seien nun n Integralgleichungen

$$\varphi_1(s) = \lambda^{(1)} \int_{a_1}^{b_1} K_1(s, t)\, \varphi_1(t)\, dt, \cdots, \varphi_n(s) = \lambda^{(n)} \int_{a_n}^{b_n} K_n(s, t)\, \varphi_n(t)\, dt$$

---

*) Für einen reellen symmetrischen Kern ist, wie Herr E. Schmidt, Math. Annalen, Bd. 63, S. 471, bewiesen hat, auch $U_2 = l_2$. Dies ist identisch mit der Aussage, daß in diesem Fall die ganze transzendente Funktion $D(x)$ höchstens vom Geschlechte 1 ist.

mit stetigen Kernen gegeben. Das Produkt

$$K_1(s_1,t_1)\,K_2(s_2,t_2)\cdots K_n(s_n,t_n) \tag{8}$$

als Funktion der beiden Variabelnsysteme $s_1, s_2, \cdots, s_n$ und $t_1, t_2, \cdots, t_n$ betrachtet, läßt sich dann als der Kern der Integralgleichung

$$\Phi(s_1,\cdots,s_n)=\mu\int\limits_{a_1}^{b_1}\cdots\int\limits_{a_n}^{b_n}K_1(s_1,t_1)\cdots K_n(s_n,t_n)\,\Phi(t_1,\cdots,t_n)\,dt_1\cdots dt_n$$

auffassen. Den Kern (8) nenne ich den durch die Kerne $K_1, K_2, \cdots, K_n$ bestimmten *Produktkern.*

Es gilt nun der Satz:

I. *Bilden die Funktionen*

$$\varphi_{\nu 1}(s),\ \varphi_{\nu 2}(s),\ \cdots$$

ein vollständiges System von Hauptfunktionen des Kerns $K_\nu(s,t)$, *so bilden die Funktionen*

$$\varphi_{1\alpha_1}(s_1)\,\varphi_{2\alpha_2}(s_2)\cdots\varphi_{n\alpha_n}(s_n) \qquad (\alpha_\nu=1,2,\cdots) \tag{9}$$

*ein vollständiges System von Hauptfunktionen des Produktkerns.**)

Hierbei denken wir uns die Funktionen (9) nach folgendem Prinzip angeordnet: von zwei Produkten

$$\varphi_{1\alpha_1}(s_1)\,\varphi_{2\alpha_2}(s_2)\cdots\varphi_{n\alpha_n}(s_n),\quad \varphi_{1\beta_1}(s_1)\,\varphi_{2\beta_2}(s_2)\cdots\varphi_{n\beta_n}(s_n)$$

soll das erste vor dem zweiten stehen, wenn entweder

$$\alpha_1+\alpha_2+\cdots+\alpha_n<\beta_1+\beta_2+\cdots+\beta_n$$

ist, oder falls

$$\alpha_1+\alpha_2+\cdots+\alpha_n=\beta_1+\beta_2+\cdots+\beta_n$$

wird, die erste nicht verschwindende unter den Differenzen

$$\alpha_1-\beta_1,\ \alpha_2-\beta_2,\ \cdots,\ \alpha_n-\beta_n$$

negativ ist. In dieser Reihenfolge mögen die Produkte (9) auch mit

$$\Phi_1(s_1,\cdots,s_n),\ \Phi_2(s_1,\cdots,s_n),\ \cdots \tag{9'}$$

bezeichnet werden.

Um den Beweis des Satzes I zu führen, haben wir nur zu zeigen, daß die Funktionen (9') in bezug auf den Kern (8) den drei im § 1 formulierten Bedingungen genügen.

Daß nun die erste Bedingung (Stetigkeit und lineare Unabhängigkeit) erfüllt ist, ist unmittelbar evident.

Ist ferner

$$\int\limits_{a_\nu}^{b_\nu}K_\nu(s,t)\,\varphi_{\nu\alpha}(t)\,dt=c_{\alpha 1}^{(\nu)}\,\varphi_{\nu 1}(s)+\cdots+c_{\alpha\alpha}^{(\nu)}\,\varphi_{\nu\alpha}(s),$$

*) Besitzt einer der Kerne $K_\nu(s,t)$ keinen Eigenwert, so besagt der Satz, daß auch der Produktkern keinen Eigenwert besitzt.

so erhält man, wenn

$$\Phi_\varrho(s_1, \cdots, s_n) = \varphi_{1\alpha_1}(s_1) \cdots \varphi_{n\alpha_n}(s_n)$$

ist,

$$\int_{a_1}^{b_1} \cdots \int_{a_n}^{b_n} K_1(s_1, t_1) \cdots K_n(s_n, t_n)\, \Phi_\varrho(t_1, \cdots, t_n)\, dt_1 \cdots dt_n$$

$$= \prod_{\nu=1}^{n} \int_{a_\nu}^{b_\nu} K_\nu(s_\nu, t_\nu)\, \varphi_{\nu\alpha_\nu}(t_\nu)\, dt_\nu$$

$$= \prod_{\nu=1}^{n} \left\{ c_{\alpha_\nu 1}^{(\nu)} \varphi_{\nu 1}(s_\nu) + \cdots + c_{\alpha_\nu \alpha_\nu}^{(\nu)} \varphi_{\nu\alpha_\nu}(s_\nu) \right\}.$$

Dies hat aber in der Tat, wie man leicht sieht, die Form

$$C_{\varrho 1}\, \Phi_1(s_1, \cdots, s_n) + \cdots + C_{\varrho\varrho} \Phi_\varrho(s_1, \cdots, s_n),$$

wobei insbesondere

$$(10) \qquad C_{\varrho\varrho} = c_{\alpha_1\alpha_1}^{(1)}\, c_{\alpha_2\alpha_2}^{(2)} \cdots c_{\alpha_n\alpha_n}^{(n)} \neq 0$$

wird.

Wir haben also nur noch die dritte Bedingung zu prüfen. Es sei also μ ein Eigenwert des Produktkerns, und es mögen die Funktionen

$$\Psi_1(s_1, \cdots, s_n), \cdots, \Psi_r(s_1, \cdots, s_n)$$

ein zu μ gehörendes vollständiges invariantes System bilden. Es sei etwa

$$(11) \qquad \int_{a_1}^{b_1} \cdots \int_{a_n}^{b_n} K_1(s_1, t_1) \cdots K_n(s_n, t_n)\, \Psi_\varrho(t_1, \cdots, t_n)\, dt_1 \cdots dt_n$$

$$= \sum_{\sigma=1}^{r} a_{\varrho\sigma}\, \Psi_\sigma(s_1, \cdots, s_n).$$

Die Determinante $|a_{\varrho\sigma}|$ ist dann gleich $\left(\frac{1}{\mu}\right)^r$, also von Null verschieden. Setzt man nun

$$\int_{a_2}^{b_2} \cdots \int_{a_n}^{b_n} K_2(s_2, t_2) \cdots K_n(s_n, t_n)\, \Psi_\varrho(s_1, t_2, \cdots, t_n)\, dt_2 \cdots dt_n = \Psi_\varrho'(s_1, s_2, \cdots, s_n),$$

so läßt sich die Gleichung (11) auch in der Form

$$(12) \qquad \int_{a_1}^{b_1} K_1(s_1, t_1)\, \Psi_\varrho'(t_1, s_2, \cdots, s_n)\, dt_1 = \sum_{\sigma=1}^{r} a_{\varrho\sigma}\, \Psi_\sigma(s_1, s_2, \cdots, s_n)$$

schreiben. Multipliziert man auf beiden Seiten mit

$$K_2(u_2, s_2) \cdots K_n(u_n, s_n)\, ds_2 \cdots ds_n$$

und integriert nach s_ν $(\nu = 2, 3, \cdots, n)$ von a_ν bis b_ν, so erhält man, da es wegen der vorausgesetzten Stetigkeit der zu betrachtenden Funktionen auf die Reihenfolge der Integrationen nicht ankommt,

$$\int_{a_1}^{b_1}\int_{a_2}^{b_2}\cdots\int_{a_n}^{b_n} K_1(s_1, t_1)\, K_2(u_2, s_2) \cdots K_n(u_n, s_n)\, \Psi'_\varrho(t_1, s_2, \cdots, s_n)\, dt_1\, ds_2 \cdots ds_n$$
$$= \sum_{\sigma=1}^{r} a_{\varrho\sigma}\, \Psi'_\sigma(s_1, u_2, \cdots, u_n).$$

Diese Formel besagt aber, daß die Gleichungen (11) richtig bleiben, wenn $\Psi_\varrho(s_1, \cdots, s_n)$ durch $\Psi'_\varrho(s_1, \cdots, s_n)$ ersetzt wird. Hieraus folgt aber (vergl. § 1), daß sich r^2 Konstanten $b_{\varrho\sigma}$ bestimmen lassen, für die

$$\Psi'_\varrho(s_1, \cdots, s_n) = \sum_{\sigma=1}^{r} b_{\varrho\sigma}\, \Psi_\sigma(s_1, \cdots, s_n)$$

wird. Es ist nun leicht zu sehen, daß die Determinante $|b_{\varrho\sigma}|$ von Null verschieden ist. Denn wäre sie Null, so wären die Funktionen Ψ'_ϱ untereinander linear abhängig, also etwa

$$c_1 \Psi_1'(s_1, \cdots, s_n) + \cdots + c_r \Psi_r'(s_1, \cdots, s_n) = 0.$$

Diese Gleichung würde richtig bleiben, wenn Ψ'_ϱ durch das in (12) links stehende Integral ersetzt wird. Diese Integrale sind aber wegen $|a_{\varrho\sigma}| \neq 0$ untereinander linear unabhängig.

Ist nun $(b'_{\varrho\sigma})$ die zu $(b_{\varrho\sigma})$ inverse Matrix, d. h. ist

$$\sum_{\tau=1}^{r} b'_{\varrho\tau}\, b_{\tau\sigma} = e_{\varrho\sigma} \qquad \left\{\begin{matrix} e_{11} = \cdots = e_{rr} = 1 \\ e_{12} = e_{13} = \cdots = e_{r,r-1} = 0 \end{matrix}\right\},$$

so erhält man aus (12)

$$(13) \qquad \int_{a_1}^{b_1} K_1(s_1, t_1)\, \Psi_\varrho(t_1, s_2, \cdots, s_n) = \sum_{\sigma=1}^{r} d_{\varrho\sigma}\, \Psi_\sigma(s_1, s_2, \cdots, s_n),$$

wo

$$d_{\varrho\sigma} = \sum_{\tau=1}^{r} b'_{\varrho\tau}\, a_{\tau\sigma}$$

zu setzen ist. Die Determinante $|d_{\varrho\sigma}|$ ist hierbei gleich dem Produkt der Determinanten $|b'_{\varrho\sigma}| = \frac{1}{|b_{\varrho\sigma}|}$ und $|a_{\varrho\sigma}|$, also von Null verschieden. Da nun die Funktionen

$$(14) \qquad \varphi_{11}(s_1),\ \varphi_{12}(s_1), \cdots$$

ein vollständiges System von Hauptfunktionen des Kerns $K_1(s_1, t_1)$ bilden, so folgt aus (13) auf Grund der am Schluß des § 1 gemachten Bemerkung, daß sich die Funktionen $\Psi_\varrho(s_1, s_2, \cdots, s_n)$ für jede Wahl der Variabeln $s_2, \cdots, s_n$ linear und homogen mit konstanten Koeffizienten durch gewisse endlich viele unter den Funktionen (14) darstellen lassen. Wählt man genauer N so groß, daß unter den Zahlen

$$c^{(1)}_{N+1,N+1},\quad c^{(1)}_{N+2,N+2},\ \cdots$$

keine Wurzeln der Gleichung

$$|d_{\varrho\sigma} - x e_{\varrho\sigma}| = 0$$

vorkommt, so läßt sich $\Psi_\varrho(s_1, s_2, \cdots, s_n)$ zunächst für feste Werte von $s_2, \cdots, s_n$ in der Form

$$\Psi_\varrho(s_1, s_2, \cdots, s_n) = \sum_{\alpha=1}^{N} f_{\varrho\alpha}\, \varphi_{1\alpha}(s_1)$$

darstellen, wo die $f_{\varrho\alpha}$ eindeutig bestimmte Konstanten sind. Da aber N von der Wahl der Werte $s_2, \cdots, s_n$ unabhängig ist, so kann auch für variable $s_2, \cdots. s_n$

$$\Psi_\varrho(s_1, s_2, \cdots, s_n) = \sum_{\alpha=1}^{N} f_{\varrho\alpha}(s_2, \cdots, s_n)\, \varphi_{1\alpha}(s_1)$$

gesetzt werden, und hier sind die $f_{\varrho\alpha}(s_2, \cdots, s_n)$ eindeutig bestimmte Funktionen von $s_2, \cdots, s_n$.

Was hier für den Kern K_1 bewiesen worden ist, gilt ebenso für jeden anderen Kern K_ν. Hieraus folgt aber leicht, daß sich $\Psi_\varrho(s_1, s_2, \cdots, s_n)$, wie zu beweisen ist, als lineare homogene Verbindung (mit konstanten Koeffizienten) gewisser endlich vieler unter den Produkten (9) darstellen läßt.*)

*) Es gilt nämlich allgemein die Regel: Hat man n Systeme von eindeutigen Funktionen

$$\begin{array}{l} \varphi_{11}(s),\ \varphi_{12}(s),\ \cdots,\ \varphi_{1m_1}(s), \\ \varphi_{21}(s),\ \varphi_{22}(s),\ \cdots,\ \varphi_{2m_2}(s), \\ \cdot\ \cdot\ \cdot\ \cdot\ \cdot\ \cdot\ \cdot\ \cdot\ \cdot \\ \varphi_{n1}(s),\ \varphi_{n2}(s),\ \cdots,\ \varphi_{nm_n}(s), \end{array}$$

und sind die Funktionen jeder Zeile untereinander linear unabhängig, so ist eine Funktion $\psi(s_1, s_2, \cdots, s_n)$ von n Variabeln, die sich für jedes ν durch die Funktionen

$$\varphi_{\nu 1}(s_\nu),\ \varphi_{\nu 2}(s_\nu),\ \cdots,\ \varphi_{\nu m_\nu}(s_\nu)$$

linear und homogen mit von s_ν unabhängigen Koeffizienten darstellen läßt, durch die Produkte

$$\varphi_{1\alpha_1}(s_1)\,\varphi_{2\alpha_2}(s_2)\cdots\varphi_{n\alpha_n}(s_n) \qquad (\alpha_\nu = 1, 2, \cdots, m_\nu)$$

linear und homogen mit konstanten Koeffizienten ausdrückbar. — Der Beweis ist leicht zu führen mit Hilfe des Satzes: m Funktionen $\varphi_1(x), \varphi_2(x), \cdots, \varphi_m(x)$ sind stets und nur dann linear unabhängig, wenn die Determinante

$$|\varphi_\alpha(x_\beta)| \qquad (\alpha, \beta = 1, 2, \cdots, m),$$

als Funktion der m Variabeln $x_1, x_2, \cdots, x_m$ betrachtet, nicht identisch verschwindet (vergl. E. Goursat, a. a. O., S. 86).

§ 4.

Die Eigenwerte des Produktkerns sind (vergl. § 1) nichts anderes als die reziproken Werte der in den Gleichungen (10) auftretenden Konstanten $C_{\varrho\varrho}$. Hieraus folgt:

II. *Sind*

$$\lambda_{\nu 1}, \lambda_{\nu 2}, \lambda_{\nu 3}, \cdots$$

die sämtlichen Eigenwerte des Kerns $K_\nu(s, t)$, *jeder Eigenwert so oft geschrieben, wie seine Ordnung angibt, so erhält man alle Eigenwerte* μ *des Produktkerns, indem man alle Produkte*

$$\lambda_{1\alpha_1}\lambda_{2\alpha_2}\cdots\lambda_{n\alpha_n} \qquad (\alpha_\nu = 1, 2, \cdots) \tag{15}$$

bildet. Hierbei ist die Ordnung eines Eigenwerts μ *genau gleich der Anzahl der Produkte* (15), *die den Wert* μ *haben.*

Für den Fall, daß einem der Kerne $K_\nu(s, t)$ kein einziger Eigenwert entspricht, besagt dieser Satz, daß auch der Produktkern keine Eigenwerte besitzt.

Aus dem Satz II kann man umgekehrt auch den Satz I ohne Mühe folgern.

§ 5.

Der Satz II läßt sich sehr leicht direkt beweisen, wenn man die Fredholmsche Theorie als bekannt voraussetzt.

Bezeichnet man nämlich unseren Produktkern (8) für den Augenblick mit

$$K\begin{bmatrix} s_1, & s_2, & \cdots, & s_n \\ t_1, & t_2, & \cdots, & t_n \end{bmatrix}$$

und den m^{ten} iterierten Kern mit

$$K^m\begin{bmatrix} s_1, & s_2, & \cdots, & s_n \\ t_1, & t_2, & \cdots, & t_n \end{bmatrix},$$

so wird z. B.

$$K^2\begin{bmatrix} s_1, & s_2, & \cdots, & s_n \\ t_1, & t_2, & \cdots, & t_n \end{bmatrix}$$

$$= \int_{a_1}^{b_1}\int_{a_2}^{b_2}\cdots\int_{a_n}^{b_n} K_1(s_1, r_1)\,K_2(s_2, r_2)\cdots K_n(s_n, r_n)\,K_1(r_1, t_1)\,K_2(r_2, t_2)\cdots K_n(r_n, t_n)\,dr_1\,dr_2\cdots dr_n.$$

Dieses Integral ist aber gleich

$$K_1^2(s_1, t_1)\,K_2^2(s_2, t_2)\cdots K_n^2(s_n, t_n).$$

Ebenso zeigt man allgemein

$$K^m\begin{bmatrix} s_1, & s_2, & \cdots, & s_n \\ t_1, & t_2, & \cdots, & t_n \end{bmatrix} = K_1^m(s_1, t_1)\,K_2^m(s_2, t_2)\cdots K_n^m(s_n, t_n).$$

Hieraus folgt

$$(16)\quad \int_{a_1}^{b_1}\int_{a_2}^{b_2}\cdots\int_{a_n}^{b_n} K^m\begin{bmatrix} s_1, & s_2, & \cdots, & s_n \\ s_1, & s_2, & \cdots, & s_n \end{bmatrix} ds_1\, ds_2 \cdots ds_n = \prod_{\nu=1}^{n}\int_{a_\nu}^{b_\nu} K_\nu^m(s_\nu, s_\nu)\, ds_\nu.$$

Für $m \geqq 3$ wird aber (vergl. § 2)

$$\int_{a_\nu}^{b_\nu} K^m(s_\nu, s_\nu)\, ds_\nu = \sum_{\alpha_\nu} \frac{1}{\lambda_{\nu\alpha_\nu}^m},$$

daher ist für $m \geqq 3$ das Integral (16) gleich der Summe der $(-m)^{\text{ten}}$ Potenzen der Zahlen (15). Hieraus folgt aber nach § 2 der zu beweisende Satz.

Bei diesem so einfachen Beweis spielt aber eine wesentliche Rolle die Tatsache, daß die Fredholmsche Funktion $D(x)$ von endlichem Geschlecht ist, was sich erst auf Grund der Poincaré-Hadamardschen Theorie der ganzen transzendenten Funktionen ergibt.

Abschnitt II.

Die assoziierten Kerne.

§ 6.

Es sei wieder eine Integralgleichung

$$\varphi(s) = \lambda\int_a^b K(s,t)\,\varphi(t)\,dt$$

mit stetigem Kern gegeben. Ich setze wie in der Einleitung

$$(17)\quad K\begin{pmatrix} s_1, & s_2, & \cdots, & s_n \\ t_1, & t_2, & \cdots, & t_n \end{pmatrix} = \frac{1}{n!}\begin{vmatrix} K(s_1,t_1), & K(s_1,t_2), & \cdots, & K(s_1,t_n) \\ K(s_2,t_1), & K(s_2,t_2), & \cdots, & K(s_2,t_n) \\ \cdot & \cdot & \cdot & \cdot \\ K(s_n,t_1), & K(s_n,t_2), & \cdots, & K(s_n,t_n) \end{vmatrix}$$

und nenne diesen Ausdruck, als Funktion der beiden Variabelnsysteme $s_1, s_2, \cdots, s_n$ und $t_1, t_2, \cdots, t_n$ betrachtet, den *zu* $K(s,t)$ *assoziierten Kern* n^{ten} *Grades*. Im folgenden bezeichne ich diesen Kern auch kurz mit K_n. Die zu studierende Integralgleichung ist die Gleichung:

$$(18)\quad \Phi(s_1,\cdots,s_n) = \mu\int_a^b\cdots\int_a^b K\begin{pmatrix} s_1, & \cdots, & s_n \\ t_1, & \cdots, & t_n \end{pmatrix}\Phi(t_1,\cdots,t_n)\,dt_1\cdots dt_n.$$

Hierbei ist zu bemerken, daß der Ausdruck K_n stets und nur dann identisch verschwindet, wenn sich $K(s)$ auf die Form

$$K(s,t) = f_1(s)\,g_1(t) + f_2(s)\,g_2(t) + \cdots + f_{n-1}(s)\,g_{n-1}(t)$$

bringen läßt.*) In diesem Fall besteht jedes vollständige System von Hauptfunktionen des Kerns aus höchstens $n-1$ Funktionen.

Ehe ich an das Studium der Integralgleichung (18) gehe, schicke ich einige vorbereitende Bemerkungen voraus.

Man habe n im Intervall $a \leqq s \leqq b$ eindeutig bestimmte (stetige) Funktionen

$$\varphi_1(s),\ \varphi_2(s),\cdots,\varphi_n(s).$$

Dann setze ich

$$\Delta\begin{pmatrix}\varphi_1, & \varphi_2, & \cdots, & \varphi_n\\ s_1, & s_2, & \cdots, & s_n\end{pmatrix} = \frac{1}{\sqrt{n!}}\begin{vmatrix}\varphi_1(s_1), & \varphi_1(s_2), & \cdots, & \varphi_1(s_n)\\ \varphi_2(s_1), & \varphi_2(s_2), & \cdots, & \varphi_2(s_2)\\ \cdot & \cdot & \cdot & \cdot\\ \varphi_n(s_1), & \varphi_n(s_2), & \cdots, & \varphi_n(s_n)\end{vmatrix}.$$

Diese Determinante ist, als Funktion von $s_1, s_2, \cdots, s_n$ betrachtet, stets und nur dann identisch Null, wenn die Funktionen untereinander linear abhängig sind.**) Allgemeiner zeigt man leicht durch den Schluß von n auf $n+1$: sind $m>n$ linear unabhängige Funktionen $\varphi_1(s), \varphi_2(s), \cdots, \varphi_m(s)$ gegeben, so sind auch die $\binom{m}{n}$ Determinanten

$$\Delta\begin{pmatrix}\varphi_{\alpha_1}, & \varphi_{\alpha_2}, & \cdots, & \varphi_{\alpha_n}\\ s_1, & s_2, & \cdots, & s_n\end{pmatrix}, \qquad (\alpha_1<\alpha_2<\cdots\alpha_n\leqq m)$$

als Funktionen von $s_1, s_2, \cdots, s_n$ betrachtet, untereinander linear unabhängig.

Sind ferner neben den Funktionen $\varphi_\alpha(s)$ noch n andere (stetige) Funktionen

$$\psi_1(s),\ \psi_2(s),\cdots,\psi_n(s)$$

gegeben, so gilt die leicht zu beweisende Formel

$$(19)\qquad \int_a^b\cdots\int_a^b \Delta\begin{pmatrix}\varphi_1, & \cdots, & \varphi_n\\ t_1, & \cdots, & t_n\end{pmatrix}\Delta\begin{pmatrix}\psi_1, & \cdots, & \psi_n\\ t_1, & \cdots, & t_n\end{pmatrix} dt_1\cdots dt_n$$

$$= \left|\int_a^b \varphi_\alpha(t)\,\psi_\beta(t)\,dt\right|. \qquad (\alpha,\beta=1,2,\cdots,n)$$

Aus dieser allgemeinen Relation ergeben sich zwei wichtige Eigenschaften der Determinante (17). Setzt man

$$\varphi_\alpha(t) = K(s_\alpha, t),$$

so wird

$$K\begin{pmatrix}s_1, & \cdots, & s_n\\ t_1, & \cdots, & t_n\end{pmatrix} = \frac{1}{\sqrt{n!}}\Delta\begin{pmatrix}\varphi_1, & \cdots, & \varphi_n\\ t_1, & \cdots, & t_n\end{pmatrix}$$

---

*) Vergl. E. Goursat, a. a. O., S. 80.

**) Vergl. die Anmerkung am Schluß des § 3.

und

$$\int_a^b \varphi_\alpha(t)\,\psi_\beta(t)\,dt = \int_a^b K(s_\alpha, t)\,\psi_\beta(t)\,dt.$$

Aus (19) folgt daher

(20) $$\int_a^b \cdots \int_a^b K\begin{pmatrix} s_1, \cdots, s_n \\ r_1, \cdots, r_n \end{pmatrix} \Delta\begin{pmatrix} \psi_1, \cdots, \psi_n \\ r_1, \cdots, r_n \end{pmatrix} dr_1 \cdots dr_n$$

$$= \frac{1}{\sqrt{n!}} \begin{vmatrix} \int_a^b K(s_1,r)\,\psi_1(r)\,dr, & \cdots, & \int_a^b K(s_1,r)\psi_n(r)\,dr \\ \cdot & \cdot\ \cdot\ \cdot & \cdot \\ \int_a^b K(s_n,r)\,\psi_1(r)\,dr, & \cdots, & \int_a^b K(s_n,r)\,\psi_n(r)\,dr \end{vmatrix}.$$

Ist ferner $L(s, t)$ ein beliebiger zweiter (stetiger) Kern und setzt man

$$\int_a^b K(s, r)\, L(r, t)\, dr = M(s, t),$$

so ergibt sich aus (20), indem man hierin $\psi_\alpha(r) = L(r, t_\alpha)$ setzt, die Formel

(21) $$\int_a^b \cdots \int_a^b K\begin{pmatrix} s_1, \cdots, s_n \\ r_1, \cdots, r_n \end{pmatrix} L\begin{pmatrix} r_1, \cdots, r_n \\ t_1, \cdots, t_n \end{pmatrix} dr_1 \cdots dr_n = M\begin{pmatrix} s_1, \cdots, s_n \\ t_1, \cdots, t_n \end{pmatrix}.$$

Hieraus folgt insbesondere die Regel:

Aus dem zu $K(s, t)$ assozierten Kern des Grades n geht durch m-fache Iteration der zu dem m^{ten} iterierten Kern $K^m(s, t)$ assoziierte Kern des Grades n hervor.

Diesen Kern bezeichne ich passend mit

$$K^m\begin{pmatrix} s_1, s_2, \cdots, s_n \\ t_1, t_2, \cdots, t_n \end{pmatrix}.$$

Ist wie früher

$$U_m = \int_a^b K^m(s, s)\, ds$$

und setzt man

$$U_m^{(n)} = \int_a^b \cdots \int_a^b K^m\begin{pmatrix} s_1, \cdots, s_n \\ s_1, \cdots, s_n \end{pmatrix} ds_1 \cdots ds_n,$$

so erhält man aus (4) die Formel

(22) $$U_m^{(n)} = \sum \frac{(-1)^{\nu_2 + \nu_4 + \cdots}}{\nu_1!\,\nu_2! \cdots \nu_n!} \left(\frac{U_m}{1}\right)^{\nu_1} \left(\frac{U_{2m}}{2}\right)^{\nu_2} \cdots \left(\frac{U_{nm}}{n}\right)^{\nu_n}$$
$$(\nu_1 + 2\nu_2 + \cdots + n\nu_n = n)$$

§ 7.

Ich beweise nun den Satz:

III. *Bilden die Funktionen:*

$$\varphi_1(s),\ \varphi_2(s),\ \cdots$$

ein vollständiges System von Hauptfunktionen des Kerns $K(s,t)$, *so erhält man ein vollständiges System von Hauptfunktionen des Kerns*

$$K\begin{pmatrix} s_1, & s_2, & \cdots, & s_n \\ t_1, & t_2, & \cdots, & t_n \end{pmatrix},$$

indem man für alle Indicessysteme $\alpha_1 < \alpha_2 < \cdots < \alpha_n$ *die Determinanten*

$$\Delta\begin{pmatrix} \varphi_{\alpha_1}, & \varphi_{\alpha_2}, & \cdots, & \varphi_{\alpha_n} \\ s_1, & s_2, & \cdots, & s_n \end{pmatrix} \tag{23}$$

aufstellt. *)

Ich setze zur Abkürzung

$$\Delta\begin{pmatrix} \varphi_{\alpha_1}, & \varphi_{\alpha_2}, & \cdots, & \varphi_{\alpha_n} \\ s_1, & s_2, & \cdots, & s_n \end{pmatrix} = \Phi_{\alpha_1, \alpha_2, \cdots, \alpha_n}$$

und denke mir diese Funktionen so angeordnet, daß $\Phi_{\alpha_1, \alpha_2, \cdots, \alpha_n}$ vor $\Phi_{\beta_1, \beta_2, \cdots, \beta_n}$ zu stehen kommt, sobald die erste nicht verschwindende unter den Differenzen

$$\alpha_n - \beta_n,\ \alpha_{n-1} - \beta_{n-1},\ \cdots,\ \alpha_2 - \beta_2,\ \alpha_1 - \beta_1$$

negativ ist. Die Anordnung ist also folgende

$$\Phi_{1,2,\cdots,n},\quad \Phi_{1,2,\cdots,n-1,n+1},\quad \Phi_{1,2,\cdots,n-2,n,n+1},\cdots$$

In dieser Reihenfolge bezeichne ich die Funktionen mit

$$\Phi^{(1)},\ \Phi^{(2)},\ \Phi^{(3)},\ \ldots.$$

Wir haben nun wieder nachzuweisen, daß diese Funktionen in Bezug auf den Kern K_n den drei für ein vollständiges System von Hauptfunktionen charakteristischen Bedingungen (vergl. § 1) genügen.

Zunächst erkennt man sofort, daß die erste Bedingung erfüllt ist.

Es sei ferner

$$\int_a^b K(s,t)\,\varphi_\alpha(t)\,dt = c_{\alpha 1}\,\varphi_1(s) + \cdots + c_{\alpha,\alpha-1}\,\varphi_{\alpha-1}(t) + c_{\alpha\alpha}\,\varphi_\alpha(s).$$

*) Ist die Anzahl der Funktionen $\varphi_\alpha(s)$ kleiner als n und der Kern K_n nicht identisch Null, so besagt der Satz, daß dieser Kern überhaupt keine Hauptfunktionen, also auch keine Eigenwerte besitzt.

Bezeichnet man für die Indiceskombination $\alpha_1 < \alpha_2 < \cdots < \alpha_n$ den größten Index α_n auch mit m und setzt

$$d_{\gamma\delta} = c_{\alpha_\gamma \delta} \qquad (\gamma = 1, 2, \cdots, n, \ \delta = 1, 2, \cdots, m)$$

für $\delta \leqq \alpha_\gamma$ und $d_{\gamma\delta} = 0$ für $\delta > \alpha_\gamma$, so erhält man

$$\int_a^b K(s, t)\, \varphi_{\alpha_\gamma}(t)\, dt = \sum_{\delta=1}^{m} d_{\gamma\delta}\, \varphi_\delta(s) \qquad (\gamma = 1, 2, \cdots, n).$$

Ist nun

$$\Phi_{\alpha_1, \alpha_2, \cdots, \alpha_n} = \Phi^{(\varrho)},$$

so ergibt sich aus (20)

$$\int_a^b \cdots \int_a^b K\begin{pmatrix} s_1, \cdots, s_n \\ t_1, \cdots, t_n \end{pmatrix} \Phi^{(\varrho)}(t_1, \cdots, t_n)\, dt_1 \cdots dt_n$$

$$= \frac{1}{\sqrt{n!}} \begin{vmatrix} \sum_{\delta=1}^{m} d_{1\delta}\, \varphi_\delta(s_1), & \cdots, & \sum_{\delta=1}^{m} d_{1\delta}\, \varphi_\delta(s_n) \\ \cdot & \cdot & \cdot \\ \sum_{\delta=1}^{m} d_{n\delta}\, \varphi_\delta(s_1), & \cdots, & \sum_{\delta=1}^{m} d_{n\delta}\, \varphi_\delta(s_n) \end{vmatrix}.$$

Nach dem verallgemeinerten Multiplikationstheorem für Determinanten läßt sich der rechtsstehende Ausdruck auf die Form

$$\frac{1}{\sqrt{n!}} \sum \begin{vmatrix} d_{1\delta_1}, & \cdots, & d_{1\delta_n} \\ \cdot & \cdot & \cdot \\ d_{n\delta_1}, & \cdots, & d_{n\delta_n} \end{vmatrix} \begin{vmatrix} \varphi_{\delta_1}(s_1), & \cdots, & \varphi_{\delta_1}(s_n) \\ \cdot & \cdot & \cdot \\ \varphi_{\delta_n}(s_1), & \cdots, & \varphi_{\delta_n}(s_n) \end{vmatrix} \qquad (\delta_1 < \cdots < \delta_n \leqq m)$$

bringen. Unter Berücksichtigung der Werte der Konstanten $d_{\gamma\delta}$ erkennt man nun sofort, daß diese Summe, wie zu beweisen ist, die Gestalt

$$C_{\varrho 1}\, \Phi^{(1)}(s_1, \cdots, s_n) + \cdots + C_{\varrho, \varrho-1}\, \Phi^{(\varrho-1)}(s_1, \cdots, s_n) + C_{\varrho\varrho}\, \Phi^{(\varrho)}(s_1, \cdots, s_n)$$

besitzt; hierbei sind die $C_{\varrho\sigma}$ gewisse Konstanten und insbesondere

$$(24) \qquad C_{\varrho\varrho} = c_{\alpha_1\alpha_1}\, c_{\alpha_2\alpha_2} \cdots c_{\alpha_n\alpha_n} \neq 0.$$

Der Nachweis, daß unsere Funktionen $\Phi^{(1)}, \Phi^{(2)}, \cdots$ auch die dritte noch zu prüfende Bedingung erfüllen, gestaltet sich nun am einfachsten folgendermaßen.

Es sei μ ein Eigenwert des assoziierten Kerns, und es mögen die r Funktionen

$$\Psi_1(s_1, \cdots, s_n), \cdots, \Psi_r(s_1, \cdots, s_n)$$

ein zu μ gehörendes vollständiges invariantes System repräsentieren. Es sei etwa

$$(25)\qquad \int_a^b\cdots\int_a^b K\binom{s_1,\cdots,s_n}{t_1,\cdots,t_n}\Psi_\varrho(t_1,\cdots,t_n)\,dt_1\cdots dt_n$$
$$=\sum_{\sigma=1}^{r} a_{\varrho\sigma}\Psi_\sigma(s_1,\cdots,s_n);$$

dann ist wieder die Determinante $|a_{\varrho\sigma}|$ gleich $\left(\frac{1}{\mu}\right)^r$, also von Null verschieden. Nun ist $K\binom{s_1,\cdots,s_n}{t_1,\cdots,t_n}$ für jedes Wertsystem $t_1,\cdots,t_n$ eine *alternierende* Funktion von $s_1,\cdots,s_n$, d. h. eine Funktion, die bei einer Vertauschung je zweier der Variabeln das Vorzeichen ändert. Folglich besitzt auch das in (25) links stehende Integral diese Eigenschaft. Da sich aber jede Funktion Ψ_ϱ wegen $|a_{\varrho\sigma}|\neq 0$ umgekehrt linear und homogen mit konstanten Koeffizienten durch diese Integrale darstellen läßt, so ist auch Ψ_ϱ eine alternierende Funktion von $s_1, s_2,\cdots,s_n$. Wird also für die Permutation $\begin{pmatrix}1 & 2 & \cdots & n\\ \nu_1 & \nu_2 & \cdots & \nu_n\end{pmatrix}$ der n Indices $1, 2,\cdots, n$ unter $\varepsilon_{\nu_1,\nu_2,\ldots,\nu_n}$ die Zahl $+1$ oder -1 verstanden, je nachdem die Permutation gerade oder ungerade ist, so kann

$$(26)\qquad \Psi_\varrho(s_{\nu_1}, s_{\nu_2},\cdots,s_{\nu_n})=\varepsilon_{\nu_1,\nu_2,\ldots,\nu_n}\Psi_\varrho(s_1,s_2,\cdots,s_n)$$

gesetzt werden. Da andererseits

$$K\binom{s_1,s_2,\cdots,s_n}{t_1,t_2,\cdots,t_n}=\frac{1}{n!}\sum\varepsilon_{\nu_1,\nu_2,\ldots,\nu_n}K(s_1,t_{\nu_1})\,K(s_2,t_{\nu_2})\cdots K(s_n,t_{\nu_n})$$

ist, so erhalten wir

$$\int_a^b\cdots\int_a^b K\binom{s_1,\cdots,s_n}{t_1,\cdots,t_n}\Psi_\varrho(t_1,\cdots,t_n)\,dt_1\cdots dt_n$$
$$=\frac{1}{n!}\sum\varepsilon_{\nu_1,\ldots,\nu_n}\int_a^b\cdots\int_a^b K(s_1,t_{\nu_1})\cdots K(s_n,t_{\nu_n})\Psi_\varrho(t_1,\cdots,t_n)\,dt_1\cdots dt_n$$
$$=\frac{1}{n!}\sum\int_a^b\cdots\int_a^b K(s_1,t_{\nu_1})\cdots K(s_n,t_{\nu_n})\Psi_\varrho(t_{\nu_1},\cdots,t_{\nu_n})\,dt_1\cdots dt_n$$
$$=\frac{1}{n!}\sum\int_a^b\cdots\int_a^b K(s_1,t_{\nu_1})\cdots K(s_n,t_{\nu_n})\Psi_\varrho(t_{\nu_1},\cdots,t_{\nu_n})\,dt_{\nu_1}\cdots dt_{\nu_n}$$
$$=\int_a^b\cdots\int_a^b K(s_1,t_1)\cdots K(s_n,t_n)\Psi_\varrho(t_1,\cdots,t_n)\,dt_1\cdots dt_n.$$

Es ist daher auch

$$(27)\qquad \int_a^b \cdots \int_a^b K(s_1, t_1) \cdots K(s_n, t_n)\, \Psi_\varrho(t_1, \cdots, t_n)\, dt_1 \cdots dt_n$$
$$= \sum_{\sigma=1}^{n} a_{\varrho\sigma}\, \Psi_\sigma(s_1, \cdots, s_n).$$

Für den hier auftretenden Produktkern

$$K(s_1, t_1)\, K(s_2, t_2) \cdots K(s_n, t_n)$$

kennen wir aber nach Satz I ein vollständiges System von Hauptfunktionen; es sind das die Produkte

$$(28)\qquad \varphi_{\beta_1}(s_1)\, \varphi_{\beta_2}(s_2) \cdots \varphi_{\beta_n}(s_n) \qquad (\beta_\nu = 1, 2, \cdots).$$

Aus der Gleichung (27) ergibt sich daher auf Grund der Ausführungen am Schluß des § 1, daß sich jede Funktion $\Psi_\varrho(s_1, s_2, \cdots, s_n)$ linear und homogen mit konstanten Koeffizienten durch gewisse endlich viele unter den Funktionen (28) ausdrücken läßt. Ist aber etwa

$$\Psi_\varrho(s_1, s_2, \cdots, s_n) = \sum A_{\beta_1, \beta_2, \cdots, \beta_n}\, \varphi_{\beta_1}(s_1)\, \varphi_{\beta_2}(s_2) \cdots \varphi_{\beta_n}(s_n),$$

so ist auch für jede Permutation $\nu_1, \nu_2, \cdots, \nu_n$ der Indices $1, 2, \cdots, n$

$$\Psi_\varrho(s_{\nu_1}, s_{\nu_2}, \cdots, s_{\nu_n}) = \sum A_{\beta_1, \beta_2, \cdots, \beta_n}\, \varphi_{\beta_1}(s_{\nu_1})\, \varphi_{\beta_2}(s_{\nu_2}) \cdots \varphi_{\beta_n}(s_{\nu_n}).$$

Multipliziert man mit $\varepsilon_{\nu_1, \nu_2, \cdots, \nu_n}$ und addiert über alle $n!$ Permutationen, so erhält man wegen (26)

$$n!\, \Psi_\varrho(s_1, s_2, \cdots, s_n)$$
$$= \sum_\beta A_{\beta_1, \beta_2, \cdots, \beta_n} \sum_\nu \varepsilon_{\nu_1, \nu_2, \cdots, \nu_n}\, \varphi_{\beta_1}(s_{\nu_1})\, \varphi_{\beta_2}(s_{\nu_2}) \cdots \varphi_{\beta_n}(s_{\nu_n}).$$

Diese Gleichung kann auch in der Form

$$\Psi_\varrho(s_1, s_2, \cdots, s_n) = \frac{1}{\sqrt{n!}} \sum A_{\beta_1, \beta_2, \cdots, \beta_n}\, \Delta \begin{pmatrix} \varphi_{\beta_1}, & \varphi_{\beta_2}, & \cdots, & \varphi_{\beta_n} \\ s_1, & s_2, & \cdots, & s_n \end{pmatrix}$$

geschrieben werden. Dies lehrt uns aber, daß Ψ_ϱ in der Tat durch gewisse endlich viele unter den Funktionen (23) linear und homogen mit konstanten Koeffizienten darstellbar ist.

§ 8.

Aus dem Satz III folgt wieder, daß die reziproken Werte der in den Gleichungen (24) auftretenden Konstanten $C_{\varrho\varrho}$ uns die sämtlichen Eigenwerte des Kerns K_n liefern. Hieraus ergibt sich:

IV. *Sind*

$$\lambda_1, \lambda_2, \ldots$$

die Eigenwerte des Kerns $K(s,t)$, jeder Eigenwert so oft geschrieben, wie seine Ordnung angibt, so erhält man in den Produkten

$$\lambda_{\alpha_1}\lambda_{\alpha_2}\cdots\lambda_{\alpha_n} \qquad (\alpha_1<\alpha_2<\cdots<\alpha_n) \tag{29}$$

die Gesamtheit der Eigenwerte μ des zu $K(s,t)$ assoziierten Kerns n^{ten} Grades. Hierbei ist die Ordnung eines Eigenwerts μ genau gleich der Anzahl der Produkte (29), *die den Wert μ haben.*

Besitzt der Kern $K(s,t)$ keinen Eigenwert oder weniger als n Eigenwerte, so besagt dieser Satz, daß der assoziierte Kern n^{ten} Grades entweder identisch Null ist oder keinen Eigenwert besitzt.

§ 9.

Im Rahmen der Fredholmschen Theorie ist der Beweis des Satzes IV folgendermaßen zu führen.

Man setze wie früher für $m \geqq 2$

$$l_m = \sum_\alpha \frac{1}{\lambda_\alpha^m},$$

ferner sei

$$l_m^{(n)} = \sum \frac{1}{(\lambda_{\alpha_1}\lambda_{\alpha_2}\cdots\lambda_{\alpha_n})^m} \quad (\alpha_1<\alpha_2<\cdots<\alpha_n;\ m\geqq 2).$$

Dann erscheint $l_m^{(n)}$ gewissermaßen als die n^{te} elementare symmetrische Funktion der Größen

$$\frac{1}{\lambda_1^m}, \frac{1}{\lambda_2^m}, \ldots.$$

Auf Grund der bekannten Waringschen Formel für die Darstellung der elementaren symmetrischen Funktionen durch die Potenzsummen, einer Formel, die auch für unendlich viele Größen mit absolut konvergenter Summe gilt, wird nun

$$l_m^{(n)} = \sum \frac{(-1)^{\nu_2+\nu_4+\cdots}}{\nu_1!\,\nu_2!\cdots\nu_n!}\left(\frac{l_m}{1}\right)^{\nu_1}\left(\frac{l_{2m}}{2}\right)^{\nu_2}\cdots\left(\frac{l_{nm}}{n}\right)^{\nu_n} \quad (\nu_1+2\nu_2+\cdots+n\nu_n=n).$$

Vergleicht man diese Formel mit der Formel (22) und beachtet, daß für $\nu \geqq 3$ die Zahl l_ν gleich U_ν wird, so erhält man für $m \geqq 3$ auch

$$l_m^{(n)} = U_m^{(n)}.$$

Aus diesen Gleichungen folgt aber auf Grund des in § 2 Gesagten der zu beweisende Satz.

§ 10.

Aus Satz IV folgt (vergl. § 1) für jedes n

$$(30)\qquad \sum_{\alpha_1<\alpha_2<\cdots<\alpha_n} \frac{1}{|\lambda_{\alpha_1}\lambda_{\alpha_2}\cdots\lambda_{\alpha_n}|^2}$$

$$\leqq \int_a^b \cdots \int_a^b \int_a^b \cdots \int_a^b \left| K\begin{pmatrix} s_1, \cdots, s_n \\ t_1, \cdots, t_n \end{pmatrix} \right|^2 ds_1 \cdots ds_n\, dt_1 \cdots dt_n.$$

Setzt man

$$\int_a^b K(s, r)\, \overline{K}(t, r)\, dr = M(s, t),^{*)}$$

so wird (vergl. Formel (21))

$$(31)\qquad \int_a^b \cdots \int_a^b K\begin{pmatrix} s_1, \cdots, s_n \\ r_1, \cdots, r_n \end{pmatrix} \overline{K}\begin{pmatrix} t_1, \cdots, t_n \\ r_1, \cdots, r_n \end{pmatrix} dr_1 \cdots dr_n = M\begin{pmatrix} s_1, \cdots, s_n \\ t_1, \cdots, t_n \end{pmatrix}.$$

Die Ungleichung (30) kann daher auch in der Form

$$(32)\qquad \sum \frac{1}{|\lambda_{\alpha_1}\lambda_{\alpha_2}\cdots\lambda_{\alpha_n}|^2} \leqq \int_a^b \cdots \int_a^b M\begin{pmatrix} s_1, \cdots, s_n \\ t_1, \cdots, t_n \end{pmatrix} ds_1 \cdots ds_n$$

geschrieben werden. — Für den Fall, daß $K(s, t)$ ein reeller (stetiger) Kern ist, hat nun Herr E. Schmidt**) bewiesen, daß der Kern $M(s, t)$ (ebenfalls im Gebiet $a \leqq s \leqq b$, $a \leqq t \leqq b$ betrachtet) nur positive Eigenwerte

$$(33)\qquad \mu_1, \mu_2, \cdots$$

besitzt, deren Anzahl mindestens gleich 1 ist; kommt ferner in (33) jeder Eigenwert so oft vor, wie seine Ordnung angibt, so ist $\sum_\alpha \frac{1}{\mu_\alpha}$ konvergent und gleich $\int_a^b M(s, s)\, ds$. Der Schmidtsche Beweis dieses Satzes läßt sich aber ohne Mühe auch auf den Fall übertragen, daß $K(s, t)$ ein nicht reeller stetiger Kern ist. Der Satz gilt ferner mutatis mutandis auch dann, wenn s und t Punkte in einem n-dimensionalen Raume bedeuten, er läßt sich daher insbesondere wegen (31) auf den Kern $M\begin{pmatrix} s_1, \cdots, s_n \\ t_1, \cdots, t_n \end{pmatrix}$ anwenden. Da nun die Eigenwerte dieses Kerns die Produkte

$$\mu_{\alpha_1}\mu_{\alpha_2}\cdots\mu_{\alpha_n} \qquad (\alpha_1 < \alpha_2 < \cdots < \alpha_n)$$

---

*) Hierin bedeutet $\overline{K}(t, r)$ die zu $K(t, r)$ konjugiert komplexe Funktion.

**) Math. Annalen, Bd. 63, S. 433 (§§ 13—19).

sind, so wird auch

$$\sum \frac{1}{\mu_1 \mu_2 \cdots \mu_n} = \int_a^b \cdots \int_a^b M\begin{pmatrix} s_1, \cdots, s_n \\ s_1, \cdots, s_n \end{pmatrix} ds_1 \cdots ds_n. ^{*)}$$

Aus (32) folgt daher die Ungleichung

$$\sum \frac{1}{|\lambda_{\alpha_1} \lambda_{\alpha_2} \cdots \lambda_{\alpha_n}|^2} \leqq \sum \frac{1}{\mu_{\alpha_1} \mu_{\alpha_2} \cdots \mu_{\alpha_n}} \quad (\alpha_1 \leqq \alpha_2 \leqq \cdots \leqq \alpha_n).$$

Hierzu sei noch bemerkt, daß, wenn die Anzahl der Eigenwerte $\mu_1, \mu_2, \cdots$ kleiner als n ist, der Kern $M\begin{pmatrix} s_1, s_2, \cdots, s_n \\ t_1, t_2, \cdots, t_n \end{pmatrix}$ identisch Null sein muß. Aus (31) folgt dann, daß auch der Kern $K\begin{pmatrix} s_1, s_2, \cdots, s_n \\ t_1, t_2, \cdots, t_n \end{pmatrix}$ identisch verschwinden muß, d. h. $K(s, t)$ hat die Form

$$K(s, t) = f_1(s)\, g_1(t) + f_2(s)\, g_2(t) + \cdots + f_{n-1}(s)\, g_{n-1}(t).$$

Man kann unser Resultat auch folgendermaßen aussprechen: entwickelt man die beiden beständig konvergenten Produkte

$$\prod_\alpha \left(1 + \frac{x}{|\lambda_\alpha|^2}\right), \quad \prod_\alpha \left(1 + \frac{x}{\mu_\alpha}\right)$$

nach Potenzen von x, so ist in der ersten Reihe jeder Koeffizient höchstens gleich dem entsprechenden Koeffizienten in der zweiten Reihe.

Abschnitt III.

Über die Berechnung der Eigenwerte und der Eigenfunktionen eines reellen symmetrischen Kerns.

§ 11.

Wir machen nun die Annahme, daß die zu betrachtende Funktion $K(s, t)$ reell, stetig und symmetrisch ist, d. h. der Gleichung $K(s, t) = K(t, s)$ genügt.

Ein solcher Kern $K(s, t)$ besitzt mindestens einen Eigenwert (vergl. Einleitung). Alle Eigenwerte

$$\lambda_1, \lambda_2, \cdots$$

---

*) Diese Gleichung läßt sich auch aus den Formeln

$$\sum_\alpha \frac{1}{\mu_\alpha^\nu} = \int_a^b M^\nu(s, s)\, ds \qquad (\nu = 1, 2, \cdots)$$

folgern.

sind reell; ferner kann man ein vollständiges System von Hauptfunktionen

$$\psi_1(s),\ \psi_2(s),\ \cdots$$

so wählen, daß jede Funktion $\psi_\alpha(s)$ eine reelle *Eigenfunktion* des Kerns wird. Es genügt daher, sich allein auf die Betrachtung der Eigenfunktionen des Kerns zu beschränken.

Die Eigenwerte des iterierten Kerns $K^2(s,t)$ sind die Größen

$$\lambda_1^2,\ \lambda_2^2,\ \cdots$$

und die Funktionen $\psi_1(s), \psi_2(s), \cdots$ bilden auch für $K^2(s,t)$ ein vollständiges System von Eigenfunktionen. Kennt man umgekehrt ein vollständiges System von Eigenfunktionen

$$(34) \qquad \varphi_1(s),\ \varphi_2(s),\ \cdots$$

des Kerns $K^2(s,t)$, so ist zwar im allgemeinen nicht jedes $\varphi_\alpha(s)$ eine Eigenfunktion von $K(s,t)$, aber jede zu einem Eigenwert λ gehörende Eigenfunktion von $K(s,t)$ läßt sich durch diejenigen unter den $\varphi_\alpha(s)$, die zum Eigenwert λ^2 von $K^2(s,t)$ gehören, linear und homogen mit konstanten Koeffizienten darstellen. Es genügt daher, die Eigenwerte und die Eigenfunktionen des Kerns $K^2(s,t)$ zu kennen, um auch die analoge Aufgabe für den Kern $K(s,t)$ als gelöst anzusehen.

Die Funktionen (34) können auch so gewählt werden, daß sie reell und *normiert orthogonal* sind, d. h. den Bedingungen

$$(35) \qquad \int_a^b \varphi_\alpha(s)\,\varphi_\beta(s)\,ds = e_{\alpha\beta} \qquad \begin{Bmatrix} e_{11} = e_{22} = \cdots = 1 \\ e_{12} = e_{13} = \cdots = 0 \end{Bmatrix}$$

genügen. Die Funktion $\varphi_\alpha(s)$ möge zum Eigenwert γ_α von $K^2(s,t)$ gehören, und es sei noch

$$\gamma_1 \leqq \gamma_2 \leqq \cdots.$$

Herr E. Schmidt*) hat nun die Existenz mindestens eines Eigenwerts des Kerns $K(s,t)$ nachgewiesen, indem er einen Eigenwert γ von $K^2(s,t)$ direkt angegeben hat. Setzt man nämlich wie früher

$$U_m = \int_a^b K^m(s,s)\,ds,$$

so werden für einen reellen symmetrischen Kern $K(s,t)$ die Größen $U_2, U_4, \cdots$ positive Zahlen. Herr Schmidt zeigt nun, daß der Ausdruck $\frac{U_{2\nu+2}}{U_{2\nu}}$ mit wachsendem ν gegen einen positiven Grenzwert c konvergiert, und daß $\gamma = \frac{1}{c}$ einen Eigenwert von $K^2(s,t)$ repräsentiert. Er beweist

*) a. a. O., § 11.

noch genauer, daß der Ausdruck $\frac{K^{2\nu}(s,t)}{c^\nu}$ mit wachsendem ν *gleichmäßig* gegen eine mithin stetige Funktion $u(s,t)$ konvergiert, die für jedes t, das $u(s,t)$ nicht identisch in s zum Verschwinden bringt, eine zum Eigenwert γ gehörende Eigenfunktion von $K^2(s,t)$ darstellt. Außerdem existiert auch der Grenzwert von $\frac{U_{2\nu}}{c^\nu}$ und es ist

$$\lim_{\nu=\infty} \frac{U_{2\nu}}{c^\nu} = \int_a^b u(s,s)\,ds = U \geqq 1. \tag{36}$$

§ 12.

Diese Resultate lassen sich noch etwas genauer fassen.

Daß γ gleich γ_1, also der kleinste Eigenwert von $K^2(s,t)$ ist, und daß U nichts anderes bedeutet, als die Ordnung des Eigenwerts γ, hat schon Herr A. Kneser in seiner in der Einleitung zitierten Arbeit durch Betrachtung der Fredholmschen Funktion $D(x)$ gezeigt. Man kann dies aber auch direkt auf elementarem Wege beweisen; zugleich erkennt man auch die eigentliche Bedeutung der Funktion $u(s,t)$.

Ist nämlich $\varphi(s)$ irgend eine Eigenfunktion von $K^2(s,t)$, die zum Eigenwert γ gehört, so ist

$$c\varphi(t) = \int_a^b K^2(s,t)\,\varphi(s)\,ds,$$

und hieraus folgt für jedes ν

$$c^\nu \varphi(t) = \int_a^b K^{2\nu}(s,t)\,\varphi(s)\,ds,$$

also

$$\varphi(t) = \int_a^b \frac{K^{2\nu}(s,t)}{c^\nu}\,\varphi(s)\,ds. \tag{37}$$

Da nun

$$\frac{K^{2\nu}(s,t)}{c^\nu}\,\varphi(s)$$

mit wachsendem ν gleichmäßig gegen $u(s,t)\,\varphi(s)$ konvergiert, so erhält man aus (37), indem man ν ins Unendliche wachsen läßt,

$$\varphi(t) = \int_a^b u(s,t)\,\varphi(s)\,ds. \tag{38}$$

Ist dagegen $\psi(s)$ eine zu einem von γ verschiedenen Eigenwert δ gehörende Eigenfunktion von $K^2(s,t)$, so ist, weil dann $\psi(s)$ und $u(s,t)$ zwei zu verschiedenen Eigenwerten eines reellen symmetrischen Kerns gehörende Eigenfunktionen sind*), für jedes t

$$\int_a^b u(s,t)\,\psi(s)\,ds = 0.$$

Bezeichnet man nun $\frac{1}{\delta}$ mit d, so wird wieder für jedes ν

$$d^\nu \psi(t) = \int_a^b K^{2\nu}(s,t)\,\psi(s)\,ds,$$

also

$$\left(\frac{d}{c}\right)^\nu \psi(t) = \int_a^b \frac{K^{2\nu}(s,t)}{c^\nu}\,\psi(s)\,ds.$$

Für genügend große Werte von ν unterscheidet sich das rechts stehende Integral beliebig wenig von $\int_a^b u(s,t)\,\psi(s)\,ds$. Da dies aber 0 ist, so erhält man $\lim_{\nu=\infty}\left(\frac{d}{c}\right)^\nu = 0$, d. h. es muß $d < c$, also $\gamma < \delta$ sein.

Ist nun $\gamma = \gamma_1$ von der Ordnung k, also $\gamma_1 = \gamma_2 = \cdots = \gamma_k < \gamma_{k+1}$, so wird jedenfalls $u(s,t)$ für jedes t eine lineare homogene Verbindung von $\varphi_1(s), \varphi_2(s), \cdots, \varphi_k(s)$ mit von s unabhängigen Koeffizienten; wir können also

$$u(s,t) = \varphi_1(s)\,\psi_1(t) + \varphi_2(s)\,\psi_2(t) + \cdots + \varphi_k(s)\,\psi_k(t)$$

setzen, wo die $\psi_\alpha(t)$ gewisse eindeutig bestimmte Funktionen von t sind. Multipliziert man aber mit $\varphi_\alpha(s)\,ds$ und integriert nach s von a bis b, so erhält man wegen (35) und (38)

$$\int_a^b u(s,t)\,\varphi_\alpha(s)\,ds = \varphi_\alpha(t) = \psi_\alpha(t).$$

Daher ist

$$u(s,t) = \varphi_1(s)\,\varphi_1(t) + \varphi_2(s)\,\varphi_2(t) + \cdots + \varphi_k(s)\,\varphi_k(t). \tag{39}$$

Setzt man noch $t = s$ und integriert nach s, so folgt aus (35) und (36)

$$U = k.$$

Ich bemerke noch folgendes: da die Funktionen $\varphi_1(t), \varphi_2(t), \cdots, \varphi_k(t)$ linear unabhängig sind, so kann man k Zahlen $t_1, t_2, \cdots, t_k$ wählen, für welche die Determinante

$$|\varphi_\alpha(t_\beta)| \qquad (\alpha, \beta = 1, 2, \cdots, k)$$

nicht verschwindet. Setzt man nun in (39) für t die Werte $t_1, t_2, \cdots, t_k$

*) Vergl. Schmidt, a. a. O., § 4.

ein, so erhält man für $\varphi_1(s), \varphi_2(s), \cdots, \varphi_k(s)$ k lineare Gleichungen, aus denen sich diese Funktionen in eindeutiger Weise bestimmen; es ergibt sich hierbei $\varphi_\alpha(s)$ in der Form

$$\varphi_\alpha(s) = c_{\alpha 1}\, u(s, t_1) + c_{\alpha 2}\, u(s, t_2) + \cdots + c_{\alpha k}\, u(s, t_k),$$

wo die $c_{\alpha\beta}$ gewisse Konstanten sind. Daher kann $u(s, t)$, als Funktion von s betrachtet, die *allgemeinste* zum Eigenwert γ gehörende Eigenfunktion von $K^2(s, t)$ genannt werden.

§ 13.

Die in den beiden letzten Paragraphen angegebenen Resultate gelten wieder auch dann, wenn s und t Punkte in einem n-dimensionalen Raume bedeuten, und an Stelle des Integrals $\int\limits_a^b$ ein n-faches Integral tritt; sie lassen sich daher auf den zu $K(s, t)$ assoziierten Kern

(40)
$$K\begin{pmatrix} s_1, & s_2, & \cdots, & s_n \\ t_1, & t_2, & \cdots, & t_n \end{pmatrix}$$

anwenden. In der Tat besitzt dieser Kern, wenn $K(s, t)$ reell, stetig und symmetrisch ist, offenbar ebenfalls diese Eigenschaften; die Symmetrie findet ihren Ausdruck darin, daß

$$K\begin{pmatrix} s_1, & s_2, & \cdots, & s_n \\ t_1, & t_2, & \cdots, & t_n \end{pmatrix} = K\begin{pmatrix} t_1, & t_2, & \cdots, & t_n \\ s_1, & s_2, & \cdots, & s_n \end{pmatrix}$$

ist. Verschwindet der Kern (40) nun nicht identisch, so muß er mindestens einen Eigenwert besitzen. Da aber nach Satz IV jeder seiner Eigenwerte die Form

$$\lambda_{\alpha_1}\, \lambda_{\alpha_2} \cdots \lambda_{\alpha_n} \qquad (\alpha_1 < \alpha_2 < \cdots < \alpha_n)$$

haben muß, so ergibt sich, daß, wenn die Anzahl der Eigenwerte $\lambda_1, \lambda_2, \cdots$ von $K(s, t)$ kleiner als n ist, der Ausdruck (40) identisch verschwinden muß, d. h. es muß $K(s, t)$ die Form

$$K(s, t) = f_1(s)\, g_1(t) + f_2(s)\, g_2(t) + \cdots + f_{n-1}(s)\, g_{n-1}(t)$$

besitzen.*)

Ist aber der Kern (40) nicht identisch Null, so verschwindet auch sein zweiter iterierter Kern

(41)
$$K^2\begin{pmatrix} s_1, & s_2, & \cdots, & s_n \\ t_1, & t_2, & \cdots, & t_n \end{pmatrix}$$

---

*) Vergl. Hilbert, a. a. O., S. 72 und Schmidt, a. a. O., § 8.

nicht identisch.*) Da dieser Kern nichts anderes ist, als der zu $K^2(s,t)$ assoziierte Kern, so erhält man seine Eigenwerte, indem man alle Produkte

$$\gamma_{\alpha_1}\gamma_{\alpha_2}\cdots\gamma_{\alpha_n} \qquad (\alpha_1<\alpha_2<\cdots<\alpha_n)$$

bildet. Die kleinste unter diesen Zahlen ist $\gamma_1\gamma_2\cdots\gamma_n$. Ist daher wie in § 6

$$U_m^{(n)}=\int_a^b\cdots\int_a^b K^m\begin{pmatrix}s_1, & \cdots, & s_n\\ s_1, & \cdots, & s_n\end{pmatrix}ds_1\cdots ds_n,$$

so erhält man durch Anwendung des Schmidtschen Satzes auf den Kern (41)

$$\lim_{\nu=\infty}\frac{U_{2\nu+2}^{(n)}}{U_{2\nu}^{(n)}}=\frac{1}{\gamma_1\gamma_2\cdots\gamma_n}.$$

Ebenso ist

$$\lim_{\nu=\infty}\frac{U_{2\nu+2}^{(n-1)}}{U_{2\nu}^{(n-1)}}=\frac{1}{\gamma_1\gamma_2\cdots\gamma_{n-1}}.$$

Hieraus ergibt sich

$$\gamma_n=\lim_{\nu=\infty}\frac{U_{2\nu+2}^{(n-1)}\,U_{2\nu}^{(n)}}{U_{2\nu}^{(n-1)}\,U_{2\nu+2}^{(n)}}. \tag{42}$$

So ist z. B. wegen der Formel (22)

$$\gamma_2=\lim_{\nu=\infty}\frac{U_{2\nu+2}\left(U_{2\nu}^2-U_{4\nu}\right)}{U_{2\nu}\left(U_{2\nu+2}^2-U_{4\nu+4}\right)},$$

$$\gamma_3=\lim_{\nu=\infty}\frac{\left(U_{2\nu+2}^2-U_{4\nu+4}\right)\left(U_{2\nu}^3-3U_{2\nu}U_{4\nu}+2U_{6\nu}\right)}{\left(U_{2\nu}^2-U_{4\nu}\right)\left(U_{2\nu+2}^3-3U_{2\nu+2}U_{4\nu+4}+2U_{6\nu+6}\right)}.$$

Auf Grund der Formeln (42) kann man also, sobald die Zahlen $U_2, U_4, \cdots$ bekannt sind, jede der Größen $\gamma_1, \gamma_2, \gamma_3, \cdots$ näherungsweise berechnen.

§ 14.

Setzt man

$$c_n=\frac{1}{\gamma_1\gamma_2\cdots\gamma_n}, \qquad (n=1,2,\cdots)$$

so konvergiert nach dem Früheren

$$\frac{1}{c_n^\nu}K^{2\nu}\begin{pmatrix}s_1, & s_2, & \cdots, & s_n\\ t_1, & t_2, & \cdots, & t_n\end{pmatrix}$$

---

*) Dies folgt (vergl. Schmidt, a. a. O., § 5) aus der Gleichung

$$K^2\begin{pmatrix}s_1, & \cdots, & s_n\\ s_1, & \cdots, & s_n\end{pmatrix}=\int_a^b\cdots\int_a^b\left[K\begin{pmatrix}s_1, & \cdots, & s_n\\ t_1, & \cdots, & t_n\end{pmatrix}\right]^2dt_1\cdots dt_n.$$

gleichmäßig gegen eine stetige Funktion

$$W\begin{pmatrix} s_1, s_2, \cdots, s_n \\ t_1, t_2, \cdots, t_n \end{pmatrix},$$

welche die allgemeinste zum Eigenwert $\gamma_1\gamma_2\cdots\gamma_n$ gehörende Eigenfunktion des Kerns (41) repräsentiert. Ebenso konvergiert der Ausdruck

$$\frac{1}{c_{n-1}^\nu} K^{2\nu}\begin{pmatrix} s_2, \cdots, s_n \\ t_2, \cdots, t_n \end{pmatrix}$$

gleichmäßig gegen die allgemeinste zum Eigenwert $\gamma_1\gamma_2\cdots\gamma_{n-1}$ gehörende Eigenfunktion

$$W\begin{pmatrix} s_2, \cdots, s_n \\ t_2, \cdots, t_n \end{pmatrix}$$

des Kerns

$$K^2\begin{pmatrix} s_2, \cdots, s_n \\ t_2, \cdots, t_n \end{pmatrix}. \tag{43}$$

Zugleich ist

$$W\begin{pmatrix} s_1, s_2, \cdots, s_n \\ t_1, t_2, \cdots, t_n \end{pmatrix} W\begin{pmatrix} s_2, \cdots, s_n \\ t_1, \cdots, t_n \end{pmatrix} = \lim_{\nu=\infty} \frac{1}{c_n^\nu c_{n-1}^\nu} K^{2\nu}\begin{pmatrix} s_1, s_2, \cdots, s_n \\ t_1, t_2, \cdots, t_n \end{pmatrix} K^{2\nu}\begin{pmatrix} s_2, \cdots, s_n \\ t_2, \cdots, t_n \end{pmatrix}.$$

Da auch hier die Konvergenz eine gleichmäßige ist, so darf unter dem Limeszeichen integriert werden. Insbesondere erhält man, indem man noch für s_1 und t_1 einfacher s und t schreibt,

$$\begin{aligned} u_n(s,t) &= \int_a^b \cdots \int_a^b \int_a^b \cdots \int_a^b W\begin{pmatrix} s, s_2, \cdots, s_n \\ t, t_2, \cdots, t_n \end{pmatrix} W\begin{pmatrix} s_2, \cdots, s_n \\ t_2, \cdots, t_n \end{pmatrix} ds_2 \cdots ds_n\, dt_2 \cdots dt_n \\ &= \lim_{\nu=\infty} \frac{1}{c_n^\nu c_{n-1}^\nu} \int_a^b \cdots \int_a^b \int_a^b \cdots \int_a^b K^{2\nu}\begin{pmatrix} s, s_2, \cdots, s_n \\ t, t_2, \cdots, t_n \end{pmatrix} K^{2\nu}\begin{pmatrix} s_2, \cdots, s_n \\ t_2, \cdots, t_n \end{pmatrix} ds_2 \cdots ds_n\, dt_2 \cdots dt_n, \end{aligned} \tag{44}$$

und auch hier ist die Konvergenz noch eine gleichmäßige.

Ich will nun zeigen, daß die hier erhaltene Funktion $u_n(s,t)$, abgesehen von einem konstanten Faktor, nichts anderes ist, als die allgemeinste zum Eigenwert γ_n gehörende Eigenfunktion des Kerns $K^2(s,t)$.

§ 15.

Es werde nämlich wie in § 7 für jedes Indicessystem $\alpha_1 < \alpha_2 < \cdots < \alpha_n$

$$\Phi_{\alpha_1,\alpha_2,\cdots,\alpha_n} = \Delta\begin{pmatrix} \varphi_{\alpha_1}, \varphi_{\alpha_2}, \cdots, \varphi_{\alpha_n} \\ s_1, s_2, \cdots, s_n \end{pmatrix} = \frac{1}{\sqrt{n!}} \begin{vmatrix} \varphi_{\alpha_1}(s_1), & \varphi_{\alpha_1}(s_2), & \cdots, & \varphi_{\alpha_1}(s_n) \\ \varphi_{\alpha_2}(s_1), & \varphi_{\alpha_2}(s_2), & \cdots, & \varphi_{\alpha_2}(s_n) \\ \cdot & \cdot & \cdot & \cdot \\ \varphi_{\alpha_n}(s_1), & \varphi_{\alpha_n}(s_2), & \cdots, & \varphi_{\alpha_n}(s_n) \end{vmatrix} \tag{45}$$

gesetzt. Dann erhält man aus (20)

$$\gamma_{\alpha_1}\gamma_{\alpha_2}\cdots\gamma_{\alpha_n}\int_a^b\cdots\int_a^b K^2\begin{pmatrix}s_1,\cdots,s_n\\ t_1,\cdots,t_n\end{pmatrix}\Phi_{\alpha_1,\alpha_2,\cdots,\alpha_n}(t_1,\cdots,t_n)\,dt_1\cdots dt_n$$
$$=\Phi_{\alpha_1,\alpha_2,\cdots,\alpha_n}(s_1,\cdots,s_n).$$

Die Funktion (45) ist daher eine zum Eigenwert $\gamma_{\alpha_1}\gamma_{\alpha_2}\cdots\gamma_{\alpha_n}$ gehörende Eigenfunktion des Kerns (41) und die Gesamtheit aller dieser Funktionen bildet nach Satz III ein vollständiges System von Eigenfunktionen des Kerns. Die Funktionen (45) sind aber wieder normiert orthogonal. Denn ist $\beta_1<\beta_2<\cdots<\beta_n$ irgend eine zweite Indiceskombination, so ergibt sich aus (19) und (35)

$$\int_a^b\cdots\int_a^b\Phi_{\alpha_1,\alpha_2,\cdots,\alpha_n}(s_1,\cdots,s_n)\,\Phi_{\beta_1,\beta_2,\cdots,\beta_n}(s_1,\cdots,s_n)\,ds_1\cdots ds_n$$
$$=\begin{vmatrix}e_{\alpha_1\beta_1}, & e_{\alpha_1\beta_2}, & \cdots, & e_{\alpha_1\beta_n}\\ e_{\alpha_2\beta_1}, & e_{\alpha_2\beta_2}, & \cdots, & e_{\alpha_2\beta_n}\\ \cdot & \cdot & \cdot & \cdot\\ e_{\alpha_n\beta_1}, & e_{\alpha_n\beta_2}, & \cdots, & e_{\alpha_n\beta_n}\end{vmatrix}.$$

Diese Determinante ist aber 1 oder 0, je nachdem die Indicessysteme $\alpha_1,\alpha_2,\cdots,\alpha_n$ und $\beta_1,\beta_2,\cdots,\beta_n$ übereinstimmen oder nicht.

Ebenso repräsentieren die Funktionen

$$(46)\qquad \Phi_{\alpha_2,\cdots,\alpha_n}=\Delta\begin{pmatrix}\varphi_{\alpha_2}, & \cdots, & \varphi_{\alpha_n}\\ s_2, & \cdots, & s_n\end{pmatrix}=\frac{1}{\sqrt{(n-1)!}}\begin{vmatrix}\varphi_{\alpha_2}(s_2), & \cdots, & \varphi_{\alpha_2}(s_n)\\ \cdot & \cdot & \cdot\\ \varphi_{\alpha_n}(s_2), & \cdots, & \varphi_{\alpha_n}(s_n)\end{vmatrix}$$

ein vollständiges System von normiert orthogonalen Eigenfunktionen des Kerns (43).

Weiter erhält man aus (45), indem man die Determinante nach den Elementen der ersten Spalte entwickelt,

$$\Phi_{\alpha_1,\alpha_2,\cdots,\alpha_n}=\frac{1}{\sqrt{n}}\{\varphi_{\alpha_1}(s_1)\,\Phi_{\alpha_2,\alpha_3,\cdots,\alpha_n}-\varphi_{\alpha_2}(s_1)\,\Phi_{\alpha_1,\alpha_3,\cdots,\alpha_n}+\cdots\}.$$

Hat man nun ein System von $n-1$ Indices $\beta_2<\beta_3<\cdots<\beta_n$, das von den n Indicessystemen

$$(47)\qquad (\alpha_2,\alpha_3,\cdots,\alpha_n),\ (\alpha_1,\alpha_3,\cdots,\alpha_n),\ \cdots,\ (\alpha_1,\alpha_2,\cdots,\alpha_{n-1})$$

verschieden ist, so wird infolge der Orthogonalität der Funktionen (46)

$$(48)\qquad \int_a^b\cdots\int_a^b\Phi_{\alpha_1,\alpha_2,\cdots,\alpha_n}(s_1,s_2,\cdots,s_n)\,\Phi_{\beta_1,\beta_2,\cdots,\beta_n}(s_2,\cdots,s_n)\,ds_2\cdots ds_n=0.$$

Bedeutet dagegen $(\beta_2, \beta_3, \cdots, \beta_n)$ die m^{te} unter den Indiceskombinationen (47), so erhält man

$$(48') \quad \int_a^b \cdots \int_a^b \Phi_{\alpha_1, \alpha_2, \ldots, \alpha_n}(s_1, s_2, \cdots, s_n)\, \Phi_{\beta_2, \beta_3, \ldots, \beta_n}(s_2, \cdots, s_n)\, ds_2 \cdots ds_n = \frac{(-1)^{m-1}}{\sqrt{n}}\, \varphi_{\alpha_m}(s_1).$$

§ 16.

Es sei nun die Ordnung des Eigenwerts γ_n von $K^2(s, t)$ gleich l. Sind ferner etwa

$$\gamma_{m+1}, \gamma_{m+2}, \cdots, \gamma_{m+l}$$

gleich γ_n, so sei $n = m + r$. Unter den Produkten $\gamma_{\alpha_1} \gamma_{\alpha_2} \cdots \gamma_{\alpha_n}$ sind dann, wie man leicht erkennt, genau $\binom{l}{r}$ gleich $\gamma_1 \gamma_2 \cdots \gamma_n$, nämlich die Produkte

$$\gamma_1 \gamma_2 \cdots \gamma_m \gamma_{m+\varkappa_1} \gamma_{m+\varkappa_2} \cdots \gamma_{n+\varkappa_r} \quad (\varkappa_1 < \varkappa_2 < \cdots < \varkappa_r;\ \varkappa_\varrho = 1, 2, \ldots, l).$$

Setzt man noch der Kürze wegen

$$\Phi_{1,2,\ldots,m,m+\varkappa_1,m+\varkappa_2,\ldots,m+\varkappa_r} = F_{\varkappa_1,\varkappa_2,\ldots,\varkappa_r},$$

so wird in Analogie mit Gleichung (39)

$$W\begin{pmatrix} s_1, s_2, \cdots, s_n \\ t_1, t_2, \cdots, t_n \end{pmatrix} = \sum_{\varkappa_1 < \varkappa_2 < \cdots < \varkappa_r} F_{\varkappa_1,\varkappa_2,\ldots,\varkappa_r}(s_1, s_2, \cdots, s_n)\, F_{\varkappa_1,\varkappa_2,\ldots,\varkappa_r}(t_1, t_2, \cdots, t_n).$$

Wird ebenso, falls $r > 1$ ist,

$$\Phi_{1,2,\ldots,m,m+\varkappa_1,m+\varkappa_2,\ldots,m+\varkappa_{r-1}} = F_{\varkappa_1,\varkappa_2,\ldots,\varkappa_{r-1}}$$

gesetzt, so ergibt sich analog für den Kern (43)

$$W\begin{pmatrix} s_2, \cdots, s_n \\ t_2, \cdots, t_n \end{pmatrix} = \sum_{\varkappa_1 < \varkappa_2 < \cdots < \varkappa_{r-1}} F_{\varkappa_1,\varkappa_2,\ldots,\varkappa_{r-1}}(s_2, \cdots, s_n)\, F_{\varkappa_1,\varkappa_2,\ldots,\varkappa_{r-1}}(t_2, \cdots, t_n).$$

Für $r = 1$ wird dagegen, da dann $\gamma_1 \gamma_2 \cdots \gamma_{n-1}$ für den Kern (43) ein Eigenwert der Ordnung 1 ist,

$$W\begin{pmatrix} s_2, \cdots, s_n \\ t_2, \cdots, t_n \end{pmatrix} = \Phi_{1,2,\ldots,n-1}(s_2, \cdots, s_n)\, \Phi_{1,2,\ldots,n-1}(t_2, \cdots, t_n).$$

Auf Grund der Formeln (48) und (48') erhält man nun durch eine einfache Rechnung

$$\int_a^b \cdots \int_a^b \int_a^b \cdots \int_n^b W\begin{pmatrix} s_1, s_2, \cdots, s_n \\ t_1, t_2, \cdots, t_n \end{pmatrix} W\begin{pmatrix} s_2, \cdots, s_n \\ t_2, \cdots, t_n \end{pmatrix} ds_2 \cdots ds_n\, dt_2 \cdots dt_n = \frac{1}{n} \binom{l-1}{r-1} \sum_{\varkappa=1}^{l} \varphi_{m+\varkappa}(s_1)\, \varphi_{m+\varkappa}(t_1),$$

oder, was dasselbe ist,

$$u_n(s,t) = \frac{1}{n}\binom{l-1}{r-1}\sum_{\varkappa=1}^{l}\varphi_{m+\varkappa}(s)\,\varphi_{m+\varkappa}(t). \tag{49}$$

Für $r = 1$ hat man hierbei $\binom{l-1}{r-1} = 1$ zu setzen, auch dann, wenn $l = 1$ ist.

Da die Funktionen $\varphi_{m+1}, \varphi_{m+2}, \cdots, \varphi_{m+l}$ nach Voraussetzung die einzigen zum Eigenwert γ_n gehörenden Eigenfunktionen des Kerns $K^2(s,t)$ sind, so erscheint in der Tat in dem früher gekennzeichneten Sinn $u_n(s,t)$ (als Funktion von s betrachtet) als die allgemeinste zu γ_n gehörende Eigenfunktion von $K^2(s,t)$.

§ 17.

Ich will noch zeigen, daß das in (44) hinter dem Limeszeichen stehende Integral sich direkt durch die Größen $U_{2\alpha}$ und die iterierten Kerne $K^{2\alpha}(s,t)$ ausdrücken läßt.

Entwickelt man nämlich, wenn $K(s,t)$ ein beliebiger Kern ist, die Determinante

$$K\begin{pmatrix} s, & s_2, & \cdots, & s_n \\ t, & t_2, & \cdots, & t_n \end{pmatrix} = \frac{1}{n!}\begin{vmatrix} K(s,t), & K(s,t_2), & \cdots, & K(s,t_n) \\ K(s_2,t), & K(s_2,t_2), & \cdots, & K(s_2,t_n) \\ \cdot & \cdot & \cdot & \cdot \\ K(s_n,t), & K(s_n,t_2), & \cdots, & K(s_n,t_n) \end{vmatrix}$$

nach den Elementen der ersten Spalte, so erhält man

$$K\begin{pmatrix} s, & s_2, & \cdots, & s_n \\ t, & t_2, & \cdots, & t_n \end{pmatrix}$$

$$= \frac{1}{n}\left[K(s,t)\,K\begin{pmatrix} s_2, & s_3, & \cdots, & s_n \\ t_2, & t_3, & \cdots, & t_n \end{pmatrix} - K(s_2,t)\,K\begin{pmatrix} s, & s_3, & \cdots, & s_n \\ t_2, & t_3, & \cdots, & t_n \end{pmatrix} + \cdots\right].$$

Auf Grund der Formel (21) folgt daher, wenn noch

$$\int_a^b\cdots\int_a^b K\begin{pmatrix} s, & s_2, & \cdots, & s_n \\ t, & t_2, & \cdots, & t_n \end{pmatrix} K\begin{pmatrix} t_2, & \cdots, & t_n \\ u_2, & \cdots, & u_n \end{pmatrix} dt_2\cdots dt_n = K\begin{Bmatrix} s, & s_2, & \cdots, & s_n \\ t, & u_2, & \cdots, & u_n \end{Bmatrix}$$

gesetzt wird,

$$K\begin{Bmatrix} s, & s_2, & \cdots, & s_n \\ t, & u_2, & \cdots, & u_n \end{Bmatrix}$$

$$= \frac{1}{n}\left[K(s,t)\,K^2\begin{pmatrix} s_2, & s_3, & \cdots, & s_n \\ u_2, & u_3, & \cdots, & u_n \end{pmatrix} - K(s_2,t)\,K^2\begin{pmatrix} s, & s_3, & \cdots, & s_n \\ u_2, & u_3, & \cdots, & u_n \end{pmatrix} + \cdots\right]$$

also

$$\begin{Bmatrix} s, & s_2, & \cdots, & s_n \\ t, & u_2, & \cdots, & u_n \end{Bmatrix} = \frac{1}{n!}\begin{vmatrix} K(s,t), & K^2(s,u_2), & \cdots, & K^2(s,u_n) \\ K(s_2,t), & K^2(s_2,u_2), & \cdots, & K^2(s_2,u_n) \\ \cdot & \cdot & \cdot & \cdot \\ K(s_n,t), & K^2(s_n,u_2), & \cdots, & K^2(s_n,u_n) \end{vmatrix}. \tag{50}$$

Ich behaupte nun, daß

$$\int_a^b \cdots \int_a^b K\begin{Bmatrix} s, & s_2, & \cdots, & s_n \\ t, & s_2, & \cdots, & s_n \end{Bmatrix} ds_2 \cdots ds_n \tag{51}$$

$$= \frac{1}{n} \sum_{p=1}^{n} (-1)^{p-1} U_2^{(n-p)} K^{(2p-1)}(s, t)$$

ist, wobei noch $U_2^{(0)} = 1$ zu setzen ist.

Man denke sich nämlich die Determinante (50) für $u_2 = s_2, \cdots, u_n = s_n$ als eine Summe von $n!$ Gliedern geschrieben, die den $n!$ Permutationen der Ziffern $1, 2, \cdots, n$ entsprechen. Diejenigen $(n-p)!$ Permutationen, in denen die Ziffer 1 in einem festen Zyklus $(1, \beta_2, \cdots, \beta_p)$ vorkommen, liefern, wie man leicht erkennt, die Teilsumme

$$(-1)^{p-1} \frac{(n-p)!}{n!} K^2(s, s_{\beta_2}) K^2(s_{\beta_2}, s_{\beta_3}) \cdots K^2(s_{\beta_{p-1}}, s_{\beta_p}) K(s_{\beta_p}, t)$$

$$K^2\begin{pmatrix} s_{\varrho_1}, & \cdots, & s_{\varrho_{n-p}} \\ s_{\varrho_1}, & \cdots, & s_{\varrho_{n-p}} \end{pmatrix},$$

wo $\varrho_1, \cdots, \varrho_{n-p}$ die von $1, \beta_2, \cdots, \beta_p$ verschiedenen Ziffern (der Größe nach geordnet) bedeuten. Integriert man nun diese Teilsummen nach $s_2, \cdots, s_n$ oder, was dasselbe ist, nach $s_{\beta_2}, \cdots, s_{\beta_p}, s_{\varrho_1}, \cdots, s_{\varrho_{n-p}}$, so erhält man

$$(-1)^{p-1} \frac{(n-p)!}{n!} K^{2p-1}(s, t) \cdot U_2^{(n-p)}.$$

Die Anzahl der Zykeln der Ordnung p, in denen die Ziffer 1 vorkommt, ist aber gleich

$$\binom{n-1}{p-1}(p-1)! = \frac{(n-1)!}{(n-p)!}.$$

Daher wird das Integral (51) in der Tat gleich

$$\sum_{p=1}^{n} (-1)^{p-1} \frac{(n-1)!}{(n-p)!} \frac{(n-p)!}{n!} K^{2p-1}(s, t) U_2^{(n-p)}$$

$$= \frac{1}{n} \sum_{p=1}^{n} (-1)^{p-1} K^{2p-1}(s, t) U_2^{(n-p)}$$

Tritt nun an Stelle von $K(s, t)$ der Kern $K^{2\nu}(s, t)$, so ist in (51) $U_2^{(n-p)}$ durch $U_{4\nu}^{(n-p)}$ und $K^{(2p-1)}(s, t)$ durch $K^{(4p-2)\nu}(s, t)$ zu ersetzen. Ist ferner $K(s, t)$, wie in unserem Fall, symmetrisch, so wird dann das Integral (51) gleich dem in (44) hinter dem Limeszeichen stehenden Integral. Die Formeln (44) und (51) liefern uns daher die merkwürdige Gleichung

$$(52)\qquad \lim_{\nu=\infty} \frac{1}{c_n^\nu\, c_{n-1}^\nu} \sum_{p=1}^{n} (-1)^{p-1}\, U_{4\nu}^{(n-p)}\, K^{(4p-2)\nu}(s,t)$$
$$= \binom{l-1}{r-1} \sum_{\varkappa=1}^{l} \varphi_{m+\varkappa}(s)\, \varphi_{m+\varkappa}(t).$$

Hierin hat man wieder $U_{4\nu}^{(0)} = 1$ und, wenn $r = 1$ ist, $\binom{l-1}{r-1} = 1$ zu setzen. Die l Funktionen $\varphi_{m+\varkappa}$ bedeuten, wie noch einmal hervorgehoben sei, die l zum Eigenwert γ_n gehörenden Eigenfunktionen; sie können deutlicher mit $\varphi_{n1}, \varphi_{n2}, \cdots, \varphi_{nl}$ bezeichnet werden. So erhält man z. B. für $n = 2$ und $n = 3$

$$\lim_{\nu=\infty} \gamma_1^{2\nu} \gamma_2^{\nu} \{ U_{4\nu} K^{2\nu}(s,t) - K^{6\nu}(s,t) \} = \binom{l-1}{r-1} \sum_{\varkappa=1}^{l} \varphi_{2\varkappa}(s)\, \varphi_{2\varkappa}(t),$$

$$\lim_{\nu=\infty} \gamma_1^{2\nu} \gamma_2^{2\nu} \gamma_3^{\nu} \left\{ \frac{1}{2} (U_{4\nu}^2 - U_{8\nu})\, K^{2\nu}(s,t) - U_{4\nu}\, K^{6\nu}(s,t) + K^{10\nu}(s,t) \right\}$$
$$= \binom{l-1}{r-1} \sum_{\varkappa=1}^{l} \varphi_{3\varkappa}(s)\, \varphi_{3\varkappa}(t).$$

Es ist noch zu bemerken, daß der in (52) links stehende Ausdruck gleichmäßig gegen seinen Grenzwert konvergiert (vergl. § 14). Setzt man nun $t = s$ und integriert nach s, so erhält man

$$(53)\qquad \lim_{\nu=\infty} \frac{1}{c_n^\nu\, c_{n-1}^\nu} \sum_{p=1}^{n} (-1)^{p-1}\, U_{4\nu}^{(n-p)}\, U_{(4p-2)\nu} = l \binom{l-1}{r-1}.$$

Andererseits ist aber, da $\binom{l}{r}$ die Ordnung des Eigenwerts $\gamma_1 \gamma_2 \cdots \gamma_n$ für den Kern (41) angibt, in Analogie mit der Formel (36)

$$(54)\qquad \lim_{\nu=\infty} \frac{U_{2\nu}^{(n)}}{c_n^\nu} = \binom{l}{r}.$$

Aus den beiden letzten Formeln ergibt sich durch Division

$$\lim_{\nu=\infty} \frac{1}{c_{n-1}^\nu\, U_{2\nu}^{(n)}} \sum_{p=1}^{n} (-1)^{p-1}\, U_{4\nu}^{(n-p)}\, U_{(4p-2)\nu} = r.$$

Man kann auch die Zahl $l - r + 1$ als einen Grenzwert darstellen. Haben nämlich für γ_{n-1} die Zahlen l' und r' dieselbe Bedeutung wie l und r für γ_n, so ist, falls $r > 1$ ist, $l' = l$, $r' = r - 1$; ist aber $r = 1$, so wird $r' = l'$. Daher wird in jedem Falle

$$\binom{l'}{r'} = \binom{l}{r-1}.$$

Es ist nun entsprechend der Formel (54)

$$\lim_{\nu=\infty} \frac{U_{2\nu}^{(n-1)}}{c_{n-1}^{\nu}} = \binom{l'}{r'} = \binom{l}{r-1}.$$

In Verbindung mit (53) ergibt sich hieraus durch Division

$$\lim_{\nu=\infty} \frac{1}{c_n^{\nu}\, U_{2\nu}^{(n-1)}} \sum_{p=1}^{n} (-1)^{p-1}\, U_{4\nu}^{(n-p)}\, U_{(4p-2)\nu} = l - r + 1.$$

Setzt man noch zur Abkürzung

$$\sum_{p=1}^{n} (-1)^{p-1}\, U_{4\nu}^{(n-p)}\, U_{(4p-2)\nu} = V_{2\nu}^{(n)},$$

so folgt aus (52) und (53)

$$(55)\quad \lim_{\nu=\infty} \frac{1}{V_{2\nu}^{(n)}} \sum_{p=1}^{n} (-1)^{p-1}\, U_{4\nu}^{(n-p)}\, K^{(4p-2)\nu}(s,t) = \frac{1}{l} \sum_{\varkappa=1}^{l} \varphi_{n\varkappa}(s)\, \varphi_{n\varkappa}(t),$$

wo wie früher $\varphi_{n\varkappa} = \varphi_{m+\varkappa}$ zu setzen ist. In dieser Formel kommen nur noch die Größen $U_{2\alpha}$ und die Funktionen $K^{2\alpha}(s,t)$ vor, die doch bei der ganzen Betrachtung als bekannt anzunehmen sind.

Zum Schluß will ich noch hervorheben, daß man die Formel (42) auch ohne Benutzung der assoziierten Kerne leicht beweisen kann, indem man die Gleichungen

$$U_{2\nu} = \sum_{\infty=1}^{\infty} \frac{1}{\gamma_\alpha^{\nu}} \qquad (\nu = 1, 2, \cdots)$$

als bekannt annimmt und die aus ihnen folgenden Gleichungen

$$U_{2\nu}^{(n)} = \sum_{\alpha_1<\alpha_2<\cdots} \frac{1}{(\gamma_{\alpha_1}\gamma_{\alpha_2}\cdots\gamma_{\alpha_n})^{\nu}} \qquad (\nu = 1, 2, \cdots)$$

betrachtet.*) Ebenso kann man die Formeln (52) und (55) aus den Schmidtschen Entwickelungsformeln**)

$$K^{2\nu}(s,t) = \sum_{\alpha=1}^{\infty} \frac{\varphi_\alpha(s)\,\varphi_\alpha(t)}{\gamma_\alpha^{\nu}} \qquad (\nu = 1, 2, \cdots)$$

ableiten. Doch erfordert dieser Beweis eine schwierige und wenig übersichtliche Rechnung.

---

*) Vergl. die Formeln und Zitate der §§ 2, 6 und 9.

**) Schmidt, a. a. O., § 8 und § 19.

16.

Über die Darstellung der symmetrischen und der alternierenden Gruppe durch gebrochene lineare Substitutionen

Journal für die reine und angewandte Mathematik 139, 155 - 250 (1911)

In der vorliegenden Arbeit behandle ich die Aufgabe, alle endlichen Gruppen von gebrochenen linearen Substitutionen (Kollineationen) zu bestimmen, die der symmetrischen oder der alternierenden Gruppe mit n Vertauschungsziffern einstufig isomorph sind. Diese Aufgabe wird so weit gefördert, daß eine genaue Übersicht über die gesuchten Kollineationsgruppen gewonnen wird.

Die symmetrische Gruppe mit n Vertauschungsziffern bezeichne ich im folgenden mit $\mathfrak{S}_n$, die alternierende Gruppe mit $\mathfrak{A}_n$.

Es genügt, unter den zu bestimmenden Kollineationsgruppen nur die *irreduziblen* (nicht zerlegbaren) zu kennen; ferner hat man zwei *äquivalente* Gruppen, d. h. zwei Gruppen, die ineinander linear transformierbar sind, als nicht verschieden anzusehen.

Unter den Gruppen gebrochener linearer Substitutionen, die der Gruppe $\mathfrak{S}_n$ oder $\mathfrak{A}_n$ isomorph sind, nehmen diejenigen eine besondere Stellung ein, die sich auch als Gruppen von $n!$ bzw. $\frac{n!}{2}$ ganzen (homogenen) linearen Substitutionen schreiben lassen. Diese Gruppen hat bereits Herr *Frobenius**) durch Berechnung der Charaktere der Gruppen $\mathfrak{S}_n$ und $\mathfrak{A}_n$

---

*) „Über die Charaktere der symmetrischen Gruppe", Sitzungsberichte der K. Preuß. Akademie, Berlin, 1900, S. 516; „Über die Charaktere der alternierenden Gruppe", ebenda, 1901, S. 303; „Über die charakteristischen Einheiten der symmetrischen Gruppe", ebenda, 1903, S. 328. — Auf anderem Wege habe ich die Charaktere der symmetrischen Gruppe in meiner Inaug.-Dissertation „Über eine Klasse von Matrizen, die sich einer gegebenen Matrix zuordnen lassen", (Berlin 1901) erhalten.

sämtlich bestimmt*). Eine einfache Methode zur wirklichen Herstellung dieser Gruppen habe ich später angegeben**).

Wir haben uns daher in dieser Arbeit nur mit denjenigen Gruppen zu beschäftigen, bei denen die Benutzung gebrochener linearer Substitutionen wesentlich ist. Eine solche Gruppe bezeichne ich mit $\mathfrak{S}_n^{(g)}$ oder $\mathfrak{A}_n^{(g)}$, je nachdem sie der Gruppe $\mathfrak{S}_n$ oder der Gruppe $\mathfrak{A}_n$ isomorph ist; entsprechend bezeichne ich die mit $\mathfrak{S}_n$ und $\mathfrak{A}_n$ isomorphen Gruppen, bei denen die gebrochenen linearen Substitutionen durch homogene lineare Substitutionen ersetzt werden können, mit $\mathfrak{S}_n^{(h)}$ und $\mathfrak{A}_n^{(h)}$.

Für $n<4$ existieren Gruppen $\mathfrak{S}_n^{(g)}$ und $\mathfrak{A}_n^{(g)}$ überhaupt nicht. Ist aber $n\geqq 4$, so ist, wie ich im folgenden zeige, die Anzahl der verschiedenen (nicht äquivalenten) irreduziblen Gruppen $\mathfrak{S}_n^{(g)}$ gleich der Anzahl v_n der Zerlegungen

$$(1.)\qquad n=\nu_1+\nu_2+\cdots+\nu_m \qquad (\nu_1>\nu_2>\cdots>\nu_m>0)$$

von n in voneinander verschiedene ganzzahlige Summanden, und zwar entspricht einer Zerlegung (1.) eine irreduzible Gruppe $\mathfrak{S}_n^{(g)}$ des Grades

$$f_{\nu_1,\nu_2,\ldots\nu_m}=2^{\left[\frac{n-m}{2}\right]}\frac{n!}{\nu_1!\,\nu_2!\cdots\nu_m!}\prod_{\alpha<\beta}\frac{\nu_\alpha-\nu_\beta}{\nu_\alpha+\nu_\beta}.$$

Hierbei bezeichne ich als den *Grad* einer Gruppe gebrochener linearer Substitutionen die um 1 vermehrte Anzahl der Variabeln, also die Anzahl der Variabeln in den zugehörigen homogenen linearen Substitutionen. Für die Zerlegung $n=n$ hat man die Zahl f_n gleich $2^{\left[\frac{n-1}{2}\right]}$ zu setzen. Bei $n=6$ sind die beiden Gruppen der Grade $f_6=4$ und $f_{3,2,1}=4$ als nicht voneinander verschieden anzusehen.

Auf die besonders interessante Gruppe $\mathfrak{S}_n^{(g)}$ des Grades $2^{\left[\frac{n-1}{2}\right]}$ hat bereits Herr *A. Wiman* in seiner wichtigen Arbeit „Über die Darstellung der symmetrischen und alternirenden Vertauschungsgruppen als Collineationsgruppen von möglichst geringer Dimensionszahl"***) hingewiesen, ohne jedoch genauer anzugeben, wie diese Gruppe für ein beliebiges n wirklich

*) Eine kurze Zusammenfassung der *Frobenius*schen Resultate findet man in § 43 dieser Arbeit.

**) „Über die Darstellung der symmetrischen Gruppe durch lineare homogene Substitutionen", Sitzungsberichte der K. Preuß. Akademie, Berlin, 1908, S. 664.

***) Math. Annalen, Bd. 52, S. 243.

herzustellen ist. Im Abschnitt VI gebe ich ein relativ einfaches Verfahren zur Bildung dieser Gruppe an.

Bei der Betrachtung der alternierenden Gruppe hat man folgendes zu beachten: Die Gruppe $\mathfrak{A}_n$ besitzt einen äußeren Automorphismus $A=\binom{P}{P'}$, wo P' aus P dadurch hervorgeht, daß man in den Zykeln der Permutation P zwei feste Ziffern, etwa die Ziffern 1 und 2 vertauscht. Aus jeder mit $\mathfrak{A}_n$ isomorphen Kollineationsgruppe $\mathfrak{K}$ erhält man daher eine zweite Gruppe $\mathfrak{K}'$ derselben Art, indem man für jede Permutation P von $\mathfrak{A}_n$ die zu P gehörende Kollineation von $\mathfrak{K}$ durch die zu P' gehörende ersetzt. Im folgenden bezeichne ich $\mathfrak{K}$ und $\mathfrak{K}'$ als *adjungierte* Gruppen.

Sieht man nun zwei adjungierte Gruppen auch dann, wenn sie nicht einander äquivalent sind, als nicht verschieden an, so wird die Anzahl der verschiedenen irreduziblen Gruppen $\mathfrak{A}_n^{(g)}$ für $n=4$ gleich Eins und für $n>4$, wie bei der symmetrischen Gruppe, gleich v_n. Die der Zerlegung (1.) entsprechende irreduzible Gruppe $\mathfrak{A}_n^{(g)}$ ist, wenn $n-m$ ungerade ist, gleich $f_{\nu_1,\nu_2,\dots\nu_n}$, und wenn $n-m$ gerade ist, gleich $\frac{1}{2}f_{\nu_1,\nu_2,\dots\nu_m}$. Diese allgemeinen Regeln erleiden jedoch eine Ausnahme für die beiden Fälle $n=6$ und $n=7$. Für $n=6$ sind zunächst von den $v_6=4$ eben erwähnten Gruppen $\mathfrak{A}_6^{(g)}$, deren Grade gleich $4, 4, 8, 20$ sind, die beiden Gruppen des Grades 4, wie bei der Gruppe $\mathfrak{S}_6$, als identisch anzusehen; außer den übrigbleibenden drei Gruppen gibt es aber noch sechs andere wesentlich verschiedene*) irreduzible Gruppen $\mathfrak{A}_6^{(g)}$ der Grade $3, 6, 6, 9, 12, 15$. Für $n=7$ kommen zu den $v_7=5$ dem allgemeinen Fall entsprechenden Gruppen $\mathfrak{A}_7^{(g)}$ noch elf andere irreduzible Gruppen der Grade $6, 6, 15, 15, 21, 21, 24, 24, 24, 24, 36$ hinzu.

Jede Gruppe $\mathfrak{S}_n^{(g)}$ und $\mathfrak{A}_n^{(g)}$ läßt sich als Gruppe von $2\cdot n!$ bzw. $2\cdot\frac{n!}{2}$ homogenen linearen Substitutionen schreiben. Diese Regel versagt nur für die beiden alternierenden Gruppen $\mathfrak{A}_6$ und $\mathfrak{A}_7$; hier kann die Mindestanzahl von homogenen linearen Substitutionen, in denen eine Gruppe $\mathfrak{A}_n^{(g)}$ $(n=6,7)$ geschrieben werden kann, auch gleich $3\cdot\frac{n!}{2}$ oder $6\cdot\frac{n!}{2}$ sein. Hierdurch erklärt sich das Auftreten der als Ausnahmefälle genannten Gruppen $\mathfrak{A}_6^{(g)}$ und $\mathfrak{A}_7^{(g)}$.

*) Hierbei werden zwei Gruppen mit konjugiert komplexen Koeffizienten als nicht voneinander verschieden betrachtet.

Besonders bemerkenswert ist das Vorhandensein zweier wesentlich verschiedener Gruppen $\mathfrak{A}_7^{(g)}$ des Grades 6, zu denen noch eine Gruppe $\mathfrak{A}_7^{(h)}$ desselben Grades hinzukommt. Die beiden Gruppen $\mathfrak{A}_7^{(g)}$ unterscheiden sich in erster Linie dadurch, daß die eine als Gruppe von $3 \cdot \frac{7!}{2}$, die andere erst als Gruppe von $6 \cdot \frac{7!}{2}$ homogenen linearen Substitutionen geschrieben werden kann. Diese beiden Gruppen hat Herr *Wiman**) bei der Untersuchung der mit der Gruppe $\mathfrak{A}_7$ isomorphen Kollineationsgruppen sechsten Grades übersehen.

Von den im vorstehenden aufgezählten Gruppen $\mathfrak{S}_n^{(g)}$ und $\mathfrak{A}_n^{(g)}$ waren bis jetzt, abgesehen von der in der Arbeit des Herrn *Wiman* erwähnten Gruppe $\mathfrak{S}_n^{(g)}$ des Grades $2^{\left[\frac{n-1}{2}\right]}$ und der zugehörigen Gruppe $\mathfrak{A}_n^{(g)}$ des Grades $2^{\left[\frac{n-2}{2}\right]}$, nur die binären, ternären und quaternären Gruppen bekannt. Die binären Gruppen $\mathfrak{A}_4^{(g)}$, $\mathfrak{S}_4^{(g)}$ und $\mathfrak{A}_5^{(g)}$ finden sich zuerst in geometrischer Einkleidung in der Arbeit des Herrn *H. A. Schwarz* „Über diejenigen Fälle, in welchen die *Gauß*ische hypergeometrische Reihe eine algebraische Funktion ihres vierten Elementes darstellt"**). Unabhängig hiervon hat diese drei Gruppen Herr *F. Klein* in seiner Arbeit „Über binäre Formen mit linearen Transformationen in sich selbst"***) aufgestellt und zugleich den Nachweis geführt, daß dies, abgesehen von zwei trivialen Fällen, die einzigen endlichen binären Substitutionsgruppen sind. Die Existenz einer ternären Gruppe $\mathfrak{A}_6^{(g)}$ hat zuerst Herr *Wiman*†) nachgewiesen, indem er gezeigt hat, daß eine schon früher von Herrn *Valentiner*††) aufgestellte ternäre Kollineationsgruppe mit der Gruppe $\mathfrak{A}_6$ isomorph ist. Unter den (irreduziblen) quaternären Kollineationsgruppen gibt es je eine Gruppe $\mathfrak{S}_4^{(g)}, \mathfrak{S}_6^{(g)}, \mathfrak{A}_6^{(g)}, \mathfrak{A}_7^{(g)}$ und je zwei Gruppen $\mathfrak{S}_5^{(g)}$ und $\mathfrak{A}_5^{(g)}$. Die Gruppen $\mathfrak{S}_6^{(g)}$ und $\mathfrak{A}_7^{(g)}$ hat zuerst Herr *F. Klein*†††) durch Betrachtungen aus der Liniengeomtrie gewonnen; jede dieser Gruppen enthält die Gruppe $\mathfrak{A}_6^{(g)}$ und je eine von den Gruppen $\mathfrak{S}_5^{(g)}$ und $\mathfrak{A}_5^{(g)}$ als Untergruppen. Die Aufzählung aller

*) a. a. O., S. 259 ff.

**) Dieses Journal, Bd. 75, S. 292.

***) Math. Annalen, Bd. 9, S. 183.

†) Math. Annalen, Bd. 47, S. 531.

††) Videnskabernes Selskabs Skrifter, 6. Raekke, Kopenhagen 1889, S. 64.

†††) Math. Annalen, Bd. 28, S. 499.

ternären und quaternären Kollineationsgruppen, die einer symmetrischen oder alternierenden Gruppe isomorph sind, hat *H. Maschke**) durchgeführt.

Im folgenden bediene ich mich der Methoden, die ich in meiner Arbeit „Über die Darstellung der endlichen Gruppen durch gebrochene lineare Substitutionen"**) auseinandergesetzt habe. Um eine genaue Übersicht über alle Gruppen $\mathfrak{S}_n^{(\varrho)}$ und $\mathfrak{A}_n^{(\varrho)}$ zu erhalten, hat man nur die Darstellungsgruppen von $\mathfrak{S}_n$ und $\mathfrak{A}_n$ aufzustellen und die ***Frobenius***schen Charaktere dieser Gruppen zu berechnen.

Die Gruppe $\mathfrak{S}_n$ besitzt für $n \geqq 4$ zwei Darstellungsgruppen $\mathfrak{T}_n$ und $\mathfrak{T}'_n$ derselben Ordnung $2 \cdot n!$, die nur für $n = 6$ einander isomorph sind. Jede dieser Gruppen hat eine invariante Untergruppe $\mathfrak{M}$ der Ordnung 2, die in dem Kommutator der Gruppe enthalten ist, und die Gruppen $\frac{\mathfrak{T}_n}{\mathfrak{M}}$ und $\frac{\mathfrak{T}'_n}{\mathfrak{M}}$ sind der Gruppe $\mathfrak{S}_n$ (einstufig) isomorph; $\mathfrak{T}_n$ und $\mathfrak{T}'_n$ unterscheiden sich dadurch, daß den Transpositionen von $\mathfrak{S}_n$ in $\mathfrak{T}_n$ Elemente der Ordnung 4, dagegen in $\mathfrak{T}'_n$ Elemente der Ordnung 2 entsprechen. Beide Gruppen lassen sich leicht auseinander ableiten; ich beschäftige mich im folgenden nur mit der Gruppe $\mathfrak{T}_n$.

Die Darstellungsgruppe von $\mathfrak{A}_n$ ist eindeutig bestimmt. Ist $n \geqq 4$, aber von 6 und 7 verschieden, so ist dies eine Gruppe $\mathfrak{B}_n$ der Ordnung $2 \cdot \frac{n!}{2}$, die in jeder der Gruppen $\mathfrak{T}_n$ und $\mathfrak{T}'_n$ als Untergruppe enthalten ist. Dagegen sind die Darstellungsgruppen von $\mathfrak{A}_6$ und $\mathfrak{A}_7$ von den Ordnungen $6 \cdot \frac{6!}{2}$ und $6 \cdot \frac{7!}{2}$.

Die Bestimmung der Darstellungsgruppen von $\mathfrak{S}_n$ und $\mathfrak{A}_n$ gelingt verhältnismäßig leicht mit Hilfe eines von Herrn *E. H. Moore* herrührenden Satzes über die Definition von $\mathfrak{S}_n$ und $\mathfrak{A}_n$ als abstrakte endliche Gruppen, der auch in den erwähnten Arbeiten von Herrn *Wiman* und *H. Maschke* eine wichtige Rolle spielt***). Wesentlich schwieriger gestaltet sich die Berechnung der Charaktere dieser Darstellungsgruppen; hierzu war insbeson-

---

*) Math. Annalen Bd. 51, S. 251.

**) Dieses Journal, Bd. 127, S. 20. Vgl. auch meine Arbeit „Untersuchungen über die Darstellung der endlichen Gruppen durch gebrochene lineare Substitutionen", ebenda, Bd. 132, S. 85. — Im folgenden zitiere ich die erste Arbeit mit D., die zweite mit U.

***) Auf anderem Wege hat Herr *de Séguier* (Comptes Rendus de l'acad. des sciences, Paris, T. 150 (1910), S. 599) die Darstellungsgruppen von $\mathfrak{S}_n$ und $\mathfrak{A}_n$ bestimmt. Bei der alternierenden Gruppe hat Herr *de Séguier* jedoch den wichtigen Ausnahmefall $n = 7$ übersehen.

dere ein genaueres Studium der Gruppe $\mathfrak{T}_n$ erforderlich, die zwar mit der symmetrischen Gruppe aufs engste verwandt, aber von wesentlich komplizierterem Aufbau ist. Die Lösung der Aufgabe gelingt mir zuletzt durch Einführung einer Klasse von symmetrischen Funktionen, die auch an und für sich von Interesse sind (Abschnitt IX).

Abschnitt I.

Die Darstellungsgruppen der Gruppen $\mathfrak{S}_n$ und $\mathfrak{A}_n$.

§. 1. Um das Verständnis des folgenden zu erleichtern, schicke ich einige Bemerkungen über die hier zur Anwendung kommenden Bezeichnungen voraus*).

Es sei $\mathfrak{H}$ eine endliche Gruppe der Ordnung h. Ordnet man den Elementen $A, B, \ldots$ von $\mathfrak{H}$ die h linearen Substitutionen (Kollineationen) von nicht verschwindenden Determinanten

$$\{A\} \qquad x_\mu = \frac{a_{\mu 1} y_1 + \cdots + a_{\mu, m-1} y_{m-1} + a_{\mu m}}{a_{m1} y_1 + \cdots + a_{m, m-1} y_{m-1} + a_{mm}}, \qquad (\mu = 1, 2, \ldots m-1)$$

$$\{B\} \qquad x_\mu = \frac{b_{\mu 1} y_1 + \cdots + b_{\mu, m-1} y_{m-1} + b_{\mu m}}{b_{m 1} y_1 + \cdots + b_{m, m-1} y_{m-1} + b_{mm}}, \ldots$$

zu, so bilden diese Substitutionen eine *Darstellung* (des Grades m) von $\mathfrak{H}$, wenn für je zwei Elemente A, B der Gruppe das Produkt $\{A\}\{B\}$ gleich wird der Substitution $\{AB\}$, die dem Produkt AB der beiden Elemente A und B entspricht. Hierbei brauchen die h Substitutionen $\{A\}, \{B\}, \ldots$ nicht sämtlich voneinander verschieden zu sein. Bezeichnet man die Koeffizientenmatrizen $(a_{\lambda\mu}), (b_{\lambda\mu}), \ldots$ mit $(A), (B), \ldots$, so besteht für je zwei Elemente A, B der Gruppe eine Gleichung

$$(2.) \qquad (A)(B) = r_{A,B}(AB),$$

wo $r_{A,B}$ eine gewisse Konstante bedeutet. Umgekehrt entspricht jedem System von h Matrizen $(A), (B), \ldots$, deren Determinanten nicht Null sind, und welche die Eigenschaft besitzen, daß für je zwei Elemente A, B von $\mathfrak{H}$ eine Gleichung der Form (2.) besteht, eine Darstellung von $\mathfrak{H}$ durch gebrochene lineare Substitutionen.

Jede der Matrizen $(A), (B), \ldots$, die den Substitutionen $\{A\}, \{B\}, \ldots$ entsprechen, ist nur bis auf einen Proportionalitätsfaktor bestimmt. Lassen sich diese Faktoren so wählen, daß die Zahlen $r_{A,B}$ sämtlich gleich Eins wer-

*) Vgl. D., Einleitung.

den, so bilden die Matrizen $(A),(B),\ldots$ selbst eine Darstellung der Gruppe $\mathfrak{H}$, die sich auch als eine Darstellung von $\mathfrak{H}$ durch die *ganzen* homogenen linearen Substitutionen

$$(A) \qquad x_\mu = a_{\mu 1} y_1 + a_{\mu 2} y_2 + \cdots + a_{\mu m} y_m,$$
$$(B) \qquad x_\mu = b_{\mu 1} y_1 + b_{\mu 2} y_2 + \cdots + b_{\mu m} y_m, \ \ldots \qquad (\mu = 1, 2, \ldots m)$$

auffassen läßt.

Zwei Darstellungen einer Gruppe durch ganze oder gebrochene lineare Substitutionen sind *äquivalent*, wenn sich die eine Darstellung in die andere durch eine ganze bzw. gebrochene lineare Transformation der Variabeln von nicht verschwindender Determinante überführen läßt. Eine Darstellung m-ten Grades durch ganze oder gebrochene lineare Substitutionen heißt ferner *irreduzibel*, wenn sich für keine ihr äquivalente Darstellung eine Zahl $k < m$ angeben läßt, so daß unter den Koeffizienten $a_{\lambda\mu}, b_{\lambda\mu}, \ldots$ ihrer Substitutionen diejenigen Null werden, bei denen $\lambda \leqq k$ und $\mu > k$ oder $\lambda > k$ und $\mu \leqq k$ ist.

Eine endliche Gruppe $\mathfrak{K}$, die eine aus invarianten Elementen von $\mathfrak{K}$ bestehende Untergruppe $\mathfrak{M}$ von der Art enthält, daß die Gruppe $\frac{\mathfrak{K}}{\mathfrak{M}}$ der Gruppe $\mathfrak{H}$ einstufig isomorph wird, nenne ich *eine durch die Gruppe $\mathfrak{M}$ ergänzte Gruppe von $\mathfrak{H}$*. Es möge, wenn

$$\mathfrak{K} = \mathfrak{M}A' + \mathfrak{M}B' + \cdots$$

ist, dem Element A von $\mathfrak{H}$ der Komplex $\mathfrak{M}A'$, dem Element B der Komplex $\mathfrak{M}B'$, usw., entsprechen. Man habe ferner irgendeine Darstellung $\varDelta'$ der Gruppe $\mathfrak{K}$ durch homogene lineare Substitutionen (Matrizen), bei der jedem Element der Untergruppe $\mathfrak{M}$ eine Matrix entspricht, die sich von der Einheitsmatrix nur um einen Zahlenfaktor unterscheidet*). Sind dann bei dieser Darstellung den Elementen $A', B', \ldots$ die Matrizen $(A), (B), \ldots$ zugeordnet, so bestehen für diese Matrizen Gleichungen der Form (2.). Daher gehört zu jeder solchen Darstellung $\varDelta'$ von $\mathfrak{K}$ durch homogene lineare Substitutionen eine Darstellung $\varDelta$ der Gruppe $\mathfrak{H}$ durch gebrochene lineare Substitutionen.

Die Gruppe $\mathfrak{K}$ läßt sich stets so wählen, daß auf diese Weise *alle* Darstellungen von $\mathfrak{H}$ durch gebrochene lineare Substitutionen erhalten werden können. Eine Gruppe $\mathfrak{K}$, die diese Eigenschaft besitzt, nenne ich *eine*

*) Diese Bedingung ist für eine irreduzible Darstellung von selbst erfüllt.

hinreichend ergänzte Gruppe von $\mathfrak{H}$. Wird die Ordnung einer solchen Gruppe möglichst klein, so bezeichne ich sie als eine *Darstellungsgruppe von* $\mathfrak{H}$. Kennt man also eine Darstellungsgruppe $\mathfrak{R}$ von $\mathfrak{H}$, so erhält man alle irreduziblen Darstellungen von $\mathfrak{H}$ durch gebrochene lineare Substitutionen, indem man alle irreduziblen Darstellungen von $\mathfrak{R}$ durch homogene lineare Substitutionen bestimmt.

Eine hinreichend ergänzte Gruppe $\mathfrak{R}$ von $\mathfrak{H}$ ist *stets und nur dann eine Darstellungsgruppe, wenn der Kommutator von* $\mathfrak{R}$ *alle Elemente der Untergruppe* $\mathfrak{M}$ *enthält.* Der Kommutator jeder Darstellungsgruppe ist ferner als abstrakte Gruppe allein durch die Gruppe $\mathfrak{H}$ eindeutig bestimmt. Dasselbe gilt für die Untergruppe $\mathfrak{M}$, die ich als den *Multiplikator* der Gruppe $\mathfrak{H}$ bezeichne. Eine Gruppe $\mathfrak{H}$, deren Multiplikator die Ordnung Eins hat, nenne ich eine *abgeschlossene* Gruppe.

§ 2. Die symmetrische Gruppe $\mathfrak{S}_n$ läßt sich durch die $n-1$ Transpositionen

$$S_1=(1,2),\ S_2=(2,3),\ \ldots\ S_{n-1}=(n-1,n)$$

erzeugen. Diese Transpositionen genügen den Gleichungen

$$\text{(I.)}\qquad S_\alpha^2=E,\ (S_\beta S_{\beta+1})^3=E,\ S_\gamma S_\delta=S_\delta S_\gamma,\qquad \left(\begin{smallmatrix}\alpha=1,2,\ldots n-1, & \beta=1,2,\ldots n-2\\ \gamma=1,2,\ldots n-3, & \delta=\gamma+2,\ldots n-1\end{smallmatrix}\right)$$

und es gilt, wie Herr *E. H. Moore**) bewiesen hat, der Satz:

„*Faßt man die Gleichungen* (I.) *als ein System von definierenden Relationen zwischen den* $n-1$ *erzeugenden Elementen* $S_1, S_2, \ldots S_{n-1}$ *auf, so ist die hierdurch definierte abstrakte Gruppe endlich und der Gruppe* $\mathfrak{S}_n$ *einstufig isomorph*“.

Man habe nun irgendeine Darstellung der Gruppe $\mathfrak{S}_n$ durch Kollineationen. Der Transposition S_α möge hierbei eine Kollineation mit der Koeffizientenmatrix A_α entsprechen; dann ist A_α also nur bis auf einen Zahlenfaktor eindeutig bestimmt. Wegen der Relationen (I.) bestehen für die A_α Gleichungen der Form

$$\text{(3.)}\qquad A_\alpha^2=a_\alpha E,$$

$$\text{(4.)}\qquad (A_\beta A_{\beta+1})^3=b_\beta E,$$

$$\text{(5.)}\qquad A_\gamma A_\delta=c_{\gamma\delta}A_\delta A_\gamma,$$

wo E die Einheitsmatrix ist, und $a_\alpha, b_\beta, c_{\gamma\delta}$ gewisse von Null verschiedene Konstanten bedeuten. Die Zahlen $c_{\gamma\delta}$ treten nur für $n>3$ auf, sie bleiben

*) Proceedings of the London Mathematical Society, Bd. 28 (1897), S. 357.

offenbar ungeändert, wenn die Matrizen A_α mit beliebigen Konstanten multipliziert werden, und sind also allein durch die betrachteten Kollineationen bestimmt.

Aus (5.) folgt

$$A_\gamma A_\delta A_\gamma^{-1} = c_{\gamma\delta} A_\delta.$$

Erhebt man auf beiden Seiten in die zweite Potenz, so ergibt sich wegen (3.)

$$(6.) \qquad c_{\gamma\delta}^2 = 1.$$

Nun sind in $S_\gamma = (\gamma, \gamma+1)$, $S_\delta = (\delta, \delta+1)$ wegen $\delta \geqq \gamma + 2$ die Ziffern $\gamma, \gamma+1, \delta, \delta+1$ voneinander verschieden. Sind weiter γ' und δ' zwei andere Indizes und ist wieder $\delta' \geqq \gamma' + 2$, so läßt sich in $\mathfrak{S}_n$ eine Permutation S angeben, welche die Indizes $\gamma, \gamma+1, \delta, \delta+1$ in die Indizes $\gamma', \gamma'+1, \delta', \delta'+1$ überführt. Es wird dann

$$S^{-1} S_\gamma S = S_{\gamma'}, \quad S^{-1} S_\delta S = S_{\delta'}.$$

Entsprechend wird, wenn bei unserer Darstellung der Permutation S eine Kollineation mit der Koeffizientenmatrix A zugeordnet ist,

$$A^{-1} A_\gamma A = c A_{\gamma'}, \quad A^{-1} A_\delta A = d A_{\delta'},$$

wo c und d gewisse von Null verschiedene Konstanten sind. Aus der Gleichung (5.) folgt nun

$$A^{-1} A_\gamma A \cdot A^{-1} A_\delta A = c_{\gamma\delta} A^{-1} A_\delta A \cdot A^{-1} A_\gamma A,$$

also

$$cd \cdot A_{\gamma'} A_{\delta'} = cd\, c_{\gamma\delta} A_{\delta'} A_{\gamma'} = cd\, c_{\gamma'\delta'} A_{\delta'} A_{\gamma'}.$$

Hieraus ergibt sich aber $c_{\gamma\delta} = c_{\gamma'\delta'}$, d. h. *alle Zahlen* $c_{\gamma\delta}$ *sind einander gleich.* Setzen wir

$$c_{\gamma\delta} = j,$$

so wird wegen (6.)

$$(7.) \qquad j = \pm 1.$$

Aus den Gleichungen (4.) folgt weiter

$$A_\beta A_{\beta+1} A_\beta = b_\beta A_{\beta+1}^{-1} A_\beta^{-1} A_{\beta+1}^{-1}.$$

Durch Erheben in die zweite Potenz erhält man leicht

$$(8.) \qquad b_\beta^2 = a_\beta^3 a_{\beta+1}^3.$$

Da wir nun die Matrizen A_α mit beliebigen Konstanten multiplizieren dürfen, so können wir die Größen a_α beliebig fixieren. Man setze zunächst

$$a_1 = a_2 = \cdots = a_{n-1} = j.$$

Dann wird wegen (7.) und (8.) $b_\beta = \pm 1$, und definiert man die Matrizen $B_1, B_2, \ldots B_{n-1}$ durch die Gleichungen

$$B_1 = A_1, \; B_2 = j b_1 \cdot A_2, \; B_3 = b_1 b_2 \cdot A_3, \; B_4 = j b_1 b_2 b_3 \cdot A_4, \ldots,$$

so genügen sie, wie man leicht erkennt, den Relationen

$$B_\alpha^2 = jE, \; (B_\beta B_{\beta+1})^3 = jE, \; B_\gamma B_\delta = j \cdot B_\delta B_\gamma.$$

Setzt man andererseits

$$a_1 = a_2 = \cdots = a_{n-1} = 1$$

und

$$C_1 = A_1, \; C_2 = b_1 \cdot A_1, \; C_3 = b_1 b_2 \cdot A_3, \; C_4 = b_1 b_2 b_3 \cdot A_4, \ldots,$$

so erhält man

$$C_\alpha^2 = E, \; (C_\beta C_{\beta+1})^3 = E, \; C_\gamma C_\delta = j \cdot C_\delta C_\gamma.$$

Ist nun $j = 1$, so werden die Relationen (I.) erfüllt, wenn man S_α durch $B_\alpha = C_\alpha$ ersetzt. Auf Grund des *Moore*schen Satzes ergibt sich daher, daß für $j = 1$ in unserer Darstellung die gebrochenen linearen Substitutionen durch homogene lineare Substitutionen ersetzt werden können. Dagegen ist dies gewiß nicht der Fall, wenn $j = -1$ ist. Für $n < 4$ kommt ferner die letztere Möglichkeit überhaupt nicht in Betracht.

§ 3. Es ist nun leicht, die Darstellungsgruppen von $\mathfrak{S}_n$ zu bestimmen. Wir bezeichnen mit $\mathfrak{T}_n$ die abstrakte endliche Gruppe, die durch das System der definierenden Relationen

$$\text{(II.)} \quad J^2 = E, \; T_\alpha^2 = J, \; (T_\beta T_{\beta+1})^3 = J, \; T_\gamma T_\delta = J T_\delta T_\gamma \quad \left(\begin{smallmatrix} \alpha = 1, 2, \ldots n-1, \; \beta = 1, 2, \ldots, n-2 \\ \gamma = 1, 2, \ldots, n-3, \; \delta = \gamma+2, \ldots n-1 \end{smallmatrix}\right)$$

zwischen den erzeugenden Elementen $J, T_1, T_2, \ldots T_{n-1}$ bestimmt ist. Ebenso sei $\mathfrak{T}_n'$ die durch die Relationen

$$\text{(II'.)} \quad J^2 = E, \; T_\alpha'^2 = E, \; (T_\beta' T_{\beta+1}')^3 = E, \; T_\gamma' T_\delta' = J T_\delta' T_\gamma', \; J T_\alpha' = T_\alpha' J$$

zwischen den erzeugenden Elementen $J, T_1', T_2', \ldots T_n'$ definierte Gruppe. In jeder der Gruppen $\mathfrak{T}_n$ und $\mathfrak{T}_n'$ ist J als invariantes Element enthalten, und bezeichnet man mit $\mathfrak{M}$ die Gruppe

$$\mathfrak{M} = E + J,$$

so werden die Gruppen $\frac{\mathfrak{T}_n}{\mathfrak{M}}$ und $\frac{\mathfrak{T}_n'}{\mathfrak{M}}$, wie ein Vergleichen der Formeln (II.) und (II'.) mit den Formeln (I.) lehrt, der Gruppe $\mathfrak{S}_n$ einstufig isomorph. Die Gruppen $\mathfrak{T}_n$ und $\mathfrak{T}_n'$ erscheinen also als zwei durch die Gruppe $\mathfrak{M}$ ergänzte Gruppen von $\mathfrak{S}_n$. Nun werden aber die Gleichungen (II.) befriedigt, wenn man das Element J durch die Matrix $j \cdot E$ und die Elemente T_α durch die Matrizen B_α ersetzt; ebenso werden die Gleichungen (II'.) erfüllt, wenn man für die Elemente J und T_α' die Matrizen $j \cdot E$ und C_α einsetzt. Aus jeder Darstellung der Gruppe $\mathfrak{S}_n$ durch gebrochene lineare Substitutionen geht daher sowohl eine Darstellung der Gruppe $\mathfrak{T}_n$, als auch eine Darstellung der Gruppe $\mathfrak{T}'_n$ durch *ganze* lineare Substitutionen hervor. Folglich sind $\mathfrak{T}_n$ und $\mathfrak{T}'_n$ als hinreichend ergänzte Gruppen von $\mathfrak{S}_n$ zu bezeichnen. Da ferner für $n \geqq 4$ wegen

$$J = T_1 T_3 T_1^{-1} T_3^{-1}, \quad J = T'_1 T'_3 T_1'^{-1} T_3'^{-1}$$

das Element J in dem Kommutator von $\mathfrak{T}_n$, bzw. $\mathfrak{T}'_n$ enthalten ist, *so sind für $n \geqq 4$ die Gruppen $\mathfrak{T}_n$ und $\mathfrak{T}'_n$ Darstellungsgruppen von $\mathfrak{S}_n$; der Multiplikator der Gruppe $\mathfrak{S}_n$ ist also für $n \geqq 4$ von der Ordnung* 2*).

Man beachte noch, daß der Kommutator von $\mathfrak{S}_n$ die alternierende Gruppe $\mathfrak{A}_n$ ist. Da der Index dieser Untergruppe gleich 2, also (für $n \geqq 4$) gleich der Ordnung des Multiplikators von $\mathfrak{S}_n$ ist, so ergibt sich (vgl. U., § 1), daß die Gruppe $\mathfrak{S}_n$ höchstens zwei einander nicht isomorphe Darstellungsgruppen besitzen kann. Wendet man aber das a. a. O. für den allgemeinen Fall einer endlichen Gruppe angegebene Verfahren an, um etwa von der Gruppe $\mathfrak{T}_n$ ausgehend zu einer zweiten Darstellungsgruppe von $\mathfrak{S}_n$ zu gelangen, so wird man gerade auf die Gruppe $\mathfrak{T}'_n$ geführt. Ist nun $n \neq 6$, so ist $\mathfrak{S}_n$ eine vollkommene Gruppe**), daher sind für $n \neq 6$ die Gruppen $\mathfrak{T}_n$ und $\mathfrak{T}'_n$ einander nicht isomorph***). Diese beiden Gruppen unterscheiden sich dadurch, daß den Transpositionen von $\mathfrak{S}_n$ in $\mathfrak{T}_n$ Elemente der Ordnung 4, dagegen in $\mathfrak{T}'_n$ Elemente der Ordnung 2 entsprechen.

Hieraus folgt auch leicht, daß $\mathfrak{T}_6$ und $\mathfrak{T}'_6$ isomorphe Gruppen sind. Denn die Gruppe $\mathfrak{S}_6$ besitzt bekanntlich einen äußeren Automorphismus, durch den jeder Transposition eine Permutation der Form $(\alpha\beta)(\gamma\delta)(\varepsilon\eta)$ zugeordnet wird. Diesen Permutationen entsprechen aber in $\mathfrak{T}_6$ Elemente der Ordnung 2, wie man durch Betrachtung der zu der Permutation (1 2) (3 4) (5 6) gehörenden Elemente $T_1 T_3 T_5$ und $J T_1 T_3 T_5$ sofort erkennt†).

Wir können folgenden Satz aussprechen:

*) Es wäre eigentlich noch zu beweisen, daß sich aus den Relationen (II.) oder (II'.) nicht etwa $J = E$ ergeben kann. Daß dies nicht der Fall ist, folgt aber daraus, daß diese Relationen, wie wir im vierten Abschnitt sehen werden, sich durch Matrizen in der Weise befriedigen lassen, daß an Stelle von E und J zwei verschiedene Matrizen treten.

**) Vgl. *O. Hölder*, „Bildung zusammengesetzter Gruppen", Math. Ann., Bd. 46, S. 321.

***) Vgl. U., S. 122.

†) Daß $\mathfrak{T}_6$ und $\mathfrak{T}'_6$ einander isomorph sind, kann man auch direkt einsehen, indem man zeigt, daß die Elemente

I. *Die Gruppen $\mathfrak{S}_2$ und $\mathfrak{S}_3$ sind abgeschlossene Gruppen. Ist aber $n > 3$, so besitzt die Gruppe $\mathfrak{S}_n$ zwei Darstellungsgruppen $\mathfrak{T}_n$ und $\mathfrak{T}'_n$ derselben Ordnung $2 \cdot n!$, die als abstrakte Gruppen durch die Relationen* (II.), *bzw.* (II'.) *definiert werden können. Nur für $n=6$ sind $\mathfrak{T}_n$ und $\mathfrak{T}'_n$ isomorphe Gruppen.*

§ 4. Ich wende mich nun zur Betrachtung der alternierenden Gruppe $\mathfrak{A}_n$. Diese Gruppe wird erzeugt durch die $n-2$ Permutationen

$$A_1 = S_2 S_1 = (1\,2\,3),\ A_2 = S_3 S_1 = (1\,2)\,(3\,4),\ \ldots\ A_{n-2} = S_{n-1} S_1 = (1\,2)\,(n-1, n),$$

die den Gleichungen

$$\text{(III.)} \begin{cases} A_1^3 = E,\ (A_1 A_2)^3 = E,\ (A_1 A_\lambda)^2 = E & (\lambda = 3, 4, \ldots n-2) \\ A_\alpha^2 = E,\ (A_\beta A_{\beta+1})^3 = E,\ A_\gamma A_\delta = A_\delta A_\gamma & \left(\begin{smallmatrix} \alpha = 2, 3, \ldots n-2,\ \beta = 2, 3, \ldots n-3 \\ \gamma = 2, 3, \ldots n-4,\ \delta = \gamma+2, \ldots n-2 \end{smallmatrix}\right) \end{cases}$$

genügen. *Durch diese Relationen wird wieder die Gruppe $\mathfrak{A}_n$ als abstrakte Gruppe eindeutig definiert*).*

Es sei wieder irgendeine Darstellung von $\mathfrak{A}_n$ durch Kollineationen gegeben. Entspricht hierbei der Permutation A_ν eine Kollineation mit der Koeffizientenmatrix P_ν, so ist P_ν nur bis auf einen Zahlenfaktor bestimmt, und es bestehen wegen (III.) Gleichungen der Form

$$(9.) \qquad P_1^3 = a_1 E,\ (P_1 P_2)^3 = b_1 E,\ (P_1 P_\lambda)^2 = c_\lambda E,$$

$$(10.) \qquad P_\alpha^2 = a_\alpha E,\ (P_\beta P_{\beta+1})^3 = b_\beta E,\ P_\gamma P_\delta = c_{\gamma\delta} P_\delta P_\gamma.$$

Die Gleichungen (10.) sind den Gleichungen (3.), (4.) und (5.) des § 2 völlig analog. Wir schließen wie dort, daß

$$(11.) \qquad c_{\gamma\delta} = c_{24} = \pm 1,\ b_\beta^2 = a_\beta^3 a_{\beta+1}^3$$

ist. Ferner erhält man aus (9.)

$$(P_2 P_1^2)^3 = (a_2 a_1 P_2^{-1} P_1^{-1})^3 = a_2^3 a_1^3 b_1^{-1} \cdot E$$

und

$$P_1 P_2 P_1 = b_1 \cdot P_2^{-1} P_1^{-1} P_2^{-1} = b_1 a_2^{-1} \cdot P_2^{-1} P_1^{-1} P_2.$$

Aus der letzten Gleichung folgt durch Erheben in die dritte Potenz

$$P_1 P_2 P_1^2 P_2 P_1^2 P_2 P_1 = P_1 (P_2 P_1^2)^3 P_1^{-1} = b_1^3 a_2^{-3} (P_2^{-1} P_1^{-1} P_2)^3.$$

$$T_1 = T'_1 T'_3 T'_5,\ T_2 = T'_3 T'_2 T'_1 T'_4 T'_3 T'_2 T'_5 T'_4 T'_3,\ T_3 = T'_1 T'_4 T'_3 T'_5 T'_4,$$
$$T_4 = T'_1 T'_2 T'_1 T'_3 T'_2 T'_1 T'_5,\ T_5 = T'_1 T'_3 T'_4 T'_3 T'_5 T'_4 T'_3$$

von $\mathfrak{T}'_6$ den die Gruppe $\mathfrak{T}_6$ definierenden Relationen genügen.

*) *E. H. Moore*, a. a. O. — Die Gruppe $\mathfrak{A}_n$ läßt sich, wie man mit Hilfe des *Moore*schen Satzes leicht zeigt, noch eleganter durch die Relationen

$$C_\alpha^3 = E,\ (C_\alpha C_\beta)^2 = E \qquad (\alpha, \beta = 1, 2, \ldots n-2,\ \beta > \alpha)$$

definieren. Doch ist diese Definition von $\mathfrak{A}_n$ für unsere Zwecke weniger geeignet.

Daher wird

$$a_2^3 a_1^3 b_1^{-1} = b_1^3 a_2^{-3} a_1^{-1}, \text{ also } b_1^4 = a_1^4 a_2^6.$$

Setzt man demnach

$$(12.) \qquad j = \frac{b_1^2}{a_1^2 a_2^3},$$

so wird $j = \pm 1$.

Aus $(P_1 P_\lambda)^2 = c_\lambda E$ erhält man ferner

$$P_\lambda P_1 P_\lambda = a_\lambda P_\lambda^{-1} P_1 P_\lambda = c_\lambda P_1^{-1}$$

und, indem man zu den dritten Potenzen übergeht,

$$a_\lambda^3 a_1 = c_\lambda^3 a_1^{-1},$$

also

$$(13.) \qquad c_\lambda^3 = a_1^2 a_\lambda^3.$$

Ist nun $n \geqq 6$ und setzt man

$$k = j \frac{a_3 c_4}{c_3 a_4},$$

so wird wegen (13.)

$$k^3 = j = \pm 1.$$

Außerdem erhält man aus den Gleichungen

$$(P_1 P_4)^2 = c_4 E, \quad P_2 P_4 = c_{24} P_4 P_2,$$

wie man leicht sieht, die Gleichung

$$P_4 P_1 P_2 = c_4 c_{24} P_1^{-1} P_2 P_4^{-1}$$

oder

$$P_4 P_1 P_2 P_4^{-1} = c_4 a_2 a_4^{-1} c_{24} \cdot P_1^{-1} P_2^{-1}.$$

Hieraus folgt, indem man auf beiden Seiten die dritte Potenz bildet,

$$b_1 = c_4^3 a_2^3 a_4^{-3} c_{24}^3 b_1^{-1}.$$

Unter Berücksichtigung der Gleichungen (11.), (12.) und (13.) schließt man hieraus, daß $c_{24} = j$ ist; daher ist auch allgemein

$$c_{\gamma\delta} = j.$$

Ist n auch $\geqq 7$, so beachte man noch die Gleichungen

$$(P_1 P_\mu)^2 = c_\mu, \quad P_3 P_\mu = j \cdot P_\mu P_3. \qquad (\mu \geqq 5)$$

Es ergibt sich hieraus

$$P_3 P_1 P_\mu = c_3 \cdot P_1^{-1} P_3^{-1} P_\mu = j c_3 \cdot P_1^{-1} P_\mu P_3^{-1},$$

also

$$P_3 P_1 P_\mu P_3^{-1} = j c_3 a_\mu a_3^{-1} \cdot P_1^{-1} P_\mu^{-1}.$$

Erhebt man auf beiden Seiten in die zweite Potenz, so erhält man

$$c_\mu = c_3^2 a_\mu^2 a_3^{-2} c_\mu^{-1},$$

also

$$c_\mu^2 a_\mu^{-2} = c_3^2 a_3^{-2}.$$

Da aber andererseits wegen (13.)

$$c_\mu^3 a_\mu^{-3} = c_3^3 a_3^{-3}$$

ist, so erhalten wir

$$\frac{c_3}{a_3} = \frac{c_5}{a_5} = \frac{c_6}{a_6} = \cdots = \frac{c_{n-2}}{a_{n-2}}.$$

Ebenso ergibt sich für $n > 7$

$$\frac{c_4}{a_4} = \frac{c_6}{a_6} = \frac{c_7}{a_7} = \cdots = \frac{c_{n-2}}{a_{n-2}}.$$

Daher wird für $n > 7$

$$\frac{c_3}{a_3} = \frac{c_4}{a_4} = \frac{c_5}{a_5} = \cdots = \frac{c_{n-2}}{a_{n-2}}.$$

Insbesondere ergibt sich für $n > 7$

$$k = j = \pm 1.$$

Man erkennt leicht, daß die hier eingeführten Größen j und k, die durch die Gleichung $k^3 = j$ verbunden sind, ungeändert bleiben, wenn die Matrizen $P_1, P_2, \ldots P_{n-2}$ mit beliebigen Konstanten multipliziert werden, *sie sind also allein durch die zu betrachtenden Kollineationen bestimmt.* Die Größe k tritt nur auf, wenn $n > 5$ ist, und wird für $n > 7$ gleich j. Also ist k nur für die beiden Fälle $n = 6$ und $n = 7$ von wesentlicher Bedeutung. *Wir werden auch später sehen, daß sich Darstellungen der Gruppen $\mathfrak{A}_6$ und $\mathfrak{A}_7$ angeben lassen, bei denen k eine primitive sechste Einheitswurzel wird.*

Um auf einfachere Formeln geführt zu werden, setzen wir zunächst für $n = 4$

$$Q_1 = \sqrt[3]{\frac{j}{a_1}} \cdot P_1, \quad Q_2 = j \cdot \frac{a_1 a_2}{b_1} \cdot P_2.$$

Dann wird

$$(14.) \qquad Q_1^3 = jE, \quad (Q_1 Q_2)^3 = jE.$$

Ist ferner $n > 4$, so sei

$$Q_1 = j \cdot \frac{c_3}{a_1 a_3} P_1, \quad Q_2 = j \cdot \frac{a_1 a_2}{b_1} \cdot P_2, \quad Q_3 = \frac{1}{a_1} \cdot \frac{b_1}{b_2} \cdot a_3 \cdot P_3, \ldots$$

und allgemein

$$Q_{2\nu} = j \cdot \frac{a_1}{b_1} \cdot \frac{b_2 b_4 \cdots b_{2\nu-2}}{b_3 b_5 \cdots b_{2\nu-1}} \cdot a_{2\nu} \cdot P_{2\nu}, \quad Q_{2\nu+1} = \frac{1}{a_1} \cdot \frac{b_1 b_3 \cdots b_{2\nu-1}}{b_2 b_4 \cdots b_{2\nu}} \cdot a_{2\nu+1} P_{2\nu+1}.$$

Dann wird, wie eine einfache Rechnung zeigt, für $n = 5$

$$(15.) \qquad Q_1^3 = Q_2^2 = Q_3^2 = (Q_1 Q_2)^3 = (Q_1 Q_3)^2 = (Q_2 Q_3)^3 = jE,$$

für $n = 6$

$$(16.)\quad \begin{cases} Q_1^3 = Q_2^2 = Q_3^2 = Q_4^2 = (Q_1 Q_2)^3 = (Q_1 Q_3)^2 = (Q_2 Q_3)^3 = (Q_3 Q_4)^3 = jE, \\ (Q_1 Q_4)^2 = kE, \ \ Q_2 Q_4 = j \cdot Q_4 Q_2, \end{cases}$$

für $n = 7$

$$(17.)\quad \begin{cases} Q_1^3 = Q_\alpha^2 = (Q_1 Q_2)^3 = (Q_1 Q_3)^2 = (Q_1 Q_5)^2 = (Q_\beta Q_{\beta+1})^3 = jE & \left(\begin{smallmatrix} \alpha=2,3,4,5, \\ \beta=2,3,4 \end{smallmatrix}\right) \\ (Q_1 Q_4)^2 = kE, \ \ Q_\gamma Q_\delta = j \cdot Q_\delta Q_\gamma, & (\gamma = 2, 3, \ \delta \geqq \gamma + 2) \end{cases}$$

und für $n > 7$

$$(18.)\quad \begin{cases} Q_1^3 = jE, \ \ (Q_1 Q_2)^3 = jE, \ \ (Q_1 Q_\lambda)^2 = jE, \\ Q_\alpha^2 = jE, \ \ (Q_\beta Q_{\beta+1})^3 = jE, \ \ Q_\gamma Q_\delta = j \cdot Q_\delta Q_\gamma. \end{cases}$$

In den Gleichungen (18.) sind die Indizes $\alpha, \beta, \gamma, \delta, \lambda$ denselben Bedingungen zu unterwerfen, wie in den Gleichungen (III.)

§ 5. Wir können nun leicht die Darstellungsgruppe von $\mathfrak{A}_n$ bestimmen*).

Man betrachte nämlich die Darstellungsgruppe $\mathfrak{T}_n$ von $\mathfrak{S}_n$. Wegen des zwischen $\mathfrak{S}_n$ und $\mathfrak{T}_n$ bestehenden (mehrstufigen) Isomorphismus entspricht der Untergruppe $\mathfrak{A}_n$ der Ordnung $\frac{n!}{2}$ von $\mathfrak{S}_n$ eine Untergruppe $\mathfrak{B}_n$ der Ordnung $2 \cdot \frac{n!}{2}$ von $\mathfrak{T}_n$. Diese Gruppe $\mathfrak{B}_n$ läßt sich durch die Elemente

$$B_1 = T_2 T_1, \ B_2 = T_3 T_1, \ B_3 = T_4 T_1, \ \ldots \ B_{n-2} = T_{n-1} T_1$$

erzeugen, und man erkennt auf Grund der Relationen (II.) unmittelbar, daß für diese Elemente die den Gleichungen (III.) analogen Gleichungen

$$(\mathrm{IV.})\quad \begin{cases} B_1^3 = J, \ (B_1 B_2)^3 = J, \ (B_1 B_\lambda)^2 = J, \\ B_\alpha^2 = J, \ (B_\beta B_{\beta+1})^3 = J, \ B_\gamma B_\delta = J \cdot B_\delta B_\gamma \end{cases}$$

bestehen. *Durch diese Gleichungen wird auch die Gruppe $\mathfrak{B}_n$ als abstrakte Gruppe eindeutig definiert.* Daß J mit den erzeugenden Elementen $B_1, B_2, \ldots B_{n-2}$ vertauschbar ist und die Ordnung 2 besitzt, ergibt sich aus (IV.) von selbst.

Die Gruppe $\mathfrak{B}_n$ ist eine durch die Gruppe $\mathfrak{M} = E + J$ ergänzte Gruppe von $\mathfrak{A}_n$ und man erkennt auch leicht, daß für $n \geqq 4$ der Kommutator von $\mathfrak{B}_n$ das Element J enthält**). Ist nun n größer als 3 und von

---

*) Daß die Gruppe $\mathfrak{A}_n$, die bekanntlich für $n > 4$ eine einfache Gruppe ist, nur eine Darstellungsgruppe besitzt, ergibt sich aus dem Satz II meiner mit U. zitierten Arbeit.

**) Dies folgt z. B. aus der Gleichung $B_1^{-1} B_2 B_1 \cdot B_2 = J B_2 \cdot B_1^{-1} B_2 B_1$.

6 und 7 verschieden, so lehren uns die Formeln (14.), (15.) und (18.), daß die Gleichungen (IV.) befriedigt werden, wenn man die Elemente J und B_ν durch die Matrizen jE und Q_ν ersetzt. Hieraus folgt aber, ähnlich wie in § 3 für die Gruppe $\mathfrak{S}_n$, daß *die Gruppe $\mathfrak{B}_n$ für $n \geqq 4$ mit Ausnahme der beiden Fälle $n=6$ und $n=7$ die Darstellungsgruppe von $\mathfrak{A}_n$ ist.*

Dagegen schließt man aus den Gleichungen (16.) und (17.), indem man noch die Gleichung $k^3 = j = \pm 1$ berücksichtigt, daß *die Darstellungsgruppen von $\mathfrak{A}_6$ und $\mathfrak{A}_7$ gewisse Gruppen der Ordnungen $6 \cdot \frac{6!}{2}$ und $6 \cdot \frac{7!}{2}$ sind.* Diese Gruppen werde ich im Abschnitt XI genauer untersuchen.

Die beiden Fälle $n=2$ und $n=3$, die wir bis jetzt außer Acht gelassen haben, sind für uns ohne Interesse. Denn $\mathfrak{A}_2$ ist von der Ordnung 1 und $\mathfrak{A}_3$ ist eine zyklische und daher gewiß eine abgeschlossene Gruppe.

Bei der Definition der Gruppe $\mathfrak{B}_n$ sind wir von der Gruppe $\mathfrak{T}_n$ ausgegangen. Man wird aber auf dieselbe Gruppe geführt, wenn man an Stelle von $\mathfrak{T}_n$ die zweite Darstellungsgruppe $\mathfrak{T}'_n$ von $\mathfrak{S}_n$ betrachtet. Man erkennt dies, indem man zeigt, daß die Elemente

$$B_1 = J T'_2 T'_1, \quad B_2 = J T'_3 T'_1, \ldots \quad B_{n-2} = J T'_{n-1} T'_1$$

von $\mathfrak{T}'_n$ den Relationen (IV.) genügen.

Für $n \geqq 4$ kann die Gruppe $\mathfrak{B}_n$ noch etwas anders charakterisiert werden. Beachtet man nämlich, daß der Kommutator von $\mathfrak{S}_n$ die Gruppe $\mathfrak{A}_n$ ist und daß der Kommutator von $\mathfrak{T}_n$ (oder $\mathfrak{T}'_n$) das Element J enthält, so ergibt sich, daß die Gruppe $\mathfrak{B}_n$ für $n \geqq 4$ nichts anderes ist, als der Kommutator von $\mathfrak{T}_n$ (oder $\mathfrak{T}'_n$). Wir können daher den Satz aussprechen:

II. *Die Darstellungsgruppe der alternierenden Gruppe $\mathfrak{A}_n$ ist, falls n größer als drei und von sechs und sieben verschieden ist, eine Gruppe der Ordnung $2 \cdot \frac{n!}{2}$, die dem Kommutator jeder Darstellungsgruppe der symmetrischen Gruppe $\mathfrak{S}_n$ einstufig isomorph ist. Dagegen sind die Darstellungsgruppen der Gruppen $\mathfrak{A}_6$ und $\mathfrak{A}_7$ von den Ordnungen $6 \cdot \frac{6!}{2}$ und $6 \cdot \frac{7!}{2}$.*

Für die Diskussion der Darstellungen der Gruppe $\mathfrak{S}_n$ durch Kollineationen ist es natürlich ganz ohne Bedeutung, welche von ihren beiden Darstellungsgruppen bevorzugt wird. Wenn nun im folgenden die Gruppe $\mathfrak{T}_n$ in den Vordergrund gestellt wird, so hat das folgenden Grund: Die Elemente $A_2, A_3, \ldots A_{n-2}$ der Gruppe $\mathfrak{A}_n$ erzeugen eine Gruppe, die der

Gruppe $\mathfrak{S}_{n-2}$ isomorph ist. In analoger Weise erzeugen die Elemente $B_2, B_3, \ldots B_{n-2}$ von $\mathfrak{B}_n$ eine durch die Gruppe $\mathfrak{M}$ ergänzte Gruppe von $\mathfrak{S}_{n-2}$. Die Gleichungen (IV.) zeigen uns aber, daß diese Gruppe der Gruppe $\mathfrak{T}_{n-2}$ und nicht der Gruppe $\mathfrak{T}'_{n-2}$ isomorph ist.

Abschnitt II.

Über die Einteilung der Elemente der Gruppen $\mathfrak{T}_n$ und $\mathfrak{B}_n$ in Klassen konjugierter Elemente.

§ 6. Ist die Permutation P der Gruppe $\mathfrak{S}_n$ gleich dem Produkt $S_\alpha S_\beta S_\gamma \cdots$ der Transpositionen $S_1 = (12)$, $S_2 = (23)$, $\ldots$, $S_{n-1} = (n-1, n)$, so entsprechen dieser Permutation in der Gruppe $\mathfrak{T}_n$ die beiden Elemente $T_\alpha T_\beta T_\gamma \cdots$ und $J T_\alpha T_\beta T_\gamma \cdots$. Wird von diesen Elementen nach irgendeiner Festsetzung das eine mit P' bezeichnet, so wird das andere gleich JP'. Wir denken uns für jede Permutation P von $\mathfrak{S}_n$ das Element P' von $\mathfrak{T}_n$ eindeutig fixiert. Dann bilden also die $n!$ Elemente P' von $\mathfrak{T}_n$ ein vollständiges Restsystem von $\mathfrak{T}_n$ mod $\mathfrak{M}$, und es ist, falls für drei Permutationen P, Q und R die Gleichung $PQ = R$ besteht, $P'Q'$ entweder gleich R' oder gleich JR'. Sind daher A und B zwei vertauschbare Permutationen, so ist $A'B'$ entweder gleich $B'A'$ oder gleich $JB'A'$. Sind ferner P und Q zwei konjugierte (ähnliche) Permutationen, so ist in $\mathfrak{T}_n$ das Element P' jedenfalls einem der Elemente Q' und JQ' konjugiert.

Eine Permutation P bezeichne ich als *eine Permutation erster oder zweiter Art, je nachdem P' und JP' konjugierte Elemente von $\mathfrak{T}_n$ sind oder nicht.* Zwei ähnliche Permutationen gehören dann entweder beide zur ersten Art oder beide zur zweiten Art.

Es seien nun

$$P, P_1, P_2, \ldots P_{h-1}$$

die sämtlichen zu einer gegebenen Permutation P ähnlichen Permutationen von $\mathfrak{S}_n$. Ist dann P von der ersten Art, so bilden die $2h$ Elemente

$$P', JP', P'_1, JP'_1, \ldots P'_{h-1}, JP'_{h-1}$$

eine Klasse konjugierter Elemente von $\mathfrak{T}_n$. Gehört dagegen P zur zweiten Art, so verteilen sich diese $2h$ Elemente auf *zwei* Klassen von je h konjugierten Elementen; hierbei geht die eine Klasse in die andere über, wenn man jedes ihrer Elemente mit J multipliziert. — Wir können diese beiden Fälle auch folgendermaßen unterscheiden: Im ersten

Falle gibt es eine Permutation Q, die mit P vertauschbar ist, ohne daß Q' mit P' vertauschbar wird, und die Anzahl $\frac{2n!}{2h}$ der mit P' vertauschbaren Elemente von $\mathfrak{T}_n$ wird gleich der Anzahl $\frac{n!}{h}$ der mit P vertauschbaren Permutationen von $\mathfrak{S}_n$. Im zweiten Fall ist dagegen für jede mit P vertauschbare Permutation Q das Element Q' mit P' vertauschbar, und die Anzahl der mit P' vertauschbaren Elemente von $\mathfrak{T}_n$ wird gleich $\frac{2n!}{h}$, also gleich der zweifachen Anzahl der mit P vertauschbaren Permutationen von $\mathfrak{S}_n$.

Man betrachte nun insbesondere zwei (vertauschbare) Permutationen A und B, von denen die erste die Ziffern $m+1, m+2, \ldots n$, die zweite die Ziffern $1, 2, \ldots m$ ungeändert läßt. Dann läßt sich A als Produkt der Transpositionen $S_1, S_2, \ldots S_{m-1}$ und B als Produkt der Transpositionen $S_{m+1}, S_{m+2}, \ldots S_{n-1}$ darstellen. Da nun aber, wenn λ einen der Indizes $1, 2, \ldots m-1$ und μ einen der Indizes $m+1, m+2, \ldots n-1$ bedeutet,

$$T_\lambda T_\mu = J T_\mu T_\lambda$$

wird, so erkennt man leicht, daß die Elemente A' und B' von $\mathfrak{T}_n$ stets und nur dann nicht vertauschbar sind, wenn die Permutationen A und B beide ungerade sind. Hieraus schließt man ohne Mühe allgemeiner:

III. *Sind A und B zwei Permutationen von $\mathfrak{S}_n$, deren Zykeln von höherer als der ersten Ordnung keine Ziffer gemeinsam haben, so sind die Elemente A' und B' von $\mathfrak{T}_n$ nur dann untereinander nicht vertauschbar, wenn die Permutationen A und B beide ungerade sind; in diesem Falle wird $A'B' = J \cdot B'A'$.*

§ 7. Auf Grund dieser Regel können wir beweisen:

IV. *Eine gerade Permutation gehört zur ersten Art, wenn unter ihren Zykeln auch solche gerader Ordnung vorkommen, und zur zweiten Art, wenn sie lauter Zykeln ungerader Ordnung enthält. Eine ungerade Permutation gehört ferner zur ersten Art, wenn sie mindestens zwei Zykeln gleicher Ordnung $m \geqq 1$ enthält, und zur zweiten Art, wenn die Ordnungen ihrer Zykeln voneinander verschieden sind.*

Wir haben, um diesen Satz zu beweisen, vier Fälle zu unterscheiden.

a) Es sei die Permutation P gerade und enthalte einen Zyklus A gerader Ordnung. Ist dann $P = AB$, so wird, da P eine gerade und A eine ungerade Permutation ist, B eine ungerade Permutation. Nun erscheinen aber A und B als zwei ungerade Permutationen, deren Zykeln (von höherer als der ersten Ordnung) keine Ziffer gemeinsam haben. Daher ist nach III

$$A'B' = JB'A',$$

oder

$$A'^{-1}(A'B')A' = JA'B'.$$

Da nun aber P' entweder gleich $A'B'$ oder gleich $JA'B'$ ist, so ergibt sich auch $A'^{-1}P'A' = JP'$; daher ist P eine Permutation erster Art.

b) Die Permutation P bestehe aus lauter Zykeln ungerader Ordnung. Dann ist die Ordnung a von P eine ungerade Zahl. Es wird daher $P'^a = J^\alpha$, wo α gleich Null oder Eins ist, und $(JP')^a = J^a J^\alpha = J^{\alpha+1}$. Folglich sind die Ordnungen von P' und JP' voneinander verschieden, und daher können P' und JP' nicht konjugierte Elemente sein, d. h. P ist eine Permutation zweiter Art.

c) Es sei P eine ungerade Permutation, die zwei Zykeln A und B derselben Ordnung $m \geq 1$ enthält. Es sei etwa

$$A = (\alpha_1, \alpha_2, \ldots \alpha_m), \quad B = (\beta_1, \beta_2, \ldots \beta_m).$$

Setzt man

$$C = (\alpha_1, \beta_1, \alpha_2, \beta_2, \ldots \alpha_m, \beta_m),$$

so wird $C^2 = AB$ und $P = C^2 D$, wo D das Produkt der von A und B verschiedenen Zykeln bedeutet. Da P eine ungerade und C^2 eine gerade Permutation ist, so wird D eine ungerade Permutation, deren Zykeln mit den Zykeln der ebenfalls ungeraden Permutation C keine Ziffer gemeinsam haben. Wir erhalten daher wieder $D'C' = JC'D'$ und

$$C'^{-1}(C'^2 D')C' = C'D'C' = JC'^2 D'.$$

Da sich nun P' von $C'^2 D'$ nur um den Faktor J unterscheiden kann, so wird auch $C'^{-1}P'C' = JP'$, folglich gehört P zur ersten Art.

d) Es bestehe nun die ungerade Permutation P aus r Zykeln $C_1, C_2, \ldots C_r$, deren Ordnungen $c_1, c_2, \ldots c_r$ untereinander verschieden sind. Dann ist $P = C_1 C_2 \cdots C_r$ nur mit den $c_1 c_2 \cdots c_r$ Permutationen

$$C_1^{\gamma_1} C_2^{\gamma_2} \cdots C_r^{\gamma_r} \qquad (\gamma_\varrho = 0, 1, \ldots c_\varrho - 1)$$

vertauschbar. Bedeutet s die Anzahl der geraden unter den Zahlen $c_1, c_2, \ldots c_r$, so ist s, weil P eine ungerade Permutation sein soll, eine ungerade Zahl. Betrachtet man nun die Elemente $C_1', C_2', \ldots C_r'$ von $\mathfrak{T}_n$, so wird für je zwei Indizes ϱ und σ entweder

$$C'_\varrho C'_\sigma = C'_\sigma C'_\varrho \quad \text{oder} \quad C'_\varrho C'_\sigma = J C'_\sigma C'_\varrho,$$

und zwar gilt hier folgende Regel: Ist für ein festes ϱ die Zahl c_ϱ ungerade, also die Permutation C_ϱ gerade, so gilt für jedes σ die erste Gleichung. Ist dagegen c_ϱ eine gerade Zahl und folglich C_ϱ eine ungerade Permutation, so besteht die zweite Gleichung nur für diejenigen $s-1$ Zahlen σ, die von ϱ verschieden sind, und für welche die Zahlen c_σ ebenfalls gerade sind. Da nun aber $s-1$ gerade ist, so erkennt man unmittelbar, daß jedes Element C'_ϱ mit dem Produkt $C'_1 C'_2 \cdots C'_r$ und folglich auch mit dem Element P', das sich von diesem Produkt nur um den Faktor J unterscheiden kann, vertauschbar ist. Folglich ist P' mit den $2c_1c_2\cdots c_r$ Elementen

$$J^\beta C_1'^{\gamma_1} C_2'^{\gamma_2} \cdots C_r'^{\gamma_r} \qquad (\beta = 0, 1;\ \gamma_\varrho = 0, 1, \ldots c_\varrho - 1)$$

vertauschbar. Es können demnach P' und JP' nicht konjugierte Elemente sein.

§ 8. Wir können nun leicht die Anzahl k'_n der Klassen konjugierter Elemente von $\mathfrak{T}_n$ bestimmen.

Eine Zerlegung

$$n = \nu_1 + \nu_2 + \cdots + \nu_m \qquad (\nu_1 \geqq \nu_2 \geqq \cdots \geqq \nu_m)$$

der Zahl n in ganzzahlige positive Summanden nenne ich *eine gerade oder eine ungerade Zerlegung, je nachdem die Anzahl der geraden unter den Zahlen* $\nu_1, \nu_2, \ldots \nu_m$ *gerade oder ungerade ist. Ferner bezeichne ich mit* k_n *die Anzahl aller Zerlegungen von* n *in gleiche oder verschiedene Summanden, mit* g_n *die Anzahl der geraden und mit* u_n *die Anzahl der ungeraden Zerlegungen von* n *in voneinander verschiedene Summanden. Außerdem verstehe ich unter* v_n *die Anzahl der Zerlegungen von* n *in gleiche oder verschiedene ungerade Summanden.* Die Zahl v_n gibt dann bekanntlich zugleich auch die Anzahl der Zerlegungen von n in voneinander verschiedene Summanden an*); daher ist

$$(19.) \qquad v_n = g_n + u_n.$$

Nun ist die Anzahl der Klassen konjugierter Permutationen von $\mathfrak{S}_n$ bekanntlich gleich der mit k_n bezeichneten Zahl. Einer Klasse von Permutationen erster Art von $\mathfrak{S}_n$ entspricht in $\mathfrak{T}_n$ nur eine Klasse konjugierter Elemente, dagegen gehören zu jeder Klasse von Permutationen zweiter Art von $\mathfrak{S}_n$ zwei Klassen konjugierter Elemente von $\mathfrak{T}_n$. Da nun die Anzahl

*) Vgl. etwa *Bachmann*, Analytische Zahlentheorie, S. 30.

der zuletzt genannten Klassen von $\mathfrak{S}_n$ nach Satz IV offenbar gleich v_n+u_n ist, so wird *die gesuchte Anzahl k'_n der Klassen von $\mathfrak{T}_n$ gleich*

$$k_n-v_n-u_n+2(v_n+u_n)=k_n+v_n+u_n.$$

Beachtet man noch die Gleichung (19.), so erhält man

$$k'_n=k_n+g_n+2u_n. \tag{20.}$$

Ich bemerke noch folgendes. Die Zahlen k_n und v_n lassen sich in bekannter Weise mit Hilfe einfacher Rekursionsformeln berechnen*). Kennt man aber die Zahl v_n, so lassen sich g_n und u_n leicht angeben. Setzt man nämlich

$$d_n=g_n-u_n,\ d_0=1,$$

so wird wegen (19.)

$$g_n=\frac{1}{2}(v_n+d_n),\ u_n=\frac{1}{2}(v_n-d_n).$$

Es ist aber, wie man leicht erkennt, für $|x|<1$

$$\sum_0^\infty d_n x^n=(1+x)(1-x^2)(1+x^3)(1-x^4)\cdots,$$

also

$$\sum_0^\infty (-1)^n d_n x^n=(1-x)(1-x^2)(1-x^3)(1-x^4)\cdots.$$

Da nun nach einer bekannten von *Euler* angegebenen Formel

$$\prod_1^\infty(1-x^\lambda)=\sum_{-\infty}^{+\infty}(-1)^\nu x^{\frac{3\nu^2+\nu}{2}}$$

ist**), so ergibt sich, daß $d_n=0$ ist, wenn n nicht die Form $\frac{3\nu^2+\nu}{2}$ hat, und daß $d_n=(-1)^{\frac{\nu^2+\nu}{2}}$ wird, wenn $n=\frac{3\nu^2+\nu}{2}$ ist.

Es mögen hier einige der Werte von g_n und u_n angeführt werden:

$g_1=1,\ g_2=0,\ g_3=1,\ g_4=1,\ g_5=1,\ g_6=2,\ g_7=2,\ g_8=3,\ g_9=4,\ g_{10}=5;$
$u_1=0,\ u_2=1,\ u_3=1,\ u_4=1,\ u_5=2,\ u_6=2,\ u_7=3,\ u_8=3,\ u_9=4\ \ u_{10}=5.$

§ 9. Ich wende mich nun zur Betrachtung der Untergruppe $\mathfrak{B}_n$ von $\mathfrak{T}_n$, die der Untergruppe $\mathfrak{A}_n$ von $\mathfrak{S}_n$ entspricht.

Die Gruppe $\mathfrak{B}_n$ wird dadurch erhalten, daß man für alle $\frac{n!}{2}$ geraden Permutationen P die Elemente P' und JP' von $\mathfrak{T}_n$ bildet. Einer Klasse $\mathfrak{C}$ von h konjugierten Permutationen der Gruppe $\mathfrak{A}_n$ entspricht wieder in der Gruppe $\mathfrak{B}_n$ entweder nur eine Klasse von $2h$ konjugierten Elementen,

*) Vgl. *Bachmann*, a. a. O., S. 28 und S. 44.
**) Vgl. *Bachmann*, a. a. O., S. 24.

oder es entsprechen ihr zwei Klassen von je h Elementen, wobei die eine Klasse in die andere übergeht, indem man jedes ihrer Elemente mit J multipliziert. Ist P eine Permutation der Klasse $\mathfrak{C}$, so tritt der erste oder der zweite Fall ein, je nachdem P' und JP' konjugierte Elemente von $\mathfrak{B}_n$ sind oder nicht.

Um also die Einteilung der Elemente von $\mathfrak{B}_n$ in Klassen konjugierter Elemente durchzuführen, haben wir vor allem zu entscheiden, für welche geraden Permutationen P die Elemente P' und JP' in bezug auf die Gruppe $\mathfrak{B}_n$ konjugiert sind. Eine solche Permutation P ist dadurch charakterisiert, daß sich eine mit P vertauschbare *gerade* Permutation Q angeben läßt, so daß P' und Q' nicht vertauschbar sind, vielmehr der Gleichung $P'Q' = JQ'P'$ genügen.

Ist nun P eine gerade Permutation zweiter Art (d. h. eine Permutation ungerader Ordnung), so sind P' und JP' bereits in $\mathfrak{T}_n$, also um so mehr in $\mathfrak{B}_n$, nicht konjugiert. Wir haben also nur die geraden Permutationen erster Art zu untersuchen, d. h. diejenigen geraden Permutationen, unter deren Zykeln auch solche gerader Ordnung vorkommen.

Ich will nun zeigen:

V. *Ist P eine gerade Permutation erster Art, so sind P' und JP' stets und nur dann konjugierte Elemente von $\mathfrak{B}_n$, wenn P mindestens zwei Zykeln gleicher Ordnung $m \geqq 1$ enthält.*

Ist nämlich A ein Zyklus gerader Ordnung von P, so ist, wie wir auf Seite 173 gesehen haben, $A'^{-1}P'A' = JP'$. Es enthalte nun P noch zwei Zykeln

$$B = (\beta_1, \beta_2, \ldots \beta_m), \quad C = (\gamma_1, \gamma_2, \ldots \gamma_m)$$

von derselben Ordnung $m \geqq 1$; hierbei kann einer der Zykeln B und C auch gleich A sein. Wir setzen

$$D = (\beta_1, \gamma_1, \beta_2, \gamma_2, \ldots \beta_m, \gamma_m),$$

so daß $D^2 = BC$ wird. Ist dann $P = BCF = D^2F$, so ist F, da P gerade ist, eine gerade Permutation; nach Satz III ist daher $D'F' = F'D'$ und also

$$D'^{-1}(D'^2F')D' = D'^2F'.$$

Folglich ist auch $D'^{-1}P'D' = P'$ und mithin

$$(A'D')^{-1}P'(A'D') = JP'.$$

Da nun A und D ungerade Permutationen sind, so ist AD in $\mathfrak{A}_n$, folglich $A'D'$ in $\mathfrak{B}_n$ enthalten. Daher sind P' und JP' in bezug auf $\mathfrak{B}_n$ konjugiert.

Es möge nun P aus r Zykeln $C_1, C_2, \ldots C_r$ bestehen, deren Ordnungen $c_1, c_2, \ldots c_r$ untereinander verschieden sind. Dann ist P in der Gruppe $\mathfrak{S}_n$ nur mit den $c_1 c_2 \cdots c_r$ Permutationen

$$C_1^{\gamma_1} C_2^{\gamma_2} \cdots C_r^{\gamma_r} \qquad (\gamma_\varrho = 0, 1, \ldots c_\varrho - 1)$$

vertauschbar. Unter diesen Permutationen sind diejenigen gerade, für welche die Summe aller γ_ϱ, denen gerade c_ϱ entsprechen, eine gerade Zahl ist. Da nun P eine gerade Permutation sein soll, so ist die Anzahl s der geraden unter den Zahlen c_ϱ eine gerade Zahl. Wir schließen nun ähnlich wie im Fall d) des § 7, daß das Element P' mit dem Element C'_ϱ von $\mathfrak{T}_n$ vertauschbar ist oder nicht, je nachdem c_ϱ eine ungerade oder eine, gerade Zahl ist. Hieraus folgt aber leicht, daß P' mit dem Element

$$C_1'^{\gamma_1} C_2'^{\gamma_2} \cdots C_r'^{\gamma_r}$$

stets vertauschbar ist, wenn die ihm entsprechende Permutation $C_1^{\gamma_1} C_2^{\gamma_2} \cdots C_r^{\gamma_r}$ gerade ist. Es gibt also keine mit P vertauschbare gerade Permutation Q für die P' und Q' nicht vertauschbare Elemente werden. Folglich können P' und JP' nicht konjugierte Elemente von $\mathfrak{B}_n$ sein, w. z. b. w.

§ 10. Es soll nun gezeigt werden, daß, *wenn l_n die Anzahl der Klassen konjugierter Elemente der Gruppe $\mathfrak{A}_n$ bedeutet, die entsprechende Anzahl für die Gruppe $\mathfrak{B}_n$ gleich wird*

$$l'_n = l_n + 2g_n + u_n, \tag{21.}$$

wo g_n und u_n dieselbe Bedeutung haben wie in § 8.

Betrachtet man nämlich eine Klasse $\mathfrak{C}$ von h konjugierten geraden Permutationen der Gruppe $\mathfrak{S}_n$, so bilden sie im allgemeinen auch in der Gruppe $\mathfrak{A}_n$ eine Klasse konjugierter Elemente. Eine Ausnahme tritt nur dann ein, wenn die Zykeln jeder Permutation von $\mathfrak{C}$ lauter voneinander verschiedene ungerade Ordnungen besitzen; in diesem Fall zerfallen die h Permutationen von $\mathfrak{C}$ in der Gruppe $\mathfrak{A}_n$ in zwei Klassen von je $\frac{1}{2}h$ konjugierten Elementen. Hierbei geht die eine Klasse aus der anderen dadurch hervor, daß man ihre Elemente mit Hilfe irgendeiner ungeraden Permutation transformiert.*)

Bedeutet daher v'_n die Anzahl der Zerlegungen von n in voneinander verschiedene ungerade Summanden, so verteilen sich die geraden Permu-

*) Vgl. *Frobenius*, „Über die Charaktere der alternierenden Gruppe", Sitzungsberichte der K. Preuß. Akademie, Berlin 1901, S. 303.

tationen *zweiter* Art in der Gruppe $\mathfrak{A}_n$ auf $v_n + v'_n$ Klassen konjugierter Elemente. Diesen Klassen entsprechen in der Gruppe $\mathfrak{B}_n$ genau $2(v_n + v'_n)$ Klassen konjugierter Elemente. Versteht man ferner unter g'_n die Anzahl der geraden Zerlegungen von n in voneinander verschiedene Summanden, unter denen auch gerade Zahlen vorkommen, so hat man in $\mathfrak{A}_n$ genau g'_n Klassen konjugierter Permutationen, die zur *ersten* Art gehören und deren Zykeln voneinander verschiedene Ordnungen besitzen. Nach Satz V entsprechen diesen g'_n Klassen in der Gruppe $\mathfrak{B}_n$ genau $2g'_n$ Klassen konjugierter Elemente. Dagegen gehört zu jeder der übrigen $l_n - (v_n + v'_n + g'_n)$ Klassen von $\mathfrak{A}_n$ in $\mathfrak{B}_n$ nur eine Klasse. Folglich ist

$$l'_n = l_n - (v_n + v'_n + g'_n) + 2(v_n + v'_n + g'_n) = l_n + v_n + v'_n + g'_n.$$

Da aber offenbar $v'_n + g'_n = g_n$ und außerdem $v_n = g_n + u_n$ ist, so erhält man die zu beweisende Formel (21.).

Die geraden Permutationen, deren Zykeln voneinander verschiedene Ordnungen besitzen, will ich im folgenden als *Permutationen dritter Art* bezeichnen. Eine solche Permutation gehört also gleichzeitig auch zur ersten Art, wenn unter den Ordnungen ihrer Zykeln gerade Zahlen vorkommen, und zur zweiten Art, wenn diese Ordnungen sämtlich ungerade sind. Nur für zwei Permutationen P und Q dritter Art kann es eintreten, daß P' und Q' in $\mathfrak{T}_n$, aber nicht in $\mathfrak{B}_n$ einander konjugiert sind. Zwei solche Elemente von $\mathfrak{B}_n$ will ich als *adjungierte Elemente* bezeichnen. Analog nenne ich zwei Permutationen von $\mathfrak{A}_n$, die in $\mathfrak{S}_n$, aber nicht in $\mathfrak{A}_n$ einander konjugiert sind, *adjungierte Permutationen.*

Abschnitt III.

Über die Zuordnung der Elemente der Gruppen $\mathfrak{S}_n$ und $\mathfrak{T}_n$.

§ 11. Wir haben bis jetzt noch keine Festsetzung darüber getroffen, welches von den beiden Elementen von $\mathfrak{T}_n$, die einer Permutation P von $\mathfrak{S}_n$ entsprechen, mit P' bezeichnet werden soll. Es ist für unsere Untersuchung von wesentlicher Bedeutung, die Bezeichnung genauer zu fixieren. Wir suchen hierbei insbesondere zu erreichen, daß *für je zwei Permutationen P und Q, die in $\mathfrak{S}_n$ oder auch in $\mathfrak{A}_n$ konjugiert sind, P' und Q' konjugierte Elemente von $\mathfrak{T}_n$ oder von $\mathfrak{B}_n$ werden.*

Ein Zyklus

$$C_{\mu,\nu} = (\mu, \mu+1, \ldots \mu+\nu-1)$$

der Ordnung ν läßt sich mit Hilfe der Transpositionen $S_\alpha = (\alpha, \alpha+1)$ in der Form

$$C_{\mu,\nu} = S_{\mu+\nu-2} S_{\mu+\nu-3} \cdots S_\mu$$

darstellen. *Unter* $C'_{\mu,\nu}$ *wollen wir dann das Element*

$$C'_{\mu,\nu} = T_{\mu+\nu-2} T_{\mu+\nu-3} \cdots T_\mu$$

verstehen. — Ist nun ν ungerade, so ist nach Satz IV von den beiden Elementen $C'_{\mu,\nu}$ und $JC'_{\mu,\nu}$ nur eins dem speziellen Element $C'_{1,\nu}$ in $\mathfrak{T}_n$ konjugiert. Es ist aber leicht zu sehen, daß dies für das Element $C'_{\mu,\nu}$ eintritt. In der Tat sind die beiden Gruppen $\mathfrak{T}_\nu$ und $\overline{\mathfrak{T}}_\nu$, die durch die Elemente $T_1, T_2, \ldots T_{\nu-1}$, bzw. $T_\mu, T_{\mu+1}, \ldots T_{\mu+\nu-2}$ erzeugt werden, wie aus den die Gruppe $\mathfrak{T}_n$ definierenden Relationen (II.) hervorgeht, isomorph, und zwar erhält man einen Isomorphismus zwischen diesen Gruppen, indem man dem erzeugenden Element $T_{1+\varrho}$ von $\mathfrak{T}_\nu$ das Element $T_{\mu+\varrho}$ von $\overline{\mathfrak{T}}_\nu$ zuordnet. Daher haben offenbar $C'_{1,\nu}$ und $C'_{\mu,\nu}$ dieselbe Ordnung ν', wobei ν' gleich ν oder gleich 2ν ist*). Wären nun $C'_{1,\nu}$ und $JC'_{\mu,\nu}$ konjugierte Elemente von $\mathfrak{T}_n$, so müßten sie von gleicher Ordnung sein, was aber, da ν ungerade ist, nicht der Fall ist.

Ist nun A irgendein Zyklus von *ungerader* Ordnung ν, so ist nur eins von den beiden Elementen von $\mathfrak{T}_n$, die zu A gehören, dem Element $C'_{1,\nu}$ konjugiert. *Das hierdurch gekennzeichnete Element bezeichne ich mit* A'. *Ist ferner*

$$P = A_1 A_2 \cdots A_m$$

eine Permutation, deren Zykeln $A_1, A_2, \ldots A_m$ *lauter ungerade Ordnungen haben, so setze ich*

$$P' = A'_1 A'_2 \cdots A'_m \text{**)}. \tag{22.}$$

Hierbei sind die Elemente $A'_1, A'_2, \ldots A'_m$, weil die Permutationen A_μ gerade sind, nach Satz III untereinander vertauschbar, daher kann in (22.) die Reihenfolge der Faktoren A'_μ beliebig abgeändert werden. Die Ordnung des Elements P' ist nichts anderes als das kleinste gemeinsame Vielfache der

---

*) Es ist, wie man zeigen kann, $\nu' = \nu$ oder $\nu' = 2\nu$, je nachdem $(-1)^{\frac{\nu^2-1}{8}}$ gleich 1 oder gleich -1 ist. Ist ferner ν eine gerade Zahl, so wird die Ordnung des Elements $C'_{\mu,\nu}$ gleich ν für $\nu = 8\lambda$ oder $\nu = 8\lambda + 6$, dagegen gleich 2ν für $\nu = 8\lambda + 2$ oder $\nu = 8\lambda + 4$.

**) Ist $P = E$, so setze ich natürlich auch $P' = E$.

Ordnungen von $A'_1, A'_2, \ldots A'_m$. Hieraus folgt leicht, daß, wenn Q eine zu P in bezug auf die Gruppe $\mathfrak{S}_n$ oder auch in bezug auf die Gruppe $\mathfrak{A}_n$ konjugierte Permutation ist, das mit Q' zu bezeichnende Element dem Element P' in $\mathfrak{T}_n$, bzw. in $\mathfrak{B}_n$ konjugiert ist.

Es sei nun C ein Zyklus von *gerader* Ordnung ν, den wir jedoch der Bedingung unterwerfen, daß er nur die Ziffern $\mu, \mu+1, \ldots \mu+\nu-1$ (in beliebiger Reihenfolge) enthalten soll. Dann läßt sich C ebenso wie der Zyklus $C_{\mu,\nu}$ als Produkt der Transpositionen $S_\mu, S_{\mu+1}, \ldots S_{\mu+\nu-2}$ darstellen. Daher sind die beiden zu C gehörenden Elemente von $\mathfrak{T}_n$ in der früher betrachteten Gruppe $\overline{\mathfrak{T}}_\nu$ enthalten. Nach Satz IV ist aber nur eins von diesen beiden Elementen in bezug auf die Gruppe $\overline{\mathfrak{T}}_\nu$ dem Element $C'_{\mu,\nu}$ konjugiert. *Das hierdurch charakterisierte Element bezeichne ich mit C'.* Dann sind also auch für zwei verschiedene Zykeln B und C der Ordnung ν, in denen nur die Ziffern $\mu, \mu+1, \ldots \mu+\nu-1$ auftreten, B' und C' in der Gruppe $\overline{\mathfrak{T}}_\nu$ einander konjugiert.

Wir betrachten jetzt die Permutationen P mit den m Zykeln

$$C_1 = C_{1,\nu_1} = (1, 2, \ldots \nu_1),\ C_2 = C_{\nu_1+1,\nu_2} = (\nu_1+1, \nu_1+2, \ldots \nu_1+\nu_2), \ldots;$$

hierbei soll $\nu_1 > \nu_2 > \cdots > \nu_m \geqq 1$ sein, und es sollen unter den ν_μ auch gerade Zahlen vorkommen, so daß P also eine Permutation zweiter oder dritter Art wird. Es wird dann

$$P = C_1 C_2 \cdots C_m.$$

Entsprechend setze ich

$$P' = C'_1 C'_2 \cdots C'_m.$$

In dieser Gleichung darf die Reihenfolge der Faktoren nicht mehr beliebig abgeändert werden. Es ist aber hervorzuheben, daß diejenigen Faktoren C'_μ, für welche die ν_μ ungerade Zahlen sind, beliebig angeordnet werden dürfen. Das Element P' bleibt also z. B. ungeändert, wenn man zuerst die Faktoren C'_μ mit ungeradem ν_μ schreibt und dann die Faktoren mit geradem ν_μ nach abnehmender Größe dieser Zahlen folgen läßt.

Ist nun Q eine beliebige Permutation, deren m Zykeln ebenso wie bei P die Ordnungen $\nu_1, \nu_2, \ldots \nu_m$ haben, so sind P und Q ähnliche Permutationen. Von den beiden Elementen der Gruppe $\mathfrak{T}_n$, die zu Q gehören, ist (auf Grund der Sätze IV und V), wenn P und Q ungerade Permutationen sind, nur eins dem Element P' in bezug auf $\mathfrak{T}_n$ konjugiert, und wenn P und Q gerade sind, nur eins in bezug auf $\mathfrak{B}_n$ mit P' konjugiert. *Dasjenige Element, das der ersten oder der zweiten Bedingung genügt, bezeichne ich mit Q'.*

Wir haben nun insbesondere für alle Permutationen P zweiter oder dritter Art festgesetzt, welches Element von $\mathfrak{T}_n$ unter P' zu verstehen ist. Für die hier noch nicht betrachteten Permutationen erster Art denken wir uns die Bezeichnung irgendwie fixiert. Beachtet man, daß für jede unter diesen Permutationen P die Elemente P' und JP' in bezug auf $\mathfrak{T}_n$, und wenn P gerade ist, auch in bezug auf $\mathfrak{B}_n$ konjugiert sind, so erkennt man, daß auf Grund unserer Festsetzungen die früher gestellte Forderung erfüllt ist: sind P und Q zwei Permutationen, die in bezug auf die Gruppen $\mathfrak{S}_n$ oder $\mathfrak{A}_n$ konjugiert sind, so sind auch P' und Q' konjugierte Elemente von $\mathfrak{T}_n$ oder $\mathfrak{B}_n$.

§ 12. Wir müssen hier noch eine Bemerkung hinzufügen, die für das spätere von Wichtigkeit ist. Sie bezieht sich auf den zuletzt behandelten Fall der Permutationen, die in Zykeln von lauter verschiedenen Ordnungen zerfallen.

Es sei insbesondere Q eine Permutation mit m Zykeln $D_1, D_2, \dots D_m$ der Ordnungen $\nu_1 > \nu_2 > \cdots > \nu_m$, und zwar möge der Zyklus D_μ nur die Ziffern

$$\nu_1 + \nu_2 + \cdots + \nu_{\mu-1} + 1, \; \dots \; \nu_1 + \nu_2 + \cdots + \nu_\mu \tag{23.}$$

(in irgendeiner Reihenfolge) enthalten. Wir haben für diesen Fall bereits festgesetzt, welche Elemente von $\mathfrak{T}_n$ wir mit $Q', D'_1, D'_2, \dots D'_m$ zu bezeichnen haben. Jedenfalls wird

$$Q' = J^\alpha D'_1 D'_2 \cdots D'_m, \tag{24.}$$

wo α einen der Werte Null oder Eins hat. *Es soll nun untersucht werden, unter welchen Bedingungen $\alpha = 0$ oder $\alpha = 1$ ist.*

Ich bezeichne mit $\mathfrak{H}_\mu$ die symmetrische Gruppe, die aus allen $\nu_\mu!$ Permutationen der Indizes (23.) besteht, und mit $\mathfrak{K}_\mu$ die der Untergruppe $\mathfrak{H}_\mu$ von $\mathfrak{S}_n$ entsprechende Untergruppe der Ordnung $2 \cdot \nu_\mu!$ von $\mathfrak{T}_n$. Hat C_μ dieselbe Bedeutung wie auf S. 180, so sind C_μ und D_μ ähnliche Permutationen von $\mathfrak{H}_\mu$; ebenso sind auf Grund der früheren Festsetzungen C'_μ und D'_μ konjugierte Elemente von $\mathfrak{K}_\mu$. Es sei H_μ eine Permutation von $\mathfrak{H}_\mu$, die der Bedingung

$$H_\mu^{-1} C_\mu H_\mu = D_\mu$$

genügt; dann ist auch

$$H_\mu'^{-1} C_\mu' H_\mu' = D_\mu'.$$

Ist nun ν_μ eine gerade Zahl, so wählen wir, was jedenfalls möglich ist, H_μ als eine gerade Permutation. Ist aber ν_μ ungerade, so ist H_μ eine gerade Permutation, wenn C_μ und D_μ in der alternierenden Gruppe mit den Vertauschungsziffern (23.) konjugiert sind, dagegen ist H_μ ungerade, wenn dies nicht der Fall ist. Es sei nun die Anzahl der Indizes μ, für die H_μ ungerade ist, gleich r, ferner sei s die Anzahl der geraden unter den Zahlen ν_μ. Für ein gerades ν_μ ist nun H_μ', da H_μ eine gerade Permutation sein soll, wie aus Satz III folgt, stets mit C_ϱ' und D_ϱ' vertauschbar, sobald nur ϱ von μ verschieden ist. Dasselbe gilt für ein ungerades ν_μ, zu dem eine gerade Permutation H_μ gehört. Ist aber ν_μ eine ungerade Zahl und H_μ eine ungerade Permutation, so ist H_μ' mit C_ϱ' und D_ϱ' vertauschbar, wenn ϱ von μ verschieden und ν_ϱ eine ungerade Zahl ist; dagegen wird, falls ν_ϱ gerade ist,

$$H_\mu'^{-1} C_\varrho' H_\mu' = J C_\varrho', \quad H_\mu'^{-1} D_\varrho' H_\mu' = J D_\varrho'.$$

Setzt man nun

$$H = H_1 H_2 \cdots H_m,$$

so wird

$$H' = J^\beta H_1' H_2' \cdots H_m',$$

wo β gleich Null oder Eins ist. Man erkennt auch ohne Mühe, daß, wenn P' wie früher das Produkt $C_1' C_2' \cdots C_m'$ bedeutet,

$$H'^{-1} P' H' = J^{rs} D_1' D_2' \cdots D_m' = J^{rs-\alpha} Q' \tag{25.}$$

wird.

Ich behaupte nun, daß *in der Gleichung* (24.) α *gleich Null oder gleich Eins ist, je nachdem* r *eine gerade oder eine ungerade Zahl ist.*

Es sei nämlich zunächst s ungerade. Dann sind P und Q ungerade Permutationen. Unter Q' ist dann das zu P' konjugierte Element von $\mathfrak{T}_n$ zu verstehen. Aus der Gleichung (25.) folgt daher

$$H'^{-1} P' H' = Q' = J^r D_1' D_2' \cdots D_m',$$

also ist $\alpha \equiv r \pmod 2$. Ist ferner s gerade, so sind P und Q gerade Permutationen dritter Art. In diesem Fall soll Q' mit P' in bezug auf die Gruppe $\mathfrak{B}_n$ konjugiert sein. Ist nun r gerade, so wird H eine gerade Permutation, folglich ist $H'^{-1} P' H' = Q'$. Die Gleichung (25.) lehrt uns daher, daß $\alpha = 0$ zu setzen ist. Ist dagegen r ungerade, so ist H eine ungerade

Permutation und man schließt leicht, daß $H'^{-1}P'H' = JQ'$ sein muß*). Wegen (25.) muß daher $\alpha = 1$ sein.

Abschnitt IV.

Allgemeine Eigenschaften der Charaktere der Gruppen $\mathfrak{T}_n$ und $\mathfrak{B}_n$.

§ 13. Betrachtet man irgendeine Darstellung einer endlichen Gruppe $\mathfrak{H}$ durch homogene lineare Substitutionen in f Variabeln (Matrizen f-ten Grades) und ist $\chi(R)$ die Spur der dem Element R von $\mathfrak{H}$ entsprechenden Substitution, so bezeichnet man nach Herrn *Frobenius* das System der Zahlen $\chi(R)$ als einen *Charakter f-ten Grades der Gruppe* $\mathfrak{H}$**). Ist die Darstellung irreduzibel, so nennt man $\chi(R)$ einen *einfachen* Charakter. Zwei Darstellungen sind stets und nur dann äquivalent, wenn sie denselben Charakter besitzen. Die Anzahl der einfachen Charaktere $\chi^{(0)}(R)$, $\chi^{(1)}(R), \ldots$ ist gleich der Anzahl der Klassen konjugierter Elemente, in die die Elemente von $\mathfrak{H}$ zerfallen, und zwischen diesen Charakteren bestehen die Relationen

$$(26.) \qquad \Sigma\chi^{(\alpha)}(R)\chi^{(\alpha)}(R^{-1}) = h, \quad \Sigma\chi^{(\alpha)}(R)\chi^{(\beta)}(R^{-1}) = 0,$$

wo R alle Elemente von $\mathfrak{H}$ durchläuft und h die Ordnung von $\mathfrak{H}$ bedeutet***).

Man bezeichnet ferner, wenn $r_0, r_1, \ldots$ beliebige ganze (nicht notwendig positive) Zahlen sind, das System der Zahlen

$$\zeta(R) = r_0\chi^{(0)}(R) + r_1\chi^{(1)}(R) + \cdots$$

als einen *zusammengesetzten Charakter* von $\mathfrak{H}$. Es ist dann wegen (26.)

$$(27.) \qquad \Sigma\zeta(R)\zeta(R^{-1}) = h(r_0^2 + r_1^2 + \cdots).$$

Nur dann, wenn diese Summe gleich h und $\zeta(E) > 0$ wird, ist $\zeta(R)$ ein einfacher Charakter. Ist keine der Zahlen $r_0, r_1, \ldots$ negativ, so gehört zu

---

*) Dies folgt daraus, daß Q' und JQ' in $\mathfrak{T}_n$, aber nicht in $\mathfrak{B}_n$ einander konjugiert sind.

**) Hieraus folgt unmittelbar, daß für zwei konjugierte Elemente R und R' von $\mathfrak{H}$ $\chi(R) = \chi(R')$ wird.

***) Einfache Beweise dieser Sätze, die zuerst Herr *Frobenius* in einer Reihe von Abhandlungen (Sitzungsberichte der K. Preuß. Akademie, Berlin, Jahrgänge 1896—1899) aufgestellt hat, findet man in zwei Arbeiten des Herrn *W. Burnside* (Acta Mathematica, Bd. 28, S. 369, und Proceedings of the London Mathematical Society, Serie 2, Bd. 1 (1904), S. 117), ferner in meiner Arbeit „Neue Begründung der Theorie der Gruppencharaktere", Sitzungsberichte der K. Preuß. Akademie, Berlin, 1905, S. 406.

$\zeta(R)$ eine Darstellung von $\mathfrak{H}$ durch Matrizen des Grades $\zeta(E)$; in diesem Fall sagt man auch, $\zeta(R)$ sei ein *eigentlicher* Charakter.

Es bedeute nun $\mathfrak{H}$ eine der Gruppen $\mathfrak{T}_n$ oder $\mathfrak{B}_n$ und entsprechend sei $\mathfrak{G}$ die Gruppe $\mathfrak{S}_n$ oder $\mathfrak{A}_n$. Ist insbesondere für einen (eigentlichen) Charakter f-ten Grades $\chi(R)$ von $\mathfrak{H}$

$$\chi(J) = j\chi(E), \quad j = \pm 1\text{*}),$$

so unterscheiden sich in der zu $\chi(R)$ gehörenden Darstellung von $\mathfrak{H}$ die den Elementen R und JR entsprechenden Matrizen (R) und (JR) nur um den Zahlenfaktor j, so daß auch

$$\chi(JR) = j\chi(R) \tag{28.}$$

wird. Durch die beiden Matrizen (R) und (JR) ist dann nur *eine* gebrochene lineare Substitution bestimmt, und die Gesamtheit dieser Substitutionen bildet eine mit der Gruppe $\mathfrak{G}$ isomorphe Gruppe $\mathfrak{K}$, die ich *die zu dem Charakter* $\chi(R)$ *gehörende Kollineationsgruppe* nenne. Ist k die Ordnung von $\mathfrak{K}$, so läßt sich $\mathfrak{K}$ stets und nur dann als Gruppe von k homogenen linearen Substitutionen schreiben, wenn $j = 1$ oder $n \leqq 3$ ist (vgl. § 2).

Einen Charakter $\chi(R)$ von $\mathfrak{H}$, für den die Gleichungen (28.) bestehen, bezeichne ich als einen *Charakter erster oder zweiter Art*, je nachdem $j = +1$ oder $j = -1$ ist.

Ist $\chi(R)$ ein einfacher Charakter erster Art von $\mathfrak{H}$, so bilden die Zahlen

$$\bar{\chi}(P) = \chi(P') = \chi(JP')$$

einen einfachen Charakter von $\mathfrak{G}$. Hierbei soll P' wie früher das der Permutation P von $\mathfrak{S}_n$ oder $\mathfrak{A}_n$ zugeordnete Element von $\mathfrak{T}_n$ oder $\mathfrak{B}_n$ bedeuten. Umgekehrt erhält man aus jedem Charakter $\bar{\chi}(P)$ von $\mathfrak{G}$ einen einfachen Charakter erster Art $\chi(R)$ von $\mathfrak{H}$, indem man die Zahlen $\chi(P')$ und $\chi(JP')$ gleich $\bar{\chi}(P)$ setzt. Daher ist die Anzahl der einfachen Charaktere erster Art von $\mathfrak{H}$ gleich der Anzahl der einfachen Charaktere von $\mathfrak{G}$, also gleich der Anzahl der Klassen konjugierter Elemente in der Gruppe $\mathfrak{G}$. Haben nun die Zahlen $k_n, k'_n, l_n, l'_n, v_n, g_n$ und u_n dieselbe Bedeutung wie in §§ 8 und 10, so erhalten wir folgendes Resultat:

Die Anzahl der einfachen Charaktere zweiter Art ist für die Gruppe $\mathfrak{T}_n$ *gleich*

*) Für einen einfachen Charakter ist diese Bedingung von selbst erfüllt.

$$k'_n - k_n = g_n + 2u_n = v_n + u_n$$

und für die Gruppe $\mathfrak{B}_n$ *gleich*

$$l'_n - l_n = 2g_n + u_n = v_n + g_n.$$

Da die Charaktere der Gruppen $\mathfrak{S}_n$ und $\mathfrak{A}_n$ bereits bekannt sind (vgl. Einleitung), können wir die Charaktere erster Art von $\mathfrak{T}_n$ und $\mathfrak{B}_n$ beiseite lassen und haben uns nur mit den Charakteren zweiter Art zu beschäftigen.

§ 14. Aus jeder Darstellung $\varDelta$ der Gruppe $\mathfrak{T}_n$ durch homogene lineare Substitutionen (Matrizen) geht eine zweite Darstellung $\varDelta'$ hervor, indem man die den Elementen von $\mathfrak{B}_n$ entsprechenden Matrizen von $\varDelta$ ungeändert läßt, dagegen die übrig bleibenden negativ nimmt. Ich bezeichne $\varDelta$ und $\varDelta'$ als *assoziierte Darstellungen* und entsprechend die zugehörigen Charaktere als *assoziierte Charaktere* von $\mathfrak{T}_n$. Zwei assoziierte Charaktere $\chi(T)$ und $\chi'(T)$ von $\mathfrak{T}_n$ sind also dadurch gekennzeichnet, daß

$$\chi'(T) = (-1)^\tau \chi(T)$$

wird, wo τ gleich Null oder Eins zu setzen ist, je nachdem T in $\mathfrak{B}_n$ enthalten ist oder nicht. Wird insbesondere $\chi(T) = \chi'(T)$, also $\chi(T) = 0$ für alle in $\mathfrak{B}_n$ nicht vorkommenden Elemente T von $\mathfrak{T}_n$, so bezeichne ich $\chi(T)$ als *sich selbst assoziiert* oder auch als einen *zweiseitigen Charakter*.

Für einen einfachen nicht zweiseitigen Charakter $\chi(T)$ von $\mathfrak{T}_n$ wird auf Grund der Gleichungen (26.)

$$\text{(29.)} \qquad \Sigma\chi(T)\chi(T^{-1}) = 2n!, \quad \Sigma(-1)^\tau\chi(T)\chi(T^{-1}) = 0,$$

also

$$\text{(30.)} \qquad \Sigma\chi(B)\chi(B^{-1}) = n!.$$

Hierbei soll T alle Elemente von $\mathfrak{T}_n$ und B alle Elemente von $\mathfrak{B}_n$ durchlaufen. Dagegen wird für einen einfachen zweiseitigen Charakter $\chi(T)$

$$\text{(31.)} \qquad \Sigma\chi(B)\chi(B^{-1}) = 2n!.$$

Hieraus folgt, daß im ersten Falle die Zahlen $\chi(B) = \varphi(B)$ einen einfachen Charakter der Gruppe $\mathfrak{B}_n$ repräsentieren; dagegen wird im zweiten Falle

$$\chi(B) = \psi(B) + \overline{\psi}(B),$$

wo $\psi(B)$ und $\overline{\psi}(B)$ voneinander verschiedene einfache Charaktere von $\mathfrak{B}_n$ sind (vgl. die Formel (27.)). Man erkennt ferner leicht, daß zwei assoziierte Charaktere von $\mathfrak{T}_n$ entweder beide von der ersten Art oder beide von der zweiten Art sind. Ebenso gehören die Charaktere $\varphi(B)$, $\psi(B)$ und $\overline{\psi}(B)$ von $\mathfrak{B}_n$ zur ersten oder zur zweiten Art, je nachdem der Charakter $\chi(T)$ von $\mathfrak{T}_n$ ein Charakter erster oder zweiter Art ist.

Unter den $g_n + 2u_n$ einfachen Charakteren zweiter Art von $\mathfrak{T}_n$ seien nun r zweiseitig und $2s$ nicht zweiseitig. Da die letzteren Charaktere paarweise auftreten, so ist s eine ganze Zahl. Beachtet man nun, daß zu jedem Paar einander assoziierter Charaktere von $\mathfrak{T}_n$ nur ein einfacher Charakter von $\mathfrak{B}_n$ gehört, dagegen jedem zweiseitigen Charakter von $\mathfrak{T}_n$ zwei Charaktere von $\mathfrak{B}_n$ entsprechen, so erhält man im ganzen $2r+s$ einfache Charaktere zweiter Art von $\mathfrak{B}_n$. Man überzeugt sich auch durch Zuhilfenahme der Gleichungen (26.), daß diese $2r+s$ Charaktere voneinander verschieden sind; ferner ergibt sich aus einem Satze des Herrn *Frobenius**), daß dies die sämtlichen einfachen Charaktere zweiter Art von $\mathfrak{B}_n$ sind. Da nun die Anzahl dieser Charaktere gleich $2g_n + u_n$ ist, so ergibt sich

$$2r + s = 2g_n + u_n.$$

Andererseits ist aber

$$r + 2s = g_n + 2u_n,$$

folglich ist

$$r = g_n, \quad s = u_n, \quad r + s = g_n + u_n = v_n.$$

Die Anzahl der zweiseitigen (einfachen) Charaktere zweiter Art von $\mathfrak{T}_n$ ist also gleich der Anzahl der geraden Zerlegungen von n in lauter verschiedene Summanden.

Ich bemerke noch, daß die Anzahl der zweiseitigen Charaktere erster Art von $\mathfrak{T}_n$ gleich ist der Anzahl der Zerlegungen von n in lauter verschiedene ungerade Summanden**).

§ 15. Ist C irgendein Element von $\mathfrak{T}_n$, das nicht in $\mathfrak{B}_n$ vorkommt, z. B. das der Transposition $S_1 = (1,2)$ entsprechende Element T_1, so erhält man einen äußeren Automorphismus A von $\mathfrak{B}_n$, indem man dem Element B von $\mathfrak{B}_n$ das Element $\overline{B} = C^{-1}BC$ zuordnet. Aus jedem Charakter $\vartheta(B)$ von $\mathfrak{B}_n$ geht daher, wie man unmittelbar erkennt, ein zweiter Charakter $\overline{\vartheta}(B)$ hervor, indem man

$$\overline{\vartheta}(B) = \vartheta(\overline{B})$$

setzt. Zwei solche Charaktere $\vartheta(B)$ und $\overline{\vartheta}(B)$ bezeichne ich als *adjungierte*

*) „Über die Relationen zwischen den Charakteren einer Gruppe und denen ihrer Untergruppen", Sitzungsberichte der K. Preuß. Akademie, Berlin, 1898, S. 501.

**) Vgl. *Frobenius*, „Über die Charaktere der symmetrischen Gruppe", § 6, und die Inaugural-Dissertation des Verfassers, § 23.

*Charaktere**). Man schließt auch sofort, daß, wenn $\vartheta(B)$ ein einfacher Charakter erster bzw. zweiter Art ist, $\overline{\vartheta}(B)$ dieselbe Eigenschaft besitzt.

Ich will nun zeigen, daß *die beiden aus einem zweiseitigen (einfachen) Charakter $\chi(T)$ von $\mathfrak{T}_n$ hervorgehenden Charaktere $\psi(B)$ und $\overline{\psi}(B)$ von $\mathfrak{B}_n$ adjungierte Charaktere sind.*

Man betrachte nämlich eine zu $\chi(T)$ gehörende (irreduzible) Darstellung $\varDelta$ von $\mathfrak{T}_n$ durch Matrizen des Grades $f=\chi(E)$. Die dem Element T entsprechende Matrix möge der Kürze wegen ebenfalls mit T bezeichnet werden. Da nun hier die zu $\varDelta$ assoziierte Darstellung mit $\varDelta$ äquivalent ist, so läßt sich eine Matrix H von nicht verschwindender Determinante angeben, so daß

$$H^{-1}TH=(-1)^{\tau}T \tag{32.}$$

wird; hierbei soll τ dieselbe Bedeutung haben wie auf S. 185. Hieraus ergibt sich, daß die Matrix H^2 mit allen Matrizen von $\varDelta$ vertauschbar ist. Da nun $\varDelta$ irreduzibel ist, so muß $H^2=aE_f$ sein, wo a eine Konstante und E_α allgemein die Einheitsmatrix α-ten Grades bedeuten soll. Wir können ohne Beschränkung der Allgemeinheit $a=1$ annehmen, so daß $H^2=E_f$ wird. Man kann daher eine Matrix M von nicht verschwindender Determinante wählen, so daß

$$M^{-1}HM=\begin{pmatrix}E_p & 0\\ 0 & -E_q\end{pmatrix}$$

wird, wo p und q gewisse positive ganze Zahlen bedeuten, deren Summe f ist. Ersetzt man nun die Matrizen T durch die Matrizen $M^{-1}TM$, so erhält man eine der Darstellung $\varDelta$ äquivalente Darstellung, für welche die Matrix $M^{-1}HM$ dieselbe Rolle spielt wie die Matrix H bei der Darstellung $\varDelta$. Wir können daher annehmen, daß bereits

$$H=\begin{pmatrix}E_p & 0\\ 0 & -E_q\end{pmatrix}$$

sei. Aus den Gleichungen (32.) ergibt sich dann, daß, wenn die Elemente von $\mathfrak{T}_n$, die der Untergruppe $\mathfrak{B}_n$ angehören, mit B, die übrigen mit C bezeichnet werden, in unserer Darstellung $\varDelta$ die Matrizen B und C die Form haben

$$B=\begin{pmatrix}P & 0\\ 0 & \overline{P}\end{pmatrix},\ C=\begin{pmatrix}0 & Q\\ \overline{Q} & 0\end{pmatrix},$$

*) Der Charakter $\overline{\vartheta}(B)$ ist von der Wahl des Elements C unabhängig.

wo P und $\overline{P}$ quadratische Matrizen der Grade p und q bedeuten, Q eine Matrix mit p Zeilen und q Spalten und $\overline{Q}$ eine Matrix mit q Zeilen und p Spalten ist. Wäre nun $p \neq q$, so würden die Determinanten der Matrizen C verschwinden, was nicht der Fall ist. Daher ist $p = q$ und also $f = 2p$.

Die Matrizen P und $\overline{P}$ bilden offenbar zwei Darstellungen der Gruppe $\mathfrak{B}_n$. Da wir aber wissen, daß $\chi(B)$ als Summe der beiden einfachen Charaktere $\psi(B)$ und $\overline{\psi}(B)$ von $\mathfrak{B}_n$ erscheint, so müssen diese Darstellungen irreduzibel sein. Wir können annehmen, daß $\psi(B)$ die Spur der Matrix P und $\overline{\psi}(B)$ die Spur der Matrix $\overline{P}$ sei.

Es möge nun C insbesondere ein Element von $\mathfrak{T}_n$ sein, dem in $\mathfrak{S}_n$ eine Transposition, etwa die Transposition $S_1 = (1, 2)$ entspricht. Dann wird das Element C^2 gleich J, also die Matrix C^2 gleich jE_f, wo $j = \pm 1$ ist. Wir erhalten daher $Q\overline{Q} = \overline{Q}Q = jE_p$. Man erkennt nun leicht, daß sich die Darstellung Δ durch eine äquivalente Darstellung ersetzen läßt, in der

$$C = \begin{pmatrix} 0 & E_p \\ jE_p & 0 \end{pmatrix}$$

wird, während H ungeändert bleibt. Hat nun aber C diese Form, so wird

$$C^{-1}BC = \begin{pmatrix} \overline{P} & 0 \\ 0 & P \end{pmatrix}.$$

Hieraus folgt aber, wie zu beweisen ist, daß

$$\psi(C^{-1}BC) = \overline{\psi}(B), \quad \overline{\psi}(C^{-1}BC) = \psi(B) \tag{33.}$$

wird.

§ 16. Ich setze noch

$$\delta(B) = \psi(B) - \overline{\psi}(B)$$

und bezeichne das System der $n!$ Zahlen $\delta(B)$ als das *Komplement des zweiseitigen Charakters* $\chi(T)$. Da man ψ und $\overline{\psi}$ vertauschen kann, so ist das Komplement $\delta(B)$ durch den Charakter $\chi(T)$ nur bis auf ein Vorzeichen bestimmt. Dieses Vorzeichen ist aber für uns ohne Bedeutung. Denn, um die beiden Charaktere $\psi(B)$ und $\overline{\psi}(B)$ von $\mathfrak{B}_n$ angeben zu können, genügt es, außer den Zahlen $\chi(J)$ entweder die Zahlen $\delta(B)$ oder die Zahlen $-\delta(B)$ zu kennen.

Die Zahl $\delta(B)$ ist offenbar nichts anderes als die Spur der Matrix HB. *Ist daher eine zu dem Charakter $\chi(T)$ gehörende Darstellung Δ bekannt, und ist H eine Matrix, die der Gleichung $H^2 = E_f$ und außerdem den Be-*

dingungen (32.) *genügt, so hat man, um das Komplement des Charakters* $\chi(T)$ *zu bestimmen, nur die Spuren der Matrizen HB anzugeben.*

Ist $\chi(J) = jf$, so genügen die Zahlen $\delta(B)$ den Gleichungen

(34.) $$\delta(JB) = j\,\delta(B),$$

ferner ist, wie aus der Formel (27.) folgt,

(35.) $$\sum_B \delta(B)\,\delta(B^{-1}) = 2n!.$$

Für jedes Element C von $\mathfrak{T}_n$, das nicht in $\mathfrak{B}_n$ vorkommt, wird wegen (33.)

(36.) $$\delta(C^{-1}BC) = -\delta(B).$$

Sind insbesondere $\overline{B} = C^{-1}BC$ und B auch in $\mathfrak{B}_n$ einander konjugiert, so wird

(37.) $$\delta(\overline{B}) = \delta(B) = 0.$$

Nur dann, wenn B und $\overline{B}$ in dem in § 10 fixierten Sinne als adjungierte Elemente von $\mathfrak{B}_n$ zu bezeichnen sind, kann $\delta(B)$ nicht Null sein. *Daher ist das Komplement* $\delta(B)$ *nur für solche Elemente B von* $\mathfrak{B}_n$ *anzugeben, denen in der Gruppe* $\mathfrak{A}_n$ *Permutationen dritter Art entsprechen.*

Allgemeiner sei $\xi(T)$ ein beliebiger zusammengesetzter Charakter von $\mathfrak{T}_n$, der sich selbst assoziiert ist, für den also $\xi(T) = 0$ wird, sobald T nicht in $\mathfrak{B}_n$ enthalten ist. Es kann dann auf unendlich viele verschiedene Arten

$$\xi(T) = r_\alpha \chi^{(\alpha)}(T) + r_\beta \chi^{(\beta)}(T) + \cdots + r_\nu \chi^{(\nu)}(T)$$

gesetzt werden, wo

(38.) $$\chi^{(\alpha)}(T),\ \chi^{(\beta)}(T), \ldots$$

einfache, nicht notwendig voneinander verschiedene Charaktere von $\mathfrak{T}_n$ bedeuten, und $r_\alpha, r_\beta, \ldots r_\nu$ ganze Zahlen sind. Hierbei soll jedoch, wenn für $\xi(T)$ die Bedingung

$$\xi(J) = j\,\xi(E), \quad j = \pm 1$$

erfüllt ist, auch

$$\chi^{(\alpha)}(J) = j\chi^{(\alpha)}(E),\ \chi^{(\beta)}(J) = j\chi^{(\beta)}(E),\ \ldots\ \chi^{(\nu)}(J) = j\chi^{(\nu)}(E)$$

sein. — Sind nun etwa

$$\chi^{(\alpha)}(T),\ \chi^{(\beta)}(T),\ \ldots\ \chi^{(\varkappa)}(T)$$

die sämtlichen zweiseitigen unter den Charakteren (38.) und kennt man zu ihnen gehörende Komplemente,

$$\delta^{(\alpha)}(B),\ \delta^{(\beta)}(B),\ \ldots\ \delta^{(\varkappa)}(B),$$

so bezeichne ich *als ein Komplement des zweiseitigen Charakters* $\xi(T)$ *jedes System von Zahlen*

$$\delta(B) = \varepsilon_\alpha r_\alpha \delta^{(\alpha)}(B) + \varepsilon_\beta r_\beta \delta^{(\beta)}(B) + \cdots + \varepsilon_\varkappa r_\varkappa \delta^{(\varkappa)}(B),$$

wo $\varepsilon_\alpha, \varepsilon_\beta, \ldots \varepsilon_\varkappa$ die Werte ± 1 haben*).

Zu jedem zweiseitigen Charakter $\xi(T)$ gehören also unendlich viele Komplemente. In jedem Fall genügen aber die Zahlen $\delta(B)$ den Bedingungen (36.)—(37.); ferner erhält man, wie leicht zu sehen ist, zwei adjungierte (zusammengesetzte) Charaktere $\vartheta(B)$ und $\bar{\vartheta}(B)$ von $\mathfrak{B}_n$, deren Summe $\xi(B)$ ist, indem man

$$\vartheta(B) = \frac{1}{2}[\xi(B) + \delta(B)], \quad \bar{\vartheta}(B) = \frac{1}{2}[\xi(B) - \delta(B)]$$

setzt. Ist insbesondere $\xi(T) = \chi(T)$ ein einfacher Charakter, so werden $\vartheta(B)$ und $\bar{\vartheta}(B)$ nur dann *eigentliche* Charaktere von $\mathfrak{B}_n$, wenn

$$\delta(B) = \pm[\psi(B) - \bar{\psi}(B)]$$

wird, wo $\psi(B)$ und $\bar{\psi}(B)$ dieselbe Bedeutung haben wie früher. Diese beiden ausgezeichneten Komplemente von $\chi(T)$ betrachten wir, wie schon vorhin erwähnt, als nicht wesentlich voneinander verschieden. Wenn wir im folgenden von *dem* Komplement eines einfachen zweiseitigen Charakters sprechen, so verstehen wir hierunter eins von diesen beiden Komplementen.

§ 17. Ist $\chi(T)$ ein beliebiger Charakter zweiter Art von $\mathfrak{T}_n$, so wird für jede Permutation P von $\mathfrak{S}_n$, wenn P' das nach den Vorschriften des § 11 zu bestimmende Element von $\mathfrak{T}_n$ ist,

$$\chi(JP') = -\chi(P').$$

Da ferner für eine Permutation P, die zur ersten Art gehört, P' und JP' konjugierte Elemente von $\mathfrak{T}_n$ sind und also $\chi(JP') = \chi(P')$ wird, so ist für jede Permutation erster Art

$$\chi(JP') = \chi(P') = 0.$$

Es genügt daher, wenn $\chi(T)$ ein Charakter zweiter Art ist, allein die Zahlen $\chi(P')$ und zwar nur für die Permutationen zweiter Art anzugeben.

Da wir ferner die Elemente P' so gewählt haben, daß zwei konjugierten Permutationen von $\mathfrak{S}_n$ auch zwei konjugierte Elemente von $\mathfrak{T}_n$ entsprechen, so ist die Zahl $\chi(P')$ allein durch die Klasse ähnlicher Permutationen von $\mathfrak{S}_n$, der P angehört, bestimmt.

Eine solche Klasse nenne ich gerade oder ungerade, je nachdem ihre Permutationen gerade oder ungerade sind. Mit $[\alpha]$ bezeichne ich eine

*) Kommt unter den Charakteren (38.) kein zweiseitiger Charakter vor, so sage ich, *das Komplement von $\xi(T)$ sei Null.*

Klasse, deren Permutationen in lauter Zykeln ungerader Ordnung zerfallen. Kommen unter den Zykeln einer Permutation P von $[\alpha]$ genau α_1 Zykeln der Ordnung 1, ferner α_3 Zykeln der Ordnung 3 vor usw., so setze ich

$$[\alpha]=[\alpha_1, \alpha_3, \ldots] \text{ und } \chi(P')=\chi_\alpha=\chi_{\alpha_1,\alpha_3,\ldots}.$$

Die Klasse $[\alpha]$ enthält

$$h_\alpha=\frac{n!}{1^{\alpha_1}\alpha_1!\,3^{\alpha_3}\alpha_3!\cdots}$$

Permutationen, und diese Zahl gibt zugleich die Anzahl der mit P' konjugierten Elemente von $\mathfrak{T}_n$ an. Die Anzahl der Klassen $[\alpha]$ ist gleich der von uns mit v_n bezeichneten Zahl. Beachtet man noch, daß P und P^{-1} ähnliche Permutationen sind und daß die Ordnung von P eine ungerade Zahl ist, so erkennt man, daß auch P' und P'^{-1} konjugierte Elemente von $\mathfrak{T}_n$ sind. Daher ist in unserem Fall $\chi(P')=\chi(P'^{-1})$, und *folglich sind die Zahlen χ_α sämtlich reell**). Ist insbesondere $\chi(T)$ ein *einfacher* Charakter zweiter Art, so folgt aus den Gleichungen (30.) und (31.)

$$\Sigma h_\alpha \chi_\alpha^2=\frac{n!}{2^\varepsilon}, \tag{39.}$$

wo die Summe über alle v_n Klassen $[\alpha]$ zu erstrecken ist und ε den Wert Null oder Eins hat, je nachdem $\chi(T)$ ein zweiseitiger Charakter ist oder nicht. Ebenso ergibt sich aus den Formeln (26.), daß für je zwei verschiedene einfache Charaktere $\chi(T)$ und $\chi'(T)$, die nicht einander assoziiert sind,

$$\Sigma h_\alpha \chi_\alpha \chi'_\alpha=0 \tag{40.}$$

ist.

Neben den Klassen $[\alpha]$, die alle geraden Permutationen zweiter Art umfassen, haben wir nur noch diejenigen Klassen von $\mathfrak{S}_n$ zu betrachten, deren Permutationen in Zykeln von lauter verschiedenen Ordnungen zerfallen. Eine solche Klasse bezeichne ich mit $(\nu), (\varrho), \ldots$ und setze, wenn eine Permutation P von (ν) genau m Zykeln der Ordnungen

$$\nu_1, \nu_2, \ldots \nu_m \qquad (\nu_1>\nu_2>\cdots>\nu_m\geqq 1)$$

enthält,

$$(\nu)=(\nu_1, \nu_2, \ldots \nu_m) \tag{41.}$$

und

---

*) Allgemein sind für jeden Charakter $\chi(T)$ und $\chi(T^{-1})$ konjugiert komplexe Zahlen. — Man kann auch leicht schließen, daß die Zahlen χ_α sämtlich ganze rationale Zahlen sind; doch werden wir dies später auf anderem Wege erkennen.

$$\chi(P') = \chi_{(\nu)} = \chi_{(\nu_1, \nu_2, \ldots \nu_m)}.$$

Die Anzahl der Klassen (ν) ist ebenfalls gleich v_n, doch sind diejenigen unter ihnen, für welche $\nu_1, \nu_2, \ldots \nu_m$ ungerade Zahlen sind, zugleich auch unter den Klassen $[\alpha]$ enthalten. Die Zahlen $\chi_{(\nu)}$ sind nur für die u_n *ungeraden* Klassen (ν) anzugeben, da die übrigen entweder unter den Zahlen χ_α vorkommen oder von selbst Null sind. Ist $\chi(T)$ ein zweiseitiger Charakter, so wird $\chi_{(\nu)}$ auch für jede ungerade Klasse (ν) gleich Null. In diesem Fall werden wir im folgenden zumeist wenigstens ein Komplement $\delta(B)$ von $\chi(T)$ anzugeben haben. Bedeutet P die Permutation

$$(1, 2, \ldots \nu_1)(\nu_1 + 1, \nu_1 + 2, \ldots \nu_1 + \nu_2) \cdots$$

der Klasse (41.), und ist P' das auf S. 180 fixierte Element von $\mathfrak{T}_n$, so setze ich, wenn (ν) eine *gerade* Klasse ist,

$$\delta(P') = \delta_{(\nu)} = \delta_{(\nu_1, \nu_2, \ldots \nu_m)}.$$

Kennt man für alle g_n geraden Klassen (ν) die Zahlen $\delta_{(\nu)}$, so kann man auf Grund der Formeln (34.)—(37.) *auch alle übrigen Zahlen $\delta(B)$ angeben.* Hierbei ist in unserem Fall in (34.) für j die Zahl -1 zu setzen.

Bezeichnet man die Anzahl $\dfrac{n!}{\nu_1 \nu_2 \cdots \nu_m}$ der Permutationen der Klasse (41.) mit $h_{(\nu)}$, so erhält man insbesondere für einen *einfachen* Charakter zweiter Art, der nicht zweiseitig ist, aus (29.) die Gleichung

$$\Sigma h_{(\nu)} \chi_{(\nu)} \overline{\chi}_{(\nu)} = \frac{n!}{2}; \tag{42.}$$

ebenso ergibt sich wegen (35.) für das Komplement eines zweiseitigen Charakters zweiter Art

$$\Sigma h_{(\nu)} \delta_{(\nu)} \overline{\delta}_{(\nu)} = n!. \tag{43.}$$

Hierbei ist in der ersten Formel die Summe über alle ungeraden Klassen (ν) und in der zweiten Formel über alle geraden Klassen (ν) zu erstrecken, ferner sind unter $\overline{\chi}_{(\nu)}$ und $\overline{\delta}_{(\nu)}$ die zu $\chi_{(\nu)}$ und $\delta_{(\nu)}$ konjugiert komplexen Zahlen zu verstehen.

Ich leite noch zwei Formeln ab, die für das folgende von Wichtigkeit sind.

Allgemein ist für jede Permutation P von $\mathfrak{S}_n$

$$\sum_\chi \chi(P') \chi(P'^{-1}) = \frac{2n!}{h_P},$$

wo die Summe über alle einfachen Charaktere erster und zweiter Art von $\mathfrak{T}_n$ zu erstrecken ist und h_P die Anzahl der mit P' konjugierten

Elemente von $\mathfrak{T}_n$ bedeutet*). Ist nun P eine Permutation zweiter Art, so ist h_P zugleich die Anzahl der mit P ähnlichen Permutationen von $\mathfrak{S}_n$. Da nun für jeden Charakter erster Art die $n!$ Zahlen $\chi(P') = \chi(P)$ einen Charakter von $\mathfrak{S}_n$ bilden, so wird

$$\sum_{\chi}{}' \chi(P')\chi(P'^{-1}) = \frac{n!}{h_P},$$

wo hier die Summe über alle Charaktere erster Art zu erstrecken ist. Daher wird auch

$$\sum_{\chi}{}'' \chi(P')\chi(P'^{-1}) = \frac{n!}{h_P},$$

wo χ alle $v_n + u_n$ Charaktere zweiter Art durchläuft. Man bezeichne nun mit

$$\chi^{(1)}(T),\ \chi^{(2)}(T), \ldots\ \chi^{(v_n)}(T)$$

v_n einfache Charaktere zweiter Art, von denen nicht zwei einander assoziiert sind; ferner sei ε_ϱ gleich Null oder Eins, je nachdem $\chi^{(\varrho)}(T)$ ein zweiseitiger oder ein nicht zweiseitiger Charakter ist. Dann kann die zuletzt erhaltene Formel für den Fall, daß P der Klasse $[\alpha]$ angehört, in der Form

(44.) $$\sum_{\varrho} 2^{\varepsilon_\varrho} \chi_\alpha^{(\varrho)^2} = \frac{n!}{h_\alpha}$$

geschrieben werden. Sind dagegen $[\alpha]$ und $[\beta]$ zwei verschiedene Klassen, so erhält man in ähnlicher Weise

(45.) $$\sum_{\varrho} 2^{\varepsilon_\varrho} \chi_\alpha^{(\varrho)} \chi_\beta^{(\varrho)} = 0.$$

Abschnitt V.

Über die zu den Charakteren der Gruppen $\mathfrak{T}_n$ und $\mathfrak{B}_n$ gehörenden Kollineationsgruppen.

§ 18. Es bedeute, wie in § 13, $\mathfrak{H}$ eine der Gruppen $\mathfrak{T}_n$ oder $\mathfrak{B}_n$ und entsprechend sei $\mathfrak{G}$ die Gruppe $\mathfrak{S}_n$ oder $\mathfrak{A}_n$. Ist g die Ordnung von $\mathfrak{G}$, so wird also die Ordnung h von $\mathfrak{H}$ gleich $2g$.

Man betrachte wieder einen einfachen Charakter $\chi(R)$ von $\mathfrak{H}$ und eine zu $\chi(R)$ gehörende Darstellung Δ von $\mathfrak{H}$ durch Matrizen (R). Ist P irgendeine Permutation von $\mathfrak{G}$, so bezeichne man die durch die Matrizen

*) Dies ist eine der Grundformeln der Theorie der Gruppencharaktere. Vgl. etwa meine auf S. 183 zitierte Arbeit „Neue Begründung der Theorie der Gruppencharaktere", Formel (XIV.).

(P') und (JP') bestimmte Kollineation mit $|P|$ und die durch diese Kollineationen erzeugte Gruppe mit $\mathfrak{K}$.

Es ist nun zunächst zu beachten, daß, wenn $n \geqq 4$ und $\chi(R)$ ein Charakter *zweiter* Art ist, die g Kollineationen $|P|$ voneinander verschieden sein müssen. Denn wäre dies nicht der Fall, so müßte es mindestens eine von E verschiedene Permutation P geben, für die $|P| = |E|$ wird, und diese Permutationen würden eine invariante Untergruppe $\mathfrak{F}$ von $\mathfrak{G}$ bilden. Ist nun zunächst $n > 4$, so müßte, da $\mathfrak{A}_n$ alsdann eine einfache Gruppe ist und $\mathfrak{S}_n$ nur die eine invariante Untergruppe $\mathfrak{A}_n$ enthält, entweder $\mathfrak{F} = \mathfrak{G}$ oder $\mathfrak{G} = \mathfrak{S}_n$ und $\mathfrak{F} = \mathfrak{A}_n$ sein. Für $n = 4$ käme für $\mathfrak{F}$ aber noch die durch die Permutationen

$$E, \; A = (1,2)\,(3,4), \; B = (1,3)\,(2,4), \; C = (1,4)\,(2,3)$$

gebildete Vierergruppe in Betracht. In jedem Fall würde also $\mathfrak{F}$ die Permutationen A und B enthalten. Für die zugehörigen Elemente A' und B' von $\mathfrak{H}$ müßten sich dann in unserer Darstellung Δ die Matrizen (A') und (B') nur um einen Zahlenfaktor unterscheiden und folglich untereinander vertauschbar sein. Es ist aber, wenn $T_1, T_2, \ldots$ die die Gruppe $\mathfrak{T}_n$ erzeugenden Elemente sind,

$$A' = J^{\alpha}\, T_1 T_3 , \quad B' = J^{\beta}\, T_2 T_1 T_3 T_2 ,$$

und es wird $A'B' = JB'A'$. Da nun auf Grund unserer Voraussetzung über den Charakter $\chi(R)$ jedenfalls $(J) = -(E)$ wird, so ergibt sich $(A')(B') = -(B')(A')$, was auf einen Widerspruch führt.

Daher ist die zu einem Charakter zweiter Art von $\mathfrak{H}$ gehörende Kollineationsgruppe $\mathfrak{K}$ für $n \geqq 4$ der Gruppe $\mathfrak{G}$ stets einstufig isomorph.

In ähnlicher Weise schließt man, daß für einen einfachen Charakter erster Art von $\mathfrak{H}$ die Gruppe $\mathfrak{K}$ nur dann der Gruppe $\mathfrak{G}$ nicht einstufig isomorph ist, wenn entweder der Grad f des Charakters gleich Eins oder $\mathfrak{G} = \mathfrak{S}_4$ und $f = 2$ ist.

§ 19. Es sei noch $\bar{\chi}(R)$ ein von $\chi(R)$ verschiedener einfacher Charakter von $\mathfrak{H}$, dem die Darstellung $\bar{\Delta}$ von $\mathfrak{H}$ durch die Matrizen $(\bar{R})$ entsprechen möge. Die zugehörige Kollineationsgruppe möge mit $\bar{\mathfrak{K}}$ bezeichnet werden, ferner sei $|\bar{P}|$ die der Permutation P von $\mathfrak{G}$ entsprechende Substitution von $\bar{\mathfrak{K}}$.

Wir wollen nun untersuchen, unter welchen Bedingungen es eintreten kann, daß die Gruppen $\mathfrak{K}$ und $\bar{\mathfrak{K}}$, abgesehen von der Reihenfolge ihrer Elemente, übereinstimmen.

Es sind hier zwei Fälle zu unterscheiden.

a) Es sei zunächst $|\bar{P}| = |P|$ für jede Permutation P von $\mathfrak{G}$. Dann unterscheiden sich die Koeffizientenmatrizen dieser beiden Kollineationen nur um einen Zahlenfaktor, und hieraus folgt, daß für jedes Element R von $\mathfrak{H}$

$$(\bar{R}) = \zeta_R \cdot (R)$$

also auch

$$\bar{\chi}(R) = \zeta_R \cdot \chi(R) \tag{46.}$$

wird, wo die ζ_R gewisse Zahlen bedeuten. Aus der ersten Gleichung folgt für je zwei Elemente R und S von $\mathfrak{H}$

$$\zeta_R \zeta_S = \zeta_{RS},$$

d. h. die Zahlen ζ_R bilden einen linearen Charakter von $\mathfrak{H}$*). Ist nun $\mathfrak{H} = \mathfrak{T}_n$, so ist der Kommutator von $\mathfrak{H}$ die Untergruppe $\mathfrak{B}_n$, deren Index gleich 2 ist. Es gibt daher außer dem Hauptcharakter $\zeta_R = 1$, der für uns nicht in Betracht kommt, nur noch einen linearen Charakter, den man dadurch erhält, daß man $\zeta_R = 1$ oder $\zeta_R = -1$ setzt, je nachdem R in $\mathfrak{B}_n$ vorkommt oder nicht. Die Gleichung (46.) lehrt uns dann, daß χ *und* $\bar{\chi}$ *assoziierte Charaktere werden.* Umgekehrt schließt man unmittelbar, *daß die zu zwei assoziierten Charakteren von* $\mathfrak{T}_n$ *gehörenden Kollineationsgruppen als nicht voneinander verschieden anzusehen sind.*

Es sei jetzt $\mathfrak{H}$ die Gruppe $\mathfrak{B}_n$. Ist dann $n > 4$, so umfaßt der Kommutator von $\mathfrak{B}_n$ alle Elemente der Gruppe, und folglich hat $\mathfrak{B}_n$ nur den einen linearen Charakter $\zeta_R = 1$, der für uns wieder auszuschließen ist. Ist aber $n = 4$, so besitzt $\mathfrak{B}_n$ drei lineare Charaktere $\zeta_0(R)$, $\zeta_1(R)$, $\zeta_2(R)$, die dadurch bestimmt sind, daß

$$\zeta_\alpha(T_2 T_1) = \varrho^\alpha, \quad \zeta_\alpha(T_3 T_1) = 1$$

wird, wo ϱ eine primitive dritte Einheitswurzel ist. Die Gruppe $\mathfrak{B}_4$ bildet also hier eine Ausnahme, was im folgenden zu berücksichtigen sein wird.

b) Es möge nun die Substitution $|\bar{P}|$ von $\bar{\mathfrak{K}}$ der Substitution $|P_1|$ von $\mathfrak{K}$ gleich sein, wo P_1 eine Permutation von $\mathfrak{G}$ bedeutet, die nicht immer gleich P ist. Offenbar erhalten wir dann einen Automorphismus A von $\mathfrak{G}$, wenn wir der Permutation P die Permutation P_1 zuordnen. Ist nun zunächst A ein innerer Automorphismus von $\mathfrak{G}$, so gibt es in $\mathfrak{G}$ eine

*) Vgl. *Frobenius* „Über die Primfaktoren der Gruppendeterminante", Sitzungsberichte der K. Preuß. Akademie, Berlin, 1896, S. 1343.

Permutation H, für die $H^{-1}PH = P_1$ wird. Dann wird man aber auf den zuerst behandelten Fall geführt, indem man $\mathfrak{K}$ durch die ihr äquivalente Gruppe ersetzt, die aus $\mathfrak{K}$ durch die lineare Transformation $|H|$ hervorgeht.

Es sei also A ein äußerer Automorphismus von $\mathfrak{G}$. Ist $\mathfrak{G} = \mathfrak{S}_n$, so kommt, da $\mathfrak{S}_n$ für $n \neq 6$ eine vollkommene Gruppe ist, nur der Fall $n = 6$ in Betracht, auf den ich später zurückkommen werde. Es sei demnach $\mathfrak{G} = \mathfrak{A}_n$. Wenn dann wieder $n \neq 6$ ist, so kann A nur ein Automorphismus sein, der dadurch erhalten wird, daß man alle Permutationen von $\mathfrak{A}_n$ mit Hilfe einer ungeraden Permutation U transformiert. Dann wird also $P_1 = U^{-1}PU$; nach Voraussetzung stimmen also die Kollineationen $|U^{-1}PU|$ und $|\overline{P}|$ überein. Bezeichnet man nun das zu U gehörende Element U' von $\mathfrak{T}_n$ mit C, so erkennt man, daß die Darstellungen $\varDelta$ und $\overline{\varDelta}$ der Gruppe $\mathfrak{H} = \mathfrak{B}_n$ in der Weise zusammenhängen, daß für jedes Element R von $\mathfrak{B}_n$

$$(C^{-1}RC) = \zeta_R \cdot (\overline{R}) \tag{47.}$$

wird, wo ζ_R eine Konstante bedeutet. Diese Zahlen ζ_R müssen dann wieder einen linearen Charakter von $\mathfrak{B}_n$ bilden. Läßt man aber den Fall $n = 4$, der sich leicht direkt erledigen läßt, beiseite, so ergibt sich nach dem früheren, daß $\zeta_R = 1$ sein muß. Aus der Gleichung (47.) folgt dann

$$\chi(C^{-1}RC) = \overline{\chi}(R),$$

d. h. χ und $\overline{\chi}$ sind adjungierte Charaktere von $\mathfrak{B}_n$ (vgl. § 15). Umgekehrt schließt man leicht, *daß, wenn χ und $\overline{\chi}$ adjungierte Charaktere von $\mathfrak{B}_n$ sind, die Gruppe $\mathfrak{K}$ in die Gruppe $\overline{\mathfrak{K}}$ oder in eine mit $\overline{\mathfrak{K}}$ äquivalente Gruppe übergeht, wenn man die Elemente von $\mathfrak{K}$ auf Grund des Automorphismus A von $\mathfrak{A}_n$ permutiert.* — Ich bezeichne in diesem Fall die Gruppen $\mathfrak{K}$ und $\overline{\mathfrak{K}}$ als *adjungierte Gruppen* (vgl. Einleitung).

In dem im vorhergehenden ausgeschlossenen Fall $n = 6$ kommt sowohl bei der Gruppe $\mathfrak{S}_6$ als auch bei der Gruppe $\mathfrak{A}_6$ noch der bekannte Automorphismus A in Betracht, durch den jedem Zyklus dritter Ordnung eine Permutation der Form $(\alpha\beta\gamma)(\delta\varepsilon\eta)$ zugeordnet wird. Wir werden auch später sehen, daß bei jeder der Gruppen $\mathfrak{T}_6$ und $\mathfrak{B}_6$ gewisse Paare von Charakteren χ und $\overline{\chi}$ vorhanden sind, für welche die zugehörigen Kollineationsgruppen vermittelst des Automorphismus A ineinander übergehen.

§ 20. Will man nur diejenigen irreduziblen Kollineationsgruppen kennen, die den Gruppen $\mathfrak{S}_n$ oder $\mathfrak{A}_n$ isomorph sind und sich nicht als

Gruppen von $n!$, bzw. $\frac{n!}{2}$ homogenen linearen Substitutionen schreiben lassen, so hat man nur die einfachen Charaktere zweiter Art von $\mathfrak{T}_n$ oder $\mathfrak{B}_n$ zu berücksichtigen. Außerdem sind nach den Ergebnissen des vorigen Paragraphen bei der Gruppe $\mathfrak{T}_n$ zwei assoziierte und bei der Gruppe $\mathfrak{B}_n$ zwei adjungierte Charaktere als nicht wesentlich verschieden anzusehen. Beachtet man noch die im vorigen Abschnitt gewonnenen Resultate über die Anzahl der Charaktere zweiter Art bei den Gruppen $\mathfrak{T}_n$ und $\mathfrak{B}_n$, so erhält man den in der Einleitung angekündigten Satz:

VI. *Ist n größer als drei und von sechs verschieden, so ist die Anzahl der wesentlich verschiedenen irreduziblen Kollineationsgruppen, die der Gruppe $\mathfrak{S}_n$ isomorph sind und nicht als Gruppen von n! homogenen linearen Substitutionen geschrieben werden können, gleich der Anzahl v_n der Zerlegungen von n in lauter verschiedene Summanden. Ist n größer als vier und von sechs und sieben verschieden, so ist die entsprechende Anzahl für die Gruppe $\mathfrak{A}_n$ ebenfalls gleich der Zahl v_n.*

Die Gruppe $\mathfrak{S}_6$ tritt wegen des Bestehens des oben erwähnten äußeren Automorphismus als Ausnahme auf. Für diese Gruppe sind, wie ich schon in der Einleitung hervorgehoben habe, von den $v_6 = 4$ dem allgemeinen Fall entsprechenden Kollineationsgruppen nur drei wesentlich verschieden. Bei der Gruppe $\mathfrak{A}_4$ reduzieren sich die $v_4 = 2$ Kollineationsgruppen (des Grades 2) infolge des Auftretens linearer Charaktere bei der Gruppe $\mathfrak{B}_4$ auf nur eine Gruppe. Die Fälle $n=6$ und $n=7$ spielen aber für die Gruppe $\mathfrak{A}_n$ eine ausgezeichnete Rolle, weil die Gruppen $\mathfrak{B}_6$ und $\mathfrak{B}_7$ nicht mehr als die Darstellungsgruppen von $\mathfrak{A}_6$ und $\mathfrak{A}_7$ erscheinen.

Abschnitt VI.

Die Hauptdarstellung zweiter Art der Gruppe $\mathfrak{T}_n$.

§ 21. In diesem Abschnitt soll die bereits in der Einleitung erwähnte mit der Gruppe $\mathfrak{S}_n$ isomorphe Kollineationsgruppe des Grades $2^{\left[\frac{n-1}{2}\right]}$ aufgestellt und genauer untersucht werden.

Sind

$$A = (a_{\varkappa\lambda}),\ B = (b_{\mu\nu})$$

zwei Matrizen der Grade p und q, so soll die Matrix des Grades pq

$$\begin{pmatrix} a_{11}B, & a_{12}B, & \dots a_{1p}B \\ a_{21}B, & a_{22}B, & \dots a_{2p}B \\ \cdot & \cdot & \cdot \\ a_{p1}B, & a_{p2}B, & \dots a_{pp}B \end{pmatrix}$$

mit $A\times B$ bezeichnet werden. Ist C eine dritte Matrix des Grades r, so ist

$$(A\times B)\times C = A\times(B\times C).$$

Diese Matrix des Grades pqr bezeichnet man mit $A\times B\times C$. In analoger Weise ist für m Matrizen $A_1, A_2, \dots A_m$, deren Grade beliebig sein können, das Zeichen

$$A_1\times A_2\times\cdots\times A_m \tag{48.}$$

zu definieren. Sind $A_1, A_2, \dots A_m$ gleich einer Matrix A, so bezeichne ich die Matrix (48.) mit $\Pi_m A$. Die Spur der Matrix (48.) ist gleich dem Produkt der Spuren der Matrizen $A_1, A_2, \dots A_m$. Sind ferner $B_1, B_2, \dots B_m$ beliebige m andere Matrizen und ist hierbei B_μ von demselben Grade wie A_μ, so wird

$$(A_1\times A_2\times\cdots\times A_m)(B_1\times B_2\times\cdots\times B_m) = A_1B_1\times A_2B_2\times\cdots\times A_mB_m \text{*)}. \tag{49.}$$

Man betrachte nun die vier Matrizen zweiten Grades

$$F=\begin{pmatrix}1&0\\0&1\end{pmatrix},\quad A=\begin{pmatrix}0&1\\1&0\end{pmatrix},\quad B=\begin{pmatrix}0&1\\-1&0\end{pmatrix},\quad C=\begin{pmatrix}1&0\\0&-1\end{pmatrix}.$$

Zwischen diesen Matrizen bestehen folgende Beziehungen:

$$\begin{cases} A^2=F,\ B^2=-F,\ C^2=F, \\ AB=-BA=-C,\ BC=-CB=-A,\ CA=-AC=B, \\ CBA=F. \end{cases} \tag{50.}$$

Man bilde nun für ein beliebiges m die Matrizen des Grades 2^m

$$M_1=\Pi_m A,\ M_2=\Pi_{m-1}A\times B,\ M_3=\Pi_{m-1}A\times C,\dots$$
$$M_{2\nu}=\Pi_{m-\nu}A\times B\times\Pi_{\nu-1}F,\ M_{2\nu+1}=\Pi_{m-\nu}A\times C\times\Pi_{\nu-1}F,\dots$$
$$M_{2m}=B\times\Pi_{m-1}F,\ M_{2m+1}=C\times\Pi_{m-1}F.$$

Diese $2m+1$ Matrizen genügen, wie man auf Grund der Formeln (49.) und (50.) leicht erkennt, den Gleichungen

$$M_{2\nu}^2=-E,\ M_{2\nu+1}^2=E, \tag{51.}$$
$$M_\varkappa M_\lambda=-M_\lambda M_\varkappa, \qquad (\varkappa\neq\lambda) \tag{52.}$$
$$M_{2m+1}M_{2m}\cdots M_2M_1=E, \tag{53.}$$

wo $E=\Pi_m F$ die Einheitsmatrix des Grades 2^m bedeutet.

---

*) Vgl. *A. Hurwitz*, „Zur Invariantentheorie", Math. Ann. Bd. 45, S. 381, und meine Dissertation, § 6.

Aus den Gleichungen (51.) und (52.) ergibt sich, daß jedes Produkt der $2m$ Matrizen $M_1, M_2, \ldots M_{2m}$, abgesehen vom Vorzeichen, gleich wird einer der

$$1+\binom{2m}{1}+\binom{2m}{2}+\cdots+\binom{2m}{2m}=2^{2m}$$

Matrizen

$$(54.)\qquad E, M_1, M_2, \ldots M_{2m}, M_1 M_2, M_1 M_3, \ldots M_{2m-1} M_{2m}, \ldots M_1 M_2 \ldots M_{2m}.$$

Man erkennt auch ohne Mühe, daß diese Matrizen, abgesehen von den Vorzeichen, mit den 4^m Matrizen übereinstimmen, die man dadurch erhält, daß man in (48.) für $A_1, A_2, \ldots A_m$ auf alle möglichen Arten die vier Matrizen F, A, B, C einsetzt. Da nun die Spuren von A, B, C gleich Null sind, so ergibt sich, daß unter den Matrizen (54.), die auch kurz mit $X_0, X_1, X_2, \ldots$ bezeichnet werden mögen, nur die erste eine von Null verschiedene Spur besitzt; die Spur von $X_0 = E$ ist aber gleich 2^m. Die Matrizen $X_0, X_1, X_2, \ldots$ reproduzieren sich ferner, abgesehen von den Vorzeichen, durch Multiplikation, und zwar ist

$$X_\varkappa^2 = \pm E, \quad X_\varkappa X_\lambda = \pm X_\mu,$$

wo μ von Null verschieden ist.

Hieraus folgt leicht, daß $X_0, X_1, X_2, \ldots$ untereinander linear unabhängig sind. Denn ist etwa

$$a_0 X_0 + a_1 X_1 + a_2 X_2 + \cdots = 0,$$

so erhält man nach Multiplikation mit $X_\varkappa$

$$a_0 X_0 X_\varkappa + \cdots + a_{\varkappa-1} X_{\varkappa-1} X_\varkappa \pm a_\varkappa E + a_{\varkappa+1} X_{\varkappa+1} X_\varkappa + \cdots = 0.$$

Da nun die Spur der links stehenden Matrix offenbar gleich $\pm 2^m a_\varkappa$ ist, so muß $a_\varkappa = 0$ sein. Beachtet man noch, daß die Anzahl der Matrizen $X_0, X_1, X_2, \ldots$ gleich ist dem Quadrat ihres Grades, so erkennt man, daß *jede* Matrix des Grades 2^m als lineare homogene Verbindung von $X_0, X_1, X_2, \ldots$ darstellbar ist. *Daher erzeugen die Matrizen* $M_1, M_2, \ldots M_{2m}$ *eine irreduzible Gruppe* *).

§ 22. Mit Hülfe der Matrizen $M_1, M_2, \ldots M_{2m+1}$ gelingt es nun, eine Darstellung zweiter Art der Gruppe $\mathfrak{T}_n$ zu bilden.

Ich setze unter der Annahme, daß $m = \left[\frac{n-1}{2}\right]$, also $n = 2m+1$ oder $n = 2m+2$ ist,

$$(55.)\qquad T_\lambda = a_{\lambda-1} M_{\lambda-1} + b_\lambda M_\lambda, \qquad (\lambda = 1, 2, \ldots n-1)$$

*) Diese Gruppe ist eine endliche Gruppe der Ordnung 2^{2m+1}.

wo

$$a_{2\nu} = -\frac{\sqrt{\nu}}{\sqrt{2\nu+1}}, \quad b_{2\nu+1} = \frac{i\sqrt{\nu+1}}{\sqrt{2\nu+1}}, \qquad (\nu=0,1,2,\ldots)$$
$$a_{2\nu+1} = -\frac{i\sqrt{2\nu+1}}{2\sqrt{\nu+1}}, \quad b_{2\nu+2} = \frac{\sqrt{2\nu+3}}{2\sqrt{\nu+1}}$$

sein soll; hierbei sind alle Quadratwurzeln positiv zu nehmen. Es wird also

$$T_1 = iM_1, \; T_2 = -\frac{i}{2}M_1 + \frac{\sqrt{3}}{2}M_2, \; T_3 = -\frac{1}{\sqrt{3}}M_2 + \frac{i\sqrt{2}}{\sqrt{3}}M_3, \; \ldots .$$

Die Größen a_λ und b_λ genügen den Gleichungen

$$b_\lambda^2 - a_{\lambda-1}^2 = (-1)^\lambda, \; a_\lambda b_\lambda = \frac{(-1)^{\lambda-1}}{2}.$$

Mit Hilfe dieser Gleichungen und der Formeln (51.) und (52.) verifiziert man ohne Mühe die Relationen

$$(56.) \quad T_\alpha^2 = -E, \quad T_\beta T_{\beta+1} + T_{\beta+1} T_\beta = E, \quad T_\gamma T_\delta = -T_\delta T_\gamma, \quad \left\{\begin{array}{l} \alpha=1,2,\ldots n-1,\\ \beta=1,2,\ldots n-2,\\ \gamma=1,2,\ldots n-3,\\ \delta \geqq \gamma+2 \end{array}\right\}$$

und hieraus folgt leicht

$$(T_\beta T_{\beta+1})^3 = -E.$$

Setzt man noch $J = -E$, so erkennt man, daß die Matrizen $J, T_1, T_2, \ldots T_{n-1}$ den die Gruppe $\mathfrak{T}_n$ definierenden Relationen (II.) genügen. Folglich erzeugen sie eine Darstellung von $\mathfrak{T}_n$ durch Matrizen des Grades $2^{\left[\frac{n-1}{2}\right]}$, und da hier $J = -E$ ist, so ist dies eine Darstellung zweiter Art. Ich bezeichne diese Darstellung im folgenden mit Δ_n und nenne sie die *Hauptdarstellung zweiter Art der Gruppe* $\mathfrak{T}_n$.

Daß die Darstellung Δ_n irreduzibel ist, erkennt man folgendermaßen: Da $T_1 = iM_1$ ist, so läßt sich offenbar jede der Matrizen $M_1, M_2, \ldots M_{2m}$ als lineare homogene Verbindung von $T_1, T_2, \ldots T_{n-1}$ darstellen. Wäre nun die durch die T_α erzeugte Gruppe reduzibel, so müßte auch die durch die M_α erzeugte Gruppe reduzibel sein. Dies ist aber, wie wir gesehen haben, nicht der Fall.

Die Gleichungen (56.) geben noch Anlaß zu folgender Betrachtung. Allein unter Benutzung dieser Gleichungen kann man offenbar jedes Produkt $T_\alpha T_\beta T_\gamma \cdots$ als lineare homogene Verbindung der 2^{n-1} speziellen Produkte

$$(57.) \quad E, T_1, \ldots T_{n-1}, \; T_1 T_2, T_1 T_3, \ldots T_{n-2} T_{n-1}, \; T_1 T_2 T_3, \ldots T_1 T_2 \cdots T_{n-1}$$

darstellen, wobei als Koeffizienten nur ganze Zahlen in Betracht kommen. Aus den Gleichungen (56.) läßt sich ferner keine lineare homogene Relation mit konstanten Koeffizienten zwischen den Produkten (57.) ableiten.

In der Tat lassen sich doch die Gleichungen (56.) durch die Matrizen (55.) befriedigen. Unter den linearen Verbindungen der Produkte dieser Matrizen kommen aber, wie wir gesehen haben, die M_a und folglich auch die 2^{2m} linear unabhängigen Matrizen (54.) vor. Ist nun n ungerade, so wird $2^{2m} = 2^{n-1}$, und daher können in diesem Fall die 2^{n-1} Produkte (57.) gewiß nicht linear abhängig sein. Ist n aber gerade, so denke man sich zu den Elementen $T_1, T_2, \ldots T_{n-1}$ noch ein weiteres Element T_n hinzugefügt und nehme zu den Gleichungen (56.) noch die Gleichungen

$$(58.) \qquad T_n^2 = -E, \quad T_{n-1} T_n + T_n T_{n-1} = E, \quad T_\beta T_n = -T_n T_\beta \quad {\scriptstyle (\beta=1,2,\ldots n-2)}$$

hinzu. Da nun, weil $n+1$ ungerade ist, aus den Gleichungen (56.) und (58.) keine lineare homogene Relation zwischen den 2^n Produkten

$$E, \; T_1, \ldots T_n, \; T_1 T_2, \ldots T_{n-1} T_n, \; T_1 T_2 T_3, \ldots T_1 T_2 \cdots T_n$$

folgen kann, so läßt sich a fortiori aus (56.) keine Relation zwischen den Produkten (57.) ableiten.

Bezeichnet man nun die Produkte (57.) mit

$$A_1, A_2, \ldots A_{2^{n-1}},$$

so ergeben sich speziell aus (56.) Gleichungen der Form

$$A_\varkappa T_a = \sum_{\lambda=1}^{2^{n-1}} t_{\varkappa\lambda}^{(a)} A_\lambda,$$

wo die $t_{\varkappa\lambda}^{(a)}$ gewisse ganze Zahlen bedeuten. Die Matrizen

$$\overline{T}_a = (t_{\varkappa\lambda}^{(a)})$$

des Grades 2^{n-1} erzeugen dann offenbar eine mit der Gruppe $\mathfrak{T}_n$ (einstufig) isomorphe Gruppe:

Die Gruppe $\mathfrak{T}_n$ läßt sich als lineare homogene Gruppe des Grades 2^{n-1} mit ganzzahligen Koeffizienten darstellen.

§ 23. Es soll nun der zu der Darstellung Δ_n gehörende (einfache) Charakter $\chi(T)$ von $\mathfrak{T}_n$, den ich als den *Hauptcharakter zweiter Art* bezeichne, berechnet werden.

Ich schicke folgende Bemerkung voraus: Es sei X eine Matrix der Form $xE + \Sigma x_a P_a$, wo die P_a Produkte gewisser k unter den Matrizen $M_1, M_2, \ldots M_{2m}$ sind. Ebenso sei $Y = yE + \Sigma y_\beta Q_\beta$, wo jedes Q_β ein Produkt von gewissen l anderen unter diesen Matrizen ist. Hierbei soll keines der Produkte P_a und Q_β gleich $\pm E$ sein. Dann sind auch die Produkte $P_a Q_\beta$ sämtlich von $\pm E$ verschieden. Daher sind nach den Ausführungen des § 21 die Spuren aller Matrizen P_a, Q_β, $P_a Q_\beta$ gleich Null. Da nun die

Spur der Einheitsmatrix E des Grades 2^m gleich 2^m ist, so ergibt sich, *daß die Spuren der Matrizen X, Y, XY die Werte $2^m x$, $2^m y$, $2^m xy$ besitzen.*

Es sei nun P eine aus ϱ Zykeln bestehende Permutation (zweiter Art) der Form

$$(59.)\qquad (1, 2, \dots \lambda_1)(\lambda_1+1, \lambda_1+2, \dots \lambda_1+\lambda_2)\cdots,$$

wobei wir $\lambda_1 \geqq \lambda_2 \geqq \cdots$ annehmen. Dieser Permutation von $\mathfrak{S}_n$ entspricht auf Grund der in § 11 getroffenen Festsetzungen in der Gruppe $\mathfrak{T}_n$ das Element

$$P' = C_{\lambda_1} C_{\lambda_2} \cdots,$$

wo

$$C_{\lambda_1} = T_{\lambda_1-1} T_{\lambda_1-2} \cdots T_1, \quad C_{\lambda_2} = T_{\lambda_1+\lambda_2-1} T_{\lambda_1+\lambda_2-2} \cdots T_{\lambda_1+1}, \quad \dots$$

zu setzen und insbesondere für $\lambda_\alpha = 1$ unter C_{λ_α} das Einheitselement E von $\mathfrak{T}_n$ zu verstehen ist. Die Matrizen, die in der Darstellung Δ_n den Elementen P' und C_{λ_α} von $\mathfrak{T}_n$ entsprechen, mögen mit denselben Buchstaben bezeichnet werden. Ist nun n ungerade, also gleich $2m+1$, so kommt M_{2m+1} in keiner der Matrizen (55.) vor und ist also, wenn die C_{λ_α} durch die $M_\varkappa$ ausgedrückt werden, überhaupt nicht zu berücksichtigen. Ist aber $n = 2m+2$, so kommt M_{2m+1} nur in T_{n-1} vor und ist daher nur dann in Betracht zu ziehen, wenn P der Zyklus $(1, 2, \dots n)$, also

$$P' = T_{n-1} T_{n-2} \cdots T_1$$

wird. Ferner drückt sich C_{λ_1} durch die Produkte der Matrizen $M_1, M_2, \dots M_{\lambda_1-1}$ aus, ebenso C_{λ_2} durch die Produkte der Matrizen $M_{\lambda_1}, M_{\lambda_1+1}, \dots M_{\lambda_1+\lambda_2-1}$ usw. Auf Grund der oben gemachten Bemerkung erkennt man daher, *daß, wenn die Spur von C_{λ_α} gleich $2^m c_\alpha$ wird, die Spur von P' den Wert $2^m c_1 c_2 \cdots$ erhält.*

Wir haben demnach nur die Spur eines Elements der Form

$$C = T_\beta T_{\beta-1} \cdots T_{\beta-\alpha}$$

zu berechnen. Es wird dann

$$C = (a_{\beta-1} M_{\beta-1} + b_\beta M_\beta)(a_{\beta-2} M_{\beta-2} + b_{\beta-1} M_{\beta-1}) \cdots (a_{\beta-\alpha-1} M_{\beta-\alpha-1} + b_{\beta-\alpha} M_{\beta-\alpha}).$$

Führt man hier die Multiplikation aus, so hat man, da es sich für uns nur um die Bestimmung der Spur von C handelt, allein diejenigen Glieder zu berücksichtigen, welche von der Form cE sind. Ist nun die Anzahl $\alpha+1$ der Faktoren T_λ von C gerade, so hat die verlangte Form nur das eine Glied

$$a_{\beta-1} b_{\beta-1} M_{\beta-1}^2 \cdot a_{\beta-3} b_{\beta-3} M_{\beta-3}^2 \cdots a_{\beta-\alpha} b_{\beta-\alpha} M_{\beta-\alpha}^2.$$

Die Spur von C wird dann wegen

$$M_\lambda^2=(-1)^{\lambda-1}E,\quad a_\lambda b_\lambda=\frac{(-1)^{\lambda-1}}{2}$$

gleich $2^{m-\frac{\alpha+1}{2}}$. Ist dagegen $\alpha+1$ ungerade, so wird die Spur von C gleich Null, es sei denn, daß $n=2m+2$ und

$$C=T_{n-1}T_{n-2}\cdots T_1$$
$$=(a_{2m}M_{2m}+b_{2m+1}M_{2m+1})(a_{2m-1}M_{2m-1}+b_{2m}M_{2m})\cdots(a_1M_1+b_2M_2)\cdot b_1M_1$$

wird. In diesem Fall enthält die Entwicklung von C das Glied

$$b_{2m+1}M_{2m+1}\cdot b_{2m}M_{2m}\cdots b_2M_2\cdot b_1M_1,$$

das auf Grund der Gleichung (53.) gleich wird

$$b_1b_2\cdots b_{2m+1}\cdot E.$$

Der hier auftretende Faktor hat den Wert

$$i\cdot\frac{\sqrt{3}}{2}\cdot\frac{i\sqrt{2}}{\sqrt{3}}\cdot\frac{\sqrt{5}}{2\sqrt{2}}\cdots\frac{\sqrt{2m+1}}{2\sqrt{m}}\cdot\frac{i\sqrt{m+1}}{\sqrt{2m+1}}=\frac{i^{m+1}\sqrt{m+1}}{2^m}.$$

Daher wird die Spur von C gleich

$$i^{m+1}\sqrt{m+1}=\sqrt{(-1)^{\frac{n}{2}}\cdot\frac{n}{2}}.$$

Sind nun die Ordnungen $\lambda_1,\lambda_2,\ldots\lambda_\varrho$ der Zykeln der früher betrachteten Permutation P sämtlich ungerade, so wird die Spur $2^m c_\alpha$ von C_{λ_α} gleich $2^{m-\frac{\lambda_\alpha-1}{2}}$, und daher wird die Spur $\chi(P')$ der Matrix P gleich

$$2^{m-\frac{\lambda_1-1}{2}-\frac{\lambda_2-1}{2}-\cdots-\frac{\lambda_\varrho-1}{2}}=2^{\frac{2m-n+\varrho}{2}}=2^{\left[\frac{\varrho-1}{2}\right]}.$$

Ist dagegen auch nur eine der Zahlen λ_α gerade, so wird $\chi(P')$ gleich Null mit Ausnahme des Falles, daß n gerade und $P=(1,2,\ldots n)$ ist. Dann wird

$$\chi(P')=\sqrt{(-1)^{\frac{n}{2}}\cdot\frac{n}{2}}.$$

Benutzt man nun die in § 17 eingeführten Bezeichnungen, so ergibt sich:

VII. *Ist* $[\alpha]$ *eine Klasse ähnlicher Permutationen von* $\mathfrak{S}_n$, *die in* σ_α *Zykeln ungerader Ordnung zerfallen, so wird für den Hauptcharakter zweiter Art* $\chi(T)$ *von* $\mathfrak{T}_n$

$$\chi_\alpha=2^{\left[\frac{\sigma_\alpha-1}{2}\right]}.$$

Für ein ungerades n *ist* $\chi(T)$ *ein zweiseitiger Charakter. Ist aber* n *gerade, so ist* $\chi(T)$ *ein nicht zweiseitiger Charakter, und es wird für die Klasse* (n) *der Zykeln* n*-ter Ordnung*

$$\chi_{(n)} = \sqrt{(-1)^{\frac{n}{2}} \cdot \frac{n}{2}},$$

für alle übrigen ungeraden Klassen (ν) *aber*

$$\chi_{(\nu)} = 0.$$

§ 24. Ist n ungerade, so haben wir noch, da $\chi(T)$ ein zweiseitiger Charakter wird, die zu $\chi(T)$ gehörenden einfachen Charaktere $\psi(B)$ und $\overline{\psi}(B)$ von $\mathfrak{B}_n$ zu bestimmen. Es genügt hierzu nach dem früheren, das Komplement $\delta(B) = \psi(B) - \overline{\psi}(B)$ von $\chi(T)$ anzugeben.

Um diese Aufgabe zu lösen, beachte man, daß für $n = 2m+1$ die Elemente $T_1, T_2, \ldots T_{n-1}$ unserer Darstellung Δ_n die Matrix M_{2m+1} nicht enthalten. Da

$$M_{2m+1} M_\lambda = -M_\lambda M_{2m+1} \qquad (\lambda = 1, 2, \ldots 2m)$$

ist, so ergibt sich daher

$$M_{2m+1}^{-1} T_\lambda M_{2m+1} = -T_\lambda.$$

Beachtet man noch, daß $M_{2m+1}^2 = E$ ist, so erkennt man, daß M_{2m+1} für die Darstellung Δ_n genau dieselbe Rolle spielt, wie die Matrix H für die in § 16 betrachtete zweiseitige Darstellung. Um daher $\delta(B)$ zu bestimmen, hat man nur für alle geraden Permutationen P die Spur der Matrix $M_{2m+1} P'$ anzugeben. Wir können uns hierbei wieder auf die Permutationen P der Form (59.) beschränken.

Da nun auf Grund der Gleichungen (51.)—(53.)

$$M_{2m+1} = \pm M_1 M_2 \cdots M_{2m}$$

ist, so wird die Spur eines Produktes

$$M_{2m+1} M_\alpha M_\beta \cdots$$

stets Null, wenn $\alpha, \beta, \ldots$ irgendwelche Indizes aus der Reihe $1, 2, \ldots 2m$ bedeuten, deren Anzahl kleiner als $2m$ ist. Hieraus folgt leicht, daß $M_{2m+1} P'$ die Spur Null hat, wenn P nicht der Zyklus $(1, 2, \ldots n)$ ist. In diesem Fall wird aber wieder

$$P' = T_{n-1} T_{n-2} \cdots T_1$$
$$= (a_{2m-1} M_{2m-1} + b_{2m} M_{2m})(a_{2m-2} M_{2m-2} + b_{2m-1} M_{2m-1}) \cdots (a_1 M_1 + b_2 M_2) \cdot b_1 M_1,$$

und die Spur $\delta(P')$ von $M_{2m+1} P'$ wird gleich der Spur von

$$M_{2m+1} \cdot b_{2m} M_{2m} \cdot b_{2m-1} M_{2m-1} \cdots b_2 M_2 \cdot b_1 M_1 = b_1 b_2 \cdots b_{2m} E,$$

also gleich

$$2^m b_1 b_2 \cdots b_{2m} = i^m \sqrt{2m+1} = \sqrt{(-1)^{\frac{n-1}{2}} n}.$$

Benutzt man wieder die Bezeichnungen des § 17, so kann man folgende Regel aufstellen:

VII*. *Ist n eine ungerade Zahl, so wird für das Komplement $\delta(B)$ des Hauptcharakters zweiter Art*

$$\delta_{(n)} = \sqrt{(-1)^{\frac{n-1}{2}} n}, \quad \delta_{(\nu)} = 0.$$

Hierbei bedeutet (n) wieder die Klasse der Zykeln n-ter Ordnung von $\mathfrak{S}_n$ und (ν) irgendeine von (n) verschiedene (gerade) Klasse.

§ 25. Der in § 22 gegebene Beweis für die Irreduzibilität der Darstellung Δ_n stützt sich auf die Betrachtung der Matrizen (54.). Es soll hier noch ein anderer Beweis angegeben werden, der nur von dem uns jetzt bekannten Charakter $\chi(T)$ der Darstellung Δ_n Gebrauch macht. Wir werden hierbei auf eine Klasse von symmetrischen Funktionen geführt werden, die für unsere ganze Untersuchung von wesentlicher Bedeutung ist.

Man hat, um die Irreduzibilität von Δ_n nachzuweisen, nur zu zeigen, daß $\chi(T)$ ein einfacher Charakter ist, und hierzu genügt es, sich zu überzeugen, daß für die Zahlen χ_a und $\chi_{(\nu)}$ die Gleichungen (39.) und (42.) des § 17 erfüllt sind. Die Gleichung (42.) kommt nur für ein gerades n in Betracht und ist unmittelbar zu verifizieren. Es handelt sich also nur um die Gleichung

$$\sum_a h_a \chi_a^2 = \frac{n!}{2^\varepsilon},$$

wo hier ε gleich Null oder Eins zu setzen ist, je nachdem n ungerade oder gerade ist. Wenden wir dieselben Bezeichnungen wie im § 17 an, so wird nach Satz VII für ein gerades n

$$\chi_a = \chi_{\alpha_1, \alpha_3, \ldots} = 2^{\frac{\alpha_1 + \alpha_3 + \cdots - 2}{2}}$$

und für ein ungerades n

$$\chi_a = \chi_{\alpha_1, \alpha_3, \ldots} = 2^{\frac{\alpha_1 + \alpha_3 + \cdots - 1}{2}}$$

Die zu beweisende Gleichung lautet daher

$$\sum \frac{n!}{1^{\alpha_1} \alpha_1! \, 3^{\alpha_3} \alpha_3! \cdots} 2^{\alpha_1 + \alpha_3 + \cdots - \varepsilon - 1} = \frac{n!}{2^\varepsilon}$$

oder einfacher

$$\sum \frac{2^{\alpha_1 + \alpha_3 + \cdots}}{1^{\alpha_1} \alpha_1! \, 3^{\alpha_3} \alpha_3! \cdots} = 2, \tag{60.}$$

wo die Summen über alle ganzzahligen (nicht negativen) Lösungen $\alpha_1, \alpha_3, \ldots$

der Gleichung

$$\alpha_1 + 3\alpha_3 + \cdots = n$$

zu erstrecken sind.

Um die Richtigkeit der Formel (60.) nachzuweisen, betrachte ich die rationale Funktion

$$f(z) = \frac{(1+x_1 z)(1+x_2 z)\cdots(1+x_r z)}{(1-x_1 z)(1-x_2 z)\cdots(1-x_r z)}$$

der Variabeln z; hierbei sollen auch $x_1, x_2, \ldots x_r$ Variable bedeuten. Entwickelt man $f(z)$ nach steigenden Potenzen von z, so möge sich

$$f(z) = 1 + q_1 z + q_2 z^2 + \cdots$$

ergeben. Dann lassen sich $q_1, q_2, \ldots$ folgendermaßen berechnen: Man setze

$$s_\lambda = x_1^\lambda + x_2^\lambda + \cdots + x_r^\lambda,$$

dann wird

$$\log f(z) = \sum_\varkappa \log \frac{1+zx_\varkappa}{1-zx_\varkappa} = 2\sum_\varkappa \left(\frac{x_\varkappa z}{1} + \frac{x_\varkappa^3 z^3}{3} + \cdots\right)$$

$$= \frac{2s_1}{1} z + \frac{2s_3}{3} z^3 + \cdots$$

und folglich

$$f(z) = e^{\log f(z)} = \sum_{\nu=0}^{\infty} \frac{1}{\nu!}\left(\frac{2s_1}{1} z + \frac{2s_3}{3} z^3 + \cdots\right)^\nu$$

$$= \sum_{\nu=0}^{\infty} \frac{1}{\nu!} \sum_{\alpha_1+\alpha_3+\cdots=\nu} \frac{\nu!}{\alpha_1!\,\alpha_3!\cdots}\left(\frac{2s_1}{1}\right)^{\alpha_1}\left(\frac{2s_3}{3}\right)^{\alpha_3}\cdots z^{\alpha_1+3\alpha_3+\cdots}.$$

Hieraus ergibt sich aber

$$(61.)\qquad q_n = \sum \frac{2^{\alpha_1+\alpha_3+\cdots}}{1^{\alpha_1}\alpha_1!\,3^{\alpha_3}\alpha_3!\cdots} s_1^{\alpha_1} s_3^{\alpha_3}\cdots. \qquad (\alpha_1+3\alpha_3+\cdots=n)$$

Wird nun speziell

$$x_1 = 1,\ x_2 = 0,\ \ldots\ x_r = 0,$$

so erhält man

$$s_1 = s_3 = \cdots = 1$$

und wegen

$$\frac{1+z}{1-z} = 1 + 2z + 2z^2 + \cdots$$

noch $q_n = 2$. Setzt man diese Werte für $s_1, s_3, \ldots$ und q_n in (61.) ein, so ergibt sich die zu beweisende Formel (60.).

Abschnitt VII.

Über die Gruppen $\mathfrak{T}_{\nu_1, \nu_2, \ldots, \nu_m}$.

§ 26. Um mit Hilfe des im vorigen Abschnitt gewonnenen Charakters auch die übrigen einfachen Charaktere zweiter Art der Gruppe $\mathfrak{T}_n$

zu berechnen, müssen wir eine spezielle Klasse von Untergruppen dieser Gruppe genauer studieren.

Es seien $\nu_1, \nu_2, \ldots \nu_m$ positive ganze Zahlen, deren Summe gleich n ist. Man bezeichne mit S_α eine Permutation von $\mathfrak{S}_n$, die nur die ν_α Ziffern

$$\nu_1 + \cdots + \nu_{\alpha-1} + 1,\ \nu_1 + \cdots + \nu_{\alpha-1} + 2,\ \ldots\ \nu_1 + \cdots + \nu_{\alpha-1} + \nu_\alpha$$

untereinander vertauscht. Dann bilden die $\nu_1!\,\nu_2!\cdots\nu_m!$ Permutationen der Form $S_1 S_2 \ldots S_m$ eine Untergruppe $\mathfrak{S}_{\nu_1,\nu_2,\ldots\nu_m}$ von $\mathfrak{S}_n$, ferner bilden die geraden unter ihnen eine Untergruppe $\mathfrak{A}_{\nu_1,\nu_2,\ldots,\nu_m}$ der Ordnung $\frac{1}{2}\nu_1!\,\nu_2!\cdots\nu_m!$*). Bezeichnet man mit $\mathfrak{S}_{\nu_\alpha}$ die symmetrische Gruppe der $\nu_\alpha!$ Permutationen S_α, so erscheint $\mathfrak{S}_{\nu_1,\nu_2,\ldots,\nu_m}$ als das direkte Produkt der m Gruppen $\mathfrak{S}_{\nu_1}, \mathfrak{S}_{\nu_2}, \ldots \mathfrak{S}_{\nu_m}$.

Zu jeder Permutation P von $\mathfrak{S}_{\nu_1,\nu_2,\ldots\nu_m}$ gehören in der Gruppe $\mathfrak{T}_n$ zwei Elemente P' und JP', und die sich auf diese Weise ergebenden $2\cdot\nu_1!\,\nu_2!\cdots\nu_m!$ Elemente bilden eine Untergruppe $\mathfrak{T}_{\nu_1,\nu_2,\ldots,\nu_m}$ von $\mathfrak{T}_n$. Ebenso entsprechen den Untergruppen $\mathfrak{A}_{\nu_1,\nu_2,\ldots,\nu_m}$ und $\mathfrak{S}_{\nu_\alpha}$ von $\mathfrak{S}_n$ in der Gruppe $\mathfrak{T}_n$ Untergruppen $\mathfrak{B}_{\nu_1,\nu_2,\ldots\nu_m}$ und $\mathfrak{T}_{\nu_\alpha}$, deren Ordnungen die Zahlen $\nu_1!\,\nu_2!\cdots\nu_m!$ und $2\cdot\nu_\alpha!$ sind. Die Gruppe $\mathfrak{T}_{\nu_\alpha}$ ist hierbei nichts anderes als die von uns studierte Darstellungsgruppe der symmetrischen Gruppe mit ν_α Vertauschungsziffern.

Die Gruppe $\mathfrak{T}_{\nu_1,\nu_2,\ldots,\nu_m}$ kann als abstrakte Gruppe folgendermaßen definiert werden. Man betrachte die m (abstrakten) Gruppen $\mathfrak{T}_{\nu_1}, \mathfrak{T}_{\nu_2}, \ldots \mathfrak{T}_{\nu_m}$ und mache die Festsetzungen:

a) Die in diesen m Gruppen enthaltenen (dem Element J von $\mathfrak{T}_n$ entsprechenden) invarianten Elemente sollen einander gleich sein und mit J bezeichnet werden;

b) sind R_α und R_β zwei Elemente der Gruppen $\mathfrak{T}_{\nu_\alpha}$ und $\mathfrak{T}_{\nu_\beta}$, so soll $R_\alpha R_\beta = R_\beta R_\alpha$ sein, wenn R_α in der Untergruppe $\mathfrak{B}_{\nu_\alpha}$ von $\mathfrak{T}_{\nu_\alpha}$ oder R_β in der Untergruppe $\mathfrak{B}_{\nu_\beta}$ von $\mathfrak{T}_{\nu_\beta}$ enthalten ist, dagegen soll, wenn weder das eine noch das andere der Fall ist, $R_\alpha R_\beta = J R_\beta R_\alpha$ sein.

Dann erzeugen die Elemente von $\mathfrak{T}_{\nu_1}, \mathfrak{T}_{\nu_2}, \ldots \mathfrak{T}_{\nu_m}$ die Gruppe $\mathfrak{T}_{\nu_1,\nu_2,\ldots,\nu_m}$, und jedes Element dieser Gruppe läßt sich als Produkt $R_1 R_2 \cdots R_m$ darstellen, wo R_α ein Element von $\mathfrak{T}_{\nu_\alpha}$ bedeutet. Hierbei werden aber je 2^{m-1} dieser Produkte einander gleich. Die Untergruppe $\mathfrak{B}_{\nu_1,\nu_2,\ldots\nu_m}$ von

*) Der Fall $\nu_1 = \nu_2 = \cdots = \nu_m = 1$ ist hierbei auszuschließen.

$\mathfrak{T}_{\nu_1, \nu_2, \ldots \nu_m}$ wird ferner dadurch erhalten, daß man nur diejenigen Produkte $R_1 R_2 \cdots R_m$ betrachtet, in denen die Anzahl der Faktoren R_α, die nicht in den zugehörigen Gruppen $\mathfrak{B}_{\nu_\alpha}$ vorkommen, gerade ist.

Im folgenden schreibe ich für $\mathfrak{T}_{\nu_1, \nu_2, \ldots \nu_m}$ und $\mathfrak{B}_{\nu_1, \nu_2, \ldots \nu_m}$ kürzer $\mathfrak{T}$ und $\mathfrak{B}$. Die Elemente von $\mathfrak{B}$ bezeichne ich mit B und die Elemente von $\mathfrak{T}$, die nicht in $\mathfrak{B}$ enthalten sind, mit C. Ebenso soll ein Element von $\mathfrak{T}_{\nu_\alpha}$ mit B_α oder C_α bezeichnet werden, je nachdem es in $\mathfrak{B}_{\nu_\alpha}$ vorkommt oder nicht.

§ 27. Ich will nun die Darstellungen der Gruppe $\mathfrak{T}$ durch homogene lineare Substitutionen (Matrizen) genauer untersuchen.

Entspricht in einer solchen Darstellung $\varDelta$ dem Element J von $\mathfrak{T}$ die Matrix $\pm E$, so bezeichne ich wieder $\varDelta$ als eine Darstellung erster oder zweiter Art, je nachdem hier das obere oder das untere Zeichen auftritt. Aus jeder Darstellung geht auch hier, wie früher bei der Gruppe $\mathfrak{T}_n$, eine zweite Darstellung $\varDelta'$ hervor, indem man die Matrizen von $\varDelta$, die den Elementen von $\mathfrak{B}$ entsprechen, ungeändert läßt, dagegen die übrigen mit -1 mutipliziert. Die Darstellungen $\varDelta$ und $\varDelta'$ nennen wir wieder *assoziierte* Darstellungen; sind insbesondere $\varDelta$ und $\varDelta'$ einander äquivalent, so soll $\varDelta$ eine *zweiseitige* Darstellung heißen. Man zeigt nun ebenso, wie in § 16 bei der Betrachtung der Gruppe $\mathfrak{T}_n$, daß sich für jede irreduzible zweiseitige Darstellung $\varDelta$ von $\mathfrak{T}$ eine Matrix H angeben läßt, welche die Darstellung $\varDelta$ in die ihr assoziierte Darstellung transformiert und außerdem der Bedingung $H^2 = E$ genügt und die Spur Null besitzt. Entspricht in $\varDelta$ die Matrix B einem Element der Untergruppe $\mathfrak{B}$ von $\mathfrak{T}$ und sind $\zeta(B)$ und $\eta(B)$ die Spuren der Matrizen B und HB, so erhält man wieder zwei einfache Charaktere $\psi(B)$ und $\overline{\psi}(B)$ von $\mathfrak{B}$, indem man setzt

$$\psi(B) = \frac{1}{2}[\zeta(B) + \eta(B)], \quad \overline{\psi}(B) = \frac{1}{2}[\zeta(B) - \eta(B)].$$

Die Zahlen $\eta(B) = \psi(B) - \overline{\psi}(B)$ nenne ich wieder das *Komplement* des Charakters $\zeta(R)$.

Es soll nun gezeigt werden, *daß man eine irreduzible Darstellung zweiter Art der Gruppe $\mathfrak{T}$ bestimmen kann, sobald man nur eine solche Darstellung für jede der die Gruppe $\mathfrak{T}$ erzeugenden Gruppen $\mathfrak{T}_{\nu_1}, \mathfrak{T}_{\nu_2}, \ldots \mathfrak{T}_{\nu_m}$ kennt.*

Man habe nämlich für die Gruppe $\mathfrak{T}_{\nu_\alpha}$ eine irreduzible Darstellung zweiter Art $\varDelta_\alpha$ des Grades f_α, zu der der Charakter $\chi^{(\alpha)}(R_\alpha)$ von $\mathfrak{T}_{\nu_\alpha}$ gehört. Die Matrizen, die in $\varDelta_\alpha$ den Elementen B_α und C_α von $\mathfrak{T}_{\nu_\alpha}$ entsprechen, mögen mit $\overline{B}_a$ und $\overline{C}_a$ bezeichnet werden. Ist Δ_a insbesondere eine zweiseitige Darstellung, so sei H_a eine Matrix, die Δ_a in die assoziierte Darstellung transformiert und der Gleichung $H_a^2 = E_a$ genügt, wo E_a die Einheitsmatrix des Grades f_a bedeutet. Die Spur von H_a ist dann Null und es ist

$$(62.)\qquad H_a \overline{B}_a = \overline{B}_a H_a, \; H_a \overline{C}_a = -\overline{C}_a H_a.$$

Ist noch $\delta^{(a)}(B_a)$ die Spur der Matrix $H_a \overline{B}_a$, so bilden die Zahlen $\delta^{(a)}(B_a)$ das Komplement des Charakters $\chi^{(a)}(R_a)$ (vgl. § 16).

Unter den m Darstellungen Δ_a seien r zweiseitig und $m-r=s$ nicht zweiseitig. Der Einfachheit wegen will ich annehmen, daß $\Delta_1, \Delta_2, \ldots \Delta_r$ die r zweiseitigen Darstellungen seien. Man setze nun $t = \left[\frac{s}{2}\right]$, so daß also $s = 2t$ oder $s = 2t+1$ wird, und bestimme nach den Vorschriften des § 21 die Matrizen $M_1, M_2, \ldots M_{2t+1}$ des Grades 2^t. Ist dann

$$F_1 = iM_1, \; F_2 = M_2, \; F_3 = iM_3, \; \ldots \; F_{2t} = iM_{2t}, \; F_{2t+1} = M_{2t+1}$$

und bedeutet F_0 die Einheitsmatrix des Grades 2^t, so wird auf Grund der Formeln des § 21

$$F_\varkappa^2 = F_0, \; F_\varkappa F_\lambda = -F_\lambda F_\varkappa, \; F_{2t+1} F_{2t} \cdots F_2 F_1 = i^t F_0.$$

Es wird also insbesondere

$$(63.)\qquad F_{s+1} F_s \cdots F_2 F_1 = i^{\frac{s}{2}} F_0$$

für ein gerades s und

$$(64.)\qquad F_s F_{s-1} \cdots F_2 F_1 = i^{\frac{s-1}{2}} F_0$$

für ein ungerades s.

Um nun eine Darstellung zweiter Art der Gruppe $\mathfrak{T}$ zu erhalten, setze ich unter Benutzung der in § 21 erläuterten Bezeichnungen

$$B_a = F_0 \times E_1 \times \cdots \times E_{a-1} \times \overline{B}_a \times E_{a+1} \times \cdots \times E_{r+s} \qquad {\scriptstyle (a=1,2,\ldots m)}$$

und ferner

$$\begin{aligned}
\Gamma_1 &= F_0 \times \overline{C}_1 \times E_2 \times \cdots \times E_{r-1} \times E_r \times E_{r+1} \times E_{r+2} \times \cdots \times E_{r+s},\\
\Gamma_2 &= F_0 \times H_1 \times \overline{C}_2 \times \cdots \times E_{r-1} \times E_r \times E_{r+1} \times E_{r+2} \times \cdots \times E_{r+s},\\
&\cdots\cdots\cdots\cdots\cdots\cdots\cdots\cdots\cdots\cdots\cdots\cdots\cdots\cdots\\
\Gamma_r &= F_0 \times H_1 \times H_2 \times \cdots \times H_{r-1} \times \overline{C}_r \times E_{r+1} \times E_{r+2} \times \cdots \times E_{r+s},\\
\Gamma_{r+1} &= F_s \times H_1 \times H_2 \times \cdots \times H_{r-1} \times H_r \times \overline{C}_{r+1} \times E_{r+2} \times \cdots \times E_{r+s},\\
\Gamma_{r+2} &= F_{s-1} \times H_1 \times H_2 \times \cdots \times H_{r-1} \times H_r \times E_{r+1} \times \overline{C}_{r+2} \times \cdots \times E_{r+s},\\
&\cdots\cdots\cdots\cdots\cdots\cdots\cdots\cdots\cdots\cdots\cdots\cdots\cdots\cdots\\
\Gamma_{r+s} &= F_1 \times H_1 \times H_2 \times \cdots \times H_{r-1} \times H_r \times E_{r+1} \times E_{r+2} \times \cdots \times \overline{C}_{r+s}.
\end{aligned}$$

Man erkennt nun auf Grund der Gleichungen $F_\varkappa^2 = F_0$ und $H_a^2 = E_a$ zunächst, daß für jedes α die Matrizen B_α und Γ_α, ebenso wie das für die Matrizen $\overline{B}_\alpha$ und $\overline{C}_\alpha$ der Fall ist, eine Darstellung zweiter Art von $\mathfrak{T}_{\nu_\alpha}$ bilden. Wegen der Gleichungen $F_\varkappa F_\lambda = -F_\lambda F_\varkappa$ und der Gleichungen (62.) ergibt sich ferner für $\alpha \neq \beta$

$$B_\alpha B_\beta = B_\alpha B_\beta, \quad B_\alpha \Gamma_\beta = \Gamma_\beta B_\alpha, \quad \Gamma_\alpha \Gamma_\beta = -\Gamma_\beta \Gamma_\alpha.$$

Hieraus folgt aber leicht, daß die Matrizen

$$B_1, \Gamma_1, B_2, \Gamma_2, \ldots B_m, \Gamma_m$$

eine Darstellung zweiter Art der Gruppe $\mathfrak{T}$ erzeugen. Der Grad dieser Darstellung ist gleich $2^t f_1 f_2 \cdots f_m$.

§ 28. Es soll nun der zu der eben gewonnenen Darstellung gehörende Charakter $\zeta(R)$ von $\mathfrak{T}$ berechnet werden, d. h. es soll, wenn dem Element R von $\mathfrak{T}$ die Matrix P entspricht, die Spur $\zeta(R)$ von P angegeben werden.

Man beachte hierzu, daß die Spur einer Matrix $X_1 \times X_2 \times \cdots$ gleich ist dem Produkt der Spuren der Matrizen $X_1, X_2, \ldots$ und daß die Spuren der Matrizen

$$F_1, F_2, \ldots F_s, \quad F_2F_1, \ldots F_sF_1, \quad F_3F_2, \ldots F_3F_2F_1, \ldots F_sF_{s-1}\cdots F_1$$

für ein gerades s sämtlich verschwinden und für ein ungerades s alle mit Ausnahme der letzten gleich Null sind (vgl. § 21). Die Spur von $F_sF_{s-1}\cdots F_1$ wird aber (für ein ungerades s) wegen (64.) gleich $(2i)^{\frac{s-1}{2}}$. Außerdem sind die Spuren von $\overline{B}_\alpha, \overline{C}_\alpha, H_\alpha\overline{C}_\alpha$ nach Voraussetzung gleich $\chi^{(\alpha)}(B_\alpha), \chi^{(\alpha)}(C_\alpha)$, $\delta^{(\alpha)}(B_\alpha)$ und für $\alpha = 1, 2, \ldots r$ die Spuren von $H_\alpha, \overline{C}_\alpha$ und $H_\alpha\overline{C}_\alpha$ gleich Null.

Auf Grund dieser Bemerkungen beweist man leicht:

a) *Ist s ungerade, so wird*

$$\zeta(B_1B_2\cdots B_m) = 2^{\frac{s-1}{2}}\chi^{(1)}(B_1)\chi^{(2)}(B_2)\cdots\chi^{(m)}(B_m), \tag{65.}$$

$$\zeta(B_1\cdots B_rC_{r+1}\cdots C_{r+s}) = (2i)^{\frac{s-1}{2}}\delta^{(1)}(B_1)\cdots\delta^{(r)}(B_r)\chi^{(r+1)}(C_{r+1})\cdots\chi^{(r+s)}(C_{r+s}) \tag{66.}$$

und $\zeta(R) = 0$, wenn das Element R von $\mathfrak{T}$ nicht von der Form $B_1B_2\cdots B_m$ oder $B_1\cdots B_rC_{r+1}\cdots C_{r+s}$ ist.

b) *Ist s gerade, so wird*

$$\zeta(B_1B_2\cdots B_m) = 2^{\frac{s}{2}}\chi^{(1)}(B_1)\chi^{(2)}(B_2)\cdots\chi^{(m)}(B_m) \tag{67.}$$

und $\zeta(R) = 0$, wenn R nicht die Form $B_1B_2\cdots B_m$ hat.

Hieraus folgt insbesondere, daß $\zeta(R)$ für ein gerades s ein zweiseitiger und für ein ungerades s ein nicht zweiseitiger Charakter ist.

Berücksichtigt man noch, daß für $\alpha \leqq r$

$$\sum_{B_\alpha} \chi^{(\alpha)}(B_\alpha)\chi^{(\alpha)}(B_\alpha^{-1}) = \sum_{B_\alpha} \delta^{(\alpha)}(B_\alpha)\delta^{(\alpha)}(B_\alpha^{-1}) = 2\cdot\nu_\alpha!$$

und für $\alpha > r$

$$\sum_{B_\alpha} \chi^{(\alpha)}(B_\alpha)\chi^{(\alpha)}(B_\alpha^{-1}) = \sum_{C_\alpha} \chi^{(\alpha)}(C_\alpha)\chi^{(\alpha)}(C_\alpha^{-1}) = \nu_\alpha!$$

wird*), so beweist man ohne Mühe die Gleichung

$$\sum_R \zeta(R)\zeta(R^{-1}) = 2\cdot\nu_1!\,\nu_2!\cdots\nu_m!.$$

Daher ist $\zeta(R)$ ein einfacher Charakter, die von uns hergestellte Darstellung der Gruppe $\mathfrak{T}$ also irreduzibel.

Ist s eine gerade Zahl, also $\zeta(R)$ ein zweiseitiger Charakter von $\mathfrak{T}$, so haben wir noch das Komplement $\eta(B)$ dieses Charakters anzugeben.

Um diese Aufgabe zu lösen, müssen wir nach dem früheren eine Matrix H des Grades $2^{\frac{s}{2}} f_1 f_2 \ldots f_m$ kennen, die unsere Darstellung der Gruppe $\mathfrak{T}$ in die assoziierte Darstellung transformiert und der Gleichung $H^2 = E$ genügt. Diese Eigenschaft besitzt aber in unserem Fall die Matrix

$$H = F_{s+1} \times H_1 \times H_2 \times \cdots \times H_r \times E_{r+1} \times E_{r+2} \times \cdots \times E_{r+s}.$$

Dies folgt unmittelbar aus den leicht zu beweisenden Gleichungen

$$H^2 = E,\quad HB_\alpha = B_\alpha H,\quad H\Gamma_\alpha = -\Gamma_\alpha H.$$

Wir haben die Spur $\eta(B)$ der Matrix HB zu berechnen, wenn die Matrix B in unserer Darstellung dem Element B der Untergruppe $\mathfrak{B}$ von $\mathfrak{T}$ entspricht. Beachtet man nun, daß die Spuren der Matrizen

$$H_\varrho,\ \overline{C}_\varrho,\ H_\varrho \overline{C}_\varrho,\ F_{s+1},\ F_{s+1} F_\varkappa F_\lambda \ldots \quad (\varrho \leqq r,\ s \geqq \varkappa > \lambda > \cdots)$$

sämtlich Null sind, abgesehen von der Matrix $F_{s+1} F_s \ldots F_2 F_1$, die nach (63.) die Spur $(2i)^{\frac{s}{2}}$ hat, so erhält man ohne Mühe die Regel:

Hat B nicht die Form $B_1 \cdots B_r C_{r+1} \cdots C_{r+s}$, so ist $\eta(B) = 0$, dagegen ist

$$(68.)\quad \eta(B_1\cdots B_r C_{r+1}\cdots C_{r+s}) = (2i)^{\frac{s}{2}} \delta^{(1)}(B_1)\cdots\delta^{(r)}(B_r)\,\chi^{(r+1)}(C_{r+1})\cdots\chi^{(r+s)}(C_{r+s}).$$

§ 29. Ich behandle nun insbesondere den für das folgende wichtigen Fall, daß $\chi^{(\alpha)}$ der im vorigen Abschnitt bestimmte Hauptcharakter zweiter Art der Gruppe $\mathfrak{T}_{\nu_\alpha}$ ist**).

---

*) Vgl. die Formeln der §§ 14—16.

**) Ist $\nu_\alpha = 1$, also $\mathfrak{T}_{\nu_\alpha} = E + J$, so hat man $\chi^{(\alpha)}(E) = 1$, $\chi^{(\alpha)}(J) = -1$ zu setzen.

Verstehen wir wieder, wenn P eine Permutation von $\mathfrak{S}_n$ ist, unter P' das nach den Vorschriften des § 11 zu fixierende Element von $\mathfrak{T}_n$, so haben wir die Zahlen $\zeta(R)$ nur für die Elemente $R = P'$ anzugeben, die den Permutationen P der Untergruppe $\mathfrak{S}_{\nu_1, \nu_2, \ldots \nu_m}$ von $\mathfrak{S}_n$ entsprechen.

Ist nun P_α eine (gerade) Permutation von $\mathfrak{S}_{\nu_\alpha}$, die in σ_α Zykeln ungerader Ordnung zerfällt, so wird in unserem Falle nach Satz VII

$$\chi^{(\alpha)}(P'_\alpha) = 2^{\left[\frac{\sigma_\alpha - 1}{2}\right]}.$$

Für ein gerades ν_α wird ferner $\chi^{(\alpha)}$ ein nicht zweiseitiger Charakter, und es ist, falls Z_α einen Zyklus der Ordnung ν_α von $\mathfrak{S}_{\nu_\alpha}$ bedeutet,

$$\chi^{(\alpha)}(Z'_\alpha) = \sqrt{(-1)^{\frac{\nu_\alpha}{2}} \cdot \frac{\nu_\alpha}{2}}.$$

Ist dagegen ν_α ungerade, so ist $\chi^{(\alpha)}$ ein zweiseitiger Charakter; dann wird noch, wenn Y_α ein Zyklus der Ordnung ν_α von $\mathfrak{S}_{\nu_\alpha}$ ist, nach VII*

$$\delta^{(\alpha)}(Y'_\alpha) = \pm \sqrt{(-1)^{\frac{\nu_\alpha - 1}{2}} \nu_\alpha},$$

wobei das obere oder das untere Vorzeichen zu nehmen ist, je nachdem Y_α in der alternierenden Gruppe $\mathfrak{A}_{\nu_\alpha}$ dem Zyklus

$$(69.)\quad (\nu_1 + \cdots + \nu_{\alpha-1} + 1,\ \nu_1 + \cdots + \nu_{\alpha-1} + 2,\ \ldots\ \nu_1 + \cdots + \nu_{\alpha-1} + \nu_\alpha)$$

konjugiert ist oder nicht (vgl. §§ 17 und 24). Für alle hier nicht erwähnten Permutationen S_α von $\mathfrak{S}_{\nu_\alpha}$ ist $\chi^{(\alpha)}(S'_\alpha) = 0$, bzw. $\delta^{(\alpha)}(S'_\alpha) = 0$ zu setzen.

Ich nehme nun an, es seien unter den m Zahlen $\nu_1, \nu_2, \ldots \nu_m$ die ersten r ungerade, die folgenden $m - r = s$ gerade, und zwar sollen die s geraden Zahlen nach abnehmender Größe geordnet sein.

Ist dann s ungerade, so wird auf Grund der Formeln (65.) und (66.)

$$\zeta(P'_1 P'_2 \ldots P'_m) = 2^{\frac{s-1}{2}} \cdot 2^{\frac{\sigma_1 - 1}{2}} \cdots 2^{\frac{\sigma_r - 1}{2}} \cdot 2^{\frac{\sigma_{r+1} - 2}{2}} \cdots 2^{\frac{\sigma_{r+s} - 2}{2}}$$
$$= 2^{\frac{1}{2}[\sigma_1 + \sigma_2 + \cdots + \sigma_m - m - 1]}$$

und

$$\zeta(Y'_1 \cdots Y'_r Z'_{r+1} \cdots Z'_{r+s}) = \pm (2i)^{\frac{s-1}{2}} \sqrt{(-1)^{\frac{\nu_1 - 1}{2} + \cdots + \frac{\nu_r - 1}{2} + \frac{\nu_{r+1}}{2} + \cdots + \frac{\nu_{r+s}}{2}} \cdot \frac{\nu_1 \nu_2 \ldots \nu_m}{2^s}}$$
$$= \pm \sqrt{(-1)^{\frac{n-m+1}{2}} \frac{\nu_1 \nu_2 \cdots \nu_m}{2}}.$$

Ebenso wird für ein gerades s wegen (67.) und (68.)

$$\zeta(P_1' P_2' \cdots P_m') = 2^{\frac{s}{2}} \cdot 2^{\frac{\sigma_1 - 1}{2}} \cdots 2^{\frac{\sigma_r - 1}{2}} \cdot 2^{\frac{\sigma_{r+1} - 2}{2}} \cdots 2^{\frac{\sigma_{r+s} - 2}{2}}$$
$$= 2^{\frac{1}{2}[\sigma_1 + \sigma_2 + \cdots + \sigma_m - m]}$$

und

$$\eta(Y_1' \cdots Y_r' Z_{r+1}' \cdots Z_{r+s}') = \pm (2i)^{\frac{s}{2}} \sqrt{(-1)^{\frac{\nu_1 - 1}{2} + \cdots + \frac{\nu_r - 1}{2} + \frac{\nu_{r+1}}{2} + \cdots + \frac{\nu_{r+s}}{2}} \cdot \frac{\nu_1 \nu_2 \ldots \nu_m}{2^s}}$$
$$= \pm \sqrt{(-1)^{\frac{n-m}{2}} \cdot \nu_1 \nu_2 \cdots \nu_m}.$$

In beiden Fällen sind die Vorzeichen bei den Quadratwurzeln*) so zu fixieren, daß das obere oder das untere Zeichen gewählt wird, je nachdem die Anzahl der Indizes ϱ, für welche der Zyklus Y_ϱ in $\mathfrak{A}_{\nu_\varrho}$ dem Zyklus (69.) *nicht* konjugiert ist, gerade oder ungerade ist. Ferner ist für alle Elemente R von $\mathfrak{T}$, für die weder R noch JR von der Form $P_1' P_2' \cdots P_m'$ oder von der Form $Y_1' \cdots Y_r' Z_{r+1}' \cdots Z_{r+s}'$ ist, $\zeta(R) = 0$ zu setzen. Ebenso wird $\eta(B) = 0$, wenn weder B noch JB die Form $Y_1' \cdots Y_r' Z_{r+1}' \cdots Z_{r+s}'$ hat.

Beachtet man noch, daß stets

$$\zeta(JR) = -\zeta(R), \quad \eta(JB) = -\eta(B)$$

sein soll, so erscheint der Charakter $\zeta(R)$ und für ein gerades s auch sein Komplement $\eta(B)$ als vollständig bestimmt.

§ 30. Unser Resultat kann, soweit es für das folgende von Wichtigkeit ist, noch wesentlich einfacher ausgesprochen werden.

Man setze μ gleich Null oder Eins, je nachdem s oder, was dasselbe ist, $n - m$ gerade oder ungerade ist. Dann ist für jede Permutation P von $\mathfrak{S}_{\nu_1, \nu_2, \ldots \nu_m}$, die in σ Zykeln ungerader Ordnungen zerfällt

$$\zeta(P') = 2^{\frac{\sigma - m - \mu}{2}}.$$

Sind ferner die Zahlen $\nu_1, \nu_2, \ldots \nu_m$ nicht sämtlich voneinander verschieden, so ist für jede Permutation Q von $\mathfrak{S}_{\nu_1, \nu_2, \ldots \nu_m}$, die in Zykeln von untereinander verschiedenen Ordnungen zerfällt, falls Q ungerade ist, $\zeta(Q') = 0$ und, falls $\mu = 0$ und Q gerade ist, $\eta(Q') = 0$. Sind aber die Zahlen $\nu_1, \nu_2, \ldots \nu_m$ voneinander verschieden, so tritt eine Ausnahme ein für diejenigen Permutationen Q von $\mathfrak{S}_{\nu_1, \nu_2, \ldots \nu_m}$, die in m Zykeln der Ordnungen $\nu_1, \nu_2, \ldots \nu_m$ zerfallen, und zwar wird dann, wenn $\mu = 1$ und Q ungerade ist,

*) Unter $\sqrt{(-1)^\lambda p}$ verstehe ich, wenn p reell und positiv ist, die Zahl $i^\lambda \sqrt{p}$.

$$\zeta(Q') = \sqrt{(-1)^{\frac{n-m+1}{2}} \cdot \frac{\nu_1 \nu_2 \cdots \nu_m}{2}}$$

und, wenn $\mu = 0$ und Q gerade ist,

$$\eta(Q') = \pm \sqrt{(-1)^{\frac{n-m}{2}} \cdot \nu_1 \nu_2 \cdots \nu_m}.$$

In der letzten Formel ist das Vorzeichen folgendermaßen zu fixieren: Es sei S die Permutation

(70.) $(1, 2, \ldots \nu_1)\ (\nu_1 + 1, \nu_1 + 2, \ldots \nu_1 + \nu_2) \cdots$
$(\nu_1 + \cdots + \nu_{m-1} + 1,\ \nu_1 + \cdots + \nu_{m-1} + 2,\ \ldots\ \nu_1 + \cdots + \nu_{m-1} + \nu_m)$

und S' das zugehörige Element von $\mathfrak{T}_n$; dann hat man das obere oder das untere Vorzeichen zu wählen, je nachdem Q' und S' in bezug auf die Gruppe $\mathfrak{B}_n$ einander konjugiert sind oder nicht.

Die Richtigkeit dieser Regeln ergibt sich ohne Mühe, wenn man die Festsetzungen und Ergebnisse des Abschnitts III genau berücksichtigt.

Insbesondere erhält man das für den nächsten Abschnitt wichtige Resultat:

Sind R und S zwei Elemente von $\mathfrak{T}_{\nu_1, \nu_2, \ldots \nu_m}$, denen in der Gruppe $\mathfrak{S}_{\nu_1, \nu_2, \ldots \nu_m}$ Permutationen zweiter Art entsprechen, so ist $\zeta(R) = \zeta(S)$, sobald nur R und S konjugierte Elemente der Gruppe $\mathfrak{T}_n$ sind. Ebenso ist $\eta(R) = \eta(S)$, wenn den Elementen R und S Permutationen dritter Art entsprechen und die Elemente R und S in bezug auf die Gruppe $\mathfrak{B}_n$ einander konjugiert sind.

Abschnitt VIII.

Einführung des Begriffs der Charakteristik.

§ 31. Mit Hilfe des in § 29 bestimmten Charakters $\zeta(R)$ von $\mathfrak{T}_{\nu_1, \nu_2, \ldots \nu_m}$ können wir einen Charakter zweiter Art für die Gruppe $\mathfrak{T}_n$ angeben*).

Ich benutze hierbei folgenden Satz des Herrn *Frobenius***):

„Es sei $\mathfrak{H}$ eine endliche Gruppe der Ordnung h, und es sei das Element R von $\mathfrak{H}$ mit h_R Elementen konjugiert. Ist dann $\mathfrak{G}$ eine Untergruppe der Ordnung g von $\mathfrak{H}$ und ist $\vartheta(R)$ ein (einfacher oder zusammengesetzter) Charakter von $\mathfrak{G}$, so erhält man einen Charakter $\gamma(R)$ von $\mathfrak{H}$, indem man

*) Die in diesem Paragraphen durchgeführte Überlegung schließt sich aufs engste an die von Herrn *Frobenius* bei der Bestimmung der Charaktere der symmetrischen Gruppe benutzte Methode an.

**) „Über die Relationen zwischen den Charakteren einer Gruppe und denen ihrer Untergruppen“, Sitzungsberichte der Kgl. Preuß. Akademie, Berlin, 1908, S. 501.

(71.) $$\gamma(R)=\frac{h}{g h_R}\sum_S \vartheta(S)$$

setzt; hierbei ist die Summe über alle h_R Elemente S von $\mathfrak{H}$ zu erstrecken, die dem Element R konjugiert sind, und $\vartheta(S)$ gleich Null zu setzen, wenn S nicht in $\mathfrak{G}$ enthalten ist.“

Es sei nun hier $\mathfrak{H}$ die Gruppe $\mathfrak{T}_n$ und $\mathfrak{G}$ die Gruppe $\mathfrak{T}_{\nu_1,\nu_2,\ldots\nu_m}$, ferner bedeute $\vartheta(R)$ den oben erwähnten Charakter $\zeta(R)$ von $\mathfrak{T}_{\nu_1,\nu_2,\ldots\nu_m}$. Wird dann der zugehörige Charakter $\gamma(R)$ von $\mathfrak{T}_n$ noch mit $\xi(R)$ bezeichnet, so ist zunächst, da $\zeta(J)=-\zeta(E)$ ist,

$$\xi(J)=\frac{h}{g}\zeta(J)=-\frac{h}{g}\zeta(E)=-\xi(E).$$

Daher ist $\xi(R)$ ein Charakter zweiter Art von $\mathfrak{T}_n$. Wir haben also die Zahlen $\xi(R)$ nur für diejenigen Elemente R von $\mathfrak{T}_n$ anzugeben, denen in $\mathfrak{S}_n$ Permutationen zweiter Art entsprechen. Ist nun aber P eine solche Permutation, so nimmt, wie wir in § 30 gesehen haben, $\zeta(S)$ für alle Elemente S von $\mathfrak{T}_{\nu_1,\nu_2,\ldots\nu_m}$, die in $\mathfrak{T}_n$ dem Element $R=P'$ konjugiert sind, denselben Wert an. Bedeutet daher $g_{P'}$ die Anzahl dieser Elemente von $\mathfrak{T}_{\nu_1,\nu_2,\ldots\nu_m}$, so wird, wenn S ein beliebiges unter ihnen ist,

$$\xi(P')=\frac{h g_{P'}}{g h_{P'}}\zeta(S).$$

Hierbei ist, wenn $g_{P'}=0$ wird, auch $\xi(P')=0$ zu setzen.

Nun soll doch aber P eine Permutation zweiter Art sein. Daher gibt (vgl. § 6) $h_{P'}=h_P$ zugleich die Anzahl der zu der Permutation P ähnlichen Permutationen von $\mathfrak{S}_n$ an; ferner ist $g_{P'}=g_P$ die Anzahl derjenigen unter diesen Permutationen, die in $\mathfrak{S}_{\nu_1,\nu_2,\ldots\nu_m}$ vorkommen, und außerdem wird hier

$$\frac{h}{g}=\frac{2n!}{2\nu_1!\,\nu_2!\cdots\nu_m!}=\frac{n!}{\nu_1!\,\nu_2!\cdots\nu_m!}.$$

Die Zahlen

$$c_P=\frac{n!}{\nu_1!\,\nu_2!\cdots\nu_m!}\,\frac{g_P}{h_P}$$

hat aber bereits Herr *Frobenius**) bestimmt: Besteht die Permutation P aus λ_1 Zykeln der Ordnung 1, ferner λ_2 Zykeln der Ordnung 2, usw., so wird

$$c_P=\Sigma\frac{\lambda_1!}{\lambda_{11}!\,\lambda_{12}!\cdots}\cdot\frac{\lambda_2!}{\lambda_{21}!\,\lambda_{22}!\cdots}\cdots,$$

*) „Über die Charaktere der symmetrischen Gruppe“, Sitzungsberichte der Kgl. Preuß. Akademie, Berlin, 1900, S. 516.

wo die Summe über alle nicht negativen ganzen Zahlen $\lambda_{11}, \lambda_{12}, \ldots$ zu erstrecken ist, die den Bedingungen

$$\lambda_1 = \lambda_{11} + \lambda_{12} + \cdots, \quad \lambda_2 = \lambda_{21} + \lambda_{22} + \cdots, \ \ldots$$
$$\nu_1 = \lambda_{11} + 2\lambda_{21} + \cdots, \quad \nu_2 = \lambda_{12} + 2\lambda_{22} + \cdots, \ \ldots$$

genügen. Haben diese Gleichungen keine Lösung, so ist $c_P = 0$ zu setzen.

Unter Benutzung der in § 17 eingeführten Bezeichnungen können wir nun den Charakter $\xi(T)$ von $\mathfrak{T}_n$ folgendermaßen kennzeichnen: *Gehört die Permutation P der Klasse*

$$[\alpha] = [\alpha_1, \alpha_3, \ldots]$$

an, so wird

$$\xi(P') = \xi_\alpha = c_\alpha \cdot 2^{\frac{\sigma_\alpha - m - \mu}{2}}, \qquad (\sigma_\alpha = \alpha_1 + \alpha_3 + \cdots) \tag{72.}$$

wo μ wie früher Null oder Eins ist, je nachdem $n-m$ gerade oder ungerade ist, und c_α die Zahl

$$c_\alpha = c_P = \Sigma \frac{\alpha_1!}{\alpha_{11}!\,\alpha_{12}!\cdots} \cdot \frac{\alpha_3!}{\alpha_{31}!\,\alpha_{32}!\cdots} \cdots$$

bedeutet. Die Summation ist hier über alle Lösungen der Gleichungen

$$\alpha_1 = \alpha_{11} + \alpha_{12} + \cdots, \quad \alpha_3 = \alpha_{31} + \alpha_{32} + \cdots, \ \ldots$$
$$\nu_1 = \alpha_{11} + 3\alpha_{31} + \cdots, \quad \nu_2 = \alpha_{12} + 3\alpha_{32} + \cdots, \ \ldots$$

zu erstrecken. Lassen diese Gleichungen keine Lösung zu, so ist $c_\alpha = 0$ zu setzen. Ist ferner $\mu = 0$ oder sind $\nu_1, \nu_2, \ldots \nu_m$ nicht untereinander verschieden, so ist für alle ungeraden Permutationen P die Zahl $\xi(P')$ gleich Null. Ist aber $\mu = 1$ und sind außerdem $\nu_1, \nu_2, \ldots \nu_m$ voneinander verschieden, so ist noch, wenn Q eine Permutation der (ungeraden) Klasse

$$(\nu) = (\nu_1, \nu_2, \ldots \nu_m)$$

ist,

$$\xi(Q') = \xi_{(\nu)} = \sqrt{(-1)^{\frac{n-m+1}{2}} \cdot \frac{\nu_1 \nu_2 \cdots \nu_m}{2}}.$$

Für alle übrigen ungeraden Klassen (λ) hat man wieder $\xi_{(\lambda)} = 0$.

Wir erkennen zugleich, daß $\xi(T)$ für $\mu = 0$ ein zweiseitiger Charakter ist. In diesem Fall haben wir noch ein Komplement $\delta(B)$ von $\xi(T)$ anzugeben. Hierzu genügt es, zwei adjungierte (einfache oder zusammengesetzte) Charaktere $\psi(B)$ und $\overline{\psi}(B)$ von $\mathfrak{B}_n$ zu kennen, deren Summe der Charakter $\xi(B)$ von $\mathfrak{B}_n$ ist, und $\delta(B) = \psi(B) - \overline{\psi}(B)$ zu setzen.

Um diese Aufgabe zu lösen, wähle man in dem oben erwähnten Satz des Herrn *Frobenius* für $\mathfrak{H}$ die Gruppe $\mathfrak{B}_n$ und für $\mathfrak{G}$ die Gruppe

$\mathfrak{B}_{\nu_1, \nu_2, \ldots \nu_m}$ und setze in (71.) für ϑ die beiden dem Komplement η des von uns betrachteten Charakters ζ von $\mathfrak{T}_{\nu_1, \nu_2, \ldots \nu_m}$ entsprechenden Charaktere

$$\frac{1}{2}[\zeta(B)+\eta(B)], \quad \frac{1}{2}[\zeta(B)-\eta(B)]$$

von $\mathfrak{B}_{\nu_1, \nu_2, \ldots, \nu_m}$. Dann erhält man, wie leicht zu sehen ist, in den zugehörigen Charakteren γ zwei adjungierte Charaktere $\psi(B)$ und $\overline{\psi}(B)$ von $\mathfrak{B}_n$, welche die verlangte Eigenschaft besitzen. Die Differenz $\delta(B) = \psi(B) - \overline{\psi}(B)$ kann auch direkt dadurch erhalten werden, daß man in (71.) für ϑ den zusammengesetzten Charakter η von $\mathfrak{B}_{\nu_1, \nu_2, \ldots \nu_m}$ einsetzt.

Auf Grund der in § 30 erwähnten Eigenschaften der Zahlen $\eta(B)$ ergeben sich ohne Mühe folgende Bestimmungsregeln für die Zahlen $\delta(B)$: *Zunächst ist stets* $\delta(JB) = -\delta(B)$. *Sind ferner die Zahlen* $\nu_1, \nu_2, \ldots \nu_m$ *nicht sämtlich voneinander verschieden, so ist* $\delta(B) = 0$ *für alle Elemente* B *von* $\mathfrak{B}_n$. *Sind dagegen* $\nu_1, \nu_2, \ldots \nu_m$ *voneinander verschieden, so ist, wenn* Q *eine Permutation der (geraden) Klasse*

$$(\nu) = (\nu_1, \nu_2, \ldots \nu_m)$$

von $\mathfrak{S}_n$ *bedeutet,*

$$\delta(Q') = \pm\sqrt{(-1)^{\frac{n-m}{2}} \cdot \nu_1 \nu_2 \cdots \nu_m}\,;$$

hierbei hat man das obere oder das untere Vorzeichen zu wählen, je nachdem Q *in der alternierenden Gruppe* $\mathfrak{A}_n$ *der Permutation* (70.) *des* § 30 *konjugiert ist oder nicht. Für alle übrigen (geraden) Permutationen* P *ist wieder* $\delta(P') = 0$ *zu setzen.*

§ 32. Der im vorigen Paragraphen bestimmte Charakter $\xi(T)$ von $\mathfrak{T}_n$ läßt sich wesentlich einfacher kennzeichnen, wenn man von den in § 25 definierten symmetrischen Funktionen q_n Gebrauch macht. Ich führe zu diesem Zweck einen neuen Begriff ein.

Man bezeichne wie früher mit s_λ die λ-te Potenzsumme der r Variabeln $x_1, x_2, \ldots x_r$ $(r \geqq n)$. Ist $\chi(T)$ ein beliebiger (einfacher oder zusammengesetzter) Charakter zweiter Art von $\mathfrak{T}_n$ und setzt man, wie in § 17, wenn P eine Permutation der Klasse $[\alpha]$ von $\mathfrak{S}_n$ ist, die in α_1 Zykeln der Ordnung 1, ferner α_3 Zykeln der Ordnung 3 usw. zerfällt,

$$\chi(P') = \chi_\alpha = \chi_{\alpha_1, \alpha_3, \ldots},$$

so nenne ich den Ausdruck

$$\Phi = \Sigma \frac{\chi_{\alpha_1, \alpha_3, \ldots}}{1^{\alpha_1}\alpha_1!\, 3^{\alpha_3}\alpha_3! \cdots} \cdot 2^{\frac{\alpha_1 + \alpha_3 + \cdots}{2}} \cdot s_1^{\alpha_1} s_3^{\alpha_3} \cdots$$

die zu dem Charakter $\chi(T)$ gehörende Charakteristik. Hierbei ist die Summe über alle Lösungen $\alpha_1, \alpha_3, \ldots$ der Gleichung

$$n = \alpha_1 + 3\alpha_3 + \cdots$$

zu erstrecken. Ich schreibe auch, indem ich die Bezeichnungen

$$\sigma_\alpha = \alpha_1 + \alpha_3 + \cdots, \quad A_\alpha = \frac{1}{1^{\alpha_1}\alpha_1!\,3^{\alpha_3}\alpha_3!\cdots}, \quad s^{(\alpha)} = s_1^{\alpha_1} s_3^{\alpha_3}\cdots$$

einführe, kurz

$$\Phi = \Sigma 2^{\frac{\sigma_\alpha}{2}} A_\alpha \chi_\alpha s^{(\alpha)},$$

wo die Summation also über alle v_n Klassen $[\alpha]$ zu erstrecken ist. Insbesondere wird, wie noch bemerkt sei, $h_\alpha = n!\,A_\alpha$ die Anzahl der Permutationen der Klasse $[\alpha]$.

Ist $\chi(T)$ ein zweiseitiger Charakter, so nenne ich Φ eine *zweiseitige Charakteristik. Ebenso soll Φ eine einfache Charakteristik heißen, wenn $\chi(T)$ ein einfacher Charakter ist.*

Durch den Ausdruck Φ ist umgekehrt der Charakter $\chi(T)$ nur dann vollständig bestimmt, wenn $\chi(T)$ ein zweiseitiger Charakter ist. Im andern Falle hat man außer den Zahlen χ_α, die durch die Koeffizienten der Funktion Φ von $s_1, s_3, \ldots$ bestimmt werden, noch die Zahlen $\chi_{(\nu)}$ zu kennen, die den *ungeraden* Klassen (ν) entsprechen (vgl. § 17). Ist ferner $\chi(T)$ ein zweiseitiger Charakter und will man noch ein Komplement $\delta(B)$ von $\chi(T)$ angeben, so hat man noch die Zahlen $\delta_{(\nu)}$ zu kennen, die zu den *geraden* Klassen (ν) gehören. *In beiden Fällen wird also der Charakteristik Φ ein System von gewissen Zahlen $\gamma_{(\nu)}$ zugeordnet, die entweder den ungeraden oder den geraden unter den v_n Klassen (ν) entsprechen.* Das System der Zahlen $\gamma_{(\nu)}$ bezeichne ich als ein *Komplement der Charakteristik Φ.* Läßt man, wenn $\chi(T)$ ein nicht zweiseitiger Charakter ist, an Stelle von $\chi(T)$ den assoziierten Charakter treten, oder wählt man, wenn $\chi(T)$ ein zweiseitiger Charakter ist, an Stelle des Komplements $\delta(B)$ das Komplement $-\delta(B)$, so bleibt Φ ungeändert, während $\gamma_{(\nu)}$ durch $-\gamma_{(\nu)}$ ersetzt wird. Die beiden Komplemente $\gamma_{(\nu)}$ und $-\gamma_{(\nu)}$ bezeichne ich als *entgegengesetzte Komplemente.*

Kennt man zwei Charakteristiken Φ und Φ' von $\mathfrak{T}_n$ und sind a und a' zwei beliebige ganze Zahlen, so ist auch $\Psi = a\Phi + a'\Phi'$ eine Charakteristik. Sind noch zwei Komplemente $\gamma_{(\nu)}$ und $\gamma'_{(\nu)}$ von Φ und Φ' bekannt, so kann man, wenn Φ und Φ' beide zweiseitig oder beide nicht zweiseitig sind, die

Zahlen $a\gamma_{(\nu)}+a'\gamma'_{(\nu)}$ als ein Komplement der Charakteristik Ψ ansehen. Ist dagegen etwa Φ nicht zweiseitig und Φ' zweiseitig, so ist Ψ eine nicht zweiseitige Charakteristik, für welche die Zahlen $a\gamma_{(\nu)}$ ein Komplement bilden.

Besteht ein Komplement einer zweiseitigen Charakteristik Φ aus lauter Nullen, so sage ich kurz, *Φ habe das Komplement Null.* Offenbar kann, wenn Φ eine beliebige Charakteristik ist, *der Ausdruck* $2\Phi=\Phi+\Phi$ *als eine zweiseitige Charakteristik mit dem Komplement Null angesehen werden.*

Da man insbesondere, wenn Φ irgendeine Charakteristik mit einem Komplement $\gamma_{(\nu)}$ ist, auch $\Phi-\Phi=0$ als eine Charakteristik auffassen kann, für welche die Zahlen $\gamma_{(\nu)}-(-\gamma_{(\nu)})=2\gamma_{(\nu)}$ ein Komplement bilden, so lassen sich für jede Charakteristik unendlich viele Komplemente angeben. Mit Hilfe der Formeln des Abschnitts IV beweist man aber leicht: *Ist Φ eine einfache Charakteristik, so gehören zu Φ nur zwei entgegengesetzte Komplemente, für die*

$$\Sigma h_{(\nu)}\gamma_{(\nu)}\overline{\gamma}_{(\nu)}=\frac{n!}{2^{\varepsilon}}$$

wird. Hierbei bedeutet $\overline{\gamma}_{(\nu)}$ die zu $\gamma_{(\nu)}$ konjugiert komplexe Zahl, ferner ist unter ε die Zahl Null oder Eins zu verstehen, je nachdem Φ eine zweiseitige Charakteristik ist oder nicht; $h_{(\nu)}$ ist die Anzahl der Permutationen der Klasse (ν)*).

§ 33. Bedeutet insbesondere $\chi(T)$ den Hauptcharakter zweiter Art der Gruppe $\mathfrak{T}_n$, so wird, wenn jetzt ε, wie in § 25, gleich Null oder Eins gesetzt wird, je nachdem n ungerade oder gerade ist,

$$\chi_a=2^{\frac{\sigma_a-1-\varepsilon}{2}}$$

Ferner läßt sich die in § 25 eingeführte Funktion q_n in der Form

$$q_n=\sum_a 2^{\sigma_a}A_a s^{(a)}=\sum_a 2^{\frac{\sigma_a+1+\varepsilon}{2}}A_a\chi_a s^{(a)} \tag{73.}$$

schreiben. Hieraus folgt aber, daß *der Ausdruck* $2^{-\frac{1+\varepsilon}{2}}q_n$ *die zu dem Hauptcharakter zweiter Art gehörende Charakteristik darstellt.*

Allgemeiner betrachte man das Produkt

$$q_{\nu_1}q_{\nu_2}\cdots q_{\nu_m}. \qquad (\nu_1+\nu_2+\cdots+\nu_m=n)$$

Dieser Ausdruck ist gleich

*) Vgl. die Formeln (42.) und (43.) des § 17.

$$\left\{\sum \frac{1}{\alpha_{11}!\,\alpha_{31}!\cdots}\left(\frac{2s_1}{1}\right)^{\alpha_{11}}\left(\frac{2s_3}{3}\right)^{\alpha_{31}}\cdots\right\}\left\{\sum \frac{1}{\alpha_{12}!\,\alpha_{32}!\cdots}\left(\frac{2s_1}{1}\right)^{\alpha_{12}}\left(\frac{2s_3}{3}\right)^{\alpha_{32}}\cdots\right\}\cdots,$$

wo die Summationsindizes den Bedingungen

$$\nu_1 = \alpha_{11} + 3\alpha_{31} + \cdots, \quad \nu_2 = \alpha_{12} + 3\alpha_{32} + \cdots, \quad \ldots$$

zu unterwerfen sind. Wird nun

$$\alpha_1 = \alpha_{11} + \alpha_{12} + \cdots, \quad \alpha_3 = \alpha_{31} + \alpha_{32} + \cdots$$

gesetzt, so läßt sich der Ausdruck auch in der Form

$$\sum \frac{2^{\alpha_1+\alpha_3+\cdots} s_1^{\alpha_1} s_3^{\alpha_3}\cdots}{1^{\alpha_1}\alpha_1!\,3^{\alpha_3}\alpha_3!\cdots} \sum \frac{\alpha_1!}{\alpha_{11}!\,\alpha_{12}!\cdots}\cdot\frac{\alpha_3!}{\alpha_{31}!\,\alpha_{32}!\cdots}\cdots$$

schreiben. Hat daher c_α dieselbe Bedeutung wie in § 31, so wird

$$q_{\nu_1} q_{\nu_2} \cdots q_{\nu_m} = \sum_\alpha 2^{\sigma_\alpha} A_\alpha c_\alpha s^{(\alpha)}.$$

Beachtet man noch die Gleichung (72.), so erhält man

$$q_{\nu_1} q_{\nu_2} \cdots q_{\nu_m} = \sum_\alpha 2^{\frac{\sigma_\alpha + m + \mu}{2}} A_\alpha \xi_\alpha s^{(\alpha)}.$$

Daher stellt der Ausdruck

$$(74.)\qquad 2^{-\frac{m+\mu}{2}} q_{\nu_1} q_{\nu_2} \cdots q_{\nu_m}$$

die zu dem in § 31 *bestimmten Charakter* $\xi(T)$ *von* $\mathfrak{T}_n$ *gehörende Charakteristik dar.* Hierbei ist μ wie früher gleich Null oder Eins zu setzen, je nachdem $n-m$ gerade oder ungerade ist.

Wir können auch unmittelbar ein Komplement $\gamma_{(\nu)}$ der Charakteristik (74.) angeben. Unter Berücksichtigung der für den Charakter $\xi(T)$ gewonnenen Resultate ergibt sich nämlich folgende Regel:

Sind zwei der Zahlen $\nu_1, \nu_2, \ldots \nu_m$ *einander gleich, so ist der Ausdruck* (74.) *eine zweiseitige Charakteristik mit dem Komplement Null. Sind dagegen diese Zahlen voneinander verschieden, so stellt der Ausdruck für* $\mu=0$ *eine zweiseitige, für* $\mu=1$ *eine nicht zweiseitige Charakteristik dar. Ein Komplement der Charakteristik* (74.) *erhält man, indem man*

$$\gamma_{(\nu)} = \sqrt{(-1)^{\frac{n-m+\mu}{2}} \cdot \frac{\nu_1 \nu_2 \cdots \nu_m}{2^\mu}}, \quad \gamma_{(\lambda)} = 0$$

setzt. Hierbei bedeutet (ν) *die Klasse der Permutationen von* $\mathfrak{S}_n$, *die in genau* m *Zykeln der Ordnungen* $\nu_1, \nu_2, \ldots \nu_m$ *zerfallen, während* (λ) *irgendeine von* (ν) *verschiedene Klasse ist.*

§ 34. Ich wende mich nun zur Betrachtung der *einfachen* Charakteristiken der Gruppe $\mathfrak{T}_n$.

Man bezeichne wie in § 17 mit

$$\chi^{(1)}(T),\ \chi^{(2)}(T),\ \ldots\ \chi^{(v_n)}(T)$$

v_n verschiedene einfache Charaktere zweiter Art von $\mathfrak{T}_n$, unter denen nicht zwei einander assoziiert sind. Dann ist jeder einfache Charakter zweiter Art von $\mathfrak{T}_n$ einem dieser v_n Charaktere entweder gleich oder assoziiert. Die zu $\chi^{(\varrho)}(T)$ gehörende Charakteristik sei

$$\Phi^{(\varrho)} = \sum_\alpha 2^{\frac{\sigma_\alpha}{2}} A_\alpha \chi_\alpha^{(\varrho)} s^{(\alpha)}.$$

Da nun zwei assoziierten Charakteren dieselbe Charakteristik entspricht, so *stellen die v_n Ausdrücke $\Phi^{(\varrho)}$ die einzigen einfachen Charakteristiken von $\mathfrak{T}_n$ dar.* Es sei noch bemerkt, daß der Koeffizient von s_1^n in $\Phi^{(\varrho)}$ gleich ist $\frac{2^{\frac{n}{2}} f_\varrho}{n!}$, wo f_ϱ den Grad des Charakters $\chi^{(\varrho)}(T)$ bedeutet; *dieser Koeffizient hat also einen positiven Wert.*

Wird wie in § 17 unter ε_ϱ die Zahl Null oder Eins verstanden, je nachdem der Charakter $\chi^{(\varrho)}(T)$ zweiseitig ist oder nicht, und bedeutet $e_{\varkappa\lambda}$ die Zahl Null oder Eins, je nachdem $\varkappa$ von λ verschieden oder gleich λ ist, so lassen sich die Formeln (39.), (40.), (44.) und (45.) des § 17 folgendermaßen schreiben:

$$\sum_\alpha A_\alpha \chi_\alpha^{(\varrho)} \chi_\alpha^{(\sigma)} = \frac{e_{\varrho\sigma}}{2^{\varepsilon_\varrho}}, \tag{75.}$$

$$\sum_\varrho 2^{\varepsilon_\varrho} \chi_\alpha^{(\varrho)} \chi_\beta^{(\varrho)} = \frac{e_{\alpha\beta}}{A_\alpha}. \tag{76.}$$

Man führe nun neben den Variablen $x_1, x_2, \ldots x_r$, als deren Potenzsummen die Größen s_λ definiert worden sind, r neue Variable $x_1', x_2', \ldots x_r'$ ein und setze entsprechend

$$t_\lambda = x_1'^\lambda + x_2'^\lambda + \cdots + x_r'^\lambda, \quad t^{(\alpha)} = t_1^{\alpha_1} t_3^{\alpha_3} \cdots$$

und

$$\Phi_1^{(\varrho)} = \sum_\alpha 2^{\frac{\sigma_\alpha}{2}} A_\alpha \chi_\alpha^{(\varrho)} t^{(\alpha)}.$$

Dann wird

$$\begin{aligned}\sum_\varrho 2^{\varepsilon_\varrho} \Phi^{(\varrho)} \Phi_1^{(\varrho)} &= \sum_\varrho 2^{\varepsilon_\varrho} \Big(\sum_{\alpha,\beta} 2^{\frac{\sigma_\alpha+\sigma_\beta}{2}} A_\alpha A_\beta \chi_\alpha^{(\varrho)} \chi_\beta^{(\varrho)} s^{(\alpha)} t^{(\beta)}\Big) \\ &= \sum_{\alpha,\beta} 2^{\frac{\sigma_\alpha+\sigma_\beta}{2}} A_\alpha A_\beta s^{(\alpha)} t^{(\beta)} \Big(\sum_\varrho 2^{\varepsilon_\varrho} \chi_\alpha^{(\varrho)} \chi_\beta^{(\varrho)}\Big).\end{aligned}$$

Auf Grund der Formel (76.) ergibt sich hieraus

$$\sum_\varrho 2^{\varepsilon_\varrho} \Phi^{(\varrho)} \Phi_1^{(\varrho)} = \sum_\alpha 2^{\sigma_\alpha} A_\alpha s^{(\alpha)} t^{(\alpha)}.$$

Der rechts stehende Ausdruck wird aber (vgl. Formel (73.)) mit Hilfe der r^2 Produkte

$$x_1 x_1', x_1 x_2', \ldots x_1 x_r', \ \ldots \ x_r x_1', x_r x_2', \ldots x_r x_r'$$

ebenso gebildet, wie der Ausdruck q_n mit Hilfe der r Größen $x_1, x_2, \ldots x_r$, und kann daher passend mit $q_n(xx')$ bezeichnet werden. Es besteht also die Gleichung

$$\sum_\varrho 2^{\varepsilon_\varrho} \Phi^{(\varrho)} \Phi_1^{(\varrho)} = q_n(xx'). \tag{77.}$$

Die Wichtigkeit dieser Formel geht aus folgender Überlegung hervor. Es seien uns v_n (nicht verschwindende) Funktionen

$$\Psi^{(\varrho)} = \sum_\alpha 2^{\frac{\sigma_\alpha}{2}} A_\alpha \zeta_\alpha s^{(\alpha)}$$

gegeben, von denen bekannt ist, daß sie als (einfache oder zusammengesetzte) Charakteristiken der Gruppe $\mathfrak{T}_n$ betrachtet werden dürfen. Man setze η_ϱ gleich Null oder Eins, je nachdem $\Psi^{(\varrho)}$ eine zweiseitige oder nicht zweiseitige Charakteristik ist; ferner sei

$$\Psi_1^{(\varrho)} = \sum_\alpha 2^{\frac{\sigma_\alpha}{2}} A_\alpha \zeta_\alpha t^{(\alpha)}.$$

Es gilt dann der Satz:

Besteht für die v_n Funktionen $\Psi^{(\varrho)}$ die der Gleichung (77.) *analoge Gleichung*

$$\sum_\varrho 2^{\eta_\varrho} \Psi^{(\varrho)} \Psi_1^{(\varrho)} = q_n(xx'), \tag{78.}$$

so stimmen diese Funktionen, abgesehen von der Reihenfolge und von den Vorzeichen, mit den v_n einfachen Charakteristiken $\Phi^{(\varrho)}$ überein.

Es sei nämlich $\zeta^{(\varrho)}(T)$ der Charakter zweiter Art von $\mathfrak{T}_n$, zu dem die Charakteristik $\Psi^{(\varrho)}$ gehört. Dann ist $\zeta^{(\varrho)}(T)$ eine ganzzahlige lineare homogene Verbindung der $v_n + u_n$ einfachen Charaktere zweiter Art, die sich für die Gruppe $\mathfrak{T}_n$ angeben lassen. Ist hierbei $p_{\varrho\sigma}$ der Koeffizient von $\chi^{(\sigma)}(T)$, so kommt, falls $\zeta^{(\varrho)}(T)$ ein zweiseitiger und $\chi^{(\sigma)}(T)$ ein nicht zweiseitiger Charakter ist, auch dem zu $\chi^{(\sigma)}(T)$ assoziierten Charakter derselbe Koeffizient $p_{\varrho\sigma}$ zu. Hieraus folgt leicht, daß

$$\zeta_\alpha^{(\varrho)} = \sum_\sigma r_{\varrho\sigma} \chi_\alpha^{(\sigma)} \qquad (\varrho, \sigma = 1, 2, \ldots v_n)$$

oder, was dasselbe ist,

$$\Psi^{(\varrho)} = \sum_\sigma r_{\varrho\sigma} \Phi^{(\sigma)}$$

gesetzt werden darf, wo die $r_{\varrho\sigma}$ gewisse ganze Zahlen sind, welche noch die Eigenschaft besitzen, daß auch $2^{\eta_\varrho-\varepsilon_\sigma} r_{\varrho\sigma}$ stets ganzzahlig ist. Die Summe

$$r_\varrho = \sum_\sigma 2^{\eta_\varrho-\varepsilon_\sigma} r_{\varrho\sigma}^2$$

ist daher eine positive ganze Zahl, die nur dann gleich Eins wird, wenn unter den Zahlen $r_{\varrho 1}, r_{\varrho 2}, \ldots$ eine gleich ± 1, die übrigen gleich Null sind.

Um also zu zeigen, daß $\pm \Psi^{(\varrho)}$ für jedes ϱ einer der Funktionen $\Phi^{(\sigma)}$ gleich wird, haben wir nur nachzuweisen, daß jede der v_n Zahlen r_ϱ gleich Eins ist. Dies ergibt sich aber wie folgt: Setzt man die Koeffizienten von $s^{(\alpha)} t^{(\alpha)}$ auf beiden Seiten der Gleichung (78.) einander gleich, so erhält man

$$\sum_\varrho 2^{\eta_\varrho} \zeta_\alpha^{(\varrho)2} = \sum_\varrho 2^{\eta_\varrho} \sum_{\sigma,\tau} r_{\varrho\sigma} r_{\varrho\tau} \chi_\alpha^{(\sigma)} \chi_\alpha^{(\tau)} = \frac{1}{A_\alpha}.$$

Man multipliziere hier mit A_α und summiere über alle v_n Klassen $[\alpha]$. Dann ergibt sich wegen (75.)

$$\sum_\varrho 2^{\eta_\varrho} \sum_\sigma \frac{r_{\varrho\sigma}^2}{2^{\varepsilon_\sigma}} = \sum_\varrho r_\varrho = v_n.$$

Wäre nun eine der Zahlen r_ϱ von Eins verschieden, also größer als Eins, so müßte ihre Summe größer als v_n ausfallen, was auf einen Widerspruch führt.

Wir haben uns noch zu vergewissern, daß die v_n Ausdrücke $\pm \Psi^{(\varrho)}$, von denen wir bereits nachgewiesen haben, daß sie einfache Charakteristiken sind, untereinander verschieden sein müssen. Wäre dies aber nicht der Fall, so würde aus

$$q_n(xx') = \sum_\varrho 2^{\varepsilon_\varrho} \Phi^{(\varrho)} \Phi_1^{(\varrho)} = \sum_\varrho 2^{\eta_\varrho} \Psi^{(\varrho)} \Psi_1^{(\varrho)}$$

offenbar eine Gleichung der Form

$$\sum_\varrho c_\varrho \Phi^{(\varrho)} \Phi_1^{(\varrho)} = 0$$

folgen, wo nicht alle c_ϱ verschwinden. Eine solche Beziehung ist aber nicht möglich. Denn setzt man in dem links stehenden Ausdruck den Koeffizienten von $s^{(\alpha)} t^{(\beta)}$ gleich Null, so ergibt sich

$$\sum_\varrho c_\varrho A_\alpha A_\beta \chi_\alpha^{(\varrho)} \chi_\beta^{(\varrho)} = 0,$$

und hieraus folgt, indem man mit $\chi_\alpha^{(\sigma)} \chi_\beta^{(\sigma)}$ multipliziert und nach α und β addiert, auf Grund der Gleichungen (75.)

$$\sum_\varrho c_\varrho \frac{e_{\varrho\sigma} e_{\varrho\sigma}}{2^{2\varepsilon_\varrho}} = \frac{c_\sigma}{2^{2\varepsilon_\sigma}} = 0.$$

Dies widerspricht aber der über die Zahlen c_ϱ gemachten Annahme.

Abschnitt IX.
Über die Funktionen $Q_{\nu_1, \nu_2, \dots \nu_m}$.

§ 35. In diesem Abschnitt soll eine Klasse von symmetrischen Funktionen der Variabeln $x_1, x_2, \dots x_r$ studiert werden, von denen ich im nächsten Abschnitt nachweisen werde, daß sie, abgesehen von konstanten Faktoren, mit den einfachen Charakteristiken der Gruppe $\mathfrak{T}_n$ übereinstimmen.

Ich betrachte wieder die Gleichung

$$f(z) = \frac{(1+x_1 z)(1+x_2 z)\cdots(1+x_r z)}{(1-x_1 z)(1-x_2 z)\cdots(1-x_r z)} = q_0 + q_1 z + q_2 z^2 + \cdots, \quad (q_0 = 1)$$

die zur Definition der Funktionen q_ν dient. Hierbei soll r nicht kleiner sein als die von uns betrachtete Zahl n. — Beachtet man, daß

$$\frac{1+x_\varkappa z}{1-x_\varkappa z} = 1 + 2x_\varkappa z + 2x_\varkappa^2 z^2 + \cdots$$

ist, so erhält man leicht

$$q_\nu = \Sigma 2^\lambda x_1^{\alpha_1} x_2^{\alpha_2} \cdots x_r^{\alpha_r}, \tag{79.}$$

wo die Summe über alle nicht negativen ganzen Zahlen $\alpha_1, \alpha_2, \dots \alpha_r$ zu erstrecken ist, die der Bedingung $\alpha_1 + \alpha_2 + \cdots + \alpha_r = \nu$ genügen, und der Exponent λ angibt, wie viele unter den Zahlen $\alpha_1, \alpha_2, \dots \alpha_r$ von Null verschieden sind.

Da $f(z) f(-z) = 1$, also

$$\left(\sum_{\nu=0}^{\infty} q_\nu z^\nu\right)\left(\sum_{\nu=0}^{\infty} (-1)^\nu q_\nu z^\nu\right) = 1$$

ist, so ergibt sich für $\nu > 0$

$$q_\nu - q_1 q_{\nu-1} + q_2 q_{\nu-2} - \cdots + (-1) q_\nu = 0. \tag{80.}$$

Setzt man daher

$$Q_{\varkappa, \lambda} = q_\varkappa q_\lambda - 2 q_{\varkappa+1} q_{\lambda-1} + 2 q_{\varkappa+2} q_{\lambda-2} - \cdots + (-1)^\lambda 2 q_{\varkappa+\lambda}, \tag{81.}$$

so wird

$$Q_{\varkappa, \lambda} = - Q_{\lambda, \varkappa};$$

die Ausdrücke $Q_{\varkappa, \lambda}$ bilden also ein alternierendes System.

Ich definiere nun, wenn $\nu_1, \nu_2, \dots \nu_m$ beliebige positive ganze Zahlen sind, den Ausdruck $Q_{\nu_1, \nu_2, \dots, \nu_m}$ durch folgende Festsetzungen: *Ist m gerade, so soll $Q_{\nu_1, \nu_2, \dots \nu_m}$ den mit Hilfe der m^2 Größen $Q_{\nu_\alpha, \nu_\beta}$ gebildeten* Pfaff*schen Ausdruck bedeuten, der in bekannter Weise aus der Rekursionsformel*

$$Q_{\nu_1, \nu_2, \dots \nu_m} = Q_{\nu_1, \nu_2} Q_{\nu_3, \nu_4, \dots \nu_m} - Q_{\nu_1, \nu_3} Q_{\nu_2, \nu_4, \dots \nu_m} + \cdots + Q_{\nu_1, \nu_m} Q_{\nu_2, \nu_3, \dots \nu_{m-1}} \tag{82.}$$

zu bestimmen ist. Für ein ungerades m soll aber

(83.) $Q_{\nu_1,\nu_2,\ldots\nu_m} = q_{\nu_1} Q_{\nu_2,\nu_3,\ldots\nu_m} - q_{\nu_2} Q_{\nu_1,\nu_3,\ldots\nu_m} + \cdots + q_{\nu_m} Q_{\nu_1,\nu_2,\ldots\nu_{m-1}}$

sein. Außerdem setze ich noch

$$Q_{\nu_1} = q_{\nu_1}.$$

Man erkennt leicht, daß $Q_{\nu_1,\nu_2,\ldots\nu_m}$ in $-Q_{\nu_1,\nu_2,\ldots\nu_m}$ übergeht, wenn zwei der Indizes vertauscht werden; daher wird $Q_{\nu_1,\nu_2,\ldots\nu_m}=0$, wenn zwei Indizes einander gleich sind. Offenbar ist in jedem Fall $Q_{\nu_1,\nu_2,\ldots\nu_m}$ ein ganzzahliges Polynom der Variabeln $x_1, x_2, \ldots x_r$, das symmetrisch und homogen vom Grade $\nu_1+\nu_2+\cdots\nu_m$ ist.

§ 36. Die Partialbruchzerlegung der Funktion $f(z)$ liefert die Gleichung

$$f(z) = 1 + 2\sum_{\beta=1}^{r} \frac{x_\beta z}{p_\beta(1-x_\beta z)}, \tag{84.}$$

wo

$$p_\beta = \frac{(x_\beta-x_1)\cdots(x_\beta-x_{\beta-1})(x_\beta-x_{\beta+1})\cdots(x_\beta-x_r)}{(x_\beta+x_1)\cdots(x_\beta+x_{\beta-1})(x_\beta+x_{\beta+1})\cdots(x_\beta+x_r)} \tag{85.}$$

zu setzen ist, und hieraus folgt für $\nu>0$

$$q_\nu = 2\sum_{\beta=1}^{r}\frac{x_\beta^\nu}{p_\beta}.$$

Setzt man in (84.) $z=-\frac{1}{x_\alpha}$, so ergibt sich, da $f\left(-\frac{1}{x_\alpha}\right)=0$ ist,

$$\sum_{\beta=1}^{r}\frac{2x_\beta}{p_\beta(x_\alpha+x_\beta)} = 1.$$

Daher ist

$$\begin{aligned} Q_{\varkappa,1} = q_\varkappa q_1 - 2q_{\varkappa+1} &= 4\sum_{\alpha,\beta=1}^{r}\frac{x_\alpha^\varkappa x_\beta}{p_\alpha p_\beta} - 4\sum_{\alpha=1}^{r}\frac{x_\alpha^{\varkappa+1}}{p_\alpha} \\ &= 4\sum_{\alpha,\beta=1}^{r}\frac{x_\alpha^\varkappa x_\beta}{p_\alpha p_\beta} - 4\sum_{\alpha,\beta=1}^{r}\frac{x_\alpha^{\varkappa+1}}{p_\alpha p_\beta}\cdot\frac{2x_\beta}{x_\alpha+x_\beta} \\ &= 4\sum_{\alpha,\beta=1}^{r}\frac{x_\alpha^\varkappa x_\beta}{p_\alpha p_\beta}\,\frac{x_\beta-x_\alpha}{x_\beta+x_\alpha}. \end{aligned}$$

Mit Hilfe der sich aus (81.) ergebenden Gleichung

$$Q_{\varkappa,\lambda} = q_\varkappa q_\lambda - q_{\varkappa+1}q_{\lambda-1} - Q_{\varkappa+1,\lambda-1}$$

zeigt man leicht, daß auch für $\lambda>1$

$$Q_{\varkappa,\lambda} = 4\sum_{\alpha,\beta=1}^{r}\frac{x_\alpha^\varkappa x_\beta^\lambda}{p_\alpha p_\beta}\,\frac{x_\beta-x_\alpha}{x_\beta+x_\alpha}$$

ist. Ich will nun allgemeiner zeigen, daß, wenn

$$A(u_1, u_2, \ldots u_m) = \prod_{\varkappa<\lambda}\frac{u_\varkappa-u_\lambda}{u_\varkappa+u_\lambda} \qquad (\varkappa, \lambda = 1, 2, \ldots m)$$

gesetzt wird,

$$Q_{\nu_1,\nu_2,\ldots\nu_m} = 2^m \sum_{\alpha_1,\alpha_2,\ldots\alpha_m=1}^{r}\frac{x_{\alpha_1}^{\nu_1}x_{\alpha_2}^{\nu_2}\cdots x_{\alpha_m}^{\nu_m}}{p_{\alpha_1}p_{\alpha_2}\cdots p_{\alpha_m}}A(x_{\alpha_m}, x_{\alpha_{m-1}}, \ldots x_{\alpha_2}, x_{\alpha_1}) \tag{86.}$$

wird.

Diese Formel läßt sich leicht mit Hilfe der Gleichungen (82.) und (83.) verifizieren, wenn wir zeigen können, daß für ein gerades m die Gleichung

$$\begin{aligned}(87.)\qquad A(u_1, u_2, \ldots u_m) = {} & A(u_1, u_2)\, A(u_3, u_4, \ldots u_m) \\ & - A(u_1, u_3)\, A(u_2, u_4, \ldots u_m) + \cdots + A(u_1, u_m)\, A(u_2, \ldots u_{m-1})\end{aligned}$$

und für ein ungerades m die Gleichung

$$\begin{aligned}(88.)\qquad A(u_1, u_2, \ldots u_m) = {} & A(u_2, u_3, \ldots u_m) \\ & - A(u_1, u_3, \ldots u_m) + \cdots + A(u_1, u_2, \ldots u_{m-1})\end{aligned}$$

besteht.

Der Beweis der Formeln (87.) und (88.) läßt sich nun folgendermaßen führen. Man bezeichne, wenn m gerade ist, mit $A'(u_1, u_2, \ldots u_m)$ den mit Hilfe des alternierenden Systems

$$\frac{u_\varrho - u_\sigma}{u_\varrho + u_\sigma} \qquad (\varrho, \sigma = 1, 2, \ldots m)$$

gebildeten *Pfaff*schen Ausdruck. Dann ist also

$$A'(u_1, u_2) = \frac{u_1 - u_2}{u_1 + u_2}$$

und

$$\begin{aligned}(89.)\qquad A'(u_1, u_2, \ldots u_m) = {} & A'(u_1, u_2)\, A'(u_3, u_4, \ldots u_m) \\ & - A'(u_1, u_3)\, A'(u_2, u_4, \ldots u_m) + \cdots + A'(u_1, u_m)\, A'(u_2, u_3, \ldots u_{m-1}).\end{aligned}$$

Nun ist aber bekanntlich

$$A'^2(u_1, u_2, \ldots u_m) = \left|\frac{u_\varrho - u_\sigma}{u_\varrho + u_\sigma}\right|.$$

Folglich verschwindet $A'(u_1, u_2, \ldots u_m)$, wenn zwei der Variabeln $u_1, u_2, \ldots u_m$ einander gleich werden. Da außerdem beide Ausdrücke

$$A'(u_1, u_2, \ldots u_m) \prod_{\varrho<\sigma}(u_\varrho + u_\sigma) \quad \text{und} \quad \prod_{\varrho<\sigma}(u_\varrho - u_\sigma)$$

ganze rationale Funktionen sind, die homogen und von demselben Grade $\frac{m(m-1)}{2}$ sind, so muß, wenn c_m der Koeffizient von $u_1^{m-1} u_2^{m-2} \ldots u_m$ in dem ersten der beiden Ausdrücke ist,

$$A'(u_1, u_2, \ldots u_m) \prod_{\varrho<\sigma}(u_\varrho + u_\sigma) = c_m \prod_{\varrho<\sigma}(u_\varrho - u_\sigma)$$

sein. Multipliziert man aber die Gleichung (89.) mit $\prod_{\varrho<\sigma}(u_\varrho + u_\sigma)$ und vergleicht auf beiden Seiten die Koeffizienten von $u_1^{m-1} u_2^{m-2} \ldots u_m$, so erhält man leicht

$$c_m = c_{m-2} - c_{m-2} + \cdots + c_{m-2} = c_{m-2}.$$

Da nun $c_2 = 1$ ist, so ist stets $c_m = 1$, und folglich ergibt sich

$$A'(u_1, u_2, \ldots u_m) = A(u_1, u_2, \ldots u_m).$$

Schreibt man nun in (89.) A für A', so erhält man die zu beweisende Formel (87.). Setzt man ferner in dieser Gleichung $u_1 = 0$ und beachtet, daß

$$A(0, u_2, \ldots u_m) = (-1)^{m-1} A(u_2, u_3, \ldots u_m) = -A(u_2, u_3, \ldots u_m)$$

ist, so erhält man eine Formel, aus der die Gleichung (88.) hervorgeht, indem man $u_2, u_3, \ldots u_m$ durch $u_1, u_2, \ldots u_{m-1}$ ersetzt und für $m-1$ wieder m schreibt.

§ 37. Von zwei Potenzprodukten

$$A x_1^{\varkappa_1} x_2^{\varkappa_2} \cdots x_r^{\varkappa_r}, \quad B x_1^{\lambda_1} x_2^{\lambda_2} \cdots x_r^{\lambda_r}$$

soll, wie üblich, das erste Produkt von höherer Ordnung heißen als das zweite, wenn die erste nicht verschwindende unter den Differenzen

$$\varkappa_1 - \lambda_1, \quad \varkappa_2 - \lambda_2, \ldots \varkappa_r - \lambda_r$$

positiv ist. Unter dem *Leitglied* einer ganzen rationalen Funktion $\varphi(x_1, x_2, \ldots x_r)$ verstehe ich das Produkt höchster Ordnung, das in φ auftritt, und unter dem Leitglied einer gebrochenen rationalen Funktion $\psi = \frac{\varphi_1}{\varphi_2}$ den Quotienten der Leitglieder von φ_1 und φ_2. Sind dann $\psi_1, \psi_2, \ldots \psi_k$ beliebige ganze oder gebrochene rationale Funktionen von $x_1, x_2, \ldots x_r$, deren Leitglieder sich nicht allein um konstante Faktoren voneinander unterscheiden, so ist das Leitglied von

$$\psi = \psi_1 + \psi_2 + \cdots + \psi_k,$$

wie man leicht schließt, gleich demjenigen der Leitglieder von $\psi_1, \psi_2, \ldots \psi_k$, dem die höchste Ordnung zukommt.

Das Leitglied der Funktion $Q_\nu = q_\nu$ der Variabeln $x_1, x_2, \ldots x_r$ ist, wie aus (79.) folgt, gleich $2x_1^\nu$. Allgemeiner beweise ich:

VIII. *Ist* $\nu_1 > \nu_2 > \cdots \nu_m > 0$ *und* $m \leqq r$*), *so ist das Leitglied von* $Q_{\nu_1, \nu_2, \ldots \nu_m}$ *gleich* $2^m x_1^{\nu_1} x_2^{\nu_2} \cdots x_m^{\nu_m}$.

Dies folgt aus der Formel (86.). In der Tat ist das Leitglied des durch die Gleichung (85.) definierten Ausdrucks p_β gleich $(-1)^{\beta-1}$, das Leitglied von $A(x_{\alpha_m}, x_{\alpha_{m-1}}, \ldots x_{\alpha_1})$ ist gleich ± 1 und speziell das von $A(x_m, x_{m-1}, \ldots x_1)$ gleich $(-1)^{\frac{m(m-1)}{2}}$. In der in (86.) rechts stehenden Summe hat daher der Summand

$$2^m \frac{x_1^{\nu_1} x_2^{\nu_2} \cdots x_m^{\nu_m}}{p_1 p_2 \cdots p_m} A(x_m, x_{m-1}, \ldots x_1)$$

*) Für $m > r$ wird $Q_{\nu_1, \nu_2, \ldots \nu_m} = 0$.

das Leitglied

$$2^m(-1)^{\frac{m(m-1)}{2}-1-2-\cdots-(m-1)} x_1^{\nu_1} x_2^{\nu_2} \cdots x_m^{\nu_m} = 2^m x_1^{\nu_1} x_2^{\nu_2} \cdots x_m^{\nu_m},$$

während die Leitglieder der übrigen Summanden für $\nu_1 > \nu_2 > \cdots > \nu_m$ jedenfalls von niedrigerer Ordnung sind.

Es bedeute nun, wie früher, k_n die Anzahl der Zerlegungen

$$n = \nu_1 + \nu_2 + \cdots \qquad (\nu_1 \geqq \nu_2 \geqq \cdots)$$

der ganzen Zahl n in positive ganzzahlige Summanden. Von zwei Zerlegungen

$$n = \lambda_1 + \lambda_2 + \cdots, \quad n = \mu_1 + \mu_2 + \cdots$$

soll die erste von *höherer Ordnung heißen als die zweite, wenn die erste nicht verschwindende unter den Differenzen* $\lambda_1 - \mu_1, \lambda_2 - \mu_2, \ldots$ *positiv ist.* Ich denke mir die $k = k_n$ Zerlegungen von n derart angeordnet, daß zuerst diejenigen kommen, bei denen die Summanden untereinander verschieden sind. Diese $v = v_n$ Zerlegungen sollen, nach abnehmender Ordnung geschrieben, mit

$$(1), (2), \ldots (v)$$

bezeichnet werden, so daß also insbesondere (1) die Zerlegung $n = n$ bedeutet. Unter

$$(v+1), (v+2), \ldots (k)$$

verstehe ich die übrigen $k - v$ Zerlegungen von n, irgendwie angeordnet. Die Anzahl der Summanden in der Zerlegung (ν) sei gleich m_ν.

Bedeutet ferner (ν) die Zerlegung $n = \nu_1 + \nu_2 + \cdots$, so bezeichne ich mit $q^{(\nu)}$ das Produkt $q_{\nu_1} q_{\nu_2} \cdots$, mit $Q^{(\nu)}$ die Funktion $Q_{\nu_1, \nu_2, \ldots}$ und mit ξ_ν die monogene symmetrische Funktion der Variabeln $x_1, x_2, \ldots x_r$, die als Summe aller Produkte der Form $x_{\alpha_1}^{\nu_1} x_{\alpha_2}^{\nu_2} \cdots$ erscheint. Das Leitglied von ξ_ν ist dann gleich $x_1^{\nu_1} x_2^{\nu_2} \cdots$. Ist nun $\nu \leqq v$, so ist $Q^{(\nu)}$ nach Satz VIII eine symmetrische Funktion von $x_1, x_2, \ldots x_r$, deren Leitglied mit dem Leitglied von $2^{m_\nu} \xi_\nu$ übereinstimmt, dagegen sind die Ausdrücke $Q^{(v+1)}, Q^{(v+2)}, \ldots Q^{(k)}$ sämtlich Null.

Ist die Zerlegung (λ) der Zahl n von höherer Ordnung als die Zerlegung (μ), *so soll auch das Produkt* $q^{(\lambda)}$ *von höherer Ordnung heißen als das Produkt* $q^{(\mu)}$. Da nun wegen (80.)

$$q_\alpha^2 = 2 q_{\alpha+1} q_{\alpha-1} - 2 q_{\alpha+2} q_{\alpha-2} + \cdots + (-1)^{\alpha-1} 2 q_{2\alpha}$$

ist, so ergibt sich leicht, daß jedes Produkt $q^{(\lambda)}$ für $\lambda > v$ in der Form

$$q^{(\lambda)} = a_1 q^{(1)} + a_2 q^{(2)} + \cdots + a_v q^{(v)}$$

darstellbar ist, wo $a_1, a_2, \ldots a_v$ gewisse ganze durch 2 teilbare Zahlen sind, und zwar können hierin nur solche unter den Produkten $q^{(1)}, q^{(2)}, \dots q^{(v)}$ wirklich vorkommen, die von höherer Ordnung sind als das Produkt $q^{(\lambda)}$. Aus den Formeln (81.), (82.) und (83.), die zur Berechnung der Ausdrücke $Q^{(\nu)}$ dienen, geht ferner leicht hervor, daß $Q^{(\nu)}$ für $\nu \leqq v$ die Form

$$q^{(\nu)} + b_1 q^{(\nu')} + b_2 q^{(\nu'')} + \cdots$$

hat, wo $q^{(\nu')}, q^{(\nu'')}, \dots$ von höherer Ordnung sind als das Produkt $q^{(\nu)}$, und $b_1, b_2, \dots$ ganze Zahlen bedeuten, die sämtlich gerade sind. Infolge der erwähnten Eigenschaft der Produkte $q^{(v+1)}, q^{(v+2)}, \dots q^{(k)}$ ergibt sich daher für $\nu \leqq v$

$$Q^{(\nu)} = q^{(\nu)} + c_1 q^{(\nu-1)} + c_2 q^{(\nu-2)} + \cdots + c_{\nu-1} q^{(1)};$$

hierbei bedeuten $c_1, c_2, \dots c_{\nu-1}$ wieder gewisse gerade ganze Zahlen. Da nun insbesondere $q^{(1)} = q_n = Q^{(1)}$ ist, so läßt sich umgekehrt $q^{(\nu)}$ in der Form

$$q^{(\nu)} = Q^{(\nu)} + d_1 Q^{(\nu-1)} + d_2 Q^{(\nu-2)} + \cdots + d_{\nu-1} Q^{(1)}$$

darstellen, wo auch $d_1, d_2, \dots d_{\nu-1}$ gerade ganze Zahlen sind.

Hieraus folgt, daß *jede homogene symmetrische ganze rationale Funktion n-ten Grades von $x_1, x_2, \dots x_r$, die sich als ganze rationale Funktion von $q_1, q_2, \dots$ (oder, was dasselbe ist, als ganze rationale Funktion der Potenzsummen $s_1, s_3, s_5, \dots$) ausdrücken läßt, zugleich als lineare homogene Verbindung von $Q^{(1)}, Q^{(2)}, \dots Q^{(v)}$ darstellbar ist**). Diese Darstellung ist eindeutig bestimmt, denn die Ausdrücke $Q^{(1)}, Q^{(2)}, \dots Q^{(v)}$ sind als Funktionen mit untereinander verschiedenen Leitgliedern linear unabhängig. Es ergibt sich zugleich, daß unter den k Produkten $q^{(\lambda)}$ nur v linear unabhängig sind; insbesondere besitzen die v Produkte $q^{(1)}, q^{(2)}, \dots q^{(v)}$ diese Eigenschaft. Ich füge noch folgende Bemerkung hinzu: Man habe einen Ausdruck der Form

$$F = e_0 q^{(\nu)} + e_1 q^{(\nu-1)} + \cdots + e_{\nu-1} q^{(1)}, \qquad (\nu \leqq v)$$

der, als lineare Verbindung der monogenen symmetrischen Funktionen $\xi_1, \xi_2, \dots$ dargestellt, die Glieder $\xi_1, \xi_2, \dots \xi_{\nu-1}$ nicht enthält. *Dann muß, wie ich behaupte, $F = e_0 Q^{(\nu)}$ sein.* Denn man denke sich F in der Form

$$F = g_1 Q^{(1)} + g_2 Q^{(2)} + \cdots + g_v Q^{(v)}$$

dargestellt. Wären nun die Zahlen $g_1, g_2, \dots g_{\nu-1}$ nicht sämtlich Null und ist $g_{\nu-\lambda}$ die erste nicht verschwindende unter ihnen, so müßte F, durch

*) Hierin ist ein bemerkenswerter Satz aus der Theorie der symmetrischen Funktionen enthalten: Ist S eine symmetrische Funktion von $x_1, x_2, \dots x_r$, die als ganze rationale Funktion der Potenzsummen $s_1, s_3, s_5, \dots$ darstellbar ist, so sind in dem Leitglied $A x_1^{\nu_1} x_2^{\nu_2} \cdots$ von S die Exponenten $\nu_1, \nu_2, \dots$ untereinander verschieden.

$x_1, x_2, \ldots x_r$ ausgedrückt, das Glied $2^{m_\nu - \lambda} g_{\nu-\lambda} \xi_{\nu-\lambda}$ enthalten, was unserer Annahme widerspricht. Würden ferner die Zahlen $g_{\nu+1}, g_{\nu+2}, \ldots$ nicht sämtlich verschwinden, so sei $g_{\nu+\mu}$ die letzte, die nicht Null ist. Drückt man dann alle $Q^{(\lambda)}$ durch $q^{(1)}, q^{(2)}, \ldots q^{(v)}$ aus, so müßte in F das Glied $g_{\nu+\mu} q^{(\nu+\mu)}$ vorkommen, was nicht der Fall ist. Daher ist $F = g_\nu Q^{(\nu)}$; da aber $Q^{(\nu)}$ das Glied $q^{(\nu)}$ mit dem Koeffizienten 1 enthält, so ergibt sich $g_\nu = e_0$, also $F = e_0 Q^{(\nu)}$.

Ist außerdem noch bekannt, daß der Koeffizient von ξ_ν in der Entwicklung von F gleich e_0' ist, so muß nach VIII $e_0' = 2^{m_\nu} e_0$ sein.

§ 38. Stellt man das Produkt $q^{(\lambda)}$ als Funktion von $x_1, x_2, \ldots x_r$ dar, so erhält man eine Gleichung der Form

$$q^{(\lambda)} = \sum_{\mu=1}^{k} a_{\lambda\mu} \xi_\mu,$$

wo die $a_{\lambda\mu}$ gewisse ganze Zahlen sind. Speziell wird, da $q^{(1)} = q_n$ und $\xi_1 = x_1^n + x_2^n + \cdots + x_r^n$ ist, wegen (79.) $a_{11} = 2$. Da die Ausdrücke $\xi_1, \xi_2, \ldots \xi_k$ linear unabhängig sind und unter den k Produkten $q^{(\lambda)}$ nur v linear unabhängige vorkommen, so ist die Determinante k-ten Grades $|a_{\lambda\mu}|$ genau vom Range v. Ich setze nun

$$D_\nu = \begin{vmatrix} a_{11}, a_{12}, \ldots a_{1\nu} \\ a_{21}, a_{22}, \ldots a_{2\nu} \\ \cdot \; \cdot \; \cdot \; \cdot \; \cdot \; \cdot \; \cdot \; \cdot \\ a_{\nu 1}, a_{\nu 2}, \ldots a_{\nu\nu} \end{vmatrix}, \quad F^{(\nu)} = \begin{vmatrix} a_{11}, a_{12}, \ldots a_{1,\nu-1}, q^{(1)} \\ a_{21}, a_{22}, \ldots a_{2,\nu-1}, q^{(2)} \\ \cdot \; \cdot \; \cdot \; \cdot \; \cdot \; \cdot \; \cdot \; \cdot \; \cdot \; \cdot \\ a_{\nu 1}, a_{\nu 2}, \ldots a_{\nu,\nu-1}, q^{(\nu)} \end{vmatrix},$$

wobei noch $F^{(1)} = q^{(1)}$ sein soll. Dann wird offenbar einerseits

$$F^{(\nu)} = D_{\nu-1} q^{(\nu)} + D'_{\nu-1} q^{(\nu-1)} + \cdots + D_{\nu-1}^{(\nu-1)} q^{(1)},$$

wo $D'_{\nu-1}, \ldots, D_{\nu-1}^{(\nu-1)}$ gewisse Determinanten des Grades $\nu-1$ bedeuten, andererseits

$$F^{(\nu)} = D_\nu \xi_\nu + \Delta'_\nu \xi_{\nu+1} + \Delta''_\nu \xi_{\nu+2} + \cdots,$$

wo $\Delta'_\nu, \Delta''_\nu, \ldots$ Determinanten ν-ten Grades sind. Auf Grund der am Schluß des vorigen Paragraphen gemachten Bemerkung ergibt sich daher für $\nu = 1, 2, \ldots v$

$$F^{(\nu)} = D_{\nu-1} Q^{(\nu)}, \quad D_\nu = 2^{m_\nu} D_{\nu-1}, \tag{90.}$$

wobei noch $D_0 = 1$ zu setzen ist; hieraus folgt

$$D_\nu = 2^{m_1 + m_2 + \cdots + m_\nu}. \qquad (\nu = 1, 2, \ldots v) \tag{91.}$$

Sind nun $x_1', x_2', \ldots x_r'$ neue r Variable, so mögen die Ausdrücke ξ_ν, $Q^{(\nu)}$ und $F^{(\nu)}$ in ξ_ν', $Q_1^{(\nu)}$ und $F_1^{(\nu)}$ übergehen, wenn an Stelle der Variabeln x_α die Variabeln x_α' treten. Hat dann $q_n(xx')$ dieselbe Bedeutung wie in § 34, so wird

$$\frac{(1+x_1x_1'z)(1+x_1x_2'z)\cdots(1+x_rx_r'z)}{(1-x_1x_1'z)(1-x_1x_2'z)\cdots(1-x_rx_r'z)}=\sum_{\nu=0}^{\infty}q_\nu(xx')z^\nu.$$

Da sich aber der links stehende Ausdruck in der Form

$$\Big(\sum_{\nu=0}^{\infty}q_\nu x_1^{\prime\nu}z^\nu\Big)\Big(\sum_{\nu=0}^{\infty}q_\nu x_2^{\prime\nu}z^\nu\Big)\cdots\Big(\sum_{\nu=0}^{\infty}q_\nu x_r^{\prime\nu}z^\nu\Big)$$

entwickeln läßt, so erhält man offenbar

$$q_n(xx')=\sum_{\lambda=1}^{k}q^{(\lambda)}\xi_\lambda'=\sum_{\lambda,\mu}a_{\lambda\mu}\xi_\mu\xi_\lambda'.$$

Beachtet man noch, daß $q_n(xx')$ ungeändert bleibt, wenn x_a und x_a' vertauscht werden, so ergibt sich $a_{\lambda\mu}=a_{\mu\lambda}$.

Die Funktion $q_n(xx')$ läßt sich also als symmetrische Bilinearform der Variabeln $\xi_1, \xi_2, \ldots \xi_k$ und $\xi_1', \xi_2', \ldots \xi_k'$ auffassen. Da nun der Rang der Determinante $|a_{\lambda\mu}|$ gleich v ist und die Hauptunterdeterminanten $D_1, D_2, \ldots D_v$ von Null verschieden sind, so wird nach einer bekannten von *Jacobi* herrührenden Formel

$$q_n(xx')=\frac{1}{D_1}F^{(1)}F_1^{(1)}+\frac{1}{D_1D_2}F^{(2)}F_1^{(2)}+\cdots+\frac{1}{D_{v-1}D_v}F^{(v)}F_1^{(v)},$$

und *hieraus folgt wegen* (90.) *und* (91.) *die für das folgende wichtige Gleichung*

$$(92.)\qquad q_n(xx')=\frac{1}{2^{m_1}}Q^{(1)}Q_1^{(1)}+\frac{1}{2^{m_2}}Q^{(2)}Q_1^{(2)}+\cdots+\frac{1}{2^{m_v}}Q^{(v)}Q_1^{(v)}.$$

Abschnitt X.

Bestimmung der einfachen Charaktere der Gruppen $\mathfrak{T}_n$ und $\mathfrak{B}_n$.

§ 39. Mit Hilfe der im letzten Abschnitt gewonnenen Resultate gelingt es nun leicht, die früher eingeführten einfachen Charakteristiken der Gruppe $\mathfrak{T}_n$ explizite darzustellen.

Es sei

$$n=\nu_1+\nu_2+\cdots+\nu_m \qquad (\nu_1>\nu_2>\cdots>\nu_m>0)$$

und $Q^{(\nu)}=Q_{\nu_1,\nu_2,\ldots\nu_m}$. Dann gilt folgende Regel, die auf Grund der Formeln (81.), (82.) und (83.) leicht zu beweisen ist:

Ist

$$q^{(\lambda)}=q_{\lambda_1}q_{\lambda_2}\cdots q_{\lambda_l} \qquad (\lambda_1>\lambda_2>\cdots>\lambda_l>0)$$

ein von $q^{(\nu)}=q_{\nu_1}q_{\nu_2}\cdots q_{\nu_m}$ verschiedenes Produkt, das in der Darstellung von $Q^{(\nu)}$ als lineare homogene Verbindung von $q^{(1)}, q^{(2)}, \ldots q^{(v)}$ wirklich vorkommt, so kann l nur einen der Werte

$$m, m-1, \ldots m-\left[\frac{m}{2}\right]$$

annehmen. Der Koeffizient c_λ von $q^{(\lambda)}$ ist ferner eine ganze Zahl, die für $l<m$ durch 2^{m-l} und für $l=m$ durch 2 teilbar ist.

Man setze nun μ, bzw. μ' gleich Null oder Eins, je nachdem $n-m$, bzw. $n-l$ gerade oder ungerade ist. Dann stellen, wie wir wissen (vgl. § 33), die Ausdrücke

$$\psi = \frac{1}{2^{\frac{m+\mu}{2}}} q^{(\nu)}, \quad \psi' = \frac{1}{2^{\frac{l+\mu'}{2}}} q^{(\lambda)}$$

zwei Charakteristiken der Gruppe $\mathfrak{T}_n$ dar. Ich beweise nun:

Der Ausdruck

$$\Psi^{(\nu)} = \frac{1}{2^{\frac{m+\mu}{2}}} Q^{(\nu)}$$

ist ebenso wie der Ausdruck ψ für $\mu=0$ eine zweiseitige und für $\mu=1$ eine nicht zweiseitige Charakteristik von $\mathfrak{T}_n$. Ferner läßt sich jedes Komplement der Charakteristik ψ zugleich auch als Komplement der Charakteristik $\Psi^{(\nu)}$ auffassen.

Aus

$$Q^{(\nu)} = q^{(\nu)} + c_\lambda q^{(\lambda)} + \cdots$$

folgt nämlich

$$\Psi^{(\nu)} = \psi + 2^{\frac{l-m+\mu'-\mu}{2}} c_\lambda \psi' + \cdots.$$

Ist nun $l=m$, also $\mu'=\mu$, so erhält man im zweiten Glied rechts $c_\lambda \psi'$, und dies ist, da c_λ gerade ist, das Zweifache einer Charakteristik, also eine Charakteristik mit dem Komplement Null. Ist ferner $l<m$, so wird, da c_λ alsdann durch 2^{m-l} teilbar ist,

$$2^{\frac{l-m+\mu'-\mu}{2}} c_\lambda \psi' = 2^{\frac{m-l+\mu'-\mu}{2}} \frac{c_\lambda}{2^{m-l}} \psi'$$

im allgemeinen wieder das Zweifache einer Charakteristik. Eine Ausnahme kann nur im Falle

$$l = m-1, \quad \mu = 1$$

eintreten; dann wird aber $\mu'=0$, und folglich ist ψ', also auch

$$2^{\frac{l-m+\mu'-\mu}{2}} c_\lambda \psi' = \frac{c_\lambda}{2} \psi'$$

eine zweiseitige Charakteristik. Jedenfalls ergibt sich, daß $\Psi^{(\nu)}$ aus ψ dadurch hervorgeht, daß zu ψ für $\mu=1$ gewisse zweiseitige Charakteristiken und für $\mu=0$ nur zweiseitige Charakteristiken mit dem Komplement Null hinzuaddiert werden. Hieraus folgt aber mit Rücksicht auf die Ausführungen des § 32 die Richtigkeit unserer Behauptung.

Setzt man nun noch $\mu=\eta_\nu$, so folgt aus (92.)

$$q_n(xx') = \sum_{\nu=1}^{v} \frac{1}{2^{m_\nu}} Q^{(\nu)} Q_1^{(\nu)} = \sum_{\nu=1}^{v} 2^{\eta_\nu} \Psi^{(\nu)} \Psi_1^{(\nu)}.$$

Auf Grund des in § 34 Bewiesenen ergibt sich daher, daß *die* $v = v_n$ *Ausdrücke* $\Psi^{(\nu)}$, *abgesehen von den Vorzeichen, mit den gesuchten einfachen Charakteristiken* $\Phi^{(\nu)}$ *der Gruppe* $\mathfrak{T}_n$ *übereinstimmen.* Wir können also $\Phi^{(\nu)} = \pm \Psi^{(\nu)}$ setzen.

§ 40. Um noch zu zeigen, daß $\Phi^{(\nu)} = + \Psi^{(\nu)}$ ist, haben wir nur nachzuweisen, daß die Entwicklung von $\Psi^{(\nu)}$ oder, was dasselbe ist, von $Q^{(\nu)}$ nach Potenzen von $s_1, s_3, s_5, \ldots$ das Glied s_1^n mit positivem Koeffizienten enthält.

Es sei dieser Koeffizient gleich $\frac{2^n g^{(\nu)}}{n!}$; wir setzen noch, falls $Q^{(\nu)} = Q_{\nu_1, \nu_2, \ldots \nu_m}$ ist,

$$g^{(\nu)} = g_{\nu_1, \nu_2, \ldots \nu_m}.$$

Dann wird zunächst, da

$$Q^{(1)} = q_n = \frac{(2s_1)^n}{n!} + \frac{(2s_1)^{n-3}(2s_3)}{(n-3)!\,3} + \cdots$$

ist,

$$g^{(1)} = g_n = 1.$$

Ferner erhält man aus (81.)

$$\frac{g_{\varkappa,\lambda}}{(\varkappa+\lambda)!} = \frac{1}{\varkappa!\,\lambda!} - \frac{2}{(\varkappa+1)!\,(\lambda-1)!} + \frac{2}{(\varkappa+2)!\,(\lambda-2)!} - \cdots + (-1)^\lambda \frac{2}{(\varkappa+\lambda)!},$$

und hieraus folgt leicht

$$g_{\varkappa,\lambda} = \frac{(\varkappa+\lambda)!}{\varkappa!\,\lambda!} \cdot \frac{\varkappa-\lambda}{\varkappa+\lambda}.$$

Allgemeiner ist, wie ich zeigen will,

$$(93.) \qquad g_{\nu_1, \nu_2, \ldots \nu_m} = \frac{(\nu_1 + \nu_2 + \cdots + \nu_m)!}{\nu_1!\,\nu_2!\cdots\nu_m!} \prod_{\alpha<\beta} \frac{\nu_\alpha - \nu_\beta}{\nu_\alpha + \nu_\beta}.$$

Unter Benutzung der in § 36 eingeführten Bezeichnung läßt sich diese Formel in der Gestalt

$$(93'.) \qquad g_{\nu_1, \nu_2, \ldots \nu_m} = \frac{(\nu_1 + \nu_2 + \cdots + \nu_m)!}{\nu_1!\,\nu_2!\cdots\nu_m!} A(\nu_1, \nu_2, \ldots \nu_m)$$

schreiben. Es sei dies bereits bewiesen, wenn an Stelle von m eine kleinere Zahl tritt. Dann erhält man, falls m gerade ist, aus der Gleichung (82.) durch Gleichsetzen der Koeffizienten von $(2s_1)^{\nu_1+\nu_2+\cdots+\nu_m}$ wegen (87.)

$$\frac{g_{\nu_1,\nu_2,\ldots\nu_m}}{(\nu_1+\nu_2+\cdots+\nu_m)!} = \frac{g_{\nu_1,\nu_2}}{(\nu_1+\nu_2)!} \cdot \frac{g_{\nu_3,\nu_4,\ldots\nu_m}}{(\nu_3+\nu_4+\cdots+\nu_m)!} - \frac{g_{\nu_1,\nu_3}}{(\nu_1+\nu_3)!} \cdot \frac{g_{\nu_2,\nu_4,\ldots\nu_m}}{(\nu_2+\nu_4+\cdots+\nu_m)!} + \cdots$$

$$= \frac{1}{\nu_1!\,\nu_2!\cdots\nu_m!} \{A(\nu_1,\nu_2)\,A(\nu_3,\nu_4,\ldots\nu_m) - A(\nu_1,\nu_3)\,A(\nu_2,\nu_4,\ldots\nu_m) + \cdots\}$$

$$= \frac{1}{\nu_1!\,\nu_2!\cdots\nu_m!} A(\nu_1,\nu_2,\ldots\nu_m).$$

Ebenso folgt für ein ungerades m aus (83.) und (88.)

$$\frac{g_{\nu_1,\nu_2,\ldots\nu_m}}{(\nu_1+\nu_2+\cdots+\nu_m)!}=\frac{1}{\nu_1!}\frac{g_{\nu_2,\nu_3,\ldots\nu_m}}{(\nu_2+\nu_3+\cdots+\nu_m)!}-\frac{1}{\nu_2!}\cdot\frac{g_{\nu_1,\nu_3,\ldots\nu_m}}{(\nu_1+\nu_3+\cdots+\nu_m)!}+\cdots$$

$$=\frac{1}{\nu_1!\,\nu_2!\cdots\nu_m!}\{A(\nu_2,\nu_3,\ldots\nu_m)-A(\nu_1,\nu_3,\ldots\nu_m)+\cdots\}$$

$$=\frac{1}{\nu_1!\,\nu_2!\cdots\nu_m!}A(\nu_1,\nu_2,\ldots\nu_m).$$

Aus der nun bewiesenen Gleichung (93.) erkennt man, daß die Zahlen $g_{\nu_1,\nu_2,\ldots\nu_m}$ für den uns allein interessierenden Fall $\nu_1>\nu_2>\cdots>\nu_m$ in der Tat positiv sind, und daß daher *die v Ausdrücke $\Psi^{(\nu)}$ die einfachen Charakteristiken der Gruppe $\mathfrak{T}_n$ repräsentieren.*

Es ist noch zu bemerken, daß die Zahlen $g_{\nu_1,\nu_2,\ldots\nu_m}$ ganze Zahlen sind. Dies ergibt sich, indem man die oben benutzten Rekursionsformeln in der Gestalt

$$g_{\nu_1,\nu_2,\ldots\nu_m}=\binom{\nu_1+\nu_2+\cdots+\nu_m}{\nu_1+\nu_2}g_{\nu_1,\nu_2}g_{\nu_3,\nu_4,\ldots\nu_m}-\binom{\nu_1+\nu_2+\cdots+\nu_m}{\nu_1+\nu_3}g_{\nu_1,\nu_3}g_{\nu_2,\nu_4,\ldots\nu_m}+\cdots$$

bzw.

$$g_{\nu_1,\nu_2,\ldots\nu_m}=\binom{\nu_1+\nu_2+\cdots+\nu_m}{\nu_1}g_{\nu_2,\nu_3,\ldots\nu_m}-\binom{\nu_1+\nu_2+\cdots+\nu_m}{\nu_2}g_{\nu_1,\nu_3,\ldots\nu_m}+\cdots$$

schreibt und außerdem noch beachtet, daß

$$g_{\varkappa,\lambda}=\frac{(\varkappa+\lambda)!}{\varkappa!\,\lambda!}\,\frac{\varkappa-\lambda}{\varkappa+\lambda}=\binom{\varkappa+\lambda}{\varkappa}-2\binom{\varkappa+\lambda-1}{\varkappa}$$

eine ganze Zahl ist.

§ 41. Setzt man wie früher

$$\Psi^{(\nu)}=\Phi^{(\nu)}=\sum_\alpha 2^{\frac{\sigma_\alpha}{2}}A_\alpha\chi_\alpha^{(\nu)}s^{(\alpha)},$$

so wird

$$Q^{(\nu)}=Q_{\nu_1,\nu_2,\ldots\nu_m}=2^{\frac{m+\mu}{2}}\Phi^{(\nu)}=\sum_\alpha 2^{\frac{\sigma_\alpha+m+\mu}{2}}A_\alpha\chi_\alpha^{(\nu)}s^{(\alpha)}.$$

Da $n-\sigma_\alpha$ und $n-m-\mu$ gerade Zahlen sind, so wird hier jeder der Faktoren $2^{\frac{\sigma_\alpha+m+\mu}{2}}$ eine ganze Zahl, und hieraus folgt auch, daß *die $\chi_\alpha^{(\nu)}$ rationale und mithin (als Summen von Einheitswurzeln) ganze rationale Zahlen sind.* Für den Grad

$$f^{(\nu)}=f_{\nu_1,\nu_2,\ldots\nu_m}$$

des einfachen Charakters $\chi^{(\nu)}(T)$ von $\mathfrak{T}_n$, zu dem die Charakteristik $\Phi^{(\nu)}$ gehört, erhält man durch Betrachtung des Koeffizienten von s_1^n in $Q^{(\nu)}$

$$f_{\nu_1,\nu_2,\ldots\nu_m}=2^{\frac{n-m-\mu}{2}}g_{\nu_1,\nu_2,\ldots\nu_m}=2^{\frac{n-m-\mu}{2}}\frac{n!}{\nu_1!\,\nu_2!\cdots\nu_m!}\prod_{\alpha<\beta}\frac{\nu_\alpha-\nu_\beta}{\nu_\alpha+\nu_\beta}.$$

Ist $\mu=0$, so ist $\chi^{(\nu)}(T)$ ein zweiseitiger Charakter und daher durch die Charakteristik $\Phi^{(\nu)}$ vollständig bestimmt. Ist aber $\mu=1$, so haben wir noch für die ungeraden Klassen (λ) von $\mathfrak{S}_n$ die Zahlen $\chi^{(\nu)}_{(\lambda)}$ anzugeben. Ebenso haben wir für $\mu=0$ noch das Komplement $\delta^{(\nu)}_{(\lambda)}$ von $\chi^{(\nu)}(T)$ zu bestimmen. — Es genügt aber hierzu, das in § 33 bestimmte Komplement $\gamma_{(\varrho)}$ der Charakteristik

$$\psi=\frac{1}{2^{\frac{m+\mu}{2}}}q_{\nu_1}q_{\nu_2}\cdots q_{\nu_m}$$

zu betrachten, das, wie wir wissen, zugleich als ein Komplement von $\Psi^{(\nu)}=\Phi^{(\nu)}$ angesehen werden kann. Es war

$$\gamma_{(\nu)}=\sqrt{(-1)^{\frac{n-m+\mu}{2}}\frac{\nu_1\nu_2\cdots\nu_m}{2^\mu}}$$

für die Klasse (ν), deren Permutationen in genau m Zykeln der Ordnungen $\nu_1,\nu_2,\ldots\nu_m$ zerfallen, und $\gamma_{(\lambda)}=0$ für alle von (ν) verschiedenen Klassen (λ). Ist nun $\overline{\gamma}_{(\varrho)}$ die zu $\gamma_{(\varrho)}$ konjugiert komplexe Zahl, so wird, wenn $h_{(\varrho)}$ die Anzahl der Permutationen der Klasse (ϱ) bedeutet, insbesondere

$$h_{(\nu)}=\frac{n!}{\nu_1\nu_2\cdots\nu_m}$$

und folglich

$$\sum_\varrho h_{(\varrho)}\gamma_{(\varrho)}\overline{\gamma}_{(\varrho)}=\frac{n!}{2^\mu}.$$

Hieraus folgt aber auf Grund des in § 32 Gesagten, daß wir für $\mu=1$

$$\chi^{(\nu)}_{(\varrho)}=\gamma_{(\varrho)}$$

und für $\mu=0$

$$\delta^{(\nu)}_{(\varrho)}=\gamma_{(\varrho)}$$

setzen dürfen.

Ich will noch die gewonnenen Resultate zusammenfassen.

IX. *Jeder Zerlegung*

$$n=\nu_1+\nu_2+\cdots+\nu_m \qquad (\nu_1>\nu_2>\cdots>\nu_m>0)$$

der Zahl n in lauter verschiedene Summanden entspricht ein einfacher Charakter zweiter Art $\chi(T)$ der Gruppe $\mathfrak{T}_n$, dessen Grad gleich

$$f_{\nu_1,\nu_2,\ldots\nu_m}=2^{\left[\frac{n-m}{2}\right]}\frac{n!}{\nu_1!\,\nu_2!\cdots\nu_m!}\prod_{\alpha<\beta}\frac{\nu_\alpha-\nu_\beta}{\nu_\alpha+\nu_\beta}$$

ist. Um die Zahlen $\chi(T)$ sämtlich zu kennen, hat man nur die zugehörigen (in § 17 gekennzeichneten) Größen $\chi_{\alpha_1,\alpha_3,\ldots}$ und $\chi_{(\varrho)}$ anzugeben. Die Zahlen $\chi_{\alpha_1,\alpha_3,\ldots}$ werden dadurch erhalten, daß man den Ausdruck $Q_{\nu_1,\nu_2,\ldots\nu_m}$ als Funktion der Potenzsummen $s_1,s_3,\ldots$ berechnet; setzt man nämlich μ gleich Null *oder Eins, je nachdem $n-m$ gerade oder ungerade ist, so wird*

$$Q_{\nu_1,\nu_2,\dots\nu_m} = \sum_{\alpha_1+3\alpha_3+\cdots=n} 2^{\frac{m+\mu+\alpha_1+\alpha_3+\cdots}{2}} \frac{\chi_{\alpha_1,\alpha_3,\dots}}{1^{\alpha_1}\alpha_1!\,3^{\alpha_3}\alpha_3!\dots} s_1^{\alpha_1} s_3^{\alpha_3}\dots.$$

Ist $\mu=1$, so wird $\chi(T)$ ein nicht zweiseitiger Charakter; für jede ungerade Klasse (λ) von $\mathfrak{S}_n$, die von der Klasse (ν), deren Permutationen in genau m Zykeln der Ordnungen $\nu_1,\nu_2,\dots\nu_m$ zerfallen, verschieden ist, wird dann $\chi_{(\lambda)}=0$, dagegen ist

$$\chi_{(\nu)} = \sqrt{(-1)^{\frac{n-m+1}{2}} \frac{\nu_1\nu_2\cdots\nu_m}{2}}.$$

Durchläuft ferner B die Elemente der Untergruppe $\mathfrak{B}_n$ von $\mathfrak{T}_n$, so repräsentieren die Zahlen $\chi(B)$ einen einfachen Charakter zweiter Art $\varphi(B)$ von $\mathfrak{B}_n$.

Ist dagegen $\mu=0$, so ist $\chi(T)$ ein zweiseitiger Charakter. Man erhält dann in den Zahlen $\chi(B)$ einen Charakter von $\mathfrak{B}_n$, der als die Summe zweier adjungierter einfacher Charaktere $\psi(B)$ und $\overline{\psi}(B)$ erscheint. Die Differenz $\delta(B)=\psi(B)-\overline{\psi}(B)$ ist im allgemeinen Null; nur dann, wenn B einer Permutation von $\mathfrak{S}_n$ entspricht, die in genau m Zykeln der Ordnungen $\nu_1,\nu_2,\dots\nu_m$ zerfällt, wird

$$\delta(B) = \pm\sqrt{(-1)^{\frac{n-m}{2}} \nu_1\nu_2\cdots\nu_m}.$$

Sieht man zwei assoziierte Charaktere von $\mathfrak{T}_n$ als nicht voneinander verschieden an, so erhält man auf diese Weise, indem man alle v_n Zerlegungen von n in lauter verschiedene Summanden in Betracht zieht, die sämtlichen einfachen Charaktere zweiter Art der Gruppen $\mathfrak{T}_n$ und $\mathfrak{B}_n$.

§ 42. Zu dem Charakter $\chi(T)$, der durch die Zerlegung $n=\nu_1+\nu_2+\cdots+\nu_m$ bestimmt ist, gehört eine irreduzible Kollineationsgruppe $\mathfrak{K}_{\nu_1,\nu_2,\dots\nu_m}$ des Grades $f_{\nu_1,\nu_2,\dots\nu_m}$, die der Gruppe $\mathfrak{S}_n$ isomorph ist, und die sich, wenn $n\geqq 4$ ist, nicht als Gruppe von $n!$ linearen homogenen Substitutionen schreiben läßt. Ist $n-m$ ungerade, so ist die der Gruppe $\mathfrak{A}_n$ isomorphe Untergruppe $\mathfrak{L}_{\nu_1,\nu_2,\dots\nu_m}$ von $\mathfrak{K}_{\nu_1,\nu_2,\dots\nu_m}$ wieder irreduzibel. Ist aber $n-m$ gerade, so zerfällt diese Untergruppe in zwei der Gruppe $\mathfrak{A}_n$ isomorphe irreduzible Kollineationsgruppen $\mathfrak{L}'_{\nu_1,\nu_2,\dots\nu_m}$ und $\mathfrak{L}''_{\nu_1,\nu_2,\dots\nu_m}$ des Grades $\frac{1}{2}f_{\nu_1,\nu_2,\dots\nu_m}$; diese beiden Gruppen sind jedoch, da sie als adjungierte Gruppen erscheinen, nach den Ergebnissen des § 19 als nicht wesentlich verschieden anzusehen.

Jede irreduzible Kollineationsgruppe, die der Gruppe $\mathfrak{S}_n$ isomorph ist und nicht als Gruppe von $n!$ linearen homogenen Substitutionen geschrieben werden kann, ist einer der v_n Gruppen $\mathfrak{K}_{\nu_1, \nu_2, \ldots \nu_m}$ äquivalent. Dasselbe gilt, wenn n von 6 und 7 verschieden ist, auch für die Gruppe $\mathfrak{A}_n$ in bezug auf die Kollineationsgruppen $\mathfrak{L}_{\nu_1, \nu_2, \ldots \nu_m}$, $\mathfrak{L}'_{\nu_1, \nu_2, \ldots \nu_m}$ und $\mathfrak{L}''_{\nu_1, \nu_2, \ldots \nu_m}$.

Ich stelle hier noch für $n = 4, 5, \ldots 9$ die Werte der Zahlen $f_{\nu_1, \nu_2, \ldots \nu_m}$ zusammen. Hierbei beziehen sich die überstrichenen Zahlen $\bar{f}_{\nu_1, \nu_2, \ldots \nu_m}$ auf diejenigen Zerlegungen von n, für die $n - m$ gerade ist, zu denen also eine Darstellung der Gruppe $\mathfrak{A}_n$ als Kollineationsgruppe des Grades $\frac{1}{2}\bar{f}_{\nu_1, \nu_2, \ldots \nu_m}$ gehört:

$$n = 4, \; f_4 = 2, \; \bar{f}_{31} = 4;$$
$$n = 5, \; \bar{f}_5 = 4, \; f_{41} = 6, \; f_{32} = 4;$$
$$n = 6, \; f_6 = 4, \; \bar{f}_{51} = 16, \; \bar{f}_{42} = 20, \; f_{321} = 4;$$
$$n = 7, \; \bar{f}_7 = 8, \; f_{61} = 20, \; f_{52} = 36, \; f_{43} = 20, \; \bar{f}_{421} = 28;$$
$$n = 8, \; f_8 = 8, \; \bar{f}_{71} = 48, \; \bar{f}_{62} = 112, \; \bar{f}_{53} = 112, \; f_{521} = 64, \; f_{431} = 48;$$
$$n = 9, \; \bar{f}_9 = 16, \; f_{81} = 56, \; f_{72} = 160, \; f_{63} = 224, \; \bar{f}_{621} = 240, \; f_{54} = 112,$$
$$\bar{f}_{531} = 336, \; \bar{f}_{432} = 96.$$

Die zu verschiedenen Zerlegungen der Zahl n gehörenden Darstellungen der Gruppen $\mathfrak{S}_n$ und $\mathfrak{A}_n$ sind im allgemeinen wesentlich voneinander verschieden (vgl. § 19). Eine Ausnahme kann nur für $n = 4$ und $n = 6$ eintreten. Für $n = 4$ sind die drei Gruppen $\mathfrak{L}_4$, $\mathfrak{L}'_{3,1}$ und $\mathfrak{L}''_{3,1}$ des Grades 2 als nicht verschieden anzusehen; dies erklärt sich aus dem Vorhandensein dreier linearer Charaktere bei der Gruppe $\mathfrak{B}_4$. Für $n = 6$ können, wie aus unserer Tabelle zu erkennen ist, nur die quaternären Gruppen $\mathfrak{K}_6$ und $\mathfrak{K}_{3,2,1}$, bzw. $\mathfrak{L}_6$ und $\mathfrak{L}_{3,2,1}$, übereinstimmen. Diese Gruppen gehen in der Tat vermittelst des bekannten äußeren Automorphismus der Gruppe $\mathfrak{S}_6$, bzw. $\mathfrak{A}_6$, durch den dem Zyklus $(1\,2\,3)$ die Permutation $(1\,3\,5)(2\,4\,6)$ zugeordnet wird, auseinander hervor (vgl. § 19). Man kann dies durch Betrachtung der zugehörigen Charaktere der Gruppen $\mathfrak{T}_6$ und $\mathfrak{B}_6$ ohne Mühe erkennen. Doch will ich hierauf nicht weiter eingehen, da bereits *H. Maschke**) gezeigt hat, daß jede der Gruppen $\mathfrak{S}_6$ und $\mathfrak{A}_6$ im wesentlichen nur einer quaternären Kollineationsgruppe isomorph ist.

§ 43. Der besseren Übersicht wegen will ich hier noch die von Herrn *Frobenius***) gewonnenen Resultate über die Darstellung der Gruppen

*) Math. Annalen, Bd. 51, S. 251.

**) Vgl. die in der Einleitung zitierten Arbeiten.

$\mathfrak{S}_n$ und $\mathfrak{A}_n$ durch homogene lineare Substitutionen anführen: Jeder Zerlegung

(94.) $$n = \lambda_1 + \lambda_2 + \cdots + \lambda_l \qquad (\lambda_1 \geqq \lambda_2 \geqq \cdots \geqq \lambda_l > 0)$$

der Zahl n in gleiche oder verschiedene Summanden entspricht eine der Gruppe $\mathfrak{S}_n$ isomorphe irreduzible Gruppe $\mathfrak{G}_{\lambda_1, \lambda_2, \ldots \lambda_l}$ linearer homogener Substitutionen in

$$\varphi_{\lambda_1, \lambda_2, \ldots \lambda_l} = \frac{n!}{(\lambda_1 + l - 1)!\,(\lambda_2 + l - 2)! \cdots \lambda_l!} \prod_{\alpha < \beta} (\lambda_\alpha - \lambda_\beta + \beta - \alpha)$$

Variabeln; hierbei sind jedoch die beiden Zerlegungen

$$n = n, \quad n = 1 + 1 + \cdots + 1$$

nicht mitzuzählen*). Sind μ_α unter den Zahlen $\lambda_1, \lambda_2, \ldots \lambda_l$ größer oder gleich α, so ist auch

(94'.) $$n = \mu_1 + \mu_2 + \cdots + \mu_m. \qquad (m = \lambda_1)$$

Stimmen nun die Zerlegungen (94.) und (94.'), die man als *assoziierte* Zerlegungen bezeichnet, nicht überein, so geht die Gruppe $\mathfrak{G}_{\mu_1, \mu_2, \ldots \mu_m}$ aus der Gruppe $\mathfrak{G}_{\lambda_1, \lambda_2, \ldots \lambda_l}$ dadurch hervor, daß man jede ihrer Substitutionen mit $+1$ oder -1 multipliziert, je nachdem die zugehörige Permutation von $\mathfrak{S}_n$ gerade oder ungerade ist; ferner ist die der Gruppe $\mathfrak{A}_n$ isomorphe Untergruppe von $\mathfrak{G}_{\lambda_1, \lambda_2, \ldots \lambda_l}$ wieder irreduzibel. Ist dagegen die Zerlegung (94.) mit der assoziierten Zerlegung (94'.) identisch, so zerfällt diese Untergruppe von $\mathfrak{G}_{\lambda_1, \lambda_2, \ldots \lambda_l}$ in zwei irreduzible Substitutionsgruppen in je $\frac{1}{2}\varphi_{\lambda_1, \lambda_2, \ldots \lambda_l}$ Variabeln. Die beiden so entstehenden mit $\mathfrak{A}_n$ isomorphen Gruppen sind aber nicht als wesentlich voneinander verschieden anzusehen: bedeutet nämlich T die Transposition (12), so kann die eine Gruppe aus der andern dadurch erhalten werden, daß man die der Permutation P von $\mathfrak{A}_n$ entsprechende lineare Substitution durch die der Permutation $T^{-1}PT$ entsprechende ersetzt**). — Betrachtet man wieder zwei äquivalente Substitutionsgruppen als nicht voneinander verschieden, so erhält man in der geschilderten Weise die sämtlichen irreduziblen Gruppen homogener linearer Substitutionen, die den Gruppen $\mathfrak{S}_n$ oder $\mathfrak{A}_n$ isomorph sind.

Die folgende Tabelle enthält für $n = 4, 5, \ldots 9$ die zugehörigen Zahlen $\varphi_{\lambda_1, \lambda_2, \ldots \lambda_l}$; hierbei wird von zwei assoziierten Zerlegungen der Zahl n nur

*) Diesen Zerlegungen entsprechen die trivialen Darstellungen von $\mathfrak{S}_n$ durch die Zahlen 1, bzw. 1 und -1.

**) Die beiden Gruppen sind also als adjungierte Gruppen zu bezeichnen (vgl. Einleitung).

eine angeführt. Die überstrichenen Zahlen beziehen sich auf diejenigen Zerlegungen, die sich selbst assoziiert sind, denen also eine Darstellung der Gruppe $\mathfrak{A}_n$ als lineare homogene Substitutionsgruppe in $\frac{1}{2}\bar{\varphi}_{\lambda_1, \lambda_2, \dots \lambda_l}$ Variabeln entspricht:

$n=4$, $\varphi_{31}=3$, $\bar{\varphi}_{22}=2$;*)

$n=5$, $\varphi_{41}=4$, $\bar{\varphi}_{311}=6$;

$n=6$, $\varphi_{51}=5$, $\varphi_{42}=9$, $\varphi_{411}=10$, $\varphi_{33}=5$, $\bar{\varphi}_{321}=16$;

$n=7$, $\varphi_{61}=6$, $\varphi_{52}=14$, $\varphi_{511}=15$, $\varphi_{43}=14$, $\varphi_{421}=35$, $\bar{\varphi}_{4111}=20$, $\varphi_{331}=21$;

$n=8$, $\varphi_{71}=7$, $\varphi_{62}=20$, $\varphi_{611}=21$, $\varphi_{53}=28$, $\varphi_{521}=64$, $\varphi_{5111}=35$, $\varphi_{44}=14$, $\varphi_{431}=70$, $\varphi_{422}=56$, $\bar{\varphi}_{4211}=90$, $\bar{\varphi}_{332}=42$;

$n=9$, $\varphi_{81}=8$, $\varphi_{72}=27$, $\varphi_{711}=28$, $\varphi_{63}=48$, $\varphi_{621}=105$, $\varphi_{6111}=56$, $\varphi_{54}=42$, $\varphi_{531}=162$, $\varphi_{522}=120$, $\varphi_{5211}=189$, $\bar{\varphi}_{51111}=70$, $\varphi_{441}=84$, $\varphi_{432}=168$, $\varphi_{4311}=216$, $\bar{\varphi}_{333}=42$.

§ 44. Aus den obigen Tabellen erkennt man, daß für kleinere Werte von n unter den Zahlen $f_{\nu_1, \nu_2, \dots \nu_m}$ die Zahl $f_n = 2^{\left[\frac{n-1}{2}\right]}$ und unter den Zahlen $\varphi_{\lambda_1, \lambda_2, \dots \lambda_l}$ (für $n>4$) die Zahl $\varphi_{n-1,1}=n-1$ am kleinsten ist. Ich will zeigen, daß diese Regel allgemein richtig ist; genauer ist für $n>6$

(95.) $$f_{\nu_1, \nu_2, \dots \nu_m} > 2 f_n \qquad (m>1)$$

und, falls nicht

$$\lambda_1 = n-1,\ \lambda_2 = 1 \text{ oder } \lambda_1 = 2,\ \lambda_2 = \lambda_3 = \dots = 1 \text{**)}$$

ist,

(96.) $$\varphi_{\lambda_1, \lambda_2, \dots \lambda_l} > 2(n-1).$$

Um die Ungleichung (95.) zu beweisen, genügt es zu zeigen, daß für die in § 40 eingeführten Zahlen $g_{\nu_1, \nu_2, \dots \nu_m}$, falls

$$n = \nu_1 + \nu_2 + \dots + \nu_m > 6,\ \nu_1 > \nu_2 > \dots > \nu_m > 0,\ m > 1$$

ist, die Ungleichung

(97.) $$g_{\nu_1, \nu_2, \dots \nu_m} > 2^{\left[\frac{m}{2}\right]+1}$$

besteht. — Dies kann man folgendermaßen erkennen. Setzt man, wie in § 36,

$$A(u_1, u_2, \dots u_m) = \prod_{\varkappa<\lambda} \frac{u_\varkappa - u_\lambda}{u_\varkappa + u_\lambda},$$

so ergibt sich mit Hilfe der Formeln (87.) und (88.) ohne Mühe für jedes u

---

*) Die zugehörige Gruppe $\mathfrak{G}_{2,2}$ ist der Gruppe $\mathfrak{S}_4$ *nicht* einstufig isomorph.

**) Diese beiden Zerlegungen von n sind einander assoziiert, daher ist $\varphi_{2,1,1,\dots 1} = \varphi_{n-1,1} = n-1$.

$$(u_1+u_2+\cdots+u_m)A(u_1,u_2,\ldots u_m)=\sum_{\varkappa=1}^{m}u_\varkappa A(u_1,\ldots u_{\varkappa-1},u_\varkappa-u,u_{\varkappa+1},\ldots u_m).$$

Hieraus folgt wegen (93'.)

$$(98.)\qquad g_{\nu_1,\nu_2,\ldots\nu_m}=g_{\nu_1-1,\nu_2,\ldots\nu_m}+g_{\nu_1,\nu_2-1,\ldots\nu_m}+\cdots+g_{\nu_1,\nu_2,\ldots\nu_{m-1},\nu_m-1},$$

wo auf der rechten Seite für $\nu_m=1$ unter $g_{\nu_1,\nu_2,\ldots\nu_{m-1},0}$ die Zahl $g_{\nu_1,\nu_2,\ldots\nu_{m-1}}$ zu verstehen ist; außerdem hat man, falls $\nu_\alpha=\nu_{\alpha+1}+1$ ist, den α-ten Summanden gleich Null zu setzen.

Man nehme nun die Ungleichung (97.), die für $n=7,8,9$ mit Hilfe der Tabelle der $f_{\nu_1,\nu_2,\ldots\nu_m}$ sofort zu verifizieren ist, als richtig an, wenn die Summe der Indizes $\nu_1,\nu_2,\ldots\nu_m$ kleiner als n ist. Ist dann nicht

$$\nu_1=m,\quad \nu_2=m-1,\ldots\nu_{m-1}=2,\quad \nu_m=1,$$

so steht in (98.) auf der rechten Seite mindestens eine Zahl $g_{\varkappa_1,\varkappa_2,\ldots\varkappa_m}$, bei der

$$\varkappa_1>\varkappa_2>\cdots>\varkappa_m>0$$

ist. Da aber die Summe der $\varkappa$ gleich $n-1$ ist, so ergibt sich

$$g_{\nu_1,\nu_2,\ldots\nu_m}\geqq g_{\varkappa_1,\varkappa_2,\ldots\varkappa_m}>2^{\left[\frac{m}{2}\right]+1}.$$

In dem ausgeschlossenen Falle muß aber, da $m+m-1+\cdots+2+1>9$ angenommen werden darf, $m\geqq 4$ sein. Man erhält dann durch wiederholte Anwendung der Formel (98.)

$$g_{m,m-1,\ldots 2,1}=g_{m,m-1,\ldots 4,2,1}+g_{m,m-1,\ldots 4,3}$$
$$>2^{\left[\frac{m-1}{2}\right]+1}+2^{\left[\frac{m-2}{2}\right]+1}\geqq 2^{\left[\frac{m}{2}\right]+1}.$$

In ähnlicher Weise beweist man die Ungleichung (96.) mit Hilfe der für die Zahlen $\varphi_{\lambda_1,\lambda_2,\ldots\lambda_l}$ bestehenden Rekursionsformel

$$\varphi_{\lambda_1,\lambda_2,\ldots\lambda_l}=\varphi_{\lambda_1-1,\lambda_2,\ldots\lambda_l}+\varphi_{\lambda_1,\lambda_2-1,\ldots\lambda_l}+\cdots+\varphi_{\lambda_1,\lambda_2,\ldots\lambda_{l-1},\lambda_l-1},$$

wo auf der rechten Seite für $\lambda_l=1$ unter $\varphi_{\lambda_1,\lambda_2,\ldots\lambda_{l-1},0}$ die Zahl $\varphi_{\lambda_1,\lambda_2,\ldots\lambda_{l-1}}$ zu verstehen und für $\lambda_\alpha=\lambda_{\alpha+1}$ der α-te Summand gleich Null zu setzen ist*).

Abschnitt XI.

Die alternierenden Gruppen mit sechs und sieben Vertauschungsziffern.

§ 45. Während im allgemeinen die sämtlichen mit der Gruppe $\mathfrak{A}_n$ isomorphen Kollineationsgruppen erhalten werden können, indem man die Darstellungen der Gruppe $\mathfrak{B}_n$ durch homogene lineare Substitutionen be-

*) Vgl. die Formel (3.) meiner Arbeit „Über die Darstellung der symmetrischen Gruppe durch lineare homogene Substitutionen", Sitzungsberichte der K. Preuß. Akademie, Berlin, 1908, S. 664.

trachtet, gilt dies nicht, wenn $n=6$ oder $n=7$ wird. Es sollen nunmehr diese beiden Ausnahmefälle erledigt werden.

In jeder Gruppe $\mathfrak{K}$ gebrochener linearer Substitutionen, die der Gruppe $\mathfrak{A}_6$ isomorph ist, lassen sich, wie ich in § 4 gezeigt habe, die Proportionalitätsfaktoren bei den Koeffizienten der einzelnen Substitutionen so wählen, daß die Koeffizientenmatrizen Q_1, Q_2, Q_3, Q_4 der den Permutationen

$$A_1=(123),\quad A_2=(12)(34),\quad A_3=(12)(45),\quad A_4=(12)(56)$$

entsprechenden Substitutionen folgenden Gleichungen genügen:

$$(99.)\quad \begin{cases} Q_1^3=Q_2^2=Q_3^2=Q_4^2=(Q_1Q_2)^3=(Q_1Q_3)^2=(Q_2Q_3)^3=(Q_3Q_4)^3=jE, \\ (Q_1Q_4)^2=kE,\quad Q_2Q_4=jQ_4Q_2, \end{cases}$$

wo j und k durch die Gruppe $\mathfrak{K}$ eindeutig bestimmte Einheitswurzeln sind, und zwar ist

$$k^6=1,\quad j=k^3.$$

Hat man umgekehrt 4 Matrizen (homogene lineare Substitutionen) Q_1, Q_2, Q_3, Q_4, für die Gleichungen der Form (99.) bestehen, wo j und k Konstanten sind, so muß von selbst $k^6=1$, $j=k^3$ werden. Ist m die kleinste positive Zahl, für die $k^m=1$ wird, so erzeugen die Q_α eine endliche Gruppe $\mathfrak{Q}$ der Ordnung $m\cdot 360$ und die zugehörigen 360 gebrochenen linearen Substitutionen bilden eine mit $\mathfrak{A}_6$ isomorphe Gruppe $\mathfrak{K}$, die sich nicht in weniger als $m\cdot 360$ homogenen linearen Substitutionen schreiben läßt. Wählt man die Q_α so, daß $\mathfrak{Q}$ der Gruppe $\mathfrak{A}_6$ oder der Gruppe $\mathfrak{B}_6$ isomorph wird, so wird $k=1$, $m=1$, bzw. $k=-1$, $m=2$. Es existieren hier aber auch Gruppen $\mathfrak{K}$, für die $m=3$ wird. Die einfachste dieser Gruppen ist die zuerst von Herrn *Valentiner* aufgestellte ternäre Kollineationsgruppe der Ordnung 360, die bekanntlich der Gruppe $\mathfrak{A}_6$ isomorph ist. Um die *Valentiner*gruppe zu erhalten, hat man nämlich für Q_1, Q_2, Q_3, Q_4 folgende lineare Substitutionen zu wählen:

(100.)

	Q_1	Q_2	Q_3	Q_4
$x_1'=$	x_1	$\frac{1}{3}(-x_1+2x_2+2x_3)$	$-x_1$	$-x_2$
$x_2'=$	ϱx_2	$\frac{1}{3}(2x_1-x_2+x_3)$	$\bar{\varepsilon} x_3$	$-x_1$
$x_3'=$	$\varrho^2 x_3$	$\frac{1}{3}(2x_1+2x_2-x_3)$	εx_2	$-x_3$,

wo

$$\varrho = \frac{-1+i\sqrt{3}}{2}, \quad \varepsilon = \frac{1}{4}(-1+i\sqrt{15}), \quad \varepsilon = \frac{1}{4}(-1-i\sqrt{15})$$

zu setzen ist. Diese vier Substitutionen genügen den Gleichungen (99.), wobei $j=1$, $k=\varrho$, also $m=3$ wird*).

Hieraus folgt zugleich:

Die Darstellungsgruppe $\mathfrak{C}_6$ *der Gruppe* $\mathfrak{A}_6$ *ist von der Ordnung* 6·360 *und kann als abstrakte Gruppe durch folgende Gleichungen definiert werden:*

$$C_1^3 = C_2^2 = C_3^2 = C_4^2 = (C_1C_2)^3 = (C_1C_3)^2 = (C_2C_3)^3 = (C_3C_4)^3 = K^3,$$
$$(C_1C_4)^2 = K, \quad C_2C_4 = K^3C_4C_2, \quad KC_a = C_aK, \quad K^6 = E.$$

§ 46. Um nun eine Übersicht über alle mit $\mathfrak{A}_6$ isomorphe Kollineationsgruppen zu erhalten, hat man jetzt nur die Charaktere der Gruppe $\mathfrak{C}_6$ zu berechnen. Hierzu haben wir vor allem die Klassen konjugierter Elemente von $\mathfrak{C}_6$ zu bestimmen.

Die 360 Permutationen von $\mathfrak{A}_6$ zerfallen in 7 Klassen ähnlicher Elemente, die durch die Permutationen

$$E, \; A=(123), \; B=(123)(456), \; C=(12)(34), \; D=(12)(3456),$$
$$F=(12345), \; F^2=(13524)$$

repräsentiert werden. Drückt man diese Permutationen durch die die Gruppe $\mathfrak{A}_6$ erzeugenden Elemente A_1, A_2, A_3, A_4 aus, so wird

$$A=A_1, \; B=A_1A_4A_3, \; C=A_2, \; D=A_4A_3A_2, \; F=A_3A_2A_1.$$

Setzt man entsprechend in der Gruppe $\mathfrak{C}_6$

$$A'=C_1, \; B'=C_1C_4C_3, \; C'=C_2, \; D'=C_4C_3C_2, \; F'=C_3C_2C_1$$

und bezeichnet die durch das Element R einer Gruppe definierte Klasse konjugierter Elemente mit (R), so verteilen sich, wie eine elementare Betrachtung lehrt, die 6·360 Elemente von $\mathfrak{C}_6$ auf die 31 Klassen konjugierter Elemente

$(K^a), (A'), (KA'), (B'), (KB'), (C'), (KC'), (K^2C'), (K^aD'), (K^aF'), (K^aF'^7)$**),

wo α die Werte 0, 1, 2, 3, 4, 5 zu durchlaufen hat. Ferner ist

$$(A') = (K^2A') = (K^4A'), \; (B') = (K^2B') = (K^4B'), \; (C') = (K^3C').$$

*) Daß die *Valentiner*gruppe sich nicht in weniger als 3·360 linearen homogenen Substitutionen schreiben läßt, hat zuerst Herr *Wiman* (Math. Annalen, Bd. 47, S. 531) nachgewiesen.

**) Das Element F' der Gruppe $\mathfrak{C}_6$ ist von der Ordnung 2·5, daher ist nicht F'^2, sondern F'^7 von derselben Ordnung wie F'.

Bedeutet σ eine primitive sechste Einheitswurzel, etwa die Einheitswurzel

$$\sigma = -\varrho = \frac{1 - i\sqrt{3}}{2},$$

so gehört zu jedem einfachen Charakter $\chi(R)$ von $\mathfrak{C}_6$ eine Zahl γ aus der Reihe 0, 1, 2, 3, 4, 5, so daß für jedes R

$$\chi(K^a R) = \sigma^{a\gamma} \chi(R) \tag{101.}$$

wird. *Dem Charakter* $\chi(R)$ *von* $\mathfrak{C}_6$ *entspricht dann eine irreduzible mit* $\mathfrak{A}_6$ *isomorphe Kollineationsgruppe des Grades* $f = \chi(E)$, *für welche die früher betrachtete Zahl* k *gleich* σ^γ *wird. Diese Zahl* $k = \sigma^\gamma$ *ist bereits dadurch vollständig bestimmt, daß* $\chi(K) = k\chi(E)$ *ist.* Bezeichnet man nun mit l_γ die Anzahl der verschiedenen einfachen Charaktere von $\mathfrak{C}_6$, für die k gleich σ^γ wird, so erhält man*)

$$l_0 = 7, \; l_1 = l_5 = 4, \; l_2 = l_4 = 5, \; l_3 = 6.$$

Für jeden Charakter $\chi(R)$ genügt es, neben der Zahl

$$k = \frac{\chi(K)}{\chi(E)}$$

nur noch die Zahlen

$$\chi(E),\ \chi(A'),\ \chi(B'),\ \chi(C'),\ \chi(D'),\ \chi(F'),\ \chi(F'^7)$$

anzugeben, da sich die übrigen Zahlen $\chi(R)$ alsdann auf Grund der Formel (101.) und der Gleichung

$$\chi(S^{-1}RS) = \chi(R)$$

berechnen lassen. Da ferner für den zu $\chi(R)$ konjugiert komplexen Charakter an Stelle der Zahl k die Zahl $\bar{k} = \sigma^{-\gamma}$ tritt, so kann man sich auf die Betrachtung der Fälle

$$k = 1, \; k = -1, \; k = \varrho, \; k = -\varrho$$

beschränken. In der Tabelle werden die diesen vier Fällen entsprechenden Charaktere durch die Zeichen $\xi, \eta, \zeta, \vartheta$ unterschieden. Zu den sechs in der Tabelle angeführten Charakteren ξ ist noch der Hauptcharakter $\xi_0(R) = 1$ hinzuzunehmen. Unter $\varkappa_1$ und $\varkappa_2$ sind die Zahlen

$$\varkappa_1 = \frac{1 + \sqrt{5}}{2}, \quad \varkappa_2 = \frac{1 - \sqrt{5}}{2}$$

zu verstehen. Die Charaktere

$$\xi_3, \xi_4; \; \eta_3, \eta_4; \; \eta_5, \eta_6; \; \zeta_1, \zeta_2; \; \vartheta_1, \vartheta_2; \; \vartheta_3, \vartheta_4$$

*) Dies folgt aus dem Satz VI meiner mit D. zitierten Arbeit (vgl. das Zitat auf S. 159).

gehen auseinander hervor, indem man $\varkappa_1$ und $\varkappa_2$, bzw. $\sqrt{2}$ und $-\sqrt{2}$ vertauscht, was in der Tabelle in abgekürzter Weise angedeutet wird.

	ξ_1	ξ_2	$\xi_{3/4}$	ξ_5	ξ_6	η_1	η_2	$\eta_{3/4}$	$\eta_{5/6}$	$\zeta_{1/2}$	ζ_3	ζ_4	ζ_5	$\vartheta_{1/2}$	$\vartheta_{3/4}$
E	5	5	8	9	10	4	4	8	10	3	6	9	15	6	12
A'	2	-1	-1	0	1	2	-1	1	-1	0	0	0	0	0	0
B'	-1	2	-1	0	1	1	-2	-1	1	0	0	0	0	0	0
C'	1	1	0	1	-2	0	0	0	0	-1	2	1	-1	0	0
D'	-1	-1	0	1	0	0	0	0	$\pm\sqrt{2}$	1	0	1	-1	$\pm\sqrt{2}$	0
F'	0	0	$\varkappa_{1/2}$	-1	0	1	1	$-\varkappa_{1/2}$	0	$\varkappa_{1/2}$	1	-1	0	-1	$\varkappa_{1/2}$
F'^7	0	0	$\varkappa_{2/1}$	-1	0	1	1	$-\varkappa_{2/1}$	0	$\varkappa_{2/1}$	1	-1	0	-1	$\varkappa_{2/1}$

Die Charaktere ξ und η ergeben sich unmittelbar durch Betrachtung der von Herrn *Frobenius* bestimmten Charaktere der Gruppe $\mathfrak{A}_6$ und der hier früher bestimmten Charaktere zweiter Art der Gruppe $\mathfrak{B}_6$ (vgl. Satz IX). Die Charaktere ζ_1 und ζ_2 entsprechen der durch die Substitutionen (100.) erzeugten Darstellung der Gruppe $\mathfrak{C}_6$ und der Darstellung, die aus dieser hervorgeht, indem man in (100.) ε und $\bar{\varepsilon}$ vertauscht. Die übrigen Charaktere habe ich folgendermaßen erhalten:

$$\zeta_3 = \bar{\zeta}_1^2 - \zeta_1, \quad \zeta_4 = \bar{\zeta}_1 \bar{\zeta}_2, \quad \zeta_5 = \zeta_1 \xi_1$$
$$\vartheta_3 = \zeta_1 \eta_1, \quad \vartheta_4 = \zeta_2 \eta_1, \quad \vartheta_1 = \zeta_1 \eta_5 - \vartheta_3 - \vartheta_4, \quad \vartheta_2 = \zeta_1 \eta_6 - \vartheta_3 - \vartheta_4.$$

Hierbei bedeuten $\bar{\zeta}_1$ und $\bar{\zeta}_2$ die zu ζ_1 und ζ_2 konjugiert komplexen Charaktere.

§ 47. Die alternierende Gruppe $\mathfrak{A}_7$ wird durch die Permutationen $A_1 = (1\,2\,3)$, $A_2 = (1\,2)(3\,4)$, $A_3 = (1\,2)(4\,5)$, $A_4 = (1\,2)(5\,6)$, $A_5 = (1\,2)(6\,7)$ erzeugt. Nach den Ergebnissen des § 4 lassen sich in jeder Gruppe $\mathfrak{K}$ gebrochener linearer Substitutionen, die mit $\mathfrak{A}_7$ isomorph ist, die diesen 5 Permutationen entsprechenden Substitutionen (bei passender Wahl der in ihnen enthaltenen Proportionalitätsfaktoren) so schreiben, daß ihre Koeffizientenmatrizen $Q_1, \ldots Q_5$ folgenden Gleichungen genügen:

$$(102.)\quad \begin{cases} Q_1^3 = Q_2^2 = Q_3^2 = Q_4^2 = Q_5^2 = jE, \\ (Q_1 Q_2)^3 = (Q_1 Q_3)^2 = (Q_1 Q_5)^2 = (Q_2 Q_3)^3 = (Q_3 Q_4)^3 = (Q_4 Q_5)^3 = jE, \\ (Q_1 Q_4)^2 = kE, \; Q_2 Q_4 = j Q_4 Q_2, \; Q_2 Q_5 = j Q_5 Q_2, \; Q_3 Q_5 = j Q_5 Q_3, \end{cases}$$

wobei wieder $k^6 = 1$, $j = k^3$ wird. Für diese Gleichungen gilt in bezug auf die Gruppe $\mathfrak{A}_7$ genau dasselbe, wie bei der Gruppe $\mathfrak{A}_6$ für die Gleichungen (99.). Auch hier lassen sich Gruppen $\mathfrak{K}$ angeben, für die k eine primitive dritte Einheitswurzel wird. Die Gleichungen (102.) werden nämlich befriedigt, wenn man für $Q_1, \ldots Q_5$ folgende lineare homogene Substitutionen in 6 Variabeln wählt, wobei $k = \varrho = \frac{-1+i\sqrt{3}}{2}$ wird:

(103.)

	Q_1	Q_2	Q_3	Q_4	Q_5
$x_1' =$	x_1	x_1	x_1	x_2	x_1
$x_2' =$	ϱx_2	x_2	$\frac{1}{2}(x_3 + 3x_6)$	x_1	x_3
$x_3' =$	$\varrho^2 x_3$	x_3	$\frac{1}{2}(x_2 + 3x_5)$	x_3	x_2
$x_4' =$	x_4	$\frac{1}{3}(-x_4 + 2x_5 + 2x_6)$	x_4	x_5	x_4
$x_5' =$	ϱx_5	$\frac{1}{3}(2x_4 - x_5 + 2x_6)$	$\frac{1}{2}(x_3 - x_6)$	x_4	x_6
$x_6' =$	$\varrho^2 x_6$	$\frac{1}{3}(2x_4 + 2x_5 - x_6)$	$\frac{1}{2}(x_2 - x_5)$	x_6	x_5.

Ich will noch eine andere Auflösung der Gleichungen (102.) durch lineare homogene Substitutionen in 6 Variabeln angeben, für die $k = -\varrho$ wird:

(104.)

	Q_1	Q_2	Q_3	Q_4	Q_5
$x_1' =$	$-x_1$	$\frac{1}{\sqrt{3}}(i x_1 + \sqrt{2}\, x_2)$	$\frac{1}{\sqrt{6}}(-i\sqrt{2}\, x_1 - 2x_4)$	$i x_2$	$i x_4$
$x_2' =$	$-\varrho x_2$	$\frac{1}{\sqrt{3}}(-\sqrt{2}\, x_1 - i x_2)$	$\frac{1}{\sqrt{6}}(-x_3 + \delta x_6)$	$i x_1$	$i x_3$
$x_3' =$	$-\varrho^2 x_3$	$\frac{1}{\sqrt{3}}(i x_3 + \sqrt{2}\, x_4)$	$\frac{1}{\sqrt{6}}(x_2 - \delta x_5)$	$i x_6$	$i x_2$
$x_4' =$	$-x_4$	$\frac{1}{\sqrt{3}}(-\sqrt{2}\, x_1 - i x_4)$	$\frac{1}{\sqrt{6}}(2x_1 + i\sqrt{2}\, x_4)$	$i x_5$	$i x_1$
$x_5' =$	$-\varrho x_5$	$\frac{1}{\sqrt{3}}(i x_5 + \sqrt{2}\, x_6)$	$\frac{1}{\sqrt{6}}(\bar{\delta} x_3 + x_6)$	$i x_4$	$i x_6$
$x_6' =$	$-\varrho^2 x_6$	$\frac{1}{\sqrt{3}}(-\sqrt{2}\, x_5 - i x_6)$	$\frac{1}{\sqrt{6}}(-\bar{\delta} x_2 - x_5)$	$i x_3$	$i x_5$.

Hierin ist

$$\delta = \sqrt{3} + i\sqrt{2}, \quad \bar{\delta} = \sqrt{3} - i\sqrt{2}$$

zu setzen.

Die durch die Substitutionen (103.) und (104.) erzeugten Gruppen $\mathfrak{D}_1$ und $\mathfrak{D}_2$ sind von den Ordnungen $3 \cdot \frac{7!}{2}$ und $6 \cdot \frac{7!}{2}$; die zugehörigen mit der Gruppe $\mathfrak{A}_7$ isomorphen Kollineationsgruppen lassen sich nicht in weniger als $3 \cdot \frac{7!}{2}$, bzw. $6 \cdot \frac{7!}{2}$ homogenen linearen Substitutionen schreiben.

Es ergibt sich zugleich der Satz:

Die Darstellungsgruppe $\mathfrak{C}_7$ der Gruppe $\mathfrak{A}_7$ ist von der Ordnung $6 \cdot \frac{7!}{2}$ und kann als abstrakte Gruppe durch folgende Gleichungen definiert werden:

$$C_1^3=C_2^2=C_3^2=C_4^2=C_5^2=K^3,\ K^6=E,\ KC_\alpha=C_\alpha K,$$
$$(C_1C_2)^3=(C_1C_3)^2=(C_1C_5)^2=(C_2C_3)^3=(C_3C_4)^3=(C_4C_5)^3=K^3,$$
$$(C_1C_4)^2=K,\ C_2C_4=K^3C_4C_2,\ C_2C_5=K^3C_5C_2,\ C_3C_5=K^3C_5C_3.$$

Man erkennt auch, daß $\mathfrak{C}_7$ die früher betrachtete Gruppe $\mathfrak{C}_6$ als Untergruppe enthält.

§ 48. Es sollen nun die Charaktere der Gruppe $\mathfrak{C}_7$ berechnet werden. Die $\frac{7!}{2}=2520$ Permutationen von $\mathfrak{A}_7$ zerfallen in 9 Klassen ähnlicher Elemente, die durch die Permutationen

$$E,\ A=(123),\ B=(123)(456),\ C=(12)(34),\ D=(12)(3456),\ F=(12345),$$
$$G=(12)(34)(567),\ H=(1234567),\ H^{-1}=(1765432)$$

repräsentiert werden. Es ist ferner

$$A=A_1,\ B=A_1A_4A_3,\ C=A_2,\ D=A_4A_3A_2,\ F=A_3A_2A_1,$$
$$G=A_2A_5A_4,\ H=A_5A_4A_3A_2A_1.$$

Setzt man entsprechend in $\mathfrak{C}_7$

$$A'=C_1,\ B'=C_1C_4C_3,\ C'=C_2,\ D'=C_4C_3C_2,\ F'=C_3C_2C_1,$$
$$G'=C_2C_5C_4,\ H'=C_5C_4C_3C_2C_1,$$

so verteilen sich die $6\cdot 2520$ Elemente von $\mathfrak{C}_7$ auf die 40 Klassen konjugierter Elemente

$$(K^\alpha),(A'),(KA'),(B'),(KB'),(C'),(KC'),(K^2C'),(K^\alpha D'),(K^\alpha F'),$$
$$(G'),(KG'),(K^2G'),(K^\alpha H'),(K^\alpha H'^{-1}),$$

wo α die Werte $0, 1, \dots 5$ durchläuft. Ferner ist

$$(A')=(K^2A')=(K^4A'),\ (B')=(K^2B')=(K^4B').$$
$$(C')=(K^3C'),\ (G')=(K^3G').$$

Für die einfachen Charaktere $\chi(R)$ von $\mathfrak{C}_7$ gilt genau dasselbe wie für die Gruppe $\mathfrak{C}_6$ (vgl. § 45). Bedeutet wieder l_γ die Anzahl der Charaktere $\chi(R)$, für die

$$\chi(K^\alpha R)=\sigma^{\alpha\gamma}\chi(R) \qquad \left(\sigma=-\varrho=\frac{1-i\sqrt{3}}{2}\right)$$

wird, so erhält man hier

$$l_0=9,\ l_1=l_5=5,\ l_2=l_4=7,\ l_3=7.$$

In der nachstehenden Tabelle werden wieder die den vier allein zu betrachtenden Fällen

$$\chi(K^\alpha R)=\chi(R),\ \chi(K^\alpha R)=(-1)^\alpha\chi(R),$$
$$\chi(K^\alpha R)=\varrho^\alpha\chi(R),\ \chi(K^\alpha R)=(-\varrho)^\alpha\chi(R)$$

entsprechenden Charaktere mit $\xi, \eta, \zeta, \vartheta$ bezeichnet. Auch hier wird der Hauptcharakter $\xi_0(R)=1$ nicht besonders angeführt. Für λ_1 und λ_2 hat man zu setzen

$$\lambda_1 = \frac{-1+i\sqrt{7}}{2}, \quad \lambda_2 = \frac{-1-i\sqrt{7}}{2}.$$

	ξ_1	$\xi_{2/3}$	ξ_4	ξ_5	ξ_6	ξ_7	ξ_8	$\eta_{1/2}$	$\eta_{3/4}$	η_5	η_6	η_7	ζ_1	ζ_2	ζ_3	ζ_4	ζ_5	$\zeta_{6/7}$	$\vartheta_{1/2}$	$\vartheta_{3/4}$	ϑ_5
E	6	10	14	14	15	21	35	4	14	20	20	36	6	15	15	21	21	24	6	24	36
A'	3	1	2	-1	3	-3	-1	2	-2	4	-2	0	0	0	0	0	0	0	0	0	0
B'	0	1	-1	2	0	0	-1	1	-1	-1	2	0	0	0	0	0	0	0	0	0	0
C'	2	-2	2	2	-1	1	-1	0	0	0	0	0	2	-1	3	1	-3	0	0	0	0
D'	0	0	0	0	-1	-1	1	0	$\pm\sqrt{2}$	0	0	0	0	-1	1	-1	1	0	$\pm\sqrt{2}$	0	0
F'	1	0	-1	-1	0	1	0	1	1	0	0	-1	1	0	0	1	1	-1	-1	1	-1
G'	-1	1	2	-1	-1	1	-1	0	0	0	0	0	2	2	0	-2	0	0	0	0	0
H'	-1	$\lambda_{1/2}$	0	0	1	0	0	$-\lambda_{1/2}$	0	-1	-1	1	-1	1	1	0	0	$\lambda_{1/2}$	-1	$\lambda_{1/2}$	1
H'^{-1}	-1	$\lambda_{2/1}$	0	0	1	0	0	$-\lambda_{2/1}$	0	-1	-1	1	-1	1	1	0	0	$\lambda_{2/1}$	-1	$\lambda_{2/1}$	1

Die Charaktere ξ und η ergeben sich wieder aus den Charakteren der Gruppe $\mathfrak{A}_7$ und den Charakteren zweiter Art der Gruppe $\mathfrak{B}_7$. Der Charakter ζ_1 entspricht der durch die Substitutionen (103.) erzeugten Darstellung der Gruppe $\mathfrak{C}_7$. Zur Berechnung der übrigen Charaktere kann man sich folgender Gleichungen bedienen:

$$\zeta_2(R) = \frac{1}{2}[\bar{\zeta}_1^2(R) - \bar{\zeta}_1(R^2)], \quad \zeta_3 = \bar{\zeta}_1^2 - \zeta_1 - \zeta_2, \quad \zeta_4 = \zeta_1\xi_1 - \zeta_3, \quad \zeta_5 = \zeta_1\xi_6 - \zeta_1\xi_5 + \zeta_3,$$

$$\zeta_6 = \zeta_1\xi_3 - \zeta_2 - \zeta_5, \quad \zeta_7 = \zeta_1\xi_2 - \zeta_2 - \zeta_5; \quad \vartheta_3 = \zeta_1\eta_1, \quad \vartheta_4 = \zeta_1\eta_2, \quad \vartheta_5 = \zeta_1\eta_3 - \vartheta_3 - \vartheta_4,$$

$$\vartheta_1 = \zeta_3\eta_3 - 2\vartheta_3 - 2\vartheta_4 - 3\vartheta_5, \quad \vartheta_2 = \zeta_3\eta_4 - 2\vartheta_3 - 2\vartheta_4 - 3\vartheta_5.$$

Hierbei bedeutet $\bar{\zeta}_1$ den zu ζ_1 konjugiert komplexen Charakter. Dem Charakter ϑ_1 entspricht die durch die Substitutionen (104.) erzeugte Darstellung der Gruppe $\mathfrak{C}_7$.

§ 49. Zu jedem einfachen Charakter χ von $\mathfrak{C}_\nu (\nu = 6,7)$ gehört eine mit der Gruppe $\mathfrak{A}_\nu$ isomorphe irreduzible Kollineationsgruppe des Grades $\chi(E)$, die mit $\mathfrak{K}(\chi)$ bezeichnet werden möge. Ist $\chi(K) = k\chi(E)$ und gehört die Einheitswurzel k zum Exponenten m, so läßt sich $\mathfrak{K}(\chi)$ in $m\frac{\nu!}{2}$, aber nicht in weniger als $m\frac{\nu!}{2}$ homogenen linearen Substitutionen schreiben. Es soll nun untersucht werden, welche unter den Gruppen $\mathfrak{K}(\chi)$ als wesentlich voneinander verschieden anzusehen sind, wenn man zwei konjugiert komplexe Gruppen und ferner zwei adjungierte Gruppen nicht voneinander unterscheidet (vgl. Einleitung).

Dem äußeren Automorphismus A der Gruppe $\mathfrak{A}_\nu$, der dadurch entsteht, daß man in den Zykeln der Permutationen von $\mathfrak{A}_\nu$ die Ziffern 1 und 2 vertauscht, entspricht, wie man leicht erkennt, ein Automorphismus $\Gamma = \binom{R}{R^*}$ der Gruppe $\mathfrak{C}_\nu$, der dadurch charakterisiert ist, daß

$$K^* = K^{-1}, \quad C_1^* = K^3 C_1^2, \quad C_\alpha^* = K^3 C_\alpha \qquad (\alpha > 1)$$

wird. Ferner führt Γ bei der Gruppe $\mathfrak{C}_6$ die Klassen konjugierter Elemente (D') und (F') in $(K^3 D')$ und (F'^7) über, ebenso bei der Gruppe $\mathfrak{C}_7$ die Klassen (D') und (H') in $(K^3 D')$ und (H'^{-1}). Dagegen bleiben für $\nu = 6$ die Klassen

$$(E), (A'), (B'), (C')$$

und für $\nu = 7$ die Klassen

$$(E), (A'), (B'), (C'), (F'), (G')$$

ungeändert.

Ist nun $\chi(R)$ ein einfacher Charakter von $\mathfrak{C}_\nu$, so bilden auch die Zahlen

$$\chi^*(R) = \chi(R^*)$$

einen einfachen Charakter von $\mathfrak{C}_\nu$. Bezeichnet man noch die zu χ und χ^* konjugiert komplexen Charaktere mit $\bar{\chi}$ und χ', so wird, da $\chi^*(K) = \chi(K^{-1})$ und $\chi(K)$ konjugiert komplexe Zahlen sind,

$$\chi^*(K) = \bar{\chi}(K), \quad \chi'(K) = \chi(K),$$

so daß also für χ und χ' die Einheitswurzel k denselben Wert erhält. Ferner wird für die Gruppe $\mathfrak{C}_6$ insbesondere

$$\chi^*(D') = \chi'(D') = k^3 \chi(D'), \quad \chi^*(F') = \chi'(F') = \chi(F'^7)$$

und für die Gruppe $\mathfrak{C}_7$

$$\chi^*(D') = \chi'(D') = k^3 \chi(D'), \quad \chi^*(H') = \chi'(H'^{-1}) = \chi(H'^{-1}).$$

Wenden wir uns nun zur Betrachtung der Kollineationsgruppen $\mathfrak{K}(\chi)$, so können wir sagen: *die zu* $\mathfrak{K}(\chi)$ *adjungierte Gruppe ist* $\mathfrak{K}(\chi^*)$, *die zu* $\mathfrak{K}(\chi)$ *und* $\mathfrak{K}(\chi^*)$ *konjugiert komplexen Gruppen sind* $\mathfrak{K}(\bar{\chi})$ *und* $\mathfrak{K}(\chi')$. Wir bezeichnen $\mathfrak{K}(\chi')$ als die zu $\mathfrak{K}(\chi)$ *koadjungierte* Gruppe. Bezüglich der in unseren Tabellen angegebenen Charaktere der beiden Gruppen $\mathfrak{C}_\nu$ ergibt sich insbesondere folgendes: Für $\nu = 6$ sind

$$\mathfrak{K}(\xi_3), \mathfrak{K}(\xi_4); \quad \mathfrak{K}(\eta_3), \mathfrak{K}(\eta_4); \quad \mathfrak{K}(\eta_5), \mathfrak{K}(\eta_6)$$

adjungierte und

$$\mathfrak{K}(\zeta_1), \mathfrak{K}(\zeta_2); \quad \mathfrak{K}(\vartheta_1), \mathfrak{K}(\vartheta_2); \quad \mathfrak{K}(\vartheta_3), \mathfrak{K}(\vartheta_4)$$

koadjungierte Gruppen. Ebenso sind für $\nu = 7$

$$\mathfrak{K}(\xi_2), \mathfrak{K}(\xi_3); \ \mathfrak{K}(\eta_1), \mathfrak{K}(\eta_2); \ \mathfrak{K}(\eta_3), \mathfrak{K}(\eta_4)$$

adjungierte und

$$\mathfrak{K}(\vartheta_1), \mathfrak{K}(\vartheta_2)$$

koadjungierte Gruppen.

Bemerkenswert ist, daß die zu den algebraisch konjugierten Charakteren ζ_6 und ζ_7, bzw. ϑ_3 und ϑ_4 von $\mathfrak{C}_7$ gehörenden Kollineationsgruppen $\mathfrak{K}(\zeta_6)$ und $\mathfrak{K}(\zeta_7)$, bzw. $\mathfrak{K}(\vartheta_3)$ und $\mathfrak{K}(\vartheta_4)$ weder als konjugiert komplexe, noch als adjungierte oder koadjungierte Gruppen erscheinen, also in dem von uns festgesetzten Sinn als wesentlich voneinander verschieden zu betrachten sind. Diese Gruppen lassen sich aber jedenfalls als Gruppen mit konjugiert algebraischen Koeffizienten schreiben.

Bei der Gruppe $\mathfrak{A}_6$ hat man noch den Automorphismus zu berücksichtigen, der dem Zyklus (123) die Permutation (123) (456) zuordnet. Durch diesen Automorphismus werden, wie aus der Tabelle der Charaktere von $\mathfrak{C}_6$ leicht geschlossen werden kann, die Gruppen $\mathfrak{K}(\xi_1)$ und $\mathfrak{K}(\eta_1)$ in die Gruppen $\mathfrak{K}(\xi_2)$ und $\mathfrak{K}(\eta_2)$ übergeführt. Die Gruppen $\mathfrak{K}(\xi_1)$ und $\mathfrak{K}(\xi_2)$, bzw. $\mathfrak{K}(\eta_1)$ und $\mathfrak{K}(\eta_2)$ können daher so geschrieben werden, daß sie dieselben Substitutionen enthalten, und werden ebenfalls als nicht voneinander verschieden angesehen (vgl. § 42).

Wir gelangen zu folgendem Schlußresultat:

Jede mit der alternierenden Gruppe $\mathfrak{A}_\nu$ $(\nu = 6, 7)$ isomorphe Kollineationsgruppe $\mathfrak{K}$ läßt sich als Gruppe von $m\frac{\nu!}{2}$ homogenen linearen Substitutionen schreiben, wo m eine der Zahlen 1, 2, 3, 6 *ist. Unter den irreduziblen Gruppen $\mathfrak{K}$ sind für $\nu = 6$ nur* 13 *und für $\nu = 7$ nur* 23 *wesentlich voneinander verschieden. Die Grade dieser Kollineationsgruppen sind in den folgenden Tabellen enthalten:*

Gruppe $\mathfrak{A}_6$

$m=1$	$m=2$	$m=3$	$m=6$
5, 8, 9, 10	4, 8, 10	3, 6, 9, 15	6, 12

Gruppe $\mathfrak{A}_7$

$m=1$	$m=2$	$m=3$	$m=6$
6, 10, 14, 14, 15, 21, 35	4, 14, 20, 20, 36	6, 15, 15, 21, 21, 24, 24	6, 24, 24, 36 .

§ 50. Vergleicht man die vorstehenden Tabellen mit den in § 42 und § 43 angegebenen Tabellen der Zahlen $f_{\nu_1, \nu_2, \ldots}$ und $\varphi_{\lambda_1, \lambda_2, \ldots}$ und berücksichtigt man die Resultate, die wir in § 44 über das Größenverhältnis dieser Zahlen gewonnen haben, so erhält man den Satz:

Man bezeichne eine mit der symmetrischen Gruppe $\mathfrak{S}_n$ isomorphe Kollineationsgruppe mit $\mathfrak{S}_n^{(h)}$ oder $\mathfrak{S}_n^{(g)}$, je nachdem sie sich in $n!$ homogenen linearen Substitutionen schreiben läßt oder nicht. Unter $s_n^{(h)}$ und $s_n^{(g)}$ verstehe man die kleinsten unter den Graden der Gruppen $\mathfrak{S}_n^{(h)}$, bzw. $\mathfrak{S}_n^{(g)}$. Die entsprechenden Zahlen für die alternierende Gruppe $\mathfrak{A}_n$ bezeichne man mit $a_n^{(h)}$ und $a_n^{(g)}$. Dann ist für $n \geqq 4$)*

$$s_n^{(h)} = n-1, \quad s_n^{(g)} = 2^{\left[\frac{n-1}{2}\right]},$$

$$a_n^{(h)} = n-1, \quad a_n^{(g)} = 2^{\left[\frac{n-2}{2}\right]}.$$

Eine Ausnahme bilden bei der alternierenden Gruppe die beiden Fälle $n=5$ und $n=6$; es ist nämlich

$$a_5^{(h)} = 3, \quad a_6^{(g)} = 3.$$

Hierin ist zugleich der zuerst von Herrn *A. Wiman***) bewiesene Satz enthalten, der besagt, daß für $n > 7$ weder die Gruppe $\mathfrak{S}_n$ noch die Gruppe $\mathfrak{A}_n$ einer Kollineationsgruppe isomorph sein kann, deren Grad kleiner als $n-1$ ist.

---

*) Die Zahlen $s_n^{(g)}$ und $a_n^{(g)}$ haben nur für $n \geqq 4$ einen Sinn. Für $n < 4$ kommt noch hinzu

$$s_2^{(h)} = a_2^{(h)} = 1, \quad s_3^{(h)} = 2, \quad a_3^{(h)} = 1.$$

**) Math. Annalen, Bd. 52, S. 243.

Inhaltsübersicht.

17.

Über Gruppen periodischer linearer Substitutionen

Sitzungsberichte der Preussischen Akademie der Wissenschaften 1911, Physikalisch-Mathematische Klasse, 619 - 627

Eine lineare homogene Substitution

$$(A) \qquad x'_k = a_{k1} x_1 + a_{k2} x_2 + \cdots + a_{kn} x_n \qquad (k = 1, 2, \cdots, n)$$

nennt man *periodisch*, wenn unter ihren Potenzen $A, A^2, A^3, \cdots$ die identische Substitution E vorkommt. Der kleinste Exponent m, für den $A^m = E$ wird, heißt die *Ordnung* von A. Notwendig und hinreichend für die Periodizität einer Substitution A ist, daß die charakteristische Determinante $|A - xE|$ von A nur für Einheitswurzeln $\rho_1, \rho_2, \cdots, \rho_n$ verschwindet und lauter lineare Elementarteiler besitzt. Die Ordnung von A ist gleich dem kleinsten gemeinsamen Vielfachen der Exponenten, zu denen die Einheitswurzeln $\rho_1, \rho_2, \cdots, \rho_n$ gehören.

Unter einer *periodischen Substitutionsgruppe* verstehen wir im folgenden eine Gruppe linearer homogener Substitutionen, die sämtlich periodisch sind. Zu diesen Gruppen gehören insbesondere alle endlichen Gruppen linearer Substitutionen von nicht verschwindenden Determinanten. Es gibt aber auch unendliche Gruppen dieser Art. Das einfachste Beispiel bildet die Gesamtheit der Substitutionen $x' = \rho x$, wo ρ alle Einheitswurzeln durchläuft. Ferner erzeugt jedes unendliche System von Substitutionen der Form

$$x'_1 = \rho_1 x_{\alpha_1}, \qquad x'_2 = \rho_2 x_{\alpha_2}, \qquad \cdots, \qquad x'_n = \rho_n x_{\alpha_n},$$

wo $\rho_1, \rho_2, \cdots, \rho_n$ Einheitswurzeln sind und $\alpha_1, \alpha_2, \cdots, \alpha_n$ bis auf die Reihenfolge die Zahlen $1, 2, \cdots, n$ bedeuten, eine unendliche periodische Substitutionsgruppe.

Ein einfaches Kriterium für die Endlichkeit einer periodischen Substitutionsgruppe verdankt man Hrn. W. Burnside[1], der gezeigt hat,

[1] *On criteria for the finiteness of the order of a group of linear substitutions*, Proceedings of the London Mathematical Society, Ser. 2, Vol. 3 (1905), S. 435. Vgl. auch A. Loewy, *Über die Gruppen linearer homogener Substitutionen vom Typus einer endlichen Gruppe*, Math. Annalen, Bd. 64, S. 264.

daß *eine solche Gruppe stets und nur dann endlich ist, wenn die Ordnungen aller Substitutionen der Gruppe unterhalb einer endlichen Schranke liegen.*

Unter Benutzung dieses Kriteriums soll hier gezeigt werden, daß für jede periodische Substitutionsgruppe $\mathfrak{G}$ in n Variabeln folgende Sätze bestehen:

I. *Jedes System von endlich vielen Substitutionen der Gruppe $\mathfrak{G}$ erzeugt eine endliche Gruppe.*

II. *Die Gruppe $\mathfrak{G}$ ist eine* Hermitesche *Gruppe, d. h. es gibt mindestens eine positiv definite* Hermitesche *Form (von nicht verschwindender Determinante), die durch jede Substitution von $\mathfrak{G}$ in sich transformiert wird.*

III. *Die Gruppe $\mathfrak{G}$ enthält eine invariante* Abelsche *Untergruppe $\mathfrak{A}$, deren Index endlich ist und unterhalb einer allein von n abhängigen Schranke liegt, sowie eine endliche Untergruppe $\mathfrak{H}$, deren Elemente zusammen mit den Elementen von $\mathfrak{A}$ die ganze Gruppe $\mathfrak{G}$ erzeugen.*

Durch diese Sätze wird die enge Verwandtschaft der allgemeinen periodischen Substitutionsgruppen mit den endlichen Gruppen dargetan. Der Satz III bildet ein Analogon zu dem bekannten von Hrn. C. Jordan herrührenden Theorem über endliche Gruppen und liefert ein Verfahren zur Aufstellung aller periodischen Substitutionsgruppen mit gegebener Variabelnanzahl. Der Beweis dieses Satzes wird hier geführt mit Hilfe einer von Hrn. L. Bieberbach[1] angegebenen und von Hrn. G. Frobenius[2] vereinfachten Methode.

§ 1.

Der Beweis des Satzes I stützt sich auf folgenden Hilfssatz:

Es seien $\omega_1, \omega_2, \cdots, \omega_p$ beliebige reelle oder komplexe Zahlen und es sei $K = P(\omega_1, \omega_2, \cdots, \omega_p)$ derjenige Zahlkörper, der aus dem Körper P der rationalen Zahlen durch Adjunktion von $\omega_1, \omega_2, \cdots, \omega_p$ hervorgeht. Dann ist der Teilkörper A von K, der von den in K enthaltenen (in bezug auf P) algebraischen Zahlen gebildet wird, ein endlicher algebraischer Zahlkörper über P. Ist ferner n eine ganze rationale Zahl, so gibt es nur endlich viele Einheitswurzeln ρ, die Gleichungen $n$ten Grades mit Koeffizienten aus dem Körper K genügen.

Wir können ohne Beschränkung der Allgemeinheit annehmen, daß unter den p Größen $\omega_1, \omega_2, \cdots, \omega_p$ die ersten q transzendente Zahlen sind, zwischen denen keine algebraische Gleichung

$$(1.)\qquad \sum A_{\nu_1, \nu_2, \ldots, \nu_q}\, \omega_1^{\nu_1} \omega_2^{\nu_2} \cdots \omega_q^{\nu_q} = 0$$

[1] *Über einen Satz des Hrn. C. Jordan in der Theorie der endlichen Gruppen linearer Substitutionen,* Sitzungsberichte 1911, S. 231.

[2] *Über den von L. Bieberbach gefundenen Beweis eines Satzes von C. Jordan,* Sitzungsberichte 1911, S. 241.

mit rationalen Koeffizienten besteht, während für $\nu > q$ die Zahlen $\omega_1, \omega_2, \cdots, \omega_q, \omega_\nu$ durch eine Gleichung dieser Art verbunden sind. Es gibt dann auch keine ganze rationale Funktion $f(x_1, x_2, \cdots, x_q)$ mit algebraischen Koeffizienten, die für $x_1 = \omega_1, x_2 = \omega_2, \cdots, x_q = \omega_q$ verschwindet. Denn ersetzt man in f die Koeffizienten auf alle möglichen Arten durch die konjugiert algebraischen Zahlen, so würde das Produkt $F(x_1, x_2, \cdots, x_q)$ der so entstehenden Funktionen $f, f', f'', \cdots$ eine ganze rationale Funktion mit rationalen Koeffizienten werden. Da aber auch $F(\omega_1, \omega_2, \cdots, \omega_q) = 0$ wird, so würden wir eine Gleichung der Form (1.) erhalten.

Bezeichnet man nun den Körper $\mathrm{P}(\omega_1, \omega_2, \cdots, \omega_q)$ mit Ω, so sind $\omega_{q+1}, \omega_{q+2}, \cdots, \omega_p$ als algebraische Zahlen in bezug auf Ω anzusehen, daher wird

$$\mathrm{K} = \Omega(\omega_{q+1}, \omega_{q+2}, \cdots, \omega_p)$$

ein endlicher algebraischer Zahlkörper über Ω. Ist k der Grad dieses Körpers, so besteht für je $k+1$ Zahlen $\alpha_1, \alpha_2, \cdots, \alpha_{k+1}$ von K eine Gleichung der Form

$$\varphi_1\alpha_1 + \varphi_2\alpha_2 + \cdots + \varphi_{k+1}\alpha_{k+1} = 0,$$

wo $\varphi_1, \varphi_2, \cdots, \varphi_{k+1}$ gewisse ganze rationale Funktionen von $\omega_1, \omega_2, \cdots, \omega_q$ mit rationalen Koeffizienten sind. Die linke Seite dieser Gleichung läßt sich als ganze rationale Funktion von $\omega_1, \omega_2, \cdots, \omega_q$ schreiben, deren Koeffizienten die Form

$$(2.) \qquad a_1\alpha_1 + a_2\alpha_2 + \cdots + a_{k+1}\alpha_{k+1}$$

besitzen, wo $a_1, a_2, \cdots, a_{k+1}$ rationale Zahlen bedeuten. Sind aber speziell $\alpha_1, \alpha_2, \cdots, \alpha_{k+1}$ (in bezug auf P) algebraische Zahlen, so müssen alle Koeffizienten (2.) verschwinden, denn andernfalls würde sich für $\omega_1, \omega_2, \cdots, \omega_q$ eine Gleichung mit algebraischen Koeffizienten ergeben. Der Teilkörper A von K besitzt also die Eigenschaft, daß je $k+1$ der in ihm enthaltenen Zahlen in bezug auf P linear abhängig sind. Hieraus folgt aber, daß A ein endlicher algebraischer Zahlkörper über P ist, dessen Grad höchstens gleich k wird.

Ist nun ρ eine primitive $m$te Einheitswurzel, für die eine Gleichung $n$ten Grades mit Koeffizienten aus dem Körper K besteht, so denken wir uns die Gleichung

$$(3.) \qquad f(x) = x^{n'} + k_1 x^{n'-1} + \cdots + k_{n'} = 0 \qquad (n' \leqq n)$$

niedrigsten Grades gebildet, der ρ im Körper K genügt. Dann muß $f(x)$ ein Divisor von $x^m - 1$ sein. Die Wurzeln der Gleichung (3.) sind daher sämtlich $m$te Einheitswurzeln, und demnach die Koeffizienten $k_1, k_2, \cdots, k_{n'}$ algebraische Zahlen, die als Größen von K im Körper A enthalten sein müssen. Bildet man nun aber, wenn l der Grad des algebraischen Zahlkörpers A ist, das Produkt $g(x)$ der l zu $f(x)$ konjugiert algebraischen Funktionen $f, f', \cdots, f^{(l-1)}$, so wird $g(x)$ eine ganze rationale Funktion des Grades $n'l$ mit rationalen Koeffizienten, die für $x = \rho$ verschwindet. Die Gleichung niedrigsten Grades, der ρ im Gebiete der rationalen Zahlen genügt, ist aber die $m$te Kreisteilungsgleichung, deren Grad gleich $\varphi(m)$ ist. Daher ist

$$\varphi(m) \leqq n'l \leqq nl.$$

Hieraus folgt aber, daß m bei festgehaltenem n eine gewisse endliche Schranke nicht überschreiten kann. Daher kommen für die Einheitswurzel ρ nur endlich viele Werte in Betracht.

Es sei nun $\mathfrak{G}$ eine beliebige periodische Substitutionsgruppe in n Variabeln. Man wähle in $\mathfrak{G}$ irgendwelche endlich viele Elemente $H_1, H_2, \cdots, H_r$ und betrachte die durch sie erzeugte Untergruppe $\mathfrak{H}$ von $\mathfrak{G}$. Die Koeffizienten einer Substitution H von $\mathfrak{H}$ sind dann gewisse ganze rationale Funktionen der $p = n^2 r$ Koeffizienten $\omega_1, \omega_2, \cdots, \omega_p$ von $H_1, H_2, \cdots, H_r$, also in dem Körper $K = P(\omega_1, \omega_2, \cdots, \omega_p)$ enthalten. Die charakteristische Gleichung $|H - xE| = 0$ von H ist daher eine Gleichung $n$ten Grades, deren Koeffizienten dem Körper K angehören. Die Wurzeln dieser Gleichung sind aber, da H als Element von $\mathfrak{G}$ periodisch ist, Einheitswurzeln. Aus dem oben bewiesenen Hilfssatz ergibt sich daher, daß unter den charakteristischen Wurzeln aller Substitutionen H von $\mathfrak{H}$ nur endlich viele voneinander verschiedene Größen vorkommen. Folglich kommen auch für die Ordnungen der Substitutionen H nur endlich viele Werte in Betracht. Nach dem Burnsideschen Kriterium ist daher $\mathfrak{H}$ eine endliche Gruppe.

§ 2.

Eine Gruppe $\mathfrak{G}$ linearer homogener Substitutionen in n Variabeln wird als *irreduzibel* bezeichnet, wenn sich kein System von $m < n$ Linearformen $y_1, y_2, \cdots, y_m$ angeben läßt, die durch alle Substitutionen von $\mathfrak{G}$ untereinander linear transformiert werden. Die Gruppe $\mathfrak{G}$ heißt ferner *vollständig reduzibel*, wenn sie sich durch eine lineare Transformation P der Variabeln in eine mit ihr ähnliche Gruppe $\mathfrak{G}' = P\mathfrak{G}P^{-1}$ überführen läßt, welche die Form

$$\mathfrak{G}' = \begin{pmatrix} \mathfrak{G}_1 & 0 & \cdots & 0 \\ 0 & \mathfrak{G}_2 & \cdots & 0 \\ . & . & \cdots & \\ 0 & 0 & \cdots & \mathfrak{G}_k \end{pmatrix}$$

besitzt, wo $\mathfrak{G}_1, \mathfrak{G}_2, \cdots, \mathfrak{G}_k$ irreduzible Gruppen sind. Diese k Gruppen sind, wenn ähnliche Gruppen als nicht voneinander verschieden angesehen werden, durch die Gruppe $\mathfrak{G}$ bis auf die Reihenfolge eindeutig bestimmt und werden als die *irreduziblen Bestandteile* von $\mathfrak{G}$ bezeichnet. Zu den vollständig reduziblen Gruppen gehören insbesondere die endlichen Gruppen und allgemeiner alle Hermiteschen Gruppen. Umgekehrt ist jede vollständig reduzible Gruppe, deren irreduzible Bestandteile Hermitesche Gruppen sind, selbst eine Hermitesche Gruppe[1].

Um nun den Satz II zu beweisen, genügt es zu zeigen:

1. Jede periodische Substitutionsgruppe $\mathfrak{G}$ ist vollständig reduzibel.

2. Eine irreduzible periodische Substitutionsgruppe ist eine Hermitesche Gruppe.

Es sei nämlich r die Anzahl der linear unabhängigen Substitutionen in der Gruppe $\mathfrak{G}$. Besitzen dann die Substitutionen $H_1, H_2, \cdots, H_r$ diese Eigenschaft, so läßt sich jedes Element G von $\mathfrak{G}$ in der Form

$$G = a_1 H_1 + a_2 H_2 + \cdots + a_r H_r \tag{4.}$$

darstellen. Nun ist aber die durch $H_1, H_2, \cdots, H_r$ erzeugte Untergruppe $\mathfrak{H}$ von $\mathfrak{G}$ nach Satz I eine endliche Gruppe, also vollständig reduzibel. Ist insbesondere $\mathfrak{H}$ eine irreduzible Gruppe, so ist a fortiori auch $\mathfrak{G}$ irreduzibel. Im anderen Falle bestimme man die lineare Transformation P der Variabeln, so daß $P\mathfrak{H}P^{-1}$ vollständig zerfällt. Dann lehrt uns die aus (4.) hervorgehende Gleichung

$$PGP^{-1} = a_1 PH_1P^{-1} + a_2 PH_2 P^{-1} + \cdots + a_r PH_r P^{-1},$$

daß PGP^{-1} in derselben Weise zerfällt wie die Substitutionen der Gruppe $P\mathfrak{H}P^{-1}$. Da dies für jedes Element G von $\mathfrak{G}$ gilt, so ist $\mathfrak{G}$ eine vollständig reduzible Gruppe. Zugleich ergibt sich, daß $\mathfrak{G}$ nur dann irreduzibel ist, wenn unter den endlichen Untergruppen von $\mathfrak{G}$ auch irreduzible Gruppen vorkommen.

Es sei nun $\mathfrak{G}$ eine irreduzible periodische Substitutionsgruppe, $\mathfrak{H}$ eine irreduzible endliche Untergruppe der Ordnung h von $\mathfrak{G}$. Wir können dann jedenfalls eine positiv definite Hermitesche Form F angeben, die durch alle Substitutionen von $\mathfrak{H}$ in sich transformiert wird. Diese Form F ist ferner, da $\mathfrak{H}$ irreduzibel ist, bis auf einen konstanten Faktor eindeutig bestimmt[2]. Ich behaupte nun, daß auch jede beliebige Substitution G von $\mathfrak{G}$ die Form F ungeändert läßt. In der Tat sei $\mathfrak{H}'$ die durch die h Elemente von $\mathfrak{H}$ und das Element G erzeugte Gruppe. Da $\mathfrak{H}'$ nach Satz I wieder eine endliche Gruppe ist,

---

[1] Vgl. den Artikel von Hrn. A. Loewy in Pascals Repertorium der höheren Mathematik, 2. Auflage, Bd. I, Kap. III, § 9.

[2] Vgl. W. Burnside, *On the reduction of a group of homogeneous linear substitutions of finite order*, Acta Mathematica, Bd. 28, S. 369; ferner G. Frobenius und I. Schur, *Über die reellen Darstellungen der endlichen Gruppen*, Sitzungsberichte 1906, S. 186.

so gibt es eine positiv definite Hermitesche Form F', die durch alle Substitutionen von $\mathfrak{H}'$ in sich übergeführt wird. Diese Form wird aber speziell auch durch die Substitutionen von $\mathfrak{H}$ nicht geändert und muß sich daher von F nur um einen konstanten Faktor unterscheiden. Hieraus folgt aber, daß die Substitution G die Form F in sich transformiert. Hiermit ist der Satz II vollständig bewiesen.

Bestimmt man, wenn F eine positiv definite Hermitesche Form der Variabeln $x_1, x_2, \cdots, x_n$ ist, die durch alle Substitutionen einer Gruppe $\mathfrak{G}$ nicht geändert wird, die lineare Transformation

$$(P) \qquad y_k = p_{k1}x_1 + p_{k2}x_2 + \cdots + p_{kn}x_n, \qquad (k = 1, 2, \cdots, n)$$

so daß

$$F = |y_1|^2 + |y_2|^2 + \cdots + |y_n|^2$$

wird, so führen die Substitutionen der mit $\mathfrak{G}$ ähnlichen Gruppe $P\mathfrak{G}P^{-1}$ die Hermitesche Einheitsform in sich über. Eine solche lineare Substitution bezeichnet man als *unitär*. Der Satz II läßt sich daher auch folgendermaßen aussprechen:

II*. *Jede periodische Substitutionsgruppe ist einer Gruppe unitärer Substitutionen ähnlich.*

Eine unendliche Gruppe unitärer Substitutionen besitzt stets infinitesimale Operationen, d. h. es läßt sich zu jedem positiven ε eine von der identischen Substitution E verschiedene Substitution A der Gruppe angeben, deren Koeffizienten sich dem absoluten Betrage nach von den Koeffizienten von E um weniger als ε unterscheiden[1]. Aus II* ergibt sich daher:

Eine periodische Substitutionsgruppe, die keine infinitesimalen Operationen enthält, ist eine endliche Gruppe.

§ 3.

Nach dem Vorgange von Hrn. Frobenius (vgl. die in der Einleitung zitierte Arbeit) soll, wenn S eine beliebige lineare Substitution in n Variabeln ist, die Quadratsumme der absoluten Beträge der n^2 Koeffizienten von S mit $\vartheta(S)$ bezeichnet werden. Für jede unitäre Substitution U ist dann

$$\vartheta(S) = \vartheta(US) = \vartheta(SU) = \vartheta(U^{-1}SU).$$

Es gilt ferner, wie Hr. Frobenius gezeigt hat, folgender Satz:

Sind A und B zwei unitäre Substitutionen einer endlichen Gruppe und ist

$$(5.) \qquad \vartheta(E-A) < \frac{1}{2}, \qquad \vartheta(E-B) < 4,$$

so ist A mit B vertauschbar.

[1] Vgl. L. Bieberbach, *Über die Bewegungsgruppen der Euklidischen Räume* (§ 9), Math. Annalen, Bd. 70, S. 297.

Der Beweis des Satzes III ergibt sich nun, indem man die von Hrn. Frobenius für den Fall einer endlichen Gruppe durchgeführte Betrachtung fast wörtlich wiederholt. Wir können wegen II* annehmen, daß die Substitutionen der zu betrachtenden periodischen Substitutionsgruppe $\mathfrak{G}$ sämtlich unitär sind. Je zwei Substitutionen A, B von $\mathfrak{G}$, die den Bedingungen (5.) genügen, sind dann, da sie in einer endlichen Gruppe, nämlich in der durch sie erzeugten Gruppe, enthalten sind, untereinander vertauschbar. Diejenigen Elemente A von $\mathfrak{G}$, für die

$$\mathfrak{S}(E-A) < \frac{1}{2} \tag{6.}$$

ist, erzeugen daher eine Abelsche Untergruppe $\mathfrak{A}$ von $\mathfrak{G}$. Da ferner für jedes Element G von $\mathfrak{G}$ aus (6.) auch

$$\mathfrak{S}(E-G^{-1}AG) = \mathfrak{S}[G^{-1}(E-A)G] = \mathfrak{S}(E-A) < \frac{1}{2}$$

folgt, so ist $\mathfrak{A}$ eine invariante Untergruppe von $\mathfrak{G}$. Zwei Elemente R und S von $\mathfrak{G}$, für welche die Komplexe $\mathfrak{A}R$ und $\mathfrak{A}S$ voneinander verschieden sind, müssen der Bedingung $\mathfrak{S}(R-S) \geqq \frac{1}{2}$ genügen. Denn andernfalls würde sich

$$\mathfrak{S}(E-SR^{-1}) = \mathfrak{S}[(R-S)R^{-1}] = \mathfrak{S}(R-S) < \frac{1}{2}$$

ergeben, d. h. SR^{-1} müßte in $\mathfrak{A}$ enthalten sein; hieraus würde aber $\mathfrak{A}R = \mathfrak{A}S$ folgen. Die Anzahl der unitären Substitutionen $R_1, R_2, \cdots$ in n Variabeln, für die

$$\mathfrak{S}(R_\alpha - R_\beta) \geqq \frac{1}{2}$$

wird, ist aber endlich und zwar kleiner als $\lambda_n = (\sqrt{8n}+1)^{2n^2}$. Daher ist der Index p der Untergruppe $\mathfrak{A}$ von $\mathfrak{G}$ endlich und kleiner als λ_n.

Wird nun

$$\mathfrak{G} = \mathfrak{A}R_1 + \mathfrak{A}R_2 + \cdots + \mathfrak{A}R_p,$$

so erzeugen $R_1, R_2, \cdots R_p$ eine endliche Untergruppe $\mathfrak{H}$ von $\mathfrak{G}$. Die Elemente dieser endlichen Gruppe erzeugen dann zusammen mit den Elementen von $\mathfrak{A}$ die ganze Gruppe $\mathfrak{G}$.

§ 4.

Nimmt man die endlichen Substitutionsgruppen als bekannt an, so läßt sich auf Grund des Satzes III ein Verfahren angeben, auch alle unendlichen periodischen Substitutionsgruppen aufzustellen.

Es sei nämlich $\mathfrak{K}$ eine beliebige endliche Gruppe linearer homogener Substitutionen in n Variabeln, $\mathfrak{B}$ irgendeine invariante Abelsche Untergruppe von $\mathfrak{K}$; hierbei kann $\mathfrak{B}$ auch die Ordnung 1 besitzen, d. h. nur die identische Substitution E enthalten. Wir können die Gruppe $\mathfrak{K}$ durch eine mit ihr ähnliche Gruppe $\mathfrak{K}' = P\mathfrak{K}P^{-1}$ ersetzen, in der jedes Element B der Untergruppe $\mathfrak{B}' = P\mathfrak{B}P^{-1}$ die Normalform

$$(7.)\qquad x'_1 = \beta_1 x_1, \quad x'_2 = \beta_2 x_2, \quad \cdots, \quad x'_n = \beta_n x_n$$

besitzt. Man wähle nun irgendwelche Substitutionen $R, S, \cdots$, von denen jede die Form

$$(8.)\qquad x'_1 = \rho_1 x_1, \quad x'_2 = \rho_2 x_2, \quad \cdots, \quad x'_n = \rho_n x_n$$

hat, wo $\rho_1, \rho_2, \cdots, \rho_n$ beliebige Einheitswurzeln sind, die nur der Bedingung unterliegen, daß stets $\rho_\varkappa = \rho_\lambda$ sein soll, wenn in allen Substitutionen B von $\mathfrak{B}'$ die Zahl $\beta_\varkappa$ gleich β_λ ist. Es soll nun gezeigt werden:

1. Die durch die Substitutionen $R, S, \cdots$ und die Elemente von $\mathfrak{K}'$ erzeugte Gruppe $\mathfrak{G}'$ ist eine periodische Substitutionsgruppe.

2. Jede periodische Substitutionsgruppe $\mathfrak{G}$ ist einer Gruppe $\mathfrak{G}'$ ähnlich, die in der geschilderten Weise aus einer endlichen Gruppe $\mathfrak{K}$ hervorgeht.

Man betrachte nämlich die durch die Substitutionen $R, S, \cdots$ und die Elemente von $\mathfrak{B}'$ erzeugte Abelsche Gruppe $\mathfrak{C}$. Diese Gruppe ist vollständig zerfallend und besitzt, wegen der über die Substitutionen $R, S, \cdots$ gemachten Voraussetzung, genau ebenso viele voneinander verschiedene irreduzible Bestandteile wie die Gruppe $\mathfrak{B}'$. In einer vollständig reduziblen Abelschen Gruppe ist aber bekanntlich die Anzahl der linear unabhängigen Substitutionen gleich der Anzahl der voneinander verschiedenen irreduziblen Bestandteile der Gruppe. Daher ist jede der Substitutionen $R, S, \cdots$ von den Elementen der Gruppe $\mathfrak{B}'$ linear abhängig. Da ferner $\mathfrak{B}'$ eine invariante Untergruppe von $\mathfrak{K}$ sein soll, so sind für jedes Element K von $\mathfrak{K}$ auch die Substitutionen

$$(9.)\qquad K^{-1}RK, \quad K^{-1}SK, \quad \cdots$$

als lineare homogene Verbindungen der Elemente von $\mathfrak{B}'$ darstellbar; daher besitzen sie sämtlich die Normalform (8.). Läßt man nun in (9.) das Element K alle Substitutionen der Gruppe $\mathfrak{K}$ durchlaufen, so erzeugen die so entstehenden Substitutionen, da sie periodisch und untereinander vertauschbar sind, eine periodische Substitutionsgruppe $\mathfrak{D}$, die durch jedes Element von $\mathfrak{K}$ in sich transformiert wird. Beachtet man, daß jedes Element Q der zu betrachtenden Gruppe $\mathfrak{G}'$ die Form DK hat, wo D in der Gruppe $\mathfrak{D}$ und K in der Gruppe $\mathfrak{K}$ enthalten ist, so ergibt sich unmittelbar, daß $\mathfrak{G}'$ eine periodische Substitutionsgruppe ist.

Es sei nun $\mathfrak{G}$ eine beliebige periodische Substitutionsgruppe, für welche die Untergruppen $\mathfrak{A}$ und $\mathfrak{H}$ dieselbe Bedeutung haben mögen wie in § 3. In der Abelschen Gruppe $\mathfrak{A}$ bestimme man m Elemente $B_1, B_2, \cdots, B_m$, durch die sich alle übrigen Elemente von $\mathfrak{A}$ linear und homogen darstellen lassen. Es bedeute $\mathfrak{K}$ die durch $B_1, B_2, \cdots, B_m$ und die Elemente von $\mathfrak{H}$ erzeugte Gruppe, ferner sei $\mathfrak{B}$ der größte gemeinsame Teiler von $\mathfrak{A}$ und $\mathfrak{K}$. Dann ist $\mathfrak{K}$ eine endliche Gruppe, in der $\mathfrak{B}$ als invariante Abelsche Untergruppe enthalten ist. Man bestimme nun eine lineare Transformation P der Variabeln, so daß die Substitutionen B der Gruppe $P\mathfrak{B}P^{-1}$ die Normalform (7.) erhalten, und betrachte die Gruppe $\mathfrak{G}' = P\mathfrak{G}P^{-1}$. Diese Gruppe wird durch die von den Elementen B von $\mathfrak{B}'$ verschiedenen Substitutionen $R, S, \cdots$ der Untergruppe $\mathfrak{A}' = P\mathfrak{A}P^{-1}$ und die Substitutionen der endlichen Gruppe $\mathfrak{K}' = P\mathfrak{K}P^{-1}$ erzeugt. Da aber $R, S, \cdots$ als lineare homogene Verbindungen der Elemente B von $\mathfrak{B}'$ darstellbar sind, so zerfallen sie in derselben Weise wie die Substitutionen B. Dies lehrt uns aber, daß die zu $\mathfrak{G}$ ähnliche Gruppe $\mathfrak{G}'$ in der vorhin angegebenen Weise aus einer endlichen Gruppe $\mathfrak{K}$ abgeleitet werden kann.

18.
Über Gruppen linearer Substitutionen mit Koeffizienten aus einem algebraischen Zahlkörper

Mathematische Annalen 71, 355 - 367 (1911)

In seiner Arbeit „On the arithmetical nature of the coefficients in a group of linear substitutions (Third Paper)"*) hat W. Burnside folgenden wichtigen Satz bewiesen: *Jede endliche Gruppe linearer Substitutionen in n Variabeln, deren Koeffizienten einem algebraischen Zahlkörper K angehören, in dem die Anzahl der Idealklassen gleich* 1 *ist, läßt sich durch eine lineare Transformation der Variabeln in eine Gruppe überführen, deren Koeffizienten ganze Zahlen des Körpers K sind.* Insbesondere kann also jede endliche Substitutionsgruppe mit rationalen Koeffizienten bei passender Wahl der Variabeln auch als Gruppe mit ganzen rationalen Koeffizienten dargestellt werden.

Die schöne und weittragende Theorie, die mein Freund E. Steinitz in der vorstehenden Arbeit entwickelt hat, gestattet nun, wie im folgenden gezeigt werden soll, den Burnsideschen Satz in bemerkenswerter Weise zu verallgemeinern. Ich beweise insbesondere, daß dieser Satz auch dann gilt, wenn nur bekannt ist, daß die Anzahl n der Variabeln zu der Anzahl der Idealklassen des Körpers K teilerfremd ist. Für jedes n läßt sich ein Relativkörper K_n in bezug auf K bestimmen, sodaß jede endliche Substitutionsgruppe in n Variabeln, deren Koeffizienten dem Körper K angehören, im Körper K_n in eine Gruppe mit ganzzahligen Koeffizienten linear transformiert werden kann.

---

*) London Math. Soc. Proc. (2) 7 (1908), S. 8.

Abschnitt I.

Über Moduln von Linearformen in einem algebraischen Zahlkörper.

§ 1.

Es sei K ein gegebener (endlicher) algebraischer Zahlkörper. Ein System $\mathfrak{M}$ von Linearformen in n unabhängigen Variabeln nenne ich einen *Modul vom Range n im Körper K,* wenn folgende Bedingungen erfüllt sind:

a) Die Koeffizienten jeder Linearform von $\mathfrak{M}$ sind ganze Zahlen des Körpers K.

b) Sind y_1, y_2 zwei Linearformen von $\mathfrak{M}$ und α_1, α_2 zwei beliebige ganze Zahlen von K, so ist auch die Linearform $\alpha_1 y_1 + \alpha_2 y_2$ in $\mathfrak{M}$ enthalten.

c) Unter den Linearformen von $\mathfrak{M}$ lassen sich n linear unabhängige angeben.

Denkt man sich für je n Linearformen von $\mathfrak{M}$ die Koeffizientendeterminante gebildet und bestimmt den größten gemeinsamen Idealteiler aller dieser Determinanten, so erhält man ein durch den Modul $\mathfrak{M}$ eindeutig charakterisiertes Ideal $\mathfrak{d}(\mathfrak{M})$ des Körpers K, das ich als den *Determinantenteiler von* $\mathfrak{M}$ bezeichne.

Ist $\mathfrak{M}_1$ irgend ein System von Linearformen, die sämtlich in $\mathfrak{M}$ enthalten sind, so soll, wie in der Theorie der Dedekindschen Moduln, $\mathfrak{M}_1$ *durch* $\mathfrak{M}$ *teilbar* heißen, was auch durch das Zeichen

$$\mathfrak{M}_1 > \mathfrak{M}$$

ausgedrückt wird.

Von Wichtigkeit sind für das folgende einige spezielle Arten von Moduln.

Sind $\mathfrak{a}_1, \mathfrak{a}_2, \cdots, \mathfrak{a}_n$ beliebige n Ideale des Körpers K und läßt man in

$$\alpha_1 x_1 + \alpha_2 x_2 + \cdots + \alpha_n x_n$$

α_1 alle durch $\mathfrak{a}_1$, ebenso α_2 alle durch $\mathfrak{a}_2$ teilbaren Zahlen durchlaufen usw., so bilden diese Linearformen einen Modul n^{ten} Ranges in K, der mit

$$[\mathfrak{a}_1, \mathfrak{a}_2, \cdots, \mathfrak{a}_n]$$

bezeichnet werden soll. Der Determinantenteiler dieses Moduls ist, wie man leicht erkennt, gleich $\mathfrak{a}_1 \mathfrak{a}_2 \cdots \mathfrak{a}_n$. Wird speziell

$$\mathfrak{a}_1 = 1,\ \mathfrak{a}_2 = 1,\ \cdots,\ \mathfrak{a}_{n-1} = 1,\ \mathfrak{a}_n = \mathfrak{a},$$

so bezeichne ich den zugehörigen Modul auch kürzer mit $\mathfrak{M}_n(\mathfrak{a})$. Für $\mathfrak{a} = 1$ wird

$$\mathfrak{E} = \mathfrak{M}_n(1) = [1, 1, \cdots, 1]$$

der die Gesamtheit aller Linearformen mit ganzzahligen Koeffizienten aus dem Körper K umfassende *Einheitsmodul* n^{ten} *Ranges.*

Allgemeiner habe man m Moduln $\mathfrak{M}_1, \mathfrak{M}_2, \cdots, \mathfrak{M}_m$ im Körper K, deren Rangzahlen $n_1, n_2, \cdots, n_m$ seien. Die $n = n_1 + n_2 + \cdots + n_m$ in den Linearformen der m Moduln auftretenden Variabeln setzen wir als voneinander unabhängig voraus. Das System der Linearformen

$$y = y_1 + y_2 + \cdots + y_m,$$

die man erhält, indem man y_1 alle Linearformen von $\mathfrak{M}_1$, y_2 alle Linearformen von $\mathfrak{M}_2$, usw., durchlaufen läßt, repräsentiert dann einen Modul $\mathfrak{M}$ vom Range n. Diesen *zerfallenden* Modul bezeichne ich mit

$$\mathfrak{M} = [\mathfrak{M}_1, \mathfrak{M}_2, \cdots, \mathfrak{M}_m].$$

Aus elementaren Sätzen der Determinantentheorie ergibt sich leicht, daß

$$\mathfrak{d}(\mathfrak{M}) = \mathfrak{d}(\mathfrak{M}_1)\,\mathfrak{d}(\mathfrak{M}_2) \cdots \mathfrak{d}(\mathfrak{M}_m)$$

wird.

Es sei ferner $\mathfrak{M}$ ein beliebiger Modul n^{ten} Ranges und $\mathfrak{q}$ ein Ideal des Körpers K. Man denke sich alle Produkte $\varkappa y$ der Zahlen $\varkappa$ von $\mathfrak{q}$ und der Linearformen y von $\mathfrak{M}$ und alle Summen von je endlich vielen dieser Produkte gebildet. Die Gesamtheit der so entstehenden Linearformen ist dann wieder ein Modul n^{ten} Ranges in K, den ich mit $\mathfrak{q}\mathfrak{M}$ bezeichne. Man erkennt ohne Mühe, daß

$$\mathfrak{d}(\mathfrak{q}\mathfrak{M}) = \mathfrak{q}^n\,\mathfrak{d}(\mathfrak{M})$$

ist.

§ 2.

Gegeben seien endlich oder unendlich viele Linearformen $y_1, y_2, \cdots$ der Variabeln $x_1, x_2, \cdots, x_n$, deren Koeffizienten ganze Zahlen des Körpers K sind; unter diesen Formen sollen n linear unabhängige vorkommen. Man denke sich alle Linearformen aufgestellt, die sich als lineare Verbindungen von irgendwelchen endlich vielen unter den Formen y_ν darstellen lassen, wobei als Koeffizienten wieder nur ganze Zahlen aus dem Körper K in betracht kommen sollen. Die Gesamtheit dieser Linearformen bildet dann einen Modul vom Range n in K, *den durch die Formen* $y_1, y_2, \cdots$ *erzeugten Modul.* Der Determinantenteiler dieses Moduls ist offenbar nichts anderes als der größte gemeinsame Idealteiler der Koeffizientendeterminanten von je n unter den Linearformen y_ν.

In Nr. 31 seiner Arbeit hat nun E. Steinitz einen fundamentalen Satz bewiesen, aus dem sich speziell für die von uns betrachteten Moduln folgendes ergibt:

In jedem Modul $\mathfrak{M}$ vom Range n im Körper K lassen sich $n + 1$ Linearformen

$$y_1, y_2, \cdots, y_{n-1}, y'_n, y''_n$$

angeben, durch die der Modul erzeugt wird. Diese Linearformen können *stets so gewählt werden, daß y'_n und y''_n sich voneinander nur um einen konstanten Faktor unterscheiden.*

Ist nun etwa

$$\alpha' y''_n = \alpha'' y'_n ,$$

wo α' und α'' ganze Zahlen von K sind, und nimmt man, was keine Beschränkung der Allgemeinheit bedeutet, α' als von Null verschieden an, so werde $y_n = \frac{1}{\alpha'} y'_n$, also

$$y'_n = \alpha' y_n , \quad y''_n = \alpha'' y_n$$

gesetzt. Dann werden alle Linearformen y von $\mathfrak{M}$ erhalten, indem man in

$$y = \xi_1 y_1 + \cdots + \xi_{n-1} y_{n-1} + (\alpha' \xi' + \alpha'' \xi'') y_n$$

$\xi_1, \cdots, \xi_{n-1}, \xi', \xi''$ alle ganzen Zahlen von K durchlaufen läßt. Oder einfacher: man erhält sämtliche Linearformen y von $\mathfrak{M}$, indem man in

$$y = \xi_1 y_1 + \xi_2 y_2 + \cdots + \xi_n y_n$$

für $\xi_1, \cdots, \xi_{n-1}$ alle ganzen Zahlen von K und für ξ_n alle durch den größten gemeinsamen Idealteiler $\mathfrak{a}$ von α' und α'' teilbaren Zahlen einsetzt. Das Ideal $\mathfrak{a}$ ist hierbei dem Ideal $\mathfrak{d}(\mathfrak{M})$ äquivalent; denn ist η die Koeffizientendeterminante der Formen $y_1, y_2, \cdots, y_n$, so wird, wie man ohne Mühe erkennt, $\mathfrak{d}(\mathfrak{M}) = \mathfrak{a}\eta$. Da ferner α' und α'' nur bis auf einen Proportionalitätsfaktor bestimmt sind, so kann $\mathfrak{a}$ durch jedes äquivalente Ideal ersetzt werden. Der Steinitzsche Satz besagt also, anders ausgedrückt:

I. *Ist $\mathfrak{M}$ ein Modul vom Range n im Körper K und ist $\mathfrak{a}$ ein beliebiges Ideal der durch den Determinantenteiler $\mathfrak{d}(\mathfrak{M})$ von $\mathfrak{M}$ bestimmten Idealklasse, so können n (linear unabhängige) Linearformen $y_1, y_2, \cdots, y_n$ mit Koeffizienten aus dem Körper K bestimmt werden, sodaß alle Linearformen y von $\mathfrak{M}$ erhalten werden, indem man in*

$$y = \xi_1 y_1 + \cdots + \xi_{n-1} y_{n-1} + \xi_n y_n$$

$\xi_1, \cdots, \xi_{n-1}$ alle ganzen Zahlen und ξ_n alle durch $\mathfrak{a}$ teilbaren ganzen Zahlen von K durchlaufen läßt.

§ 3.

Der Satz I läßt sich noch einfacher formulieren, wenn man den Begriff der Äquivalenz zweier Moduln einführt.

Es sei $\mathfrak{S}$ ein beliebiges System von Linearformen

$$y = s_1 x_1 + s_2 x_2 + \cdots + s_n x_n$$

und $P = (p_{\varkappa\lambda})$ eine gegebene lineare Substitution. Das System $\mathfrak{S}'$ der Linearformen

$$y' = s_1 \sum p_{1\lambda} x_\lambda + s_2 \sum p_{2\lambda} x_\lambda + \cdots + s_n \sum p_{n\lambda} x_\lambda ,$$

das aus $\mathfrak{S}$ durch Anwendung der Substitution P hervorgeht, bezeichne ich kurz mit $\mathfrak{S}P$. Ist die Determinante von P nicht Null, so ist umgekehrt $\mathfrak{S} = \mathfrak{S}' P^{-1}$.

Sind nun $\mathfrak{M}$ und $\mathfrak{M}'$ zwei Moduln n^{ten} Ranges im Körper K, so kann es eintreten, daß sich eine lineare Substitution mit ganzen oder gebrochenen Koeffizienten aus K angeben läßt, sodaß $\mathfrak{M}' = \mathfrak{M}P$ wird.*) Ich bezeichne dann $\mathfrak{M}$ und $\mathfrak{M}'$ als *äquivalente Moduln*. Da aus $\mathfrak{M}' = \mathfrak{M}P$, $\mathfrak{M}'' = \mathfrak{M}Q$

$$\mathfrak{M}'' = \mathfrak{M}'(P^{-1}Q)$$

folgt, so sind zwei Moduln, die einem dritten äquivalent sind, auch untereinander äquivalent. Die Gesamtheit der einander äquivalenten Moduln bildet eine *Modulklasse*.

Der Satz I ist nun, wie man unmittelbar erkennt, seinem Inhalt nach völlig identisch mit folgendem Satz:

I*. *Ist $\mathfrak{M}$ ein Modul vom Range n im Körper K und ist $\mathfrak{a}$ ein beliebiges Ideal der durch den Determinantenteiler $\mathfrak{d}(\mathfrak{M})$ von $\mathfrak{M}$ bestimmten Idealklasse, so ist $\mathfrak{M}$ dem Modul $\mathfrak{M}_n(\mathfrak{a})$ äquivalent.*

Auf Grund dieses Satzes läßt sich ein einfaches Kriterium für die Äquivalenz zweier Moduln angeben und zugleich die Anzahl der Modulklassen bestimmen:

II. *Zwei Moduln n^{ten} Ranges im Körper K sind dann und nur dann einander äquivalent, wenn ihre Determinantenteiler derselben Idealklasse angehören. Die Anzahl der Klassen, in welche die Moduln n^{ten} Ranges zerfallen, ist für jedes n gleich der Anzahl h der Idealklassen des Körpers K.*

Geht nämlich $\mathfrak{M}'$ aus $\mathfrak{M}$ durch lineare Substitution P hervor, so ist offenbar

$$\mathfrak{d}(\mathfrak{M}') = \mathfrak{d}(\mathfrak{M}) \cdot |P|$$

wo $|P|$ die Determinante von P bedeutet. Daher sind $\mathfrak{d}(\mathfrak{M})$ und $\mathfrak{d}(\mathfrak{M}')$ äquivalente Ideale. Ist aber diese Bedingung erfüllt, so sei $\mathfrak{a}$ irgend ein Ideal der durch $\mathfrak{d}(\mathfrak{M})$ und $\mathfrak{d}(\mathfrak{M}')$ bestimmten Idealklasse. Dann ist nach Satz I* jeder der Moduln $\mathfrak{M}$ und $\mathfrak{M}'$ dem Modul $\mathfrak{M}_n(\mathfrak{a})$ äquivalent, folglich sind sie auch einander äquivalent.

Man denke sich nun, wenn $\mathfrak{a}_1, \mathfrak{a}_2, \cdots, \mathfrak{a}_h$ Repräsentanten der h Idealklassen von K sind, die h Moduln

$$\mathfrak{M}_n(\mathfrak{a}_1), \mathfrak{M}_n(\mathfrak{a}_2), \cdots, \mathfrak{M}_n(\mathfrak{a}_h)$$

gebildet. Da der Determinantenteiler von $\mathfrak{M}_n(\mathfrak{a}_\lambda)$ gleich $\mathfrak{a}_\lambda$ ist, so sind unter diesen Moduln nicht zwei einander äquivalent; ferner ist wieder nach Satz I* jeder Modul n^{ten} Ranges einem unter ihnen äquivalent. Es gibt also in der Tat h und nicht mehr als h Modulklassen.

Daß die Anzahl der Modulklassen für jedes n endlich ist, war mir bereits seit langer Zeit bekannt. Auf die überraschende Tatsache, daß

*) Die Determinante der Substitution P ist dann von selbst nicht Null.

für die Äquivalenz zweier Moduln die Äquivalenz ihrer Determinantenteiler hinreichend und daher die Anzahl der Modulklassen von n unabhängig ist, hat mich jedoch erst E. Steinitz aufmerksam gemacht. Im Einverständnis mit ihm publiziere ich den Satz mit der von mir vorgeschlagenen Beweisanordnung.

Abschnitt II.

Anwendung auf die Theorie der Gruppen linearer Substitutionen.

§ 4.

Unter einer *Substitutionsgruppe n^{ten} Grades* verstehe ich im folgenden ein endliches oder unendliches System $\mathfrak{G}$ von linearen homogenen Substitutionen in n Variabeln, welches die Eigenschaft besitzt, daß das Produkt von je zwei Substitutionen von $\mathfrak{G}$ wieder in $\mathfrak{G}$ enthalten ist. Es wird also insbesondere nicht vorausgesetzt, daß die Determinanten der Substitutionen von $\mathfrak{G}$ nicht verschwinden sollen. Ist P eine lineare Substitution von nicht verschwindender Determinante und ersetzt man jede Substitution G von $\mathfrak{G}$ durch $P^{-1}GP$, so bilden diese Substitutionen eine mit $\mathfrak{G}$ ähnliche Gruppe, die ich kurz mit $P^{-1}\mathfrak{G}P$ bezeichne.

Eine Substitutionsgruppe $\mathfrak{G}$, deren Koeffizienten Zahlen des algebraischen Zahlkörpers K sind, nenne ich eine *in K rationale Gruppe.* Läßt sich insbesondere eine (ganze) Zahl μ von K angeben, sodaß für jeden Koeffizienten γ der Gruppe $\mathfrak{G}$ das Produkt $\mu\gamma$ eine ganze Zahl wird, so soll $\mathfrak{G}$ eine *Gruppe mit beschränkten Nennern* heißen. Man kann diese Eigenschaft von $\mathfrak{G}$ noch anders ausdrücken: denkt man sich jeden Koeffizienten γ als reduzierten Idealbruch $\frac{\mathfrak{m}}{\mathfrak{n}}$ dargestellt, so sollen unter den Nennern $\mathfrak{n}$ nur endlich viele Ideale vorkommen. Das kleinste gemeinsame Vielfache dieser Ideale bezeichne ich als den *Generalnenner der Gruppe.*

Zu den Gruppen mit beschränkten Nennern gehören insbesondere die endlichen Gruppen. Ein anderes wichtiges Beispiel erhält man folgendermaßen: Es sei $\mathfrak{a}$ ein (ganzes) Ideal des Körpers K; man bilde dann alle Substitutionen

$$A = \begin{pmatrix} a_{11}, & a_{12}, & \cdots, & a_{1n} \\ a_{21}, & a_{22}, & \cdots, & a_{2n} \\ \cdot & \cdot & \cdot & \cdot \\ a_{n1}, & a_{n2}, & \cdots, & a_{nn} \end{pmatrix},$$

in denen die $(n-1)^2+1$ Koeffizienten

$$a_{11}, \cdots, a_{1\,n-1}, a_{21}, \cdots, a_{2,n-1}, \cdots, a_{n-1,1}, \cdots, a_{n-1\,n-1}, a_{nn}$$

ganze Zahlen des Körpers K, ferner

$$a_{1,n},\ a_{2,n},\ \cdots,\ a_{n-1,n}$$

durch $\mathfrak{a}$ teilbare ganze Zahlen und endlich

$$a_{n1},\ a_{n2},\ \cdots,\ a_{n,n-1}$$

ganze oder gebrochene Zahlen von K sind, die sich als Idealbrüche mit dem Nenner $\mathfrak{a}$ schreiben lassen. Es sollen also, anders ausgedrückt, für *jede* Zahl α des Ideals $\mathfrak{a}$ die Produkte

$$\alpha a_{n1},\ \alpha a_{n2},\ \cdots,\ \alpha a_{n,n-1}$$

ganze Zahlen von K sein. Die Gesamtheit dieser Substitutionen A bildet offenbar eine Gruppe. Diese durch das Ideal $\mathfrak{a}$ und die Zahl n eindeutig bestimmte (in K rationale) Gruppe bezeichne ich mit $\mathfrak{G}_n(\mathfrak{a})$. Speziell ist $\mathfrak{G}_n(1)$ die durch die Gesamtheit aller Substitutionen mit ganzzahligen Koeffizienten aus dem Körper K gebildete Gruppe.

§ 5.

Ist $\mathfrak{G}$ eine beliebige in K rationale Substitutionsgruppe n^{ten} Grades, deren Nenner beschränkt sind, so gibt es jedenfalls Linearformen

$$y = \mu_1 x_1 + \mu_2 x_2 + \cdots + \mu_n x_n$$

mit ganzzahligen Koeffizienten aus K, die bei Anwendung irgend einer Substitution von $\mathfrak{G}$ wieder in ganzzahlige Linearformen übergehen. Denn ist μ irgend eine ganze Zahl von K, die durch den Generalnenner von $\mathfrak{G}$ teilbar ist, so besitzen die Linearformen

$$\mu x_1,\ \mu x_2,\ \cdots,\ \mu x_n$$

diese Eigenschaft.

Ich betrachte nun nach dem Vorgange von W. Burnside die Gesamtheit $\mathfrak{L}$ der ganzzahligen Linearformen y, die der erwähnten Bedingung genügen. Das System $\mathfrak{L}$ repräsentiert offenbar einen Modul vom Range n im Körper K. Geht nun eine Form y von $\mathfrak{L}$ durch Anwendung einer Substitution G von $\mathfrak{G}$ in $z = y_G$ über, so gehört auch z dem Modul $\mathfrak{L}$ an. Denn zunächst ist z nach Voraussetzung eine ganzzahlige Linearform. Wendet man aber auf z eine beliebige Substitution G' von $\mathfrak{G}$ an, so geht die so entstehende Form $z_{G'}$ aus y durch Anwendung der Substitution GG' hervor. Da GG' wieder in $\mathfrak{G}$ enthalten ist, so ist auch $z_{G'} = y_{GG'}$ eine ganzzahlige Linearform, folglich ist z eine Form von $\mathfrak{L}$.

Unter Benutzung der in § 3 eingeführten Bezeichnungen können wir diese Eigenschaft des Moduls $\mathfrak{L}$ folgendermaßen ausdrücken: Für jede Substitution G von $\mathfrak{G}$ ist das System von Linearformen $\mathfrak{L}G$ in $\mathfrak{L}$ enthalten, also $\mathfrak{L} G > \mathfrak{L}$.' Neben dem Modul $\mathfrak{L}$ gibt es noch andere Moduln n^{ten} Ranges in K, welche der Bedingung genügen, daß für jede Substitution G von $\mathfrak{G}$

$$\mathfrak{M} G > \mathfrak{M}$$

wird. Ich nenne dann $\mathfrak{M}$ einen *zu der Gruppe* $\mathfrak{G}$ *gehörenden Modul.* Die Anzahl dieser Moduln ist unendlich groß; denn ist $\mathfrak{q}$ ein beliebiges Ideal von K, so ist, wie man leicht erkennt, zugleich mit $\mathfrak{M}$ auch $\mathfrak{q}\mathfrak{M}$ ein zu $\mathfrak{G}$ gehörender Modul (vgl. § 1).

Für die Anwendungen ist noch folgende Bemerkung von Wichtigkeit: Man setze, wenn $G = (g_{\varkappa\lambda})$ eine Substitution von $\mathfrak{G}$ ist,

$$y_\varkappa^G = g_{\varkappa 1} x_1 + g_{\varkappa 2} x_2 + \cdots + g_{\varkappa n} x_n.$$

Diese Linearform geht durch Anwendung irgendeiner Substitution G' von $\mathfrak{G}$ in die Linearform $y_\varkappa^{GG'}$ über. Sind daher insbesondere für ein festes $\varkappa$ unter den Linearformen $y_\varkappa^G$, die den verschiedenen Substitutionen G von $\mathfrak{G}$ entsprechen, n linear unabhängige vorhanden, so erzeugen für jede ganze Zahl μ von K, die durch den Generalnenner der Gruppe teilbar ist, die ganzzahligen Linearformen $\mu y_\varkappa^G$ einen zu $\mathfrak{G}$ gehörenden Modul $\mathfrak{M}_\varkappa$. Diese Bedingung ist insbesondere für jedes $\varkappa$ erfüllt, wenn $\mathfrak{G}$ in bezug auf den Körper K irreduzibel ist. Bei einer irreduzibeln Gruppe $\mathfrak{G}$ kann man demnach unmittelbar n zu ihr gehörende Moduln angeben.

Es sei umgekehrt $\mathfrak{M}$ ein gegebener Modul n^{ten} Ranges im Körper K. Die Gesamtheit der linearen Substitutionen G, die der Bedingung $\mathfrak{M} G > \mathfrak{M}$ genügen, bildet dann eine Gruppe $\mathfrak{G}$ n^{ten} Grades, die ich *die zu dem Modul gehörende Gruppe* nenne. Es ist leicht zu sehen, daß $\mathfrak{G}$ eine in K rationale Gruppe mit beschränkten Nennern ist. Denn wählt man in $\mathfrak{M}$ irgendwelche n linear unabhängige Formen

$$z_\varkappa = a_{\varkappa 1} x_1 + a_{\varkappa 2} x_2 + \cdots + a_{\varkappa n} x_n, \qquad (\varkappa = 1, 2, \cdots, n),$$

so geht $z_\varkappa$ durch Anwendung einer Substitution $G = (g_{\varkappa\lambda})$ von $\mathfrak{G}$ in eine Linearform $z'_\varkappa$ mit den Koeffizienten

$$b_{\varkappa\lambda} = \sum_{\nu=1}^{n} a_{\varkappa\nu} g_{\nu\lambda}$$

über. Da $z'_\varkappa$ in $\mathfrak{M}$ enthalten sein soll, so sind die $b_{\varkappa\lambda}$ ganze Zahlen des Körpers K; folglich ist auch, wenn Δ die Determinante $|a_{\varkappa\lambda}|$ bedeutet, jede der Zahlen $\Delta g_{\nu\lambda}$ eine ganze Zahl von K. Hieraus ergibt sich zugleich, daß der Generalnenner der Gruppe $\mathfrak{G}$ ein Divisor des Determinantenteilers $\mathfrak{d}(\mathfrak{M})$ von $\mathfrak{M}$ ist.

Ist $\mathfrak{M}$ speziell der Modul

$$\mathfrak{M}_n(\mathfrak{a}) = [1, \cdots, 1, \mathfrak{a}],$$

so wird $\mathfrak{G}$, wie man leicht erkennt, nichts anderes als die früher betrachtete Gruppe $\mathfrak{G}_n(\mathfrak{a})$. Die Gruppe $\mathfrak{G}_n(1)$ kann also als die zu dem Einheitsmodul $\mathfrak{E}$ gehörende Gruppe charakterisiert werden.

§ 6.

Ich beweise nun folgenden Satz:

III. *Es sei $\mathfrak{H}$ die Gruppe der h Idealklassen des Körpers K und $\mathfrak{H}_n$ diejenige Untergruppe von $\mathfrak{H}$, deren Idealklassen sich als n^{te} Potenzen darstellen lassen. Ist dann, wenn N den Index der Untergruppe $\mathfrak{H}_n$ von $\mathfrak{H}$ bedeutet,*

$$\mathfrak{H} = \mathfrak{H}_n A_1 + \mathfrak{H}_n A_2 + \cdots + \mathfrak{H}_n A_N$$

und sind $\mathfrak{a}_1, \mathfrak{a}_2, \cdots, \mathfrak{a}_N$ beliebige Ideale aus den Idealklassen $A_1, A_2, \cdots, A_N$, so läßt sich jede in K rationale Gruppe $\mathfrak{G}$ n^{ten} Grades, deren Nenner beschränkt sind, durch eine lineare Transformation der Variabeln in eine Untergruppe einer der N Gruppen

$$\mathfrak{G}_n(\mathfrak{a}_1),\ \mathfrak{G}_n(\mathfrak{a}_2),\ \cdots,\ \mathfrak{G}_n(\mathfrak{a}_N)$$

*überführen.**)

Die N Ideale $\mathfrak{a}_1, \mathfrak{a}_2, \cdots, \mathfrak{a}_N$ sind dadurch charakterisiert, daß jedes Ideal $\mathfrak{a}$, mit der n^{ten} Potenz eines passend gewählten Ideals $\mathfrak{q}$ multipliziert, einem unter ihnen äquivalent wird, und daß, wenn $\varkappa$ von λ verschieden ist, $\mathfrak{a}_\varkappa$ und $\mathfrak{a}_\lambda \mathfrak{q}^n$ für kein Ideal $\mathfrak{q}$ äquivalente Ideale sind. — Es sei nun $\mathfrak{M}$ ein zur Gruppe $\mathfrak{G}$ gehörender Modul mit dem Determinantenteiler $\mathfrak{d}$; ferner sei $\mathfrak{a}_\varkappa$ dasjenige Ideal aus der Reihe $\mathfrak{a}_1, \mathfrak{a}_2, \cdots, \mathfrak{a}_N$, für das sich $\mathfrak{q}$ so bestimmen läßt, daß $\mathfrak{a}_\varkappa$ und $\mathfrak{d}\mathfrak{q}^n$ äquivalente Ideale werden. Man bilde dann den zu $\mathfrak{G}$ gehörenden Modul $\overline{\mathfrak{M}} = \mathfrak{q}\mathfrak{M}$. Da der Determinantenteiler dieses Moduls gleich $\mathfrak{d}\mathfrak{q}^n$ ist, so ist $\overline{\mathfrak{M}}$ nach Satz I* dem Modul

$$\mathfrak{M}' = \mathfrak{M}_n(\mathfrak{a}_\varkappa)$$

äquivalent. Daher läßt sich eine lineare Substitution P von nicht verschwindender Determinante, deren Koeffizienten dem Körper K angehören, so bestimmen, daß $\mathfrak{M}' = \overline{\mathfrak{M}} P$ wird. Für jede Substitution G von $\mathfrak{G}$ folgt dann aus $\overline{\mathfrak{M}} G > \overline{\mathfrak{M}}$

$$\overline{\mathfrak{M}} G P > \overline{\mathfrak{M}} P$$

oder, was dasselbe ist,

$$\mathfrak{M}'(P^{-1} G P) > \mathfrak{M}'.$$

Da nun die zu $\mathfrak{M}'$ gehörende Gruppe die Gruppe $\mathfrak{G}_n(\mathfrak{a}_\varkappa)$ ist, so ist $P^{-1}GP$ in $\mathfrak{G}_n(\mathfrak{a}_\varkappa)$ enthalten. Die aus $\mathfrak{G}$ durch die lineare Transformation P der Variabeln hervorgehende Gruppe $\mathfrak{G}' = P^{-1}\mathfrak{G}P$ ist daher eine Untergruppe von $\mathfrak{G}_n(\mathfrak{a}_\varkappa)$.

*) Die Zahl N gibt bekanntlich zugleich die Anzahl der Idealklassen an, deren n^{te} Potenz die Hauptklasse ist.

Wir können ohne Beschränkung der Allgemeinheit annehmen, daß A_1 die Hauptklasse und $\mathfrak{a}_1$ das Hauptideal 1 ist. Ist dann insbesondere der Determinantenteiler $\mathfrak{d}$ des Moduls $\mathfrak{M}$ der n^{ten} Potenz eines Ideals äquivalent, so wird das vorhin betrachtete Ideal $\mathfrak{a}_\varkappa$ gleich $\mathfrak{a}_1 = 1$ und die Gruppe $\mathfrak{G}'$ wird als Untergruppe von $\mathfrak{G}_n(1)$ eine Gruppe mit ganzzahligen Koeffizienten aus dem Körper K. Ist umgekehrt bekannt, daß $\mathfrak{G}$ einer Gruppe $\mathfrak{G}'$ ähnlich ist, deren Koeffizienten ganze Zahlen von K sind, so läßt sich auch eine lineare Transformation Q mit ganzzahligen Koeffizienten aus dem Körper K bestimmen, sodaß $\mathfrak{G}' = Q\mathfrak{G}Q^{-1}$ wird. Dann ist, wenn $\mathfrak{E}$ wie früher den Einheitsmodul bedeutet, für jede Substitution G von $\mathfrak{G}$

$$\mathfrak{E}(QGQ^{-1}) > \mathfrak{E},$$

und folglich

$$\mathfrak{E}Q(G) > \mathfrak{E}Q.$$

Daher ist der Modul $\mathfrak{E}Q$, dessen Determinantenteiler das durch die Determinante von Q bestimmte Hauptideal ist, ein zu $\mathfrak{G}$ gehörender Modul. Wir erhalten also den Satz:

IV. *Eine in K rationale Substitutionsgruppe $\mathfrak{G}$ n^{ten} Grades, deren Nenner beschränkt sind, läßt sich dann und nur dann in eine Gruppe mit ganzzahligen Koeffizienten aus dem Körper K linear transformieren, wenn sich ein zu $\mathfrak{G}$ gehörender Modul angeben läßt, dessen Determinantenteiler der n^{ten} Potenz eines Ideals von K äquivalent ist.*

Ist insbesondere n zu h teilerfremd, so ist jedes Ideal von K der n^{ten} Potenz eines gewissen Ideals äquivalent. Aus IV ergibt sich daher:

V. *Jede in K rationale Substitutionsgruppe mit beschränkten Nennern, in der die Anzahl der Variabeln zu der Anzahl h der Idealklassen von K teilerfremd ist, läßt sich durch eine lineare Transformation der Variabeln in eine Gruppe mit ganzzahligen Koeffizienten aus dem Körper K überführen.**)

Der Satz IV läßt noch eine wichtige Folgerung zu. Ist nämlich K' ein algebraischer Körper, der K als Teilkörper enthält, so erzeugen die Linearformen des Moduls $\mathfrak{M}_n(\mathfrak{a}_\varkappa)$ in K' einen Modul $\mathfrak{M}'_n(\mathfrak{a}_\varkappa)$, dessen Determinantenteiler das durch $\mathfrak{a}_\varkappa$ bestimmte Ideal von K' ist. Betrachtet man nun $\mathfrak{G}_n(\mathfrak{a}_\varkappa)$ als eine im Körper K' rationale Gruppe, so erscheint $\mathfrak{M}'_n(\mathfrak{a}_\varkappa)$ als ein zu dieser Gruppe gehörender Modul. Wählt man nun insbesondere, was jedenfalls möglich ist, für K' einen Körper, in dem die Ideale $\mathfrak{a}_1, \mathfrak{a}_2, \cdots, \mathfrak{a}_N$ Hauptideale werden, so folgt aus Satz IV, daß jede der N Gruppen $\mathfrak{G}_n(\mathfrak{a}_\varkappa)$ einer Gruppe mit ganzzahligen Koeffizienten aus dem Körper K' ähnlich wird. Unter Berücksichtigung des Satzes III erhalten wir daher:

*) Für den Fall $h = 1$ hat diesen Satz bereits W. Burnside bewiesen.

VI. *Für jede Zahl n läßt sich ein Relativkörper K_n in bezug auf den Körper K bestimmen, sodaß jede in K rationale Substitutionsgruppe n^{ten} Grades mit beschränkten Nennern einer Gruppe ähnlich wird, deren Koeffizienten ganze Zahlen des Körpers K_n sind.*

Eine endliche Gruppe linearer Substitutionen mit beliebigen Koeffizienten läßt sich, wie G. Frobenius*) bewiesen hat, stets auch als Gruppe mit algebraischen Koeffizienten darstellen.**) Aus VI folgt aber genauer:

VII. *Jede endliche Gruppe linearer Substitutionen läßt sich in eine Gruppe mit ganzen algebraischen Koeffizienten linear transformieren.*

Es sei noch bemerkt, daß man diesen Satz nach der Burnsideschen Methode auch ohne Benutzung der Steinitzschen Resultate beweisen kann.

Der Satz V läßt sich noch etwas verallgemeinern: Es sei $\mathfrak{G}$ eine beliebige in K rationale Gruppe n^{ten} Grades mit beschränkten Nennern und d der größte gemeinsame Teiler der Zahlen n und h. Man bilde in bekannter Weise die Substitutionsgruppe

$$\mathfrak{G}^{(d)} = \begin{pmatrix} \mathfrak{G} & 0 & \cdots & 0 \\ 0 & \mathfrak{G} & \cdots & 0 \\ \cdot & \cdot & \cdots & \cdot \\ 0 & 0 & \cdots & \mathfrak{G} \end{pmatrix}$$

des Grades dn, die durch d-maliges „Aneinanderreihen" aus $\mathfrak{G}$ hervorgeht. Sind nun $\mathfrak{M}_1, \mathfrak{M}_2, \cdots, \mathfrak{M}_d$ irgendwelche d zu $\mathfrak{G}$ gehörende Moduln, so gehört, wie man leicht erkennt, der aus ihnen zusammengesetzte Modul

$$\mathfrak{N} = [\mathfrak{M}_1, \mathfrak{M}_2, \cdots, \mathfrak{M}_d]$$

des Ranges dn zu der Gruppe $\mathfrak{G}^{(d)}$ (vgl. § 1). Wählt man insbesondere

$$\mathfrak{M}_1 = \mathfrak{M}_2 = \cdots = \mathfrak{M}_{d-1} = \mathfrak{M}, \quad \mathfrak{M}_d = \mathfrak{q}\,\mathfrak{M},$$

wo $\mathfrak{q}$ ein Ideal von K ist, so wird, falls $\mathfrak{d}$ der Determinantenteiler von $\mathfrak{M}$ ist, der Determinantenteiler von $\mathfrak{N}$ gleich $\mathfrak{q}^n \mathfrak{d}^d$. Ist aber $d = hx - ny$, so wird für $\mathfrak{q} = \mathfrak{d}^y$

$$\mathfrak{q}^n \mathfrak{d}^d = \mathfrak{d}^{hx}$$

ein Hauptideal. Aus Satz IV folgt daher, daß *die Gruppe $\mathfrak{G}^{(d)}$ in jedem Fall einer Gruppe mit ganzzahligen Koeffizienten aus dem Körper K ähnlich ist.*

---

*) „Über die Darstellung der endlichen Gruppen durch lineare Substitutionen. II", Berlin Ber. 1899, S. 482; vgl. auch meine Arbeit „Arithmetische Untersuchungen über endliche Gruppen linearer Substitutionen", ebenda 1906, S. 164.

**) Unter einer endlichen Gruppe linearer Substitutionen ist hierbei nicht etwa eine beliebige Gruppe mit nur endlich vielen Substitutionen, sondern eine Gruppe zu verstehen, die einer abstrakten endlichen Gruppe einstufig isomorph ist.

§ 7.

Eine in K rationale Substitutionsgruppe $\mathfrak{G}$ des Grades n, deren Nenner beschränkt sind, bezeichne man als *Gruppe vom Typus* $(\mathfrak{a})$, wenn $\mathfrak{G}$ in eine Untergruppe der Gruppe $\mathfrak{G}_n(\mathfrak{a})$ linear transformierbar ist. Man schließt nun in ähnlicher Weise wie beim Beweis der Sätze III und IV, daß $\mathfrak{G}$ *dann und nur dann vom Typus* $(\mathfrak{a})$ *ist, wenn sich ein zu* $\mathfrak{G}$ *gehörender Modul angeben läßt, dessen Determinantenteiler einem Ideal der Form* $\mathfrak{a}\mathfrak{q}^n$ *äquivalent ist.* Eine Gruppe vom Typus $(\mathfrak{b})$ ist daher zugleich auch vom Typus $(\mathfrak{a})$, wenn sich das Ideal $\mathfrak{q}$ so bestimmen läßt, daß $\mathfrak{a}\mathfrak{q}^n$ und $\mathfrak{b}$ äquivalente Ideale werden.

Der Satz III besagt nun, daß unter den sämtlichen Gruppen $\mathfrak{G}$ nur höchstens N verschiedene Typen vertreten sind, nämlich die Typen $(\mathfrak{a}_1)$, $(\mathfrak{a}_2), \cdots, (\mathfrak{a}_N)$. Der Vollständigkeit wegen will ich noch nachweisen, daß diese N Typen insofern wesentlich voneinander verschieden sind, als nicht *jede* Gruppe vom Typus $(\mathfrak{a}_\varkappa)$ zugleich auch vom Typus $(\mathfrak{a}_\lambda)$ ist. Es genügt hierzu zu zeigen:

Sind $\mathfrak{a}$ *und* $\mathfrak{b}$ *zwei Ideale von* K *und ist* $\mathfrak{b}$ *nicht einem Ideal der Form* $\mathfrak{a}\mathfrak{q}^n$ *äquivalent, so ist die Gruppe* $\mathfrak{G}_n(\mathfrak{a})$ *nicht vom Typus* $(\mathfrak{b})$.

Dieser Satz ist als bewiesen anzusehen, wenn wir zeigen können, daß jeder zu $\mathfrak{G}_n(\mathfrak{a})$ gehörende Modul $\mathfrak{M}$ einen Determinantenteiler der Form $\mathfrak{a}\mathfrak{q}^n$ besitzen muß. Dies erkennt man aber folgendermaßen: Es sei

$$y = \mu_1 x_1 + \mu_2 x_2 + \cdots + \mu_n x_n$$

eine beliebige Linearform von $\mathfrak{M}$. Die Koeffizienten $\mu_\varkappa$, die in den verschiedenen Formen y auftreten, bilden offenbar für jedes $\varkappa$ ein Ideal $\mathfrak{m}_\varkappa$. Ist A irgendeine Substitution von $\mathfrak{G}_n(\mathfrak{a})$, so enthält nach Voraussetzung $\mathfrak{M}$ auch die Form y_A, die aus y durch Anwendung der Substitution A hervorgeht. Nun kommt aber in $\mathfrak{G}_n(\mathfrak{a})$, wenn die Indizes $\varkappa$ und λ beide kleiner als n oder beide gleich n sind, eine Substitution A vor, für die $y_A = \mu_\varkappa x_\lambda$ wird. Daher muß zunächst

$$\mathfrak{m}_1 = \mathfrak{m}_2 = \cdots = \mathfrak{m}_{n-1}$$

sein. Ist ferner α eine beliebige durch $\mathfrak{a}$ teilbare ganze Zahl und α' eine beliebige gebrochene Zahl von K, die als Idealbruch mit dem Nenner $\mathfrak{a}$ darstellbar ist, so können in $\mathfrak{G}_n(\mathfrak{a})$ zwei Substitutionen A und A' gewählt werden, für die

$$y_A = \mu_1 \alpha x_n, \quad y_{A'} = \mu_n \alpha' x_1$$

wird. Daher sind die Ideale $\mathfrak{m}_1\mathfrak{a}$ und $\mathfrak{m}_n$ durcheinander teilbar, also $\mathfrak{m}_n = \mathfrak{m}_1\mathfrak{a}$. Zugleich erkennen wir aber, daß die Linearformen von $\mathfrak{M}$ dadurch erhalten werden, daß man in y die Koeffizienten $\mu_1, \mu_2, \cdots, \mu_n$

unabhängig voneinander alle Zahlen der Ideale $\mathfrak{m}_1, \mathfrak{m}_2, \cdots, \mathfrak{m}_n$ durchlaufen läßt. Daher wird (vgl. § 1)

$$\mathfrak{M} = [\mathfrak{m}_1, \cdots, \mathfrak{m}_{n-1}, \mathfrak{m}_n] = [\mathfrak{m}_1, \cdots, \mathfrak{m}_1, \mathfrak{m}_1\mathfrak{a}] = \mathfrak{m}_1 \mathfrak{M}_n'(\mathfrak{a}).$$

Da der Determinantenteiler dieses Moduls gleich $\mathfrak{m}_1^n\mathfrak{a}$ ist, so ist unsere Behauptung bewiesen.

Ist n nicht teilerfremd zu h, so läßt sich in K ein Ideal $\mathfrak{a}$ wählen, für das $\mathfrak{a}\mathfrak{q}^n$ niemals ein Hauptideal wird. Die Gruppe $\mathfrak{G}_n(\mathfrak{a})$ ist dann gewiß nicht vom Typus (1), also keiner Gruppe mit ganzzahligen Koeffizienten aus K ähnlich. Der Satz V läßt sich daher in folgender Weise ergänzen:

V.* *Ist die Zahl n nicht teilerfremd zu der Anzahl der Idealklassen des Körpers K, so läßt sich nicht jede in K rationale Substitutionsgruppe n^{ten} Grades mit beschränkten Nennern in eine Gruppe mit ganzzahligen Koeffizienten aus dem Körper K linear transformieren.*

19.

Bemerkungen zur Theorie der beschränkten Bilinearformen mit unendlich vielen Veränderlichen

Journal für die reine und angewandte Mathematik 140, 1 - 28 (1911)

In der vorliegenden Arbeit will ich eine Reihe von Einzelfragen behandeln, die sich auf die von Herrn *Hilbert**) begründete Theorie der beschränkten Bilinearformen beziehen.

Nach Zusammenstellung einiger der bis jetzt bekannten Resultate (§ 1) leite ich in §§ 2 und 3 mehrere neue (hinreichende) Kriterien für die Beschränktheit einer vorgelegten Bilinearform ab. Die Kriterien des § 2 beziehen sich vornehmlich auf absolut beschränkte, die des § 3 auf nicht absolut beschränkte Bilinearformen. Den Inhalt des § 4 bildet der Beweis eines elementaren Satzes über definite *Hermite*sche Formen und eines analogen Satzes über definite *Hermite*sche Kerne. In § 5 teile ich einen neuen einfachen Beweis für die Beschränktheit der zuerst von Herrn *Hilbert* eingeführten Bilinearformen

$$A = \sum_{\substack{p,q=1 \\ (p \neq q)}}^{\infty} \frac{x_p y_q}{p-q}, \quad S = \sum_{p,q=1}^{\infty} \frac{x_p y_q}{p+q}$$

mit und betrachte einige mit ihnen nahe verwandte Formen. In § 6 gebe ich ein sehr einfaches Beispiel für eine reelle quadratische Form an, die beschränkt, aber nicht absolut beschränkt ist. Als Anwendung ergibt sich

*) „Grundzüge einer allgemeinen Theorie der linearen Integralgleichungen", Vierte Mitteilung, Nachr. d. K. Ges. d. Wiss., Göttingen, math.-phys. Kl. 1906, S. 157—227. — Vgl. auch *E. Hellinger*, „Neue Begründung der Theorie quadratischer Formen von unendlichvielen Veränderlichen", dieses Journal, Bd. 136 (1909), S. 210—271, und *E. Hellinger* und *O. Toeplitz*, „Grundlagen für eine Theorie der unendlichen Matrizen", Math. Annalen, Bd. 69, S. 289—330.

ein durchaus elementarer Beweis für die in den neueren Untersuchungen über *Fourier*sche Reihen mehrfach benutzte Tatsache, daß der Ausdruck

$$\left|\frac{\sin t}{1}+\frac{\sin 2t}{2}+\cdots+\frac{\sin nt}{n}\right|$$

für alle n und alle reellen t unterhalb einer festen Schranke liegt.

Im Schlußparagraphen leite ich mit Hilfe einer Integralformel eine allgemeine Klasse von beschränkten Bilinearformen mit reellen, nicht negativen Koeffizienten ab, in der die vorhin erwähnte Form S als ganz spezieller Fall enthalten ist. Ich erhalte zugleich ein System von unendlich vielen beschränkten Bilinearformen, die dadurch bemerkenswert sind, daß sich ihre oberen Grenzen genau angeben lassen. Insbesondere ergibt sich, daß die obere Grenze der Form S genau gleich π ist.

Ich bediene mich im folgenden der Bezeichungen, die *E. Hellinger* und *O. Toeplitz* in ihrer oben zitierten Arbeit eingeführt haben.

§ 1.

Es sollen hier einige bekannte Definitionen und Resultate zusammengestellt werden, von denen in der Folge Gebrauch gemacht wird*).

Die Bilinearform

$$(1.)\qquad A = A(x, y) = \sum_{p,q=1}^{\infty} a_{pq} x_p y_q \text{**)},$$

deren Koeffizienten a_{pq} reelle oder komplexe Zahlen sein können, wird als *beschränkt* bezeichnet, wenn sich eine positive Zahl M angeben läßt, so daß für alle n und für alle reellen oder komplexen Zahlen $x_1, \dots x_n$, $y_1, \dots y_n$, die den Bedingungen

$$(2.)\qquad |x_1|^2+\cdots+|x_n|^2 \leqq 1,\quad |y_1|^2+\cdots+|y_n|^2 \leqq 1$$

genügen, der absolute Betrag des „n-ten Abschnittes"

$$A_n = A_n(x, y) = \sum_{p,q=1}^{n} a_{pq} x_p y_q$$

von A nicht größer als M, also

$$(3.)\qquad \left|\sum_{p,q=1}^{n} a_{pq} x_p y_q\right| \leqq M$$

*) Vgl. insbesondere die Arbeit von *E. Hellinger* und *O. Toeplitz*.

**) Solange über die Konvergenz dieser Doppelreihe nichts bekannt ist, soll das Summenzeichen nur zur Zusammenfassung des Koeffizientensystems (a_{pq}) dienen.

wird. Oder anders ausgedrückt: Ist μ_n das Maximum des absoluten Betrages von $|A_n(x, y)|$ unter den Nebenbedingungen (2.), so soll die obere Grenze μ der Zahlen $\mu_1, \mu_2, \ldots$ endlich sein.

Die Zahl μ_n läßt sich folgendermaßen berechnen: Man bezeichne allgemein die zu einer Zahl u konjugiert komplexe Größe mit $\bar{u}$ und setze

$$h_{pq} = \bar{h}_{qp} = a_{1p}\bar{a}_{1q} + a_{2p}\bar{a}_{2q} + \cdots + a_{np}\bar{a}_{nq}.$$

Dann ist μ_n gleich der Quadratwurzel aus der größten unter den (sämtlich reellen, nicht negativen) Wurzeln der Gleichung

$$\begin{vmatrix} h_{11}-x, & h_{12}, & \ldots & h_{1n} \\ h_{21}, & h_{22}-x, & \ldots & h_{2n} \\ \cdot & \cdot & \cdot & \cdot \\ h_{n1}, & h_{n2}, & \ldots & h_{nn}-x \end{vmatrix} = 0.$$

Zugleich ist μ_n^2 das Maximum der definiten *Hermite*schen Form

$$\sum_{p,q=1}^{n} h_{pq} x_p \bar{x}_q = \sum_{r=1}^{n} |a_{r1}x_1 + a_{r2}x_2 + \cdots + a_{rn}x_n|^2$$

unter der Nebenbedingung

$$(4.) \qquad |x_1|^2 + |x_2|^2 + \cdots + |x_n|^2 \leqq 1.$$

Hieraus folgt leicht: Ist $a_{pq} = \bar{a}_{qp}$, so ist die Bilinearform $A(x, y)$ dann und nur dann beschränkt, wenn die *Hermite*sche Form

$$A(x, \bar{x}) = \sum_{p,q=1}^{\infty} a_{pq} x_p \bar{x}_q$$

beschränkt ist, d. h. wenn sich eine Zahl M angeben läßt, so daß für alle n und für alle Werte $x_1, x_2, \ldots x_n$, die der Bedingung (4.) genügen,

$$|A_n(x, \bar{x})| = \left| \sum_{p,q=1}^{n} a_{pq} x_p \bar{x}_q \right| \leqq M$$

wird.

Ist A beschränkt, so *konvergiert die Reihe* (1.) *im Sinne der Konvergenz der Doppelreihen, sowie auch bei zeilen- und kolonnenweiser Summation für alle Wertsysteme* x_p, y_q, *für welche die Reihen*

$$X = |x_1|^2 + |x_2|^2 + \cdots, \quad Y = |y_1|^2 + |y_2|^2 + \cdots$$

konvergent sind. Außerdem ist für jede Zahl M, die den Bedingungen (3.) genügt, auch

$$(5.) \qquad \left| \sum_{p,q=1}^{\infty} a_{pq} x_p y_q \right| \leqq M,$$

sobald $X \leqq 1$, $Y \leqq 1$ wird, und daher für beliebige Werte von X und Y

$$\left| \sum_{p,q=1}^{\infty} a_{pq} x_p y_q \right| \leqq M\sqrt{XY} \leqq \frac{M}{2}(X+Y).$$

Umgekehrt folgen aus der Ungleichung (5.) auch die Ungleichungen (3.). Jede Zahl M, die diese Eigenschaft besitzt, heißt eine *obere Schranke* der Form A; die kleinste unter diesen Zahlen nennt man die *obere Grenze* von A. Man kann diese Zahl auch als die obere Grenze der früher betrachteten Zahlen $\mu_1, \mu_2, \ldots$ definieren.

Die Konvergenz der Doppelreihe (1.) *für alle Wertsysteme* x_p, y_q, *für welche die Reihen* X, Y *konvergent sind, ist nicht nur eine notwendige, sondern auch hinreichende Bedingung für die Beschränktheit der Form* A. *Dasselbe gilt für die Reihen*

$$\sum_{p=1}^{\infty}\Big(\sum_{q=1}^{\infty} a_{pq}x_p y_q\Big), \text{ bzw. } \sum_{q=1}^{\infty}\Big(\sum_{p=1}^{\infty} a_{pq}x_p y_q\Big).^{*)}$$

Für jede obere Schranke M von A wird

$$(6.)\qquad \sum_{r=1}^{n}|a_{pr}|^2 \leqq M^2, \quad \sum_{r=1}^{n}|a_{rq}|^2 \leqq M^2.$$

Eine notwendige Bedingung für die Beschränktheit der Form A *ist daher, daß die Reihen*

$$|a_{p1}|^2+|a_{p2}|^2+\cdots, \quad |a_{1q}|^2+|a_{2q}|^2+\cdots \qquad (p, q=1, 2, \ldots)$$

konvergieren und daß ihre Summen unterhalb einer endlichen Grenze liegen. Eine hinreichende Bedingung ist die Konvergenz der Doppelreihe $\sum\limits_{p,q=1}^{\infty}|a_{pq}|^2$.

Sind

$$A=\sum_{p,q=1}^{\infty} a_{pq}x_p y_q, \quad B=\sum_{p,q=1}^{\infty} b_{pq}x_p y_q$$

zwei nicht notwendig beschränkte Bilinearformen, für welche die Reihen

$$\sum_{r=1}^{\infty}|a_{pr}|^2, \quad \sum_{r=1}^{\infty}|b_{rq}|^2 \qquad (p, q=1, 2, \ldots)$$

konvergent sind, so konvergieren auch die Reihen

$$f_{pq}=\sum_{r=1}^{\infty} a_{pr}b_{rq}.$$

Die mit Hilfe dieser Zahlen gebildete Bilinearform

$$F=\sum_{p,q=1}^{\infty} f_{pq}x_p y_q$$

heißt dann die aus A und B durch *Komposition* oder *Faltung* hervorgehende Form und wird mit AB bezeichnet. Sind insbesondere A und B

*) Während die anderen hier angeführten Resultate sich schon in der *Hilbert*schen Arbeit finden, ist dieser Satz erst von *E. Hellinger* und *O. Toeplitz* (a. a. O. § 10) ausgesprochen und bewiesen worden.

beschränkte Formen, so besitzt auch F diese Eigenschaft. *Die obere Grenze von $F = AB$ ist höchstens gleich dem Produkt der oberen Grenzen der Formen A und B.*

Sind insbesondere für die Form A die Reihen $\sum_r |a_{rq}|^2$ sämtlich konvergent, so konvergieren auch die Reihen

$$c_{pq} = \bar{c}_{qp} = \sum_{r=1}^{\infty} a_{rp}\,\bar{a}_{rq}. \qquad (p, q = 1, 2, \ldots)$$

Die *Hermite*sche Form

$$(7.) \qquad C = \sum_{p,q=1}^{\infty} c_{pq}\, x_p\, \bar{x}_q,$$

die durch Faltung der beiden Formen

$$A' = \sum_{p,q=1}^{\infty} a_{qp}\, x_p y_q, \quad \bar{A} = \sum_{p,q=1}^{\infty} \bar{a}_{pq}\, x_p y_q$$

entsteht, nenne ich *die Norm* $N(A)$ von A*). Von einer Bilinearform A, für welche die Reihen $\sum_r |a_{rq}|^2$ konvergieren, sage ich auch, *A sei eine Form, deren Norm existiert.* Jede beschränkte Form besitzt diese Eigenschaft. Für jedes Wertsystem $x_1, x_2, \ldots$ mit konvergenter Summe $\sum_p |x_p|^2$ ist dann auch die Reihe (7.) konvergent, und es wird zugleich

$$C = N(A) = \sum_{p=1}^{\infty} \left| \sum_{q=1}^{\infty} a_{pq}\, x_q \right|^2.$$

Von besonderer Wichtigkeit für das Folgende ist der (übrigens sehr leicht zu beweisende) Satz (vgl. *E. Hellinger* und *O. Toeplitz*, a. a. O., S. 304):

Hilfssatz I. *Die Bilinearform A ist dann und nur dann beschränkt, wenn die Norm von A existiert und beschränkt ist. Die obere Grenze von $N(A)$ ist genau gleich dem Quadrat der oberen Grenze von A.*

Da jeder Abschnitt der Form $N(A)$ eine nicht negativ definite *Hermite*sche Form (in endlich vielen Variabeln) ist, so erkennt man leicht auch die Richtigkeit des folgenden Satzes:

Hilfssatz II. *Sind $A, B, \ldots$ endlich viele Bilinearformen, deren Normen existieren, und ist die Hermitesche Form*

$$H = N(A) + N(B) + \cdots$$

---

*) Hat man, was vielfach der Fall ist, neben der Norm von A noch die Norm von A' zu betrachten, so kann man passend $N(A)$ als die *Zeilennorm* und $N(A')$ als die *Kolonnennorm* von A bezeichnen.

beschränkt, so ist auch jede der Formen $A, B, \ldots$ beschränkt. Ist η die obere Grenze von H, so ist die obere Grenze jeder der Formen $A, B, \ldots$ höchstens gleich $\sqrt{\eta}$.

Es sei noch bemerkt, daß, wenn die Koeffizienten a_{pq} der beschränkten Form A reell sind, auch die Variabeln x_p, y_q als reell angenommen werden dürfen, ohne daß sich hierbei der Wert der oberen Grenze von A ändert. An Stelle der *Hermite*schen Form $N(A) = C(x, \bar{x})$ kann dann die quadratische Form $C(x, x)$ gesetzt werden.

§ 2.

Die Bilinearform

$$A = \sum_{p,q=1}^{\infty} a_{pq} x_p y_q$$

ist jedenfalls beschränkt, wenn die Form

$$\sum_{p,q=1}^{\infty} |a_{pq}| x_p y_q$$

beschränkt ist. In diesem Fall sagt man, A sei *absolut beschränkt.*

Ich will nun beweisen:

I. *Sind in einer Bilinearform A die Reihen*

$$z_p = \sum_{r=1}^{\infty} |a_{pr}|, \quad k_q = \sum_{r=1}^{\infty} |a_{rq}| \qquad (p, q = 1, 2, \ldots)$$

sämtlich konvergent und unterhalb einer endlichen Schranke gelegen, so ist A absolut beschränkt. Ist $z_p \leqq \zeta$, $k_q \leqq \varkappa$, so ist die obere Grenze von A höchstens gleich $\sqrt{\zeta\varkappa}$.

Da hier nur die absoluten Beträge der Zahlen a_{pq} eine Rolle spielen, so können wir annehmen, daß alle a_{pq} reell und nicht negativ sind. Aus der Konvergenz der Reihen k_q folgt insbesondere, daß auch die Reihen $\sum_r a_{rq}^2$ konvergieren; wir können daher die Norm

$$C = N(A) = \sum_{p,q} c_{pq} x_p x_q$$

bilden. Nun wird hier

$$c_{pq} = \sum_{r=1}^{\infty} a_{rp} a_{rq} \quad \text{und}$$

$$c_p = \sum_{q=1}^{\infty} c_{pq} = \sum_{r=1}^{\infty} a_{rp} \sum_{q=1}^{\infty} a_{rq} = \sum_{r=1}^{\infty} a_{rp} z_r \leqq \zeta \sum_{r=1}^{\infty} a_{rp} \leqq \zeta\varkappa.$$

Ist nun ω_n das Maximum von

$$C_n = \sum_{p,q=1}^{n} c_{pq} x_p x_q$$

unter der Nebenbedingung (4.) des § 1, so wird ω_n gleich der größten Wurzel der Gleichung

$$\begin{vmatrix} c_{11}-x, & c_{12}, & \dots & c_{1n} \\ c_{21}, & c_{22}-x, & \dots & c_{2n} \\ \dots & \dots & \dots & \dots \\ c_{n1}, & c_{n2}, & \dots & c_{nn}-x \end{vmatrix} = 0.$$

Daher lassen sich n reelle Zahlen $\xi_1, \xi_2, \dots, \xi_n$ bestimmen, die nicht sämtlich Null sind und den Gleichungen

$$\omega_n \xi_p = \sum_{q=1}^{n} c_{pq}\xi_q \qquad (p=1,2,\dots n)$$

genügen. Ist $|\xi_p|$ insbesondere die größte unter den Zahlen $|\xi_1|, |\xi_2|, \dots |\xi_n|$, so wird*)

$$\omega_n |\xi_p| \leqq \sum_{q=1}^{n} c_{pq}|\xi_q| \leqq |\xi_p| \sum_{q=1}^{n} c_{pq} \leqq |\xi_p|\cdot\zeta\varkappa,$$

also $\omega_n \leqq \zeta\varkappa$. Folglich ist $N(A)$ beschränkt und die obere Grenze von $N(A)$ höchstens gleich $\zeta\varkappa$. Hieraus folgt aber auf Grund des Hilfssatzes I der zu beweisende Satz.

Es verdient noch bemerkt zu werden, daß die Form A *nicht beschränkt zu sein braucht, wenn sie nur die Eigenschaft besitzt, daß die Reihen*

$$\sum_{r=1}^{\infty} |a_{pr}|^{1+\varrho}, \quad \sum_{r=1}^{\infty} |a_{rq}|^{1+\varrho}$$

für jedes positive ϱ konvergent sind und unterhalb einer (von ϱ abhängigen) Grenze liegen. Dies erkennt man an dem Beispiel

$$A = \sum_{p,q=1}^{\infty} \frac{\log p \log q}{p+q} x_p y_q.$$

Die erwähnte Bedingung ist hier, wie man leicht schließt, erfüllt. Setzt man aber

$$x_1 = y_1 = 0, \quad x_p = y_p = \frac{1}{\sqrt{p}\log p}, \qquad (p \geqq 2)$$

so sind die Reihen $\sum_p x_p^2$, $\sum_p y_p^2$ konvergent, dagegen wird die Doppelreihe

$$\sum_{p,q=2}^{\infty} \frac{1}{\sqrt{pq}(p+q)}$$

divergent. Die Form A ist also nicht beschränkt.

Die Gesamtheit $\mathfrak{A}$ der Bilinearformen, die den Bedingungen des Satzes I genügen, besitzt offenbar die Eigenschaft, daß die aus zwei Formen

*) Vgl. *G. Frobenius*, „Über Matrizen aus positiven Elementen, II.", Sitzungsberichte der K. Preuß. Akademie, Berlin, 1909, S. 514—518.

von $\mathfrak{A}$ durch Faltung hervorgehende Form wieder in $\mathfrak{A}$ enthalten ist. Man kann daher $\mathfrak{A}$ als eine Untergruppe in der Gruppe $\mathfrak{B}$ aller beschränkten Formen bezeichnen.

Aus dem Satz I ergibt sich leicht:

II. *Sind*

$$(u_{pq}),\ (v_{pq}) \qquad (p, q = 1, 2, \ldots)$$

zwei Koeffizientensysteme, für welche die Reihen

$$u_p = \sum_{r=1}^{\infty} |u_{pr}|^2,\quad u'_q = \sum_{r=1}^{\infty} |u_{rq}|^2,$$
$$(p, q = 1, 2, \ldots)$$
$$v_p = \sum_{r=1}^{\infty} |v_{pr}|^2,\quad v'_q = \sum_{r=1}^{\infty} |v_{rq}|^2,$$

konvergent und sämtlich unterhalb einer endlichen Schranke gelegen sind, so ist die Bilinearform

$$A = \sum_{p,q=1}^{\infty} u_{pq} v_{pq} x_p y_q$$

absolut beschränkt. Ist insbesondere

$$u_p \leqq u,\quad u'_q \leqq u,\quad v_p \leqq v,\quad v'_q \leqq v,$$

so wird die obere Schranke von A höchstens gleich $\sqrt{uv}$.

Auf Grund der bekannten Ungleichung

$$|x_1 y_1 + x_2 y_2 + \cdots|^2 \leqq [|x_1|^2 + |x_2|^2 + \cdots][|y_1|^2 + |y_2|^2 + \cdots]$$

erhält man nämlich

$$\sum_{r=1}^{\infty} |u_{pr} v_{pr}| \leqq \sqrt{\sum_{r=1}^{\infty} |u_{pr}|^2 \cdot \sum_{r=1}^{\infty} |v_{pr}|^2} \leqq \sqrt{uv}$$

und ebenso

$$\sum_{r=1}^{\infty} |u_{rq} v_{rq}| \leqq \sqrt{uv}.$$

Daher genügt A den Bedingungen des Satzes I, wobei $\zeta = \varkappa = \sqrt{uv}$, also auch $\sqrt{\zeta\varkappa} = \sqrt{uv}$ wird.

Insbesondere ergibt sich auf Grund der Ungleichungen (6.) des § 1:

III. *Sind*

$$A = \sum_{p,q=1}^{\infty} a_{pq} x_p y_q,\quad B = \sum_{p,q=1}^{\infty} b_{pq} x_p y_q$$

zwei beschränkte Bilinearformen, so ist die Form

$$P = \sum_{p,q=1}^{\infty} a_{pq} b_{pq} x_p y_q$$

absolut beschränkt. Die obere Grenze von P ist, wenn α *und* β *die oberen Grenzen von A und B bedeuten, höchstens gleich* $\alpha\beta$.

Es ist noch zu bemerken, daß der erste Teil unseres Satzes als trivial erscheint, wenn eine der Formen A oder B absolut beschränkt ist. Denn ist z. B. A absolut beschränkt und ist b die obere Grenze der Zahlen $|b_{pq}|$, so ergibt sich für alle Wertsysteme $x_1, \ldots x_n$, $y_1, \ldots y_n$, die den Bedingungen (2.) des § 1 genügen,

$$\left|\sum_{p,q=1}^{n} a_{pq} b_{pq} x_p y_q\right| \leqq b \sum_{p,q=1}^{n} |a_{pq} x_p y_q| \leqq b\alpha',$$

wenn α' die obere Grenze der Form $\sum_{p,q} |a_{pq}| x_p y_q$ bedeutet. Die in unserem Satz gegebene Abschätzung der oberen Grenze von P kann aber unter Umständen auch in diesem Fall dem wahren Wert näher kommen als die sich auf dem zweiten Wege ergebende. Ist z. B.

$$A = B = 2x_1 y_1 + x_1 y_2 + x_2 y_1 - 2x_2 y_2,$$

so wird $\alpha = \beta = \sqrt{5}$, $b = 2$, $\alpha' = 3$; hier ist also $\alpha\beta < b\alpha'$.

Insbesondere ergibt sich, daß, wenn $A = \sum_{p,q} a_{pq} x_p y_q$ eine beschränkte Bilinearform mit der oberen Grenze α ist, für jedes $\nu > 1$

$$A^{(\nu)} = \sum_{p,q=1}^{\infty} a_{pq}^{\nu} x_p y_q$$

eine absolut beschränkte Form wird, deren obere Grenze höchstens gleich α^{ν} ist. Ist ferner α_2 die obere Grenze von $A^{(2)}$ und $a \geqq |a_{pq}|$, so wird für $\nu \geqq 2$ die obere Grenze von $A^{(\nu)}$ höchstens gleich $a^{\nu-2}\alpha_2$. Sind daher $c_1, c_2, \ldots$ beliebige Zahlen, für welche die Reihe

$$c_1 a + c_2 a^2 + \cdots$$

absolut konvergiert, so wird, wenn $A_n^{(\nu)}$ der n-te Abschnitt der Form $A^{(\nu)}$ ist,

$$|c_1 A_n^{(1)} + c_2 A_n^{(2)} + \cdots| \leqq |c_1|\alpha + \alpha_2(|c_2| + |c_3| a + \cdots).$$

Hieraus folgt:

IV. *Ist*

$$A = \sum_{p,q=1}^{\infty} a_{pq} x \, y_q$$

eine beschränkte Bilinearform, in der $|a_{pq}| \leqq a$ *ist, und bedeutet*

$$f(x) = c_1 x + c_2 x^2 + \cdots$$

eine beliebige Potenzreihe, die für $x = a$ *absolut konvergiert, so ist auch die Form*

$$\sum_{p,q=1}^{\infty} f(a_{pq}) x_p y_q$$

beschränkt.

Z. B. sind zugleich mit A auch die Bilinearformen

$$\sum_{p,q=1}^{\infty} \frac{a_{pq}}{\xi - a_{pq}} x_p y_q, \quad \sum_{p,q=1}^{\infty} \log\left(\frac{\xi}{\xi - a_{pq}}\right) x_p y_q$$

beschränkt, sobald nur der absolute Betrag von ξ größer ist als die obere Grenze der Zahlen $|a_{pq}|$.

Der Satz IV läßt sich in naheliegender Weise verallgemeinern, indem man an Stelle einer beschränkten Bilinearform A ein System von endlich vielen beschränkten Formen $A, B, C, \ldots$ betrachtet.

§ 3.

Ähnlich wie bei einer analogen Betrachtung in der Theorie der unendlichen Reihen kann man sich die Aufgabe stellen, diejenigen Koeffizientensysteme

$$u_{pq} \qquad (p, q = 1, 2, \ldots)$$

zu studieren, welche die Eigenschaft besitzen, daß für *jede* beschränkte Bilinearform $\sum_{p,q} a_{pq} x_p y_q$ auch die Form $\sum_{p,q} a_{pq} u_{pq} x_p y_q$ beschränkt ist. Ein solches System u_{pq} will ich im folgenden als *ein Multiplikatorensystem für beschränkte Bilinearformen* bezeichnen.

Sind u_{pq} und v_{pq} zwei Multiplikatorensysteme, so besitzen offenbar auch die Systeme

$$au_{pq} + bv_{pq}, \quad u_{pq} v_{pq}$$

dieselbe Eigenschaft; hierbei können a und b zwei beliebige reelle oder komplexe Zahlen bedeuten. Ferner bilden zugleich mit den Zahlen u_{pq} auch die Zahlen u_{qp} ein Multiplikatorensystem.

Das einfachste Multiplikatorensystem u_{pq} erhält man, indem man, wenn

$$a_1, a_2, \ldots; \quad b_1, b_2, \ldots$$

beliebige Größen sind, deren absolute Beträge unterhalb einer endlichen Schranke liegen, $u_{pq} = a_p b_q$ setzt. Aus dem Satz III folgt ferner, daß die Koeffizienten u_{pq} einer beliebigen beschränkten Bilinearform ein Multiplikatorensystem bilden. Allgemeiner genügt es, daß die Reihen

$$\sum_{r=1}^{\infty} |u_{pr}|^2, \quad \sum_{r=1}^{\infty} |u_{rq}|^2 \qquad (p, q = 1, 2, \ldots)$$

konvergent und sämtlich unterhalb einer endlichen Schranke gelegen sind (vgl. Satz II).

Auf Grund des auf S. 4 angeführten Satzes von *E. Hellinger* und *O. Toeplitz* erhält man ferner ein Multiplikatorensystem für beschränkte Bilinearformen, indem man ein System u_{pq} betrachtet, welches die Eigenschaft besitzt, daß für *jede* konvergente Doppelreihe $\sum_{p,q} a_{pq}$ auch die Doppelreihe $\sum_{p,q} a_{pq} u_{pq}$ konvergiert. Diese Bedingung ist z. B., wie *G. H. Hardy**) gezeigt hat, erfüllt, wenn die Zahlen

$$u_{pq}, \quad u_{pq} - u_{p+1,q}, \quad u_{pq} - u_{p,q+1}, \quad u_{pq} - u_{p+1,q} - u_{p,q+1} + u_{p+1,q+1}$$

reell und nicht negativ sind und

$$\lim_{q=\infty} u_{1q} = \lim_{p=\infty} u_{p1} = 0$$

ist.

Ich möchte hier aber auf zwei ganz andere Klassen von Multiplikatorensystemen aufmerksam machen, die für die Anwendungen von größerer Bedeutung zu sein scheinen.

Ist $h_{pq} = \bar{h}_{qp}$, so bezeichnet man die (formal gebildete) *Hermite*sche Form

$$H = \sum_{p,q=1}^{\infty} h_{pq} x_p \bar{x}_q$$

als *(nichtnegativ) definit*, wenn für jedes n und jedes Wertsystem $x_1, x_2, \ldots x_n$

$$H_n = \sum_{p,q=1}^{n} h_{pq} x_p \bar{x}_q \geqq 0$$

wird. Es genügt hierzu bekanntlich, daß die (sämtlich reellen) Hauptunterdeterminanten der Matrix

$$\begin{pmatrix} h_{11}, h_{12}, \ldots h_{1n} \\ h_{21}, h_{22}, \ldots h_{2n} \\ \cdot\;\cdot\;\cdot\;\cdot\;\cdot\;\cdot\;\cdot\;\cdot \\ h_{n1}, h_{n2}, \ldots h_{nn} \end{pmatrix}$$

nicht negative Werte haben. Läßt sich insbesondere eine positive Zahl μ angeben, so daß für alle n

$$H_n \geqq \mu \left[|x_1|^2 + |x_2|^2 + \cdots + |x_n|^2 \right]$$

wird, so nennt man H *positiv definit*. In einer definiten *Hermite*schen Form ist die obere Grenze h der (nichtnegativen) Zahlen $h_{11}, h_{22}, \ldots$ zugleich auch die obere Grenze der absoluten Beträge aller Zahlen h_{pq}.

Es gilt nun der Satz:

V. *Die Koeffizienten h_{pq} jeder definiten Hermiteschen Form, für welche die obere Grenze h der Zahlen $h_{11}, h_{22}, \ldots$ endlich ist, bilden ein Multiplikatorensystem für beschränkte Bilinearformen. Ist die obere Grenze der beschränkten Bilinearform*

---

*) Proceedings of the London Mathematical Society, Serie 2, Bd. 1 (1903), S. 124—128, und Bd. 2 (1904), S. 190—191.

$$A = \sum_{p,q=1}^{\infty} a_{pq} x_p y_q$$

gleich α, so ist die obere Grenze der Bilinearform

$$B = \sum_{p,q=1}^{\infty} a_{pq} h_{pq} x_p y_q$$

höchstens gleich αh.

Man bestimme nämlich, was bekanntlich stets möglich ist, n Linearformen

$$z_p = \sum_{q=1}^{n} l_{pq} x_q \qquad (p=1, 2, \ldots n)$$

so, daß

$$H_n = \sum_{p,q=1}^{n} h_{pq} x_p \bar{x}_q = \sum_{r=1}^{n} z_r \bar{z}_r$$

wird. Dann erhält man

$$h_{pq} = \sum_{r=1}^{n} l_{rp} \bar{l}_{rq}.$$

Setzt man nun

$$A_n(x_p, y_q) = \sum_{p,q=1}^{n} a_{pq} x_p y_q,$$

so wird

$$B_n(x_p, y_q) = \sum_{p,q=1}^{n} a_{pq} h_{pq} x_p y_q = \sum_{p,q,r=1}^{n} a_{pq} l_{rp} \bar{l}_{rq} x_p y_q.$$

Dies läßt sich in der Form

$$B_n(x_p, y_q) = \sum_{r=1}^{n} A_n(l_{rp} x_p, \bar{l}_{rq} y_q)$$

schreiben. Nun ist aber, wenn α die obere Grenze von A ist,

$$|A_n(l_{rp} x_p, \bar{l}_{rq} y_q)| \leqq \alpha \sqrt{\sum_{p=1}^{n} |l_{rp} x_p|^2 \cdot \sum_{p=1}^{n} |\bar{l}_{rp} y_p|^2}$$
$$\leqq \frac{\alpha}{2} \sum_{p=1}^{n} |l_{rp}|^2 (|x_p|^2 + |y_p|^2).$$

Daher wird

$$|B_n(x_p, y_q)| \leqq \frac{\alpha}{2} \sum_{p,r=1}^{n} |l_{rp}|^2 (|x_p|^2 + |y_p|^2)$$
$$\leqq \frac{\alpha}{2} \sum_{p=1}^{n} h_{pp} (|x_p|^2 + |y_p|^2)$$
$$\leqq \frac{\alpha h}{2} \sum_{p=1}^{n} (|x_p|^2 + |y_p|^2).$$

Für alle Wertsysteme x_p, y_q, die den Bedingungen (2.) des § 1 genügen, ist also in der Tat, wie zu beweisen ist,

$$|B_n(x_p, y_q)| \leqq \alpha h.$$

Eine andere Klasse von Multiplikatorensystemen, die aber mit den eben betrachteten eng verwandt sind, liefert der folgende Satz:

VI. *Es seien*

$$f_p(t),\ g_p(t) \qquad (p=1,2,\ldots)$$

beliebige reelle oder komplexe Funktionen der reellen Variabeln t, und es mögen die absoluten Beträge der Produkte von je zweien dieser Funktionen im Intervall $a \leqq t \leqq b$ *integrabel sein*). Lassen sich dann zwei Zahlen* μ *und* ν *angeben, so daß*

$$\int_a^b |f_p|^2\,dt \leqq \mu, \quad \int_a^b |g_p|^2\,dt \leqq \nu \qquad (p=1,2,\ldots)$$

wird, so bilden die Zahlen

$$v_{pq} = \int_a^b f_p g_q\,dt \qquad (p,q=1,2,\ldots)$$

ein Multiplikatorensystem für beschränkte Bilinearformen. Ist

$$A = \sum_{p,q=1}^{\infty} a_{pq} x_p y_q$$

eine beschränkte Bilinearform mit der oberen Grenze α, *so ist die obere Grenze* β *der Bilinearform*

$$B = \sum_{p,q=1}^{\infty} a_{pq} v_{pq} x_p y_q$$

höchstens gleich $\frac{1}{2}\alpha(\mu+\nu)$ **).

Für den n-ten Abschnitt B_n der Form B erhält man nämlich, wenn $\sum_{p=1}^{n} |x_p|^2 \leqq 1$, $\sum_{p=1}^{n} |y_p|^2 \leqq 1$ ist, ähnlich wie vorhin

$$\begin{aligned} |B_n| &= \left| \int_a^b \Big(\sum_{p,q=1}^{n} a_{pq} f_p g_q x_p y_q \Big) dt \right| \\ &\leqq \int_a^b \left| \sum_{p,q=1}^{n} a_{pq} f_p g_q x_p y_q \right| dt \\ &\leqq \frac{\alpha}{2} \int_a^b \sum_{p=1}^{n} (|f_p x_p|^2 + |g_p y_p|^2)\,dt \\ &\leqq \frac{\alpha(\mu+\nu)}{2}. \end{aligned}$$

Aus dem Satz VI folgt z. B., daß die Zahlen

$$\frac{1}{a_p + b_q} \qquad (p,q=1,2,\ldots)$$

ein Multiplikatorensystem für beschränkte Bilinearformen bilden, wenn nur die reellen Bestandteile aller Größen a_p, b_p positiv sind und eine positive

*) Es kann sich hierbei auch um uneigentliche Integrale handeln.

**) Genauer kann gezeigt werden, daß $\beta \leqq \alpha\sqrt{\mu\nu}$ ist.

untere Grenze besitzen. Denn setzt man

$$f_p(t) = e^{-a_p t},\quad g_p(t) = e^{-b_p t},\quad a = 0,\quad b = \infty,$$

so sind unter den über die Zahlen a_p, b_p gemachten Voraussetzungen die Bedingungen unseres Satzes erfüllt; es wird aber dann

$$v_{pq} = \frac{1}{a_p + b_q}.$$

§ 4.

In enger Beziehung zu dem Satze V steht folgender Satz, der trotz seiner Einfachheit nicht bekannt zu sein scheint:

VII. *Sind*

$$A = \sum_{p,q=1}^{n} a_{pq} x_p \overline{x}_q,\quad B = \sum_{p,q=1}^{n} b_{pq} x_p \overline{x}_q \quad (a_{pq} = \overline{a}_{qp},\ b_{pq} = \overline{b}_{qp})$$

zwei (nichtnegativ) definite Hermitesche Formen, so besitzt auch die Form

$$C = \sum_{p,q=1}^{n} a_{pq} b_{pq} x_p \overline{x}_q$$

dieselbe Eigenschaft. Bezeichnet man mit α und α' die größte und die kleinste charakteristische Wurzel der Form A, ferner mit b und b' die größte und die kleinste unter den Zahlen $b_{11}, b_{22}, \ldots, b_{nn}$, so liegt jede charakteristische Wurzel der Form C zwischen $\alpha' b'$ und αb.

Man bestimme nämlich wie beim Beweise des Satzes V n^2 Zahlen l_{pq} so, daß

$$b_{pq} = \sum_{r=1}^{n} l_{rp} \overline{l}_{rq}$$

wird. Dann wird

$$C = \sum_{r=1}^{n} A(l_{r1} x_1,\ l_{r2} x_2, \ldots l_{rn} x_n).$$

Dies zeigt schon, daß C nichtnegativ definit ist. Ferner ist bekanntlich stets

$$\alpha' \sum_{p=1}^{n} |u_p|^2 \leqq A(u_1, u_2, \ldots, u_n) \leqq \alpha \sum_{p=1}^{n} |u_p|^2.$$

Daher ergibt sich

$$C \leqq \alpha \sum_{r,p=1}^{n} |l_{rp} x_p|^2 = \alpha \sum_{p=1}^{n} b_{pp} |x_p|^2 \leqq \alpha b \sum_{p=1}^{n} |x_p|^2.$$

Ebenso erhält man

$$C \geqq \alpha' b' \sum_{p=1}^{n} |x_p|^2.$$

Hieraus folgt aber die Richtigkeit des zweiten Teils unseres Satzes.

Ist insbesondere A eine positive *Hermite*sche Form, d. h. $\alpha' > 0$, so erkennt man, daß auch C diese Eigenschaft besitzt, es sei denn, daß $b' = 0$ wird. Dies tritt aber nur dann ein, wenn B eine der n Variabeln überhaupt nicht enthält, d. h. wenn in dem Koeffizientenschema von B eine Zeile und die zugehörige Kolonne lauter Nullen enthalten.

Ich will noch eine Anwendung des Satzes VII angeben.

Man bezeichne nach dem Vorgange von *Hilbert**) eine für $a \leqq s \leqq b$, $a \leqq t \leqq b$ definierte, reelle oder komplexe *stetige* Funktion $K(s,t)$ der reellen Variabeln s und t als einen *nichtnegativ definiten Hermiteschen Kern*, wenn $K(s,t)$ und $K(t,s)$ konjugiert komplexe Größen sind und das Doppelintegral

$$I(x) = \int_a^b \int_a^b K(s,t)\, x(s)\, \overline{x}(t)\, ds\, dt$$

einen nichtnegativen Wert erhält, wie auch die stetige Funktion $x(s)$ gewählt wird**). — Man beweist ohne Mühe, wie *W. H. Young****) gezeigt hat, daß der Kern $K(s,t)$ stets und nur dann nichtnegativ definit wird, wenn für jedes n und jedes System von n Zahlen $s_1 < s_2 < \cdots < s_n$ des Intervalls $a \leqq s \leqq b$ die *Hermite*sche Form

$$H = \sum_{p,q=1}^{n} K(s_p, s_q)\, x_p\, \overline{x}_q$$

nichtnegativ definit ist. Auf Grund dieses Kriteriums ergibt sich aus dem Satz VII unmittelbar der Satz:

VII*. *Sind $K(s,t)$ und $L(s,t)$ zwei nichtnegativ definite (stetige) Hermitesche Kerne, so besitzt das Produkt $K(s,t)\, L(s,t)$ dieselbe Eigenschaft.*

---

*) Grundzüge einer allgemeinen Theorie der linearen Integralgleichungen, Erste Mitteilung, Nachr. d. K. Ges. d. Wiss., Göttingen, math.-phys. Kl. 1904, S. 79.

**) *Hilbert* betrachtet a. a. O. nur *positiv* definite Kerne, d. h. solche, für die das Integral $I(x)$ positiv und nur dann Null wird, wenn $x(s) = 0$ ist.

***) „A note on a class of symmetric functions and on a theorem required in the theory of integral equations“, Messenger of Mathematics, Vol. XL (1910), S. 37—43. — *Young* beschränkt sich auf die Betrachtung reeller symmetrischer Kerne, doch läßt sich sein Beweis unmittelbar auch auf *Hermite*sche Kerne übertragen. Die Voraussetzung, daß $K(s,t)$ eine stetige Funktion sein soll, scheint von wesentlicher Bedeutung zu sein. Es genügt ferner keineswegs zu wissen, daß die Formen H positiv sind, um behaupten zu können, daß $K(s,t)$ ein positiv definiter Kern ist.

§ 5.

Mit Hilfe der Integralformel

$$\sum_{p,q=1}^{n}\left(\frac{1}{p+q}+\frac{1}{p-q}\right)x_p y_q = \frac{1}{\pi}\int_{-\pi}^{\pi} t\,[\Sigma(x_p \sin pt - y_p \cos pt)]^2\,dt,$$

in der für $p=q$ das Zeichen $\frac{1}{p-q}$ durch die Null zu ersetzen ist, hat *Hilbert**) bewiesen, daß die beiden Bilinearformen

$$A = \sum_{\substack{p,q=1\\(p\neq q)}}^{\infty}\frac{x_p y_q}{p-q},\quad S = \sum_{p,q=1}^{\infty}\frac{x_p y_q}{p+q}$$

beschränkt sind. Als obere Schranken für A und S ergeben sich hierbei die Zahlen 4π und 2π.

Ich will nun zeigen, daß *man die Beschränktheit der Formen A und S sehr einfach aus dem Hilfssatz* II *des* § 1 *folgern kann.* Man setze nämlich

$$S_0 = \sum_{p,q=1}^{\infty}\frac{x_p y_q}{p+q-1}$$

und

$$N(A) = A'A = \sum_{p,q=1}^{\infty} c_{pq} x_p x_q,\quad N(S_0) = S_0' S_0 = \sum_{p,q=1}^{\infty} d_{pq} x_p x_q.$$

Dann wird

$$c_{pp} + d_{pp} = {\sum_{r=1}^{\infty}}'\frac{1}{(p-r)^2} + \sum_{s=1}^{\infty}\frac{1}{(p+s-1)^2} = {\sum_{r=-\infty}^{+\infty}}'\frac{1}{(p-r)^2} = 2\sum_{s=1}^{\infty}\frac{1}{s^2} = \frac{\pi^2}{3}\qquad (r\neq p)$$

und für $p \neq q$

$$c_{pq} + d_{pq} = {\sum_{r=1}^{\infty}}'\frac{1}{(p-r)(q-r)} + \sum_{s=1}^{\infty}\frac{1}{(p+s-1)(q+s-1)}$$

$$= {\sum_{r=-\infty}^{\infty}}'\frac{1}{(p-r)(q-r)} = \frac{1}{q-p}{\sum_{r=-\infty}^{\infty}}'\left[\frac{1}{p-r}-\frac{1}{q-r}\right].\qquad (r\neq p,\ r\neq q)$$

Dies ist aber, wie man leicht erkennt, gleich $\frac{2}{(q-p)^2}$. Also ergibt sich

$$(1.)\qquad N(A) + N(S_0) = \frac{\pi^2}{3}\sum_{p=1}^{\infty} x_p^2 + \sum_{\substack{p,q=1\\(p\neq q)}}^{\infty}\frac{2}{(q-p)^2}x_p x_q$$

Die erste rechts auftretende Form ist gewiß beschränkt, und zwar ist ihre obere Grenze gleich $\frac{\pi^2}{3}$. Die zweite Form genügt aber den Bedingungen des Satzes I, und da hier die Reihen

*) Vgl. *H. Weyl*, „Singuläre Integralgleichungen mit besonderer Berücksichtigung des *Fourier*schen Integraltheorems", Inaugural-Dissertation, Göttingen 1908, S. 83. — Einen elementaren Beweis für die Beschränktheit der Form S hat *F. Wiener*, Math. Annalen, Bd. 68 (1910), S. 361—366, angegeben. Die Formen A und S sind, wie *O. Toeplitz*, „Zur Theorie der quadratischen Formen von unendlichvielen Veränderlichen" (Nachr. d. K. Ges. d. Wiss., Göttingen, math.-phys. Kl., 1910), gezeigt hat, in einer allgemeinen Klasse von beschränkten Bilinearformen als spezielle Fälle enthalten.

$$\sum_{r=1}^{\infty}{}' \frac{2}{(r-p)^2}, \quad \sum_{s=1}^{\infty}{}' \frac{2}{(q-s)^2} \qquad (r \neq p,\ s \neq q)$$

kleiner sind als $4 \sum\limits_{p=1}^{\infty} \frac{1}{p^2}$, so ist ihre obere Grenze höchstens gleich $\frac{2\pi^2}{3}$. Daher ist die obere Grenze von (1.) höchstens gleich $\frac{\pi^2}{3} + \frac{2\pi^2}{3} = \pi^2$. *Die oberen Grenzen von A und S_0, und folglich auch von S, sind also höchstens gleich π.*

Allgemeiner setze man, wenn λ eine beliebige reelle Zahl ist,

$$A_\lambda = \sum_{p,q=1}^{\infty} \frac{x_p y_q}{p-q+\lambda}, \quad S_\lambda = \sum_{p,q=1}^{\infty} \frac{x_p y_q}{p+q-1+\lambda};$$

hierbei sind für ein ganzzahliges λ diejenigen Glieder, deren Nenner 0 sind, nicht mitzuschreiben. Man erhält dann leicht für ein ganzzahliges λ

$$N(A_\lambda) + N(S_\lambda) = N(A_0) + N(S_0)$$

und für ein nicht ganzzahliges λ

$$N(A_\lambda) + N(S_\lambda) = \sum_{r=-\infty}^{+\infty} \frac{1}{(r+\lambda)^2} \cdot \sum_{p=1}^{\infty} x_p{}^2 = \frac{\pi^2}{\sin^2 \lambda\pi} \sum_{p=1}^{\infty} x_p^2.$$

Daher sind die Formen A_λ und S_λ für jedes λ beschränkt; ihre oberen Grenzen sind für ein ganzzahliges λ höchstens gleich π und für ein nicht ganzzahliges λ höchstens gleich $\frac{\pi}{|\sin \lambda\pi|}$.

Um eine Anwendung des Satzes V zu geben, will ich zeigen, daß die Bilinearform

$$A^{(\mu)} = \sum_{\substack{p,q=1 \\ (p \neq q)}}^{\infty} \frac{(pq)^{\frac{\mu-1}{2}}}{p^\mu - q^\mu} x_p y_q,$$

die für $\mu = 1$ in die vorhin betrachtete Form A übergeht, auch *für jedes positive $\mu > 1$ beschränkt ist, und daß ihre obere Grenze höchstens gleich $\frac{\pi}{\mu}$ wird.* — Es genügt hierzu, zu beweisen, daß die quadratische (reelle *Hermite*sche) Form

$$(2.) \qquad \sum_{p,q=1}^{\infty} \frac{p-q}{p^\mu - q^\mu} (pq)^{\frac{\mu-1}{2}} x_p x_q,$$

in der für $p = q$ der in der Form $\frac{0}{0}$ erscheinende Koeffizient gleich $\frac{1}{\mu}$ zu setzen ist, nichtnegativ definit ist.

Dies ergibt sich folgendermaßen. Sind $u_1, u_2, \ldots$ beliebige komplexe Zahlen mit positiven reellen Bestandteilen, so folgt aus der Formel

$$\frac{1}{u_p + \bar{u}_q} = \int_0^{\infty} e^{-(u_p + \bar{u}_q)t}\, dt,$$

daß die *Hermite*sche Form

$$\sum_{p,q=1}^{\infty} \frac{x_p \overline{x}_q}{u_p + \overline{u}_q}$$

nichtnegativ definit ist*). Sind daher $s_1, s_2, \ldots$ beliebige positive Zahlen und bedeutet α eine beliebige komplexe Zahl, so bilden die Zahlen

$$\frac{\alpha + \overline{\alpha}}{\alpha s_p + \overline{\alpha} s_q}$$

und folglich auch, wenn ε eine Größe vom absoluten Betrage 1 ist, die Zahlen

$$\frac{1-\varepsilon}{s_p - \varepsilon s_q}$$

das Koeffizientensystem einer definiten *Hermite*schen Form. Durchläuft nun, wenn m eine positive ganze Zahl ist, ε die von 1 verschiedenen m-ten Einheitswurzeln, so wird

$$\frac{1}{m} \sum_{\varepsilon} \frac{1-\varepsilon}{s_p - \varepsilon s_q} = \frac{s_p^{m-1} - s_q^{m-1}}{s_p^m - s_q^m} \quad \text{oder} \quad \frac{1}{m} \sum_{\varepsilon} \frac{1-\varepsilon}{s_p - \varepsilon s_q} = \frac{m-1}{m s_p},$$

je nachdem s_q von s_p verschieden oder gleich s_p ist. Für jede positive ganze Zahl $l < m$ erhält man daher nach Satz VII das Koeffizientensystem (h_{pq}) einer definiten quadratischen Form, indem für $s_p \neq s_q$

$$h_{pq} = \frac{s_p^l - s_q^l}{s_p^m - s_q^m} = \prod_{r=l+1}^{m} \frac{s_p^{r-1} - s_q^{r-1}}{s_p^r - s_q^r}$$

und für $s_p = s_q$

$$h_{pq} = \frac{l}{m s_p^{m-l}} = \prod_{r=l+1}^{m} \frac{r-1}{r s_p}$$

setzt. Schreibt man nun $s_p^{\frac{1}{l}}$ für s_p und μ für $\frac{m}{l}$, so erkennt man, daß der Ausdruck

$$\sum_{p,q=1}^{\infty} \frac{s_p - s_q}{s_p^{\mu} - s_q^{\mu}} x_p x_q, \tag{3.}$$

in dem für $s_p = s_q$ der Koeffizient von $x_p x_q$ gleich $\frac{1}{\mu s_p^{\mu-1}}$ zu setzen ist, für jedes System von positiven Zahlen $s_1, s_2, \ldots$ und für jedes rationale $\mu > 1$ eine nichtnegativ definite quadratische Form repräsentiert. Durch

---

*) Auf rein algebraischem Wege folgt dies aus der leicht zu beweisenden Determinantenrelation

$$\left| \frac{1}{u_p + v_q} \right| = \frac{\prod\limits_{p<q} (u_p - u_q)(v_p - v_q)}{\prod\limits_{p,q} (u_p + v_q)} \qquad (p, q = 1, 2, \ldots n)$$

Grenzübergang ergibt sich sofort, daß dies auch für ein irrationales $\mu > 1$ richtig bleibt.

Daß nun die quadratische Form (2.) definit ist, folgt hieraus als ganz spezieller Fall, indem man in (3.) für $s_1, s_2, \ldots$ die Zahlen $1, 2, \ldots$ wählt und x_r durch $r^{\frac{\mu-1}{2}} x_r$ ersetzt.

§ 6.

Schreibt man den n-ten Abschnitt A_n der alternierenden Form $A = \sum_{p \neq q} \frac{x_p y_q}{p-q}$ in der Gestalt

$$A_n = \sum_{\substack{p,q=1 \\ (q>p)}}^{n} \frac{x_p y_q - x_q y_p}{p-q}$$

und ersetzt, wenn t eine reelle Größe bedeutet, x_p und y_p durch $x_p \sin pt$, bzw. $x_p \cos pt$, so erhält man, da π eine obere Schranke von A ist, für reelle Werte x_p

$$\left| \sum_{\substack{p,q=1 \\ (q>p)}}^{n} \frac{\sin(p-q)t}{p-q} x_p x_q \right| \leq \pi \sqrt{\sum_{p=1}^{n} (x_p \sin pt)^2 \cdot \sum_{p=1}^{n} (x_p \cos pt)^2}$$
$$\leq \frac{\pi}{2} \sum_{p=1}^{n} x_p^2 (\sin^2 pt + \cos^2 pt) = \frac{\pi}{2} \sum_{p=1}^{n} x_p^2.$$

Hieraus folgt, daß der Ausdruck

$$F(t) = \sum_{\substack{p,q=1 \\ (p \neq q)}}^{\infty} \frac{\sin(p-q)t}{p-q} x_p x_q$$

eine reelle quadratische Form darstellt, deren obere Grenze höchstens gleich π ist.

Einen einfacheren Beweis für die Beschränktheit der Form $F(t)$ und eine genauere Abschätzung der oberen Grenze liefert die folgende Methode: Es genügt offenbar $0 \leq t \leq \pi$ anzunehmen. Setzt man

$$\Phi(t) = t \sum_{p=1}^{n} x_p^2 + \sum_{\substack{p,q=1 \\ (p \neq q)}}^{n} \frac{\sin(p-q)t}{p-q} x_p x_q = t \sum_{p=1}^{n} x_p^2 + F_n(t),$$

so wird

$$\frac{d\Phi(t)}{dt} = \sum_{p,q=1}^{n} x_p x_q \cdot \cos(p-q)t = \left(\sum_{p=1}^{n} x_p \cos pt\right)^2 + \left(\sum_{p=1}^{n} x_p \sin pt\right)^2.$$

Daher ist für jedes System reeller Zahlen $x_1, x_2, \ldots x_n$ die Funktion $\Phi(t)$ von t monoton wachsend, also

$$\Phi(0) \leq \Phi(t) \leq \Phi(\pi).$$

Dies gibt aber

$$(1.) \qquad -t \sum_{p=1}^{n} x_p^2 \leq F_n(t) \leq (\pi - t) \sum_{p=1}^{n} x_p^2.$$

Folglich ist die Form $F(t)$ beschränkt und ihre obere Grenze (für $0 \leqq t \leqq \pi$) höchstens gleich der größeren der beiden Zahlen t und $\pi - t$.

Die Form $F(t)$ ist deshalb interessant, weil sie wohl das einfachste Beispiel für eine *reelle quadratische Form liefert, die beschränkt, aber (für jedes t, das nicht ein ganzzahliges Multiplum von π ist) nicht absolut beschränkt ist**). Dies erkennt man folgendermaßen: Wäre $F(t)$ für ein gegebenes t absolut beschränkt, so müßte sich eine Zahl M angeben lassen, so daß

$$\sum_{\substack{p,q=1\\(p\neq q)}}^{n} \left| \frac{\sin(p-q)t}{p-q} \right| \leqq M(1^2+1^2+\cdots+1^2) = Mn$$

oder, ausführlicher geschrieben,

$$2\sum_{\nu=1}^{n-1}(n-\nu)\left|\frac{\sin \nu t}{\nu}\right| \leqq Mn$$

wird. Dies gibt aber

$$\sum_{\nu=1}^{n-1}\left|\frac{\sin \nu t}{\nu}\right| \leqq \frac{M}{2} + \frac{1}{n}\sum_{\nu=1}^{n-1}|\sin \nu t| < \frac{M}{2}+1.$$

Eine solche Zahl M kann aber, wenn t nicht ein ganzzahliges Multiplum von π ist, nicht existieren, da die Reihe

$$\sum_{\nu=1}^{\infty}\left|\frac{\sin \nu t}{\nu}\right|$$

alsdann bekanntlich divergent ist.

Setzt man noch in (1.) $x_1 = x_2 = \cdots = x_n = 1$, so erhält man (für $0 \leqq t \leqq \pi$)

$$-tn \leqq 2\sum_{\nu=1}^{n-1}(n-\nu)\frac{\sin \nu t}{\nu} \leqq (\pi - t)n,$$

also

$$(2.) \qquad -\frac{t}{2} + \frac{1}{n}\sum_{\nu=1}^{n-1}\sin \nu t \leqq \sum_{\nu=1}^{n-1}\frac{\sin \nu t}{\nu} \leqq \frac{\pi - t}{2} + \frac{1}{n}\sum_{\nu=1}^{n-1}\sin \nu t.$$

Hieraus folgt speziell, daß für alle n und alle reellen t

$$(3.) \qquad S_n(t) = \left|\sum_{\nu=1}^{n-1}\frac{\sin \nu t}{\nu}\right| < \frac{\pi}{2} + 1 = 2{,}570\ldots$$

ist.

*) Ein allgemeines Verfahren zur Herstellung von quadratischen Formen, die diese Eigenschaft besitzen, hat *O. Toeplitz* in der auf S. 16 zitierten Abhandlung angegeben. — Das einfachste Beispiel einer reellen alternierenden Bilinearform, die beschränkt, aber nicht absolut beschränkt ist, repräsentiert die Form $\sum_{p \neq q} \frac{x_p y_q}{p-q}$; vgl. *E. Hellinger* und *O. Toeplitz*, a. a. O., S. 308.

Daß der Ausdruck $S_n(t)$ unterhalb einer von n und t unabhängigen Zahl M liegen muß, hat zuerst ***A. Kneser****) bewiesen. Diese Tatsache spielt insbesondere in den neueren Arbeiten von ***L. Fejér*** über überall stetige Funktionen, deren *Fourier*sche Reihen an einzelnen Stellen divergieren, eine nicht unwichtige Rolle**). Die durch (3.) gelieferte Abschätzung von $S_n(t)$ hat ***Fejér*** in seiner Note „Sur les sommes partielles de la série de *Fourier*"***) mit Hilfe eines interessanten allgemeinen Satzes über *Fourier*sche Reihen gewonnen.

Schreibt man in (2.) $\pi - t$ für t, so erhält man durch Addition beider Ungleichungen (für $n = 2m$)

$$-\frac{\pi}{2} + \frac{1}{m}\sum_{\mu=1}^{m} \sin(2\mu-1)t \leqq 2\sum_{\mu=1}^{m} \frac{\sin(2\mu-1)t}{2\mu-1} \leqq \frac{\pi}{2} + \frac{1}{m}\sum_{\mu=1}^{m} \sin(2\mu-1)t.$$

Daher ist für alle m und alle reellen t

$$(4.) \qquad \left|\sum_{\mu=1}^{m} \frac{\sin(2\mu-1)t}{2\mu-1}\right| < \frac{\pi}{4} + \frac{1}{2}.$$

Beachtet man noch, daß für $0 < t < \pi$

$$\sum_{\mu=1}^{k} \sin(2\mu-1)t = \frac{\sin^2 kt}{\sin t} \geqq 0$$

ist, so erhält man wegen (4.) und

$$2\sum_{\mu=1}^{m} \frac{\sin(2\mu-1)t}{2\mu-1} - \sum_{\mu=1}^{m} \frac{\sin(2\mu-1)}{\mu} = \sum_{\mu=1}^{m} \frac{\sin(2\mu-1)t}{\mu(2\mu-1)} \geqq 0$$

die für $0 \leqq t \leqq \pi$ geltende Formel

$$0 \leqq \sum_{\mu=1}^{m} \frac{\sin(2\mu-1)t}{\mu} < 2\left(\frac{\pi}{4} + \frac{1}{2}\right).$$

Hieraus folgt zugleich, daß für alle m und alle reellen t

$$\left|\sum_{\mu=1}^{m} \frac{\sin(2\mu-1)t}{\mu}\right| < \frac{\pi}{2} + 1$$

---

*) „Beiträge zur Theorie der *Sturm-Liouville*schen Darstellung willkürlicher Funktionen", Math. Annalen, Bd. 60 (1905), S. 402—423, § 1.

**) Vgl. vor allem die Arbeit von *L. Fejér*, „*Lebesgue*sche Konstanten und divergente *Fourier*reihen", dieses Journal, Bd. 138 (1910), S. 22—53. — Siehe auch *A. Haar*, Zur Theorie der orthogonalen Funktionensysteme", Inaug.-Dissertation, Göttingen, 1909, S. 17.

***) Comptes Rendus de l' Acad. des Sciences, Paris, Mai 1910, S. 1299. — Es wird hier noch gezeigt, daß die obere Grenze der Ausdrücke $S_n(t)$ nicht kleiner sein kann als $\int_0^\pi \frac{\sin x}{x} dx = 1{,}85\ldots.$

wird. — Auch diese Ungleichung findet sich in der erwähnten Note von *L. Fejér.*

Ich will noch zeigen, daß man die Ungleichung (3.) durch die präzisere Ungleichung

$$\left|\sum_{\nu=1}^{n-1}\frac{\sin\nu t}{\nu}\right| < \frac{\pi}{2}+\frac{1}{\pi}+\frac{1}{2}=2{,}389\ldots$$

ersetzen kann. Wegen (2.) genügt es hierzu, zu zeigen, daß

$$\frac{1}{n}\left|\sum_{\nu=1}^{n-1}\sin\nu t\right| < \frac{1}{\pi}+\frac{1}{2} \tag{5.}$$

ist*). — Dies beweise ich folgendermaßen. Man betrachte die alternierende Bilinearform

$$D=\sum_{\substack{p,q=1\\(p<q)}}^{n}(x_p y_q - x_q y_p).$$

Die charakteristische Determinante dieser Form ist, wie man leicht zeigt,

$$\varDelta(x)=\begin{vmatrix} -x, & 1, & 1,\ldots & 1\\ -1, & -x, & 1,\ldots & 1\\ -1, & -1, & -x,\ldots & 1\\ \cdot & \cdot & \cdot & \cdot\\ -1, & -1, & -1,\ldots & -x\end{vmatrix}=(-1)^n\frac{(x+1)^n+(x-1)^n}{2}.$$

Die Wurzeln der Gleichung $\varDelta(x)=0$ sind die Größen

$$i\operatorname{cotg}\frac{(2\nu+1)\pi}{2n}. \qquad (\nu=0,1,\ldots n-1)$$

Für $\nu=0$ erhält man die dem absoluten Betrage nach größte Wurzel. Da aber

$$\operatorname{cotg}\frac{\pi}{2n}<\frac{2n}{\pi}$$

ist, so wird für reelle $x_p y_q$

$$|D|\leqq\frac{2n}{\pi}\sqrt{\sum_{p=1}^{n}x_p^2\cdot\sum_{p=1}^{n}y_p^2}\leqq\frac{n}{\pi}\sum_{p=1}^{n}(x_p^2+y_p^2).$$

Setzt man speziell $x_p=\sin pt$, $y_p=\cos pt$, so erhält man

$$\left|\sum_{\substack{p,q=1\\(p<q)}}^{n}\sin(p-q)t\right|\leqq\frac{n^2}{\pi},$$

*) Aus der Formel

$$\frac{2}{\pi}=\frac{1}{\pi}\int_0^{\pi}\sin x\,dx=\lim_{n=\infty}\frac{1}{n}\sum_{\nu=0}^{n-1}\sin\frac{\nu\pi}{n}$$

geht hervor, daß die genaue obere Grenze des in (5.) links stehenden Ausdrucks nicht kleiner als $\frac{2}{\pi}$ sein kann.

oder

$$\left|\sum_{\nu=1}^{n-1}(n-\nu)\sin\nu t\right| \leqq \frac{n^2}{\pi},$$

also

$$\frac{1}{n}\left|\sum_{\nu=1}^{n-1}\sin\nu t\right| \leqq \frac{1}{\pi}+\frac{1}{n^2}\left|\sum_{\nu=1}^{n-1}\nu\sin\nu t\right| < \frac{1}{\pi}+\frac{n(n-1)}{2n^2} < \frac{1}{\pi}+\frac{1}{2}.$$

§ 7.

Es sei $K(s, t)$ eine reelle Funktion der positiven Argumente s, t, die folgenden Bedingungen genügt:

1. $K(s, t)$ wird niemals negativ und ist homogen vom Grade -1.

2. In jedem endlichen Gebiet

$$0 < a \leqq s \leqq b,\quad 0 < a \leqq t \leqq b$$

ist $K(s, t)$ (im Riemannschen Sinne) integrabel.

3. Das Integral

$$k = \int_0^\infty \frac{K(s,1)}{\sqrt{s}}\,ds = \int_0^\infty \frac{K(1,t)}{\sqrt{t}}\,dt = \int_1^\infty \frac{K(s,1)+K(1,s)}{\sqrt{s}}\,ds\,^{*)}$$

hat einen endlichen Wert.

Sind dann $f(s)$ und $g(s)$ zwei nicht negative Funktionen des positiven Arguments s, die in jedem Intervall $0 < a \leqq s \leqq b$ integrabel sind und für die die Integrale

$$F = \int_0^\infty f^2(s)\,ds,\quad G = \int_0^\infty g^2(s)\,ds$$

endliche Werte haben, so existiert, wie ich behaupte, auch das Doppelintegral

$$I = \int_0^\infty\int_0^\infty K(s,t)\,f(s)\,g(t)\,ds\,dt,$$

und zwar ist stets

$$(1.)\qquad I \leqq k\sqrt{FG}\,^{**)}.$$

Es genügt offenbar, zu zeigen, daß für je zwei positive Zahlen $a < b$

$$I_{a,b} = \int_a^b\int_a^b K(s,t)\,f(s)\,g(t)\,ds\,dt \leqq k\sqrt{FG}$$

---

*) Das Übereinstimmen dieser drei Integrale folgt aus der vorausgesetzten Homogeneitätseigenschaft der Funktion $K(s, t)$.

**) Für $K(s,t) = \frac{1}{s+t}$ hat eine ähnliche Ungleichung bereits *H. Weyl* (a. a. O., S. 85) bewiesen.

ist. Denn die Formel (1.) wird dann erhalten, indem man a gegen 0 und b gegen ∞ konvergieren läßt. Diese Schlußweise ist gestattet, weil die zu integrierende Funktion nicht negativ ist.

Man setze nun in $I_{a,b}$

$$s = u, \; t = uv.$$

Dann wird wegen der Homogeneität von $K(s,t)$

$$I_{a,b} = \iint K(1,v)\, f(u)\, g(uv)\, du\, dv.$$

Das Integrationsgebiet ist hierbei durch die Ungleichungen

$$a \leqq u \leqq b, \quad \frac{a}{u} \leqq v \leqq \frac{b}{u}$$

bestimmt. Da nun

$$\frac{a}{b} \leqq \frac{a}{u}, \quad \frac{b}{u} \leqq \frac{b}{a}$$

wird, so vergrößern wir das Integrationsgebiet, wenn wir es durch das Rechteck

$$a \leqq u \leqq b, \quad \frac{a}{b} \leqq v \leqq \frac{b}{a}$$

ersetzen. Folglich ist

$$I_{a,b} \leqq \int_a^b \int_{\frac{a}{b}}^{\frac{b}{a}} K(1,v)\, f(u)\, g(uv)\, du\, dv = \int_{\frac{a}{b}}^{\frac{b}{a}} K(1,v)\, dv \int_a^b f(u)\, g(uv)\, du.$$

Nun ist auf Grund einer bekannten Ungleichung

$$\left[\int_a^b f(u)\, g(uv)\, du\right]^2 \leqq \int_a^b f^2(u)\, du \cdot \int_a^b g^2(uv)\, du.$$

Ferner ist

$$\int_a^b f^2(u)\, du \leqq \int_0^\infty f^2(u)\, du = F$$

und

$$\int_a^b g^2(uv)\, du = \frac{1}{v} \int_{av}^{bv} g^2(u)\, du \leqq \frac{1}{v} \int_0^\infty g^2(u)\, du = \frac{G}{v}.$$

Daher wird

$$\int_a^b f(u)\, g(uv)\, du \leqq \frac{1}{\sqrt{v}} \sqrt{FG}$$

und also, wie zu beweisen ist,

$$I_{a,b} \leqq \sqrt{FG} \int_{\frac{a}{b}}^{\frac{b}{a}} \frac{K(1,v)}{\sqrt{v}}\, dv \leqq k\sqrt{FG}.$$

Man setze nun, wenn

$$a_1, a_2, \dots; b_1, b_2, \dots$$

beliebige positive Zahlen sind,

$$s_n = a_1 + a_2 + \dots + a_n, \quad t_n = b_1 + b_2 + \dots + b_n.$$

Sind ferner $x_1, x_2, \dots$ und $y_1, y_2, \dots$ zwei Systeme nicht negativer Zahlen mit konvergenten Quadratsummen

$$X = \sum_{p=1}^{\infty} x_p^2, \quad Y = \sum_{p=1}^{\infty} y_p^2, \tag{2.}$$

so bestimme man die Funktionen $f(s)$ und $g(t)$ folgendermaßen: für

$$s_{n-1} < s \leqq s_n, \quad t_{n-1} < t \leqq t_n \qquad (s_0 = t_0 = 0)$$

soll

$$f(s) = \frac{x_n}{\sqrt{a_n}}, \quad g(t) = \frac{y_n}{\sqrt{b_n}}$$

sein. Ist ferner $\lim\limits_{n=\infty} s_n = \sigma$ oder $\lim\limits_{n=\infty} t_n = \tau$ eine endliche Zahl, so verstehe man für $s \geqq \sigma$, bzw. für $t \geqq \tau$ unter $f(s)$ oder $g(t)$ den Wert 0. Durch diese Festsetzungen sind dann $f(s)$ und $g(t)$ für alle positiven Werte des Arguments eindeutig definiert. Diese Funktionen sind jedenfalls in jedem endlichen Intervall integrabel; ferner wird

$$F = \int_0^{\infty} f^2(s)\,ds = X, \quad G = \int_0^{\infty} g^2(s)\,ds = Y.$$

Die Formel (1.) geht aber dann über in die Formel

$$\sum_{p,q=1}^{\infty} \frac{x_p y_q}{\sqrt{a_p b_q}} \int_{s_{p-1}}^{s_p} \int_{t_{q-1}}^{t_q} K(s,t)\,ds\,dt \leqq k\sqrt{XY}, \tag{3.}$$

d. h. *die links stehende Bilinearform ist beschränkt und ihre obere Grenze ist höchstens gleich der Zahl k.*

Ein einfacheres Resultat erhält man, wenn man $K(s,t)$ noch folgender Bedingung unterwirft:

4. Bei festem s soll $K(s,t)$ mit wachsendem t und bei festem t mit wachsendem s nicht zunehmen.

Dann wird offenbar

$$\int_{s_{p-1}}^{s_p} \int_{t_{q-1}}^{t_q} K(s,t)\,ds\,dt \geqq K(s_p, t_q)(s_p - s_{p-1})(t_q - t_{q-1})$$
$$\geqq K(s_p, t_q)\, a_p b_q.$$

Aus der Formel (3.) ergibt sich daher, daß auch *der Ausdruck*

$$\sum_{p,q=1}^{\infty} K(s_p, t_q)\sqrt{a_p b_q}\, x_p y_q \tag{4.}$$

eine beschränkte Bilinearform darstellt, deren obere Grenze höchstens gleich k ist.

Setzt man z. B., wenn m irgend eine positive Zahl bedeutet,

$$K(s,t)=\frac{1}{\sqrt[m]{s^m+t^m}},$$

so wird

$$k=\int_0^\infty \frac{dt}{\sqrt[m]{1+t^m}.\ \sqrt{t}}=\frac{1}{m}\int_0^\infty (1+u)^{-\frac{1}{m}}\, u^{\frac{1}{2m}-1}\,du=\frac{\Gamma^2\left(\frac{1}{2m}\right)}{m\Gamma\left(\frac{1}{m}\right)}$$

und speziell $k=\pi$ für $m=1$. *Daher ist für alle Wertsysteme x_p, y_q, für welche die Reihen* (2.) *konvergieren, und für alle Zahlen s_n, t_n, die den Bedingungen*

$$0<s_1<s_2<\cdots,\quad 0<t_1<t_2<\cdots$$

genügen:

$$\left|\sum_{p,q=1}^{n}\frac{\sqrt{(s_p-s_{p-1})(t_q-t_{q-1})}}{\sqrt[m]{s_p^m+t_q^m}}\,x_p y_q\right| \leqq \frac{\Gamma^2\left(\frac{1}{2m}\right)}{m\Gamma\left(\frac{1}{m}\right)}\sqrt{\sum_{p=1}^{\infty}x_p^2\cdot\sum_{p=1}^{\infty}y_p^2}\,.$$

Ein anderes interessantes Beispiel erhält man, wenn man unter $K(s,t)$ den reziproken Wert der größeren der beiden Zahlen s und t versteht. Dann wird, wie man leicht erkennt, $k=4$. Für $s_n=t_n$, $x_n=y_n$ ergibt sich daher

$$\left|\sum_{p=1}^{\infty}\frac{s_p-s_{p-1}}{s_p}x_p^2+2\sum_{\substack{p,q=1\\(q>p)}}^{\infty}\frac{\sqrt{(s_p-s_{p-1})(s_q-s_{q-1})}}{s_q}x_p x_q\right|\leqq 4\sum_{p=1}^{\infty}x_p^2.$$

Wählt man in (4.) speziell $a_n=b_n=1$, also $s_n=t_n=n$, so erhält man die Bilinearform

$$A(x,y)=\sum_{p,q=1}^{\infty}K(p,q)\,x_p y_q. \tag{5.}$$

Ich will nun zeigen, *daß die obere Grenze α dieser Form für jede Funktion $K(s,t)$, die den Bedingungen 1.—4. genügt, genau gleich k ist.*

Da nämlich auf Grund unseres allgemeineren Resultats $\alpha\leqq k$ sein muß, so wird für alle positiven Werte $x_1, x_2, \ldots$, für welche die Reihe $X=\sum_p x_p^2$ konvergiert,

$$Q=\frac{A(x,x)}{X}\leqq k. \tag{6.}$$

Man setze nun speziell, wenn $\varrho>\frac{1}{2}$ ist, $x_n=\frac{1}{n^\varrho}$. Dann wird $X=\zeta(2\varrho)$ und, wie man auf Grund der Gleichung

$$K(p,p) = \frac{K(1,1)}{p}$$

leicht erkennt,

$$A(x,x) = -K(1,1)\zeta(2\varrho+1) + \sum_{p=1}^{\infty}\sum_{q=p}^{\infty}\frac{K(p,q)}{(pq)^\varrho} + \sum_{q=1}^{\infty}\sum_{p=q}^{\infty}\frac{K(p,q)}{(pq)^\varrho}.$$

Nun ist aber wegen der Eigenschaften *1.* und *4.* der Funktion $K(s,t)$

$$\sum_{q=p}^{\infty}\frac{K(p,q)}{(pq)^\varrho} \geqq \int_p^\infty \frac{K(p,t)}{(pt)^\varrho}dt = \frac{1}{p^{2\varrho}}\int_1^\infty \frac{K(1,t)}{t^\varrho}dt$$

und ebenso

$$\sum_{p=q}^{\infty}\frac{K(p,q)}{(pq)^\varrho} \geqq \frac{1}{q^{2\varrho}}\int_1^\infty \frac{K(s,1)}{s^\varrho}ds.$$

Daher wird

$$A(x,x) \geqq -K(1,1)\zeta(2\varrho+1) + \zeta(2\varrho)\int_1^\infty \frac{K(u,1)+K(1,u)}{u^\varrho}du.$$

Bezeichnet man den Quotienten (6.) in unserem Fall mit Q_ϱ, so ergibt sich also

$$(7.)\qquad k \geqq Q_\varrho \geqq -K(1,1)\frac{\zeta(2\varrho+1)}{\zeta(2\varrho)} + \int_1^\infty \frac{K(u,1)+K(1,u)}{u^\varrho}du.$$

Nun ist allgemein, wenn $\varphi(u)$ eine nicht negative, in jedem endlichen Intervall $1 \leqq a \leqq u \leqq b$ integrable Funktion ist, für die das Integral

$$\int_1^\infty \frac{\varphi(u)}{\sqrt{u}}du$$

existiert,

$$\lim_{\varrho=\frac{1}{2}+0}\int_1^\infty \frac{\varphi(u)}{u^\varrho}du = \int_1^\infty \frac{\varphi(u)}{\sqrt{u}}du.$$

Läßt man daher in (7.) ϱ gegen $\frac{1}{2}$ konvergieren, so erhält man, da

$$\lim_{\varrho=\frac{1}{2}}\zeta(2\varrho+1) = \zeta(2) = \frac{\pi^2}{6},\quad \lim_{\varrho=\frac{1}{2}}\zeta(2\varrho) = \infty$$

und

$$\int_1^\infty \frac{K(u,1)+K(1,u)}{\sqrt{u}}du = k$$

ist,

$$\lim_{\varrho=\frac{1}{2}+0} Q_\varrho = k,$$

d. h. man kann, wenn ε eine beliebige positive Größe ist, ϱ so nahe bei $\frac{1}{2}$ wählen, daß $Q_\varrho > k-\varepsilon$ wird. Dies zeigt deutlich, daß die obere Grenze α der Form (5.) nicht kleiner als k sein kann. Da nun $\alpha \leqq k$ war, so muß $\alpha = k$ sein.

Der Vollständigkeit wegen will ich noch erwähnen, daß die Resultate dieses Paragraphen sich auch auf Funktionen von mehr als zwei Variabeln übertragen lassen. Es gilt nämlich folgender Satz:

Es sei $K(s_1, s_2, \dots s_n)$ *eine nichtnegative Funktion der positiven Variabeln* $s_1, s_2, \dots s_n$, *die homogen vom Grade* $1-n$, *in jedem endlichen Gebiet*

$$0 < a \leqq s_\nu \leqq b \qquad (\nu = 1, 2, \dots n)$$

integrabel ist, und für die das $(n-1)$*-fache Integral*

$$k_n = \int_0^\infty \int_0^\infty \cdots \int_0^\infty \frac{K(1, s_2, s_3, \dots s_n)}{\sqrt[n]{s_2 s_3 \cdots s_n}} ds_2 ds_3 \cdots ds_n$$

einen endlichen Wert hat. Sind dann $f_1(s), f_2(s), \dots f_n(s)$ *beliebige nichtnegative Funktionen des positiven Arguments* s, *die in jedem endlichen Intervall* $0 < a \leqq s \leqq b$ *integrabel sind, und für die die Integrale*

$$F_\nu = \int_0^\infty f_\nu^n(s)\, ds \qquad (\nu = 1, 2, \dots n)$$

endliche Werte haben, so existiert das n*-fache Integral*

$$I_n = \int_0^\infty \int_0^\infty \cdots \int_0^\infty K(s_1, s_2, \dots s_n) f_1(s_1) f_2(s_2) \cdots f_n(s_n)\, ds_1 ds_2 \cdots ds_n,$$

und zwar ist

$$I_n \leqq k_n \sqrt[n]{F_1 F_2 \cdots F_n}.$$

Setzt man noch voraus, daß K *für jedes* ν *bei festen* $s_1, \dots s_{\nu-1}, s_{\nu+1}, \dots s_n$ *und wachsendem* s_ν *nicht zunehmen soll, so ist, wenn*

$$a_1^{(\nu)}, a_2^{(\nu)}, \dots \qquad (\nu = 1, 2, \dots n)$$

beliebige positive Zahlen sind und

$$s_n^{(\nu)} = a_1^{(\nu)} + a_2^{(\nu)} + \cdots + a_n^{(\nu)} \qquad (\nu = 1, 2, \dots n)$$

gesetzt wird, die n*-fach unendliche Reihe*

$$U = \sum_{p_1, \dots p_n = 1}^{\infty} K(s_{p_1}^{(1)}, s_{p_2}^{(2)}, \dots s_{p_n}^{(n)}) \left[a_{p_1}^{(1)} a_{p_2}^{(2)} \cdots a_{p_n}^{(n)}\right]^{\frac{n-1}{n}} x_{p_1}^{(1)} x_{p_2}^{(2)} \cdots x_{p_n}^{(n)}$$

für alle nichtnegativen Wertsysteme $x_p^{(1)}, x_p^{(2)}, \dots x_p^{(n)}$ *konvergent, für welche die Reihen*

$$X_\nu = \sum_{p=1}^{\infty} (x_p^{(\nu)})^n \qquad (\nu = 1, 2, \dots n)$$

konvergent sind, und zwar ist stets

$$U \leqq k_n \sqrt[n]{X_1 X_2 \cdots X_n}.$$

In dieser Ungleichung kann, wenn speziell alle $a_\mu^{(\nu)}$ *gleich* 1 *gesetzt werden, die Zahl* k_n *durch keine kleinere Zahl ersetzt werden.*

Printed in the United States
By Bookmasters